KT-557-979

THIS ITEM CANNOT BE
ISSUED FOR LOAN

B.S.U.C - LIBRARY
00242997

The Environment Encyclopedia and Directory

2005

The Environment Encyclopedia and Directory 2005

4th edition

LONDON AND NEW YORK

First published 1994
Fourth Edition 2005

Haines House, 21 John Street, London WC1N 2BP
(A member of the Taylor & Francis Group)

ISBN 1 85743 224 X
ISSN 1352-6480

Printed on Ocean White Recycled paper manufactured from 100% recycled fibres of which 75% are derived from post-consumer waste

Typeset by AJS Solutions, Huddersfield and Dundee
Printed and bound in the United Kingdom by
Polestar Wheatons, Exeter

FOREWORD

The publication of the fourth edition of THE ENVIRONMENT ENCYCLOPEDIA AND DIRECTORY, the first edition of which appeared in 1994, reflects the enduring importance of environmental issues and the continuing demand for accurate and up-to-date information on the subject. Indeed, as this volume went to press, the Kyoto Protocol, passed in 1997, eventually entered into force in the wake of its ratification by the Russian Federation, but still with the USA not committed to its terms. The book brings together cartographic, informative, directory, bibliographic and biographic data concerning the environment in a single volume.

The book is divided into five main sections. Firstly, there is an introductory essay, new to this edition, which examines the topical issue of emissions trading in the context of international climate change, and discusses its evolution as the response of nations and corporations to the greenhouse gas emission targets set by the Kyoto Protocol.

A series of 26 global and regional maps, dealing with climate, vegetation, population, rainforests, energy, protected areas, fishing, pollution and other subjects of relevance to the environment, provide a graphical counterpart to the other sections of the volume.

The A–Z of the Environment provides approximately 1,000 definitions and descriptions of terms, concepts, events and programmes relating to the environment. Cross-references within the definitions are given in bold type. As well as a large number of extensively updated and rewritten entries, there are many entries new to this edition, including Biosphere 2, Ecotourism, International Coral Reef Initiative, Kola Peninsula, Lake Nyos, Natura 2000, World Summit on Sustainable Development and Yangzi River.

The Directory of Environmental Organizations lists the leading international environmental organizations, followed by a chapter on every country in the world. Each country chapter begins with the principal environmental ministry or agency, which is followed by other relevant national ministries, then alphabetical listings of other governmental and non-governmental organizations. Subsidiary organizations can be found indented below the appropriate parent institution. There is a total of almost 4,500 entries from around the world, giving, where appropriate, contact names, addresses, telephone and fax numbers, e-mail and internet details, dates of foundation, affiliations and a brief outline of each organization's activities. The Index of Environmental Organizations, which can be found at the end of the book, lists entrants in the Directory section according to 11 fields of activity.

The Bibliography of Environmental Periodicals lists more than 1,200 relevant titles alphabetically, giving addresses, telephone and fax numbers, e-mail and internet information, names of editors, foundation dates, and details of frequency, subject matter, language and circulation.

Finally, the Who's Who of the Environment provides biographical details, addresses and telephone, fax and e-mail contact information for almost 800 leading personalities involved in environmental affairs around the world.

February 2005

ACKNOWLEDGEMENTS

The Editors would like to thank all the organizations, periodicals and individuals listed in *The Environment Encyclopedia and Directory 2005* who completed questionnaires and returned them to us.

Special acknowledgements are due to the following: Heidi Bachram for the introductory essay; Edward Oliver, for the maps; Gareth Wyn Jones for revising and updating the A–Z of the Environment; Michael Chapman for co-ordinating the updating of the Directory of Environmental Organizations, the Bibliography of Environmental Periodicals and the Who's Who of the Environment.

With regard to the maps on pages 13–38 of this book, we are particularly grateful for permission to make use of material from the following sources: *The State of the Environment Atlas*, by Joni Seager (London, Penguin Books; New York, Simon & Schuster, 1995, US title: *The State of the Earth Atlas*); *The State of the World Atlas*, by Dan Smith (London, Penguin Books, 1999; map illustrations and graphics copyright (c) Myriad Editions Limited, 1999); *World Bank Atlas 2000* and *World Development Indicators 2000*, (c) 2000, maps republished and statistical data used with permission of The World Bank, permission conveyed through Copyright Clearance Center, Inc; *Collins Concise Atlas*, *Collins Pocket Planet Earth Factfile* and *Times Atlas of the World*, (c) Bartholomew Ltd 2000, reproduced by permission of HarperCollins Publishers; *Environment Atlas*, by David R Wright, and *Atlas of the World*, copyright (c) George Philip Limited; *World Resources 1998–99*, World Resources Institute/Oxford University Press; *Statistical Yearbook*, 43rd issue, United Nations, 1999; and *World Population Prospects: The 1998 Revision, Vol I: Comprehensive Tables*, United Nations.

CONTENTS

LIST OF ABBREVIATIONS

AAA	Agricultural Adjustment Administration
AAAS	American Association for the Advancement of Science
AAF	Army Air Force
AASA	Associate of the Australian Society of Accountants
AB	Aktiebolag; Alberta; Bachelor of Arts
ABA	American Bar Association
AC	Companion of the Order of Australia
ACA	Associate of the Institute of Chartered Accountants
ACCA	Associate of the Association of Certified Accountants
Acad	Academy, Académie
Accad	Accademia
Accred	Accredited
ACIS	Associate of the Chartered Institute of Secretaries
ACP	African, Caribbean and Pacific Group of States; American College of Physicians
ACS	American Chemical Society
ACT	Australian Capital Territory
ADC	Aide-de-camp
Adm	Admiral
Admin	Administration, Administrative, Administrator
AERE	Atomic Energy Research Establishment
AF	Air Force
AFC	Air Force Cross
AFESD	Arab Fund for Economic and Social Development
affil	affiliated
AFL	American Federation of Labor
AG	Aktiengesellschaft (Joint Stock Company)
Agric	Agriculture
Ags	Aguascalientes
ai	*ad interim*
AIA	American Institute of Architects; Associate of Institute of Actuaries
AIAA	American Institute of Aeronautics and Astronautics
AIB	Associate of the Institute of Bankers
AICE	Associate of the Institute of Civil Engineers
AIChE	American Institute of Chemical Engineers
AID	(US) Agency for International Development
AIDS	Acquired Immune Deficiency Syndrome
AIEE	American Institute of Electrical Engineers
AIME	American Institute of Mining Engineers; Associate of the Institution of Mining Engineers
AIMechE	Associate of the Institution of Mechanical Engineers
AK	Alaska; Knight of the Order of Australia
Akad	Akademie
AL	Alabama
ALS	Associate of the Linnaean Society
Alt	Alternate
AM	Albert Medal; Alpes Maritimes; Master of Arts; Member of the Order of Australia
Amb	Ambassador
AMICE	Associate Member of the Institution of Civil Engineers
AMIEE	Associate Member of the Institution of Electrical Engineers
AMIMechE	Associate Member of the Institution of Mechanical Engineers
ANU	Australian National University
AO	Officer of the Order of Australia
Apdo	Apartado
approx	approximately
Apptd	Appointed
Apt	Apartment
AR	Arkansas
ARA	Associate of the Royal Academy
ARC	Agriculture Research Council
ARCS	Associate of the Royal College of Science
ARIBA	Associate of the Royal Institute of British Architects
ARSA	Associate of the Royal Scottish Academy; Associate of the Royal Society of Arts
ASEAN	Association of South East Asian Nations
ASLIB	Association of Special Libraries and Information Bureaux
ASME	American Society of Mechanical Engineers
Asoc	Asociación
Ass	Assembly
Asscn	Association
Assoc	Associate
ASSR	Autonomous Soviet Socialist Republic
Asst	Assistant
avda	avenida
Ave	Avenue
Aug	August
AZ	Arizona
b	born
B AGR	Bachelor of Agriculture
B AGR SC	Bachelor of Agricultural Science
B ARCH	Bachelor of Architecture
B CH, B CHIR	Bachelor of Surgery
B COM(M)	Bachelor of Commerce
B ECONS	Bachelor of Economics
B ED	Bachelor of Education
B ENG	Bachelor of Engineering
B LIT(T)	Bachelor of Letters
B LL	Bachelor of Laws
B MUS	Bachelor of Music
B PHIL	Bachelor of Philosophy
B SC	Bachelor of Science
BA	Bachelor of Arts; British Airways
BAAS	British Association for the Advancement of Science
BAFTA	British Academy of Film and Television Awards
BAO	Bachelor of Obstetrics
BAS	Bachelor in Agricultural Science
BBC	British Broadcasting Corporation
BC	British Columbia
BCE	Bachelor of Civil Engineering
BCL	Bachelor of Canon Law; Bachelor of Civil Law
BCN	Baja California Norte
BCS	Bachelor of Commercial Sciences; Baja California Sur
BD	Bachelor of Divinity
Bd	Board
BDS	Bachelor of Dental Surgery
BE	Bachelor of Education; Bachelor of Engineering
Beds	Bedfordshire
BEE	Bachelor of Electrical Engineering
Berks	Berkshire
BFA	Bachelor of Fine Arts
BFI	British Film Institute
BIM	British Institute of Management
biog	biography
BIPM	Bureau International des Poids et Mesures
BIS	Bank for International Settlements
Bl	Block
BL	Bachelor of Laws
Bldg	Building

LIST OF ABBREVIATIONS

BLS	Bachelor in Library Science
blvd	boulevard
BM	Bachelor of Medicine
BMA	British Medical Association
Bn	Battalion
BP	Boîte postale (Post Office Box); British Petroleum
BPA	Bachelor of Public Administration
Brig	Brigadier
BS	Bachelor of Science; Bachelor of Surgery
BSA	Bachelor of Scientific Agriculture
BT	Baronet
Bucks	Buckinghamshire
c	child, children; cuadra (block)
c	*circa*
C	Celsius
C BIOL	Chartered Biologist
C CHEM	Chartered Chemist
C ENG	Chartered Engineer
C LIT	Companion of Literature
C PHYS	Chartered Physicist
C SC	Candidate of Sciences
CA	California; Chartered Accountant
CABI	Commonwealth Agricultural Bureaux International
Cambs	Cambridgeshire
Camp	Campeche
Cand	Candidate, Candidature
Capt	Captain
Cards	Cardiganshire
CARICOM	Caribbean Community and Common Market
CB	Companion of the (Order of the) Bath
CBC	Canadian Broadcasting Corporation
CBE	Commander of (the Order of) the British Empire
CBI	Confederation of British Industry
CBIM	Companion of the British Institute of Management
CBS	Columbia Broadcasting System
CC	Companion of the Order of Canada
CCP	Chinese Communist Party
Cdre	Commodore
CE	Chartered Engineer, Civil Engineer
CEAO	Communauté Economique de l'Afrique de l'Ouest
Cen	Central
CEO	Chief Executive Officer
CERN	Conseil (*now* Organisation) Européen(ne) pour la Recherche Nucléaire
CFC	chlorofluorocarbon
CGT	Confédération Général du Travail
CH	Companion of Honour
CH B	Bachelor of Surgery
CH M	Master of Surgery
Chair	Chairman, Chairperson, Chairwoman
Chem	Chemistry
Chih	Chihuahua
Chis	Chiapas
CIA	Central Intelligence Agency
Cia	Compagnia, Compañia (Company)
Cie	Compagnie (Company)
CIEE	Companion of the Institution of Electrical Engineers
C-in-C	Commander-in-Chief
circ	circulation
CIS	Commonwealth of Independent States
CLD	Doctor of Civil Law (USA)
CM	Master in Surgery
CMG	Companion of (the Order of) St Michael and St George
CNAA	Council for National Academic Awards
CNRS	Centre National de la Recherche Scientifique
Co	Company; County
CO	Colorado; Commanding Officer
Coah	Coahuila
Col	Colima; Colonel; Colonia
Coll	College
Comm	Commission
Commd	Commissioned
Commdg	Commanding
Commdr	Commander, Commandeur
Commdt	Commandant
Commr	Commissioner
Conf	Conference
Confed	Confederation
Cons	Conservative; Conservative and Unionist Party
Contrib	Contribution, Contributor
Corp	Corporate
Corpn	Corporation
Corresp	Correspondent, Corresponding
CP	Communist Party; Caixa Postal/Case Postale (Post Office Box)
CPA	Certified Public Accountant; Commonwealth Parliamentary Association
CPSU	Communist Party of the Soviet Union
cr	created
CSCE	Conference on Security and Co-operation in Europe
CSIRO	Commonwealth Scientific and Industrial Research Organization
CT	Connecticut
Cttee	Committee
CVO	Commander of the Royal Victorian Order
d	daughter(s)
D ARCH	Doctor of Architecture
D CN L	Doctor of Canon Law
D COMM	Doctor of Commerce
D ECON	Doctor of Economics
D EN D	Docteur en Droit
D EN MED	Docteur en Médecine
D ENG	Doctor of Engineering
D ÈS L	Docteur ès Lettres
D ÈS SC	Docteur ès Sciences
D HIST	Doctor of History
D HUM LITT	Doctor of Humane Letters
D IUR	Doctor of Law
D IUR UTR	Doctor of both Civil and Canon Law
D JUR	Doctor of Law
D LIT(T)	Doctor of Letters; Doctor of Literature
D(R) MED	Doctor of Medicine
D MIN SCI	Doctor of Municipal Science
D MUS	Doctor of Music
D PHIL	Doctor of Philosophy
D SC	Doctor of Science
D SC S	Doctor of Social Science
D TECH	Doctor of Technology
D THEOL	Doctor of Theology
DB	Bachelor of Divinity
DBA	Doctor of Business Administration
DC	District of Columbia
DCL	Doctor of Civil Law
DCS	Doctor of Commercial Sciences
DCT	Doctor of Christian Theology
DD	Doctor of Divinity
DDS	Doctor of Dental Surgery
DE	Delaware
Dec	December
Del	Delegate, delegation
Denbighs	Denbighshire
Dept	Department
Desig	Designate
Devt	Development
DF	Distrito Federal
DFA	Doctor of Fine Arts
Dgo	Durango

DH Doctor of Humanities
DHL Doctor of Hebrew Literature
DIC Diploma of Imperial College
DIP AD Diploma in Art and Design
DIP AGR Diploma in Agriculture
DIP ED Diploma in Education
DIP(L) ENG Diploma in Engineering
DIP GEOG Diploma in Geography
Dir Director
Dist District
Div Division; Divisional
DLS Doctor of Library Science
DM Doctor of Medicine (Oxford)
DMD Doctor of Dental Medicine
DMS Director of Medical Services
DMV Doctor of Veterinary Medicine
DN Distrito National
DO Doctor of Ophthalmology
DPH Diploma in Public Health
DPM Diploma in Psychological Medicine
Dr Doctor
DR AGR Doctor of Agriculture
DR ING Doctor of Engineering
DR IUR Doctor of Laws
DR JUR Doctor of Laws
DR OEC (PUBL) Doctor of (Public) Economy
DR RER NAT Doctor of Natural Sciences
DR RER POL Doctor of Political Science
DR SC Doctor of Sciences
DR SC NAT Doctor of Natural Sciences
DS Doctor of Science
DST Doctor of Sacred Theology
DTM (& H) Diploma in Tropical Medicine (and Hygiene)
DUP Diploma of the University of Paris

E East, Eastern
EC European Communities
ECA Economic Co-operation Administration; Economic Commission for Africa
ECAFE Economic Commission for Asia and the Far East
ECE Economic Commission for Europe
ECLAC Economic Commission for Latin America and the Caribbean
Econs(s) Economic(s)
ECOSOC Economic and Social Council
ed edited, editor
Ed Editor
ED Doctor of Engineering (USA)
ED D Doctor of Education
ED M Master of Education
Edif Edifice, Edificio (Building)
Edn Edition
Educ Education
EEC European Economic Community
EFTA European Free Trade Association
e.g. exempli gratia (for example)
EIA Environmental Impact Assessment
EIB European Investment Bank
EM Master of Engineering (USA)
Emer Emeritus
Eng Engineering
ENG D Doctor of Engineering
EP European Parliament
EPA Environmental Protection Agency
ESCAP Economic and Social Commission for Asia and the Pacific
ESCWA Economic and Social Commission for Western Asia
esq esquina (corner)
est established
ETH Eidgenössische Technische Hochschule (Swiss Federal Institute of Technology)
EU European Union
EURATOM European Atomic Energy Community
Exec Executive
Exhbn Exhibition
Ext Extension

f founded, founder
F ENG Fellow, Fellowship of Engineering
FAA Fellow of Australian Academy of Science
FAATS Fellow of Australian Academy of Technological Sciences
FACC Fellow of the American College of Cardiology
FACCA Fellow of the Association of Certified and Corporate Accountants
FACE Fellow of the Australian College of Education
FACP Fellow of American College of Physicians
FACS Fellow of the American College of Surgeons
FAHA Fellow Australian Academy of the Humanities
FAIA Fellow of the American Institute of Architects
FAIAS Fellow of the Australian Institute of Agricultural Science
FAIM Fellow of the Australian Institute of Management
FAO Food and Agriculture Organization
FASE Fellow of Antiquarian Society, Edinburgh
FASSA Fellow Academy of Social Sciences of Australia
Fax Facsimile
FBA Fellow of the British Academy
FBIM Fellow of the British Institute of Management
FBIP Fellow of the British Institute of Physics
FCA Fellow of the Institute of Chartered Accountants
FCAE Fellow Canadian Academy of Engineering
FCGI Fellow of the City and Guilds of London Institute
FCIA Fellow Chartered Institute of Arbitrators
FCIB Fellow Chartered Institute of Bankers
FCIC Fellow of the Chemical Institute of Canada
FCIS Fellow of the Chartered Institute of Secretaries
FCIWEM Fellow of the Chartered Institution of Water Environmental Management
FCO Foreign and Commonwealth Office
FCSD Fellow Chartered Society of Designers
FCT Federal Capital Territory
FCWA Fellow of the Institute of Cost and Works Accountants
Feb February
Fed Federal, Federation
FFCM Fellow of Faculty of Community Medicine
FGS Fellow of the Geological Society
FGSM Fellow of the Guildhall School of Music
FIA Fellow of the Institute of Actuaries
FIAL Fellow of the International Institute of Arts and Letters
FIAM Fellow of the International Academy of Management
FIAMS Fellow of the Indian Academy of Medical Sciences
FIARB Fellow of the Institute of Arbitrators
FIB Fellow of the Institute of Bankers
FIBA Fellow of the Institute of Banking Associations
FIBIOL Fellow of the Institution of Biologists
FICE Fellow of the Institution of Civil Engineers
FICHEME Fellow of the Institute of Chemical Engineers
FID Fellow of the Institute of Directors
FIE Fellow of the Institute of Engineers
FIEE Fellow of the Institution of Electrical Engineers
FIEEE Fellow of the Institute of Electrical and Electronics Engineers
FIJ Fellow of the Institute of Journalists
FIL LIC Licentiate in Philosophy

FIM Fellow of the Institute of Metallurgists
FIME Fellow of the Institute of Mining Engineers
FIMECHE Fellow of the Institute of Mechanical Engineers
FIMI Fellow of the Institute of the Motor Industry
FINSTF Fellow of the Institute of Fuel
FINSTM Fellow of the Institute of Marketing
FINSTP Fellow of the Institute of Physics
FINSTPET Fellow of the Institute of Petroleum
FIPM Fellow of the Institute of Personnel Management
FIRE Fellow of the Institution of Radio Engineers
FITD Fellow of the Institute of Training and Development
FL Florida
FLA Fellow of the Library Association
FLS Fellow of the Linnaean Society
fmr former
fmrly formerly
FNI Fellow of the National Institute of Sciences of India
FNZIA Fellow of the New Zealand Institute of Architects
FoE Friends of the Earth
FRACO Fellow of the Royal Australian College of Ophthalmologists
FRACP Fellow of the Royal Australasian College of Physicians
FRACS Fellow of the Royal Australasian College of Surgeons
FRAES Fellow of the Royal Aeronautical Society
FRAI Fellow of the Royal Anthropological Institute
FRAIA Fellow of the Royal Australian Institute of Architects
FRAIC Fellow of the Royal Architectural Institute of Canada
FRAM Fellow of the Royal Academy of Music
FRAS Fellow of the Royal Asiatic Society; Fellow of the Royal Astronomical Society
FRBS Fellow of the Royal Society of British Sculptors
FRCM Fellow of the Royal College of Music
FRCO Fellow of the Royal College of Organists
FRCOG Fellow of the Royal College of Obstetricians and Gynaecologists
FRCP(E) Fellow of the Royal College of Physicians (Edinburgh)
FRCPI Fellow of the Royal College of Physicians of Ireland
FRCP(UK) Fellow of the Royal College of Physicians (United Kingdom)
FRCS(E) Fellow of the Royal College of Surgeons (Edinburgh)
FRECONS Fellow of the Royal Economic Society
FRES Fellow of the Royal Entomological Society
FRFPS Fellow of the Royal Faculty of Physicians and Surgeons
FRGS Fellow of the Royal Geographical Society
FRHISTS Fellow of the Royal Historical Society
FRHORTS Fellow of the Royal Horticultural Society
FRIBA Fellow of the Royal Institute of British Architects
FRIC Fellow of the Royal Institute of Chemists
FRICS Fellow of the Royal Institute of Chartered Surveyors
FRMETSOC Fellow of the Royal Meteorological Society
FRPS Fellow of the Royal Photographic Society
FRS Fellow of the Royal Society
FRSA Fellow of the Royal Society of Arts
FRSAMD Fellow of the Royal Scottish Academy of Music and Drama
FRSC Fellow of the Royal Society of Canada; Fellow of the Royal Society of Chemistry
FRSE Fellow of the Royal Society of Edinburgh
FRSL Fellow of the Royal Society of Literature
FRSM Fellow of the Royal Society of Medicine
FRSNZ Fellow of the Royal Society of New Zealand
FRSS Fellow of the Royal Statistical Society
FRSSA Fellow of the Royal Society of South Africa
FSA Fellow of the Society of Antiquaries
FSIAD Fellow of the Society of Industrial Artists and Designers
FTI Fellow of the Textile Institute
FTS Fellow of Technological Sciences
FWAAS Fellow of the World Academy of Arts and Sciences
FZS Fellow of the Zoological Society

GA Georgia
GATT General Agreement on Tariffs and Trade
GB Great Britain
Gen General
GIS Geographic Information System
Glam Glamorganshire
Glos Gloucestershire
GmbH Gesellschaft mit beschränkter Haftung (Limited Liability Company)
GOC (in C) General Officer Commanding (in Chief)
Gov Governor
Govt Government
GPO General Post Office
Grad Graduate
Gro Guerrero
GRSM Graduate of the Royal School of Music
GSO General Staff Officer
Gto Guanajuato

ha hectare(s)
HABITAT United Nations Centre for Human Settlements
Hants Hampshire
hc *honoris causa*
HE His Eminence; His (or Her) Excellency
Herefords Herefordshire
Herts Hertfordshire
Hgo Hidalgo
HH His (or Her) Highness
HI Hawaii
HLD Doctor of Humane Letters
HM His (or Her) Majesty
Hon Honorary; Honourable
Hosp Hospital
HQ Headquarters
HRH His (or Her) Royal Highness
Hunts Huntingdonshire

IA Iowa
IAEA International Atomic Energy Agency
IATA International Air Transport Association
IBRD International Bank for Reconstruction and Development (World Bank)
ICAO International Civil Aviation Organization
ICC International Chamber of Commerce; International Computing Centre
ICE Institution of Civil Engineers
ICFTU International Confederation of Free Trade Unions
ICI Imperial Chemical Industries
ICJ International Court of Justice
ICOM International Council of Museums
ICSC International Civil Service Commission
ICSU International Council of Scientific Unions, International Council for Science
ID Idaho
IDA International Development Association
IDB Inter-American Development Bank
IEE Institution of Electrical Engineers

IEEE	Institution of Electrical and Electronic Engineers
IFAD	International Fund for Agricultural Development
IFC	International Finance Corporation
IFRB	International Frequency Registration Board
IL	Illinois
ILO	International Labour Organization
IMCO	Inter-Governmental Maritime Consultative Organization
IMECHE	Institution of Mechanical Engineers
IMF	International Monetary Fund
IMO	International Maritime Organization
IN	Indiana
Inc	Incorporated
Ind	Independent
Insp	Inspector
Inst	Institut; Institute; Institution; Instituto
INSTRAW	International Research and Training Institute for the Advancement of Women
Int	International
INTELSAT	International Telecommunications Satellite Organization
IOC	International Olympic Committee
IPU	Inter-Parliamentary Union
ITU	International Telecommunications Union
IUCN	World Conservation Union
IUPAC	International Union of Pure and Applied Chemistry
IUPAP	International Union of Pure and Applied Physics
Jal	Jalisco
Jan	January
JCB	Bachelor of Canon Law
JCD	Doctor of Canon Law
JD	Doctor of Jurisprudence
JP	Justice of the Peace
Jr	Junior
Jt	Joint
jtly	jointly
JU D	Doctor of Law
JU DR	Doctor of Law, *Juris utriusque Doctor* (Doctor of both Civil and Canon Law)
JUD	*Juris utriusque Doctor* (Doctor of both Civil and Canon Law)
KBE	Knight of the British Empire
KCB	Knight Commander of (the Order of) the Bath
kg	kilograms(s)
KG	Knight of (the Order of) the Garter
km	kilometre(s)
KS	Kansas
KT	Knight of (the Order of) the Thistle
KY	Kentucky
L EN D	Licencié en Droit
L ÈS L	Licencié ès Lettres
L ÈS SC	Licencié ès Sciences
L PH	Licentiate of Philosophy
L TH	Licentiate in Theology
LA	Los Angeles; Louisiana
Lab	Laboratory; Labour; Labour Party
Lancs	Lancashire
LDS	Licentiate in Dental Surgery
Legis	Legislative
Leics	Leicestershire
LHD	Doctor of Humane Letters
LIC EN DER	Licenciado en Derecho
LIC EN FIL	Licenciado en Filosofía
LIC MED	Licentiate in Medicine
Lieut	Lieutenant
Lincs	Lincolnshire
LITT D	Doctor of Letters
LL B	Bachelor of Laws
LL D	Doctor of Laws
LL L	Licentiate of Laws
LL M	Master of Laws
LM	Licentiate of Medicine, or Midwifery
LORCS	League of Red Cross and Crescent Societies
LRAM	Licentiate of the Royal Academy of Music
LRCP	Licentiate of the Royal College of Physicians
LSE	London School of Economics
Ltd(a)	Limited; Limitada
m	marriage, married; metre(s)
m.	million
M AGR	Master of Agriculture (USA)
M ARCH	Master of Architecture
M CH	Master of Surgery
M CH D	Master of Dental Surgery
M COM(M)	Master of Commerce
M DIV	Master of Divinity
M ED	Master of Education
M ENG	Master of Engineering (Dublin)
M PH	Master of Philosophy (USA)
M SC	Master of Science
M SC S	Master of Social Science
MA	Massachusetts; Master of Arts
Maj	Major
Man	Management, Manager, Managing
Math	Mathematical, Mathematics
MB	Bachelor of Medicine; Manitoba
MBA	Master of Business Administration
MBE	Member of (the Order of) the British Empire
MCE	Master of Civil Engineering
MCL	Master of Civil Law
MDS	Doctor of Medicine; Maryland
MDS	Master of Dental Surgery
ME	Maine
mem	member
MEP	Member of the European Parliament
Méx	México
MFA	Master of Fine Arts
Mfg	Manufacturing
Mfrs	Manufacturers
MI	Michigan
MIA	Master of International Affairs
MICE	Member of the Institution of Civil Engineers
Mich	Michoacán
MICHEME	Member of the Institution of Chemical Engineers
Middx	Middlesex
MIEE	Member of the Institution of Electrical Engineers
MIGA	Multilateral Investment Guarantee Agency
Mil	Military
MIMARE	Member of the Institute of Marine Engineers
MIMECHE	Member of the Institution of Mechanical Engineers
MIMINE	Member of the Institution of Mining Engineers
MINSTT	Member of the Institute of Transport
MISTRUCTE	Member of the Institution of Structural Engineers
MIT	Massachusetts Institute of Technology
MJ	Master of Jurisprudence
MN	Minnesota
MO	Missouri
Mon	Monmouthshire
Mor	Morelos
Movt	Movement
MP	Member of Parliament
MPA	Master of Public Administration (Harvard)
MRAS	Member of the Royal Asiatic Society
MRC	Medical Research Council
MRCP(E)	Member of the Royal College of Physicians (Edinburgh)

MRCP(UK)	Member of the Royal College of Physicians (United Kingdom)
MRCS(E)	Member of the Royal College of Surgeons (Edinburgh)
MRCVS	Member of the Royal College of Veterinary Surgeons
MRI	Member of the Royal Institution
MRIA	Member of the Royal Irish Academy
MRIC	Member of the Royal Institute of Chemistry
MS	Master of Science; Master of Surgery; Mississippi
MT	Montana
MTS	Master of Theological Studies
MU DR	Doctor of Medicine
MUS BAC *or* B	Bachelor of Music
MUS DOC *or* D	Doctor of Music
MUS M	Master of Music (Cambridge)
MVD	Master of Veterinary Medicine
N	North, Northern
n/a	not available
NAS	National Academy of Sciences (USA)
NASA	National Aeronautical and Space Administration
Nat	National
NATO	North Atlantic Treaty Organization
Nay	Nayarit
NB	New Brunswick
NBC	National Broadcasting Corporation
NC	North Carolina
NCD	National Capital District (Papua New Guinea)
ND	North Dakota
NE	Nebraska; North East
NF	Newfoundland
NGO	Non-Governmental Organization
NH	New Hampshire
NJ	New Jersey
NL	Nuevo León
NM	New Mexico
no	number, numero
Northants	Northamptonshire
Notts	Nottinghamshire
Nov	November
nr	near
NS	Nova Scotia
NSW	New South Wales
NT	Northern Territory
NV	Nevada
NW	North West
NWT	North West Territories
NY	New York
NZ	New Zealand
NZIC	New Zealand Institute of Chemistry
OAPEC	Organization of Arab Petroleum Exporting Countries
OAS	Organization of American States
OAU	Organization of African Unity
Oax	Oaxaca
OBE	Officer of (the Order of) the British Empire
Oct	October
OECD	Organisation for Economic Co-operation and Development
Of	Oficina (Office)
OH	Ohio
OK	Oklahoma
OM	Member, Order of Merit
ON	Ontario
OPEC	Organization of the Petroleum Exporting Countries
OR	Oregon
Org	Organization
Oxon	Oxfordshire

PA	Pennsylvania
Parl	Parliament, Parliamentary
PC	Privy Councillor
PD B	Bachelor of Pedagogy
PD D	Doctor of Pedagogy
PD M	Master of Pedagogy
PDF	portable document format
PE	Prince Edward Island
Pembs	Pembrokeshire
per	pereulok (lane, alley)
Perm	Permanent
PH B	Bachelor of Philosophy
PH D	Doctor of Philosophy
PH DR	Doctor of Philosophy
PH L	Licentiate of Philosophy
PHARM D	Docteur en Pharmacie
pl	place, platz, ploschad (square)
PLC	Public Limited Company
PMB	Private Mail Bag
PO(B)	Post Office (Box)
POW	Prisoner of War
PQ	Province of Quebec
pr	prospekt (avenue)
Pres	President
Prin	Principal
Prof	Professor
Prov	Province, Provincial
Pty	Proprietary
Publ(s)	Publication(s)
Publr	Publisher
Pue	Puebla
Pvt	Private
Q Roo	Quintana Roo
QC	Queen's Counsel
Qro	Querétaro
RA	Royal Academy
RACP	Royal Australasian College of Physicians
RAF	Royal Air Force
RAM	Royal Academy of Music
RCA	Radio Corporation of America; Royal Canadian Academy; Royal College of Art
RCRA	Resource Conservation and Recovery Act
Rd	Road
Regt	Regiment
Rep	Representative, Represented
Repub	Republic
retd	retired
Rev	Reverend
RI	Rhode Island
RIBA	Royal Institute of British Architects
RNZAF	Royal New Zealand Air Force
RS DR	Doctor of Social Sciences
RSA	Royal Scottish Academy; Royal Society of Arts
RSL	Royal Society of Literature
RSPB	Royal Society for the Protection of Birds
Rt Hon	Right Honourable
Rt Rev	Right Reverend
s	son(s)
S	South, Southern
SA	Sociedad Anónima, Société Anonyme; South Africa
SAE	Society of Aeronautical Engineers
Salop	Shropshire
SALT	Strategic Arms Limitation Treaty
SB	Bachelor of Science (USA)
SC	South Carolina
SC B	Bachelor of Science
SC D	Doctor of Science
SD	South Dakota
SE	South East
SEATO	South East Asia Treaty Organization

Sec	Secretary
Secr	Secretariat
Sept	September
Sin	Sinaloa
SIPRI	Stockholm International Peace Research Institute
SJD	Doctor of Juristic Science
SK	Saskatchewan
SLP	San Luis Potosi
SM	Master of Science
Soc	Société, Society
Sok	Sokak (street)
Son	Sonora
SPA	Società per Azioni
Sq	Square
Sr	Senior
St	Saint; Street
Staffs	Staffordshire
STB	Bachelor of Sacred Theology
STD	Doctor of Sacred Theology
STL	Licentiate of Sacred Theology
STM	Master of Sacred Theology
str	strasse
Supt	Superintendent
SW	South West
Tab	Tabasco
Tamps	Tamaulipas
Tech	Technical, Technology
Tel	Telephone
Temp	Temporary
TH B	Bachelor of Theology
TH D	Doctor of Theology
TH DR	Doctor of Theology
TH M	Master of Theology
Tlax	Tlaxcala
TN	Tennessee
Trans	Translation, Translator
Treas	Treasurer
TU(C)	Trades Union (Congress)
TV	Television
TX	Texas
u	utca (street)
UAE	United Arab Emirates
UED	University Education Diploma
UK	United Kingdom (of Great Britain and Northern Ireland)
UKAEA	United Kingdom Atomic Energy Authority
ul	ulitsa (street)
UMIST	University of Manchester Institute of Science and Technology
UN(O)	United Nations (Organization)
UNA	United Nations Association
UNCED	United Nations Conference on Environment and Development (Earth Summit)
UNCITRAL	United Nations Commission on International Trade Law
UNCTAD	United Nations Conference on Trade and Development
UNDP	United Nations Development Programme
UNDRO	United Nations Disaster Relief Office
UNEP	United Nations Environment Programme
UNESCO	United Nations Educational, Scientific and Cultural Organization
UNFPA	United Nations Fund for Population Activities
UNHCR	United Nations High Commissioner for Refugees
UNICEF	United Nations International Children's Emergency Fund
UNIDIR	United Nations Institute for Disarmament Research
UNIDO	United Nations Industrial Development Organization
UNIFEM	United Nations Development Fund for Women
UNITAR	United Nations Institute for Training and Research
Univ	University
UNPF	United Nations Population Fund
UNRISD	United Nations Research Institute for Social Development
UNRRA	United Nations Relief and Rehabilitation Administration
UNRWA	United Nations Relief and Works Agency
UNSCEAR	Secretariat of the United Nations Scientific Committee on the Effects of Atomic Radiation
UNU	United Nations University
UNV	United Nations Volunteers
UPU	Universal Postal Union
US(A)	United States (of America)
USAF	United States Air Force
USSR	Union of Soviet Socialist Republics
UT	Utah
VA	Virginia
Ver	Veracruz
Vic	Victoria
Vol(s)	Volume(s)
VT	Vermont
vul	street (Ukraine)
W	West, Western
WA	Washington (State); Western Australia
Warwicks	Warwickshire
WEU	Western European Union
WFC	World Food Council
WFP	World Food Programme
WFTU	World Federation of Trade Unions
WHO	World Health Organization
WI	Wisconsin
WIDER	World Institute for Development Economics Research
Wilts	Wiltshire
WIPO	World Intellectual Property Organization
WMO	World Meteorological Organization
Worcs	Worcestershire
WTO	World Tourism Organization; World Trade Organization
WV	West Virginia
WWF	World Wide Fund for Nature, World Wildlife Fund (USA)
WY	Wyoming
Yorks	Yorkshire
YT	Yukon Territory
Yuc	Yucatán
Zac	Zacatecas

INTERNATIONAL TELEPHONE CODES

To make international calls to telephone and fax numbers listed in the book, dial the international code of the country from which you are calling, followed by the appropriate code for the country you wish to call (listed below), followed by the area code (if applicable) and telephone or fax number listed in the entry.

	Country code	+ or – GMT*
Afghanistan	93	+4½
Albania	355	+1
Algeria	213	+1
Andorra	376	+1
Angola	244	+1
Antigua and Barbuda	1 268	–4
Argentina	54	–3
Armenia	374	+4
Australia	61	+8 to +10
Australian External Territories:		
Australian Antarctic Territory	672	+3 to +10
Christmas Island	61	+7
Cocos (Keeling) Islands	61	+6½
Norfolk Island	672	+11½
Austria	43	+1
Azerbaijan	994	+5
The Bahamas	1 242	–5
Bahrain	973	+3
Bangladesh	880	+6
Barbados	1 246	–4
Belarus	375	+2
Belgium	32	+1
Belize	501	–6
Benin	229	+1
Bhutan	975	+6
Bolivia	591	–4
Bosnia and Herzegovina	387	+1
Botswana	267	+2
Brazil	55	–3 to –4
Brunei	673	+8
Bulgaria	359	+2
Burkina Faso	226	0
Burundi	257	+2
Cambodia	855	+7
Cameroon	237	+1
Canada	1	–3 to –8
Cape Verde	238	–1
The Central African Republic	236	+1
Chad	235	+1
Chile	56	–4
China, People's Republic	86	+8
Special Administrative Regions:		
Hong Kong	852	+8
Macao	853	+8
China (Taiwan)	886	+8
Colombia	57	–5
The Comoros	269	+3
Congo, Democratic Republic	243	+1
Congo, Republic	242	+1
Costa Rica	506	–6
Côte d'Ivoire	225	0
Croatia	385	+1
Cuba	53	–5
Cyprus	357	+2
'Turkish Republic of Northern Cyprus'	90 392	+2
Czech Republic	420	+1
Denmark	45	+1
Danish External Territories:		
Faroe Islands	298	0
Greenland	299	–1 to –4
Djibouti	253	+3
Dominica	1 767	–4
Dominican Republic	1 809	–4
Ecuador	593	–5
Egypt	20	+2
El Salvador	503	–6
Equatorial Guinea	240	+1
Eritrea	291	+3
Estonia	372	+2

	Country code	+ or – GMT*
Ethiopia	251	+3
Fiji	679	+12
Finland	358	+2
Finnish External Territory:		
Åland Islands	358	+2
France	33	+1
French Overseas Departments:		
French Guiana	594	–3
Guadeloupe	590	–4
Martinique	596	–4
Réunion	262	+4
French Overseas Collectivité Départementale:		
Mayotte	269	+3
Overseas Collectivité Territoriale:		
Saint Pierre and Miquelon	508	–3
French Overseas Countries:		
French Polynesia	689	–9 to –10
New Caledonia	687	+11
Gabon	241	+1
Gambia	220	0
Georgia	995	+4
Germany	49	+1
Ghana	233	0
Greece	30	+2
Grenada	1 473	–4
Guatemala	502	–6
Guinea	224	0
Guinea-Bissau	245	0
Guyana	592	–4
Haiti	509	–5
Honduras	504	–6
Hungary	36	+1
Iceland	354	0
India	91	+5½
Indonesia	62	+7 to +9
Iran	98	+3½
Iraq	964	+3
Ireland	353	0
Israel	972	+2
Italy	39	+1
Jamaica	1 876	–5
Japan	81	+9
Jordan	962	+2
Kazakhstan	7	+6
Kenya	254	+3
Kiribati	686	+12 to +13
Korea, Democratic People's Republic (North Korea)	850	+9
Korea, Republic (South Korea)	82	+9
Kuwait	965	+3
Kyrgyzstan	996	+5
Laos	856	+7
Latvia	371	+2
Lebanon	961	+2
Lesotho	266	+2
Liberia	231	0
Libya	218	+1
Liechtenstein	423	+1
Lithuania	370	+2
Luxembourg	352	+1
Macedonia, former Yugoslav republic	389	+1
Madagascar	261	+3
Malawi	265	+2
Malaysia	60	+8
Maldives	960	+5
Mali	223	0
Malta	356	+1
Marshall Islands	692	+12
Mauritania	222	0
Mauritius	230	+4
Mexico	52	–6 to –7
Micronesia, Federated States	691	+10 to +11
Moldova	373	+2
Monaco	377	+1
Mongolia	976	+7 to +9

	Country code	+ or – GMT*
Morocco	212	0
Mozambique	258	+2
Myanmar	95	$+6\frac{1}{2}$
Namibia	264	+2
Nauru	674	+12
Nepal	977	$+5\frac{3}{4}$
Netherlands	31	+1
Netherlands Dependencies:		
Aruba	297	–4
Netherlands Antilles	599	–4
New Zealand	64	+12
New Zealand's Dependent and Associated Territories:		
Tokelau	690	–10
Cook Islands	682	–10
Niue	683	–11
Nicaragua	505	–6
Niger	227	+1
Nigeria	234	+1
Norway	47	+1
Norwegian External Territory:		
Svalbard	47	+1
Oman	968	+4
Pakistan	92	+5
Palau	680	+9
Palestinian Autonomous Areas	970	+2
Panama	507	–5
Papua New Guinea	675	+10
Paraguay	595	–4
Peru	51	–5
The Philippines	63	+8
Poland	48	+1
Portugal	351	0
Qatar	974	+3
Romania	40	+2
Russian Federation	7	+2 to +12
Rwanda	250	+2
Saint Christopher and Nevis	1 869	–4
Saint Lucia	1 758	–4
Saint Vincent and the Grenadines	1 784	–4
Samoa	685	–11
San Marino	378	+1
São Tomé and Príncipe	239	0
Saudi Arabia	966	+3
Senegal	221	0
Serbia and Montenegro	381	+1
Seychelles	248	+4
Sierra Leone	232	0
Singapore	65	+8
Slovakia	421	+1
Slovenia	386	+1
Solomon Islands	677	+11
Somalia	252	+3
South Africa	27	+2
Spain	34	+1
Sri Lanka	94	+6
Sudan	249	+2
Suriname	597	–3
Swaziland	268	+2
Sweden	46	+1
Switzerland	41	+1
Syria	963	+2
Tajikistan	992	+5
Tanzania	255	+3
Thailand	66	+7
Timor-Leste	670	+9
Togo	228	0
Tonga	676	+13
Trinidad and Tobago	1 868	–4
Tunisia	216	+1
Turkey	90	+2
Turkmenistan	993	+5
Tuvalu	688	+12
Uganda	256	+3
Ukraine	380	+2
United Arab Emirates	971	+4
United Kingdom	44	0
United Kingdom Crown Dependencies	44	0
United Kingdom Overseas Territories:		
Anguilla	1 264	–4
Ascension Island	247	0
Bermuda	1 441	–4
British Virgin Islands	1 284	–4
Cayman Islands	1 345	–5
Diego Garcia (British Indian Ocean Territory)	246	+5
Falkland Islands	500	–4
Gibraltar	350	+1
Montserrat	1 664	–4
Pitcairn Islands	872	–8
Saint Helena	290	0
Tristan da Cunha	2 897	0
Turks and Caicos Islands	1 649	–5
United States of America	1	–5 to –10
United States Commonwealth Territories:		
Northern Mariana Islands	1 670	+10
Puerto Rico	1 787	–4
United States External Territories:		
American Samoa	1 684	–11
Guam	1 671	+10
United States Virgin Islands	1 340	–4
Uruguay	598	–3
Uzbekistan	998	+5
Vanuatu	678	+11
Vatican City	39	+1
Venezuela	58	–4
Viet Nam	84	+7
Yemen	967	+3
Zambia	260	+2
Zimbabwe	263	+2

* The Times listed compare the standard (winter) times in the various countries. Some countries adopt Summer (Daylight Saving) Time—i.e. +1 hour—for part of the year.

World Ozone Depletion

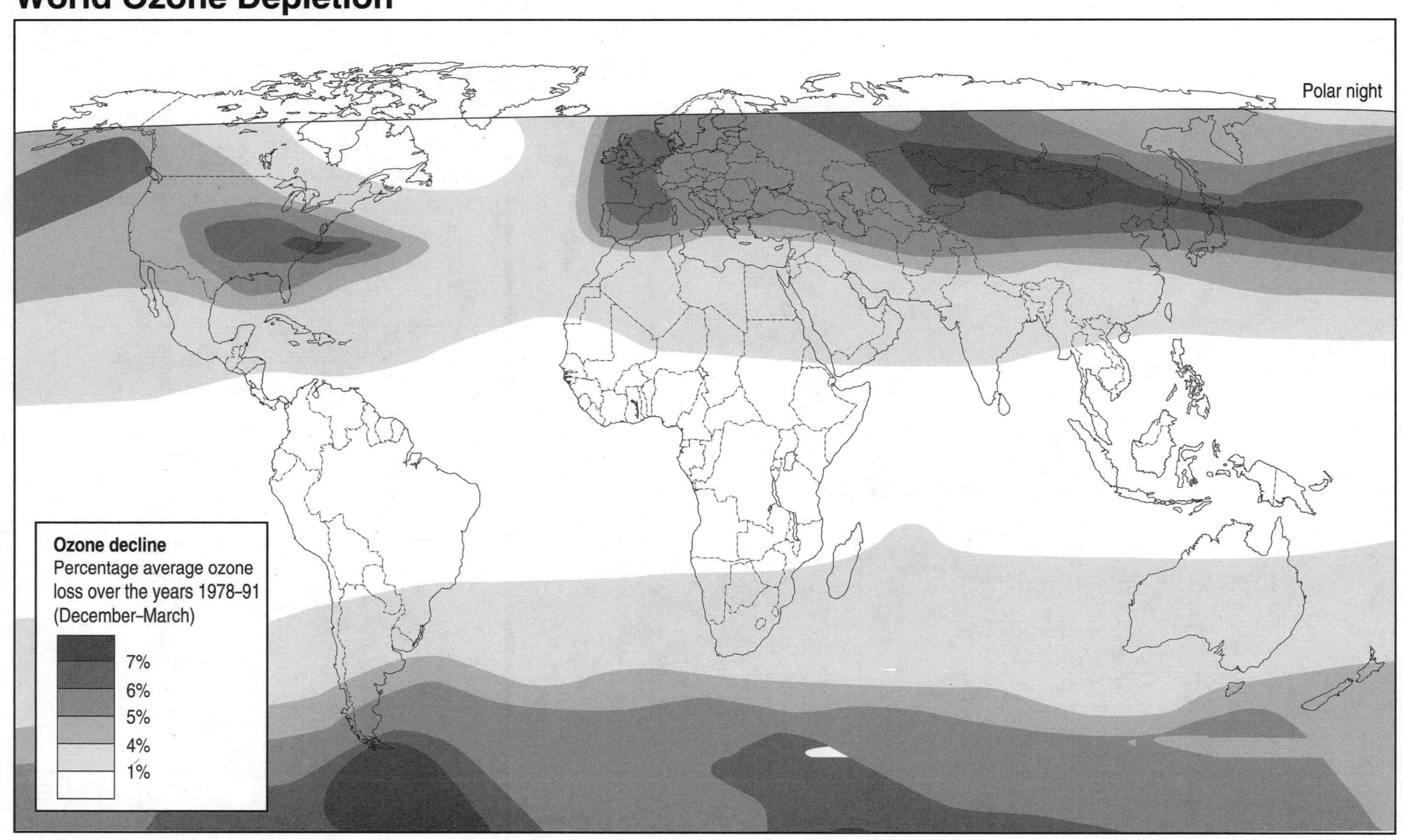

Source: *The State of the Environment Atlas,* Penguin Books, 1995

Vegetation

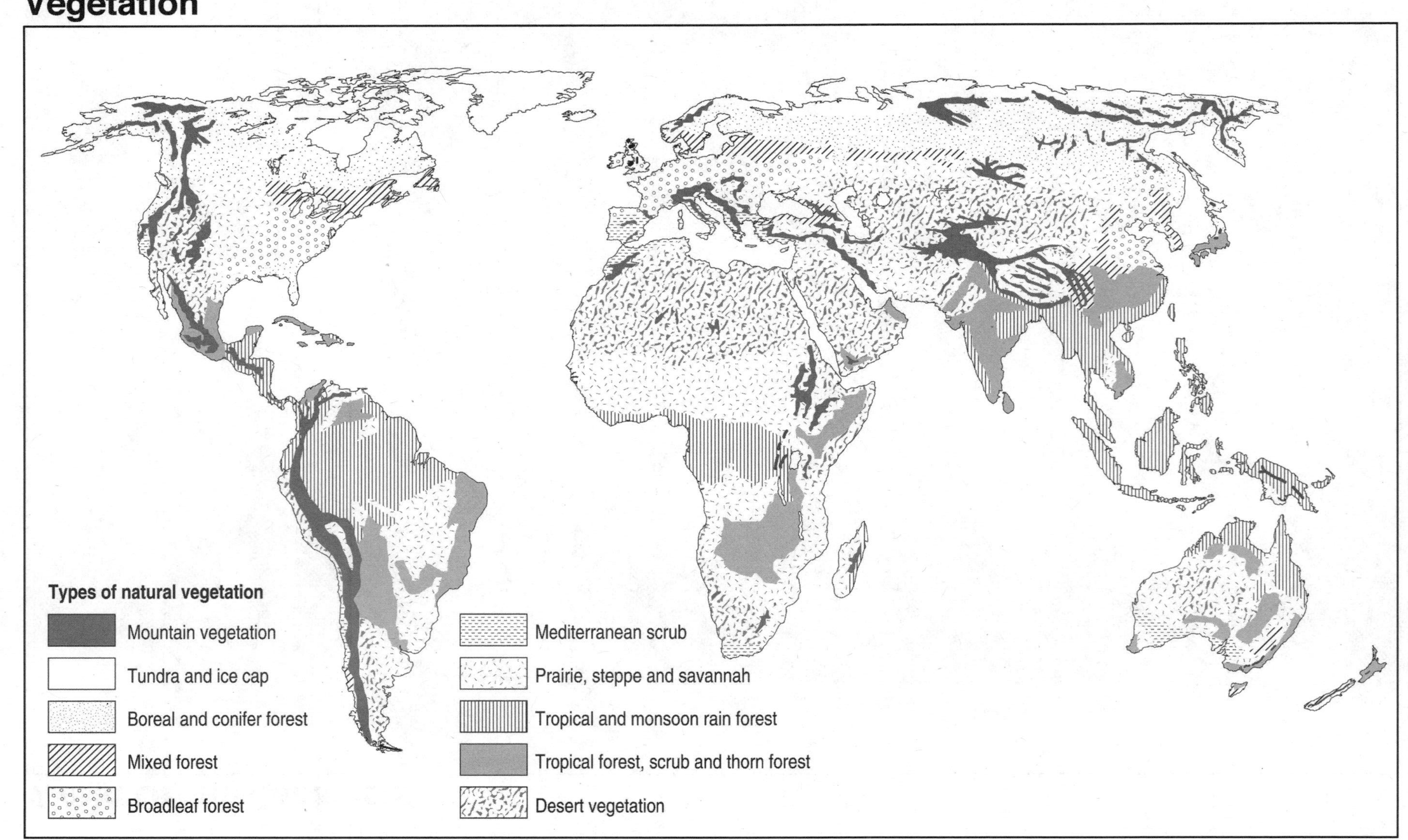

Source: *Times Atlas of the World,* © Bartholomew Ltd 2000; reproduced by permission of HarperCollins Publishers

Forest Area

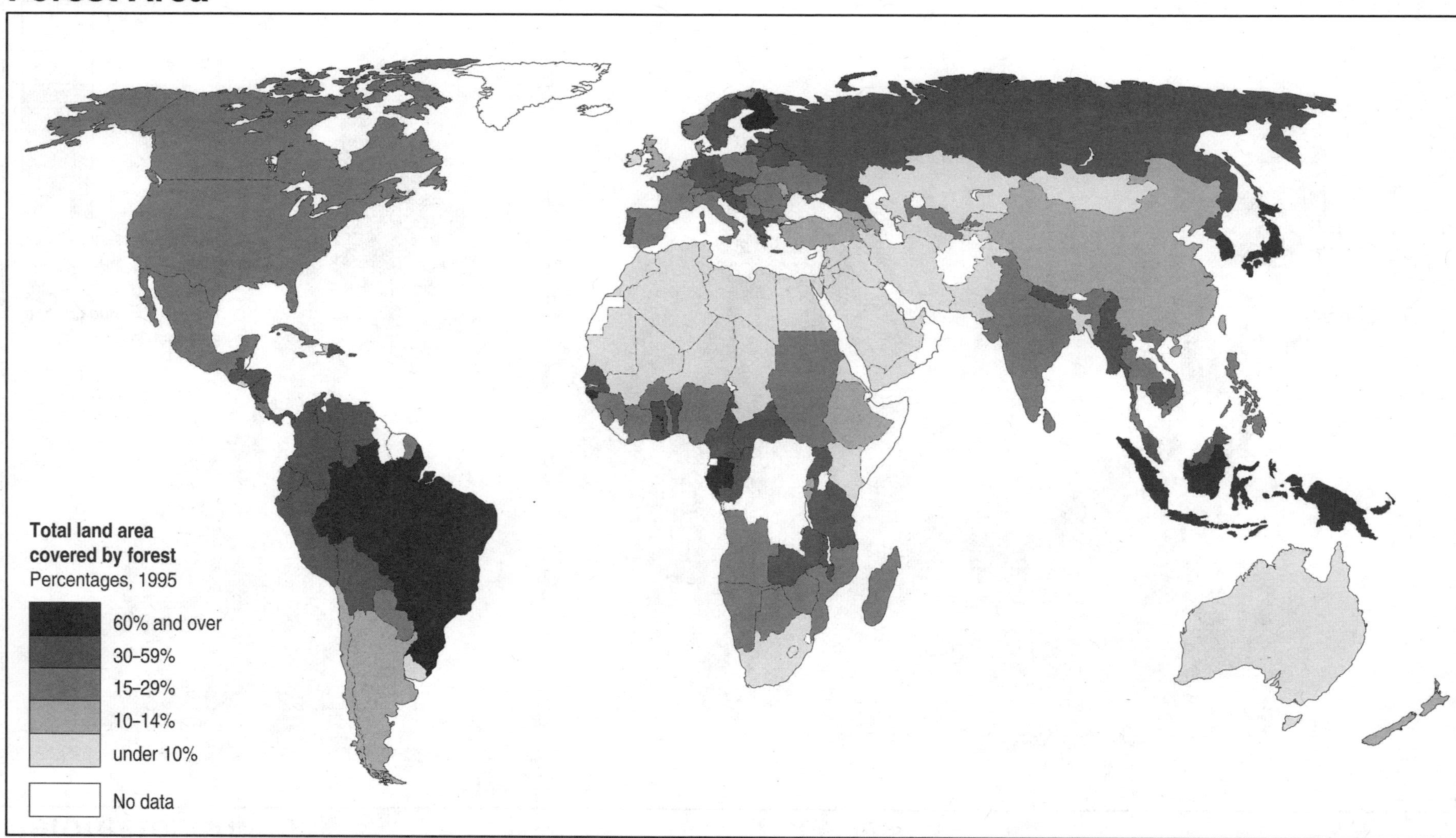
Total land area
covered by forest
Percentages, 1995
60% and over
30–59%
15–29%
10–14%
under 10%
No data

Deforestation

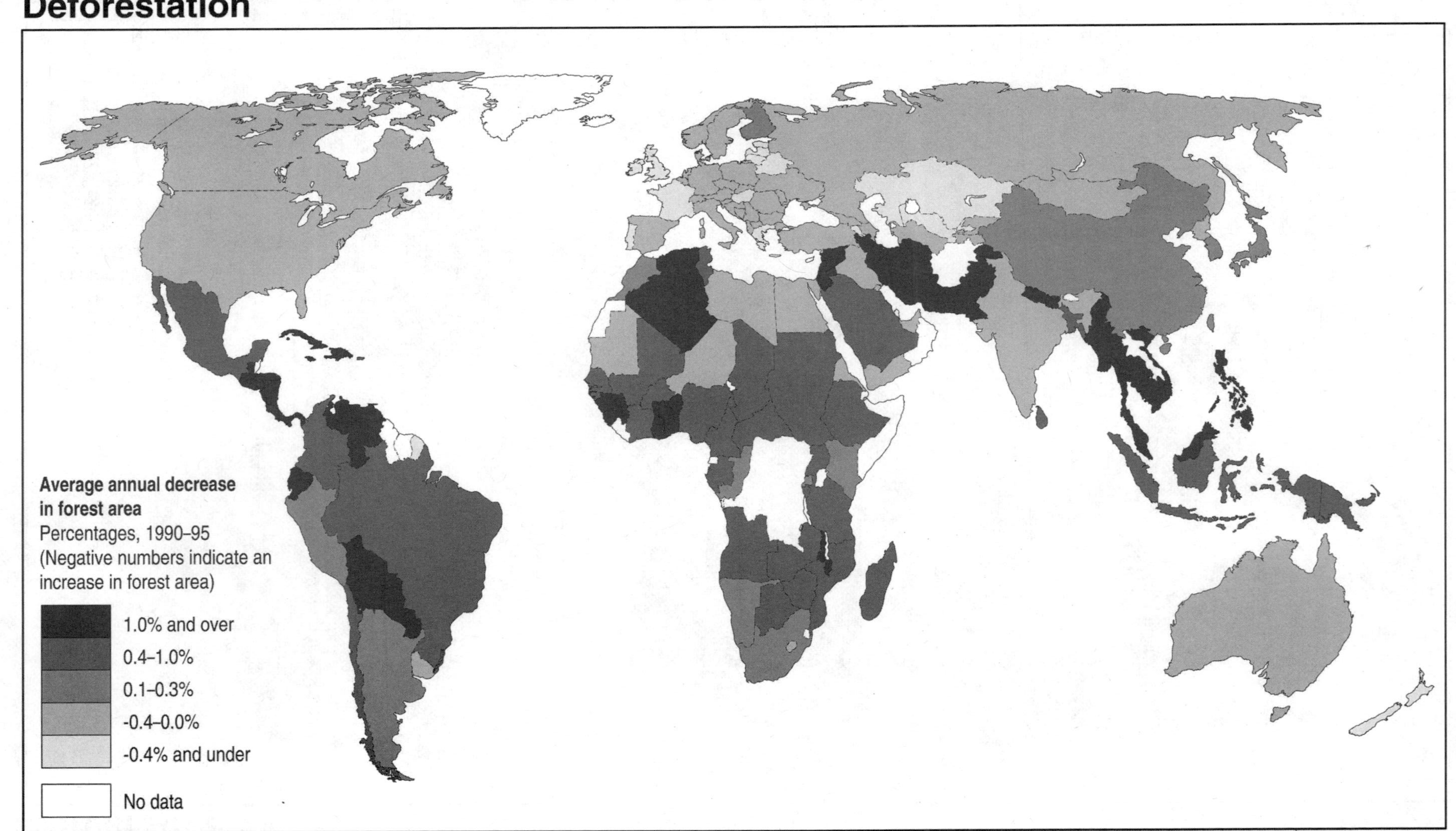
Average annual decrease
in forest area
Percentages, 1990–95
(Negative numbers indicate an
increase in forest area)
1.0% and over
0.4–1.0%
0.1–0.3%
-0.4–0.0%
-0.4% and under
No data

Rainforest Destruction

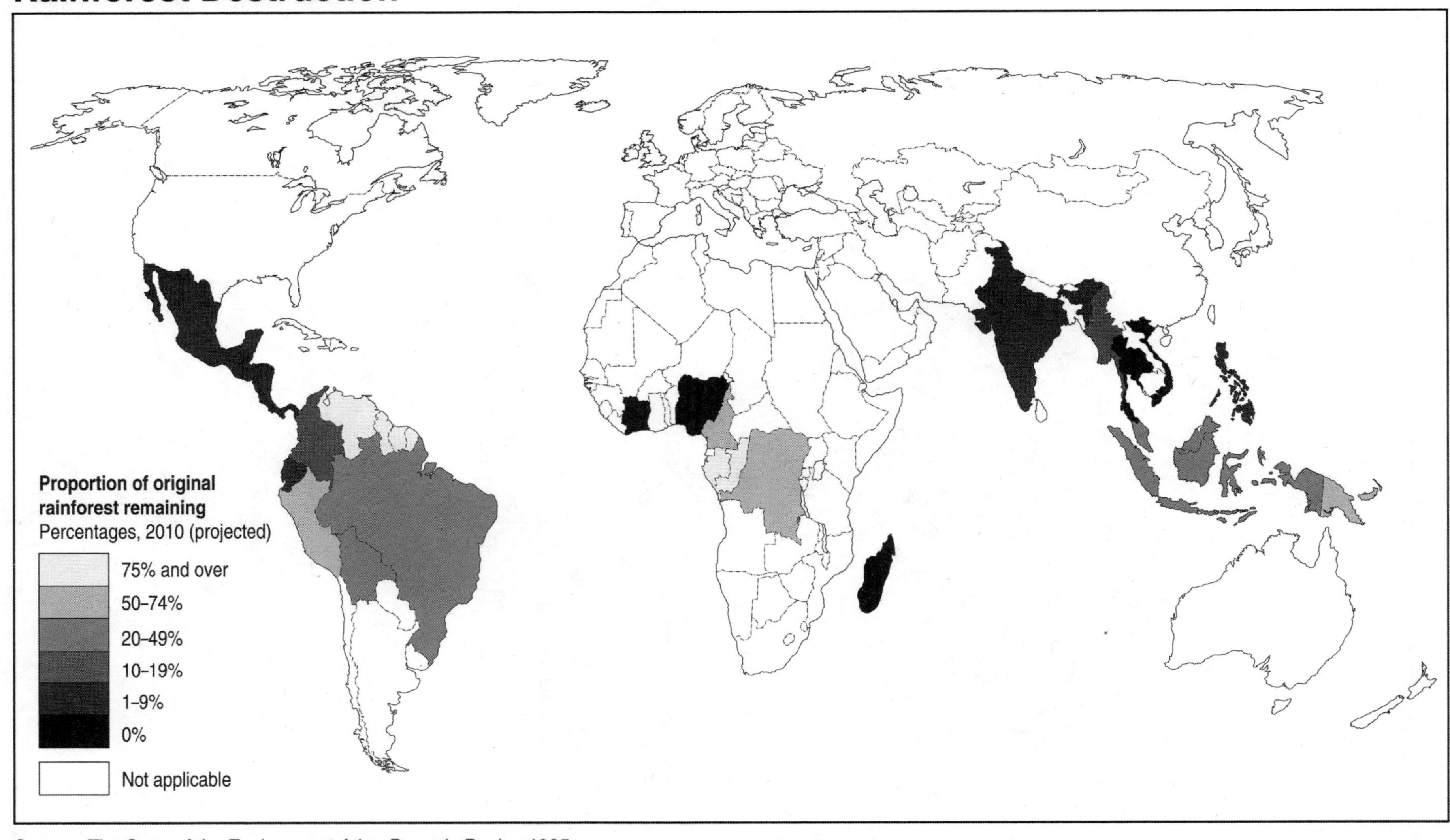

Source: *The State of the Environment Atlas,* Penguin Books, 1995

Deforestation in the Amazon

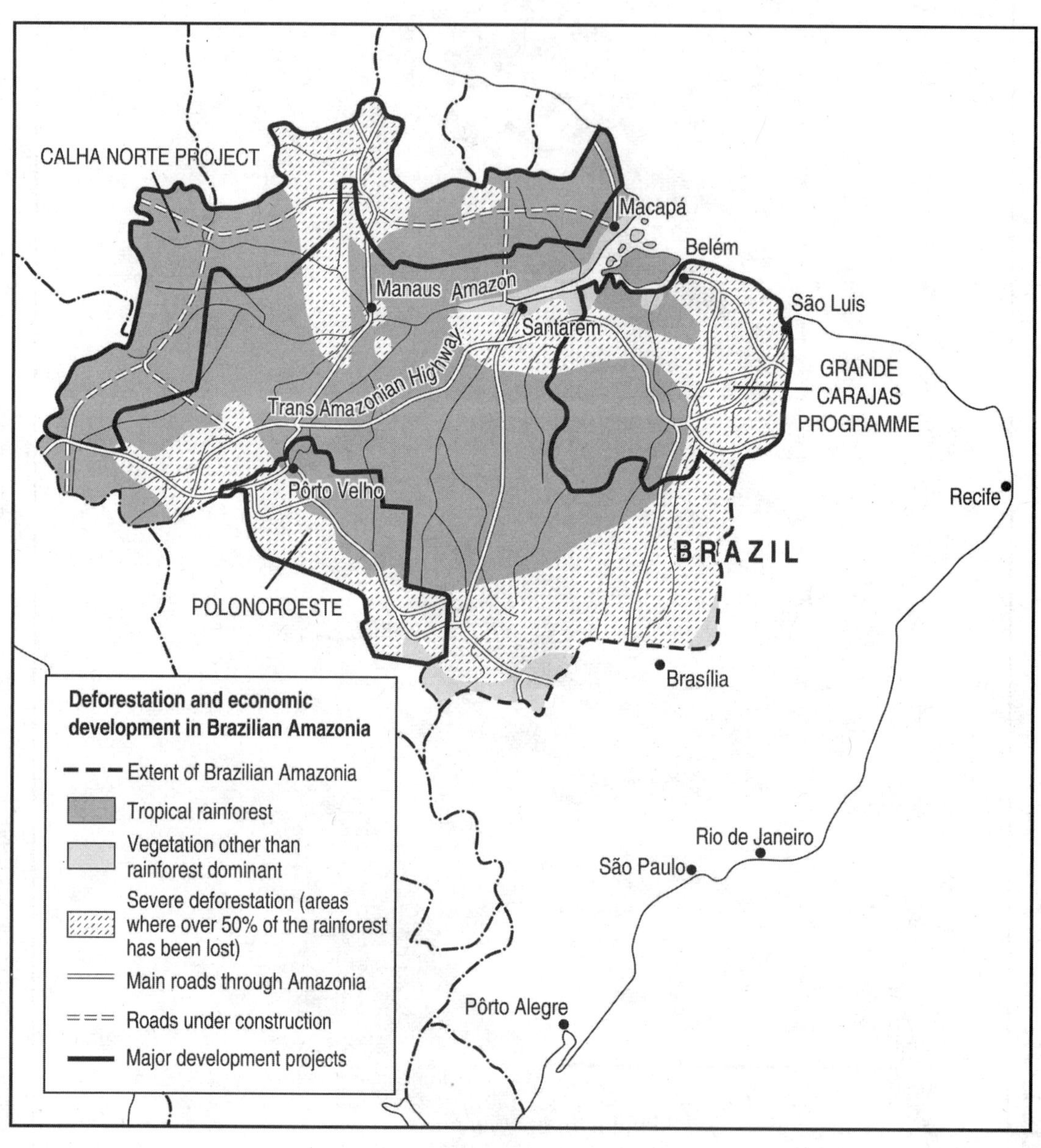

Source: *Collins Concise Atlas*, © Bartholomew Ltd 2000; reproduced by permission of HarperCollins Publishers

Soil Degradation

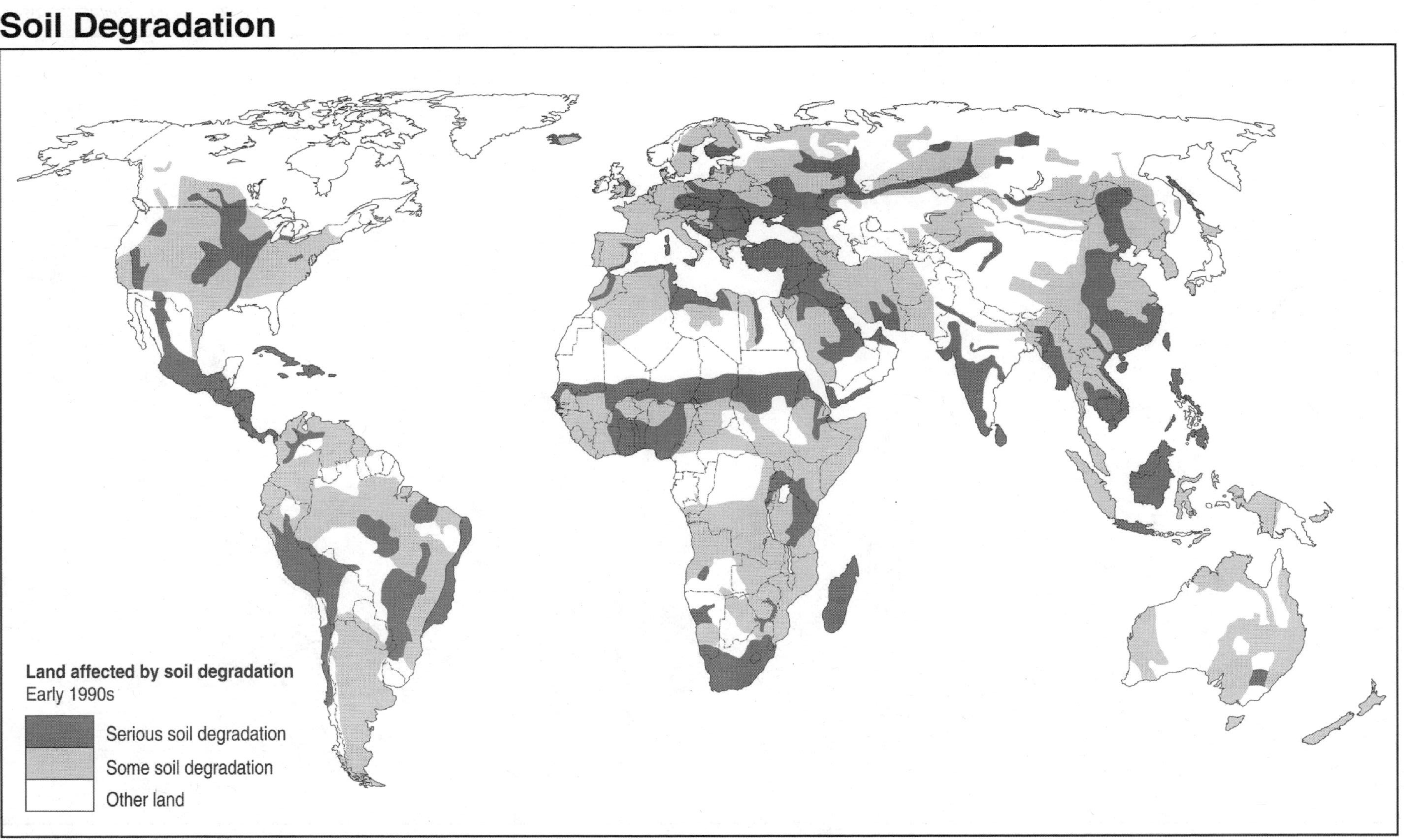

Source: *The State of the Environment Atlas,* Penguin Books, 1995

Water Pollution

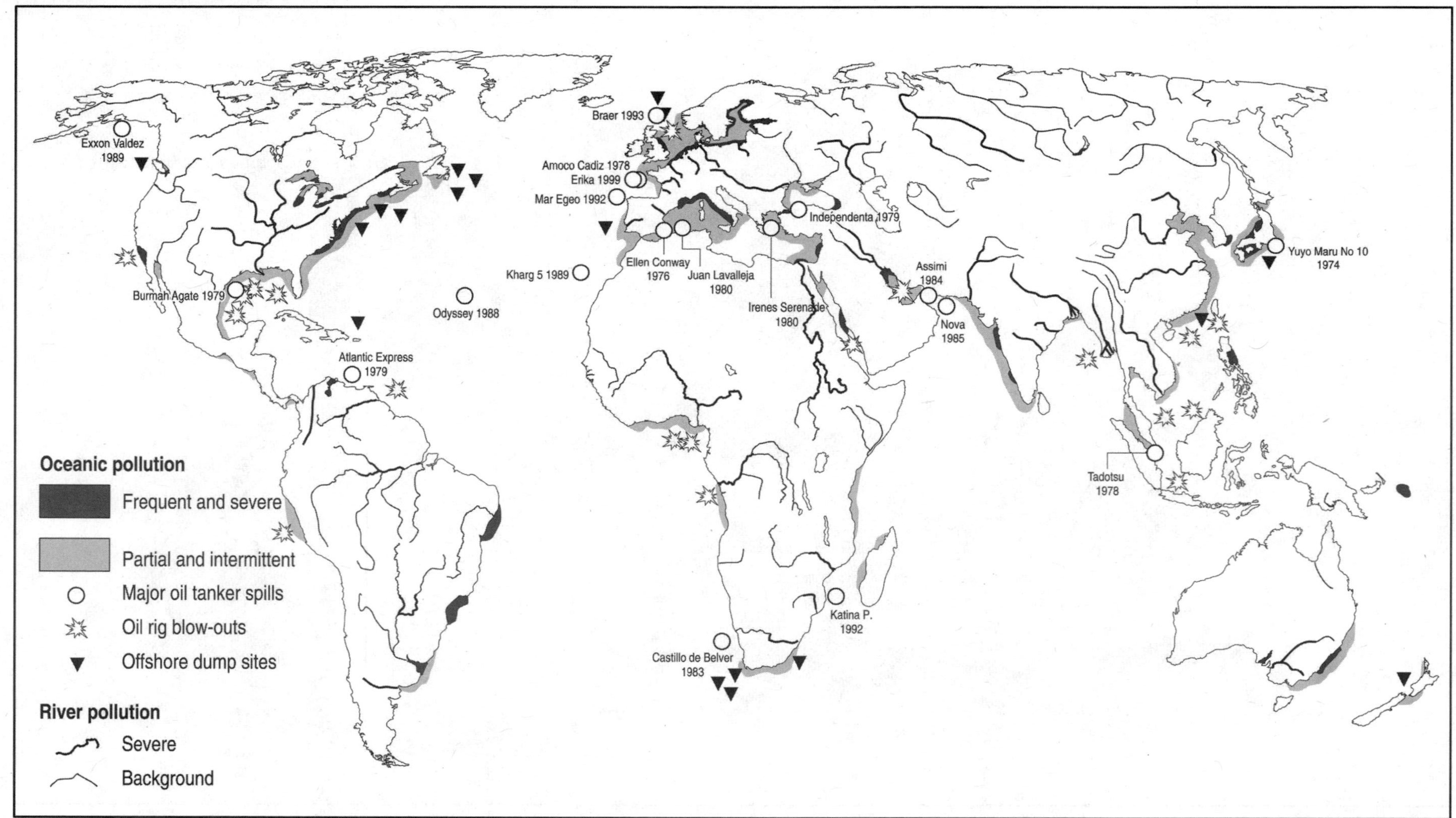

Sources: *Collins Pocket Planet Earth Factfile,* © Bartholomew Ltd 2000; reproduced by permission of HarperCollins Publishers; *The State of the Environment Atlas,* Penguin Books, 1995

Access to Drinking Water

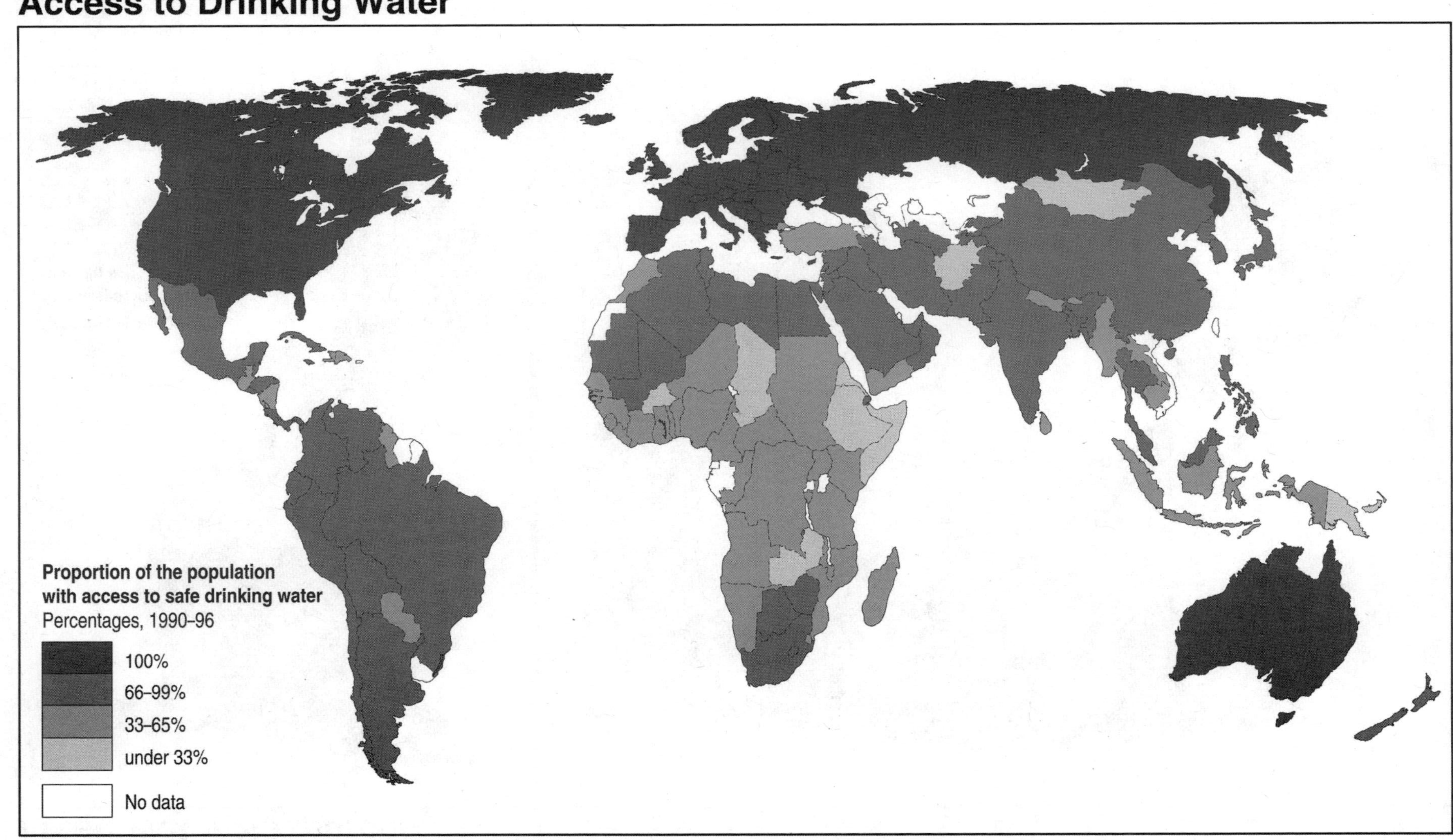

Source: *The State of the World Atlas,* Penguin Books, 1999

Fresh Water Resources

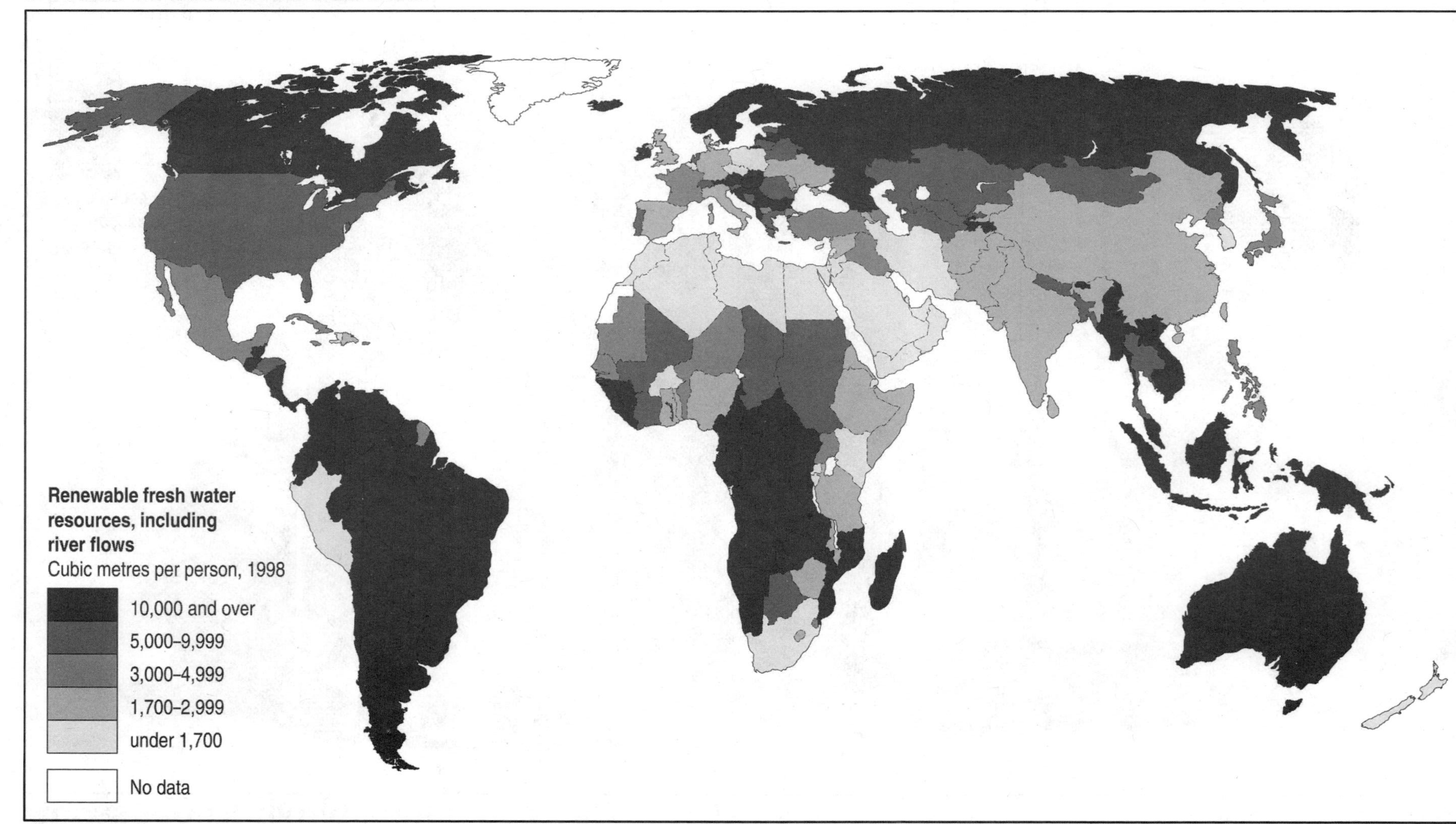

Water Use

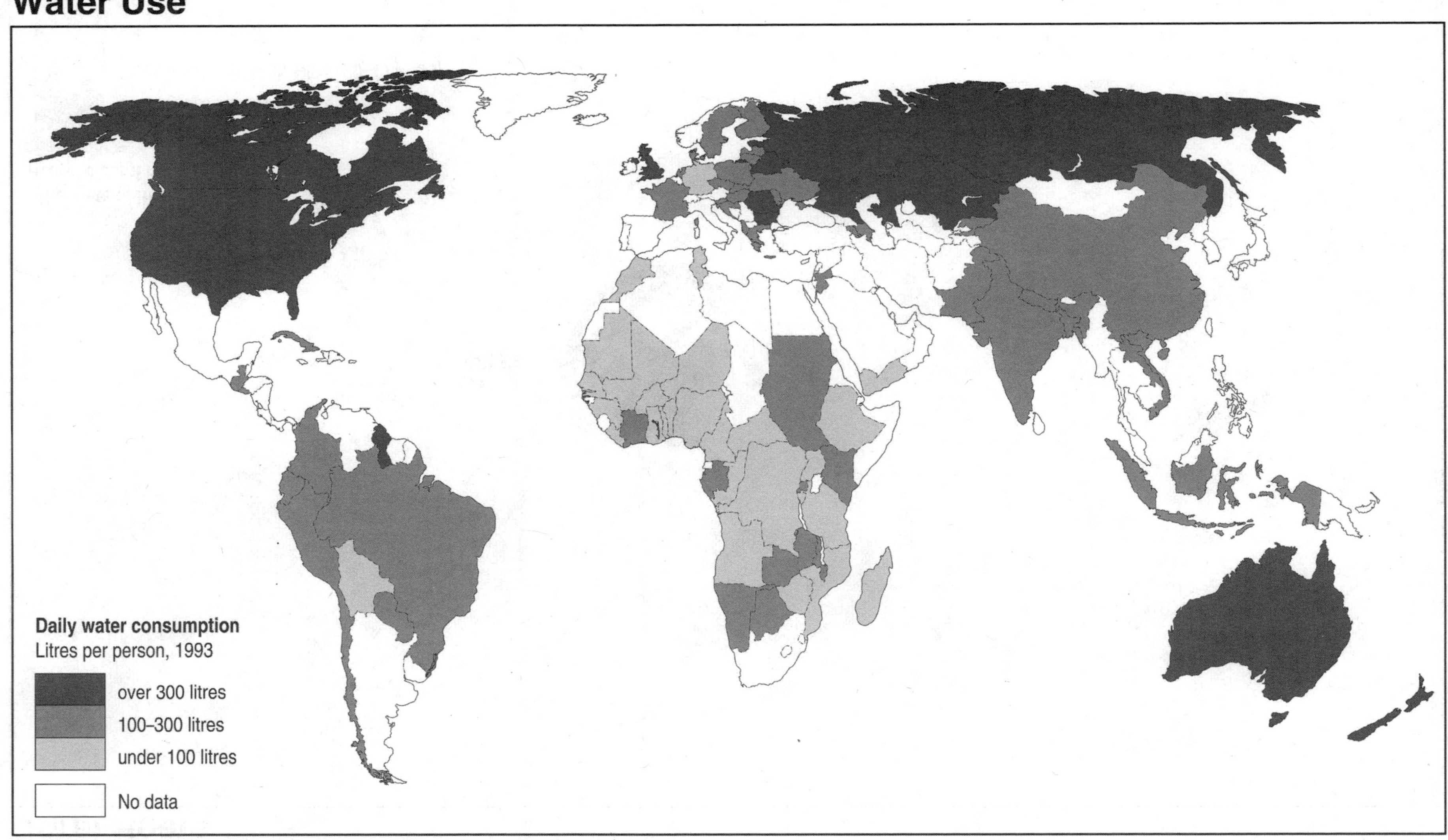

Source: *The State of the World Atlas,* Penguin Books, 1999

Coral Reefs

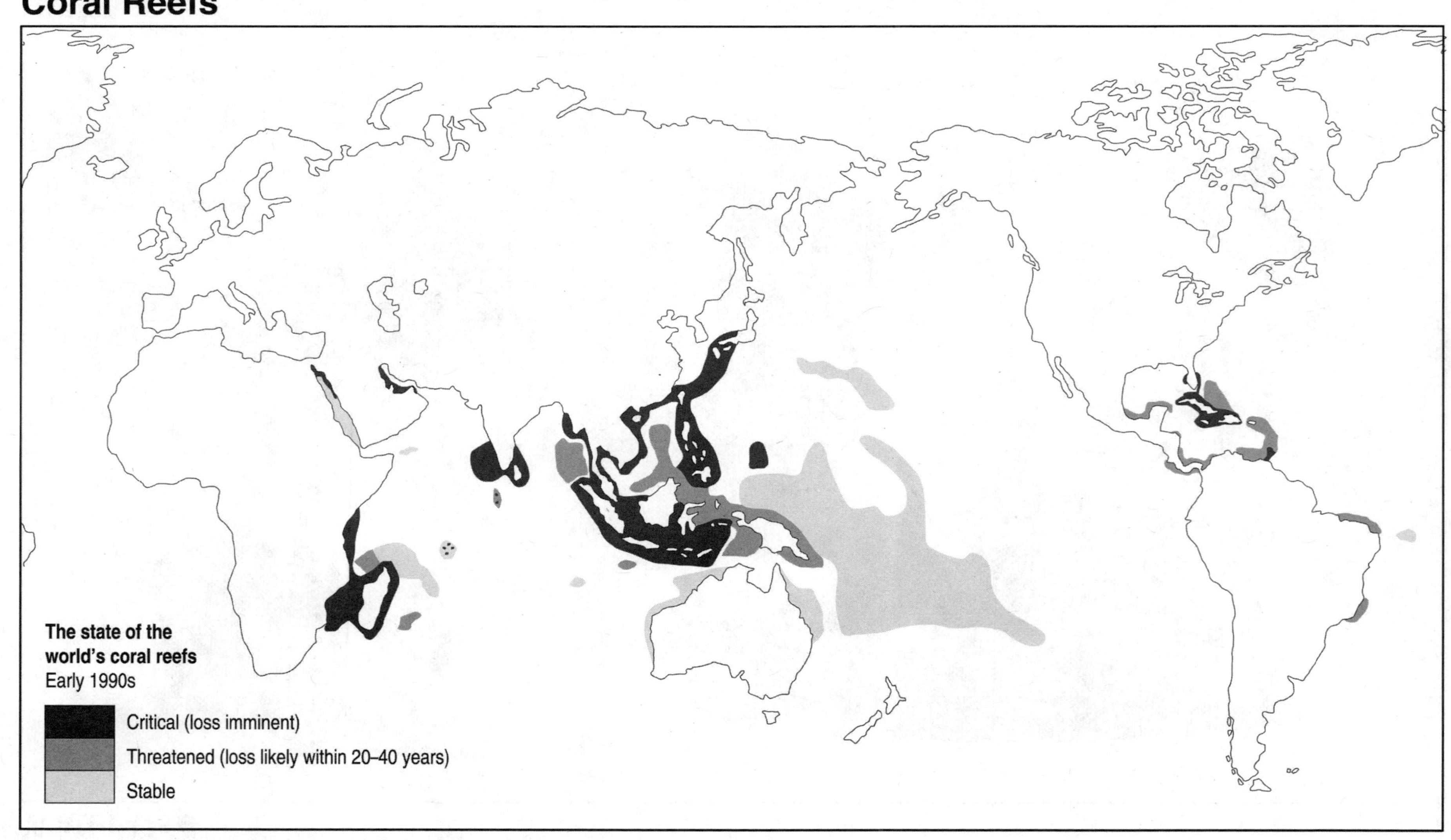

Source: *The State of the Environment Atlas,* Penguin Books, 1995

Wetlands

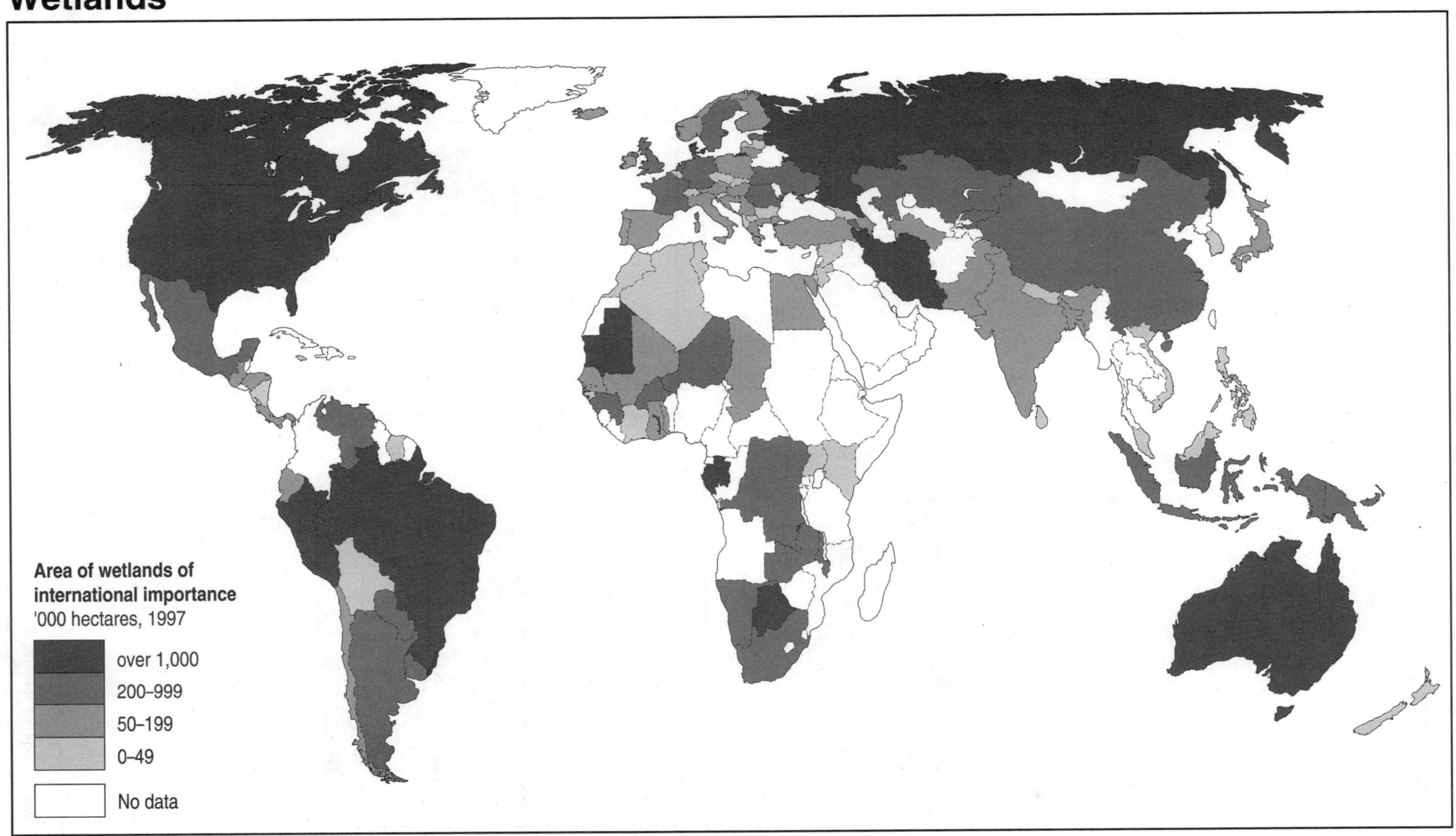

Source: *World Resources 1998–99,* World Resources Institute/Oxford University Press

The Aral Sea

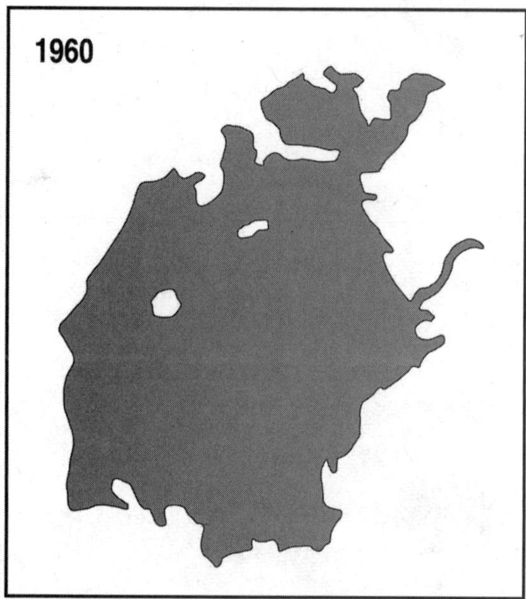

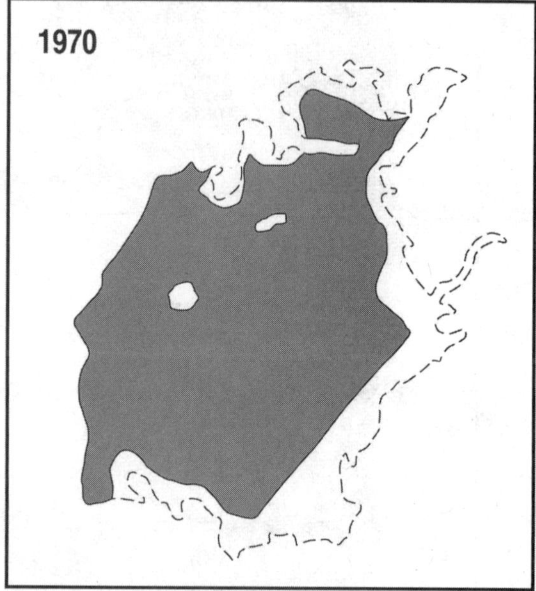

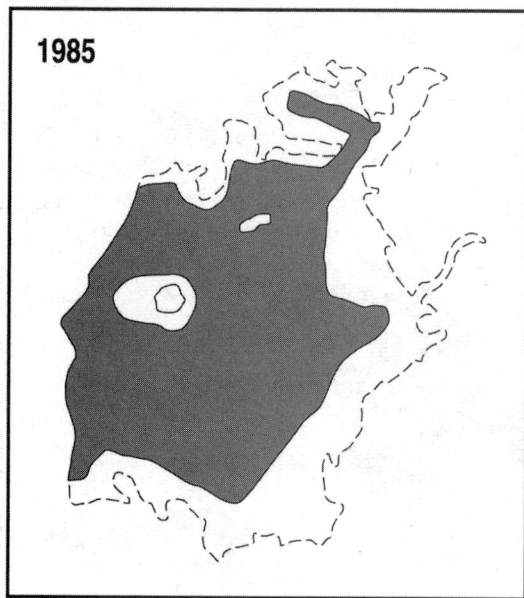

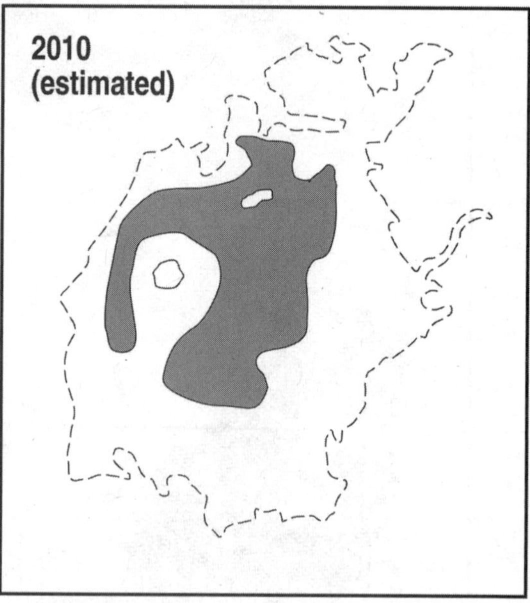

Source: *Collins Pocket Planet Earth Factfile,* © Bartholomew Ltd 2000; reproduced by permission of HarperCollins Publishers

Fishing

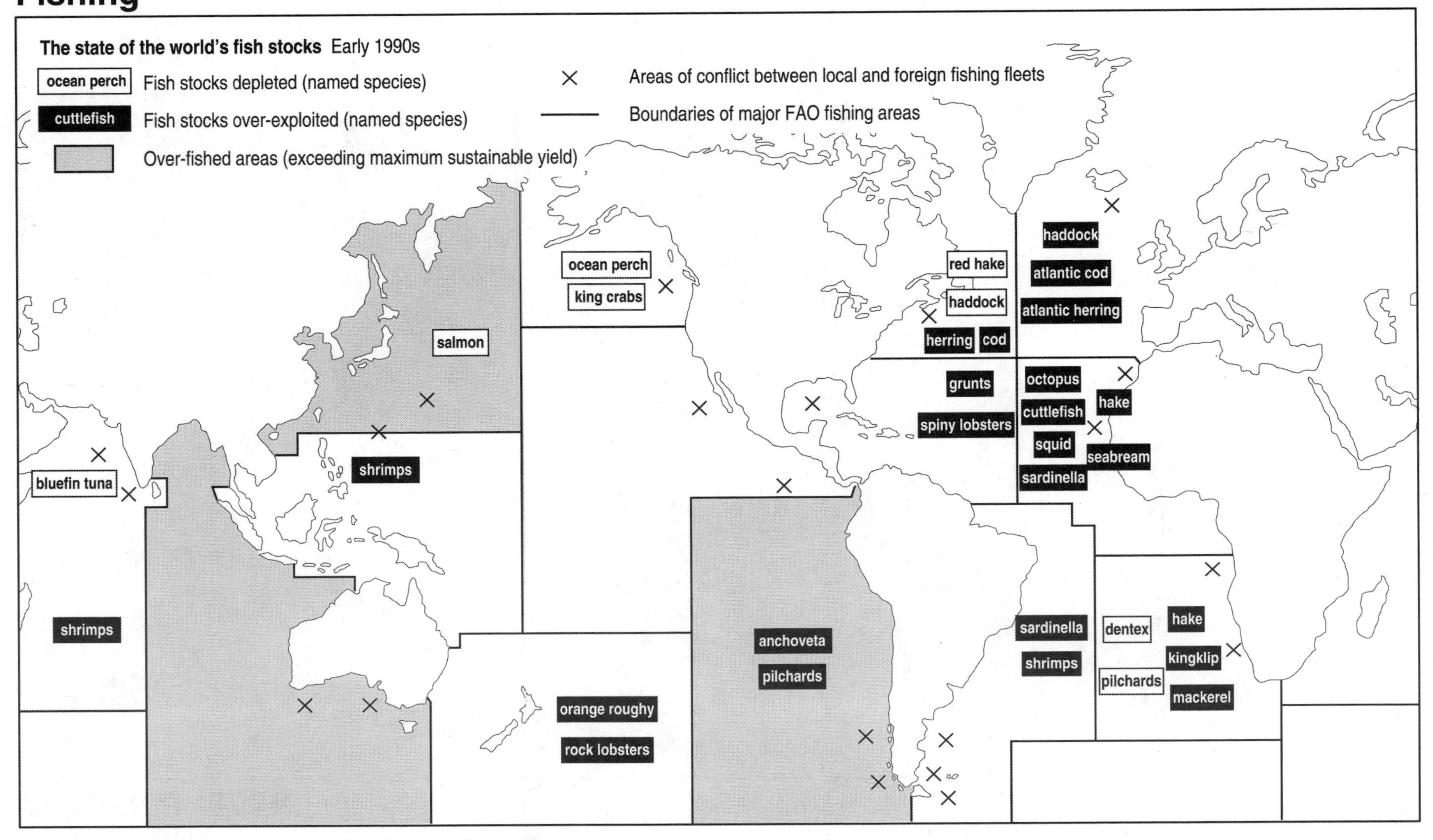

Source: *The State of the Environment Atlas,* Penguin Books, 1995

Protected Areas

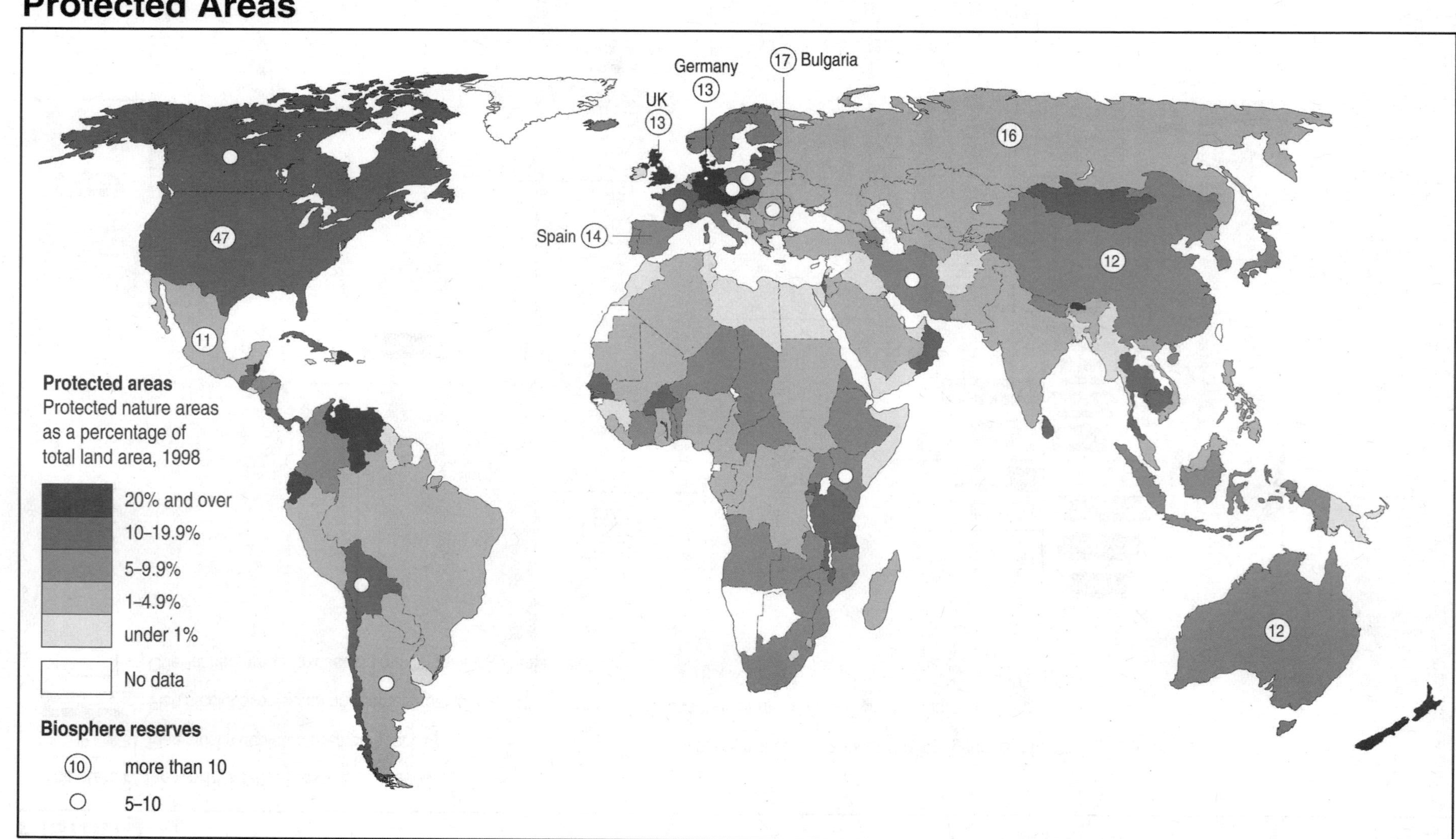

Source: *The State of the World Atlas,* Penguin Books, 1999

Energy Production

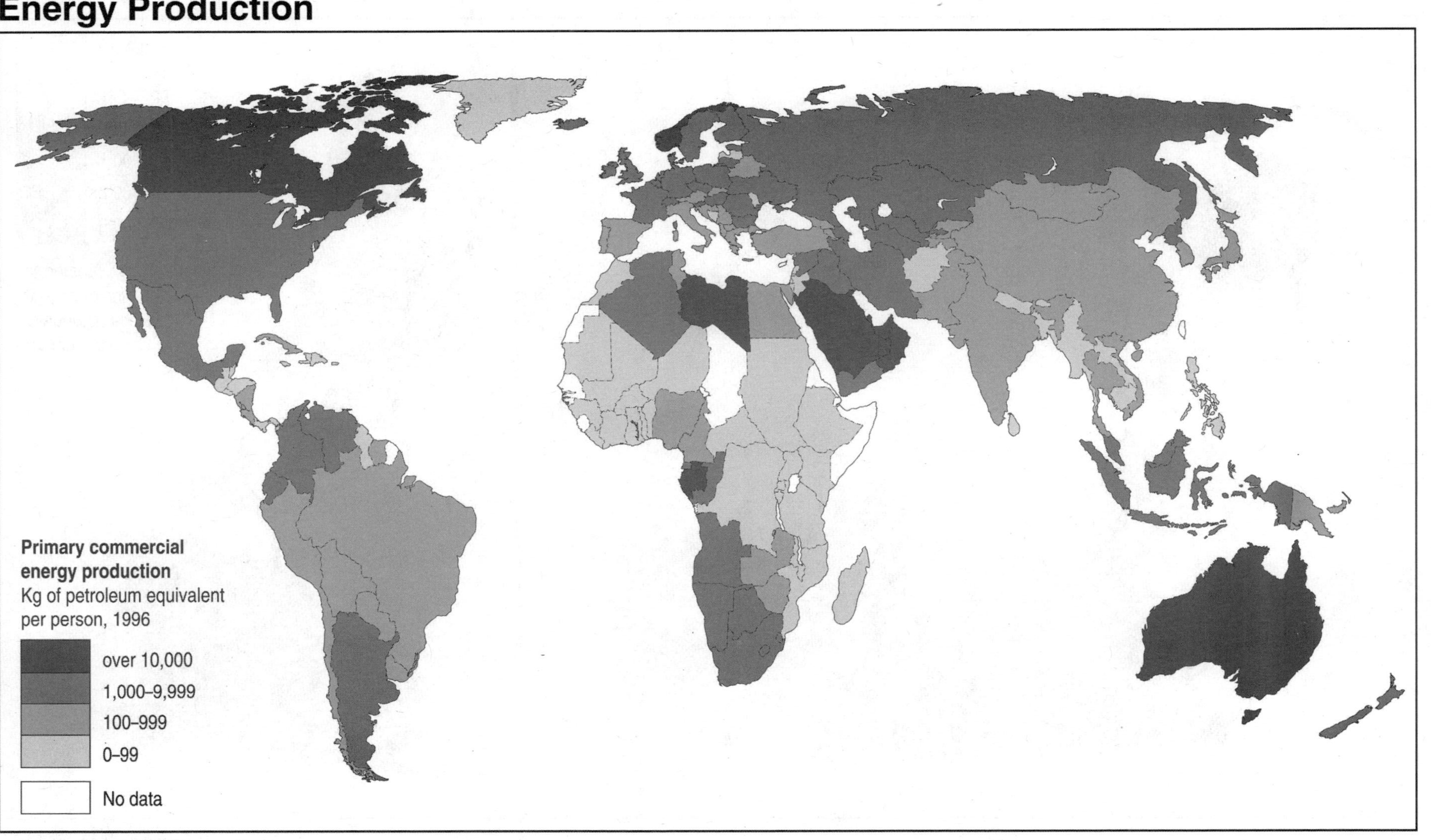

Source: United Nations Statistics Division

Energy Use

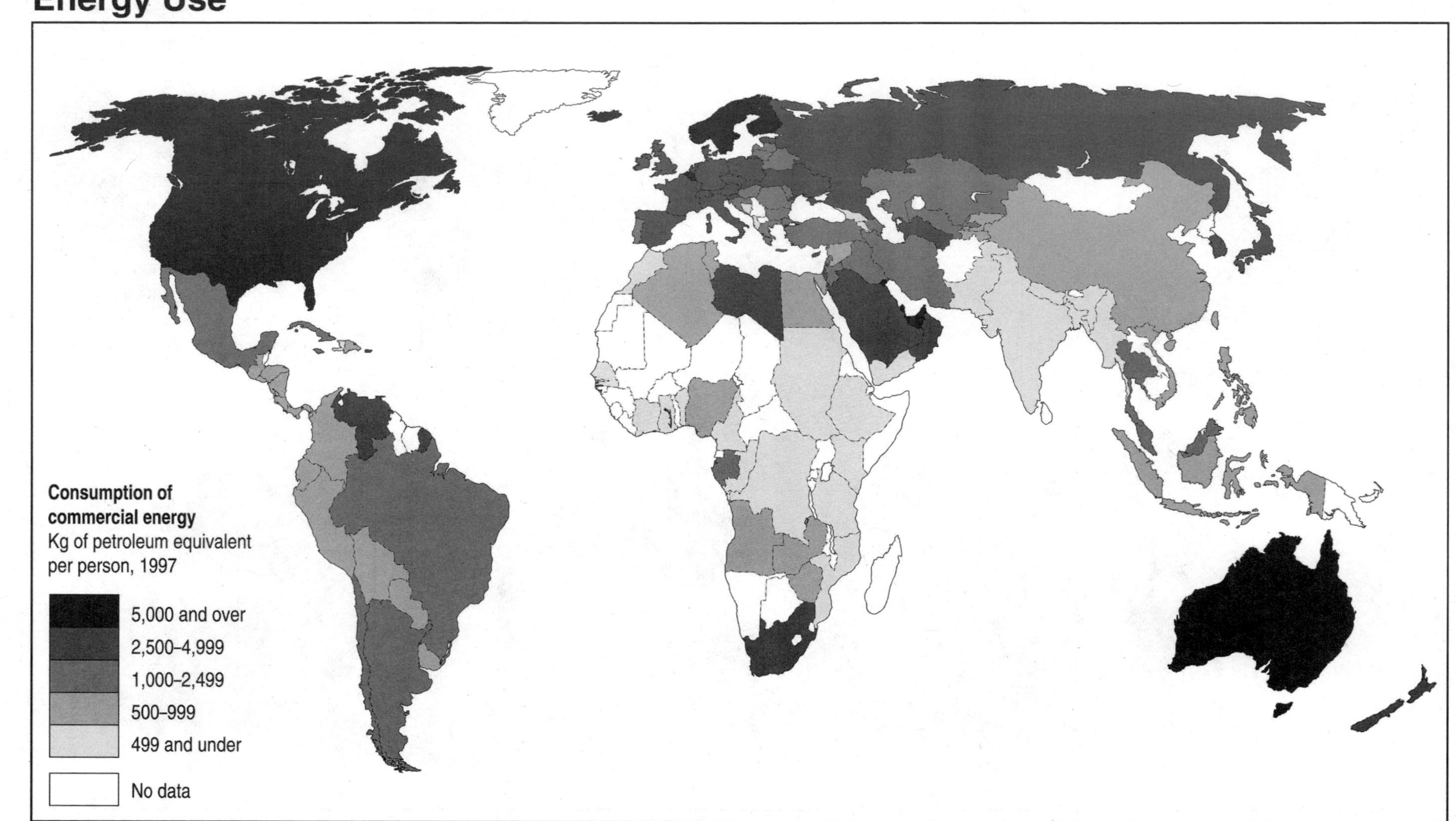
Consumption of
commercial energy
Kg of petroleum equivalent
per person, 1997
5,000 and over
2,500–4,999
1,000–2,499
500–999
499 and under
No data

Energy Efficiency

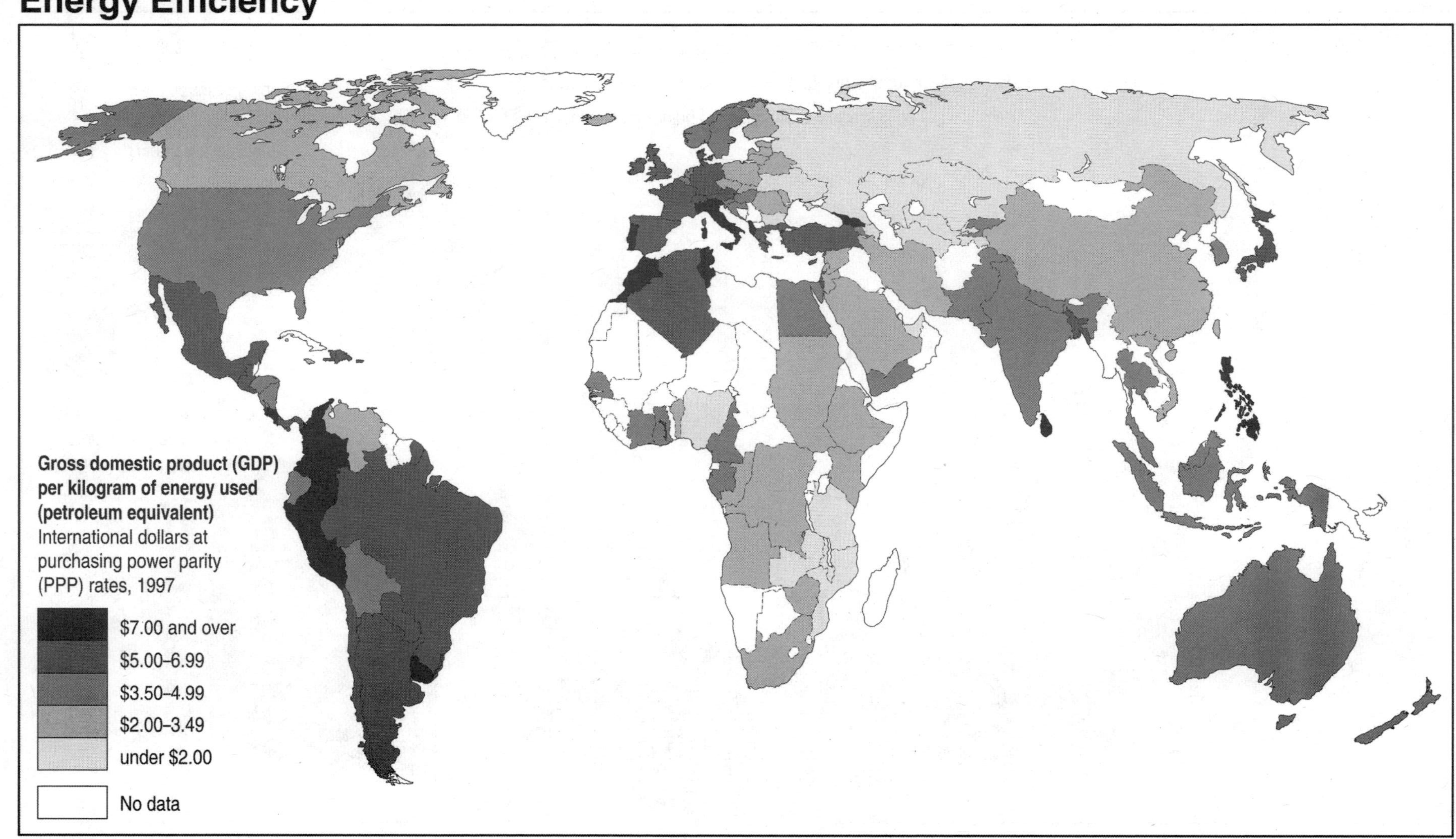

Republished with permission of The World Bank, from *World Bank Atlas 2000*, © 2000; permission conveyed through Copyright Clearance Centre, Inc

Nuclear Power

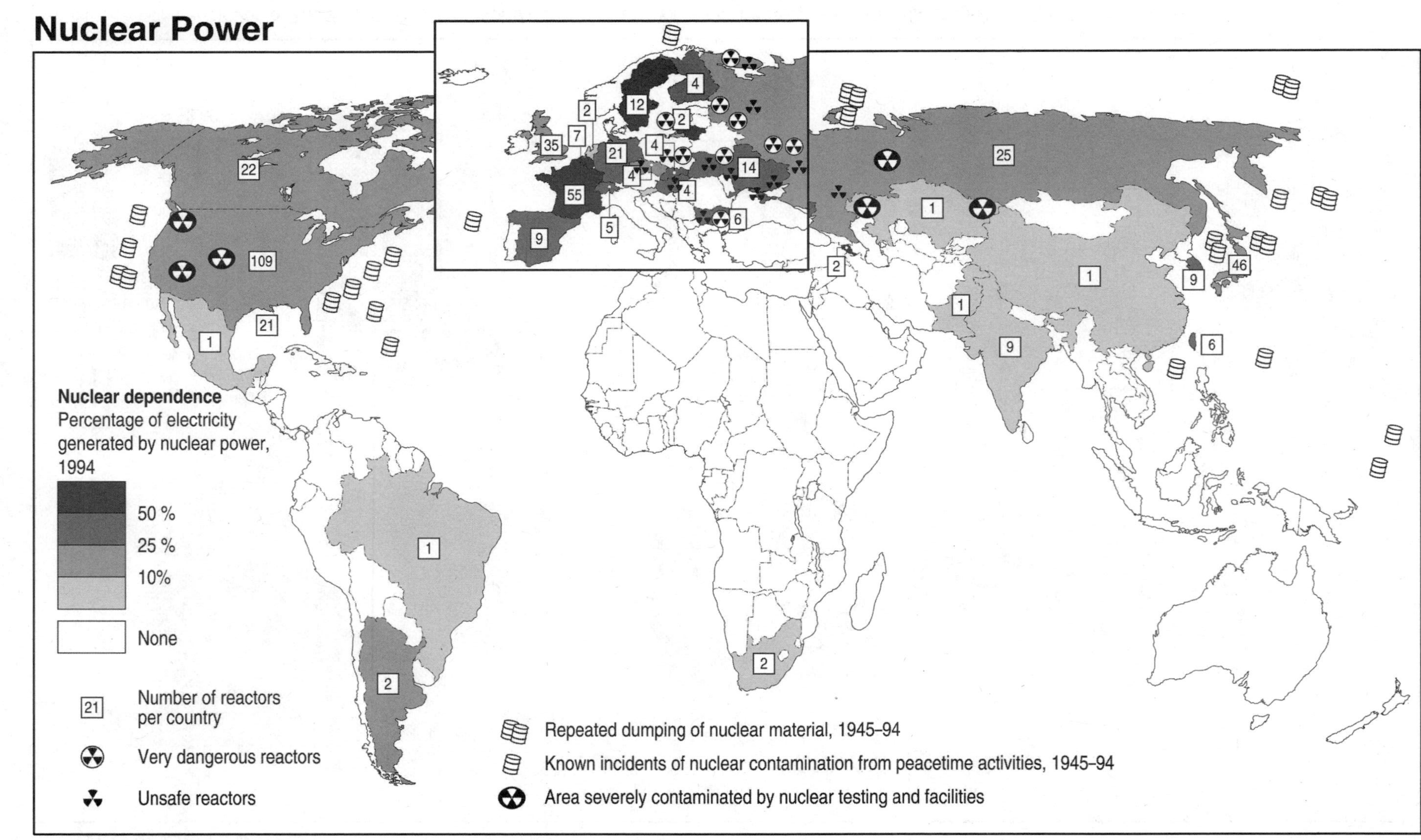

Source: *The State of the Environment Atlas,* Penguin Books, 1995

A–Z of the ENVIRONMENT

Pearce Wright

Updated by William R Moomaw and Kelly Sims Gallagher

Revised for this edition by Gareth Wyn Jones

William R Moomaw is Professor of International Environmental Policy and Director of the International Environment and Resource Policy Program at the Fletcher School of Law and Diplomacy, Tufts University. He is also Director of the Institute of the Environment and Co-Director of the Global Development and Environment Institute at Tufts University, and Vice-Chairman of the Board of Earthwatch. Professor Moomaw specializes in global environmental issues, particularly climate change.

Kelly Sims Gallagher completed her doctorate in international energy policy at the Fletcher School of Law and Diplomacy, specializing in US–Chinese co-operation for energy development in the People's Republic of China, and is now Director of the Energy Technology Innovation Project at the Belfer Center for Science and International Affairs (BCSIA), Harvard University. She was formerly Science Policy Director with responsibility for atmospheric protection at Ozone Action, Washington, DC, and a Truman Scholar in the office of US Vice-President Albert A Gore, Jr, and has participated in numerous international negotiations on global climate change.

Pearce Wright was formerly Science Editor of *The Times* and is now a freelance writer on science, medicine and the environment.

Gareth Wyn Jones is a freelance writer and editor.

Note: Words printed in **bold** type indicate that an entry on the subject so marked is to be found elsewhere in the A–Z of the Environment.

A

ABSOLUTE DROUGHT is the British definition of a period of at least 15 consecutive days in which less than 0.01 inches of rainfall is recorded. In the USA **dry spell** has a similar meaning. (See also **drought**.)

ABSORPTION is the process by which a substance retains radiant energy, heat and light waves, instead of reflecting, refracting or transmitting them. Minor gases, such as **ozone**, nitrous oxide, **water vapour** and **carbon dioxide** (see also **greenhouse gases**), have a significant effect on atmospheric absorption. Oxygen and nitrogen, the main atmospheric gases, do not absorb very much radiation. Another definition of absorption is the process in which the energy of **electromagnetic radiation** is taken up by a molecule and transformed into a different form of energy.

ACCEPTABLE DAILY INTAKE (ADI) is the measurement of the amount of any chemical substance that can be safely consumed by a human being in a day. Calculations are usually based on the maximum level of a substance that can be fed to animals without producing any harmful effects. This is divided by a 'safety factor' to allow for the differences between animals and humans and to take account of the variation in human diets. The ADI is set by international agencies: the Joint Expert Committee on Food Additives (which first introduced the term in 1956), a body of scientists from the **Food and Agriculture Organization** and the **World Health Organization**; the European Union's Scientific Committee for Food; and the USA's Food and Drug Administration (FDA). The levels are set and discussed independently by each of the agencies.

ACCLIMATIZATION refers to the process by which plants and animals adjust to changes in climate. For example, plants may undergo changes that mean they have to synthesize new types of protein and membranes that will be more tolerant of cold weather. Although acclimatization is generally a natural process, it is projected that, with the advent of rapid global **climate change**, plants and animals will be forced into a rapid acclimatization process in order to adjust to new climatic conditions. Under the existing global climate change scenarios for the 21st century, scientists estimate that many plants will be unable to acclimatize fast enough to keep up with the changing climate. For example, the **Intergovernmental Panel on Climate Change** notes that if levels of **greenhouse gases** in the atmosphere continue to increase, approximately one-third of the world's forested areas will undergo major changes in broad vegetation types. In other words, some species will not survive, but will be completely replaced by other species in a given area, accompanied by a severe reduction in **biological diversity**.

ACID is a substance that contains hydrogen and splits up in water, liberating hydrogen ions and creating conditions that are chemically reactive. An acidic substance has a **pH** value of less than 7.0 (neutral), alkaline substances more.

ACID RAIN can more usefully be included in the broader term acid deposition, which includes acidic snow, mist, sleet, rain, fog, gas, as well as the dispersion of dry particles. It is a regional rather than a global phenomenon and results from the emission of air pollutants from particular industrial activities, especially power production, in combination with certain meteorological conditions. Normal precipitation has a **pH** of 5.0, while acid rain has pH values lower than 5.0. Acid rain upsets the balance of nature, disrupting **ecosystems** and destroying **forests, woodlands**, plants and crops. It can kill aquatic life by altering the chemical balance of lakes and rivers; and it corrodes building materials and fabrics. The pollution is caused principally by discharges of **sulphur dioxide** and **nitrogen oxides** released by burning **fossil fuels** such as **coal** and petroleum. Other by-products emitted by industry and vehicles to the air can add to the mixture. The airborne pollutants often interact chemically with **water vapour**. Some of the pollutants drift back to earth within a few miles of the source of emission. Others are lofted high into the atmosphere and carried by prevailing winds for hundreds of miles before falling back to earth as dust, rain, snow or mist. This problem is experienced all over the world. Discharges from very tall chimneys of modern fossil fuel power stations in the United Kingdom have been detected in Scandinavia. A small proportion of acid rain in Europe is believed to come from the USA, where the phenomenon helped prompt reform of the **Clean Air Act**. About two-thirds of sulphur dioxide and one-quarter of nitrous oxide produced in the USA come from electric power generation (where it is reliant on the burning of fossil fuels). Acid rain now falls on about 30% of the total land area of the People's Republic of China, polluting water supplies, harming agriculture and damaging forests; it is a particular problem in the ecologically rich valley of the Chang Jiang (River Yangzi or Yangtze). After a debate on acid rain at the first **UN Conference on the Human Environment**, held in Stockholm, Sweden, in 1972, international pressure grew for controls. Thirty-three countries signed the Geneva Convention on the Long-Range Transport of Air Pollutants, one of the main aims of which was to combat acid rain. Other regional and international co-operation agreements have followed, as well as research into prevention. Many of the pollutants that cause acid rain can be removed by employing technologies such as scrubbers on power stations and **catalytic converters** on cars. Acid rain was a term first coined in 1872, in a book of that name by Robert Angus Smith, a British chemist with the **Alkali Inspectorate**.

ACTINIDES are a group of 15 radioactive elements produced when **uranium** fuel is irradiated in a nuclear reactor. They include **plutonium**, **americium** and **neptunium**. All of the group have atomic numbers higher than uranium, and so are also referred to as the **transuranic elements**. The health hazard presented by the actinides, if they are released into the environment, comes from the potency of their radioactive characteristics. They are **alpha-emitters**, and therefore can cause intense localized damage in tissues if absorbed into the body.

ACTINOMETER is the general name for any instrument which measures radiant energy, particularly **solar energy**.

ACTION PLAN FOR THE HUMAN ENVIRONMENT: See **UN Conference on the Human Environment**

ACTIVATION PRODUCTS are produced when normal metal and concrete construction materials and components employed in building a nuclear power station come under constant irradiation by neutrons. The materials absorb neutrons to become radioactive, and strong emitters of **gamma radiation**, creating difficulties in dismantling nuclear reactors and disposal of the pieces.

ADAPTABILITY refers to the ability of an organism to change its mode of behaviour and even its physiology in order to survive under new conditions that would otherwise be too stressful. In ecological terms, when something is adapted, it means that its structure is well suited to its function. New conditions that organisms must adapt to include environmental stresses such as **acid rain**, land-use changes like **deforestation**, and changes in the **climate**.

ADDITIVE is a substance mixed in small quantities with another product to modify the chemical or physical state of the commodity. Additives are used to make food look visually more attractive, in the case of colouring agents, as well as to preserve and extend the life of the product. Additives may also become environmental pollutants, such as tetra-ethyl **lead**, introduced into petrol as an **anti-knock agent**. It was developed as a cheaper way of blending high-octane fuels for use in high compression automobile engines. It improves the process of combustion as well as eliminating 'knocking' or 'pinking'. Unfortunately, tetra-ethyl lead is a very toxic form of lead when discharged into the air from car exhausts. It is especially dangerous for young children in whom it causes brain damage. For regulation on additives to food, see **E-numbers**.

ADENOSINE TRIPHOSPHATE (ATP) is one of the two fundamental power units in animals and plants. ATP is the principal energy-carrying compound in the cells of all living organisms. It is found wherever critical energy conversion processes are taking place such as in the conversion of sunlight by **photosynthesis** to plant tissue or in extracting the energy from sugar at the instant it is demanded by the muscle of an animal. It is used to bring about such biological processes as muscle contraction, the transport of ions and molecules across cell membranes, and the synthesis of biomolecules. ATP is regenerated using the chemical energy obtained from the oxidization of food.

ADVANCE INFORMED AGREEMENT (AIA): See **Biosafety Protocol**

ADVANCED GAS-COOLED REACTOR (AGR) was the name of the British-designed nuclear reactor that the British Government chose for its second nuclear power programme. The AGR followed the first generation of **Magnox** nuclear power stations. The plan was to increase the United Kingdom's nuclear electricity capacity from the 5,500 MW provided by Magnox stations to 11,000 MW. The new plan was announced in 1964, just three years after the first two Magnox power stations had been switched on. The aim of the AGR was to achieve higher efficiency by reaching higher temperatures. The AGR, with a nuclear core of 520 metric tons of fuel, is more compact than the Magnox. However, it has a slightly higher percentage, 2%, of enriched **uranium-235**. The AGR programme encountered many design difficulties, delays and excessive costs. It was the subject of successive parliamentary and government inquiries. However, the first two AGRs eventually came into operation in February 1976, at **Hinkley Point**, in Somerset, and Hunterston, in Scotland. Some remain in operation. All such gas-cooled reactors are now considered obsolete.

AEDES AEGYPTI is the most important of the *Aedes* family of **mosquitoes** found in the tropics and subtropics because it transmits serious illnesses such as dengue fever and yellow fever. It is found in estuaries and river areas, usually near human settlements. The female will lay her eggs in any available stagnant water: puddles, discarded tin cans, old tyres, coconut shells, even holes in trees. Other species of *Aedes* are carriers of filariasis and other viruses that cause various forms of encephalitis. Much of this problem is caused by poverty, which produces an unsanitary environment. Attempts to eliminate pesticide sprays have only added to the environmental depreciation.

AEROBIC RESPIRATION: See **respiration**

AEROSOLS are small particles in the atmosphere that can affect **climate**, depending on how they reflect or absorb radiation. Natural events like volcanic eruptions can produce sulphate aerosol clouds, which tend to reflect sunlight back into the atmosphere. **Particulate** aerosols can also attract moisture in the atmosphere. Human-produced aerosols are used, for instance, in hairsprays, deodorants, polishes and other consumer goods, which formerly used **chlorofluorocarbons (CFCs)** as propellants.

AFFORESTATION is the planting of trees on land that has not been forested for a long time, if ever. Large-scale afforestation would help reduce global **climate change** because new forest would absorb heat-trapping **carbon dioxide** out of the atmosphere. It is also believed to help prevent **desertification**, although that remains a problem in northern China, where the People's Republic has been carrying out possibly the biggest afforestation programme in history—according to government statistics, 42,000m. trees were planted between 1982 and 2003, with 560m. people involved in the effort in the last year alone. Afforestation should be distinguished from **reforestation**. (See also **forests and woodlands**.)

AFLATOXINS are a group of poisonous substances produced by the fungus mould, *Aspergillus flavus*, which grows on seeds and grain. They were first identified in the 1960s, when a large flock of birds died on a poultry farm in the United Kingdom in strange circumstances. It was found that they had been fed with meal contaminated with a poisonous mould. Aflatoxins can damage the immune system and cause liver cancer. Most of the reported cases in people have come from the Indian sub-continent and East Africa, where environmental conditions suit aflatoxin growth. Undernourished children are at most risk. Aflatoxins are a major problem in the tropics and subtropics because they thrive in moist and warm conditions and also because stored cereals and nuts, especially groundnuts (peanuts), are often contaminated. The use of chemical fertilizers is thought to play a major part in the spread of aflatoxin contamination because they tend to increase the water content of the crop and encourage the growth of mould.

AGENDA 21, the Rio Declaration on Environment and Development, is a comprehensive programme of action to be achieved by governments, development agencies, UN organizations and independent sector groups in every area where human activity affects the environment. Agenda 21 was adopted in Rio de Janeiro, Brazil, by the UN Conference on Environment and Development, the **Earth Summit**, in 1992. Agenda 21 is the response to the UN General Assembly's call, in December 1989, for a global meeting to devise strategies to halt and reverse the effects of **environmental degradation** 'in the context of increased national and international efforts to promote sustainable and environmentally sound development in all countries'. Behind Agenda 21 is the belief that humanity has reached a defining moment in its history. Present policies deepen the economic divisions within and between countries and this increases poverty, hunger, sickness and illiteracy world-wide and is causing the continued deterioration of the ecosystem on which we depend for life on earth. Agenda 21 concentrates on ways to change course and on how to improve living standards for those in need with the implementation of a global partnership for **sustainable development**.

AGENT ORANGE was the herbicide that the US military sprayed in Viet Nam in the 1960s and early 1970s to defoliate vast tracts of jungle that provided cover for the Viet Cong, whom they were fighting. The ecological impact of the action caused concern among many scientists. However, there was even greater concern about the health of the rural populations in Viet Nam. By the time military use of Agent Orange (codenamed for the orange bands identifying the drums it was stored in) ceased in 1971, an estimated 19m. gallons (72m. litres) had been used in Viet Nam. Agent Orange is an organochlorine **herbicide**, containing traces of an impurity known by the initials TCDD (tetrachlorodibenzo-paradioxin), which is a member of the **dioxin** family of chemicals, regarded as among the most poisonous ever made.

AGRICULTURAL HABITATS are those areas created by humans to cultivate crops and graze domestic animals in order to provide food. Domestication of plants and animals has resulted in complete alteration of natural landscapes in many parts of the world. In creating agricultural habitats, humans usually remove almost everything that has a detrimental effect on farming, such as **pests**, weeds and indigenous plants and animals, leaving only those organisms that are helpful to crop cultivation. The result is that the diversity of plant and animal species has been greatly reduced and the number of **monoculture** habitats has increased. A large increase in the demand for food over the centuries has led to an increase in **intensive agriculture** and the mechanization of farming methods. Industrialization in Europe, from the end of the 18th century, had a major impact on agriculture, both in the provision of new machinery that could improve the efficiency of agriculture and in the production of synthetic **fertilizers** to improve the fertility of the soil. Modern agricultural methods are energy-intensive and require extensive use of **pesticides** and **fungicides**, which further alter the agricultural habitats, although in Europe at least from the late 20th century there was some attempt to encourage farmers to maintain land with non-agricultural flora and fauna in mind.

AGROCHEMICALS are the man-made **fertilizers** and **pesticides** that were increasingly produced in industrialized countries following the development of **intensive agriculture** in the 1950s. They are the products of the modern chemical and petrochemical industries. Manufactured nitrate fertilizers have produced huge rises in the yields per hectare of cereals, vegetables, grazing pasture and fruit. However, the high crop yields have become dependent on very heavy doses of agrochemicals. World fertilizer use rose from 15m. metric tons in 1950 to over 150m. tons in 1990 (levels of use remained fairly constant from the late 1980s into the early 2000s). Nitrogenous fertilizers accounted for 37% of the total at the beginning of the 1960s, but for 60% by 2002. One estimate suggests that if agrochemical use stopped overnight, the world's harvest would be halved. On the debit side, the high use of agrochemicals has resulted in the leaching of fertilizers into waterways where they cause **eutrophication** and **algal blooms**. Nitrates from fertilizers are also leaking into **ground-water aquifers**, which supply drinking water. The worrying problem for water engineers is that it may take decades for nitrates to work their way through the overlying rocks before reaching the ground-water's natural reservoirs. Pesticide poisoning comes from direct contamination or from residues that have accumulated in the food chain, causing extensive deaths among wildlife and raising anxiety about human health. An estimate by the **World Health Organization** suggests that nearly 3m. people suffer from acute pesticide poisoning every year; of these, 200,000 are workers in the developing countries who die because they are not trained properly in the use of protective masks and coveralls when spraying fields.

AIR POLLUTION can be caused by natural events and human activities. It usually refers to the emission of ash, **particulate** matter, soot and harmful gases from factories and power stations, as well as to automobile emissions. Examples of air pollution include **carbon monoxide** and **sulphur dioxide**. It can also be caused in rural areas by **pesticide** sprays from fields, or dust from mines and quarrying. Natural sources, such as dust blowing from deserts, the ash and dust from volcanic eruptions and salt from the sea being blown inland by the wind, are other contributory factors. The gases rising from bogs, marshes, **wetlands**, forest fires and decomposing material can also cause air pollution. Larger particles fall to the ground near the source, but smaller particles can be carried long distances by the wind, often resulting in acid deposition (see **acid rain**). Air pollution has a damaging effect on health. Apart from irritated eyes, respiratory diseases such as **asthma** and other breathing problems, studies have shown that high **lead** content in the air in cities and towns probably results in brain damage and a reduction in the intelligence of children. Air pollution causes severe environmental degradation: in towns the acidity of the pollution causes the fabric of buildings to deteriorate and metals to tarnish. Throughout the world, air pollution is a serious cause for concern and great efforts are being made to reduce it through implementation of clean air legislation (see **Clean Air Act**), the use of smokeless fuels and the fitting of **catalytic converters** to cars. Related to air pollution are the emissions of **ozone**-depleting and **greenhouse gases**, which often have the same sources as air pollution. In 2004 it was reported that the People's Republic of China intended to impose fuel economy standards even more strict than those in many parts of the USA, in an attempt to limit pollution and encourage the development of so-called **hybrid vehicles**.

AIR QUALITY is an environmentally important issue because clean air is essential to the health of humans and to the general well-being of all **flora and fauna**. It is measured in various ways with the aim of monitoring the effect of traffic and industry, most often in urban areas. In order to maintain clean air, many countries have adopted air quality standards. The **World Health Organization** established a set of guidelines, revised in 1999, that countries should adhere to in order to protect their populations from severe air pollution.

AIR QUALITY STANDARDS are maximum limits placed on the concentration of specified air pollutants during a defined time. They are also known as ambient standards. The standards vary from city to city and country to country. If the targeted concentration level is exceeded at any time, then the region is in non-compliance with the standard at that time. In fact, the standards are difficult to enforce because polluting gases are sometimes invisible and are quickly dispersed. The first national standards were set in the United Kingdom after London was seriously affected by smog in 1952. After a government inquiry, the **Clean Air Act** was passed. In the USA, **photochemical smog** experienced in Los Angeles, California, led to the passing of similar legislation. Emission standards for hazardous air pollutants from factories, automobiles and aircraft were set, and the federal Clean Air Act eventually allowed private citizens to sue anyone who violates an emission standard.

AIRBORNE DISEASES describes illnesses that can be caused by airborne infectious organisms and by **air pollution**. Infections include the common cold, flu, influenza and tuberculosis, in which **bacteria** are transmitted through the air from an infected person, usually by coughs and sneezes. Allergic complaints, such as **asthma** and **hay fever**, are often caused by airborne pollen, rusts, moulds and dust. Car emissions contribute to increases in the number of people who suffer from chest diseases and other ailments, and include **nitrogen oxides**, **particulates** and the products of **lead** additives in petrol.

ALBEDO is the proportion of sunlight reflected from a surface. It is an important factor in the regulation of the temperature of the **atmosphere** and the general circulation of air and water around the planet. The albedo of the earth can affect the climate and may be changed by alterations in forest cover, the size of deserts, and the amount of the surface that is covered by ice or snow. Observations by satellites have provided more precise measurements of the earth's albedo effect. This is useful for providing more precise projections of the potential impacts of global **climate change**, for example. The value of any albedo varies, depending on the nature and properties of the terrestrial surface. The incident angle of the sun's radiation is also very important. Water has a low albedo of around 5% when the sun is high in the sky, but acts like a mirror and reflects about 70% of solar radiation when the sun's rays strike the surface obliquely.

ALDEHYDES are widely used in the chemical and pharmaceutical industry. They are also an **air pollution** hazard when they are among the **volatile organic compounds (VOCs)** emitted from vehicles. VOCs pose a risk to health because they are toxic or carcinogenic (see **carcinogen**). Increasing attention has focused on aldehydes. They are likely to be produced in greater amounts as they are now used as additives to **biofuels** or **alternative fuels** such as methanol and **ethanol**. Safety recommendations of the **World Health Organization** suggest that the concentration of aldehydes in the atmosphere over a 30-minute period should be less than 30 parts per 1,000m.

ALDRIN is a not a substance that appears naturally. It is a toxic chlorinated hydrocarbon **insecticide**. It disappears very quickly from the soil and was therefore thought to be a 'safe' insecticide, after it was first produced in 1948. However, once in the atmosphere it is converted into dieldrin, which accumulates in the bodies of animals and enters the **food chain**. The numbers of birds of prey, in particular, were dramatically reduced. Since strict regulations on its use were introduced in many countries during the 1960s and 1970s, the numbers of most afflicted species have gradually improved. In 1979 the European Communities (now the European Union) issued a directive to all member states banning the marketing and sale of aldrin in plant protection products, but permitting its use in agriculture, horticulture and food storage, but subsequent directives and legislation outlawed it completely. Between 1972 and 1987 in the USA aldrin was licensed for killing **termites**, but the manufacturer then voluntarily abandoned the chemical's registry for that use, and it is no longer used in North America. Aldrin is regulated under international treaties as a **persistent organic pollutant**.

ALGAE are simple green, aquatic plants without stems, roots or leaves. They are among the microscopic organisms that form the start of the **food chain**. Algae are found floating in the sea (see **phytoplankton**) and fresh water, but they also grow on the surface of damp walls, rocks, the bark of trees and on soil. They contain chlorophyll and other pigments that let them grow by **photosynthesis**, which is the process that uses sunlight to convert **carbon dioxide**, water and salts into a source of food. During the summer algae flourish near the surface of the oceans and they are referred to as a **carbon sink**. On land, algae can be useful in improving the fertility of soil by **nitrogen** fixation, the process by which nitrogen in the atmosphere is converted into a type of **nitrate** that is essential for crop plants. Algae can, however, be troublesome in streams and lakes polluted by fertilizer run-off or sewage wastes. Nitrates in the effluent cause algae to grow in thick layers, and far too quickly to be handled by the natural cycle of growth and decomposition: such an **algal bloom** can render a body of water unable to sustain life. At the beginning of the 21st century this was increasingly a problem in the **oceans**, creating so-called dead zones.

ALGAL BLOOM refers to excessive and rapid growth of **algae** and other aquatic plants when they are stimulated to grow too quickly by pollution. It takes place when there are too many **nutrients** in the water and is aggravated when accompanied by a rise in temperature. Although the algae grow quickly they soon die because they have swallowed up all the water's nutrients. As they decompose they tend to rise to the surface and form a green slime. The natural processes that decompose decaying plants depend on the oxygen available in the water, and when the algae becomes too prolific the process of decay is said to impose too great a biochemical oxygen demand (**BOD**). Because of the reduction of oxygen in the water, **eutrophication**, decomposition cannot take place and the stream or lake becomes lifeless. Algal blooms have increased because higher levels of **nitrates** and **phosphates** from agricultural areas have leached from the fields into watercourses. Not all algal blooms have this great an effect, particularly if occurring at sea, but they could still be classed as harmful algal blooms if the algae release toxins that may not harm shellfish, for instance, but could have severe effects further up the **food chain**.

The **ALKALI INSPECTORATE** was the first agency established in the world to control emissions from the chemical industry. It was established under the Alkali Act 1863 by the British Government. One of the first inspectors was Robert Angus Smith, who first used the phrase **acid rain**. In the 1960s the Inspectorate was given wider duties for monitoring of industrial **air pollution** in the United Kingdom. However, environmentalists were dissatisfied on several counts. First, the Inspectorate's reports on individual companies were kept private. In fact, they were covered by the Official Secrets Act. Secondly, the criteria for accepting emissions were not based on absolute pollution control standards, but on the principle of the so-called best practical means. In other words, if an industrialist persuaded the alkali inspector that a chemical works was employing the best practical means of operation, the activities of the works would continue. (This kind of problem remains widespread globally.) The Inspectorate later became part of Her Majesty's Inspectorate of Pollution, an organization which had a far wider remit and which was, in turn, incorporated within the Environment Agency in 1996.

ALP ACTION was established in 1990 by the Bellerive Foundation. The Foundation, which was founded by Prince Sadruddin Aga Khan (and had him as its president until his death in 2003), is a group of environmentalists, scientists, industrialists and businessmen whose aim is to raise funds to co-ordinate grass-roots **conservation** efforts, preserve unique **habitats**, encourage **reforestation**, revive traditional agriculture, and control the excesses of tourism in the Alps. Alp Action also has a programme of initiatives designed to increase public awareness of these problems.

ALPHA-EMITTERS are radioactive elements that, in the process of **decay**, release **alpha radiation**.

ALPHA RADIATION is one of the three principal forms of **ionizing radiation** to which people on earth are exposed. The radiation is only produced by some radioactive elements as they go through their process of **decay**, but it applies particularly to the radiation emitted by the **actinides**. Alpha radiation consists of relatively large fragments of an atomic nucleus with two protons and two neutrons, which is the same number found in the nucleus of a helium atom. Consequently, alpha particles are relatively cumbersome compared with **beta** and **gamma radiation**. However, alpha particles carry a lot of energy that can easily be absorbed by a wafer-thin layer of tissue. If alpha particles are being emitted by a radioactive element that has entered the lungs or the bone marrow, the damage done to the tissue can trigger a radiation-induced cancer. The damage is caused by the ionizing radiation effect within the cells of tissue, producing changes in the chemistry of the substance that can bring about the death of a cell or its transformation into a malignant one.

The **ALPS** link seven European countries: Austria, France, Germany, Italy, Liechtenstein, Slovenia and Switzerland. These ranges are being damaged by pollution smog that comes from European industrial centres and drifts across the mountains. The Alps are a valuable **ecosystem**. They are the natural habitat of a wide variety of flora and fauna, some of it unique to the area. The Alpine forests protect houses and winter sports centres against rock falls and landslides, and also stabilize the banks of its rivers. However, they have been badly affected by **acid rain** and **air pollution**. Large dams have destroyed the ecological balance of rivers and **wetlands**, and many rivers are polluted to such an extent that the fish are contaminated. Several governments, especially that of Switzerland, have brought in strict legislation designed to control the emission of pollutants. Some animal species, including the **otter**, lynx, brown bear and wolf, have virtually disappeared from the Alps. Bearded vultures, bred in captivity, have been returned to the area. **Conservation** groups have been very active in the Alps, particularly on issues like acid rain and the pressure of **tourism**. **Alp Action** has been established by the Bellerive Foundation to co-ordinate conservation programmes. An Alpine Convention (Convention on the Protection of the Alps) was signed by the Alpine countries and the European Communities (now known as the European Union) in November 1991.

ALTERNATIVE FARMING SYSTEMS are the methods of agriculture developed in response to the growing awareness of environmental damage caused by **intensive agriculture** and farming practices relying on heavy doses of **agrochemicals**. In Europe, the development of intensive farming was encouraged by the inducements and subsidies offered under the **Common Agricultural Policy** of the European Communities (now the European Union). However, there is increasing evidence that many farmers are turning to alternative systems. These do not necessarily entail complete cessation of the use of chemical **fertilizers** or **pesticides**, but only decreased dependence on them. Nor are such farmers returning to old-fashioned farming methods: they still use modern machinery and hybrid seeds, and try to concentrate on practices that conserve resources and make for a healthier soil. They use animal manures that add nutrients to the soil and use new methods such as **integrated pest management** to reduce plant diseases and increase crop yields.

ALTERNATIVE FUELS is a term that refers to several types of fuel advocated for use in cars and lorries because they are less polluting than petrol or diesel. The refining of petrol and diesel from crude petroleum is not a sustainable technology over the long term. The principal alternative fuels for road vehicles are: **natural gas**; **liquefied petroleum gas (LPG)**; **ethanol** derived from wood, corn or sugar; rape methyl ester; **hydrogen**, preferably produced using electricity from **solar energy**; and methanol from natural gas or **coal**. Some produce lower carbon dioxide emissions than conventional fuels but are more expensive. There is no perfect fuel in environmental terms: the best choice depends on which of the environmental problems associated with burning fuel is most severe in each case. For example, the use of **methane** in the form of **compressed natural gas (CNG)** has significant environmental advantages over diesel for large vehicles such as buses, road sweepers and refuse collection lorries that make frequent stops in urban areas. CNG-powered vehicles produce few **particulates** such as **PM10s** and the **hydrocarbons** they emit have less than half the **ozone**-forming potential of those from diesel fuel. Since the courts imposed the introduction of CNG use for public transport in 1998 there has been some improvement in the **air quality** of the Indian city of Delhi. There are other advantages of developing processes to produce **biofuels** from crops: biofuels can also be used as feedstock, or raw materials, for the industries that currently manufacture plastics, agrochemicals and pharmaceutical products from petrochemicals. The biggest biofuel project hitherto has been the promotion of ethanol in Brazil, notably in the 1980s.

ALUMINIUM (Al) became implicated as an **environmental health** hazard in the 1980s on two counts. Biomedical scientists looking for possible causes of Alzheimer's disease, the premature senility indicated by loss of memory and confusion, found a circumstantial link with aluminium. The theory is a controversial one. Post-mortem examinations compared two groups of people. One group, with a higher incidence of Alzheimer's, had lived in an area of the United Kingdom where public drinking water supplies were dosed heavily with aluminium sulphate to remove peaty materials. The analyses of brain cells showed microscopic particles of aluminium, together with other changes in tissue used as the definitive diagnosis of the disease. The second event occurred at **Camelford**, in Cornwall, the United Kingdom, when a bulk delivery of aluminium sulphate was tipped directly into the local drinking water supply instead of into storage tanks. The pollution caused serious intestinal and dermatological illnesses in the local population.

The **AMAZON** is the second longest river in the world (over 6,400 km) and the largest by the size of its watershed, the number of tributaries and the volume of water discharged into the sea. The Amazon and its tributaries pass through five countries before entering the heart of the basin in Brazil. Much of the basin is dominated by the vast tracts of the world's largest **tropical rainforest**, which is huge enough to have a significant effect on the **climate** of the globe, as well as being home to a huge diversity of plant and animal species. Significant species in the river itself include the boto, one of two Amazon river **dolphin** species, and the carnivorous piranha fish, one of more than 1,000 freshwater species at least found in the Amazon basin, while the jungle, particularly the rainforest canopy, is a rich habitat for insects, birds and various other forms of wildlife. There is a minimum of over 300 mammal species in the Amazon, a bird population boasting over 1,600 species and 500 species of amphibians. Such estimates may not do full justice to the wealth of **biological diversity**, and they are vague at best when claiming a further 100,000 invertebrate species, of which 70% are insects. Despite extensive and continuing **deforestation** (estimated at some 30,000 sq km annually at the beginning of the 2000s), there are still 3.3m. sq km of this unique resource remaining. Government attempts to prevent illegal logging, unregulated mineral prospecting or farmers practising **slash-and-burn agriculture** are constrained by a lack of resources, corruption and an imperative for economic development in a country suffering massive poverty among much of its population.

AMBIENT STANDARDS: See **air quality standards**

AMERICIUM (Am) is one of the radioactive elements that did not exist before 1942 and which were first created by nuclear **fission** in the first nuclear reactor. The element is one of the **high-level waste** products from the nuclear industry that have caused controversy because there is no way to dispose of them. Other highly active wastes include **plutonium** and **neptunium**, which remain in the waste effluents produced after reprocessing spent **nuclear fuel** (see **nuclear reprocessing**); americium belongs to a group of elements that are also called the **actinides** or **transuranic elements**. They get their group name from the fact that they have **atomic numbers** greater than 92, and so lie beyond uranium in the periodic table. Before nuclear fission and its by-products were discovered, **uranium** was the heaviest natural element. Americium is particularly hazardous because it is an **alpha-emitter** and has a **half-life** of over 450 years.

AMINO ACIDS are the building blocks of life. They are the basic components from which proteins are assembled. Proteins are formed by various combinations and proportions of the 20 most common amino acids. Plants and many micro-organisms synthesize amino acids from simple inorganic compounds. Animals have to rely on their diet to obtain theirs. They can also be produced deliberately in **genetically modified organisms**—strains of bacteria and fungi—in fermentation tanks. All living organisms, from the simplest of life forms, the viruses, to the human animal, are composed mainly of proteins. Amino acids themselves consist of many combinations of just four elements: carbon, hydrogen, oxygen and nitrogen. There are also amino acids that never occur in protein and which are essential nutrients.

AMMONIA (NH_3) plays a critical part in the **nitrogen cycle**, which is a natural process through which plants get nitrogen from the atmosphere. Nitrogen-fixing bacteria are able to achieve a similar reaction and were used by scientists before the **Haber process** was invented. If the temperature and pressure are normal, these bacteria release ammonium ions, which are converted by nitrifying bacteria into nitrite and nitrate ions. Ammonia is a colourless gas with a strong smell and is very soluble in water. It is also a principal ingredient of many chemical **fertilizers**. It is produced by the Haber process, and more than 80m. metric tons are used every year, in the manufacture of nitric acid, fertilizers, explosives, dyes and resins.

AMOCO CADIZ was the name of the petroleum tanker wrecked off the coast of Brittany, France, in March 1978, spilling its cargo of 220,000 metric tons. Over 320 km of coastline were affected. More than 4,500 birds of 33 species were found contaminated with petroleum on nearby beaches. Extensive areas of **salt marsh** were damaged, oyster fishing grounds were disrupted and some animal species, such as sea urchins, were almost completely exterminated. It is estimated that some 30,000 seabirds were killed, along with some 240,000 tons of fish and shellfish. At the time, it was the largest loss of marine life from an oil spill ever recorded. It was also the first spill to affect estuarine tidal rivers. Clean-up costs totalled ECU 199m. The disaster has been one of the most extensively studied pollution spills.

AMPHIBIAN POPULATIONS have become important indicators of the planet's ecological health. A decline in the size or health of an amphibian population reflects damage of some kind to its **ecosystem**. Amphibian populations are good monitors of local conditions because they remain in fairly confined areas for all their lives. The number of frogs, toads and salamanders is declining in many parts of the world where they are victims of **vanishing habitats**, chemical contamination and depletion of the **ozone layer**. Destruction of the ozone layer is harmful because **ultraviolet radiation (UV-B)** from the sun reach the earth's surface instead of being filtered out. Since the late 1980s many scientists have believed that declining amphibian populations may be a phenomenon with a global cause as much as a local one.

ANIMAL SOCIETIES refer to the particular way different species of animals group together, if only for sexual reproduction. When animals live in tight-knit groups for all or part of their lives, the groups are those commonly known as a herd, shoal, school, pack, swarm, etc. The value of such groups may be mutual protection, or for defence of a feeding area in order to restrict the numbers of animals to a level that a habitat can support. Some groupings, such as ants, bees and termites, have rigid caste systems. In these cases members of the group are kept together by passing food from one to another (trophallaxis) and by **pheromones**, which give a common odour to the colony. Thus, a strange bee attempting to enter a hive is stopped, investigated by guard bees and repulsed. Of course, bees play a crucial role in an ecosystem as one of the main agents of pollination of plants.

ANOPHELES is a type of mosquito that seriously affects human health. The malarial protozoal parasites (four species of the *Plasmodium* genus) have a life cycle split between a vertebrate host (almost exclusively human, but in some circumstances other higher primates) and an insect vector or carrier. The vector can only come from 60 of about 380 Anopheline mosquitoes. The parasite is transmitted into the host body by the bite of female *Anopheles*. Moreover, some species of the *Anopheles* genus transmit the parasites that cause Bancroftian filariasis, which can result in blindness.

ANOXIC describes organisms that are capable of living in the absence of oxygen. Anoxic conditions can arise in polluted lakes, rivers and estuaries. They occur when, for example, heavy **algal blooms** generated by **eutrophication** eventually die. When algae fall to the bottom of a lake or to the sea-bed, their decomposition results in massive deoxygenation of the water, killing fish and other aquatic flora and fauna. **Waterlogging**, produced by accidental or deliberate flooding, is another cause of anoxic conditions when the air is driven from the soil. This has an effect on many biochemical processes in the soil and may even trigger a phenomenon known as a **chemical time bomb**, or the release of harmful substances previously stored safely in the soil.

The **ANTARCTIC ICE SHEET** is the huge blanket of ice that covers the Antarctic continent. It forms the southern ice cap, which is kept in a constant balance by snow accumulation and the release of icebergs into the Antarctic Ocean. The Western Antarctic ice sheet (containing about 10% of the world's ice), in particular, is highly vulnerable to global **climate change** and scientists are not sure how the ice sheet will respond (the Eastern Antarctic sheet, which is vastly larger, is more stable, as it rests within a bowl of mountains). A warmer world could produce more precipitation, including increased snowfall, making the ice sheet become thicker. However, increased air and ocean temperatures are also likely to have a melting effect, which could cause **sea levels** to rise dramatically, flooding many **small island states** and other low-lying areas. Sea levels from global climate change could rise by as much as 1 m by 2100. There is evidence that the Western Antarctic ice sheet has melted significantly in the geologically recent past, having a dramatic impact on sea levels and global climate. Alarming signs of instability are already evident in the west, particularly on the **ice shelves** of the Antarctic peninsula. Such ice shelves are important in maintaining the stability of the ice sheet itself, as they shield the land-based ice from direct contact with the melting influence of the ocean.

The **ANTARCTIC OZONE HOLE** was the first known instance of thinning of the protective layer of **ozone** in the earth's upper atmosphere. The ozone shield stops harmful **ultraviolet radiation (UV-B)** from the sun reaching the ground. Many experts say that any sustained increase in the amount of UV-B radiation reaching the earth's surface could lead to an increase in the incidence of cancer in humans and animals, and lower crop yields. It could also affect the marine **food chain**, starting at the level of microscopic **plankton** in the oceans. The ozone layer is depleted by certain man-made chemicals, including **chlorofluorocarbons (CFCs)** used in **aerosols** and refrigeration, the pesticide methyl bromide, organic **solvents** and **methane** (natural gas). Destructive changes were first reported by scientists monitoring the amount of ozone in the **stratosphere** over the South Pole. In 1985 a team led by Joseph Farman of the **British Antarctic Survey** reported in the scientific journal *Nature* that they had detected, over a number of seasons, a 40% decrease in the usual springtime regeneration of **ozone** over **Antarctica**. The team, using equipment developed in the 1920s, discovered the lowest concentration of ozone ever recorded. Eleven years before, in 1974, the chemists F Sherwood Rowland and Mario J Molina, of the University of California, had already predicted a decline in global ozone. They had demonstrated mechanisms involving man-made chemicals in the laboratory that, if repeated in the natural environment, would destroy ozone. The international scientific community was sceptical, particularly as modern monitoring equipment on sophisticated satellites had not detected ozone depletion in the upper atmosphere. It was later found that measurements made by satellites had in fact indicated the depletion of ozone, but that scientists had failed to interpret the data when processing the extensive records. Faced with the evidence of Farman's group, scientists began work to determine whether the depletion would spread beyond Antarctica. Arguments continued about whether the phenomenon observed by the Antarctic team could have been caused by natural forces or by synthetic chemicals. Tests to confirm the potential sources of chemical destruction were inconclusive, but stated that 'a chemical mechanism is fundamentally responsible for the hole'. A larger expedition to the Antarctic was organized, comprising 150 scientists and support teams, and representing 19 organizations and four nations. It was discovered, using satellite, aircraft, balloon and ground measurements, that, compared with long-term records, the average ozone concentration in a region twice as large as the USA declined by nearly one-half between August and October, and in some areas decreased to negligible levels. The substances found to be responsible for ozone depletion were to be phased out under the **MontrTal Protocol**. Ozone depletion continued to increase each year after Farman's discovery. In October 1998 the New Zealand Government reported that the Antarctic ozone hole covered 27.2m. sq km, more than the combined areas of Russia and Canada (the two largest countries in the world). Despite some shrinkage in 1999, in September 2000 the hole covered a record 28.3m. sq km. The ozone hole shrank again in 2001, was slightly larger in 2002 and again reached over 27m. sq km in 2003. By 2000 it was hoped that some of the ozone-depleting gases had reached their maximum concentration in the atmosphere and that, over the decades, reduced emissions would allow the recovery of stratospheric ozone.

ANTARCTICA, one of the earth's last great wildernesses, has a strong relationship with the global ecosystem and an important role in the global climate system. The continent acts as a 'refrigerator', which affects the global atmosphere and ocean circulation. Its importance means that many countries carry out research in Antarctica, with 19 countries operating 37 wintering stations (mid-2002), although seven countries also maintain territorial claims on the continent (neither the USA nor Russia, like the USSR before it, recognize these claims). The 1961 Antarctic Treaty was designed to ensure that Antarctica would be used exclusively for peaceful purposes. Nuclear explosions and the disposal of radioactive wastes were forbidden. Twelve nations with an interest in Antarctica (Argentina, Australia, Belgium, Chile, France, Japan, New Zealand, Norway, South Africa, the USSR, the United Kingdom and the USA) signed the Treaty, agreeing to the establishment of joint scientific research projects to encourage co-operative endeavours between the parties. Sovereignty claims were frozen by the Treaty and some developing countries contend that the membership conditions are exclusionary, although 32 other states have acceded (of which 15 have acquired consultative status by virtue of their scientific activities in Antarctica). The Treaty forbids the exploitation of living resources in Antarctica or any activity that would cause deterioration of those resources, which has resulted in a degree of protection for wild animals and the sparse plant life. Numerous measures under the Treaty have enhanced this protection, notably the Agreed Measures for the Conservation of Antarctic Flora and Fauna, Conventions for the Conservation of Antarctic Seals and of Antarctic Marine Living Resources, and the designation of Specially Protected Areas and of Sites of Special Scientific Interest. However, there is concern about the **environmental degradation** of the continent. This has especially been brought about by depletion of the **ozone layer**, by the onset of global **climate change**, by industrial pollutants originating at lower altitudes, and by damage caused by humans, in the form of construction, waste, disturbance to plants and animals, and fuel spillages. Proposals in 1988 that would have allowed limited mining on the continent were strongly opposed by France and Spain, and eventually the other parties to the Treaty (lastly the United Kingdom and the USA) reached agreement to reverse the policy. In October 1991 a draft agreement was signed by 24 countries in Madrid, Spain, to ban mineral and petroleum exploration in Antarctica for at least 50 years. The agreement, the Protocol on Environmental Protection, calls for the Antarctic to be declared a natural reserve devoted to peace and science. It includes new regulations for wildlife protection, waste disposal, marine pollution and continued monitoring. It entered into force on 14 January 1998. However, no controls have been imposed on **tourism**, which has grown dramatically since the Protocol was signed. Tourism, albeit **ecotourism**, already brought 4,800 people to Antarctica in 1991, but the number of visitors had reached 15,325 by 2001 and was expected to be around 27,000 in 2003. In that last year the Antarctic and Southern Ocean Coalition voiced the growing concerns about the impact of relatively large numbers of visitors on such a delicate ecosystem and urged the introduction of tourist quotas. Meanwhile, since December 1994 there has been a whaling sanctuary around Antarctica, established below 40°S by the **International Whaling Commission.**

ANTHRAX is an infectious disease of wild and domesticated animals caused by the organism *Bacillus anthracis*. Under certain conditions this organism forms highly resistant spores which persist and keep their virulence in contaminated soil or other material for many years. Almost all animals, especially herbivores, are vulnerable to the disease, but usually it occurs in cattle, sheep, goats and horses, all of which are infected as they graze on contaminated land. Domestic pets or animals kept in captivity usually catch it by eating contaminated food. Anthrax can be transmitted to man by direct contact, inhalation or ingestion of infected material, such as wool, hair, hides, bones or the carcasses of infected animals. It is also occasionally transmitted by wearing furs or leather goods or using brushes that are contaminated by spores. Anthrax is one of the oldest recorded diseases in the world, dating back to a reference by Moses in the book of Exodus. In 1863 anthrax became the first disease in which a specific micro-organism was identified as the cause. It was also the first infectious disease against which a bacterial vaccine, developed by Louis Pasteur in 1881, was found to be effective, and which subsequently led to the development of bacteriology and immunology. To control the disease, strict quarantine regulations are used during outbreaks, including the burning of diseased carcasses. Anthrax was considered highly suitable for those researching biological weapons. Its power was demonstrated in the Second World War when British scientists, fearing a biological attack, tested the disease on a flock of sheep on a small Scottish island. Gruinard remained quarantined for 48 years, until 1990 and the end of an intensive, four-year, £0.5m. decontamination project. Since the terrorist attacks on the USA in September 2001 and the increased sensitivity of governments to such materials being used by 'rogue' states or terrorists, interest in anthrax and its vaccines has been revived. In 2001 the **World Health Organization** issued draft guidelines on the Health Aspects of Biological and Chemical Weapons.

ANTHROPOGENIC describes an object or disturbance to the environment that is created by humans or that follows from emissions of harmful substances as a result of human activities.

ANTI-KNOCK AGENTS are substances that are added to petrol (gasoline) to make sure the fuel burns more efficiently. Engine knock, a metallic knocking sound and loss of power caused by inefficient **combustion** of fuel, is produced by the petrol-air mixture burning too quickly. It results in lots of explosions in too many parts of the cylinder at once instead of a single smooth detonation that pushes the piston down. The **diesel engine**, invented in 1892 by Rudolf Diesel, avoided engine knock by putting the fuel-air mixture under high pressure, so the heat of compression alone was enough to ignite it. Petrol is a complex mixture of **hydrocarbons**, some of which burn more quickly than others. The amount of knock produced by a given petrol is measured and known as its **octane rating**. One of the principal functions of petroleum refining is to produce a hydrocarbon mixture with a high octane rating. The object is to meet the demands of car manufacturers, who through the years have produced new engines with an increasingly high **compression ratio**. This means the petrol-air mixture is compressed to greater densities before ignition. The task of producing high octane ratings was made easier by the use of chemicals that, when added in small quantities, reduce knock. The most effective of these was tetra-ethyl **lead**. Leaded petrol was introduced in 1925.

The **APHELION** is the point at which the orbit of a planet is at its farthest point from the sun. In the case of the earth, it occurs early in July. It has relevance in the theory of **climate change** proposed by **Milankovitch**.

APHIDS are part of a world-wide family of over 3,600 species, which includes the greenfly. They are about the size of a pinhead and are serious plant pests. They stunt the growth of plants, transmit plant viruses and also deform leaves, buds and flowers. They are very prolific. Wingless females produce living young without fertilization during the summer. When the plant containing mother and offspring dies, some aphids grow two pairs of wings and fly to another plant. They are kept under control with chemical sprays and by natural predators, including ladybird beetles.

AQUACULTURE is the cultivation and harvest of freshwater or marine animals and plants, in ponds, tanks, cages or on protected beds. This is usually done in inland waters, estuaries or coastal waters. It is estimated that commercial fish farming accounts for more than 10% of the world's fish needs. Fish farming usually concentrates on molluscs, including oysters, mussels and clams, because they are usually immobile and fetch high prices. Shrimps and salmon are also farmed, but the stock have to be caught in the wild first, so that they can be brought up to commercial standards in pens. Aquaculture is not new. In Asia freshwater fish have been farmed for some 4,000 years, usually on small farms. The People's Republic of China remains by far the largest aquaculture producer. At the end of the 20th century more than one-half of the fish eaten in Israel and one-quarter in the People's Republic of China and in India came from fish farms. In other parts of the world new forms of aquaculture are being developed and encouraged as a result of fears of a world shortage of fish in the next century. For example, **seaweed** is being grown as food for humans and livestock, and for industrial materials. There are, however, some concerns about the likelihood that farmed fish might be more exposed to toxins and impurities in the water.

AQUEOUS PHASE CONVERSION: See **wet (aqueous) phase conversion**

AQUIFERS are layers of rock, sand or gravel that can absorb water and allow it to flow. An aquifer acts as a **ground-water** reservoir when the underlying rock is impermeable. This may be tapped by wells for domestic, agricultural or industrial use. A serious environmental problem arises when the aquifer is contaminated by the seepage of sewage or toxins from waste dumps (see **leachate**). If ground-water in aquifers is withdrawn more quickly than it is regenerated, a fresh water supply for animals and humans can be depleted. In coastal areas, excess extraction can cause salt water to seep into the aquifer, ruining the supply of fresh water. The Ogallala aquifer—the largest in the world—which stretches from southern South Dakota to north-western Texas in the USA, is severely depleted. The overall rate of withdrawal from this aquifer is eight times its natural recharge rate. Aquifer depletion is also a serious problem in the north of the People's Republic of China and in Mexico City, Bangkok (Thailand) and in many parts of India. Quetta, the city of Baluchistan in Pakistan, was a city of 50,000, but at the beginning of the 21st century had 1m. inhabitants, all plumbing a non-replenishable or fossil aquifer, which scientists reckoned would be exhausted soon after 2010.

The **ARAL SEA** in Central Asia was once one of the world's largest inland bodies of water. Over a period of 25 years it shrank dramatically and, by the end of the 20th century, it had lost 70% of its water, more than 68,000 sq km (26,250 sq miles) having been turned into salt flats and desert. Its surface area shrank by over one-half, its depth fell by 19 m and its salinity tripled between 1960 and 1965 alone. A scheme designed to irrigate a large area of desert in the former USSR diverted water from the lake's major tributaries, the Amu-Dar'ya and the Syr-Dar'ya. The former, the more southerly of the two rivers, is the most severely affected, generally drying up far short of the Aral Sea. The environment of the surrounding area suffered severe damage. There are violent dust storms along the shores and the Sea's waters are too saline for fish and too shallow for ships (once 69 m at its deepest, the Aral Sea could only reach as far as 37 m by 1997). In the irrigated desert area, water in the irrigation channels is badly polluted, drinking water is of a very low quality and **soil erosion** is a rapidly increasing problem. The decline of the Aral Sea made life much more difficult for the surrounding populations. Efforts are under way to restore the Aral Sea, but have met with little success. Uzbekistan and Turkmenistan are the worst affected states, but the other former Soviet, Central Asian countries are also concerned.

The **ARCTIC** generally refers to the north polar regions, but it has also become used to denote the territory lying within the Arctic Circle and to the landscape, climate and plant and animal life found in those regions. Both the Arctic and Antarctic play a very large part in maintaining the earth's heat balance. Since a large part of the earth's water is frozen in the ice fields, snow and glaciers of the polar regions, the possible melting caused by global warming could have world-wide repercussions. Melting could cause catastrophic **sea level** rises and add to **climate change**. Both polar areas only absorb energy during their respective summer period and radiate heat throughout the year. Each year the outflow of energy must be balanced, more or less, by the inflow, otherwise the polar regions would become progressively colder. The vastly different geography of the two polar regions means that there are equally large differences in the way they influence the global **climate**. Unlike Antarctica, which is a huge land mass with bordering oceans and covers about three times the area of the Arctic, the northern region is an ocean, largely or almost entirely covered with **sea ice**, depending on the season (save the less fluctuating Greenland **ice sheet**). The Arctic Ocean has the land of three continents at its perimeter: North America, Asia and Europe. However, the continental shelves that extend into the Arctic Ocean make it one of the world's most nutrient-rich seas and most fertile fishing grounds. Another difference between the polar regions is that the earth is also nearest the sun during the Antarctic summer, so its **atmosphere** receives 7% more solar radiation than reaches the northern atmosphere during the Arctic summer. The southern limits of the Arctic are more difficult to define and include areas of Scandinavia warm enough to support reptiles. Some ecologists set the limit as the boundary where the sparse scrub of the **polar tundra** and the conifer-dominated **taiga** forest meet. The taiga covers more than one-half of Canada and much of Alaska and Scandinavia. Only the top 50 cms of the tundra thaw each summer, but many species of migratory birds use it as their breeding ground. The protection of **flora and fauna** and the exploitation of the mineral resources of the region are the subject of the **Arctic Treaty** and other international agreements. Many scientists believe global warming could melt the tundra's **permafrost**. Aside from destroying the tundra's ecosystem, they fear melting would release vast amounts of **methane**, which would accelerate the **greenhouse effect**. In June 2004, for the first time, UNESCO added two Arctic sites to its list of **World Heritage sites**. Denmark submitted Greenland's Ilulissat Icefjord, where the fast-moving glacier, Sermeq Kujalleq, meets the sea. Russia proposed the natural system of the Wrangel Island reserve in Chukotka; the island has the most **biodiversity** of anywhere in the high Arctic, hosting the highest densities of Pacific walrus and of ancestral polar bear dens, as well as important nesting grounds and plant species (it was also believed to have had a unique species of dwarf mammoth, that survived to as recently as around 2,000 BC).

The **ARCTIC COUNCIL** was established in 1996 by the eight countries bordering the **Arctic** Ocean. The purpose of its foundation was to protect the fragile **environment** of the region. The Council comprises Canada, where the secretariat was initially based, and Denmark, Finland, Iceland, Norway, the Russian Federation, Sweden and the USA. Land within the Arctic Circle is sparsely populated, but environmentalists believe it suffers disproportionately from pollution and other human activities. The Council was to provide a permanent forum for governments to formulate common guidelines to protect the environment from the effects of development. Environmental impact issues were to be given prime consideration when any future decisions were made about mining or petroleum exploration. Various programmes and research projects have allowed the organization to develop the Arctic Council Action Plan.

The **ARCTIC MILITARY AND ENVIRONMENTAL CO-OPERATION (AMEC) PACT** was signed by the USA, Norway and the Russian Federation in 1996. The agreement was to handle jointly the disposal of nuclear wastes from disused and scrapped submarines that are accumulating on the **Kola peninsula** in Russia. Projects include the development of containers for interim storage of liquid and solid **nuclear wastes**, and deciding how to deal with nuclear reactors that were previously discarded simply by dumping them at sea. Environmental groups are concerned that spent fuel and other nuclear waste from vessels withdrawn from service under disarmament agreements could result in **radiation** leaking into the Barents Sea. Norway, which shares a 200 km **Arctic** border with the Russian Federation, had encouraged its partners in the AMEC Pact to draw international attention to the environmental hazards on the Kola peninsula. Critics maintained that, although the disarmament negotiations were a step in the right direction, they failed to address the issue of spent nuclear fuel properly.

The **ARCTIC MONITORING AND ASSESSMENT PROGRAMME (AMAP)** was agreed in 1991 by environment ministers from the eight countries bordering the Arctic Ocean. A task force was formed two years later. Its brief was, first, to discover the extent and sources of pollution from **persistent organic pollutants (POPs)**, **heavy metals** and other elements, and **radionuclides**; and, second, to prepare a plan for the cleaning up and prevention of pollution. Its reports to the **Arctic Council** have helped formulate environmental strategies.

The **ARCTIC NATIONAL WILDLIFE REFUGE, ALASKA** is not only the largest of the USA's 530 wildlife sanctuaries, but was its most pristine. No humans live here and in winter the territory is frozen and lifeless, but in summer the land thunders with the passing of 130,000 caribou.

ARCTIC OZONE DEPLETION is a relatively new phenomenon. The worst **ozone** depletion is over **Antarctica** in the southern hemisphere, but there was also significant ozone depletion over the Arctic during the 1990s. About 15% of ozone is lost during the Arctic spring, corresponding to a 22% increase in radiation. The Arctic will continue to be vulnerable to large ozone losses, a situation that could be made worse by global **climate change**.

The **ARCTIC TREATY** was an agreement made between the then USSR (Russia is the relevant successor state), Canada, the People's Republic of China and Japan in 1990. The four countries pledged to collaborate on the protection of the environment and scientific research in the **Arctic**. The treaty covers the monitoring of pollution, exchanges of technical information and funding for environmental protection programmes. The countries also agreed to co-ordinate research into timber harvesting, wildlife management and offshore petroleum exploration.

AROMATICS IN FUEL is a term describing a category of compounds found in petrol (gasoline) and which includes environmentally unacceptable substances like **benzene**. However, the proportion of aromatics in **hydrocarbon** fuels affects the **octane rating**. Unleaded petrol frequently has higher levels of aromatics in order to increase the octane rating of the fuel.

ARSENIC (As) is used mainly in **pesticides**, plant desiccants and wood preservatives in the form of arsenous oxide, copper acetoarsenite (Paris green) and calcium or lead arsenite. Arsenic is a **carcinogen**. Fruits sprayed with such substances may contain enough arsenic to be toxic to humans. Arsenic may also be emitted by metal smelters and **coal** combustion. Arsenic is a phytotoxic agent that accumulates in aquatic organisms. The level of arsenic in drinking water as a result of **ground-water** contamination is also a major concern. The most serious arsenic contamination of ground-water in the world is in **Bangladesh**, where much of the population is exposed to arsenic, especially in southern and eastern Bangladesh. A similar situation exists in parts of neighbouring West Bengal, an Indian state.

ARTESIAN WELLS are bore holes sunk into **aquifers** to extract the **ground-water**. The term is applied loosely to any well in which water rises under pressure, but not necessarily to the surface. If the water level is higher than the outlet of the well then the water will flow upwards under pressure. If the well-head is above the water level in the aquifer the water has to be pumped up and this is known as a sub-artesian well.

ASBESTOS is the generic name for a group of fibrous mineral silicates. It includes blue asbestos (crocidolite), white asbestos (chrisotile) and brown asbestos (amosite). After they are mined the asbestos fibres are separated from the rock and are spun into a cloth. When inhaled the fibres penetrate the lungs and the tissues of the bronchial tubes, resulting in asbestosis, a crippling lung disease. Asbestos also causes cancer of the lung, the gastro-intestinal tract and mesothelioma, a malignant cancer of the inner lining of the chest. However, because it is a poor conductor of electricity and highly resistant to heat, asbestos has been widely used over the years in fire-fighting suits, and building and insulating materials. In 1968 crocidolite and amosite (both amphibole asbestos, with the more dangerous fibre type) were banned in the United Kingdom, and most countries followed suit. However, there are still many buildings and products still containing asbestos, and many people continue to sicken, as the consequences of asbestos-fibre inhalation can be many years in appearing. Chrisotile or white asbestos (the most common serpentine fibre) continues to be sold. Most asbestos uses were finally banned in the USA in 1989, but a court ruling in 1991 permitted some established usages of chrisotile and declared mass removal impractical and unnecessary. Ten mainly northern European countries have the most restrictive bans on usage. The biggest asbestos mines are in Québec, in Canada.

ASTHMA is a chronic respiratory disease which is increasingly common. The **World Health Organization (WHO)** reckons that anywhere between 100m. and 150m. people world-wide suffer from asthma and that incidence is increasing by about roughly 50% every decade. Developed countries have been affected most, but developing countries too show increases, albeit varying widely. According to Belgian scientists, asthma in Western Europe as a whole doubled in 10 years, from the late 1980s to the late 1990s. In parts of Latin America and Africa, and in India, there is a relatively high incidence of asthma, but in the Western Pacific it ranged from over 50% among children of the Caroline Islands to virtually nil in Papua New Guinea. Asthma can occur when **particulate** material irritates the membranes lining the respiratory tract and gives rise to breathing difficulties. While there are concerns that asthma is linked to **air pollution**, because the growth in the number of cases has coincided with large increases in vehicle traffic, no causal links with levels of air pollution have been demonstrated. A study of six urban areas in the USA showed a statistical relationship between air pollutants and deaths, although the factor most strongly implicated was cigarette smoking. After allowing for smoking and other risk factors, the death rate was 25% higher in the most-polluted city than in the least-polluted city. The link was stronger in the case of tiny particulates, or **PM10s**, in the atmosphere. There are also suggestions that centrally heated, carpeted houses provide an ideal habitat for the house mite, a potential trigger, and possibly even a primary cause, of asthma. A European Union project was established to bring together 14 European research centres to form the European Network for Understanding Mechanisms of Severe Asthma (ENFUMOSA). The network was to study the difference between severe and mild asthma, and generate and exchange information about asthma in each country in order to compare the types, patterns and underlying causes of severe asthma. The WHO participates in a number of initiatives on the issue.

The **ATMOSPHERE** is the envelope of colourless gases surrounding the planet in a 700 km-thick layer. It consists primarily of **nitrogen** (about 78%), **oxygen** (20.9%), argon (almost 1%) and **carbon dioxide** (0.03%), as well as traces of krypton, xenon, neon and helium, plus **water vapour**, traces of **ammonia**, organic matter, **ozone** and various salts. The proportion of most gases in the lower atmosphere is relatively stable, although the level of water vapour, which is a major **greenhouse gas**, can vary greatly. Weather and climate conditions are contained in the lowest zone of the atmosphere, the **troposphere**. The atmosphere is particularly efficient at filtering potentially harmful radiation, particularly **ultraviolet radiation**. It also acts as a shield against solid matter from space, by burning up any particles with heat generated by air friction. Above the troposphere is the **stratosphere**, which extends up to 48 km above the surface of the earth and is where aircraft tend to fly. The **ozone layer** is between the stratosphere and the mesosphere, the latter being the coldest part of the atmosphere and extends up to 80 km from the ground. Beyond that is the thermosphere, where the air is thin and which is itself divided into the lower ionosphere (now important for reflecting radio waves back to earth) and the outermost layer or exosphere.

ATMOSPHERIC CHEMISTRY is the study of chemical processes in the atmosphere. The discovery of the **ozone hole** over the **Antarctic** was one example of atmospheric chemistry. Three chemists, Paul J Crutzen, F Sherwood Rowland and Mario J Molina, won the Nobel Prize for this discovery in 1995.

ATMOSPHERIC CIRCULATION describes the general movement and circulation of air which transfers energy between different levels of the **atmosphere**. The mechanisms of circulation are very complicated. They involve the transfer of energy between the oceans and the atmosphere, the land and the atmosphere as well as the different levels of the atmosphere. Some atmospheric movements occur frequently enough to be recognized, for example the trade winds and the all-important **jet stream**.

ATMOSPHERIC DUST is the quantity of fine particles discharged into the atmosphere which can cause a dust veil and **climate change**. The particles include emission from volcanic activity and **aerosol** clouds from agricultural and industrial activities. The major sources of natural aerosols are dust from drought-stricken soils, sea-salt spray, volcanic eruptions and soot from natural forest and brushwood fires. Major sources of **anthropogenic** aerosols are the burning of **fossil fuels** and **slash-and-burn agriculture** and intensive land development in the developing world. Although the heaviest particles are soon washed out of the atmosphere by rainfall, or simply by gravity, they still affect the **radiation** balance in the atmosphere. Depending on the nature of the materials creating the dust veil, the effect may add to or counteract warming trends attributed to climate change. The effect on warming depends on the quantity and types of the particles and the nature of the land or ocean below. A method of calculating the severity of volcanic dust in terms of the mass of material ejected, and the duration and maximum extent of the spread, was devised by Hubert H Lamb by way of a **dust veil index**. Regional effects of particular atmospheric dusts may be more significant than their global effect. **Sulphur** aerosols in the northern hemisphere may be causing more low cloud cover, because the aerosols increase the condensation rates in clouds.

ATMOSPHERIC GASES include the natural and human-induced vapours and gases in the atmosphere like **water vapour**, **carbon dioxide**, **methane** and **chlorofluorocarbons (CFCs)**, which are responsible for global warming (see **climate change**). The composition of dry air is 78% nitrogen and 21% oxygen. Among the smaller quantities of other gases there is about 0.03% of carbon dioxide, which is one of the **greenhouse gases**. The moisture content of air is critical to calculations about global warming, because water vapour is the principal greenhouse gas. However, the level of water vapour varies enormously by season and region.

ATMOSPHERIC POLLUTION: See **air pollution**

The **ATOMIC NUMBER** defines an element by the number of protons it contains in the nucleus of a single atom. Elements are classified by the number of protons they have. The first element in the periodic table, **hydrogen**, has one proton. Helium has two, and **carbon** has six protons. Among natural elements, **uranium** has the most protons, with 92.

AUDUBON is a name inextricably linked to the origins and rise of the modern conservation movement. John Audubon was a naturalist who in the early 19th century painted all known species of North American birds. His books, illustrated with coloured plates, remain classic works on ornithology. Audubon's publications were among a growing number of writings and lectures by naturalists on wildlife and conservation that led to the development of the modern environmental movement, which was marked by the foundation of the **Sierra Club** in 1892. Other groups, particularly those concerned with birds and their habitats, took Audubon's name as their own, in tribute to his seminal role. The fact that the modern **conservation** movement had its beginnings in the New World has been attributed to the fact that within the memory of a single generation there had been extreme man-made changes in the landscape, as well as disruption and destruction of wildlife.

AVAILABLE NUTRIENTS are the elements in a soil solution that can be readily taken up by plants.

B

BACKGROUND RADIATION is the radiation resulting from natural sources, as opposed to man-made sources, and to which people are exposed in everyday, normal life. It comes from the sun and terrestrial origins, such as rocks and soil. On average, every person receives two millisieverts (mSv) (see **sievert**) of background radiation each year. Amounts vary depending on location. At high altitudes there is more cosmic radiation than at sea level. Other parts of the world have rocks containing small amounts of uranium and thorium that release **radon** gas into the air.

BACTERIA are a group of single-cell micro-organisms, the smallest of the living organisms. Some are vital to sustain life, while others are responsible for causing highly dangerous human diseases, such as **anthrax**, **tetanus** and **tuberculosis**. Bacteria are found everywhere, in the soil, water and air. For instance, fleas on black rats in the holds of Italian ships brought a bacterium to Messina in 1347 which caused the Black Death (bubonic plague) and ravaged Europe's population. In the first scientific use of **biological control** for pest management, a beetle was introduced from Australia to California to destroy cottony cushion scale, a pest that was destroying the state's citrus groves.

A **BACULOVIRUS** is a class of insect virus which has been used as a **genetically modified organism (GMO)** to create a **biopesticide**, an agent that kills insect pests by **biological control**. The baculovirus is used as a carrier, or vector, replacing healthy genes with another carrying the virus which will kill an insect pest.

The **BALTIC SEA** is the partially landlocked, fairly shallow sea that is bordered by Sweden, Finland, Russia, Estonia, Latvia, Lithuania, Poland, Germany and Denmark. It is the largest expanse of **brackish** water in the world and stretches from southern Denmark almost to the **Arctic** circle, a total of 160,000 square miles. The Baltic is a haven for wildlife. It is also tideless, sluggish and has been a victim of pollution from the countries on its shores over many years, from a variety of sources: toxic wastes, **fertilizer** run-offs, petroleum spills and discharges, and domestic wastes. Fish are contaminated and the Baltic seal is practically extinct. In 1974 the **Helsinki Convention** was the first regional agreement drawn up to protect an enclosed sea from land-based pollutants. Originally signed by seven states (including the USSR), the revised Convention, which came into effect in 2000, had 10 contracting parties, including all nine countries and the European Union (to which all but Russia belonged). In 2001 the Council of Baltic Sea States, which consists of all nine countries and the European Union, established a unit to pursue the Baltic 21 agenda, a regional version of **Agenda 21**. Of particular concern was the need to reduce the incidence of **eutrophication** as a result of excessive nutrients being washed into the Baltic. Since the 1990s pollution from **polychlorinated biphenyls (PCBs)**, **DDT** and **mercury** has shown signs of decreasing.

BANGLADESH could be one of the countries most vulnerable to flooding as a result of rises in **sea levels** caused by global **climate change**. The country is periodically inundated by monsoon floods, which, on average, cover 20% of the territory every other year and 37% one year in every 10. In 1987 and 1988 floods affected 40% and 60% of the territory, respectively. As the Bangladeshi people are used to repeated flooding, they have historically lived and coped with the problem. However, the country's **population** doubles in size every 30 years, which puts ever-growing pressure on limited land resources and increases the demand for food. The potential impact of climatic change from global warming could be disastrous. Much of the country is low-lying delta (hence it is prone to flooding), vulnerable to sea surges. This would be exacerbated if sea levels rose, and the country would lose much territory outright. (It has been calculated that merely a 1 m rise in sea level would deprive Bangladesh of 18% of its land area.) Bangladesh established a Flood Plan Co-ordination Organization to conduct studies with the aim of finding a lasting solution to the problem. It also asked the **World Bank** for help in developing a five-year Flood Action Plan. Bangladesh is the site of the most serious **arsenic** contamination of **ground-water** in the world. Much of the population is exposed to arsenic, especially in southern and eastern Bangladesh.

The **BASLE CONVENTION ON THE CONTROL OF TRANSBOUNDARY MOVEMENTS OF HAZARDOUS WASTES AND THEIR DISPOSAL** was an agreement drawn up under the auspices of the **UN Environment Programme (UNEP)**. It came into effect in May 1992 and was adopted by 116 governments and the European Communities (from November 1993 known as the European Union). Its main aim was to impose strict controls on the trans-border transport of hazardous wastes, and on their disposal. It gives developing countries the right to turn away shipments of hazardous waste from their shores. It means that a country exporting substances that it considers hazardous has to get permission for their delivery from the proposed importer, who then has 60 days to issue a 'prior informed consent'. The ultimate goal of the Convention is to cut the generation of hazardous wastes to a minimum. The **recycling** industry in the developed countries considers that the Convention is too strict. Before the Basle Convention came into force, the **Organisation for Economic Co-operation and Development (OECD)** approved new rules to protect the recycling industry which threaten to undermine the agreement. The rules place hazardous waste substances into three categories. Those on the red list, dangerous hazardous wastes, will still remain under the control of the Basle Convention. Amber substances on the OECD's list of rules give the importer 30 days, not the Convention's 60 days, to accept the load and once permission is given it is valid for a year. The European Communities included the lists in their waste regulation, which was officially adopted in 1993. Critics are alarmed because some substances on the OECD's green list, including **lead**, **cadmium** and plastics, are not considered dangerous by member states and could, therefore, be exported anywhere in the world. The Basle Convention (160 parties by 2004) and its organization continued to develop regulations to help protection of human health and the environment. On 1 July, for instance, the secretariats of the Basle Convention and the Convention for the Protection and Development of the Marine Environment of the Wider Caribbean Region (the 1983 Cartagena Convention) signed an understanding on co-operating in achieving their aims.

BATTERIES provide the electricity for the starter motors, ignition and lighting systems of vehicles. The use of batteries as the power source for electrically driven transport was for many years limited to vehicles like fork-lift trucks and some retailers' delivery vans, including milk-floats. Electric power is attractive because it avoids the emission of air **pollutants** at the point of use. Hence **electric vehicles** could also be described as **zero-emission vehicles**. The disadvantages of batteries relate principally to their weight, cost and durability, and the need for frequent recharging for long-distance journeys. **Hybrid vehicles** offer an alternative to complete dependence on batteries: a gas turbine or small internal combustion engine drives a generator that in turn produces electrical energy to charge the storage system. In the long term, the preferred technology for electric propulsion may be **fuel cells**.

BECQUEREL (Bq) is the modern unit for expressing the radioactivity of a substance. It is superseding the **curie**. One becquerel is equal to the decay, or disintegration, of one atomic nucleus per second. Antoine Henri Becquerel (1852–1908) was the discoverer of natural radioactivity, in 1896. His name was originally given to the rays emitted by uranium, and he shared the 1903 Nobel Prize for Physics with Pierre and Marie Curie, for their study of the 'Becquerel radiation'.

BELS: See **decibels**

BENEFIT–COST ANALYSIS: See **cost–benefit analysis**

BENTHOS is the highly varied collection of plant and animal life that inhabits the sea bottom, the Benthic zone. Most are found in relatively shallow seas less than 200 metres deep. The type of life found in the zone depends on the proportion of sand and rock, the amount of light, and the temperature of the water. Organisms have been able to colonize inhospitable regions of the zone because of extreme specialization. Benthos, an important part of the **food chain**, occurs in many forms: some are fixed to the sea floor, while others are capable of active movement on or within the sediment on the bottom. In shallow waters, the Benthic zone is also home to molluscs, such as oysters and mussels, crustaceans, such as crabs and lobsters, and worms, starfish and bottom-dwelling fish, including plaice and cod, which are important human food sources.

BENZENE (C_6H_6) forms 3%–5% of petrol (gasoline), depending on the national regulations that limit its use. Benzene is a colourless, liquid **hydrocarbon**, discovered in 1825 by Michael Faraday, the British chemist, as an extract from coal tar. However, in 1865 the German chemist, Frederick Kekule, recognized that the benzene molecule was a ring of six carbon atoms, and it was this that made benzene so industrially important. The benzene ring is the building block of a vast family of organic chemical compounds known as the aromatics (see **aromatics in**

fuel). It was on this discovery that the creation of the modern petrochemical and organic chemical industries was based. Benzene is also well known as a **carcinogen**, or cancer-causing agent. New limits on industrial exposure of 100 parts per 1,000 million, proposed in the USA in 1993, meant the air in US factories would have benzene levels less than those commonly found in European cars and in congested urban areas. However, new European Union regulations in 2000 marked an intensification of the process of treating benzene far more strictly.

BENZENE HEXACHLORIDE (BHC): See **organochlorine compounds**

BENZO(A)PYRENE (BaP) is the most-hazardous isomer in the benzopyrene group of **carcinogens** contained in car fumes, industrial solvents and the smoke from tobacco and wood stoves. It belongs to a large family of compounds called **polycyclic aromatic hydrocarbons** produced by the incomplete combustion of petrol and fuel oil.

The **BERNE CONVENTION** is the name given to the Convention on the Conservation of European Wildlife and Natural Habitats. Originally negotiated in Switzerland and adopted in 1979, the Convention aims to protect all forms of endangered wildlife in specified parts of Europe, mainly through its appendices, which prescribe different levels of protection for species of flora and fauna, as well proscribing certain forms of use and certain methods of catching and killing.

BEST AVAILABLE TECHNIQUES NOT ENTAILING EXCESSIVE COST (BATNEEC) is a principle of environmental protection according to which expenditure to control pollution should be proportional to the environmental benefit it brings. BATNEEC should also be used to direct releases of waste into the environmental medium—land, air or water—that is best able to absorb it, with minimum pollution of the environment as a whole.

BEST AVAILABLE TECHNOLOGY (BAT) is the term the European Commission adopted in 1993 to signify the latest, most-advanced techniques that are in use or should be used to reduce emissions of pollutants to the environment. Where appropriate, **European Communities directives** on pollution control specify the best available technology that will achieve the required standards. The use of BAT is intended to ensure that 'end-of-pipe solutions', when untreated pollutants are emitted directly to the air, water or land, are replaced by clean, low-waste technologies.

BEST PRACTICES are those environmental or social practices that are worth serving as models for others due to their success in achieving sustainable development. For example, at the **Habitat II** conference in Istanbul, Turkey, in 1996, a CD-ROM describing model housing and shelter projects was distributed so that other countries could emulate them.

BETA RADIATION is the name given to the **ionizing radiation** which is produced as a stream of high-speed electrons emitted by certain types of radioactive substance when they **decay**. The intensity of **radiation** energy produced in human tissue by a beta particle is a hundred times less than that produced by an **alpha radiation** particle, but it travels slightly deeper into tissue.

BHOPAL is the capital of Madhya Pradesh in central India and site of one of the worst environmental disasters in world history. In the early hours of 3 December 1984 an estimated 45 metric tons of the gas methyl isocyanate, a highly volatile and toxic chemical used in the manufacture of **pesticides**, escaped from a tank at a plant owned by Union Carbide India Ltd (UCIL), a subsidiary of Union Carbide, an international company based in the USA. The gas drifted over the densely populated area surrounding the plant, killing many inhabitants instantly. There was widespread panic as thousands of others tried to escape from the city. The official death toll was 2,500, although a subsequent state government investigation attributed some 3,800 deaths to the incident, and some claim that up to 5,000 died immediately. As many as 50,000 other people were treated for respiratory problems, temporary blindness and eye irritations, lung damage and severe vomiting, which were caused by exposure to the gas. The health of an estimated 200,000 people was affected and years after the leak thousands were still being treated for side-effects. A report on the disaster concluded that it was caused by a combination of badly designed equipment and sub-standard operating and safety procedures. By 1987 more than 500,000 local residents, more than two-thirds of Bhopal's population, had put in claims for compensation (this had risen to 600,000 by 2004). In 1989 the Indian Supreme Court awarded damages of a mere US $470m. against Union Carbide. The criminal case entered its 15th year in 2004, with many having not testified, the Government seeming reluctant to pursue the case and the US-based Union Carbide having severed its links and investment in UCIL in 1994. There were also fears about a second disaster in the making, from the slow pollution of **ground-water** by toxic chemicals dumped or badly stored at the site. A lawsuit against the parent company was initiated in New York state, USA, in 1999. Although the case suffered a reverse in 2003, a superior court reinstated much of the action in March 2004.

BIOACCUMULATION is a process that can be both environmentally useful and environmentally hazardous. It can also occur naturally or be formed as part of a planned biotechnology process. Bioaccumulation is the accumulation by an organism of materials that are not an essential component or nutrient of that organism. Usually it refers to the accumulation of metals, but it can apply to bioaccumulation of persistent synthetic substances such as **organochlorine compounds**. Many organisms, such as plants, **fungi** and **bacteria**, will accumulate metals when grown in solutions containing them. The process can be employed usefully as a purification process to remove toxic **heavy metals** from waste water and contaminated land. The hazard arises when crops are grown unwittingly on polluted land, and the plants become the route for poisons to enter the **food chain**.

The **BIOCENOSE** is a community of all living things (plants, animals and other organisms), which, together with the environment or **biotope** in which it exists, constitutes an **ecosystem**.

The **BIOCHEMICAL ENERGY CYCLE** involves the absorption of **solar energy** by plants and its conversion into **carbohydrate** compounds, which are then available as food for animals.

BIODEGRADABLE refers to waste substances that can be decomposed by bacteria or other biological means. Commonly, this suggests that the decomposed remains will be non-toxic and will not accumulate in the **food chain**. Paper, wood, wool and leather are biodegradable.

BIODIVERSITY: See **biological diversity**

BIOFUELS are the bulk biological materials that can be converted into liquid fuels, such as **ethanol** and methanol, either for use as fuel or as the raw material feedstock of the chemical industry. The process may begin with crops being deliberately grown for this purpose, and these range from wood pulp and sugar cane, to seeds like maize and oil-seed rape that can be converted into methanol. However, the initial **biomass** can be wet materials such as starch, sugar, sewage waste water and other effluent used in **fermentation**. Processing of sugar cane, by fermentation and distillation, was developed in Brazil in response to the 1973 energy crisis in order to produce ethanol. Brazil developed a major fuel called Proalcool, producing over 3,000m. gallons of it by 1990. In the USA an ethanol-based fuel called Gasohol was developed. Methanol is less easy to make. Another gaseous biofuel, or **biogas**, is hydrogen, which is made by **photolysis** of water, which almost mimics the process of **photosynthesis**. However, in photosynthesis in plants the hydrogen is not created in the form of a gas, but is used to make sugars for growth of the plant.

BIOGAS is a gas, rich in **methane**, which is produced by the **fermentation** of animal dung, human sewage or crop residues in an airtight container. It is used as a fuel, to heat stoves, lamps, run small machines and to generate electricity. The residues of biogas production are used as a low-grade organic fertilizer. Biogas fuels do not usually cause any pollution to the atmosphere, and because they come from **renewable energy** resources they have great potential for future use.

The **BIOGEOCHEMICAL CYCLE** is the movement of chemical elements in a circular pathway, from organism to physical environment, back to organisms. The process is termed a nutrient cycle if the elements concerned are **trace elements**, which are essential to life. A biogeochemical cycle occurs when vegetation decomposes and minerals are incorporated naturally in the **humus** for future plant growth.

BIOINVASIVE SPECIES are foreign organisms—animal or plant species—that disturb the ecological balance of an environment into which they have been accidentally or deliberately introduced. Japanese knotweed, which spread uncontrollably in the United Kingdom, and the escape of mink are examples. In the USA a major example is the sea lamprey fish, which has been devastatingly destructive for commercial fisheries because it preys on trout and other stocks. It gained access to the **Great Lakes** through the Welland Ship Canal built in 1829 to bypass Niagara Falls. The problems of bioinvasive species imported through ship ballast is a particular problem in the **Black Sea**, and in 2001 the countries of that region urged international action on ballast regulations.

BIOLOGICAL CONCENTRATION, or biomagnification, relates to the behaviour of certain contaminants, especially **chlorinated hydrocarbons**, when absorbed into the body. These chemicals get into the **food chain** and are hazardous because they are very stable and highly soluble in the fatty tissue. They resist the processes of **metabolism** and accumulate in the body. Substances that possess the characteristics of biomagnification include the pesticides **DDT**, dieldrin, **aldrin**, endrin, heptachlor and mirex. They also include the **polychlorinated biphenyls**, a family of chemicals used extensively in industry since the mid-1930s.

BIOLOGICAL CONTROL is a form of pest control that is an alternative to using synthetic chemicals. It is the control of one species by another. Australia can boast involvement in both the first scientific use of biological control for pest management (a beetle was introduced from Australia to California, in the USA, to destroy cottony cushion scale, a pest that was destroying the state's citrus groves) and the most famous example (the introduction of myxomatosis into Australia in order to control rabbits). Another organism used for years as a biological control agent has been the bacteria *Bacillus thuringiensis*, which produces a protein that kills caterpillars. *Bacillus thuringiensis* has been used as a spray. Recently scientists isolated the protein and have used it to make a **biopesticide**.

BIOLOGICAL DIVERSITY describes the incalculable natural variety of the world's flora and fauna and their **habitats**. Although the record of all living species is still incomplete, the concern about a reduction in biological diversity rests on reports produced by environmental and scientific groups, which point to the threat of extinction faced by plants and animals as a result of man-made phenomena, such as **acid rain** and **deforestation**. Biodiversity is also a possible casualty of **climate change**, which could also cause further loss of forests, and the disruption of **wetlands** and the **polar tundra**, complex **ecosystems** that have existed for millennia. Many species might be unable to migrate rapidly enough to cope with climate change in order to find a suitable habitat. Since species of flora and fauna are tied together in a web of interdependence, there is concern that the whole ecosystem could unravel as a result of the destruction of a vital link. Forests, for example, have adapted geographically to a narrow temperature and moisture range. In northern temperate zones there are profound changes in the belts of vegetation, coinciding with a shift of 4°C (7.2°F) in the average temperature.

BIOLOGICAL DIVERSITY CONVENTION: See **Convention on Biological Diversity (CBD)**

The **BIOLOGICAL EFFECTS of IONIZING RADIATION (BEIR)** is a special committee, established by the US National Academy of Sciences, to investigate the way **ionizing radiation** affects the human body. As a result of its findings, it also proposes safety limits. The committee said, in a report published in 1972, that there is a risk in any dose of radiation, but contradicted this in another report eight years later. The world's scientific community is uncertain about the effects of exposure to low-level radiation over a long period because there is little information on this issue. Most observations are based on information gained from observing survivors of the atomic bomb explosions in Japan at the end of the Second World War. The BEIR committee continues to produce reports periodically.

BIOMASS refers, strictly speaking, to the total weight of all the living things in an ecosystem. However, it has come to refer to the amount of plant and crop material that could be produced in an ecosystem for making **biofuels** and other raw materials used in industry. Biomass can be divided into several areas of activity. Algal biomass refers to single-celled plants that can be grown in ponds, by **aquaculture**, to make food materials. Plant biomass refers to crops like sugar cane.

A **BIOME** is a large geographic area, and more precisely one of the 14 major ecological regions into which the world can be divided. Each biome comprises a very complex **biotic** community of a myriad of plants and animals. Biomes are defined in terms of their animals and plant life, which in turn reflect the major climatic regime of the region: **polar tundra**, **desert**, **mangroves** and a variety of **forest** and woodland, and various **savannah**, **grasslands** and **steppelands**, as well as **oceans** and **ice sheets**.

BIOPESTICIDES are naturally occurring biological agents used to kill pests by causing specific biological effects rather than by inducing chemical poisoning. The idea is based on mimicking processes that arise naturally, such as protection of the coffee bean by its caffeine content, or use of a bacterial substance, known as the BTK protein, that interferes with the absorption of food by some insects. The argument for biopesticides, and against conventional chemical pesticides, is that they are more easily **biodegradable** and more target specific.

BIOREMEDIATION describes the use of biological processes, using micro-organisms, to clean up pollutants. It ranges from sewage treatment to methods of destroying or removing toxic materials in polluted soil and waste tips. Contaminated sites can be inoculated with appropriate **microbes**.

The **BIOSAFETY PROTOCOL** to the 1992 UN **Convention on Biological Diversity (CBD)** has the objective of ensuring the safe transfer, handling and use of living modified organisms (LMOs—also known as **genetically modified organisms (GMOs)** that may have an adverse effect on biological diversity and its components. Negotiations for a biosafety measure were launched in 1994. The negotiations proved extremely contentious because six grain-exporting nations known as the Miami Group (Argentina, Australia, Canada, Chile, Uruguay and the USA) feared that such a protocol would inhibit trade. The participating governments were unable to adopt the Cartagena Protocol on Biosafety (named after the Colombian city in which the original conference of parties had been convened one year previously) until January 2000. Even now the USA remains uncommitted in any way to the agreement. The Protocol establishes a Biosafety Clearing-House enabling countries to share information about genetically modified organisms. Exporters are required to obtain an importing country's approval, through a procedure known as advance informed agreement (AIA), for initial shipments of certain GMOs intended for release into the environment, for example seeds and trees. GMOs intended for food, feed and processing—in other words, most commodities—are exempted from the AIA requirement. However, their labelling must indicate the presence of GMOs, so that countries can decide whether to import those commodities based on a scientific risk assessment. New negotiations, scheduled to end by 2004, were to be initiated to address the issue of liability for any damage resulting from cross-border movements of GMOs. Palau became the 50th country or regional economic integration organization party to the CBD to ratify the Protocol, in June 2003, and it duly came into force on 11 September. One year after Palau, Gambian ratification gave the Protocol exactly 100 parties.

The **BIOSPHERE** is the part of the earth that supports life and consists of the soil (lithosphere), oceans and waterways (hydrosphere) and the **atmosphere** up to 15 km above the earth's surface. It has three main functions for plants and animals: providing a safe habitat so that individual organisms can complete their life cycle; providing a stable habitat to allow evolution of species; and providing a system in which energy is supplied by the sun and the materials essential to life are recycled. The biosphere depends on a complex series of inter-relationships between the inorganic (soil, water and air) and living organisms. The biosphere supports a multitude of different **ecosystems**, each one interacting with its neighbour. Hence, a change in one ecosystem triggers a knock-on effect, changing neighbouring systems. The word biosphere was coined by Eduard Suess, an Austrian geologist, but deepened into its present meaning in the 1920s by Vladimir Ivanovich Vernadskii, the Russian scientist who first popularized the concept of the noosphere, a sort of intellectual equivalent of the physical biosphere.

BIOSPHERE 2 was a 1.3-ha structure designed to contain an artificial environment. It was built near Oracle, Arizona, USA, by Edward P Bass, Biosphere Space Ventures and others, at a cost of some US $150m. The project was designed to see if people could live in a closed, artificial **biosphere** and if experiments could be conducted without harming the earth ('Biosphere 1'). The project conducted two sealed experiments, the first for two years from September 1991 and the second for six months in 1994. The project lost some credibility when, in the first experiment, **oxygen** and **carbon dioxide** levels fell and had to be augmented from outside. It was an engineering rather than a scientific achievement, but the project demonstrates the complexity and scale of the natural environment and the dangers inherent in attempting to emulate such a system (part of the science of biospherics). Management was transferred to Columbia University (based in New York, USA) in 1995, which between 1996 and 2003 put 1,200 graduate students per year in Biosphere 2.

BIOSPHERE RESERVES are protected land and coastal areas that are approved under the **Man and the Biosphere (MAB)** programme of **UNESCO** in conjunction with the **Convention on International Trade in Endangered Species of Wild Flora and Fauna (CITES)**. Each reserve has to have an **ecosystem** that is recognized for its diversity and usefulness as a **conservation** unit. The reserves have at least one core area where there can be no interference with the natural ecosystem. A transition zone surrounds this and within it scientific research is allowed. Beyond this is a buffer zone that protects the whole reserve from agricultural, industrial and urban development. Biosphere reserves and buffer zones were regarded as examples of a new generation of conservation techniques. A 1992 **World Resources Institute** report, published in collaboration with the **UN**

Environment Programme and the **UN Development Programme**, said that many of the reserves were created simply by giving the designation to existing **national parks** or **nature reserves**, with no new land or separate regulations to accompany the designation. Agencies have also not received enough money to make the complicated zoning system work. The report also said that buffer zones were meant to enhance the park or reserve's conservation aims and to do this there needs to be control over human activities within it, while at the same time encouraging economic incentives to people living near the park, so that the area of the natural **habitat** in question can be widened. In practice this only works if the management has the authority and ability to carry out the regulations. The Network of World Biosphere Reserves was established at an international conference in Seville, Spain, in 1995. By July 2003 there were 440 biosphere reserves in 97 countries.

BIOTA is the general description of all the **flora and fauna** of the planet. It covers all living things from the tiniest submicroscopic organism to the elephant and giant redwood tree, all of which are supported by the **biosphere**.

BIOTECH or **BIOTECHNOLOGY** is a combination of biology and technology. The word first appeared in print in 1919 and is used to describe developments in the application of biological organisms for commercial and scientific purposes. So 'bio' stands for biology and the science of life, and 'tech' stands for technology, or the tools and techniques that the biotechnologists have in their workbox. Those tools and techniques include micro-organisms and a range of methods for manipulating them, such as **genetic engineering**. Historically, the first use of biotechnology took place centuries ago with the discovery of **fermentation** of wine and beer, and later bread. Fermentation still remains at the centre of biotechnology. However, modern biotechnology came of age with genetic engineering, an advance that has given scientists the ability to alter nature, and take advantage of the processes and substances that **microbes**, and anything else with genes, produce naturally without any prompting. The new technologies, developed in the early 1970s, allowed scientists, for the first time, to work in detail on the genetic changes of cells, genes and DNA (deoxyribonucleic acid), the chemical chains containing the genetic information of most organisms. This meant that genes could be transplanted to another plant or animal. This has led to long debates about the implications of the release of genetically engineered microbes into the environment. It also pioneered the way forward into **gene therapy**. The first commercial product of biotechnology was synthetic insulin (first licensed for use in the USA in 1982). Biotechnology (Bt) crops such as soybeans and corn were first grown commercially in 1996 (99% of such crops are still soybeans and corn, together with cotton and canola). By the beginning of the 21st century biotech crops were fairly widely used in some parts of the world, especially North America, but restrained elsewhere by fears of environmental contamination by **genetically modified organisms**. The first field test of a genetically modified vertebrate, a trout, was approved in 1990, and the first cloned animals were born in the mid-1990s, while at the beginning of the 2000s rice was the first food plant to have a map of its genome to be completed.

BIOTIC refers to the living organisms, plants and animals in an **ecosystem**.

A **BIOTOPE** is an environment in which a community of organisms exists, and together they constitute an **ecosystem**.

The **BIRDS DIRECTIVE**, the Directive on the Conservation of Wild Birds, was issued by the Council of the European Communities (now the European Union) in April 1979 and obliged member states to create protected areas, to ensure their upkeep and management and to re-establish destroyed **biotopes**. This system became part of the wider initiative enshrined in the **Habitats Directive** of 1992. Nevertheless, a European Commission report on wild bird species in Europe published in March 2002 identified a decline of almost 25% in their population, despite the efforts of just over 20 years.

The **BLACK SEA** is the large inland sea bordered by Bulgaria, Romania, Ukraine, Russia, Georgia and Turkey. Over 16m. people live in the coastal regions around the Black Sea, an inland water basin that reaches a maximum depth of over 2,200 m (average depth is 1,300 m) and covers 440,000 sq km (including the 18,000 sq km of the shallow Azov Sea). Seventeen countries lie in the main drainage basin of the Black Sea. Among the great rivers flowing into the Black Sea are the Danube (the second largest in Europe), the Dneiper (third), the Dneister, the Don (fourth) and the Kuban. A severely limited outflow to the Mediterranean is through the narrow Bosphorus Straits (thence to the Sea of Marmara and through the Dardanelles to the Aegean), which also allows an undercurrent of seawater (less than one-half the volume of the outflow) back into the Sea. This water does not mix and the Black Sea is characterized by a shallow surface layer of fresher water over a much colder, more saline, and hence denser, lower layer. The lack of mixing means that there is insufficient oxygen for decomposition, and what there is is used entirely by bacteria, leaving 87% of the Black Sea otherwise devoid of marine life (i.e. below about 180 m beneath the surface). However, the surface of warmer, more hospitable waters is thinning, as the **anoxic** layer rises. The Black Sea has some of the world's largest deposits of several mineral gases, including **ammonia**, **methane** and ethane. Tons of highly toxic hydrogen sulphide, the debris of decomposed remains of animals and vegetation, are contained in the deep lower waters. There are plans to build a power station to use the deposits of mineral gas for electricity generation. The plan also includes the development of a pumping station designed to separate water saturated with the gases and the hydrogen sulphide, which will be used to produce **fertilizers**, with clean water being returned to the sea. It is estimated that money spent on purifying the sea will amount to only 20% of the resources accruing from the power that could be generated.

The Black Sea is also severely polluted by the dumping of waste and by waste released from the factories on its shores. This threatens the life that flourishes in the top layer of water, on the sea shelf, and in the deltas and wetlands along the coast. Protection measures were slow in coming to the Black Sea region, and have not been helped by instability in some of the countries, but progress has been made since the 1990s. By early 1994 the six littoral countries had ratified their 1992 Convention for the Protection of the Black Sea against Pollution, signed in Bucharest, Romania, in 1992. The implementation of the Bucharest Convention is entrusted to a secretariat known as the Istanbul Commission after its home, the Turkish city on the Bosphorus. Further ministerial co-operation and international assistance also enabled the 1993 Black Sea Environmental Programme (BSEP) to co-ordinate action and information exchange between about 40 institutions in the six countries. Among the BSEP initiatives later in the decade was the Black Sea Strategic Action Plan (BS-SAP), which was adopted by all the countries in Istanbul in 1996. This support was reaffirmed and revised in June 2002. In 2002–04 the Global Environmental Facility and the UN Development Fund were co-operating in an Ecosystem Recovery Project, focusing on the problems of **eutrophication**. The main problem has been fertilizer run-off and the resulting excess of nutrients entering the Black Sea (mostly from the coastal countries themselves, but 30% from other countries, via the Danube), but also solid-waste, sewage and toxic-waste discharges, petroleum pollution and the introduction of **bioinvasive species**, mainly dumped from ship ballast.

BLEACHING: See **chemical bleaches** and **coral reefs**

The **BLOOD–BRAIN BARRIER** is the mechanism by which the body usually keeps poisonous substances that have invaded the blood stream from getting into the brain and central nervous system. There are substances that manage to breach the barrier, such as the toxic **heavy metals**.

BOD, standing for biochemical (or biological) oxygen demand, is an official measure of how polluting a liquid is when discharged into a river, stream, brook or lake. Sewage is treated with the main aim of reducing the BOD of the liquid, so that it can be discharged without causing serious pollution. Those authorities concerned with water pollution usually set limits on the volume and BOD of effluent before it can be discharged. The organic matter in healthy streams and lakes gives the water a BOD of a few parts per million, which the system can handle easily. Strict pollution controls, mostly on discharges of sewage, have greatly reduced the BOD levels in some of the principal rivers in Western Europe and North America in recent years. The River Thames in the United Kingdom, for example, was so polluted in the late 1950s that few fish could survive in it. In the 1960s new sewage works were built and others modernized, and the fish began to return to the river. The number of birds, swans, teal, shelduck, dunlin and tern has also increased since the 1960s. However, BOD cannot measure the extent to which a waterway may be polluted with harmful chemicals used in **pesticides**, **nitrates** from farmland or **leachates** from toxic waste dumps unless they raise the oxygen demand of the water.

A **BOILING WATER REACTOR (BWR)** is a thermal or slow reactor similar to the **pressurized water reactor**. The expense of turbine shielding and radiological protection tends to counteract the costs of a simpler design. An advanced boiling water reactor (ABWR) has also been developed. BWRs provide about 4% of the world's nuclear-generated electricity (in 2001 there were 92 BWR nuclear plants, of which 32 were in the USA and 28 in Japan).

The **BONN CONVENTION** is the Convention on the Conservation of Migratory Species of Wild Animals (CMS), which was drawn up to protect all migratory species considered to be endangered through all or part of their range. Signed in 1979, the CMS came into effect in 1983 and imposes strict conservation responsibilities. Hunting or trapping of 51 species threatened with **extinction**, half of which are birds, is strictly prohibited. The Convention also aims to encourage countries along migration routes to draw up agreements to protect another 2,000 threatened species. Animals have always migrated, either to find food, an agreeable climate or to spawn. However, as they migrate along established routes, across land, through the air or the sea, many species are seriously threatened, either during the journey or on arrival at their destination. Apart from hunting, trapping and netting, many long-established habitats have been destroyed or damaged by pollution. **Petroleum spills** pose a serious threat for migrating birds, and, at the end of their journey, breeding, resting and feeding places are often threatened because marshes, **wetlands** and estuaries have been drained to make way for agriculture or industry, or for recreational facilities. Four regional conservation agreements have been concluded under the Bonn Convention: on seals in the Wadden Sea (a part of the North Sea along a stretch of the Dutch, German and Danish coasts); on bats in Europe (Eurobat); on small **cetaceans** in the **Baltic** and North Seas (ASCOBANS); and on African-Eurasian migratory water birds (AEWA). The Bonn Convention encompasses more than 40 countries and four continents.

BOREAL FOREST, also known as **taiga**, describes the largely coniferous forest areas in the northern hemisphere between the latitudes of 50° and 70°. Large areas of boreal forest are found in Scandinavia, Siberia and North America. Severe winters and heavy snowfalls mean that the growing season is short, sometimes as little as three months. As well as the spruce, pine, larch and fir trees, a few hardwoods also grow among the conifers and make up a valuable timber resource. Boreal forests are home to the beaver, sable, mink and fox, as well as larger mammals, such as the grey wolf and its prey, caribou, reindeer and moose. Bears also have ranges that spread northwards into the boreal forests. Species that do not hibernate during the winter have adapted to make it easier to move over snow. For example, the reindeer and moose have evolved splayed-out feet to support their weight. Boreal forests also provide industrial resources, such as coal, natural gas, petroleum and mineral ores. The needle-like leaves of the conifers retain heat as they do not radiate heat quickly or lose it to the wind. The conical shape of the trees makes it easier for them to shed snow before its weight causes structural damage. There are frequent frosts in the boreal forests and, in fact, only some two months in the year are free of frost. While some boreal forest has been over-exploited by man for timber and wood pulp, there are, because of their enormous economic value, many areas that have strict forest management schemes, with major replanting programmes designed to replace trees lost to forestry.

BORON (B) is an unusual element in that it occurs in a non-metallic and a metallic form. The element is never found free in nature, but is present in volcanic springs and rock formations of volcanic origin. In one form it is regarded as a mineral element, commonly found as boric acid in soil, providing an essential **trace element** for plant growth, but not, apparently, essential to animals.

BOVINE SOMATOTROPHIN: See **BST**

BOVINE SPONGIFORM ENCEPHALOPATHY (BSE) is a disease in cattle that threatens the **food chain**. It is also known informally as mad cow disease because of the erratic behaviour of cattle suffering from the brain damage caused by the disease. The risk to humans lies in the hypothetical transmission of the illness to trigger its human equivalent, Creutzfeldt-Jakob Disease (CJD). A variant of CJD (vCJD) is thought sometimes to be caused by consuming contaminated beef products made from diseased cattle. The longer established, equivalent disease in sheep and goats is scrapie, and in deer and elk chronic wasting disease. Analysis of brain proteins linked with both diseases has shown genetic similarities between the animal and human forms of the disease. The precise cause of either disease is surrounded by uncertainties, but an abnormal form of a naturally occurring protein called a prion is known to be involved. Studies in the mid-1990s indicated that normal prions are important to the function of brain cells. However, in diseases such as BSE and CJD, the prions mutate and begin to destroy brain cells, resulting in a deterioration of the structure of brain tissue. Scientists comparing cow prion genes and human prions to those found in animals such as sheep, rodents and monkeys found two significant similarities in the genetic sequences for cows, humans, chimpanzees and gorillas that no other animal shared. An outbreak of BSE in British cattle in the 1990s led to a ban on British beef exports and the instigation of a culling programme (the disease was first reported in the United Kingdom in 1986, but the scale of the problem did not emerge at once). The disease was subsequently found in other countries, mainly in Europe (by the start of 2004 Sweden was the only European Union country to be completely free of BSE), but also in Israel and Oman. There have been occurrences elsewhere, notably, recently, one case in the USA in December 2003, which caused considerable alarm, but the infected cow was claimed to have arrived from Canada.

BP is the abbreviated form of the expression 'before present' day. It is widely used in geochronology, the study of dating various substances and layers from the earth, especially with regard to **carbon dating**.

BRACKISH is the term used to describe water which is seawater mixed with **fresh water**. It is any water that contains too much salt for drinking, but not enough to be seawater. The largest brackish body of water in the world is the **Baltic Sea**.

The ***BRAER*** was the Liberian-registered supertanker that went aground on the coast of the Shetland Islands in the United Kingdom in January 1993. It was carrying 85,000 metric tons of Norwegian light crude petroleum, making it, potentially, one of the largest petroleum spills in history, and having the capability of causing more environmental damage than the ***Exxon Valdez***, which went aground in Alaska, USA, in 1989. However, the petroleum was dispersed by one of the longest storms in many years. Nevertheless, hundreds of birds died and thousands of salmon from Shetland's fish farms were contaminated, eventually resulting in compensation claims of some ú50m. Most of the pollution dispersed into the sea, but a large amount was blown onto the land, contaminating fields where sheep grazed.

The **BRANDT COMMISSION** is the popular name for the Independent Commission on International Development Issues (ICIDI). It was named after the former Chancellor of the Federal Republic of Germany (West Germany), Willy Brandt, who chaired the body. The 18-member Commission, which sat in 1977–79, was given the brief to examine economic and social inequalities and find solutions to them. Its report, *North–South: A Programme for Survival*, published in 1980, recommended urgent action to improve relations between the developed and the developing worlds, covering aid, food, trade and energy. The successor body was disbanded after publishing a second report, *Common Crisis*, in 1983. In 2001 the Brandt 21 Forum was formed to revive interest in, update the proposals of and advance the aims of the old Brandt Commission.

BRENT SPAR is the disused North Sea petroleum-storage platform that became the focus of a bitter battle, in June 1996, between the Shell petroleum company and the **Greenpeace** environmental group. The petroleum company initially intended to scuttle the rusting 14,500 metric ton *Brent Spar* in 2,000 m of water in the North Atlantic. Instead the platform and its 90 tons of sludge residue were to be disposed of in some other way. In addition to preventing the platform from being dumped at sea, the confrontation opened a formal scientific debate among European Union countries about the best way of **decommissioning** large structures used by the off-shore petroleum and gas industries. There were, however, also anxieties about decommissioning processes, as evidenced, for instance, by the widespread protests about plans in 2003 to dismantle in the United Kingdom a toxic 'ghost fleet' of old ships from the USA.

BRINE is any water which is strongly impregnated with salt.

The **BRITISH ANTARCTIC SURVEY (BAS)** began conducting scientific research in Antarctica in 1962 and is responsible for nearly all the United Kingdom's research in the region. It has five permanent research stations strategically placed in **Antarctica** and the South Atlantic, which carry out year-round research and monitoring. The BAS is administered from its headquarters in Cambridge, United Kingdom, which is responsible for the maintenance and operation of the research stations and for transporting more than 100 expedition field personnel and 1,100 metric tons of equipment and essential supplies to the polar continent. Two ice-strengthened research ships are used, one of which provides the deep-sea platform for marine projects. The BAS collaborates with about 30 other countries active in Antarctic research, as part of the Antarctic Treaty. The Survey's research covers the interdependent physical, chemical and biological processes of the entire Antarctic system. The BAS investigates, for example, pattern and change in the physical environment of Antarctica, the geological evolution, and the dynamics of **freshwater** and land **ecosystems**. Research into polar **ice sheets** aims to understand their potential to raise global sea levels in a warmer climate brought about by the **greenhouse effect**. A scientist from the BAS discovered the **Antarctic ozone hole**.

The **BRITISH ECOLOGICAL SOCIETY** is the oldest ecological society in the world. It was formed in 1913 and since 1980 it has operated as a charitable company, limited by guarantee. The membership is open to all who are genuinely interested in **ecology**, in the United Kingdom or abroad. There are some 4,500 members, about one-half of them from overseas. The Society holds a number of meetings every year, including the winter meeting at which a wide range of papers are delivered, often by research students. The spring symposium and other smaller meetings are more specialized and feature lectures by world-renowned speakers. The Society produces four scientific journals and a bulletin free to its members. It also supports expeditions, small projects and ecological education. The Society is becoming increasingly involved in policy matters, and in the promotion of ecology in education and government.

BRITISH NUCLEAR FUELS (BNFL) was formed in 1971 when the British Government decided to transfer the production activities of the **United Kingdom Atomic Energy Authority** to a separate company. BNFL manufactures nuclear fuel at Springfields, conducts uranium **enrichment** at Capenhurst, and undertakes reprocessing and high-level waste storage at **Sellafield**. It owns and controls the only low-level radioactive waste dump in the United Kingdom, at Drigg, near Sellafield. It is an international business, employing some 23,000 people in 16 countries and controlling about 12% of the global nuclear market (2003).

The **BROADS** are the freshwater lakes of East Anglia in the United Kingdom. Broads are usually edged with reeds and are connected to a slow-flowing river near its estuary. It is thought that the Norfolk and the Suffolk Broads were created primarily by the removal of **peat** in medieval times and subsequent flooding by the neighbouring rivers, which often had linking channels.

BROMINE (Br) is a reddish-brown liquid element that gives off suffocating vapours.

BROWN COAL is a brown fibrous material that is at the intermediate stage between **peat** and bituminous coal, which is good quality soft coal. The term 'brown' is sometimes used as an alternative name for **lignite**. In practice, brown coal is nearer bituminous or hard coal, and the plant debris that has been compressed over thousands of years is unrecognizable as such and has changed into an amorphous mass. Brown coal is usually extracted in open cast operations to yield a heavy soft fuel. It is usually used in power stations located near the coal head. It tends to have a high sulphur content, and so is dirty coal to burn if used in a power station without **flue gas desulphurization** equipment.

The **BRUNDTLAND REPORT** is the report of the World Commission on Environment and Development, set up by the General Assembly of the UN in 1983. It was named after Gro Harlem Brundtland, the former Norwegian Prime Minister who chaired the body. The brief of the Commission was to draw up a global agenda for change and, in particular, to devise a strategy for achieving **sustainable development** for the year 2000 and beyond. It urged the replacement of chemically dependent agriculture by **sustainable agriculture** and that there be a shift from export crops to food crops for local consumption. It recommended that 'soft' energy systems (**renewable energy** plus energy conservation measures) should be adopted, describing them as the best way towards a sustainable future. The Report also urged industry to assess the potential impact of new technologies before their employment, so that their production, use and disposal do not put too much stress on environmental resources. The report set out seven goals for the future: to revive economic growth; to change the quality of growth; to meet essential needs for jobs, food, energy, water and sanitation; to ensure a sustainable level of **population**; to conserve and enhance the resource base; to reorientate technology and management risk; and to merge environment and economics in decision-making.

BST (bovine somatotrophin) is the name of a hormone found in cattle that is responsible for growth and the quality of milk yields. BST was one of the substances to be produced by **biotech**. Animal welfare groups are opposed to the introduction of BST as an additive in the cow's diet, arguing that some BST might be passed on to children through milk.

BURN-UP is the amount of energy that can be extracted from **nuclear fuel** before it is considered spent.

The **BUSINESS CHARTER FOR SUSTAINABLE DEVELOPMENT** is a list of 16 principles of environmental management published in 1991 by the International Chamber of Commerce in Paris, France. The list provides a code of behaviour to encourage **environmental auditing** and good management practices that conserve raw materials, devise clean production technologies and minimize waste.

BUSINESS COUNCIL FOR SUSTAINABLE DEVELOPMENT: See **World Business Council for Sustainable Development**

C

C3 PLANTS are a group of plants including soyabean, wheat and cotton. They are classified as C3 because in the process of carbon fixation, during **photosynthesis**, they produce a compound essential for growth that consists of molecules each containing three atoms of carbon. The photosynthesizing productivity of C3 plants is expected to increase if the **carbon dioxide** concentrations in the atmosphere continue to grow.

C4 PLANTS are a group of plants that include maize and sorghum. The classification refers to a process of carbon fixation usually found in tropical plants with high growth rates. These plants have adapted to high temperatures, strong light, low water supply and low **carbon dioxide** levels. They are classified as C4 because in the process of carbon fixation, during **photosynthesis**, they produce a compound essential for growth that consists of molecules that each contain four atoms of carbon. Compared with C3 plants, C4 plants show little change in their rate of photosynthesis as a result of an increase of carbon dioxide concentrations in the atmosphere.

CADMIUM (Cd) is one of the toxic **heavy metals** that caused deaths and permanent illness in a series of major pollution incidents around the world. Cadmium has no useful biological purpose, but it has wide industrial applications, and is obtained from the compound known as greenockite. It is used in metal plating to prevent corrosion, in rechargeable batteries and as a pigment in certain plastics and paints. Special care is taken in the industrial smelting of ores and subsequent handling of cadmium because it is a category 1 **carcinogen**. Occupational exposure is also known to have caused heart, chest and kidney disorders. **Environmental health** problems have resulted from exposure to various sources of cadmium pollution of the air, soil and water. The worst recorded incident was in Japan, where an entire village was affected by prolonged exposure to cadmium. Polluted effluent water had contaminated rice crops over the long term, causing painful brittle bones and kidney failure in those affected.

CAESIUM (Cs) is soft, silvery-white and highly reactive and belongs to the alkali group of metals. It is a radiation hazard, because it can occur in two radioactive forms. Caesium-134 is produced in nuclear reactors, not directly by **fission**, but by the reaction. It emits **beta** and **gamma radiation** and has a **half-life** of 2.06 years. Caesium-137 is a fission product of **uranium** and occurs in the fall-out from nuclear weapons. It emits beta- and gamma-rays and has a half-life of 30 years. Caesium-137 was the principal product released into the atmosphere, and hence the **food chain**, from atmospheric testing of nuclear weapons and from the **Windscale fire** and **Chornobyl** nuclear accidents. After the Chornobyl accident, which spread a radiation cloud across much of Europe, the European Commission proposed new and more restrictive limits on levels of caesium in food and drinking water.

CAM PLANTS include plants such as cacti and other succulents. They are distinctive in that, unlike C3 and C4 plants, they are able to carry out carbon fixation in the dark, as well as being able to **photosynthesize** in sunlight. CAM stands for crassulacean acid metabolism.

CAMELFORD is the town in Cornwall, in the United Kingdom, where in 1989 20 metric tons of **aluminium** sulphate were accidentally unloaded into the public drinking-water supply. The pollution caused prolonged cases of stomach illness. There was concern about the possible long-term effects because the accident coincided with medical research that suggested that aluminium was a cause of Alzheimer's disease. A later concern was that the contaminated water was a cause of an increase in childhood leukaemia. In 1996 a British Department of Health report stated that there was no evidence to suggest that short-term exposure to the contamination in the water was related to three cases of leukaemia.

CANDU reactors or pressurized heavy-water reactors (PHWRs) are designed to react natural uranium, and so need a more efficient moderator (**heavy water**). With no expensive fuel-enrichment facilities or large pressure vessels necessary, this type of reactor is popular in less developed countries. Other advantages include the least down-time and the most efficient fuel use of any type of reactor, but disadvantages are the cost of the heavy water and that the reactors can be used for the production of **plutonium**, which is used in weapons programmes. The CANDU reactor was developed in the 1960s by a government-private consortium in Canada (CANDU is a registered trademark, an acronym for Canada Deuterium Uranium). All of Canada's nuclear reactors are of this type.

CAPTIVE BREEDING is a method of breeding in zoos or reserves those species of animals, plants, birds and fish that are close to **extinction**. Animals and plants in the wild are threatened with extinction as their natural habitats are destroyed, or fragmented, to clear land for agriculture and to build houses for a growing population. Hunting is also a threat to habitats, as is pollution. It is recognized that captive breeding is not the ideal way of preserving a species, but, if habitats cannot be protected, it is argued it is at least a way to save and protect certain species until they can be safely released back into the wild. The breeding programmes take a great deal of management, particularly during the transition back to the wild, and there have been some notable successes, including the Arabian **oryx**, the red wolf, the black-footed ferret, the **rhinoceros**, the Californian condor, the **golden lion tamarin** monkey and the bearded vultures of the Alps. Great care is taken while the animals are in captivity to keep them as isolated as possible in surroundings that resemble as much as possible their original habitats. Efforts are made to ensure that they do not become too trusting of humans. The bald eagle, which has been in decline since the 1940s, is also being bred in captivity, from eggs recovered in the wild, and some have been released back into the wild. In Oklahoma, USA, at the Sutton Avian Research Centre, elaborate steps are taken to ensure that the chicks do not 'imprint' humans as parents, which would mean they would not be able to survive alone in the wild. The chicks are fed via a glove puppet with an eagle head. Programmes to prepare animals for their release, include public meetings to inform and reassure the local human population, and training for the animals in how to go about hunting live prey in readiness for the wild. Once released animals are kept in reserves and a strict monitoring programme is maintained, using radio collars in some cases so that straying animals can be returned to their reserve, or taken back into captivity if they are thought to be in danger.

CARBOHYDRATES are one of three major components from which all organic matter is formed. The other two are **proteins** and lipids. Carbohydrates are a vast family of compounds made from various arrangements of carbon, hydrogen and oxygen. They get their name (watered carbon) because the hydrogen and oxygen are in the correct proportions to form water: that is, two atoms of hydrogen to one of oxygen. The most important carbohydrates are starches, sugars, cellulose and gums.

CARBON (C) is a non-metallic element which is unique in the number of compounds it can form and the way in which they are formed. Its molecules form part of many other elements. Chemically, carbon is unique because it can form many compounds containing chains and rings of carbon atoms. Pure carbon occurs naturally as diamond or graphite. However, it is carbon's role as the universal constituent of all living things—**microbes**, plants and animals—which makes it so important. The primary deposits of carbon compounds, such as the carbonates (chalk and limestone) and the **fossil fuels** (coal, petroleum and natural gas), are derived from living things. Only tiny quantities occur as graphite and diamond. Carbon burns in oxygen to form **carbon dioxide**. Before the extensive burning of fossil fuels by industrialized countries and the massive burning of trees, bringing about **deforestation**, both of which are primary **carbon sinks**, the **carbon budget** was in balance; meaning the annual exchange of carbon between the atmosphere and the oceans and flora and fauna was equal. There are two stable **isotopes** of carbon and four radioactive ones. One of the latter, carbon-14, is used in **carbon dating**.

The **CARBON BUDGET** is the balance in the exchange of **carbon** between the **atmosphere** and living things in the **biosphere**. **Deforestation** and the burning of **fossil fuels** upset this balance because both activities release carbon into the atmosphere. This means that more carbon is entering the atmosphere than is being absorbed by the biosphere, which is the cause of the **greenhouse effect**.

The **CARBON CYCLE** explains how every living thing, from the simplest microbe to human beings, needs **carbon** to survive. Carbon is the element that occurs in all the compounds that make up living organisms. It is the basic constituent of the **carbohydrates**, fats, proteins and **nucleic acids** such as DNA and RNA. The carbon cycle is the flow of carbon from the **atmosphere** to the **biosphere** and back. Green plants are the primary producers of food and are at the base of the **food chain**; animals consume plants, and **bacteria** and **fungi** are the principal decomposers. The chain begins when plants absorb **carbon dioxide** from the air in the process of **photosynthesis**, in order to make starch, sugars, proteins and cellulose. When eaten by animals, the organic plant material is converted into molecules for building animal-type tissues. However, in making the molecules necessary for the growth of tissues, both plants and animals use energy derived from breaking down particular carbohydrates. This process, aerobic **respiration**, converts the carbohydrates back into carbon dioxide and water, which are then excreted back into the environment. Even a small disruption to a food chain, by pollution or a natural phenomenon, can have catastrophic effects on plants and animals. As well as absorbing carbon dioxide, plants also release carbon dioxide during respiration and burning. Scientists calculate that 100,000m. metric tons of carbon dioxide per year are removed from the atmosphere by green plants during photosynthesis, 50,000m. tons are put back into the air by respiration, and 2,000m. tons are released when trees are burned. Another 50,000m. tons of carbon dioxide are released through decomposition of vegetation of all types. It is feared that **climate change** might alter this balance and a warmer earth might see forests becoming carbon sources rather than sinks. Marine life depends on the carbon cycle, while the oceans absorb an estimated 104,000m. tons of carbon dioxide per year and release 100,000m. tons per year by biological and chemical processes. Humans now release 6,000m. tons of carbon into the atmosphere each year through the burning of **fossil fuels**.

CARBON DATING, or radiocarbon dating, was a method of determining the age of organic remains such as wood, bone and shells, which was devised for archaeological research. However, it has become an important technique for studies of the environment, especially for investigating objects that contain evidence of past **climate change**. The objects include terrestrial fossils, sediments taken from the ocean and ice cores drilled from glaciers and the polar ice fields. Because of the steady bombardment of the earth by **cosmic radiation**, a small proportion of **nitrogen** atoms in the atmosphere are transformed by neutron bombardment into the radioactive nuclei of carbon-14, which has a **half-life** of 5,600 years. The atmosphere also contains normal carbon-12. Both forms of carbon atoms are absorbed by living trees, or other plants or animals, in the form of **carbon dioxide**, and during **photosynthesis** or **respiration** they are converted into plant tissue or bone. When the plant or animal dies, the carbon-14 becomes locked in any buried, preserved material, and the **radioactivity** of the buried carbon-14 continues to decay. The carbon-12 is stable and therefore remains in its original concentrations. An analysis of the proportions of carbon-14 to carbon-12 is used to calculate the age of an object.

CARBON DIOXIDE (CO_2) is a natural, essential component of the air and is the vital ingredient of the **carbon cycle**. However, it cannot fuel **respiration** and in sufficient quantity can suffocate (as in the 1986 disaster at **Lake Nyos** in Cameroon). It also becomes a form of pollution, one of the **greenhouse gases**, when excessive amounts are discharged into the air, primarily from burning **coal**, petroleum and **natural gas**, and by **deforestation**. Measurements made of gases trapped in air bubbles in the polar **ice sheets** and **glaciers** show that, prior to the intensive coal-burning that fuelled the industrial revolution, the concentration of carbon dioxide in the atmosphere was 280 parts per million. This had increased by more than 25% by 1990. Calculations by the **Intergovernmental Panel on Climate Change** indicated that a doubling of carbon dioxide in the atmosphere produced by human activity was likely to occur by 2050, compared to pre-industrial levels before 1800. According to the same forecasts, by 2030 carbon dioxide levels would have caused global temperatures to increase by an average of 1.8°C (3.2°F).

CARBON MONOXIDE (CO) is a colourless, poisonous gas that is slightly lighter than air. The main source of carbon monoxide is from the **combustion** processes of industrial furnaces and car engines, when the carbon in the coal or petrol is burned incompletely. Consequently, some of the carbon combines with oxygen in the air to form carbon monoxide. Carbon monoxide is a source of pollution in traffic-congested urban streets. When carbon monoxide is inhaled, it combines with red blood cells by replacing the useful oxygen in the molecules that give healthy blood its bright red colour. The tissues of the body are then deprived of oxygen.

CARBON SINK refers to any part of the environment, from trees to the oceans, that acts as a reservoir or store for carbon. Some carbon sinks, like those in the shells of former marine organisms or in **fossil fuel** deposits, can be many millions of years old. Others may be seasonal, like the bloom of **phytoplankton** in the ocean. (See also **sequestration**.)

CARBON TAX is a policy measure to limit **carbon dioxide** emissions into the atmosphere. It is often discussed in the context of vehicle exhausts or power station emissions, and achieving improvements in fuel efficiency by levying an energy tax on petrol. Some advocates of a carbon tax believe it should be applied at a uniform rate across all sectors of the economy that generate carbon dioxide by burning **fossil fuels** such as petroleum, **coal** and **natural gas**. In October 1991 a carbon tax was proposed to the European Communities (EC) ministers of the environment in a recommendation for action that could take effect by the year 2000. The EC considered phasing in a carbon tax of US $10 on each barrel of petroleum by that date. The scheme originated from concern about the impact of carbon dioxide and other gaseous emissions on global **climate change**, and from fears that, if not controlled in time, the human-induced phenomenon of global warming would be irreversible. The proposal of an energy tax was part of a 'no regret' strategy adopted by the European Commission, which also included proposals to curb other forms of air, land and water **pollution**. The idea of a carbon tax met with much resistance in the USA, but was widely suggested as a means of achieving the aims of the **Kyoto Protocol**.

CARBON TETRACHLORIDE $_{(CCl4)}$ is an industrial solvent made from chlorinating methane. Moist carbon tetrachloride breaks down to the poisonous gas, phosgene, and hydrogen chloride. Safer alternatives are being developed.

A **CARCINOGEN** is a substance that causes cancer in humans and animals. The first carcinogen to be discovered was **arsenic**, followed later by **cadmium**, cobalt, chromium and **asbestos**. A wide range of substances are now suspected or have been proved to be carcinogenic. They include the chemicals in **tobacco** and tobacco smoke, synthetic **polychlorinated biphenyls** used in plastics, electrical components and brake fluids in automobiles, and many of the synthetic compounds used in industry, such as resins and adhesives.

The **CARIBBEAN** is believed to be one of the **regional seas** most polluted by **pesticide** run-off and industrial discharges. Industrial pollution, from sugar refineries, mining, offshore petroleum drilling and industrial waste all contribute to the pollution. The sea also contains some of the busiest petroleum tanker routes in the world. The increased volume of **tourism** is another threat to the Caribbean. Fish in the **coral reefs** in the area are being depleted and coral is being taken to sell to tourists. It is subject to a convention signed under the UN Environment Programme's Regional Seas Programme, the Cartagena Convention (the Convention for the Protection and Development of the Marine Environment of the Wider Caribbean Region), which was named for the Colombian city where the original agreement was signed in 1983. It has three protocols, on combating oil spills (also signed in Cartagena in 1983), on wildlife and specially protected areas (Jamaica, 1990) and on the consequences of land-based pollution (Aruba, 1999). The responsible organization is the Caribbean Environment Programme (CEP). The CEP was made the responsible regional body for action on coral reefs under the **International Coral Reef Initiative**, established in 1994. The Caribbean includes some of the largest reef systems in the world, be they the barrier reefs of Belize or Andros (Bahamas) or the reef system of the Turks and Caicos Islands.

CARING FOR THE EARTH, A STRATEGY FOR SUSTAINABLE LIVING, is the revised and more practical version of the **World Conservation Strategy**, published in 1980, which outlined a single, united approach to global problems. The updated version, like the previous document, was sponsored by major international conservation organizations. It was launched in 1991 in more than 60 countries.

The **CARRYING CAPACITY** is the maximum population that a habitat can support indefinitely within a given set of environmental conditions. When carrying capacity is exceeded, problems associated with **overpopulation** are experienced.

CARTAGENA PROTOCOL ON BIOSAFETY: See **Biosafety Protocol**

CASH CROPS are, literally, any crops that are grown for sale in markets or for export. They include coffee, cocoa, sugar, bananas, vegetables, peanuts and non-foods, like **tobacco** and cotton. Huge areas of countries in the developing world have been turned over to cash crops. Those countries with no mineral or petroleum resources depend on cash crops for foreign money, so that they can import materials to develop roads, for construction, or to buy Western consumer goods and, indeed, food. However, critics argue that cash crops are planted on land that would otherwise be used to grow food for the local community and say this is a cause of world famine, as well as of economic instability. Cash crops, such as peanuts, can ruin the land if it is not left fallow after six years of harvests. Moreover, if the best agricultural land is used for cash crops, local farmers are forced to use marginal land to grow food for local consumption, and this has a further dramatic effect on the environment.

CATALYTIC CONVERTERS are designed to clean up the exhaust fumes from petrol-driven vehicles, which are otherwise the major threat to **air quality** standards in congested urban streets and on motorways. Converters remove **carbon monoxide**, the unburned **hydrocarbons** and the oxides of nitrogen. These compounds are damaging to human health and the environment in a variety of ways. Scientists estimated that in the British capital of London alone, in 1990, 1.3m. metric tons of pollutants were being pumped into the air each year. More than 80% of this came from cars. Carbon monoxide is poisonous to humans and also contributes to the **greenhouse effect**. Hydrocarbons irritate the lungs and react with sunlight to produce **photochemical smog**. **Nitrogen oxides** react with water vapour in the atmosphere to produce **acid rain**, and also help form photochemical smog. Catalytic converters, sometimes referred to simply as 'cats', can remove over 90% of the pollutants if they are a 'full three-way system'. Although catalytic converters are contained in a cylindrical tube about 30 cm long and 23 cm wide, the technology is an extension of the industrial type of **catalytic cracking**, which is used to break down crude petroleum into its constituent parts: petrol, paraffin, diesel and heavy engine oil. The converter is attached to the vehicle's exhaust near the engine. Exhaust gases pass through the cellular ceramic substrate, a honeycomb-like filter. While compact, the intricate honeycomb structure provides a surface area of 23,000 sq m. This is coated with a thin layer of platinum, palladium and rhodium metals, which act as catalysts that stimulate a reaction to changes in the chemical composition of the gases. Platinum and palladium convert **hydrocarbons** and **carbon monoxide** into **carbon dioxide** and water vapour. Rhodium changes **nitrogen oxides** and **hydrocarbons** into nitrogen and water, which are harmless.

CATALYTIC CRACKING is the process for treating thick, heavy crude petroleum in order to separate out the light fractions, like petrol and those which provide the raw material from which the petrochemical industry makes plastics, paints and pesticides. The basic process of cracking, or separation of the various fractions, by heating crude petroleum at a high pressure and temperature is made more efficient by the use of a catalyst; hence the name catalytic cracking.

CATALYTIC EXTRACTION PROCESSING (CEP) is a method of recycling waste metals in which a catalyst is used to maintain a particular chemical reaction. Waste substances are injected into a bath of molten metal. The bonds holding the elements together are severed and the liberated atoms recombine in a different manner to make marketable, reusable materials. For example, a waste stream of chlorobenzene contaminated with cobalt oxide can produce a cobalt metal alloy, a ceramic product and hydrogen chloride gas, the last of which is easily converted to calcium chloride, to be used for road de-icing or industrial refrigeration. A CEP reactor has been used to treat **isocyanate** wastes from polyurethane manufacture and the tars and reactor clean-out wastes from chlorinated hydrocarbon production. In another example of CEP, plastic and metal components from a missile guidance system were converted to reusable alloys and ceramic products. Thousands of patents exist for different CEPs. An improved catalyst can make an established chemical manufacturing process more efficient, leading to cheaper products, while a wholly new catalyst can facilitate a completely new process.

CDM: See **Clean Development Mechanism**

CETACEANS are aquatic mammals that include **whales**, porpoises and **dolphins**. About 80 species are found in practically every marine environment, as well as some in freshwater rivers and lakes. Generally, cetaceans are divided between two sub-orders, the Odontoceti or toothed species and the Mysticeti, which have plates of baleen (whalebone) to filter nutrients from the seawater. The latter group consists of fewer species, but includes the largest cetaceans, all of which have double blowholes. Dolphins and porpoises are all toothed. Porpoises actually include killer whales, pilot whales, harbour porpoises and the very endangered vaquitas. In 1990, during the third International Conference on the Protection of the North Sea, the ministers of the countries adjoining the North Sea signed a memorandum of understanding owing to concern about the effects of pollution and habitat deterioration on the small-cetacean population. An agreement under the **Bonn Convention** was signed in 1991. Research projects by British and other coastal European nations aimed to increase knowledge about diseases in marine mammals and the effects of pollution and other human influences and, in co-ordination with other North Sea countries, to establish a system for the long-term monitoring of the mortality of cetaceans. However, Europe was considered behind the USA in dealing with small cetaceans. Most issues dealing with the large cetaceans, whales, came under the emit of the **International Whaling Commission**.

CHANG JIANG: See **Yangzi river**

CFCs: See **chlorofluorocarbons (CFCs)**

CHAIN REACTION is the process by which energy is released by splitting atoms of **uranium-235**, by the process of **fission**, either in a controlled form in a nuclear reactor or in the uncontrolled destructive form of an atomic bomb. The process of fission can be produced in **fissile** substances. As more free neutrons are released from a uranium fission event than are needed to initiate the event, the reaction can become self-sustaining.

CHEMICAL BLEACHES are used together with heat, steam and large amounts of water to produce paper from pulp. Bleaching is used to give brightness and strength, but the **chlorine** compounds commonly used for that purpose are also one of the primary concerns about the content of waste water discharges from the paper and pulp industry. **Unbleached paper** is increasingly being used where whiteness is unnecessary.

CHEMICAL TIME BOMB is a term used to describe a delayed environmental disaster caused by the accumulation of a mixture of chemicals in **soil**, **sediments** or **waste disposal** dumps. In time, the interactions between the substances in question reach a state in which they outweigh any natural neutralizing process. The full significance of the phenomenon only began to be recognised in the late 1980s as ecologists discovered how soils hold harmful substances, preventing them from reaching **ground-water** supplies or the **food chain**. An example of a chemical time bomb was the disaster at **Love Canal**.

CHEMICAL TRANSFORMATIONS are the processes by which primary, or directly emitted, pollutants such as **sulphur dioxide** and **nitrogen oxides** become converted into **secondary pollutants** such as acids, and **particulates** of **sulphates** and **nitrates**.

CHEMISTERILANTS are chemical compounds that cause sterilization or prevent reproduction in insects or birds and rodents.

CHERNOBYL: See **Chornobyl**

CHLORDANE is a toxic chlorinated hydrocarbon **insecticide**, banned in the United Kingdom, France and Germany, and then proscribed throughout the European Union. Its use is also severely restricted in the USA. However, it is still in agricultural use in some parts of the developing world. Chlordane was mostly used to protect telephone and electricity poles, fences and other wood in close contact with the ground, from fungal decay micro-organisms and insects, particularly **termites**. When absorbed through the skin it affects the nervous and the respiratory systems. It can also lead to liver and kidney damage. Chlordane accumulates in the **food chain**.

CHLORINATED HYDROCARBONS are toxic chemicals that are almost entirely synthetic and do not occur in nature. Because of this the micro-organisms that bring about natural decay can not break down the compounds, and they persist in the environment. Products made from chlorinated hydrocarbons are now largely banned in industrialized countries, but are still manufactured and used in developing countries. Hydrocarbons are organic chemicals containing only **carbon** and **hydrogen** in their molecules. It is when the hydrogen is replaced with **chlorine** that chlorinated hydrocarbons are produced. The most infamous chlorinated hydrocarbon is the **insecticide**, **DDT**, which caused widespread environmental pollution and destroyed wildlife. Chlorinated hydrocarbons are stored in body fat and move up the **food chain**. Small amounts ingested regularly build to high levels in the tissues. A predator eating several small animals with large amounts of chlorinated hydrocarbons could well absorb a fatal dose. As well as insecticides, the contamination of water by **chlorinated solvents** is a growing problem as many of these substances are toxic and suspected **carcinogens**.

CHLORINATED SOLVENTS are the **chlorinated hydrocarbons** that are widely used in industry and domestic products. Solvents are liquids capable of dissolving materials for many purposes. Chlorinated solvents are causing a growing environmental problem because most are poisonous to some extent. Some are used in dry-cleaning processes to remove grease. Chloroform was used as an industrial degreaser and **carbon tetrachloride**, which is closely related to chloroform, has been used in fire extinguishers. These solvents have also been used in paint removers and thinners for typewriter correcting fluid. Inhaling chlorinated solvents can cause drowsiness or unconsciousness. Putting newly dry-cleaned clothes in a warm car or confined space can lead to a damaging build up of fumes. Chloroform has been banned as an ingredient of medicines in North America and some European countries. Chlorinated solvents linger in the environment because they are not easily broken down by micro-organisms. If they are dumped in **landfill** sites they may evaporate and cause damage to the **ozone layer**, or penetrate **ground-water**. Chlorinated solvents do not burn easily. If their vapours are exposed to a naked flame, phosphene, a poisonous gas, forms. If they are burned at too low a temperature, **dioxin** is created.

CHLORINE (Cl) is a very reactive and highly toxic green, gaseous element, belonging to the **halogen** family of substances. It is one of the most widespread elements, as it occurs naturally in seawater, salt lakes and underground deposits, but usually occurs in a safe form as common salt (NaCl). It accounts for about 0.055% of the earth's crust. It is used commercially in large quantities by the chemical industry both as an element to produce chlorinated organic solvents such as **polychlorinated biphenyls** and for the manufacture of polyvinyl chloride or PVC plastics, thermoplastics and hypochlorite bleaches. Chlorine was the basis for the organochlorine pesticides like **DDT** and other agricultural chemicals that killed wildlife.

CHLORINE LOADING POTENTIAL (CLP) is the simplest of the system of indices developed to enable policy-makers to assess and compare the impact of various **halogens**, or halocarbon gases, as **ozone**-depleting substances. The chlorine loading potential compares the amount of **chlorine** released by a particular gas with that released by the same mass of CFC-11 (see **chlorofluorocarbons (CFCs)**) when each gas breaks down into its constituents. CLP calculations became increasingly important in the search for beneficial alternatives to the most-damaging ozone-depleting substances.

CHLOROFLUOROCARBONS (CFCs) are a family of chemicals, discovered in 1930 by a US chemist, Thomas Midgley, which revolutionized industrial and domestic **refrigeration**. Aside from their thermodynamic properties as refrigerant gases, CFCs were non-toxic, non-flammable, non-corrosive and stable. Development of the convenience-food business relied heavily on deep freezers cooled by CFCs. Soon after, other applications followed with CFC propellants for **aerosol** spray cans and the 'blowing' of foam plastics from which the lightweight, heat-insulating packaging associated with fast foods is made. CFCs mainly comprise methyl chloroform and **carbon tetrachloride**, the solvent cleaner. A combination of the stability of CFC compounds and the effect of the chlorine molecules they contained, once they were released in the upper atmosphere and came into contact with the earth's protective **ozone layer**, became the source of the environmental problem. CFCs are also potent **greenhouse gases**. Concern about destruction of the ozone layer led to the **MontrTal Protocol**, agreed in 1987, to phase out CFC production and consumption. From 1996 CFCs could no longer be produced in industrialized countries, although developing countries were allowed to continue producing CFCs until 2010. This exemption led to a 'black' market (illegal trade) in CFCs in some parts of Europe and the USA, although the US authorities dealt severely with this practice. By the end of the 20th century CFCs in the atmosphere had been reduced by 85% since the peak of world production. Alternatives to CFCs include hydrochlorofluorocarbons (HCFCs), which are much less ozone-depleting, and hydrofluorocarbons (HFCs), which are less suitable because they are potent **greenhouse gases**.

CHOLERA is a disease transmitted primarily through contaminated water and food, especially raw vegetables and seafood. If drinking water supplies become contaminated, particularly in overpopulated areas with bad sanitation, infection spreads rapidly. The disease-causing organism is a bacterium called *Vibrio cholerae*.

CHORNOBYL (or Chernobyl) is a small town in Ukraine that was the site of the world's worst accident at a civil nuclear power station. Nearly 20 years later the damage remained Europe's single biggest environmental problem. On 26 April 1986 one of four nuclear power reactors at the Soviet power station exploded, an event which experts in the then USSR had claimed was impossible, when an uncontrolled reaction known as a **prompt criticality** took place in a rod of nuclear fuel. The accident blew the reinforced concrete top from the graphite-moderated channel-tube reactor (**RBMK**), releasing a cloud of radioactive waste that spread across the whole of northern Europe. As a result of the far-reaching health, social, political and economic implications of the accident, Soviet experts presented a report on the accident to an open international meeting, known as the Chernobyl Post-Accident Review, which was held on 25–29 August 1986 at the headquarters of the **International Atomic Energy Agency** in Vienna, Austria. According to official figures, of the 300 firefighters and rescuers who went to the emergency, 31 died within one week and more than 200 others suffered acute **radiation** sickness. Over 135,000 people had to be evacuated from the immediate area and large amounts of fresh food had to be destroyed. Between 300,000 and 600,000 people formed teams to clean up the 30-km evacuation zone around the reactor, but many of them entered the zone two years after the accident. Following the accident, thousands of people were exposed to doses of radiation that were expected to cause long-term radiation-related illnesses, such as leukaemia. Within 10 years of the incident more than 1,000 people developed thyroid cancer alone, the most notable increase occurring among children who were less than 15 years old at the time of the explosion. An additional 3,000 cases were expected between 1997 and 2001. The First International Conference on the Biological and Radiological Aspects of the Chernobyl Accident, held in September 1990, was told that the greatest concentration of cases occurred in an area 200 km north of Chornobyl, where rain washed radioactive **iodine** from the atmosphere, causing severe contamination. Rural areas near Chornobyl would remain in quarantine for decades, even centuries. Eighteen years on, exact casualty figures remained under debate. Some claimed that 39,000 people world-wide would die from cancer as a direct consequence of the accident, while birth defects and growth problems among children in Ukraine have increased by 230%. About 3m. children require medical treatment, and 3.5m. still live on those areas most affected by the accident, in Ukraine, Belarus and Russia. There may be worse consequences for over 7m. people yet to emerge. Hitherto, Ukraine has spent US $5,000m. on decontamination efforts, but the overall costs of the accident may amount to $140,000m. Resealing the reactor was considered necessary in the first decade of the 2000s. Once the fire and radioactive emissions had been halted, the reactor was sealed in concrete, but the ruins contained 40,000 cubic metres of long-lived highly active waste, including 100 metric tons of the reactor's original fuel. Since the 1990s there have been various international aid initiatives to help in decontamination and offers to help decommission the power station.

CITES: See **Convention on International Trade in Endangered Species of Wild Flora and Fauna**

The **CLEAN AIR ACT** was first introduced in the United Kingdom in 1956, to reduce **air pollution**, following the death of 4,000 Londoners during a smog in 1952. The Act controlled the emissions of smoke, grit and dust into the atmosphere, but not **sulphur dioxide**. It was later reinforced by the Clean Air Act 1968. As a result of the legislation, smokeless zones were established and there was a dramatic improvement in general **air quality standards**. City-based power stations were closed and larger replacements, with tall chimneys, were built in more-remote areas, often close to coal-mining regions. This, of course, had its own problems, but in the cities the main pollution problems came to be from motor vehicles and from sewerage, industrial and household waste. Western European countries soon followed suit, although the rest of Europe lagged behind. In the USA some cities had introduced clean air regulations in the late 19th century, but only from the 1940s were general concerns expressed about air quality. After severe smog incidents on both the west and east coasts of the mainland USA, the federal authorities introduced the Air Pollution Control Act 1955 and a full-fledged Clean Air Act in 1964. Although the legislation was revised continually through the late 1960s, it was soon deemed inadequate, and the Clean Air Act 1970 was a major revision of the previous one. Revisions throughout the 1970s were followed by a hiatus, until the reforms of the Clean Air Act 1990. Developments in the early 2000s were believed to be undermining much environmental legislation in the USA. Most countries have equivalent legislation, although it is not always sufficient to address their problems and often neglected vehicular traffic. The People's Republic of China, for example, passed an Air Pollution and Control Law that was amended in 1995. In 2004 a number of new Chinese laws indicated that the vast country was introducing further regulation and standards in an effort to improve air quality standards in its polluted cities.

The **CLEAN DEVELOPMENT MECHANISM (CDM)** is a unique policy instrument that was an integral part of the **Kyoto Protocol** on global **climate change** in 1997. It is one of three so-called flexible mechanisms under the Protocol designed to help industrialized countries reach their targets for emission reductions more cheaply and easily. This mechanism is intended to help finance cleaner development in less-developed countries by means of credit received by industrialized countries for pollution control measures that they pay for in developing countries. For example, if one country paid to upgrade a power station in a second country, in order to make it more energy-efficient, the first country would receive credit under the Kyoto Protocol for **greenhouse gas** emissions reductions. The CDM was applauded as a major contribution to sustainable development, despite some constraints placed on it in the detailed negotiations on its operation, details which were finally resolved in December 2003 under the **Framework Convention on Climate Change**.

CLEAN ENERGY SOURCES: See **renewable energy**

CLEAR-CUTTING is a method of timber harvesting in which all trees in a forested area are completely removed in a single cutting. A clear-cut site can be reforested through natural regeneration or deliberate planting. Almost two-thirds of annual US timber production is harvested by clear-cutting. It can increase the volume of harvested wood and shortens the time for establishing a new stand of trees, but it can also lead to severe **soil erosion**, flooding and landslides. It leaves unsightly swaths of denuded land that take decades to regenerate, and often adversely affects the **biological diversity** of a forest.

CLEARING-HOUSE was launched in 1982 by the **UN Environment Programme (UNEP)** to help developing countries to manage their own environmental problems. The programme also helps to work out, monitor and manage priority programmes and projects, and finds funding and other resources. It also acts as a 'broker', matching up potential donors with identified projects and forging links between developing countries and donors. Clearing-house operates in projects throughout the world, including a study of the impact of **sea level** rise in the Maldives, a cleaning-up programme for the heavily populated Manila Bay in the Philippines, and reducing water pollution in the Chinese canal city of Shaoxing. It also organizes regional workshops on hazardous **waste disposal** and management. Clearing-house relies on voluntary contributions and support from the international community. Several groups now operate similar systems, such as that under the **Biosafety Protocol**.

CLIMATE is the average weather conditions in a particular region of the world, taken over a 30-year period.

Many aspects of the earth's geography affect the climate. Equatorial, or low, latitudes are hotter than the polar latitudes because of the angle at which the sun's rays arrive at the earth's surface. The difference in temperature at the Equator and at the poles has an influence on the **global circulation** of huge masses of air. Cool air at the poles sinks and spreads along the surface of the earth towards the Equator. Cool air forces its way under the lower-density warmer air in the lower regions, pushing the lighter air up and toward the poles, where it will cool and descend. These movements of air are, in effect, conveyor belts carrying prodigious quantities of energy. They also have a dominant effect on the climate. Equatorial regions with rising air have low-pressure conditions, and the hot humid air, cooling as it rises, sheds a lot of rain. Human activity is largely responsible for the problem known as global **climate change**.

The **CLIMATE ACTION NETWORK (CAN)** is an international consortium of non-governmental organizations (NGOs), with regional and national nodes, that work on the issue of global **climate change**. CAN is credited for focusing international attention on the issue of climate change, and for helping to bring about the **Kyoto Protocol** by exerting strong international pressure on governments world-wide. It also serves as a model for international NGO collaboration on other issues.

CLIMATE CHANGE once referred to natural short-term and long-term changes in weather patterns. Later the term came to be used primarily in the context of human-induced changes to the climate associated with the **greenhouse effect**. The emission of heat-trapping gases like **carbon dioxide** from the burning of **fossil fuels** is causing the atmosphere to warm. There are now considered to be six main human-emitted **greenhouse gases**: carbon dioxide, **methane**, nitrous oxide, hydrofluorocarbons (HFCs), sulphur hexafluoride and perfluorocarbons (PFCs). These are to be regulated under the **Kyoto Protocol** on climate change, which was adopted in 1997 but not in force by mid-2004. The first greenhouse gases to be regulated were the **chlorofluorocarbons (CFCs)**, which were to be phased out under an agreement of 1987 (later commitments were by 1996 for the main users, the industrialized countries, and by 2010 for all countries).

The earth has known many climates. These have left their prehistoric marks on the shape of the land by erosion of rocks, in fossils in the sediment of deep oceans showing changes in the habitats of plant and animal species, and in centuries-old ice sheets containing trapped gas bubbles and particles from ancient atmospheres. Early historical records contain evidence of **droughts**, **floods** and severe winters, indicating the different climates experienced by our ancestors. The first systematic meteorological observations, a relatively new source of data, have their origins in records started in the first half of the 18th century. The definitive climatic history of the planet has yet to be written. However, speculation that the discharge of carbon dioxide gases from industrial activities might cause climatic change by global warming was made by a US scientist, J Tyndall, as long ago as 1863. In 1896 the Swedish chemist, Svante Arrhenius, calculated that if the carbon dioxide levels in the atmosphere were doubled, as a result of **coal**-burning, temperatures would increase by 6°C (10.8°F). In the 1930s and 1950s scientists recognized the importance of **water vapour** as an even more abundant greenhouse gas, which could physically interact with carbon dioxide in a warming process that constituted an example of a **feedback cycle**. Greater attention was focused on the issue in the mid-1960s, prompted by publication of the first long-term records of changing carbon dioxide concentrations. These were amassed at an observatory at Mauna Loa, in Hawaii, USA, by Charles Keeling and are still being measured; between 1958 and 1988 Keeling measured an increase in atmospheric carbon dioxide of 11%, from 315 parts per million to 351 parts per million.

The **Intergovernmental Panel on Climate Change (IPCC)**, a group of more than 2,500 scientists, calculated that the earth had already warmed by about 0.6°C (1°F) during the 20th century and projected that it would increase in temperature by another 1.1°–3.6°C (2°–6.5°F) by the end of the 21st century. The difference in temperature from the last ice age to the end of the 20th century was about 5°C (9°F). If the earth were to warm by even 1.9°C (3.5°F) by 2100, it would be the fastest rate of warming experienced during the previous 10,000 years, the period in which modern civilization developed. Moreover, in 2001 the IPCC revised its predictions upwards. However, the danger of climate change is its unpredictability. It is not the averages but the extremes, such as severe droughts and storms, that cause the greatest damage. The predicted impacts of global climate change include: increased **desertification**, more **extreme weather** such as heat waves and flooding, a greater spread of infectious diseases like **malaria** and **cholera**, an increase in **sea level** of 15–94 cm (6–37 inches), which would inundate most low-lying areas and small islands, increased scarcity of **fresh water**, shifting and transformation of many forested areas (see **forests and woodlands**), some forest loss, and challenges to agriculture and food supply.

Other human activities have the potential to affect the climate. A change in the **albedo** (reflectivity) of the land brought about by **desertification** and **deforestation** affects the amount of **solar energy** that is absorbed at the earth's surface. Man-made **aerosols** produced by the sulphur released from power stations can have a cooling effect and also modify clouds. Changes in **ozone** levels in the stratosphere owing to CFCs may influence climate, and vice versa.

CLIMATE MODELS are attempts to describe mathematically how the climate system functions and how the various factors driving it interact. The models are converted into immensely complex computer programmes, which can only be analysed with super-computers. The purpose is to attempt to predict the chain reaction that would happen as a consequence of changing one factor, such as the level of one of the **greenhouse gases**, and the change it would cause in terms of global warming. Climate models are complex because, in addition to predicting the interaction of the greenhouse gases, they have to analyse the cumulative effects of other factors: the extent of sea and ice on the earth and the effect this has on the level of **water vapour** in the atmosphere; what happens when the amount, types and height of **clouds** change, because clouds differ in their effect on global warming; how different parts of the ocean absorb and release **carbon dioxide** according to the seasons and the temperature of the water. Since the physics and chemistry of some of these processes are not fully understood, climate modelling is clouded by uncertainties, in much the same way as economic modelling.

CLONING: See **genetic engineering**

CLOUDS are visible masses of condensed water vapour floating in the lower **atmosphere**. They are one of the largest areas of uncertainty for scientists devising **climate models** with which to simulate the ways in which global warming might increase cumulatively. For example, a rise in air temperature will accelerate evaporation from the **ocean**. The **water vapour** produced may form high-altitude cirrus clouds. They have a significant **greenhouse effect**, trapping heat radiated from the earth. Other clouds, such as the stratocumuli that cover about one-third of the ocean, have a predominantly cooling effect, because they reflect a good deal of sunlight back into space. Much research time has been spent trying to decide how to define the realistic cloud effects that are to be built into the models. Oceans also have an enormous thermal inertia. While scientists accept that the way oceans absorb and release their heat has a profound influence on weather patterns, some experts believe there might be a delay of at least 20 years between the time when **greenhouse gases** reach a particular concentration and the time at which the total increase in temperature is achieved.

The **CLUB OF ROME** is an association formed in 1968 by the Italian industrialist Aurelio Peccei, which united people concerned at the state of the world, which was drawing too heavily on finite resources, and who thought alternative policies should be adopted. Its members came from business, politics and the world of social and environmental sciences. The Club commissioned research into the problems of pollution, urban planning, inflation, unemployment, and the increasing gap between the developed and developing countries. In a 1972 report, ***Limits to Growth***, the Club of Rome predicted that if the then present trends continued, the limit to economic growth could be reached within 100 years and would result in over-development and collapse. Ecologists were on the whole in favour of the report, but it was criticized by economists and technologists. The Club of Rome was also part of the movement that recommended a New International Economic Order (NIEO), to lessen the inequality between the developed and the developing world. This concept was included in proposals adopted by the UN in 1974. By the 1980s the possibility of a NIEO had faded, primarily because of the economic situation in the world.

COAL tends to be regarded as one of the major sources of **carbon dioxide** pollution, which is believed to be enhancing the **greenhouse effect** and, consequently, increasing the likelihood of global **climate change**. The carbon, released in combination with oxygen with which it combines when burned in power stations, was trapped as coal aeons ago, when trees and plants were buried in swamps. Coal is part of a family of hydrocarbon **fossil fuels**. The lowest form, both in terms of its carbon content and use as an energy source, is **peat**, which contains the highest amounts of water and other non-combustible materials. **Lignite**, or brown coal, comes next, and is in plentiful supply in opencast sites; then comes hard or bituminous coal and, finally, the highest ranking by carbon energy content, anthracite. When burned, one ton of hard coal yields three times the energy of a ton of lignite; they are the two most widely used grades world-wide. The additional environmental problem comes from coal with a high sulphur content. When this coal is burned, a chemical reaction occurs with oxygen to produce **sulphur dioxide**, the principal source of **acid rain**. These emissions can be greatly reduced at existing types of coal burning stations by filters and **flue gas desulphurization** technology, such as electron beam desulphurization. However, a more direct method lies in newer coal-burning processes such as **fluidized bed combustion** and by energy conservation measures, like **combined heat and power** energy systems, that drastically reduce the amount of coal burnt for a given amount of energy.

COALBED METHANE is the methane generated in **coal** seams, which would benefit the environment and improve safety if it was extracted and used as a fuel. Otherwise, methane leaking into the air presents an environmental hazard, because it is a powerful **greenhouse gas**. Methane absorbs about 21 times as much heat energy as carbon dioxide.

COASTAL AREAS describe the areas where the land masses meet the seas. Some 60% of the world's population live within 100 km of the coast; coastal waters include many essential ecosystems and provide breeding areas for an estimated 90% of the world's fish catch. Coastal areas throughout the world are under enormous environmental stress, which is caused by a wide range of factors, including pollution and the destruction and deterioration of marine habitats. Coastal pollution, from land-based industry, logging or mining, agricultural run-off and human sewage, is a major problem. Untreated sewage can lead to increased growth of poisonous **algal bloom**, which often flows back with the tide to pollute the beaches and shorelines. There is also a threat to human health from the **bacteria** and viruses in sewage that cause stomach upsets, hepatitis, cholera and **typhoid**. These are caught either by swimming in contaminated waters or by eating seafood caught in them. The Group of Experts on the Scientific Aspects of Marine Pollution (GESAMP), an advisory panel of the UN, reported in 1990 that pathogenic organisms that are discharged in domestic sewage from estuaries, drainage canals and rivers into the sea are 'the principal problem for human health on a world-wide scale'. Accidental petroleum spills also threaten marine organisms and can cause serious damage in coastal areas. Another serious problem is the devastation of coastal habitats, in particular the **wetlands**, **mangrove forests**, **salt marshes** and **sea-grasses** in order to make way for industrial, urban and recreational developments. **Coral reefs**, home to one-third of the world's fish species, are also gradually being eroded by pollution and are being over-exploited. The reefs, together with the mangrove forests, provide valuable protection against coastal erosion. Another threat is global warming (see **climate change**), which could lead to a rise in **sea levels**. Many countries would then face the danger of their coastal areas disappearing under the rising sea.

COCKROACHES are among the most primitive of the living winged insects. It is thought they have been unchanged for more than 300m. years, and are among the oldest fossil insects. Cockroaches are usually found in tropical climates, but a few species, out of the total 3,500 known species, have become pests. They are common household pests in many countries, and are often imported by ship and accidentally taken into homes in groceries. Cockroaches eat plant and animal products, including food, paper, clothing and soiled hospital waste, and contaminate their environment with droppings and an unpleasant odor, to which many people are allergic. They are a major health hazard and carry harmful bacteria, protozoan parasites and faunal pathogens, including those that cause **typhoid**, leprosy and **salmonella**. In dark, humid places like the corners of kitchens, bakeries and hospitals, where the conditions are right, they breed prolifically all year round. Huge amounts have been spent throughout the world in an attempt to eradicate cockroaches, but cockroaches have survived because they breed so quickly and rapidly develop resistance to most of the pesticides that have been used. They are also highly resistant to **radiation**.

COLIFORM BACTERIA are a group of **bacteria** that inhabit the intestines of humans and other mammals, including cattle. Their presence is taken as a measure of the degree to which rivers, lakes and drinking water supplies are contaminated by human waste. Public health standards often include the maximum acceptable concentration of these bacteria in water supplies.

COMBINED HEAT AND POWER (CHP) systems are power stations that use the heat that would be wasted by conventional power stations to provide hot water and space heating in domestic and industrial premises. Conventional power stations produce electricity with about 35% efficiency. The remainder is discharged into the environment as waste hot water. With CHPs the energy efficiency level can be as high as 75% or more. Many US and European cities, for example New York and Boston, MA (USA), Copenhagen (Denmark) and Hamburg (Germany), have CHP schemes for **district heating**, which by improving energy use saves fuel and reduces energy costs. An early CHP system was established at Battersea power station in London (United Kingdom) in 1950. The heat was piped under the River Thames to provide heat and hot water for blocks of flats on the opposite side of the river until the power station was closed down in 1984. Combined heat and power is also known as cogeneration.

COMBUSTION, or burning, occurs when **carbon**-containing fuels, such as wood, **coal**, petroleum and **natural gas**, are burned and the carbon is oxidized to form **carbon dioxide**, which is widely regarded as the primary greenhouse gas. The **hydrocarbon** fuels, like coal and petroleum, come from ancient plants, which grew by absorbing carbon from the air before being decomposed and were buried millions of years ago.

COMMAND AND CONTROL is the name for a type of environmental regulation that tends to set an absolute limit on the amount of pollution control, and also sometimes legislates how the pollution control is actually to be achieved. Many economists believe command and control regulations are too rigid and that there are more flexible policies that would cost less to industry. However, one advantage of command and control policies is that they are relatively transparent and usually easy to enforce. In the USA especially, it is more commonplace and politically acceptable to use market-based policy mechanisms such as **emissions trading** in environmental regulation.

COMMENSAL RELATIONSHIP describes that between two living organisms of different species, in which one is unaffected by the relationship, while the other benefits. (See also **symbiosis** and **parasitism**.)

The **COMMISSION ON HEALTH AND ENVIRONMENT** was established by Dr Hiroshi Nakajima, Director-General of the **World Health Organization (WHO)**, as an independent body in 1990 to respond to the concern about the impact that environmental change can have on human health. The WHO Commission, chaired by Simone Veil, a French Member of the European Parliament, was supported by four expert panels to cover the areas of energy, food and agriculture, industry, and **urbanization**. In March 1993 a draft Global Strategy for Health and Environment was published by the WHO, based on the findings and recommendations of the Commission and the outcome of the **Earth Summit**. The draft strategy outlines the WHO's objectives and proposed action programmes to achieve them. The priority areas decided upon were: a broader programme to promote environmental health; an expanded programme for the promotion of chemical safety; wider action and collaboration throughout the WHO on issues related to health and the environment; and stronger partnerships with other international and non-governmental organizations. The focus was to be on three main areas of **environmental health**: rural, urban and global. The strategy for global matters will place particular emphasis on urban areas where the expanding population and centres of modern economic development result in special health and environmental problems.

The **COMMON AGRICULTURAL POLICY (CAP)** was adopted by the European Communities (known as the European Union from 1993) to protect agricultural production and farm prices. It took effect in 1962. Incentives and subsidies, paid during the 1970s and 1980s, led to food 'mountains'. Many small farms were encouraged to switch to **intensive agriculture** and this led to the exhaustion of the soil and widespread environmental damage throughout Europe, causing loss of habitats for wildlife. **Grasslands** and **wetlands** have also been put under threat. In the late 1980s several measures were taken to redress some of the environmental and financial problems caused by the CAP. The CAP now funds Environmentally Sensitive Areas, where farmers are paid to farm in ways that benefit the environment and its wildlife. The **set-aside policy** was a measure, introduced in 1988, that paid, the farmer to take land out of production and use it for more environmentally beneficial purposes. This only began to be introduced properly with the more serious reform efforts of 1992, while in May 2000, in preparation for enlargement of the European Union, it was decided to continue the process of reform begun in 1992, with emphasis on environmental concerns.

COMPOST is a mixture of partially decomposed plant and animal substances that can be used as a soil conditioner or fertilizer. Some people recycle their leftover food and plant cuttings in order to make compost (see **recycling (domestic)**).

COMPRESSED NATURAL GAS (CNG) is the form in which **methane** can be stored as an **alternative fuel** for use in less-polluting vehicles. CNG is very suitable for large urban vehicles that make frequent stops, such as buses, road sweepers and refuse collection lorries. CNG emits very little **particulate** material into the air when burnt, and a relatively small amount of the **hydrocarbons** that form the basis of **photochemical smog** and ground-level **ozone**. It is suitable as a substitute for diesel and can be used as one of the fuels for **hybrid vehicles**.

The **COMPRESSION RATIO** is the characteristic of car engines that has led to the use of **anti-knock agents** to improve the **octane rating** of petrol.

CONSERVATION means protection from destructive influences, decay or waste. It applies equally to the protection of valuable paintings and antiquities in museums as it does to old buildings and all forms of animal and plant wildlife. As a result, the conservation movement consists of groups that focus on separate issues. For some, the protection of nature is most important and is achieved by creating **national parks** and **nature reserves**. For others, it is the sustained production of valuable materials from the living resources of the land and sea, or the prudent use of fuels and minerals. Conservation has become synonymous with actions to improve the quality of life for people by the control or elimination of environmental **pollution**. Conservation is seen as critical to our survival because life depends on the health of the **biosphere**, the air, water, soil and rock, on which all life on earth exists. The modern conservation movement has its roots in the mid-19th century in the USA, where parks such as **Yosemite National Park** were created. However, in the years after the Second World War the world population expanded rapidly, increasing pressure on land and resources. Synthetic **pesticides** began to have a severe effect on the environment and became a target for conservationists. By 1970 the problems of pollution of the air, land, rivers and sea became a global concern. However, no single country could contain the situation. It was obvious that to solve the problems there had to be an international authority with powers to co-ordinate environmental protection activities. Many recommendations were suggested for conserving wildlife and halting pollution. Many of them were eventually incorporated in the Action Plan for the Human Environment produced by the **UN Conference on the Human Environment**, which was held in 1972. However, critics argue that unless countries are willing to delegate more authority to international organizations, progress towards the solution of global problems will be slow. A conservable resource can usually be used for more than one purpose; for example, a forest can also be used for recreation. However, preservationists (see **preservation**) usually argue that land should be left untouched rather than being managed for multiple purposes.

CONSUMER PRODUCTS IN ENVIRONMENTALLY RISKY CATEGORIES is a term used in European Union discussion papers to describes goods containing hazardous substances which can present environmental waste-disposal problems. Such goods include plastics like polyvinyl chloride (PVC), which contain **organochlorine compounds** and organic solvents; pesticides containing organochlorine and **organophosphorous compounds**; paints containing **heavy metals**, pigments, **solvents** and organic residues; and **batteries** and other components of used vehicles, which can also contain heavy metals.

The **CONTAINMENT** is the reinforced steel or concrete vessel that encloses a nuclear reactor. It is designed to withstand minor explosions in the core, to keep **radionuclides** from escaping into the environment, and to be safe against terrorist attack.

The **CONTINENTAL SHELF** is the gently sloping seabed of the shallow water nearest to a continent, covering about 45 miles from the shore and deepening over the sloping sea floor to an average depth of 400 ft. It continues until it reaches the **continental slope**. The continental shelf contains most of the important fishing grounds and a range of resources, including gas and petroleum, sand and gravel. However, the shelf is, in general, a structural extension of the continent, and so may also be a source of minerals found in that region, such as tin, gold and platinum.

The **CONTINENTAL SLOPE** is the steep slope which descends from the edge of the continental shelf until it reaches the abyssal depths.

The **CONVENTION ON BIOLOGICAL DIVERSITY (CBD)** was adopted in May 1992 after four years of negotiations under the sponsorship of the **UN Environment Programme (UNEP)**. It was presented to the UN Conference on Environment and Development (the **Earth Summit**) and adopted there by 153 states and the European Communities (now the European Union). At mid-2004 188 parties had ratified, acceded to, approved or agreed to the CBD. Although the USA had signed the CBD in 1993, it had persisted in refusing to pursue any form of ratification, owing to concerns with the convention's intellectual property rights, technology transfer and finance provisions. The CBD contains provisions intended to ensure effective national action to curb the destruction of biological species, **habitats** and **ecosystems**. Among them are: that countries regulate to conserve their biological resources; the imposition of legal responsibility upon nations for the environmental impact of their private companies in other countries; technology transfer on preferential and concessional terms, where it does not prejudice property rights or patents; the regulation of biotechnology firms; access and ownership of genetic material; and compensation to developing countries for extraction of their genetic materials. The convention also embraces the idea that industrialized countries must help developing countries financially and with know-how. It also accepts that the responsibility for setting up a network of protected areas rests first with each country, and that the first beneficiaries of the conservation and sustainable use of wild plant and animal species should be the rural communities and indigenous peoples whose traditional knowledge and respect for those resources has preserved them for centuries. The convention does not specifically refer to **genetically modified organisms (GMOs)** and the ecosystems of which they are a part. Such organisms, however, are understood to be included in the broader form referring to living modified organisms (LMOs) resulting from biotechnology (see **biotech**), which has led to the **Biosafety Protocol**, which, upon coming into effect in 2003, became the first internationally binding agreement governing the transboundary movement of LMOs (i.e. GMOs). The convention does not include the idea of drawing up a global list of protected areas. Instead, every country which is a party to it will put together its own list of protected areas in the expectation that, eventually, they will become part of a global list.

CONVENTION ON CLIMATE CHANGE: See **Framework Convention on Climate Change**

CONVENTION ON THE CONSERVATION OF EUROPEAN WILDLIFE AND NATURAL HABITATS: See **Berne Convention**

The **CONVENTION ON DESERTIFICATION (CCD)** is formally known as the Convention to Combat Desertification in those Countries Experiencing Serious Drought and/or Desertification, particularly in Africa. It had its origins in the **Earth Summit** in Rio de Janeiro, Brazil, and was adopted in 1994, coming into force in 1996. By 1997 113 countries had ratified the CCD. The emphasis of the Convention is on improving processes and enhancing local participation in decision-making. It requires states to decentralize authority and resources to the local level and encourages the formation of partnerships and collaborations. Parties are supposed to address the underlying causes of desertification through National Action Plans, using them to enable citizens to fight the causes of desertification.

The **CONVENTION ON THE ESTABLISHMENT OF AN INTERNATIONAL FUND FOR COMPENSATION FOR OIL POLLUTION DAMAGE** was adopted at an **International Maritime Organization** conference in 1971 in order to provide compensation for victims of pollution damage caused by petroleum. It came into effect in 1978. (See also **Torrey Canyon**.)

The **CONVENTION ON INTERNATIONAL TRADE IN ENDANGERED SPECIES OF WILD FLORA AND FAUNA (CITES)** emerged from meetings started in 1973 between a number of countries that were becoming increasingly concerned about the exploitation of wildlife. Negotiations led to a treaty to prevent international trade from threatening **endangered species** with **extinction**. The treaty came into force in July 1975. At mid-2004 there were 166 parties to CITES (all European Union countries are required to comply with CITES under a directive adopted in 1984 and updated in 1997). In 1976 the **World Conservation Union (IUCN)** established a system called TRAFFIC to help the CITES Secretariat monitor and regulate trade in wildlife. There are over 13,000 known species of mammals and birds in existence, as well as thousands of reptiles, amphibians and fish, millions of invertebrates and some 250,000 flowering plants. Extinction is a natural feature of the evolution of life on earth, but in modern times humans have been responsible for the loss of most of the animals and plants that have disappeared. These include 17 species of bear, five species of wolf and fox, four species of cat, 10 species of cattle, sheep, goat or antelope, five species of horse, zebra and ass, and three species of deer. The last dodo, a large flightless bird, was killed in Mauritius in 1681, while the passenger pigeon, which had previously been extremely common in North America, was also exterminated for food early in the 20th century. Many species decline in numbers because of **vanishing habitats** and increased exploitation as human populations grow. Trade in animals and plants continues because transport systems have made live shipments easier anywhere in the world. Millions of live animals and plants are shipped around the world every year to supply the trade in domestic animals, furs, leather, ivory, timber and ornamental plants. CITES established a world-wide system of controls on international trade in wildlife and wildlife products. The treaty divided species traded into two categories. Animals and plants listed for protection in Appendix I are so endangered that trade in those species is banned. They include apes, the **giant panda**, great **whales**, **tigers**, the **rhinoceros**, sea turtles, some species of **mussel**, orchids and cacti. Appendix I also lists those species which would be at risk of becoming endangered if trade in them was not controlled and monitored, including birds of prey, **dolphins** and porpoises, African **elephants**, tortoises and some species of snail. Species listed in Appendix II, such as certain parrots and snakes are subject to strict trade controls. A third annexe, Appendix III, includes any species listed by a party to CITES as having special protection within its jurisdiction, as many countries enforce even stricter controls than those required by CITES when they wish to give extra protection to a listed species. Countries may also ban trade in all their wildlife. Customs officers in most member states enforce the CITES regulations. Governments have to submit reports, including trade records, to the CITES Secretariat in Switzerland.

The **CONVENTION ON THE LAW OF THE NON-NAVIGATIONAL USES OF INTERNATIONAL WATERCOURSES** was adopted on May 21 1997 by 103 nations and 27 abstentions. Notably, three nations voted against adoption: the People's Republic of China, Turkey and Burundi. The treaty would enter into force on the 90th day following the deposit of the 35th instrument of ratification at the UN. At the beginning of 2004 it remained open for signature. The Convention provides the first strong guiding principles for use of **fresh water** and for future agreements on shared watercourses. Signatories have a non-binding commitment to prevent harm to other states, co-operate in good faith to use and protect shared watercourses and peacefully settle disputes over watercourses, including the use of arbitration as necessary. Adoption of the treaty after 27 years of negotiation was a significant achievement. This development may indicate a new international recognition that the allocation of freshwater resources requires a more comprehensive approach. In 1992 four similar principles with respect to fresh water were agreed in Dublin, Ireland, that laid the foundation for the Convention. The principles were: fresh water is a finite and vulnerable resource, essential to sustain life, development and the environment; water development and management should be based on a participatory approach, involving users, planners and policy-makers at all levels; women play a central part in the provision, management and safeguarding of water; and water has an economic value in all its competing uses and should be recognized as an economic good.

CONVENTION ON LONG-RANGE TRANSBOUNDARY AIR POLLUTION (LRTAP): See **Long-Range Transboundary Air Pollution Convention**

CONVENTION ON THE CONSERVATION OF MIGRATORY SPECIES OF WILD ANIMALS (CMS): See **Bonn Convention**

The **CONVENTION ON PERSISTENT ORGANIC POLLUTANTS** was signed by about 150 countries in May 2001, in Stockholm (Sweden), and came into effect on 17 May 2004. Its objective is the eventual cessation of intentional production and use of all **persistent organic pollutants (POPs)**: pesticides such as **aldrin**, **chlordane**, **DDT**, dieldrin, endrin, heptachlor, mirex and toxahene, and industrial chemicals such as **polychlorinated biphenyls (PCBs)** and hexachlorine. These synthetic organic pollutants are known informally as the **Dirty Dozen**. It has been debated whether the continued use of DDT for malaria prevention should be permitted. POPs **bioaccumulate** in the fatty tissue of humans and animals, with the greatest degree of accumulation occurring at the top of the **food chain**. It is thought that POPs may be **endocrine disrupters**.

CONVENTION ON THE PROTECTION OF THE ENVIRONMENT OF THE BALTIC SEA AREA: See **Helsinki Convention**

CONVENTION FOR THE PROTECTION OF THE WORLD CULTURAL AND NATURAL HERITAGE: See **World Heritage Site**

CONVENTION ON WETLANDS OF INTERNATIONAL IMPORTANCE ESPECIALLY AS WATERFOWL HABITAT (RAMSAR): See **Ramsar Convention**

COOLING PONDS are the large pools at nuclear power stations in which spent **nuclear fuel** is stored for one to two years after its removal from the nuclear reactor. One-third of the nuclear fuel rods in a nuclear power station are replaced each year. The used fuel is stored in water to cool down thermally and radioactively before it can be transported for **nuclear reprocessing** or permanent storage. Cooling ponds in the USA are dangerously full because a long-term storage facility has not been approved. A site was identified in Yucca Mountain, Nevada, but this site proved to be very controversial for geological reasons and because many Nevadans were opposed to the establishment of such a facility in their state. Building works were not scheduled to begin until the early 2010s, enhancing worries about the safety of the cooling pools.

COPPER (Cu) is very toxic in high concentrations, but is also an essential **trace element**, because it is a component of many proteins. A deficiency of copper in the body leads to anaemia, since iron can not be used without it.

CORAL BLEACHING is a potential new threat to **coral reefs**. Widespread bleaching (resulting from the loss of the coral's zooxanthellae symbionts or a reduction in the level of their photosynthetic pigment concentrations) was reported during late September 1987 by divers in Puerto Rico, Jamaica, the Bahamas, Haiti, the Cayman Islands, the US Virgin Islands, the British Virgin Islands, the Dominican Republic and southern Florida, USA. During January—summertime in the southern hemisphere—severe bleaching was also reported on Australia's **Great Barrier Reef**, one of the largest and most established reef systems in the world, and one of only two double barrier reefs. Marine biologists began to search for a stress that could cause such a widespread outbreak of bleaching. **Pollution** and **sediments** are known to cause coral bleaching, but some observers think that water temperature rises of one or two degrees could be a cause. It is believed that, as temperatures rise, increasingly smaller fluctuations from the mean yearly temperature would be necessary to cause bleaching outbreaks. An increase in bleaching could also possibly increase the rate of accumulation of **carbon dioxide** in the atmosphere, because coral reefs act as large **carbon sinks**.

CORAL REEFS have been built up from the skeletons of reef-building coral, a small primitive marine animal, and other marine animals and **algae** over thousands of years. Reefs depend, in the first place, on the **symbiosis** between scleractinian corals and an algae known as zooxanthellae. Coral reefs tend to occur in clear, shallow and sunlit seas. Coral reefs are one of the most productive and diverse ecosystems and are estimated to yield about 12% of the world's fish catch. They are very vulnerable to any change in their environment, especially pollution, because it makes the water opaque. They must have light in order that **photosynthesis** by the algae can take place. Like trees, corals reflect the environmental conditions in which they grow, indicating marine pollution, sea-surface temperature and other aquatic conditions. Corals can also indicate previous sea levels. For example, raised reefs off Bermuda, Australia, the Bahamas and Hawaii indicate that the sea level was some six metres higher during the warmest time between the last two ice ages than it is today. Even without the possible effects of **climate change**, it is feared that the inexorable rise of the human population will have an adverse effect on coral. Many reefs are being destroyed by sewage pollution and by the proliferation of the crown of thorns starfish, which feeds on the living coral. Some reefs have been mined for limestone to provide materials for increased building programmes on tropical islands. Moreover, as a result of tourism, some reefs have been depleted to make jewellery or decorations for aquaria. At the time of the **Earth Summit** in 1992 about 10% of the world's coral reefs had been damaged. By the 2002 **World Summit on Sustainable Development** this had risen to 27%. Other reports claimed that if immediate and dramatic action was not taken to lessen human impacts, fully one-third of the reefs could be destroyed within 30 years (adding to the 11% already destroyed and one-fifth damaged). (See also the **Great Barrier Reef**.)

CORINAIR is one of the **CORINE** projects and generates an inventory of atmospheric emissions based on the levels of discharges from the 12 European Communities member states in 1985. CORINAIR-85 covered the three pollutants that were targeted by an earlier project, initiated by the countries of the **Organisation for Economic Co-operation and Development (OECD)** in 1980. CORINAIR, which is co-ordinated by the **European Environment Agency**, produces an inventory of emissions of **sulphur dioxide**, **nitrogen oxides** and **volatile organic compounds (VOCs)** in Europe. Nine principal categories of emission sources were initially identified, but the number was subsequently revised. The change was needed to meet the aims of the 1986 European Monitoring and Evaluation Programme (EMEP), a co-operative programme for the monitoring and evaluation of the long-range transmission of air pollutants in Europe. In 1990 the number of principal sources was extended to cover: public power, CIR combustion, industrial combustion processes, fuels, solvent use, road transport and other waste, agricultural and natural sources. By the mid-1990s the CORINAIR project covered 250 activities and eight pollutants within the 11 categories of source. The system is used by 30 countries in Europe for producing national pollution inventories and for making submissions to the **Long-Range Transboundary Air Pollution Convention**.

CORINE is the abbreviation for Co-ordination of Information on the Environment. It was established in the mid-1980s by a European Communities decision (85/338/EEC) to devise a system for gathering and comparing information on the state of the environment. One of the purposes of the programme is to ensure that the information contained in thousands of databases is stored in a standard format. CORINE Land Cover is one of a series of inventories and data bases that generate easily comprehensible maps and statistical overviews. The project has been run by the **European Environmental Agency** since 1995.

COSMIC RADIATION is the radiation that comes from outer space, much of which is blocked by the upper layers of the **atmosphere**. Some cosmic radiation is very energetic and is able to penetrate a mile or more into the earth.

COST–BENEFIT ANALYSIS is a policy tool for evaluating the economic advantages and disadvantages of enacting an environmental policy. In order to measure the benefits and costs of a proposed policy, numerical values must be assigned to various goods, a task which can be difficult in the case of ecological goods. For example, in order to study a potential **air pollution** policy, one would need to know the extent to which the public valued cleaner air, and how much each person would be willing to pay in order to obtain it. The pollution damages, such as the health costs associated with respiratory diseases caused by air pollution, could be calculated by estimating visits to the doctor or aggregating the cost of asthma medicines. Against the potential benefits, one would calculate the cost to society of implementing the policy: for example, how much it would cost power generating companies to install additional pollution control equipment in their plants, and whether these costs would be passed on to consumers. There is often insufficient information to make informed decisions based on cost–benefit analysis. Nevertheless, this method is regularly used in the USA to assess environmental legislation.

The **CURIE** (Ci) is the unit of measurement of the amount of **radioactivity** of a substance. One curie is the amount of a substance that it takes to produce 37,000m. atomic disintegrations per second. There are significant differences between the physical quantities of a substance and its radioactivity. For example, 10,000 Ci of newly produced, lethal, radioactive **iodine**-130 would be scarcely noticeable, consisting of only a few milligrams of material. If the 10,000 Ci were newly produced **caesium**, the substance would weigh about 150 grams and be about the size of a tangerine. However, iodine has a short **half-life**, and every 12 hours it loses one-half its radioactivity. Whereas an exposure of three minutes to newly produced iodine-130 would be lethal, three minutes' exposure to the same substance 40 days later would be within acceptable recommended dose limits. In contrast, 10 minutes' exposure to caesium, which has a half-life of 30 years, would be lethal even 10 years after it was produced. There are also other factors that make some substances more biologically dangerous. One curie, and in some cases a small fraction of one curie, may cause genetic abnormalities, cancer or death, depending on the length and circumstances of exposure and whether the substance is an emitter of **alpha**, **beta** or **gamma radiation**. The curie is being replaced by the **becquerel**, which is equal to the decay, or atomic disintegration, of just one nucleus per second. Marie Skodlowska Curie (1867–1934) coined the term radiation and it was in her honour that the unit was named in 1910. She had continued the early research on uranium 'rays' done with her husband, Pierre Curie (1859–1906). For this work the couple had shared the 1903 Nobel Prize for Physics with Henri Becquerel, but Marie Curie also became the first person to win two Nobel Prizes, when she was awarded the Nobel Prize for Chemistry in 1911, for the discovery of **radium** and polonium (see **polonium-210**).

CURITIBA is a city in south-west Brazil that has been described as being 'designed with nature'. Its **population** had increased to over 2m. by 1990, but since 1965 its development has been planned with environmental factors as priorities. One priority was to establish an efficient bus service and, consequently, three-quarters of the population commute by bus. There is also a network of cycle lanes and the main street is closed to all vehicles. Riverbanks and flood plains have been converted to parks and disused factories and riverside buildings have become sports and leisure centres. The city's former waste dump is now a botanical garden. More than 70% of the population sort materials for recycling and in low-income areas families can exchange filled rubbish bags for bus tokens, parcels of surplus food and school notebooks.

CYANOBACTERIA are a large and varied group of bacteria, and are remarkable in that they possess chlorophyll, carry out **photosynthesis** in the presence of sunlight and air, and release oxygen. Fossil evidence suggests cyanobacteria existed over 3,000 million years ago. They are believed to have been the first oxygen-producing organisms and to have been responsible for generating the oxygen in the atmosphere on which life depends. Cyanobacteria can also carry out **nitrogen** fixation. They are the only organisms known to fix both carbon and nitrogen. The organisms are widely distributed in **fresh water**, the marine environment, rocks and soil.

D

DAMS are one of the most environmentally controversial of major man-made developments. Dams are built to provide **irrigation** and **hydroelectric power**, and to control flooding. Dams have been built throughout history, the oldest known being in Egypt (c. 2,700 BC). In the modern era, for example, the USA has built about 5,000 dams, Japan 2,000 and India 1,000. Since the Second World War several ambitious and extremely large dams have been constructed in a relatively short period of time, primarily because at the time they offered solutions to two great needs, food and energy. However, sometimes dams create problems that are more severe than those they are intended to remedy. For instance, millions of people have been uprooted from their homes to make way for the reservoirs, and **water-borne diseases** have increased in many areas around large dams. Often the flooded land is fertile and agriculturally productive. Impoundment of the waters also stops river silt, which generally has a high nutrient value, from being carried to the sea. In the short term the building of a dam and reservoir often dramatically increases fish yields. However, there is evidence that, in later years, because submerged vegetation and soils soon rot down and the amount of nutrient is reduced, the fish stocks are depleted. There is also an adverse effect on fish life within the river basin and in the seas immediately beyond its estuary. Aquatic weeds, which often invade reservoirs, and the pesticides subsequently used to destroy them, also affect fish populations. The flooding of forests, agricultural land and bush has also led to a loss of habitats for thousands of species, some of them already endangered. Where dams have been built on waters that cross territorial boundaries, there are often bitter international disputes. There is also a suspicion that very large dams could trigger **earthquakes**. In June 2004 the **World Wide Fund for Nature (WWF)** released a report bemoaning the threat to some of the world's largest rivers from dams, citing the threat to freshwater species' reproduction and migration and the loss of nutrients and oxygen. The report particularly mentioned the Yangzi (Yangtze or Chang Jiang) in the People's Republic of China, the basin of the River Plate in South America and the Tigris and Euphrates in Turkey and the Middle East. Sixty per cent of the world's 227 largest rivers were reported to be fragmented by dams. In the same year there were reports of problems on the Mekong and at its delta as a result of Chinese dams built in Yunnan since the mid-1990s, but causing problems with water levels and fish stocks in countries downstream (where the river provides 80% of protein). At the end of the 20th century the largest completed dam in the world was the **Itaipú Dam** between Brazil and Paraguay, which had 12,600 MW of generating capacity. However, once completed, the **Three Gorges Dam** in China would become the largest dam and have generating capacity of some 18,000 MW, but it will displace over 1m. people and irreversibly alter an entire watershed and **ecosystem**. Other dams can make claim to particular records, such as: the highest dam, Rogun, on the Vakhsh river in Tajikistan (335 m); the longest dam, the Hirakud, on the Mahanadi in Orissa, India (the dyke is some 25 km in length); or the biggest dam by volume, the Syncrude Tailings barrage in Alberta, Canada (540m. cu m of materials). The reservoirs can cover a large area, but some are shallow, others deep and long, so capacity varies enormously; among the biggest, apart from that filling behind the Three Gorges since 2003, are the Owen Falls reservoir in Uganda, Lake Nasser in Egypt (behind the Aswan High Dam) or Lake Mead in the USA (behind the Hoover Dam).

The **DANUBE** is the second largest river in Europe (some 2,860 km in length), following only the Volga in length and volume. It flows through Germany, Austria, Slovakia, Hungary, Croatia, Serbia and Montenegro, Romania, Bulgaria and Ukraine, but in total it drains from 17 European countries. The Danube delta forms the second largest **wetlands** area in Europe. Almost one-third of the pollution of the **Black Sea** has its source in the Danube. A European Union (EU) clean-up project of the river basin was announced in 1993. Funds were to be used in a long-term work plan involving the EU, those countries with land drained by the Danube, and other countries. The river, on its upper reaches, has a hydroelectric power project at Gabčikovo-Nagymarós that is seen as the largest water diversion scheme in Europe. From the 1980s the project has aroused widespread opposition and disagreement between Slovakia (part of Czechoslovakia until 1 January 1993) and Hungary about water use. Water pollution had prompted increased negotiation between the countries of the Danube basin from the 1980s, resulting in the Danube Environmental Programme and its Strategic Action Plan, which was formulated in the mid-1990s alongside the Convention on Co-operation for the Protection and Sustainable Use of the Danube River. The latter, signed in Bulgaria by the riparian states and the EU in 1994, is usually known as the Danube River Protection Convention and it came into force in 1998, allowing the establishment of the International Commission for the Protection of the Danube River. A Joint Action Programme for 2001–05, which emphasized the reduction of nutrients running into the watercourse, was then produced.

DDT was banned as a pesticide in the USA in 1973 and in much of Europe in 1984, after a long campaign by environmentalists that began with the publication of ***Silent Spring*** in 1962. DDT is still used in many developing countries for spraying swamps that are the breeding ground of the larvae of the malaria-carrying **Anopheles** mosquitoes. The pesticide was declared an environmental hazard because of its harm to wildlife and because it was accumulating in the **food chain**. As DDT is not readily metabolized into harmless waste products by animals, including humans, it becomes stored in body fat. Wildlife such as worms, snails and voles absorb DDT from eating plants and seeds. The DDT then accumulates in their predators, eventually reaching animals that prey on other animals, and also in animals that eat plant tissue. However, from the post-Second World War years until the late 1960s it was accepted as the miracle pesticide. DDT was cheap to produce. It killed body lice, which can carry typhus, and the mosquito that transmits **malaria** and yellow fever. It was seen as both a public health benefit for combating malaria and other insect-borne diseases, and in boosting crop production by killing the Colorado beetle and hosts of other plant-destroying insects. In the industrialized world it became an integral part of **intensive agriculture**. Scarcely a home was without a DDT product to deal with flies, clothes moths, pet fleas and garden insects. DDT was first synthesized in 1873 by a German graduate student, but he had no idea of its insecticidal value, so it was forgotten. It was rediscovered by the Swiss entomologist, Dr Paul Müller, when he was looking for a long-lasting insecticide against the clothes moth and found that it was extremely effective against disease-carrying flies and mosquitoes. This discovery earned him the Nobel Prize for Medicine in 1948. By the beginning of the 21st century the insecticide was in such infrequent use that the acronym generally applied to drug discovery technology.

DEAD ZONES: See **oceans**

DEADTIME is the time it takes waste to decompose, provided it is **biodegradable**. The description is employed when assessing the types of waste that could be disposed of in a **landfill** site. Depending on the material, calculations of deadtime hinge on the amount of water and air there is seeping through the site. Decomposition takes longer when water and air are scarce.

DEBT-FOR-NATURE SWAPS are innovative schemes designed to help alleviate the debt burden of developing countries and, at the same time, provide financial support for nature conservation in countries with environmental problems. Several banks began to accept that many of the loans they had made could never be repaid and so conservation organizations turned the debt crisis to ecological good by buying the debt at a discount so that valuable natural resources under pressure could be saved. Under the debt-for-nature scheme, the debt of a developing country is bought by a conservation organization, at a discount, on the secondary debt market. In return, the debtor country has an obligation to carry out an agreed conservation programme. For the developing country it means a reduction of the debt they have to service and, instead of having to pay an outside source in hard currency, it can use local currency to invest in conservation projects. The funds of the purchasing organization are, in effect, converted at a more favourable exchange rate. As a result of these schemes, which were launched in 1987, environmental programmes are in place in several countries, including Madagascar, Zambia, Costa Rica, Ecuador and Poland. For example, the US environmental group Conservation International bought US $650,000 of Bolivia's debt in return for 15,000 sq km of **tropical rainforest**. Between 1989 and 2003 the **World Wide Fund for Nature (WWF)** was responsible for 16 debt-for-nature swaps with a number of countries, including Ecuador (the first) and Madagascar (the most recent), with a total face value of US $84m. In Zambia, for instance, the local currency equivalent, among other conservation programmes, went towards protection of **rhinoceros** and African **elephant** populations. Other programmes include conserving wetlands and forests, establishing nature reserves, and halting soil erosion and habitat destruction. There are provisions, too, for education and training programmes.

DECAY refers both to biological decomposition of organic matter to create natural fertilizer and **humus**, and to the very important feature of all radioactive materials determining how their **radioactivity** declines with time. Radioactive nuclei change spontaneously into stable non-radioactive materials in one or two stages. Each step is accompanied by the emission of radiation. The decay process is defined by the time it takes for the activity of a substance to drop by 50%. Each radioactive species has its own fixed **half-life**, which varies from fractions of a second to millions of years.

DECIBELS (dBs) are the units commonly used as measurements of sound power or loudness. The decibel equals one-10th of a bel; the bel got its name from Alexander Graham Bell, inventor of the telephone. Laws to control **noise pollution** set limits on decibels. However, they are not always easy to interpret, because a change of 1 dB is equivalent to a change of 26% in the intensity of sound, and a doubling in loudness or power of a sound adds 3 dBs to its rating. Hence, a sound of 21 dBs is twice as loud as a sound of 18 dBs.

DECIDUOUS FORESTS are the temperate forests comprised of trees that seasonally shed their leaves, located in the east of the USA, in Western Europe from the **Alps** to Scandinavia, and in eastern Asia. The hardwoods of these forests have been exploited extensively since the 16th century. The trees of deciduous forests usually produce nuts and winged seeds. A relatively small number of mammals live in the forests, including bears, badgers, deer, wild pigs and squirrels, but their seed and nut-gathering activities have a considerable effect on the distribution of seeds and, therefore, the survival of deciduous forest trees. In spring and summer migrant birds return from their winter quarters in warmer climates.

The **DECLARATION OF THE HAGUE** was the first document drawn up by an intergovernmental meeting in which national leaders expressed a willingness to diminish their sovereignty in specific areas for the sake of the planet. The meeting was arranged in March 1989 by the Prime Ministers of the Netherlands (the host), Norway and France. As the implications of the twin problems of global warming (see **climate change**) and **ozone** depletion emerged from the latest research, the Governments of these countries were concerned about the difficulties of obtaining the unprecedented degree of international co-operation needed to avert ecological disaster. The declaration was a statement of principles, not an action plan. However, it proposed new institutions that would be needed to finance and facilitate activities such as the transfer of energy-efficient technologies between countries and provide **chlorofluorocarbon (CFC)** substitutes to developing countries in exchange for their agreement to limit carbon dioxide and CFC emissions.

DECOMMISSIONING is a term often used to describe the dismantling of obsolete nuclear power stations and the disposal of the associated radioactive wastes. Similar problems were increasingly encountered following the end of the Cold War, with the need to decommission many of the nuclear-powered submarines of the US, former Soviet and British navies. The same principles apply to the decommissioning of both nuclear power stations and nuclear submarines. At the end of a nuclear power station's useful life, the last batch of spent **nuclear fuel** is removed and the cooling fluids are drained. When a reactor has been retired, there are three options for decommissioning. The first, known as 'immediate dismantlement' (in the USA this is referred to as DECON), involves prompt decontamination and disassembly of the plant. The second, 'mothballing' or 'safe enclosure' (SAFSTOR), entails some initial clean-up, followed by several decades of quarantine, before the parts of the nuclear plant are dismantled and sent to a disposal site. The third, known as 'entombment' (ENTOMB), involves encasing the reactor in concrete. Although nuclear power provided over 15% of the world's electricity by 1990, not one large commercial station had been dismantled completely, so costs were uncertain. However, work had begun on the development of technologies for the decontamination, dismantling and transportation of **activation products**, and metal and concrete structures. The first major nuclear power station completely decommissioned in the USA was Fort St Vrain, in 1992. By January 2004 three others had also achieved this status, while 19 others were in the process of decommissioning (but most in SAFSTOR). Valuable lessons had been learnt from Germany, which had opted for immediate decommissioning of some former East German reactors in the 1990s. Elsewhere in Europe the favoured method is to start with a period of mothballing: 10 reactors in the United Kingdom, three in France (which was also developing recycling facilities), for instance. In Japan the DECON of a **Magnox** reactor from 1998 was expected to take up to 10 years and cost 25,000m. yen (over US $200m.). Increasing experience of decommissioning was expected to lower costs and improve technique. By 2004 almost 100 commercial power reactors, over 250 research reactors, a number of fuel cycle facilities and about 100 mines had ceased operations. Estimates suggested that decommissioning a typical **pressurized water reactor (PWR)**, the most commonly used nuclear power system, would create 18,000 cu m of **low-level waste** alone: 70 times as much as was produced each year by all the world's operating nuclear reactors. The greatest problem for nations with nuclear plant, therefore, is the lack of permanent disposal facilities, especially for **intermediate-level waste** and **high-level waste**. More than 350 civil nuclear reactors were expected to reach the end of their lives between 2005 and 2035.

DEFOLIANTS are **herbicides** used to remove leaves from plants and trees. (See also **Agent Orange**.)

DEFORESTATION began thousands of years ago when humans first cut down trees for fuel and cleared space for agriculture. At the beginning of the 21st century the activity remained concentrated in **tropical rainforests**. It was reckoned that 16m. ha of **forest and woodland** are lost annually. Technically, deforestation is the conversion of forested land to non-forested land, or the reduction of forest cover within a forest. The potential impact on climate comes about through altering the **carbon cycle** and **nitrogen cycle**. Destruction of the forest releases **carbon dioxide** into the air. Forest resources are essential to both development and the preservation of the global environment. Mismanagement of forests—inadequate fire control, unsustainable commercial logging, overgrazing and the harmful effects of airborne pollutants—linked to degradation of soil and water sources, loss of wildlife and **biological diversity**, and the aggravation of global warming. **Agenda 21** stated that urgent action was necessary to conserve and expand existing forests, and proposed the creation of national forestry action programmes for **sustainable development** of forests by governments. Deforestation is different from **harvesting**, which is the term used for commercial logging.

DEGRADABLE PLASTICS were developed as an answer to criticisms of the pollution problem created by disposable plastic bags, cups, cartons and bottles. The assumed virtue of degradable plastics is that they are formulated either to break down in sunlight (photodegradable) or to be decomposed by bacteria (**biodegradable**). The biodegradable variety is made from a chemical polymer that contains 6%–10% starch, binding it together. Bacteria feed on the starch, breaking the plastic into small pieces. Some environmentalists regard degradable plastic made from traditional petrochemical raw materials as a sleight of hand, rather than the development of a genuinely 'greener' product. They are said to work at cross purposes to **recycling**.

DELIBERATE RELEASE describes the process of intentionally releasing something into the environment. One example would be the release of **genetically modified organisms (GMOs)**, which may be **biopesticides** or genetically modified seeds of, say, the potato, tomato or wheat plants, which have been designed for better pest-resistance or improved growth. The first GMO was released in 1986 and was a genetically engineered frost-proofing bacterial strain called ice-minus. An international treaty known as the **Biosafety Protocol** was agreed in January 2000 to address issues related to the release of GMOs.

A **DELTA** is a vast, fan-shaped creation of land, or low-lying plain, formed from successive layers of sediment washed from uplands to the mouth of some rivers, such as the **Danube**, the **Ganges**, the Nile, the Mississippi and the **Yangzi** (Chang Jiang). The nutrient-rich sediment is deposited by rivers at the point where, or before which, the river flows into the sea. Deltas are formed when rivers supply and deposit sediments more quickly than they can be removed by waves or ocean currents. The importance of deltas was first discovered by prehistoric man, who was attracted to them because of their abundant animal and plant life. Connecting waterways through the deltas later provided natural routes for navigation and trade, and opened up access to the interior. Deltas are highly fertile and often highly populated areas. They would be under serious threat of flooding from any **sea level** rise and are also affected by **dams** upstream.

DEMERSAL describes those fish that live on or near the seabed. They include round-fish, such as cod, haddock, hake and dogfish, but also the flatfish that lie on the seabed—plaice, sole, turbot, brill, skate and ray, for example. Most of these demersal fish, which are the main product of fishing fleets, lay huge numbers of eggs, which float to the surface. The subsequent fry add to the spring **plankton**. The demersal zone is the sea or ocean near the seabed or the coast, whereas the open ocean and its denizens are described as pelagic.

DENITRIFICATION is the loss of **nitrogen** from soil by biological or chemical means. It is a gaseous loss, unrelated to loss by physical processes such as through **leachates**.

DESALINATION is a process for converting salt or **brackish** water into a source of safe drinking water. Parts of the world with severe water shortages are looking to desalination plants to solve their problems. Already use of desalination plants is increasing in the richer countries, such as Saudi Arabia, Kuwait, the United Arab Emirates and the USA. Desalination of water is still nearly four times more expensive than obtaining water from conventional sources. However, technology is improving and costs are likely to decrease slightly in the future. As **fresh water** becomes more scarce, it is likely that desalination plants will gradually become more popular. There is also now more interest in building distillation plants beside electric installations, so that the waste heat from power generation can be used to drive the desalination process. The largest proportion of desalination plants treat sea water, while a smaller number are devoted to treating brackish water. There are other applications: the treatment of effluent water; the treatment of river water for boilers; and the removal of pollutants, like **nitrates** and pesticides, from **ground-water**.

DESERT VEGETATION has to develop ingenious ways to protect itself from the extremes of almost permanent drought and high daytime temperatures. Some **xerophytes** have deep and widespread roots, others develop a fibrous network in the surface soil to catch the dew, like many cacti which store the moisture in their fleshy stems. Many desert plants are able to grow very quickly if it suddenly rains, and often seeds can remain dormant, sometimes for many years, waiting for rain to provide the right conditions for germination. **Lichen**, which grow on rocks, can play a major role because they can capture and fix **nitrogen** from the atmosphere. The nitrogen is passed back into the soil by snails, which absorb the lichen from the rocks. As they escape to the shade of the rocks to avoid the daytime heat they excrete the nitrogen into the soil with their faeces, which enriches the soil to encourage plant growth.

DESERTIFICATION is the process of land damage that allows the soil to spread like a desert in arid and semi-arid regions. There is a loss of vegetative cover and the soil deteriorates in texture, nutrient content and fertility. Desertification affects the lives of three-quarters of the world's population, 70% of all drylands and one quarter of the total land area of the planet. The results of it include poverty, a decline in soil fertility and the degradation of rangeland, rain-fed crops and irrigated land. There are many reasons for desertification, but the majority are caused by human activities such as **overgrazing**, **deforestation**, poor land management and over-exploitation. Overuse of pesticides and herbicides can deprive the soil of nutrients, rendering it sterile desert, while **waterlogging** can trigger **salinization**. Of the agricultural land in arid regions, 80% suffers from moderate to severe degrees of desertification. If humans cause the destruction they are also its victims. They have to leave the countryside for the cities and, as refugees, face intolerable living conditions and are threatened by disease, disability, malnutrition and death. There are some remedial and preventive actions against desertification, among them **irrigation**, planting of trees and grasses, and the erection of fences to secure sand dunes. **Agenda 21** states that the priority in combating desertification should be establishing preventive measures for lands that are not yet, or are only slightly, degraded. At the same time, severely degraded land should not be neglected. For regions prone to desertification and drought, better information and monitoring systems are needed to identify the priority areas for action. To address the problems associated with desertification, a number of governments negotiated a **Convention on Desertification**, which came into force in 1996. However, in 2004, 10 years after it was negotiated, it was estimated that Central Asia still lost between 8,000 and 10,000 sq km per year to desertification. In the People's Republic of China sand dunes have reached within 100 km of the capital, Beijing, and the dust storms affect even the Korean peninsula and Japan; the Gobi is reported to be increasing in size by about 10,400 sq km per year. In sub-Saharan Africa it was reckoned that some 60m. people might be forced to move from newly desert areas, if current trends continued. Even as far south as Nigeria, 3,500 sq km become desert annually. Major cities are also threatened, with scientists predicting that both Sana, the capital of Yemen, and Quetta, the capital of Pakistan's Baluchistan province, will have exhausted their water supplies by about 2010.

DESERTS cover between one-fifth and one-third of the world's land surface. They are regions formed as a result of an extreme lack of water, either because there is almost no rainfall, or because they are isolated from the sea, or cut off by high mountains so that moisture in the atmosphere is lost before it can reach the area. The main concern now is that desert regions are spreading across wider areas of the world, because of the process of **desertification**, in which land becomes desert-like because of a combination of: over-cultivation of poor soils; **overgrazing** by sheep, camels, cattle and goats on impoverished land; cutting down too much **fuelwood**; or because of large irrigation schemes that lead to **salinization** of the land. Traditional deserts can be 'hot', like the Sahara, or 'cold' like the deserts north of the **Himalayas** or in Afghanistan. There are also semi-arid deserts, such as those in northern Eurasia or the uplands of Utah and Montana in the USA, and coastal deserts, one example of such, the Atacama in northern Chile, being the driest in the world. Some deserts support no living matter at all, in others the **desert vegetation** is sparse. Most desert animals are nocturnal or active during the cooler periods of dawn and dusk to avoid the daytime heat. They have developed a remarkable adaptation to the arid conditions. The camel and the Arabian **oryx**, for example, survive on very little water and small mammals get their water solely from the foliage they eat. At the edges of many of the world's deserts there are stretches of less extreme climate and some woody scrub vegetation survives, but they are now under increasing pressure as a result of overgrazing by local populations.

DESICCATION is a gradual increase in dryness, primarily because of a change in **climate**. It can be caused by natural changes, like a reduction in rainfall, or alterations to a river's flow. It is also caused by human actions, such as **overgrazing**, **irrigation** schemes or **deforestation**, which all result in a drying-out of the soil and a deterioration in the amount and quality of vegetation.

DETERGENTS have been used for decades to remove dirt and grease from clothes, dishes and surfaces. Until 1918 soap was the main product used, but then synthetic chemicals were added to improve performance. It was in the 1960s when these additives caused recognizable problems, associated with the huge increase in the volume of detergents used by industry and in washing machines. They had ingredients that failed to break down in water. Effluent containing these chemicals disrupted sewage treatment processes, and the release of partially treated effluent into rivers caused the detergents to produce large quantities of foam. The foam was dispersed on the wind and spread **bacteria** that caused infection in humans and animals. In the 1960s the ingredients of detergents were altered to make them **biodegradable**. However, these contain a great amount of **phosphate** materials, which can lead to **eutrophication**.

DIATOMS are single cell organisms which are abundant in marine and freshwater **plankton** and **benthos**, seabed-dwelling organisms.

DIELDRIN: See **aldrin**

DIESEL ENGINES are internal **combustion** engines that operate without an electrical spark. The mixture of air and fuel vapour is rapidly heated by the high degree of compression in the cylinder and spontaneously ignites. The diesel engine has greater fuel efficiency than a petrol-powered engine, but releases more **particulates**, as a dark smoke, and produces more **nitrogen oxides** and odour. Diesel engines are used throughout the world in larger land and sea vehicles. European countries tend to use them in automobiles to a far greater extent than in North America, because of such engines' high efficiency.

DIMETHYLSULPHIDE (DMS) is a **sulphur**-containing compound released by **phytoplankton**, the microscopic marine plants at the bottom of the **food chain**. DMS has become a major factor in calculating the importance of ocean-derived sulphur on **climate change**, particularly global warming. DMS is released into the water when phytoplankton dies and decays. It evaporates to the air, where it reacts with oxygen to create **sulphur dioxide**. This is then converted into **aerosols** of **particulate** sulphate, which increases the formation of clouds. In turn, clouds reflect the sun's radiation back into space. DMS could thus indirectly counteract the effect of **greenhouse gases**, which cause global warming.

DIOXINS are a family of 75 different compounds, or of 210 compounds if, as is often argued, a group of closely related agents called **furans** is included. Dioxins are never deliberately manufactured, but are referred to as **intermediate chemicals** because they are temporary substances formed during one of the manufacturing stages of other more-complex chemical materials, usually pesticides. However, dioxins are very poisonous and traces often remain in the end-product. They can also be released as by-products during the incineration of electrical insulating materials and other plastic substances that contain **polychlorinated biphenyls (PCBs)**. Dioxin came to prominence as a serious environmental hazard in 1976, when it was released in an accident at a chemical plant at **Seveso**, in Italy. The explosion happened at the critical stage in the production process of the herbicide 2,4,5-T (trichlorophenol). An uncontrollable exothermic reaction occurred inside a process vessel. A large **aerosol** cloud containing neat dioxin was released into the atmosphere and was spread by the wind, contaminating 1,800 ha of populated land. Plants and animals were severely affected and humans exposed to the toxic cloud developed a severe skin condition called chloracne. Contamination of the soil with dioxin was detectable as much as 10 years later. Fatal accidents involving dioxin had previously taken place in the United Kingdom and the Netherlands, but had been contained within factory sites. Dioxin was also the pollutant in the famous **Love Canal** incident in the USA. The most notorious form of dioxin is tetrachlorodibenzo-para-dioxin (TCDD), the herbicide, more commonly known as **Agent Orange**, that was sprayed as a defoliant by the US forces in Viet Nam.

DIRECT TOXICITY ASSESSMENT is a modern technique for measuring and controlling the toxicity of effluent discharges. It involves testing the effect of effluent on organisms at different levels of the **food chain**, from **algae** and macro-invertebrates such as water-fleas, to fish. Such tests determine the level of effluent treatment and dilution that is needed to protect the aquatic environment. Previously, water **pollution** monitoring relied on chemical analyses to set limits on effluent discharges into coastal and inland waters. Such a technique can be inadequate for complex discharges that contain a variety of substances in very small quantities. Some of these substances may be present in quantities too small to be detected, even though they may be highly toxic to aquatic life when part of a combination of chemicals. Such combinations of chemicals can react together to become more toxic than could be predicted using knowledge of individual substances. Direct toxicity assessment, however, provides what is called a real-effect measure of the potency of an effluent.

The **DIRTY DOZEN** is a name given to a group of, in fact, 18 chemicals identified by the **Pesticide Action Network (PAN)** as examples of the most dangerous pesticides that threaten human health. The Dirty Dozen includes campheclor (toxaphene), **chlordane**/heptachlor, chlordimeform, **DDT**, dibromochloropropane, the 'drins' (**aldrin**, dieldrin and endrin), **lindane**, ethyl parathion, **paraquat**, pentachlorophenol and 2,4,5-T. All of them accumulate in the **food chain** and cause death and disease in many species of wildlife. Traces of some of these chemicals have been found in Antarctica, many miles from sites where the pesticides were used. By 1990 the Dirty Dozen had been banned or severely restricted in at least 60 countries, but there is evidence that some are still being used in developing countries. They are to be regulated under the **Convention on Persistent Organic Pollutants**, which took effect in 2004.

DISASTERS include natural phenomena like **earthquakes**, volcanic eruptions, hurricanes, typhoons and tidal waves in which there are heavy fatalities or destruction of vast areas of habitation. Disasters are also caused by humans and there has been a steady increase in the number of man-made disasters over recent years. **Environmental degradation**, particularly in countries where there is heavy **deforestation** or **desertification**, or where marginal land has been overcultivated or overgrazed, can trigger disasters like floods or drought, landslides or mudslides, or make the effects of a natural disaster worse. For example, deforestation in the Himalayas has led to increased flooding in **Bangladesh**, a country that is already a victim of floods. In the Sahel region of Africa widespread desertification has increased the prolonged drought. Other man-made disasters include petroleum spills, chemical plant leakages or explosions, and nuclear radiation. A growth in population means that, should an accident occur in a crowded city, many more people are killed or injured because they are forced to live closer to the locations of potentially dangerous sites, as was the case in **Bhopal** in India when a chemical plant exploded.

DISINFECTANTS are chemicals, or other agents, that kill or render inactive micro-organisms in animals, seeds or other plant parts that cause disease. The term also refers to chemicals used to clean or sterilize inanimate objects.

DISTILLATION is the start of the petroleum-refining process, when crude petroleum is split into a number of parts or cuts. The separation is made on the basis of the boiling points of the various **hydrocarbons** in petroleum, enabling them to be divided at specific temperatures.

DISTRICT HEATING schemes use the surplus hot water from power stations by channelling it to provide space heating for buildings in the vicinity. District heating systems are also capable of burning virtually any type of fuel, including municipal refuse. There are many district heating schemes in mainland Europe, Scandinavia, the former USSR, the United Kingdom and the USA. Some power stations generate electricity and hot water, and are referred to as **combined heat and power**. Others just generate heat.

DNA: See **nucleic acids**

DNA-DAMAGE WEIGHTED DOSE is a measurement for assessing the potential risk to human health of the

ultraviolet radiation in the sun's rays. It is a calculation of the amount of biologically active ultraviolet radiation—so-called because it can damage deoxyribonucleic acid (DNA) in humans, animals and plants—reaching the earth's surface. A few hours' exposure to UV radiation may result in sunburn and snow-blindness in humans. Repeated exposure can cause skin ageing, skin cancer, depression of the body's immune system and corneal cataracts.

DNOC (4,6-dinitro-o-cresol) was the first chemical insecticide, discovered in 1892. Later, it was found it could kill annual weeds in cereal crops without damaging the crop. DNOC did not appear to leave toxin residues in the soil, so did not enter the **food chain**. However, it was later found to be harmful to wildlife, animals and humans, if inhaled or absorbed through the skin. It was replaced by other **herbicides**, including **2,4-D**.

The **DOBŘÍŠ ASSESSMENT** is a report named after Dobříš Castle in the Czech Republic (then part of Czechoslovakia). This was the venue at which the first pan-European conference of 36 environmental ministers or their deputies took place, in June 1991. The meeting was also attended by representatives of many international institutions and European non-governmental environmental organizations. The delegates requested a detailed report on the state of the European environment, the full title of which was *Europe's Environment: The Dobříš Assessment*. It was prepared by the **European Environment Agency**, published by the European Union in 1995.

DOBSON UNITS (DU) are used for measuring the total amount of **ozone** in a vertical column of air above the earth's surface; the method was perfected in 1956 by George Dobson, a British pioneer of stratospheric ozone and first president of the International Ozone Commission. He had long suspected seasonal changes occurred in the **ozone layer**, but he made his first measurements at Halley Bay, in **Antarctica**, almost 25 years before the **ozone hole** was observed by his colleagues. When Dobson made his first measurements the ozone level was 150 DU less than calculations suggested it should have been. He initially suspected faulty equipment. However, readings made the following year confirmed the pattern. Measurements made elsewhere in the world did not repeat the pattern. However, throughout the 1960s the seasonal decrease in ozone levels at Halley Bay were monitored. Then the depletion, now known as the ozone hole, was confirmed by Joe Farman and Jonathan Shanklin of the **British Antarctic Survey**, between 1982 and 1985. A reassessment of satellite photographs revealed the development of a hole every spring since 1970. Serious depletion was then confirmed over the Arctic, showing that ozone depletion was not restricted to the southern hemisphere.

DOLPHINS are the most agile and intelligent of the **cetaceans**. They are able to communicate both with other members of their group and with humans, with high-pitched barks and squeaks. They appear to have some affinity with humans, which has led to legendary tales of dolphins saving shipwrecked sailors. Most often seen are the common and the bottle-nosed dolphins. They move in herds, following herring or sardine shoals. There is evidence that dolphins live up to 35 years in the wild. However, they are under threat, because they swim and feed with tuna, and frequently get trapped in the **drift-nets** used to catch tuna. These were at the centre of the **tuna-dolphin dispute** between the USA and Mexico. Pollution is another hazard. For example, a bottle-nose dolphin found stranded in Cardigan Bay in the United Kingdom in 1988 had levels of dieldrin, **DDT** and **polychlorinated biphenyls** in its blubber more than 10 times higher than those found in other cetaceans sampled around the British coasts in previous years. There are also river dolphins, notably those of the **Ganges** and the **Yangzi**, which are particularly endangered.

DRIFT-NETS were formerly made with natural fibres such as hemp, cotton or flax, and were easily visible and acoustically detected by marine mammals. In the 1950s fishing nets began to be made with cheaper, very thin, but extremely strong plastic. They are now made from a synthetic material that is invisible to marine mammals and is not **biodegradable**. One net can be up to 50 km long. They are primarily used by fishing fleets in Japan, China (Taiwan) and the Republic of Korea (South Korea). It is estimated that 7,000 km of net are used every year in the Pacific alone. An estimated 1,200 metric tons or more of commercial drift-nets are lost or discarded at sea every year. Drift-nets, or broken-off pieces of net, continue to be a hazard for many years because they can trap and kill marine animals, such as turtles, **dolphins** and porpoises. In 1989 the UN General Assembly adopted a resolution calling for the banning of ocean drift-nets. The International Convention for the Prevention of Pollution from Ships (MARPOL), which came into effect in 1988, aimed to cut down the number of nets lost or discarded, by banning the disposal at sea of all plastics including, but not limited to, synthetic ropes, synthetic fishing nets and plastic garbage bags. In 1989 a Convention for the Prohibition of Fishing with Long Driftnets in the South Pacific was signed in Wellington, New Zealand, and two years later the UN passed a resolution for a moratorium on drift-net fishing on the high seas. In 2004 the European Union was moving towards the complete banning of drift-nets in the **Baltic Sea**. (See also the **tuna-dolphin dispute**.)

DROUGHT can be the result of natural weather conditions or of a change in the environment produced when trees are cut down or vegetation is destroyed. Droughts can have a disastrous and long-term effect. For example, the drought in Africa's Sahel region, which started in 1968, caused some 100,000 deaths over five years and irreversibly changed the lives of people in the region. The Ethiopian drought of 1973–74 was responsible for the deaths of an estimated 200,000 people. Five continents had serious droughts in 1982. As well as the droughts in northern and southern Africa, Australia, Italy and Spain had their worst droughts for over 50 years. Crops were also destroyed in Nicaragua, when a dry spell came in the middle of what is normally the rainy season. Moreover, in India, a delay in the annual **monsoon** set back crop production. Sometimes droughts are connected to the occurrence of the **El Niño** weather phenomenon. In hot and cold desert regions there is usually an almost permanent drought. Plants adapt to the harsh environment and crops can be produced only by continual **irrigation**, using **ground-water** resources.

A **DRY SPELL** is regarded as 14 consecutive days without measurable rainfall in the USA. In the United Kingdom, a similar meaning is signified by **absolute drought**.

DRY/WET DEPOSITION is pollution on the ground resulting from **acid** formed in the atmosphere. Dry deposition is the result of dry phase conversion of **sulphur dioxide** or **nitrogen oxides** into their respective acids. Wet deposition is the acid content of rain, snow, sleet or mist (see **acid rain**) formed by the **wet (aqueous) phase conversion** of sulphur dioxide or nitrogen oxides to the respective acids.

DUNG FUEL is dried animal excrement that is used as fuel in developing countries, especially when **fuelwood** is scarce or not available. Millions of tons are burned each year, mostly for cooking. However, dung is a valuable natural **fertilizer**, especially in those countries where there are already food shortages, and the over-use of it for burning deprives the soil of vital nourishment.

A **DUST DEVIL** is a short-lived whirl of dust particles which are lifted to a considerable height above the ground. It is caused by intense solar heat on the ground surface in arid lands and can reach speeds of over 30 km per hour, but dies away without causing any serious damage.

DUST STORMS happen in arid or semi-arid regions, when a large volume of dust is whipped up by the wind and causes a lack of visibility, or sometimes zero visibility. Dust can be carried to great heights and over great distances. Storms usually occur when there are excessively high temperatures, low humidity, and a high electric force. Sometimes they appear as a wall of dust or as a whirlwind. Dust storms can cause havoc to crops in areas that are dependent on cultivating fragile **ecosystems**. They also pose a serious danger to aircraft.

The **DUST VEIL INDEX** was the scale devised by the British climatologist, Hubert H Lamb, to rank the severity of volcanic dust in terms of the mass of material thrown into the **atmosphere**, the length of time it takes to be washed out by rain and the extent of its spread round the planet in the meantime.

DUSTBOWL was the term used to describe a vast area of the Great Plains in the USA, which is primarily arable land dedicated to large-scale cereal farming. Agriculture spread into areas where rainfall was frequently low and this, combined with extensive droughts in the 1930s, severely damaged the soil. There was massive **soil erosion** and large quantities of precious topsoil were blown away, creating frequent dust storms. Millions of hectares of land were laid to waste across five states. Huge quantities of agricultural chemicals and water were applied to the land and restored much of it to a state where it was able to support crops again.

DUTCH ELM DISEASE is a serious tree disease caused by a fungus carried by two species of bark beetles. It was first recognized in the Netherlands in 1919 and spread rapidly through Europe. Ten years later it appeared in the USA and Canada. The beetles use the bark of trees killed by the disease as breeding sites. Flying from tree to tree they carry spores of the fungus with them, and these block the vessels that carry water to the leaves of the tree. The leaves wilt and eventually die. Dutch elm disease was found in the United Kingdom in 1931 and, over six weeks, 20% of elm trees in the south of England were killed. In the 1960s a more virulent strain of the fungus developed in the northern USA and Canada, and this strain was believed to have been introduced into the United Kingdom by beetles carried on imported logs from Canada in 1964. Within two years 9m. British elms had died. Almost all the elms in southern England were wiped out.

E

E-NUMBERS are an internationally accepted coding system to identify substances added to food during its manufacture and processing. Food additives have been suspected of provoking symptoms of hyperactivity, **asthma**, eczema and migraine. The numbers cover six categories of food additives: preservatives; colourings; emulsifiers and stabilizers; antioxidants; sweeteners; and other miscellaneous additives. E-numbers were introduced by the European Communities (now the European Union—EU) in order to reassure consumers that any substance added to food has been thoroughly tested and pronounced safe. To get an E-number, the EU's seal of approval, a food additive must pass stringent safety tests and be approved by all member countries. Synthetic flavourings were not initially regulated by the EU, but various countries within Europe, Australia and New Zealand, Japan and the USA had lists of permitted flavourings, and further regulation was introduced during the 1990s.

The **EARTH OBSERVING SYSTEM (EOS)** is the network of satellites which now scan the surface of the earth and its **atmosphere** from space. Interpreting the information gathered by one of the **Earth Resources Technology Satellites** involves very advanced computer analysis on the ground. There are immense advantages, from the environmental point of view, in exploring from space with a combination of **infrared radiation** sensors, radar scanners, and photographic and television cameras. These instruments show seasonal and longer-term events. They also provide an early warning of potential short-term disasters: developing hurricanes, volcanic eruptions, locust swarms, forest fires. Earth observation began with the first meteorological satellite, **Tiros-1**, which was launched in 1960. An international network of spacecraft for meteorological forecasting now forms a permanent World Weather Watch system. A second group of satellites was developed in parallel with the weather system. It began with the US family of specialist reconnaissance Landsats, and now includes the French SPOT satellites and the European Space Agency's ERS (Earth Research Satellite) series. These specialist spacecraft can read the signature of the reflected radiation from objects on the ground. They can tell the difference between various rocks, soil, plants and trees, and, from the colour of the vegetation, show whether a crop or forest is suffering from drought or disease. The European Remote Sensing satellite ERS-1 has unique instruments for surveying oceans and sea ice.

EARTH RESOURCES TECHNOLOGY SATELLITES are the family of spacecraft that monitor atmospheric, terrestrial and ocean conditions. They have become the world's primary early warning systems of global scale pollution and ecological changes, such as the impact of deforestation. The first of the satellites launched between 1972 and 1984 by the USA's National Aeronautics and Space Administration from 1972 to 1978 were called Landsat. They were designed to collect information about the earth's natural resources, particularly the location of mineral deposits, and to conduct inventories of crops. From the mid-1980s the US earth resources satellite programme was taken over by the Earth Observation Satellite Corpn (Eosat). A parallel European programme is known as the European Remote Sensing (ERS) satellite.

The **EARTH SUMMIT**, the UN Conference on Environment and Development (UNCED), was the biggest and most ambitious world environmental conference. It was held over 12 days in June 1992 in Rio de Janeiro, Brazil. Sixty world leaders gathered to formulate a series of agreements and treaties. The 27-point Rio Declaration was signed as a commitment to pursue **sustainable development** and the eradication of poverty. It ranges over major environmental issues, such as land, air, water, living things and management of toxic waste. It covers the principles of **polluter-must-pay**, free trade and environmental protection, and upholds the rights of women, children and indigenous peoples. The key to this is that countries have sovereign rights to exploit their resources, but an equal responsibility to make sure that their actions do not harm the environment for others. Two international conventions were also signed at the end of the Summit: the non-binding UN **Framework Convention on Climate Change**, which was intended to reduce **greenhouse gas** emissions in industrialized countries (to their 1990 levels by 2000); and a **Convention on Biological Diversity**, to protect the world's animal and plant life. (See also **Agenda 21**.) In 1997 a conference known as Rio Plus 5 was held in New York, USA, to review progress on implementing Agenda 21 and the so-called Rio Conventions. There was general agreement that not enough progress had been made, and the renewed sense of urgency instilled by Rio Plus 5 helped to motivate the **Kyoto Protocol** to the Framework Convention on Climate Change, which was adopted later in the same year. Another 'Earth Summit' was held in South Africa in 2002, the **World Summit on Sustainable Development**.

EARTHQUAKES have been responsible for some of the worst disasters ever to occur in urban environments. The earthquake in Shaanxi (Shensi) province (now in the People's Republic of China) in 1556 is believed to have been the most deadly ever and killed 830,000 people, destroying entire towns and villages. The most deadly earthquake of the 20th century was also in China, in Tangshan, when well over 250,000 people were killed in 1976. The largest earthquake was in Chile in 1960. Improved construction techniques have helped reduce death tolls, but many cities in the developed world have simply not been tested recently. Thus, the earthquake in **Kobe**, Japan, on 17 January 1995 led to 6,336 deaths and totally destroyed more than 100,000 buildings. Iran has suffered two very deadly earthquakes recently, one in 1990 killing some 50,000, while at Bam, in the south-east, about 40,000 died in December 2003. Estimates suggest that during the 20th century there were 1,120 fatal earthquakes, causing a total of 1.53m. deaths. The activity of an earthquake may resume after a period of days, years or even thousands of years. Major earthquakes are natural phenomena, caused when energy is released by a slippage of the continental plates into which the earth's crust is divided. However, minor earthquakes can be triggered by human activities like underground nuclear testing, the construction of **dams** and reservoirs and the use of deep wells to pump liquid wastes into the earth, if such activities disturb the rock layers beneath the surface. There is no way precisely to predict an earthquake, or its ferocity, but enough is now known about the design of earthquake-resistant buildings to reduce fatalities in all but the most extreme cases.

EARTHWATCH is a global environmental assessment programme, which was included in the Action Plan for the Human Environment adopted by the **UN Conference on the Human Environment** in 1972. It has four main aims: to provide a basis on which to identify specific knowledge and to take the steps to find it; to generate the new knowledge needed to guide decision-making; to gather and evaluate specific data in order to discern and predict important environmental conditions and trends; and to distribute knowledge to scientists and technologists and provide useful and up-to-date information to decision-makers at all levels. The Earthwatch programme of the **UN Environment Programme** has four main arms: the **Global Environment Monitoring System (GEMS)**; the **Global Resource Information Database (GRID)**; the **International Register of Potentially Toxic Chemicals (IRPTC)**; and **INFOTERRA**, the global information system.

ECO-LABELLING is a European Union (EU) initiative to encourage the promotion of environmentally friendly products. The scheme came into operation in late 1992 and was designed to identify products that are less harmful to the environment than equivalent brands. It was hoped that by buying labelled goods, consumers would be able to put pressure on manufacturers and retailers to make and to stock 'greener' products. This includes the effect they have on the environment at all stages. For example, eco-labels will be awarded to products that do not contain **chlorofluorocarbons** damaging to the **ozone layer**, to those products that can be, or are, recycled, and to those that are energy efficient. The labels are awarded on environmental criteria set by the EU. These cover the whole life cycle of a product, from the extraction of raw materials, through manufacture, distribution, use and disposal of the product. The first products to carry the EU eco-labels were washing machines, paper towels, writing paper, light bulbs and hairsprays.

The **ECO-MANAGEMENT AND AUDIT SCHEME (EMAS)** was devised by the European Union and came into operation in 1995. It was designed to encourage a more rapid adoption of good environmental management practice in industry, including techniques such as **life-cycle analyses**, **integrated pollution control**, materials and product substitution, **recycling** and **environmental impact assessment**. The EMAS requires a company to conduct an environmental review of its operations, set a policy, establish a programme, implement comprehensive environmental management systems and conduct regular environmental audits of its activities.

ECOLOGICAL ECONOMICS is an increasingly widespread interdisciplinary field that draws together both natural and social scientists. Ecological economists view the human economy as one integrated part of the earth's entire ecological system. In terms of the laws of thermodynamics, they see the human economy as drawing on the earth's matter and energy to meet human needs, then depositing the residual back into the earth as waste. This process is referred to as throughput. It is widely held among ecological economists that total resource throughput is occurring at a rate that may jeopardize the earth's ability to supply the human economy with the resources it needs, and its ability to absorb the human economy's waste.

ECOLOGICAL NICHE refers to the characteristics of an environment that provides all the essential food and protection for the continued survival of a particular species of flora or fauna. In addition to food and shelter, there is no long-term threat to existence in that place from potential predators, parasites and competitors. The concept of the ecological niche goes a long way beyond the idea of the species habitat.

ECOLOGIST refers both to a specialist who studies the relationship between plants and animals and their environments, that is **ecology**, and to the magazine founded in the United Kingdom in 1970 to report on ecological and environmental science, and to defend the rights of those people most affected by environmentally damaging development projects. The magazine dealt with difficult and unfashionable issues, many of which were borne out during its years of publication. The principal founders of the magazine were Edward Goldsmith, one of the pioneers of the modern Green movement and co-author of the *Blueprint for Survival*, Peter Bunyard, Robert Allen and Michael Allaby. In 1984 Goldsmith and Nicholas Hildyard produced *Social and Environmental Effects of Large Dams*, a book that highlighted the environmental effects that large **dams** were having, particularly in developing countries. With other environmentalists throughout the world, *The Ecologist* was also active in campaigning against the vast complex of dams planned for the Narmada river in India (see **Sardar Sarovar Dam**).

ECOLOGY is the study of living things as they exist in their own natural surroundings, or habitats. The term was adopted by the German naturalist, Ernst Haeckel, in 1869, and comes from the Greek *oikos*, meaning home or dwelling. He spelt it oecology, and it covered work in biology in which the observations and experiments of the habitats of plants and animals were treated as separate studies. The word was originally meant to suggest the meaning 'the economy of nature'. Ecology has come to be more broadly defined as the interrelationship between all living things and their environments. Vladimir Ivanovich Vernadskii in the 1920s defined it as 'the science of the **biosphere**'. The **ecologist** needs to have a command of chemistry, physics and statistics, as well as biology. Hence, the **environmental sciences**, which depend on ecological research, were the first interdisciplinary sciences.

ECONET is the name shared by several environmental projects. One of the largest is a programme to map and model the environment of the South China Sea. It is an ambitious plan devised by the Hong Kong University of Science and Technology. The scheme was expected to last 15 years, at an estimated cost of US $150m. The almost totally enclosed sea is bordered by 10 rapidly developing countries and, although by the mid-1990s it was not badly polluted, there were fears that it soon would be. Dangers would be increased if there was a resolution to territorial disputes over the Spratly Islands and other islets, as the exploitation of mineral resources could then begin in earnest. Scientists insist that there is an urgent need for the project to proceed before development destroys the sea's **ecosystem**. Increasing coastal development has led to large amounts of untreated sewage and industrial wastes being pumped into the sea. The speed of development has also caused the destruction of large areas of **mangrove forests**, making coastlines susceptible to flooding. A major obstacle to the designers of the project is the difficulty of ensuring the collaboration of countries with poor foreign political relations. However, when the People's Republic of China and Viet Nam indicated that the project would receive support from their environmental protection agencies, the **UN Development Programme** asked that a detailed proposal for EcoNet be submitted.

The name EcoNet is also shared by one of the computer networks provided by the Institute for Global Communications on the internet. It is dedicated solely to environmental preservation, conservation and sustainability and is intended to help organizations and communities concerned about local and world-wide environmental issues to co-operate more effectively and efficiently using the internet. Information sources and on-line discussion groups on topics like environmental justice, global warming (see **climate change**), energy policy, rainforest preservation, legislative activities, water quality, toxic substances and environmental education are available through EcoNet.

ECONOMIC BOTANY is the study of plants that could one day be useful to humanity. It was started by Sir William Hooker, the first director of the **Royal Botanic Gardens** at Kew, then on the outskirts of the British capital. It was studies at Kew that led to the development of the viscose-rayon and acetate fibre industries. Vital investigations into cures for AIDS, finding plant sources for the creation of safer **insecticides**, and developments of new programmes in molecular biology are among the many research projects that are being carried out at the Royal Botanic Gardens, which has also developed the **World Seed Bank**.

ECONOMIC INSTRUMENTS, in the context of the environment, are fiscal measures implemented in order to benefit the environment. Such instruments include a **carbon tax** or **energy tax** on petrol to curb its use and reduce **carbon dioxide** emissions; **emissions trading**; incentives to use **solar energy** rather than non-renewable **fossil fuels**; and **road pricing** policies. The use of economic instruments to meet environmental objectives has become more widespread, especially in the USA, where **command and control** regulatory measures are viewed as excessively expensive. There are various economic approaches to achieving environmental objectives. At a global level they include negotiation of international trade agreements to take account of the environmental health of the planet and to encourage **sustainable development**. Demands by environmental groups for improvements in environmental standards are often countered by the argument on the part of industrial groups and governments, that environmental controls would damage trade. One solution to this problem would be for governments to offer compensation for loss of trade to industries that accepted more stringent environmental controls, if those controls led to more expensive products or the cessation of certain manufacturing operations. There are few international agreements that take account of both environmental and trade objectives, but examples include the environmental side agreement to NAFTA (see **North American Agreement on Environmental Cooperation**); the **Montréal Protocol**, which was designed first to restrict, then to ban, the trade in **ozone**-depleting substances like the **chlorofluorocarbons** used in **aerosols**, refrigerators and air-conditioning systems; the **Convention on International Trade in Endangered Species of Wild Flora and Fauna (CITES)**; and the **Basle Convention on the Control of Transboundary Movements of Hazardous Wastes and their Disposal**, which was signed to prevent the export of such wastes to developing countries.

ECOSYSTEMS are resources that are formed by a combination of interacting living and inanimate, but biologically active, components. An example of an ecosystem is a particular type of freshwater habitat in which it is ideal for trout populations to flourish, because of the availability and quality of food. The list of ecosystems is almost endless, since the term refers to the particular biological and physical attributes that characterize every large or small woodland, forest, hedgerow, pasture, heathland, wetland, beach, desert or pond, lake, river or ocean. An ecosystem cannot be separated into its living and non-living parts because it is an interdependent, dynamic unit, although the community of organisms is sometimes described as a **biocenose** and its physical environment as the **biotope**. The term first appeared in 1935, in a publication of the British ecologist, Sir Arthur Tansley, although the word had been invented by his colleague, Roy Clapham, in 1930. The term **biosphere** was already in use, but an ecosystem became popular because it could have a more local and specific application.

ECOTOURISM or sustainable **tourism** is ecologically sustainable travel to natural areas that conserves the environment and improves the well-being of local populations and cultures. The principles of ecotourism involve minimizing human impact, fostering environmental awareness and respect, financially benefiting both conservation efforts and local people, and increasing sensitivity to a country's particular environmental, political and social climate. Ecotourism has evolved from a growing awareness of the damage even well-meaning tourism can do, as well as the need of many countries to exploit their natural environments without destroying them. It is a concept that evolved in the last decades of the 20th century, the oldest organization being the International Ecotourism Society (TIES), which was founded in 1990. The **UN Environment Programme** soon realized the importance of ecotourism to the aims of the 1992 **Convention on Biological Diversity**, for instance. Indeed, the UN designated 2002 the International Year of Ecotourism, in an effort to encourage sustainable tourism development and propagate the emerging principles and best practice of ecotourism. A summit was held in Québec, Canada, in May of that year, and a Declaration adopted. A Center of Ecotourism and Sustainable Development was created in the US capital in 2003.

ECOTOXICOLOGY is the study of the harmful biological effects of specific pollutants or combinations of pollutants that are discharged into the environment. It refers especially to the effect caused by the discharge of 'cocktails' of contaminants into rivers, estuaries and coastal waters. Common pollutants that are toxic to aquatic life in **fresh water** include **ammonia**, chromium, **copper**, cyanide, nickel, **nitrite**, **phenol** and **zinc**. Studies in ecotoxicology have led to a new approach to the control of effluent. Ammonia, for instance, is one of the most frequent pollutants of fresh water in European countries, but its toxicity to aquatic life is very variable and depends on many factors. Local water quality conditions have a considerable effect on its toxicity. Extensive research has focused on the complex chemical and biological processes influencing the toxicity of ammonia to fish and other aquatic life in order to produce standards of practical control for their protection. The information on lethal and sub-lethal effects of pollutants is used in computer models employed to predict the impact of particular concentrations of contaminants on fish. The development of ecotoxicology methods for setting water quality standards has depended on development of new biological assessment techniques, both for laboratory screening of water samples and for routine monitoring of effluent.

EFFLUENT is the waste liquid from domestic sewage, industrial sites or from agricultural processes. Effluents are harmful when they enter the environment, especially in **fresh water**, because of their polluting chemical composition.

EIA: See **environmental impact assessment**

EL NIÑO or El Niño-Southern Oscillation (ENSO) is a climatic phenomenon in which the waters of the Eastern Pacific retain more warmth than normal and the trade winds and ocean currents that normally flow across the tropical Pacific Ocean from the Americas to Asia abruptly reverse (see **ocean circulation**). El Niño has been blamed for many disasters much further afield, such as bush fires in Australia arising from **drought**, rain storms in California, USA, storms in normally calm Pacific Islands, and also for interrupting the Indian and East African monsoons. The abnormal retention of water of a high temperature along the South American coast has led to the demise of **plankton**, squid and fish that prefer to live in cold waters. With the death of the anchovy and sardine, hake, herring and mackerel disappear or die. Sea-birds and **seals** that feed on them abandon their nests or lose their young. The decaying fish are washed on to the beaches producing large quantities of sulphurous gas. It was thought that El Niño was predictable, arriving normally every three years, and lasting for about 18 months. However, the 1990 episode reached maturity during early 1992 and began to weaken in mid-1992, only to increase unexpectedly in November 1992. It only disappeared completely early in 1995. The 1997 episode was more normal in its duration, as was the 2002–03 occurrence, when Australia suffered its worst drought in a century. Some researchers warn that El Niño could be affected by global warming (see **climate change**) and, if this should increase, it could have an influence on this destructive climatic force. The opposite and alternate phenomenon to El Niño is called La Niña (or El Viejo), which is characterized by unusually cold ocean temperatures in the equatorial Pacific. A Tropical Ocean Global Atmosphere programme has evolved a specific ENSO observation system.

ELECTRIC VEHICLES are recommended by some environmentalists because they constitute a method of road transport that can reduce the **air pollution** by **nitrogen oxides**, **carbon dioxide** and **particulates** associated with conventional car exhausts. Electric vehicles are also sometimes referred to as **zero-emission vehicles**. There are other advantages: electric motors require less maintenance than internal combustion engines since they have fewer moving parts and are more reliable and quieter. However, since most electricity is generated by polluting power stations, the use of electric vehicles may just displace the air pollution that would otherwise be generated by car exhaust. Therefore the environmental benefits of electric vehicles depend on whether the generation of the electricity used to charge the vehicles is clean. Manufacturers in Japan, the USA and Europe have started to sell electric cars. France is particularly interested in electric vehicles because of the plentiful supply of electricity for recharging **batteries** that is available from **hydroelectric power** and nuclear energy. The main technical problem associated with electric vehicles is to develop new types of batteries or **fuel cells** that are cheaper, last longer, weigh less and have faster recharge times than those previously available. In the long term, the preferred technology for electric propulsion could be fuel cells, which are based on an electrochemical process in which hydrogen, or a hydrogen-rich fuel such as a **hydrocarbon** or one of the aliphatic alcohols, reacts with oxygen to produce electricity, heat and water. Fuel cells invented by Cambridge university scientists in the United Kingdom provided the water supply for the moon voyages of the US Apollo spacecraft. A major problem in developing fuel cells for road vehicles is the limited life of the 'stack', the inner part of the cell in which the reaction takes place. Certain Japanese car companies began marketing a **hybrid vehicle** that is a combination of an electric vehicle and a conventional one and that achieves much greater **energy efficiency** than conventional vehicles.

ELECTROMAGNETIC RADIATION usually refers to the low-level electromagnetic radiation emitted from high-voltage power lines, microwave ovens, radar dishes and radios. All living organisms generate minute electrical currents. Indeed, in animals they are important regulators of the central nervous system. Magnetic fields develop when an electric current is passed. The brain, and all other organs, produces very low-level magnetic fields, which can be detected outside the body. Low-frequency electromagnetic radiation (microwaves) was thought to be harmless because it is non-ionizing, but in the 1980s its effects began to be more fully understood. The **World Health Organization**, in a report on the effects on health of microwaves, said that exposure led to the appearance of autonomic and central nervous disturbances, which produced damaging effects. In 1982 it was reported that cattle on a dairy farm located near a high-voltage electricity power line had given birth to stillborn or defective calves, and that chickens were laying abnormal eggs.

ELECTRON BEAM DESULPHURIZATION is a high technology device for extracting **sulphur dioxide** from gases produced in a power station or factory furnace before they are released into the atmosphere.

ELEPHANTS were traditionally hunted for their valuable **ivory**, which can be used to make piano keys, jewellery and ornaments. They are the largest land mammal and live in rainforests or **savannah**. In the second half of the 20th century, in most of the countries of Africa, the elephant population rapidly declined, sometimes because of illegal poaching, but mostly because its habitat disappeared as a result of human encroachment (see **vanishing habitats**). The **Convention on International Trade in Endangered Species of Wild Flora and Fauna (CITES)** placed the African elephant on Appendix 1, the highest protection designated under the Convention, in 1989. In 1985 CITES had set up the ivory quota system in an attempt to maintain the ivory trade within realistic limits. Most of the 103 member countries agreed not to allow the sale of ivory. Subsequently, several countries in southern Africa reported that their elephant populations had increased. South Africa and Zimbabwe carried out culling programmes. The Kruger National Park in South Africa has to reduce stocks by 600 animals per year to avoid overcrowding. Some are killed, others given to private reserves and contraception was also considered as a means of limiting numbers. Given adequate space, African elephants make a valuable contribution to a stable **ecosystem**. As they only digest about one-half of the vast amount of leaves, grass, fruits and bark they eat, their dung provides essential nutrients to restore the soil of the savannah. When humans take over parts of the elephants' range, elephants eat crops and become regarded as pests. Asian elephants are also considered to be an **endangered species**. Vanishing habitats, rather than poaching, are the greatest threats to their survival. The Asian elephant is largely domesticated and widely used in the logging industry.

EMISSIONS TRADING is a term used to describe an **economic instrument** used to control pollution. It is most commonly used in the context of regulation of **air pollution**. The concept is that, when an overall pollution limit is set, permits or credits are issued or sold to emitters, which they are allowed to trade with third parties who must also comply with environmental laws. For example, if a power station decided to control its emission of air pollution to a higher degree than was necessary to fulfil its legal obligations, it could sell a credit certifying that the emissions control took place to another company. The purchasing company might wish to buy such a credit if it had not successfully reduced its own emissions or if such emissions controls were prohibitively expensive. Proponents of emissions trading argue that such policies are more economically efficient, because the companies that can reduce emissions most cheaply would theoretically achieve emissions reductions first. Others have expressed ethical concerns about creating a market for pollution. In accordance with European Commission proposals made in 2000 for compliance with the **Kyoto Protocol**, the European Union was scheduled to begin an emissions-trading scheme by 2005.

ENDANGERED SPECIES are those of the planet's **flora and fauna** that are threatened with **extinction**. Hunting and poaching to fuel the trade in **ivory**, horn, skins, fur and feathers are a long-standing threat to already-endangered species like **elephants**. Pollution, agricultural expansion, loss of **wetlands**, **deforestation** and other erosion of habitats have been added to the hazards. Breaches of the **Convention on International Trade in Endangered Species of Wild Flora and Fauna** pose another threat, brought about by the illegal, direct trade in live animals by collectors. Extinction, like the dinosaurs' demise, is also a feature of natural environmental change or catastrophe. However, human activity was responsible for most of the animals and plants known to have been lost in the past two centuries. They include antelope, 17 species or subspecies of bears, four of cats, three of deer, foxes, goats, horses, wolves and zebras. The last dodo was killed in Mauritius in 1681. In the early 20th century the passenger pigeon, which gathered in flocks by the hundred in North America, was killed off for food. Threats to large animals like the African **elephant**, **rhinoceros**, gorilla, **tiger**, **giant panda**, **polar bear**, **otter** and **whale**, have come to public attention because of the work of conservation groups, such as the **World Wide Fund for Nature** and the campaigning of the environmental groups like **Greenpeace** and **Friends of the Earth**. Species have been rescued from the brink of extinction by successful, controlled breeding of managed herds, as in the case of the **oryx**, and by creating protected reserves, as in the case of the Indian **tiger**. The annually updated **Red Data Books** or Red List of the **World Conservation Union (IUCN)** provide a definitive register of endangered species.

ENDOCRINE DISRUPTORS are **persistent organic pollutants (POPs)** that mimic hormones under certain circumstances. Scientific claims regarding this phenomenon are still considered controversial, but the theory is that such synthetic organic chemicals—typically used as pesticides and industrial chemicals—can disrupt endocrine systems in humans and animals, which could affect reproductive systems, early development and immune systems. The decline in numbers of beluga whales in the St Lawrence River in Canada seems likely to be related to the high levels of **polychlorinated biphenyls (PCBs)** in their bloodstream. One whale was discovered to have contained levels of PCBs 10 times higher than that necessary for its entire body to have qualified as hazardous waste under Canadian law. Consequently, such corpses are treated as toxic waste to prevent the re-release of PCBs and other industrial chemicals into the environment. International negotiations on the issue of POPs resulted in the **Convention on Persistent Organic Pollutants** in the early 2000s.

The **ENERGY BALANCE** is the way the earth and **atmosphere** interact to absorb incoming **solar energy**. The earth–atmosphere system absorbs about 68% of the solar energy that reaches the outer layer of the atmosphere. Some of the solar energy is bounced directly back into space rather than penetrating to lower layers of the atmosphere. Some is absorbed by the various gases in the air and turned into heat. Some is absorbed by the ground and some is reflected back into the atmosphere. Sunlight stimulates **photosynthesis** and so some is stored temporarily in vegetation. Solar energy is stored for longer periods in the trees and plants that were turned into **fossil fuels**. Objects respond differently. Dark rocks readily absorb solar energy, because they have a low **albedo**. Ice and snow reflect most of it back. Water is better at absorbing heat than land because of convection currents. These exchanges between the earth and atmosphere balance out through the year to give the earth a fairly consistent environment, season by season. Consequently, global temperatures have varied very little over centuries. Human beings may be changing this balance by burning fossil fuels and enhancing the **greenhouse effect**. While there is an overall balance between energy absorbed and energy lost, the processes by which it is maintained are fairly turbulent. The turbulence is not restricted to the atmosphere. The oceans carry immense quantities of energy in currents from one part of the globe to another. (See also **global circulation**.)

ENERGY CONSERVATION is a far broader concept than just the realization of the benefits of saving **fossil fuel** by greater **energy efficiency**. Energy conservation covers the relationships between energy, **ecology** and the environment. For example, the methods of extracting and using the sources of energy, from chopping trees for fuelwood to coal, petroleum and nuclear power generation, can adversely affect the ecology and environment of the world. Energy conservation includes the development of the renewable, non-polluting sources of power like **solar energy**, **wind power** and tidal energy. Energy conservation also embraces developments like the **genetic engineering** of crops, so that plants can flourish without the need for high levels of manufactured fertilizer, which is an energy-intensive process.

ENERGY CROPS are those plants grown wholly or partly for their energy content, and in effect providing a means of storing solar energy in the form of **carbohydrate**. Both cellulose and starch can then be fermented to produce **ethanol** and **methane**. **Biomass** and **biofuels** are also examples of energy crops.

ENERGY EFFICIENCY refers to actions to reduce the use of energy. Examples of how to conserve energy include using compact fluorescent light bulbs, better building design, the modification of production processes, better selection of road vehicles and transport policies, the adoption of **district heating** schemes in conjunction with electrical power generation, and the use of domestic insulation and double glazing in homes.

ENERGY IMPACTS is a term that describes the **pollution** of the **atmosphere**, land and water caused by the extraction, transport and burning of **fossil fuels**. Any energy installation, including those based on renewable sources, has an environmental impact, which includes the energy consumption involved in building it. An assessment of the total environmental implications of an installation is produced by a **life-cycle analysis**, a technique that measures the sum of the environmental impacts of the resources used to provide a product or service, plus the amount of emissions and wastes. In the case of burning fossil fuels, the emissions into the air include **carbon dioxide**, **sulphur dioxide**, **nitrogen oxides** and **particulates**. Carbon dioxide is the main **greenhouse gas** contributing to global warming (see **climate change**). Sulphur dioxide and nitrogen oxides are the principal contributors to **acid rain**. During the extraction and processing stage of all fossil fuels, another potent greenhouse gas, **methane**, is liberated. Other energy impacts include the polluting effects of **petroleum spills** and the costs involved in the safe disposal of **nuclear wastes**.

ENERGY TAXES are one of the **economic instruments** that are available to encourage consumers to transfer from use of **fossil fuels**, which emit polluting gases like **carbon dioxide**, **sulphur dioxide**, **nitrogen oxides** and **particulates**, to more environmentally friendly energy sources. Energy taxes were levied for many years purely for revenue-gathering purposes, but taxes levied for environmental reasons remain controversial and are the subject of disputes between governments and institutions like the European Union on one side, and industry on the other. One example of an environmentally beneficial energy tax is the premium tax on leaded petrol in many countries, which is designed to encourage use of **lead-free petrol**. Taxation systems can also be used to encourage the development of **renewable energy sources**.

ENRICHMENT is the process of taking natural uranium ore, which consists of uranium-238 (99.3%) and **uranium-235** (0.7%), and changing these proportions. The amount of U-235 is increased to 3.5%–4% in order to provide uranium fuel for a nuclear reactor, and to almost 100% for the fuel for an atomic bomb. For 30 years enrichment was done by the difficult and expensive gaseous diffusion method, based on a discovery by Lars Onsager, a Norwegian-US chemist working at Yale University in the USA. He received the Nobel Prize for his research. Gaseous diffusion is both difficult and expensive. The method depends first on making uranium hexafluoride gas and pumping it through thousands of special filters. Owing to the tiny difference in the nucleus of the atoms, slightly more U-235 gets through each filter and the proportion of U-235 steadily increases. In the 1970s scientists at the United Kingdom Atomic Energy Authority's research establishment at Harwell invented a gas centrifuge method of enrichment that was both quicker and cheaper.

ENTROPY refers to the amount of disorder in a system. It involves the factors that define the energy state of a substance, and entropy relates to the unavailability of the energy because of the disorder in a system.

The word **ENVIRONMENT** suggests different things to different people. To the meteorologist and climatologist, for instance, it usually means the **atmosphere**. To the environmental engineer, it usually means the atmosphere in an enclosed space, such as a factory, office block or hospital, which the engineer has to maintain in a safe, healthy and comfortable state. To the **ecologist**, the environment is synonymous with the term habitat, within which plants and animals live. In the context of this publication environment has the broadest possible connotation. It refers to the entire range of external influences, natural and man-made, that can impinge on the **life-support systems** essential for health and survival.

ENVIRONMENT 2010: OUR FUTURE OUR CHOICE: See **Towards Sustainability**

ENVIRONMENTAL ACCOUNTING: See **green accounting**

ENVIRONMENTAL ACTION PROGRAMMES: See **Towards Sustainability**

The **ENVIRONMENTAL CHEMICALS DATA AND INFORMATION NETWORK (ECDIN)** is one of the principal sources of toxicological data concerning any of the 100,000 substances marketed in the European Union, and the 200–300 new substances that appear each year. ECDIN is maintained by the European Commission's Joint Research Centre.

ENVIRONMENTAL CRIME refers to the infringement of laws such as those that ban the use or sale of ecologically dangerous substances like **chlorofluorocarbons (CFCs)**, or the trade in endangered species of animals. Many illegal traders in CFCs, for which there was a thriving 'black' market, were arrested and prosecuted in the USA in the late 1990s. European authorities were relatively slow to take decisive action against such criminals, but had introduced measures by the early 2000s.

ENVIRONMENTAL DEGRADATION is the process by which the **soil**, a rich but fragile ecosystem, containing millions of micro-organisms to sustain fertility, is destroyed. It can be sterilized by salts, contaminated by chemicals, under-fertilized, buried under buildings or swamps, damaged by changes in the traditional patterns of grazing and shifting cultivation which gave soil time to recover between crops, or even blown away by the wind. In arid, semi-arid and dry sub-humid areas the process of **desertification** begins. People eking out a living from the land overgraze and cut down trees and shrubs and, when **fuelwood** runs out, they are forced to burn dung, one of the only fertilizers that is affordable. In dry areas productive land turns into desert and in more humid areas it is debased and becomes unproductive wasteland. Environmental degradation is also caused by air- or water-borne pollution to rivers, seas and forests, and contributes to the corrosion of buildings. Historically, a dramatic form of environmental degradation on Easter Island (an isolated Chilean territory in the Pacific) had virtually wiped out the Polynesian population and culture by the time Europeans first landed there (completing the process) in the 18th century. More recently, and on a greater scale, the Sahel drought in Africa between 1968 and 1974 impelled the UN to convene the Conference on Desertification. It adopted a plan of action to combat desertification. Ninety-four countries pledged support at the Conference, but progress has been slow. (See also **soil erosion**.)

ENVIRONMENTAL HEALTH refers to causes of ill health and diseases in people that are produced or exacerbated by environmental conditions. The impact on health of weather, climate, water supply, eating habits and mode of life generally is not a modern-day discovery. In the 4th century BC they were taken to be essential considerations for 'whoever wishes to investigate medicine properly'. Nowadays a multiplicity of additional environmental factors must be added to the list. One is **overpopulation** in big cities and the tempo and tensions of life in them that cause stress. Others include: **air pollution** from cars, industrial emissions and nuclear radiation; pollution of land by industrial wastes and pesticides; and pollution of streams, rivers, lakes and seas by sewage, industrial effluent and petroleum. Some of the changes are obvious because of blatant large-scale pollution, others are being wrought by imperceptible alterations in the environment such as **low-level radiation** releases and accumulation of low levels of toxic **heavy metals** in the **food chain**. The traditional approach to medicine has always been that the primary role of the doctor is the treatment of the individual patient. Now the practice of medicine has moved into an era where an understanding of the social and environmental factors in relation to disease has become of paramount importance.

The **ENVIRONMENTAL HEALTH ACTION PLAN FOR EUROPE (EHAPE)** was agreed by 50 countries at the second European Conference on Environment and Health, held in Helsinki, Finland, in 1994 and organized by the **World Health Organization (WHO)** Regional Office for Europe (WHO/EURO). The action plan is based on the following definition: '**Environmental health** comprises those aspects of human health, including quality of life, that are determined by physical, biological, social and psychosocial factors in the environment. It also refers to the theory and practice of assessing, correcting, controlling and preventing those factors in the environment that can potentially affect adversely the health of present and future generations.' Progress on the Action Plan was reviewed by the third Conference, held in London in 1999. The fourth Conference on European Environment and Health was held in Budapest, Hungary, in June 2004.

An **ENVIRONMENTAL IMPACT ASSESSMENT (EIA)** is an audit of the environmental consequences of any planned major development before it leaves the drawing board stage. It should provide both an ecological and amenity balance sheet. The idea of EIA techniques was adopted originally in the 1970s by petroleum companies, to cope with tougher legislation on all their industrial and shipping activities. An EIA directive from the European Commission lists 35 obvious developments that require an audit, ranging from power stations and airports to scrapyards and toxic waste dumps. However, there are more than 200 European Union directives that involve environmental controls. EIAs pose questions such as whether a development would affect the local topography, whether it would sterilize agricultural land, which pollutants would be unleashed and what would be the potential threats to flora and fauna, and to **environmental health**.

ENVIRONMENTAL INDICATORS is a term that refers to agreed sets of measurements needed to monitor environmental change. Such data is only useful if it is assembled in a form that meets several needs. One is to ensure the policing of specific international and national regulations. Another is to organize indicators into a descriptive list that provides a profile of environmental quality. Social and economic indicators are used along with environmental data to give a comprehensive picture of the state of the environment either in a particular country or region, or globally. Information relevant to policy-makers tends to be based on environmental indicators that help explain changes in the environment and their relationship with human activities. There are three main groups of environmental indicators: state-of-the-environment indicators, stress indicators and pressure indicators.

ENVIRONMENTAL LAW covers a wide spectrum of options from binding 'hard' laws, such as national legislation, to 'soft' laws, covering guiding principles, recommended practices and procedures, and standards. Environmental law also attempts to reconcile international considerations with concerns that focus on very specific problems such as soil degradation, marine pollution or the depletion of non-renewable resources. It embraces **environmental impact assessment**, which in some countries is becoming increasingly binding. Until 1972 there were 58 international treaties and other agreements on the environment. By 1992 close to 900 more agreements had been adopted regionally or internationally. The **Earth Summit** in that year prompted a massive increase in the number of conventions, treaties and protocols negotiated thereafter, and this process was deepened further in the 2000s.

ENVIRONMENTAL LIABILITY is the penalty to be paid by an organization for the damage caused by pollution and restoration necessary as a result of that damage, whether by accidental spillages from tankers, industrial waste discharges into waterways or land, or deliberate or accidental release of radioactive materials. The concept is part of the **polluter-must-pay** principle. However, the philosophy is difficult to put into practice. Ultimately, the cost of pollution is met by the consumer through the price of goods or taxes for public services. Progress in enforcing liability rests on the adoption of an increasing number of environmental laws and negotiation of international agreements. The potential scale of petroleum pollution, which became apparent with the ***Torrey Canyon*** petroleum tanker spillage in 1967, led to the first major understandings on liability. These resulted in the establishment of the 1969 International Convention on Civil Liability for Oil Pollution Damage and, in 1971, to the International Fund for Compensation for Oil Pollution Damage. There are now over 200 European Union directives alone governing the environment.

The **ENVIRONMENTAL PROTECTION AGENCY (EPA)** is the US Government's agency responsible for controlling the pollution of air and water, pesticides, radiation hazards and noise pollution. The EPA is also involved in research to examine the effects of pollution. Its head was elevated to cabinet level during the administration of President Bill Clinton (1993–2001), but suffered a loss of influence under George W. Bush.

ENVIRONMENTAL SCIENCE is the general label given to the vast range of research into the natural environment.

ENVIRONMENTALLY FRIENDLY: See **green consumerism**

ENVIRONMENTALLY SENSITIVE AREAS (ESAs) are places that can be officially defined under European Union (EU) regulations as being important for the **conservation** of **biological diversity**, landscapes or cultural features, where these depend upon the survival of traditional forms of farming. Recognized ESAs are eligible for EU grants to ensure that traditional agriculture is maintained, with appropriate safeguards for environmental protection. ESAs are of particular importance for the protection of **national parks** and national **nature reserves**.

ERTS: See **Earth Resources Technology Satellite**

ESSENTIAL ELEMENTS are the basic elements, sometimes referred to as the macronutrients, that are needed to produce the basic molecules of life. There are 10: **carbon**, **oxygen**, **hydrogen**, **nitrogen**, **phosphorous**, **sulphur**, potassium, magnesium, calcium and iron. In addition, there are other vital minerals that are needed in small amounts, referred to as the trace elements. They include **boron**, **manganese**, **zinc**, **copper** and **molybdenum**. Excess amounts of trace elements can also be toxic.

An **ESTUARY** is the area at the mouth of a river where it broadens into the sea, and where fresh and sea water intermingle to produce **brackish** water. The estuarine environment is very rich in wildlife, particularly aquatic, but it is very vulnerable to damage as a result of the actions of humans. Experience has shown these ecosystems are at risk if there is an artificial change in the deposition of **sediments** and salinity caused by the abstraction of water upstream, or by the construction and operation of flood-control barrages on the estuary. Chemical pollution from domestic and industrial sources can cause serious problems, as can pollution from ships or the discharges of coolant water from power stations. (See also **delta**.)

ETHANOL is the alcohol best known for its intoxicating qualities, and for centuries it has been made from grain and other crops. However, from the mid-1970s it has attracted the interest of environmentalists, as the fuel of the future for cars and lorries (see **biofuels**). It can be blended with motor fuel or used on its own in modified engines. The possibilities of using ethanol to replace petroleum-derived motor fuel have been demonstrated extensively in Brazil. A huge government-subsidized industry was set up to produce enough ethanol to fuel one-half the country's cars, although falling petroleum prices had undermined the effort by the 1990s. It can be seen that the question of expanding the global use of ethanol, or any **renewable energy source**, was not a technical but an economic one. Crop-derived ethanol had to compete with petroleum, and fuel crops must be grown on the planet's limited amount of fertile land at a time when many people are starving. Scientific research in 1989 showed ethanol could be produced from wood and cellulose wastes, instead of valuable food crops like grain.

EUROENVIRON is an extension of the 19-nation European Research Co-ordination Agency (EUREKA) research programme, created to establish a European programme of research and development to counterbalance technological development in the USA and Japan. Euroenviron extends this to green technologies for halting pollution emissions, reducing waste discharges and saving energy.

The **EUROPEAN ARCTIC STRATOSPHERIC OZONE EXPERIMENT (EASOE)** was an extensive series of chemistry studies by European countries of the **stratosphere** over the northern hemisphere conducted during the winter of 1991/92. They showed levels of man-made chlorine compounds in the stratosphere that were capable of destroying **ozone**.

EUROPEAN COMMUNITIES (EC) DIRECTIVES adopted in 1980 and 1985 established legal limits for emissions of **sulphur dioxide**, **particulates** and **nitrogen dioxide**. Directives are one of several forms of EC legislation and are binding upon member states of the European Union, while leaving the precise means by which a goal is to be achieved to the discretion of each state. Other environmentally relevant directives cover issues such as **noise pollution**, water quality in rivers, drinking water quality, beaches, **eco-labelling**, waste dumps, industrial discharges and nuclear **radiation** levels.

EC DIRECTIVE ON THE CONSERVATION OF WILD BIRDS: See **Birds Directive**

EC DIRECTIVE ON THE CONSERVATION OF NATURAL HABITATS AND OF WILD FAUNA AND FLORA: See **Habitats Directive**

The **EUROPEAN CONVENTION ON CIVIL LIABILITY FOR DAMAGE RESULTING FROM ACTIVITIES DANGEROUS TO THE ENVIRONMENT** is an agreement concluded in 1993 between the member countries of the Council of Europe that defines legal liability for environmental damage. It represents the first international legal instrument to set out the principle of civil liability for all damage caused in the environment, with the exception of damage resulting from the transport of dangerous products and nuclear accidents, which are covered by other international conventions. The convention requires that each signatory 'shall ensure that, where appropriate, taking account of the risks of the activity, operators conducting a dangerous activity on its territory, be required to participate in a financial security scheme or to have and maintain a financial guarantee up to a certain limit, of such type and terms specified by international law, to cover the liability under this Convention'. This is interpreted by the Council as a requirement that operators must 'take out compulsory insurance as soon as the environment insurance market is sufficiently developed'. Ultimately, the Convention aims to make insurance cover or payments into national compensation funds compulsory in order to pay for damage.

The Convention is open to other states, as well as members of the Council of Europe. It deals with compensation for damage to the environment, to persons and property, as well as compensation for economic loss following impairment of the environment; for example, the loss of earnings to a tourist industry following pollution of a beach. The phrase 'dangerous activity' covers actions that produce or make use of a provisional list of over 100 dangerous substances. It also concerns the use of **genetically modified organisms (GMOs)** or micro-organisms, and involves the treatment of waste and waste dumps. Liability rests on the person who controls the dangerous activity. This may be an individual or a public authority, so that a local authority operating a refuse dump, an industrialist manufacturing **fertilizers**, a farmer using them, and a laboratory making GMOs, will be liable for any damage caused by their activities. The Convention will extend to creeping pollution, such as harmful substances that are deposited in a waste dump and then seep into the ground contaminating **ground-water** and drinking water sources.

The **EUROPEAN ENVIRONMENT AGENCY (EEA)** is a European Union (EU) institution established in Copenhagen, Denmark, in November 1994. Non-EU members were permitted to join and it was the first EU institution to admit the accession states of 2004, including some with a legacy of severe environmental damage. It was set up to produce reliable, objective and comparable information for those involved in developing European environmental policy and to give early warning of impending environmental problems. The EEA is supported by the national information systems incorporated into the European Environment Information and Observations Network (EIONET). Specific tasks are carried out under contract by European Topic Centres in priority fields, which include inland waters, marine and coastal environments, **air quality**, air emissions, nature conservation, **soil**, forests (see **forests and woodlands**), land cover, and catalogue and data sources. One of the Agency's aims is to reform the existing environmental information system, in such a way that Europe can benefit from the results of years of measurements and monitoring. The Agency also aims to incorporate European information into international environmental monitoring programmes, for example those of the UN agencies, and to co-operate with other EU and international bodies and programmes. In February 2004 the EEA and the European Commission launched the European Pollutant Emission Register, the first Europe-wide record of industrial discharges into air and water. The EEA publishes a report on the state of the European environment every three years.

The **EUROPEAN ENVIRONMENTAL BUREAU (EEB)** is a **non-governmental organization (NGO)** formed in 1974 to co-ordinate the activities of European NGOs in order to maximize their impact on the environmental projects and policies of the European Communities (now known as the European Union). Based in Brussels, Belgium, along with many of the principal EU institutions, its main areas of interest are environmental policy, education, nature protection, pollution control, land use, planning, transport, energy and agriculture. Another area is the environmental aspect of European relations with developing countries. It operates through lobbying, general and specific publications, and also runs a series of seminars and conferences.

The **EUROPEAN INVENTORY OF EXISTING COMMERCIAL CHEMICAL SUBSTANCES (EINECS)** is a database that lists more than 100,000 chemical substances marketed in the European Union (EU) and is one of several measures to control chemical risks. About 1,000 of the substances listed account for 95% of the total production of chemicals in Europe. EU legislation distinguishes between 'existing' and 'new' chemicals, existing chemicals being defined as those that were commercially available before September 1981. A chemical substance must be registered as new if it cannot be found among those listed in EINECS. EINECS is supplemented by the European List of Notified Chemical Substances (ELINCS). At least 13m. synthetic compounds have been produced in laboratories, and either reproduce the structure of substances found in nature or have a completely new chemical structure. The speed of industrial development in the late 20th century and into the 21st century has not allowed knowledge of the environmental effects of chemical products to keep pace with the use of such substances. In the USA, the **Toxic Release Inventory** fulfils a similar role to EINECS.

The **EUROPEAN WASTE CATALOGUE (EWC)** is a directory established by the European Union (EU) in order to provide a common nomenclature for cross-referencing national lists of waste categories and methods of disposal. EU countries have an agreed order of preference for waste management, starting with reduction of the generation of wastes by means of more efficient manufacturing, reducing the amount of disposable material in consumer goods and increasing the durability of products. Next in the order of preference is the separation of usable components in waste at source, which is achieved by means of more efficient control of effluent from production processes, separation of paper, glass, plastic and metals by householders, and concentration of used **tyres** and petroleum at collection centres. Third in the order of preference is direct reuse of waste products if possible, such as the return of materials to the production process in steel-making or cement kiln operations, and the **incineration** of household wastes to recover energy. Fourth, treatment to recycle usable materials from mixed industrial waste, by means of magnetic separation of ferrous metals and thermal processes for non-ferrous materials. Fifth in order of preference is destruction of useless waste by incineration, or neutralizing **acid** wastes with alkalis. Sixth is permanent storage of waste in or on land. Finally, dumping at sea, which is to be avoided as far as possible.

The **EUROPEAN UNION (EU) WASTE STREAMS PROGRAMME** is an EU-supported project to study the best way to recycle or dispose of a number of selected categories of waste: used **tyres** and cars, **chlorinated solvents**, electronic, construction and hospital waste, and sewage sludge. Another purpose of the Programme is to change the behaviour of the people and organizations responsible for the generation, handling and management of waste. The Programme evolved gradually in the late 1980s and 1990s as harmonization measures were implemented. According to the annual report for 1993 of the European Commission's Directorate-General XI, the EU produced an estimated 250m. metric tons of municipal waste and 850m. tons of industrial waste per year. **Landfill**, which is the most common method of **waste disposal**, carries the risk, if not properly managed, of contaminants such as **heavy metals** (including **arsenic**, **cadmium**, **lead**, **copper**, **manganese** and **zinc**), **ammonia** and organic compounds like **polychlorinated biphenyls (PCBs)** leaching into **soil** and **ground-water** (see **leachate**). **Incineration**, unless properly regulated, discharges a number of toxic substances into the atmosphere. **Recycling** has the advantage of reducing emissions and saving reusable materials, but involves a considerable amount of sorting and treatment.

EUROTRAC/TOR is an **air pollution** monitoring network for recording the concentration of **ozone** and **photochemical smog** at ground level throughout Europe.

EUTROPHICATION is a process of pollution that occurs when a lake or stream becomes over-rich in plant nutrients. It becomes overgrown in **algae** and other aquatic plants. The plants die and decompose. In decomposing the plants rob the water of oxygen and the lake, river or stream becomes lifeless. **Nitrate** fertilizers which drain from the fields, nutrients from animal wastes and human sewage are the primary causes of eutrophication. They have a high biological oxygen demand (**BOD**). This is the figure showing how much oxygen is needed to purify water affected by a particular form of contamination.

The **EVERGLADES** are the subtropical marshy region of southern Florida, USA, covering some 4,000 square miles (10,400 sq km), which stretch down to the **mangrove forest** swamps that border the Gulf of Mexico and Florida Bay. The Everglades contain a variety of ecosystems, providing habitats for a range of wildlife and flora and fauna. Much of the area is covered with saw grass and, on low-lying islands, trees and shrubs, including cypresses, palms and other tropical species, which form a natural habitat for deer, panthers and a variety of reptiles. The mild climate, combined with open areas of water, makes an ideal climate for wading birds, and for many kinds of snake, turtle and alligator. The Everglades also house a wide variety of insects. In the 19th century regulations were introduced to protect birds that were being hunted for their feathers. Subsequently, similar restrictions were placed on the killing of alligators for their skins. In 1947 the central and southern regions of the glades were designated as the Everglades **National Park** to protect wildlife habitats. However, a great part of the Everglades has been damaged by the fertilizers that are discharged from nearby agricultural areas. They have led to the growth of cat-tails and furry grasses, which have taken over some 6,000 acres (2,400 ha), depleting the oxygen in the water and therefore the number of wading birds. The Everglades are also under constant pressure from a large increase in population and from the growth of **tourism**.

EXTERNALITIES, within the context of environmental economics, are benefits and costs that are not included in the market price of an economic good. For example, the cost to society of air pollution from power stations, in the form of increased doctors' visits or incidences of asthma, for example, is not incorporated into the price of electricity. This reduces the economic efficiency of electricity markets because, if the costs of air pollution were included in the price, demand for electricity might decrease as consumers had found ways to enhance their **energy efficiency**. In this example, the cost of the air pollution is the environmental externality.

EXTINCTION describes the complete disappearance of a species of plant or animal from the planet. It can be a natural process, part of the general evolutionary pattern, but it can also be caused by humans. There have been mass extinctions throughout history, but these were lengthy, mostly gradual processes, although some are associated with dramatic, but natural, events. The current period is reckoned to be the sixth mass extinction to have taken place. Species are now being lost at a much more rapid pace than is usual, entirely owing to human activities, such as the growing levels of pollution and hunting, excessive fishing, and the destruction of habitats. Species are lost because of **deforestation** of the tropical forests and by the damage being done to **coral reefs** and **wetlands**. It is estimated that wildlife species are being lost at a rate over 1,000 times faster than ever before. The International Union for the Conservation of Nature and Natural Resources (see **World Conservation Union (IUCN)**) estimated in 1988 that about 800 species of mammals, reptiles and amphibians were threatened with extinction. Another 1,073 birds, 596 species of fish and more than 2,000 species of invertebrates were also in danger of being lost forever. The IUCN regularly produces its **Red Data Books** listing endangered species, noting those in imminent danger of extinction.

EXTREME WEATHER describes all unusually strong weather phenomena, such as major tornadoes, storms that cause severe flooding, heatwaves and blizzards. Scientists have projected that incidences of extreme weather are likely to increase as a result of global **climate change**.

The ***EXXON VALDEZ*** was the petroleum tanker responsible for one of the worst petroleum pollution disasters. It devastated the ecology of hundreds of miles of pristine Alaskan coastline and the breeding grounds of marine life, fish and mammals centred on Smith Island. It became grounded off the island in 1989. The petroleum slick spread over some 3,000 square miles (7,800 sq km), contaminating 350 miles (560 km) of beaches. It is estimated that the **petroleum spill** was responsible for the deaths of some 33,000 birds and 1,000 **otters**. The pollution of the shores of Prince William Sound destroyed marine organisms and vegetation and disturbed the local fishing industry. Four years later scientists discovered brain-damaged seals, birds that do not breed, fish with malformed jaws and a dwindling number of **whales** in the area. It is believed the whales inhaled toxic vapours from the slick and died. Controversy continues as, although compensation has been agreed, court actions are delaying much restoration work.

F

FALL-OUT refers to both the airborne and waterborne **radioactivity** from nuclear weapons testing and emissions, in both accidental and planned circumstances, and from nuclear power stations. The most intensive testing of nuclear weapons in the atmosphere occurred during the periods 1955–58 and 1961–62. After the **Nuclear Test Ban Treaty** on atmospheric testing was agreed between the USA, the former USSR and the United Kingdom in 1963, only France and the People's Republic of China continued with atmospheric tests. The most important **radionuclides** in weapons' fall-out are **caesium**-90 and strontium-90, which are both fissile products of uranium and plutonium. The two most serious cases of accidental fall-out were from the **Windscale fire** in 1957 and, particularly, the explosion of the **Chornobyl** nuclear reactor in 1986. The **fission products** released from Windscale were **iodine**-131, caesium-137, strontium-89, strontium-90 and **polonium-210**. A large area around Windscale (United Kingdom) became contaminated with iodine and the soil had to be removed. In the Chornobyl accident, the fall-out also included aerosol particles of plutonium. The area round Chornobyl could remain contaminated for centuries.

FAO: See **Food and Agriculture Organization**

The **FAST BREEDER REACTOR (FBR)** is a type of nuclear reactor that breeds more **fissile** fuel while it is consuming **nuclear fuel** to generate power. The first FBR was built by the United Kingdom Atomic Energy Authority at Dounreay in Scotland. The reactor is called fast because it has no moderator to slow down the neutrons that are generated once the chain reaction has begun. This allows some neutrons to escape from the fuel core to be absorbed by a blanket of **uranium**-238. Some of the U-238 is then converted to **plutonium**, which is extracted by the **nuclear reprocessing** of the uranium. Typically, a fast breeder relies on liquid metal cooling. FBRs for commercial power generation have been built in France, Japan, the former USSR and the USA, and in 2003 one was planned for India and another for the People's Republic of China (using Soviet technology). However, there is greater risk of the proliferation of nuclear weapons with FBRs than with the thermal **pressurized water reactors (PWRs)**.

FAUNA: See **flora and fauna**

The **FAUNA AND FLORA PRESERVATION SOCIETY (FFPS)** is the oldest international **conservation** organization. It is a voluntary organization, with its headquarters in London, United Kingdom, which promotes international conservation of wildlife, especially **endangered species** of plants and animals. The Society, an independent body, increases public awareness about endangered wildlife by holding meetings and exhibitions and publishing information. It also promotes the establishment and management of wildlife reserves and the implementation of laws covering the protection and conservation of wild animals and plants.

FEEDBACK CYCLES or feedback loops take place within a given system where there is a continuous cycle of sensing, evaluating and reacting to changes in environmental conditions. A feedback cycle can occur within many different types of systems, including physical and ecological ones. A negative feedback is where a flow of information counteracts the effects of a change in the external conditions on the system, reducing or neutralizing its impact. A positive feedback is when the information compounds the effects of a change in the same direction, making it either worse or better. An example of a physical feedback loop is global **climate change**. When there are increased levels of carbon dioxide in the air owing to human burning of fossil fuels, trees and plants respond in an instance of negative feedback by absorbing the excess carbon dioxide. However, if temperatures were to increase enough to cause forests to die, then the trees would release more carbon dioxide into the air, which would cause the atmosphere to warm even more, causing more forest death, which would in turn contribute to the acceleration of this positive feedback loop.

FERMENTATION is a process of breaking down plant material and food without the use of oxygen. In other words, it is an anaerobic process. The most well-known fermentation process involves adding yeast to the raw materials from which alcohol is derived, in the making of wine and brewing of beer. It is also used to cause the dough to rise in baking. However, fermentation is also the basis of the **biotech** industry. Many pharmaceutical products, like antibiotics, employ fermentation at the preliminary stage to make the active ingredient that is eventually packaged as a tablet, capsule or liquid. It is the process by which waste materials, such as cellulose and straw, and organic refuse, are deliberately decomposed to generate **biogas**.

FERRUGINOUS DISCHARGES are a type of effluent, like that in the floodwater from abandoned coal mines, which is contaminated with iron.

FERTILE nuclear materials, such as **uranium**-238 and thorium, are those that can be converted into **fissile** material in a nuclear reactor.

FERTILIZERS are any material that is added to the **soil** in order to provide and replace vital nutrients for crop growth. That material can be natural or organic, such as **compost** and manure. The term is more commonly associated today with manufactured fertilizers, made from substances derived from chemical processes or mining. Most manufactured fertilizers are made of a compound of salts of nitrogen, phosphorus and potassium. The use of artificial fertilizers increased rapidly throughout the world from the 1950s when farmers realized that fertilizers meant they no longer had to leave fields fallow to regain fertility or, indeed, keep farmyard animals just for their manure.

FILTER FEEDERS are a variety of animal, critical to the **food chain**, which live mostly on the ocean floor or in **plankton**. They feed by creating a current running through a filter, a scraping device to move food from the filter, and a water exit. The filter absorbs particles of an acceptable size and rejects those that are too large. In sponges and sea cucumbers, the entire body acts as the filter, water inlet and water exit. Mussels and oysters are filter feeders. However, the baleen **whales** are the largest filter feeders. They have blow-holes that expel the water that has been ingested along with the **krill** that they feed on.

FINITE FUELS are the deposits of **coal**, petroleum, gas (see **fossil fuels**) and **uranium** ores that were formed millions of years ago by geological and biogeological processes. As such they are non-renewable reserves, and are finite as it is possible to calculate their date of exhaustion according to certain rates of exploitation.

FIRE is one of the greatest tools of mankind. Once early man had learned to control it, he was able to provide warmth and light, to cook, to kill insects and to clear forests of undergrowth so that game could be more easily hunted. Later, he found that this burning produced more richly abundant **grasslands**. Agriculture brought a new urgency to clear brush and trees, and used ash as a fertilizer. This procedure, **slash-and-burn agriculture**, continues in many tropical and temperate areas throughout the world today. Fire also permitted the opening up of industrial and manufacturing processes. However, uncontrolled and spontaneous fire is destructive. Fires in forests, grassland or brushland are a constant problem, causing loss of vegetation, trees and crops, and sometimes the lives of wildlife or humans. Dried grass, leaves and light branches ignite easily and this often generates enough heat to set fire to tree stumps and the matted undergrowth of the forest floor. Evergreens, such as pine, cedar, fir and spruce, have flammable fuels that burst into flames if they are heated sufficiently by a forest fire. In some ecosystems, such as in the Australian bush, fire can be part of the natural cycle.

FISH FARMING: See **aquaculture**

FISHING QUOTAS: See **total allowable catches (TACs)** and **straddling stocks**

FISSILE describes a substance which is capable of **fission**.

FISSION is the process of atom-splitting employed in a nuclear reactor and an atomic bomb. Splitting an atom is accompanied by the release of an immense burst of raw energy. Fission occurs when a neutron strikes the nucleus of an atom of a fissile substance, such as **uranium-235** or **plutonium**-239, causing it to split into more or less equal halves. In addition to the release of energy, many more neutrons are also produced. They strike other uranium or plutonium atoms, generating further neutrons and, under special circumstances, sustaining the fission process in a **chain reaction**. In a nuclear reactor the chain reaction is controlled to generate heat. In a bomb, energy is released as an uncontrolled destructive force.

FISSION PRODUCTS are the substances formed by splitting atoms of **uranium-235** or any other fissile element. Fission products are contained in the fall-out from nuclear weapons and in the various forms of **nuclear waste** from nuclear power stations and from **nuclear reprocessing** of spent fuel. Fission products cover a wide range of radioactive elements, or radionuclides, from **americium**, **caesium** and **iodine** to strontium and **zinc**. One year's operation of an average nuclear power station produces waste fission products of 5,000m. **curies** of radioactivity. Many of these products have an extremely short **half-life**. A significant proportion of them are rendered safe simply by storing spent nuclear fuel for one to two years in cooling ponds at the nuclear power station, during which time a large amount of **radioactivity** decays. Hence, the 180m. curies of radioactivity in each metric ton of spent nuclear fuel reduces to one-260th of its initial level, to 693,000 curies, within one year. The remaining fission products take much longer to disintegrate. Some will still be dangerous in 10,000 years. Accidents such as the **Windscale fire** in 1957 and the **Chornobyl** explosion were disastrous because they pumped fission products in their most radioactive state into the atmosphere.

FLATFISH is a generic term for any of the 600 species of oval-shaped, flattened fish. Flatfish include halibut, plaice, sole, turbot, dab and flounder, and they live in tropical and cold waters. Most are **demersal**, marine fish and are found along the **continental shelf**, but some are found in **fresh water**. Flatfish are important to environmental monitoring because they can give early warning signs when chemical contamination is becoming toxic. As well as studying the effects of contamination on the flatfish immune system, scientists have been investigating the impact on liver cells. The fish act as a continuous monitor, because it is possible to link subtle changes in their liver cell function, and health, to specific groups of contaminants. These techniques are being used by scientists in Europe and the USA. They are likely to be much more cost-effective than conventional analysis, and demonstrate whether the contaminants are actually harming fish.

FLOODS are caused either by tidal waves (or **tsunami**), when rivers overflow their natural or artificial banks on to dry land, or by excessively high rainfall over a short space of time, or when large amounts of ice formed on rivers melt in the spring. Sometimes this is a natural process, such as when a river submerges the surrounding floodplain. Egypt's Nile basin is an example of flooding being beneficial to the environment, because agricultural productivity is maintained by the deposits of fertile sediment left on the land after flooding. Severe and sudden flooding from **extreme weather** can cause large-scale loss of life and damage to property and land. **Environmental degradation**, especially where there is large-scale **deforestation** or urbanization, also contributes to a cycle of severe drought followed by equally disastrous floods, which have been experienced in some regions in recent years. Forests control the run-off to rivers and most of the annual rainfall in the forested watershed of rivers in the tropics is trapped in the sponge-like network of roots beneath the ground. If the forest is uprooted, and therefore the 'sponge' is taken away, only a small amount of water remains in the soil and the greatest proportion goes into local rivers in a short space of time, resulting in serious flooding as well as **soil erosion**. Another problem is that global **climate change** could lead to a rise in **sea levels** and flood low-lying **coastal areas**. A rise of more than 10 cm in sea levels would have a serious impact on humans, plants and animals in low-lying areas like **Bangladesh**, the Maldives or Florida (USA).

FLORA AND FAUNA are, respectively, the plants or plant life in a region or at a point of time, and a collective name for animals or animal life of any region or point of time.

FLORAS are full catalogues and checklists of known plant species. Many of the world's plants are disappearing at a rapid rate, particularly in the tropical and subtropical forests, where habitats are being destroyed. It is now a major priority to explore these regions and bring back for cultivation the most rare and scientifically interesting species, some of which are potentially of great value for food sources or medicines. The **Royal Botanic Gardens** in the United Kingdom have been involved in the production of floras for Africa, Australia, Cyprus, India and Iraq. New studies are concentrating on regions where the vegetation is particularly diverse, or of special concern for the conservation of genetic resources. Small samples are collected by plant-hunters from the Gardens and brought back to be cultivated and studied, and eventually to be used as a reservoir to support wild populations so that they do not become extinct (see **World Seed Bank**).

FLUE GAS DESULPHURIZATION is the necessary, but somewhat costly, method of removing **sulphur dioxide** and **nitrogen oxides** from the gas in power stations when **fossil fuels** are being burned. The control technologies include scrubbers and electrostatic precipitators, which remove emissions from flue gases before they are discharged into the atmosphere.

FLUIDIZED BED COMBUSTION is the most efficient way of burning all fuels, but particularly coal. The furnace contains a layer or bed of fine inert particles such as sand or ash. When hot air is blown up from below the fuel and inert particules behave like a fluid. The fuel burns very efficiently and traps waste products in the bed instead of in gas discharges through the chimney.

FLUORIDES are substances containing fluorine that may be added to water supplies as a way of preventing dental problems. Individuals growing up in areas where the natural water is deficient in fluoride minerals tend to have poorer teeth than those in areas of abundant fluoride. The addition of fluorides to water supplies was the cause of considerable controversy. The anti-fluoride lobby, one of the first really vocal health and environmental movements, continues to object to the treatment of water supplies with fluoride.

A **FLUX** is a general state of change. For example, it refers to processes by which **carbon** is exchanged between the atmosphere and oceans.

FLYWHEELS are an alternative to batteries as a way of storing energy for propulsion purposes. A flywheel is set spinning in a vacuum at up to 200,000 revolutions per minute, and can be recharged in minutes.

FOG is a cloud-like body of water droplets suspended in the atmosphere. It reduces visibility to less than one kilometre. It is formed when the lower layers of the atmosphere cool and **water vapour** condenses.

The **FOOD AND AGRICULTURE ORGANIZATION (FAO)** is the oldest of the permanent specialized UN agencies. It was established in 1945 to co-ordinate food, agriculture, forestry and fisheries development programmes, to improve standards of living in rural communities and to fight malnutrition and hunger at the end of the Second World War. It advises governments on policy and agricultural planning (giving technical assistance to individual countries on a project basis), on nutrition, improved food storage and distribution, fertilizer production, pest and disease control, **soil** fertility and **irrigation**. It also runs educational programmes, keeps statistics on world production, trade and consumption of agricultural products, and publishes yearbooks, periodicals and research bulletins. Its headquarters are in Rome, Italy, and there are other offices throughout the world. It is governed by a council elected by the biennial FAO conference of member countries. In 1985 it began to concentrate on the problems of tropical forestry, with the introduction of measures to promote conservation and the increased use of **afforestation** in agricultural and industrial planning.

The **FOOD CHAIN** is a process that begins with the simplest green plants and ends in a large animal that has no predators relying on it for food. Examples of food chains could include: green plant, aphids, ladybird, insectivorous bird, hawk. Or marine **algae**, copepod, small fish, large fish, man. Or again, green plant, rabbit, fox. The lower members of the food chain have to support the higher members. Consequently, they have to reproduce at a rate greater than that necessary to maintain their species, because they provide a margin to feed those above them in the chain. Any serious fluctuations in the numbers at any point in the chain, either as a result of natural causes or contamination from pollution, can have disastrous repercussions for every member. Of course, feeding is the way animals get the energy they need to exist. In the process the energy moving along a food chain is described as passing from one trophic level to the next (see **trophic structure**). Plants form the first, producer, trophic level. Then come the consumer levels of herbivores, carnivores and parasites. Ecologists have shown that the number of trophic levels is limited to five or six. Once the consumers die, they are recycled back to the bottom of the feed chain by decomposers like **fungi** and **bacteria**.

FOOD POISONING is the general term to describe illness caused by eating food that is infected with bacteria or contaminated with toxins. In recent years outbreaks of certain types of infection have increased, particularly poultry-borne **bacteria** such as **salmonella**, and also **listeria**, which is responsible for much of the contamination of salads and dairy produce. Precise figures for the incidence of food poisoning are not known, because most people have it in such a mild form that they do not report it. In some cases, however, it can be serious and even life-threatening. Salmonella and the campylobacter group of bacteria are killed when they are thoroughly heated. Outbreaks of food poisoning usually occur when there is inadequate hygiene in the preparation and handling of food, or if food is stored at the wrong temperature.

FORAMINIFERA are an abundant family of simple, mostly marine, organisms, which are important to scientists studying **climate change**. Most have a shell that falls to the bottom of the sea when the organisms die and that becomes incorporated into rock, appearing as chalk or limestone. About one-third of all sediment at the bottom of the sea is made up of discarded shells. The living foraminifera either crawl around or attach themselves to rocks, feeding on **bacteria**, **diatoms** and other algae.

A **FOREST CONVENTION** was the objective of sustained efforts by governments and environmental organizations after the 1992 **Earth Summit**. By 2004 there was as yet no international convention for the protection of forests, the only comparable agreement being the non-binding **Statement of Principles on the Management, Conservation and Sustainable Development of all Types of Forest** of 1992. As a result of the controversy and lack of agreement among countries and non-governmental organizations concerning the need for, and eventual character of, a forest convention, the UN Commission on Sustainable Development (CSD) established the **Intergovernmental Panel on Forests** at its third session in 1995. A UN Forests Forum was established to meet annually for five years from 2000, but no progress seemed likely towards a forest convention until the end of that process.

FOREST PRINCIPLES: See **Statement of Principles on the Management, Conservation and Sustainable Development of all Types of Forest**

A **FOREST STAND** is a community of trees, including above-ground and below-ground **biomass** and soils, sufficiently uniform in species, composition, age, arrangement and condition to be managed or viewed as a unit.

FORESTS AS CARBON SINKS: See **sequestration**

FORESTS AND WOODLANDS cover about one-third of the land area on earth. In 1995 it was reckoned that only 30 countries have more than one-half of their area wooded, and that 85% of the world's forest cover is located in just 25 countries. In fact, only seven countries (Brazil, Canada, the People's Republic of China, the Democratic Republic of Congo—formerly Zaire—Indonesia, Russia and the USA) contain more than three-fifths of the global total. Forests are indispensable to humanity but they are at the heart of environmental concern. Forests and woodlands play many roles in the social and material well-being of mankind in general and particularly of rural populations. Pine forests in the north, the **tropical rainforests** on the Equator and the temperate forests of the southern hemisphere regulate the climate, protect water resources and provide forest products. One-half the world's population depends on them for fuel, they are home to millions of plant and animal species and they secure water supplies by withholding rainfall. The forests are also a source of new medicines, chemicals, foods, commercial products and services. Trees recycle moisture through their leaves, absorb the sun and soak up carbon dioxide, but as more and more trees are cut down **deforestation** contributes an estimated 10% to global warming. Loss of forestland is particularly rapid in developing countries, and many species are being driven towards **extinction** with the removal of their habitat. The forests are depleted by logging for timber, firewood and charcoal, or cleared for cattle. Another threat is the hundreds of millions of people who, driven by the shortage of more suitable land, cut down the forests to build houses and to farm. They then move on because the land loses its fertility. The **Earth Summit** adopted and recommended for endorsement the **Statement of Principles on the Management, Conservation and Sustainable Development of all Types of Forest**. In 1985 the UN Conference on Tropical Timber ratified the **International Tropical Timber Agreement (ITTA)**, to regulate the tropical timber trade. It encourages co-operation between producing and consuming countries.

FOSSIL FUELS are the energy-containing materials which were converted over many thousands of years from their original form of trees, plants and other organisms after being buried in the ground. Physical and chemical processes occurred in the earth's crust that changed them into **coal**, **peat**, petroleum or **natural gas**. Fossil fuel use, when expressed for clarity in terms of coal-equivalent energy, increased from 1,000m. metric tons per year at the beginning of the 20th century to 12,000m. tons in 1990. Use increased thereafter. If fossil fuels are completely burned, or oxidized, they are converted into **carbon dioxide**, water and heat energy. Therefore, burning fossil fuels destroys the energy source. Fossil fuels are used for other purposes. Coal and petroleum are used for the manufacture of a wide variety of carbon-containing compounds such as plastics, synthetic fibres, medicines, food and pesticides. Although **nuclear fuels** are inorganic substances, unlike the biological origins of fossil fuels, they are also destroyed when they are used in the generation of heat in nuclear reactors or in nuclear weapons.

The **FRAMEWORK CONVENTION ON CLIMATE CHANGE (UNFCCC)** (see also **climate change**) began with a meeting of government representatives convened in Geneva, Switzerland, in September 1990 by the **World Meteorological Organization** and the **UN Environment Programme**. The purpose of the meeting was to prepare for negotiations based on the work of the **Intergovernmental Panel on Climate Change**, which had begun in November 1988. At the **Earth Summit** in Rio de Janeiro, Brazil, in 1992 the UNFCCC was adopted by all nations present, and it came into force two years later. The Convention calls for all countries to inventory their **greenhouse gas** emissions, to take all measures possible to reduce emissions, and to review regularly the adequacy of the Convention. Industrialized countries made a commitment to try to reduce greenhouse gas emissions to their 1990 levels by 2000. Very few of them were expected to meet that goal, which made negotiations for the **Kyoto Protocol** to the UNFCCC difficult, since many environmentalists and developing countries doubted the sincerity of industrialized countries' commitment to reducing emissions. In 1995, at a meeting in Berlin, Germany, participating governments decided that there was a need for a legally binding treaty to reduce emissions and adopted the Berlin Mandate, which committed them to negotiating a binding treaty by 1997. As a result of these negotiations, the Kyoto Protocol was adopted in 1997, but to take effect it needed a large economy like either the Russian Federation or the USA to ratify it. In late 2004 Russia ratified the Protocol, which came into effect in February 2005.

FRESH KILLS LANDFILL is the world's largest **landfill** waste disposal site and is located on Staten Island, off New York, USA. It also became the subject in 1993 of a major research project, applicable to all such sites suitable for **methane** generation, into why even readily degradable items, such as food, decompose very slowly in solid waste landfills. Converting solid wastes to methane is an attractive way to degrade refuse and collect a **renewable energy** source. The researchers' report, entitled *Environmental Factors Influencing Methanogenesis from Refuse in Landfill Samples*, shows that microbial activity is inhibited wherever the landfill's moisture content is low, the pH is either very acidic or alkaline (low or high), and, surprisingly, where sulphate ion levels are high. Analysis of the site revealed that paper and textiles, which compose over 40% of the landfill's volume, contain high levels of sulphate. However, this ion is not found in fresh paper and textile wastes. Instead, the authors suspect that construction and demolition debris that contain gypsum and constitute up to 14% of the landfill may be the source of the sulphate. As sulphate is leached from the gypsum the paper and textiles may soak up the water and serve as a reservoir.

FRESH WATER is an essential element for life. Some 70% of the earth is covered by water, but only a fraction of that is readily available as fresh water. Some 94% of the water is salty and is found in **oceans** as **seawater** and, of the rest, 99% is out of reach, frozen in icecaps and **glaciers**, or buried deep underground. Only about 8% of freshwater supplies are withdrawn as drinking water or for domestic or municipal use; the rest is used in agriculture and industry. Freshwater supplies are under intense environmental pressure from two main sources: pollution of supplies and the fact that in the second half of the 20th century world water use more than trebled, and was predicted to increase into the 21st century. Although the annual global rainfall is more than adequate to satisfy the need for fresh water, its distribution throughout the world is uneven, leading to severe **droughts** in several regions. Pollution of fresh water stems from industrial and human wastes, agricultural fertilizers, **acid rain**, and seepage from toxic waste dumps. Excess nutrients from sewage and soil erosion cause **algal blooms** that deplete the oxygen in the water; **heavy metals** and synthetic organic compounds from industry, mining and agriculture **bioaccumulate** in aquatic organisms; and pathogens from sewage spread disease. The **Commission on Health and Environment**, set up by the World Health Organization, reported in 1992 that fresh water should be considered as a scarce, valuable and finite resource. An estimated 2,500m. people, the report said, suffer from illnesses linked to insufficient or contaminated water and lack of sanitation. The **UN Environment Programme (UNEP)** made the 1980s the **International Drinking Water Supply and Sanitation Decade** because of global concerns about water supplies. At the end of the 1980s there were still about 1,232m. people without safe water, and about 1,600m. without sanitation. In 2000, according to UNEP, over 1,100m. were still without access to safe drinking water and another 2,400m. without access to basic sanitation. The agency warned that one-third of the world's population would suffer chronic water shortages within a few decades if concerted action was not taken soon. In 1997 the **Convention on Non-Navigational Uses of International Watercourses** was adopted in response to the problems associated with freshwater resources. The problems of access to fresh drinking water and basic sanitation were at the heart of the aims set by the UN Millennium Declaration of 2000 and by the 2002 **World Summit on Sustainable**

Development. It was proposed to halve the current proportions by 2015.

FRIENDS OF THE EARTH (FoE) is an international pressure group that operates in dozens of countries. It started in the USA as an off-shoot of the conservation group the **Sierra Club**. It is a non-party group that lobbies governments and politicians on all environmental issues and works to increase public awareness. FoE is very active in campaigning against the construction of new nuclear power stations and in establishing safety levels for existing nuclear facilities. The organization campaigns against: air, sea and land pollution, ozone depletion, acid rain and carbon dioxide build-up; and the use of whale products, damaging aerosol propellants, leaded petrol, food additives and non-returnable bottles. FoE's activities include campaigns for recycling, increased use of bicycles and public transport, alternatives to tropical hardwoods, better energy conservation and home insulation and the adoption of catalytic converters in cars to reduce emissions that cause photochemical smog.

FUEL ADDITIVES AND DETERGENTS: See **additives** and **detergents**

FUEL CELLS are a source of power for **electric vehicles** and are expected to replace conventional **batteries** in the 21st century. The attraction of the fuel cell is that it is a way of generating electric power that has water as its only exhaust. The technology was invented in 1839 by a Briton, Sir William Grove, but remained without practical applications for more than a century. Fuel cells convert chemical energy into electrical energy and differ from batteries in that they must be fed continuously with a chemical fuel but do not require frequent recharging. The first practical device was the Bacon Fuel Cell, a six-kilowatt hydrogen-oxygen device perfected by Francis Bacon, an engineer at Cambridge University in the United Kingdom, and which provided the electricity and water for the journey to the moon of the Apollo spacecraft. A similar design supplies electricity for astronauts on the space shuttle. Other fuels such as methanol, **ethanol**, **natural gas** and **liquefied petroleum gas** will also react with oxygen to generate electricity.

FUELWOOD accounts for the greater part of the energy consumption of many developing countries. Supplies of wood and charcoal are running dangerously low in some countries, and fuelwood is being consumed at a faster rate than it can be replanted. The **Food and Agriculture Organization** had predicted that by 2000 some 1,000m. people faced chronic fuelwood shortages. Often there are no alternatives to fuelwood, or if there are they are too expensive. There are many reasons for the shortage of fuelwood: **deforestation**, as forests have been over-exploited; populations have increased; and the need for energy, to heat and cook, has increased. However, as the shortages reach crisis point, this sets in motion a chain of other events. Without wood water cannot be boiled, so disease spreads, people are forced to eat only one cooked meal per day, and they burn **dung fuel** and crop residues that would be better used to fertilize the soil. In rural communities village women often take a whole day to collect wood to last only a day or two; time that used to be spent growing food. The lack of other energy resources, therefore, causes the depletion of fuelwood reserves, which can have devastating impact on the environment and perpetuate poverty, as can be seen in heavily populated parts of Haiti, which have been denuded of trees or brush.

FUNGI are a group of plants without chlorophyll, leaves, actual stems and roots. They include mould, mildews, rusts, yeasts and mushrooms, and are remarkably effective at breaking down, or decomposing, almost anything from wood to leather, and bread to cloth, and thus are important elements of the **food chain**. The by-products of this decomposition process are sometimes toxic, causing great problems to the environment, while others, at the same time, are a boon, producing some cancer-curing toxins. Others have been the basis of beneficial agents like cyclosporin, the immune suppressant 'miracle drug' that has made organ transplants possible.

FUNGICIDES are chemicals used to kill or halt the development of fungi that cause plant disease, such as: storage rot; seedling diseases; root rots; vascular wilts; leaf blights, rusts, smuts and mildews, and viral diseases. These can be controlled by the early and continued application of selected fungicides that either kill the pathogens or restrict their development. Fungal diseases are more difficult to control than insects because the fungus lives at close quarters with its host. It is difficult to find chemicals that will kill the fungus without harming the plant and, because **fungi** may also go through secondary cycles rapidly during a three-month growing season, repeated applications of fungicides may be necessary. They must also be applied to plants during stages when they are vulnerable to inoculation by pathogens, before there is any evidence of disease. There are about 225 fungicidal materials, most of which are recently discovered organic compounds. **Sulphur** is probably the oldest effective fungicide and it is still used in gardens.

FURANS, like **dioxins**, form a group of highly toxic contaminants found in some **herbicides** and **pesticides**. Neither furans nor dioxins are produced for their own sake, but are known as **intermediate chemicals** or intermediaries, compounds formed during the manufacture of other substances. There are a total of 285 furans and dioxins.

FUSION is a potential clean source of energy using technology that, it is believed, will harness the sort of nuclear reactions that take place within the sun and in the hydrogen bomb. However, even at the earliest estimate, the commercial generation of fusion energy is at least 50 years away. The attraction of fusion is that it avoids the pollution associated with nuclear **fission** and the burning of **fossil fuels**. Whereas nuclear fission, the principle on which nuclear power stations operate, splits heavy atoms like **uranium** to create lighter ones, fusion begins with hydrogen and forms heavier ones. The atoms of hydrogen, in the form of tritium, are squeezed by immense electromagnetic fields until two of them fuse to form helium, and in the process release large amounts of energy. Experimental fusion reactors are operating in Europe, the USA and the Russian Federation. The drawback is the immense research cost.

FUSION TORCH is the most efficient incinerator for neutralizing even the most hazardous and toxic compounds. It uses a burner called a plasma torch which heats the wastes to thousands of degrees Celsius to reconvert them to their natural constituent elements.

G

G7, the Group of Seven, is a group of leading industrialized countries (it was originally the Group of Five or G5); it held the world's first Green Summit, in Paris, France, in July 1989. The seven countries that participate in annual summits accounted for 12% of the world's population but contributed nearly one-half of the pollutants that cause global warming, the environmental issue singled out for the agenda of the Paris gathering. While omitting specific targets, the meeting, which included the leaders of Canada, France, Germany, Italy, Japan, the United Kingdom and the USA, stated that it 'strongly advocated common efforts to limit emissions of carbon dioxide and other greenhouse gases that threaten to induce climate change'. The ideas prepared the foundation for the **Framework Convention on Climate Change**. During the 1990s the Russian Federation gradually began to participate in the annual meetings and consequently the group became commonly known as the G8. Before every meeting of the G8, a meeting of environmental ministers is held to prepare recommendations for the heads of government.

G15 is the group of 15 semi-industrialized countries that meets to discuss development and other concerns for the developing world. It was formed in 1990 and consisted of: Algeria, Argentina, Brazil, Egypt, India, Indonesia, Jamaica, Malaysia, Mexico, Nigeria, Peru, Senegal, Venezuela, the former Yugoslavia and Zimbabwe.

G-77, the Group of 77, was formed by 77 developing countries at the end of the first session of the **UN Conference on Trade and Development** in 1964. The number of member states subsequently increased and stood at 132 in mid-2004, on its 40th anniversary, but the name of the group remained unchanged. The G-77 often negotiates as a bloc in **multilateral environmental agreement** negotiations.

GAIA is the name of the ancient Greeks' all-pervading earth goddess. The name was suggested by the novelist William Golding to describe the theory expounded by the scientist James Lovelock and now shared by many other biologists and chemists, which depicts the whole biological, physical and chemical world as a self-adjusting living organism. *Gaia* was used as the title of a book by Lovelock. In 2004 Lovelock outraged many in the environmental movement when he suggested that nuclear power, hitherto strongly opposed, was the best means of dramatically lowering the use of **fossil fuels** and so minimizing the effects of **climate change**. He was worried that **renewable energy** sources would take too long and be too expensive to introduce on the scale necessary.

GAMMA HEXACHLOROCYCLOHEXANE (HCH), the active ingredient in **lindane**, is a synthetic organic compound used as a **pesticide** substitute for **DDT**. Hexachlorocyclohexane was discovered in the 19th century, but its insecticidal properties were only recognized in 1940 when French and British entomologists found it to be effective against all insects tested. HCH was also found to be a mixture of five isomers, or different forms, but only the gamma isomer has insecticidal properties.

GAMMA RADIATION is the most intense form of nuclear **radiation**; the rays can penetrate deep into the body, or even pass through it, without losing their energy.

The **GANGES** (Ganga) river rises in the Himalayas in Nepal and northern India and flows into the Bay of Bengal, passing through the world's largest **delta** in **Bangladesh**. It irrigates the Ganges plain, which contains several of India's principal cities, and is the world's second largest agricultural area, supporting more than 300m. people. However, it is also a major depository for sewage and industrial discharges. The Ganges is a sacred river for the Hindus, who bathe in its waters and are cremated on its banks. **Topsoil** has been eroded and the silt levels in the Ganges have risen. This has largely been caused by **deforestation** in the Himalayas and on the river banks, which has caused flooding in the low-lying areas of Bangladesh. There are also fears about the effects of dams on its headwaters and tributaries, particularly the threat to the endangered Ganges dolphin.

GARP is the acronym for Global Atmospheric Research Project, an international group created to map and follow meteorological phenomena. It also monitors their effects on the world weather system.

GASOHOL: See **biofuels**

GATT: See **World Trade Organization (WTO)**

GEF: See **Global Environment Facility**

GENE BANKS are storehouses of seeds or vegetative tissue, kept in low humidity and temperature, to help maintain genetic diversity. Sometimes known as seed banks or germ plasm banks, their contents mostly originate from a wide range of primitive strains and wild crop varieties. The International Board for Plant Genetic Resources (IBPGR), which was established in 1974, promotes the collection, documentation, evaluation, conservation and eventual use of genetic resources of significant plant species. Gene banks are the subject of international controversy because they contain seeds that have mostly been acquired from the developing countries by the industrially rich countries, where they have been used in breeding programmes to develop new strains. Instead of taking decades over a traditional plant breeding programme by fertilization, it is now possible to manipulate directly the genes of plants, creating **genetically modified organisms**, which are plants modified to give a higher resistance to disease and improved growth and yields and, therefore, increase the profit of the plant breeder and farmer. Often this results in the patenting of material that exists in nature, usually in a developing country.

GENE SYNTHESIS is the complete manufacture of a gene using chemical reagents from the laboratory shelf to create a segment of DNA (see **nucleic acids**). The synthesis is done in a machine and is an alternative to using biochemical scissors, **restriction enzymes**, to snip a gene from a cell and transfer it to a micro-organism, which is then grown in fermentation tanks.

GENE THERAPY is genetic manipulation that specifically relates to changing the genetic make-up of a cell by deleting or inserting genetic material. There are two approaches: germ-line gene therapy and somatic cell gene therapy. The former changes the 'germ cells' of the sperm and ova, and would have a permanent effect on all descendants of whoever had the therapy. It is banned by most countries. Somatic cells are all the other cells of the body. Changing these does not affect the germ cells, only the individual who has been engineered.

GENETIC ENGINEERING is an important element of the biotechnology industry. It entails the modification of the genetic characteristics of a microbe, plant or animal by inserting a gene from another variety or species. The many purposes of genetic engineering include producing fruit and vegetables that are more disease-resistant, bigger or stay fresh for longer; extending the colour range of flowers; breeding cows that produce extra vitamins in their milk; and altering yeast or other cells so they can be used to produce antibiotics and other pharmaceuticals in **fermentation** tanks. Studies into the potential environmental risks of **genetically modified organisms (GMOs)** make it difficult for EU advisers to negotiate European Union-wide regulations. The perception of the risks posed by GMO technology varies greatly, according to the proposed products and their applications. Many products are expected to offer environmental benefits, with no risk. For example, the impact of an established ecologically damaging pesticide might be avoided by growing genetically modified pest-resistant plants, or by using biological control rather than chemical control of pests. Another form of genetic engineering has been cloning, which is also controversial, particularly as the clones are often defective. There have been several successful cloning ventures of animals since the first in 1952 (a tadpole), but the birth of Dolly, a sheep, in the United Kingdom in 1997 raised the possibility of human cloning and, therefore, the controversy. Dolly was the first mammal to be successfully cloned from an adult cell. The type of cloning used was reproductive cloning, and this has since been used as part of efforts to save endangered species (in 2001 a wild ox or gaur was cloned—although it died soon after birth—as was, with more longevity, a wild sheep or mouflon, in Italy). Recombinant DNA or gene cloning has been used for many years, but the most controversial method is therapeutic cloning, which involves the harvesting of stem cells from embryos.

GENETIC PEST RESISTANCE IN PLANTS is a potential alternative to using chemical **pesticides**. Genetic engineers are working toward identifying and introducing into plants the genes that give resistance to **pests**. There are two possible ways of doing it. The first is to locate existing genes that give protection and seem to be more abundant in hardy wild varieties of flora. The object is then to transfer them to valuable crop plants otherwise susceptible to attack and disease. This approach seems more promising in the search for resistance to bacterial and fungal **pathogens**. The trick is to compare the genes of the resistant plant and the pathogen, and to identify in the pathogen the 'avirulence genes', which make it so infectious, and the corresponding plant genes evolved to stop them. The second approach is to introduce a completely new type of gene into the plant. For example, the bacterial organism *Bacillus thuringiensis* produces a substance that stops insects absorbing food when it gets into their gut. The gene for the *Bacillus thuringiensis* toxin, called BTK protein, has been inserted into plants. Consequently, when the insect nibbles the leaf, the insect dies.

GENETICALLY MODIFIED ORGANISMS (GMOs) are genetically engineered organisms released into the environment. They may be micro-organisms designed for use as **biopesticides**, or seeds that have been altered genetically to give a plant better disease resistance or growth. Creatures too can be genetically engineered, with the first modified insect, a pink bollworm, which harms cotton plants, beginning trials in 2001 (the moth has a defective gene, designed to be spread among the wild population). (See **Biosafety Protocol**, where GMOs are called living modified organisms.) In July 2001 the Codex Alimentarius Commission, the main food standards body of the **World Health Organization**, adopted the first global standards for assessing the safety of genetically modified (GM) foods. In March 2002 further work produced basic risk-analysis principles for all food derived from **biotechnology**. Most battles about GMOs have been fought over the first generation of biotech crops, mainly soybeans, corn and cotton. Since the first commercial planting in 1996, the area covered by biotech crops had reached about 53m. ha in 2001, an increase of 19% on the previous year. In 2001 genetically modified plants reportedly accounted for 46% of the global soybean crop, and 20% of the cotton crop. Many scientists urge the benefits of biotechnology in commercial crops, such as greater yields from existing acreages or the less insecticide necessary with pest-resistant cotton, but others fear contamination of wild species, a loss of biodiversity and sometimes the concentrated use of herbicides. GMOs are widely cultivated in North America, but not in the European Union (EU), which has created some tension between the world's two major trading blocs. In December 2003 the USA and Canada challenged at the World Trade Organization the EU's de facto moratorium on the import of GMOS since 1998 (introduced after widespread public concern about the environmental impact of the plants). The EU, which was about to admit some countries already growing GMOs, such as Poland, claimed that its stringent new rules (finalized in January 2004) were not an infringement of trade. Meanwhile, in February 2004 the USA had succeeded in persuading the People's Republic of China to allow the import of genetically modified crops.

GEOCHEMISTRY is the study of the chemistry of the earth, particularly its crust, waters and atmosphere. The chemistry of the land affects that of the oceans, because a number of chemical particles are present in both sea and **fresh water**. The particles in **seawater** are often the result of run-off from the land, while others come from volcanic activity.

A **GEOLOGICAL PERIOD** is the measure used as a geological timescale to divide the ages of the rock formations into epochs.

GEOTHERMAL ENERGY taps the vast reservoir of heat that lies beneath the earth's surface and is created by the energy welling up from beneath the earth's crust. Some of this energy can be captured directly, when it emerges at the surface in the form of **geysers** or hot springs, or is pumped from natural underground reservoirs of hot water. Other ways are being tested of extracting energy from hot rocks by drilling deep boreholes, pumping water down one and extracting it from another once it has been heated in the rock.

GEYSERS are hot springs that spout jets of steam and water. Geysers, a name derived from the Icelandic word for 'to rush forth', are produced when underground waters are heated by coming into contact with, or being very close to, molten volcanic rock. They are commonly associated with recent volcanic activity and often make a dramatic spectacle as they throw a column of steam and boiling water high above the surface of the ground. Some geysers discharge continuously, others may erupt violently for a few minutes and then remain inactive for hours or days. Sometimes a geyser, like Old Faithful in the **Yellowstone National Park** in the USA, has a regular and predictable pattern and erupts for a few minutes every hour. It is possible in some areas to harness and control the geyser steam to generate electricity, as with the geysers near San Francisco, USA. However, the disadvantage is that often the power station has to be built on a site near the geyser that is prone to damage either from volcanic activity or **earthquakes**.

GIANT PANDAS, despite legal protection, special reserves, harsh penalties for poaching and a well-funded conservation programme, continue to decline in number. It has been estimated that there are only about 1,000 in the wild, with another 100 in captivity. They have become an **endangered species** for three reasons. As the human population expands, more and more of the pandas' habitat has been cleared and they have been forced to live in isolated areas, cut off from each other and from other sources of bamboo, their only food. When one variety of bamboo dies off, the pandas die if they cannot find another kind to eat. Secondly, although the People's Republic of China has severe penalties for killing pandas and almost every country prohibits the sale of panda skins, poaching still goes on. Some pandas have also been killed accidentally in traps set for other animals. Thirdly, the panda does not breed very well in captivity. However, there were signs that some efforts were being rewarded, and by 2002 the Chinese Government had established 40 panda reserves, protecting over 1m. ha and about 45% of the total panda habitat. Moreover, a comprehensive survey in 2004 discovered that there were rather more pandas surviving in the wild than previously thought (nearly 1,600). The panda is the internationally recognized symbol of the **World Wide Fund for Nature**, an environmental conservation non-governmental organization.

GLACIERS are slow-moving masses of ice that accumulate on mountains or in polar regions. They are found where warm, moist air or warm water meets cold air or water. They move, influenced by the force of gravity and the pressure of the ice, above layers of slush and slide downhill, eventually melting at lower levels to form rivers or reaching sea level, where they form **ice shelves** or fall into the water as icebergs. Glacier ice covers about 10% of the earth's surface. At the height of the **Pleistocene** ice age it probably covered about 29% of the earth. As a result of global **climate change**, glaciers are in retreat throughout the world and are projected to decline dramatically during the 21st century. Many are predicted to disappear by 2020, according to a 2001 report, which reported that the ice-fields of Mt Kilimanjaro in Tanzania had retracted by 80% since 1912, while in South America all but two of the six glaciers Venezuela had in 1972 had disappeared and the Quelccaya glacier in Peru retreated 32 times faster in 1999–2000 than in 1963–83. Even in the great mountain ranges of central Eurasia a significant contraction of glaciers has been recorded. Geological exploration indicates the effect of the ice in shaping land forms, grinding rocks and moving **sediments** over large distances.

GLASOD (Global Assessment of Soil Degradation) was a three-year global study sponsored by the **UN Environment Programme (UNEP)** and co-ordinated by the International Soil Reference and Information Centre (ISRIC) in the Netherlands. More than 250 soil scientists and 21 regional co-ordinators were asked to estimate the degree, type and causes of soil degradation caused by human intervention since 1945. Parts of the survey were to be combined with information on population, climatology and vegetation loss in world desertification. GLASOD was also involved with a more detailed global soils project, the World Soils and Terrain Digital Database (SOTER), which was also co-ordinated by ISRIC for the International Society of Soil Science (ISSS). This project was due to be completed within 15–20 years. UNEP sponsored the GLASOD study because it is politically important to have an assessment of good quality now, rather than an assessment of very good quality in 15 or 20 years' time.

GLASSPHALT is glass recovered from the ash of waste incinerators and other furnaces. It provides an alternative to gravel as an aggregate and has been used in road construction materials in the USA.

GLOBAL 2000 was the study commissioned by US President Jimmy Carter in 1977, in order to gather information about what life would be like in the run-up to the year 2000. A report published in 1980, *The Global Report to the President*, concluded that by the beginning of the 21st century the world would be more crowded, more polluted, less ecologically stable and more vulnerable to destruction than in 1980, and that for hundreds of millions of the desperately poor, the outlook for food and other necessities would be no better, and for many it would be worse. It attracted a lot of support, but President Carter was defeated in the 1980 election by Ronald Reagan, who did not take the project further. Jimmy Carter subsequently started the Global 2000 project, an independent centre where steps are taken to act on implementing some of the report's recommendations.

GLOBAL ATMOSPHERIC RESEARCH PROJECT: See **GARP**

GLOBAL CIRCULATION is the key factor in maintaining the different climatic zones of the earth: the desert, temperate agricultural areas, forest, prairie or ice-field. Day-to-day climate changes in middle latitudes are often so great that it seems unlikely that there is an overall pattern to the weather. Yet there are large-scale wind circulations over most of the earth that are persistent and predictable, and they drive the weather machine. They form the general global circulation of the **atmosphere**. Although everyone is familiar with the life-giving sunshine and warmth from the sun, it is not so widely recognized that the earth and its atmosphere radiate heat back into space. If they did not the planet would become increasingly hot and barren. However, the incoming solar heating of the earth's surface is largely concentrated in the lower latitudes, while the outgoing radiation from the earth is more uniform. This imbalance is the underlying source of the major systems of winds and ocean currents, which take excess heat from the tropical regions to the poles (see **atmospheric circulation**, **ocean circulation**, **energy balance** and the **Hadley Cell**). The pattern of the circulating winds is influenced by the rotation of the earth, the distribution of the oceans, continents, mountain ranges, deserts, forests, snow and ice. In addition, there are fluctuations in the sun's energy output. All these things interact to make a vastly complicated engine pushing the atmosphere around the globe, as surface winds and high-level winds, or **jet streams**. Weather and climate forecasting depends on portraying these processes mathematically, creating so-called global circulation models, or descriptions of the **global climate system**, for making computer predictions.

GLOBAL CLIMATE CHANGE: See **climate change**

The **GLOBAL CLIMATE COALITION (GCC)** is a group of primarily US commercial and industrial interests, formed after President Bill Clinton's election in the USA in 1992, with the aim of representing the interests of business on economic and social issues arising from **climate change** and global warming. It worked to discredit the study of climate change, paying for large public relations campaigns that portray the science as being full of uncertainty. It also commissioned many economic studies that usually concluded that it would be prohibitively expensive to reduce **greenhouse gas** emissions. Its declared goal is to ensure that plans for reductions of greenhouse gas emissions in the USA are 'cost-effective'. The coalition supported President Clinton's declared policy on global warming in April 1993, calling 'not for more bureaucracy or regulation or unnecessary costs, but instead for American ingenuity and creativity to produce the best and most energy-efficient technology'. However, the coalition also made clear its opposition to action beyond the **Framework Convention on Climate Change** for reductions in emissions, which was signed at the UN conference in Rio de Janeiro in 1992, and its opposition to the **Kyoto Protocol**. In late 1999 and early 2000 several significant members of the coalition, including Ford Motor Company, British Petroleum and Texaco, withdrew their membership as a result of intense public pressure, citing concerns about the tactics of the group. However, with the election of George W. Bush to the presidency in 2000, the group was assured that the USA would not sign the Kyoto Protocol. Indeed, the Bush presidency was widely to be perceived to favour the energy industry and, in many cases, opposed to environmentalist interests.

GLOBAL CLIMATE OBSERVING SYSTEM: See **World Meteorological Organization**

The **GLOBAL CLIMATE SYSTEM** is the complete web of physical, chemical and biological processes that happen within the earth–atmosphere system in response to the energy received from the sun.

The **GLOBAL ENERGY BALANCE** is the overall balance between the heat energy received by the earth and atmosphere and that lost into space. The process starts with the amount of energy absorbed from the sun. The earth and the atmosphere absorb 68% of **insolation**, energy from the sun, and the rest is reflected back into space as the E-A (earth-atmosphere) **albedo**. However, there are numerous energy transfers between different layers of the atmosphere and the earth. (See also **energy balance** and **global circulation**.)

The **GLOBAL ENVIRONMENT FACILITY (GEF)** is the largest source of multinational funds in the form of grants for environmental protection. It is a programme of the **World Bank** and has four focal areas: **climate change**; **biological diversity**; protection of international waters; and **ozone** depletion. It was formed in 1990 by the **UN Environment Programme**, the **UN Development Programme** and the World Bank, and was subsequently restructured.

GLOBAL ENVIRONMENT FUNDS were first officially proposed by Gro Harlem Brundtland, the former Norwegian Prime Minister who chaired the commission responsible for the **Brundtland Report**. Such Funds would provide money for a global programme to reduce pollution and improve the environment. The concept led to the creation of the **Global Environment Facility**.

The **GLOBAL ENVIRONMENT MONITORING SYSTEM (GEMS)** is a world-wide collective effort to monitor the global environment and assess the health of its constituents. The GEMS Programme Activity Centre (PAC) works with UN bodies, governments and national and international institutions to monitor atmosphere and climate, environmental pollution, renewable resources and environmental data. It was established in 1975 by the **UN Environment Programme (UNEP)**. Data on climate gathered by GEMS has encouraged international action to save the ozone layer and to mitigate the effects of climate change. The World Glacier Monitoring Service, backed by UNEP, UNESCO and the Swiss Federal Institute of Technology, incorporates over 750 glacier stations in 21 countries. It has published the *World Glacier Inventory*, the first survey of all the world's glaciers and permanent ice-fields. GEMS also monitors environmental pollutants, including chemical pollutants in the air, water and even food, which pose a threat to human health.

GLOBAL ENVIRONMENTAL CHANGE refers to environmental changes that affect the total global system, and those that occur at local or regional levels but have consequences for the total earth system. It includes any alteration of the climate system as a result of the build-up of **greenhouse gases**, depletion of the stratospheric **ozone layer**, changing **sea levels**, loss of **biological diversity**, **desertification**, and depletion of **tropical rainforest**. It is accepted that human activities have caused, added to, or accelerated global environmental change. However, throughout history a process of continual change has been recorded. Current research, being coordinated by the Inter-Agency Committee on Global Environmental Change, and involving government agencies responsible for the global environment, is concentrating on the likely consequences of the present more rapid change on our weather, the natural environment, crops, and the way we live. By marrying past environmental change with the current acceleration in global environmental change, scientists are working out how best to predict, for example, the probable future extent of global warming (see **climate change**), and how quickly these changes will come about. The best ways to respond to these changes can then be worked out.

The **GLOBAL ENVIRONMENTAL RADIATION MONITORING NETWORK (GERMON)** was set up after the Chornobyl disaster, in order to strengthen national and international monitoring systems and to arrange for information to be exchanged between countries should such a disaster happen again. It was established by the **UN Environment Programme** and the World Health Organization, in co-operation with other UN bodies and national institutions. It provides information on radiation levels under normal conditions. If GERMON detects abnormally high levels it circulates reports and any other details of a nuclear accident to countries that are affected.

The **GLOBAL OCEAN OBSERVING SYSTEM (GOOS)** is a scheme which was established by the Intergovernmental Oceanographic Commission and the **World Meteorological Organization**. It is a permanent monitoring network, which feeds data about the oceans into research programmes on **climate change**.

GLOBAL RESOURCE INFORMATION DATABASE (GRID), which comes within the framework of the **Global Environment Monitoring System (GEMS)**, uses computer technology to process and analyse information in order to produce environmental maps and print-outs for planners. Based on environmental geographic information systems (GIS), these can be used in mainframe computers to study global issues or with microcomputers to study national or local ones. GRID compiles geographically-referenced environmental data compiled by other organizations, and supplies it to other GRID users. It also helps other countries and institutions to receive GIS, with related image-analysis technology, in order to develop a global environmental information exchange network. It operates through seven regional centres, each acting as a kind of environmental switchboard to send data to users in its region. GRID has helped to set up an environmental database for Uganda to assist development planning and to estimate the number of African **elephants**. It has identified suitable sites to develop aquaculture in Costa Rica and evaluated new methods to use satellite data to map and monitor the tropical forests.

GLOBAL STRATEGY FOR HEALTH AND ENVIRONMENT: See **Commission on Health and Environment**

GLOBAL WARMING: See **climate change**

GLOBAL WARMING POTENTIAL (GWP) refers to the relative effect of a substance that behaves as a **greenhouse gas** and, hence, its potential effect on the **climate**. The concept of GWPs has developed into an index that can be used to compare the overall climatic effect of a particular greenhouse gas with that of the same amount of **carbon dioxide**. For instance, carbon dioxide has a GWP of 1, whereas **methane** has a GWP of 63, **carbon tetrachloride** has a GWP of 470 and CFC-12, one of the **chlorofluorocarbons (CFCs)**, has a GWP of 7,100. Substances with higher GWPs are more-powerful greenhouse gases.

GLOSS is the acronym for Global Sea-Level Observing System, a world-wide network of stations that measures the level of the sea. There are more than 300 tide gauges around the world, set up either on and around **coastal areas**, or in the middle of the **oceans**. Average **sea level** figures are relayed monthly to the Permanent Service for Mean Sea Level at the Proudman Oceanographic Laboratory (POL) on Merseyside, in the United Kingdom. These ground measurements are combined with satellite images of the surface of the sea in order to produce a rapid global monitoring system.

The **GOLDEN LION TAMARIN** is a small monkey that lives in the **tropical rainforest** of Brazil, close to Rio de Janeiro. However, numbers have declined, along with those of many other species, as all but a tiny fraction of the forest has been cleared for agriculture and to allow for the expansion of urbanization north and south of Rio de Janeiro. There are now an estimated 500 golden lion tamarins left in the wild. The majority of these are in a reserve near Rio, but that is threatened by the expansion of cattle-farming on adjoining ranches and by squatters. A **captive breeding** programme has ensured that there are still enough tamarins to make a viable population.

GRASSLANDS cover nearly one-fifth of the earth's land surface. They include **savannah** in Africa, the prairies of North America, the Pampas of South America and the steppes of Russia and Central Asia. Grassland ecosystems support thousands of different species, above and below the ground, and have a vital part to play in maintaining the ecological balance of the world. Ruminants supported by grasslands—beef and dairy cattle, water buffalo, goats and sheep—supply most of the world's meat, milk and leather. As the world's population and the demand for food grows, the grasslands have come under environmental pressure from **overgrazing**, and because many grasslands are being converted into agricultural croplands.

A **GRAY** (Gy) is the standard unit for an absorbed dose of radiation, i.e. it measures how much energy radiation deposits in a body, enabling a comparison of biological effect. It was defined in 1940 by Louis Harold Gray, a British physicist, for whom it was named. His work on the effect of radiation on biological systems effectively founded the science of radiobiology.

The **GREAT BARRIER REEF** spreads for some 2,300 km off the north-east coast of Australia. It is the largest reef system in the world and consists of thousands of individual reefs, shoals and small islands. There are at least 350 species of coral, and living among them are anemones, worms, gastropods, lobsters, crayfish, prawns, crabs and numerous fish and birds. In the 1960s, when there were proposals to drill for petroleum and minerals in the reef, environmentalists brought the beauty and biological significance of the Great Barrier Reef to the attention of the Australian people. In 1975 the Great Barrier Reef Marine Park Act was passed, banning mining from the reef. The reef is now managed by a joint federal and state government authority and is zoned to allow various groups to visit the reef for commercial fishing, shell collecting, **tourism** and recreational and scientific activities. In 1981 it became a **World Heritage Site**. However, in common with other **coral reefs**, it remains in danger from pollution and other forms of human interference, and it is uncertain to what extent **climate change** will impact on it. The most immediate challenge has been protecting the reef system from the 2m. visitors annually, with the Government introducing measures to sustain both the industry and the reef from the late 1990s. In a further effort to protect this unique ecosystem the Australian Government introduced new legislation, which took effect in July 2004, banning fishing in one-third of the national park, making it the largest stretch of protected ocean in the world. This national park surrounding the Great Barrier Reef covers 334,000 sq km (roughly the size of Japan).

The **GREAT LAKES**, in North America, contain 18% of the world's fresh, liquid surface water and stretch across 1,200 km. Some 30m. people in the USA and Canada live in the Great Lakes basin. The lakes, Superior, Michigan, Huron, Erie and Ontario, were in danger of becoming dead lakes—especially Lake Erie—until strict pollution controls were brought in to remove sources of pollution, particularly sewage. There is still concern, however, about the level of toxic chemicals in the lakes and **acid rain** damage to the streams that feed the lakes. In 1985 the eight US states and two Canadian provinces that border the lakes signed the Great Lakes Charter. Signatories agreed to consult each other over any plans to divert water from the lakes. A year later the Toxic Substances Control Agreement was signed. This aims to reduce toxic discharges as much as possible. It was agreed that the Niagara river discharges into Lake Superior were to be cut by one-half by 1995. The Agreement also bans petroleum drilling in all the lakes. In 1987 the Great Lakes Water Quality Agreement was signed. In May 2004 the US Government created the Great Lakes Interagency Task Force to assist in the co-ordination of agencies concerned with the region.

GREEN ACCOUNTING takes into account the real value of natural resources, such as clean air and water, before they become degraded. For example, when a forest is cut down, a company counts timber sales as income, which later appears as growth in industry and national accounts, but no one, in strict accountancy terms, pays the environmental costs that might ensue, except those who suffer because of increased **soil erosion** and flooding, loss of **fuelwood** and wildlife habitats, damaged watersheds and increased global **climate change**. Green accounting calls for assessments by governments or individual companies of the existence and effectiveness of policies to prevent damage to the environment from their processes, products and services. The UN and other organizations are working to produce a manual for environmentally adjusted accounting. Proponents of green accounting argue that the gross domestic product (GDP) of a country should be adjusted for environmental change. An article published in *Nature* magazine in 1998 estimated the global value of ecosystem services at US $16,000,000m.–$54,000,000m. each year. (See also **polluter-must-pay**.)

GREEN CONSUMERISM is public awareness of environmental issues and the wish to replace hazardous products, such as conventional **detergents**, bleaches, leaded petrol (see **lead-free petrol**) and **insecticides** with more acceptable, or 'friendlier', alternatives. 'Green' consumers tend to buy goods that have minimum impact on the environment in terms of electricity use, **waste disposal** and use of scarce resources such as hardwoods from **tropical rainforests**, and avoid any product that threatens wildlife or health. Such consumers choose to buy, where possible, products that have been, or can be, recycled or reused (see **recycling (domestic)**), cosmetics and toiletries that have not been tested on animals, and energy-efficient household goods, such as long-life light bulbs. The 'green' consumer also avoids any product that contains **chlorofluorocarbons (CFCs)**, which harm the **ozone layer**, and selects foods that are organically grown.

GREEN PLANS are long-term strategies for taking account of environmentally sensitive issues in national development policies.

GREEN PLANTS have the unique ability to transform **solar energy** into food by **photosynthesis**, thereby providing the food that sustains all other forms of life.

GREEN POLITICS describes any political activity that is concerned with conservation and promotes the protection of the environment. It is thought that the first 'Green' party was the Values Party, launched in New Zealand in 1972. It did not win any seats in national elections, but in 1979 the West German ecology party (Die Grünen—the Greens) won seats in local elections and went on to win seats in the federal parliament. The movement grew through the 1980s and into the 1990s, particularly in the European Union. Green politicians were successful in pushing serious environmental issues into mainstream politics. However, their levels of representation were increasingly challenged by the main political parties realizing the significance of environmental issues and the number of voters that were concerned about them. Environmental protection and awareness began to appear in the manifestos of other parties, although in some countries Greens became part of the political mainstream, even entering government, albeit in coalitions. In the former Communist states too environmental movements had flourished from the 1980s, particularly after the **Chornobyl** disaster, but mainly in protest at the legacy of ruined environments left by the command economy. In the USA the Greens made powerful local challenges to the two established parties, and were credited with depriving the Democrats of a presidential election victory in 2000.

The **GREEN REVOLUTION** was the name given to the widespread development of high-yield strains of wheat, corn and rice during the 1960s and early 1970s. It was more formally known as the Indicative World Plan for Agricultural Development. The revolution came after the Food and Agriculture Organization held the World Food Congress in 1963. A Freedom from Hunger campaign was set up with the goal of increasing food supplies and solving the world's hunger problems. The hybrid cereals were short-stemmed, so they were not blown over by the wind, and were quick to mature, making it possible to grow three crops on the same land every year. Norman Borlaug, the US agricultural scientist and plant pathologist, developed the high-yield strains that greatly increased crop yields. He laid the groundwork for the green revolution and was awarded the Nobel Prize for Peace in 1970. In spite of its success in increasing yields, the programme came in for a great deal of criticism for a variety of reasons: in order to get the high yields, seeds were used with a high addition of **fertilizers** and **pesticides**, and they also needed a great deal of water, which often led to large **irrigation** schemes; storage of the high-yield strains was also a problem, and because of their high water content they were susceptible to moulds, such as **aflatoxins**; as a result of changing from traditional agriculture patterns, particularly in the developing world, there was a loss of genetic diversity in many areas, which was followed by pest infestations; the hybrids were also vulnerable to pests and diseases, which led to an increase in the use of pesticides, which has had an effect on those countries which depend on fish as a major source of animal protein.

GREEN TAX is a general term for the use of taxation to reduce **pollution**. More formally, it is a method of using **economic instruments** to increase the cost of more polluting activities, thereby encouraging industrialists and consumers to choose more environmentally friendly manufacturing processes, consumer products and habits. While the US Government tends to choose regulation to reduce pollution, the European Union countries tend to prefer green taxation. For example, whereas the USA used regulations to introduce **catalytic converters** on cars, most European countries use excise duty on vehicles to favour environmentally cleaner models. Taxation was used to create a differential between the price of leaded petrol and the less-polluting unleaded fuel.

The **GREEN TURTLE** is a large quadruped sea reptile that is in danger of **extinction**. It has been hunted extensively in the Caribbean Sea for meat and for its shell, which is used to make ornaments and jewellery. The female does not produce eggs every year, but when she does she always returns to her native beach to lay her eggs. Eggs are often removed from the beach because they are thought to have aphrodisiac qualities, and when eggs are allowed to hatch they are prey to shore birds. However, the increased use of the beaches for **tourism** and tourist facilities has now destroyed many of the turtle's nesting sites. This destruction of habitat led to dwindling populations of the green turtle throughout the 20th century and threatened the turtle's survival. One project to protect the turtle's future involved taking eggs from nests, which are scooped out of the sand by the mother turtle, and moving some of them to another beach where species of green turtles have been seen in the past. It is hoped that the hatched female turtles will subsequently return to lay their eggs at the new 'native' beach.

The **GREENHOUSE EFFECT** is a normal phenomenon that has been altered by human activities. Life as we know it on earth is possible owing to a natural greenhouse effect that keeps the planet 33°C (60°F) warmer than it would otherwise be. **Water vapour**, **carbon dioxide** and other trace gases such as **methane** and nitrous oxide trap solar heat and slow its loss by re-radiation into space, making the earth warm enough for plants and animals to survive. As a result of industrialization and population growth, greenhouse gas emissions from human activities have consistently increased. These steady additions have begun to upset a delicate balance, significantly increasing the amount of greenhouse gases in the atmosphere and enhancing their insulating effect. A wide variety of activities contribute to greenhouse gas emissions. Burning of **coal**, petroleum and **natural gas** releases about 6,000m. metric tons of carbon into the atmosphere annually. Burning and logging of forests contributes another 1,000m.–2,000m. tons each year. In addition, the use of certain chemicals releases a substantial amount of greenhouse gases such as hydrofluorocarbons (HFCs), sulphur hexafluoride and perfluorocarbons (PFCs). The atmospheric level of carbon dioxide, the most abundant greenhouse gas, increased by 30% between 1860 and the end of the 20th century. Many greenhouse gases will stay in the atmosphere for periods of time ranging from decades to centuries. If nothing is done to reduce these concentrations of greenhouse gases in the atmosphere, they are expected to double by 2050, compared to pre-industrial levels. This is one of the principal motors of **climate change**.

GREENHOUSE GASES are so-called because of their heat-absorbing properties, which vary enormously. While most attention is focused on **carbon dioxide**, other gases are far more potent even if present in far smaller quantities: **methane** is many times more potent than carbon dioxide, as are **chlorofluorocarbons (CFCs)** and nitrous oxide, owing to the higher **global warming potentials (GWP)** of these gases. The other main human-generated greenhouse gases are sulphur hexafluoride, the hydrofluorocarbons (HFCs) and the perfluorocarbons (PFCs). **Water vapour** is the most abundant of all greenhouse gases, but is naturally present, on the whole. Between the start of the industrial revolution in the 18th century and the end of the 20th century, the total amount of greenhouse gases in the atmosphere increased with an effect equivalent to an increase of 50% in carbon dioxide alone.

GREENPEACE, one of the first of the environmental pressure groups, was founded in 1971, dedicated to campaigning against abuse of the natural world by non-violent direct action protests. All of its protests are backed by scientific research. Greenpeace began when a group of US citizens and Canadians chartered a ship and went to Amchitka Island, Alaska, where a series of nuclear tests was planned. As a result of their presence only one test was carried out. The island subsequently became a bird sanctuary. Greenpeace continued in this active style, protesting about the slaughter of **seals**, against whaling (see **whales**) and the dumping of waste at sea. Their methods, which often put campaigners in great personal danger, have made headlines around the world, bringing remote issues directly to the public's attention. Its action against the slaughter of seals contributed to a European Communities ban on imports of seal pelts and products while the number of seals slaughtered dropped to one-sixth of its previous level. In 1987 it raised its World Park Antarctica flag in an attempt to prevent exploitation of the continent. Greenpeace also carried out successful campaigns to prevent the dumping of radioactive and chemical waste, and campaigned against French nuclear tests in the Pacific, dumping of used oil rigs in the deep sea and on many other issues.

GROUND-WATER is perhaps the world's most valuable resource. It represents 97% of total fresh water reserves, excluding that stored in the polar ice-caps. Most countries draw heavily on ground-water for daily supplies: in the USA over 50% of the population relies on ground-water for domestic water supplies and in the Russian Federation the figure is almost 70%. In the European Union the proportion of the population varies from 30% to almost 60%. Ground-water accumulates below the earth's surface and is a valuable part of the **hydrological cycle**. The traditional benefits of ground-water were its cheapness to develop, and the fact that it was good quality water that needed minimal treatment, but that is no longer necessarily the case. Ground-water has two sources: mineral springs rising from deep underground, and rain. **Artesian wells** are sometimes used to bring ground-water to the surface. In many parts of the world a large proportion of ground-water is brought to the surface for domestic, industrial and agricultural use, including **irrigation** in arid areas. Much ground-water is stored in **aquifers**. There was a significant increase in the use of ground-water in the late 20th century, and water supplies throughout the world are being depleted and polluted at a disturbing rate. Significant lowerings of the level of the water table have a serious effect on the environment. A lowering in coastal areas can lead to the incursion of seawater into ground-water and, inland, it can cause subsidence (as, for example, in Mexico City, increasingly in the early 2000s). Ground-water resources throughout the world are also polluted by a variety of sources associated with urban and industrial development, and modern agricultural practices. A particular concern is that ground-water is becoming polluted by effluent in rivers and leaks from **landfill** sites, waste dumps and nuclear waste sites.

GUANO is a deposit of droppings of fish-eating birds, such as the brown pelicans of Peru. On many islands and coasts of the world, guano deposits have built up to considerable depths. Guano is rich in **nitrogen** and **phosphorous** and in many areas it is mined and used in the manufacture of **fertilizer**. On the Pacific island of Nauru extensive mining of guano deposits has rendered 80% of the island uninhabitable. The guano of bats is also prized as a fertilizer.

H

The **HABER PROCESS** is a chemical manufacturing process on which the modern **agrochemical** nitrogen-based fertilizer industry was founded. The Haber process makes **ammonia** by taking **nitrogen** from the air and combining it with hydrogen (usually from **methane** in natural gas) under pressure. Prior to its invention by the German chemist, Fritz Haber, in 1908, there was no industrially useful way of extracting atmospheric nitrogen as a raw material, even though the air around us is composed of 78% nitrogen. The best commercial sources of nitrogen at the beginning of the century were **nitrate** deposits in the Atacama **desert** of northern Chile. Production of ammonia by the Haber process provided a molecule containing nitrogen in a suitable form for making fertilizer. All plant life needs nitrogen, which is necessary to build up **amino acids** and **proteins**. Animal life also requires proteins, but acquires them by eating plants. Although there is plenty of nitrogen in the atmosphere, plants and animals have no means of making use of gaseous nitrogen, or fixing it to form compounds, because it is chemically inert. Plant life obtains nitrogen from nitrates in the soil. However, as nitrates are inorganic salts that are soluble in water, they are liable to be washed rapidly from the soil in the form of **leachates** after rainfall. However, nitrogen-fixing bacteria are capable of converting atmospheric nitrogen into ammonia, either alone or co-operatively with leguminous plants. Once ammonia is formed, it is converted into nitrates by nitrifying bacteria. Without such activity, life on earth would be impossible. However, the natural rate of conversion of atmospheric nitrogen to nitrates by bacteria in the soil was not enough to support the level of crop-growing farmers implemented under the **intensive agriculture** schemes of the **green revolution** that were started in the 1950s and then encouraged by the European Union's **Common Agricultural Policy**. Industrial fixation of nitrogen has grown at such a pace that there is now concern about whether another natural bacterial process, denitrification or reconversion of surplus nitrates back to nitrogen, can keep up. The consequence has been the over-accumulation of nitrates, which have leached into lakes and rivers causing **eutrophication**, and increasingly into the **ocean**.

A **HABITAT** is the geographical locality and physical and biological context in which a plant or animal naturally grows or lives. It can be either the geographical area over which it extends, or the particular station in which a specimen is found. It is the term mainly used to indicate the type of locality, such as the sea, the mountains, forest, etc.

HABITAT II was the second UN conference to concentrate on problems in the urban environment. It was held in Turkey in June 1996 and focused on the public health hazard of **lead** in petrol, an issue that the European Union, Canada and the USA dealt with aggressively. Children exposed to lead at an early age are at risk of suffering from various conditions, including learning difficulties, hearing loss, reduced attention span and behavioural abnormalities. Habitat II compiled a list of the major developing cities around the world where millions of people are exposed to high lead levels. As a result, the **World Bank** offered support for a world-wide programme to phase out leaded petrol. The Bank was prepared to give countries economic help to make lead-free petrol one of their highest environmental priorities. It lent Thailand US $90m. for its programme of refinery conversion and of public education to end the use of leaded petrol, which was blamed for 90% of airborne lead pollution in congested cities such as Bangkok. Thailand's four-year programme of withdrawal of leaded petrol, concluded in 1996, is claimed to have increased the average intelligence quotient (IQ) of children in Bangkok by four points. Cities listed at the Habitat II conference as still being exposed to high lead levels include: Algiers, Algeria; Cairo, Egypt; Cape Town, South Africa; Hong Kong; Jakarta, Indonesia; Jeddah, Saudi Arabia; Karachi, Pakistan; Lagos, Nigeria; Lima, Peru; Mexico City; and Nairobi, Kenya. A conference known as Istanbul Plus 5 was held in 2001 to review progress since HABITAT II, but the proposed HABITAT III conference was not scheduled until 2016.

The **HABITATS DIRECTIVE** was adopted by the European Communities in May 1992, about 18 months before the organization came to be called the European Union (EU). The directive establishes a legal framework for the conservation and preservation of Europe's most threatened species. It commits members of the EU to maintain wildlife and habitats with a favourable conservation status. The Habitats Directive requires nations to set aside about 10% of their land and sea for conservation. Besides nominating important habitats for protection, governments must also protect sites containing named species, including, at sea, **dolphins**, **otters** and common and grey **seals**. The all-EU protected area system of Special Areas of Conservation under the Habitats Directive and of

Special Protected Areas under the Birds Directive were together known as **Natura 2000**.

The **HADLEY CELL** is a system of **atmospheric circulation** that ultimately distributes air from the tropics to the poles. It is sustained by large-scale convection currents in which hot air is replaced by cooled air. The process was first described by George Hadley, in 1735, to explain the trade winds. (See also **global circulation**.)

HAGUE DECLARATION: See **Declaration of the Hague**

HALF-LIFE is the period of time it takes one-half of the number of atoms of a radionuclide to disintegrate. Over each successive half-life, one-half of the atoms remaining decay, so that after two half-lives one-quarter of the original amount is left. The half-life of radionuclides varies enormously from microseconds to millions of years. The concept can also be applied to the amount of time that other non-radioactive natural and artificial substances remain in the environment.

HALOGENATED COMPOUNDS are the vast family of carbon substances containing **halogens** (fluorine, **chlorine, bromine** or **iodine**). They include **chlorofluorocarbons**, **DDT** and a wide range of other organochlorine **pesticides** and **herbicides**, as well as the **polychlorinated biphenyl** stabilizers that were added to plastics and lubricating oils from the 1930s.

HALOGENS are a group of naturally occurring elements which include fluorine, chlorine, **bromine** and **iodine**. They are important in much synthetic carbon-based industrial chemistry, and chemical substances containing them often cause lasting damage when released into the environment because of their specific chemical reactivity with other molecules in the atmosphere or with tissues if they are ingested by animals.

HARMFUL ALGAL BLOOMS (HABs): See **algal blooms**

HARVESTING relates to agriculture and forestry. A crop is harvested after it has fully grown and is ready to be sold or traded. In the context of forestry, harvesting is the logging of trees for commercial purposes. It is considered to be a different activity from **deforestation** since the cutting of the trees is managed and new trees are usually grown to create a future crop for commercial loggers.

HAY FEVER, or allergic rhinitis, causes irritation of the eyes, ears, nose and palate. It is an allergy triggered, usually in the spring and summer, by airborne pollen from trees, flowers, grasses and weeds. Allergic rhinitis can also be caused by any inhaled allergen, including house dust, feathers, fungus spores and animal dandruff (see also **indoor allergens**). There is now growing medical concern about the role of **air pollution**, which is thought to be aggravating the symptoms of allergic respiratory diseases, such as hay fever.

The **HAZARDOUS AND NOXIOUS SUBSTANCES (HNS) CONVENTION** is an international compensation scheme for victims of maritime disasters. It was agreed in 1995 by members of the **International Maritime Organization**. The Convention demands compensation against maritime incidents that involve loss or damage as a result of **fire**, explosion or **pollution**, including **petroleum spills** and the release of noxious substances. It was an important step towards the introduction of a compensation fund, improvement of the safety of life at sea and the prevention of marine pollution. The fund would be a two-tier system under which compensation would first be paid by ship owners, who would be required to have insurance to cover their liability for damage. Further compensation would be available, if needed, from an international compensation fund financed by contributions from receivers of cargoes of hazardous and noxious substances. The United Kingdom was the first country to sign the Convention, following two major petroleum tanker accidents off the British coast in the space of three years, the first of which was the **Braer**. In 1996, after the **Sea Empress** ran aground off the Welsh coast, compensation to local businessmen and fishermen was expected to be £76m., most of which was to be paid by the British Government from a fund financed by a tax on petroleum importers.

The **HAZARDOUS WASTE REDUCTION AND COMBUSTION STRATEGY** was a major programme announced by the US **Environmental Protection Agency (EPA)** in the spring of 1993, which was designed to protect public health and the environment by reducing the amount of hazardous waste produced in the country and by strengthening federal controls governing hazardous waste incinerators and industrial furnaces. It included the first-ever hazardous waste reduction guidelines the EPA had ever published. The purpose was to change the national approach to hazardous waste management. A temporary freeze was introduced on the expansion of waste handling capacity in order to examine the country's waste strategy. The strategy's goals were: to overhaul the federal rules governing waste **incineration**; to introduce immediate use of full risk assessments, including those for indirect exposure, in the scientific evidence backing applications for permits; to specify tougher **dioxin** emission standards and more stringent control for metals with new **waste disposal** permits; and to conduct a national review of the relative roles of waste combustion and waste reduction in hazardous waste management. Under the federal hazardous waste law, the Resource Conservation and Recovery Act (RCRA), hazardous waste producers had been required since 1984 to have waste reduction programmes in place, but no specific guidance had been issued to define those programmes. According to the EPA, nearly 5m. metric tons of hazardous waste were burned each year, in 184 incinerators and 171 industrial furnaces, including 34 cement kilns. About one-half of the 5m. tons are burned each year at 15 large commercial incinerators and the 34 cement kilns, which also take commercial waste. The remaining incinerators and industrial furnaces are 'on-site' facilities, permitted for non-commercial use only.

HAZDAT is an abbreviation of the Hazardous Substance Release and Health Effects Database of the US Agency for Toxic Substance and Disease Registry (ATSDR). HazDat is a scientific and administrative database developed to provide access to information on the release of hazardous substances from Superfund sites and from emergency events, and on the effects of hazardous substances on the health of the population. HazDat information includes site characteristics, activities and site events, contaminants found, contaminant media and maximum concentration levels, impact on population, community health concerns, ATSDR public health threat categorization, ATSDR recommendations, the environmental fate of hazardous substances, exposure routes and physical hazards at the site or event.

HEALTH FOR ALL 2000 was a project of the **World Health Organization (WHO)** to raise the standard of health for the world's population by 2000. After WHO was established in 1948 there was a marked general improvement in health, and life expectancy continued to increase. However, in 1990 it was estimated that some 1,000m. people in the world were diseased, in poor health or malnourished. In developing countries, millions suffer from a range of diseases, from **malaria**, **tuberculosis (TB)** and respiratory infections, to **schistosomiasis**, onchocerciasis and leishmaniasis. Despite the formation of WHO, other diseases significantly increased, such as heart disease from unhealthy diets, smoking-related illnesses and AIDS. One of the most important factors, and a focus both of the **World Summit on Sustainable Development** in 2002 and WHO policy, is access to **fresh water**, for drinking and for sanitation.

HEAT ISLANDS describe the sprawling urban and industrial areas where air and surface temperatures are higher than in remote places because of the amount of surplus heat energy discharged from buildings and vehicles. Although they are called islands, these areas may cover hundreds of square kilometres in industrialized countries. These regions also include some of the major meteorological monitoring stations from which data is collected to look for signs of global **climate change**. Some experts believe that heat islands distort the overall picture of global warming, although adjustments are made to temperature measurements made at such stations.

HEAVY METAL POISONING refers to the health hazard represented by a number of **heavy metals** that are similar in the effects they cause at low levels of contamination. The main ones are antimony, **arsenic**, **lead**, **mercury** and **thallium**, which all interfere with one of the normal enzyme activities of the body. The group also includes **cadmium**, which damages the joints and kidneys.

HEAVY METALS are a group of metals that have no useful biological function if absorbed into the body. However, when ingested even in small amounts from food, air or water, they may cause irreversible damage to the brain and the central nervous system, or to the kidneys and liver. They include **mercury**, **lead** and **cadmium**. One of the most important natural safeguards against this form of **heavy metal poisoning** is usually provided by the **blood–brain barrier**. It is a remarkable filtering system evolved by the body to protect its most vital organ, allowing the blood to carry essential nutrients to the brain while blocking harmful molecules. Unfortunately, when heavy metals are transformed into certain complex compounds, like the tetra-ethyl lead that is added to petrol to increase its octane number, the metal can cross the blood-brain barrier. Conversion of metals into the more toxic forms can occur in the environment or in the body. Metal pollution first came to prominence in the **Minamata Bay** incident in Japan in the 1950s. Since the 1930s the Chisso chemical plant had been discharging methyl mercury into the bay. Instead of dispersing to sea as expected, it slowly accumulated in the **food chain**, until the fish on which the local population depended largely for food were contaminated.

HEAVY WATER is dideuterium oxide (D_2O or ^2H_2O), which is chemically the same as ordinary or light water (H_2O), but the hydrogen atoms consist of the heavy isotope, deuterium, in which the nucleus contains a neutron as well as the proton normally found in a hydrogen atom. Heavy water is used in some types of nuclear reactor as a neutron moderator, allowing a reaction with uranium. Allowing the use of natural uranium means that heavy water is subject to official regulation in many countries, owing to its implication in nuclear proliferation (heavy-water reactors have been used for weapons programmes in India, Israel, the Democratic People's Republic of Korea (North Korea) and Pakistan).

HEDGEROWS have an important role in the environment, providing protection against erosion by wind or rain and giving shelter to livestock and crops. They control the drainage of water from fields and provide a valuable habitat for wildlife. Some insects and fungi found in hedgerows harm crops but, on the whole, hedgerows house wildlife that plays a valuable part in the environment, such as the bumblebee, which pollinates flowering crops, and birds of prey and insect predators, which keep farm **pests** under control. In the post-Second World War period thousands of kilometres of hedgerows disappeared when fields were cleared for more extensive farming or because there was a lack of long-term management. In the United Kingdom alone, government research showed that between 1984 and 1990 more than 85,000 km of hedgerows were lost. There is evidence that this trend was halted in the 1990s, when new plantings exceeded the removal of hedgerows. In 1996 the British Government drafted proposals for the protection of hedgerows which are of significant historic, wildlife or landscape value.

The **HELSINKI CONVENTION** is named for the Finnish capital, where the Convention on the Protection of the Marine Environment of the Baltic Sea Area was first signed. The Convention was adopted in 1974 by the then seven states with a **Baltic** littoral and was concerned with environmental protection issues in the marine zones. There was a major revision in 1992, and the new Helsinki Convention came into effect in 2000, now with 10 signatories (nine countries and the European Union) with a wider remit. The Baltic Marine Environment Protection Commission, the so-called Helsinki Commission (HELCOM), was established to enforce the protection requirements of the Convention.

HEPATITIS, inflammation of the liver, is a major cause of human ill health and death. Hepatitis A is caused by a virus that is usually contracted by consuming food or drink (especially water, milk and shellfish) that is contaminated with faeces. However, it is also an **environmental health** problem associated with poor sanitation, particularly as a result of overcrowding in human settlements, which frequently leads to epidemics, especially among the young. Adults who have not previously been exposed to the virus are particularly at risk when they travel in, or move to, an area of high infection where there is poor sanitation. There is evidence that the Hepatitis A virus is transmitted not only from man to primates, but also from infected primates to animal handlers. Hepatitis B is transmitted in blood and other body fluids. Hypodermic needles, dental and surgical instruments, tattoo needles and razors have been implicated in the transmission of infection. Additional forms of hepatitis virus and disease are designated as C and D.

HERBICIDES are chemicals used to kill weeds and other undesirable plants. They have played a part in the increase of agricultural productivity, allowing a greatly improved quality and quantity of food. They are also used on industrial sites, roadsides, ditches, irrigation canals, railway embankments and power lines to remove plants that might cause damage or a fire risk. Some, known as contact herbicides, destroy only the plant tissue they touch. Others, translocated herbicides, are absorbed by the leaves and carried within the plant so that it dies. Herbicides are not usually acutely toxic to humans and other mammals, although some, such as **paraquat**, dinoseb and **DNOC**, are poisonous.

HETEROTROPHIC RESPIRATION is the release of **carbon dioxide** by the decomposition of organic matter such as **biomass** or trees.

HIGH-LEVEL WASTE refers to the **fission products** that contain relatively large quantities of **alpha-emitters** and have a long **half-life**, and are left behind after treating spent nuclear fuel. Owing to the intensity of its radioactivity, high-level waste also generates a lot of heat. It is often left in either liquid form in water-cooled tanks or cast into glass ingots, in order to cool for 50–100 years before disposal into some form of underground repository.

The **HIMALAYAS** form the highest mountain range in the world, generally characterized by snowy peaks and **glaciers**. However, the Himalayas also contain rich arable land, extensive grasslands and forest, as well as mineral deposits. The mountains act as a climatic divide and influence air and water circulation. Owing to its height the range prevents the cold continental air from the north getting to India in winter; to the south-west, the rain-bearing monsoon winds give up most of their moisture before crossing the mountains, so there is extensive desert in the rain shadow to the north. The Himalayas are prone to erosion. A rapidly increasing population in the Himalayas, **deforestation**, and unsustainable farming on the mountains' steep slopes mean that the land cannot absorb water. Part of the major flood problems of **Bangladesh** is a direct effect of water flooding from the Himalayas. Moreover, a report at the beginning of the 2000s cited evidence for global warming (see **climate change**) even among the high mountains of the Himalayas and its flanking ranges, with losses of **glacier** ice ranging between one-fifth and two-fifths in places.

HINKLEY POINT, on the Somerset coast in the United Kingdom, is the site of both a **Magnox** and an **advanced gas-cooled reactor** nuclear power station. In 1989 a public inquiry was held into a planning application for a third station, a US-type **pressurized water reactor**, but it was not built. In most of Europe, except France, the **Chornobyl** accident had ensured the loss of public support for a nuclear power programme.

HOLOCENE EPOCH: See **Quaternary**

HOMEOSTASIS is the maintenance of conditions within a living system. One example is body temperature in humans, which has to be maintained at about 37°C (98°F). Homeostasis is not a static, unchanging condition but rather a dynamic state where internal conditions are constantly changing to keep the overall environment constant. Homeostasis is maintained by positive and negative feedbacks (see **feedback cycles**) from external conditions.

HORMONE-LIKE POLLUTANTS and, more informally, sex-change chemicals, are terms applied to a wide range of man-made compounds in the environment. As they are similar to the natural family of oestrogen hormones produced by the body, they can interfere with human reproduction. They are the subject of one of the most controversial debates about environmental pollution. The agents are used in a wide range of activities, including agriculture and manufacturing industries. The synthetic substances in question include well-known pollutants such as **DDT**, the plastics ingredient D"X-A and some of the **polychlorinated biphenyls**, which are suspected of contributing to such problems as breast cancer, a reduction in sperm count among men in Europe and North America, and an increase in testicular cancer. Later suspects include such chemicals as nonylphenol and phthalates, which are believed to be present in effluent from European factories in quantities that are in breach of environmental safety limits. The Scottish Environment Protection Agency in the United Kingdom estimated that the amounts of nonylphenol and phthalates present in 11 rivers in southern and central Scotland could be as much as 10 times the recommended limits. Researchers have linked accidental releases of hormone-like chemicals with reproductive abnormalities in wildlife. When discharged from factories into rivers, the substances in question have been known to cause fish to change sex. The debate over pollution control centres on whether such substances are capable of harming humans when present in the environment at very low levels. Discussions have been held between environmental authorities and industry about the possibility of substituting the compounds with safer alternatives that, in many cases, are already available. World-wide this issue was reviewed extensively in the late 1990s and early 2000s, and increasing regulation was imposed. (See also **endocrine disruptors** and **persistent organic pollutants (POPs)**.)

The **HUMAN ENVIRONMENT** refers to any of the conditions in which people live. However, human environment has come to mean all human settlements in villages, towns or major cities, which require environmental management to provide water and public spaces, remove public wastes, and protect water and **air quality**. At the start of the 20th century 14% of the world's population were urban dwellers. By 1960 this figure had increased to 34%, by 1985 to 42% and by 2000 to 47%. This figure was expected to reach 60% by 2030. While urbanization has been the most significant aspect of changes in the distribution of the global population, the statistics can obscure the tremendous diversity in the patterns of human settlement.

HUMUS is the decomposed organic matter that gives **soil** both its dark brown colour and its nutrient richness. High quality soils rich in humus have often taken centuries to build up, only to be depleted of their goodness in a few years by **intensive agriculture**.

HYBRID VEHICLES or hybrid-electric vehicles are essentially electrically driven cars, buses and delivery vans that are, however, not completely dependent on **batteries**. Such vehicles use a gas turbine or small internal combustion engine to recharge their batteries, a design which increases the fuel efficiency of vehicles by a factor of two or three. In the early 2000s government regulation, tax incentives and celebrity endorsement had begun to make them more popular in California, in the USA, in particular. In 2004 four models were available commercially, and at least 10 other models were due to be introduced to the market within the next four years.

A **HYDROCARBON** is a gaseous, liquid or solid organic compound consisting of **carbon** and **hydrogen**. Hydrocarbons are the most important constituents of the **fossil fuels coal**, petroleum and **natural gas**.

HYDROELECTRIC POWER describes the production of electricity by generators driven by water. The generators can be contained in barrages built across rivers or in the wall of a **dam** constructed specifically to provide hydroelectric power. In some cases, dams are constructed for the combined purpose of regulating river flows for agricultural development, reducing flooding and providing energy. However, such schemes were increasingly acknowledged to have their own environmental costs, and the larger projects certainly were viewed with less favour by the beginning of the 21st century (at which time hydropower accounted for 17% of the world's generated electricity).

HYDROGEN is a major component of water and **hydrocarbons** which, on initial consideration, seems an ideal source of alternative or **renewable energy**. Its energy density—the amount of energy released when it is burned—is three times that of the equivalent weight of petrol. It has no carbon, so the only chemical by-product of combustion is water. It is renewable because it can be extracted from water or plant material. Experimental hydrogen-fuelled cars are in existence. The disadvantage of hydrogen is storage: at normal temperatures it is a gas and must be stored under pressure in heavy cylinders. Scientists are experimenting with methods of storing hydrogen in devices which can absorb it like a sponge.

HYDROGRAPHIC AND CHEMICALS TRACER STUDY is the name of a research project that is part of the **World Ocean Circulation Experiment**. The research is designed to show exactly how the vast quantity of energy absorbed by the oceans from the sun regulates the world's climate (see **ocean circulation**). The hydrographic part of the study involves cruise ships and satellites to monitor hundreds of different types of instruments, some radio-controlled and some sonar devices, which record the pattern of currents flowing across the oceans and their temperatures. In practice, these ocean currents are vast conveyor belts of energy. The hydrographic experts are employing the latest computer and microchip technology for studies that the pioneers, 100 years ago, did by dropping plumb lines and towing buoys behind sailing ships. However, the oceanographers are also using a new technique of trace analysis. They are measuring the chemical constituents of **seawater** because they contain telltale markers of the recent source of the water. **Oxygen** is a standard indicator together with the nutrients silicate, **phosphate** and **nitrate**. For example, the cold water recently discharged from polar waters into the southern oceans is high in oxygen and nutrients, particularly silicates. In addition, the man-made **chlorofluorocarbon** pollutant gases also provide chemical tracers. As it is known when they must have entered the ocean from the atmosphere, scientists can calculate the flow of some ocean currents that carried them to the places where they were detected. These measurements are particularly important in calculating the time it takes, in the mixing process, for surface water to get carried to the ocean bed. Oceanographers can 'age' a particular parcel of water and work out its recent history.

HYDROLOGICAL CYCLE is the way that water transfers from its gaseous, liquid and solid states from the sea to land and back again. Water evaporates from the **oceans** into the **atmosphere** and is carried by the wind to the land where it condenses and falls as rain, snow, sleet, hail or dew. Some evaporates as it falls and returns to the oceans by winds moving in the opposite direction. Water that reaches the ground goes back to the sea when it evaporates from the land, or as **water vapour** from plants and vegetation, or as the run-off from rivers, or **groundwater** movements.

HYDROLOGY is the study of the occurrence, properties, movement and use of water and ice on or under the earth's surface. It follows the **hydrological cycle** from precipitation, as rainfall and snow, to the discharge of water by rivers into the sea, or its return by evaporation back into the **atmosphere**.

HYDROPONICS is the method of cultivating plants by growing them without soil, using a **nutrient** solution. The plants' roots are held in gravel or sand. Crops can be grown hydroponically in arid regions, where soil is infertile or toxic. Large-scale cultivation is a very expensive operation, so its use is restricted. It is a method, though, that can be used on a small scale, at low cost, in areas of the world where there are frequent water shortages, including cities. Hydroponics is being used in research experiments by scientists from the National Aeronautics and Space Administration (NASA) in Florida, USA, to prepare for a future colonization of Mars, when space crews will have to grow their own food and recycle air and water in order to survive. Crops, such as potatoes, wheat, soyabean and strawberries, are grown in an airtight chamber. The experiment will help scientists to understand how plants would survive in a sealed environment on Mars. The Biomass Production Chamber is designed to provide food for future pioneering astronauts and also to monitor the changes in temperature, humidity, light, carbon dioxide and oxygen. Scientists are also looking at how plants cope with increased carbon dioxide, which contributes to the **greenhouse effect**. Another experiment, **Biosphere 2**, in the Arizona desert in the USA, has been studying miniature **tropical rainforest**, **savannah** and **coral reefs** in a large airtight greenhouse.

HYDROSPHERE refers to all the water on the planet, whether it is in the oceans, lakes, rivers or frozen in glaciers and the ice sheets. It includes the **water vapour** in the atmosphere. The movement of moisture between the hydrosphere, the **atmosphere**, and the **lithosphere** is the **hydrological cycle**.

I

ICE SHEETS form over the polar landmasses. The two great ice sheets, Greenland and **Antarctica**, between them hold more than 90% of the earth's **fresh water**. The thickness of the ice sheets continually changes as ice is redistributed within the sheet. Falling snow forms ice on the central plateau of the ice sheets. This is then balanced by loss at the coastal region in the formation of **ice shelves** and the calving of icebergs. This can take many thousands of years. Monitoring with radar altimeter data is crucially important in providing very precise and repeated measures of the topography of the ice sheets. This data will provide answers to a fundamental question, which is still unanswered after decades of fieldwork: are the ice sheets shrinking or growing? If they are shrinking this could be one of the most sensitive indicators of global **climate change**. More than 2m. measurements made with the Geosat altimeter over the Antarctic ice sheet show surface flow features and reflect the underlying topography of the Antarctic landmass. Data from **Earth Resources Technology Satellites**, which have a wider range, will provide a greatly improved coverage of Antarctica. The increasing rate of ice calving off the main ice sheets and shelves since the last decades of the 1900s has heightened fears global warming is having an irreversible effect on the ice sheets, particularly on the more vulnerable Greenland and Western Antarctica ones, as well as the Arctic **sea ice**, which is reckoned to have thinned by about 40% between the 1960s and the end of the 20th century.

ICE SHELVES are formed over the open seas, as opposed to **ice sheets** which form over polar landmasses. When ice sheets and **glaciers** reach the edge of the land, ice shelves extending hundreds of kilometres into the ocean may be formed. Much of the northern polar ice cap is **sea ice**, effectively a massive ice shelf attached to the Greenland ice sheet. Ice shelves are subject to dramatic changes. For example, the Wordie ice shelf, off the western side of the Antarctic peninsula, over 10–15 years, rapidly disintegrated because of a rise in mean air temperature. The increased amounts of melt-water, combined with underlying fracture processes, led to the shelf breaking up. Even more catastrophic, in 1986, at the edge of the much larger Larsen ice shelf, which is at the most northerly edge of the Antarctic peninsula, a large promontory, weighing some 1,000,000m. metric tons, became detached as an enormous iceberg. The iceberg, thereafter slowly melting, equalled the total annual loss of ice from Antarctica during a normal summer. The rate of calving continued to increase, especially from the vulnerable Larsen and, on the other side of the peninsula, Wilkins **ice shelves**. The sector known as the Larsen A ice shelf lost 1,300 sq km in 1995 and Larsen B and Wilkins together lost over 3,000 sq km in 1998. However, what would alarm scientists more is if the massive Ross and Filchner-Ronne ice shelves were affected (these effectively barricade the Western Antarctic ice sheet and its five great ice streams from entering the ocean directly, which, if it happened, could accelerate the rate of melting and possibly disrupt the great currents of the **ocean circulation** system). Indeed, in 2000 the so-called B15 iceberg broke off the Ross ice shelf. At some 11,000 sq km it was the largest iceberg hitherto (it, in turn, broke into numerous smaller bergs in 2003), but in 2002 the largest overall ice event ever recorded was the breaking off and disintegration of the entire Larsen B (3,250 sq km in extent and 200 m thick) ice shelf in a period of about one month. There is evidence that previous meltings from the Western Antarctic ice sheet have had a dramatic impact on sea levels and global climate.

INCINERATION of hazardous wastes in an incinerator provides, in theory, an alternative option for disposal to burial in the ground or dumping at sea. In practice, however, many substances undergo transformations into more harmful agents, such as **dioxin**, and are discharged into the air if the furnace is not controlled properly. Incinerators are usually operated by local authorities and private waste contractors where land area for waste disposal is limited, as in Japan and Europe.

INDEPENDENT COMMISSION ON INTERNATIONAL DEVELOPMENT ISSUES (ICIDI): See **Brandt Commission**

INDICATOR SPECIES are species that serve as early warnings that a community or **ecosystem** is being degraded. One example would be the current phenomenon of **amphibian** decline. In the late 20th century hundreds of the world's amphibian species began to vanish or rapidly decline in number in different places around the world. Many amphibians are also born with extreme deformities. It is not clear that there is a single cause for this problem, but scientists have speculated that the decline is caused by a variety of environmental stresses, such as increased **ultraviolet radiation (UV-B)** resulting from **ozone** depletion, and toxic **pollution** in water supplies. As amphibians live both in water and on land, they are exposed to all kinds of pollutants. Their soft, permeable skin makes them extremely vulnerable to changes in the environment. There is concern that amphibians are one of the species affected early by modern environmental stresses and that other species may soon be similarly affected.

INDICATORS: See **environmental indicators**

INDOOR ALLERGENS are those allergens that play a key role in triggering some cases of **asthma**. Sources of indoor allergens include house dust mites, **fungi**, domestic pets and **cockroaches**. The dust mite, a microscopic organism that lives primarily in carpeting, pillows, mattresses and upholstery, is responsible for significant residential allergen exposure. The allergens are contained in faecal balls that accumulate in bedding and other surfaces. Cat dandruff, another common allergen, becomes airborne in very small particles and can linger in buildings long after the cat has gone. Environmental control is important in reducing disease severity, which may mean removing carpeting and replacing it with polished floors, reducing clutter, and lowering the relative humidity on which allergens, such as fungi, thrive. Covering mattresses and pillows with impermeable materials and washing bed linens weekly in hot water will help to reduce allergic illnesses, and is particularly useful in limiting exposure during childhood. Concern about indoor allergens as a public health threat coincides not only with the trend towards spending more time indoors, but also with the move to construct more energy-efficient offices and residences.

Buildings are increasingly sealed tightly and weather-stripped, and with central heating and air-conditioning, there may be a reduction in the exchange of air from the outside. That, plus a relatively high level of interior humidity, particularly in homes with excess moisture in carpets and carpet padding, combines to encourage the growth of airborne allergens or even **sick building syndrome**.

INDUSTRIAL COMPETITIVENESS AND PROTECTION OF THE ENVIRONMENT was a proposition in a communiqué made in November 1992 by the European Commission to the Council of Ministers of the European Communities (renamed the Council of the European Union after 1993) and the European Parliament. It asserted that 'what is good business can also be good for the environment'. The object of the message was to raise awareness that environmental protection had become a pre-condition for industrial expansion, and had to be incorporated in measures for stimulating industrial development. In asserting that industrial development and environmental protection were not necessarily incompatible, industries that made environmental concerns an integral part of their strategy could derive significant competitive advantages from developing lean technologies that cut energy costs or space requirements. Beyond the production process, there was greater consumer awareness of ecologically sound products like unleaded petrol and **phosphate-free detergents**. By 1990 the European Communities environmental protection market had expanded to over ECU 50,000m. in value and was expected to grow by between 50% and 100% before the end of the 20th century.

INDUSTRIAL ECOLOGY is the term that is given to the analysis of industrial process and the flow of raw materials and energy used during manufacturing and product use, along with the associated heat, **pollution** and waste materials. The concept attempts to identify opportunities to reduce or eliminate waste by recycling or reusing materials (see **recycling (industrial)**). The analysis is based upon the model of a natural **ecosystem** in which all substances move in closed cycles during which they are transformed by different organisms and physical processes. In its broadest definition, industrial ecology measures the consequences for the natural world of resource withdrawal, manufacturing, transport, use and disposal at the end of a product's useful life. Sometimes such analysis is referred to as industrial metabolism, but this term seems better suited for describing the flows of energy, materials and waste in a specific manufacturing or industrial process, rather than for describing the total impact of the life-cycle industrial process on the ecosystem in which it is embedded.

INDUSTRY IMPACTS is a term that refers to the environmental disruption and **pollution** caused by the extraction and processing of raw materials and their subsequent manufacture into intermediate and finished products. The main impacts arise directly from emissions to air or water, or from indirect effects on land. Such impacts, and their implications for human health, may be local, transboundary (see **transboundary pollution**) or global. Manufacturing industry can play a part in providing solutions to environmental problems by means of clean technology—the development of new processes and machinery necessary for pollution abatement—and modified products. Industry's share of Europe's total energy consumption declined from 49% to 41% between 1970 and 1989, principally as a result of greater energy efficiency. Industrial water-use accounts for over 50% of water-use in Europe. However, companies in the pulp and paper-manufacturing sector and in some sectors of the chemical industry have made significant reductions in water consumption and waste water discharges while maintaining product quality.

INFOTERRA was formed as a result of the **UN Conference on the Human Environment**, held in Stockholm in 1972, and operates as an exchange of environmental information, tapping a vast reservoir of information and experience that has been accumulated over the years. Governments, industry and researchers across the world have access to the information. It is now the largest environmental information system in the world, linking national and international institutions and experts in 176 (2004) countries. It works through a network of institutions, which are designated by governments as a focal point for the scheme. Each one prepares a listing of environmental expertise within its borders and chooses the best sources for inclusion in INFOTERRA's International Directory, which lists more than 7,000 institutions. INFOTERRA deals with well over 20,000 queries on the environment each year, and it has proved a very effective way of supplying environmental information to users world-wide.

INFRARED RADIATION is sometimes referred to as the long-wave heat radiation that is emitted by any hot object, including people. It accounts for 60% of the sun's radiation. Infrared is also a significant factor in the process of global warming (see **climate change**). In practice, infrared refers to the band of radiation energy with wavelengths ranging from 3.0 micrometres to 100 micrometres. Like any other hot object, when the earth's surface has been warmed by solar radiation during the day it emits long-wave infrared radiation. Most of it is absorbed by the various gases in the atmosphere. However, there is a band of infrared of 8–11 micrometres that passes through a cloudless sky and escapes into space. This is called the **radiation window**. However, this window is partially closed by **clouds** and atmospheric pollutants, which means the atmosphere absorbs the heat rather than letting it escape back into space.

INSECTICIDES are a class of **pesticides** used to destroy insects. A variety of substances have been used over the years: hot water, soapy water, whitewash, vinegar, turpentine, brine, lye and fish oil. Before the Second World War arsenicals, nicotine, **sulphur**, cryolite and hydrogen cyanide gas were used as insecticides, but then new chemicals, the synthetic organic insecticides, were discovered, the first of which was **DDT**.

INSOLATION refers to the energy from the sun that is received and trapped by the **atmosphere** and earth, and which comes mostly in the form of radiation ranging from **ultraviolet radiation** wavelengths, through visible light to the **infrared radiation**.

INTEGRATED PEST MANAGEMENT (IPM) describes a method that combines the judicious use of **pesticides** with **biological control** to control pests. It is a strategy being worked out by the UN Environment Programme, the Food and Agricultural Organization, the UN Development Programme and others, using the pests' natural predators, pest-resistant crop varieties, crop rotation and traditional means of control. Studies are in progress in regional programmes in some 40 countries and include cotton production in North and East Africa, Latin America and the Near East, rice fields in South East and South Asia, and small farmers' crops in West Africa.

INTEGRATED POLLUTION CONTROL (IPC) refers to control measures for industries with a particularly high pollution potential. IPC provides measures and procedures to prevent or to minimize emissions from industrial installations, using the best practicable means to achieve a high level of protection for the environment as a whole. Since these activities pose a risk of increasing exposure to harmful substances in all media, including air, soil, water and food, the IPC concept evolved when it became clear that approaches to controlling emissions in one medium alone might encourage redirecting the impact of pollution into other environmental media. For many years there was separate legislation on the control of releases into each medium. The implementation of one set of control measures was sometimes at the expense of an increase in releases to another medium, in which case the result did not represent an overall improvement in environmental quality. IPC operates under a guiding principle that the **best available techniques not entailing excessive cost** should be used to prevent releases or, where this is not practicable, to minimize and render harmless any substances released.

INTELLIGENT VEHICLE HIGHWAY SYSTEMS (IVHS) are large-scale developments in the European Union, the USA and Japan using electronic technology in the management of road traffic flow. IVHS can take various forms. In industrialized countries traffic signals on motorways and traffic lights in cities are often controlled by computers that monitor the flow and congestion of traffic using road sensors. Experiments have been conducted with 'intelligent' cars to test electronic systems for automated navigation, which can be used to ensure that cars are separated by specified distances and travel at more-uniform speeds, reducing fuel consumption. In the early 2000s the IVHS systems were being used in the European Union to help map **noise pollution**.

INTENSIVE AGRICULTURE is a farming technique which uses high levels of capital, labour and **agrochemicals** to achieve high outputs and yields from a unit of land.

The **INTERGOVERNMENTAL PANEL ON CLIMATE CHANGE (IPCC)** is sponsored by two UN agencies: the **World Meteorological Organization** and the **UN Environment Programme**. It was established to assess the potential threat of global warming as a result of pollution of the air by **greenhouse gases** and other human activities that might alter the climate. The IPCC first met in November 1988. Three expert groups were set up: one to provide a scientific assessment of the current knowledge of **climate change**; another to assess the potential impacts of climate change such as changes in **sea level**; and a third to consider response strategies. All three groups reported to the meeting of the IPCC in Sweden in August 1990, motivating the negotiation of the UN **Framework Convention on Climate Change (UNFCCC)**. The results of this first study were published as three reports in 1991. The reports are updated every five years, the Second Assessment Report (1995) being released in 1996. The Third Assessment Report, published in 2001, included reports from three working groups about the science and impacts of climate change, as well as the economic, technological and social issues related to global warming. In it the IPCC warned that climate change could be far worse than previously feared, and beyond the ability of humankind to adapt, unless greenhouse gas emissions were cut substantially. The IPCC also produces technical reports, which are requested by governments and negotiators and usually focus on a single issue, and special reports, which are more detailed studies of given issues. The IPCC reports are consensus documents that review the state of science on global climate change using only peer-reviewed literature. They are rigorously reviewed and approved by both scientists and governments. The reports are frequently instrumental in climate change negotiations; for example, the First Assessment Report informed the negotiations at the **Earth Summit** and the Second Assessment Report helped prompt the negotiation of the **Kyoto Protocol**.

The **INTERGOVERNMENTAL PANEL ON FORESTS (IPF)** was established at the third session of the UN Commission on Sustainable Development (CSD) in 1995. Its broad mandate includes implementation of the decisions relating to forestry that arose from the **Earth Summit**, as well as scientific research and forest assessment to provide more information for international negotiators. A related group, also established by the CSD, is the Intergovernmental Forum on Forests, which aims to foster continuing dialogue between non-governmental organizations, governments and industry on **forests and woodland** issues.

INTERMEDIATE CHEMICALS, often referred to simply as intermediates, are substances that are created temporarily during the synthesis of complicated chemical compounds. If the chemical reaction goes wrong, or is interrupted, the intermediate may be released as a pollutant. Public attention was first drawn to the hazards of intermediates, and the existence of **dioxin** as an intermediate, by the **Seveso** chemical works explosion in Italy. It was not the first industrial accident involving dioxin. However, the earlier ones had been restricted to contamination of workers, and so were listed as occupational accidents and not environmental pollution accidents. Seveso was the first major environmental disaster attributed to an intermediate organic chemical. Other environmental pollution accidents have happened with other intermediates at chemical works. Manufacturing of products like pesticides have gone wrong, and the resulting **effluent** containing toxic intermediates has poured into rivers. The problem is not limited to the manufacture of new stocks of chemicals. Destruction of long-established substances, such as **polychlorinated biphenyls (PCBs)**, by **incineration**, must be carried out meticulously, to make sure intermediates, including dioxin, are properly neutralized. However, the PCB family of substances, which were used extensively in industry for 50 years after the Second World War, are very stable and highly soluble in fat once in the food chain. Since the PCBs are so stable, rather than being metabolized they accumulate in fat. This resistance to **metabolism** which allows a pollutant to accumulate, and includes many other organochlorine agents such as **DDT**, is referred to as **biological concentration**.

INTERMEDIATE-LEVEL WASTE is the radioactive scrap produced by the nuclear industry in the form of contaminated metal, discarded components and pipework from nuclear reactors and the cladding stripped from spent **nuclear fuel** elements.

The **INTERNATIONAL ATOMIC ENERGY AGENCY (IAEA)** is an autonomous inter-governmental agency (although it is administratively part of the UN, and reports to the UN General Assembly) set up to develop the civil and peaceful uses of atomic energy. However, it is also the repository of the **Non-Proliferation Treaty (NPT)**. The agency was established in 1957 and its work includes research on the applications of nuclear energy to medicine, agriculture, water location and industry. It runs conferences and training programmes, sends out publications to promote the exchange of technical information and skills, and gives technical assistance, in particular to developing countries. It establishes and administers radiation safeguards. Its headquarters are in Vienna, Austria. At the beginning of the 21st century it was involved with attempts led by the USA to monitor the nuclear industry in Iran, a member country.

The **INTERNATIONAL BANK FOR RECONSTRUCTION AND DEVELOPMENT (IBRD)** is known as the **World Bank**, but it is, in fact, one arm of the World Bank Group. It has 184 (2004) member states and operates as an independent organization, but with the status of a UN specialized agency. Since 1948 most of its loans have been to projects that secure economic growth in developing countries, foster educational and social schemes, and resettlement programmes. It advises on development programmes in developing countries, sponsors consultative groups and meetings, and publishes annual statistics.

The **INTERNATIONAL CODE OF CONDUCT ON THE DISTRIBUTION AND USE OF PESTICIDES** aims to establish voluntary standards of conduct for governments and all those involved in the distribution and use of pesticides. The Code was published in 1986 by the Food and Agriculture Organization and developed with the UN Environment Programme (UNEP), the World Health Organization and other agencies. It is aimed especially for use in countries where national controls are weak or do not exist. UNEP backs up the Code with training programmes.

INTERNATIONAL COMMISSION FOR THE PROTECTION OF THE DANUBE RIVER: See **Danube**

INTERNATIONAL CONVENTION FOR THE PREVENTION OF POLLUTION FROM SHIPS (MARPOL): See **International Maritime Organization (IMO)**

INTERNATIONAL CONVENTION FOR THE REGULATION OF WHALING, which came into force on 10 November 1948, is one of the earliest international environmental treaties, although the Convention was not established for conservation purposes. Parties are required to implement regulations and decisions arising from the annual meeting of the **International Whaling Commission (IWC)**. Complete catch reports are required from members engaged in subsistence and scientific whaling. Whale harvests declined dramatically from 38,977 in 1970 to 688 in 1990. In 1982 an amendment was created to phase out commercial whaling over a three-year period. The ban became effective in 1986. Iceland, Japan and Norway remained the three countries still committed to whaling at the beginning of the 21st century, although only Norway still sanctioned overtly commercial hunts (i.e. not for 'scientific reasons').

The **INTERNATIONAL CORAL REEF INITIATIVE (ICRI)** is a partnership of governments, non-governmental organizations and international bodies founded in 1994 to address the rapid global decline of **coral reefs**, an issue given a high priority by the 1992 **Agenda 21** declaration. There were various regional initiatives, and ICRI itself produced its Framework for Action plan. Coral reef management and conservation was likely to benefit, at least with information about best practice, from the 2000 establishment of the International Coral Reef Action Network, described as a global partnership for coral reefs.

The **INTERNATIONAL COUNCIL FOR SCIENCE (ICSU)**, formerly known as the International Council of Scientific Unions, is a non-governmental group of scientific organizations representing 80 national academies of science, 20 international unions and 26 other bodies called scientific associates. The object of ICSU is to achieve the free exchange of scientific information and co-operation between the world's scientific communities. To accommodate the spread of activities in which many of the member unions are involved, ICSU has established 13 scientific committees with experts from a multiplicity of scientific disciplines. One of them is the Scientific Committee on Problems of the Environment (SCOPE), which directs attention to developing countries. SCOPE was formed in 1969 in response to environmental concerns emerging at the time. Its first task was to prepare a report on global environment monitoring for the **UN Conference on the Human Environment** in Stockholm, Sweden. The mandate of SCOPE is: to assemble, review and assess the information available on human-induced environmental changes and the effects of these changes on humans; to assess and evaluate the methods of measuring environmental parameters; and to provide an intelligence service on current research and a body of informed advice.

INTERNATIONAL COUNCIL OF SCIENTIFIC UNIONS (ICSU): See **International Council for Science (ICSU)**

The **INTERNATIONAL DRINKING WATER SUPPLY AND SANITATION DECADE** began in 1980. It was launched by the UN with the aim of achieving 'clean water and adequate sanitation for all by 1990'. Hundreds of millions of people did receive safe water and proper sanitation for the first time, in many parts of the developing world. However, it was always recognized to be an ambitious programme and by the end of the decade there were almost as many people without access to safe water and sanitation as there were when it began. **Population** increases, **droughts** and economic crises all contributed to its lack of overall success, together with the fact that the millions of dollars that the programme required every day to make it work was not forthcoming from the industrialized and developed countries. Some initiatives did emerge from the decade: national action plans were drawn up, even though many had to be shelved because of lack of funds. As technologies become cheaper there is hope that some schemes could be completed at a fraction of the previous costs for items like hand pumps and borehole equipment. The UN's International Conference on Water and the Environment, held in Dublin, Ireland, in 1992, reported that the cost of keeping the original decade target would be US $36,000m. per year. The **World Summit on Sustainable Development (WSSD)**, held in South Africa in 2002, focused on the issue of clean **fresh water** for drinking and sanitation, and urged immediate action. The WSSD contributed to the initiative for international targets of halving the proportion of people without access to clean drinking water and basic sanitation by 2015.

The **INTERNATIONAL FUND FOR PLANT GENETIC RESOURCES** was set up in 1989 under the auspices of the **Food and Agriculture Organization (FAO)**. It is administered by the FAO and was launched to conserve and promote the use of plant genetic resources at a global level. The Fund was intended to assist developing countries in the identification and survey of plant resources, as well as to assist in plant breeding and seed multiplication and distribution. It provides a way for countries, inter-governmental and **non-governmental organizations**, and private industries and individuals to maintain the genetic diversity of the world's plants.

INTERNATIONAL GEOPHYSICAL YEAR (IGY) was the international geophysical research programme in which more than 70 countries took part. IGY began in July 1957 and ended in December 1958, and during that time several important projects were carried out simultaneously. These projects included the exploration of the variability of solar radiation and **sunspots**, and gathering information on **cosmic radiation**, the nature of the upper atmosphere, and ice in **glaciers** and **ice sheets**. The programme's data was interpolated with weather fluctuations and used orbiting satellites for the first time. Ocean currents were mapped, including the subsurface currents that have profound effects on all life on earth (see **ocean circulation**). A variety of charting devices were used during IGY to study undersea landslides and currents, volcanic eruptions and their output. Much of the work carried out during IGY is still the subject for on-going research.

The **INTERNATIONAL HYDROLOGICAL DECADE** was a research programme on water problems. It began on 1 January 1965, following a resolution adopted at the general conference of **UNESCO**. The scientific programme covered all aspects of water: its characteristics; its movement; and the use of water on and below the earth and in the atmosphere (see **hydrology**). The information acquired over the period of the Decade was important in preparing development projects of widely different kinds. National committees were set up in 96 countries, bringing together, in many cases for the first time, representatives of different national organizations concerned with water problems.

The **INTERNATIONAL INSTITUTE FOR ENVIRONMENT AND DEVELOPMENT (IIED)** is an international, independent policy research institute. It works mostly in the developing world, undertaking research and analysis on environmental and development topics, including environmental planning, human settlements, forestry, and the use and production of energy. It is a charitable, non-profit-making organization based in London, United Kingdom, and in Buenos Aires, Argentina. It is funded by international organizations, governments, private and corporate foundations, and from public donations. IIED also co-operates with other environmental development agencies, including the **UN Environment Programme**, the **World Conservation Union (IUCN)** and the **World Resources Institute**.

The **INTERNATIONAL MARITIME ORGANIZATION (IMO)** is one of the specialized agencies of the UN and was the first international body devoted exclusively to maritime matters. Its prime objectives are safety at sea and prevention of marine pollution. The Inter-governmental Maritime Consultative Organization, as the IMO was formerly known, began operations in 1959 and the IMO subsequently promoted 30 conventions and protocols, and adopted more than 700 codes and recommendations on safety at sea. In the 1970s the IMO became the agency responsible for negotiating international agreements controlling pollution from ships. Its members have adopted the international convention on Oil Pollution Preparedness, Response and Co-operation (OPRC), designed to improve the ability of nations to cope with emergencies, such as tanker accidents. It does not yet have mandatory status, but many of its provisions were used as the basis for the IMO's response to the massive pollution of the Persian (Arabian) Gulf in 1991. The organization was able to provide assistance that saved many ecologically important and sensitive areas from major damage. The first important IMO agreement on controlling pollution from ships was the International Convention for the Prevention of Pollution from Ships (MARPOL), which deals not only with pollution by petroleum, but also chemicals, garbage and sewage. More recently the IMO has been the forum for international negotiations on dumping at sea, which has been proposed as a means of routine disposal of industrial wastes. The negotiations led to the Convention on Prevention of Marine Pollution by Dumping of Wastes and Other Matter, better known as the **London Convention**, which halted dumping of radioactive waste at sea.

The **INTERNATIONAL MONETARY FUND (IMF)** is a specialized agency of the UN. It was established in 1947 to promote international monetary stability, the growth of world trade and the stability of foreign exchange. It also gives financial assistance to the 184 member countries when they have balance of payments problems and smoothes multilateral payments and financial arrangements among member states. The IMF works in close collaboration with the **World Bank**.

The **INTERNATIONAL NUCLEAR EVENTS SCALE (INES)** is a measure of the severity of an uncontrolled release of radioactivity into the environment and grades nuclear accidents in terms of public health and environmental impact. It was devised to make a distinction between accidents with consequences limited to a nuclear plant and events causing off-site radioactive pollution, like **Chornobyl** and the **Windscale fire**. The scale was proposed by the **International Atomic Energy Agency** on a trial basis in 1989, partly in response to the Chornobyl accident, and was formally adopted in 1992. Its purpose is to enable nuclear events to be graded according to seven levels of severity so that the threat to public health and safety can be communicated promptly to all the relevant public services. It covers abnormal events during the operation and decommissioning of nuclear power stations and **nuclear fuel** reprocessing plants, the management of **nuclear waste** storage sites and the transport of nuclear fuel and nuclear wastes. The scale begins with anomalies (level 1) and incidents (levels 2 and 3) that do not necessitate off-site protective measures and are of little environmental concern. Accidents (levels 4 to 7) cause off-site contamination and are likely to require protective countermeasures, including the administration of **iodine** tablets to humans, imposition of local food controls and decontamination of **soils** and other materials. The Chornobyl accident was graded at level 7 on the INES scale because of the scale of the international emergency, mass evacuation, continuing land quarantine and long-term treatment for health effects that it caused.

The **INTERNATIONAL PROGRAMME ON CHEMICAL SAFETY (IPCS)** was formed in 1980 to assess the risks that specific chemicals pose to human health and the environment. It was set up by the World Health Organization (WHO), the UN Environment Programme and the International Labour Organization. It liaises with scientists to ensure that international research methods are comparable and reliable. The organizations work with the **International Register of Potentially Toxic Chemicals (IRPTC)**, which provides the basis for chemical safety decisions. IPCS, which is based at WHO's headquarters in Geneva, Switzerland, processes the information and publishes evaluations of health and environmental risks, and advice on control and exposure levels. IPCS produces detailed environmental information for scientists and non-technical health and safety guides for administrators, managers and decision-makers. It also provides international chemical safety cards for easy reference in the workplace and poisons information booklets for medical use. IPCS also works with the IRPTC on a training programme on chemical control for developing countries. A computerized poisons information package (INTOX) is also available, in three languages, to countries without the necessary facilities for coping with poisonings.

The **INTERNATIONAL REGISTER OF POTENTIALLY TOXIC CHEMICALS (IRPTC)** operates through a network of national and international organizations, industries and external contractors. Its computerized central data files have profiles of more than 1,000 chemicals, as well as special files on **waste disposal** and management, on chemicals currently being tested for toxic effects and a file of national regulations covering over 9,000 substances. The IRPTC is working to make the files available on-line. The organization plays an important part in the drive to ensure that substances that are banned or severely restricted in developed countries are not dumped in developing countries. Between 1950 and the mid-1990s world production of organic chemicals increased from 7,000m. metric tons to around 250,000m. tons. With over 70,000 chemicals in common use, the greatest threat is ignorance about their use or dangers. By 1992 a total of 149 national authorities had been nominated to act as channels for a procedure known as a **prior informed consent (PIC)**. Initially, PIC was applied to all chemicals banned or severely restricted in 10 or more countries. Eventually it was to apply to all chemicals banned or severely restricted in any country. The Rotterdam Convention on PIC entered into force in February 2004.

The **INTERNATIONAL TROPICAL TIMBER AGREEMENT (ITTA)** came into force in 1985 and is managed by the International Tropical Timber Organization. The agreement's main role is in marketing and promotion, but it also includes the goal that producer countries should be helped to develop better techniques for reforestation and forest management. It also works to encourage more timber processing in producing countries, so adding to the value of the products, and supports research and development programmes. Environmental organizations, among them the **UN Environment Programme**, the **World Wide Fund for Nature (WWF)** and the **World Conservation Union (IUCN)**, co-operated with ITTA to build ecological considerations into the agreement. Some countries decided only to import wood from sustainable sources from 1995. There has been some criticism that ITTA achievements are uneven and restricted to tropical forests, but it is felt the agreement could form the model for a revised agreement to include all types of **forests and woodlands**.

The **INTERNATIONAL UNDERTAKING ON PLANT GENETIC RESOURCES** was adopted by the member states of the **Food and Agriculture Organization (FAO)** in 1984. The agreement proposed that all germ plasm, including seeds improved by **biotech**, should be freely available to all. However, there is considerable world-wide disagreement about ownership and property rights, and therefore the rights to commercial use, of the seeds contained in **gene banks**. The dispute centres around the fact that more than 90% of the genetic material held in gene banks around the world, almost all of it in developed countries, came originally from farmers in the developing world. In 1986 Brazil, China (Taiwan), Ethiopia and India banned the export of germ plasm from specific crops. The USA also refused to allow genetic material held in a gene bank to be sent to Nicaragua. There was increasing pressure to allow farmers in the developing world who have developed strains after years of cultivation to share in the rewards. The Undertaking was reaffirmed in 1989 and a resolution endorsed the concept of farmers' rights. The Undertaking was drawn up to ensure that plant genetic resources are preserved and made as widely available as possible for plant breeding for the benefit of present and future objectives, and to conserve plant genetic resources of economic and/or social interest, particularly for agriculture. Another objective focuses on cultivated plants, plants or varieties that have been in cultivation in the past, primitive versions of cultivated plants, wild relatives of such plants and certain special genetic stocks. The agreement is not legally binding on the signatories. There is also no mechanism to settle disputes. In 1989, at the FAO conference, the **International Fund for Plant Genetic Resources** was set up.

INTERNATIONAL UNION FOR CONSERVATION OF NATURE AND NATURAL RESOURCES (IUCN): See **World Conservation Union (IUCN)**

The **INTERNATIONAL WASTE IDENTIFICATION CODE (IWIC)** constitutes a uniform classification system to describe waste considered to be hazardous. The code was developed by the **Organisation for Economic Co-operation and Development (OECD)** in 1988, immediately adopted by the European Communities (now the European Union—EU) and has been formally incorporated into EU environmental legislation since 1983. It is not applied world-wide. It means that different countries have compatible standards by which to compare their progress in dealing with waste. The hazard of a particular type of waste can be judged by comparing it with a list that describes wastes from six different perspectives: the reason why the material is intended for disposal; disposal operations; generic types of potentially hazardous wastes; their constituents; the hazardous characteristics; and the activities generating the wastes. The IWIC facilitates the compilation of a comprehensive 'cradle-to-grave' dossier for a consignment of waste. Any batch of waste being transported from one country to another is accompanied by such a description.

The **INTERNATIONAL WHALING COMMISSION (IWC)** was set up in 1946 after the signing of the **International Convention for the Regulation of Whaling**. This was the first effort to control the hunting of **whales**. However, because the Commission has no regulatory powers to enforce its decisions, whales continued to be hunted, some to near-**extinction**. By the end of the 1960s scientists and large numbers of the general public concerned about conservation of wildlife were becoming alarmed at the depletion of whale stocks. A series of campaigns to save the whale were conducted against the IWC, co-ordinated principally by **Greenpeace**. Whaling was banned in 1980, but Iceland, Japan and Norway continue to assert hunting rights, sometimes under the guise of scientific research, which is permitted by IWC rules. Most members of the IWC are now mainly concerned with the preservation of the great cetaceans. A whale reserve was declared by the IWC in the southern oceans in 1994, between latitude 40°S and the continent of **Antarctica**.

The **INTERTIDAL MARINE ENVIRONMENT**, sometimes known as the littoral zone, describes the area between land and sea which is regularly exposed to the air by the tidal movement of the sea. Sand beaches, rocky bottoms, protected bays, **salt marshes**, **mangrove forests** and **coral reefs** are all part of the intertidal marine environment. Marine organisms that inhabit the intertidal marine environment have to adapt to periods of exposure to air and to the waves created by wind, which makes it the most physically demanding of the marine habitats. In this semi-terrestrial region, plants and animals have to adjust their breathing to survive for most of the time spent out of the water. Different species survive at various levels within the intertidal region, depending on their adaptability and whether the area is sandy, muddy or rocky. Species found within these levels include limpets, molluscs and barnacles, winkles, snails, and burrowing molluscs, like clams, and also sea urchins and crustaceans, as well as predators such as starfish, dog whelks and crabs.

INVERSION: See **temperature inversion**

IODINE (I) was the first essential **trace element** to be discovered as such. It is indispensable to life because it is essential for the synthesis of thyroid hormones, which regulate the metabolic rate in all cells, it controls growth and development, and helps burn excess fat. It is present in seawater and was previously extracted from seaweed. It is still generally available in kelp tablets from chemists. In addition to the natural stable form, there are 14 radioactive isotopes of iodine. Radioactive iodine-131, a **fission product** released in the fall-out from weapons testing, and in nuclear accidents such as the **Windscale fire** and **Chornobyl** explosion, is indistinguishable to our bodies from stable iodine. As it is a very volatile element, iodine is one of the radioactive **isotopes** most likely to be released in large quantities in an accident. If it contaminates pasture land, it is absorbed rapidly into the **food chain** and the milk supply. As was discovered after the Chornobyl disaster, cattle, sheep and goats grazing over large areas of ground will concentrate radioactive fall-out in their bodies. So, an element that is normally essential to our health becomes concentrated in the thyroid gland. In children iodine contamination can cause serious developmental problems.

IONIZATION is the process of producing ions when an atom or molecule temporarily loses some of its electrons. There are several ways in which ions may be formed from atoms. Ionization occurs in certain chemical reactions and when some substances are in a dissolved state in solution. Ions are produced by **ionizing radiation** if the rays carry enough energy to break up a molecule or detach electrons from an atom.

IONIZING RADIATION refers particularly to **radiation** that causes electrons to be stripped from an atom, and so converts an atom from its neutral state into an electrically charged state. The process temporarily alters the chemistry of the atom. If a lot of ionization happens to atoms that form organic molecules in the cells of the body, the change can lead to disruption of the normal biochemistry and the death of a cell or its transformation into a cancerous one. Ionizing radiation includes X-rays, **alpha**, **beta** and **gamma radiation**.

IRRIGATION is one of the most effective ways known of increasing agricultural productivity, and irrigated land produces more than 30% of the world's food. Irrigation has been practised for centuries, especially in parts of Africa, Asia and South America. In arid areas, traditional methods of irrigation draw on centuries of experience of the best ways to ensure the proper distribution of water, the clearing of drainage ditches, and observing fallow periods. Some modern irrigation schemes, encouraged because they were seen as a way to ease world hunger, are built on a vast scale and many are now criticized. Apart from the huge expense, they mostly disregard fallow seasons and are organized to grow three crops on the same ground every year. Consequently, the mismanagement of some schemes has led to waterlogged land, the leaching of soil nutrients and **salinization**, which sterilizes the land. There is also concern that a large number of crops are grown as **cash crops** for export to earn the money to pay for the installation of irrigation schemes. The worst fear is that irrigation projects create new and permanent places for the carriers of diseases such as **malaria**, filariasis and **schistosomiasis**. Agricultural pests, insects and rats thrive in newly irrigated areas because of the year-round food supplies.

ISLAND BIOGEOGRAPHY is the study of the relationship between the size of **ecosystems** and the quality and quantity of their **biological diversity**. Although originally such studies were conducted on islands of land surrounded by water, the concept has been extended to include small patches of forest, grassland or other types of ecosystem that are surrounded by a different kind of ecosystem. The general principle is that the smaller the ecosystem, the fewer species it can support.

ISLANDS are unique but fragile and vulnerable **ecosystems** because the evolution of their flora and fauna has taken place in relative isolation. Many remote islands have some of the most unique flora in the world; some have species of plants and animals that are not found anywhere else, which have evolved in a specialized way, sheltered from the fierce competition that species face on mainland. Endemic island animals have little defence when predators, such as cats, dogs or pigs, are introduced into an island, because they have developed without any natural predators. It is thought that the dodo became extinct because during its long evolution on the island of Mauritius it lost its ability to fly, and its sense of fear, because these things had become unnecessary during its evolution. **Vanishing habitats**, lost because of growing populations, island development, agriculture, **deforestation**, mining and quarrying, the introduction of non-native species, and **tourism** are a big threat to many islands. Most of the recorded **extinctions** of species have occurred on islands: 75% of all mammals driven to extinction have been island dwellers, as have 97 out of 108 extinct bird species. The tropical rainforests of Madagascar, in the Indian Ocean, have an estimated 12,000 different plant species and some 70% of them are unique to the island, including the Madagascar rosy periwinkle, which is a valuable component of a drug used to fight leukaemia and Hodgkin's disease. Madagascar's remaining rainforest is under constant pressure to provide land for its growing population and to accommodate tourists, thus threatening the survival of these species. Some islands are now deliberately protected from human contact, such as the **World Heritage Site** that protects Henderson, Ducie and Oeno, which are attached to the British dependency of Pitcairn. Another threat to island ecosystems, whether they are inhabited by people or not, is **climate change** and the projected rise in **sea level** that particularly concerns **small island states**.

ISO 14000 is a series of voluntary international standards for environmental management developed for use by private companies by the International Organization for Standardization (ISO), which formed a Strategic Advisory Group on the Environment in 1991 to consider the need for such standards. For example, ISO 14011, one of the standards in the series, provides guidelines for **environmental auditing**. It was hoped that companies would be certified by ISO 14000 in order to be viewed as good corporate environmentalists, which would improve their corporate image, potentially save costs, create regulatory relief and meet consumer demand.

ISOCYANATE is a member of the family of poisonous cyanide compounds. It can be produced when certain types of plastic catch fire.

ISOTOPE refers to a species of a chemical element with a different atomic mass, but the same **atomic number**, or characteristic, which determines which element it is. The chemical nature of a particular element is a property of the number of protons in the nucleus. However, the other particles that make up the nucleus of an atom, the neutrons, may vary and so change the atomic mass. Atoms of elements with differing neutron numbers are known as isotopes. Uranium-238, for example, has 92 protons and 146 neutrons, whereas **uranium-235** has 92 protons but only 143 neutrons. Some isotopes are stable: carbon-12, for instance. Others, such as carbon-14, are unstable.

ISOTOPIC ANALYSIS is a primary method of dating the origin of samples of material that provide evidence of the state of the earth's climate hundreds or thousands of years ago. The samples may be fossils of primitive marine organisms retrieved from sediments on the sea bed or bubbles of atmospheric gas trapped for thousands of years in a glacier. The technique depends on the fact that there are two forms, or isotopes, of **oxygen**. One is stable and the other slowly decays. A measurement of the proportion of the two forms of oxygen from a specimen can be used to calculate when it became buried or trapped in an ice sheet.

The **ITAIPÚ DAM** is the largest hydroelectric dam in the world, with a capacity of 12,600 MW, although the **Three Gorges Dam** in the People's Republic of China will surpass Itaipú upon completion (scheduled for 2009). Itaipú is located on the River Paraná, on the border between Brazil and Paraguay. It took 18 years to build, at a cost of about US $18,000m. The magnitude of the project is also demonstrated by the fact that in 1995 Itaipú alone accounted for 25% of the energy supply in Brazil and 78% in Paraguay. Each country is entitled to one-half of the power generated, but Paraguay exports much of its share. The Dam is administered by Itaipú Binacional, a joint Paraguayan-Brazilian government commission.

Two innovative programmes implemented by the Itaipú authorities are the Mymba Kuera and Gralha Azul projects. The former programme attempted to minimize the effects of reservoir flooding on the fauna of the region by catching animals and releasing them in biological reserves. The Gralha Azul project was designed to create and afforest a protective zone around the reservoir on the Brazilian side. Despite the inclusion of forest protection in the initial planning stages of the project, over 700 sq km of forests were negatively affected by the Dam's construction. Major effects were the complete loss of forest lands, particularly on the Paraguayan side, the general reduction in the amount of forest lands and the extinction of some plant types, including a rare orchid. The majority of the damage occurred in the first few years of construction. However, this process was retarded when binational efforts were made to study and minimize any further damage.

IVORY is the raw material for which humans have hunted **elephants** in order to make piano keys, jewellery and ornaments. In 1989 the **Convention on International Trade in Endangered Species of Wild Flora and Fauna (CITES)** placed the African elephant on Appendix 1, a list of species afforded the highest protection designated under the Convention. In 1985 CITES set up the ivory quota system in an attempt to maintain the ivory trade within realistic limits. Most of the 103 member countries agreed not to allow the sale of ivory. Subsequently, several countries in southern Africa reported that their elephant populations had increased.

The **IXTOC 1** oil well blow-out, in the Gulf of Mexico in 1979, was the world's biggest recorded spill, releasing over 500,000 metric tons of petroleum into the environment over a period of nine months. Since the disaster happened a long distance from the nearest coast, the discernable environmental impact was remarkably small compared, for instance, with the ecological havoc caused by the inshore wreck of the ***Amoco Cadiz*** or the ***Torrey Canyon***.

J

The **JET STREAM** is a strong and enduring belt of westerly winds that travels at an altitude of 12,000 m in the **troposphere**. Its typical speed varies from 110 km per hour in summer to 180 km per hour in winter.

The **JOINT GLOBAL OCEAN FLUX STUDY (JGOFS)** is an international investigation that scientists believe will help them understand the links between the ocean **carbon cycle** and **climate change**. The world's **oceans** have been referred to as the 'sleeping giant' of global carbon dioxide control. There is an estimated 20 times more carbon dissolved in **seawater** than occurs on land in plants, animals and soil. Release of just 2% of the carbon stored in the ocean would more than double the level of carbon dioxide in the atmosphere. Each year about 15 times more carbon dioxide is taken up and released by natural marine processes than the total produced by the burning of fossil fuel, deforestation and other human activities. The amounts of carbon dioxide entering and leaving the oceans, over an annual cycle, are estimated to be in balance. Research suggests that balance was reached not long after the last ice age, 10,000 years ago. Nevertheless, the exchanges of carbon dioxide between the air and sea are controlled by many different processes. They are influenced by regional and seasonal factors, with direct and indirect links to climatic factors. Consequently, scientists are concerned that the effects of climate change could have a major influence on some of the processes involved in carbon dioxide exchange, either accelerating or slowing down temperature changes.

The JGOFS study investigated the physical conditions of the oceans and the atmosphere that govern the rate at which carbon dioxide is dissolved from or released into the air, and the biological processes whereby marine plants absorb and release carbon dioxide. Carbon dioxide is extracted from the air by **photosynthesis** of the **phytoplankton**, which flourish in the top layers of the sea, and is released by **respiration**. Uncertainties about global warming arise because scientists do not know exactly what happens to the carbon being absorbed and released from many sources. Human activities add about 5,000m.–6,000m. metric tons of carbon to the atmosphere per year, but the annual increase in atmospheric carbon is only 3,000m. tons. It was thought the balance must be absorbed by the oceans. Recent estimates of the processes for absorbing carbon by the oceans leave over 1,000m. tons unaccounted for. Improvements in understanding the mechanisms of processing carbon are urgently needed for reliable predictions of future increases in atmospheric carbon dioxide, and hence, global warming.

JOHANNESBURG DECLARATION ON SUSTAINABLE DEVELOPMENT: See **World Summit on Sustainable Development**

JOINT IMPLEMENTATION (JI) is one of three mechanisms put in place by the **Kyoto Protocol** on global **climate change** in order to promote **sustainable development**. This policy instrument would allow industrialized countries to collaborate on projects for the reduction of **greenhouse gas** emissions and to share the credit for any such reductions. For example, if one country were to assist another in **retrofitting** a power station in order to make it more efficient, then both countries would share the credit for reducing emissions. Although the Protocol had not been implemented at mid-2004, the procedure was informing other international agreements. (See also the **Clean Development Mechanism**.)

K

KARIN B was a ship that was refused entry to European ports with its cargo of hazardous chemical waste, containing **polychlorinated biphenyls**, in 1988. The cargo had been dumped illegally in Nigeria by an Italian chemical company, and the *Karin B* was trying to take it for **incineration** in a chemical disposal plant. The waste cargo was eventually returned to Italy. The incident exposed the trade in toxic waste from rich countries to the poor, developing countries. It paved the way for the **Basle Convention on the Control of Transboundary Movements of Hazardous Wastes and their Disposal**, which came into effect in 1992.

KELP is brown **seaweed** that grows below low-level tides, usually anchored to the seabed. Kelp, when burned, provides ash that is used as a **fertilizer**. It is also a valuable source of **iodine**, potash and soda.

KINETIC ENERGY is the energy of a moving object owing to its motion. A fast-flowing river or a waterfall possess kinetic energy. This energy can be harnessed to drive electric generators for **hydroelectric power**. The kinetic energy of wind and waves can be harnessed as **renewable energy sources**.

KOBE, Japan's largest port, suffered one of the most disastrous **earthquakes** of modern times in January 1995. The quake measured 7.2 on the Richter scale and the environmental devastation it caused led to the death of more than 6,300 people. It destroyed an industrial hub that over the previous 130 years had played a principal role in each step of Japan's development into a modern economy. The quake caused over 100,000 buildings to collapse and damaged electricity and gas supplies and railway lines. It also set off a fire-storm that obliterated Kobe's worst-hit district, Nagata ward, a warren of mostly wooden shops and small factories in the centre of the city.

The **KOLA PENINSULA** lies on Russia's border with Norway and Finland, on the Barents Sea and the **Arctic** Circle. The regional capital is Murmansk. The peninsula is highly polluted and regarded with anxiety by neighbouring countries. Problems began with the massive development of heavy industry after the Second World War, specifically the vast nickel-smelting works, the sulphurous emissions of which affected tracts of Finland and Norway. Moreover, the concentration of nuclear reactors on the peninsula, considered to be the world's most hazardous, caused further concern. Advisers to the Norwegian Government believed the Kola peninsula had the highest concentration of nuclear reactors, both active and derelict, in the world. At least 70 nuclear submarines of the former Soviet navy were withdrawn from service after the collapse of the USSR. Nuclear fuel was removed from only 20 of them, partly because of a lack of storage sites for the remainder. Environmental groups were concerned that spent fuel and other nuclear waste from vessels withdrawn from service under disarmament agreements could result in **radiation** leaking into the Barents Sea. Greater fears about terrorists obtaining such materials after 2001 accelerated efforts to address the problems, although agreements such as the **Arctic Military and Co-operation Pact** in the mid-1990s had begun the process.

KRILL is the collective name for some 80 species of shrimp-like crustaceans, about 5 cm long, which live exclusively in the cold waters around the poles, particularly in **Antarctica**. They feed on **plankton** and form the second level of the marine **food chain**. Krill have a high level of protein and are an important food source for fish, five species of baleen **whales**, some species of seal and several species of birds, including penguins. Krill shoals weighing as much as 2m. metric tons are not unknown. Whales consume several million tons of krill every year, but when the moratorium on whale hunting was declared, krill became more abundant as a food source for sea birds and seals. However, krill are now being caught in vast quantities by Russian and Japanese fishing fleets, for use as animal feedstock and, in some cases, as a nutritious food source for humans. Pollution of the oceans presents a threat to krill, because it could destroy the plankton on which they feed.

The **KYOTO PROTOCOL** on global **climate change** is a protocol to the UN **Framework Convention on Climate Change (UNFCCC)**. The Protocol was adopted in 1997 in Kyoto, Japan, and was the first international treaty to mandate reductions in **greenhouse gas** emissions for industrialized countries. These reductions are legally enforceable, not voluntary, as they were under the UNFCC. The Protocol required industrialized countries to reduce their **greenhouse gas** emissions by an average of 5.2% below 1990 levels. Thus, the European Union (EU) was to reduce its emissions by 8% below 1990 levels, but the Russian Federation only had to reduce its emissions to 1990 levels. All countries were to inventory their emissions and report their activities for the monitoring of compliance. The Protocol regulated the six main greenhouse gases: **carbon dioxide**, nitrous oxide, **methane**, hydrofluorocarbons, sulphur hexafluoride and perfluorocarbons. Three innovative, flexible mechanisms were established by the Protocol: the **Clean Development Mechanism**, **Joint Implementation**, and **emissions trading**. All are designed to support **sustainable development** and minimize the costs of reducing emissions, but have been criticized in some quarters for creating potential loopholes in the Protocol as they provide countries with so much flexibility in reducing their emissions. The Kyoto Protocol had not entered into force by the time its details were to be finalized at a meeting convened in The Hague, Netherlands, in 2000. That this meeting failed to reach agreement turned out not to be as disastrous a development as the election of George W. Bush to the US presidency in November. The USA, which is responsible for about one-quarter of the world's greenhouse gas emissions (but has only one-20th of the population), had already announced that it would not ratify the Protocol until there was greater participation by developing countries (although their potential for adding to the problem is much greater than their actual, minimal responsibility), but President Bush, who assumed office in January 2001, subsequently refused to consider signing the Kyoto Protocol. Under the complicated arrangements of the Protocol, this meant that it could only enter into force if ratified by Russia (the EU countries, Japan and Canada had already signed). However, in late 2003 and early 2004 it seemed that Russia's commitment was unreliable. The country used the issue as a bargaining point on other matters with the EU and Japan, but finally ratified the Protocol in November 2004, thus allowing it to come into force in February 2005. Meanwhile, the Kyoto proposals were already having an effect on domestic legislation in various countries and on other international agreements.

L

LA NIÑA: See **El Niño**

LAKE BAIKAL, in Siberia, is the oldest lake in the world. Its depth of 1,600 m also makes it the world's deepest lake. More than 1,500 species of animals and 1,000 plants have been found in Baikal and some 1,300 of these have not been recorded anywhere else in the world. The local community, many of them Buddhists, regard the lake as holy and, until the 1950s, its water was noted for its purity. Large-scale development after the Second World War brought factories to the shores of the lake, including the Baikalski paper mill, which has poured massive amounts of industrial effluent into the lake every day for more than 30 years. Large areas of the surrounding forests have been destroyed by pollution from the emissions of the factories' chimneys. Since the fall of the Communist regime in the former USSR, environmental campaigns have been set in motion to clean up the lake and some remedial programmes have been started, including a new pipeline from the paper mill taking discharges to a nearby river. Lake Baikal is favoured by scientists because of its wealth of unique flora and fauna. The Russian Academy of Science negotiated new collaborative agreements with overseas organizations. For example, the US Geological Survey co-operated with Soviet researchers to collect sediment from the floor of the lake. The lake was placed on the list of **World Heritage Sites** in 1996, but pollution has continued to such an extent that in 2004 UNESCO debated putting Baikal onto the endangered list.

LAKE NYOS in Cameroon was the site of a major natural environmental disaster on 21 August 1986. Lake Nyos is the water-filled throat of an old volcano. Although some of the details are disputed, the main event was the sudden release of **carbon dioxide**, which had built up under pressure in the deeper waters. It is most likely that a landslide, perhaps some sort of volcanic activity, disrupted the lake and permitted the deadly release of the carbon dioxide. The heavier-than-air cloud of carbon dioxide spread as far as 25 km from the lake, moving fast enough to flatten vegetation, including some trees. About 1,700 were suffocated to death and 845 were hospitalized.

LAND–OCEAN INTERACTION STUDY (LOIS) is the name of a research project to collect data so that scientists will be able to predict the likely impact on the coastal zone in the United Kingdom of natural and man-induced environmental changes over the next 50 to 100 years. The possible changes under investigation are fluctuations in **sea level**, tidal patterns and the wind–wave climate. The researchers are examining the possible effects of local influences such as altered patterns of agriculture, the use of waterways and water pollution; the impact of trends in **climate** attributed to global warming (see **climate change**); and the results of changes in land-use for coastal defence, or for development and leisure. LOIS came about because the United Kingdom lies on a wide **continental shelf** bordering the Atlantic Ocean. It possesses an extremely long and intricate coastal zone of great physical and ecological diversity, on which lie many centres of large population, commerce and industry. Interaction between human and natural factors is causing the physical and biological environments of the coastal zone to change rapidly. In many places cliffs and dunes are being eroded. In others, the coast is being built out and estuaries are being filled up as a result of accumulation of sediment. Some of the **sediment** may be supplied by rivers and carried along the coastline, or carried shoreward from the sea. Human activities not only influence the physical character of this zone, they have an especially strong influence on its ecosystems. These have been significantly modified over the last 200 years by an increasing flow to the coastal zone of an ever more diverse range of substances connected with agriculture, urban growth, the burning of fossil fuels and industrial development. While some of these topics have been explored separately, the Natural Environment Research Council decided that a piecemeal approach was no longer a satisfactory way to attempt to understand how best to manage coastal zones. Consequently, a large number of research groups from universities and other institutes were invited to conduct the LOIS project. However, LOIS itself forms only one piece in the jigsaw of a particular area of environmental research. LOIS will obtain information on which to work from larger studies such as BOFS (Biogeochemical Ocean Flux Study) and TIGER (**Terrestrial Initiative in Global Environmental Research**). (See also **coastal areas**.)

LANDFILL is the oldest method of **waste disposal** for the solid matter discarded in the domestic dustbin, along with packaging material and paper from shops and offices. Landfill sites are usually disused quarries and gravel pits. When they were filled, previous practice was to cover them up with soil and forget about them. Housing estates have been built, often with disastrous consequences, on old landfill dumps. Waste burial has now become a serious technology and a potential source of energy. Landfill sites can be designed to be bioreactors, which deliberately produce **methane** gas as a source of **biofuel** or alternative energy. Traditionally, waste tips remained exposed to air and aerobic microbes—those that thrive in air—in order to turn some of the waste into compost. However, open tips also encourage vermin, smell in hot weather and disfigure the landscape. In the 1960s, as a tidier and safer option, landfill operators began to seal each day's waste in a clay cell. While excluding vermin, the clay also excluded air. Decomposition relied on anaerobic microbes, which die in air. However, the process produced methane (natural gas), which was a safety hazard. Methane can be extracted by sinking a network of perforated pipes into the site. (See also **Fresh Kills Landfill** and **waste disposal**.)

LANDSAT: See **Earth Resources Technology Satellite**

The **LAW OF THE SEA CONFERENCES** are a series of UN conferences which began in 1958. The first UN Conference on the Law of the Sea (UNCLOS I) was convened to standardize the national limits of territorial waters. Traditionally, the limit of territorial waters was three nautical miles, because this had been the longest range of naval artillery in the 17th century. Chile was the first nation to claim 200 miles along its shoreline and other nations soon followed with similar claims. The conference tried to negotiate claims to territorial waters and to investigate questions of national rights to coastal mineral deposits. The second conference, UNCLOS II, took place in 1960 and was unable to reach agreement. In 1973 the third conference began and continued until 1982 when the **Law of the Sea** Treaty was agreed.

The **LAW OF THE SEA (UNCLOS III)** was the treaty agreed in 1982 at the end of the third of the UN **Law of the Sea Conferences**, which lasted from 1973 to 1982. It was an attempt to establish controls over a wide range of maritime problems, including fisheries, navigation, **continental shelves**, the deep seabed, scientific research and marine pollution. Out of 150 nations who took part in the conference, only a few have ratified the Treaty. Thirty-three countries rejected it, including the USA, the United Kingdom and the Federal Republic of Germany (then just West Germany). Others have signed but not ratified the Treaty. However, many of the Treaty's provisions to regulate maritime activities within defined zones and boundaries have been widely accepted as customary international law.

The Treaty was criticized because the question of areas of overlap between adjoining countries was not resolved, and because the arrangements could have meant that six countries would have a claim to more than one-half of desirable coastal and continental shelf areas of the entire world. Under the terms of the Treaty, 40% of the seas come under the control of countries with **coastal areas**. The treaty proposes: a territorial zone, extending 12 nautical miles from land, with a universal right to total sovereignty given to coastal states; a contiguous zone of 24 miles, with restricted rights; an exclusive economic zone (EEZ), a 200-mile area, with rights over scientific research, environmental preservation and economic operations; and a continental shelf over which states are allowed to explore and exploit resources as long as they do not infringe the legal status of water and air above. Sixty per cent of the seas, the 'high seas', are outside the jurisdiction of any one nation, but there is a proposal that there should be no restrictions on fishing or passage, although countries are bound by the Treaty to co-operate in preventive measures to control pollution. It is proposed that 42% of the 'high seas' are designated the 'common heritage of mankind', the resources of which would be controlled by the UN. To this end, the International Seabed Authority (ISA) was established in Jamaica in 1994.

LEACHATE is the liquid that drains from a waste dump when water percolates through the waste and gathers up polluting materials. The water may be from rain or streams flowing over or through the surface. A poorly managed **landfill** site can produce large amounts of leachate. Fresh waste produces the strongest leachate, with a very high biochemical oxygen demand (**BOD**). If there are residues of **ammonia** or **heavy metals** in the dump, the leachate could cause severe pollution if a large quantity were to get into rivers. Older leachates have lower BODs, but still contain a large amount of dissolved ammonia, which is especially toxic to fish.

LEAD (Pb) is a poisonous metal, but one that has been used widely for thousands of years—in **batteries**, alloys, shot, paints, petrol, in construction and in plumbing. Lead in paint, especially the older high-lead forms, has caused many deaths. Adding tetra-ethyl lead to petrol as an **anti-knock agent** to stop the engine knocking has led to distribution of lead into the atmosphere, and high concentrations appear in urban dust and air. This also increases levels of lead in food. The problem is exacerbated when lead solder is used in tins for food, but this is being phased out in many countries. **Lead poisoning** causes stomach pains, headaches, tremors and irritability. In severe cases it causes coma and death. There is controversy about the effects of low levels of lead on the development of a child's brain. It affects the nerves and the brain at very low concentrations. Research in the USA has found that the mental development of babies is affected by lead levels in the blood as low as 10 micrograms per decilitre.

LEAD-FREE PETROL is petrol free of organic lead compound, which was traditionally added during blending as the most convenient way to obtain a desired **octane rating**.

LEAD POISONING can occur at any age, but it refers particularly to the problem for children younger than six years old. Accumulated exposure can affect both the growth and intellectual development of an infant. Contamination by very tiny levels appears to be enough to impair intelligence if it happens in the very young. The primary hazard is from lead-based paint, ancient lead plumbing and airborne emission from exhausts of cars using leaded petrol. However, cases of contamination have also been attributed to lead workers bringing home lead dust on their clothes.

LEAN-BURN ENGINES are a type of more fuel-efficient road vehicle engine designed in order to reduce the emission of **nitrogen oxides** and **hydrocarbons**. Lean-burn engines use a higher air-to-fuel ratio than traditional engines and the fuel mixture passing through the injection unit is controlled by a microprocessor.

LEGUME INOCULANTS, one of the most straightforward applications of **biotech**, involve injecting **nitrogen**-fixing bacteria into the **soil** around the roots of legumes like peas and beans, in order to increase the soil population of the naturally occurring **rhizobia**, the nitrogen-fixing bacteria that live symbiotically with legumes (see **symbiosis**). Nitrogen is an essential macronutrient, which are substances people need in their diet. Though 80% of the air is nitrogen, plants and animals cannot harness it to convert into **amino acids** and protein. They rely on other sources of nitrogen, such as natural or man-made sources of **ammonia** and **nitrate** fertilizers for plants, which in turn provide amino acids and protein for animals. Only a few organisms can extract nitrogen from the air and convert it into a form that plants can assimilate. The rate at which this process of nitrogen fixation can supply nitrogen to a plant puts a limit on plant growth. In addition to inoculating soils, biologists are experimenting with ways of improving the efficiency with which plants process nitrogen.

LICHENS are composite organisms formed by the **symbiosis** between species of **fungi** and an **algae**. They are either crusty patches or bushy growths on tree trunks, stone walls, roofs or garden paths. As they have no actual roots, they get their sustenance from the atmosphere and rainwater. Lichens play an important role in the detection and monitoring of pollution, especially **sulphur dioxide**, as they are highly sensitive to pollution and different species disappear if pollution reaches specific levels.

LIFE-CYCLE ANALYSES (LCAs) are among the array of new methods under development by which companies can reduce the environmental effects of their activities through attention to the design of their products and production methods. The environmental impact (see **environmental impact assessment**) of a product is the sum of the resources needed to make it, plus the amount of waste created in its manufacture, during its use and in its disposal. An LCA scrutinizes the environmental costs of using particular materials and the energy requirements to make a commodity, as well as the waste released at each stage of a product's life. The LCA method can assist governments in putting policies like **eco-labelling** schemes into practice. However, different approaches to life-cycle analysis can lead to contradictory conclusions. A Swedish study concluded that milk cartons were preferable to refillable glass bottles, whereas a Swiss analysis found the opposite to be the case. A German study decided the two containers were equally acceptable, finding milk cartons better in terms of low atmospheric emissions and glass bottles a better option when considering discharges to water.

LIFE-SUPPORT SYSTEMS are those environmental conditions that are essential for human life: air, water, food and shelter. The environment can also present a hazard to health. According to the **World Health Organization (WHO)**, 'good health is a state of complete physical, mental and social well-being and not merely an absence of disease or infirmity'. It represents a balanced relationship of the body and the mind and complete adjustment to the total environment. The WHO says, conversely, that disease is maladjustment or maladaptation in an environment. It is a reaction for the worse between a person and the hazards or adverse influences in that individual's external environment. For the purposes of **environmental health**, WHO puts environmental influences into three broad categories: the physical environment, the biological environment and the human environment. The atmosphere is the primary part of the physical environment. A person suffocates in seconds without the oxygen in air. By contrast, exposure to excess solar radiation can cause sunburn, skin cancer and genetic mutations. While water is essential for life—a person can survive at most for a week without drinking water—it also presents a range of health hazards. Many diseases are caused by water-borne organisms. Bacteria causing **cholera**, dysentery, paratyphoid and typhoid are transmitted by contaminated water. Water-borne parasites cause **malaria**, tapeworm, yellow fever and **schistosomiasis**. Viruses which cause infectious **hepatitis** and poliomyelitis are other **water-borne diseases**. In addition to pathogens, there are other impurities in water. They include pesticides, herbicides, lead, zinc, mercury, arsenic, nitrates, fluorides, selenium, molybdenum, cadmium and sodium.

LIGNITE is a low-grade brown-black **coal** which gives off the most smoke and least heat. It is used mostly for generating electricity. It is an intermediate stage of formation between **peat** and bituminous or hard coal.

LIMITS TO GROWTH was a study commissioned by the **Club of Rome** and published in 1972 by a team of economists and futurologists working with Prof Dennis Meadows at the Massachusetts Institute of Technology, in the USA. The Club of Rome was formed by a group of international businessmen who were concerned by speculation about the impact on society and the environment of emerging trends in science and technology. The study employed elaborate computer calculations to process the most complex mathematical model so far devised to predict future developments from current trends. It projected a **Malthusian** vision that predicted the collapse of world order if **population** growth, industrial expansion, increasing pollution, insufficient food production and exploitation of mineral resources continued at current rates. *Limits to Growth* proposed a series of alternative developments to offset disaster. It called for a 'Copernican revolution of the mind' that rejected the belief in endless growth and the tacit acceptance of waste in a throw-away society. It proposed **zero population growth**, a levelling-off on industrial production, pollution control, **recycling** of materials, the manufacture of more durable and repairable goods, and a shift from consumer goods to a more service-oriented economy.

LINDANE is an odourless and highly volatile **insecticide**. It was popular for use as a household fumigant and was sold in pellet form to attach to light bulbs or electric wall vaporizers. Tests later proved that it was dangerous to humans and pets.

LIQUEFIED PETROLEUM GAS (LPG) used for cooking and heating is believed to contribute to dangerous levels of **air pollution** and of problematic **photochemical smog** in Mexico City, for instance. The form of LPG used in the city consists of one-half propane combined with a mixture of butane and butenes. The latter produce **ozone**, the main component of smog. It is estimated that leaking unburnt LPG could be responsible for one-third of the smog in Mexico City.

LISTERIA is a bacterium, known properly as *Listeria monocytogenes*, that causes the fever- and flu-like illness listeriosis. The infection is often transmitted by contaminated dairy products, such as soft cheese and yoghurt. Normally the symptoms are so mild that the sufferer is unaware of the infection. However, in pregnancy it can lead to miscarriage, still birth, neonatal death and serious neurological damage, including mental retardation.

The **LITHOSPHERE** is the earth's solid upper shell on which all species live. It is also the source of raw materials and energy. Increased industry and agriculture puts a strain on the lithosphere and can lead to landslides, mudflows, **ground-water** pollution and other geological disasters.

LIVING MODIFIED ORGANISMS (LMOs): See **genetically modified organisms (GMOs)**

LOKI is the volcano that erupted beneath the Vatnajokull glacier in Iceland in November 1996. Weeks after the eruption, torrents of melted ice from a vast frozen lake under the glacier began to pour onto the coastal plains, sweeping away bridges and power lines and causing other environmental devastation. It was estimated that the lake, the Grimsvotn, held some 3,000m. cu m of melt-water.

The **LONDON CONVENTION** is the familiar name for the Convention on Prevention of Marine Pollution by Dumping of Wastes and Other Matter. It was adopted in 1972 at a conference held under the auspices of the United Kingdom and came into force in 1975. The Convention, previously the London Dumping Convention, controls and regulates on a global scale the disposal at sea of wastes and other material of any kind, including ships and platforms. It covers all the world's oceans, aiming to control the dumping of waste from ships and aircraft. There is a specific blacklist of substances most dangerous to the environment: **mercury** and mercury compounds, **cadmium** and cadmium compounds, organohalogen compounds (see **halogenated compounds**), persistent plastics, crude petroleum, fuel oil, heavy diesel oil, hydraulic fluids and lubricating oils (if taken on board to be dumped) and high-level radioactive waste and other radioactive materials if the **International Atomic Energy Agency** rules they are unsuitable. Materials produced for chemical or biological warfare are also on the blacklist. A 'grey' list of substances that can only be dumped with a special permit includes: wastes with significant amounts of **arsenic**, **lead**, **copper** and zinc, organosilicon compounds, cyanides, fluorides and **pesticides**. The Convention has specific regulations on the dumping of several other materials that may present a risk to the marine environment and human health. It also controls the **incineration** of wastes on board ship and, in 1988, contracting parties agreed to take steps to minimize or substantially reduce the use of incineration of noxious liquid waste by January 1994.

LONDON DUMPING CONVENTION: See **London Convention**

LONDON SMOG: See **Clean Air Act**

The **LONG-RANGE TRANSBOUNDARY AIR POLLUTION (LRTAP) CONVENTION** was the first environmental treaty signed by both the East and West and is one of the most significant international treaties to address **air pollution** and **acid rain**. The International Institute for Applied Systems Analysis (IIASA), an organization based in Vienna, Austria, was created to maintain discussions between the West and the former Eastern bloc during the Cold War on issues like defence, water, trade and air. IIASA created a model demonstrating acid deposition that was part of the motivation for LRTAP. The UN Economic Commission for Europe (ECE) convened negotiations to develop the 1979 Convention in response to the problems with **transboundary pollution** and acid deposition in Europe. Five protocols were negotiated between 1979 and the end of the 20th century to amend the LRTAP Convention to address **nitrogen oxides**, **volatile organic compounds** and **sulphur dioxide**.

LOVE CANAL was a housing estate in Niagara City, upper New York state, USA. It was built in the late 1950s on a **landfill** site used during the 1940s and early 1950s to dump thousands of tons of chemical waste. In 1977 toxic gases released from buried and corroding drums of chemical waste began to bubble through the earth. The waste, mainly **dioxin**, seeped into the houses. The site was declared a Federal Disaster Area, the first time that a national emergency had been declared as a result of chemical pollution. It was more than 12 years later that the site was finally declared free from contamination. The incident ultimately led to the formation of a **non-governmental organization** called the Citizens Clearinghouse for Hazardous Waste (subsequently renamed the Center for Health, Environment and Justice) by Lois Gibbs

LOW-LEVEL RADIATION can refer to the radiation from **low-level wastes**, or to a long-standing controversy over whether such a thing as a safe threshold of **radiation** exists. Most studies of the response of people to radiation have been conducted on individuals known to have received doses of 100 rems, the equivalent of one **sievert** in the new standard units. Information about the effect of doses of radiation below 10 rems, or 0.1 of a sievert, is less reliable. First, there are no radiation dose records for people who live near nuclear power stations or **nuclear reprocessing** and waste handling plants (see **Three Mile Island**). Secondly, researchers believe a cancer induced by low-level radiation may not appear for 30 or more years. The main source of information on the effects of low-level radiation are from the continuing studies of the survivors of the nuclear bombs dropped on Japan at the end of the Second World War.

LOW-LEVEL WASTE is the most bulky kind of radioactive waste. It ranges from laboratory rubbish, such as bits of broken glassware, rubber gloves and pieces of plastic, to the cleaning materials used in the nuclear industry. Much of it was formerly sealed in concrete drums and dumped at sea, a practice that was stopped under the conditions of the **London Convention**.

LRTAP: See **Long-Range Transboundary Air Pollution Convention**

M

MACHYNLLETH, in mid-Wales and previously noted only as the site of the only independent Welsh parliament (1402), is the site of the Centre for Alternative Technology, where a community of 30 to 40 people provide a living example of 'environmentally friendly' behaviour. The community uses **wind power**, water turbines and **solar energy** to provide energy for their washing and cleaning, and to drive the usual household devices. Wood stoves are used to heat houses and for cooking.

MAD COW DISEASE: See **bovine spongiform encephalopathy (BSE)**

MAGNETICALLY LEVITATED SURFACE VEHICLES (MAGLEVs) use advanced transport technology and could become widespread in the 21st century in the form of high-speed trains. Superconducting magnets are used to levitate such vehicles. This eliminates the rolling resistance from track experienced by the wheels of a conventional train, and linear motors provide the forward propulsion. The potential environmental advantages of MAGLEV vehicles are that they do not emit the **pollutants** associated with petrol- and diesel-driven vehicles, and that their speed would reduce dependence on air transport, which causes serious environmental damage.

MAGNOX REACTORS were the nuclear reactors that powered the United Kingdom's first nuclear power programme. They used natural **uranium** as fuel encased in magnesium alloy, or Magnox, canisters. The reactors were cooled by carbon dioxide. The versions built for the electricity utilities were based on the Calder Hall reactor, built at Windscale (see **Sellafield**). Calder Hall was called the first commercial nuclear power station when it was opened in 1956. In fact, it was built to produce **plutonium** for the British atomic weapons programme. The electricity it generated was a by-product that was sold to the national electricity grid. In all, the British built 26 such reactors, and one was built in Italy and one in Japan (in 2003 only eight remained operational, at four sites in the United Kingdom, and all were due to shut down by 2010). Similar gas-cooled reactors, apart from the second-generation **advanced gas-cooled reactor**, were the **UNGG** reactors in France and some in the Democratic People's Republic of Korea (North Korea).

MAJOR ACCIDENT REPORTING SYSTEM (MARS) is a database of accidents with catastrophic consequences at industrial installations or during the transport of hazardous materials. A European Union directive to register major accidents was introduced because of the increasing quantities of toxic, flammable and explosive substances that are used, produced or processed at high temperatures or pressures. The MARS database lists over 150 accidents, analysis of which identifies three major points: most accidents occurred in petroleum refineries and in the petroleum industry, whereas the ceramic, cement, surface-coating and dyes industries were those least susceptible to accidents; highly flammable gases were the most frequent cause of accidents, and **chlorine** the most commonly released agent; and, statistically, most accidents happened during normal operations, although the frequency of accidents was actually higher during short periods of maintenance, start-up, shut-down and other non-standard plant conditions.

MALARIA is a disease associated with bad environmental conditions producing stagnant water near homes. The stagnant water may be the result of schemes intended to improve irrigation by a network of ditches or water storage impoundments. Control calls for elimination of the breeding sites of the **mosquitoes** that carry the disease-causing parasite. Hopes of eliminating the mosquito with **DDT** and other pesticides have faded. The insects appear to have developed resistance to the pesticides and this has further aggravated the problem of infestations.

MALTHUSIAN refers to the theory of the classical economist, Thomas Malthus (1766–1834), who maintained that population growth would always outrun food supply and that in the end human numbers would have to be kept down by famine, disease or war. He was the first person in the United Kingdom to be appointed a professor of political economy and he first formulated what is now called the law of diminishing returns.

MAN AND THE BIOSPHERE (MAB) was a **UNESCO** project started in October 1970. It was launched to further long-term conservation of the earth's **ecosystems**, based on sound scientific foundations. MAB results have increased scientific understanding of ecosystems and their inter-relationships. The detailed research opened up new avenues for tropical ecosystems in particular. New species were discovered, and many forest plant species were seen to have potential medicinal benefits. By the 1980s about 25% of Western medicines contained at least one ingredient supplied by the **tropical rainforests**. Under the MAB programme, by 2003, 440 **biosphere reserves** had been established world-wide.

MAN-MADE HABITATS are those areas developed and influenced by man over the years to provide shelter and food for the population. (See **agricultural habitats**.)

MANFREDONIA is the site of a petrochemical works explosion in Italy in 1976. Ten metric tons of **arsenic** trioxide, 18 tons of potassium oxide and 60 tons of water escaped from an **ammonia** cooling column. One thousand people were evacuated from an area of 4 sq km around the plant. Olive groves and almond plantations were among the crops contaminated in an area of 15 sq km. Several herds of sheep were poisoned by arsenic, and abnormally high levels of arsenic were measured in sea water and fish.

MANGANESE (Mn) is an essential **trace element**, required both by animals and plants. In plants it is involved in the process of **photosynthesis**. The leaves of plants deficient in manganese become pale green or yellow. In animals it is a growth factor necessary to encourage bone development. Critical enzymes need manganese. However, inhalation of manganese dust as a pollutant also causes poisoning, including brain damage.

MANGROVE FORESTS or swamps form one of the most diverse **ecosystems** throughout the tropics. They are also one of the most endangered. Mangroves grow at the edge of the sea and provide a unique habitat for some 2,000 species of fish, invertebrates and plants. They are a valuable source of timber, pulpwood, fuel and charcoal, and of raw materials used in dyes, glues, rayon and tannin. They are also the breeding ground for a variety of fish, shrimps, crabs and molluscs. Mangrove forests play an important part in the desalination of **seawater**. As the roots of the trees are in the seawater they protect shorelines from erosion and act as a bulwark against storms. However, over the years many acres of mangrove forests have been destroyed. They have either been cut down for charcoal production, fish farming development, or to clear areas for agricultural or urban expansion. They have also been affected by **pesticides** leached from agricultural areas and choked by sediment from coastal development and mining operations. The mangroves of the Philippines, for example, have been reduced to only 380 sq km from an estimated 5,000 sq km in the 1920s. Most of the lost area has been cut for its timber, or converted into fish farms or rice fields. In many areas throughout the tropics the destruction of the mangrove forests has started a chain of coastal degradation. However, in some countries, including the USA and Thailand, mangrove forests are being cultivated to protect coastlines from erosion, to trap the silt draining off the land and to filter out pollution.

MARPOL: See **International Maritime Organization (IMO)**

MASS EXTINCTION: See **Extinction**

MAUNA LOA OBSERVATORY, in Hawaii, USA, was the world's first observation station to measure long-term changes in **carbon dioxide** concentrations in the **atmosphere**. It was established by a US scientist, who began work in 1957. Subsequently, carbon dioxide measurements have been made by scores of stations across the world. The carbon dioxide content of the atmosphere has increased by about 25% since the burning of petroleum and coal on a large scale began during the industrial revolution, before which the carbon dioxide concentration was 280 parts per million (ppm). When observations began at Mauna Loa, carbon dioxide concentrations had reached 315 ppm, rising to 360 ppm by 2000. The increase in carbon dioxide concentrations results from the total annual emissions into the atmosphere, which are estimated at about 7,000m. tons and are brought about by the burning of **fossil fuels** and **deforestation**. Only 50% of **anthropogenic** carbon dioxide remains in the atmosphere. The other human-induced emissions are a small part of the large transfers of carbon dioxide conducted between the atmosphere, the ocean and the terrestrial **biosphere** during seasonal growing cycles. Relatively small changes in the world **ocean circulation** and chemistry, the temperature of the oceans, or in the life cycle of terrestrial vegetation, could dramatically affect the levels of carbon dioxide in the atmosphere.

MEAs: See **multilateral environmental agreements**

MEDEA is a committee of scientists formed to study satellite data relating to **Antarctica**, in order to measure changes in the extent and shape of the ice-cap that might indicate **climate change**. Some of the data took the form of pictures originally taken in the 1960s and 1970s by a series of US spy satellites placed in polar orbits in order to monitor the former USSR. MEDEA was to compare these now-declassified pictures with images from Radarsat, a satellite launched in 1995 as a joint Canadian–US operation.

The **MEDITERRANEAN SEA** has been the dumping ground for millions of metric tons of waste, mostly emanating from land sources, for several decades. The Mediterranean has a coastal population of over 100m. Another 100m. tourists visit the area every year. A high proportion of raw sewage is dumped into the sea. There are 60 petroleum refineries on the coastline, producing substances that are very destructive to the marine environment. The direct dumping of wastes from ships is banned, but the Mediterranean is still one of the most polluted seas in the world. To compound the situation, the sea has a high rate of evaporation and a slow rate of water renewal, and, consequently pollutants accumulate but do not break down. It is believed that it takes about 80 years for the Mediterranean to renew its water. Shellfish from the sea are unsafe for human consumption, except for those from a few recognized growing areas. A Mediterranean Action Plan has been drawn up in an attempt to halt pollution and clean up the sea. The action plan, part of the **UN Environment Programme (UNEP)** Regional Seas Programme, involves 16 Mediterranean states that border the sea. These states have initiated water improvement projects. In 2004, however, UNEP reported that parts of the Mediterranean, such as the Adriatic, were suffering from so-called dead zones, stretches of **anoxic** waters that were increasingly appearing in the world's seas and **oceans** as a result of **eutrophication**.

MERCURY (Hg) is widely used in industry, but it has long been known that it is highly poisonous and, hence, an occupational hazard, causing brain damage and death if as little as one gram is ingested. It is a highly toxic **heavy metal** which attacks the nervous system. Small doses cause irritability and headaches, while large doses lead to convulsions, coma and death. Like all heavy metals it accumulates through **food chains**. The increased toxicity of mercury was recognized relatively recently once it was realized that it could be transformed in the environment into an organic compound and then absorbed in food, such as fish.

The **METABOLISM** is the sum of all the physical and chemical routes by which the body processes foodstuffs and other compounds. As foodstuffs and other substances are broken, the new compounds are referred to as metabolites. Sometimes metabolites are poisonous. For example, the processes of metabolism bind strontium to bone and allow **lead**, **mercury** and **cadmium** to cross the **blood–brain barrier**.

METHANE (CH_4) is the main constituent of natural gas. It is the simplest **hydrocarbon** and a valuable raw material for making organic chemical compounds. Methane is produced by decaying organic matter. Consequently, as well as being found in deep natural gas wells, methane also bubbles up in marshes and bogs, is produced in the flatulence of ruminant animals like cows and in paddy-field cultivation, and is generated by termites as they eat their way through forest wastes. Methane is also a **greenhouse gas** and an **ozone layer** destroyer. As a greenhouse gas, atmospheric methane has a high capacity for absorbing heat, a property which makes methane a significant contributor to **climate change**. Molecule for molecule, methane is 20 times more effective as a greenhouse gas than carbon dioxide, which is widely regarded as the most critical greenhouse gas.

METHYL-MERCURY is an **organometallic compound**, which is a substance produced when a metal atom, like **mercury**, is combined with an **organic group** by the process of **methylation**. Organometallic compounds may be man-made or produced naturally in the environment when a metal pollutant is caused to react with an organic group. If an environmentally produced agent, like methyl-mercury, gets into the **food chain**, it can be highly toxic to people. The **Minamata Bay** pollution disaster was caused by factory effluent containing mercury, which was converted on the seabed into methyl-mercury and then entered the food chain. The methylation process was performed by marine micro-organisms, which in turn were eaten by other marine life. The toxin accumulated in shellfish eaten by the local population, with tragic results.

METHYL TERT-BUTYL ETHER (MTBE) is the chemical name of an additive that can be put into petrol as an alternative to the environmentally hazardous **organometallic compound** tetra-ethyl **lead**. Additives are used in petrol as a cheaper means of increasing the **octane rating** of automobile fuel in preference to more expensive, additional refining. In other words, the performance of low octane-rated blends of petrol can be boosted by additives like MTBE. These additives are also referred to as oxygenates, which have the purpose of ensuring better combustion of the **hydrocarbon** fuel. They also reduce the amount of carbon dioxide in the exhaust gases. There are still by-products of combustion, such as **nitrogen oxides**, that can cause **photochemical smog**. The by-products can be removed by devices like **catalytic converters**. However, converters are damaged by lead. The impetus for MTBE-type additives began in California, USA, and they were then adopted in the US **Clean Air Act**. Demand for MTBE was 8m. metric tons in 1991, with demand forecast for 35m. tons by 2001.

METHYLATION is a process by which compounds are modified by the addition of a methyl group, usually by a reaction involving methyl alcohol. It is a vital process, which is exploited by the chemical and pharmaceutical industry and in research. However, methylation can happen naturally, and a relatively innocuous substance can be converted in the environment into a substance that is highly toxic if it gets into the food chain, as happened in the **Minamata Bay** tragedy. Besides being the product of important industrial processes, many essential methyl compounds are made by animals and plants.

MICROBES are micro-organisms, single cells, or clusters of cells, that can survive on their own if they have nourishment. They include **bacteria**, **algae**, **fungi** and yeasts. They are independent cells, unlike brain cells, for example, which can only survive in the brain, or leaf cells, which can only survive in the plant leaf. They can adapt to any environment around the world. They have in fact been used over the centuries, albeit sometimes unknowingly, by farmers, animal and plant breeders, vintners, brewers and bakers.

The **MICROBIOLOGICAL RESOURCES CENTRES (MIRCENs)** are a network of regional institutions which maintain regional centres for **microbiology** research and development of micro-organisms. The six centres were started by the UN Environment Programme in collaboration with the UNESCO in 1974. It was the first attempt to extend **biotech** to the developing world, and maximize the potential of using micro-organisms to solve regional and national environmental problems. Biotechnology offers developing countries a way to tap their enormous **biological diversity** for economic development. The centres were established in Brazil, Guatemala, Egypt, Kenya, Thailand and Senegal. Subsequently, other centres were established or accredited. Each MIRCEN is a network of collaborating laboratories and institutions, maintaining a regional **gene bank** of microbial resources, and serves as a centre of excellence for conservation, research and training. In Guatemala, MIRCEN concentrates on the use of microbes to convert coffee wastes into **biogas** and organic fertilizer; in Thailand research is carried out on turning cassava wastes into power alcohol; in Cairo, Egypt, they are examining the use of micro-organisms in pest and vector control and in degrading persistent pollutants. Other MIRCENs are exploring biological fertilizers, based on nitrogen-fixing bacteria.

MICROBIOLOGY is a branch of biology that deals with micro-organisms, and the major scientific discipline from which principles are applied for use in **biotech**, which produces vaccines, hormones, vitamins, organic solvents, dairy products, wine and beer, and compost. Microbiology has contributed in various ways to such advances as the identification, in the 1940s, of DNA (see **nucleic acids**) as the genetic material, the revelation of the universal genetic code in the 1950s, and the production of human insulin from genetically engineered bacteria in the 1980s.

MIGRATION: See **Bonn Convention**

Milutin **MILANKOVITCH** and James Croll conceived an astronomical theory of **climate change**. They analysed the pattern of the last five ice ages, the **Pleistocene** ice ages, and concluded that the climate had been influenced by the variation in the earth's orbit of the sun. The earth does not revolve in a circular orbit round the sun at a constant velocity, but over a period of years changes occur that affect the amount and distribution of solar radiation received by the earth. First, the tilt of the earth's axis of rotation relative to the plane of its orbit varies between 21.80° and 24.40° over a period of 40,000 years. This effects the seasonality or thermal range between summer and winter. Secondly, the ellipticity or eccentricity of the earth's orbit varies slightly, and in the 1930s Milankovitch, a Yugoslavian (Serbian) mathematician, calculated a periodicity of 96,000 years for this variation. In other words, the orbit becomes alternately more circular and more elliptical. Thirdly, the season when the earth is nearest the sun (perihelion) changes over a period of 21,000 years. At present it is nearest in the northern hemisphere winter, on 7 January, and farthest in the northern hemisphere summer, on 7 July. Milankovitch analysed these cycles and computed the times when he argued they conspired to trigger the start of ice ages.

MILLENNIUM SEED BANK: See **World Seed Bank**

The **MILLIMETRE-WAVE ATMOSPHERIC SOUNDER (MAS)** is a special instrument developed as part of a long-term, international environmental monitoring programme for analysing the **atmosphere** from the US space shuttle. The MAS searches the atmosphere for global changes in levels of **ozone**, chlorine monoxide, **water vapour**, temperature and pressure. The MAS is, in effect, a receiver that 'listens' to the emissions from molecules in the atmosphere for their unique signatures. Measurements began on space flight STS-45, on 24 March 1992, at the start of a 10-year programme. The same instrument was due to go into orbit every year for 10 years, covering a solar cycle, in order to sort out natural variations, such as **sunspot** fluctuations, from man-induced ones. Since scientists have yet to devise a long-term method of permanently measuring anything from orbit whose accuracy did not soon degrade, the MAS instrument has to be flown back and forth in the space shuttle. As little as a 1% change in accuracy invalidates any readings. By retrieving the MAS after every flight and recalibrating it, scientists can be assured that readings are comparable year after year. The researchers are looking for changes of a fraction of 1% in ozone, chlorine monoxide and water vapour levels for indications of global **climate change**. They are also looking for temperature changes because as the earth's surface heats up, the upper atmosphere cools down. The atmosphere should cool more than the surface heats up, and it might be easier to spot changes in the atmosphere because that change occurs at a faster rate. It was initially decided to go into orbit because ground-based instruments have to look through all the layers of the atmosphere and it is difficult to separate them out and interpret the measurements. In orbit, the MAS looks tangentially across the edge of the atmosphere and can scan much thinner layers of it. Measurements can also be made on a global scale, rather than just from one location.

MINAMATA BAY in Japan had the misfortune to bring the health hazards of **heavy metal poisoning** to public attention. The danger of **mercury** poisoning was well-known during the 19th century as an occupational hazard in industries handling the substance. The accident that shocked the world into recognizing its dangers if discharged into the environment happened at Minamata. Since the 1930s methyl mercury had been discharged from the Chisso chemical plant into the bay, which was the source of shellfish for local villagers. Mercury discharges were thought to be diluted and dispersed to sea. In practice, the mercury settled on the seabed and was absorbed into the **food chain**, especially by **filter feeders** such as shellfish. Once in the food chain, mercury gradually accumulated in the bodies of local people, causing a life-threatening and painful disease. By the 1950s it was causing mental retardation, seizures, loss of co-ordination and partial or total loss of vision and hearing. At first, the mystery illness, which came to be known as Minamata disease, was simply called itai-itai, resulting from the cry of pain from its victims when they tried to move their limbs. Mothers who were apparently healthy gave birth to children with severe physical and mental disabilities. The last report on the incident, made by the **Worldwatch Institute**, recorded 649 deaths and 1,385 other sufferers from Minamata disease.

MIXING LAYER refers to the lowest layer of the atmosphere, close to the earth's surface, into which most airborne pollutants are emitted. Until the early 1960s it was believed that air pollutants, once released, were eventually dispersed and diluted through the atmosphere until they had reached negligibly low concentrations. However, measurements made in the development of a new field of research in atmospheric chemistry showed otherwise.

Three observations were made concerning the behaviour of air pollutants. First, under certain meteorological conditions, namely a **temperature inversion**, pollutants are trapped in high concentrations close to the ground and form **photochemical smog**. Secondly, there can be interaction between different man-made and natural pollutants, which can happen near the source of pollution or after it has dispersed. The results of this interaction are returned to the ground in forms like **acid rain** that are damaging to **ecosystems**, health and buildings. Thirdly, individual pollutants stay in the atmosphere for differing lengths of time. This duration, or residence time, is determined by the processes of deposition and chemical conversion. If a pollutant remains in the atmosphere for 30 days, it is likely to spread by the process of vertical mixing between the highest and lowest levels of the **troposphere**. When resident times approach six to 12 months, pollutants can be carried from one hemisphere to another. After 12 months, pollutants start spreading from the troposphere up to the **stratosphere**, which occurs in the case of European and North American emissions of **chlorofluorocarbons**, which deplete the **ozone layer**, and of the so-called **greenhouse gases** responsible for **climate change**.

MOLLUSCICIDES are used to control snail populations, both of the type which inhabit gardens, fields or greenhouses, and those which are the intermediate hosts of parasites that cause disease in humans. Some species of snail, especially freshwater snails, are intermediate hosts to organisms which cause **schistosomiasis** in humans, and liver and lung fluke in humans, dogs, cats, and domestic animals. One molluscicide, metaldehyde, has been used as a bait for slugs and snails since it was discovered in 1936. (See also **pesticides**.)

MOLYBDENUM (Mo) is an essential **trace element** that is a component of a number of enzymes. Excess amounts of it are injurious to the human body.

MONOCULTURE is the cultivation of a single crop, usually on a large area of land. This can result in **soil erosion**, loss of **biological diversity** and pest infestation.

The **MONSOON** is the intense seasonal rainfall which falls largely in the Indian subcontinent, the People's Republic of China and South-East Asia. It is caused by a large-scale change of wind direction in the tropics, largely brought about because of changes in the heating of **oceans**.

The **MONTRÉAL PROTOCOL** was signed in 1987 by 24 countries, which agreed to reduce **chlorofluorocarbon (CFC)** production by 50% by 1999. By the end of the 20th century most countries in the world were parties to the Protocol, which was a product of the **Vienna Convention for the Protection of the Ozone Layer**. In a series of amendments to the Convention in the early 1980s, known as the London Amendments and Copenhagen Amendments, countries agreed to phase out CFC production completely by 1996 in the industrialized world. Developing countries were allowed over 10 years longer to phase out their production of **ozone**-depleting substances. In part because developing countries were still producing CFCs in the late 1990s, a 'black' market (illegal trade) in CFCs appeared in the industrialized countries. Many developing countries started to phase out their production with the assistance of a Multilateral Fund contributed to mainly by industrialized countries. At the end of the 1990s the People's Republic of China was the largest producer and consumer of CFCs. One of the main alternatives to CFCs, the hydrofluorocarbons (HFCs), are major **greenhouse gases** and are now meant to be regulated under the **Kyoto Protocol**. The other main alternative is the hydrochlorofluorocarbons (HCFCs), which are also ozone-depleting substances, albeit much less damaging than CFCs. Even so, HCFCs were also to be phased out eventually. The other main ozone-depleting substance regulated under the Montréal Protocol is methyl bromide, a potent greenhouse gas and dangerous **pesticide** used mainly on grapes, strawberries, cut flowers and tomatoes. The USA is the largest producer and consumer of methyl bromide, closely followed by Israel, in terms of production. As a result of the Montréal Protocol, by 2000 some forms of CFC appeared to have reached their maximum concentration in the atmosphere, but the ozone layer had not yet started to recover. Ozone depletion was expected to continue long into the 21st century because CFCs persist for many decades in the atmosphere.

MOSQUITOES are among the worst disease-carrying pests known to mankind. There are some 3,000 species, distributed almost world-wide. In the tropics they are particularly abundant. The males of most species feed on plant juices, but the females need blood to reproduce. They get this by biting and sucking the blood of humans and animals and, at the same time, often transmit serious diseases to both humans and livestock. The ***Aedes aegypti*** breeds in rivers and coastal areas, near human settlements. It is the principal carrier responsible for transmitting dengue fever and yellow fever. The female lays her eggs in stagnant water. Other species of *Aedes* act as vectors of filariasis and other viruses that cause forms of encephalitis. ***Anopheles*** mosquitoes also inhabit both tropical and temperate regions. The **malaria** parasite, *Anopheles* plasmodium, is transmitted only through the bite of a female. Other species of *Anopheles* transmit the parasites that cause Bancroftian filariasis, which can lead to blindness. It was projected that, as a result of **climate change**, some species of mosquitoes would be able to move into new areas of the world where they were not previously present, which could increase the spread of mosquito-borne infectious diseases.

MOUNT PINATUBO is the **volcano** that erupted in the Philippines in 1991 with serious environmental consequences, having previously been dormant for 600 years. The eruption caused the immediate deaths of more than 300 people, many of whom were trapped in buildings that collapsed under the weight of volcanic ash. More than 200,000 people were evacuated. The volcanic plume reached an altitude of more than 30 km. The finest **aerosols** reached the upper atmosphere and eventually covered much of the planet with a layer of stratospheric **particulates**. These particles reflected so much sunlight back into space that meteorologists took it into account in their calculations of global warming. Mount Pinatubo was the most intensely observed volcanic eruption on record. Scientists studied the volcanic cloud, and its effect on **climate** and stratospheric **ozone**, in great detail, using equipment based both on the ground and in the air, including solar photometers, airborne aerosol counters and satellite instruments.

MOUNTAINS are an important source of water, energy, minerals, forest and agricultural products, and recreation. They are storehouses of **biological diversity** and **endangered species** and an essential part of the global ecosystem. About 10% of the world's population depend on mountain resources and nearly one-half of these people are affected by the degradation of mountain watershed areas. Mountain ecosystems are very vulnerable. They are increasingly sensitive to **soil erosion**, landslides and rapid loss of habitat and genetic diversity. Widespread poverty and an increase in the numbers of mountain inhabitants lead to **deforestation**, cultivation of marginal lands, excessive livestock grazing, loss of **biomass** cover and other forms of **environmental degradation**. As little is known about mountain ecosystems, **Agenda 21** proposed the establishment of a global mountain database. This is essential for the launch of programmes that would contribute to the **sustainable development** of mountain ecosystems. The proposals also focus on promoting watershed development and alternative employment for people whose livelihoods are linked to practices that degrade mountains.

Negotiations on a **MULTILATERAL AGREEMENT ON INVESTMENT (MAI)** were initiated by members of the **Organisation for Economic Co-operation and Development (OECD)** in May 1995. The goal of the negotiations was to reach agreement on a new set of regulations for the treatment of foreign investors, in order to increase their market access, provide them with greater protection and devise a dispute-resolution mechanism for investors and host countries. The early drafts of the agreement had almost no binding environmental, labour or social provisions to ensure that foreign investors would behave responsibly. Moreover, the agreement could have exempted foreign investors from certain domestic regulations. Therefore, many environmentalists and other social activists were strongly opposed to the MAI. Negotiations failed as a result of widespread controversy and opposition, but the issue was taken up again by both the OECD and the **World Trade Organization (WTO)**.

MULTILATERAL ENVIRONMENTAL AGREEMENTS (MEAs) are internationally negotiated agreements designed to protect the environment. Sometimes they include some form of trade restriction to achieve their aims, and there is often some conflict or contradiction between the goals of multilateral environmental agreements and trade agreements. MEAs are often used to resolve issues related to **transboundary pollution**. On a global scale they include agreements like the **Montréal Protocol** on **ozone**-depleting substances, the **Convention on International Trade in Endangered Species of Wild Flora and Fauna** and the **Basle Convention on the Control of Transboundary Movements of Hazardous Wastes and their Disposal** on hazardous waste shipments.

MURUROA ATOLL is a coral island in the South Pacific Ocean used by the French Atomic Energy Commission (CEA) for underground nuclear tests from 1966. It was the site of a series of confrontations between environmentalists and the French authorities. Tests were conducted by drilling an 800 m shaft below the atoll lagoon. French officials insist that the explosions posed no danger to the regional populations or to the environment. The closest inhabited island is 200 km away and the explosions were believed to cause no radioactive contamination or significant ecological impact. The last programme of tests began in September 1995 and, when it was completed, the French Government ceased testing and signed the UN Comprehensive Test Ban Treaty. A total of 140 underground nuclear tests were conducted at Mururoa and the neighbouring Fangataufa atoll over 30 years, despite regional objections. Although the site was cleared, the environmental group **Greenpeace** and countries in the southern hemisphere are concerned about long-term pollution. They fear that the underground tests may have cracked the basalt rock beneath the coral atolls, causing **radioactive** materials to leak out into the ocean. Consequently, at the request of the French ministry of foreign affairs, specialist teams were organized by the **International Atomic Energy Agency** to investigate the environmental situation and prepare an **environmental impact assessment** covering the next 100 years. International teams of experts visited the site to collect samples of **plankton**, fish, **sea-water**, lagoon **sediments**, coral, **soil**, coconuts and vegetation for analysis. The US $1.5m. research project was completed in 1998, but extended into a long-term monitoring programme.

MUSSELS are bivalve marine molluscs found in most waters, but more especially in cool seas. There are some 1,000 types of freshwater mussels in streams, lakes and ponds throughout most of the world. Some species, such as the blue mussel, are an important food source in Europe and are often farmed. The mussel is also extremely important as a biological indicator of water quality. Scientists have now developed new, sensitive techniques that indicate when human activity is causing damage to marine life. Using mussels, researchers are currently working on a programme, which uses as a yardstick the amount of energy available for growth. The presence of toxins in the water may also influence the animal's internal defence mechanisms. As a result of the use of specific blood cells from mussels and fish, there are now simple tests to assess the immune system efficiently. Mussel populations studied from 26 sites along the east coast of the United Kingdom show that there is considerable variation in the mussels' ability to resist infection between sites. Those from the generally less polluted Scottish coast, for example, tend to show higher resistance than those from the more polluted Thames and Teesmouth areas.

A **MUTAGEN** is an agent that causes changes to plants and animals, particularly to their genetic material and especially at the time of reproduction. Certain chemicals and forms of radiation are power mutagens that damage the DNA, or genetic material (see **nucleic acids**), in the centre of every cell of a living organism.

MUTATIONS come about as a result of changes in the DNA (see **nucleic acids**) which forms the genes in a nucleus of a cell. Mutations in sperm or egg cells may lead to inherited defects in children. Mutations in the non-germ (or somatic) cells that make up the body may lead to disease in an individual. The DNA of a person contains about 50,000 genes. A chemical pollutant or exposure to radiation may damage any gene.

N

NADPH: See **photosynthesis** and **respiration**

NAFTA ENVIRONMENTAL SIDE AGREEMENT: See **North American Agreement on Environmental Co-operation (NAAEC)**

NARMADA DAMS: See **Sardar Sarovar Dam**

NASA: See **National Aeronautics and Space Administration (NASA)**

The US **NATIONAL AERONAUTICS AND SPACE ADMINISTRATION (NASA)** was created in 1958 in response to the launch of the USSR's Sputnik rocket. NASA's main objective for many years was to explore options for human space flight. However, the massive research and development effort led to many inventions that benefit the environment, such as **photovoltaic solar power systems** and efficient engine technologies. NASA's Landsat and **Earth Observing System** have helped provide comprehensive data about the surface of the earth. NASA satellites—in particular the **Total Ozone Mapping Spectrometer** Satellite—continue to monitor **ozone** depletion and other climatic factors.

The US **NATIONAL OCEANS AND ATMOSPHERE ADMINISTRATION (NOAA)** was established to predict environmental changes, protect life and property, and provide policy-makers with reliable scientific information. NOAA is concerned with both short- and long-term environmental change and is one of the most important collectors of environmental data in the world. The National Climatic Data Center, for example, co-ordinates the collection and analysis of climatic data for the USA and much of the rest of the world, providing an important data set for the study of global **climate change**. NOAA also works to prepare for emergencies caused by extreme weather events and earthquakes by providing timely and reliable observations, assessments and information to policy-makers. A new field of activity for NOAA was the issuing of seasonal outlooks of climate variability based on research into the **El Niño** events.

NATIONAL PARKS are areas of outstanding natural beauty, set aside for the **conservation** of flora and fauna, and scenery, and for recreation, if this does not conflict with the conservation objectives of the parks and their landscapes. Hunting, logging, mining, commercial fishing, agriculture and livestock grazing are all controlled within national parks, as is industrial activity. The first such area was **Yellowstone National Park** in Wyoming, USA, established in 1872. It was regarded as one of the first national efforts to protect wild nature. National parks are usually selected for their unique characteristics: outstanding beauty, unusual geological factors, exceptional wild animals or plant life. National parks tend to be found in more densely populated countries, such as the United Kingdom (which declared its smallest national park, and the first in 15 years, in the New Forest in 2004), and are usually areas of natural beauty that have been changed by long periods of human occupancy. More emphasis is placed on the protection of traditional landscapes, which also include facilities for sailing, walking, climbing and fishing. National parks have had amazing success in restoring plant and animal species. The **World Conservation Union (IUCN)** has encouraged all countries to set aside at least 10% of national land and water resources for conservation, preferably as national parks. In fact, although they vary in size and character, and have different styles of management, more than 100 countries have already established national parks.

NATURA 2000 is the European Union-wide system of conservation areas established by the **Habitats Directive** of 1992 and the **Birds Directive** of 1979. The purpose is to ensure that the overall network safeguards the species and habitats listed in the annexes to the Habitats Directive. The Birds Directive also included the injunction to re-establish and create habitats.

NATURAL GAS is an attractive fuel for small- to medium-sized power stations from an environmental viewpoint, because it almost eliminates emissions of **sulphur dioxide**, reduces **nitrogen oxides** by about 45%, and discharges about 50% less **carbon dioxide** than **coal** when burned.

NATURAL GAS VEHICLES: See **alternative fuels** and **compressed natural gas (CNG)**

NATURAL RESOURCES used to refer to the living resources such as forests, wildlife and plants that were able to regenerate through reproduction. They were regarded by the early conservationists as separate from the inanimate, non-renewable resources such as mineral and **fossil fuel** deposits that, once used, were irreplaceable. Anxiety about the environment, and particularly concern generated by ***Limits to Growth***, which was published in 1972, focused attention on the need to conserve the inanimate natural resources. **Recycling** is one of the most attractive methods of reusing or refining waste materials instead of producing new materials. Consequently, recycling can reduce the volume of mineral extraction. This is also applicable to products made from living resources, such as wood and paper, and to the use of sewage concentrate as a fertilizer, as an alternative to having it manufactured in a chemical plant.

NATURE RESERVES are areas allocated to preserve and protect certain animals or plants. They differ from **national parks**, which are largely places for public recreation, because they are provided exclusively to protect species for their own sake. **Endangered species** are increasingly being kept in nature reserves to prevent them from extinction, particularly in India, Indonesia and some African countries. Nature reserves were used once to preserve the animals that landowners hunted, but, in the 19th century, they became places where animals were kept to prevent them from dying out. Special refuges and sanctuaries are also often designated to protect certain species or groups of wild animals or plants, especially if their numbers and distribution have been significantly reduced. They also serve as a place for more plentiful species to rest, breed or winter. Many parts of the world also have marine and aquatic reserves to protect different species of sea or freshwater plant and animal life, as, for example, on Australia's **Great Barrier Reef**.

NEKTON describes animals that swim powerfully enough to make them independent of tides and currents. Fish predominate, but there are also aquatic mammals and other animals.

NEPTUNIUM (Np) is a radioactive, silvery element of the **actinide** series of substances. In 1940 it became the first **transuranic element** to be synthesized. It forms part of the **high-level waste** products that are produced in spent nuclear fuel. Neptunium is particularly hazardous because it is an **alpha-emitter**.

NGANA is a parasitic disease found in cattle in tropical Africa and spread by the **tsetse** fly. As a consequence of the disease millions of square kilometres in the sub-Saharan region cannot safely be used for rearing livestock.

NIMBUS-7 is a meteorological satellite, operated by NASA, the US space agency, which found in the second half of 1992 that global **ozone** levels were 2%–3% lower than in any previous year and 4% lower than normal. The very lowest levels were observed in December 1992, when the global average of ozone levels was approximately 280 **Dobson units**. By comparison, a normal December value is about 293 Dobson units. Previously, the lowest level, of 286 Dobson units, had been observed in December 1987. The 1992 ozone levels were especially low in the mid-latitudes of the northern hemisphere. The December 1992 mid-latitude ozone levels were 9% below normal. The low mid-latitude ozone values continued into 1993. The January 1993 ozone levels were 13%–14% below normal. Scientists can only speculate on the cause of the low ozone values in 1992. While the exact cause is unknown, the low ozone may be related to the continuing presence of particles produced in the upper atmosphere following the eruption of **Mount Pinatubo** in the Philippines in June 1991. Nimbus-7 has measured ozone levels since November 1978 and was the sole monitor of global ozone levels until 1991 and the primary one into the mid-1990s. Its measurements were backed by the NASA TOMS instrument orbiting on the Russian Meteor-3 satellite, which was launched in August 1991, and the **Shuttle Solar Backscatter Ultraviolet** instruments flown on the space shuttle and the NOAA-11 satellite.

NIMBY and NIMBYism are rather pejorative terms often encountered in the context of the environment. They are derived by abbreviation of the phrase 'not in my backyard', ironically attributed to individuals who object to a proposed development, such as a road bypass, nuclear power station or rail link, on the grounds that it will affect the amenity of the area in which they live, although they have no objection in principle to a similar development taking place elsewhere.

NIÑA: See **El Niño**

NIÑO: See **El Niño**

A **NITRATE** or **NITRATE RELEASE FERTILIZER** is a man-made fertilizer, based on ammonium nitrate, that has been spread intensively in the years since the Second World War to provide the **nitrogen** needed for rapid growth of crops. Surplus nitrates are washed into rivers and threaten water supplies, causing a potential health hazard for both children and adults. Chemically, a nitrate is a salt of nitric acid, a compound containing NO_3.

NITRATE-SENSITIVE AREAS have water sources with high levels of **nitrates**, which are thought to have an affect on health. High nitrate levels are a problem and add to the costs of providing drinking water. Research has identified adjustments that could solve the problem of nitrate **leachates**, but solutions vary from area to area, owing to different hydrogeological, land-use and climatic conditions.

NITRIC ACID (HNO_3) is one of the most widely used reagents within the chemical laboratory and industry. It is formed in the atmosphere by chemical reactions involving the **nitrogen oxides** discharged from coal- and petroleum-fired power stations and petrol-driven vehicles, to produce a damaging and corrosive environmental pollutant.

NITRIFICATION is the process by which **ammonia** compounds, including man-made **fertilizers** and the **humus** provided by organic matter of plant and animal origin, are converted into **nitrites** and then **nitrates**, which are then absorbed as a nutrient by crops. Excess nitrate can be leached into surface waters and groundwaters, causing pollution. Excess nitrate may also be converted by microbes back into gaseous **nitrogen**, which is an important greenhouse gas, and released back into the atmosphere.

A **NITRITE** is a compound which is formed as an **intermediate chemical** when **ammonia**-based fertilizers, whether man-made or of natural origin, are converted by microbes in the soil to nitrate. On heavily fertilized land, the rate of formation of nitrite may be too rapid and some of it gets leached into waterways as pollution. Chemically, a nitrite is the radical, nitrogen dioxide, or any compound containing it (a salt or ester of nitrous acid).

NITROGEN (N) is an essential **nutrient** in the food supply of plants and the diets of animals. Animals obtain it in nitrogen-containing compounds, particularly **amino acids**. Although the atmosphere is nearly 80% gaseous nitrogen, very few organisms have the ability to use it in this form. The higher plants normally obtain it from the **soil** after micro-organisms have converted the nitrogen into **ammonia** or **nitrates**, which they can then absorb. This conversion of nitrogen, known as nitrogen fixation, is essential for the formation of amino acids which, in turn, are the building blocks of **proteins**. Nitrogen starvation of a plant is indicated by yellowness of its green leaves and shoots, and retarded flowering and fruiting. Some plants have special root nodules that are better at absorbing the nitrogen substances. In modern intensive agriculture, farmers have increased the production of crops by spreading nitrate-based **agrochemicals** on the land. While productivity has risen, the excess fertilizers have become environmentally damaging. They drain into waterways and can cause **eutrophication**.

The **NITROGEN CYCLE** is the complex set of processes by which crops acquire the large amount of **nitrogen** they need to make **proteins**, **nucleic acids** and other biochemicals of which they are composed, and how the nitrogen returns to the atmosphere. The ultimate source of nitrogen in the global ecosystem is the molecular nitrogen in the atmosphere. To a very limited extent, some dissolves in water. However, there is none in rock. Yet many arable crops will remove at least one metric ton of nitrogen, in the form of **nitrate**, from each five hectares of land during the growing season. Molecular nitrogen from the atmosphere first enters biological pathways by assimilation, or nitrogen fixation, by a variety of micro-organisms which live freely in **soil**. Once the nitrogen has been captured by one of the micro-organisms, there are many different routes, or processes, by which it is handled. Thus, the nitrogen cycle can be divided broadly into processes that add nitrogen in the form of nitrate to the soil and the processes that remove it. Since the recycling of nitrogen has been going on for millions of years, there are, in fact, various places where nitrogen is temporarily stored. Some is retained in the soil as decomposing plant matter, waiting to be released and taken up by new crops as nitrate. That nitrate is produced by nitrifying bacteria living in the soil that thrive on **ammonia**, which is produced by decaying plant and animal material. Nitrogen is an important element in ammonia, and in processing it the nitrifying bacteria produce nitrate that can be absorbed by the roots of plants. In addition to the nutrient-rich **humus** of decomposed plant matter or animal urea, man-made fertilizer is added to the soil in manufactured forms of nitrates and urea. The absorption of nitrates by plants is the major process of removing nitrogen from the soil. However, some nitrogen is released from nitrates by denitrifying bacteria, and escapes in gaseous form from the soil back into the atmosphere. Some drains away in the form of **leachates** of nitrates into waterways. Leaching of nitrates into the surface and ground-waters is of particular concern. European Union directives set a limit of 50 mg per litre for nitrate in drinking water or in any water where there is a risk of **eutrophication**.

NITROGEN DIOXIDE (NO_2) is a gas found in the exhaust fumes of cars and lorries. It forms **photochemical smog** in bright sunshine and has been suggested as a possible cause of **asthma** and other breathing difficulties.

NITROGEN OXIDES (NO_x) are a group of gases that include two of environmental importance formed directly or indirectly by the combustion of fuel in furnaces and vehicle engines: **nitrogen dioxide** (NO_2) and nitric oxide (NO). Nitrogen oxides can inhibit plant growth and contribute to **acid rain**. Nitric oxide undergoes a number of complex chemical reactions in the **atmosphere** to produce nitrogen dioxide, which in turn can lead to the formation of ground-level or **tropospheric ozone**. Ozone and nitrogen dioxide are the principal **secondary pollutants** attributable to transport (although small amounts of the latter pollutant are emitted directly in vehicle exhausts). Nitrous oxide (N_2O) is a potent **greenhouse gas** that is regulated under the **Kyoto Protocol**.

NOAA: See **National Oceans and Atmosphere Administration**

NOISE POLLUTION is perhaps the most obvious example of environmental pollution, caused when noise from traffic, aircraft, factories, loud radios, or neighbouring houses and gardens, becomes a nuisance or detrimental to health. Noise can be a constant intrusion or merely an irritant. It can also cause severe damage to health and hearing, and at a high level can rupture eardrums. The level of noise is measured in **decibels (dBs)**, which is the unit used for classifying the intensity of sound and allows different sources of noise to be compared. A report in the *New England Journal of Medicine* suggested that for 25 years noise levels had risen by about 1 dB per year. Noise is rapidly becoming a serious polluting factor. An increase in the amount of aircraft, road and rail traffic, portable radios and tape recorders, and household gadgets all contribute to growing noise pollution. There are laws covering the control of noise for road traffic, aircraft and airports. For example, the European Communities (shortly to become known as the European Union—EU) issued a directive on aircraft noise that came into force in 1993. New regulations also covered a new noise limit for buses, coaches and heavy goods vehicles. In 2003 the EU required 'noise maps' for every major city, in preparation for further regulation within five years.

NON-GOVERNMENTAL ORGANIZATIONS (NGOs) are private, voluntary, non-profit organizations that act in the public interest, educate the public, or act as advocates on behalf of a given issue. Throughout the world there are more than 5,000 international NGOs that are concerned with the environment and development, with millions of supporters. In all, it is estimated that globally there are some 12,000 such NGOs, ranging from small grassroots agencies to influential international groups like **Greenpeace** and **Friends of the Earth**. The development of such NGOs started primarily during the post-Second World War period, when movements were formed because of growing concern for the environment. After the **UN Conference on the Human Environment** in Stockholm in 1972, these movements expanded into developed and also developing countries. The NGOs operate on a regional, national or international level as an independent voice, sometimes acting as the focus of opposition to government policies and at other times acting in support of those policies. They are growing, both in numbers and influence, and are increasingly involved in decision-making processes at both the national and international level.

NON-LINEARITY refers to the fact that ecological systems do not always respond to **anthropogenic** pressure in a predictable and linear way. Ecological thresholds are often unknown by scientists because they are only discovered as a result of being crossed by accident. If such thresholds are passed, an unexpectedly large amount of damage can occur as a result of positive feedbacks (see **feedback cycles**). For example, the problem of **ozone** depletion demonstrated that the stratospheric **ozone layer** did not respond to increased concentrations of ozone-depleting substances such as **chlorofluorocarbons** in a linear way. At a certain level of concentration, ozone depletion became much worse and resulted in the **ozone hole** over **Antarctica**.

The **NON-PROLIFERATION TREATY (NPT)** is an arrangement by which countries that signed the treaty allow their nuclear power stations, and other nuclear sites, to be inspected in order to demonstrate that no materials are being diverted for weapons purposes. The Treaty is supervised by the **International Atomic Energy Agency (IAEA)**, which is based in Vienna, Austria. India, Israel and Pakistan do not subscribe to the NPT, but have developed nuclear weapons. Since India and Pakistan openly tested these weapons in 1998 there has been pressure for them to subscribe to the international system. The Democratic People's Republic of Korea (North Korea) was found to be in violation of its treaty obligations in 2002, having revived its nuclear-weapons programme, but declared it had withdrawn from the NPT in the following year. It has been under international pressure to submit to IAEA supervision ever since. The only other country believed to be ambitious to develop nuclear weapons is Iran, which officially still subscribes to the NPT. Any clandestine programme must, therefore, be in breach of international law. There has been a long-running dispute with the USA, which has been strong in its claims that Iran had a weapons programme (not merely a civil nuclear-power one) and in its condemnation of Russian help for any nuclear programme. The European Union attempted to mediate and urged Iran to submit to IAEA inspections, but their efforts seemed to have stalled in mid-2004.

The **NORTH AMERICAN AGREEMENT ON ENVIRONMENTAL CO-OPERATION (NAAEC)** is a 'side agreement' to the North American Free Trade Agreement (NAFTA). Under the side agreement, each country must prepare public reports on the state of the environment, strive for high levels of environmental protection and enforce domestic environmental laws and regulations. New organizations formed by this agreement are: the Commission for Environmental Cooperation (CEC); a Joint Public Advisory Council, consisting of business, government and environmental representatives; and a Council of Ministers, comprising the environment ministers of Canada, Mexico and the USA. One interesting provision of the Agreement is that any citizen who believes that a member state is 'failing effectively to enforce its environmental law' may submit its concerns to the CEC Secretariat, which will create a factual record of the complaint.

NORTH SEA: See **Oslo Treaty**

The **NUCLEAR FUEL CYCLE** describes a complex chain of events which begins with the mining and refining of **uranium** ore, followed by uranium enrichment. Rods of uranium fuel are then fabricated and placed in a reactor. When removed from a reactor where they have been subjected to months of bombardment by neutrons, the fuel rods are still predominantly uranium. In addition, they also contain potentially valuable **plutonium**, potentially hazardous long-lived radioactive wastes called **actinides** and many short- and medium-term radioactive wastes. Reactors were built originally to produce plutonium for weapons. Only a small proportion of the uranium fuel was converted into plutonium, which was then extracted from the spent fuel by chemical **nuclear reprocessing**. When the first nuclear power stations came into operation in the mid-1950s, reserves of uranium ore were still in short supply. Therefore, it made sense to continue the practice of reprocessing in order to recover the useful uranium left in the fuel. There was also the possibility of using plutonium in a new type of reactor. However, the disadvantage of reprocessing lies in the problems of handling and disposing of a wide variety of **low-level**, **intermediate-level** and **high-level waste** streams. All the activities from mining of uranium ore and fuel fabrication to reprocessing and waste disposal constitute the nuclear fuel cycle.

NUCLEAR FUELS are obtained from inorganic minerals extracted by mining. Although they are at least partially consumed when used in nuclear reactors for the production of heat, they differ from **fossil fuels** in the way they release energy. Burning of fossil fuels, such as **coal**, petroleum and **natural gas**, is a chemical reaction. Nuclear fuels, such as **uranium**, are destroyed by a process of spontaneous disintegration, called **fission**, and prompted by natural radioactivity. If the process is left to occur naturally in **uranium**-bearing rock, the rate of change is imperceptibly small. In a man-made nuclear reactor the energy-releasing processes of disintegration, which in the natural state happen slowly over thousands of millions of years, are compressed into minutes. The release of energy is harnessed to generate steam, which drives electricity generators. Fears of contamination from the fuel and its wastes, as well as of the potentially catastrophic accidents possible, traditionally motivated opposition by environmentalists to the nuclear industry. However, the scale of the threat of environmental disaster from **fossil fuels** always presented a counter argument. In May 2004 James Lovelock, a prominent British scientist and author of the **Gaia** hypothesis, claimed that the urgency of the global situation demanded the use of nuclear power as the most practical and efficient means to prevent any further generation of **greenhouse gases**. His suggestion created much controversy.

NUCLEAR MELTDOWN is when the core of a nuclear reactor overheats and permits **radioactivity** to escape. In **pressurized water reactors**, **boiling water reactors** and breeder reactors this could happen in an accident involving a loss of coolant, with emergency cooling systems failing. **Fast breeder reactors** are considered more susceptible to meltdown than thermal or slow reactors, as the nuclear reactions are less easy to control. The only known large-scale nuclear meltdown took place at **Chornobyl**, Ukraine (then part of the USSR), in 1986, although there have been some partial meltdown accidents, notably at **Three Mile Island** in the USA in 1979, and lesser incidents at Chalk River, in Ontario, Canada, in 1952 and Lagoona Beach, near Detroit, in Michigan, USA, in 1966.

NUCLEAR RADIATION: See **radiation**

NUCLEAR REPROCESSING is the method by which unused **uranium** and newly created **plutonium** can be recovered from spent nuclear fuel rods after they are removed from a nuclear reactor. Nuclear power technology is distinguished from all other energy technologies by one particular feature. Unlike the ash of residues from coal or petroleum burning, the spent **nuclear fuel** from a nuclear power station contains both potentially valuable material and uniquely hazardous wastes. The first nuclear reactors and reprocessing plants were not built as sources of electrical energy. The reactors were built expressly for the 'transmutation' of **fertile** uranium-238 into weapons-grade plutonium-239, which is a fissile material, meaning it is able to undergo nuclear fission, in order to achieve the splitting of the atom. However, only a small proportion of the uranium-238 was converted into plutonium, and so plutonium had to be extracted chemically. The method became called reprocessing.

In principle, the idea is straightforward. Fuel rods are dissolved in a vat of nitric acid, creating a mixture of uranium nitrate, plutonium nitrate and comparable salts of the hundreds of other by-products in spent fuel. Separation of these is the crucial stage in the process. Although the activity is a chemical engineering task, reprocessing works are far from ordinary chemical plants. As the materials are so radioactive, they must be kept behind heavy shielding at all times and every operation has to be carried out by remote control. Plutonium is still created automatically in the uranium fuel used in the reactors of commercial power stations. One-third of the nuclear fuel rods in a nuclear reactor are replaced each year. The old rods are stored for over a year in **cooling ponds** at nuclear power stations, to allow the bulk of the short-lived radioactive substances to disintegrate or **decay**. The medium- and long-lived radioactive substances remain in the fuel rods that are then, ideally, transported in protected heavy-duty, water-filled nuclear fuel flasks to a reprocessing or waste storage plant. In the United Kingdom they are taken to the **Sellafield** site operated by **British Nuclear Fuels (BNFL)**, and in France to a site at Cap la Hague operated by Cogema. Reprocessing is a controversial activity, because of the wastes, especially the **high-level waste**, generated in recovering the uranium and plutonium. There are several objections by environmentalists against creating stockpiles of plutonium. It is a potential weapons material, a **carcinogen** and highly dangerous, and it presents unusual risks when transported.

The **NUCLEAR TEST BAN TREATY** is an agreement between the nuclear superpowers to stop the testing of nuclear weapons. Its formal name is the Treaty Banning Nuclear Weapons Tests in the Atmosphere, in Outer Space and Underwater. The Treaty came into force in 1963. The agreement bans all above-ground testing and does not cover underground testing. A Comprehensive Test Ban Treaty was negotiated in the 1990s, but has not been signed by the USA and, therefore, nor by the People's Republic of China, India and Pakistan. Russia, France and the United Kingdom did sign the 1996 agreement, but, controversially, only after some final tests had been conducted in the Pacific.

NUCLEAR WASTE is any unwanted by-products of controlled nuclear reactions, particularly the waste products of nuclear power stations and nuclear reprocessing. Controlled nuclear reactions involve the **fission** of **uranium** and **plutonium**, which gives rise to a host of fission products that must be kept isolated from the environment until they have decayed to a harmless state. (See also **low-level**, **intermediate-level** and **high-level waste**.)

NUCLEAR WINTER was the phrase used to describe the potential atmospheric and climatic consequences of the dust, gases, smoke and **radioactivity** that would be generated by a nuclear war. The discovery by researchers, in the 1970s, that dense clouds of soil particles may have played a role in previous mass extinctions of fauna and flora on earth refocused attention on the possible consequences of nuclear war. In the 1980s leading members of the international scientific community compiled a report, *Environmental Consequences of Nuclear War*, describing the effects of widespread nuclear explosions on climate, ecosystems and food supply. Sponsored by the International Council of Scientific Unions (see **International Council for Science (ICSU)**), the scientists held workshops in a number of European, Soviet, American and Asian cities. They assessed the biological effects resulting from likely changes in sunlight, temperature, rainfall, atmospheric chemistry, **ionizing radiation**, **ultraviolet radiation**, plant and animal growth, and the resultant agricultural productivity.

NUCLEIC ACIDS are sometimes referred to as the molecules of life. There are two types, DNA (deoxyribonucleic acid), which carries the genetic code, and RNA (ribonucleic acid). They consist of long chains of nucleotides.

NUMERICAL FORECASTING is the forecasting of the behaviour of the weather using mathemetical models.

NUTRIENTS have a particular meaning for the ecologist. They refer to those chemical elements that are involved in the construction of living tissue and that are needed by both plant and animal. The most important, in terms of bulk, are **carbon**, hydrogen and **oxygen**, with other essential ones including **nitrogen**, potassium, calcium, sulphur and **phosphorous**. Plants get their nutrients either from the atmosphere (carbon and oxygen) or by extracting them from the soil with their roots (nitrogen, potassium, calcium and other salts). Animals depend on their food and water intake for their nutrient supply.

NUTRITION is the process by which an organism provides itself with, or is provided with, all the materials and nutrients needed for healthy growth and survival. The raw materials needed by green plants are water, carbon dioxide and a variety of inorganic substances, but particularly the so-called **essential elements** or macronutrients. There are 10 of these: **nitrogen**, **phosphorous**, **sulphur**, potassium, magnesium, calcium and iron, together with **carbon**, **hydrogen** and **oxygen**. In addition, plants require other elements in such tiny amounts that they are referred to as **trace elements**, or micronutrients. They include **boron**, **manganese**, **zinc**, **copper**, **molybdenum** and other elements. A concentration in water of a few parts per million of the trace elements is nutritionally sufficient. The sources of all these materials vary with the environment of the plant, so survival can be threatened by disruption of its ecosystem or habitat. All these substances, except carbon dioxide, enter from the soil solution surrounding the roots of the plant. Carbon enters the plant, in combination with oxygen, as carbon dioxide molecules. Water is the source of hydrogen, which, like carbon, forms part of the structural materials of plant tissue, which are synthesized in growth. Oxygen enters the plant in two ways: as part of the carbon dioxide molecule and as part of the water molecule. It is also used in making plant tissue, but in its gaseous form it is also needed for aerobic respiration.

NYOS: See **Lake Nyos**

O

OBNINSK: See **RBMK**

OCEAN CIRCULATION or thermohaline circulation is the way energy from the sun, stored in the sea, is transported around the world. The currents explain, for example, why the United Kingdom has ice-free ports in winter, while St Petersburg, in Russia, but at the same latitude as the Shetland Islands, needs ice breakers. Evidence is growing that the world's ocean circulation was very different during the last ice age and has changed several times in the distant past, with dramatic effects on climate. Scientific research programmes, such as that of the **World Ocean**

Circulation Experiment, are now working to tie down the critical links between circulation and climate. The long-term aim is to forecast the **climate change** caused by the **greenhouse effect** up to 100 years in the future. The oceans are vital as storehouses, as they absorb more than one-half of the sun's heat reaching the earth. This heat, which is primarily absorbed near the Equator, is carried around the world and released elsewhere, creating currents that last up to 1,000 years. Oceans warm slowly. As the earth rotates and the wind acts upon the surface, currents carry warm tropical water to the cooler parts of the world. The strength and direction of the currents are affected by landmasses, bottlenecks through narrow straits, and even the shape of the seabed. When the warm water reaches polar regions its heat evaporates into the atmosphere, reducing its temperature and increasing its density. When **seawater** freezes it leaves salt behind in the unfrozen water and this cold water sinks into the ocean and begins to flow back to the tropics. Eventually it is heated and begins the cycle all over again. As the current of warm tropical water in the North Atlantic heads towards the **Arctic** it passes north-west Europe, keeping the ports ice-free and the winters warm and mild. However, the warm current of this Gulf Stream misses the west side of the Atlantic, so that Canada and Greenland have cold, icy winters. The influence of these currents can be seen to dramatic effect in the Pacific, with the periodic **El Niño** weather phenomenon, which is caused by the reversal of the flow of the cold, usually northerly Humboldt or Peru current and its accompanying trade winds. There are often serious climate consequences to such an event, for places widely separated on the globe. Furthermore, environmentalists fear that global warming may trigger a melting of the **ice shelves** and **sea ice**, or even of the **ice sheets**, and this influx of fresh water could disrupt the thermohaline circulation, the great ocean conveyer, significantly.

OCEAN CIRCULATION MODELLING refers to a major international research effort designed to produce methods and data for predicting what is happening in the oceans, and is comparable to methods used for weather forecasting. The ocean modelling work is the ultimate goal of the many projects that form the **World Ocean Circulation Experiment (WOCE)**. The WOCE experiments have been designed to meet the needs of the World Climate Research Programme, intended to improve understanding of the pattern of the currents that move vast amounts of energy, known as the global **ocean circulation**. Earth is unique in having two separate but interacting fluid envelopes: the **atmosphere** and the **oceans**. Both influence the climate through the way they interact with each other. The effect of the winds on the surface layer of the ocean is referred to as one of the atmospheric forces driving the oceans; but the strength of the winds depend on how much energy the atmosphere has obtained from the ocean. Thus, while atmospheric 'forcing' drives the oceans, the ocean in turn regulates the climate of the atmosphere, until a balance is reached. The traditional way of dividing the world into three broad regions, tropical, temperate and polar zones, is possible because of that balance. This balance was reached 10,000 years ago, at the end of the last ice age. **Climate change** seems likely to disrupt it.

Against this background, WOCE is the largest investigation ever undertaken into understanding the physical characteristics of the oceans. However, in order to produce a computerized mathematical model for forecasting the future behaviour of the oceans, it is necessary to have an accurate knowledge of its present state. The purpose of WOCE is to understand precisely how the vast amount of the sun's energy, which is absorbed into the oceans, is then released to regulate the climate. Although the great ocean currents act as conveyor belts carrying energy round the globe, the process of energy transfer is done unevenly. The way warm near-surface water is transported up the Atlantic to the high northern latitudes is not repeated in the same way in the southern Pacific with a comparable flow towards **Antarctic** latitudes. The monitoring of the oceans' conveyor belts is done by measuring physical characteristics, such as the heat and density of the various layers of water at different depths. However, changes to these measurements, which could come from climate change, could in turn be linked to biological characteristics of the oceans. The marine **biota** are intimately involved in the cycles of nutrients and carbon within the oceans. Concern about man-made emissions of carbon dioxide has focused attention on the role of biology in the global **carbon cycle**. Two biological processes are of key importance. First, **photosynthesis** by **phytoplankton** (marine plants), which consumes carbon dioxide, thus reducing the partial pressure of the gas in seawater and increasing its transfer (flux) from atmosphere to ocean. Secondly, living organisms die and sink to the ocean depths, a process that can be thought of as a pump maintaining a high carbon dioxide content in deep water, and a low content in surface waters and the atmosphere. The marine ecosystem is, however, not that simple. Growth of phytoplankton is limited by nutrients other than carbon; the interaction between different elements within the **biogeochemical cycle** causes a great deal of complexity. Phytoplankton compete with **bacteria** for inorganic nutrients, and are themselves consumed by higher predators, such as fish and mammals. Biological models are aimed at predicting carbon fluxes between the atmosphere and deep ocean.

The **OCEAN DRILLING PROGRAMME (ODP)** is the name of the largest multinational research programme for ocean exploration. It was started in 1968 by the world's largest oceanographic research centre, the Scripps Institution for Oceanography, which is based at La Jolla, California, USA. It became international in 1975 and provides a continuous flow of discoveries about the earth's past climate, which have relevance in predicting its future climate. A temperature history of the ocean's deep waters between 70m. and 20m. years ago, based on measurements of benthic foraminifera (see **benthos**) from deep-sea drilling cores, shows a prominent warming phase and two cooling phases in the earth's climate. Changes in climate leave their mark in the contents of individual organic compounds and species locked in **sediments**.

OCEANS cover 70% of the earth's surface and are very important to the human population as a source of minerals, fish, shellfish, sand and gravel, and petroleum and gas reserves. The oceans contain 97% of the world's water and also account for 84% of the moisture from evaporation. However, the marine environment is under threat from rising pollution, from agricultural chemicals, litter, plastic, chemical and bacterial pollution, petroleum spills and untreated human sewage. Shipping and dumping at sea contribute about 10% of the oceans' pollution. This pollution, in turn, enters the **food chain** causing serious food poisoning. Over-fishing has dangerously depleted fish stocks. Fishing gear, such as large-scale **drift-nets** that are sometimes 60 km long and not selective in the fish that are trapped in them, have caused particular concern. There are fears that annual fish catches may soon fall short of demand. Marine mammals have been threatened because their habitats have been damaged by pollution and competition over food supplies. **Agenda 21** called for improved monitoring of pollution levels and compliance with shipping protocols and agreements, and stricter international regulations to reduce the risk of accidents. It suggested that ocean spill response centres and a global database on marine pollution should be set up and that ports should have facilities for collecting rubbish, petroleum and chemical residues from ships. It also said that marine species should be restored and maintained at sustainable levels and that selective fishing gear should be used to minimize waste. (See also **London Convention** and **seawater**.) Early in 2004 it was reported by the **UN Environment Programme** that the emerging threat to the world's oceans was so-called 'dead zones', oxygen-starved areas devoid of life, which had doubled in size since 1990. They mainly affected coastal waters, vital to fish stocks, and some stretched over 70,000 sq km. The product of **eutrophication**, from the inflow of excess nutrients, dead zones had long existed in the Gulf of Mexico and Chesapeake Bay (USA), but had spread to the **Baltic Sea**, the **Black Sea**, the Adriatic Sea and now even the Gulf of Thailand and the Yellow Sea. They were also beginning to appear off South America, Japan, Australia and New Zealand.

The **OCTANE RATING** or octane number is a measurement that indicates the properties of a particular blend of petrol (motor fuel or gasoline). The primary purpose is to show that a fuel will behave efficiently in a car engine with a high **compression ratio**: hence, the regular and premium grades on sale at filling stations. Ideally, the combustion of the fuel-air mixture in an engine proceeds smoothly from the spark plug across the cylinder chamber. Since the fuel is highly compressed before ignition, the mixture may explode in several places to produce a distinctive knocking sound and loss of power. Knock resistance is achieved by a careful blending of fuels. The perfect anti-knock fuel comes by judicious mixing of a blend of iso-octane and normal-heptane, both pure **hydrocarbons**. A blend of 90% iso-octane and 10% heptane would have an octane number of 90. In practice, petrol with the equivalent properties is obtained by producing more complex blends at the refinery and use of **anti-knock agents** additives like tetra-ethyl-lead and tetra-methyl-lead (see **organometallic compounds**). The additives are cheap, as well as being effective as knock depressants. Leaded petrol is also an environmental hazard and is being phased out under environmental laws.

OIL POLLUTION PREPAREDNESS, RESPONSE AND CO-OPERATION CONVENTION (OPRC): See **International Maritime Organization (IMO)**

OIL SPILLS: See **petroleum spills**

OLD-GROWTH FORESTS are those that have remained mostly untouched by humans and are usually highly diverse **ecosystems**. They are hundreds of years old in the case of species like the giant sequoias (see **redwood tree**) and are virtually irreplaceable, because they have taken so long to develop. As a result, they are unique and increasingly rare and protected resources. More than 95% of old-growth forests have been cleared in the USA, but surviving examples such as the giant redwood forests in California are protected. Forests that replace old-growth forests tend to have less biological diversity. Old-growth forests store a large quantity of **carbon**, which helps reduce global warming. They are rich **habitats** for many species.

1,3 BUTADIENE is a **hydrocarbon** solvent found in vehicle emissions. It has attracted concern as a **carcinogen** because it is known from laboratory research to cause leukaemia and lymphomas.

OPEN-CAST MINING, sometimes called strip-mining, is a method of excavating large quantities of materials such as **brown coal**, peat, iron and copper ores, clay, sand, gravel, quartz or limestone by removing topsoil to expose mineral deposits near the surface. Open-cast mining is estimated to occupy 0.1% of land in Europe. In the Czech Republic, near the town of Most in Bohemia, a single concentration of various open-cast operations alone covers 10,000 ha. The local impact of such mining can be dramatic. For example, the open lignite mines in the former German Democratic Republic (East Germany), near Leipzig, spoil the landscape and contribute to high levels of **air pollution**.

The **ORGANIC** method is a way of growing food and husbanding eggs and livestock naturally, without the use of any manufactured additives, **fertilizers**, or **pesticides**. Organic fertilizers, such as **compost** and **humus**, which are produced entirely from plant sources, are used as materials for building soil fertility on farms, allotments and gardens.

The **ORGANIC GROUP** are compounds containing **carbon** and, usually, hydrogen, which are the starting point of the branch of chemistry classed as organic chemistry. When an organic group combines with elements such as, for example, the metals **lead** or **mercury**, they form **organometallic compounds** like tetra-ethyl lead and **methyl-mercury**. Organometallic substances include some of the most hazardous environmental toxins. They can be very poisonous if they are absorbed into the body by inhaling polluted air or eating contaminated food, because they tend to react with the central nervous system. Originally, organic chemicals were those chemical compounds based on carbon that were found in living organisms. Subsequently, the modern chemical and pharmaceutical industry was largely created by making complex carbon-based compounds from petrochemicals, such as plastics, synthetic fibres and drugs.

ORGANIC RECYCLING relates to the use of crop residues and manure in low-cost methods of non-intensive agriculture, which are employed in developing countries.

The **ORGANISATION FOR ECONOMIC CO-OPERATION AND DEVELOPMENT (OECD)** is a group of 26 industrialized nations that constitutes a forum where representatives of member governments discuss and seek to co-ordinate all aspects of their economic and social policies. The basic structure of OECD's work in the field of the environment is outlined by its Environmental Strategy for the 1990s, which aims to integrate environmental and economic decision-making, to improve environmental performance, including the reduction of **pollution**, and to increase international co-operation. Responsibility for implementing the programme lies principally with the OECD's Environment Policy Committee and several inter-committee working groups. Another of OECD's specialized bodies, the Group on Urban Affairs, is concerned with economic, social and administrative issues in cities, as well as ecological aspects of the built environment. OECD has also supported international efforts to implement **Agenda 21**. In May 2001, in Paris, France, a meeting of OECD environment minister adopted a new programme, the Environmental Strategy for the First Decade of the 21st Century. Based on criteria for environmental sustainability, the Environmental Strategy identified five linked objectives for enhancing its policies: maintaining ecosystems through efficient management of natural resources; separating environmental pressures from economic growth; improving information for decision-making and assessing progress indicator measurements; enhancing the quality of life with the interaction of environmental and social policies; and improving global governance and co-operation.

An example of OECD's environmental initiatives is the rules formulated to protect the **recycling** industry. OECD estimated that in 1990 40,000m. metric tons of waste were exported for recycling throughout the world, a large part of it to developing countries. OECD's rules divide hazardous wastes into three categories, red, amber and green, for trade between its member countries. On the red list are dangerous wastes, such as **asbestos**, polychlorinated biphenyls (PCBs) and chlorinated **dioxin**. Ash from iron and steel manufacture, petrol, coke and bitumen, lead acid **batteries**, waste oils and petroleum/water mixtures, **phenols**, **thallium** compounds and hydrogen peroxide solutions are on the amber list. The green list—regarded as non-hazardous—includes scrap metals, gold, platinum, silver, iron and steel, copper, nickel, lead, **cadmium**, plastics, ceramics, mining wastes, glass, rubber wastes, textile and tannery wastes, and certain ashes from coal-fired power stations. Some accused the OECD of using its rules to undermine the provisions of the **Basle Convention on the Control of Transboundary Movements of Hazardous Wastes and their Disposal**.

ORGANOCHLORINE COMPOUNDS are a large group of synthetic substances that contain **carbon**, **chlorine** and **hydrogen**. They are among the category of **persistent organic pollutants (POPs)**. Once released, their stability enables them to remain for decades in the environment, where they become cause for concern, because of their toxicity. Organochlorine compounds and other POPs accumulate in the **food chain** and threaten the health of humans and of wildlife. **Benzene** hexachloride (usually known by the acronym BHC) was first prepared in the 19th century and exists in many forms, including the pesticide **lindane**. Other particularly pernicious examples of organochlorine compounds include **polychlorinated biphenyls (PCBs)**, chlorinated dibenzodioxins, polychlorinated dibenzodioxins (PCDD) and polychlorinated dibenzofurans (PCDFs), and **pesticides** like **DDT** and **chlordane**.

ORGANOCHLORINE SOLVENTS, also known as chlorinated solvents, are used for cleaning and degreasing metals, dry-cleaning, paint-stripping and in industry as an **intermediate chemical**. They pose threats to human health and the environment in several ways. For instance, exposure over long periods may cause cancer and birth defects. Exposure occurs as a result of **air pollution** and, to a lesser extent, drinking contaminated water. The most common compounds are methylene chloride (acronym MC), trichloroethane, trichloroethylene, perchloroethylene, CFC-11 and CFC-113 (see **chlorofluorocarbons—CFCs**).

An **ORGANOMETALLIC COMPOUND** is a compound which consists of a metal atom combined with an organic group substance. For example, tetra-ethyl lead is an organometallic compound. Industrially, organometallic compounds have an increasingly important role as catalysts. In the environment, organometallic compounds tend to be hazardous substances. Tetra-ethyl lead is used as an additive to petrol to improve its **octane rating**. However, tetra-ethyl lead emitted from car exhausts is a particularly dangerous contributor to **air pollution**. If absorbed into the body, it gets into the bloodstream and can then cross the **blood–brain barrier** to cause brain damage.

ORGANOPHOSPHOROUS COMPOUNDS achieved notoriety in the 1990s as the hazardous ingredient of the **pesticides** blamed for causing so-called Gulf War syndrome. The pesticides in question were sprayed on tents and equipment to keep them free of flies and other potential disease-carrying insects. Commercial organophosphorous pesticide and **herbicide** compounds for agricultural use were derived from substances developed as possible chemical-warfare nerve agents during the Second World War, by the United Kingdom and the USA. The organophosphorous agricultural sprays had the advantage over **organochlorine compounds** that they decomposed very quickly after use.

The **ORYX** once ranged over the Arabian peninsula, but they had been hunted to near-**extinction** by the early 1960s. Herds of the Arabian oryx have been returned to Oman after what is recognized as one of the most successful results of **captive breeding**. What were thought to be the last few surviving animals were rounded up and taken to the Phoenix Zoo in Arizona, USA. The animals bred well and were distributed to European zoos. In 1980, 18 animals were returned to Oman. The animals were penned so that they could gradually acclimatize to their new surroundings and be studied. The Sultan of Oman and the Harasis tribe welcomed the oryx back and became their protectors, wardens and observers. They were released from the pens after a few months and settled down into coherent herds. By 1991 the Omani herd was still thriving, while other herds, both from other captive breeding programmes and from the small native population, which was unknown to the original rescuers, continued to thrive in parks in Qatar, Abu Dhabi, Jordan, Saudi Arabia, Bahrain and Dubai.

The **OSLO TREATY** came into force in 1974 and banned the dumping of certain wastes, such as **mercury**, **cadmium** and persistent plastics, in the North Sea. The Oslo Treaty has similar black lists and grey lists to the **London Convention** of substances that are banned from dumping. The Treaty was signed by the Governments of Belgium, Denmark, France, the Federal Republic of Germany (then just West Germany), the Netherlands, Norway, Sweden and the United Kingdom. Meetings called the North Sea Conferences were subsequently organized. At the first, in 1984, it was agreed that the phasing out of the use and discharge of **polychlorinated biphenyls (PCBs)** should be stepped up, that wastes, including sewage sludge, should not be dumped in the North Sea and that specified hazardous waste should come under more stringent control. In 1987, at the second Conference, it was agreed to reduce by one-half discharges of 'substances that are toxic, persistent and liable to bioaccumulate' by 1995 (see **bioaccumulation**). Restrictions on the use of marine **incineration** and the dumping of sewage sludge were also agreed. At the third North Sea Conference, in 1990, the eight signatories to the original Treaty and Switzerland had agreed to stop dumping sewage sludge at sea, but the United Kingdom said it would only phase this out by 1998. The dumping of PCBs was causing some disagreement among the signatories of the Oslo Treaty. Denmark, Sweden and Norway wanted this to be prohibited by 1995, Germany by 1998, Belgium, the United Kingdom and the Netherlands by 2005, and France by 2010. Such disagreements were echoed at the fourth (1995) and fifth (2002) Conferences, the latter being held in Bergen, Norway. The sixth North Sea Conference is scheduled to take place in Sweden in 2007.

OTTERS are semi-aquatic mammals that prefer water, but can also travel across land faster than a man can run. Their fur is highly prized and because of it some species have been hunted to near-extinction. The fur is critical to the survival of the otter because, unlike other northern marine mammals, which rely on blubber to protect them from the cold, the fur is its only insulation. Otters are particularly vulnerable to **petroleum spills** because when petroleum penetrates their fur, cold water gets to the skin and the otter freezes to death. During the 1989 petroleum spill from the ***Exxon Valdez***, many hundreds of sea otters died. Otters are also seen as enemies of fishermen, because they eat abalone, clams and mussels, a popular food of many humans. The giant sea otter population of North America had dwindled by 1910, and few thought it would survive. However, in the mid-1930s the animals were sighted again, and they had re-established colonies along much of their traditional range by the end of the century. River otters were among Europe's most endangered mammals, badly affected by the pollution of rivers that has contaminated the eels and fish on which otters depend. In the United Kingdom they have been protected since the 1981 Wildlife and Countryside Act was passed. In most European countries otters enjoy similar protection, except in a few cases of seasonal or traditional exception. Much of Western Europe, however, has few, fragmented otter populations, with conservation efforts including reintroduction as well as protection. In central and eastern Europe there are determined efforts to ensure that the existing populations are not so reduced.

OUR PLANET, OUR HEALTH was the name of the report of the World Health Organization (WHO) Commission on Health and Environment, prepared for the **Earth Summit** in Brazil in 1992. The report said that the deterioration of the environment, with its huge daily toll in human lives, was one of the most serious threats facing the world. Its key message was that the environmental well-being of the planet and the health of humanity depend on each other. It argued that health rarely receives high priority in environmental policies and development plans, despite the fact that the quality of the environment and the nature of development are major determinants of health.

OVERGRAZING is intensive grazing by animals, for example cattle, sheep or goats, on an area of pasture. It has become a serious threat to the world's rangelands and **grasslands**. Several factors have led to overgrazing, which leads to the soil being degraded and becoming liable to erosion by wind and rain, and even to **desertification**. The main pressures leading to widespread overgrazing have been the need to increase the size and numbers of herds to produce more food for an increasing human **population**, and the transformation of traditional pasture land into plantations to grow **cash crops**. Throughout the dry tropics, where traditionally herds ranged over vast areas, intensive livestock-rearing schemes have taken over, mostly to provide meat for the export market. Well-digging operations have also led to heavy concentrations of animals in small areas.

OVERPOPULATION is a term used to describe the situation where a species has exceeded the **carrying capacity** of a given area, causing **environmental degradation** and reducing the quality of life for the resident population. When a population exceeds the carrying capacity of a region, it often experiences some form of dieback until it reaches a more sustainable size.

OXYGEN (O) comprises about two-thirds of the body weight of plants and animals. It is an essential constituent of carbohydrates, lipids (fatty and oily substances) and most **proteins**. It is necessary, either directly or indirectly, for **respiration** in all forms of life. Some eminent biologists have said that most, if not all, the oxygen in the **atmosphere** (O_2) is derived from the **photosynthesis** of green plants. In its natural state it is an odourless, colourless gas, which occurs in the atmosphere, dissolved in the oceans and other water (the **hydrosphere**) and is chemically combined in the earth's strata (the **lithosphere**). The composition of dry air is over 20% oxygen.

OXYGEN/PEROXIDE-BLEACHED products are an environmentally softer alternative to the synthetic bleaches, which are found in conventional washing powders or used in industrial processes for bleaching paper products, tampons and disposable nappies, for instance, and can leave traces of **dioxin** contamination.

OXYGENATES are the additives put into petrol, like **methyl tert-butyl ether (MTBE)** and tert-amyl ether, and which are preferred as **anti-knock agents** to lead additives (see **organometallic compounds**).

OZONE (O_3) is a molecule that consists of three atoms of oxygen stuck together, in contrast to the oxygen in the atmosphere which only has two atoms of oxygen. The ozone form is very reactive and is toxic to humans and living matter in the **troposphere**, near the surface of the earth. However, in the **stratosphere**, ozone provides a protective layer that blocks harmful **ultraviolet radiation** from the sun. A fall in the level of stratospheric ozone allows radiation to reach the ground, causing skin cancers and killing sensitive plants. Ozone destruction occurs when man-made gases, particularly chlorinated chemicals, **chlorofluorocarbons (CFCs)** and halons, are discharged into the atmosphere. When they reach the stratosphere they decompose and release **chlorine** atoms that trigger an ozone-destroying reaction. The chlorine remains unchanged and can cause 100,000 destructive reactions before it eventually leaves the atmosphere. So

100 g of CFCs emitted from aerosol sprays, in which they are used as propellants, can destroy over three metric tons of ozone. Ozone is also formed at street level by the action of sunlight on the **nitrogen oxides** and unburned **hydrocarbons** released from vehicle exhausts. Low-level ozone is a nose and chest irritant that can precipitate severe bouts of **hay fever** and provoke bronchitis among the elderly.

The **OZONE HOLE** is the description given to the discovery by Dr Joe Farman, a British scientist, of an alarming deterioration in the **ozone layer**. This was formally confirmed and published in the scientific magazine, *Nature*, in 1984. Measurements of ozone had been routinely made for decades at the start of spring at the **British Antarctic Survey (BAS)** base, at Halley Bay. Regular seasonal records are kept of **Dobson units**. The units are measurements made with a special ground-based instrument showing the total amount of ozone in a vertical column of the atmosphere directly overhead. Each subsequent spring the hole in the ozone layer, which forms a protective screen in the sky to filter harmful **ultraviolet radiation** from the sun, increased in size and duration. The hole in Antarctica normally first appears in August, reaches its maximum extent in September or early October and disappears by December. In 1999 the **Antarctica ozone hole** was more than twice the size of the People's Republic of China, while in September 2000 it covered a record area. Towards the end of the 20th century Arctic ozone depletion in the northern hemisphere also became increasingly severe. Over the Arctic, about 15% of ozone is lost during the Arctic spring, corresponding to a 22% increase in radiation. Ozone loss over the more densely populated mid-latitudes was also significant. In 1979–1997 estimated ozone loss was more than 5% during the winter and spring months in mid-latitudes, meaning that a proportionally larger amount of UV-B reached the earth's surface. Threats to the ozone layer remained abundant. The 1998 Scientific Assessment warned that the Arctic would continue to be vulnerable to large ozone losses. This situation could be made worse by global **climate change**. Estimates of the date by which the ozone layer would recover were extended to at least 2050, even assuming full compliance with the **Montréal Protocol** and its amendments.

The **OZONE LAYER**, or ozonosphere, is the protective layer of ozone in the **stratosphere**, found at a level between 15 and 30 km above the earth, and which prevents harmful ultraviolet (UV-B) radiation, a potential cause of skin cancer and a threat to plant life, from reaching the ground. The fragile shield is being damaged by chemicals released on earth. The main chemicals depleting stratospheric ozone were **chlorofluorocarbons (CFCs)**, used in refrigerators, **aerosols**, and as cleaners in many industries, and halons, used in fire extinguishers. The damage is caused when these chemicals release highly reactive forms of **chlorine** and **bromine**. It is hoped that the international phasing out of such chemicals will allow the ozone layer to regenerate by the middle of the 21st century.

P

PALAEOCLIMATOLOGY is the study of climate conditions in the geological past. While most scientists are sure that an increase in the levels of carbon dioxide and other **greenhouse gases** in the atmosphere would result in increased **climate change**, they are still uncertain about the significance for the effects on the climate of natural events: **sunspot** activity, volcanic dust and phenomena like the **Milankovitch** cycle and the onset of ice ages. Consequently, studies of past climates are intended to help remove some of the uncertainties about the factors influencing climatic change. Although there are few simple and accurate indicators of very ancient climates, proxy data is available from which to assemble a picture of the environment. The remains of pollens, beetles and molluscs are valuable because of their profusion. Knowledge of the habitats and **ecosystems** in which today's flora and fauna flourish, and of how the various species are distributed, provides a means of comparison. The occurrence of the fossil remains of those species similar to today's species of animals and plants is indirect evidence of what the temperature and seasonal conditions must have been like at the time those remains were formed. However, the biological data is often inconclusive and needs confirmation by lithological, geochemical and palaeomagnetic indicators. The lithological evidence is the record contained in sediments and rocks, which show the type of weathering that created those formations. Geochemical indicators include evidence of the composition of the calcium carbonate taken from marine fossil shells in ocean cores. The technique is to measure the ratio of the **isotopes** of oxygen in the fossil shells. This reflects the temperature of the water when the shell was formed during the lifetime of the animal, and the chemical composition of the **seawater** at the time. Palaeomagnetism is a study of the earth's geomagnetic field in the geological past.

PALAEONTOLOGY is the study of fossil plants and animals and of the rock formations where they are found. The information is used increasingly in **environmental science** to build up a picture of past environments, the patterns of **climate change**, the species of plants and animals that existed, and those that may already have disappeared through **extinction**. As concern grows about the increasing loss of species because of human activities, it is essential to compare past data with the present to assess what damage is being done to the world's **biological diversity**. The effect of losing species is not known, but past studies should help to identify the danger signs when a species or population drops dangerously low, or is killed off by pollution, for example, or by predators.

The **PANEL OF EXPERTS ON ENVIRONMENTAL MANAGEMENT FOR VECTOR CONTROL (PEEM)** is the organization that promotes alternatives to **pesticides** in controlling **vector-borne diseases**. Since it was formed in 1980 it has focused specifically on the various aspects of water resources development associated with **water-borne diseases**. The **UN Environment**

Programme, the **Food and Agriculture Organization** and the **World Health Organization** use the PEEM to promote environmental means of controlling the vectors of diseases such as **malaria**, **schistosomiasis** and sleeping sickness, which affect millions of people every year.

PARAQUAT is a **herbicide** and **defoliant**. Unlike most other herbicides it is highly toxic to mammals and has been responsible for poisonings, some of them fatal. Less than a teaspoonful of paraquat is likely to be lethal if swallowed, while it can also be absorbed through the skin. There is no known antidote. Concentrated paraquat is listed as a poison in the United Kingdom and much of Europe, and is only sold to professionals. The USA has restricted its use to licensed operators and it is also severely restricted in many other parts of the world, but is widely used in developing countries. When used properly as a spray in agriculture or horticulture it is regarded as fairly harmless, because it sticks to the soil and cannot poison other plants. However, there is some evidence that some soils do not hold it so tightly and that it can drain into rivers, where it becomes extremely toxic.

PARASITISM describes the relationship between two organisms in which one, the parasite, lives on or within the other, the host. Unlike **symbiosis**, where two organisms have relationships that are mutually beneficial, parasites depend on the host in a variety of ways, and usually inflict harm by causing severe illness and even death. Humans are infected by diseases transmitted or caused by almost every kind of parasitic organism, from viral infections, such as the common cold and AIDS, to skin diseases contracted from **fungi** and ringworm, to **malaria** and sleeping sickness, in which the parasites (respectively, the **mosquito** and the **tsetse** fly) invade many hosts during their life cycles. Malaria, for example, is spread when an already infected mosquito bites a person to feed on their blood and passes on the disease they have picked up after biting an infected person. Several of the most serious tropical illnesses are caused by macroparasites, microscopic organisms that include parasitic viruses, **bacteria** and fungi. Larger organisms, such as worms, insects, ticks, vertebrates and plants, often give birth to offspring on a host and leave it to establish new infections. The tropical disease filariasis is caused by nematode roundworms that are spread from person to person by mosquitoes. The worms cause inflammation and eventually block the lymph vessels which, in severe cases, results in elephantiasis, an extremely disfiguring disease.

PARTICULATE TRAPS are devices attached to diesel vehicles to prevent **particulates** from the engine being discharged into the **atmosphere**. (See **retrofitting**.) Such traps were initially too bulky to be fitted to smaller vehicles. Engineers have experimented with designs for burning off the trapped particulates, but this solution was complicated by the difficulty of destroying the particulates without increasing the amounts of **nitrogen oxides** produced.

PARTICULATES are tiny flecks of soot and similar substances in the **atmosphere**. A principal source of particulates is black exhaust smoke from road vehicles. Among road vehicles, 77% of particulates are produced by heavy goods vehicles, 10% by buses, 7% by light delivery vehicles and 6% by cars. A **diesel engine** produces far more particulates than a petrol engine. Particulates in vehicle exhausts mainly consist of carbon and unburnt compounds from the fuel or lubricating oil. Particulates are also produced by **tyre** wear, the chimneys of power stations, industrial process plant and even such activities as barbecues and fireworks. Secondary particulates may form in the atmosphere and include nitrates and sulphates created from **nitrogen oxides** and **sulphur dioxide**. Small particles of solid or liquid may form **aerosols**. Particulate material can irritate the membranes lining the respiratory tract, causing breathing difficulties, and is suspected of triggering heart attacks and **asthma**. Different types of particle are described in various ways that indicate their risk to health, the size of the particle being an important factor. (See **PM10**.)

A **PATHOGEN** is any disease-producing organism.

PEAT is a dense accumulation of only partially decomposed plant materials, which has formed in **wetlands** over a period of thousands of years. It collects in still water, which does not have enough oxygen to complete the total process of decay. Peat, in fact, is the first stage in the formation of **lignite** and **coal**. It is used as fuel in Ireland, Scotland (United Kingdom), the Netherlands and the Nordic countries, and traditionally rural communities have cut the peat over the years without causing any environmental harm. However, mining the peatlands destroys them faster than the rate at which they can regenerate. There is concern now that too many peatlands, a vital element in the **water cycle**, are being destroyed. Peatlands are home to leeches, worms, molluscs, arthropods, fish, amphibians and wildfowl. They are important as habitats for birds and for many **endangered species**. **Air pollution** is degrading the peatlands and the danger is that when peatlands are oxidized they release tons of carbon dioxide, a major contributory factor in the **greenhouse effect**.

The **PEBBLE BED MODULAR REACTOR (PBMR)** is an advance design of nuclear reactor being developed by South African and US interests. It claims a significantly higher level of both safety and efficiency, using helium as a coolant, at very high temperatures, to drive the turbine directly. Complex steam management systems are unnecessary and transfer efficiency is high. A demonstration reactor operated for over two decades in Germany (until after the **Chornobyl** accident) and a prototype was planned in South Africa in the early 21st century.

PEDOLOGY is the scientific study of the origin, structure, composition, distribution and classification of **soil** types.

PENTACHLOROPHENOL (PCP), one of the universally toxic phenolic compounds, is a general purpose agent that is used as a **fungicide**, **herbicide** and **molluscicide**, particularly in Egypt where it is used to control snails that carry the larval human blood flukes that cause **schistosomiasis**. It is also used in wood preservatives and is very poisonous.

PERMAFROST is permanently frozen surface ground that has been below freezing point, 0°C (32°F), for more than two years. It is estimated to cover 20% of the world's land surface, including glaciers. It is found in the **Arctic**, sub-Arctic and **Antarctica**. It can penetrate to depths of up to 700 m. In the summer months the upper layers of the ground may thaw, but there is concern now that if the temperatures rise as a result of **climate change**, this could have a dramatic impact on the Arctic permafrost. The permafrost is a vast storehouse of rotten organic matter that would produce vast quantities of the greenhouse gas **methane**, if defrosted. The fear is that climate change and global warming would thaw the permafrost and the huge increase in atmospheric levels of methane would in turn accelerate climate change in an example of a positive **feedback cycle**. (See also **polar tundra**.)

PERSISTENT ORGANIC POLLUTANTS (POPs) are a large group of synthetic **chlorine**-containing organic chemicals that include **polychlorinated biphenyls (PCBs)** and **chlorofluorocarbons (CFCs)**, and are often used as **pesticides**, **herbicides** and agents added to paints and plastics to improve their characteristics. After their invention in the 1930s, the use of **organochlorine compounds** developed rapidly because they were highly stable substances. Just over 30 years later, in 1962, the book ***Silent Spring*** was published, warning of the threat to health and the environment of pollution by organochlorine agents that were toxic at concentrations of a few parts per million. The **Convention on Persistent Organic Pollutants** came into effect in May 2004. POPs bioaccumulate (see **bioaccumulation**) in the fatty tissue of humans and animals, with the greatest degree of accumulation occurring at the top of the **food chain**. It is thought that POPs may be **endocrine disrupters**, because they can mimic hormones under certain circumstances.

PEST is the term used to describe an animal, insect, plant or other organism that is considered dangerous or inconvenient to man, because it either spreads disease or destroys crops. To the farmer, pests include insects and mites that damage crops, weeds that compete with field crops for **nutrients** and water, plants that choke **irrigation** channels or drainage systems, rodents that eat young plants and grain, and birds that eat seedlings or stored foodstuffs. In the home, flies, **cockroaches**, moths, slugs, snails, **termites**, beetles and bugs, rats and mice are all regarded as destructive pests. In tropical regions **mosquitoes** and the **tsetse** fly carry a threat because of the serious diseases they transmit. The development of a wide range of **pesticides** led to the eradication or control of many pests, but several brands have now been withdrawn or restricted because they were found to cause cancer, or because they were persistent or non-selective and caused widespread pollution. Many species have also developed **pesticide resistance**. Other methods of pest control are now being developed or expanded. **Integrated pest management** uses a variety of methods to control pests, including the introduction of natural predators, crop rotations and the use of pest-resistant varieties of plants; **biological control** relies on nature's own sets of checks and balances, by using natural predators or releasing sterilized males.

A **PESTICIDE** is a synthetic chemical compound produced to kill insects, vegetation, fungi and rodents that are regarded as a nuisance or health threat, or as an agricultural **pest** or weed. In practice, attempts to control pests with chemicals are centuries old. Three hundred years ago **arsenic** was used in honey as an ant bait. Various substances like Paris green (copper acetoarsenite—see **arsenic**), **mercury** formulations and other compounds, such as kerosene mixed with soap, were developed to protect crops. The progress of the synthetic chemical industry during the 1930s, particularly following the belated recognition of **DDT**, which had been discovered 60 years earlier, gave a new dimension to synthetic pesticide development. Modern pesticides are divided into categories that describe their main targets, such as **insecticides**, **herbicides**, **fungicides**, **rodenticides** and **molluscicides**. Less obvious are: acaricides, that kill mites; pediculicides, that kill lice; silvicides, that kill trees and bushes; and slimicides, that kill slimes. Agricultural and industrial experts who favour their use argue that pesticides are essential to prevent crops from being destroyed by insects. Environmentalists claim they cause an unacceptable level of environmental pollution and that not enough is known about their effect on humans. Their use steadily increased in the post-Second World War years, but by the 1970s many, including **aldrin**, **benzene** hexachloride, **chlordane**, **DDT**, dieldrin, dinitrophenol, endrin and heptachlor, had been banned or restricted. (See also **chemisterilants**, **defoliants**, **disinfectants**, **pheromones** and **repellents**.)

The **PESTICIDE ACTION NETWORK (PAN)** is a world-wide affiliation of groups and individuals opposed to the misuse and spread of **pesticides**. It has campaigned to have the use of the chemicals known as the **Dirty Dozen** phased out.

PESTICIDE POISONING is caused by the widespread and indiscriminate use or abuse of **pesticides**. The World Health Organization estimates that 40,000 people, mostly in developing countries, die every year and another 2m. suffer some sort of injury as a result of pesticide poisoning. A number of factors have led to the high incidence of pesticide poisoning. Many pesticides are persistent and, even where they have been proscribed, people still ingest minute amounts of the chemicals. In some countries people do not always understand the instructions giving the safety warnings or the recommended dosage on pesticide containers, either because the labels are misleading, or simply because the operator cannot read. Some people do not wear protective clothing when using pesticides, others wash the application equipment in irrigation channels or rivers. Since many homes do not have running water, workers and their families often wash in streams contaminated with pesticides. Careless and profligate aerial spraying has exacerbated the problem. (See also the **International Programme on Chemical Safety**.)

PESTICIDE RESISTANCE is the ability of a **pest** to develop an immunity to **pesticides**. The Food and Agricultural Organization (FAO) reported in 1990 that 1,600 species had developed resistance to the most widely used pesticides. Insects, mites, fungi and rodents have all developed some resistance to both chemical and biological methods of control. In the 1950s rats developed resistance to **warfarin**. Thirty years later, there was evidence that they were becoming resistant to the second-generation **rodenticides**. The **UN Environment Programme** stated in 1989 that there were signs that chemicals were losing their edge in the fight against pests. Natural predators, as well as the pests they were intended for, were also being wiped out by pesticides, leaving the field open to new pest species.

PETROLEUM SPILLS, caused when tankers collide at sea or break up, disgorging their load in coastal waters, are a major source of pollution, causing death to seabirds, fish and wildlife. However, these accidental spillages, dramatic though they are, are not the major cause of pollution by petroleum at sea. Most marine pollution from ships is the result of routine operations by tankers at sea, such as washing out crude petroleum tanks. An estimated 1m. metric tons of petroleum also seep into the seas from freighters, tankers and offshore drilling rigs each year, as do millions more tons of crude petroleum products, such as gasoline solvents and waste engine oil. The International Convention for the Prevention of Pollution from Ships (MARPOL) was the first comprehensive anti-pollution convention adopted by the **International Maritime Organization**, and came into effect in 1983. It also deals with pollution from chemicals, other harmful substances, garbage and sewage. The MARPOL convention greatly reduces the amount of petroleum that can be discharged into the sea by ships. It bans discharges in already polluted areas, such as the **Black Sea** and the Red Sea. Residues from tank cleaning operations must be discharged in port reception areas and not the sea. The Convention also includes a number of modified construction requirements, such as protective and segregated ballast tanks.

pH is a measure of acidity—the lower the number, the greater the acidity. For example, battery acid has a pH of about 1.0, normal rainfall has a pH of 5.0 and a neutral solution has a value of 7.0. The opposite of acidity is alkalinity, where the pH can range up to 14.0. **Acid rain** has pH values lower than around 4.0.

PHENOL, or carbolic acid, is perhaps best known as an antiseptic. However, it is the basis of a variety of chemical industry products. Phenol was formerly obtained by the distillation of coal tar, which was a source of river pollution. Other manufacturing processes are now used, such as **benzene**, for example.

A **PHEROMONE** is a very powerful chemical produced by insects and many other animals for communication. Pheromones have been synthesized in the laboratory to trap some species.

PHOSPHATE is the form of mineral compound in which plants obtain **phosphorous**, which is an essential nutrient. It also refers to any of a large group of chemical compounds derived either naturally as phosphate rocks, including sediments containing the calcium phosphate shells of former marine invertebrates, bird droppings or **guano**, and artificial fertilizers made by an industrial process from **phosphoric acid**, an acid derivative of phosphorous.

PHOSPHATE-FREE DETERGENTS are detergents in which a **phosphate** ingredient has been replaced by an environmentally safer substance, such as zeolite. The substitute will not act as a nutrient, and so will not cause **eutrophication** as a result of the accelerated growth of plants and micro-organisms if it is released into waterways.

PHOSPHORIC ACID (H_3PO_4) is the most important acid that is derived from **phosphorous** and it is used to make phosphate salts for fertilizers. It is also used in the food processing and textile industries.

PHOSPHOROUS (P) is a vital mineral that is a major constituent of bones and teeth, **nucleic acids** (DNA and RNA), cell membranes, and energy transfer systems in the body involving **adenosine triphosphate**. In animals, 80%–85% of phosphorous is found in bone and teeth, in combination with calcium. The phosphorous cycle is less complicated than the **nitrogen** and **carbon cycles**. Phosphorous does not enter the atmosphere in any form other than occasional dust. Hence the phosphorous cycle involves only the soil and aquatic parts of the natural environment. The influx of phosphorous to rivers and lakes, together with the formation of sewage and run-off from fertilized land, stimulates the growth of small aquatic plants and micro-organisms, blue-green **algae**, with disastrous consequences for the ecosystem. Plants take up phosphorous as phosphate from the soil or water. Animals eliminate excess organic phosphorous in their diets by excreting phosphorous salts in their urine.

PHOTOCHEMICAL REACTIONS are the chemical reactions that need light or **ultraviolet radiation** to occur. The reactions range from those that are absolutely essential to life, such as sunlight-driven **photosynthesis**, and those that are a threat to life when ultraviolet radiation reacts with **nitrogen oxides** and **hydrocarbons** to produce **photochemical smog**.

PHOTOCHEMICAL SMOG is the **air pollution** caused by the action of sunlight on unburned **hydrocarbons** and **nitrogen oxides**, mostly from car exhausts. **Tropospheric ozone**, a **secondary pollutant**, plays an important part in its formation. Photochemical smog is associated with large industrial areas and urban complexes, and causes eye irritations, breathing problems and damage to plant life. It was first reported in Los Angeles County, USA, in 1942, and later throughout California. Strict controls were established in the late 1960s in California, including the compulsory installation of **catalytic converters** and the use of **lead-free petrol**. The measures did ease the problem, but by 1989 there had been such an increase in the number of motor vehicles that incidences of smog were back to pre-1967 levels. In 1989 more stringent regulations were introduced, to be implemented in three phases. The regulations included a proposal that all cars would have to be powered by electricity or other 'clean' technologies by 2007. Free parking for petrol-driven vehicles would be abolished and, by 1994, a limit would be set on the numbers of cars per family. Petrol-powered lawnmowers and solid-fuel barbecues would be banned. Photochemical smog has been recorded in one-half the states in the USA and also in Australia, India, Belgium and the United Kingdom. In the northern hemisphere the problem occurs mainly between April and September and has its origin in areas of dense traffic, although the highest levels of tropospheric ozone are frequently measured in rural or suburban areas, in air masses that have drifted away from urban centres. In anticyclonic weather several days may pass before an inflow of cleaner air dilutes a polluted air mass.

PHOTOCHEMISTRY is the branch of chemistry dealing with photochemical reactions.

PHOTOLYSIS is a chemical reaction produced by exposure to light or **ultraviolet radiation**. Research scientists specializing in this branch of photochemistry are experimenting with photolysis as a way of storing **solar energy** in a novel kind of **renewable energy** system.

PHOTOSYNTHESIS is the process by which plants manage to tap the sun's energy. It is a complex process, which essentially consists of two stages. First, the green pigment chlorophyll absorbs some of the energy of sunlight, which activates it in such a way that part of the energy is captured chemically in the formation of a substance known as **adenosine triphosphate (ATP)**. ATP becomes a temporary store of energy and, consequently, is known as an energy carrier. The reaction also produces a chemical-reducing agent called NADPH. Both of these compounds are needed at a later stage. A second stage begins that is not dependent on sunlight. **Carbon dioxide** is drawn into the plant and converted into organic compounds in a reaction which uses the energy supplied by the ATP and the chemical-reducing properties of NADPH. The organic compounds produced are various types of sugars, needed for building new plant tissues, food storage and the **respiration** needed to generate the energy to keep the plant cells operating.

PHOTOVOLTAIC SOLAR POWER SYSTEMS use semiconductors, similar to the materials used for microchips such as silicon, for the direct conversion of sunlight into electricity (see **solar energy**). The crystals can be made into solar cells smaller than a postage stamp, in order to power pocket calculators or large roof-top **solar panels** of several square metres. The electricity generated by solar panels may be used to charge a car-type battery, which may, in turn, provide power for household lighting, refrigeration, telephones, radio and television, and perhaps water heating. Solar power systems are environmentally clean sources of energy and have the potential to provide an inexhaustible supply of electricity, provided it can be stored in some form of battery. Although photovoltaic solar power systems were invented a long time ago, the prohibitive costs of the semiconductor materials restricted their use. So, initially, their main use was in space technology, producing electricity for satellites and spacecraft. In the 1980s manufacturers developed cheaper technologies for mass-producing solar cells, similar to printing on rolls of paper. Small photovoltaic solar power systems could be very important for families in rural parts of the developing world that are a long way from an electricity grid. **Dams** and **hydroelectric power** once seemed to be the most promising developments for bringing irrigation agriculture and electricity to developing countries. However, experience shows such schemes usually mean flooding fertile valleys and the displacement of large numbers of people in parts of the world where there are no alternative settlements available. So, although it was thought that the advent of hydroelectric power would make rural populations more productive, in practice rural economies could not afford to build the networks that would link them into the hydroelectric distribution grid. Solar power will, perhaps, offer a better alternative.

PHYTOPLANKTON are drifting microscopic plants that, with an animal component, **zooplankton**, make up **plankton**. They are so small that in a teaspoonful of seawater there could be thousands of phytoplankton plants. They are at the root of the ocean **food chain** and play an important part in the regulation of the climate by capturing heat gases in the atmosphere, such as **carbon dioxide**, **methane** and water vapour. However, if they grow excessively, in **algal blooms** fed by run-off nutrients, they can begin a process resulting in the problem of dead zones in the **oceans**. (See also **ocean circulation modelling**.)

PIPER ALPHA was the offshore petroleum platform that exploded on 6 July 1988 in the North Sea, causing the death of 167 men. The initial explosion was caused by the ignition of a low-lying cloud of gas condensate, which had leaked into the compression module. This caused extensive damage and led to a huge petroleum fire. It was followed by three other explosions when gas risers from the connecting platforms were ruptured. Many of the crew died from smoke inhalation in the accommodation section, where they had assembled to wait for helicopters. The public inquiry held after the destruction of the platform ruled that it would become a legal requirement for operators of such platforms to compile a Safety Case identifying the risks to which platform workers are exposed and quantifying them to show that they are not too high.

PLANKTON form one of the most diverse assemblies of organisms and are found at the start of the **food chain** in the **oceans** and in **fresh water**. Having little or no power of locomotion, plankton drift with the tides and currents, but in incredible numbers. There are **phytoplankton** (plants) and **zooplankton** (animals). The most important of the organisms are **diatoms**, which are microscopic green plants with beautiful siliceous skeletons. Diatoms and other tiny green plants form the marine pastures for the herbivores of the sea, which in turn fall prey to the carnivores. There are great seasonal blooms of minute green plants in spring. By early summer, vast numbers of larvae hatch from their eggs. In early autumn, the diatoms flourish once again. Plankton are vulnerable to **ultraviolet radiation**, which is increasingly prevalent as a result of the destruction of the **ozone layer** and the growth of chemical pollutants. The phytoplankton, which need sunlight for photosynthesis, flourish down to depths of 80 m, and survive, although in diminishing numbers, down to 180 m, below which there is insufficient light for **photosynthesis**.

PLEISTOCENE describes the geological period that included the last four major ice ages, and which covered the interval from 1.5m. to 10,000 years ago. The subsequent period, which includes the present day, is the Holocene. (See also **Quatenary**.)

PLUTONIUM (Pu) is a man-made **transuranic element** created for the first time at the University of California, Berkeley, in the USA, in February 1941. Only a few grams were made. Four years later several kilograms had been produced to fuel two atomic bombs. The first was tested over the desert at Alamogordo, in New Mexico, and exploded with a force equivalent to 18,500 metric tons of trinitrotoluene (TNT). The second was dropped over Nagasaki, Japan, on 9 August 1945, the fissile material used in the bombing of Hiroshima three days before this having been **uranium-235**. As with the first test, a mushroom cloud of radioactive **fall-out** rose some 12,000 m into the atmosphere, and over a period of months it spread round the earth. Plutonium is created in a nuclear reactor. It is made when fuel rods of uranium are bombarded with neutrons. A fission **chain reaction** follows, in which some of the uranium atoms are split into atoms of lower **atomic**

number, accompanied by an immense release of energy. Other atoms, such as plutonium, are produced by a transmutation of uranium to a higher atomic number by the absorption of a neutron into its nucleus. Plutonium is extracted from the reactor fuel by a chemical separation process known as **nuclear reprocessing**. It is started by dissolving uranium fuel rods in nitric acid after their removal from the reactor. As the plutonium is extracted, great care is needed to keep the material in small quantities to avoid creation of a critical mass. Plutonium is heavily regulated for a second reason. It is regarded as the most poisonous substance on earth. One millionth of a gram is capable of causing cancer of the lung or bone marrow if absorbed into the body.

PM10 is the term used to describe a particularly hazardous category of air pollutants. It refers to the microscopic **particulate** materials (PMs), that are most likely to cause breathing difficulties in humans. PM10s have an aerodynamic diameter of 10 microns or less. They are easily inhaled and small enough to penetrate into the deep lung where they become lodged. Particles that are small enough to penetrate deep into the lung and stay there are referred to as thoracic or respirable particles. PM10s were regulated in the USA from 1987, when the **Environmental Protection Agency** replaced the existing air quality standard.

POLAR BEARS have no natural enemies except humans, but there are nevertheless threats to the survival of the species. They need a large area in which to hunt, and also reproduce at a very slow rate. They have also been hunted for many years for food, fur and sport. There are other major threats to their survival, including **vanishing habitat** and **petroleum spills**. Even minor changes in the **food chain** have a dramatic effect in a fragile environment like that of the far north. Thus, when the ***Exxon Valdez*** spilled 11m. metric tons of petroleum into Alaska's Prince William Sound in 1989, severe damage was caused to the breeding grounds of the polar bear. The former USSR banned the hunting of polar bears in 1956, while the USA, Canada, Norway and Greenland took measures to protect them some years later. However, in Canada and Alaska the native Inuit (Eskimos) are allowed to hunt and kill polar bears for food and fur. There are debates as to whether this practice should be allowed to continue. In the early 1990s there were only an estimated 25,000 wild polar bears left in the world, in spite of the measures employed to protect them, although studies conducted for the newly established **Arctic Council** in 1996 showed that the number of bears had increased slightly. By that time it was feared that the increasing prevalence of **polychlorinated biphenyls (PCBs)** in the environment threatened the fertility of polar bears. (See **endangered species**.)

The **POLAR ICE SHEET PROGRAMME (PISP)** is an international project involving 23 countries. It is the largest comprehensive research study of the polar **ice sheets** ever carried out.

The **POLAR ORBITING SATELLITE** is one of the most important developments space technology brought to meteorology and **numerical forecasting**. The satellite, which circles the earth over the poles, at right angles to the Equator, provides scientists with observations and measurements of parts of the globe that were inaccessible for routine monitoring. The polar satellites enabled more accurate global measurements to be made of the amount of snow cover and likely changes that could be caused from human activity.

The **POLAR TUNDRA** are the treeless regions south of the **Arctic** polar ice—in the northern areas of Siberia, the Nordic region, Canada and Alaska. The extreme cold of the long winters, with temperatures sometimes as low as -36°C (-32.8°F), limits the vegetation in the tundra to mosses, **lichen**, sedges and rushes, with some grasses and flowering herbs. Cold, drying winds and **permafrost** just below the ground surface also restrict plant growth. The tundra are particularly fragile environments and the flat, almost featureless expanse of the tundra leaves animals exposed to the harsh climatic conditions; many avoid the worst tundra conditions in the winter. The only large mammal that stays in the tundra throughout the winter is the musk ox, which is protected from the cold by long, matted, windproof hair over thick fat layers. They huddle together for warmth and for protection against predators. During the long daylight hours of the brief summer plant growth is sufficient to attract insects, migrant birds and migrating herbivores, such as moose, caribou and elk, moving northwards from the **boreal forests**. When petroleum was discovered in the North American tundra, the number of people arriving as a result put a great strain on the region. Driving vehicles over the frozen ground resulted in erosion and damaged the vegetation, which can only be restored slowly. Pipelines disrupted the migratory routes of the caribou because the animals were afraid of the alien structures. The possibility of **climate change** presents a real threat to the tundra. In a warmer climate it is feared that the **boreal forest** could encroach into the region and deplete the present tundra.

A **POLLUTANT** is any biologically harmful material or agent that contaminates or degrades a natural environment, causing **pollution** of air, land or water.

The **POLLUTER-MUST-PAY** principle is the notion that those who cause industrial **pollution** should pay for its effects, either by finding ways to avoid pollution in the first place, or by providing compensation for its effects. The principle has not been easy to enforce, particularly where atmospheric pollutants are concerned, because there are so many sources and it is not always possible to trace the guilty parties. Pollution can also be blown away from its source at a great speed. However, there has been more success following disasters, such as **petroleum spills** from tankers or nuclear incidents. **Air quality standards**, usually fixed regionally, are often more effective in controlling pollution levels. There is concern that the cost of preventing, repairing, or compensating for pollution would, eventually, be paid for by the consumer.

POLLUTION is the act of discharging a substance, either accidentally or deliberately, that in any way damages or poses a threat to the environment.

POLONIUM-210 (Po-210) is a radioactive element and an **alpha-emitter**. It is produced by irradiating bismuth-209 in a nuclear reactor. The polonium is then used in combination with beryllium as the neutron trigger for atomic bombs. Although it was 30 years before the truth was admitted, polonium was released in the radioactive **fallout** from the **Windscale fire** in 1957. As the British Government withheld the truth that bismuth was being irradiated in one section of the reactor, the cancer-causing risks of **fall-out** containing polonium from the Windscale fire were underestimated. Polonium and **radium** were discovered in the late 1890s, as a consequence of the research into **uranium** and **radiation** conducted by Pierre and Marie Curie (see **curie**).

POLYCHLORINATED BIPHENYLS (PCBs) are a family of some 60 synthetic compounds that were invented in the 1930s and widely used by industry for 40 years because of their chemical stability. They were regarded as being among the safest man-made materials. Their principal uses were as the insulating material in electrical capacitors, transformers and cables; in heat-exchange and hydraulic systems; in plastics, paints and adhesives; and as additives in lubricating oils. The toxicity and **carcinogenic** properties of PCBs released into the environment were recognized after two major accidents in Japan. PCBs are **organochlorine compounds** and are poisonous to animals in the same way as **DDT**. PCBs also have harmful effects on the fertility of mammalian species. For instance, PCBs are suspected of causing sterility among **polar bears** in the **Arctic** and grey **seals** in the Baltic. The stability of PCBs allowed them to accumulate in the **food chain** and spread throughout the global environment before their danger became apparent. Disposal of PCBs presents further risks as they release **dioxin** when incinerated.

POLYCYCLIC AROMATIC HYDROCARBONS (PAHs) are a particularly persistent form of **air pollution**. They are carcinogenic (see **carcinogen**) and are produced by industrial solvents, wood stoves and the incomplete combustion of organic compounds like vehicle and fuel oil. Several hundred PAHs exist, the best known of which is **benzo(a)pyrene (BaP)**.

POLYMERS are man-made plastics derived from the petrochemical industry. In effect, polymers are long-chain molecules. They are made by linking together a relatively simple chemical compound, a monomer, to form a complex one, in a process known as polymerization. Essential parts of animals and plants, such as **proteins**, cellulose and **nucleic acids**, are polymers. In general, synthetic polymers, such as nylon, polyethylene, polypropylene and polyurethane, are stable compounds, hence their attraction for use in textiles and fabrics, and in engineering components and packaging materials, among other applications. Polymers are also the basis of concrete, glass, rubber and paper. As they are lighter than metals, polymer-reinforced materials are used increasingly in boats, vehicles and aircraft. The disadvantages of polymers include the problem of **waste disposal**. Scientists are devising polymers that will decompose harmlessly in sunlight. However, the bulk of discarded plastics raises disposal problems.

POPULATION growth forecasts indicate that by 2020 there will be nearly 8,000m. people living on earth. By 2050 the total is estimated to reach just under 10,000m., assuming that there are no major global disasters, such as wars, famines or epidemics of disease that might bring about drastic reductions in human numbers. A growing population puts a great strain on environmental resources, food, energy, **transport** and land, and can result in **overpopulation**. As more and more people compete for these resources there is a danger that they will move into previously uninhabited areas, that more trees will be cut down to provide fuel, causing **deforestation**, and that more **fossil fuels** will be used up. In the late 1990s, considered globally, most people still lived in rural areas, but the pattern of human settlements is changing. It was estimated that by 2000 45% of the population of developing countries (1,972m.) and 75% of that of developed countries (945m.) would live in cities. Services, communications, transport and water supplies may break down under this strain. About one-third of city dwellers in developing countries live in slums and shanty towns, where clean water is scarce, fuel is expensive and refuse is rarely collected. Disease becomes endemic and violence common. The **UN Environment Programme**, in co-operation with the UN Centre for Human Settlements (Habitat), developed environmental guidelines for urban development that were to be applied in cities in order to make urban management more considerate of the environment. **Agenda 21** made many proposals for improving the social, economic and environmental quality of human settlements and the living and working environments of all people, particularly the urban and rural poor, through planning and management. It advocated that developing countries should promote **reforestation**, **afforestation** and **biomass** energy, and make more use of **solar energy**, **hydroelectric power** and **wind power** as energy sources. It also stated that all countries should adopt transport plans that encourage high-occupancy public transport, safe cycle paths and footways, and development patterns that reduce the demand on transport.

POROUS ASPHALT reduces the **noise pollution** caused by vehicle **tyres** on road surfaces, especially when moving at high speed. European Union legislation sets limits on the noise that may be produced by road vehicles. Porous asphalt surfaces are up to four decibels quieter under dry conditions than conventional asphalt surfaces, and up to eight decibels quieter in wet conditions. Porous asphalt is more expensive either than conventional asphalt or than concrete, which is the noisiest surface of the three, but the cheapest and most durable. Civil engineers are experimenting with an even quieter material known as whisper concrete.

POVERTY is linked in a vicious circle with **environmental degradation** and **population** growth. The majority of people who are unable to meet their own basic needs for nutrition, shelter and health (see **life-support systems**) live in environmentally vulnerable areas, where even minor changes in climate or land use can have a dramatic effect on the quality of the local environment and its ability to support the local population. Rapid population growth adds to the problem.

The **PRECAUTIONARY PRINCIPLE** is one of the basic concepts for environmental protection. The principle demands that if an activity or use of a substance carries a significant risk of causing environmental damage, then it should not continue. In essential circumstances, an activity or use of a product may be kept to a minimum provided the maximum practicable safeguards are adopted. The precautionary principle was incorporated in one of the environmental articles of the Treaty on European Union in 1992 (Article 38, amending Article 130r of the Treaty establishing the European Economic Community). It was also a guiding principle of the **Framework Convention on Climate Change** and the subsequent **Kyoto Protocol**.

PRESERVATION is the concept of protecting a resource like an **old-growth forest** so that it remains untouched by humans. By contrast, advocates of **conservation** believe that a given resource can be managed in a sustainable manner so that it can have multiple uses.

PRESSURIZED HEAVY-WATER REACTORS (PHWRs): See **CANDU reactors**

PRESSURIZED WATER REACTORS (PWRs) are the most commonly built version of the light water reactor (LWR). The PWR was a direct offshoot of the development of nuclear propulsion by the US Navy for submarines. The PWR uses ordinary water both as a moderator and coolant, and it uses enriched uranium in which the more **fissile** isotope of uranium, **uranium-235**, has been increased to form 3% of the uranium fuel. The reactor is operated under pressure to prevent the water from boiling. The reactors have a relatively small, compact core and a high power density compared with the British **Magnox** and **advanced gas-cooled reactors (AGRs)**. However, the design is a source of controversy among environmentalists, who believe that an accident resulting in a sudden loss of coolant could degenerate rapidly into a meltdown of the nuclear fuel. Their fears were reinforced by the **Three Mile Island** accident. Nevertheless, over one-half of the world's commercial nuclear generating units (438 in 2000) are PWRs, with over 250 in the USA and nearly 60 in France.

PREVENTION OF MARINE POLLUTION BY DUMPING OF WASTES AND OTHER MATTER: See **London Convention**

PRIOR INFORMED CONSENT (PIC) is a concept that allows potential importing countries to refuse shipments of dangerous substances, hazardous waste or **pesticides**. In 1989 the **Food and Agriculture Organization (FAO)** of the UN amended its **International Code of Conduct on the Distribution and Use of Pesticides** to adopt the principle of prior informed consent, but both this and a later set of London Guidelines are non-binding. The first round of negotiations on a legally binding international agreement on PIC was held in Belgium in 1996. The agreement was concluded in the Netherlands, as the Rotterdam Convention on the Prior Informed Consent Procedure for Certain Hazardous Chemicals and Pesticides in International Trade. This allowed countries to refuse imports only of banned or severely restricted chemicals and dangerous pesticides.

PRINCE WILLIAM SOUND, ALASKA: See ***Exxon Valdez***

PROALCOOL: See **biofuels**

PROMPT CRITICALITY is the type of explosion in a nuclear reactor that caused the **Chornobyl** disaster, but which the nuclear industry had believed impossible. The nuclear industry had always claimed that the physical conditions of all types of nuclear reactor would make a nuclear explosion impossible.

PROPERTY RIGHTS are entitlements defining an owner's privileges and limitations for the use of a resource. Property rights are vested with individuals or states, or can be held commonly. They are seen to be essential to making economic transactions efficient. Economic transactions are difficult to make in an efficient manner when there is a lack of clearly defined property rights. When property rights are commonly held (not exclusively controlled by a single agent), as is the case with air and the **atmosphere**, such resources can be more easily overexploited.

PROTEINS are one of the three major compounds of which all organisms are formed. Thus, any change in the environment or pollutant that affects the synthesis of proteins is a threat to the survival of animals or plants.

PROTOZOA are single-cell animals, and the smallest and most profuse of all micro-organisms. They often feed on other **microbes**, primarily **bacteria** and some small **algae**. They are best known as insect and animal parasites, and some can cause **malaria** and other debilitating diseases. This, scientists think, could be turned to advantage by **genetic engineering** to develop non-chemical pesticides. The first protozoan pesticide, which was approved by the **Environmental Protection Agency** in the USA, kills grasshoppers, which can destroy millions of dollars' worth of crops every year.

PURE AIR only exists in laboratory conditions. In the natural environment the atmosphere is laden with dust particles which come from volcanic ash, fine sand from **dust storms** and sea salt, plus the particles and gases from chimneys and cars. A laboratory sample of pure, dry air comprises 78.08% **nitrogen**, 20.94% **oxygen**, 0.03% **carbon dioxide**, an absolutely vital 0.0006% of **ozone** and, less importantly, 0.093% argon, 0.0018% neon, 0.00005% hydrogen and even smaller traces of krypton and xenon.

PYRITE is the most widespread sulphide mineral, combining a metal and **sulphur**. When combined with iron, forming iron pyrites, it superficially resembles gold in appearance and is known as 'fool's gold'. If wetland regions are drained artificially and the land aerated, pyrites locked in the soil can oxidize to create **sulphuric acid** that pollutes the land.

Q

The **QUATERNARY** is the geological period, still in progress, that has lasted less than 2m. years. In geological terms it spans very little time, but it includes **climate change** of an unparalleled abruptness and magnitude in the earth's history. The hallmark of the quaternary is a series of episodes of dramatic cooling and spread of **ice sheets**, interspersed with a retreat of the ice sheets, or interglacial periods. The period covering the past 10,000 years is referred to as the Holocene epoch, or the postglacial time. Some experts suggest this may be a misnomer because there is no evidence to show that we are not simply in an interglacial period. The question has obvious implications in studying climatic change and separating natural and man-made influences on the climate. Consequently, great effort is devoted to the study of quaternary climates. One of the most common techniques is that of pollen analysis, which involves analysing the proportion of different types of pollen in non-glacial sediments, to show whether the plant species was characteristic of a warm or cold climate. Faunal remains, ranging from large mammals to beetles, have also been used frequently in establishing patterns of climate change. Since 1980, **foraminifera**, fossils of primitive marine organisms from the **sediment** of the seabed, have been used widely to divine ocean temperatures. They have been used with **isotopic analysis** of oxygen from sediments and from glacier ice from cores taken from the continental ice sheets of Greenland and Antarctica. Isotopic analysis from ice sheets have dated back more than 100,000 years, although in 2004 it was announced that a 3-km ice core had been drilled recording more than twice as long ago as any previous one and could take scientists back about 740,000 years. **Carbon dating** of organic material is effective for analysing material between 30,000 and 50,000 years old. For timescales of a few thousand years, or less, environmental scientists study **lichens**, tree rings and varves. The study of deep-sea sediment shows there have been 17 glacials (ice ages) and 17 interglacials in Europe in the last 1.7m. years. The climatic pattern shown by these analyses closely follows the fluctuations in the **insolation** curve, the solar energy received per unit of area of the earth.

R

RABIES is one of the oldest diseases known to man. In the 16th century it was described in the Americas as a bat-transmitted disease that was fatal to man. Dogs, cats, rats, squirrels, foxes, jackals, cattle and bats are reported as the main source of rabies in several countries. It is an acutely infectious disease of the central nervous system to which all warm-blooded animals, including humans, are susceptible. It is usually transmitted by bites and licks from an infected animal. It has been known to be an **airborne disease** in caves inhabited by rabid bats.

RADIATION refers, in general, to the emission of any rays from either natural or man-made origins, such as radio waves, the sun's rays, medical X-rays and the **fall-out** and **nuclear wastes** produced by nuclear weapons and nuclear energy production. Radiation is usually divided between non-ionizing radiation, such as thermal radiation (heat) and light, and nuclear radiation. Non-ionizing radiation includes **ultraviolet radiation** from the sun which, although it can damage cells and tissues, does not involve the **ionization** events of nuclear radiation. Nuclear radiations only became a cause for concern during the 1950s and 1960s. Routine monitoring of the air found that radioactive materials from atmospheric tests of atomic and hydrogen bombs by the nuclear powers spread radioactive fall-out across the whole northern hemisphere. Radiation exposure is taken to mean exposure to **alpha**, **beta** and **gamma radiation** , X-rays and neutrons emitted by nuclear power stations or the wastes they release. Absorbed dose is a term employed when measuring the effect on the body of exposure to nuclear radiation. The unit of activity of a radioactive substance indicates how many atomic disintegrations are happening each second as it decays (**becquerel**). However, it says very little about the biological effects of the substance, because it does not indicate whether the radiation it produces, when its atoms disintegrate, is in the form of alpha, beta or gamma radiation. Each of these behaves differently on the body. One measure of the effect of radiation is the amount of energy it deposits in a piece of tissue. This quantity is the absorbed dose, and the unit of absorbed dose is the **gray**.

The **RADIATION WINDOW** is the condition that exists when the energy emitted from the earth is in the form of radiation that passes through a cloudless sky and escapes into space. The radiation window is partially closed by **clouds** and atmospheric pollutants, which absorb its energy. As a rule, a clear night sky means a colder night, because energy escapes into space instead of warming the **atmosphere**. Since the gases in the atmosphere control the loss of energy from earth to space, changes in the concentration of atmospheric gases can alter the **global energy balance**. The surface of the earth absorbs short wavelength solar radiation and gets warm. Like any object that gets hot, the earth then radiates some energy, but as radiation of a longer wavelength which, in the lower atmosphere, is readily absorbed by water vapour and water droplets, forming clouds and carbon dioxide gas. These in turn heat up. In stopping the energy escaping into space, the clouds and gases are said to close the radiation window. The result is what is called the **greenhouse effect**.

RADIOACTIVITY is a phenomenon associated with the nucleus of an element that happens to be in an unstable form. Radioactivity is emitted when an atom tries to convert from the radioactive to the stable form or when, in special circumstances, it undergoes **fission**. The radioactivity of a substance is determined by the number of nuclei that decay per second. The unit of activity was formerly the **curie** (Ci), one of which is equal to 37,000m. decays per second. Since it is a very large unit, scientists often use smaller measurements such as the millicurie (mCi, one-1,000th of a curie). The curie is also being replaced by the **becquerel** (Bq). One becquerel is equal to just one decay per second. The nuclei of most naturally occurring elements are stable, with the exception of **uranium-235**, **radium**, **radon** and thorium. Most artificially produced nuclei are radioactive.

RADIOCARBON DATING: See **carbon dating**

A **RADIONUCLIDE** is any element that is radioactive.

RADIUM (Ra) is a naturally occurring radioactive element obtained from pitchblende, uranium ore. It was discovered by Pierre and Marie **Curie** in 1898 and named radium by them. It is pure white until exposed to air, when it blackens. The origin of radium is **uranium**-238 that is dispersed throughout soil and rock. Uranium-238 slowly

decays at the start of the creation of a long series of radionuclides that result in the production of lead, in a process measured in a **half-life** of 4,500m. years. One of the decay products is **radon**. Radium was used as the luminous additive in paint for the dials of clocks and watches, until it was discovered workers using it contracted radiation-induced cancers.

RADON (Rn) is a naturally occurring, radioactive, noble gas that presents a definable cancer risk in some regions, especially those where there is a lot of granite. It is one of the heaviest gases. Radon is a decay product of radium and can accumulate in poorly ventilated buildings.

RAINFORESTS: See **tropical rainforests**

The **RAMSAR CONVENTION** is the Convention on Wetlands of International Importance, Especially as Waterfowl Habitat. It is named after the town in Iran (now known as Sakht-Sar) where it was signed in 1971. The Convention, which came into force in 1975, was one of the first international agreements to deal with environmental problems. It was also unique because it protects a specific type of ecosystem, **wetlands**, on a world-wide basis. Wetlands play an important role in purifying water and as a source of food and fuel. They are important breeding grounds and nursery areas for a great proportion of the world's fish catch. They are under threat from pollution, both from industry and agricultural fertilizers and pesticides, and by many human activities, including the construction of dams and canals, the removal of **peat** and forestry. By 2004 138 parties had signed the Convention, each agreeing to specify at least one wetland that would either be protected or replaced with one of equal worth if the original site was destroyed. There were 1,368 wetland sites throughout the world designated as of international importance. The work of the Ramsar Convention is shared by the **World Wide Fund for Nature**, the **World Conservation Union (IUCN)** and the International Wildfowl Research Bureau. There is a permanent, funded secretariat in Switzerland, hosted by the IUCN.

RBMK, an acronym derived from the Russian term for a pressure-tube reactor (reaktor bolshoi moshchnostyi kanalnyi), is the type of nuclear reactor that exploded at the **Chornobyl** atomic power station in Ukraine (then part of the USSR) in 1986. The accident was caused by an uncontrollable power surge when the operators of the plant disengaged several safety devices during a test of electrical equipment. The Chornobyl reactor lacked the pre-stressed concrete containment structure that surrounds nuclear power stations elsewhere in the world. Without this protection, radioactive material escaped to the environment. The RBMK is a light-water cooled, graphite-moderated reactor, which did not use such expensive materials as other models, but was less stable as a result. The first example supplied power to Obninsk, Russia (then part of the USSR), in the 1950s and was the first nuclear power plant in the world, but few now remain operational, and all those that do have enhanced safety features.

RECYCLING (DOMESTIC) of glass, paper and cans has three benefits for the environment: it saves valuable raw materials; it saves energy that would otherwise have been consumed making original products; and it cuts down on litter and waste refuse. Collection points for waste glass, paper and cans are becoming familiar fixtures in Australia, much of Europe, Japan, New Zealand and North America. Glass-making is very energy-intensive and uses a great deal of sand, limestone and soda ash, but, once made, glass can be recycled virtually forever. In Switzerland, for example, more than 40% of glass is recycled and this accounts for 60% of glass production. It is estimated that every year each citizen of an industrialized country will consume paper or cardboard to the equivalent of two trees. It takes about two tons of wood to produce one ton of paper and the manufacture of paper products requires a great deal of energy, water and chemicals. Over 30% of paper products are made from recycled paper, including stationery, paper towels, toilet paper, egg boxes and newspapers. Making paper from recycled fibre uses 40% less energy than producing it from virgin pulp. The recycled paper industry has also found ways to improve the quality of its products. The smelting of **aluminium** cans saves 95% of the energy used to make aluminium from raw materials. The **Worldwatch Institute** reported in 1987 that if recovery rates of aluminium were doubled world-wide, 'over 1m. tons of air pollutants, including toxic fluoride, would be eliminated'.

RECYCLING (INDUSTRIAL) is a long-established practice, in that scrap metal, glass and paper have been mixed with virgin raw materials for decades. One estimate suggests steel produced from scrap reduces air pollution by 85%, cuts water pollution by 75% and eliminates mining wastes altogether. Similarly, paper made from recycled pulp reduces air pollutants by 75% and water effluent by 35%, and minimizes pressure on the forests. Only 5% as much energy is needed to produce **aluminium** from scrap, compared to production from bauxite ore. Of course, the classic example of recycling is the car. In 1990 Europeans sent 15m. old cars to the breaker's yard, and by the first years of the 2000s the figure had passed 20m. While scrap metal merchants thrived long before the environmental appreciation of the need for recycling, it was only with the planning of cars for the 1990s that designers of vehicles began thinking about the most economic way of dismantling as well as assembling them. A car has about 20,000 components. A traditional breaker's yard would reclaim 50%–75% of the weight of materials in a car, depending on how much of the vehicle could be stripped. An average pre-1985 car comprises 40% sheet metal, 13% cast iron, 13% machine-forged steel parts, 10% electrical and mechanical equipment, 9% plastic, 4% rubber, 3.5% glass, 1.5% fabrics, 1% petroleum and grease, 0.5% copper and zinc, and miscellaneous ceramics and non-ferrous metal components, etc. Since 1960, the proportion of plastic has risen from 2%, and by the year 2000 exceeded 20%.

RED DATA BOOKS (or the **RED LIST**) are a series of publications produced by the **World Conservation Union (IUCN)**. They constitute an inventory of rare plant and animal species. The information listed includes the status, geographical distribution, population size, habitat and breeding rate of each species. The books also contain the **conservation** measures, if any, that have been taken to protect the species. There are five categories of rarity status: **endangered species**; vulnerable organisms, which are those unlikely to adapt to major environmental changes; rare organisms, which are at risk only because there are already few of them in the world; out-of-danger species, which were formerly in one of the preceding categories, but were removed from danger by conservation actions; and indeterminate species, which are plants and animals thought likely to be at risk, but about which not enough is known to assess their status. The IUCN collected data on endangered species for 35 years and in 1996 initiated a thorough scientific assessment of 5,205 species, carried out by more than 500 scientists world-wide. The IUCN's most comprehensive report to date warned that one-quarter of all mammal species were at risk of extinction. It blamed human **population** growth and economic development for reducing, fragmenting and degrading the habitats of the species in question. The introduction of foreign species also contributed to the threat, as well as exploitation, **pollution** and **climate change**. The report also found that 11% of known bird species, 20% of reptiles, 25% of amphibians (see **amphibian populations**) and 34% of fish, particularly **freshwater** species, are threatened with extinction. Of the 26 orders of mammals, 24 included species under threat. The largest proportion of threatened species was in the order that includes monkeys and apes, where 46% were under threat. By 2002, in two years, the IUCN had increased the extinction risk for primate species from one-fifth to one-third. The 2003 Red List put the extinction risk for critically endangered mammals at 16% of the 1,130 threatened species, and 15% of the 1,194 threatened bird species.

RED TIDE is **seawater** that is covered or discoloured by the sudden growth of **algal bloom** or by a great increase in single-celled organisms, dinoflagellates. Red tides are often fatal to many forms of marine life and, in some cases, can result in human deaths, because the dinoflagellates are eaten by clams and mussels, which concentrate the paralysing toxins that they produce.

The **REDWOOD TREE**, *Sequoia sempervirens*, is the tallest living coniferous evergreen tree, and is known to reach a height of over 90 m. The redwood tree is found from south-west Oregon down to central California, in the USA. Their reddish brown bark is resistant to threats such as insects, **fungi** and **fire**, and can be more than 30 cm thick on an old tree. When a tree is cut, it sprouts and natural reproduction occurs through seed production. Redwood timber is used in carpentry and for furniture, fence posts and panelling. Knots cut from the trunk are used for bowls, trays and veneers. Some of the redwood forests in the north-western USA are endangered **old-growth forests**.

REFORESTATION is the activity of replanting land with trees in order to regenerate a forest, soon after the previous crop of trees has been cut down or harvested (see **harvesting**). The term is used when no change in land use occurs. (See also **afforestation**.)

REFRIGERATION was the discovery that did much to make food safer, enabling it to be kept longer, and also kept fresh during international export journeys. However, refrigerators formerly used **chlorofluorocarbons (CFCs)**, which contribute to the depletion of the **ozone layer**. Refrigeration is responsible for the largest and fastest-growing use of CFCs in the developing world. The industrial countries, and some developing countries, have taken exceptional steps to control and, eventually, ban the production of CFCs under the **Montréal Protocol**.

REGIONAL SEAS are most affected by pollution because they are either enclosed or partly enclosed and have a slow rate of water renewal. Owing to this, they do not have the self-cleansing and restorative properties of oceans. Many of these seas support dense human populations and cater for an enormous number of tourists. The **Mediterranean**, **Caribbean** and **Baltic Seas** are especially vulnerable. The Regional Seas Programme, co-ordinated by the **UN Environment Programme (UNEP)**, was started in the mid-1970s to bring together coastal nations in order to prevent further decline in 10 of the world's **coastal areas**, inshore waters and open **oceans**. The programme drew up a series of individual action plans, which set a strategy for co-operation on research, monitoring, pollution control and protection, and the rehabilitation and development of coastal and marine resources. Each action plan contained a legally binding convention covering general commitments and detailed protocols dealing with specific issues, such as **petroleum spills**, dumping, emergency co-operation and protected areas. The Regional Seas Programme is funded by UNEP and trust funds set up by the governments involved. Although the programme has had some success in taking steps to reduce pollution—from industrial and agricultural discharges, direct dumping from ships, petroleum pollution and agricultural run-off—the scale of the problem is enormous.

RENEWABLE ENERGY, or alternative energy, refers to those sources of power which produce energy without depleting the earth's non-renewable resources, the **finite fuels** such as **coal**, petroleum, **natural gas** and **uranium**. Alternative energies include **solar energy**, examples of which are solar architecture, involving the design of energy-efficient buildings, and solar cells that convert the sun's energy directly into electricity (see **photovoltaic solar power systems**); **wind power**, often in remote areas or offshore sites; tidal, **wave** and **hydroelectric power**; **biomass**, which uses wood chips, agricultural residues and high-energy plants to produce energy; and **geothermal energy**. Hydroelectricity, which provided 17% of the world's power supply at the beginning of the 21st century (utilizing only one-third of its economically exploitable potential), is the only renewable source in general use, but is increasingly seen as having negative environmental impacts as well, mainly because of the building of **dams**. Renewable forms of energy are regarded as clean sources because they do not involve the emission of **sulphur dioxide**, which causes **acid rain**, or of **greenhouse gases**, like **carbon dioxide**. The lack of greenhouse gas emissions means that renewable energy is regarded as being essential to making the world's energy system less **carbon**-intensive. Renewable energy also has the advantage of being appropriate for remote or inaccessible places that might be hard to service by conventional means. The cost of renewable energy, especially wind technologies, began to be reduced significantly towards the end of the 20th century, increasing the likelihood of its adoption. The potential of wave power was even conceded by conservative voices in the energy industry, which predicted that the seas and oceans could provide 10% of the world's power needs. The waves have an estimated reserve of energy of two terawatts (the equivalent of about 1,000 power stations). Cost, however, made it unlikely to be commercially feasible until the end of the first decade of the 21st century. Other, more remote, marine potential includes tidal, ocean thermal conversion and ocean-current power. The potential of wind and solar energy is perhaps even greater: a windy country like the United Kingdom could potentially supply its energy needs three times over with this resource, while even solar energy in a country not noted for hours of sunshine, could provide two-thirds of demand. However, there were still issues of cost, as well as **NIMBY** objections and the fears of some conservationists about the effect on landscape and, particularly, on bird populations (in danger from the giant rotors driving wind turbogenerators). Nevertheless, progress from the late 1990s, particularly if government support was involved (as in Denmark, Germany, the United Kingdom and parts of the USA), indicated that globally installed wind capacity might reach 150 gigawatts by 2010. Solar energy also had massive potential, although the technology still needed significant investment. A revolution in the energy industry, including massive changes in infrastructure, and considerably greater research and development resources would be necessary if the world was even to approach one-half its total primary energy supply coming from solar and other renewable sources by 2050 (according to a report by Shell). In 2004 the eminent scientist, James Lovelock (author of the **Gaia** theory), expressed doubts about whether it was possible to introduce renewable energy sources quickly enough or on sufficient scale to counter the imminence of environmental catastrophe. He provoked considerable controversy by suggesting that **nuclear fuels** were the best short-term replacement for **fossil fuels**.

REPELLENTS are substances used to repel ticks, gnats, **mosquitoes** and flies, birds, dogs, deer, moles, rabbits and rodents. They act because they are either foul-smelling or unpleasant to touch or taste. They are, therefore, less environmentally invasive than the various '-cides' which kill the irritant or **pest**.

RESIDENCE TIME is the length of time individual **pollutants** remain in the environment. The residence time of pollutants in the **atmosphere** is determined by the processes of deposition and chemical conversion. If a pollutant remains in the atmosphere for 30 days, it is likely to spread by the process of vertical mixing between the highest and lowest levels of the **troposphere**. When resident times approach six to 12 months, pollutants can be carried from one hemisphere to another. After 12 months, pollutants start spreading from the troposphere up to the **stratosphere**, which occurs in the case of European and North American emissions of **chlorofluorocarbons**, which deplete the **ozone layer**, and of the so-called **greenhouse gases** responsible for **climate change**.

RESPIRATION has two meanings. It is a vital process by which plants and animals obtain their biological energy, by breaking down organic molecules created in earlier chemical reactions. For example, in a primary process like **photosynthesis**, a reaction driven by sunlight involving **carbon** and **oxygen** creates complex molecules, such as **carbohydrates** and glucose, which then provide a store of energy. When this energy is needed for other purposes, both plants and animals can undo the results of photosynthesis to unlock the stored energy by respiration. During this transformation of a complex molecule, the molecule returns to a simple form like carbon dioxide when the energy is released. The essence of respiration is that the energy stored in compounds like starch and sugar is available on demand when needed by the organism's fundamental power units, known as **adenosine triphosphate** and NADPH. They need the energy for fuelling the movement of muscles or for making some other molecule. The other meaning of respiration is a physical-chemical process of transporting oxygen to, and carbon dioxide from, tissues, and is part of breathing.

RESTRICTION ENZYMES are the special agents which, in 1971, two US microbiologists, Daniel Nathans and Hamilton Othanel Smith, discovered could be used as biochemical scissors to snip genes from DNA (see **nucleic acids**). This discovery became the foundation of **genetic engineering** and the modern **biotechnology** industry.

RETROFITTING is the renovation of a building or alteration of a system to take advantage of newly improved technologies. Examples include the installation of equipment to control discharges of **pollutants**, like **sulphur dioxide** and **nitrogen oxides** from power stations, or to ensure more efficient combustion in vehicles with **lean-burn engines**. Retrofitting is often a cost-effective way of reducing air pollution as it avoids the necessity of buying new industrial plant or vehicles before the useful life of existing equipment has ended.

The **RHINE** is the largest, and reputedly the most polluted, Western European river. It has its source in Switzerland and flows through France, Germany, the Netherlands and into the North Sea. From Reichenau in Switzerland, where the anterior and posterior Rhine streams (Voerderrhein and Hinterrhein, respectively) unite, the river flows for 1,320 km to the North Sea and the Rhine delta, where it is joined by the Meuse. The Rhine basin covers about 250,000 sq km, an area containing more than 51m. people split between six countries. The river passes through major urban and industrial areas and for many years it has suffered from severe pollution from the large volumes of industrial waste discharged into it from chemical plants on its banks. Concern about the increased salinity of the river goes back to the 1950s, although as early as 1885 the riparian states signed a Salmon Treaty, in an effort to protect the declining fish population. This was a result not only of pollution, but also of **dam** building. Canals, dams and drainage all affected river levels and water quality, and little was organized multilaterally. By 1935 salmon had disappeared from the Rhine, although it was only after the Second World War that efforts were made to co-operate on environmental issues. In 1963 France, the Federal Republic of Germany (then just West Germany), Luxembourg, the Netherlands and Switzerland formed the International Commission for the Protection of the Rhine against Pollution (ICPRP) and, joined by the European Communities (now known as the European Union), in 1976 signed the Convention for the Protection of the Rhine against Chemical Pollution. These arrangements were subsumed into and displaced by a new Convention for the Protection of the Rhine, which had been negotiated by January 1998 and was signed in April 1999. Purification plants along the river, built at a cost of US $38,500m. between 1970 and 1990, had helped, but did not tackle the causes of pollution. Efforts to deal with this were motivated by the shock of the **Sandoz** disaster, and the aims of the ICPRP's 1986 Rhine Action Programme could be increased. Although many experts believed the targets set in the 1987 Programme (to halve levels of hazardous pollutants by 1995) could not be achieved, in fact between 1985 and 1992 testing showed many of the metallic and other pollutants had been reduced by almost this level or sometimes more. Thus, in 1991 the German chemical industry federation agreed to reduce the flow of toxic chemicals into the Rhine. At the same time, domestic pollution controls were strengthened. These international initiatives have already produced some improvements in the water quality of the Rhine: the concentration of **heavy**

metals has fallen and the biological treatment of organic waste has reduced depletion of oxygen and the number of fish killed. In the 1990s salmon were reintroduced to the Rhine and in 1996 the first one was caught. In 2004 the closure of the potash mines in Alsace, France, would be of great benefit to continuing attempts to clean the Rhine and restore its damaged habitats.

RHINOCEROS populations in the last 25 years of the 20th century decreased by as much as 85%. The number of black rhinoceroses, for example, dwindled from 65,000 in the 1970s to less than 2,500. It became a critically **endangered species** and a losing battle was fought against poachers who kill for the rhinoceros' horn. Several African countries brought in strict measures to prevent the killing of rhinoceros, but still the rhinoceros was hunted down, mostly because these measures have increased the price of the scarce rhinoceros horn. Trade in all rhinoceros products is banned in most countries. Education programmes to reduce the demand for rhinoceros horn have been implemented, as have **captive breeding** programmes. Rhinoceroses have been moved to protected reserves and ranches in Africa and 10 black rhinoceroses were taken to the USA. A count of black rhinoceroses in Zimbabwe in 1992 revealed that there were only 250 left in the wild; another 160 were in special game ranches. Zimbabwe also tried to discourage poaching by de-horning black rhinoceroses. The horns grow back at a rate of about three inches per year. Throughout Africa there were only about 2,400 black rhinoceroses by the mid-1990s, but by 2002 there were about 3,100 and by 2004 over 3,600. The white rhinoceros, which had been reduced to a population of just 50 in the wild at the beginning of the 20th century, had a population of some 11,000 one century later, but there remain at least two threatened sub-species. The northern white rhinoceros, for example, has only about 20 animals in a single unit left, in the Democratic Republic of Congo (formerly Zaire).

RHIZOBIA are a genus of bacteria that live in the roots of leguminous plants, such as the pea family, in a symbiotic relationship (see **symbiosis**).

RIO CONVENTION: See **Convention on Biological Diversity (CBD)**

RIO PLUS 5: See **Earth Summit**

RIO SUMMIT: See **Earth Summit**

RISK MANAGEMENT is the process by which early efforts and assessments are taken to prevent environmental risks or accidents. It was a major part of **Towards Sustainability**, the fifth Environmental Action Programme of the European Communities (now the European Union), and emphasized four key areas of concern. One was industrial accidents and hazards, and another covered chemicals. The number of chemicals has greatly increased, so that now there are some 100,000 used in manufacturing industries. Most of these can be hazardous if they are incorrectly applied or released in large quantities, and their effects are either unknown or long-lasting. The European Commission has been assessing the potential risks of existing chemicals and has targeted 50 specific chemicals in comprehensive programmes to reduce risks. Another area of concern was product-labelling, so that dangerous substances could be identified quickly in an emergency. The fourth area of concern was **biotech**, because it is not always known what effect certain **genetically modified organisms** could have on the environment in the long term. The fifth Environmental Action Programme planned to set up a standardized set of environmental risk assessment requirements and safety measures for biotechnology products. This process was carried forward in the sixth Programme, and also takes place in other countries.

RNA: See **nucleic acids**

ROAD PRICING, or selling road space, is one of the **economic instruments** under consideration by transport policy-makers as a means of reducing congestion and the environmental impact of increasing numbers of cars and lorries. There are several ways of selling road space. In France, Norway and other countries, tolls are collected from vehicles as they enter certain stretches of road. In Singapore an area licensing scheme operates during the morning and evening rush-hours: licences can be bought for around $3 per day at kiosks in the central area and must then be displayed alongside the motor taxation certificate. London, the capital of the United Kingdom, took this a stage further in February 2003, when a congestion charge was levied on almost all motor vehicles (except motorcycles) entering the centre of the city during the normal working week. The unexpected success of the scheme in reducing traffic levels was likely to encourage similar schemes in other crowded city centres.

RODENTICIDES are **pesticides** used to destroy several small mammals, especially rodents, which damage either houses, stored products or cultivated crops. Rodents breed prolifically and are constantly competing with people for food. In the developing countries, where grain is stored in the open, rats are known to eat as much as 20% of grain crops. There are other ways to destroy rodents, such as shooting or trapping, but poisoning is the most widely used, because it is both economical and effective. Rodenticides differ in their chemical make-up and also in the way they affect animals. The most successful group are the coumarins, or anticoagulants, a classic example of which is **warfarin**. They require repeated ingestion over several days, making the rodent weaker by the day. Coumarins inhibit the material in the blood that causes blood to clot, finally resulting in internal bleeding.

The **ROYAL BOTANIC GARDENS**, at Kew, Surrey, in the United Kingdom, has been at the forefront of botanical research and plant conservation for some 200 years. The gardens house the largest and most diverse collection of living plants, and the most comprehensive research collection of preserved plant material in the world. Kew's botanists and horticulturalists advise on environmental management, preserve **endangered species** as seeds or living plants, and carry out research programmes to identify alternative crops as sources of medicines, food, and fuel. Kew Gardens' first director, Sir William Hooker, started the programme of '**economic botany**', the study of plants that were 'either eminently curious or in any way serviceable to mankind'. He established Kew as a centre of experimental scientific research. Regular expeditions are mounted to remote regions of the world by botanists from Kew, working in close co-operation with scientists from the host country, to collect seeds, seedlings or cuttings of plants that are endangered, or thought to be near **extinction**, and some that have not been discovered before. They are grown in conditions as much like their native habitat as possible. The plant collection is used for research, and cultivated plants are exchanged with other botanic gardens, scientific institutes and universities for a whole range of studies. Many are then reintroduced into the wild. Today, as natural plant habitats throughout the world are being destroyed and plants are becoming extinct, a priority of the research at Kew is to assess the value of what is being lost, so that a way can be found to reduce and reverse the rate of destruction of the world's plant species. The **World Seed Bank** or Millennium Seed Bank, which houses the seeds of over 5,500 species of flowering plants—one of the world's largest collections—is at the Royal Botanic Gardens' annexe at Wakehurst Place in Sussex.

S

SALINIZATION occurs when salt concentrations in the **soil** build up and leave it unable to support plant life. All soils contain a small amount of salt, but if there is regular rainfall it is flushed out of the soil into the **ground-water** or flows through streams and rivers to the sea and so does not accumulate in the soil. However, when rainfall is low and evaporation rates high, as they are in arid areas, the soil tends to have a naturally high salt content. Salinization occurs when the delicate salt balance is upset. The salt builds up in the root zone of crops, or sometimes forms a salt crust on the surface. A major cause of salinization in arid zones is **irrigation** schemes that are continued year after year without the land being left fallow. If the land is not well drained, the **water table** rises and brings the salt to the surface. The UN conference on **desertification** reported that one-10th of the world's irrigated land is waterlogged. Salinization and **waterlogging** affect many countries badly, including Argentina, Central Asia, the People's Republic of China, Egypt, India, Iraq, Pakistan, Peru and the USA.

SALMONELLA is the general name for a family of micro-organisms, one of the largest groups of bacteria, that includes those most frequently implicated in **food poisoning** and gastroenteritis. Unhygienic handling and inadequate cooking of poultry and meat, improper storage of cold meats and, more recently, contamination of battery-reared hen eggs, are the most common sources of salmonella infections.

SALT MARSHES are areas of **brackish**, shallow water usually found in **coastal areas** and in **deltas**. There are also inland marshes in arid areas where the water has a high salt level owing to evaporation. They were once considered to be wastelands, but recently the importance of the salt marsh ecosystem has been recognized. They are environmentally delicate areas, extremely vulnerable to pollution by industrial or agricultural chemicals, or to **thermal pollution**, which often results when river water has been used as the coolant in power stations and industrial plants. The water temperature of the salt marsh is usually higher than that of the adjacent sea, and the water is usually shallower. These factors contribute to increased **photosynthesis** and, therefore, a flourishing plant growth that supports a large number of animals. Soil brought to the marshes by rivers and the droppings of animals that feed there also increase the fertility of the marsh. Many animals that live in the continental shelf region, especially fish and crustaceans, spend part of their lives in salt marshes, either as larvae or as feeding adults. Moreover, salt marshes are now considered to be efficient **carbon sinks**, owing to their high rate of soil deposit (owing to tides) and that the **bacteria** in this habitat do not emit **methane**.

The **SANDOZ** chemical factory, near Basle in Switzerland, caught fire in the early hours of 1 November 1986. More than 30 metric tons of **herbicides** and **insecticides** from exploding drums were washed into the River **Rhine**. The fire started with an electrical fault in a warehouse that contained 1,246 tons of chemicals, principally insecticides. A slick spread 48 km down the river, destroying hundreds of thousands of fish and threatening the drinking water supply for towns on the river. Two hundred miles of the upper Rhine lost most or all of its marine life and the Government of the Federal Republic of Germany (then just West Germany) ordered that all wells along the Rhine should be closed. People in towns near the West German capital of Bonn had to have water ferried in by tanker. Fish, birds, insects and river plants were killed between Basle and Karlsruhe, West Germany. The disaster on a river that had begun a slow improvement since the 1970s encouraged an accelerated programme of recovery.

The **SARDAR SAROVAR DAM** is the largest of the 30 controversial large **dams** (and over 3,000 smaller ones) under construction on India's Narmada river. The Sardar Sarovar Dam is in Gujarat, the upstream states affected being Maharashtra and Madhya Pradesh. The Narmada scheme was planned as the world's largest **irrigation** project when work started in 1979. The aim was to use river water to irrigate millions of hectares of farmland in western India. An inevitable consequence of the project was that 250 villages would be submerged and thousands of people displaced from their homes. This was one of the main reasons why environmentalists described the dam as an ecological disaster. Work on the dam, which was meant to reach 130 m in height but was only one-quarter complete by the mid-1990s, was halted many times because of protests and legal challenges. In 1995 India's Supreme Court ordered the Government to stop construction and review the environmental and human impact of the dam. In 1993 the **World Bank**, after paying US $280m. of its original offer of a $450m. loan to build the dam, had withheld the remainder until India agreed to adhere to environmental guidelines. India refused to accept the balance of the loan, on the grounds that it preferred to complete the project using its own resources. Officials vowed to complete the project and in October 2000 the Supreme Court allowed construction to continue as long as proper resettlement had take place before flooding. The ruling allowed the Sardar Sarovar Dam to rise to 90 m (anti-dam activists had favoured 88 m), and only to go higher when approved by an environmental authority. The Narmada Control

Authority allowed the Government of Gujarat to have the dam raised to 103 m in 2003 and 110 m in March 2004. Opposition shifted to detailed arguments over compliance with that ruling and to politics again. Meanwhile, there was also controversy about whether the cost of the project, even with the hydroelectric scheme enabled, was justifiable, and whether the irrigation canals to spread the dammed waters had been built.

SATELLITE TRACKING TECHNOLOGY has been used to track sperm **whales** in the Gulf of Mexico, among other roving and migratory species. This venture started in the summer of 1993. Twelve sperm whales were followed with the aid of space satellites by a team of marine biologists and marine mammal specialists from Oregon State University in the USA. Concern about the sperm whale's status as an **endangered species** persuaded the US Department of the Interior's Mineral Management Service to fund the research. The department is also responsible for leasing offshore gas and petroleum exploration areas. With petroleum and gas activities in the Gulf of Mexico moving into deeper water, it has been accepted that there is a need to know more about sperm whale activities in these deep water areas in order to prevent any conflicts. The presence of sperm whales in the Gulf of Mexico had been confirmed in 1990 during aerial surveys. For the satellite tracking of whales, a radio transmitter is attached to the mammal. Two types of radio tags were being used. One, about the size of a flashlight, records the depth and duration of each dive, and water temperature. The other tag, a miniaturized version one-fifth the size of the first, allowed the researchers to track the whale's location for a much longer period of time. Similar tracking technology was used to monitor other migrations, with examples in the early 2000s being geese from Ireland to western Canada and swans from Siberia to Western Europe.

The **SATELLITES AND SURFACE METEOROLOGY PROJECT** is an enormous investigation that forms part of an even wider research venture, the **World Ocean Circulation Experiment (WOCE)**. The overall purpose of WOCE is to understand exactly how the **oceans** regulate the world's climate and to accumulate data in order to produce computer models for predicting the effects of **climate change** on the oceans, and vice versa. The oceans contain vast currents that are, in effect, huge conveyor belts, carrying energy absorbed from the sun from the Equator to the polar regions (see **ocean circulation**). However, the oceans also release large quantities of energy into the atmosphere, and these releases are primarily responsible for the world's weather patterns (see **atmospheric circulation**). The purpose of the Satellites and Surface Meteorology Project is to understand more fully the interaction between the sea surface and the atmosphere. Remote sensing of the sea surface from space has become an established technique for regular, large-scale observations. Measurements from satellites can include sea surface temperatures, wind speeds, wave height and slopes in sea level, which are all related to ocean currents. The main spacecraft used for satellite measurements are the European Space Agency's Earth Research Satellite (ERS) series, the joint US/French TOPEX/POSEIDON programme and the US National Aeronautic and Space Administration's **Earth Observing System (EOS)** programme. The most important measurements are the transfers (or fluxes) of heat, water and momentum between the air and sea, which are needed for driving ocean models and for verifying climate models of the 'coupled' ocean-atmosphere system. Measurements of these fluxes include the use of a special instrument called the scatterometer on the ERS-1.

A **SAVANNAH** is typically an area of tropical **grassland** in Africa, with a combination of abundant high grass and low tree growth. Savannahs have a seasonal climate with wet and dry seasons, and they extend across a wider area than **tropical rainforest**. They are the grazing ground for wild animals and, more recently, for domesticated species, such as cattle and goats.

SCHISTOSOMIASIS, or bilharzia, is a debilitating parasitic **water-borne disease**. It is transmitted by a species of water snail, which acts as host to the first larval stage of flukes, schistosoma. It is estimated that more than 200m. people in 74 tropical countries suffer from the disease and at least 600m. more are at risk of contracting it. Approximately 200,000 people die from schistosomiasis every year. Most at risk are those who come into contact every day with water, through swimming, fishing or washing clothes. When the flukes leave the snail, they pass through human skin and produce large quantities of eggs, which pass into the intestine or bladder. These cause anaemia and inflammation. Cirrhosis of the liver and enlargement of the spleen can follow. A large number of the eggs also pass out of the body in urine or faeces and, consequently, the cycle continues. **Molluscicides** in the water can control the snail hosts, while a vaccine is in the development stage.

SEA EMPRESS was the name of the tanker which ran aground in February 1996, spilling 72,000 metric tons of petroleum into the sea near Milford Haven, Wales, in the United Kingdom. Compensation to local businessmen and fishermen amounted to some £76m. (approximately US $119m.).

SEA-GRASSES grow close to the shores in the shallow waters of temperate or tropical seas. They play a valuable role in trapping **sediment**, slowing down erosion and helping to keep waters clear. Many species of fish and shellfish are dependent on sea-grass beds at some stage in their lives, either as nurseries or as a life-long habitat. Sea-grasses are very vulnerable to pollution, especially from sedimentation, when shallow coastal waters are affected by the erosion of agricultural land. They are also under threat because of the depletion of **mangrove forests**, and as a consequence of dredging, the development of **tourism** and petroleum production, and the release of industrial wastes.

SEA ICE plays an important role in substantially modifying interactions between the ocean and the atmosphere (see **atmospheric circulation**). Ice disrupts the normal exchange of energy and momentum between the two. It can have a very significant effect on the climate because it effectively insulates the sea from the cold atmosphere. The melting and freezing of sea ice influences the structure of the ocean, which can affect **ocean circulation** and the ability of oceans to remove carbon dioxide from the atmosphere. Data from a range of remote sensing systems can now determine the expanse, age and, to some degree, the thickness of sea ice. Altimeter data and microwave imagers are used to map the extent of ice cover. The main areas of sea ice are in the polar regions, around the continent of Antarctica and, notably, the Arctic sea ice that extends from the Greenland ice sheet and over the North Pole. A 2002 study concluded that the permanent Arctic sea ice was retreating at an increasing rate of about 14% per year. In Antarctica the calving from **ice shelves** and the disintegration of sea ice was reportedly picking up pace from the 1980s. Although ice shelves and sea ice in general do not have the same direct impact on **sea levels** and global temperatures, they are important in regulating thermohaline or **ocean circulation**.

Rises in **SEA LEVEL** are a possible consequence of **climate change**. During the 20th century the mean sea level rose by 15 cm (six inches) and, during the 21st century, it could rise by a further 15 cm to one metre as a result of global warming. Revised estimates by the **Intergovernmental Panel on Climate Change**, published in 2001, expected the rise in sea level during the 21st century to be nearer 1 m, however, even without the contribution of melting polar ice. Contrary to common belief, projected sea-level rises would be primarily a result of the thermal expansion of oceans owing to higher oceanic temperatures. It was not known for sure whether the polar ice caps and **ice sheets** would melt, releasing their enormous store of fresh water. As one of the projected impacts of climate change is increased precipitation, it is possible that the polar ice caps could increase somewhat, owing to more snowfall. If one of the land-based ice sheets were to break off as a result of climate change, it could cause a catastrophic rise in sea level. However, even the more conservative projected rise could endanger many low-lying regions and **small island states**. This is because there are many regions that would actually be inundated by such a rise and also because storm surges would reach further inland as a result of the higher sea level. This concern prompted the formation of the Alliance of Small Island States (AOSIS) as a negotiating bloc in the **Kyoto Protocol** negotiations. Another low-lying country with few resources to combat rising sea levels is the heavily populated delta country of **Bangladesh**. Even a one-metre rise would inundate one-half of the country's rice-growing land and force the relocation of 40m. people, quite apart from the terminal threat to the important habitat of the Sunderbans (see **tigers**).

The **SEA TURTLE DISPUTE**, also known as the Shrimp–Turtle Dispute, was one of several major issues that raised concerns about the impact of the **World Trade Organization (WTO)** on the environment. The dispute arose over a US law designed to protect endangered sea turtles, thousands of which are killed each year in shrimp trawl nets. The law required nations catching wild shrimp and exporting them to the USA to be certified as having adopted specific conservation measures. These measures require shrimp trawls to be equipped with 'turtle-excluder devices' (TEDs), which allow turtles to escape unharmed. The law was challenged at the WTO by India, Malaysia, Pakistan and Thailand, which argued that it was an illegal restriction on their shrimp exports. A WTO dispute-settlement panel ruled that the law presented an arbitrary and unjustifiable restriction to international trade. In early 1998 the USA appealed against the panel's decision, but it was upheld by the WTO Appellate Body, with some important reservations. Environmental non-governmental organizations had submitted an *amicus curiae* brief in support of the USA that was not considered at first, but was later accepted by the Appellate Body, indicating that submissions by civil society could be considered by the WTO. Regulatory reforms and domestic court action resulted in the USA, in April 2000, producing a list of 41 countries from which shrimp exports were certified (16 use TEDs and 25 do not put sea turtles at risk).

SEAWATER accounts for more than 98% of the mass of the **hydrosphere** and covers just over 70% of the globe. Owing to the composition and stability of the **oceans**, and the way they are controlled, they are of great importance to the climate, and great attention has been given to studying the effects of pollution. Long-term experiments include the **Joint Global Ocean Flux Study**, the **Ocean Drilling Programme** and the **World Ocean Circulation Experiment**. Man's activities are believed to be accelerating the change in the composition of seawater. Although the effects are mainly obvious in estuaries and coastal regions, there are signs of the global consequences of pollution. These consequences have begun to be found by the ocean studies that are analysing tens of thousands of samples of seawater, from which marine chemists have been compiling a global map of the chemical composition of seawater. As there are no specific preserved records to provide direct evidence about the composition of seawater in earlier geological periods, a picture of how seawater evolved to its present-day composition has been assembled by indirect evidence. The primary sources of information have come from the analysis of preserved ocean **sediments**, which contain fossils and can be investigated using **isotopic analysis**. The sediments can usually be accurately dated. Interpreting the relationships between the characteristics of seawater and features contained in the sedimentary record requires very complex analysis.

SEALS are marine mammals found in most of the world's marine environments, but more often in cool or cold coastal waters. Their bodies contain a great deal of fat which insulates them and acts as a pressure regulator as they dive rapidly to depths of over 250 m and stay submerged for up to 30 minutes at a time. One group of seals, the eared seals, includes the sea lions and fur seals that are often seen in circuses and aquaria. Elephant seals, named because they have large noses, were hunted for their fur to near **extinction** in the 19th century, but a rescue operation carried out by the USA and Mexico resulted in a large population of elephant seals being established on the islands off California. In 1982, 25,000 pups were born, compared with only six in 1911. In the **Baltic Sea**, increasing sterility among another species of seal, the grey seal, is attributed to the prevalence of **polychlorinated biphenyls (PCBs)** in the environment. However, more grey seals (38% of the world population) live around the United Kingdom, mainly in Scotland, where they flourish. Indeed, this is the case for many seals (40% of all Europe's common seals live in British waters) and the local fishing industry is suspicious of the effect of seals on fish stocks. For this reason culling, banned in the United Kingdom since 1978, is periodically permitted in Canada, where the biggest such exercise in 25 years was allowed in 2004, amid much controversy (in April the Government announced that hunters could kill up to 350,000 seals).

SEAWEED is the red, brown or green **algae** growing in or by the sea. Many forms are edible and are used as food, particularly in the People's Republic of China and Japan, and some of its extracts are used in medical products. It is a source of many substances, such as **iodine** and the gelling agents in food processing. Combined with straw it can provide an alternative to burning **peat**, which is scarce. Experts have stated that seaweed has a far greater potential for health and nutritional purposes. The European Communities (now the European Union) launched a four-year project in April 1993 in which researchers across Europe were carrying out a variety of tests on the possible future uses of seaweed. These range from an extract to reduce blood cholesterol levels to the development of a variety of foodstuffs. Currently, large-scale seaweed cultivation is still carried out only in Asia, mainly the People's Republic of China. Intensive seaweed cultivation began in China in 1951, and production increased from 62 metric tons of kelp in 1952 to over 3m. tons in 1998. Most, worldwide, is for food, although some Asian countries cultivate seaweed for carageenan and crude agar production.

SECOND WORLD CLIMATE CONFERENCE: See **World Meteorological Organization**

SECONDARY POLLUTANTS are formed as a result of chemical reactions in the **atmosphere** between primary pollutants. Major secondary pollutants include **nitrogen dioxide** and **ozone**, which are produced by reactions involving the fumes from road vehicles. Although small amounts of nitrogen dioxide are released directly in vehicle emissions, much more is produced when the nitric oxide discharged from vehicles reacts with oxidants in the atmosphere. In the presence of strong sunlight, nitrogen dioxide reacts with the **volatile organic compounds (VOCs)** in vehicle exhausts to create an excess of **tropospheric ozone**. In polluted atmospheres, other oxidation reactions may take place involving **hydrocarbons**, **aldehydes** or other compounds.

SEDIMENTS are the particles of soil and rock fragments that are washed from mountains and uplands to be deposited in rivers, pushed along by **glaciers**, carried out to sea and blown by wind. Sediments were first seriously studied by scientists on the HMS *Challenger* expedition between 1872 and 1876, the first completely scientific circumnavigation of the world, which collected specimens of plants and animal life. Sediments are divided into two groups: those that are usually land-based, consisting of rock, gravel, sand and mud; and those that drop from all levels of water and finally land on the bottom of the sea, such as the shells or skeletal remains of **plankton**.

SELLAFIELD is the name of the spent nuclear fuel reprocessing plant operated by British Nuclear Fuels, in Cumbria, the United Kingdom. The plant was known for 30 years as Windscale, the site in 1957 of the first acknowledged nuclear power accident. A nuclear pile producing **plutonium** for the United Kingdom's nuclear weapons programme caught fire, discharging a cloud of radioactive **iodine**, strontium and **caesium**. Surrounding grassland was contaminated and milk from dairy herds had to be thrown away. Spent nuclear fuel from the United Kingdom's first generation of **Magnox** nuclear power stations is stored at Sellafield. When it is 'cool' enough, it is reprocessed to extract the plutonium and remaining **uranium**. The process also creates three batches of waste classed as **low-level**, **intermediate-level** and **high-level waste**. In 1977 the first public inquiry, called the **Windscale Inquiry**, into the use of nuclear energy was forced by environmental organizations, led by **Friends of the Earth (FoE)**. The Windscale Inquiry was concerned with a plan to build a new type of plant for handling the waste from spent nuclear fuel, called THORP (thermal oxide reprocessing plant). Controversy centred on claims by opponents of Sellafield, who alleged that discharges from the plant caused leukaemia among children in the region and were responsible for radiation-induced cancers among the work-force. The Sellafield plant manufactures nuclear fuel rods, reprocesses spent nuclear fuel from nine countries and treats and stores radioactive wastes. The Irish Government criticized this operation over radioactive emissions to the Irish Sea and the Norwegian Government expressed concern over discharges of the radioactive element technetium-99. In 2000 the facility was also accused by the British Government of falsifying data since 1996. Sellafield was named Windscale until 1988, and before that the site was the original Calder Hall base of the British nuclear programme.

SEQUESTRATION in environmental terms refers to **carbon** sequestration, which is the process of increasing the carbon content of carbon pools (other than the **atmosphere**). Carbon is sequestered by the planting of trees and other organic matter that absorbs carbon as it grows. **Oceans** are also potential places for carbon to be sequestered. A colloquial term for a place where carbon is sequestered is a 'sink' (see **carbon sink**). The opposite of a sink—something that generates carbon emissions—is a 'source'. Forest fires, such as the major one that occurred in Indonesia during 1997, are good examples of carbon sources. Carbon sequestration could be an important way to combat global **climate change**, because high concentrations of carbon in the atmosphere are a major cause of such change. The management of sequestration was a contentious aspect of the **Kyoto Protocol** negotiations for a number of reasons. For example, it was disputed whether carbon sequestration has benefits as permanent as closing a **coal**-fired power station, and consequently whether it is an acceptable alternative. However, the fact that forests and oceans have another value because they help regulate climate may create an additional incentive for humanity to protect them.

The **SET-ASIDE POLICY** was introduced in 1988 by the European Communities (now the European Union), both to reduce cereal output and as an alternative to subsidized storage and the accumulation of grain 'mountains' in Europe. It is also designed to improve the environment, which had been damaged during years of **intensive agriculture**. Under the policy, farmers are paid to take at least 20% of their land out of production and use it in one of four ways to improve the environment. It can be set aside for permanent fallow, allowing gamebirds to flourish and attracting seed-eating birds, such as finches, to eat any thistle, teasel or cereal heads left after mowing. Small mammals, mice, voles and shrews, also thrive, leading to increased numbers of birds of prey and predatory animals. Rotational fallow allows some of the rare wild flowers of arable ground to restore depleted seed banks in the soil. In some areas it may also encourage ground-nesting birds to return. The set-aside land can also be used for woodland, which has long-term value to wildlife and the landscape. The fourth option is for the farmer to use the land for non-agricultural purposes. In 1992 this process of reform of the **Common Agricultural Policy** was intensified further, and in 2000 more measures to favour the environment were introduced.

SEVESO, in northern Italy, was the place that alerted the world to the dangers of **dioxin**. A vat used in the manufacture of the bacteriocide hexachlorophene exploded at a chemical plant in Seveso in 1976, spreading a huge cloud of lethal dioxin over the countryside. About 18 sq km of surrounding countryside were contaminated and more than 900 people had to be evacuated, once the outbreak of skin disorders and dying wildlife was attributed to the incident at the chemical plant. About 4% of local farm animals died. In the year following the accident, the rate of birth defects increased by more than 40%. The soil in the most heavily contaminated area was so polluted that the top 40 cm had to be removed and buried in two massive plastic-lined pits. The factory was dismantled and some two metric tons of dioxin-contaminated waste were removed for disposal. Many years later soil samples from the area were still 'dioxin-positive'.

SEWAGE TREATMENT comes in two stages—primary and secondary treatment. The primary stage involves a process of screening solids from sewage, leaving a sludge and relatively clear water for further treatment or for disposal into rivers, the sea or on to the land. In the secondary stage the sludge is stirred constantly in vast tanks to get more oxygen into the mixture, allowing bacteria to break down the organic matter and leave a harmless residue that falls as a sediment to the bottom of the tank. After processing, the clear water on top of the tank is discharged into rivers and the sediment is used as **landfill** or discharged at sea.

SEX-CHANGE CHEMICALS: See **hormone-like pollutants** and **endocrine disruptors**

The **SHUTTLE SOLAR BACKSCATTER ULTRAVIOLET (SSBUV)** instrument was an instrument used as part of a US space shuttle mission, in the spring of 1993, which confirmed the decline in **ozone** levels in the northern hemisphere, first found by the **Nimbus-7** meteorological satellite. Only in the equatorial region were ozone values well within the range of the previous year's data. The SSBUV flew eight times on the space shuttle between 1989 and 1996, and the readings became the basis for future measurements. The shuttle-borne instrument complements measurements from the one on the NOAA-11 SBUV/2 satellite, which also monitored ozone from January 1989.

SICK BUILDING SYNDROME describes a situation where, for no apparent reason, a large number of people working in a particular building feel unwell. Although not universally recognized by the medical establishment, the syndrome has been investigated by a number of leading research journals, such as the *New England Journal of Medicine*. The symptoms associated with the syndrome include dry or itchy eyes, nose or throat, skin rashes, headaches, irritability, lethargy and lack of concentration. There are specific diseases that can be attributed to the workplace, like **asthma** and toner's lung. The environmental factors that have been blamed for the syndrome include dampness, dust and dust mites, organisms that thrive where humidifiers or air-conditioning are used, and carpets that are not properly cleaned. Some studies suggest that symptoms are linked to the stress of going to work. Some symptoms were found to be worst at the beginning of the week but had disappeared by Saturday evening. Other symptoms could be attributed directly to office buildings affected by high levels of paper dust, chemical smog from photocopiers, automatic vending machines and other equipment, **ozone**, **carbon monoxide**, **electromagnetic radiation** and cigarette smoke.

The **SIERRA CLUB** is one of the world's most influential conservation groups. The organization was created in 1892, in the USA, by the naturalist John Muir, and can claim to be the founder of the modern environmental movement. Muir was a leading advocate of wilderness preservation. The ideas of conservation of the environment and appreciation of the nature of ecosystems on which the Club is based have been adopted by environmental activists like **Friends of the Earth (FoE)**, **Greenpeace** and the parties that launched the **Green politics** movement in Europe. These ideas are enshrined in Muir's book *My First Summer in the Sierra*, a diary of camping and exploration in 1869, which is regarded as one of the classics of American geographical writing. The Club gradually extended its activities to encompass its present concern for preservation of the scenic and amenity resources and the wildlife of both the sierras and of other comparable areas in North America.

The **SIEVERT** (Sv) is the modern unit used for expressing the dose of radiation to which a person has been exposed. One sievert is equivalent to 100 rem (roentgen equivalent man), an older measure. The average dose to which a person is exposed through **background radiation** is only two millisieverts (1,000ths of a sievert). The unit was named in 1979 for Prof. Rolf Sievert (1896–1966), a Swedish scientist.

SILENT SPRING was the book that gave the first warnings about the harmful effects of synthetic chemical **pesticides** on wildlife, particularly on species of birds. Written by Rachel Carson, a trained biologist and ecologist, and first published in the USA in 1962, *Silent Spring* cited more than 600 research papers showing the destruction of ecosystems as a result of human activity. In particular, Carson indicated 23 substances belonging to the families of the organochlorine and organophosphate pesticide compounds, the use of which had become commonplace in intensive agriculture. All but nine of these substances have since been banned in most of the industrialized countries. For many years, pesticides were very successful in killing or controlling insects and mites that damage crops, and in saving human lives by killing off pests such as the larvae of the malaria-carrying mosquito. Carson pointed to the evidence of accumulation of pesticides in the **food chain**. As an example, she described how, in the 1950s, robins were dying in the grounds of Michigan State University. The culprit was **DDT**, which had been sprayed on US elm trees to protect them from **Dutch elm disease**, imported by a fungus-carrying beetle. Earthworms ate the leaves when they fell to the ground, and were then eaten by birds and other predators. The clue came from research in which earthworms were fed to crayfish, which subsequently died. The poison was shown to be DDT. Subsequent scientific tests showed that 11 contaminated worms was a lethal dose for a robin. Carson documented other deaths: of kestrels, hawks and larks in Norfolk, Virginia; of migrating salmon in a Canadian river, killed by forestry spraying; and of thousands of fish on the Colorado river.

SILT is the fine mineral material formed from the erosion of rock fragments and deposited by rivers and lakes. Its particles are the intermediate form between sand and clay. The particles can range in size from 0.01–0.05 mm in diameter.

SINKS: See **sequestration**

SLASH-AND-BURN AGRICULTURE is a traditional farming system, also known as shifting cultivation or swidden agriculture, that has been used by generations of farmers in tropical forests and the **savannah** of north and east Africa. In north-east India it is known as *jhum* agriculture. It is known to be an ecologically sound form of cultivation, and because the soil is poor in **tropical rainforests** it is a sustainable method of farming. It is still practised today, primarily in the developing countries. Small areas of bush or forest are cleared and the smaller trees burned. This unlocks the nutrients in the vegetation and gives the soil fertilizer that is easily taken up by plants. A few years later the soil is degraded and the farmer moves on to do the same at another site. The original ground is left fallow for anything up to 20 years so that the forest can regenerate. With the growth in **population** and in the subsequent need for more farming land to produce food, the method is increasingly being used today to clear large areas of tropical forests for cattle ranching, and in most cases the ground is not left fallow for long enough and, with modern mechanized farming systems, not enough tree stumps or suitable habitats for plant life are left to start the regeneration process.

SMALL ISLAND STATES refers to the group of nations that are vulnerable to rises in **sea level** caused by **climate change**. Within the UN system, these countries are usually referred to as small island developing states. These nations formed a coalition, within the context of the **Framework Convention on Climate Change (UNFCCC)** and the **Kyoto Protocol**, called the Alliance of Small Island States (AOSIS—which included 43 members and observers by 2004). The Maldives hosted the first Small States Conference on Sea Level Rise. As a result of the Conference, the 1989 Malé Declaration on Global Warming and Sea Level Rise was made, setting out the steps small island states should take to combat the effects of a potential rise in **sea level**. (After the **Earth Summit** a Programme of Action for the Sustainable Development of Small Island States was announced in 1994, in Barbados. This was scheduled to be reviewed at a conference in Mauritius early in 2005.) The Maldives, like many other countries, faces a serious threat from any rise in sea levels. The average height of the islands is about two metres above mean sea level and a rise of 50 cm would have serious consequences. Some islands would be flooded while others could be submerged. As a result of stronger wave action, the infrastructure at some resorts could also be affected. As the Maldivian economy relies mainly on tourism and fisheries, climate change could have a damaging effect. Some low-lying states are facing the possibility that they cannot sustain an existence beyond the mid-21st century, such the small Pacific country of Tuvalu, the highest point of which is only four metres above sea level. Countries like Tuvalu and Kiribati, which have refuges in New Zealand offered to their populations, could disappear completely, while other nations could lose much territory. Meanwhile, the small island states find that there are many other environmental issues that affect their vital interests.

SMOG: See **photochemical smog**

SOIL means very specific but different things to different people. Farmers and gardeners regard it as the top 25–30 cm of the earth's surface. Good soil is dark brown, a sign that it is nutrient-rich in **humus** resulting from decades, or even centuries, of decomposing organic matter. To geologists and soil scientists (pedologists) it consists of layers of variable depth, any of which may be affected physically or chemically by the atmosphere. The various layers are referred to as soil horizons. Early soil scientists regarded soil simply as fragmented rock mixed with organic compounds from decomposed plants. With the development of **ecology** in the 1930s, awareness grew of the influence of the living components in the soil like micro-organisms, insects and worms as well as plants, and of the large variations in soil formations resulting from climatic differences and human activities. The pioneers of modern soil science identified seven major factors in soil formation: climate; organisms; parent material; relief or physical shape; **hydrology**; human factors; and time. Soil temperatures are fairly constant below one metre, approximating to the mean annual air temperature. However, in higher levels the temperatures are more variable and the rates of many processes, like the decomposition of organic matter, are closely linked to maximum air temperature. Consequently, soil fertility could be altered significantly by climate change. The effect of human activity on soils has been significant since at least the Neolithic period. Changes vary from the beneficial, increasing crop yields, to the destructive, leading to pollution and land degradation. Artificial drainage of **wetlands** can improve aeration and let crops grow on previously unproductive swamps. Conversely, unwholesome common minerals may be locked in the waterlogged soil, such as **pyrite**. Pyrite reacts with the oxygen from the air to produce **sulphuric acid**, increasing acidity of the soil. Similarly, **deforestation** for **slash-and-burn agriculture**, followed by soil cultivation and growth of arable crops, can result in the gradual loss of organic matter and nutrient reserves accumulated in soil during the forested period. Cultivation may be followed by weakening of the soil structure and susceptibility to erosion by run-off and wind, and a steady decline in soil fertility because of leaching.

SOIL ACIDIFICATION occurs naturally, but the process is intensified by the emission of **sulphur**, **nitrogen** and **ammonia** compounds from burning **fossil fuels**, industrial processes and agricultural activity. Acidification changes the chemical balance of the soil and causes the release of elements such as **aluminium**, iron, calcium, magnesium and other metals. The Scandinavian countries are among those most severely affected by **transboundary pollution**. The main sources of the acidic emissions that fall there are from the northern countries of mainland Europe and from central Europe, but also from the United Kingdom and Russia (notably the **Kola peninsula**).

SOIL EROSION can be the result of any of a multiplicity of factors that are slowly degrading a large proportion of the world's arable and growing land. Every year the world's **population** increases by over 80m., but there is less **topsoil** (over 20,000m. metric tons less) on which to produce food. **Soil** is a rich and fertile, but fragile, ecosystem. A layer of soil one centimetre in depth can take centuries to develop but be lost in a season: blown away by wind, scoured from deforested slopes by rain or **environmental**

degradation, sterilized by salts (see **salinization**), poisoned by chemicals, exhausted of its nutrients or buried under swamps and buildings. The problem afflicts industrialized countries as well as the developing countries, but for different reasons. The human disaster provoked when soil erosion strikes was first described in the 1930s by John Steinbeck, in his novel *The Grapes of Wrath*. It was a tragedy of families trapped by circumstances when drought turned the Great Plains of the USA into a **dustbowl**. It also happens as a product of **deforestation** or **desertification**.

The **SOLAR ENERGY** falling just on the earth's surface each year is equal to about four times the total estimated **fossil fuel** reserves of the planet. Solar energy is harnessed in many ways, including as forms of **renewable energy** that are derivatives of solar power. In addition, there are **biofuels**, which are dependent on the sun for growing wood for burning, and crops like sugar cane that can be converted into **ethanol** and alcohol fuels for cars and lorries. However, passive solar energy systems form the method that humankind has used for 2,000 years to harness the sun's rays. Passive solar energy refers to the way houses and buildings are designed and built to take the most advantage of the sun. Features such as a patio, verandah and conservatory, which are built as sun traps, are taken for granted. Most of the world's architectural styles have been shaped by the prevailing pattern of sunshine. Traditional designs of buildings in hot regions position the openings for windows and doors with great care to allow a breeze to cool the inhabitants. Highly reflective colours help to reduce the solar heating of the buildings during the day and the dissipation of the stored heat of the building at night to the low temperatures of the cloudless sky. In the higher latitudes, large areas of glass on the south-facing side of buildings capture the sun. Active solar energy systems go beyond architecture and judicious choice of building materials, by employing pumps, valves, motors and electronic control mechanisms for applications like water heating, space heating and providing heat for industrial processes. **Photovoltaic solar power systems** convert sunlight directly to electricity and are used for remote navigation lights and beacons, automatic weather stations and remote telecommunications links.

SOLAR PANELS are one of various devices for collecting **solar energy**, either by direct heating of water or direct conversion of sunlight to electricity.

SOLVENTS are capable of dissolving other liquids, solids and gases and form the bulk of chemical compounds, such as paint, plastic and **pesticides**. By their very nature, solvents tend to be damaging to the environment.

SOUTHERN OSCILLATION: See **El Niño**

SSBUV: See **Shuttle Solar Backscatter Ultraviolet**

STAND: See **forest stand**

The **STATEMENT OF PRINCIPLES ON THE MANAGEMENT, CONSERVATION AND SUSTAINABLE DEVELOPMENT OF ALL TYPES OF FOREST** was a document agreed at the **Earth Summit** and regarded as the first global consensus on the management, conservation and **sustainable development** of forests. This document calls for a recognition of the vital role of all types of forests in maintaining the ecological processes and balance at local, national, regional and global levels, in protecting fragile ecosystems, watersheds and freshwater resources, and as sources of genetic material for **biotech** products, as well as **photosynthesis**. According to the statement, all types of forests embody complex and unique ecological processes that are the basis for their present and potential capacity to provide resources to satisfy human needs. Their sound management and conservation is of concern to the governments of the countries to which they belong and they are of value to local communities and to the environment as a whole. The Principles state that the costs of achieving benefits associated with forest conservation needs and increased international co-operation should be shared by the international community. They also call for appropriate measures to protect forests against harmful effects of pollution, including air-borne pollution, fires, pests and diseases. All countries, especially the developed countries, should take positive action towards **reforestation**, **afforestation** and forest conservation.

The **STEPPELANDS** are the steppes of Eurasia and the prairies of North America, which occupy the heartland of the two great land masses of the northern hemisphere and are the granaries of these two parts of the world. Their climatic characteristics are light rainfall in the spring and early summer, which are the growing seasons, and dry, sunny conditions in the rest of the summer, which are ideal for ripening and harvesting. Since they are inland and far away from the moderating influences of the **oceans**, the steppelands' climate is one of large diurnal, or 24-hour, variation and some large annual ranges of temperature and precipitation. During the warm summers, monthly mean temperatures vary between 17°C (63°F) and 20°C (68°F). Winter is long, with freezing temperatures well below 0°C (32°F) for months. However, the vulnerability of the agriculture of the steppes was demonstrated more than once during the 20th century. Climatic fluctuations have caused a decrease in the moisture-bearing westerly winds that nourish crops.

STOMATA are the tiny pores or holes in the leaves of plants that let in the **carbon dioxide** used in **photosynthesis**, and release oxygen and water vapour. Comparisons of preserved specimens with today's plants suggest that leaves now have fewer stomata than those of the same but earlier species. The investigations began with plants of 200 years ago, but have been extended to cover intervals over the past 160,000 years. Long-term environmental studies are examining whether these changes might be related to changes in the level of carbon dioxide in the atmosphere. The investigations include experiments in growing trees under differing carbon dioxide levels.

STRADDLING STOCKS are those fish whose habitat falls within both the Exclusive Economic Zone created by the **Law of the Sea (UNCLOS III)** (up to 200 nautical miles off the coast of any sovereign country) and the high seas. These fish can be overfished and driven to extinction. The problems associated with straddling stocks and highly migratory fish species, driven by a very public dispute over the turbot between Canada and Spain, led to the Agreement on Conservation and Management of Straddling Fish Stocks and Highly Migratory Stocks, which was adopted in 1995. It limits freedom of fishing on the high seas, subjecting it to the conservation mandates of

ANGOLA

Governmental Organizations

Ministry of Fisheries: Avda 4 de Fevereiro 30, Edif Atlântico, CP 83, Luanda; tel (2) 311420; fax (2) 310199; e-mail geral@angola-minpescas.com; internet www.angola-minpescas.com; Minister: Salomão Luheto Xirimbimbi; Vice-Minister: Dr Victória Francisco Lopes doe Barros Neto.

Ministry of Agriculture and Rural Development: Avda Comandante Gika 2, CP 527, Luanda; tel (2) 322694; fax (2) 323217; e-mail gabminander@netangola.com; Minister: Gilberto Buta Lutukuta.

Ministry of Energy and Water: Avda 4 de Fevereiro 105, CP 2229, Luanda; tel (2) 393681; fax (2) 298687; Minister: José Maria Botelho de Vasconcelos.

Ministry of Urban Affairs and the Environment: Luanda; tel (2) 336717; Minister: Diakumpuna Sita José.

Direcção Nacional da Conservação de Natureza (National Nature Conservation Department): BP 74, Luanda.

Non-Governmental Organizations

Instituto de Investigação Agronómica (Agronomic Research Institute): CP 406, Huambo; f 1962; incorporates agrarian documentation centre; publications.

National Ecological Party of Angola—PNEA: Luanda; Leader: Sukawa Dizizeko Ricardo; member of Democratic Civilian Opposition alliance.

ANTIGUA AND BARBUDA

Governmental Organizations

Ministry of Agriculture, Lands, Marine Resources and the Environment: Queen Elizabeth Highway, St John's; tel 462-1543; fax 462-6104; Minister: Charlesworth Samuel.

National Parks Authority: POB 1283, St John's; tel 460-1379; fax 460-1514; e-mail natpark@candew.ag; Parks Commr: Ann Marie Martin; f 1985; restoration and conservation of heritage sites; museums; archaeological research centre; guided tours; interpretation centre.

ARGENTINA

Governmental Organizations

Ministry of Social Development: Avda 9 de Julio 1925, 13°, 1332 Buenos Aires; tel (11) 4379-3648; e-mail desarrollosocial@desarrollosocial.gov.ar; internet www.desarrollosocial.gov.ar; Minister: Dr Alicia Margarita Kirchner de Mercado; f 1999.

Secretaría de Desarrollo Sustentable y Política Ambiental (Secretariat of Sustainable Development and Environmental Policy): San Martín 459, 1°, 1004 Buenos Aires; tel (11) 4348-8200; fax (11) 4348-8300; e-mail omassei@sernah.gov.ar; internet www.medioambiente.gov.ar; Sec: Dr Atilio Armando Savino; fmrly Secretaría de Recursos Naturales y Desarrollo Sustentable (Secretariat of Natural Resources and Sustainable Development); adopted current name 1999.

Dirección de Fauna y Flora Silvestres—DFYFS (Wildlife Department): San Martín 459, 2°, 1004 Buenos Aires; tel (11) 4348-8551; fax (11) 4348-8554; e-mail vlichtsc@sernah.gov.ar; Dir: Victoria Lichtschein.

Subsecretaría de Ordenamiento y Política Ambiental (Undersecretariat for Planning and Environmental Policy): San Martín 459, 2°, 1004 Buenos Aires; tel (11) 4348-8435; e-mail rpatrouilleau@sernah.gov.ar; internet www.desar-.desarrollosocial.gov.ar.

Ministry of Health and Environment: Avda 9 de Julio 1925, 1332 Buenos Aires; tel (11) 4381-8911; fax (11) 4381-2182; e-mail consultas@msal.gov.ar; internet www.msal.gov.ar; Minister: Dr Ginés González García.

Administración de Parques Nacionales (National Parks Administration): Avda Santa Fe 690, 1059 Buenos Aires; tel (11) 4311-0303; fax (11) 4311-6633; e-mail parquesn@teletel.com.ar; internet siaa.sernah.gov.ar/apn; Pres: Dr Felipe R Lariviere; f 1936; affiliated to the Department of Tourism; management of national parks, national reserves and natural monuments; scientific research; monitoring the planning and use of parks and recreational areas.

Centro de Ecología Aplicada del Litoral—CECOAL (Centre for Applied Ecology): Ruta Provincial No 5, Km 2,5, 3400 Corrientes; tel (3783) 54418; fax (3783) 54421; Dir: Prof Dr Juan José Neiff; f 1973; research on applied ecology, with emphasis on evaluation and diagnostics in relation to natural resources; environmental impact assessment; areas of interest include large tropical rivers, flood-plain wetlands and biodiversity; affiliated to the Consejo Nacional de Investigaciones Científicas y Técnicas—CONICET.

Comision Nacional de Energía Atómica—CNEA (National Commission for Atomic Energy): Avda del Libertador 8250, 1429 Buenos Aires; tel (11) 4704-1000; fax (11) 4704-1154; e-mail comunicacion@cnea.gov.ar; internet www.cnea.gov.ar; Pres: Dr José Pablo Abriata; Vice-Pres: Ing Jorge Fabián Calzoni; f 1950.

Centro Atómico Ezeiza (Ezeiza Atomic Centre): Presbitero Juan Gonzalez y Aragon 15, 1802 Ezeiza, Buenos Aires; tel (11) 6779-8100; fax (11) 6779-8500; e-mail gusalvar@cae.cnea.gov.ar; internet caebis.cnea.gov.ar; research into nuclear energy and radioactive waste and their effects on the environment.

Dirección de Recursos Naturales Renovables (Office of Renewable Natural Resources): Boulogne sur Mer esq Avda El Libertador, Parque General, San Martín, 5500 Mendoza; tel and fax (261) 424-5863; Dir: Eduardo Torres; management of fauna, native flora, forests and protected natural areas.

Instituto Nacional de Tecnología Agropecuaria—INTA (National Institute of Agricultural Technology): Rivadavia 1439, 1033 Buenos Aires; tel (11) 383-5095; fax (1) 383-5090; e-mail postmaster@inta.gov.ar; internet www.inta.gov.ar; Pres: Ing Agr Carlos Chepi; Dir Nacional: Ing Agr Roberto Bochetto; f 1956; agricultural research and extension; maintains 42 experimental stations and 220 extension agencies; affiliated to the Secretariat of Agriculture.

Centro de Investigación en Ciencias Agropecuarias—CICA (Agricultural and Sciences Research Centre): Complejo de Investigación Castelar, Las Cabañas y de Los Reseros, 1712 Castelar, Buenos Aires; tel (11) 4621-1683; fax (11) 4481-1316; e-mail dircica@inta.gov.ar; internet cica.inta.gov.ar; Prin Officer: Oscar Grau; research programmes leading to improved agricultural and fisheries production, in terms of increased yields and reduced environmental impact.

Instituto de Clima y Agua (Climate and Water Institute): Los Reseros y Las Cabañas s/n, 1712 Castelar, Buenos Aires; tel (11) 4621-1684; fax (11) 4621-5663; e-mail cdibella@inta.gov.ar; internet www.inta.gov.ar/cya; Prin Officer: César Rebella; research into climate and water resources, including the effects of agriculture on climate and of climate on agriculture.

Instituto de Ingeniería Rural (Institute of Rural Engineering): Avda Pedro Díaz 1798 y Avda Williams, Casilla 25, 1712 Castelar, Buenos Aires; tel and fax (11) 4665-0495; e-mail ingrura@inta.gov.ar; internet www.inta.gov.ar/cica/iir/iir.htm; Prin Officer: Roberto Delafosse; development of machinery and systems which improve agricultural yields and the protection of the environment.

Instituto de Microbiología y Zoología Agrícola—IMYZA (Institute of Microbiology and Agricultural Zoology): Las Cabañas y de Los Reseros, Villa Udaondo CC25, 1712 Castelar, Buenos Aires; tel (11) 4481-4320; fax (11) 4621-1701; e-mail imiza@cnia.inta.gov.ar; internet www.inta.gov.ar/cnia/imyza/imyza.htm; Dir: Roberto Lecuona; f 1958; research into the use of micro-organisms to improve agricultural yields, to reduce the impact of pollution on yields and to reduce the impact of agriculture on the environment; biological pest control.

Instituto Tecnológico de Chascomús (Chascomús Technological Institute): Camino Circunvalación Lag Km 6, CC 164, 7130 Chascomús, Buenos Aires; tel (2241) 430323; fax (2241) 424048; e-mail iglesias@criba.edu.ar; Dir: Alberto Alvari Iglesias; studies into aquatic ecosystems, water quality, soils.

Museo Provincial de Ciencias Naturales 'Florentino Ameghino' (Florentino Ameghino Provincial Museum of Natural Sciences): Primera Junta 2859, 3000 Santa Fé; tel and fax (342) 4573770; e-mail ameghino@ceride.gov.ar; internet www.unl.edu.ar/santafe/museocn.htm; Dir, Biology: Carlos A Virasoro; Library and Documentation Centre: Andrea Van Perdeck; Botanic area: Lic Andea Bosisio; Invertebrate Zoology: Lic Monica Paris de Bassi; Vertebrate Zoology: Prof Edelvita Fioramonti: Museology area: Tac Rodolfo Ferreyra; f 1914; affiliated to the Ministry of Education; study and management of national fauna and flora; maintenance of zoological, botanical and other collections; ecological education; preservation of natural and cultural heritage; includes a Library and Documentation Centre, Zoological Laboratory, Vascular Herbarium and Cultural Centre.

Secretaría de Agricultura, Ganadería, Pesca y Alimentación (Department of Agriculture, Livestock, Fisheries and Food): Avda Paseo Colón 922, 1°, Of 146, 1063 Buenos Aires; tel (11) 4349-2291; fax (11) 4349-2292; e-mail prensa@mecon.gov.ar; internet www.sagpya.mecon.gov.ar; Sec: Ing Agr Miguel Santiago Campos; f 1871; part of the Ministry of Economy; undertakes regulatory, promotional, advisory and administrative responsibilities on behalf of the meat, livestock and fisheries industries.

Centro de Documentación e Información Forestal 'Ingeniero Agrónomo Lucas A Tortorelli' (Lucas A Tortorelli Centre for Forestry Documentation and Information): Avda Paseo Colón 982, Anexo Jardin, 1063 Buenos Aires; tel (11) 4349-2124 / 2125; fax (11) 4349-2102; e-mail bfores@mecon.gov.ar; internet www.sagpya.mecon.gov.ar/new/0-0/forestacion/index.php; Library: Nilda Fernandez; f 1948; the most important forestry library in Argentina.

Non-Governmental Organizations

Agencía de Cooperación en Ingeniería Ambiental (Environmental Engineering Co-operation Agency): Quintana 2351, 3400 Corrientes, Corrientes; tel (3783) 42-0411; Pres: Carlos Patiño.

Amigos de la Tierra Argentina (Friends of the Earth—FoE Argentina): Córdoba 5051, 1426 Buenos Aires; tel (11) 4773-5947; fax (11) 4777-9837; e-mail postmaster@amigos.org.ar; Prin Officers: Osvaldo B Corvalán, María Belén Villa, Roberto Reisinger, Maximo Winkler, María del C Cogurno, Ana Ganopol, Anahi Magdaleno; f 1984; environmental education; research into flora and fauna; protection of natural areas; environmental management; processing of household waste.

Asociación Ambiental (Environmental Association): 25 de Mayo 114, 3280 Colón, Entre Ríos; tel (3447) 42-1940; fax (3447) 42-2282; Pres: Marta Catalina Uechi.

Asociación Ambientalista Verde X Gris (Green/Grey Environmental Association): Marmol 177, 1888 Florencio Varela, Buenos Aires; tel (11) 4255-2459; Pres: María Virginia Demattei.

Asociación Amigos de la Ecología y Protección del Medio Ambiente Humano (Association of Friends of Nature and Protection of the Human Environment): San Martín 451, 1°, 4000 Tucumán, Tucumán; Pres: Gloria Carrizo.

Asociación Amigos del Arbol y la Tierra—Patagonia Verde—AAyT (Association of Friends of the Earth and Trees—Green Patagonia): Doctor Baraja 479, 8504 Carmen de Patagones, Comarca Patagones-Viedma; tel and fax (2920) 46-1418; e-mail larranaga@arnet.com.ar; Pres: Dr. Carlos Alberto Larrañaga; Vice-Pres: Carlos Fioraventi Strassner; conservation of flora and fauna; study of pollution; agricultural and fisheries programmes; mem, Clean Up World.

Asociación Argentina contra la Contaminación Ambiental (Argentine Association against Environmental Pollution): Junín 956, 4°, 1113 Buenos Aires; tel (11) 4963-3414; Pres: Juan Moretton.

Asociación Argentina de Ingeniería Sanitaria y Ciencias del Ambiente—AIDISAR (Argentine Environmental Engineering and Science Association): Belgrano 1580, 3°, 1093 Buenos Aires; tel (11) 4381-5832; fax (11) 4381-5903; e-mail aidisar@aidisar.org.ar; internet www.aidisar.org.ar; Pres: Enrique Arntsen; programmes in health, environment and waste management; publishes a list of environmental engineering and science institutions.

Asociación Argentina de Medios por el Medio Ambiente (Argentine Association for Environmental Methods): Irigoyen

Freyre 2670, 3000 Santa Fe; tel (342) 453-5339; e-mail proteger@aamisde.satlink.net; Pres: Liliana A Corra.

Asociación Argentina para la Naturaleza (Argentine Nature Association): Belgrano 571, 1°, 1708 Morón, Buenos Aires; Pres: María José Toledo; (11) 4483-4190.

Asociación Austral para la Vida Silvestre—AAViS (Southern Association for Tree Life): Intevu 12b, 58, Casilla 20, 9420 Río Grande, Tierra del Fuego; tel (2964) 42-3080; Pres: Fabio Mackowiak; campaigns in southern Argentina on issues including conservation of flora and fauna, sustainable development, preservation of natural resources, energy, soils, waste, health and demography, environmental education and improvement of agriculture.

Asociación Civil de Ecoamigos (Civil Association of Friends of the Environment): Patricias Mendocinas 602, 1714 Ituzaingo, Buenos Aires; tel (11) 4450-4761; fax (11) 4624-3538; Pres: Alejandra Casella.

Asociación Civil Los Algarrobos—Desarrollo Sustentable (Los Algorrobos Civil Association for Sustainable Development): 25 de Mayo 376, Casilla 96, 5176 La Cumbre, Córdoba; tel (3548) 45-2199; fax (3548) 45-1756; e-mail acla@acla.org.ar; internet www.acla.org.ar; Pres: Andrés Hamilton Joseph; Vice-Pres: Daniel Tomasini; Programme Dir: Margarita Astralaga; f 1989; environmental education and promotion of sustainable agriculture and development.

Asociación Civil Pro Vicente López (Vicente López Civil Association): Avda Maipú 169, 1°, Departamento J, 1638 Vicente López, Buenos Aires; tel (11) 4797-2859; Pres: Santiago Jelenic.

Asociación Cultural Mariano Moreno—ACMM (Mariano Moreno Cultural Association): Belgrano 450, 1876 Bernal, Buenos Aires; tel (11) 4252-2292; fax (1) 4259-1039; Pres: Ruben Ravera; environmental education, promotion of use of renewable natural resources, improvement of agriculture, health and welfare.

Asociación de Protección al Ambiente Serrano La Calera (La Calera Association for the Protection of the Mountain Environment): Uriburu 461, Casilla 40, 5151 La Calera, Córdoba; tel (3543) 46-7327; fax (3543) 46-6291; Pres: Gabriel Rolfi.

Asociación de Protección del Ambiente Serrano (Association for the Protection of the Mountain Environment): 25 de Mayo 376, 5178 La Cumbre, Córdoba; tel (3548) 45-2073; fax (3548) 45-1737; Pres: José Velez.

Asociación Ecologista Nueva Tierra (New Earth Ecology Association): Francisco González 2016e, 1650 San Martín, Buenos Aires; tel and fax (11) 4754-6438; e-mail nuevatierra@hotmail.com; Pres: Susana Papale.

Asociación para la Protección del Medio Ambiente y Desarrollo Humano—ECOVIDA (Association for Environmental Protection and Human Development—ECOVIDA): Marcos A Zar 174, Casilla 44, 9120 Puerto Madryn, Chubut; tel and fax (2965) 45-2169; Pres: Federico Yacianci.

Asociación para la Protección del Medio Ambiente y Educación Ecológica de la República Argentina—APRODEMA (Argentine Association for Environmental Protection and Ecological Education): San Martín 933, 5°, Departamento Q, 1004 Buenos Aires; tel and fax (11) 4311-1263; Pres: Cayetano Umana Minetto; research programmes on ecosystems in Argentina and pollution in Buenos Aires.

Asociación Turismo y Ambiente (Tourism and Environment Association): Calle 9 1583, 1900 La Plata, Buenos Aires; tel (221) 452-2135; Pres: Angel Hugo Merlo.

Aves Argentinas—Asociación Ornitológica del Plata (River Plate Ornithological Association): 25 de Mayo 749, 2°, Departamento 6, 1002 Buenos Aires; tel and fax (11) 4312-1015; e-mail aop@aorpla.org.ar; Pres: Juan Carlos Chebez; Exec Dir: Andrés Bosso; f 1916; scientific research and conservation of wild birds and their habitats; publication of journal and magazine; courses for ornithologists; Argentinian Naturalists School; camps; outings; library; affiliated with BirdLife International.

Capacitando: Avda Luro 3351, 7600 Mar del Plata, Buenos Aires; tel (223) 475-2278; fax (223) 475-6272; Pres: Dora Beatriz Fontanes; areas of interest include conservation of plant and animal species, forestry and environmental education.

Centro Ambiental Argentino (Argentine Environmental Centre): Ayacucho 242, 8°, Departamento D, 1000 Buenos Aires; tel (11) 4953-2060; fax (11) 4784-2427; Pres: Yolanda Ortiz.

Centro de Estudios Ambientales para la Planificación y el Desarrollo (Centre for Environmental Studies for Planning and Development): Sarmiento 230, 5°, Departamento A, 8000 Bahía Blanca, Buenos Aires; tel and fax (291) 453-5247; e-mail jecalo@criba.edu.ar; Pres: Daniel Rabano.

Centro de Estudios para el Desarrollo de Proyectos Ambientales (Research Centre for the Development of Environmental Projects): Corrientes 982, 1640 Martínez, Buenos Aires; tel (11) 4798-2361; Pres: Máximo Ullmann.

Centro de Estudios y Promoción Agraria—CEPA (Centre for the Promotion and Study of Agriculture): Olleros 3877, 1427 Buenos Aires; tel (11) 4553-6810; promotion of sustainable agriculture, rural development and protection of soils from deterioration and erosion.

Centro de Investigación y Extensión Forestal Andino Patagónico (Andes–Patagonia Forestry Expansion and Research Centre): Ruta 259, Km 4, Casilla 238, 9200 Esquel, Chubut; tel and fax (2945) 44-3948; conducts programmes aimed at expanding forest cover and reducing damage caused to forest areas by fire.

Centro de Protección de la Naturaleza (Nature Protection Centre): 9 de Julio 3563, Casilla 538, 3000 Santa Fe; tel and fax (342) 456-2609; Pres: Héctor Pertout.

Centro Interdisciplinario de Estudios sobre el Desarrollo Latinoamericano—CIEDLA (Interdisciplinary Research Centre on Latin American Development): Fundación Konrad Adenauer, L N Alem 690, 20°, 1001 Buenos Aires; tel (11) 4313-3522; fax (11) 4311-2902; e-mail kas-ciedla@kas-ciedla.org.ar; internet www.kas-ciedla.org.ar; Dir: Dr Dieter W Benecke; f 1981; research projects; seminars and conferences on Latin America; library of approx 5,000 volumes containing documentation on Latin America and Europe.

Centro Latinoamericano de Estudios Ambientales—CLEA (Latin American Environmental Research Centre): Callao 1103, 10°, Departamento I, 1023 Buenos Aires; tel (11) 4813-0968; fax (11) 4322-3950; Pres: Pablo Quiroga; f 1989; environmental impact assessment, education, training and legislation; regional planning.

Centro Mocoví 'Ialek Lav'a': Avda Sarmiento s/n, Barrio Sanurbano, 2728 Melincue, Santa Fe; tel (3465) 49-9015; fax (3465) 49-9197; Co-ordinator: Ariel Araujo; concerned with environmental issues affecting indigenous communities, including deforestation and land-use.

Centro Regional para el Desarrollo y el Medio Ambiente (Regional Centre for Development and the Environment): Los Almendros 1003, 5881 Merlo, San Luis; tel (2656) 47-5482; fax (2656) 47-5365; Pres: Claudio Huberman.

Club Andino Tucumán (Tucumán Andean Club): Las Heras 50, 4000 San Miguel de Tucumán, Tucumán; tel (381) 4301256; Pres: Manuel Parajón.

Comisión Ambientalista Xanaes (Xanaes Environmental Commission): Reconquista 12, 5960 Río Segundo, Córdoba; tel (3572) 42-1988; fax (3572) 42-1976; e-mail coamxa_rio2@yahoo.com; Pres: Mario Edgardo Cánepa; affiliated to Consejo de Organizaciones Ambientales no Gubernamentales de la Provincia de Córdoba; environmental monitoring and protection; protection of animals; publications.

Comisión Nacional de Ecología y Desarrollo Humano (National Commission for Ecology and Human Development): Martín Fierro 1313, 1875 Wilde, Buenos Aires; tel (11) 4201-9008; Pres: Nestor González.

Comité Argentino para la Energía Solar (Argentine Solar Energy Committee): Saraza 450, 1424 Buenos Aires; tel (11) 4772-7005; fax (11) 4394-3635; Pres: Horacio Lafontaine.

Compañeros de las Americas, Patagonia-Montana (Partners of the Americas, Patagonia-Montana): Padre Milanesio 1099, Junín de los Andes, 8371 Neuquén; tel (972) 91553; fax (972) 91262; Pres: Esmeraldo Correale; regional office of Partners of the Americas (Washington, DC, USA); promotes economic and social development through technical training; development activities are originated by eight sub-committees, including those for health, natural resources, emergency-preparedness and ecotourism; brs in Bariloche, Chuele Chuel, Neuquén, Zapala and San Martín de los Andes.

EcoBios: Almirante Zar 323, Puerto Deseado, 9050 Santa Cruz; tel and fax (97) 87-1237; e-mail rqfrere@criba.edu.ar; Prin Officers: Dr Patricia Alejandra Gandini, Dr Esteban Frere; f 1982; conservation of sea-birds in Patagonia.

Eco-Geo Argentina—Asociación Ecologista de Santiago del Estero (Eco-Geo Argentina—Santiago del Estero Ecology Association): Avda Roca Sur 811, 4200 Santiago del Estero; tel (385) 421-4170; Pres: María Margarita Cambrini.

Equística—Defensa del Medio Ambiente (Ekistics—Defence of the Environment): H de la Quintana 33, Rosario 2000, Santa Fe; tel and fax (341) 63-7011; Prin Officer: Rosa Mirtha Fuentes Huisman; f 1980; research into air quality, transportation, housing and environmental law; environmental impact assessment, particularly in relation to the Paraná River; information and advisory service for industry; environmental education; research and dissemination of information on climate change; promotes Agenda 21 at local level; affiliated with Foro del Buen Ayre.

Fundación Ambiente y Recursos Naturales—FARN (Foundation for Environment and Natural Resources): Monroe 2142, 1428 Buenos Aires; tel (11) 4788-4266; e-mail info@farn.org.ar; internet www.farn.org.ar; Exec Dir: Daniel A Sabsay; f 1985; research, promotion, consultation and training in the areas of environmental policy, legislation, economics and administration.

Fundación Bariloche (Bariloche Foundation): Casilla 138, Avda 12 de Octubre 1915, 8400 San Carlos de Bariloche; tel and fax (2944) 42-2050; e-mail csuarez@criba.edu.ar; internet www.bariloche.com.ar/ideefb/spanish/fb/index.htm; Pres: Carlos E Suárez; Vice-Pres: Juan Carlos Secondi; energy economics; environment; quality of life; epistemology; environment, development and energy.

Buenos Aires Office: Fundación Bariloche, Piedras 482, 2° H, 1070 Buenos Aires; tel (11) 4331-1646; fax (11) 4334-4717; e-mail idee@bariloche.com.ar.

Fundación Cruzada Patagonica (Patagonian Crusade Foundation): Felix San Martín 678, Junín de los Andes, 8371 Neuquén; tel and fax (972) 91262; Pres: Dr Germán A Pollitzer; Treas: Eduardo Ramis; Head of Education: Sergio Rumene; f 1979; education, demography, agriculture and public transport serving Campeñiuen Araucanian communities.

Fundación del Sur (Foundation of the South): Cochabamba 449, 1150 Buenos Aires; tel (11) 4361-8549; fax (11) 4307-0545; e-mail jlmerega@fundasur.org.ar; internet www.fundasur.org.ar; Pres: Roberto Gedeón; Exec Dir: Juan Luis Mérega; f 1987; environmental programmes covering issues including desertification and municipal waste management; development of community programmes, with emphasis on support for small businesses; local development; quarterly newsletter.

Fundación Neuquén para la Conservación de la Naturaleza (Neuquén Foundation for Nature Conservation): Correo Central 25, San Martín de los Andes, 8370 Neuquén; tel (972) 27311; fax (972) 27111; Pres: Douglas Reid; Vice-Pres: Ariel Sibileau; f 1988; affiliated with Nature Conservancy (USA).

Fundación para la Conservación de las Especies y el Medio Ambiente—FUCEMA (Foundation for the Conservation of Species and the Environment): Pringles 10, 3°, 1183 Buenos Aires; tel and fax (11) 4983-7949; e-mail webmaster@fucema.org.ar; internet www.fucema.org.ar/; Pres: Rodolfo Alejandro Tecchi; f 1990; field studies; monitoring and management of protected areas; training of biologists and park rangers; compilation of biodiversity inventories; education; publications.

Fundación Vida Silvestre Argentina—FVSA (Argentinian Wildlife Foundation): Defensa 251, 6°, Departamento K, 1065 Buenos Aires; tel (11) 4331-3631; fax (11) 4331-4086; e-mail info@vidasilvestre.org.ar; internet www.vidasilvestre.org.ar; Pres: Dr Héctor Laurence; Vice-Pres: Enrique Götz; Sec: Mauricio Rumboll; f 1977; national associate organization of the World Wide Fund for Nature—WWF International; promotion of conservation and wise use of natural resources; maintenance of two wildlife reserves; environmental education.

Greenpeace Argentina/Cono Sur (Greenpeace Argentina/Southern Cone Regional Office): Mansilla 3046, 1425 Buenos Aires; tel (11) 4962-0404; fax (11) 4962-7164; e-mail greenpeace.argentina@dialb.greenpeace.org; internet www.greenpeace.org.ar; Dir of Communications: Oscar Soria; f 1987; concerned with pesticides, nuclear energy and trade in toxic materials; office responsible for Argentina, Chile and Uruguay.

Grupo de Educadores Ambientalistas—G de EA (Environmentalist Teachers Group): Laprida 1373, San Isidro 1642, Buenos Aires; tel (11) 4743-9467; fax (11) 4812-9467; Dir-Gen: Lilian Lavalle de Vignaroli; Prin Officers: Monica Ramírez de Spinardi; f 1991; training in environmental education for teachers; design of environmental education and conservation projects for small communities; workshops.

Grupo Ecologista Misionero—GEM (Misiones Ecology Group): Col Alvarez 2299, 14° F, 3300 Posadas, Misiones; tel (752) 20952; Gen Co-ordinator: A Nelci Pascual; Sec: Alberto Goñi; f 1990; establishment of parks or reserves in threatened environments or those representative of ecosystems in the province of Misiones; study and preservation of endangered animal and plant species; lectures and conferences; publication of articles in newspapers, with the aim of making the public aware of the need for environmental preservation.

Instituto Argentino de Investigaciónes de las Zonas Aridas—IADIZA (Argentinian Institute for Research into the Arid Zones): Casilla 507, CRICYT, Bajada del Cerro s/n, 5500 Mendoza; tel (261) 24-1995; fax (261) 428-7995; e-mail cricyt@planet.losandes.com.ar; Dir: Juan Carlos Guevara; Research Dir: Virgilio G Roig; f 1972; attached to the National Research Council of Argentina; involved in research and education in the areas of ecology and taxonomy of desert flora and fauna, management of Argentina's desert regions, global change, biodiversity of arid lands, and assessment and control of desertification.

Instituto de Desarrollo Tecnológico para la Industria Química—INTEC (Institute of Technological Development for the Chemical Industry): Güemes 3450, 3000 Santa Fe; tel (342) 4532965; fax (342) 4532965; e-mail director@intec.unl .edu.ar; internet intec.ceride.gov.ar/; Dir: Dr Alberto E Cassano; 1st Vice-Dir: Mario G Chiovetta; 2nd Vice-Dir: Dr Alberto Cardona; f 1975; concerned with chemistry and the environment; research and development in advanced oxidation technologies; different reacting systems employing ultraviolet light applied to air and water pollution; energy optimization in chemical plants; domestic waste management; publications.

Instituto de Estudios e Investigaciones sobre el Medio Ambiente—IEIMA (Institute of Studies and Research on the Environment): Calle Adolfo Alsina 1816, 1090 Buenos Aires; tel (11) 4374-2951; fax (11) 4372-1850; e-mail ieima@ciudad .com.ar; Pres: Elva Roulet; Dir: Elida Barreiro; f 1990; research and documentation in all areas of environmental science, technology and development.

Instituto de Investigaciones Fisiológicas y Ecológicas—IFEVA. Universidad de Buenos Aires (Institute of Physiological and Ecological Research, University of Buenos Aires): Avda San Martín 4453, 1417 Buenos Aires; tel (11) 4524-8070; fax (11) 4514-8730; e-mail sanchez@ifeva.edu.ar; internet www.ifeva.edu.ar; Dir: Rodolfo A Sánchez; crop improvement research.

Proteger Foundation for Sustainable Development and Health: Belgrano 3569, Casilla 550, 3000 Santa Fe; tel (342) 497-0298; fax (342) 455-2293; e-mail jcproteg@satlink .com; Pres: Lilian Corra; Dir: Jorge Cappato; f 1991; sustainable development projects; promotion of appropriate technology; campaigns for ecosystem protection; environmental health training; environmental education.

Sociedad Argentina para el Derecho y la Administración del Ambiente y de los Recursos Naturales—SADARN (Argentinian Society for Environmental and Natural Resources Law and Management): Uruguay 467, 10°, Of B, 1015 Buenos Aires; tel (11) 4371-5867; fax (11) 4812-1575; e-mail pgrettilex@elsitis.net; Pres: Dr Eduardo A Pigretti; Vice-Pres: Dr Pascual V Politi; Sec: Dr Graciela Berra Estrada; f 1977; participation in environmental projects and conferences; formulation of proposals for legislation; creation of Comité Argentino pro Tribunal Internacional del Ambiente (Argentinian Committee for an International Environmental Tribunal); affiliated with Comisión Interamericana de Derecho y Administración del Ambiente—CIDYAA (Inter-American Cmmission for Environmental Law and Management) and the International Court of Environmental Arbitration and Conciliation.

Sociedad Central de Arquitectos (Central Society of Architects): Montevideo 938, 1019 Buenos Aires; tel (11) 4812-3644; fax (1) 42-6629; e-mail info@socearq.org; internet www .socearq.org; Pres: Arq. Carlos Lebrero; f 1886; professional association.

TETRA TECH Argentina, SA: Reconquista 144, 6°, 1003 Buenos Aires; tel (11) 4345-5410; fax (1) 4345-5420; Chair: Dr Humberto Romero; environmental impact assessment.

TRAFFIC Sudamérica—Oficina Argentina (TRAFFIC South America—Argentina Office): Ayacucho 1477, 9° B, 1111 Buenos Aires; tel and fax (11) 4811-4348; Dir: Tomás Waller; Consultants: Juan X Gruss, Patricio A Micucci; f 1991; field surveys; statistics on trade in wildlife; concerned with sustainable management of wildlife and commercially valuable species.

Universidad Nacional del Sur, Departamento de Agronomía (National University of the South, Department of Agronomy): San Andrés 800, 8000 Bahía Blanca, Buenos Aires; tel (291) 43-0024; fax (291) 42-1942; e-mail webmaster@ criba.edu.ar; internet www.criba.edu.ar/agronomia; Dir: Miguel A Cantamutto; research on all environmental issues affecting and arising from agricultural activity, including soils, water resources, contaminated sites, pests and forestry.

ARMENIA

Governmental Organizations

Ministry of Energy: 375010 Yerevan, Republic Sq, Government Bldg 2; tel (1) 52-19-64; fax (1) 52-63-65; e-mail minenergy@minenergy.am; internet www.minenergy.am; Minister: Armen Movsissian; nature conservation and protection strategy; co-ordination of ecological research; environmental monitoring; preventive measures against pollution; natural resource management and regulation of the use and reproduction of natural resources; international co-operation in the field of environmental protection and management; responsibility for the State Environmental Inspectorate, the Environmental Monitoring Centre, Hayantar State Forestry Enterprise, the Hydrometeorological Service and the Geological Survey.

Ministry of Agriculture: 375010 Yerevan, Nalbandian St 48; tel (1) 52-46-41; fax (1) 15-10-36; internet www.minagro .am; Minister: David Lokian.

Ministry of Nature Protection: 375010 Yerevan, Government Bldg 3, Republic Sq; tel (1) 52-10-99; fax (1) 58-54-96; e-mail iac@mnpiac.am; internet www.mnpiac.am; Minister: Vardan Ayvazian.

Non-Governmental Organizations

Armenian Agricultural Institute: 375009 Yerevan, Teryana St 74; tel (1) 52-45-41; fax (1) 52-23-61; e-mail asmagric@ usda.am; Rector: A P Tarverdyan; Pro-Rector, Instruction: Yuri Marmaryan; Pro-Rector, Scientific Section: Daniel Petrosyan; f 1930; departments include those of agrochemistry and soil science, plant protection and irrigation; library of 540,000 vols; publications; faculties: agrarian, food technologies, economics, veterinary medicine, and livestock breeding/ engineering.

Armenian Scientific Research Institute for Scientific Technical Information: Yerevan; tel (2) 23-67-74; e-mail nfp@globinfo; Dir: Marat B Edilian; f 1982; serves as the INFOTERRA national focal point.

Association for Human Sustainable Development: Yerevan, Issahakian St 28; tel (2) 52-85-30.

Centre for Ecological-Noosphere Studies, Armenian National Academy of Sciences: 375025 Yerevan, Abovian St 68; tel (2) 56-93-31; fax (2) 58-02-54; e-mail ecocentr@sci.am; internet www.sci.am/about/research/18-econoosf.htm; Dir: Armen K Sagatelian; Deputy Dir: Robert H Revazyan; f 1989; research in fields including environmental geochemistry, biomonitoring, information technology in environmental protection; environmental impact assessments.

Council on Problems of the Biosphere, Armenian National Academy of Sciences: 375019 Yerevan, Marshal Baghramian St 24; tel (2) 52-07-04; Chair: G A Baghramian.

Ecological Survival: Yerevan, Marshall Baghramian St 24d, Apt 902; tel (2) 27-93-35; e-mail akhosrov@aua.am.

Ecoteam of Armenia: 375001 Yerevan, Abovian St 22[a], Apt 53; tel (2) 53-01-23; Pres: Artashes Sarkissian.

EPAC: Yerevan, Koriun St 8, Apt 8; e-mail epac@arminco.com.

Greens' Union of Armenia: 375093 Yerevan, Mamikonian St 47/13; tel (2) 25-76-34; fax (2) 15-17-95; e-mail yeranos@armgreen.arminco.com; Pres: Hakob Sanasarian; Foreign Relations Co-ordinator: Anahit Allakhverdian; f 1986; environmental education; establishment and development of ecologically clean technologies; monitoring and preservation of food, water, soil and air quality; rational use of resources; recycling; waste treatment; alternatives to nuclear power; protection of fauna and flora.

Institute of Botany and Botanical Garden, Armenian National Academy of Sciences: 375063 Yerevan, Avan; tel (2) 62-58-40; Dir: Ashot A Charchoglian; f 1944.

Institute of Hydroecology and Ichthiology: 375019 Yerevan, Marshal Baghramian St 24; tel (2) 27-92-68; e-mail rhovan@sci.am.

Scientific Research Institute of Environmental Hygiene and Occupational Toxicology: 375001 Yerevan, Tumanian St 8; tel (2) 58-24-13; fax (2) 15-10-97; research into environmental pollution and toxic wastes.

Sevan Institute of Hydroecology and Ichthyology, Armenian National Academy of Sciences: 378610 Sevan, Ul Kirova 186; tel (76) 2-34-19; Dir: R O Hovhanesyan.

AUSTRALIA

Governmental Organizations

Department of the Environment and Heritage—Environment Australia: GPOB 787, Canberra, ACT 2601; tel (2) 6274-1111; fax (2) 6274-1123; internet www.ea.gov.au; Minister: Senator Robert Hill; Parl Sec: Dr Sharman Stone; Sec: Roger Beale; department of the Federal Government, advice on and implementation of policies and programmes for environmental conservation and ecologically sustainable development; affiliated with Australian Greenhouse Office, Australian Heritage Commission, Great Barrier Reef Marine Park Authority, Australian Oceans Office.

Australian Heritage Commission: POB 787, Canberra, ACT 2601; tel (2) 6274-2111; fax (2) 6274-2095; internet www.environment.gov.au/heritage/index.html; First Asst Sec: Bruce Leaver; identification of places of natural and cultural value; management of World Heritage Sites in Australia; maintenance of the National Estate data base; advice on heritage protection.

Australian and New Zealand Environment and Conservation Council—ANZECC: c/o Department of the Environment and Heritage, GPOB 787, Canberra, ACT 2601; tel (2) 6274-1428; fax (2) 6274-1858; e-mail anzecc@ea.gov.au; internet www.environment.gov.au/anzecc; forum for the exchange of information and experience, and the development of co-ordinated policies on national and international environmental and conservation issues; members represent Australian federal, state and territory governments, and the governments of New Zealand and Papua New Guinea.

Bureau of Meteorology Research Centre—BMRC: 150 Lonsdale St, Melbourne, Vic 3001; tel (3) 9669-4444; fax (3) 9669-4660; e-mail ask_bmrc@bom.gov.au; internet www.bom.gov.au/bmrc; Chief: Dr M J Manton; f 1985.

Department of Agriculture, Fisheries and Forestry: GPOB 858, Canberra, ACT 2601; tel (2) 6272-3933; fax (2) 6272-5161; internet www.affa.gov.au; Minister: Warren Truss; Minister forFisheries, Forestry and Conservation: Ian Macdonald; department of the Federal Government.

Ministerial Council on Forestry, Fisheries and Aquaculture—MCFFA: c/o Agriculture, Fisheries and Forestry—Australia, GPOB 858, Canberra, ACT 2601; tel (2) 6272-4441; fax (2) 6272-4875; e-mail scf@affa.gov.au; internet www.affa.gov.au/ffid/mcaffa/mcaffa.html; Chair (Forestry): Wilson Tuckey; Chair (Fisheries and Aquaculture): Warren Truss; f 1994; includes New Zealand (member) and Papua New Guinea (observer); forum for development of policies consistent with objectives of federal, state and territory governments and, wherever relevant, the New Zealand Government; provides a means to achieve integration of action by governments on issues relating to forestry, fisheries and aquaculture; consultative body; affiliated with Standing Cttee on Forestry, Standing Cttee on Fisheries and Standing Cttee on Aquaculture.

Natural Resource Management Ministerial Council–NRMMC: GPOB 858, Canberra, ACT 2601; tel (2) 6272-4145; fax (2) 6272-4772; e-mail nrmmc@mincos.gov.au; internet www.mincos.gov.au; Jt Chair: Warren Truss (Australian Minister for Agriculture, Fisheries and Forestry), Ian Campbell (Australian Minster for Environment and Heritage); development of integrated and sustainable agricultural, land and water management policies, strategies and practices; members are Australian federal, state and territory ministers, and New Zealand ministers, responsible for agriculture, soil conservation, water resources and rural adjustment matters.

Australian Museum: 6 College St, Sydney S, NSW 2010; tel (2) 9320-6000; fax (2) 9320-6050; e-mail info1@austmus.gov.au; internet www.amonline.net; Dir: Frank Hawarth; f 1827; natural history and human studies; promotion of understanding of Australia's natural environment and cultural heritage through public programmes and exhibitions, and scientific research; taxonomic and environmental research based on the museum's collections and field studies; commercial work for environmental surveys.

Commonwealth Scientific and Industrial Research Organization—CSIRO: Bag 10, Clayton S, Vic 3169; tel (3) 9545-2176; fax (3) 9545-2175; e-mail enquiries@csiro.au;

internet www.csiro.au; Chair: Catherine Livingstone; Chief Exec: Dr Geoff Garrett; f 1926; independent statutory authority created by the Commonwealth of Australia; conducts research in all fields of the physical and biological sciences except defence science, nuclear energy and clinical medicine.

CSIRO Atmospheric Research: Private Bag 1, Aspendale, Vic 3195; tel (3) 9239-4400; fax (3) 9239-4444; e-mail chief@dar.csiro.au; internet www.dar.csiro.au.

CSIRO Division of Wildlife and Ecology: GPOB 284, Canberra, ACT 2601; tel (2) 6242-1600; fax (2) 6242-1555; e-mail enquiries@dwe.csiro.au; internet www.dwe.csiro.au; Chief: Dr Steve Martin; Deputy Chief: Allen Kearns; f 1949; focuses on the ecology of Australian landscapes; management of natural resources; research to develop Australia's capacity to predict and respond aptly to future environmental change; affiliated with CSIRO Divisions of Land and Water, Atmospheric Research, Entomology and Marine Research.

CSIRO Land and Water: PMB No 2, Glen Osmond, SA 5064; tel (8) 26246-5621; fax (8) 26246-5595; e-mail graham.harris@cbr.clw.csiro.au; internet www.clw.csiro.au; Chief of Div: Dr Graham Harris.

CSIRO Land and Water: Black Mountain Laboratories, GPOB 1666, Canberra, ACT 2601; tel (2) 6246-5700; fax (2) 6246-5800; e-mail enquiries@adl.clw.csiro.au; internet www.clw.csiro.au; Chief: Dr Graham Harris; component of the Commonwealth Scientific and Industrial Research Organisation—CSIRO; research on soil, water and atmospheric processes to promote understanding and sustainable management of land and water resources.

Department of Industry, Science and Resources: 20 Allara St, Canberra, ACT 2601; tel (2) 6213-6000; fax (2) 6213-7000; e-mail distpubs@dotrs.gov.au; internet www.dotrs.gov.au; Minister: Senator Nick Minchin.

Ministerial Council on Mineral & Petroleum Resources—MCMPR: c/o Resources Division, Dept. Industry, Tourism & Resources, GPOB 9839, ACT 2601; tel (2) 6213-7159; fax (2) 6213-7955; e-mail algis.kusta@industry.gov.au; internet www.industry.gov.au/mcmpr; Sec: A. Kusta; members are federal, state and territory ministers, and those of New Zealand and Papua New Guinea, with responsibility for energy and resources.

Department of Natural Resource and Mines Queensland: GPOB 2545, Brisbane, Qld 4001; tel (7) 3896-3111; internet www.nrm.qld.gov.au; Minister: Stephen Robertson; Dir-Gen: Terry Hogan.

DPI Forestry, Queensland Department of Primary Industries: POB 944, Brisbane, Qld 4001; tel (7) 3234-0001; fax (7) 3234-0271; internet www.dpi.qld.gov.au/forest; Exec Dir (acting): Gary Bacon; manages State-owned forest resources; supplies timber products from sustainably managed and renewable forests; provides scientific expertise in tropical and subtropical forest management.

Environment ACT: MacArthur House, Level 2, 12 Wattle St, POB 144, Lyneham, ACT 2602; tel (2) 6267-9777; fax (2) 6207-2227; e-mail EnvironmentACT@act.gov.au; internet www.environment.act.gov.au; Exec Dir: Dr Maxine Cooper; Dir: Elizabeth Fowler; part of the Urban Services department of Australian Capital Territory.

Great Barrier Reef Marine Park Authority—GBRMPA: POB 1379, 2–68 Flinders St, Townsville, Qld 4810; tel (7) 4750-0700; fax (7) 4772-6093; e-mail registry@gbrmpa.gov.au; internet www.gbrmpa.gov.au; Chair: Virginia Chadwick; f 1975; affiliated to the Department of the Environment and Heritage; provision for the protection, wise use, understanding and enjoyment of the Great Barrier Reef through care and development of the Marine Park; principal adviser to the Federal Government; the Great Barrier Reef Ministerial Council co-ordinates policy between the Federal and Queensland governments and comprises two ministers from each government; the Great Barrier Reef Consultative Committee is an independent advisory body to ministers and the GBRMPA.

Moreton Bay Environmental Education Centre: POB 373, Wynnum, Qld 4178; tel (7) 3906-9111; fax (7) 3906-9100; e-mail info@moretoneec.eq.edu.au; internet www.moretonbayeec.qld.edu.au; Prin: Eileen J Mithcell; f 1986; affiliated to Education Queensland; land- or water-based environmental education programmes.

New South Wales Environment Protection Authority: 59–61 Goulburn St, Sydney, NSW 2000; tel (2) 9995-5000; fax (2) 9995-5999; e-mail info@epa.nsw.gov.au; internet www.epa.nsw.gov.au; Dir-Gen: Lisa Corbyn.

New South Wales National Parks and Wildlife Service: 43 Bridge St, POB 1967, Hurstville, NSW 2220; tel (2) 9585-6444; fax (2) 9585-6527; e-mail info@npws.nsw.gov.au; internet www.npws.nsw.gov.au; Dir-Gen: Brian Gilligan; f 1967; conservation and management of the natural and cultural heritage, including national parks, nature reserves, historic sites, Aboriginal heritage and reserves; community awareness programmes.

Northern Territory Department of Infrastructure, Planning and Environment: Darwin Plaza Bldg, Level 2, 21 Smith St, Darwin, NT 0800; tel (8) 8924-4139; fax (8) 8924-4053; internet www.ipe.nt.gov.au; Minister for Transport and Infrastructure, Lands and Planning and Parks and Wildlife: Dr Chris Burns; Minister for Environment and Heritage: Marion Scrymgour.

Office of Energy Policy, South Australian Department of Mines and Energy: 30 Wakefield St, Level 19, Adelaide, SA 5000; tel (8) 8226-5500; fax (8) 8226-5523; CEO: Dr Cliff F Wong; f 1978; advice to government on energy policy; regulation of electricity and gas industries; implementation of energy conservation programmes within government.

Office of the Supervising Scientist—OSS: Tourism House, 40 Blackall St, Barton, ACT 2600; tel (6) 274-1222; fax (6) 274-1519; e-mail oss@oss.erin.gov.au; internet www.erin.gov.au/portfolio/epg/oss.html; Supervising Scientist (acting): Dr Arthur Johnston; Asst Sec: Stewart Needham; f 1978; supervision of environmental performance at uranium mines through reviews every six months; policy advice on nuclear and mining issues; programmes to identify and encourage best-practice environmental management in mining, in collaboration with industry; provision of environmental regulation services on Australia's Indian Ocean territories; agency of the Environment Protection Group of Environment Australia.

Environmental Research Institute of the Supervising Scientist—ERISS: Private Bag 2, Jabiru, NT 0886; tel (8) 8979-9711; fax (8) 8979-2499; e-mail enquiries@eriss.erin.gov.au; internet www.eriss.gov.au; formerly Alligator Rivers Region Research Institute; protection of the Alligator Rivers region of the Northern Territory from the effects of uranium mining; research into tropical wetlands management and ecosystems; monitoring of water quality.

Queensland Environmental Proection Agency—EPA: 160 Ann St, Brisbane, Qld 4000; tel (7) 3227-7111; internet www.epa.qld.gov.au; Minister: Desley Boyle; f 1998; promotion of ecologically sustainable development; protection and wise use of the natural and cultural heritage and a cleaner and safer environment.

Queensland Parks and Wildlife Service—QPWS: 160 Ann St, Brisbane, Qld 4000; tel (7) 3227-8186; internet www.epa.qld.gov.au/environmental_management; f 1998.

South Australian Department for Environment and Heritage: POB 1047, Adelaide, SA 5001; tel (8) 8204-9000; fax (8) 8226-4321; internet www.environment.sa.gov.au; Minister: Iain Evans; Chief Exec: Allan Holmes; environmental management and planning; landscape protection; conservation strategies; protected areas; resource management; land-use planning; environmental policy; regulation control; sustainable development; national parks; integration of conservation and development; pollution control.

State Forests of New South Wales: Locked Bag 23, Pennant Hills, NSW 2120; tel (2) 9980-4100; fax (2) 9980-7010; e-mail katied@brushbox.forest.nsw.gov.au; internet www.forest.nsw.gov.au; Prin Officer: Dr Bob Smith; f 1916; formerly the New South Wales Forestry Commission; management of the public forests of NSW in order to maintain a balance between timber production and recreational uses; protection of biodiversity; soil and water conservation; preservation of the historical and archaeological value of forests.

Tasmanian Department of Primary Industries, Water and Environment: Marine Board Bldg, 1 Franklin Wharf, BPOB 44A, Hobart, Tas 7001; tel (3) 6233-8011; fax (3) 6234-1335; e-mail webmaster@dpiwe.tas.gov.au; internet www.dpiwe.tas.gov.au; Minister: David Llewellyn; Sec: Kim Evans; f 1998; land information; property services; surveying; valuation; planning and environmental management; land titles office; parks and wildlife office.

Victoria Department of Natural Resources and Environment: 8 Nicholson St, E Melbourne, Vic 3002; tel (3) 9637-8000; fax (3) 9637-8148; e-mail customer.service@mre.vic.gov.au; internet www.mre.vic.gov.au; Minister for Environment and Conservation: Sheryl Garbutt; Minister for Energy and Resources: Candy Broad; Minister for Agriculture: Keith Hamilton; management of public land and natural resources in Victoria; responsibilities include national parks, forests, fisheries, fire protection, protection of land from degradation and protection of flora and fauna.

Sustainable Energy Authority: 215 Spring St, Ground Floor, Melbourne, Vic 3000; tel (3) 9655-3222; fax (3) 9655-3255; e-mail advice@sea.vic.gov.au; internet www.sea.vic.gov.au; CEO: Keith Fitzmaurice; f 1990; promotes energy efficiency and renewable and sustainable energy in all sectors of the Victorian economy.

Western Australia Department of Conservation and Land Management: Hackett Dr, Crawley, WA 6009; tel (8) 9422-0300; fax (8) 9386-1578; e-mail info@calm.wa.gov.au; internet www.calm.wa.gov.au; f 1985; conservation and management of wildlife, land and water resources; planning and management of national parks, nature reserves, marine parks, marine nature reserves, conservation parks, state forests and timber reserves.

Western Australia Department of Environmental Protection—DEP: Westralia Sq, 8th Floor, 141 St George's Terrace, Perth, WA 6000; tel (8) 9322-7000; fax (8) 9322-1598; e-mail dep_library@environ.wa.gov.au; internet www.environ.wa.gov.au; CEO: Dr Bryan Jenkins; provision of professional and technical support to the Western Australia Environmental Protection Authority; monitoring of compliance with the Environmental Protection Act; development of environmental policies and codes of practice; education and community awareness programmes; library and information service; environmental impact assessment; waste management.

Western Australia Environmental Protection Authority—EPA: Westralia Sq, 8th Floor, 141 St George's Terrace, Perth, WA 6000; tel (8) 9222-7000; fax (8) 9222-7155; e-mail dep_info@environ.wa.gov.au; internet www.environ.wa.gov.au/epa; Chair: Bernard Bowen; Deputy Chair: Dr Elizabeth Mattiske; independent statutory authority; provision of environmental advice to government; monitoring of compliance with the Environmental Protection Act; assessment of proposals and development of policies to protect the environment and to prevent, control and abate pollution.

Non-Governmental Organizations

ANU Enterprise Pty Ltd: Australian National University, GPOB 4, Canberra, ACT 2601; tel (2) 6249-3811; fax (2) 6275-1433; internet www.anuenterprise.com.au; Gen Man: Dr Neil Hamilton; Man, Forestry and Environment Div: Dr Ken Shepherd; f 1979; commercial company, known as Anutech Pty Ltd until 2004, wholly owned by Australian National University; intended to facilitate consultancies, market discoveries and contract research; consultancy services in environmental impact assessment, forestry and rural development; professional courses for overseas students; co-operation with School of Resource and Environmental Management.

Appropriate Technology for Community and Environment—APACE: Univ of Technology, Sydney, POB 123, Broadway, NSW 2007; tel (2) 9514-2554; fax (2) 9514-2611; e-mail apace@uts.edu.au; internet www.apace.uts.edu.au; Pres: Dr Paul Bryce; Programme Dir: Donnella Bryce; f 1976; design, promotion and implementation of low-impact, environmentally responsible technology in partnership with indigenous communities throughout the world.

Association for Research and Environmental Aid Ltd—AREA: POB A529, Sydney S, NSW 1235; tel and fax (2) 9282-6983; e-mail area@kcs.com.au; Chair: Peter Wolfe; f 1978; provision of environmental and development assistance to developing countries; project design, implementation, management, monitoring and evaluation; environmental impact assessment; resource assessment; forestry; use of wildlife resources; use of marine resources; international trade; pollution control; energy; agriculture; human ecology; endangered animal species; design and planning of protected areas; watershed management.

Australasian Primatological Society: POB 500, One Tree Hill, SA 5114; Prin Officer: Dr Graeme Crook; affiliated to the International Primatological Society.

Australian and New Zealand Solar Energy Society Ltd—ANZSES: POB 1140, Maroubra, NSW 2035; tel (2) 9311-0003; fax (2) 9311-0004; e-mail anzses@keystone.arch.unsw.edu.au; internet www.anzses.org; Chair: Assoc Prof John Todd; Vice-Chair: P C Thomas; Sec: Assoc Prof Jai Singh; f 1978; non-profit society to promote social and economic development through the environmentally sound utilization of solar energy; areas of interest include transport, climate change and waste management; annual international conference; quarterly magazine; serves as the Asia-Pacific Regional Office of the International Solar Energy Society—ISES.

Australian Committee for IUCN: POB 528, Sydney, NSW 2001; tel (2) 9247-6300; fax (2) 9247-8778; Sec: Pam Eiser.

Australian Conservation Foundation: 340 Gore St, Fitzroy, Vic 3065; tel (3) 9416-1166; fax (3) 9416-0767; e-mail melbourne@acfonline.org.au; internet www.acfonline.org.au; Exec Dir: Don Henry; f 1966; campaigns for an ecologically sustainable society, protection of biodiversity and management of natural resources; promotes positive global change; protection of forests and endangered species; campaigns on waste minimization and reduction of ozone layer depletion, greenhouse gas production, pollution and land degradation.

Australian Council of National Trusts: POB 1002, Civic Sq, Canberra, ACT 2608; tel (2) 6247-6766; fax (2) 6249-1395; e-mail acnt@spirit.com.au; internet www.austnattrust.com.au; Chair: Diane Weidner; Exec Officer: Alan Graham;

f 1965; national co-ordinating body of the Australian National Trust movement; conservation of Australia's indigenous, historic and natural heritage; lobbying of government agencies, private organizations and other national and international bodies to conserve Australia's cultural heritage and to promote community awareness, interest and involvement; development of educational, promotional, marketing and funding strategies.

Australian Geographic Society: POB 321, Terrey Hills, NSW 2084; tel (2) 9473-6700; fax (2) 9473-6701; e-mail editorial@ausgeo.com.au; internet www.australiangeographic.com.au; Chief Exec: Rory Scott; Managing Editor: Teri Cowley; encouragement and financial support for research into Australia's flora, fauna and the environment; fund-raising through Australian Geographic Shops and sales of Australian Geographic Journal.

Australian Greens: c/o Senate, Parliament House, Canberra, ACT 2600; tel (61) 6277-3170; fax (61) 6277-3185; e-mail senator.brown@aph.gov.au; internet www.greens.org.au; Prin Officer: Senator Bob Brown; environmental political party.

Australian Institute of Marine Sciences: Private Mail Bag 3, Townsville, Qld 4810; tel (7) 4753-4211; fax (7) 4772-5852; Dir: Dr R Reichert; f 1972; carries out research in the marine sciences, with particular attention to the tropics; maintains a fleet of research vessels; library.

Australian National Parks Council—ANPC: POB 2227, Canberra, ACT 2601; tel (2) 6257-1063; Pres: Stephen Johnston; f 1971; protection, promotion and extension of the national park systems; NGO forum.

Australian National University, School of Resource Management and Environmental Science: Canberra, ACT 0200; tel (2) 6249-2579; fax (2) 6249-0746; e-mail forestry@anu.edu.au; Prin Officer: Prof P J Kanowski; f 1990; undergraduate and postgraduate education and research in forestry, geology, resource and environmental management, and environmental science.

Centre for Resource and Environmental Studies—CRES: Australian National University, Canberra, ACT 0200; tel (2) 6125-4277; fax (2) 6125-0757; e-mail office@cres.anu.edu.au; internet cres.anu.edu.au; Dir: Prof Mike Hutchinson; Exec Officer: Adele Doust; part of the Australian National University; transdisciplinary research and postgraduate training centre; focuses on resource and environmental issues at local, national and global levels, with emphasis on problems related to use of the environment.

Centre for Resource and Environmental Studies—CRES, Australian National University: Canberra, ACT 0200; tel (2) 6249-4588; fax (2) 6249-0757; e-mail wasson@cres.anu.edu.au; internet cres.anu.edu.au; Dir: Prof Robert Wasson; applied interdisciplinary research, supported by basic research on the environment, resource use, human institutions and public policy; specialist consulting and contract research; postgraduate training; research areas include spatial modelling of climate, vegetation and terrain; ecological economics; conservation biology and landscape ecology; water resources; public policy and institutional analysis; catchment management; environmental history; sustainability analysis.

Australian Rainforest Conservation Society, Inc: 19 Colorado Ave, Bardon, Qld 4065; tel (7) 3368-1318; fax (7) 3368-3938; e-mail aila.keto@rainforest.org.au; internet www.rainforest.org.au; Pres: Dr Aila Keto; Dir: Dr Keith Scott; f 1982; rainforest conservation; ecologically sustainable management of native forests; energy issues; identification and management of world heritage sites.

Australian Trust for Conservation Volunteers—ATCV: POB 423, Ballarat, Vic 3353; tel (3) 5333-1483; fax (3) 5333-2290; e-mail info@conservationvolunteeers.com; internet www.conservationvolunteeers.com; Exec Dir: Colin Jackson; f 1982; recruitment and management of volunteers for conservation projects including seed collection, tree planting, weed control, path conservation and maintenance, flora and flora surveys, and habitat restoration; member of the World Conservation Union—IUCN.

Biodiversity Coalition: c/o Post Office, Cygnet, Tas 7112; tel (3) 6295-1745; fax (3) 6295-1964; Co-ordinator: Alistair Graham; f 1991; publication of newsletter; information networking.

Cairns and Far North Environment Centre, Inc: POB 323N, North Cairns, Qld 4870; tel (7) 4032-1746; fax (7) 4053-3779; e-mail cafnec@internetnorth.com.au; Co-ordinator: Gavan McFadzean; f 1981; agency for liaison and referral between the public, industry and government; campaigns on issues of environmental importance; public environmental education; co-ordination of local environmental bodies; research into environmental processes and effects of land-use and development on the environment; library; affiliated with Queensland Conservation Council.

Canberra and South-East Region Environment Centre: POB 1875, Canberra City, ACT 2601; tel (2) 6248-0885; Pres, Man Cttee: Bren Weatherstone; Information Officer: Susan Smith; environmental library; resource centre for the public and 30 member groups in Canberra.

Conservation Council of South Australia, Inc: tel (8) 8223-5155; fax (8) 8232-4782; e-mail general@ccsa.asn.au; internet www.ccsa.asn.au; Exec Officer: Michelle Grady; Pres: Margaret Bolster; f 1971; association of over 60 environment groups in South Australia; environmental advocacy; environmental advice, information and education services; liaison with government, industry, trade unions and community groups on environmental matters; member of the World Conservation Union—IUCN.

Conservation Council of Western Australia, JNC: City West Lotteries House, 2 Delhi St, Perth, WA 6005; tel (8) 9420-7266; fax (8) 9420-7273; e-mail conswa@iinet.net.au; internet www.iinet.net.au/~conswa; Pres: Dr Sue Graham-Taylor; f 1967; environmental conservation; information; education.

Ecological Society of Australia, Inc: POB 1564, Canberra, ACT 2601; tel (2) 6249-5092; fax (2) 6249-5095; Pres: Prof J B Kirkpatrick; Sec: Dr J Whinan; f 1963; scientific study of organisms in relation to their environment; application of ecological principles; reservation and sound management of natural areas; advice to governmental and other agencies where appropriate.

Energy Research, Development and Information Centre—ERDIC, University of New South Wales: Sydney, NSW 2052; tel (2) 9385-5555; fax (2) 9662-4265; Dir: Dr G D Sergeant; f 1978; co-ordination of research and consultancy work; information centre; publication of annual report.

Environment Centre of the Northern Territory—NT, Inc: Home Centre, 4 Cavenagh St, Darwin, NT 0800; tel (8) 8981-1984; fax (8) 8941-0387; Campaigns Co-ordinator: Jamie Pittock; promotion of environmental awareness; environmental information; forum for Northern Territory conservationists nationally and internationally; research; lobbying of governments and industry.

Environment Centre of Western Australia: 587 Wellington St, Perth, WA 6000; tel (8) 9322-3045; Prin Officers: Luci Ravi, Carina Calzoni; f 1978; information resource centre for environmental issues; contact point for a range of environmental initiatives.

Environmental Law Centre—ELC, Macquarie University: School of Law, Sydney, NSW 2109; tel (2) 9850-7099; fax (2) 9850-7682; e-mail zada@law.law.mq.edu.au; Dir: Zada Lipman; f 1984; undergraduate and postgraduate teaching, and research; consultancy in environmental law, local government, pollution, heritage and indigenous culture; participation in the Environmental Law Commission of the World Conservation Union—IUCN.

Environmental Technology Centre, Murdoch University: Murdoch, WA 6155; tel (8) 9360-7310; fax (8) 9360-7311; e-mail etc@murdoch.edu.au; internet www.etc.murdoch.edu.au; Chair: Prof Goen Ho; Dir: Dr Kuruvilla Mathew; Research Man: Dr Martin Anda; f 1978; research and development; education and training; environmental technology for sustainable development; Australian Cooperative Research Centre nodes for Desert Knowledge, environmental biotechnology and sustainable tourism.

Fraser Island Defenders Organization: 36 Kemp St, POB 71, Gladesville, NSW 2111; tel (2) 9817-4660; Project Officer: J Sinclair; f 1971; promotion of wise use of natural resources; conservation of flora and fauna.

Gippsland Waters Coalition—GWC: POB 463, Moe, Vic 3825; tel (3) 5127-6508; fax (3) 5127-7560; Sec: Neil Grigg; f 1988; organization of community groups; protection of aquatic environments; promotion of ecologically sustainable allocation of resources.

Greening Australia Ltd: POB 74, Yarralumla, ACT 2600; tel (2) 6281-8585; fax (2) 6281-8590; e-mail general@greeningaustralia.org.au; internet www.greeningaustralia.org.au; Chief Exec: Mark Thomas; national community organization; conservation of trees and vegetation; maintains more than 40 regional centres, including eight main branches in the Australian states and territories.

Greenpeace Australia Pacific: GPO Box 3307, Sydney, NSW 2001; tel (2) 9211-4666; 1800-815-151 (toll free); fax (2) 9211-4588; e-mail greenpeace@au.greenpeace.org; internet www.greenpeace.org.au; CEO: Peter Mullins; f 1978; an independent organisation campaigning to ensure a just, peaceful, sustainable environment for future generations. Aims: end the nuclear threat by global nuclear disarmament and closure of the nuclear industry, replacing it with non-radioactive alternatives; stop genetically engineered food by removing genetically engineered organisms from the environment and promoting a GE-free future; save the oceans by bringing an end to overfishing, pirate fishing and commercial whaling; eliminate toxics by abolishing POPs, PVC and chlorine and promoting alernatives, and by preventing the dumping of waste in developing nations and oceans; stop climate change by phasing out fossil fuels and replacing them with renewable energy, such as wind and solar power; save the forests by protecting Melanesia's ancient forests from wanton destruction and supporting solutions developed and run by village communities.

Griffith University, Faculty of Environmental Sciences: Nathan, Qld 4111; tel (7) 3875-7519; fax (7) 3875-7459; e-mail w.hogarth@ens.gu.edu.au; Dean: Prof Bill Hogarth; f 1975; incorporates the Australian School of Environmental Studies, the School of Environmental Engineering, the School of Environmental Planning and the Graduate School of Environmental Sciences and Engineering.

ICLEI Australia-New Zealand: 4/255 Bourke St, Melbourne, Vic 3000; tel (3) 9639-8688; fax (3) 9639-8677; e-mail anz@iclei.org; internet www.iclei.org/anz; national office of the International Council for Local Environmental Initiatives—ICLEI.

Jacobs Well Environmental Education Centre: 843 Pimpama–Jacobs Well Rd, Norwell, Qld 4208; tel and fax (7) 5546-2317; e-mail jacobeec@jacobseec.qld.edu.au; internet jacobseec.qld.edu.au; Prin Officer: Glenn Leiper; f 1974; affiliated to Education Queensland; environmental education activities for students of all ages and levels of ability in the education system; school day trips and camps; programmes focus on local environmental issues and are conducted in coastal wetlands and remnant forests.

Marine Education Society of Australasia—MESA: POB 616, Indooroopilly, Qld 4068; tel and fax (7) 3378-0128; internet www.education.monash.edu.au/peninsula/seaweek; Pres: Ian Tibbetts; Sec: Jan Oliver; f 1984; promotion of awareness and understanding of marine environments; national forum for discussion and advocacy of marine education; brs in each state and territory; br in New Zealand.

Men of the Trees, Inc (Queensland): POB 283, Clayfield, Qld 4011; tel (7) 3262-1096; Pres: N Brennan; Vice-Pres: Noel Dawson; f 1981; tree-planting; creating awareness of the importance of trees; affiliated with International Tree Foundation (United Kingdom).

Monarto Zoological Park: Princes Hwy, Monarto South, SA 5254; tel (08) 8534-4100; fax (08) 8534-4077; e-mail MZPadmin@monartozp.com.au; internet www.monartozp.com.au; Dir: Chris Hannacks; Tourism and Marketing: Andrew Brown; Admin: Karyn Biggs; Education: Ian Walton; Curator: Peter Clark; Vet: Ian Smith; f 1993; Monarto Zoological Park is a 1,000 hectare fauna and flora sanctuary renowned for the breeding of rare and endangered species; visit the multi-award winning interactive visitor centre featuring a café style bistro and gift shop; the one and a quarter hour guided safari bus tour features a drive through habitats including cheetah, Southern White rhinoceros and giraffe.

Murray Darling Association, Inc: POB 359, Albury, NSW 2640; tel (2) 6021-3655; fax (2) 6021-2025; Pres: Max Moor; Gen Man: Leon Broster; f 1944; association of 100 local governments, groups, corporations and individuals with an interest in maintaining the environmental interests of the basin of the Murray and Darling Rivers while also encouraging sustainable development of the river valleys' resources.

National Environmental Law Association—NELA: Unit 4, Wollongong St, Fyshwick, ACT 2609; tel (2) 6239-3715; fax (2) 6239-3717; e-mail nela@effect.net.au; f 1982; maintains branches in each of the Australian states and the ACT.

National Parks Association of New South Wales, Inc: POB A96, Sydney S, NSW 1235; tel (2) 9299-0000; fax (2) 9290-2525; e-mail npansw@npansw.org.au; internet www.npansw.org.au; Exec Officer: Andrew Cox; Admin Off: Sam McGuinness; Marine Off: Nicky Hammond; Woodlands Off: Cecile Van der Burgh; f 1957; wilderness protection; environmental research; resource management; environmental education; lobbying government; land-use planning; management and planning of protected areas; camps and excursions.

National Trust of Australia (NSW): Observatory Hill, Watson Rd, The Rocks, NSW 2000; tel (2) 9258-0123; fax (2) 9258-1264; e-mail webmaster@nsw.nationaltrust.org.au; internet www.nsw.nationaltrust.org.au; Exec Dir: Else Atkin; Pres: Justice B S J O'Keefe; conservation of the natural and built environments.

Nature Conservation Council of New South Wales: 39 George St, The Rocks, NSW 2000; tel (2) 9247-4206; fax (2) 9247-5945; e-mail nccnsw@peg.apc.org; internet www.peg.apc.org/~nccnsw; Exec Officer: John Connor; f 1955; organization comprising over 100 environment groups in NSW; public information centre; campaigns on forests, protected areas, land use, pollution, waste management, environmental research, urban planning, national parks, wildlife, energy, urban bushland and environmental law.

North Queensland Conservation Council, Inc—NQCC: POB 364, Townsville, Qld 4810; tel (7) 4771-6226; fax (7) 4721-1713; e-mail nqcc@byte-tsv.net.au; internet www.nqcc.org.au; Co-ordinator: Jeremy Tager; f 1975; conservation of the natural and human environments; responsible for an area between the towns of Mackay and Innisfail and the border with the Northern Territory to the west; affiliated with Queensland Conservation Council.

Organic Growers' Association of Western Australia: POB 213, Wembley, Perth, WA 6913; tel (419) 968-821; e-mail orggrow@iinet.net.au; internet www.iinet.net.au/~orggrow; Sec: Sheryn Aubey; collection and dissemination of information on organic growing; demonstrations of practical alternatives to chemicals in agriculture; restoration of living soil.

Pelican Lagoon Research and Wildlife Centre: Penneshaw, Kangaroo Island, SA 5222; tel (8) 8553-7174; fax (8) 8553-7383; e-mail echidna@kln.net.au; internet www.echidna.edu.au; Dir: Mike McKelvey; Resident Biologist: Dr Peggy Rismiller; f 1982; long-term environmental and biological field research into native Australian flora and fauna; life history studies of the Australian short-beaked echidna, Rosenberg's goanna, island tiger snake and little penguin; has run Mallee woodland ecology workshops working model for sustainable research and living (implemented 1980); consultant for global community workshops in sustainable development (Expo 2000).

Project Jonah Australia: POB 328, Eastwood, NSW 2121; tel (2) 9899-2111; fax (2) 9876-8698; Pres: Dr Brendon Gooneratne; Sec: Patricia Lawson; whale conservation; environmental education; lobbying of government.

Queensland Conservation Council: Elizabeth St, POB 12046, Brisbane, Qld 4002; tel (7) 3221-0188; fax (7) 3229-7992; e-mail qccqld@powerup.com.au; internet www.qccqld.org.au; Man: Helenka King; Co-ordinator: Imogen Zethoven; co-ordination of the activities of interest groups; liaison with government and media on behalf of interest groups; environmental information centre; protecting, conserving and sustaining Queensland's natural environment and encouraging the development of an ecologically sustainable state.

Rainforest Information Centre: 13 Wotherspoon St, POB 368, Lismore, NSW 2480; tel (2) 6621-8505; e-mail rainforestinfo@ozemail.com.au; internet www.forests.org/ric; Prin Officers: John Seed, John Revington, Ruth Rosenhek, Anja Light; f 1980; concerned with global rainforest protection and reforestation; information on rainforests; Goodwood Project; campaign to ban toxic gold mining; deep ecology workshops.

Royal Zoological Society of South Australia, Inc: Adelaide Zoological Gardens, Frome Rd, Adelaide, SA 5000; tel (8) 8267-3255; fax (8) 8239-0637; internet www.adelaidezoo.com.au; CEO: E J McAlister, AO; Adelaide Zoo – Director: Mark Craig; Business Manager: Tom Reid; Monarto Zoological Park Director: Chris Hannocks; Business Manager: Karyn Biggs; f 1878; promotion of awareness and understanding of animal life; preservation and propagation of rare and endangered species.

RSPCA Australia, Inc: POB E369, Kingston, ACT 2604; tel (2) 6282-8300; fax (2) 6282-8311; e-mail rspca@rspca.org.au; internet www.rspca.org.au; Pres: Dr Hugh Wirth; Exec Officer: Jenny Hodges; Scientific Officer: Dr Bidda Jones; Government Liaison Officer: Brendan Jacomb; f 1981 as Royal Society for the Prevention of Cruelty to Animals; enforces existing laws relating to the treatment of animals and lobbies for new legislation; disseminates information and creates public awareness; operates and supports animal clinics and shelters; member of the World Society for the Protection of Animals (WSPA); maintains eight branches in the Australian states and territories.

RSPCA New South Wales: 201 Rookwood Rd, POB 34, Yagoona, Sydney, NSW 2199; tel (2) 9709-5433; fax (2) 9796-2258; f 1873.

Soil and Water Conservation Association of Australia—SAWCAA: 9 The Crest, Killare, NSW 2071; tel (2) 9484-3703; fax (2) 9980-6728; Pres: R A Farley; Exec Officer: R W Roberts; f 1985; promotion of sustainable land use, in order to protect soil and water resources for present and future generations; education and information; research; state branches; federal council; publication of quarterly journal.

Tasmanian Conservation Trust, Inc: 102 Bathurst St, Hobart, Tas 7250; tel (3) 6234-3552; fax (3) 6231-2491; e-mail tct@southern.com.au; internet www.tct.org.au; Dir: Michael Lynch; f 1968; conservation of the natural and cultural heritage of Tasmania.

Total Environment Centre, Inc—TEC: 1/88 Cumberland St, Sydney, NSW 2000; tel (2) 9247-4714; fax (2) 9247-7118; Dir: Jeff Angel; f 1972; non-profit community organization; environmental campaigning and information centre.

University of Adelaide, Department of Environmental Science and Management: Roseworthy, SA 5371; tel (8) 8303-7804; fax (8) 8303-7956; e-mail bwilliams@roseworthy.adelaide.edu.au; Prin Officers: Dr Brian D Williams, Prof Hugh Possingham; f 1979; undergraduate and postgraduate teaching; research in environmental science; consultancy in various fields, including environmental chemistry, environmental toxicology, conservation biology and geographic information systems—GIS.

University of Tasmania: School of Geography and Environmental Studies, Private Bag 78, Hobart, Tas 7001; tel (3) 6226-2834; fax (3) 6226-2989; e-mail secretary.env@geog.utas.edu.au; internet www.geol.utas.edu.au/geog; Head of School: Prof Jamie Kirkpatrick; f 1974; postgraduate teaching; research.

University of Western Australia, Department of Zoology: Nedlands, Perth, WA 6907; tel (8) 9380-2237; Head of Dept: Assoc Prof W J Bailey; Chair of Zoology: Prof S D Bradshaw; undergraduate and postgraduate teaching; research.

Water Research Foundation of Australia: Centre for Resource and Environmental Studies, Australian National University, Canberra, ACT 0200; tel (2) 6249-0660; fax (2) 6249-0757; e-mail wrfa@cres.anu.edu.au; internet cres.anu.edu.au/wrfa/index.htm; Chair: Hon J G Beale; Research Dir: Prof I White; Office Man: Kathy Hicks; f 1955; initiation and support of innovative research into water and land use; support for training in water and land resource management; dissemination of information.

Wild Life Preservation Society of Australia: 36 Diamond Rd, Pearl Beach, NSW 2256; POB 3428, Sydney, NSW 2001; tel and fax (2) 4344-4708; Pres: Vincent Serventy; f 1909; sustainable development; conservation of biodiversity; international co-operation.

Wilderness Society: POB K29, Haymarket, Sydney, NSW 2000; tel (2) 9267-7929; fax (2) 9264-2673; Dir: Karenne Jurd; national, non-profit, community-based organization dedicated to the protection and conservation of wilderness areas in Australia; local-level activism, public education and lobbying of politicians and industry; current campaigns focus on forests, arid lands, Cape York (Qld), and federal and state legislation affecting wilderness areas.

Tasmania Office: Wilderness Society, 130 Davey St, Hobart, Tas 7000; tel (3) 6234-9799; fax (3) 6224-1497.

Wildlife Preservation Society of Queensland, Inc: 133 George St, 2nd Floor, Brisbane, Qld 4000; tel (7) 3221-0194; fax (7) 3221-0701; Pres: Dr Glen Ingram; Dir: Adrian Jeffreys; f 1962; community-based volunteer organization committed to creating an ecologically sustainable future for people and wildlife through advocacy and action; environmental education; consultation; campaigning for environmental legislation; publications.

Wildlife Survival: POB R1519, Royal Exchange, NSW 2000; tel (2) 9247-6300; fax (2) 9247-8778; Prin Officer: Pam Eiser; f 1981; campaigns on kangaroo slaughter, protection of endangered species' habitats, Antarctic conservation, and preservation and management of wildlife.

World Wide Fund for Nature—WWF Australia: Level 13, 235 Jones St Ultimo NSW 2007; tel (2) 9281-5515; fax (2) 9281-1060; e-mail enquiries@wwf.org.au; internet www.wwf.org.au; Pres: Rob Purves, CEO: Dr David Butcher; f 1978; national affiliate organization of WWF International; support for the World Conservation Strategy; conservation projects for endangered species and habitats; environmental education projects.

Melbourne Office: 9 Church St, 1st Floor, Hawthorn, Vic 3122; tel (3) 9853-7244; fax (3) 9853-4156; e-mail wwfmelb@ozemail.com.au; Pres: Prof Alistair Gilmour.

Australian External Territories

NORFOLK ISLAND

Governmental Organizations

Ministry for Land and the Environment: Old Military Barracks, Quality Row, Kingston, Norfolk Island 2899; tel 22003; fax 23378; e-mail executives@assembly.gov.nf; Minister: (vacant).

Australian Nature Conservation Agency, Norfolk Island Conservancy: POB 310, Norfolk Island 2899; tel 22695; fax 23397; e-mail fred.howe@ea.gov.au; Government Conservator: Fred Howe; Deputy Park Manager: John Henderson; affiliated to Australian National Park and Wildlife Service; maintains a national park.

Non-Governmental Organizations

Norfolk Island Flora and Fauna Society: POB 305, Norfolk Island 2899; tel and fax 22800; e-mail xtian@norfolk.net.nf; Pres: Margaret Christian; Sec: John McCoy; Treas: Laigh Christian-Mitchell; f 1967; maintains a natural history museum; proceeds are used for the conservation of plant and bird species; undertakes practical fieldwork to enhance the chances of survival of endemic and native species on the island; organizes public information evenings and guided walks (by arrangement); publications.

AUSTRIA

Governmental Organizations

Ministry of Agriculture and Forestry, the Environment and Water Management: Stubenreng 5, 1012 Vienna; tel (1) 711-00-0; fax (1) 711-00-21-40; e-mail office@lebensministerium.at; internet www.lebensministerium.at; Minister: Josef Proell.

Bundesanstalt für Agrarwirtschaft—AWI (Federal Institute of Agricultural Economics): Marxerg. 2, 1030 Vienna; tel (1) 877-36-7419; fax (1) 877-36-51-7490; e-mail awi-bmlfuw.gov.at; internet www.awi-bmlfuw.gv.at; Dir: Dr Hubert Pfingstner; f 1960; central agricultural organization.

Bundesanstalt für Landtechnik (Federal Institute for Agricultural Engineering): 3250 Wieselburg, Rottenhauserstr 1; tel (7416) 52175; fax (7416) 52175-45; e-mail direktion@blt.bmlf.gv.at; internet www.blt.bmlf.gv.at; Dir: Dr Johann Schrottmaier; f 1947; research, development and testing in agricultural engineering and renewable raw materials, hillside farming, energy from biomass, sustainable development of less-favourable areas.

Umweltbundesamt (Federal Environment Agency): 1090 Vienna, Spittelauer Lände 5; tel (1) 31-30-40; fax (1) 31-30-454-00; e-mail mayer@ubavie.gv.at; internet www.ubavie.gv.at; Dir: Dr Wolfgang Struwe; f 1985; responsible for environmental monitoring and elaboration of federal environmental policies.

Österreichische Bundesforste—ÖBF (Austrian Federal Forests): Pummergasse 10–12 , 3002 Purkersdorf; tel (2231) 600; fax (2231) 600-219; e-mail bundesforste@bundesforste.at; internet www.bundesforste.at; Prin Officers: Dr Georg Erlacher, Dr Thomas Uher; f 1925/1997; national timber management; real-estate management; water, land resources and renewable energy; consulting.

Zentralanstalt für Meteorologie und Geodynamik (Central Institute for Meteorology and Geodynamics): 1190 Vienna, Hohe Warte 38, Postfach 342; tel (1) 36026-2003; fax (1) 369-12-33; e-mail dion@zamg.ac.at; internet www.zamg.ac.at; Dir: Prof D Fritz Neuwirth; f 1851; affiliated to the Federal Ministry of Education, Science and Culture; all activites of a national meteorological and geophysical service.

Non-Governmental Organizations

Ärzte für eine gesunde Umwelt (Doctors for a Healthy Environment): 6020 Innsbruck, Weinhartstr 2; tel and fax (512) 566-890; e-mail aegu@magnet.at; member of Koordinationsstelle Ökobüro.

BirdLife Österreich—Gesellschaft für Vogelkunde (BirdLife Austria—Ornithological Society): 1070 Vienna,

Museumpl 1/10/8; tel (1) 523-46-51; fax (1) 523-52-54; e-mail birdlife@blackbox.at; Pres: Dr Gerhard Loupol; Dir: Dr Andreas Ronner; f 1953; member of Umweltdachverband ÖGNU; scientific studies of bird populations in Austria; conservation action for threatened species; designates candidate sites for protected area status; lectures and excursions for members; publications on bird conservation and scientific ornithological topics.

Die Grünen—Die Grüne Alternative (Greens—Green Alternative): 1070 Vienna, Lindengasse 40; tel (1) 52-125-0; fax (1) 52-691-10; e-mail infopool@gruene.or.at; internet www.gruene.at; Chair: Prof Dr Alexander Van der Bellen; f 1986; political organization; campaigns for environmental protection, peace and social justice.

Global 2000: 1120 Vienna, Flurschützstr 13; tel (1) 812-57-30; fax (1) 812-57-28; e-mail office@global2000.at; internet www.global2000.at; Chairs: Brigid Weinzinger, Ingmar Höbarth; Media Spokesperson: Andreas Baur; f 1982; campaigning organization concerned with genetic engineering, nuclear power in Eastern Europe and renewable agricultural resources.

Greenpeace (Greenpeace in Central & Eastern Europe): 1050 Vienna, Siebenbrunnengasse 44; tel (1) 545-45-80; fax (1) 545-45-80-98; e-mail office@at.greenpeace.at; internet www.greenpeace.at; Exec Dir: Dr Bernhard Drumel; f 1983; national office of Greenpeace International.

Institut für Energieforschung (Institute of Energy Research): Joanneum Research, 8010 Graz, Elisabethstr 11; tel (316) 876-13-38; fax (316) 876-13-20; e-mail ief@joanneum.at; internet www.joanneum.at/ief; Head: Dr Jozef Spitzer; f 1988; part of Joanneum Research; research on energy issues for industry and for national and international government agencies. Main focus is on: electricity from renewable energy sources; decision-making aid for policy makers; cooling with heat; clean energy from waste; greenhouse gas balance models; regional energy strategies; regional and national carbon balances; environmental friendly space heating; energy optimization in production processes.

Koordinationsstelle ÖKOBÜRO (Ecobureau Co-ordination Centre): 1010 Vienna, Volksgartenstr 1; tel (1) 524-93-77; fax (1) 524-93-77-20; e-mail oekobuero@blackbox.net; internet www.message.at/oekobuero; 'umbrella' body co-ordinating the activities of 12 environmental organizations.

Naturfreunde-Internationale—NFI (International Friends of Nature): 1150 Vienna, Diefenbachgasse 36; tel (1) 892-38-77; fax (1) 812-97-89; e-mail nfi@nfi.at; internet www.nfi.at; Sec-Gen: Manfred Pils; f 1895; promotes sustainable regional development and tourism; environmental education; affiliated with Friends of Nature organizations in 22 countries.

Österreicher Naturschutzbund—ÖNB (Austrian Nature Protection Union): 5020 Salzburg, Arenbergstr 16; tel (662) 64-290-915; fax (662) 64-37-344; Pres: Prof Dr Eberhard Stüber; f 1913; environmental conservation; newsletter; member of Umweltdachverband ÖGNU.

Österreichische Gesellschaft für Landschaftsplanung und Landschaftsarchitektur—ÖGLA (Austrian Society for Land Planning and Landscape Architecture): 1020 Vienna, Schiffamtsgasse 18/6; tel (1) 216-584-413; fax (1) 216-584-415; e-mail buero.knoll@eunet.at; member of Umweltdachverband ÖGNU.

Österreichische Gesellschaft für Natur- und Umweltschutz (Austrian Society for the Protection of Nature and the Environment): 1080 Vienna, Alserstr 21; tel (1) 40-113-0; e-mail umweltdachverband@oegnu.or.at; Man Dir: Franz Maier; lobbying of government, aid agencies and industry; teacher training; publications; protected areas; education; conferences.

Österreichische Ökologie-Institut für angewandte Umweltforschung (Austrian Institute for Applied Ecological Research): 1070 Vienna, Seidengasse 13; tel (1) 523-61-05-0; fax (1) 523-58-43; e-mail oekoinstitut@ecology.at; internet www.ecology.at; member of Koordinationsstelle Ökobüro.

Österreichische Wasserschutzwacht (Austrian Water Protection Group): 8010 Graz, Kaiser-Franz-Jozef-Kai 50; tel (316) 83-27-32; member of Umweltdachverband ÖGNU.

Österreichischer Alpenverein—OeAV (Austrian Alpine Club): 6010 Innsbruck, Wilhelm-Greilstr 15; tel (512) 595-47; fax (512) 575-528; e-mail peter.hasslacher@alpenverein.at; internet www.alpenverein.at; Prin Officer: Peter Hasslacher; f 1862; member of Umweltdachverband ÖGNU and of Commission International pour la Protection des Alpes—CIPRA.

Österreichischer Fischereiverband (Austrian Fisheries Union): 1160 Vienna, Haberlgasse 32/13; tel (664) 1204461; fax (1) 3177418; e-mail ifis_gerhard@compuserve.com; Pres: Hans Harra; Prin Officer: Gerhard Woschitz; member of Umweltdachverband ÖGNU and Arbeitsgemeinschaft für die Fischerei der Alpenländer.

Österreichischer Wildgehege-Verband (Austrian Game Reserve Union): 1010 Vienna, pA MPA, Kärntner Ring 14; tel (1) 505-61-24; fax (1) 504-55-81; member of Umweltdachverband ÖGNU.

Rainforest Foundation: 1030 Vienna, Weyrgasse 5/2; tel (664) 357-45-75; fax (1) 713-35-94-73; e-mail rffa@magnet.at; Prin Officer: Carlos Macedo; f 1996; implements projects for the protection of rainforest; affiliated with similar organizations in Brazil, Norway, the United Kingdom and the USA.

Salzburger Nationalparkfonds—NPF (Salzburg National Park Fund): 5741 Neukirchen, Markt 306; tel (6565) 6558-0; fax (6565) 6558-18; e-mail nationalpark@salzburg.at; internet www.npht.sbg.ac.at; Prin Officer: Harald Kremser; f 1984; promotion of the National Park; management of protected areas; landscape protection; scientific research; environmental education and public relations; visitor information and management; affiliated with Carinthia and Tyrol National Park Funds.

Umweltdachverband ÖGNU (Environmental Group): 1080 Vienna, Alserstr 21; tel (1) 40-113; e-mail umweltdachverband@oegnu.or.at; umbrella body co-ordinating the activities of 35 environmental organizations.

Verband der Naturparke Österreichs (Austrian Nature Park Association): 8010 Graz, Alberstr 10; tel (316) 318-848-99; fax (316) 318-848-88; e-mail oear.stmk@oear.co.at; Man: Franz Handler; f 1995; national co-ordinating body for the 30 Austrian nature parks; member of Umweltdachverband ÖGNU.

Vereinte Grüne Österreichs—VGÖ (United Green Party of Austria): Linz; tel (732) 66-83-91; fax (732) 65-06-68; Chair: Adi Pinter; Gen Secs: Wolfgang Pelikan, Günter Ofner; f 1982; ecologist political party.

Vienna Institute for Development and Co-operation—VIDC: 1010 Vienna, Weyrgasse 5/1; tel (1) 713-35-94; fax (1) 713-35-94-73; e-mail vidc@magnet.at; Dir: Erich Andrlik; f 1985; fosters interdisciplinary research in co-operation with institutions in Africa, Asia and Latin America; advises Austrian NGOs on the implementation of development projects and provides assistance; conducts rainforest protection projects; organizes cultural exchanges; conferences and workshops.

WWF Österreich (World Wide Fund for Nature—WWF Austria): 1160 Vienna, Ottakringerstr 114–116, Postfach 1; tel (1) 488-170; fax (1) 488-17-29; e-mail wwf@wwf.at; internet www.wwf.at; Prin Officers: Winfried Walter, Karl Wagner; national affiliate organization of the World Wide Fund for Nature—WWF International; nature conservation in Austria and abroad.

AZERBAIJAN

Governmental Organizations

State Committee for Ecology and Environmental Protection: 370016 Baku, Azadlyg Sq 1, Government House; Chair: Ali Hasanov.

Ministry of Agriculture and Produce: 1016 Baku, Azadlyg Sq 1, Government House; tel (12) 493-53-55; Minister: Ismet N Aliyev.

Ministry of Ecology and Natural Resources: 370073 Baku, Bahram Agayev St 100a; tel (12) 492-59-07; fax (12) 439-84-32; internet eco.gov.az; Minister: Hussein Baghirov.

Ministry of Fuel and Energy: Baku, Gasanbek Zardabi St 88; tel (12) 447-05-84; fax (12) 431-90-05; e-mail mfe@mfe.az; Minister: Majid Karimov.

State Committee for Geology and Mineral Resouces: 370016 Baku, Azadlyg Sq 1, Government House; Chair: Islam Tagiyev.

State Committee for Hydrometeorology: 370016 Baku, Azadlyg Sq 1, Government House; Chair: Zulfugar Musayev.

State Committee for Material Resources: 370016 Baku, Azadlyg Sq 1, Government House; Chair: Hulmammad Javadov.

State Committee for Improvements of Soil and Water Economy: 370016 Baku, Azadlyg Sq 1, Government House; tel (12) 93-61-54; fax (12) 93-11-76; Chair: Ahmed Ahmedzade; f 1923.

Non-Governmental Organizations

Agricultural Research Institute: 370098 Baku, Pos Pirshagi, Sovkhoz No 2; tel (12) 24-11-50.

Azerbaijan Research Institute of Water Economy: 370012 Baku, Tbilisi Ave 69[a]; tel (12) 31-69-90; Dir: Dr Elchin Surkhoi ogly Gambarov; f 1961.

Botanical Garden, Azerbaijan Academy of Sciences: Baku, Patamdartskoye Ave 40; Dir: U M Agamirov.

Commission on Nature Conservation, Azerbaijan Academy of Sciences: Dept of Biological Sciences, 370001 Baku, Istiglaliyat St 10; Chair: G A Aliyev.

Commission on the Caspian Sea, Azerbaijan Academy of Sciences: Dept of Earth Sciences, Baku, H Javid St 31; tel (12) 39-35-41; fax (12) 92-56-99; Chair: R M Mamedov.

Institute of Geography, Azerbaijan Academy of Sciences: 370143 Baku, Hussein Javid St 31; tel (12) 38-29-00; Dir: B A Budgov; f 1945; library; publications.

Institute of Soil Science and Agrochemistry, Azerbaijan Academy of Sciences: 370073 Baku, Krylova St 5; Dir: I Sh O Iskenderov.

Institute of Zoology, Azerbaijan Academy of Sciences: 370073 Baku, Ketchid 1128, Block 504; tel (12) 39-73-71; e-mail zoology@dcacs.ab.az; Dir: M A Musaev; Deputy Dir: Prof T K Mikailov; f 1936; species composition in Azerbaijan; scientific fundamentals for the protection of useful species and the control of harmful species.

Research Institute for Plant Protection: Gyanja, Fioletova St 57; tel (222) 34781.

Society of Soil Scientists, Azerbaijan Academy of Sciences: 370601 Baku, Isteglal 10; Chair: D M Guseinov.

V L Komarov Institute of Botany, Azerbaijan Academy of Sciences: 370073 Baku, Patamdartskoye shosse 40; tel (12) 39-32-30; Dir: V D O Gajiyev.

BAHAMAS

Governmental Organizations

Ministry of Health and Environmental Services: Royal Victoria Gardens, Shirley St, POB N-3730, Nassau; tel 322-7425; fax 322-7788; Minister: Marcus C Bethel.

Department of Environmental Health Services—DEHS: POB N 3729, Nassau; tel 322-4908; fax 322-3607; e-mail lhepburn@mail.dehs.bs; internet www.dehs.bs; Dir: Dr Donald Cooper; Asst Dir: L Hepburn; Chief Health Insp: M McKenzie; solid waste management; food safety and building control; analysis of pollutants and other contaminants; control of disease-spreading pests and rodents.

Ministry of Agriculture, Fisheries and Local Government: Levy Bldg, East Bay St, POB N-3028, Nassau; tel 325-7502; fax 322-1767; Minister: V Alfred Gray; responsibilities include development of co-operatives, wildlife conservation, control of the importation of fauna and flora, and veterinary medicine.

Department of Fisheries: Levy Bldg, E Bay St, POB N-3028, Nassau; tel 393-1014; fax 393-0238; Prin Officer: Colin Higgs; management of marine resources.

Non-Governmental Organizations

Bahamas National Trust: POB N-4105, Nassau; tel 393-1317; fax 393-4978; e-mail bnt@bahamas.net.bs; internet www.bahamas.net.bs/environment; Dir: Gary E Larson; f 1959; mandated with management and development of national park and protected area system; resource conservation; wildlife and habitat protection; historic preservation; environmental education.

BAHRAIN

Governmental Organizations

Ministry of Electricity and Water: POB 2, Manama; tel 17533133; fax 17533035; Minister: Sheikh Abdullah bin Salman al-Khalifa.

Ministry of Municipalities and Agriculture: POB 26909, Manama; tel 17293693; fax 17293694; Minister: Dr Muhammad Ali as-Sitri.

Ministry of Public Works and Housing: POB 5, Muharaq Causeway Rd, Manama; tel 17535222; fax 17533095; Minister: Fahmi bin Ali al-Jouder.

Non-Governmental Organizations

Arabian Gulf University: POB 26671, Manama; tel 440044; fax 440002; Vice-Dean and Dir, Biotechnology Programme: Muhammad Nabil Alaa El-Din; UNEP Global Environment Outlook—GEO collaborating centre.

Bahrain Centre for Studies and Research: POB 496, Manama; tel 754757; fax 754678; e-mail besr@patlco.com; internet www.besr.gov.bb; Sec-Gen: Dr Salman Rashid al-Zayani; f 1981; scientific study and research in all fields of the natural sciences; library of 7,000 vols.

Murshidat Al Bahrain (Girl Guides of Bahrain): POB 43, Manama; tel 685276; fax 685336; Chair: Fawzia Amin; Int Commr: Faika Ameen; f 1970; education on family life, the environment and community service; the advancement of girls and young women through community development projects; member of the World Association of Girl Guides and Girl Scouts.

Water Sciences and Technology Association—WSTA: POBox 20018, Manama; tel 826512; fax 826513; e-mail swta@batelco.com.bh; wsta-secretary@wsta-gcc.org; internet www.wsta-gcc.org; Pres: Abdullatif I Al-Mugrin; Vice-Pres: Dr Mohammed F Al-Rashed; Sec: Dr Mahmoud Abdulraoof Mohd; Treasurer: Ebrahim A Al-Kaabi; encourages scientific studies and research; provides training and develops local capability in the field of water sciences and technology, with particular emphasis on water resources and treatment; co-operates with universities and research centres.

BANGLADESH

Governmental Organizations

Ministry of the Environment and Forests: Dhaka; tel (2) 7165436; e-mail admin1@moef.gov.bd; internet www.moef.gov.bd; Minister: Shahjahan Siraj.

Department of Environment: Paribesh Bhaban, E-16, Agargaon, Sher-e-Bangla Nagar, Dhaka 1207; tel (2) 812461; fax (2) 9118682; e-mail infobnfp@bdcom.com; Dir-Gen: A R Khan; serves as the INFOTERRA national focal point.

Ministry of Agriculture: Bangladesh Secretariat, Bhaban 4, 2nd 9-Storey Bldg, Dhaka; tel (2) 832137; internet www.bangladesh.gov.bd/moa; Minister: M K Anwar.

Ministry of Local Government, Rural Development and Co-operatives: Bangladesh Secretariat, Bhaban 7, 1st 9-Storey Bldg, 6th Floor, Dhaka; tel (2) 318682; Minister: Abdul Mannan Bhuiyan; includes Department of Environment Pollution Control.

Ministry of Power, Energy and Mineral Resources: Bangladesh Secretariat, Bhaban 6, First Floor, Dhaka; tel (2) 865918; fax (2) 861110; Prime Minister and Minister: Begum Khaleda Zia.

Ministry of Water Resources: Bangladesh Secretariat, Bldg 4, 4th Floor, Dhaka; tel (2) 868688; fax (2) 862400; Minister: Maj (retd) Hafizuddin Ahmad.

Non-Governmental Organizations

Bangladesh Centre for Advanced Studies—BCAS: 620 Rd 10[a], Dhanmondi, GPOB 3971, Dhaka 1205; tel (2) 815829; fax (2) 811344; e-mail saleem@pradeshta.net; Exec Dir: Dr Saleemul Huq; Dir: Atiq Rahman; research on the environment; UNEP Global Environment Outlook—GEO collaborating centre.

Bangladesh Poush: 43 New Eskaton Rd, Dhaka 1000; tel (2) 402801; Pres: Haroun er-Rashid; f 1985; afforestation; health and sanitation; environmental education; disaster management; environmental conservation; preservation of biodiversity; fund-raising; research; publications.

Institute for Environment and Development Studies—IEDS/FoE Bangladesh: 29 Purana Paltan, 3rd Floor, POB 3691, Dhaka 1000; tel (2) 242351; fax (2) 868746; Prin Officers: Chowdhury M Farouque, Amir Hussain Chowdhry, Mofizur Rahman; f 1985; active through research, campaigns, demonstrations, training and publications; concerned with environmental issues affecting land, water and air; issues include the Himalayan environment, banning of trade in frog legs, organic farming, tropical rainforests, mangrove swamps, global warming and climate change, waterlogged land, and air and water pollution; part of Friends of the Earth International.

IUCN Bangladesh Country Office: House 3[a], Rd 15, Dhanmondi, Dhaka 1209; tel and fax (2) 8122577; e-mail iucnbd@citechco.net; Contact: Dr Ainun Nishat; preparation with government of the National Conservation Strategy; collaboration with other IUCN branches, NGOs and government in the identification and development of projects for conservation and sustainable management of natural resources and biodiversity; implementation of projects; securing donor support; symposia and workshops to create public awareness of environment and development issues.

Multidisciplinary Action Research Centre: House 12, Rd 12, Dhanmondi R/A, Dhaka 1209; tel (2) 311462; Man Dir: Monowar Hossain; sustainable development.

Wildlife Society of Bangladesh: c/o Zoology Dept, University of Dhaka, Dhaka 2; tel (2) 506330; Pres: Prof K Z Husain; f 1981; organization of wildlife biologists and other students of wildlife in Bangladesh; environmental education.

BARBADOS

Governmental Organizations

Ministry of Energy and Public Utilities: Govt Headquarters, Bay St, St Michael; Minister: Anthony P Wood.

Environment Unit: Sir Frank Walcott Bldg, 4th Floor, Culloden Rd, St Michael; tel 431-7687; fax 437-8859; e-mail envdivn@mailhub.caribsurf.com; internet www.environment.gov.bb; main policy co-ordinating body of the Government regarding environmental policy at the national, regional and international level; advises the Government on environmental policy; develops programmes to implement policy; develops and executes environmental education and public awareness programmes; acts as a technical focal point for regional and international policy; conserves and manages coastal and marine areas; operates the Climate Change Programme and Ozone Programme; addresses the development of policy, management and assessment of biodiversity; active in sustainable development policy formulation through the National Commission on Sustainable Development; reviews progress in implementing various international conventions; distributes resources; affiliated with the Coastal Zone Management Unit—CZMU, National Conservation Commission—NCC, Environmental Special Projects Unit—ESPU and Energy Unit—EU.

Environmental Special Projects Unit—ESPU: ESPU Project House, 1 Sturges, St Thomas; tel and fax 438-7761; e-mail info@espu.gov.bb; internet www.espu.gov.bb; Project Man: Steve Devonish; Tech Officer: Mark Brathwaite; f 1996; fmrly National Conservation Commission Project Unit; redevelopment of Harrison's Cave.

Ministry of Agriculture and Rural Development: Graeme Hall, POB 505, Christ Church; tel 428-2150; fax 420-8444; internet www.barbados.gov.bb/minagri; Minister: Senator Erskine Griffith.

Ministry of the Housing, Lands and the Environment: National Housing Corpn Bldg, The Garden, Country Rd, St Michael; tel 431-7800; fax 435-0174; Minister: Elizabeth Thompson.

Non-Governmental Organizations

Barbados Environmental Association: 210 Courts Plaza, St George St, POB 132, Bridgetown; tel and fax 427-0619; Pres: Gordon Bispham; Vice-Pres: Leo Brewster; Sec: Stephen Boyce; f 1987; awareness-raising on environmental issues and the need for conservation and management of natural resources; research on environmental issues.

Barbados Museum and Historical Society: St Ann's Garrison, St Michael; tel 427-0201; fax 429-5946; e-mail museum@barbmuse.org; Dir: Alissandra Cummins; Pres: Dr Trevor Carmichael; Nat History Curator: Cherrie Parris; f 1933; zoological, botanical, geological and other scientific specimen collections; environmental education programmes; natural history data bases; affiliated with the International Conservation Centre—ICCROM, the Caribbean Conservation Association—CCA and the Museums Asscn of the Caribbean—MAC; member of the International Council of Museums—ICOM.

Barbados National Trust: Ronald Tree House, 10th Ave, Belleville, St Michael; tel 436-9033; fax 429-9055; e-mail natrust@sunbeach.net; internet www.wow.net/sunhead/bnt; Exec Dir: Penelope Hynam-Roach; Deputy Exec Dir: William Gollop; Public Relations Officer: William Cummins; f 1961; compilation of a listing and photographic record of places of natural beauty and their animal and plant life; acquisition and opening to the public of sites of historic and architectural interest; affiliated with the Heritage Canada Foundation, the National Trust for Historic Preservation (USA) and the national trusts of Australia, Fiji, Zimbabwe, England and Wales, and Scotland (United Kingdom).

Centre for Resource Management and Environmental Studies—CERMES, University of the West Indies: Cave Hall Campus, POB 64, Bridgetown; tel 417-4000; fax 424-4204; e-mail cermes@uwichill.edu.bb; internet www.cavehill.uwi.edu/cermes; Dir: Hazel A Oxenford (acting); Sr Lecturers: Robin Mahon, Leonard Nurse; f 1985; teaching: graduate MSc in natural resource and environmental management, short courses; research: policy and governance of natural resources, management systems and approaches, fisheries biology and applied fisheries science, coastal ecology; outreach: regional projects putting resource management into action.

BELARUS

Governmental Organizations

Ministry for Natural Resources and Environmental Protection: 220048 Minsk, vul Kalektarnaya 10; tel (17) 220-66-91; fax (17) 220-55-83/220-47-71; e-mail minproos@mail.belpak.by; internet www.president.gov.by/Minpriroda/index_e.htm; Minister: Lyavontsy I Kharonzhyk; First Deputy Minister: Vasily Podolyako, Deputy Ministers: Aleksandr Apatsky, Valentin Malishevsky; areas of activity include air quality, climate, conservation of wildlife, waste management and water quality; supervises the regional committees for environmental protection, hydrometeorology's work, exploring of minerals, ecological expertise, ecological information, management and nature usage, international agreements.

Belarusian Research Centre for Ecology: 220002 Minsk, Horuzhaya 31 A; tel (17) 234-70-65; fax (17) 234-80-72; e-mail eco@ecoprom.belpak.minsk.by; Dir: V I Aleshka; Technical Staff: Michail Yu Kalinin, Alezander V Yakovenko, Galina A Chalake; the Information Department serves as the INFOTERRA national focal point.

Central Research Institute for Complex Use of Water Resources—CRICUWR: Minsk, vul Slavinskaga 1; tel (17) 264-65-22; fax (17) 264-27-34; e-mail cricuwr@infonet.by; Dir: Mikhail M Cherepansky; f 1961; assessment and forecasting of natural and anthropogenic changes in water resources; hydroecological monitoring of water resources and quality; regulation of water pollution prevention and water resource use.

Ministry of Energy: 220050 Minsk, vul K Marksa 14; tel (17) 229-83-59; fax (17) 229-84-68; Minister: Alyaksandr A Aheyeu.

Ministry of Forestry: 220039 Minsk, vul Chkalova 6; tel (17) 224-47-05; fax (17) 224-41-83; e-mail infor@komleshoz.org; internet www.komleshoz.org; Minister: Pyotr Syamashka.

Minsitry of Agriculture and Food: 220050 Minsk, vul Kirava 15; tel (17) 227-37-51; fax (17) 227-42-96; e-mail kanc@mshp.minsk.by; internet mshp.minsk.by; Minister: Leonid V Rusak.

State Committee for Energy and Energy Supervision: Minsk; Chair: Lew Dubovik.

State Committee for Hydrometeorology: Minsk; Chair: Yury Pakumeyka.

State Committee for Land Resources, Geodesy and Cartography: Minsk; Chair: Georgiy Kuznyatsow.

Committee on the Consequences of the Accident at the Chornobyl Nuclear Power Plant—Chornobyl Committee: 220004 Minsk, pr Masherava 23; tel (17) 289-15-96; fax (17) 227-27-36; Chair: Ivan A Kenik; f 1990; formerly the Ministry for Emergency Situations and the Protection of the Population in the Aftermath of the Chornobyl Nuclear Power Station Disaster; responsible for state policy in the field of protection of the population from the consequences of the Chornobyl disaster; co-ordinates implementation of scientific research on health, ecology, radioecology, applied science, social protection and the rehabilitation of affected citizens, maintainance of contaminated and protected areas, decontamination and radioactive waste management.

Non-Governmental Organizations

Association of Professional Ecologists: 220004 Minsk, Oboinyi per 4; tel (17) 226-71-06.

Belarus Research Institute for Land Reclamation and Grassland, Academy of Agricultural Sciences of the Republic of Belarus: 220040 Minsk, vul M Bogdanovich 153; tel (17) 232-49-41; fax (17) 232-64-96; Dir: Vladislav Filippovich Karlovsky; f 1930; reclamation of waterlogged soils and development and maintenance of reclaimed land.

Belarus State Polytechnic Academy, Ecology Department: 220027 Minsk, Skaryny pr 65; tel (17) 239-91-29; fax (17) 231-30-49; Head: Dr Sergei Dorozhko; environmental education for engineers; ecological training programmes; scientific research; student exchange.

Bureau on Environmental Consultancy—BURENCO: 220027 Minsk, Skaryny pr 65; tel (17) 231-30-52; fax (17) 231-30-49; e-mail ecology_zone8@infra.belpak.minsk.by; Dir: Vladimir Koltunov; f 1996; implementation of Environmental Management Systems for industrial enterprises in Minsk; advice to government on environmental policy; Eco-Team project addresses the potential for energy saving in households; environmental education and teacher training; affiliated with Eindhoven University of Technology (Netherlands).

Belarusian Ecological Green Party: 220068 Minsk, vul Asipenki 19; tel (17) 220-11-16; fax (17) 256-82-72; f 1993; f 1998 by the merger of Belaruskaya Ekalagichnaya Partya (Belarusian Ecological Party—f 1993) and Belaruskaya Partya Zyalenych (Green Party of Belarus—f 1992).

Belarusian Ecological Union: 220030 Minsk, Lenina 15[a]; tel (17) 227-87-96; unites various groups concerned with environmental issues.

Belaruskaya Partya Zyaleny Mir (Belarusian Greenpeace Party): 246023 Gomel, vul Brestskaya 6; tel (23) 247-08-08; fax (23) 247-96-96; Leaders: Oleg Gromyka, Nick Lekunovich; f 1994; political organization.

Belovezhskaya Pushcha Museum: Brestskaya oblast, Belovezhskaya Pushcha Game Reserve; tel 56-1-69; fax 2-12-83; e-mail box@npbprom.belpak.brest.by; internet sity .belpak.brest.by/bp/index.shtml; Dir: Vasilij P Zhukov; f 1945; game reserve museum showing the work being done to preserve the European bison, and to acclimatize other animals in the same part of Belarus.

Chornobyl Socio-Ecological Union: 220000 Minsk, Myasnikob Str 39; tel (17) 220-39-04; fax (17) 271-58-19; e-mail sasha@by.glas.apc.org; Pres: Vasily Yakavenka; Vice-Pres: Yuri Voronzhtsev; f 1990; concerned with the social welfare of victims of the Chornobyl disaster and environmental clean-up; campaigns on environmental issues.

Ecomir Republican Scientific and Engineering Centre for Remote Sensing of Environment, Academy of Sciences of Belarus: 220012 Minsk, vul Surganava 2; tel (17) 268-53-60; fax (17) 239-31-71; Dir: A A Kovalev; f 1990.

Institute of Radioecological Problems, Academy of Sciences of Belarus: 220109 Minsk, Sosny; tel (17) 246-72-53; fax (17) 246-76-15; Dir: G A Sharovarov.

Institute of the Problems of the Use of Natural Resources and Ecology, Academy of Sciences of Belarus: 220114 Minsk, Staroborisovsky trakt 10; tel (17) 264-26-31; fax (17) 264-24-13; Dir: I I Lishtvan; f 1932; environmental research.

Maladziezhny Ekalagichny Rukh 'Belaya Rus' (White Rus Youth Ecological Movement): Minsk; tel (17) 239-91-29; fax (17) 234-53-50; Co-ordinator: Dr Sergei Dorozhko; dissemination of environmental information; co-ordination of the activities of environmental organizations in Belarus; ecological camps and trips; seminars; conferences; cultural exchange.

Ranitsa Group: 220007 Minsk, vul Zhukovskogo 9/2, kv 39; tel (17) 222-18-47; f 1991; environmental education.

Students' National Ecocenter of Belarus—RECU: 220023 Minsk, vul Makaenka, d 8; tel (17) 264-11-68; e-mail root@swta.minsk.by; Dir: Dr Lidia A Kurganova; f 1930; environmental forums and education.

BELGIUM

Governmental Organizations

Ministry of the Environment, Consumer Protection and Sustainable Development: 1 rue Marie Thérèse, 1000 Brussels; tel (2) 549-09-20; fax (2) 512-21-23; Minister: Bruno Tobback; federal institution.

Federale Diensten voor het Leefmilieu/Services Fédéraux pour les Affaires Environnementales (Federal Department of the Environment): c/o Ministry of Social Affairs, Public Health and the Environment, 19 blvd Pachéco, BP 5, 1010 Bruxelles; tel (2) 210-46-87; fax (2) 210-48-52; e-mail environment@health.fgov.be; internet www.environment.fgov.be; comprises the Studies and Coordination Office, the Product Standards Office and the Risk Management Office.

Studiecentrum voor Kernenergie/Centre d'Etude de l'Energie Nucléaire—SCK.CEN (Nuclear Research Centre): Boeretang 200, 2400 Mol; tel (14) 33-25-86; fax (14) 33-25-84; e-mail averlede@sckcen.be; internet www.sckcen.be; Chair, Bd of Govs: Prof Frank Deconinck; Gen Man: Paul Govaerts; Public Relations Officer: Anne Verledens; responsible to the Secretary of State for Energy and Sustainable Development; conducts research on nuclear safety, radioactive waste, and protection and safeguards against radiation.

Ministry of Small and Medium-Sized Businesses, the Liberal Professions and the Self Employed, and Agriculture: Brussels; Minister: Sabine Laruelle.

Administration of Natural Resources and the Environment: Brussels Region, Trierstraat 49, 1040 Brussels.

Department of the Environment and Infrastructure for the Flemish Region: Graaf de Ferrarisgebouw, Koning Albert II-laan 20, bus 2, 1000 Brussels; tel (2) 553-71-02 fax (2) 553-71-05; e-mail leefmilieu.infrastructure@lin.vlaanderen.be; Minister of the Environment and Agriculture: Vera Dua; Sec-Gen: Fernand Desmyter; regional institution.

Instituut voor Natuurbehoud (Institute of Nature Conservation): Kliniekstraat 25, 1070 Brussels; tel (2) 558-18-11; fax (2) 558-18-11; e-mail info@instnat.be; internet www.instnat.be; Dir: Eckhart Kuijken; f 1986; scientific institute of the Flemish Community; conducts scientific studies, investigations and operations in connection with nature conservation, with the aim of formulating courses of action and scientific criteria for the implementation of a nature conservation policy.

Openbare Vlaamse Afvalstoffenmaatschappij—OVAM (Flemish Public Waste Agency): Kanunnik De Deckerstraat 22–26, 2800 Mechelen; tel (15) 28-42-84; fax (15) 20-32-75; e-mail info@ovam.be; internet www.ovam.be; Admin-Gen: Frank Parent; Deputy Admin-Gen: Henny De Baets; responsible for waste management in Flanders; concerned with reduction and recycling of waste, and soil quality.

Direction Générale des Ressources Naturelles et de l'Environnement, Région Wallonne—DGRNE (Department of Natural Resources and the Environment for the Walloon Region): 15 ave Prince de Liège, 5100 Namur; tel (81) 33-50-50; fax (81) 33-51-22; e-mail dgrne@mrw.wallonie.be; internet mrw.wallonie.bedgrne; Dir-Gen: Claude Delbeuck; regional institution.

Institut Bruxellois pour la Gestion de l'Environnement/Brussels Instituut voor Milieubeheer—IBGE-BIM (Brussels Institute for Environmental Management): Gulledelle 100, 1200 Brussels; tel (2) 775-75-11; fax (2) 775-76-21; e-mail info@ibgebim.be; internet www.ibgebim.be; f 1989; regional environmental administration; gathers, analyses and disseminates environmental information; oversees environmental impact assessments.

Institut Scientifique de Service Public—ISSeP (Public Service Scientific Institute): 200 rue du Chéra, 4000 Liège; tel (4) 252-71-50; fax (4) 252-46-65; e-mail direction.gene@issep.be; internet www.issep.be; Prin Officers: Michel de Waele, Claude Michaux; f 1990; environmental pollution measurements and related activities, including water analysis for the Walloon region, air emission measurements and overall assessment, waste and landfill studies, and maintenance of networks for the measurement of air and water pollution; affiliated to the European Network of Environmental Research Organizations—ENERO.

Office of the Minister in charge of the Environment for the Brussels-Capital Region: 54 ave Louise/Louizalaan, BP 10, 1050 Brussels; tel (2) 517-12-00; fax (2) 511-94-42; e-mail dgosuin@gosuin.irisnet.be; internet www.brussels.irisnet.be/En/1en_admi/1en_2gov/1en_2gov.htm; Minister in charge of the Environment, Water Policy, Nature, Green Spaces and Nature Preservation, Public Cleanliness and Foreign Trade: Didier Gosuin; regional institution.

Services Fédéraux des Affaires Scientifiques, Techniques et Culturelles/Federale Diensten voor Wetenschappelijke, Technische en Culturele Aangelegenheden—SSTC/DWTC (Federal Office for Scientific, Technical and Cultural Affairs—OSTC): 8 rue de la Science, 1000 Brussels; tel (2) 238-34-11; fax (2) 230-59-12; e-mail jama@belspo.be; internet www.belspo.be; Mans: Ward Ziarko, Georges Jamart; Tech Staff: Hilde Van Dongen, François Vanderseypen; affiliated to the Office of the Prime Minister; serves as the INFOTERRA national focal point.

Federale Raad voor Duurzame Ontwikkeling/Conseil Fédéral du Développement Durable—FRDO-CFDD (Federal Council for Sustainable Development): Secretariat, Aduatukersstraat 71–75, 1040 Brussels; tel (2) 743-31-50; fax (2) 743-31-55; e-mail mail@frdo-cfdd.fgov.be; internet www.belspo.be/frdocfdd; Perm Secs: Jan De Smedt, Catherine Mertens; f 1997; replaced the National Council for Sustainable Development (f 1993); advisory body comprising representatives of environmental and development organizations, consumers' unions, trade unions, employers' federations, energy producers and scientific institutions; liaises with the federal authorities on matters related to sustainable development policy, particularly the implementation of international commitments such as Agenda 21, the Framework Convention on Climate Change (UNFCCC) and the Convention on Biological Diversity (CBD); acts as a forum for debate on the issue of sustainable development; organizes symposia; creates public awareness; maintains eight working groups: biological diversity and forests, energy and climate, federal planning, genetically modified organisms, international relations, product standards, scientific research on sustainable development, and the socio-economic dimension of sustainable development; publications.

Institut Royal des Sciences Naturelles de Belgique/Koninklijk Belgisch Instituut voor Natuurwetenschappen—IRScNB-KBIN (Royal Belgian Institute of Natural Sciences—RBINS): 29 rue Vautier, 1000 Brussels; tel (2) 627-45-15; fax (2) 627-41-13; e-mail willem.devos@naturalsciences.be; internet www.naturalsciences.be; Dir: Daniel Cahen; Deputy Dirs: Anouk Dubois, Jaqueline Verheyen; Public Relations Officer: Willem De Vos; Librarian: Laurent Meese; Intl Relations Manager: Olivier Retout; f 1846; hosts and provides links to various environmental websites and data bases; focal point on the convention on biological diversity.

Vlaamse Landmaatschappij—VLM (Flemish Land Agency): Gulden-Vlieslaan 72, 1060 Brussels; tel 2-5437200; fax 2-5437397; e-mail landinrichting@vlm.be; internet www.vlm.be; Admin-Gen: Roland de Paepe; Dep Admin-Gen: Guido Clerx; Gen Man: Marc Heyerick; f 1988; regional institution; maintains five provincial branches in Bruges, Ghent, Herentals, Leuven and Hasselt.

Vlaamse Milieumaatschappij—VMM (Flemish Environment Agency): A Van De Maelestraat 96, 9320 Erembodegem; tel (53) 72-62-11; fax (53) 77-71-68; e-mail info@vmm.be; internet www.vmm.be; Admin-Gen: Frank Van Sevencoten; Deputy Admin-Gen: John Huylebroeck; Information Officer: Johan Janda; regional institution; maintains six field offices, two laboratories, and an office in Brussels.

Non-Governmental Organizations

Anders Gaan Leven—Agalev (Ecologist Party): 23 Brialmontstraat, 1210 Brussels; tel (2) 219-19-19; fax (2) 223-10-90; e-mail agalev@agalev.be; internet www.agalev.be; Pres: Jos Geysels; f 1982; Flemish ecological political party with members in the Flemish and national parliaments; member of the European Federation of Green Parties—EFGP.

Belgian National Union for Conservation of Nature: 84 Italiëlei, 2000 Antwerp; tel (3) 231-26-04; fax (3) 233-64-99; Sec-Gen: Julius-Anton Smeyers; federation of the main nature conservation NGOs in Belgium; nature conservation; environmental education.

Bond Beter Leefmilieu—BBL Vlaanderen vzw (League for a Better Environment Flanders): Tweekerkenstraat 47, 1000 Brussels 1; tel (2) 282-17-20; fax (2) 230-53-89; e-mail hostmaster@bblv.be; internet www.bondbeterleefmilieu.be; Pres: Luc Hens; Gen Sec: Lieve Grauls; f 1971; one of four regional environmental federations in Belgium; comprises more than 120 environmental organizations in the Flemish-speaking region; protection of the environment from pollution and insensitive development by encouraging co-operation between the federation and its member organizations, representing member organizations where necessary, acting as legal adviser to the environmental movement, political lobbying, influencing public opinion through the media, and collaborating with other NGOs in Europe and world-wide.

Brusselse Raad voor het Leefmilieu—BRAL vzw (Brussels Council for the Environment): Zaterdagplein 13, 1000 Brussels; tel (2) 217-56-33; fax (2) 217-06-11; e-mail bral@bralvzw.be; internet www.bralvzw.be; Gen Sec: Peter Mortier; Pres Gen Ass: Albert Martens; f 1973; one of four regional environmental federations in Belgium; owing to its geographical situation, BRAL is an environmental organization focusing on urban or city-related issues—climate change is studied from the angle of city car pollution; energy—green buildings and insulation in privately owned apartments and homes.

Centre Permanent d'Education à la Conservation de la Nature—CPECN (Permanent Centre for Conservation Education): Parc de Mariemont, 29 rue du Parc, 7170 Manage; tel (64) 23-80-10; fax (64) 23-80-19; e-mail secretariat@crie-mariemont.be; internet www.ful.ac.be/hotes/crie; Dir: Jean-Pierre Cokelberghs; Librarian: Catherine Vieuxtemps; f 1986; conservation education and strategies.

Collectif d'Echanges pour la Technologie Appropriée (Exchange Group for Appropriate Technology): 18 rue de la Sablonnière, 1000 Brussels; tel (2) 218-18-96; fax (2) 223-14-95; Sec-Gen: Francis Douxchamps.

De Wielewaal, Natuurvereniging vzw (De Wielewaal Nature Association): Graatakker 11, 2300 Turnhout; tel (14) 47-29-50; fax (14) 47-29-51; e-mail info@wielewaal.be; internet www.moerbeke.be/wielewaal; Prin Officer: Peter Sterckens; Contact: Lieven Van Hooste; f 1933; study and conservation of nature; nature education.

Earth Action: 146 blvd Brand Whitlock, 1200 Brussels; tel (2) 736-80-52; fax (2) 736-99-50; Contact: Nicholas Dunlop.

Ecolo (Ecologist Party): 8 rue du Séminaire, 5000 Namur; tel (81) 22-78-71; fax (81) 23-06-03; e-mail ecolo.sf@ecolo.be; internet www.ecolo.be; Fed Secs: Jacques Bauduin, Philippe Defeyt, Brigitte Ernst; f 1980; French-speaking ecological party; affiliated with the European Federation of Green Parties—EFGP.

Friends of the Earth Belgium/Les Amis de la Terre: pl de la Vingeanne, 5100 Dave (Namur); tel (81) 40-14-78; fax (81) 40-23-54; e-mail amis.delaterre@gate71.be; internet www.ful.ac.be/hotes/amisterre; national office of Friends of the Earth International.

Greenpeace Belgium: Vooruuitgangstraat 317, 1030 Brussels; tel (2) 274-02-00; fax (2) 201-19-50; e-mail info@be.greenpeace.org; national office of Greenpeace International.

Institut E van Beneden, Université de Liège (E van Beneden Institute, Liège University): Faculté des Sciences, 22 quai van Beneden, 4020 Liège; tel (4) 366-50-81; fax (4) 366-50-10; e-mail jc.ruwet@ulg.ac.be; Prin Officer: Prof Jean-Claude Ruwet; wildlife preservation; management of natural resources.

Inter-Environnement Bruxelles—IEB asbl: 16 rue Marcq, 1000 Brussels; tel (2) 223-01-01; fax (2) 223-12-96; e-mail info@ieb.be; internet www.ieb.be; Gen Sec: Anne-France Rihoux; f 1974; one of four environmental federations in Belgium; federation of environmental organizations and local residents' associations in the Brussels-Capital region.

Inter-Environnement Wallonie—IEW asbl: 7 rue de la Révolution, 1000 Brussels; tel (2) 219-89-46; fax (2) 219-91-68; e-mail iew@ecoline.org; internet www.ful.ac.be/hotes/iew; Gen Sec: Thérèse Snoy; f 1974; one of four environmental federations in Belgium; independent, non-profit federation of over 120 environmental organizations in French-speaking Belgium; member of European Environmental Bureau—EEB.

Koninklijke Maatschappij voor Dierkunde van Antwerpen (Royal Zoological Society of Antwerp): 26 Koningin Astridplein, 2018 Antwerp; tel (3) 202-45-40; fax (3) 231-00-18; Dir: F J Daman; f 1843; environmental education; research.

Laboratoires de l'Environnement de l'ULB à Treignes, Université Libre de Bruxelles—ULB (ULB Environment Laboratories at Treignes, Free University of Brussels): 81 rue de la Gare, 5670 Treignes; tel (60) 39-96-24; fax (60) 39-94-50; e-mail jcverhae@ulb.ac.be; Prin Officer: Dr Jean-Claude Verhaeghe; f 1972; environmental research in the fields of ecology, wildlife behaviour and habitats, and agricultural techniques; the University is a member of the World Conservation Union—IUCN.

Ligue Royale Belge pour la Protection des Oiseaux—LRBPO (Royal Belgian League for the Protection of Birds): 43–45 rue de Veeweyde, 1070 Brussels; tel (2) 521-28-50; fax (2) 527-09-89; e-mail lrbpo@birdprotection.be; internet www.users.skynet.be/RNS/croisades/actions/erika/ligue_royale_po.html; Chair: Roger Arnhem; Deputy Chairs: Michel David, Jan Rodts; f 1922; active in the protection of wild birds throughout Europe by opposing hunting, trapping and trade, and through conservation of their natural habitats; member of Inter-Environnement Wallonie—IEW asbl.

Nationaal Verbond voor Natuurbescherming (Belgian Union for the Conservation of Nature): Bervoetstraat 33, 2000 Antwerpen; tel (3) 231-26-04; fax (3) 233-64-99; Gen Sec: J A Smeyers; f 1953; federation of major NGOs dealing with nature conservation in Belgium; environmental education.

Stichting Leefmilieu vzw (Environment Foundation): Kipdorp 11, 2000 Antwerp; tel (3) 231-64-48; fax (3) 232-63-98; e-mail leefmilieu@village.uunet.be; internet www.stichtingleefmilieu.be; Sec-Gen: Bavo Rombouts; information; education; forum on environmental problems and solutions; publications.

Vlaamse Jeugdbond voor Natuurstudie en Milieubehoud 'Natuur 2000' vzw (Flemish Youth Federation for the Study and Conservation of Nature): Bervoetstraat 33, 2000 Antwerp; tel (3) 231-26-04; fax (3) 233-64-99; e-mail natur2000@pandora.be; internet www.natur2000.be; Chair: Maarten Bogaert; f 1967; nature studies; conservation activities; environmental education; affiliated with Nature 2000 International.

World Blue Chain for the Protection of Animals and Nature: 39 ave de Vise, 1170 Brussels; e-mail bwk@vt4.net; Man Dir: Marleen Elsen-Verloot; protection of animals and nature.

World Wide Fund for Nature—WWF Belgium: 608 chaussée de Waterloo, 1050 Brussels; tel (2) 340-09-99; fax (2) 340-09-33; e-mail communications@wwf.be; internet www.wwf.be; Pres and Chair: Guido Ravoet; f 1966; national affiliate organization of World Wide Fund for Nature—WWF International; conservation of fauna, flora, land, waters, soil and natural resources; fund-raising.

BELIZE

Governmental Organizations

Ministry of Natural Resources, the Environment, Commerce and Industry: Market Sq, Belmopan; tel 822-2226; fax 822-2333; e-mail info@mnrei.gov.bz; internet www.mnrei.gov.bz; Minister of the Environment: Vildo Marin; Minister of Natural Resources: John Briceño.

Ministry of Agriculture, Fisheries and Co-operatives: 2nd Floor, West Block Bldg, Belmopan; tel 822-2241; fax 822-2403; e-mail mafpaeu@btl.net; Minister of Agriculture: Servulo Baeza; Minister of Fisheries and Co-operatives: Mike Espat.

Non-Governmental Organizations

Belize Audubon Society: POB 1001, Belize City; tel (2) 77369; fax (2) 78562; Pres: Therese Rath; Vice-Pres: David Craig, Lombardo Riverol; f 1969; management of reserves; environmental education; community information programme.

Belize Center for Environmental Studies—BCES: POB 1081, Belize City; tel (2) 34988; fax (2) 34985; Man Dir: Lou Nicolait; Admin: Frances Griffith; f 1988; non-profit organization; integration of natural resource conservation and environmentally sound planning with sustainable economic development in Belize, through education, research, publication of findings, technical assistance and training, policy evaluation, environmental impact assessments, dissemination of information, networking and use of a resource centre.

Monkey Bay Wildlife Sanctuary: POB 187, Belmopan; tel (8) 23180; fax (8) 23361; e-mail mbay@pobox.com; internet www.watershedbelize.org; Prin Officers: Matthew Miller, Joshua T Brown, Jr; f 1990; protection of biodiversity; ecotourism and residential placement programmes to promote understanding between different cultures; community management and monitoring of watersheds; natural history education field programmes; affiliated with the Monkey Bay Wildlife Fund (Tokyo, Japan) and Sibun Watershed Asscn (Belmopan, Belize).

Programme for Belize: POB 749, 2 South Park St, Belize City; tel (2) 75616; fax (2) 75635; e-mail pfbel@btl.net; internet www.pfbelize.org; Exec Dir: Edilberto Romero; f 1988; private conservation organization; management of 240,000 acres of forested land; linking of conservation with economic development by testing principles and approaches to sustainable land management; community outreach and environmental education.

BENIN

Governmental Organizations

Ministry of the Environment, Housing and Urban Development: BP 01-3621, Cotonou; tel 31-55-96; fax 31-50-81; e-mail sg@environnement.bj; internet www.mehubenein.net; Minister: Luc-Marie Constant Gnacadja; Gen Sec: Pascal Yaha.

Direction de l'Aménagement du Territoire (Directorate of Physical Planning): BP 01-3621, Cotonou; tel 30-07-42; Dir: Asse A Severin; serves as the INFOTERRA national focal point.

Ministry of Agriculture, Stockbreeding and Fishing: 03 BP 2900, Cotonou; tel 30-04-10; fax 30-03-26; e-mail sg@agriculture@gouv.bj; internet www.agriculture.gouv.bj; Minister: Lazare Sehoueto.

Department of Agriculture: BP 58, Porto Novo; tel 21-32-90.

Department of Forests and Natural Resources: BP 03-2900, Cotonou; tel 33-06-62; fax 33-19-56.

Ministry of Mining, Energy and Water: 04 BP 1412, Cotonou; tel 31-29-07; fax 31-35-46; e-mail sg@energie.gouv.bj; internet www.energie.gouv.bj; Minister: Kamarou Fassassi.

Non-Governmental Organizations

Les Amis de la Terre (Friends of the Earth—FoE): BP 03-1162, Cotonou; tel 30-21-05; fax 30-22-05; Prin Officer: Venance Dassi; environment; development; community health.

BHUTAN

Governmental Organizations

National Environment Commission: Royal Government of Bhutan, POB 466, Thimphu; tel (2) 323384; fax (2) 323385; e-mail rnrncc@druknet.net.bt; Deputy Minister: Dasho Nado Rinchen; Chief Communications Officer and Man, INFOTERRA Nat Focal Point: Kunzang Dorji; responsible for the formulation of Bhutan's National Environmental Strategy and for the majority of technical reports and guidelines relating to Bhutan's environment; the Communication Unit serves as the INFOTERRA National Focal Point, liaises with national and international organizations, and maintains environmental data bases and an environmental library information system.

Ministry of Agriculture: POB 252, Thimphu; tel (2) 322129; fax (2) 323153; internet www.moa.gov.bt; Minister: Lyonpo Sangye Ngedup Dorji.

Department of Forests: Thimphu; tel (2) 23055; fax (2) 22395.

Non-Governmental Organizations

Royal Society for the Protection of Nature—RSPN: POB 325, Thimphu; tel (2) 322056; fax (2) 323189; e-mail rspn@druknet.net.bt; internet rspn.cjb.net; Pres: Dasho Sangay Thinlay; Exec Dir: Lam Dorji; f 1987; environmental education programmes; integrated conservation and development programmes; membership and fund-raising activities.

World Wide Fund for Nature—WWF Bhutan Programme: POB 210, Thimphu; tel (2) 23528; fax (2) 23518; Country Rep: Mingma Norbu Sherpa; Conservation Programme Dir: Kinzang Namgay; f 1992; national affiliate organization of World Wide Fund for Nature—WWF International; surveying and management of protected areas; anti-poaching activities and tiger conservation; conservation fellowship and training; institutional strengthening of national conservation organizations; environmental education and awareness-raising.

BOLIVIA

Governmental Organizations

Ministry of Sustainable Development and Planning: Avda Arce 2147, Casilla 12814, La Paz; tel (2) 231-0860; e-mail sdnp@coord.rds.org.bo; internet www.rds.org.bo; Minister: Gustavo Pedraza Mérida; the Deputy Minister for the Environment, Natural Resources and Forest Development is responsible for the following institutions: Dirección General de Clasificación de Tierras y Cuencas (Directorate-General for Classification of Soils and River Basins), Dirección General de Impacto, Calidad y Servicios Ambientales (Directorate-General for Environmental Impact, Quality and Services), Dirección General de Biodiversidad (Directorate-General for Biodiversity), Dirección General de Desarrollo Forestal Sostenible (Directorate-General for Sustainable Forest Development) and Servicio Nacional de Areas Protegidas (National Service for Protected Areas).

Ministry of Agriculture, Livestock and Rural Development: Avda Camacho 1471, La Paz; tel (2) 2200885; fax (2) 2336041; e-mail correspondencia@maca.gov.bo; estadisticas@maca.gov.bo; internet www.agrobolivia.gov.bo; Minister: Diego Montenegro Ernst; Statistical Bureau Man: Ing José Napoleón Pacheco; f 1888; politics, strategies, programmes for agricultural, livestock and fisheries sector.

Fundación para la Defensa de la Naturaleza (Nature Defence Foundation): Casilla 128, Oruro; tel (52) 51501; Conservationist: Ana Rosa Quiroga Nogales; Ecologist: Gonzalo A Lázaro; f 1990; promotes the value of traditional knowledge for sustainable development, the importance of environmental knowledge to the conservation of diversity, the use of indigenous institutions to encourage participation in the development process and experiences in indigenous development planning; affiliated with Green Point.

Non-Governmental Organizations

Academia Nacional de Ciencias de Bolivia (Bolivian National Academy of Sciences): Avda 16 de Julio 1732, Casilla 5829, La Paz; tel (2) 36-3990; fax (2) 37-9681; e-mail aciencia@ccibo-entelnet.bo; Pres: Carlos Aguirre; Vice-Pres: Ismael Montes de Oca; research into public policies, environmental sciences, solar energy, astronomy; advisory services to private and public institutions on policy issues; serves as the INFOTERRA national focal point; affiliated with Bolivian Asscn for the Advancement of Science, Bolivian Inventors Asscn and Bolivian Asscn of Women in Science.

Apoyo para el Campesino Indígena del Oriente Boliviano—APCOB (Support for the Native Peasant of Eastern Bolivia): Calle Cuatro Ojos 80, Casilla 4213, Barrio San Luis, Santa Cruz de la Sierra; tel (3) 54-2119; fax (3) 54-2120; e-mail apcob@bibosl.scz.entelnet.bo; internet www.latinwide.com/apcob; Prin Officers: Jürgen Riester, Graciela Zolezzi; f 1980; consolidation and protection of inter-municipal territory, and sustained and integral management of natural resources; efficient development and management of the forestry industry; support to native organizations and promotion of women's issues through development programmes.

Asociación Ecológica del Oriente—ASEO (Eastern Bolivian Ecological Association): Ballivián 157, Casilla 4831, Santa Cruz de la Sierra; tel (3) 33-9252; fax (3) 33-9252; Pres: Ester Ballerstaedt; f 1987; environmental education, monitoring and awareness-raising.

Bolivian Association for Conservation—TROPICO, Conservation Data Centre: Casilla 11250, Edif El Ciprés, Dpto 5b, Calle Campos 296, Esq 6 de Augusto, La Paz; tel (2) 435005; fax (2) 435027; e-mail tropico@mail.megalink.com; Exec Dir: Zimena Aramayo; f 1987; management of protected areas; conservation and management of wildlife; maintains an information centre.

Bolivian Committee for IUCN: c/o Estación Biológica del Beni—Reserva de la Biosfera, Casilla 5829, La Paz; tel (2) 36-3990; fax (2) 36-9192; Dir: Carmen Miranda; member of the World Conservation Union—IUCN.

Catholic Relief Services, Bolivia: Jacinto Benavente 2190, 2°, Casilla 2561, La Paz; tel (2) 32-3335; fax (2) 39-2228; e-mail crsbo@caoba.entelnet.bo; Dir: Kristen Sample; Deputy Dir: Holly Inurreta; f 1943; financial and technical support for sustainable agriculture and natural resource management, preventive health care and micro-enterprise development; national office of Catholic Relief Services (Baltimore, MD, USA).

Centro de Comunicación y Desarrollo Andino—CENDA (Centre for Communication and Andean Development): Tadeo Haenke 2231, Casilla 3226, Cochabamba; tel (42) 43412; fax (42) 81502; e-mail cenda@pino.cbb.entelnet.bo; internet www.cedib.org/webs/cenda/index.html; education, research, communication and agroforestry programmes for Quechua communities.

Centro de Datos para la Conservación—CDC Bolivia (Conservation Data Centre): Calle 26 y Avda Muñoz Reyes s/n, Cota Cota, Casilla 11250, La Paz; tel and fax (2) 79-7399; Exec Dir: Maria R de Marconi; f 1986; provision of scientific and technical bases for conservation; compilation, analysis and processing of information; technical assistance.

Centro de Investigación Agrícola Tropical—CIAT (Centre for Tropical Agricultural Research): Avda Ejercito 131, Casilla 247, Santa Cruz de la Sierra; tel and fax (3) 32-3177; e-mail rhzciat@bibosi.scz.entelnet.bo; f 1976; research covers agroforestry and technology transfer, farm and community forestry, and silviculture of native tree species.

Centro de Investigación y Documentación para el Desarrollo del Beni—CIDDE-BENI (Research and Documentation Centre for the Development of Beni): Casilla 86, Cipriano Barace 732, Trinidad, Beni; tel (46) 52037; fax (46) 52038; e-mail ciddeben@sauce.ben.entelnet.bo; Dir: Carlos Navia; f 1984; forest management and silviculture; sustainable development; environmental campaigning.

Centro de Investigaciones y Estudio de la Capacidad de Uso Mayor de la Tierra—CUMAT (Land Capability Research Centre): Casilla 10387, Sucursal Miraflores, La Paz; tel (2) 41-7570; fax (2) 41-2496; e-mail cumat@datacom-bo.net; Exec Dir: Juan Carlos Quiroga; f 1984; concerned with environmental impact assessment, land use, deforestation, agricultural alternatives, image processing, watershed management and agricultural management projects; conducts research and acts as a technical consultancy.

Centro de Servicios Múltiples de Tecnologías Apropiadas—SEMTA (Appropriate Technology Centre): Alfredo Ascarrunz 2675, Casilla 15041, La Paz; tel (2) 41-0042; fax (2) 41-1497; e-mail warisata@wari.bo; Exec Dir: Juan José Castro Guzman; Financial Dir: Estela M de Lenz; Gen Co-ordinator of Programmes: Carlos Loayza; f 1980; non-profit, independent institution working for sustainable development based on agro-ecology and self-sufficiency, particularly in rural areas of Bolivia.

ECOTOP: Tiquipaya-Apote, Casilla 1836, Cochabamba; tel (42) 88820; fax (42) 88381; f 1985; formerly POLEN; consultancy in agroecology and rural development.

Estación Biológica del Beni—Reserva de la Biosfera, Academia Nacional de Ciencias de Bolivia (Beni Biological Station—Biosphere Reserve, Bolivian National Academy of Sciences): Avda 16 de julio 1732, Casilla 5829, La Paz; tel (2) 35-0612; fax (2) 35-0612; e-mail cmiranda@ebb.rds.org.bo; Exec Dir: Carmen Miranda; f 1982; maintains seven programmes: the Protection Programme, implemented by a corps of forest rangers; the Natural Resource Management Programme for conservation and production; the Scientific Research Programme on fauna and flora; the Monitoring Programme, dedicated to tropical ecosystems; the Public Use Programme of environmental education and ecotourism; the Programme of Support to Local Communities, concerned with sustainable development and traditional knowledge; and the Co-operation and Diffusion Programme, encouraging collaboration with other scientific centres and dissemination of information to the public and other interested parties.

Fundación Amigos de la Naturaleza—FAN (Friends of Nature Foundation): Casilla 8706, La Paz; tel (2) 33-3806; Exec Dir: Hermes Justiniano; f 1988; private, non-profit organization which aims to protect the biodiversity of Bolivia.

Institut de Recherche pour le Développement—IRD (Development Research Institute): Casilla 9214, La Paz; tel (2) 22-7724; fax (2) 22-5846; e-mail gibon@mail.megalink.com; internet www.ird.fr; Researchers: Bernard Pouyaud (glacial hydrology), Geneviève Bourdy (medicinal plants), Jean-Pierre Dujardin (parasitology), Charles-Eduoard de Suremain (nutrition), (vacant—hydrobiology); f 1945; a centre of Institut de Recherche pour le Développement, France; co-operative development work; trains local researchers.

Liga de Defensa del Medio Ambiente—LIDEMA (Environmental Defence League): Avda 20 de octubre 1763, Casilla 11237, La Paz; tel (2) 2419393; fax (2) 2912322; e-mail lidema@lidema.org.bo; internet www.lidema.org.bo; Exec Dir: Lic Jenny Gruenberger; f 1985; private non-profit organization which co-ordinates the activities of 30 Bolivian conservation organizations; development of national projects in education, basic and applied research, and public relations; policy and legislative initiatives, support for sustainable production.

Museo Nacional de Historia Natural—MNHN (National Museum of Natural History): Casilla 8706, La Paz; tel (2) 79-5364; e-mail mnhn@mnhn.rds.org.bo; Exec Dir: Lilián Villalba M; f 1980; biological and ecological research; conservation; education; publications.

Protección del Medio Ambiente Tarija—PROMETA (Protection of the Tarija Environment): O'Connor 559, Casilla 59, Tarija; tel (66) 33873; fax (66) 45865; e-mail prometa@uajms.bo; Institutional Development Co-ordinator: Hernán Ruiz Fournier; f 1990; conservation of biodiversity; research; environmental education, policy and lobbying.

BOSNIA AND HERZEGOVINA

Governmental Organizations

Environmental Steering Committee: 71 000 Sarajevo; f 1999; established with the aim of co-ordinating environmental activities between the two 'entities': the Federation and the Serb Republic of Bosnia and Herzegovina.

Ministry of Agriculture, Water Management and Forestry of the Federation of Bosnia and Herzegovina: Titova 15, 71 000 Sarajevo; tel (33) 443338; fax (33) 663659; e-mail info@fmpvs.gov.ba; internet www.fmpvs.gov.ba; Minister: Marinko Božić; First Sec: Faketa Begović; responsible for drafting legislation in the fields of water management, agriculture, forestry and veterinary science, as well as for monitoring of its implementation. The ministry consists of four sectors with the following tasks: Water management: management and protection of natural resources, equal share of natural resources development; Agriculture: soil, incentives, fruit growing, cattle breeding, etc; Veterinary: animal diseases, control of facilities (slaughterhouses, plants, etc; Forestry: management of forests, inspection (monitoring). Each section has its inspection with chief federal inspector.

Ministry of Foreign Trade and Economic Relations: Musala 9, 71000 Sarajevo; tel and fax (33) 206142; e-mail info@vet.gov.ba; internet www.vet.gov.ba; Minister: Dragan Doko; Sec: Slaviša Kreštalica.

Ministry of Physical Planning and the Environment of the Federation of Bosnia and Herzegovina: Maršala Tita 9a, 71000 Sarajevo; tel and fax (33) 473124; e-mail mehmedagic@fbihvlada.gov.ba; Minister: Ramiz Mehmedagić; Sec: Biljana Knežević.

Ministry of Urbanism, Civil Engineering and Ecology of the Serb Republic: Trg srpskih junaka 4, 51000 Banja Luka; tel (51) 215511; fax (51) 215548; e-mail mgr@mgr.vladars.net; internet www.mpugie.rs.ba; Minister: Hrusto Tupeković; Sec: Svjetlana Jovanović.

Commission to Preserve National Monuments: c/o High Representative of the International Community, 71 000 Sarajevo, trg Djece Sarajeva bb; tel (33) 447275; fax (33) 447420; Annexe 8 of the 1995 Dayton Accords provided for establishment of the Commission; after five years responsibility for its operation was to be transferred to the Government of Bosnia and Herzegovina; the Director-General of UNESCO was to appoint the first Chairman and one other member of the five-member Commission.

Non-Governmental Organizations

Bosnian Environmental Technologies Association—BETA: 71 000 Sarajevo, Stjepana Tomića 1; tel and fax (33) 200226; e-mail mjuric@utic.net.ba; Contact: Prof Esma Velagić-Habul.

Centar za Okolišno Odiženi Razvoj Bih-COOR (Centre for the Environmentally Sustainable Development of Bosnia and Herzegovina—CESD): 71 000 Sarajevo, Tomica 1; tel and fax (33) 207949; e-mail coorsa@bih.net.ba; internet www.coor.ba; Chair: Prof Dr Tarik Kupusović; Deputy Chair: Dragana Selmanagic; f 2000; training, education and awareness raising programmes on environmental protection and sustainable development; promotion of cleaner production towards decision makers, publishing news and information on successful case studies, in-plant demonstrations and projects for industries and training on CP; publishing leaflets and brochures about environmental protection and sustainable development.

Centre for Ecology and Natural Resources: 71 000 Sarajevo, Zmaja od Bosne 35; tel (33) 649196; fax (33) 649342; e-mail sdug@usa.net; internet www.cepres.pmf.unsa.ba; Contact: Prof Dr Sulejman Redžić.

EKOBiH—Society for Environment, Protection and Improvement of Bosnia and Herzegovina: 71 000 Sarajevo, Stari Grad, Patke bb; tel (33) 232634; fax (33) 441040; Contact: Faik Kulović.

Ekološki Pokret Zeleni (Green Eco-Movement): 77240 Bosanska Krupa, Reis-ul Dž Čauševića bb; tel (77) 472844; fax (77) 471072; Contact: Zajnil Palić.

Ekološko Društvo RS 'EKOS' (EKOS Ecological Society): 78000 Banja Luka, Stepe Stepanovića 75; tel (51) 61392; fax (51) 63024; Contact: Prof Dr Branislav Nedović.

Fondacija za Podsticanje Uravnotezenog Razvoja i Kvalitet Zivota—Fond Eko (Foundation for Sustainable Development, Stimulation and Quality of Life): 71 000 Sarajevo, Branilaca Sarajeva 47, ul Rudarska 72; tel (33) 670448; fax (33) 211354; e-mail fondeko@bih.net.ba; Contact: Prof Dr Dubravka Šoljan; activities include liaison with government on the matter of protected areas.

Hrvatska Ekološka Udruga Buna (Croatian Eco Association Buna): 88000 Mostar, Buna bb; tel (36) 480038; fax (36) 480380; Contact: Damir Brljević.

Hydro-Engineering Institute: 71 000 Sarajevo, Stjepana Tomica 1; tel (33) 212466/67; fax (33) 207949; e-mail heis@heis.com.ba; internet www.heis.com.ba; Contact: Prof Dr Tarik Kupusovic; Exec Dir's: Dalila Jabucar, Sela Cengic, Branko Vucijak, Sanda Midzic, Admir Ceric, Ranko Tica; f 1991; work scope: scientific and research activities, education, planning, consulting services, revision of technical-investment documentation for water management, hydro-engineering and environmental engineering; activities: hydrologic field measurements; hydrology of surface and ground water streams and lakes; hydrological aspects of reservoir design; measuring, observation and modelling of water runoff from urban areas.

Institute for Agropedology: 71 000 Sarajevo, Dolina 6; tel (33) 221780; fax (33) 268261; e-mail zapsaapksa.com.ba; Contact: Prof Dr Husnija Resulović; Dir: Esad Bukalo; concerned with soil contamination in industiral areas, in the vicinity of roads; maps of contaminated soils with heavy metals; soil pollution in urban areas.

Meteorological Institute: 71 000 Sarajevo, Bardakičije 12; tel (33) 663508; fax (33) 524040; e-mail martin_tais@hotmail.com; Contact: Tais Martin; f 1891.

Unija Vodoprivrednih Drustava (Union of Water Resources Societies): 71 000 Sarajevo, H Cemerlica 25; tel (33) 655759; fax (33) 664861; e-mail vodoprivreda_bih@zamir-sa.ztn.apc.org; Contact: Avdo Saric.

BOTSWANA

Governmental Organizations

National Conservation Strategy (Co-ordinating) Agency: Private Bag 0068, Gaborone; tel 302050; fax 302051; internet www.gov.bw/government; Contact: Mushanana L Nchunga; part of the Ministry of Local Government, Lands and Housing; serves as the secretariat of the National Conservation Strategy Advisory Board and co-ordinates the execution of its decisions; advises and supports ministries and government departments in the execution of their environmental responsibilities; assists local authorities in formulating local and district conservation strategies; provides advice to central and local government and the public on natural resources conservation; assesses and prepares standards for Environmental Impact Assessments; co-ordinates Botswana's implementation of international environmental agreements.

Ministry of Agriculture: Private Bag 003, Gaborone; tel 3950602; fax 3975805; Minister: Johnnie Swartz.

Ministry of Minerals, Energy and Water Affairs: Khama Crescent, Private Bag 0018, Gaborone; tel 3656600; fax 372738; Minister: Charles Tibone.

Ministry of Trade, Industry, Wildlife and Tourism: Private Bag 004, Gaborone; tel 3911233; fax 3975239; Minister of Environment, Wildlife and Tourism: Capt Kitso Mokaila.

Department of Town and Regional Planning: Private Bag 0042, Gaborone; tel 354277; fax 313280; e-mail infoterra@info.bw; Tech Officer and Man, INFOTERRA nat focal point: Tuelo Nkwane; part of the Ministry of Lands and Housing; serves as the INFOTERRA national focal point.

Department of Wildlife and National Parks: POB 131, Gaborone; tel 371404; Dir: G Seeletso.

Non-Governmental Organizations

Chobe Wildlife Trust: POB 55, Kasane; tel 650516; fax 650223; e-mail cwt@info.bw; internet www.chobewildlifetrust.com; Patron: Lt-Gen Seretse Khama Ian Khama; Chair: Stephen Griesel; Vice-Chair: Richard Randall; CEO: Tracey Draper; conservation in northern Botswana; environmental education; assistance to Department of Wildlife and National Parks; campaign against poaching; support for protection of the rhinoceros and community-based natural resource management.

IUCN Botswana Country Office: Private Bag 00300, Plot 2403, Hospital Way, Ext 9, Gaborone; tel 301584; fax 371584; e-mail iucn@iucnbot.bw; internet www.iucnbot.bw; Contact: Ruud Jansen; f 1984; national office of the World Conservation Union—IUCN.

Kalahari Conservation Society: Plot 112, Independence Ave, POB 859, Gaborone; tel 3974557; fax 3914259; e-mail admin@kcs.org.bw; internet www.kcs.org.bw; CEO: Felix Minggae; Conservation Officer: Edwin Thibedi; Public Relations Officer: Stella Chiwaya; Education Officer: Herbert Kebafetotse; Community Liaison Officer: Mercy Motladiile; f 1982; research, education and lobbying for conservation of wildlife and the environment; advocates sustainable use of natural resources, particularly where this benefits local communities.

Francistown Branch: Kalahari Conservation Society, POB 1086, Francistown; tel and fax 212766.

Khama Rhino Sanctuary Trust: POB 10, Serowe; tel and fax 430713; fax 4635808; e-mail krst@botsnet.bw; internet www.khamarhinosanctuary.org; Chief Warden: Moremi M Tjibae; Vice-Chair: Bathusi Letlhare; Contact: K Spruyt; f 1992; conservation and re-introduction of endangered rhinoceros species; education; eco-tourism.

Veld Products Research and Development—VPR&D: POB 2020, Gaborone; tel 347047; fax 347363; e-mail veldprod@info.bw; Man Dir: R Sunstrum; f 1994; independent, non-profit organization encouraging community-based natural resource management for economic growth; development of non-timber forest products; domestication of indigenous fruit trees in semi-arid areas; member of the World Conservation Union—IUCN.

BRAZIL

Governmental Organizations

Ministry of the Environment: Esplanada dos Ministérios, Bloco B, CEP 70068-900 Brasília, DF; tel (61) 322-7819; fax (61) 322-8469; e-mail marina-silva@mma.gov.br; internet www.mma.gov.br; Minister: Maria Osmarina Marina da Silva Vaz de Lima.

Ministry of Agrarian Development: Esplanada dos Ministerios, Bloco A, 8° andar, Brasília, DF; tel (61) 223-8002; fax (61) 322-0492; e-mail miguel.rossetto@mda.gov.br; internet www.mda.gov.br; Minister: Miguel Rossetto.

Ministry of Agriculture, Fisheries and Food Supply: Esplanada dos Ministérios, Bloco D, 8°, 70043-000 Brasília, DF; tel (61) 226-5161; fax (61) 321-8360; e-mail ccnagri@agricultura.gov.br; internet www.agricultura.gov.br; Minister: Roberto Rodrigues; includes the National Parks and Reserves Department.

Empresa Brasileira de Pesquisa Agropecuária—EMBRAPA (Brazilian Agricultural Research Corporation): SAIN, Parque Rural, Final da Av W-3 N, CP 04-0315, 70770-910 Brasília, DF; tel (61) 348-4433; fax (61) 347-1041; Pres: Alberto Duque Portugal; f 1973.

Instituto Brasileiro de Desenvolvimento Florestal—IBDF (Brazilian Institute for Forestry Development): Setor de Áreas Isoladas, Av L4 Norte, 70800 Brasília, DF; tel (61) 321-2324; Forestry Engineer: Dr Antonio José Costa de Freitas Guimarães.

Instituto Nacional de Colonização e Reforma Agrária—INCRA (National Institute for Land Settlement and Agrarian Reform): SBN, Palácio do Desenvolvimento, 14°, 70040 Brasília, DF; f 1970.

Superintendência do Desenvolvimento da Pesca—SUDEPE (Bureau for the Development of the Fishing Industry): Edif da Pesca, Av W-3 N, Quadra 506, Bloco C, 70040 Brasília, DF; tel (61) 272-3229; Supt: Aécio Moura da Silva.

Instituto Brasileiro do Meio Ambiente e Recursos Naturais Renováveis—IBAMA (Brazilian Institute of Environment and Renewable Natural Resources): SAIN, Av L4 Norte, Bloco C Subsolo, 70800-200 Brasília, DF; tel (61) 226-8221; fax (61) 322-1058; e-mail mmarilia@sede.ibama.gov.br; internet www.ibama.gov.br; Pres: Marilia Marreco Cerqueira; f 1967; merged with SEMA (National Environmental Agency) in 1988; responsible for the annual formulation of national and regional forestry, fishery and environmental plans; management and conservation of renewable natural resources, particularly flora and fauna; administration of national parks and nature reserves.

Centro de Pesquisas para Conservação das Aves Silvestres—CEMAVE (Research Centre for Conservation of Wild Birds): Parque Nacional de Brasília, Via Epia SMU, 70630 Brasília, DF; tel (61) 233-3251; fax (61) 233-5543; e-mail cemave@ibama.gov.br; internet www.bdt.org.br/ibama/cemave; Contact: Monica Koch; f 1977; adopted current name 1992; activities include tagging and monitoring of bird species.

Centro Nacional de Informação Ambiental—CNIA (National Environmental Information Centre): SAIN, Av L4 Norte, Bloco C Subsolo, 70800-200 Brasília, DF; tel

(61) 316-1205; fax (61) 226-5094; e-mail cnia@sede.ibama.gov.br; Prin Officer: Rita de Cassia do Vall Caribé; f 1989; diffusion of information on all aspects of the environment, forestry, pollution, waste, pisciculture and aquaculture.

Instituto Nacional de Meteorologia—INMET (National Meteorological Institute): Eixo Monumental Via S1, Brasília, DF; tel (61) 322-0033; fax (61) 226-3712; e-mail aathayde@inmet.gov.br; internet www.inmet.gov.br; Prin Officers: Augusto Cesar Vaz de Athayde, José Mauro de Rezende; f 1909; weather forecasting; climate research.

Non-Governmental Organizations

Assesoria e Servicios a Projetos em Agricultura Alternativa—AS-PTA (Consultants for Projects in Alternative Agriculture): Rua da Candalária 9, 6°, 20091-020 Rio de Janeiro, RJ; tel (21) 253-8317; fax (21) 233-8663; e-mail aspta@ax.apc.org; Exec Dir: Silvio Gomes de Almeida; conducts research and disseminates information on agroecology, with the aim of promoting sustainable agricultural development among small producers.

Associação Brasileira de Ecologia (Brazilian Association of Ecology): Avda Atlântica 734, Apto 1201, Rio de Janeiro, RJ; tel (21) 275-0111; Pres: A Monteiro da Silva; campaigns on ecology, quality of life, food and agriculture, traditional knowledge, wild flora and fauna, and health; scientific research; botanical gardens.

Associação Brasileira de Engenharia Sanitária e Ambiental—ABES (Brazilian Association of Sanitary and Environmental Engineering): Av Beira Mar 216, 13°, 20021-060 Rio de Janeiro, RJ; tel (21) 210-3221; fax (21) 262-6838; e-mail abes@ax.ibase.org.br; f 1966; development and improvement of sanitary services; raising awareness of environmental preservation issues; 23 brs throughout Brazil.

Associação Catarinense de Preservação da Natureza—ACAPRENA (Santa Catarina Nature Preservation Association): Rua Antonio da Veiga 140, CP 1507, 89010-971 Blumenau, SC; tel (47) 321-0020; fax (47) 322-8818; f 1973; environmental education and monitoring.

Associação de Defesa do Meio Ambiente, São Paulo—ADEMASP (Association for the Defence of the Environment, São Paulo): CP 832, 04531 São Paulo, SP; tel (11) 280-7244; fax (11) 280-5468; Pres: Dr Paulo Nogueira-Neto; f 1955; campaigns on wild fauna and flora, endangered animal and plant species, and ecology; environmental and scientific research.

Associação de Preservação do Meio Ambiente do Alto Vale do Itajaí—APREMAVI (Alto Vale do Itajaí Environmental Preservation Association): CP 218, 89160-000 Rio do Sul, SC; tel and fax (478) 22-0326; e-mail apremavi@ax.apc.org; Prin Officers: Miriam Prochnow, Wigold Schäffer; f 1987.

Brazilian Committee for IUCN: c/o Rua Conselheiro Carrao 640, Bela Vista, 01328 São Paulo, SP; tel (11) 289-5031; fax (11) 280-5468; Regional Councillor: Dr José Pedro de Oliveira Costa.

Centro de Estudos Alternativos e Vida Natural—CEAVN (Centre for Alternative Studies and Natural Living): Conjunto Eldorado, Quadra F, No 55, Feitosa, 57043-460 Maceió, AL; alternative technology, ecology and culture.

Comissão Pró-Índio de Acre—CPI-AC (Pro-Indian Commission): Rua Pernambuco 1.025, Bosque, 69907-580 Rio Branco, AC; tel and fax (68) 224-1426; e-mail cpi@mdnet.com.br; Education Co-ordinator: Vera O S Paiva; Environment and Agriculture Co-ordinator: Renato A Gavazzi; Health Co-ordinator: Paulo J B Alencar; f 1979; education and training.

Companhia de Tecnología de Saneamento Ambiental—CETESB: Caixa 20941, 01000 Sao Paulo, SP.

Fundação Brasileira para a Conservação da Natureza—FBCN (Brazilian Foundation for Conservation of Nature): Rua Miranda Valverde 103, 22281-000 Rio de Janeiro, RJ; tel (21) 537-7565; fax (21) 537-1343; Pres: Jairo Costa; Exec Officer: Anthenor Navarro; research into Brazilian biodiversity; protection of wildlife habitats; environmental education programmes.

Fundação Gaia (Gaia Foundation): Rua Jacinto Gomes 39, 90040-270 Porto Alegre, RS; tel (51) 331-3105; fax (51) 330-3567; e-mail fundgaia@zaz.com.br; internet www.fgaia.org.br; Pres: José A Lutzenberger; Vice-Pres: Lara Lutzenberger; f 1987; promotion of ecological consciousness, sustainable development, regenerative agriculture and democracy; defence of the cultural identity of minorities; protection of ecosystems, flora and fauna; demonstrates sustainable farming methods and offers courses on sustainable development issues at the rural headquarters, Rincão Gaia.

Fundação Grupo Esquel Brasil—FGEB (Esquel Group Foundation—EGF Brazil): Edif Belvedere, 8°, Sala 801-A, SAS, Quadra 6, Bloco L, 70070-100 Brasília, DF; tel (61) 322-2062; fax (61) 322-1063; e-mail esquelbr@ax.apc.org; internet www.esquel.org/brazil.htm; Exec Sec: Dr Sílvio Sant'Ana; national office of the Esquel Group Foundation (Washington, DC) and part of Grupo Esquel; independent, non-profit organization working for sustainable development in Brazil.

Fundação Museu do Homem Americano—FUMDHAM (Museum of American Man Foundation): Abdias Neves 551, 64770 São Raimundo Nonato, PI; tel (86) 582-1612; fax (86) 582-1656; e-mail fumdham@zaz.com.br; Scientific Dir: Dr Anne-Marie Pessis; f 1986; private scientific research institute responsible for the protection of the Serra da Capivara National Park.

Fundação Pró-Natureza—FUNATURA (Foundation for Nature): Edif Gemini Center II, Salas 201/7, Bloco B, SCLN 107 70743-520 Brasília, DF; tel (61) 274-5449; fax (61) 274-5324; e-mail funatura@essencial.com.br; Pres: Maria Tereza Jorge Pádua; Dir: César Vitor do Santo; f 1986; promotion of the role of the private sector in the protection of Brazilian biodiversity; environmental education.

Fundação SOS Mata Atlântica (Brazilian Foundation for the Conservation of the Atlantic Rainforest): Rua Manoel dâ Nóbrega 456, Paraíso, 04001-001 São Paulo, SP; tel (11) 3887-1195; fax (11) 3885-1680; e-mail smata@ax.apc.org; internet www.sosmatatlantica.org.br/index.html; Pres: Roberto L Klabin; Dir, Institutional Relations: Mario César Mantovani; f 1986.

Fundação Vitória Amazônica (Amazonian Victory Foundation): Conjunto Morada do Sol, Rua R/S, Quadra Q, Casa 7, Aleixo, 69060-080 Manaus, AM; tel (92) 642-1336; fax (92) 236-3257; e-mail fva@fva.org.br; internet www.fva.org.br; Pres: José Joaquim Marques Marinho; Vice-Pres: Rita de Cássia Guimarães Mesquita; Exec Man: Carlos César Durigan; Asst Exec Man: Ana Cristina Ramos de Oliveira; f 1990; an NGO with actions focused in rio Negro basin; aiming environmental conservation allied to the improvement of life quality of the inhabitants of this region, through sustainable uses of the natural resources and research on biodiversity and sociodiversity.

Fundo Mundial para a Natureza—WWF Brasil (World Wide Fund for Nature—WWF Brazil): SHIS EQ QL 6/8, Conjunto E, 2°, 71620-430 Brasília, DF; tel (61) 248-2899; fax (61) 248-7176; e-mail panda@wwf.org.br; Prin Officer: Garo Batmanian; affiliated to World Wide Fund for Nature—WWF International.

Greenpeace Brazil São Paulo: Rua dos Pinheiros 240, Conjunto 32, Pinheiros, 05422-000 São Paulo, SP; tel (11) 3066-1155; fax (11) 282-5500; e-mail greenpeace.brazil@dialb .greenpeace.org; affiliated to Greenpeace International.

Gritos da Terra Entidade Ecológica (Gritos da Terra Ecological Group): Conj Alfredo G de Mendonça, Bloco 16, Apto 302, Jacarecica, 57033-020 Maceió, AL; Pres: José Raimundo dos Santos; f 1987; environmental coalition of governmental and non-governmental groups; general and human ecology; opposition to large dam construction projects; environmental education; promotion of cycling and the rights of rural communities; lectures.

Institut de Recherche pour le Développement—IRD: CP 7091, 71619-970 Brasília, DF; tel (61) 248-5323; fax (61) 248-5378; e-mail orstom@apis.com.br; Delegate to Brazil: Maurice Lourd; f 1979; formerly Mission ORSTOM au Brésil (ORSTOM Mission in Brazil); headquarters of the Brazilian delegation to Latin America; missions at various universities and research institutes.

Instituto de Antropologia e Meio Ambiente—IAMA (Institute for Anthropology and Environment): Rua Afonso Vaz 454, Jd Pirajussara, 05580-001 São Paulo, SP; tel (11) 211-6724; fax (11) 210-1338; e-mail iama@ci-sp.rnp.br; Prin Officers: Mauro de Mello Leonel Jr, Betty Mindlin; f 1987; analysis of public policies and major governmental and private-sector economic projects; support for ethnological, ethnohistoric and ethnobiological studies of culturally distinct groups; establishment of multidisciplinary scientific working groups for the development of methodologies for environmental impact assessment.

Instituto de Biociencias, Departamento de Ecologia, Universidade Estadual Paulista 'Julio de Mesquita Filho' (Biosciences Institute, Ecology Department, São Paulo State University): Av 24A, No 1515, CP 199, 13506-900 Rio Claro, SP; tel (195) 34-0244; fax (195) 34-4433; internet www .unesp.br; Head, Inst of Biosciences: Prof Dr Osvaldo Aulino da Silva; education and research in pollution, biodiversity, ecosystems, restoration ecology, remote sensing and applied ecology.

Instituto de Pesquisas e Estudos Florestais—IPEF (Forest Research and Studies Institute): Av Pádua Dias 11, CP 530, 13400-970 Piracicaba, SP; tel (19) 433-6155; fax (19) 433-6081; e-mail mmpoggia@carpa.ciagri.usp.br; internet jatoba .esalq.usp.br; Scientific Dir: Prof Dr Walter de Paula Lima; Pres: Manoel de Freitas; f 1968; research areas include conservation and management of natural resources, industrial forest products and reforestation, forest ecology, biomass energy and remote sensing; library; publications; affiliated with the University of São Paulo.

Instituto de Pesquisas Ecológicas—IPE (Institute of Ecological Research): Casilla 47, 12960-000 Nazaré Paulista, SP; tel and fax (11) 7861-1327; e-mail ipe@ax.ibase.org.br; Pres: Suzana Machado Padua; Vice-Pres: Claudio V Padua; f 1992; research on rare and endangered species; conservation projects for wildlife and habitats; co-operation with landowners to establish isolated fragments of forest as protected areas or private reserves; training programmes in conservation biology for professionals; environmental education programmes in local communities; affiliated with Wildlife Preservation Trust International and the Center for Environmental Research and Conservation, Columbia University (NY, USA).

Instituto PNUMA Brasil (Brazilian UNEP Institute): Av. Nilo Peçanha, 50 - sl 1708, 20044-900 Rio de Janeiro, RJ; tel (21) 3084-1020; fax (21) 3084-4233; e-mail contato@ brasilpnuma.org.br; internet www.brasilpnuma.org.br; Prin Officer: Haroldo Mattos de Lemos; Tech Dir: Carlos Raja Gabaglia Penna; Devt Dir: José Roberto Naninho; Admin Dir: Roberto Padula; Legal Dir: O Graza Conto; Project Coordinator: Nelio Paes de Banos; promotion of aims and programmes of UNEP; distribution of translations of UNEP literature; conferences on relevant issues. Training activities, including post-graduation courses on envionmental management.

Instituto Socioambiental (Socio-Environmental Institute): Av Higienopolis 983, Higienopolis, 01238-001 São Paulo, SP; tel (11) 825-5544; fax (11) 825-7861; e-mail socioamb@ax.apc .org; internet www.socioambiental.org; Exec Sec: Nilto Tatto; Deputy Exec Sec: Marina Kahn; protects social rights and assets related to the environmental and cultural heritage; fosters socio-economic development to ensure ecologically sustainable access to and management of natural resources; research and dissemination of information.

Instituto Sul Mineiro de Estudos e Conservação da Natureza (South Minas Gerais Research and Nature Conservation Institute): Fazenda Lagoa, 37127-000 Monte Belo, MG; tel and fax (35) 571-1500; e-mail ism@ax.apc.org; Prin Officer: Maria Christina Weyland Vieira; geographical and biological research; natural resource conservation; environmental education; international relations in the areas of forest conservation and sustainable development.

International Wildlife Coalition, Brazil: CP 5087, 88040-970 Florianópolis, SC; tel (51) 982-5157; fax (48) 234-1580; Pres: José Truda Palazzo; Vice-Pres: Dr Maria do Carmo Both; research, education and consultancy in wildlife and nature protection in Brazil and abroad; projects on whales, environmental restoration, and establishment and enforcement of wildlife sanctuaries.

Movimento de Apoio à Resistência dos Waimiri Atroari—MAREWA (Support Movement for Waimiri Atroari Resistance): BR-174, Km 107, Rua Cupiuba 16, 69735-000 Presidente Figueiredo, AM; tel (92) 324-1117; fax (92) 324-1117; Prin Officers: Doroti Alice Müller, Egídio Schwade; ethnohistorical archive; support for rights of indigenous peoples.

Movimento pela Vida—MOVIDA (Movement for Life): Rua Sargento Jaime 370, CP 230, Prado, 57010-070 Maceió, AL; tel (82) 223-8791; fax (82) 231-7459; opposition to chemical pollution; campaigns in defence of rainforests in the state of Alagoas; environmental education.

Núcleo de Apoio à Pesquisa sobre Populações Humanas e Áreas Úmidas Brasileiras, Universidade de São Paulo—NUPAUB-USP (Research Centre on Human Population and Wetlands in Brazil, University of São Paulo): Rua do Anfiteatro 181, Colmeias, Favo 06, Cidade Universitária, Butantã, 05508-90 São Paulo, SP; tel (11) 211-0011; fax (11) 813-5819; e-mail nupaub@org.usp.br; Prin Officer: Prof Antônio Carlos S Diegues.

Sociedade Brasileira de Paisagismo (Brazilian Landscape Association): CP 19-178, 04599-970 São Paulo, SP; tel (11) 280-4279; fax (11) 280-4462; Prin Officers: Rodolfo Geiser, Paulo del Pichia; research, training courses and conferences in the fields of landscape and agronomy.

Sociedade de Defesa do Pantanal—SODEPAN (Society for the Defence of Marshland): Parque de Exposições Laucidio Coelho, Avda Américo Carlos da Costa 320, 79080-170 Campo Grande, MS; tel (67) 742-1891; fax (67) 742-3388; Pres: Beatriz Diacopulos Rondon; Exec Dir: Joaquim Rocha Azevedo; non-profit organization; conservation of the environment of the Paraguay river basin and tributaries; protection of flora and fauna.

União Protetora do Ambiente Natural—UPAN (Union for Protection of the Natural Environment): Lindolfo Collor 560, CP 189, 93001-970 São Leopoldo, RS; tel (51) 592-7933;

fax (51) 592-6617; Exec Sec: Carlos Cardoso Aveline; Prin Officer: Aline dos Santos; f 1992.

Universidade Federal de Viçosa, Departamento de Engenharia Florestal (Federal University of Viçosa, Forestry Engineering Department): 36571-000 Viçosa, MG; tel (31) 899-2465; fax (31) 899-2478; e-mail apsouza@mail.ufv.br; internet www.ufv.br; Head: Prof Dr Amaury Paulo de Souza; f 1960; forestry production and management; environmental impact assessment; wood and derivatives technology.

BRUNEI

Governmental Organizations

Ministry of Development: Old Airport, Jalan Berakas, Bandar Seri Begawan BB 3510; tel 2241911; e-mail info@mod.gov.bn; internet www.mod.gov.bn; Minister: Dato' Seri Paduka Dr Haji Ahmad bin Haji Jumat; f 1984; infrastructure, housing, construction, town and country planning, national development, management of natural resources, the environment, waste management, air quality, climate change, energy, water management.

Department of Environment, Parks and Recreation: Ministry of Development, Bandar Seri Begawan BB3510; tel 2383222; fax 2383644; e-mail modenv@brunet.bn; internet www.www.jastre.gov.bn; Dir: Haji Mohd Zakaria Haji Sarudin; Senior Environment Officers: Pengiran Shamhary Pengiran Dato' Paduka Haji Mustapha, Martinah Haji Tamit; f 1993; responsible for co-ordinating environmental matters in Brunei; serves as the secretariat to the National Committee on the Environment; maintains the National Environmental Resource Information Centre—NERIC; responsible for the management and maintenance of parks and recreational areas.

Ministry of Health: Jalan Menteri Besar, Bandar Seri Begawan BB 3910; tel 2226640; fax 2240980; e-mail moh2@brunet.bn; internet www.moh.gov.bn; Minister: Pehin Dato' Haji Abul Bakar bin Haji Apong.

Ministry of Industry and Primary Resources: Jalan Menteri Besar, Bandar Seri Begawan BB 39102069; tel 2382822; fax 2382807; e-mail MIPRS2@brunet.bn; internet www.industry.gov.bn; Minister: Pehin Dato' Haji Abdul Rahman Taib.

Department of Agriculture, Research and Development Division: Jalan Menteri Besar, Bandar Seri Begawan 2069; Head of Div: Haji Insanul Bakti; comprises Commodity Units, Crop Production Units, Product Development Units and Scientific Services Units (Soil Science Unit and Analytical Laboratory Service Unit).

Jabatan Perhutanan (Department of Forestry): Bandar Seri Begawan 2067; tel 2222450; fax 2241012.

Jabatan Perikanan (Department of Fisheries): Bandar Seri Begawan 2069; tel 2383068; fax 2383069; e-mail bruneifisheries@brunet.bn; internet www.fisheries.gov.bn; mem of Southeast Asian Fisheries Development Centre; development and management of fisheries and related industries.

Ministry of Culture, Youth and Sports, Museums Department: Simpang 336, Jalan Kebangsaan, Bandar Seri Begawan BC 4415; tel 2244545; fax 2242727; e-mail bmlibry@brunet.bn; internet www.kkbs.gov.bn; Dir: Pengiran Haji Hashim bin Pengiran Haji Mohd Jadid; Librarian: Hajah Fatimah Haji Aji; affiliated to Brunei Museums Dept; museum of history, culture and natural history; member of the World Conservation Union—IUCN, Royal Anthropological Inst of Great Britain, Museum Asscn, Royal Asiatic Soc Malaysian Branch, International Council of Museums, Malayan Historical Soc, Pacific Science Asscn, International Council of Archives, World Crafts Council, Convention on International Trade in Endangered Species of Flora and Fauna—CITES.

Non-Governmental Organizations

Brunei Nature Society: POB 2241, Bandar Seri Begawan 1922 Panaga.

Natural History Society: c/o PTR/3, Brunei Shell Petroleum Co, Seria 7082.

BULGARIA

Governmental Organizations

Ministry of the Environment and Water: 1000 Sofia, William Gladstone St 67; tel (2) 9406222; fax (2) 9862533; e-mail slavv@moew.government.bg; internet www.moew.government.bg; Minister: Dolorès Borisova Arsenova; environmental protection, monitoring and education; nature conservation; sustainable development; maintains regional inspectorates.

Ministry of Agriculture and Forestry: 1040 Sofia, Botev Blvd 55; tel (2) 980-99-27; fax (2) 980-62-56; internet www.mzgar.government.bg; Minister: Mekhmed Mekhmed Dikme; areas of activity include geological mapping and biological agriculture.

Ministry of Energy and Energy Resources: 1040 Sofia, Triadica St 8; tel (2) 987-84-25; e-mail pressall@doe.bg; internet www.doe.org; Minister: Milko Korachev.

Committee on Energy: 1000 Sofia, Blvd Knjaz Dondukov 1; Chair: Konstantin Rosinov; regulatory authority.

N Pushkarov Research Institute of Soil Science: PO Box 1369, Shosse Bankya St 7, 1080 Sofia; tel (2) 24-61-41; fax (2) 24-89-37; e-mail soil@mail.bg; internet www.iss-poushkarov.org; Dir: Prof Dr Sci N V Kolev; Deputy Dirs: Dr B Georgiev, Dr N Dinev; Scientific Sec: Assoc Dr E Filcheva; f 1947; monitors the pollution of soils and underground waters.

National Centre of Hygiene, Medical Ecology and Nutrition: 1431 Sofia, Blvd Dimitur Nestorov 15; tel (2)

958-18-94; fax (2) 958-12-77; e-mail nch@aster.net; Dir: Assoc Prof Nikolay Rizov; assesses the impact of environmental pollution on the health of the population.

Nuclear Regulatory Agency: 1574 Sofia, Blvd Shipchenski Prokhod 69; tel (2) 940-68-00; fax (2) 940-69-19; e-mail mail@bnsa.bas.bg; internet www.bnsa.bas.bg; Chair: Emil Vapitev; f 1957; nuclear safety; radiation protection, radioactive waste management; nuclear material and physical protection; emergency planning and response; legislation; international co-operation.

Non-Governmental Organizations

Asotsiatsiya na Ekolozite ot Obshtinite v Balgaria: (Association of Ecologists in Bulgarian Municipalities): 8000 Bourgas, Christo Botev St 13; tel and fax: (56) 42-044; Contact: Nikolai Sidjimov.

Association of Bulgarian Ecologists—ABECOL: 1000 Sofia, Slaveykov Pl 7; tel (2) 81-72-22; fax (2) 88-53-49; Pres: Prof Dr Simeon Nedialkov; f 1990; research and education in all aspects of ecological science; environmental impact assessment; landscape planning; pollution; waste management; forestry; sustainable agriculture; urban ecology; eco-recreation; federation of regional ABECOL societies.

Balgarski klub za ekologichna kultura (Bulgarian Club for Ecological Culture): 1504 Sofia, Krakra 24; tel (2) 43-21-63.

Balgarsko Prirodoizpitatelno Druzhestvo (Bulgarian Society for Nature Research): 1164 Sofia, Dragan Tsankov St 8; tel (2) 71-07-41; e-mail butsa13@hotmail.com; Pres: Dr Dimiter Vodenicharov; f 1896.

Bulgarian Society for the Protection of Birds—BSPB/BirdLife Bulgaria: 1111 Sofia, Yavorov Complex 71, 4th Floor, Apt 1; tel (2) 971-58-55; fax (2) 971-58-56; e-mail bspb_hq@bspb.org; internet www.bspb.org; Devt Dir: Sylvia Andonova; Conservation Dir: Nikolay Petkov; Exec Dir: Boris Barov; Natura 2000 and IBA Officer: Irina Kostadinova; f 1988; non-profit organization dedicated to the preservation of biodiversity through field projects, research programmes on species and habitats, and management of species populations and protected sites; advocacy and policy-related activities; education and public awareness; ecotourism and sustainable development; affiliated with BirdLife International.

Bulgarian Society of Natural Research: 1421 Sofia, Blvd Dragan Tsankov 8, POB 1136; tel (2) 66-65-94; Chair: Prof Dr Vassil Golemanski; f 1896; society of natural scientists; promotes the study and conservation of the environment; 20 regional brs and 5 scientific sections, including one on ecology.

Central Laboratory of General Ecology, Bulgarian Academy of Sciences: 1113 Sofia, Gagarin St 2; tel (2) 73-61-37; fax (2) 70-54-98; e-mail ecolab@ecolab.bas.bg; internet www.ecolab.bas.bg; Dir: Assoc Prof Georgi Hiebaum; f 1956; research and education in the fields of theoretical and applied ecology and environmental impact assessment; main research subjects include interrelationships and processes in ecosystems, fluxes of substances and energy and their transformations in marine, freshwater and terrestrial ecosystems, the carbon cycle, cycles of nutrients, the fate of toxic substances in ecosystems, natural diversity in biological systems, the functional role of ecosystems, community ecology of parasitic and free-living organisms, the biodiversity of parasitic worms and terrestrial invertebrates in natural ecosystems and urban areas, biodiversity protection, cytogenetic monitoring of mutagenic impact, aquatic monitoring.

Central Laboratory of Solar Energy and New Energy Sources: 1784 Sofia, Blvd Tzarigradsko chaussée 72; tel and fax (2) 75-40-16; e-mail solar@phys.bas.bg; Dir: Assoc Prof Petko Vitanov; research into alternative energy production.

Chernomorski Ekologichen Informatsionen Tsentar (Black Sea Environmental Information Center): 9002 Varna, Bratya Miladinovi St 68g, Apt 23; tel (52) 23-09-57; e-mail banchev@ms3.tu-varna.acad.bg; Contact: Roumyana Sabcheva Peteva.

Conservative Ecological Party: c/o 1000 Sofia, Rakovski Blvd 134; tel (2) 88-25-01; Chair: Khristo Bisserov; political party; member of the Sayuz na Demokratichni Sili—SDS (Union of Democratic Forces) alliance.

Druzhestvo Ekoenergetika (Ecoenergetic Society): 1000 Sofia, Blvd Dondukov 9; tel (2) 80-23-23; fax (2) 88-15-30; e-mail ecogl@bulnet.bg; Contact: Dimo Stoilov.

ECO-CLUB 2000: 1124 Sofia, Svetoslav Terter St 27; tel and fax (2) 46-60-62; e-mail ecoc2000@nat.bg; internet www.bulgaria.com/ecoc2000; Contact: Dimitur Vassilev.

'Ecofarm' Association for Biological Agriculture: 4000 Plovdiv, 12 Mendeleev St; tel (32) 26-59-09; fax (32) 26-59-03; e-mail vpopov@au-plovdiv.bg; Prin Officers: Stoitcho Peev Karov, Vladislav Haralamdiev Popov, Plamen Pavlov Paraskevov; f 1996; agricultural and environmental preservation; consultancy and support for farmers involved in developing organic and sustainable agriculture; promotion of organic agriculture and environmentally sound agricultural practices at a national level; demonstrations and training for farmers, researchers, agronomists and students of organic agriculture; affiliated to the Agroecological Centre and Higher Inst of Agriculture.

ECOS Foundation: 1000 Sofia; tel (2) 714-33-71; fax (2) 87-24-00; Dir-Gen: Ognian Champoev; educational foundation; aims to promote ecological awareness.

Ekologichno Dvizhenie Chista Priroda (Clean Nature Environmental Movement): 5800 Pleven, Storgozia Res Area, Block 65, Apt 19; tel (64) 67-625; Contact: Plamen Lakov.

Fondatsiya TAYM—Ecoproekti (TIME—This Is My Environment Eco-Projects Foundation): 1504 Sofia, Yanko Sakazov St 58; tel (2) 44-32-36; e-mail time@mbox.digsys.bg; Contact: Veleslava Tsakova.

Green Society Foundation: 1000 Sofia; tel and fax (2) 87-24-21; Pres: Assoc Prof Petur Gulubov; Exec Dir: Radiana Stanoeva; f 1991; co-operates in projects and campaigns with other groups; operates seven environmental programmes.

Informatsionen i Ucheben Centar po Ekologija (Center for Environmental Information and Education): 1309 Sofia, Indzhe Voivoda St, Block 9, Entrance 3, 2nd Floor; tel (2) 920-13-41; e-mail ceie@bulnet.bg; Contact: Petko Kovachev.

Institut za Gorata (Forestry Research Institute, Bulgarian Academy of Sciences): 1756 Sofia, Blvd St Kliment Ohridski 132; tel (2) 62-29-61; fax (2) 62-29-65; e-mail forestin@bulnet.bg; Dir: Prof Dr Ivan Raev; f 1928; forestry; forest inventory; forest soil science and microbiology; ecology and environmental conservation; forest conservation; forest genetics and tree-breeding; tree physiology; wildlife management; forest harvesting; library of 39,000 vols; publications.

Institute of Botany, Bulgarian Academy of Sciences: 1113 Sofia, Acad Georgi Bonchev St 23; tel (2) 72-09-51; fax (2) 71-90-32; e-mail botinst@iph.bio.bas.bg; Dir: Prof Dr Emanel Palamarev; f 1889; plant taxonomy, monitoring and conservation of rare and threatened plant species; assessment of plant resources; palaeobotany.

Institute of Oceanology, Bulgarian Academy of Sciences: 9000 Varna, Kv Asparuhovo St 40; tel (52) 77-20-38; fax (52) 77-42-56; e-mail office@io.bas.bg; Dir: Prof Asen Konsulov; research in the fields of marine geology, hydrology, chemistry, biology, ecology, geophysics, physical oceanography, coastal management and underwater technology, with emphasis on the Bulgarian Black Sea coastline and shelf.

Institute of Water Problems, Bulgarian Academy of Sciences: 1113 Sofia, Acad Georgi Bonchev St 1; tel (2) 72-25-72; fax (2) 72-25-77; e-mail santur@bgcict.acad.bg; internet www.iwp.bas.bg; Dir: Prof Dr Ohanes Santurjian; f 1963; rational planning, construction and operation of water resource systems; rehabilitation and maintenance of inland waters.

Institute of Zoology, Bulgarian Academy of Sciences: 1000 Sofia, 132 'Kl. Okhridski' Blvd.; tel (2) 88-31-63; fax (2) 88-28-97; e-mail zoology@bgcict.acad.bg; Dir: Prof Dr Vassil Grigorov Golemansky; f 1947; morphology, taxonomy, ecology and evolution of Bulgarian and Palaearctic fauna; experimental zoology and protozoology; biological monitoring and preparation of the *Red Data* series on animal and bird life in Bulgaria; publication of journals.

Mezhdunarodna Fondatsiya Ekoselishta (Eco-Settlements Foundation): 1000 Sofia, Neofit Rilski St 60; tel (2) 89-90-42; Contact: Stoyan Shivarov.

Mladezhka Ekologichna Morska Akademiya (Youth Environmental Marine Academy): 9009 Varna, Mladost, Block 102-2-1-1; tel (52) 44-70-03; Contact: Yordan Radenkov.

National Ecoglasnost Movement: 1000 Sofia, Blvd Knjaz Dondukov 39, 4th Floor, Rm 45; tel (2) 986-22-21; fax (2) 88-15-30; e-mail ekogl@bulnet.bg; Pres: Ivan Sungarsky; Sec: Hachadur Chuldjian; f 1989; campaigns against nuclear power and pollution; promotion of human rights and democracy; protection of nature in the Black Sea and the River Danube; sustainable development; renewable energy; recycling; waste management; has a political wing, the Ecoglasnost Political Club, which contests presidential and general elections in alliance with leftist parties; member of Friends of the Earth International.

National Institute of Meteorology and Hydrology, Bulgarian Academy of Sciences: 1184 Sofia, Carigradsko shosse 66; tel (2) 975-39-96; fax (2) 88-44-94; Dir: Assoc Prof V Sharov; monitors air, soil and water pollution, and radioactivity.

Natsionalen Ekologichen Klub (National Ecological Club): 1113 Sofia, Akad G Bonchev St, Block 29[a], na Ban; tel (2) 70-52-25; Prin Officer: Prof Vassil Sgurev; independent public organization; environmental conservation; co-operation with government agencies and NGOs; ecological education.

Priroden Fond (Wilderness Fund): 1113 Sofia, Gagarin St 2; fax (2) 70-54-98; Chair: Jeko Spirodonov; f 1990; assesses the best places to establish reserves for the protection of the natural environment.

Sdruzhenie EKOFORUM (ECO-FORUM Association): 1113 Sofia, Gagarin St 2; tel (2) 70-53-79; fax (2) 55-10-67; e-mail radev@mgv.bg; Gen Sec: Assoc Prof Pavel Georgiev; organization of scientists, entrepreneurs and public workers committed to promoting sustainable development and conservation, particularly in the transition to a market economy.

Sdruzhenie Priroda na Zaem (Borrowed Nature Association): 1301 Sofia, Blvd A Stamboliyski 20b; tel (2) 80-95-58; e-mail bornat@sf.cit.bg; Contact: Kliment Mindjov.

Sdruzhenie za Ekologichno Obuchenie i Upravlenie (Environmental Management Training Centre): 1000 Sofia, P Parchevich St 4; tel (2) 51-80-49; e-mail dora.iordanova@bulmail.sprint.com; Contact: Veselina Stoyanova.

SREDETSC (Bulgarian Red Cross Youth—BRCY): 1527 Sofia, Blvd Dondukov 61; tel (2) 65-61-85; fax (2) 65-61-39; e-mail youth@redcross.bg; internet www.redcross.bg; Pres: Anton Borissov; Youth Dir: Anton Valev; f 1921; part of Bulgarian Red Cross—BRC and International Federation of Red Cross and Red Crescent Societies—IFRC; training and activities in the fields of first aid, healthy living, leadership, ecology and sustainable development, social relief, disaster preparedness, disaster response and international humanitarian law.

Union for Nature Protection: 1000 Sofia; tel (2) 83-26-72; Chair: Assoc Prof Dr Svetoslav Gerasimov; f 1928 as the Union for Native Nature Protection; restructured 1991; independent public organization with local brs and international contacts.

United Front for Ecological Saving of Kardjali: 6600 Kardjali, Republikanska St 30; tel (361) 256-42; fax (361) 256-42; Pres: Dimo Bodurov; Sec: Todor Dimitrov; f 1989; protection of the people and environment of Kardjali; co-operation with the Citizen Action Committee for the Environment in developing solutions to city environmental problems.

BURKINA FASO

Governmental Organizations

Ministry of Agriculture, Water Resources and Fisheries: 03 BP 7005, Ouagadougou; tel 32-41-14; fax 31-08-70; e-mail ministere-agriculture@cenatrin.bf; internet www.agriculture.gov.bf; Minister of State: Salif Diallo.

Direction Generale de la Preservation et la Conservation de la Nature et de l'Environnement—DGPE (Directorate-General for the Preservation and Conservation of Nature and the Environment): 03 BP 7044, Ouagadougou; tel 50-30-63-97; fax 50-31-81-34; Dir: Bruno R Salo; serves as the INFOTERRA national focal point.

Ministry of Animal Resources: 03 BP 7026, Ouagadougou 03; tel 32-61-07; fax 31-84-75; internet www.mra.gov.bf; Minister: Dofinwiya Alphonse Bonou; includes Plant Protection Service.

Ministry of the Environment and Quality of Life: 565 rue Neto, Koulouba, 03 BP 7044, Ouagadougou 03; tel 32-40-74; fax 30-70-39; e-mail messp@liptinfor.bf; internet www.environnement.gov.bf; Minister: Laurent Gouindé Sedogo.

Centre National de la Recherche Scientifique et Technologique: BP 7047, Ouagadougou; tel 50-33-23-94; fax 50-30-50-03; Dir-Gen: Michel P Sedogo; f 1950; conducts basic and applied research in natural sciences, agriculture and energy; library of 20,000 vols; publications.

Non-Governmental Organizations

Association Burkinabé d'Action Communautaire/Groupe Energies Renouvelables et Environnement—ABAC-GERES: (Burkinabé Community Action Association/Renewable Energy and Environment Group): 01 BP 4071,

Ouagadougou; tel 50-36-26-30; fax 50-36-02-18; e-mail geres@fasonet.bf; federation of women's organizations, urban and rural associations and those active in crafts and light industry.

Centre d'Etudes Economiques et Sociales d'Afrique Occidentale (West African Centre for Economic and Social Studies): BP 305, Bobo-Dioulasso 01; tel 20-97-10-17; fax 20-97-08-02; e-mail cesau.bobo@fasonet.bf; Dir: Rosalie Ouoba; f 1960; areas of study include the development of rural communities, land administration and the environment.

Centre Ecologique Albert Schweitzer—CEAS (Albert Schweitzer Ecological Centre): 01 BP 3306, Ouagadougou; tel 50-30-23-65; fax 50-34-10-65; e-mail ceas-rb@fasonet.bf; appropriate technology for agriculture and the food industry.

CIRAD—Forêt: BP 1759, Ouagadougou 01; tel 50-33-40-98; fax 50-30-76-17; Head of Mission: Denis Depommier; f 1963; research in silviculture, agroforestry and tree-breeding.

Institut de Recherche pour le Développement—IRD, Ouagadougou (Development Research Institute, Ouagadougou): BP 182, Ouagadougou 01; tel 50-30-67-37; fax 50-31-03-85; internet www.ird.bf; Dir: Georges Grandin; formerly Centre ORSTOM; a regional centre of the Institute, which is based in France; fields of research include hydrology, geography, agronomy, botany and ecology.

Institut de Recherche pour le Développement—IRD, Bobo-Dioulasso (Development Research Institute, Bobo-Dioulasso): BP 171, Bobo-Dioulasso 01; tel 20-97-12-69; fax 20-97-09-42; Dir: J L Devineau; f 1947; formerly Centre ORSTOM; research areas include medical entomology, human geography, ecology, malariology; library of 1,200 vols.

Institut d'Etudes et de Recherches Agricoles: BP 7192, Ouagadougou 03; tel 50-34-02-69; fax 50-34-02-71; Dir: Célestin P Belem; f 1978.

IUCN Burkina Faso Country Office: 01 BP 3133, Ouagadougou 01; tel 50-31-31-54; fax 50-30-31-51; e-mail uicnbf@fasonet.nf; Contact: Sapré Laurent Millogo; national office of the World Conservation Union—IUCN.

BURUNDI

Governmental Organizations

Ministry of Land Management, the Environment and Tourism: Bujumbura; Minister: Albert Mbonerane.

Ministry of Agriculture and Livestock: Bujumbura; tel 222087; Minister: Pierre Nlikumagenge; includes Department of Water Resources and Forests.

Ministry of Development, Planning and Reconstruction: BP 1830, Bujumbura; tel 223988; Minister: Séraphine Wakana.

Institut National pour la Conservation de la Nature (National Institute for Nature Conservation): Présidence de la République, BP 938, Bujumbura; tel (2) 6063.

Institut National pour l'Environnement et la Conservation de la Nature—INECN (National Institute for the Environment and Nature Conservation): BP 56, Gitega; tel 402071; fax 402075; Dir-Gen: Dr Gaspard Bikwemu; responsibilities include management of national parks and nature reserves; serves as the INFOTERRA national focal point.

Non-Governmental Organizations

Institut des Sciences Agronomiques du Burundi (Agronomic Science Institute of Burundi): BP 795, Bujumbura; tel 223390; fax 225798; Dir-Gen: Dr Jean Ndikurana; f 1962; agronomic research and farm management.

CAMBODIA

Governmental Organizations

Ministry of Environment: 48 blvd Sihanouk, Tonle Bassac, Khan Chamkarmon, Phnom-Penh; tel (23) 427894; fax (23) 427844; e-mail moe-cabinet@camnet.com.kh; internet www.moe.gov.kh; Minister: Dr Mok Mareth; Sec of State: To Gary; Chief of Office: Roath Sith; the Office of Research and Editing in the Department of Education and Communication serves as the INFOTERRA national focal point.

National Environmental Action Plan Secretariat—NEAP: 48 blvd Sihanouk, Tonlo Bassac, Chamkamon, Phnom-Penh; tel (23) 721074; Co-ordinator: Pum Vicheth; f 1996; funded by the World Bank; integrates environmental considerations into national development, defines strategies, co-ordinates financing, conducts participatory national environmental planning; five main areas of activity are natural resource management of the Tonle Sap, protected areas management, energy development, urban waste management and coastal fisheries management.

Ministry of Agriculture, Forestry, Hunting and Fisheries: 200 blvd Norodom, Phnom-Penh; tel (23) 211351; fax (23) 217320; e-mail infor@maff.gov.kh; internet www.maff.gov.kh; Minister: Chan Sarun.

Ministry of Water Resources and Meteorology: 47 blvd Norodom, Phnom-Penh; tel (23) 724289; fax (23) 426345; e-mail mowram@cambodia.gov.kh; internet www.mowram.gov.kh; Minister: Lim Kean Hor.

Non-Governmental Organizations

Culture and Environment Preservation Association—CEPA: 13 St. 122, Khan Toul Kork, Phnom Penh; tel (23) 361097; Pres: Vann Piseth; f 1996; fisheries studies, reforestation, mangroves.

IUCN Cambodia: c/o UNESCO, 38 blvd Samdech Sothearos, BP 29, Phnom-Penh; tel (23) 26945; fax (23) 26163; national office of the World Conservation Union—IUCN.

CAMEROON

Governmental Organizations

Ministry of the Environment and Forestry: BP 1341, Yaoundé; tel 220-42-58; fax 222-94-87; e-mail onadef@camnet.cm; internet www.camnet.cm; Minister: Tanyi Mbianyor Oben.

Department of Wildlife and Protected Areas: c/o Central Post Office, Yaoundé; nature conservation; wildlife management.

Ecole pour la Formation de Spécialistes de Faune—Ecole de Faune Garoua (School for the Training of Specialists in Wildlife Management—Garoua Wildlife School): BP 271, Garoua; tel 27-11-25; fax 27-31-35; Prin: Ibrahim Soare Njoya; training in wildlife management for French-speaking African states; research and consultancy on environmental issues.

Office National de Développement des Forêts—ONADEF (National Office for the Development of Forests): BP 1341, Yaoundé; tel 20-42-58; fax 21-53-50; e-mail onadef@camnet.cm; internet www.camnet.cm/investir/envforet/index.htm; Dir-Gen: Jean Williams Solo; Deputy Dir-Gen: Georges Ngoue Mbangamina; f 1981; management of forest resources.

Ministry of Agriculture: Quartier Administratif, Yaoundé; tel 222-05-33; fax 222-50-91; Minister: Zacharie Perevet; Sec of State: Aboubakary Abdoulaye.

Ministry of Livestock, Fisheries and Animal Husbandry: c/o Central Post Office, Yaoundé; tel 222-33-11; Minister: Dr Hamadjoda Adjoudji.

Ministry of Mines, Water Resources and Energy: Quartier Administratif, BP 955, Yaoundé; tel 23-34-00; fax 222-61-77; e-mail minmee@camnet.cm; internet www.camnet.cm/investir/minmee; Minister: Dr Yves Mbele.

Institut de la Recherche Agricole pour le Développement—IRAD (Institute of Agricultural Research for Development): BP 2123, Yaoundé; tel 23-26-44; fax 23-35-38; Dir: Dr J A Ayuk-Takem; f 1996; affiliated to the Ministry of Higher Education; research areas include agriculture, agronomy and botany; publications; agricultural research destined for development in both rural and urban areas.

Institut des Recherches Zootechniques at Vétérinaires—IRZV (Institute for Zoological and Veterinary Research): BP 1457, Yaoundé; tel and fax 23-24-86; Dir: Dr John Tanlaka Banser; f 1974; affiliated to the Ministry of Higher Education; research on livestock, fisheries, wildlife and the environment.

Non-Governmental Organizations

Association des Clubs des Amis de la Nature du Cameroun (Association of Friends of Nature Clubs of Cameroon): BP 11261, Yaoundé; tel and fax 21-42-34; Co-ordinator: Patris Tapchem; Consultant: Célestin Modeste Mbomba; f 1985; research, and collection and dissemination of information and documentation relevant to the environment and forests; training in environmental and forest protection for local officers; rural awareness-raising; affiliated with the Centre d'Etude, de Recherche et de Documentation en Droit International et pour l'Environnement—CERDIE and the Fédération des Organisations Non-Gouvernementales pour l'Environnement au Cameroun—FONGEC.

Fédération des Organisations Non-Gouvernementales pour l'Environnement au Cameroun—FONGEC (Federation of Non-Governmental Organizations for the Environment in Cameroon): Yaoundé; Perm Sec: Flaubert Djateng.

Institut de Recherche pour le Développement—IRD (Development Research Institute): BP 1857, Yaoundé; tel 20-15-08; fax 20-18-54; Rep: Alain Valette; a regional centre of the Institute, which is based in France; research areas include sedimentology, hydrology, ecology, demography, geography and ornithology.

International Centre for Research in Agroforestry—ICRAF, Cameroon Office: IRA/ICRAF Project, BP 2067, Yaoundé; tel 23-75-60; fax 23-74-40; Regional Co-ordinator: Dr Bahine Duguma; f 1987.

IUCN Cameroon: BP 284, Maroua; tel 29-22-69; fax 29-22-68; Project Leader: Michael Allen; national office of the World Conservation Union—IUCN; manages a project for the conservation and development of the Waza-Logone region.

World Wide Fund for Nature—WWF Cameroon: BP 6776, Yaoundé; tel 21-42-41; fax 21-42-40; e-mail lsome@wwf.carpo.org; internet www.panda.org; Regional Rep: Laurent Somé; national affiliate organization of World Wide Fund for Nature—WWF International.

CANADA

Governmental Organizations

Environment Canada: 70 Crémazie St, Gatineau, QC K1A 0H3; tel (819) 997-2800; fax (819) 953-2225; e-mail enviroinfo@ec.ga.ca; internet www.ec.ga.ca; Minister: Stéphane Dion; federal institution with ministerial status.

Atmospheric Environment Service—AES: 4905 Dufferin St, Downsview, ON M3H 5T4; tel (416) 739-4392; fax (416) 739-5700; e-mail robert.crawford@ec.gc.ca; Data Man: Bob Crawford; f 1971; weather forecasting; meteorological research; data collection and archiving; policy development; climate and air quality services.

Atmospheric and Climate Science Directorate—Meteorological Service of Canada: 4905 Dufferin St, Downsview, ON M3H 5T4; tel (416) 739-4588; fax (416) 739-4265; e-mail michel.beland@ec.gc.ca; Dir-Gen: Dr Michael Béland; Dir, Climate Research: Dr Doug Whelpdale; Dir, Air Quality Research: Dr Keith Puckett; Dir, Meteorological Research: Jim Abraham; f 1871; the Atmospheric and Climate Science Directorate (ACSD) conducts research in climate, meteorology, air quality, and environmental impact and adaptation. It produces science assessments on pressing environmental issues (such as climate change, acid rain, and the depletion of the ozone layer, etc) for Canadians and government policy makers.

Canadian Association on Water Quality—CAWQ/ ACQE: 867 Lakeshore Rd, Burlington, ON L7R 4A6; tel (905) 336-6291; fax (905) 336-4420; e-mail suzanne.ponton@ cciw.ca; internet www.cawq.ca; Pres: Yves Comeau; f 1967; formerly Canadian Association on Water Pollution Research and Control; research on the scientific, technological, legal and administrative aspects of water quality and water pollution control; exchange of information; practical application of such research for public benefit; sponsors three annual regional symposia on water pollution research; publications; affiliated with the International Water Asscn.

Canadian Wildlife Service—CWS: Vincent Massey Sq, 17th Floor, Ottawa, ON K1A 0H3; tel (819) 997-1095; fax (819) 997-2756; e-mail cws-scf@ec.gc.ca; internet www.cws -scf.ec.gc.ca/cwshom_e.html; protection and management of migratory birds, nationally significant habitats and endangered species; biological research; studies on the socio-economic importance of wildlife.

Atlantic Regional Library: Canadian Wildlife Service, 17 Waterfowl Lane, POB 1590, Sackville, NB E0A 3C0; tel (506) 364-5019; fax (506) 364-5062; e-mail jean.sealy@ec .gc.ca; Librarian: Jean Sealy.

Québec Region: 1141 route de l'Eglise, CP 10100, Sainte-Foy, PQ G1V 4H5; tel (418) 648-7225.

National Laboratory for Environmental Testing: National Water Research Institute, Canada Centre for Inland Waters, 867 Lakeshore Rd, POB 5050, Burlington, ON L7R 4A6; tel (905) 336-4648; fax (905) 336-6404; e-mail robin.sampson@cciw.ca; internet www.cciw.ca/nwrie/nlet/ intro.html; Dir: R C J Sampson; f 1983; analytical, scientific and advisory services; quality management programme; technology advancement and transfer, focusing on automation in sampling and analysis; liaison with academic and industrial laboratories.

National Water Research Institute: Environment Canada, 867 Lakeshore Rd, POB 5050, Burlington, ON L7R 4A6; e-mail nwriscience.liaison@ec.gc.ca; internet www .nwri.ca; Dir-Gen: Dr John H. Carey; Dir, Science Liaison Branch: Dr Alex T Bielak; The National Water Research Institute (NWRI) is a Directorate of Environment Canada's Environmental Conservation Service. The Institute is Canada's largest freshwater research facility with over 300 staff including aquatic ecologists, hydrologists, toxicologists, physical geographers, modellers, limnologists, environmental chemists, research technicians, and experts in linking water science to environmental policy. NWRI has two main centres: the larger at the Canada Centre for Inland Waters on the shores of the Great Lakes in Burlington, Ontario; the other at the National Hydrology Research Centre, in the heart of the Canadian Prairies in Saskatoon, Saskatchewan. NWRI also has staff located in Gatineau, Québec; Fredericton, New Brunswick; and Victoria, British Columbia; working with other government departments, universities and research organizations to address a variety of water-related issues.

National Water Research Institute, National Hydrology Research Centre: 11 Innovation Blvd, Saskatoon, SK S7N 3H5; tel (306) 975-5717; fax (306) 975-5143; Dir: F J Wrona; research into climate and hydrology, land–atmospheric processes, the northern environment, nutrient impacts, ecosystem rehabilitation, and ground-water and contaminants; interdisciplinary research with the University of Saskatchewan and other national and international groups; Scientific Information Section disseminates information nationally and internationally.

Agriculture and Agri-Food Canada: Sir John Carling Bldg, 930 Carling Ave, Ottawa, ON K1A 0C7; tel (613) 759-1000; fax (613) 759-6726; e-mail info@agr.gc.ca; internet www .agr.gc.ca; Minister: Andrew Mitchell; federal institution with ministerial status.

Fisheries and Oceans Canada: 200 Kent St, 13th Floor, Station 13228, Ottawa, ON K1A 0E6; tel (613) 993-0999; fax (613) 990-1866; e-mail info@dfo-mpo.gc.ca; internet www .dfo-mpo.gc.ca; Minister: Geoff Regan; federal institution with ministerial status; conservation, development and sustainable economic use of fisheries resources.

Natural Resources Canada: 580 Booth St, Ottawa, ON K1A 0E4; tel (613) 995-0947; fax (613) 996-9094; e-mail webmaster@nrcan.gc.ca; internet www.nrcan-rncan.gc.ca; Minister: Ruben John Efford; federal institution with ministerial status.

Alberta Agriculture, Food and Rural Development: 7000 113th St, Edmonton, AB T6H 5T6; tel (403) 427-2127; fax (403) 422-2861; internet www.agric.gov.ab.ca; Minister: Shirley McClellan; provincial institution with ministerial status.

Alberta Environment: Petroleum Plaza, South Tower, 10th Floor, 9915, 108th St, Edmonton, AB T5K 2G8; tel (403) 427-2700; fax (403) 422-4086; e-mail env.infocent@gov.ab.ca; internet www3.gov.ab.ca/env; Minister of the Environment: Lorne Taylor; provincial institution with ministerial status.

British Columbia Ministry of Agriculture, Food and Fisheries: POB 9058, Station Provincial Government, Victoria, BC V8W 9E2; tel (250) 387-1023; fax (250) 387-1522; e-mail agf.webmaster@gems2.gov.bc.ca; internet www.agf .gov.bc.ca; Minister: John van Dongen; provincial ministry.

British Columbia Ministry of Forests: POB 9529, Station Provincial Government, Victoria, BC V8W 9C3; tel (250) 387-6240; fax (250) 387-1040; internet www.gov.bc.ca/for; Minister: Michael de Jong; provincial ministry.

British Columbia Ministry of Sustainable Resource Management: POB 9352, Station Provincial Government, Victoria, BC V8W 9M2; tel (250) 356-9076; fax (250) 356-8273; e-mail srm.minister@gems6.gov.bc.ca; internet www .gov.bc.ca/srm; Minister of Water, Land and Air Protection: Bill Barisoff.

Canadian Council of Ministers of the Environment—CCME: 123 Main St, Suite 360, Winnipeg, MB R3C 1A3; tel (204) 948-2090; fax (204) 948-2125; internet www.mbnet .mb.ca/ccme; Pres, Cncl of Ministers: Paul Bégin; Dir-Gen: Peter André Globensky; Chair, Man Cttee: Diane Jean; f 1964; Canadian intergovernmental forum meeting twice a year for discussion and joint action on environmental issues of national and international concern; 14 environment ministers from federal, provincial and territorial governments; two steering committees advise the Council and co-ordinate projects assigned to intergovernmental task groups; projects cover issues including air quality, packaging, waste management and integration of environment and economy.

Canadian Museum of Nature: PO 3443, Station D, Ottawa, ON K1P 6P4; tel (800) 263-4433; fax (613) 364-4020; e-mail enquiries@mus-nature.ca; internet www.nature.ca; Pres: Joanne DiCosimo; f 1842; the national natural science museum with departments of zoology, botany and earth sciences; education; includes a library and archives.

Institute for Chemical Process and Environmental Technology, National Research Council Canada: 1200 Montréal Rd, Bldg M-12, Ottawa, ON K1A 0R6; tel (613) 993-4041; fax (613) 957-8231; e-mail don.singleton@nrc-cnrc .gc.ca; internet www.icpet-iptce.nrc-cnrc.gc.ca; Dir-Gen: Dr Donald L Singleton; Program Dir: Dr Yves Delandes; Dir of

Commercialization: Kevin Jonasson; f 1916; performing innovative research and development in responsible chemistry technologies for sustainable development; the institute works closely with industry partners in sectors where chemistry is a sgnificant component of the value chain, focusing on key technology application areas: fuel cells, oil sands and bio-products.

Manitoba Agriculture, Food and Rural Initiatives: 165 Legislative Bldg, 450 Broadway Ave, Winnipeg, MB R3C 0V8; tel (204) 945-3722; fax (204) 945-3470; e-mail minagr@leg.gov.mb.ca; internet www.gov.mb.ca/agriculture; Minister: Rosann Wowchuk; provincial institution with ministerial status.

Manitoba Conservation: 200 Saulteaux Cresc., Winnipeg, MB R3J 3W3; tel (204) 945-6784; e-mail mincon@leg.gov.mb.ca; internet www.gov.mb.ca/natres; Minister: Stan Struthers; provincial institution with ministerial status; replaced Manitoba Environment and Manitoba Natural Resources 1999.

National Round Table on the Environment and the Economy/Table Ronde Nationale sur l'Environnement et l'Economie—NRTEE/TRNEE: Canada Bldg, Suite 200, 344 Slater St, Ottawa, ON K1R 7Y3; tel (613) 947-7287; fax (613) 992-7385; e-mail admin@nrtee-trnee.ca; internet www.nrtee-trnee.ca; Chair: Dr Stuart Lyon Smith; Exec Dir and CEO: David McGuinty; Information Services Officer: Edwin Smith; independent federal agency; forum for discussion of ways to incorporate environmental considerations in economic policy and practice; reports to the Prime Minister of Canada.

New Brunswick Department of Agriculture, Fisheries and Aquaculture: Agricultural Research Station, 850 Lincoln Rd, POB 6000, Fredericton, NB E3B 5H1; tel (506) 453-2666; fax (506) 453-7170; e-mail jean-francois.pelletier@gnb.ca; internet www.gnb.ca/0027/index-e.asp; Minister: David Alward; provincial institution with ministerial status.

New Brunswick Department of Natural Resources and Energy: Hugh John Flemming Forestry Complex, 1350 Regent St, POB 6000, Fredericton, NB E3B 5H1; tel (506) 453-2207; fax (506) 444-5839; e-mail brent.roy@gnb.ca; internet www.gnb.ca/0078/index-e.htm; Minister: Keith Ashfield; provincial institution with ministerial status.

New Brunswick Department of the Environment and Local Government: Marysville Pl., 20McGloin St, POB 6000, Fredericton, NB E3B 5H1; tel (506) 453-2690; fax (506) 457-4991; e-mail jason.humphrey@gnb.ca; internet www.gnb.ca/0009/index-e.asp; Minister: Brenda Fowlie; provincial institution with ministerial status.

Newfoundland and Labrador Department of Environment and Conservation: Confederation Bldg, West Block, 4th Floor, POB 8700, St John's, NL A1B 4J6; tel (709) 729-2664; fax (709) 729-6639; internet www.gov.nl.ca/env; Minister: Tom Osborne; provincial institution with ministerial status.

Newfoundland and Labrador Department of Fisheries and Aquaculture: Petten Bldg, 30 Strawberry Marsh Rd, POB 8700, St John's, NL A1B 4J6; tel (709) 729-3723; fax (709) 729-6082; internet www.gov.nl.ca/fishaq; Minister: Trevor Taylor; provincial institution with ministerial status.

Newfoundland and Labrador Department of Natural Resources: Natural Resources Bldg, 50 Elizabeth Ave, POB 8700, St John's, NL A1B 4J6; tel (709) 729-4715; fax (709) 729-2076; internet www.gov.nl.ca/nr; Minister: Ed Byrne; provincial institution with ministerial status.

Northwest Territories Department of Resources, Wildlife and Economic Development: Scotia Centre Bldg, Suite 600, 5102, 50th Ave, POB 1320, Yellowknife, NT X1A 2L9; tel (867) 669-2388; fax (867) 873-0388; e-mail brendan_bell@gov.nt.ca; internet www.gov.nt.ca/rwed/default.htm; Minister: Brendan Bell; territorial institution with ministerial status.

Nova Scotia Department of Agriculture and Fisheries: 5151 George St, POB 2223, Halifax, NS B3J 3C4; tel (902) 424-4560; fax (902) 424-4671; e-mail dentrca@gov.ns.ca; internet www.gov.ns.ca.nsaf; Minister: Chris A d'Entremont; provincial institution with ministerial status.

Nova Scotia Department of Fisheries and Aquaculture: 5151 George St, Halifax, NS B3J 1M5; tel (902) 424-4560; fax (902) 424-4671; e-mail groveske@gov.ns.ca; internet www.gov.ns.ca/fish; Minister: Ernest L Fage; provincial institution with ministerial status.

Nova Scotia Department of Natural Resources: Founders Sq, 1701 Hollis St, POB 698, Halifax, NS B3J 2T9; tel (902) 424-8633; fax (902) 424-7735; e-mail dnrweb@gov.ns.ca; internet www.gov.ns.ca/natr; Minister: Richard Hurlburt; provincial institution with ministerial status.

Nova Scotia Department of the Environment and Labour: 5151 Terminal Rd, 5th Floor, POB 697, Halifax, NS B3J 2T8; tel (902) 424-5300; fax (902) 424-0503; e-mail webster@gov.ns.ca/enla; internet www.gov.ns.ca/enla; Minister: Kerry Morash; provincial institution with ministerial status.

Nova Scotia Environmental Assessment Board: POB 2107, Halifax, NS B3J 3B7; tel (902) 424-6387; fax (902) 424-5376; e-mail hlfxterm.envi.nicholsl@gov.ns.ca; Chair: William H Charles; Dir: Shirley L Nicholson; advisory body reporting to the Nova Scotia Minister of the Environment; public consultation review process; alternate dispute resolution; environmental awards programme.

Nunavut Department of Sustainable Development: Box 1000, Station 204, Iqaluit, Nunavut Territory, NU X0A 0H0; tel (867) 975-5925; fax (867) 975-5980; e-mail dsd@gov.nu.ca; internet www.gov.nu.ca/sd.htm; Minister of the Environment: Olayuk Akesuk; Deputy Minister: Katherine Trumper; Asst Deputy Minister: Peter Ittinuar; territorial institution with ministerial status.

Ontario Ministry of Agriculture and Food: 1 Stone Rd W, Guelph, ON N1G 4Y2; tel (519) 826-3100; e-mail aboutomafra@omaf.gov.on.ca; internet www.gov.on.ca/omafra; Minister: Steve Peters; provincial ministry.

Ontario Ministry of Natural Resources: Whitney Block, Rm 6630, 99 Wellesley St W, Toronto, ON M7A 1W3; tel (416) 314-1990; fax (416) 314-1944; e-mail mnr.nric@mnr.gov.on.ca; internet www.mnr.gov.on.ca; Minister: David Ramsay; provincial ministry; provision for resource development and outdoor recreation facilities; administration, protection and conservation of public lands and waters.

Ontario Forest Research Institute: Ontario Ministry of Natural Resources, 1235 Queen St E, Sault Ste Marie, ON P6A 2ES; tel (705) 946-2981; fax (705) 946-2030; e-mail information.ofri@mnr.gov.on.ca; Gen Man: David Deyoe; f 1945; forest research and other scientific activities in support of ecological sustainability; main areas of research include ecological land classification, ecology of intensive plantations, long-term ecology, establishment and physiology, forest growth and yield, genetic resource management, hardwood and mixed-wood silviculture, forest health, landscape ecology, stock quality assessment and vegetation management alternatives.

Ontario Ministry of the Environment: 135 St Clair Ave W, Suite 100, Toronto, ON M4V 1P5; tel (416) 325-4000; fax (416)

323-4264; e-mail picemail@ene.gov.on.ca; internet www.ene .gov.on.ca; Minister: Leona Dombrowsky; f 1972; provincial ministry.

Parks Canada: 25 Eddy St, Hull, PQ K1A 0M5; e-mail parks_webmaster@pch.gc.ca; internet parkscanada.pch.gc.ca; Minister of Candian Heritage: Liza Frulla; part of Canadian Heritage; replaced the Canadian Parks Service.

Prince Edward Island Department of Agriculture, Fisheries, Aquaculture and Forestry: Jones Bldg, 5th Floor, 11 Kent St, POB 2000, Charlottetown, PE C1A 7N8; tel (902) 368-4880; fax (902) 368-4857; e-mail kjmacadam@gov.pe.ca; internet www.gov.pe.ca/af; Minister: Kevin MacAdam; Deputy Minister: Paul Jelley; provincial institution with ministerial status.

Prince Edward Island Department of the Environment and Energy: Jones Bldg, 4th and 5th Floors, 11 Kent St, POB 2000, Charlottetown, PE C1A 7N8; tel (902) 368-5000; fax (902) 368-5830; e-mail jwballem@gov.pe.ca; internet www.gov .pe.ca/enveng; Minister: Jamie Ballem; Deputy Minister: Lewie Creed; provincial institution with ministerial status.

Québec Ministry of Agriculture, Fisheries and Food: 200 chemin Sainte-Foy, 12ème étage, Québec, PQ G1R 4X6; tel (418) 380-2110; fax (418) 643-8422; e-mail info@agr .gouv.qc.ca; internet www.agr.gouv.qc.ca; Minister: Françoise Gauthier; provincial ministry.

Québec Ministry of Natural Resources, Wildlife and Parks: Edif de l'Atrium, 5700 4ème Ave Ouest, Charlesbourg, PQ G1H 6R1; tel (418) 627-8600; fax (418) 643-0720; e-mail service.citoyens@mrnfp.gouv.qc.ca; internet www.mrnfp .gouv.qc.ca; Minister: Sam Hamad; provincial ministry.

Québec Ministry of the Environment: Edif Marie-Guyart, rez-de-chausée, 675 blvd René-Lévesque est, Québec, PQ G1R 5V7; tel (418) 521-3830; fax (418) 646-5974; e-mail info@ menv.gouv.qc.ca; internet www.menv.gouv.qc.ca; Minister: Thomas J Mulcair; provincial ministry; responsible for protection of wildlife, terrestrial and freshwater environments, protected areas, human health, endangered animal and plant species, land use and development planning, pollution control, quality of life and landscape protection; environmental research; education and training.

Saskatchewan Department of Agriculture, Food and Rural Revitalization: Walter Scott Bldg, 3085 Albert St, Regina, SK S4S 0B1; tel (306) 787-5140; fax (306) 787-2393; e-mail brusnak@agr.gov.sk.ca; internet www.agr.gov.sk.ca; Minister: Mark Wartman; provincial institution with ministerial status.

Saskatchewan Department of Environment: 3211 Albert St, Regina, SK S4S 5W6; tel (306) 787-2700; fax (306) 787-2947; internet www.se.gov.sk.ca; Minister: David Forbes; provincial institution with ministerial status.

Yukon Territory Ministry of Renewable Resources: POB 2703, Whitehorse, YT Y1A 2C6; tel (403) 667-5811; fax (403) 393-6295; e-mail srm.minister@gems6.gov.bc.ca; internet www.gov.bc.ca/srm; Minister: Eric Fairclough; territorial ministry.

Non-Governmental Organizations

Alberta Environmental Network—AEN: 10511 Saskatchewan Dr, Edmonton, AB T6E 4S1; tel (403) 433-9302; fax (403) 433-9305; e-mail aen@web.net; internet www.web.net/~aen; Man Dir: Barry Breau; f 1979; support for networking, communication and resource-sharing between non-governmental environmental organizations; affiliated to the Canadian Environmental Network.

Aquatic Ecosystem Health and Management Society—AEHMS: c/o Canada Centre for Inland Waters, POB 5050, Burlington, ON L7R 4A6; tel (905) 336-4867; fax (905) 634-3516; Pres: Dr M Munawar; Sec: Dr P Ross; Treas: Dr D Malley; f 1989; concerned with the health and recovery of marine, freshwater and estuarine ecosystems; promotes and undertakes scientific initiatives to protect the global aquatic environment from toxic contamination and nutrient enrichment; environmental information through public awareness programmes and education.

Arctic Institute of North America, University of Calgary: Library Tower, 2500 University Drive NW, Calgary, AB T2N 1N4; tel (403) 220-7515; fax (403) 282-4609; Exec Dir: Michael Robinson; community-based economic planning and development; joint ventures with northern communities using participatory action research to preserve traditional knowledge, train communities and encourage self-reliance.

Association Québecoise pour l'Evaluation d'Impacts—AQEI (Quebec Association for Impact Assessment): CP 785, Succursale pl d'Armes, Montréal, PQ H2Y 3J2; tel and fax (514) 990-2193; e-mail mondor@aqei.qc.ca; internet www .cam.org/~aqei; Co-ordinator: Françoise Mondor; affiliated to the International Association for Impact Assessment—IAIA.

Banff Centre for Management, Resource and Environment Management Program: POB 1020, Banff, AB T0L 0C0; tel (403) 762-6583; fax (403) 762-6422; Sr Program Man: Felicity N Edwards; resource management; environmental management and planning; environmental education; member of the World Conservation Union—IUCN.

BC Environmental Network: 610, 207 W Hastings St, Vancouver, BC V6B 1H7; tel (604) 879-2279; fax (604) 879-2272; e-mail info@bcen.bc.ca; internet www.bcen.bc.ca; f 1979; grouping of the majority of local environmental organizations in the province of British Columbia; governed by an eight-member elected steering committee; annual conference; affiliated with the Canadian Environmental Network.

British Columbia Research Inc: 3650 Wesbrook Mall, Vancouver, BC V6S 2L2; tel (604) 224-4331; fax (604) 224-0540; internet www.bcr.bc.ca; Pres and CEO: Dr Hugh Wynne-Edwards; Dir of Environmental Sciences: Dr James Hill; f 1993; toxicity analysis and field services; environmental management and planning; treatment and processes; development and evaluation of technology; industrial hygiene and risk management.

Calgary Zoological Society: Calgary Zoo, Botanical Garden and Prehistoric Park, POB 3036, Station B, Calgary, AB T2M 4R8; tel (403) 232-9300; fax (403) 237-7582; e-mail alex@ calgaryzoo.ab.ca; internet www.alex@calgaryzoo.ab.ca; Pres and CEO: Alex Graham; Development: Brian Duclos; Animal Care and Habitat: Dr Clement Lanthier; Planning and Facility Operations: Don Peterkin; Communication, Education and Research: Dr Sherry Rainsforth; Business Operations: Keith Scott; f 1929; education; recreation; conservation; scientific study.

Canadian Arctic Resources Committee: 1276 Wellington Street, 2nd floor, Ottawa, ON K1Y 3A7; tel (613) 759-4284; fax (613) 759-4581; e-mail info@carc.org; internet www.carc .org; Exec Dir: Karen Wristen; Research Dir: Kevin O'Reilly; Communications Dir: Clive Tesar; Cumulative Effects Program Dir: Shelagh Montgomery; f 1971; a citizen's organization dedicated to the long-term environmental and social well-being of northern Canada and its peoples; believing in the application of sustainable development and the precautionary principle; policy and advocacy work is grounded in solid scientific and socio-economic research and experience; main activities are research and communications through publications such as the quarterly journal *Northern Perspectives*.

Canadian Association for Humane Trapping: 390 Bay St, Suite 1202, Toronto, ON M5H 2Y2; tel (416) 363-2614; fax (416) 363-8451; Exec Dir: Jane Vinet; liaison with trappers, the fur industry, governments and animal welfare organizations.

Canadian Circumpolar Institute—CCI, University of Alberta: Old St Stephen's College, 3rd Floor, North Wing, 8820 112th St, Edmonton, AB T6G 2E1; tel (403) 492-4512; fax (403) 492-1153; e-mail canadian.circumpolar.institute@ualberta.ca; internet www.ualberta.ca/~ccinst/polar/polar.html; Dir: Dr Clifford G Hickey; Assoc Dirs: Ken Barron, Heather Young Leslie; f 1990; community-based research, co-operative efforts, adaptive management approaches and sustainable development; research into the environment, resources, wildlife, traditional knowledge, politics, health, land use, economics, social policy and people of northern Canada and other circumpolar nations.

Canadian Coalition for Nuclear Responsibility, Inc/Regroupement pour la Surveillance du Nucléaire—CCNR: CP 236, Snowdon Station, Montréal, PQ H3X 3T4; tel (514) 489-5118; fax (514) 489-5118; e-mail ccnr@web.net; internet www.ccnr.org; Pres: Gordon Edwards; Sec and Treas: Marc Chénier; f 1975; research and educational activities relating to all aspects of nuclear technology and the alternatives, including uranium mining, reactor safety, energy policy, weapons proliferation, radioactive waste management, medical and environmental effects of radiation and standard-setting.

Canadian Committee for IUCN: c/o Canadian Museum of Nature, POB 3443, Station D, Ottawa, ON K1P 6P4; tel (613) 566-4795; fax (613) 364-4022; e-mail abreau@mus-nature.ca; Pres: William Rapley; Vice-Pres: Stephen Fuller; Sec: Jacques Prescott; f 1981; national office of the World Conservation Union—IUCN; membership includes nine federal and provincial government depts, 20 NGOs and 32 individuals across Canada.

Canadian Ecophilosophy Network: POB 5853, Station B, Victoria, BC V8R 6S8; tel (604) 598-7004; e-mail ecosophy@islandnet.com; internet www.ecostery.org; Contacts: Victoria Stevens, Alan R Drengson; f 1983; exchange of ideas and valuations relating to ecophilosophy, a philosophy based on a commitment to respect nature and live according to ecocentric values; publications.

Canadian Environmental Law Association: 517 College St, Suite 401, Toronto, ON M6G 4A2; tel (416) 960-2284; fax (416) 960-9392; Exec Dir (acting): Michelle Swenarchuk; Pres: Graham Rempe; f 1971; concerned with environmental law reform and litigation.

Canadian Environmental Network/Réseau Canadien de l'Environnement—CEN/RCE: 300-945 Wellington St, Ottawa, ON K1Y 2X5; tel (613) 728-9810; fax (613) 728-2963; e-mail cen@cen.web.net; internet www.cen.web.net; Nat Dir: Anne-Marie Cotter; Nat Caucus Co-ordinators: Chantal Bois, Derek Stack; f 1977; facilitates and advances the work of its member groups to promote ecologically sound ways of life.

Canadian Forestry Association: 185 Somerset St W, Suite 203, Ottawa, ON K2P 0J2; tel (613) 232-1815; fax (613) 232-4210; e-mail cfa@canadianforestry.com; internet www.canadianforestry.com; Pres: Barry Waito; Gen Man: Dave Lemkay; Office Admin: Sheila A Rust; f 1900; wise use and sustainable development of forests and their resources; National Forest Week, wildfire prevention campaign, National Forest Strategy, forest education, National Forest Congress and Canadian Urban Forests Conferences; Canada's Forests teaching kits; publications.

Canadian Global Change Program: Royal Society of Canada, 225 Metcalf St, Suite 308, Ottawa, ON K2P 1P9; tel (613) 991-6996; e-mail cgcp@rsc.ca; Chair, Bd of Dirs: Dr Hugh Morris; Dir: Dr Brian Bornhold; f 1985; co-ordination of research-planning activities; conferences and seminars; policy analysis related to global change issues; publications.

Canadian Institute for Environmental Law and Policy—CIELAP: 517 College St, Suite 400, Toronto, ON M6G 4A2; tel (416) 923-3529; fax (416) 923-5949; e-mail cielap@cielap.org; internet www.cielap.org; Exec Dir: Anne Mitchell; f 1970; environmental law, policy research and education; member of the World Conservation Union—IUCN.

Canadian Institute of Forestry—CIF: 151 Slater St, Suite 606, Ottawa, ON K1P 5H3; tel (613) 234-2242; fax (613) 234-6181; e-mail cif@cif-ifc.org.ca; internet www.cif-ifc.org; Pres: Dr John Barker; First Vice-Pres: Ralph Roberts; Second Vice-Pres: Evelynne Wrangler; f 1908; aims to promote the stewardship of Canada's forest resources through leadership, professional competence and public awareness; comprises 22 sections throughout Canada; 16 working groups focus on forestry issues, including silviculture and forest protection; established the Forestry Awareness in Canada Trust; serves as the secretariat of the Pest Management Caucus; continuing forestry education programme; awards; publications include a journal, newsletters and policy papers; member of the International Union of Societies of Foresters—IUSF; collaborates closely with the Society of American Foresters—SAF.

Canadian Nature Federation/Fédération Canadienne de la Nature—CNF/FCN: 1 Nicholas St, Suite 606, Ottawa, ON K1N 7B7; tel (613) 562-3447; fax (613) 562-3371; e-mail cnf@cnf.ca; internet www.cnf.ca; Exec Dir: Julie Gelfand; Marketing Dir: Ruth Catana; Dir, Conservation Programs: Caroline Schultz; Dir, Wildlands Campaign: Kevin McNamee; formerly Canadian Audubon Society; campaigns on parks, wildlife, habitat protection, endangered species and forest management; publications.

Canadian Parks and Wilderness Society—CPAWS: 401 Richmond St W, Suite 380, Toronto, ON M5V 3A8; tel (416) 979-2720; fax (416) 979-3155; Trustee: Ted Mosquin; f 1963; preservation of wild ecosystems in parks, wilderness and natural areas; preservation of biodiversity, habitats and species.

Canadian Society of Environmental Biologists: POB 962, Station F, Toronto, ON M4Y 2N9; promotion of the management of natural resources based on sound ecological principles.

Canadian Solar Industries Association—CanSIA: 2415 Holly Lane, Suite 250, Ottawa, ON K1V 7P2; tel (613) 736-9077; fax (613) 736-8938; e-mail cansia@magmacom.com; internet www.newenergy.org/newenergy/cansia.html; Pres: R K Swartman; Man: Ortrud Seelemann; f 1978; dissemination of technical and commercial information relating to applications of solar energy; represents the interests of the solar energy industry to Canadian provincial and federal governments in order to ensure favourable legislative and economic conditions; promotion of market development for solar technology; co-operation with other stake-holders in the field of renewable energy.

Canadian Wildlife Federation: 2740 Queensview Drive, Ottawa, ON K2B 1A2; tel (613) 721-2286; fax (613) 721-2902; e-mail info@cwf-fcf.org; internet www.cwf-fcf.org; Exec Vice-Pres: Colin Maxwell; Gen Man: Richard Leitch; f 1932; conservation of wildlife and wise use of Canada's renewable resources; liaison with individuals, provincial governments, municipalities and businesses; funding of research on species at risk and wildlife conservation projects; promotion of environmental legislation.

Centre Canadien d'Etude et de Coopération Internationale—CECI (Canadian Centre for International Research and Co-operation): 3185, rue Rachel Est, Montréal, Québec, H1W 1A3, Canada; tel (514) 875-9911; fax (514) 875-6469; e-mail info@ceci.ca; internet www.ceci.ca; Head: Michel Chaurette; f 1968; non-profit international development aid corporation which assigns voluntary workers to developing countries; campaigns on afforestation, wildlife, land restoration, soil conservation, resource protection and forest conservation.

Centre for Atmospheric Chemistry, York University: Steacie Science Bldg, Rm 006, Toronto, ON M3J 1P3; tel (416) 736-5410; fax (416) 736-5411; e-mail cac@yorku.ca; internet www.cac.yorku.ca; Dir: Geoffrey W Harris; f 1985; forum for transmission of information and development of new research initiatives among atmospheric chemists.

Centre for the Environment, Brock University: St Catharines, ON L2S 3A1; tel (905) 688-5550; fax (905) 688-6389; e-mail envi@spartan.ac.brocku.ca; internet www.brocku.ca/envi; affiliated with organizations in Thailand and Argentina.

Ducks Unlimited Canada: POB 1160, Stonewall, MB R0C 2Z0; tel (204) 467-3000; fax (204) 467-9028; e-mail webfoot@ducks.ca; internet www.ducks.ca; Pres: John Messer; Exec Vice-Pres: Gord Edwards; management of water-fowl breeding habitats in Canada; water-fowl and wetland research; public education.

Ecology Action Centre: 1568 Argyle Street, Suite 31, Halifax, NS B3J 2B3; tel (902) 429-2202; fax (902) 422-6410; e-mail eac@ecologyaction.ca; internet www.ecologyaction.ca; environmental education and advocacy on issues relevant to Nova Scotia; reference library; publications.

Energy Research Unit, University of Regina: 3737 Wascana, Regina, SK S4S 0A2; tel (306) 585-4269; fax (306) 585-5205; e-mail kybettbd@cas.uregina.ca; Dir: Prof B D Kybett; Assoc Dir: Prof P Tontiwachwithikul; f 1978; research into the capture and purification of carbon dioxide emissions for use in the petroleum and food industries; development and administration of renewable energy systems in Africa; environmental assessment and planning.

Environmental Engineering Laboratory, McGill University: Dept of Civil Engineering, 817 Sherbrooke St W, Montréal, PQ H3A 2K6; tel (514) 398-6861; fax (514) 398-7361; e-mail gehr@civil.lan.mcgill.ca; internet www.mcgill.ca/civil; Head: Ronald Gehr; teaching, research and contract analyses; projects include industrial waste treatment by peroxidase enzymes, modelling of odour impact from industrial air pollution sources, disinfection by ultraviolet light and other agents, fate and remediation of contaminants in soils and ground-waters.

Environmental Law Centre (Alberta) Society: 10709 Jasper Ave, Suite 204, Edmonton, AB T5J 3N3; tel (780) 424-5099; fax (780) 424-5133; e-mail elc@elc.ab.ca; internet www.elc.ab.ca; Exec Dir: Cindy Chiasson; f 1982; aims to ensure that the law serves to preserve and protect the environment; public information and assistance; legal research; monitoring and reform.

Environmental Research and Studies Centre: 3-23 Business Building, University of Alberta. T6G 2R6; tel (403) 492-5825; fax (403) 492-3325; e-mail Beverly.Levis@ualberta.ca; internet www.ualberta.ca/ESRC; Dir: Dr Ray Rasmussen; Man: Beverly Levis; f 1990; acts as a co-ordination and information agency for the extensive research and studies programs at the University of Alberta; programmes include: publication of Environmental News; an open and free interdisciplinary lecture series focused on current environmental topics; a web based information resource of environmental research; publication of a resource guide for sustainable living and learning; support for environmental research activities.

Federation of Alberta Naturalists: POB 1472, Edmonton, AB T5J 2N5; tel (403) 453-8639; fax (403) 453-8553; Pres: Jorden Johnston; Exec Dir: Glen Semenchuk.

Federation of Ontario Naturalists—FON: Conservation Centre, Moatfield Park, 355 Lesmill Rd, Don Mills, ON M3B 2W8; tel (416) 444-8419; fax (416) 444-9866; Environment Dir: Marion Taylor; campaigns on freshwater ecology, national parks, wildlife, water-fowl and other birds, migratory species, wetland ecosystems, wild flora, endangered animal species, poaching and management of protected areas; environmental education.

Forest Management and Policy, Lakehead University: Faculty of Forestry, 955 Oliver Rd, Thunder Bay, ON P7B 5E1; tel (807) 343-8508; fax (807) 343-8116; e-mail pduinker@lakeheadu.ca; Chair: Assoc Prof Dr Peter N Duinker; f 1988; research, development and applications in forest management and policy; biodiversity assessment, wildlife habitat analysis, environmental impact assessment, conflict resolution and public participation.

Friends of the Earth Canada: 206–260 St Patrick St, Ottawa, ON K1N 5K5; tel (613) 241-0085; fax (613) 241-7998; e-mail foe@intranet.ca; internet www.foecanada.org; CEO: Beatrice Olivastri; Admin: Karen Cartier; Program Man: Kathleen Kelso; f 1978; national office of Friends of the Earth International—FoEI.

Geotechnical Science Laboratories, Carleton University: Dept of Geography, 1125 Colonel By Drive, ON K1V 0A4; tel (613) 520-9004; fax (613) 520-9005; e-mail peter_williams@carleton.ca; Dir: Prof Peter J Williams; Admin: Else Groves; f 1978; fundamental and applied research into the properties and behaviour of the ground and environment; main areas of research are climate change, ground contamination, cold regions, permafrost, and gas and petroleum pipelines; other areas include soil fertility and erosion, glaciology and clay landslides.

Green Party of Canada: POB 397, London, ON N6A 4W1; tel (519) 474-3294; fax (519) 474-3294; Leader: Joan Russow; f 1983; environmental political organization.

Greenpeace Canada: 250 Dundas St W, Suite 605, Toronto, ON M5T 2Z5; tel (416) 597-8408; fax (416) 597-8422; e-mail greenpeace.toronto@dialb.greenpeace.org; internet www.greenpeacecanada.org; Exec Dir: Jeanne Moffat; f 1971; national office of Greenpeace International.

Institut National de la Recherche Scientifique—Eau, Université du Québec (National Institute for Scientific Research—Water Unit, University of Quebec): 2800 Einstein 20, POB 7500, Ste-Foy, PQ G1V 4C7; tel (418) 654-2524; fax (418) 654-2600; e-mail jpv@inrs-eau.uquebec.ca; internet www.inrs-eau.uquebec.ca; Dir: Dr Jean-Pierre Villeneuve; f 1970; postgraduate education and applied research in environmental and water sciences, including hydrology, hydrogeology, watershed management, climate changes, biogeochemistry, limnology, ecotoxicology, waste-water treatment and pollution control.

Institute for Environmental Studies—IES, University of Toronto: School of Graduate Studies, 33 Willcocks St, Suite 1016V, Toronto, ON M5S 3E8; tel (416) 978-6526; fax (416) 978-3884; e-mail ies.gradinfo@utoronto.ca; internet www.utoronto.ca/env/es.htm; Dir: Rodney R White; Assoc Dirs: Roger I C Hansell, Frances Silverman; f 1974; graduate collaborative programmes in environmental studies and toxicology; a programme in environment and health is under development; five-year partnership with Environment Canada's Environmental Adaptation Research Group; Climate Change Study Group; symposium on natural and man-made hazards; forest ecosystems forum; five-year environmental training project in Viet Nam; involved in trilateral collaborative research programme between Canada, Mexico and the USA; engaged in geographic information systems (GIS) research in the People's Republic of China; areas of interest include environmental fate, the effects and modelling of contaminants, atmospheric change and biodiversity, and the insurance industry and the environment.

Institute for Risk Research, University of Waterloo: Waterloo, ON N2L 3G1; tel (519) 885-1211; fax (519) 725-4834; e-mail irr@mail.eng.uwaterloo.ca; internet workbench .uwaterloo.ca/irr; Dir: John Shortreed; f 1982; research, education and information dissemination; areas of activity include risk management in the public interest, transportation risks, hazardous waste management and environmental health.

International Development Research Centre—IDRC: 250 Albert St, POB 8500, Ottawa, ON K1G 3H9; tel (613) 236-6163; fax (613) 563-2476; e-mail info@idrc.ca; internet www.idrc.ca; Information Officer: Lise Daoust; created and funded by the Canadian Parliament.

Inuit Tapiriiksat Kanatami (Inuit Brotherhood of Canada): 170 Laurier Ave W, Suite 510, Ottawa, ON K1P 5V5; tel (613) 238-8181; fax (613) 234-1991; Pres: Rosemarie Kuptana; Sec and Treas: Martha Flaherty; advocacy, research and consultation on issues affecting Canadian Inuit, including human rights, justice, autonomy, social policy, housing, environmental health and contaminants, environmental assessment and planning, traditional skills and wildlife management.

IUCN Canada Country Office: 380 rue St-Antoine Ouest, Bureau 3200, Montréal, PQ H2Y 3X7; tel (514) 287-9704; fax (514) 287-9057; e-mail mercer@iucn.ca; Dir: Malcolm Mercer.

Metropolitan Toronto Zoological Society: 361[a] Old Finch Ave, Scarborough, ON M1B 5K7; tel (416) 392-5900; fax (416) 392-5934; internet www.torontozoo.com; Exec Dir: William A Rapley; f 1972; environmental education; research into zoology, conservation and wildlife.

New Brunswick Environmental Network/Réseau Environnemental du Nouveau-Brunswick: 167 Creek Rd, Waterford, NB E4E 4L7; tel (506) 433-6101; fax (506) 433-6111; e-mail nben@nbnet.nb.ca; internet www.web.net/nben; Co-ordinator: Mary Ann Coleman; f 1991; organization of civic environmental groups in the province of New Brunswick; affiliated to the Canadian Environmental Network.

Newfoundland and Labrador Environmental Network: POB 944, Corner Brook, NF A2H 6J2; tel and fax (709) 634-2520; e-mail cbnlen@nfld.net; Exec Dir (acting): Greg Mitchell; affiliated to the Canadian Environmental Network.

Ocean Voice International: 2255 Carling Ave, Suite 400, Ottawa, ON K2B 1A6; tel (613) 721-4541; fax (613) 721-4562; e-mail ovi@cyberus.ca; internet www.ovi.ca; Pres: Don E McAllister; Vice-Pres: Jaime Baquero; Exec Dir: Angela Jellett; f 1987; formerly International Marinelife Alliance Canada; conservation of marine biodiversity and ecosystems; promotion of sustainable marine resource management through research and education; projects in the Atlantic, Indian and Pacific Oceans.

Ontario Affiliate of the IAIA: Environmental Management Branch, Ottawa ON K1N 5A1; tel (613) 738-0708; fax (613) 738-0721; e-mail oaia@oaia.on.ca; internet oaia.on.ca; Pres: W Paul McDonald; non-profit association of impact assessment practitioners; affiliated to the International Association for Impact Assessment—IAIA.

Ontario Federation of Anglers and Hunters—OFAH: POB 2800, Peterborough, ON K9Y 8L5; tel (705) 748-6324; fax (705) 748-9577; e-mail ofah@oncomdis.on.ca; internet www.ofah.org; Exec Vice-Pres: Richard G Morgan; f 1928; co-ordination of conservation activities of angling and hunting associations; education and lobbying; support for scientific research.

Organization for the Rehabilitation and Enhancement of the Environment Canada—OREE, Inc: 1130 Elizabeth St, St Laurent, PQ H4L 4L1; tel (509) 86-0815; fax (509) 86-0559; CEO: Arnel Gervais; f 1986; promotion and development of sustainable agricultural production in Haiti; discouragement of cash crop production; soil conservation and control of desertification.

Prince Edward Island Eco-Net: 126 Richmond St, Charlottetown, PE C1A 1H9; tel (902) 566-4170; fax (902) 566-4037; e-mail peien@isn.net; internet www.isn.ca/~network; Co-ordinator: Christine Jackson; affiliated to Canadian Environmental Network; provides information and resources to local conservationist groups; maintains a nationwide network of contacts.

Recycling Council of Ontario: 489 College St, Suite 504, Toronto, ON M6G 1A5; tel (416) 960-1025; fax (416) 960-8053; e-mail rco@web.net; internet www.web.net/rco; Exec Dir: John Hanson; Chair: Anne Mathesson; f 1978; promotion of waste reduction, reuse, recycling and composting; waste management resource centre, information software and publications; annual conference and awards.

School for Resource and Environmental Studies—SRES, Dalhousie University: 1312 Robie St, Halifax, NS B3H 3E2; tel (902) 494-3632; fax (902) 494-3728; e-mail sres@is.dal.ca; internet quasar.sba.dal.ca:2000/sres.html; Dir: Prof R P Côté; Academic Programme Co-ordinator: Prof Ann Dwire; f 1978; interdisciplinary research and education in natural resource management and environmental problems.

Sierra Club Canada: 1 Nicholas St, Suite 620, Ottawa, ON K1N 7B7; tel (613) 241-4611; fax (613) 241-2292; e-mail sierra .club.canada@sierraclub.org; internet www.sierraclub.ca; Exec Dir: Elizabeth May; Dir, Energy and Atmosphere: Louise Comean; national office of Sierra Club (USA); local-level environmental organization; local, national and international lobbying and activism; public education.

Society Promoting Environmental Conservation—SPEC: 2150 Maple St, Vancouver, BC V6J 3T3; tel (604) 732-7736; fax (604) 736-7115; e-mail enviro@spec.bc.ca; internet www.spec.bc.ca; Pres: David Capman; information and practical demonstrations on energy-saving strategies, organic gardening, composting, waste reduction and recycling; network facility for environmental organizations; proposal of alternatives to the use of hazardous pesticides and herbicides; campaigns for protection of wetlands.

Toxicology Centre, University of Saskatchewan: 44 Campus Drive, Saskatoon, SK S7N 5B3; tel (306) 966-7441; fax (306) 931-1664; e-mail karsten.liber@usask.ca; Prin Officer: Dr Karsten Liber; f 1983; conducts research on a broad range of toxicological and ecotoxicological issues.

Union Québécoise pour la Conservation de la Nature—UQCN (Quebec Union for Nature Conservation): 690 Grande Allée Est, 4e Etage, Québec, PQ G1R 2K5; tel (418) 648-0991; fax (418) 648-2104; Pres: Harvey Mead; union of societies concerned with natural sciences and the environment; environmental education; promotion of rational use of resources.

Water Environment Association of Ontario: 63 Hollyberry Trail, North York, ON M2H 2N9; tel (416) 502-1440; fax (416) 502-1786; f 1971; formerly Pollution Control Association of Ontario; member of Water Environment Federation; seminars, workshops and annual conference.

Western Canada Wilderness Committee: 27 Abbott St, Vancouver, BC V6B 2K7; tel (604) 683-8220; fax (604) 683-8229; e-mail info@wildernesscommittee.org; internet www .wildernesscommittee.org; Prin Officers: Adriane Carr, Sue

Fox; f 1980; public education; promotion of wilderness preservation.

World Wildlife Fund—WWF Canada: 245 Eglinton Ave E, Suite 410, Toronto, ON M4P 3J1; tel (416) 489-8800; fax (416) 489-3611; e-mail panda@wwfcanada.org; internet wwfcanada.org; national affiliate organization of World Wide Fund for Nature—WWF International; conservation of wild animals, plants and habitats; implementation of the World Conservation Strategy.

CAPE VERDE

Governmental Organizations

Ministry of the Environment, Agriculture and Fisheries: Ponta Belém, Praia, Santiago; tel 261-57-17; fax 261-40-54; Minister: Dr José António Pinto Monteiro.

Instituto Nacional de Investigação e Desenvolvimento Agrario (National Institute for Agricultural Research and Development): CP 84, Praia, Santiago; tel 71-11-47; fax 71-11-33; e-mail inida@cvtelecom.cv; Prin Officers: José Gabriel Levy, Carlos Pinheiro Silva, María Tereza Veracruz, Zuleika Salazar Levy; f 1979; research and training in crop production and protection, soils and natural resource management.

Centre de Documentation et Information pour le Développement—CDID (Information Centre for Development): 6 SC 0,0,300, BP 120, Praia; tel 61-39-69; fax 61-15-27; Dir: Daniel Avelino Pires; serves as the INFOTERRA national focal point.

CENTRAL AFRICAN REPUBLIC

Governmental Organizations

Ministry of Water Resources, Forests, Hunting and Fishing: BP 830, Bangui; tel 61-95-58; fax 61-57-41; Minister: Denis Kossi-Bella.

Ministry of Agriculture and Animal Husbandry: Bangui; Minister: Lt-Col Parfait M'Baye.

Ministry of Mines and Energy: Bangui; tel 61-20-54; fax 61-60-76; Minister: Maj Sylvain N'Doutingaï.

Ministry of Public Health and Population: Bangui; tel 61-29-01; Minister: Nestor Mamadou Nali.

Commission Environnement et Tourisme, Assemblée Nationale (National Assembly, Environment and Tourism Commission): Bangui; tel 61-02-34; Pres: Thierry Ignifolo van den Boss; f 1993; controls the adoption and application of environmental legislation.

Direction de la Santé Communautaire, Service d'Hygiène et de la Salubrité de l'Environnement (Public Health Directorate, Hygiene and Environmental Quality Service): BP 71, Bangui; tel 61-04-22; Dir: Makando; environmental monitoring; prevention of the spread of disease.

Direction Générale de l'Environnement (Directorate-General for the Environment): BP 830, Bangui; tel 61-95-58; fax 61-57-41; Dir-Gen: Gustave Doungoube; Dir: Clément Amarou; f 1995; development of national environmental policy; co-ordination of the environmental activities of government departments and of non-governmental organizations.

Institut de Recherches Agronomiques de Boukoko (Boukoko Agronomic Research Institute): M'Baiki, BP 44, Boukoko; Dir: M Gondjia; f 1948; research on tropical agriculture, including studies of plant diseases and entomology.

Non-Governmental Organizations

Association Centrafricaine pour le Respect de la Nature—ACRN (Central African Association for Respect for Nature): Paroisse St Jacques de Kpetene, Bangui; Pres: Abbé Jonas Clément Morouba; f 1999.

Association Cri de Forêts (Cry of the Forests Association): Mairie, Bimbo; Pres: Albert-Roger Bassa; Sec-Gen: Michel Dambeti; f 1996; subsidiary to CEDIFOD and OCDH; programmes to combat deforestation and poaching and reduce poverty in forest areas; raising of livestock and cultivation of crops; education; affiliated with CIONGCA and RONGED.

Association des Volontaires pour le Développement Economique en Centrafrique (Association of Volunteers for Economic Development in Central Africa): BP 690, Bangui; tel 61-73-21; Exec Pres: Paul Gérard Rekouane; f 1992; encourages use of appropriate technology in the protection and rational management of the country's environment.

Association Multidisciplinaire de Développement Durable (Multidisciplainary Association for Sustainable Development): BP 1185, Bangui; fax 61-35-61; Pres: Dr Liclere Dogertina; Sec-Gen: Joseph Bata; f 1997; economic and environmental research; programmes to promote sustainable development.

Association pour la Conservation de la Nature et de ses Ressources—ACNR (Association for the Conservation of Nature and Natural Resources): Paroisse St Jacques de Kpetene, BP 919, Bangui; Pres: Crépin Kabba Dali; Sec-Gen: Hubert Guelagbadanga; f 1995; active in various fields of environmental protection, especially combating deforestation and desertification.

Comité National de Lutte contre l'Incendie, les Feux de Brousse et autres Calamités (National Committee for Combating Fires and other Disasters): BP 1859, Bangui; tel 61-11-37; Prin Officers: Lucien Gonda, Alasan Abdoulaye; f 1983; protection of people, property and land from fires and other natural disasters.

Education à la Maîtrise de la Fécondité (Education for the Management of Natural Resources): BP 335, Bangui; tel 61-44-43; Prin Officer: Marie José Koulaye; f 1984; development of forestry; improvement of diet through increased use of local produce; promotion of traditional medicinal remedies.

Institut de Recherche pour le Développement—IRD (Development Research Institute): BP 893, Bangui; tel 61-20-89; fax 61-68-29; Dir: Jean-Yves Gac; f 1949; formerly Centre ORSTOM de Bangui (ORSTOM Centre in Bangui); a regional centre of the Institute, which is based in France; research areas include pedology, hydrology, geology, geophysics and demography.

Organisation Centrafricaine pour la Défense de la Nature—OCDN (Central African Organization for Nature Protection): BP 147, Km 5, Bangui; Pres: Maxim Balalou; Sec-Gen: Patrice Passe-Sanand; f 1992.

Organisation pour la Promotion des Initiatives de Développement Durable (Organization for the Promotion of Initiatives in Sustainable Development): Bangui; tel 61-13-18; Prin Officer: Patrick Zacko; f 1999.

Réseau des ONG de l'Environnement et du Développement Durable (Network of Environment and Sustainable Development NGOs): BP 1389, Bangui; Sec-Gen: Limbogo Ngakeu Msa; National Co-ordinator: Marc Karangaze; f 1998.

Station Expérimentale de la Maboké (Maboké Research Station): par M'Baiki; Dir: Roger Heim; f 1963; research areas include the protection of materials in tropical regions, zoology, botany and the protection of natural resources; under the direction of the Muséum National d'Histoire Naturelle (Paris, France).

CHAD

Governmental Organizations

Ministry of the Environment and Water: BP 905, N'Djamena; tel 52-60-12; fax 52-38-39; e-mail facdrem@intnet.td; Minister: Adoum Diar; incorporates Direction des Pêches et de l'Aquaculture—DPA (Department of Fisheries and Aquaculture) and Direction de la Protection de la Faune et des Parcs Nationaux—DPFPN (Department of Wildlife Protection and National Parks); the Division of Training, Projects, Archives and Statistics serves as the INFOTERRA national focal point.

Direction des Forêts et de la Protection de l'Environnement—DFPE (Department of Forests and Protection of the Environment): BP 447, N'Djamena; tel 51-50-32; reforestation; prevention of desertification; protection of plants and wildlife.

Division of National Parks and Wildlife Reserves: BP 905, N'Djamena; tel 51-23-05; fax 51-43-97.

Ministry of Agriculture: BP 441, N'Djamena; tel 52-69-79; fax 52-51-19; Minister: Pascal Yoadimnadji.

Ministry of Land Management, Town Planning and Housing: BP 436, N'Djamena; tel 52-31-89; fax 52-39-35; Minister: Sandjima Dounia.

Ministry of Livestock: BP 750, N'Djamena; tel 52-98-53; Minister: Mahamat Abdoulaye.

Direction des Ressources en Eau et de la Météorologie—DREM (Directorate of Water Resources and Meteorology): BP 429, N'Djamena; tel 52-30-81; fax 52-30-43; e-mail sacdrem@intnet.td; Dir: Neasmiandogo Betoloum; Head, Hydrology Service: Younane Nelngar; f 1975; hydrological and climatological monitoring; compilation of data bases.

CHILE

Governmental Organizations

Comisión Nacional del Medio Ambiente—CONAMA (National Environment Commission): Obispo Donoso 6, Providencia, Santiago; tel (2) 240-5600; fax (2) 244-1262; e-mail cpina@conama.cl; internet www.conama.cl; Pres, Chair, Governing Cncl, and Minister Sec-Gen of the Presidency: Alvaro García Hurtado; Exec Dir: Adriana Hoffman; responsible to the President of the Republic.

Ministry of Agriculture: Teatinos 40, Santiago; tel (2) 393-5000; fax (2) 393-5050; e-mail contacto@minagri.gob.cl; internet www.minagri.gob.cl; Minister: Jaime Campos Quiroga.

Corporación Nacional Forestal—CONAF (National Forestry Corporation): Avda Bulnes 285, Of 501, Santiago; tel (2) 390-0219; fax (2) 671 5881; e-mail consult@conaf.cl; internet www.conaf.cl; Exec Dir: Carlos Weber Bonte; conservation, management and optimum use of national forest resources; management of protected areas; four technical programmes are Wildlife Heritage, Forest Control, Forest Management and Development, and Fire Management.

Ministry of Housing and Urban Development: Juan Antonio Ríos 6, Santiago; tel (2) 638-0801; fax (2) 633-3892; e-mail contactenos@minvu.cl; internet www.minvu.cl; Minister: Claudio Orrego Larraín.

Ministry of Public Health: Enrique MacIver 541, 3°, Santiago; tel (2) 639-4001; fax (2) 633-5875; e-mail consulta@minsal.cl; Minister: Pedro García Aspillaga.

Instituto Antártico Chileno (Chilean Antarctic Institute): Luis Thayer Ojeda 814, Casilla 16521, Correo 9, Santiago; tel (2) 231-8195; fax (2) 232-0440; e-mail inach@inach.cl; internet www.inach.cl; Dir: Oscar Pinochet; Sub-Dir: Jorge Berguño; f 1963; co-ordinates scientific work performed by Chileans in Antarctica, in accordance with the rules of the Antarctic Treaty and especially with its Protocol of Environmental Protection.

Non-Governmental Organizations

Alianza Humanista–Verde (Humanist–Green Alliance): Santiago; Leaders: Andrés R Koryzma, José Tomás Sáenz; environmental political party.

Asociación para la Difusión de los Programas de las Naciones Unidas sobre Medio Ambiente—ADNUMA (Association for the Dissemination of United Nations

Programmes on the Environment): 2 Norte 942, Casilla 102, Talca; tel (71) 22-7046; fax (71) 22-7046; f 1994; environmental education programmes.

Centro de Estudios Agrarias y Ambientales—CEA (Agricultural and Environmental Studies Centre): Casilla 164, Valdivia; tel (63) 21-5846; fax (63) 21-3149; e-mail cea@valdivia.uca.uach.cl; Dir: Patricia Möller; environmental education; promotion of sustainable development; environmental impact studies; wildlife management.

Centro de Investigación y Planificación del Medio Ambiente—CIPMA (Centre for Environmental Research and Planning): Holanda 1109, Casilla 16362, Providencia, Santiago; tel (2) 231-0602; fax (2) 334-1095; Pres: Guillermo Geisse G; Research Co-ordinator: Michael Nelson.

Colegio de Ingenieros Forestales (College of Forestry Experts): San Isidro 22, Of 503, Casilla 9686, Santiago; tel (2) 639-3289; fax (2) 638-5280; Pres: Jorge I Correa Drubi; Sec: Leonardo Araya Valdebenito; f 1972; trade association; publications.

Comité Nacional Pro-Defensa de la Fauna y Flora—CODEFF (National Committee for the Defence of Flora and Fauna): Avda Francisco Bilbao 691, Providencia, Santiago; tel (2) 251-0262; fax (2) 251-8433; e-mail info@codeff.mic.cl; internet www.codeff.cl; Contacts: Miguel Stutzin Schottlander, Jenia Jofre Canobra; environmental education, research and legal action; improvement of environmental legislation and its enforcement; affiliated with Friends of the Earth—FoE Chile.

Greenpeace Chile: Eleodoro Flores 2424, Ñuñoa, Santiago; tel (2) 343-7788; fax (2) 204-0162; e-mail greenpeace.chile@dialb.greenpeace.org; national office of Greenpeace International.

Red Ecológica Anti-Smog (Anti-Smog Ecological Network): Casilla 4012, Santiago 21; tel (2) 682-0636; e-mail jotacea@esfera.cl; Prin Officer: Juan Carlos Araya; f 1989; public information and education on air pollution and its effects on humans; affiliated with Red Nacional de Acción Ecológica—RENACE.

Red Nacional de Acción Ecologica—RENACE (National Ecological Action Network): Seminario 774, Ñuñoa; tel (2) 223-4483; fax (2) 223-4522; e-mail renace@rdc.cl; internet www.renace.net; Dirs: Flavia Liberona Céspedes, Alvaro Gómez Concha, Nicolas Binfa Alvarado; Exec Sec: Oriana Reyes Olivares; f 1988; exchange of information and expertise; ecological development; environmental education and training publications; affiliated with 148 organizations throughout Chile.

Sociedad de Vida Silvestre de Chile (Chile Wildlife Society): Avda Alemania 0422, Casilla 1705, Temuco; tel (45) 21-0773; fax (45) 23-4126; e-mail cea@valdivia.uca.uach.cl; Pres: Darcy Ríos Leal; environmental education; publication of various environmental bulletins; International Congress on Natural Resources Management every two years.

CHINA, PEOPLE'S REPUBLIC

Governmental Organizations

State Environmental Protection Administration—SEPA: 115 Xizhimennei Nanxiaojie, Beijing 100035; tel (10) 66153366; fax (10) 66151768; e-mail ymzhao@nepa.go.cn; internet www.sepaeic.gov.cn/english/SEPA/index.htm; Minister: Xie Zhenhua; fmrly National Environmental Protection Agency—NEPA; adopted current name and assumed ministerial status 1998.

Organic Food Development Centre: State Environmental Protection Administration, 8 Jiangwangmiao St, POB 4202, Nanjing, Jiangsu; tel and fax (25) 5420606; e-mail lizf@njnet.nj.ac.cn; Prin Officers: Prof Li Zhengfang, Assoc Prof Xiao Xingji; f 1994; research on agriculture and the environment, organic food and rural development; organic inspection, certification and consultation; education and training in the fields of ecological agriculture and appropriate technology; member of the International Federation of Organic Agriculture Movements—IFOAM.

Ministry of Agriculture: 11 Nongzhanguan Nanli, Chaoyang Qu, Beijing 100026; tel (10) 64192293; fax (10) 64192468; e-mail webmaster@agri.gov.cn; internet www.agri.gov.cn; Minister: Du Qinglin.

Agro-Environment Protection Institute—AEPI: 31 Fukang Rd, Tianjin 300191; tel (22) 23367139; fax (22) 23369542; e-mail caep@public.tpt.tj.cn; Dir: Prof Tao Zhan; Deputy Dir: Zhang Hongsheng; f 1979; research into the impact of industrial wastes, agro-chemicals and unsustainable use of natural resources on the agro-ecological environment; monitoring and assessment of environmental quality; raising public awareness; publications; maintain seven specialized cttees: Farmland Eco-Environment Protection Cttee, Pesticide Pollution and Control Cttee, Animal Husbandry Environment Protection Cttee, Agro-Environment Management Cttee, Agro-Environment Monitoring Cttee, Eco-Agriculture Cttee, and Sub-Society for Fisheries Environment Protection.

Ministry of Land and Natural Resources: 3 Guanyingyuanxiqu, Xicheng Qu, Beijing 100035; tel (10) 66127001; fax (10) 66175348; e-mail bzxx@mail.mlr.gov.cn; internet www.mlr.gov.cn; Minister: Sun Wensheng.

Ministry of Water Resources: 1 Baiguang Lu, Ertiao, Xuanwu Qu, Beijing 100053; tel (10) 63203069; fax (10) 63202650; internet www.mwr.gov.cn; Minister: Wang Shucheng.

China Institute for Radiation Protection: 270 Xuefu St, POB 120, Taiyuan, Shanxi; tel (351) 7020266; fax (351) 7020407; e-mail cirpnnc@mh.ty.col.co.cn; Prin Officers: Yang Huating, Li Shushen; f 1962; study and application of radiation measurements; radiation dosimetry; diffusion of pollutants in the atmosphere, surface- and ground-water, and the ecosystem; radioactive waste treatment and disposal; affiliated to the China National Nuclear Corporation.

Environmental Protection Research Institute, China National Council for Light Industry: 11 Fucheng Rd, Haidian Dist, Beijing 100037; tel and fax (10) 68429835; Dir: Shen Zhenhuan; f 1979; research into environmental technologies for light industry; main areas of activity include biochemical treatment, physico-chemical treatment, waste cellulose hydrolysis, monitoring and analysis, engineering design, environmental impact assessment and risk analysis.

Gansu Grassland Ecological Research Institute: POB 61, Lanzhou; tel (931) 8914070; fax (931) 8663778; e-mail nanzb@public.lz.gs.cn; Dir College of Pastoral Agriculture

Science and Technology, Lanzou Uni: Zhibia Nan; Chief of Library: Liheng Li; research into grassland farming, grassland ecology and sustainable development; training and education for students and professionals; consultancy; publications; affiliated with the Ministry of Agriculture and with Lanzou University, Lanzhou.

Research Institute of Environmental Law: Wuhan University, Law School, Wuhan, Hubei 430072; tel (27) 7882712; fax (27) 7882661; e-mail caishouq@public.wh.hb.cn; Dir: Prof Cai Shouqiu; Vice-Dirs: Prof Wang Xi, Prof Li Qijia; f 1981; an institute of both the National Environmental Protection Agency—NEPA and of Wuhan University; research in the basic theories guiding the development of environmental law and policy in the People's Republic of China; participation in environmental law-making at national and local levels of government; environmental law courses, including a Master's degree programme; training programmes in environmental law, policy and management for government officials at various levels; legal consultants to NEPA and other governmental and non-governmental bodies; legal representation; academic co-operation and exchanges both nationally and internationally.

Xinjiang New Energy Research Institute: 40 Beijing S Rd, Urumqi, Xinjiang, Uygur 830011; tel (991) 3835905; fax (991) 3835920; Dir: Wang Guo Hua; f 1979; research and development of solar photoelectric sources and water heaters, and wind-driven generators; design, manufacture and installation of photothermal and photoelectric projects; institute of the Xinjiang Uygur Autonomous Region Science and Technology Committee.

Non-Governmental Organizations

China Association of the Environmental Protection Industry: 9 Sanlihe Rd, Haidian Dist, Beijing 100835; tel (10) 68393892; fax (10) 68393748; e-mail caepi@public3.bta.net.cn; Deputy Sec-Gen: Xiaoyu Jiang; f 1993; participation in trade development programme and surveying of the national environmental protection industry; research into development strategies for the industry; assistance to businesses in the introduction of foreign capital and technology; organization of international exhibitions and conferences; consultancy; promotion of international co-operation.

China Wildlife Conservation Association: 18 Hepingli Dongjie, Hepingli, Beijing 100714; tel (10) 64216343; fax (10) 64238030; e-mail cwca@public3.bta.net.cn; Gen Sec: Zhen Rende; Vice Gen Sec: Chen Runsheng; resource protection; wildlife management and research; environmental education.

Chinese Ecosystem Research Network: 3 Datun Rd, POB 9717, Beijing 100101; tel (10) 64931980; fax (10) 64931970; e-mail zhaosd@cern.ac.cn; Prin Officer: Prof Zhao Shidong; f 1988; monitoring of changes in environments and ecosystems; research into the structure, function, dynamics and management of ecosystems; contributions to the study of global change and sustainable development; affiliated with the Commission for Integrated Survey of Natural Resources—CISNAR, Chinese Academy of Sciences.

Chinese Society for Environmental Sciences: 115 Xizhimennei Nanxiaojie, Beijing 100035; tel (10) 661006; Pres: Li Jingzhao; f 1979; promotion of environmental chemistry; exchange of scientific information; advice; publications.

Commission for Integrated Survey of Natural Resources—CISNAR, Chinese Academy of Sciences—CAS: POB 9717, Beijing 100101; tel (10) 64889797; fax (10) 64914230; e-mail shkcheng@cisnar.ac.cn; Dir: Prof Cheng Shengkui; f 1956; co-ordination of survey teams for rational use of natural resources; multidisciplinary research into natural resources.

Friends of Nature—FON: Beijing; e-mail office@fon.org.cn; internet www.fon.org.cn; Pres: Prof Liang Congjie; f 1994; first environmental NGO in the People's Republic of China; protection of habitats and endangered species, including the snubnosed monkey, Tibetan antelope and wild yak; campaigns against illegal logging; liaison with government; creation of public awareness of ecological issues; educational activities for children.

Hangzhou Environmental Health Institute, Zhejiang Medical University: 353 Yan-An Rd, Hangzhou, Zhejiang 310031; tel (571) 7022700; fax (571) 7077389; Prin Officer: Dr Huang Xing-shu; f 1985; environmental toxicology studies; assessment of the impact of chemical pollution and electromagnetic radiation on human health; postgraduate education and training.

Institute of Applied Ecology: POB 417, Shenyang 110015; tel (24) 3902096; fax (24) 3843313; Dir: Sun Tie-Hang; f 1954.

Institute of Botany, Chinese Academy of Sciences—CAS: 141 Xizhimen Wai St, Beijing 100044; tel (10) 68353841; fax (10) 68319534; Dir: Prof Zhang Xinshi; multidisciplinary research organization; botanical garden; conservation of rare and endangered plant species; study of plant genetic resources.

Institute of Desert Research: Academia Sinica, Lanzhou.

Institute of Hydrobiology, Chinese Academy of Sciences—CAS: Luojiashan, Wuhan, Hubei 430072; tel (27) 87883482; fax (27) 87875132; e-mail nic@ihb.ac.cn; Dir: Dr Zhou Zouyan; f 1978; ecology; conservation; biology of rearing and breeding; ethology; includes the Department of River Dolphin Research.

Institute of Scientific and Technological Information: Chinese Academy of Forestry, Wan Sou San, Beijing 10091; integral unit of the Chinese Academy of Forestry.

Institute of Soil and Water Conservation, Chinese Academy of Sciences—CAS: Yang Ling, Shaanxi 712100; tel (910) 7012412; fax (910) 7012210; e-mail libinfo@ms.iswc.ac.cn; internet ms.iswc.ac.cn; Dir: Prof Jun Liang Tian; Deputy Dir: Prof Li Rwi; f 1956; prevention and control of soil erosion; evaluation of land and water resources and their appropriate use; study of the theories and techniques of restoring vegetation and of the sustainable farming of dry land.

Institute of Soil Science, Chinese Academy of Sciences—CAS: 71 E Beijing Rd, POB 821, Nanjing, Jiangsu 210008; tel (25) 7712572; fax (25) 3353590; e-mail zhcao@ns.issas.ac.cn; Prin Officer: Cao Zihong; f 1935; agricultural research; environmental protection; development of new techniques for fertilization.

IOI China: National Marine Data and Information Service, State Oceanic Administration of China, 93 Liuwei Rd, Hedong District, Tianjin 300171; tel (22) 24301292; fax (22) 24304408; e-mail houwf@netra.nmdis.gov.cn; Dir: Prof Hou Wenfeng; national office of the International Ocean Institute—IOI, Malta.

Kunming Institute of Zoology, Chinese Academy of Sciences—CAS: Kunming 650107, Yunnan; tel (871) 5190390; fax (871) 51918323; e-mail wanglin@mail.kiz.ac.cn; internet www.kiz.ac.cn; Dir: Weizhi Ji; research on wildlife conservation and sustainable use of animal resources.

Red Scarf Environmental Protection Action Group: Changsha Wangyue Second Primary School, c/o National Environmental Protection Agency, Beijing 100035; tel (10)

6615-1937; fax (10) 6615-1762; f 1991; attempts to improve the local environment; educational camps for children.

Research Centre for Eco-Environmental Sciences, Chinese Academy of Sciences—RCEES/CAS: POB 2871, Beijing 100085; tel (10) 62555019; fax (10) 62555381; e-mail zhunk@mail.rcees.ac.cn; internet www.rcees.ac.cn; Exec Dir, INFOTERRA national focal point, and Dir, Documentation and Information Centre: Prof Zhang Kangsheng; f 1975; formerly Institute of Environmental Chemistry; adopted current name 1986; affiliated to the State Environmental Protection Administration—SEPA; serves as the INFOTERRA national focal point; comprises six research laboratories: Laboratory of Environmental Aquatic Chemistry, Laboratory of Environmental Analytical Chemistry and Ecological Toxicology, Laboratory of Atmospheric Pollution Chemistry, Laboratory of Systems Ecology, Laboratory of Ecological Engineering, Research Group on National Status and Environmental Policy; also includes three engineering centres: Membrane Technology, Control of Water Pollution and Waste-water Recycling, and Environmental Biological Engineering; library of 30,000 vols; publications.

China, Special Administrative Regions

HONG KONG

Governmental Organizations

Environmental Protection Department: Southorn Centre, 24–28th Floors, 130 Hennessy Rd, Wan Chai; tel 28351001; fax 28382155; e-mail enquiry@epd.gov.hk; internet www.info.gov.hk/epd; Dir: R J S Law; enforcement of environmental protection legislation; environmental monitoring; strategic planning; policy development.

Agriculture, Fisheries and Conservation Department: Cheung Sha Wan Government Offices, 303 Cheung Sha Wan Rd, Kowloon; tel 21506666; fax 23113731; e-mail afdenq@afcd.gcn.gov.hk; internet www.info.govhk.afcd; Dir: Lessie Wei Chui Kit-yee; f 1946; facilitation of agricultural and fishery production and improvement of productivity; monitoring of the welfare of animals, disease control in animals and plants, and compliance with relevant legislation; conservation of flora, fauna and natural habitats, including marine habitats; management of country parks, special areas and marine reserves; control of international trade in endangered species of animals and plants; provision and efficient operation of government wholesale market facilities for fresh food produce; provision of technical services and professional expertise.

Non-Governmental Organizations

Conservancy Association: Capri Bldg, Unit B, 7th Floor, 130 Austin Rd, TST, Kowloon; tel 27286800; fax 27285538; e-mail cahk@netvigator.com; internet www.netvigator.com/~cahk; Gen Sec: Lister Cheung; f 1968; independent organization; monitors and participates in the formulation of government environmental protection policies; training and environmental protection programmes; supports other community groups in the organization of activities to promote environmental protection; environmental resource centre; information resources and educational activities for the general public.

Green Power: 2 Jordan Rd, Ground Floor, GPOB 3723, Kowloon; tel 23142662; fax 23142661; e-mail greenpow@hk.linkage.net; internet www.greenpower.org.hk; Prin Officer: Alexander Yan; f 1988; conservation of local endangered species, including the Chinese white dolphin (*Sousa chinensis*); initiation of eco-village project in Guangzhou (People's Republic of China) and eco-park project in Lungkwutan (Hong Kong); public environmental initiatives.

Tsuen Wan Environmental Resource Centre: Tak Wah Park, Area 18, Tak Wah St, Tsuen Wan; tel 29448204; fax 29448203; internet www.greenpower.org.hk; CEO: Dr Man Chi-Sum; f 1997; environmental protection and education.

Greenpeace China: 302 Mandarin Bldg, 35 Bonham Strand E, Sheung Wan; tel 28548300; fax 27452426; e-mail greenpeace.china@dialb.greenpeace.org; internet www.greenpeace-china.org.hk; Exec Dir: Ho Wai-Chi; national office of Greenpeace International.

Hong Kong Observatory: 134[a] Nathan Rd, Kowloon, Hong Kong; tel 29268200; fax 23119448; e-mail mailbox@hko.gcn.gov.hk; f 1883; Observatory is a department of the Hong Kong Government.

World Wide Fund for Nature—WWF Hong Kong: 1 Tramway Path, Central Mail, POB 12721, Hong Kong; tel 25261011; fax 28452734; Pres: The Hon Sir Kenneth Ping-fan Fung; national affiliate organization of World Wide Fund—WWF International; also operates the WWF programme for the People's Republic of China.

MACAU

Governmental Organizations

Direcção dos Serviços Meteorológicos e Geofísicos—SMG (Macau Meteorological and Geophysical Bureau): Rampa do Observatório, Taipa Grande, CP 93, Macau; tel 850522; fax 850557; e-mail meteo@smg.gov.mo; internet www.smg.gov.mo; f 1952; responsible to the Secretary for Transport and Public Works.

Gabinete Técnico do Ambiente (Technical Cabinet of the Environment): Macau.

Serviços Florestais e Agrícolas de Macau (Forestry and Agricultural Services of Macau): Rua Central 107, Macau; principal body with responsibility for management of natural resources and nature conservation.

CHINA, TAIWAN

Governmental Organizations

Environmental Protection Administration—EPA: 41 Chung Hua Rd, Sec 1, Taipei; tel (2) 23117722; fax (2) 23116071; e-mail www@sun.epa.gov.tw; internet www.epa.gov.tw; Dir-Gen: Lin Jun-yi; Admin: Dr Edgar Lin; subordinate organization of the Executive Yuan (Council of Ministers).

Atomic Energy Council: 67 Lane 144, Kee Lung Rd, Sec 4, Taipei; tel (2) 23634180; fax (2) 23635377; internet www.aec.gov.tw; Chair: Dr Hsia Der-yu; subordinate organization of the Executive Yuan.

Institute of Nuclear Energy Research: POB 3, Lung-Tan 32500; tel (2) 23651717; fax (2) 24711064; Dir: Dr Hsia Der-yu; f 1968.

Council of Agriculture—COA: 37 Nan Hai Rd, Taipei 100; tel (2) 23812991; fax (2) 23310341; e-mail webmaster@www.coa.gov.tw; internet www.coa.gov.tw; Chair: Ching-Lun Lee; Sec-Gen: Shin-hsien Chan; f 1984; subordinate organization of the Executive Yuan; policy-making body with responsibility for agriculture, forestry, fisheries, the animal industry and food administration; promotion of technology and provision of external assistance.

National Science Council—NSC: 106 Ho-Ping E Rd, 17th–22nd Floors, Sec 2, Taipei 10636; tel (2) 27377992; fax (2) 27377248; e-mail nsc@nsc.gov.tw; internet nscnt07.nsc.gov.tw/english/index.html; Chair: Weng Cheng-i; subordinate organization of the Executive Yuan; promotion of overall national scientific and technical development, support for academic research, and establishment and administration of science-based industrial parks (SIPs).

Non-Governmental Organizations

Green Party: 281 Roosevelt Rd, 11F-1, Sec 3, Taipei; tel (2) 23621362; fax (2) 23621361; e-mail gptaiwan@ms10.hinet.net; internet gptaiwan.yam.org.tw; Chair: Chen Guang-Yeu; f 1996; environmental political party; formed by a faction of the Democratic Progressive Party—DPP.

Taiwan Environmental Protection Union—TEPU: 29 Lane 128, Sec 3, Roosevelt Rd, Taipei; tel (2) 23636419; fax (2) 23623458; e-mail tepu@ms1.hinet.net; internet tepu.yam.org.tw; Contacts: Cheng-Yan Kao, Kuang-Yu Chen; f 1987; promotes environmental protection and ecological preservation; ten local chapters throughout Taiwan.

COLOMBIA

Governmental Organizations

Ministry of the Environment, Housing and Territorial Development: Calle 37, No 8-40, Santafé de Bogotá, DC; tel (1) 288-6877; fax (1) 288-9788; internet www.minambiente.gov.co; Minister: Sandra Suárez Pérez.

Instituto de Hidrología, Meteorología y Estudios Ambientales—IDEAM (Institute of Hydrology, Meteorology and Environmental Studies): Diagonal 97, No 17-60, 1°, 2°, 3° y 7°, Santafé de Bogotá, DC; tel (1) 283-6927; fax (1) 635-6218; internet www.ideam.gov.co; Dir: Pablo Leyva; Man, INFOTERRA national focal point: Flor Alba Moreno Galindo; f 1976; fmrly Instituto Colombiano de Hidrología, Meteorología y Adecuación de Tierras—HIMAT; adopted current name 1995; irrigation; flood control; drainage; hydrology; meteorology; the Documentation Centre serves as the INFOTERRA national focal point.

Ministry of Agriculture and Rural Development: Avda Jiménez, No 7-65, Santafé de Bogotá, DC; tel (1) 334-1199; fax (1) 284-1775; e-mail minagric@colomsat.net.co; internet www.minagricultura.gov.co; Minister: Carlos Gustavo Cano Sanz.

Instituto Nacional de los Recursos Naturales Renovables y del Ambiente—INDERENA (National Institute of Renewable Natural Resources and the Environment): Diagonal 34, No 5-18, 3°, Santafé de Bogotá, DC; tel (1) 285-4417; fax (1) 283-3458; f 1968; research into conservation and use of renewable natural resources; environmental protection; management of parks and natural areas.

Corporación Autónoma Regional del Valle del Cauca—CVC (Autonomous Regional Corporation of the Cauca Valley): Carrera 56, No 11-36, 4°, Apdo 2366, Santiago de Cali, Valle del Cauca; tel (92) 339-8949; fax (92) 330-4080; Deputy Dir: Oscar Libardo Campo Velasco; f 1954; execution of government policies and programmes concerning the environment and sustainable resource use; ensures compliance with government regulations; land reclamation; water resource management; conservation aid; land use.

Corporación Nacional de Investigación y Fomento Forestal—CONIF (National Corporation of Forestry Research and Development): Parque La Florida, Apdo 095153, Santafé de Bogotá, DC; tel (1) 267-6844; fax (1) 221-8624; Pres: Alberto Leguizamo Barbosa.

Instituto Colombiano de la Reforma Agraria—INCORA (Colombian Institute of Agrarian Reform): Centro Administrativo Nacional—CAN, Avda El Dorado, Apdo 151046, Santafé de Bogotá, DC; tel (1) 222-0963; Dir: Germán Bula E; f 1962; administration of public land on behalf of the Government; reclamation of land by irrigation, drainage and construction in order to increase productivity in agriculture and stock-breeding; technical assistance and loans; supervision of redistribution of land.

Non-Governmental Organizations

Centro de Investigaciones para el Desarrollo Integral—CIDI, Universidad Pontificia Bolivariana (Research Centre for Integrated Development, Bolivian Pontifical University): Circular 1, 70.01, Apdo 56006, Medellín, Antioquia; tel (1) 415-9191; fax (1) 411-2372; e-mail cidi@jonva.upb.edu.co; internet www.upb.edu.co; Dir: Dr Eduardo Dominguez; f 1969; research and consultancy in environmental pollution, energy technology, urban development, bioengineering and biophysics; information centre.

Comité Colombiano de la UICN (Colombian Committee for IUCN): Avda 13, No 87-43, Apdo 55402, Santafé de Bogotá, DC; tel (1) 616-9363; fax (1) 210-4515; Deputy Dir: Dr Elsa Matilda Escobar; national office of the World Conservation Union—IUCN.

Confederación Colombiana de Organismos no Gubernamentales (Colombian Confederation of Non-Governmental Organizations): Carrera II, No 71-40, Oficina 301, Santa Fé de Bogotá, DC; tel and fax (1) 2127363; e-mail confe@colnodo.apc.org; internet www.ourworld.compuserve.com/homepages/ongcolombia; Pres: Alberto Jiménez; Exec Dir: Inés Useche de Brili; f 1989; institutional development, information diffusion.

Corporación Colombiana para la Amazonia 'Araracuara'—COA (Araracuara Colombian Association for the Amazon): Calle 20, 5-44, Apdo 034174, Santafé de Bogotá, DC; tel (1) 283-6755; fax (1) 286-2418; Man: Dr Dario Fajardo Montaña; Tech Dir: Dr Braulio Gutiérrez; scientific research on biological, social and ecological aspects of the Amazon region of Colombia.

Corporación Integral para el Desarrollo Cultural y Social—CODECAL (Integrated Corporation for Cultural and Social Development): Carrera 38d, No 60-56, Santafé de Bogotá, DC; tel (1) 315-3185; fax (1) 222-5807; e-mail codecal@colnodo.apc.org; internet www.colnodo.apc.org/codecal; f 1972; basic environmental educational activities including courses and workshops, publications and community projects.

Fondo para la Protección del Medio Ambiente en Colombia—FEN Colombia (Colombian Environmental Protection Foundation): Calle 71ª, No 6-30, 20°, Santafé de Bogotá, DC; tel (1) 217-2100; fax (1) 211-9776; Dir: Angel Guarnizo Vásquez; Research Co-ordinator: Eduardo Guerrero; promotion of research into protection of natural resources and the environment.

Fundación Codesarrollo (Co-development Foundation): Edif Caracas 1, 5°, Calle 54, No 45-63, Medellín, Antioquia; tel (4) 513-1311; fax (4) 513-1322; e-mail codesarro@antioq.grupopro.com.co; Exec Dir: Luis Alberto Gómez Ramírez; f 1960; independent organization; marketing of recyclable wastes and facilitation of their appropriate management; provision of solid waste classification and disposal services to industry, business and the residential sector; recycling facilities; environmental assessment and consultancy; impact studies; environmental planning, management and auditing.

Fundación Herencia Verde—FHV (Green Heritage Foundation): Calle 4ª Oeste, No 3ª-32, Apdo 32802, Cali, Valle del Cauca; tel (2) 3880-8484; fax (2) 3881-3257; e-mail fhv@cali.cetcol.net.co; Exec Dir: Antonio Solarte Sánchez; f 1983; conservation of natural resources; environmental education; planning and management of nature reserves and tropical forest; sustainable development; conservation of biodiversity; research into medical use of plants.

Fundación Natura—FN (Nature Foundation): Calle 61, No 4-26, Apdo 55402, Santafé de Bogotá, DC; tel (1) 345-1216; fax (1) 249-6250; e-mail enatura@impsat.net.co; internet www.natura.org.co; Exec Dir: Dr Juan Pablo Ruiz; f 1984; management of protected area projects; co-operation with the Government and the private sector; conservation of biological resources.

Fundación para la Comunicación Popular—FUNCOP-CAUCA (Foundation for Popular Communication): Calle 10 Norte, No 29, Apdo 2096, Popayán, Cauca; tel (28) 23-5590; fax (28) 23-5590; e-mail funcop@popayan.cetcol.net.co; internet www.funcop.org.co; Exec Dir: María Cecilia López Sacconi; organization concerned with democracy, rural development projects, training of rural communities, health and ecology.

Fundación para la Defensa del Interés Público—FUNDEPUBLICO (Foundation for Defence of the Public Interest): Calle 71, No 5-83, Santafé de Bogotá, DC; tel (1) 210-4737; Pres: Dr German Sarmiento Palacio; promotion of environmental legislation.

Fundación para la Educación Superior—FES (Higher Education Foundation): Apdo 5744, Cali, Valle del Cauca; tel (2) 884-5933; fax (2) 883-4706; e-mail fesambi@ibm.net; Pres: Dr Mauricio Cabrera; f 1964; promotion of social development in Colombia; supports programmes for sustainable socio-economic development, education, health and the environment.

Fundación para la Investigación y Protección del Medio Ambiente—FIPMA (Foundation for Research and Protection of the Environment): Carrera 36ª, No 5b1-54, Apdo 2741, Cali, Valle del Cauca; tel (2) 556-7519; fax (2) 558-5092; Dir: José M Borrero; Sec: Martha C Velásquez; lobbying and activism for environmental protection; nature protection and promotion of the sustainable use of resources.

Fundación para un Mejor Ambiente—FMA (Foundation for a Better Environment): Calle 8ª, No 3-14, 1°, Apdo 1565, Cali, Valle del Cauca; tel (2) 382-3271; fax (2) 382-4627; Exec Dir: Dr Humberto Swann Barona; non-profit organization for conservation of the environment.

Fundación Pro-Sierra Nevada de Santa Marta (Foundation for the Sierra Nevada of Santa Marta): Apdo 5000, Santafé de Bogotá, DC; tel (1) 310-0571; fax (1) 217-3487; Exec Dir: Juan Mayr; conservation of the ecological and cultural heritage of the Sierra Nevada; research; sustainable development; co-operation with the Government and the Indian and Colono communities.

Fundación Puerto Rastrojo—FPR (Puerto Rastrojo Foundation): Carrera 10, No 24-76, Of 602, Apdo 241438, Santafé de Bogotá, DC; tel (1) 284-9010; fax (1) 284-3028; e-mail rastrojo@openway.com.co; Dir: Patricio von Hildebrand; evaluation and protection of Amazonian biodiversity; management and protection of threatened species' populations; guidance for indigenous communities in sustainable use of forest resources; defence of protected areas and of the territories of indigenous communities.

Instituto de Estudios Ambientales—IDEAS, Universidad Nacional de Colombia (Environmental Studies Institute, Colombian National University): Edif Camilo Torres, Bloco B-7, Carrera 50, No 27-70, Santafé de Bogotá, DC; tel (1) 368-1295; fax (1) 368-1295; e-mail idea@bacata.usc.unal.edu.co; Prin Officers: Julio Carrizosa Umaña, Margarita Pacheco; f 1991; sustainable development in urban and rural environments; sustainable transport; environmental education.

Partners of the Americas, Antioquia (Colombia): Calle 52, No 47-42, Edif Coltejer, 14°, Medellín, Antioquia; tel (4) 514-3211; fax (4) 251-6117; e-mail compamericas@epm.net.co; internet www.partners.net; Pres: Gabriel Márquez Vélez; Vice-Pres: Margarita Correa Henao; promotes public participation; mobilizes regional co-operation to deal with economic affairs and social development; aids community leaders; supports NGOs; initiates programmes on health, education, sports, culture and the environment.

Sociedad Colombiana de Ecología (Colombian Society of Ecology): Edif Camilo Torres, Bloco C, Módulo 4, 5°, Carrera 50, No 27-70, Apdo 8674, Santafé de Bogotá, DC; tel (1) 221-5504; fax (1) 221-6950; Pres: Dr J M Idrobo; promotion of conservation and rational use of natural resources; preservation of wildlife habitats and biodiversity.

COMOROS

Governmental Organizations

Ministry of Planning and Territorial Improvements: Moroni; Minister: Mamza Saïd.

Ministry of Rural Development, Fisheries, Crafts and the Environment: Moroni; Minister: Abdoulhamid Mohamed.

Société de Développement de la Pêche Artisanale des Comores—SODEPAC (Comoros Society for the Development of the Fishing Industry): Moroni; government agency.

CONGO, DEMOCRATIC REPUBLIC

Governmental Organizations

Ministry of the Environment: Kinshasa-Gombe; Minister: Anselme Enerunga.

Direction Programmation, Formation et Relations Internationales—DPFRI (Directorate of Planning, Training and International Relations): 15 ave des Cliniques, BP 12348, Kinshasa-Gombe; tel (12) 33250; Man, INFOTERRA national focal point: Ntondo Lumuka Nantole.

Ministry of Agriculture: Immeuble SOZACOM, 3rd Floor, blvd du 30 juin, BP 8722 KIN I, Kinshasa-Gombe; tel (12) 31821; Minister: Valentin Senga.

Ministry of Energy: Immeuble SNEL, 239 ave de la Justice, BP 5137 KIN I, Kinshasa-Gombe; tel (12) 22570; Minister: Kalema Lusona.

Centre d'Etude et d'Expérimentation des Technologies Appropriées—CEETA (Appropriate Technology Study and Experimentation Centre): BP 2849, Bukavu; tel 3072; Dir: Dr Georges Defour; f 1980; research areas include engineering, renewable energy, soil erosion and pesticides.

Institut National pour l'Etude et la Recherche Agronomique—INERA (National Institute for Agronomic Studies and Research): BP 2037, Kinshasa 1; tel (12) 32332; Pres: Dr Masi Mango Ndyanabo; f 1933; library of over 38,000 vols; publications.

Institut pour la Conservation de la Nature (Institute for Nature Conservation): 13 ave de la Clinique, BP 868, Kinshasa 1; tel (12) 31401; fax (12) 275647; Pres: Mankoto Ma Mbaelele; Dir and Admin: Mokwa Vankang Izmtsho; formerly Institut Zaïrois pour la Conservation de la Nature; official government agency responsible for the management, monitoring and administration of national parks, nature reserves, hunting areas and trapping stations.

Non-Governmental Organizations

Centre de Recherche pour l'Exploitation de l'Energie Renouvelable—CREER, Université de Kinshasa (Research Centre for the Exploitation of Renewable Energy, University of Kinshasa): BP 127, Kinshasa XI; tel (12) 30123.

CONGO, REPUBLIC

Governmental Organizations

Ministry of Agriculture, Livestock, Fishing and the Promotion of Women: BP 2453, Brazzaville; tel 81-41-31; fax 81-19-29; Minister: Jeanne Dambenzet.

Ministry of Forestry and the Environment: BP 98, Brazzaville; tel 81-41-37; fax 81-41-34; e-mail secretariat@minifor.com; Minister: Henri Djombo.

Service National de Reboisement—SNR (National Reforestation Department): BP 839, Pointe-Noire; tel 94-02-79; fax 94-09-05; e-mail snco10@calva.com; Man Dir: Antoine Moutanda; Head, Tech Dept: Georges Mapola; fmrly Office Congolais des Forêts (Congolese Forest Agency); development of the forestry industry; reforestation of forest and savanna; afforestation of savanna; agroforestry activities; affiliated with several NGOs and research institutions.

Ministry of Mines, Energy and Water Resources: BP 2124, Brazzaville; tel 81-02-64; fax 81-50-77; Minister: Philippe Mvouo.

Centre d'Etudes sur les Ressources Végétales—CERVE (Study Centre for Plant Resources): BP 1249, Brazzaville; tel 81-21-83; Dir: Prof Laurent Tchissambou; f 1985; cataloguing of the plant species of the Republic of the Congo; promotion of traditional phytotherapy; development of indigenous and exotic fodder plants.

Conseil Supérieur de l'Environnement (Higher Council for the Environment): BP 2499, Brazzaville; tel 83-39-99; Dir: Jean Nanga-Maniane; f 1977; formation, development and coordination of environmental policy in the Congo.

Unité de Recherche sur les Ecosystèmes Aquatiques—UREA (Research Unit on Aquatic Ecosystems): Centre de Recherche Géographique et de Production Cartographique—CERGEC, BP 125, Brazzaville; fax 83-84-12; Head of Hydrometric Network: Bienvenu Maziezoula; responsible for the national hydrometric network and for liaison with relevant interational organizations.

Non-Governmental Organizations

Centre Technique Forestier Tropical (Centre for Tropical Forestry): BP 764, Pointe-Noire; Dir: J-C Delwaulle; f 1958; forestry research.

Institut de Recherche pour le Développement—IRD (Development Research Institute): BP 181, Brazzaville; tel 83-26-80; fax 83-29-77; Dir: Claude Reichenfeld; f 1947; formerly ORSTOM-DGRST Centre; a regional centre of the Institute, which is based in France; fields of research include bioclimatology, hydrology, pedology, botany, agricultural and medical entomology, nutrition, major endemic diseases, microbiology, phytopathology and social sciences.

Pointe-Noire Centre: Institut de Recherche pour le Développement—IRD, BP 1286, Pointe-Noire; tel 94-02-38; fax 94-39-81; Dir: France Reversat; f 1950; biological and physical oceanography; pedology; botany; plant ecology and physiology.

COSTA RICA

Governmental Organizations

Ministry of the Environment and Energy: Avdas 8 y 10, Calle 25, Apdo 10.104, 1000 San José; tel 233-4533; fax 257-0697; e-mail root@ns.minae.go.cr; internet www.miniae.go.cr; Minister: Carlos Manuel Rodríguez Echandi.

Ministry of Agriculture and Livestock: Apdo 10.094, 1000 San José; tel 231-2344; fax 231-2062; internet www.mag.co.cr; Minister: Rodolfo Coto Pacheco.

Consejo Nacional de Investigaciones Científicas y Tecnológicas—CONICIT (National Council for Scientific and Technological Research): Apdo 10318, 1000 San José; tel 224-4172; fax 225-2673; e-mail farmijo@www.conicit.go.cr; internet www.conicit.go.cr; Man, INFOTERRA national focal point: Freddy Armijo; Tech Information Officer: Francisco Vargas Villabolos; the Registry and Information Center (CERICYT) serves as the INFOTERRA national focal point.

Non-Governmental Organizations

Asociación ANAI (ANAI Association): Apdo 170, 2070 Sabanilla; tel 224-3570; fax 253-7524; e-mail anaicr@correo.co.cr; Co-Dirs: William McLarney, James R Lynch, Robert B Mack; f 1983; promotion of sustainable development amongst rural communities and local organizations in the Talamanca region; preservation of biodiversity; forest management.

Asociación Comunidades Ecologistas La Ceiba/Amigos de la Tierra Costa Rica—COECOCEIBA-AT (La Ceiba Ecologist Communities Association/FoE Costa Rica): Apdo 12.423, 1000 San José; tel 223-3925; e-mail gavitza@racsa.co.cr; Pres: Gabriel Rivas Ducca; Vice-Pres: Carmen Juanes Marcos; Sec: Isaac Rojas; f 1999; community management of forest and biodiversity; campaign work to achieve sustainability (and against activities that destroy sustainability); networking with social and ecologist organizations affiliated to FoEI.

Asociación Conservacionista de Monteverde—ACM (Monteverde Conservation League): Apdo 10 581, 1000 San José; tel 645-5003; fax 645-5104; e-mail acmmcl@sol.racsa.co.cr; internet www.acmonteverde.org; Pres: Jorge Maroto; Exec Dir: Carlos L Muñoz; Dir of Environment and Human Development: Yúber Rodríguez; Dir of Protection and Management: Gerardo Céspedes; f 1986; conservation of rainforests; acquisition of land for reforestation; environmental education; forest reserve of over 50,000 acres (20,235 ha); conservation research; habitat restoration; ecotourism.

Asociación Cultural SEJEKTO de Costa Rica (SEJEKTO Cultural Association of Costa Rica): Apdo 1.293, Moravia; Pres: José Duálok Rojas; f 1985; nature conservation; scientific research.

Asociación Demográfica Costarricense (Costa Rican Demographic Association): Apdo 10.203, 1000 San José; tel 231-4361; fax 231-4430; Exec Dir: Victor H Morgan Alvarado; private non-profit organization; information, research, education and training in natural resources and the environment.

Asociación para la Conservación de los Recursos Naturales de Costa Rica—ACORENA-CR (Costa Rican Association for Conservation of Natural Resources): 150m oeste Bomberos de Guápiles, Apdo 84.330, 1000 San José; tel 287-0540; fax 287-0545; Vice-Pres: Franklin Carmiol Umaña; f 1990; private conservation group; environmental education in rural areas; protection of tropical forests.

Asociación Talamanqueña de Ecoturismo y Conservación—ATEC (Talamanca Association for Ecotourism and Conservation): Puerto Viejo de Talamanca, Limón; tel 750-0398; fax 750-0191; e-mail atecmail@racsa.co.cr; internet www.greencoast.com; Pres: Willis Rankin Gonzalez; Vice-Pres: Alex Paz Balma; Dirs: Noble Baker, Alaine Berg; f 1990; non-profit organization dedicated to the promotion of ecologically sound tourism, environmental education, conservation and local enterprise.

Centro Científico Tropical (Tropical Science Centre): B La Granja, Apdo 83.870, 1000 San José; tel 253-3267; fax 253-4963; e-mail cecitrop@sol.racsa.co.cr; internet www.cct.or.cr; Exec Dir: Julio C Calvo; f 1962; scientific research, consulting and teaching programmes concerning the biological and physical resources of the tropical environment; principal areas of interest include climate change, forestry, ecology, dendrology, land-use planning, environmental impact assessment and conservation of natural resources; owns and operates the Monteverde Cloud Forest Preserve and Los Cusingos Neotropical Bird Sanctuary.

Centro de Derecho Ambiental y de los Recursos Naturales—CEDARENA (Environmental and Natural Resources Law Center): 75m suroeste del antiguo Higuerón, Apdo 134, 2050 San Pedro; tel 253-7239; fax 225-5111; e-mail cedarena@sol.racsa.co.cr; Pres: Silvia Chaves Quesada; Treas: Robert Wells; f 1989; non-profit, independent association of lawyers, law students and volunteers concerned with environmental law, land tenure and the rights of indigenous peoples; establishment of a permanent centre for research, information exchange and education.

Corporación de Investigaciónes para el Desarrollo Socio Ambiental—CIDESA (Socio-Environmental Development Research Corporation): Apdo 414, 3000 Alajuela; tel 233-1072; fax 223-1609; Vice-Pres: Florangel Villegas Verdú; environmental management and planning; scientific research; resource management; conservation education.

Fundación de Parques Nacionales (National Parks Foundation): Apdo 1.108, 1000 San José; tel 257-2239; fax 222-4732; Pres: Dr Pedro León; protection and development of the

Protection): 1011 Budapest, Fő u 44–50; tel (1) 457-3545; fax (1) 201-4282; e-mail bacso.laszlone@ktmdom2.ktm.hu; internet www.ktm.hu/intez/fofel; Dir-Gen: Dr Katalin Ughy; Head, Secr: László Bacsó; f 1991; inspectorate for waste management and control of noise, air and water pollution; co-ordinates 12 regional environmental inspectorates and nine national park directorates.

Környezetgazdálkodási Intézet (Institute for Environmental Management): 1369 Budapest, POB 352; tel (1) 374-3500; fax (1) 311-5826; e-mail kozpont.kgi@ktm.x400gw.itb.hu; Dir: Dr E Istv; Tech Officer: Dr Julianna Tokar; promotes and sustains sound environmental management; houses the INFOTERRA national focal point, which disseminates environmental information, provides an environmental information service and manages the National Professional Library of Environmental Protection and Water Management.

Ministry of Agriculture and Rural Development: 1055 Budapest, Kossuth Lajos tér 11; tel (1) 301-4000; fax (1) 302-0402; internet www.fvm.hn; Minister: Dr Imre Németh.

Non-Governmental Organizations

Alba Kör Erőszakmentes Mozgalom a Békéért (Alba Circle, Non-Violent Movement for Peace in Hungary): 1461 Budapest, V Vadász u 29, POB 225; tel and fax (1) 332-6109; e-mail alba@albakor.hu; internet www.albakor.hu; f 1990; includes a Green Group for the protection of the environment against military and civilian damage.

Bokor Oko-Csoport (Eco Group of Bokor): 2000 Szentendre, Kovács L u 16; tel and fax (26) 318-797; e-mail di@bok.zpok.hu; internet www.bocs.hu; Sec: Istvan Döry; f 1989; promotes ecology and biodiversity.

Eco-Village Network: 7478 Bárdudvarnok, Visnyeszéplak 48; tel (82) 481-487; fax (82) 269-4460; e-mail pacsirta@vpszk.beme.hu; Contact: Róbert Fidrich.

Egyetemes Létezéós Természetvédelmi Egyesÿlet (ELTE Nature Conservation Club): 1054 Budapest, Vadász u 29; tel and fax (1) 111-7855; e-mail etk@zpok.hu; internet www.etk.hu; Chair: Veronika Móra; Sec: Ernese Balogh; f 1983; educational and conservation projects; publications.

Environmental Management and Law Association—EMLA: 1082 Budapest, Állő út 66/b VI/4; tel (1) 303-5504; fax (1) 333-2931; e-mail emla@emla.zpok.hu; Contact: Kiss Csaba.

Environmental Partnership Foundation: 1117 Budapest, Móricz Zsigmond kšrtér 15, I/1; tel and fax (1) 166-8866; e-mail okotars@pship3.zpok.hu; Contact: Zsuzsa Foltanyi.

Erdeszeti Tudomanyos Intezet (Forest Research Institute): 1023 Budapest, Frankel Leo u 42–44; tel (1) 115-0624; fax (1) 115-1806; f 1949; general research in forestry and environmental matters.

Gaja Kornyezetvedo Egyesulet (Gaia Environmental Society): 8008 Székesfehérvar, Petófi s u 5, POB 40; tel (22) 343-669; Contact: Istvan Szilli; environmental education; promotes biodiversity and ecology, health and sanitation.

Geo-Environ Nature Protection Society: 6701 Szeged, POB 653; tel and fax (62) 310-865; Contact: Tith Imre; environmental education; lobbying activities to preserve the environment.

Green Action: 3525 Miskolc, Kossuth u 13; tel (46) 349-806; fax (46) 352-010; promotes local-level environmental activity; co-operates with international NGOs and governments; conservation and protection of natural energy sources.

Green Heart Youth Nature Conservation Movement: 2013 Pomáz, Mátyás Király u 2; tel and fax (26) 325-957; e-mail zoldsziv@freemail.com; internet www.c3.hu/~zsziv; Pres: Anikó Orgoványi; Vice-Pres: Victor András; natural and environmental education; conservation of environmental resources; water quality monitoring on the Danube and Tisza rivers in five countries; publications; 15,000 members in 16 countries; member groups in London and in Manila, Philippines.

Green Party of Hungary: Budapest; Chair: Zoltan Medveczky; environmental political party.

Hungarian Environmental Education Foundation: 1113 Budapest, Zsombolyai u 6; tel (1) 132-2177; Contact: Victor András.

Hungarian Friends of Nature Alliance: 1065 Budapest, Bajcsy-Zsilinszky út 31, I/3; tel (1) 111-2467; fax (1) 153-1930; e-mail nagyand@okopr.zpok.hu; Contact: László Kalmár.

ISTER (East European Environmental Research Unit): 1054 Budapest, Vadász u 29; tel and fax (1) 153-0100; e-mail vamaa@vargham.zpok.hu; Pres: János Vargha; CEO: György Droppa; research, advisory work and publications concerning environmental protection.

IUCN Hungary: 1371 Budapest, POB 433; tel (1) 201-8964; fax (1) 202-0452; Contact: Ferenc Németh.

Környezeti Nevelési Hálózat (Environmental Education Network): 2000 Szentendre, Kovács L u 16; tel (26) 318-797; fax (26) 318-195; e-mail di@bok.zpok.hu; Contact: János Geiszler.

Levegő Munkacsoport (Clean Air Action Group): Budapest 1465, POB 1676; tel (1) 206-5599; fax (1) 165-0438; e-mail levego@levego.zpok.hu; internet www.levego.hu; Nat Sec: András Lukács; Chair, Cttee of Experts: Dr Dezső Radó; f 1988; association of 57 Hungarian environmental NGOs; environmental protection in relation to transport and energy issues; green economics; public awareness-raising activities; research; advocacy.

Living Water Network: 2600 Vác, Ilona u 3, POB 184; tel (27) 304-484; fax (27) 311-179; e-mail barabas@habakukk.zpok.hu; internet www.zpok.hu; Contact: Katalin Rozgonyi.

Magyar Környezetvédelmi Egyesÿlet—MKVE (Hungarian Society for Environmental Protection): 1387 Budapest, Szende Pál u 13, POB 42; tel (1) 277-1322; Contact: Sámuel Kántor.

Magyar Madártani és Természetvédelmi Egyesület—MME (BirdLife Hungary): 1121 Budapest, Költő u 21; tel and fax (1) 275-6247; e-mail mme@mme.hu; internet www.mme.hu; Pres: György Kállay; Vice-Pres: László Haraszthy; Dir: József Fidlóczky; f 1974; environmental education; nature conservation; conservation of birds, including migratory species; publications; affiliated to BirdLife International.

Magyar Természetvédők Szövetsége (Friends of the Earth Hungary): 1091 Budapest, Ulloi U 91/B III/21; tel and fax (1) 216-7297; e-mail mtvsz@elender.hu; internet www.mtvsz.hu; Pres: Erzsbèbet Schmuck; f 1989; promotes exchange of information, co-operation and joint programmes with other organizations; environmental policy development; creates public awareness of environmental issues; has 76 member groups; affiliated with Friends of the Earth International—FOEI, the European Environmental Bureau—EEB, CEEWEB, the World Conservation Union—IUCN and ANPED.

Magyar Tudományos Akadémia Ökológiai és Botanikai Kutatóintézete (Ecological and Botanical Research Institute, Hungarian Academy of Sciences): 2163 Vácrátót, Alkotmány u 2–4; tel (28) 360-122; fax (28) 360-110; Dir: Dr Attila

Borhidi; f 1952; terrestrial and aquatic ecology; nature conservation.

Naturfilm Ltd: 1221 Budapest, Leanyka u 3; tel and fax (1) 229-1062; Contact: Rácz Gábor; production of nature and wildlife films and books; mem, UNEP 500 Roll of Honour 1987.

Ökoszolgálat Alapítvány, Környezetvédelmi Tanácsadó Szolgálat (Ecoservice Foundation, Environmental Counselling Service): 1054 Budapest, Vadász u 29; tel and fax (1) 269-4460; e-mail okosz@okoszolagat.hu; internet www.okoszolagat.hu; Head: Andrea Hágen; f 1990; information service; data base of environmental organizations, publications and courses.

Okotars Alapítvany (Environmental Partnership for Central Europe): 1519 Budapest, POB 411; tel (1) 166-8866; fax (1) 181-3393; e-mail foltanyi@pship.zpok.hu; Country Dir: Zsuzsa Foltanyi; f 1991; provides small grants and technical and organizational assistance to NGOs and local authorities in Hungary and other Central European countries; part of a network of similar organizations in the Czech Republic, Poland and Slovakia.

Országos Erdészeti Egyesület (Hungarian Forestry Association): 1027 Budapest, Fő u 68; tel (1) 201-6293; fax (1) 201-7737; Pres: András Schmotzer; Sec-Gen: Gábor Barátossy; f 1866; forestry and forest industries; environmental protection; 5,000 mems.

Pangea Environmental and Cultural Association: 2601 Vác, Béke út 57; tel (27) 304-484; fax (27) 304-483; e-mail postmast@goncol.zpok.hu; internet www.zpok.hu; Contact: László Breuer.

Parnasszus Cultural Association, Environmental Section: 4032 Debrecen, Doberdó út 13; tel (52) 329-949; Contact: Klára Iveti.

Reflex Kornyezetvédő Egyesület (Reflex Environmental Protection Society): 9024 Győr, Bartók Béla u 7; tel (96) 316-192; fax (96) 310-988; e-mail reflex@fnet.gyor.hu; Pres: József Lajtmann; f 1987; renewable energy, energy efficiency consultancy, environmental law and legal consultancy, nature conservation, environmental conflict resolution and involvement by the public and shareholders in environmental decision- and policy-making; solid waste management and recycling; environmental planning; affiliated with Reflex Energy Office.

Tree of Life Environmental Alliance: 3300 Eger, Bajcsy-Zsilinszky u 9; tel and fax (36) 411-036; e-mail eletfa@mail.sprinternet.hu; Contact: Ferenc Bárdos.

Vízgazdálkodási Tudományos Kutató Rt—VITUKI (Water Resources Research Centre): 1095 Budapest, Kvassay Jeno út 1; tel (1) 215-6140; fax (1) 216-1514; e-mail vitukirt@vituki.hu; internet www.datanet.hu/hydroinfo/vituki; Dir-Gen: P Bakonyi; Deputy Dir-Gen: I Mayer; f 1952; comprises three institutes; provides research, consulting and engineering services, including water quality surveys, formulation of monitoring strategies, sampling programmes for conventional and trace pollutants, toxicological and hydrobiological analyses, environmental base-line surveys, impact reports and assessments; incorporates the National Water Quality Central Laboratory; library and documentation centre; training courses and conferences.

ICELAND

Governmental Organizations

Ministry for the Environment: Vonarstræti 4, 150 Reykjavík; tel 5458600; fax 5624566; e-mail postur@environment.is; internet www.environment.is; Minister: Sigriður Anna Þorðardóttir; Man, Dept of Public Relations and Information: Hugi Olafson; the Department of Public Relations and Information serves as the INFOTERRA national focal point.

Hollustuvernd Ríkisins (Environment and Food Agency): Ármúla 1[a], POB 8080, 105 Reykjavík; tel 5851000; fax 5851010; internet www.hollver.is.

Náttúruvernd Ríkisins (Nature Conservation Agency): Hlemmur 3, POB 5324, 125 Reykjavík; tel 5627855; fax 5627790; e-mail nr@ni.is; internet www.ni.is/nr; f 1997; formerly the Nature Conservation Council; protection of Iceland's natural landscapes; the Nature Conservation Act aims to encourage the interaction of man and nature in such a way as to prevent unnecessary damage to wildlife and land and to prevent the pollution of sea, fresh water and air.

Veiðistjóraembættið (Wildlife Management Institute): Hafnarstræti 97, POB 465, 602 Akureyri; tel 4622820; fax 4622325; e-mail veidist@nattfs.is; internet www.nattfs.is/veidi; Dir: Asbjorn Dagbjartsson; f 1957; wildlife management, wildlife management research, regulation of hunting and animal damage control.

Ministry of Agriculture: Sölvhólsgata 7, 150 Reykjavík; tel 5459750; fax 5521160; e-mail postur@lan.stjr.is; internet www.stjr.is/lan; Minister: Guðni Ágústsson.

Ministry of Fisheries: Skúlagötu 4, 150 Reykjavík; tel 5458300; fax 5621853; e-mail postur@hafro.is; internet www.stjr.is/sjr; Minister: Árni M Mathiesen.

Landgræðsla Ríkisins (Soil Conservation Service): Gunnarsholt, 851 Hella; tel 4883000; fax 4883010; e-mail lgr@landgr.is; internet www.landgr.is; Dirs: Sveinn Runólfsson, Dr Andrés Arnalds; f 1907; erosion control; reclamation of areas of degraded soil; protection and improvement of vegetation; seed production (*Leymus arenarius, Deschampsia berengensis* and *Lupinus noothatensis*).

Orkustofnun (National Energy Authority of Iceland): Grensásvegur 9, 108 Reykjavík; tel 5696000; fax 5688896; e-mail os@os.is; internet www.os.is; Dir Gen: Thorkell Helgesson; f 1967; independent organization reponsible to the Ministry of Industry and Commerce; offers advice on matters concerning energy; studies Icelandic energy resources and their use.

Non-Governmental Organizations

Institute for Environmental Issues and Development: Hedi, Rangarvallahreppi, 851 Hella; tel 9875179.

Samband Islenskra Kristnibodsfélaga (Icelandic Mission Society): POB 4060, 124 Reykjavík; tel 5888899; fax 5888840; e-mail skuli@sik.is; internet www.sik.is; Gen Sec: Skuli Svavarsson; f 1929; activities in developing countries include promoting reforestation.

Skógraektarfélag Íslands (Icelandic Forestry Association): Ránargata 18, 101 Reykjavík; tel 5518150; fax 5627131; e-mail skog@skog.is; internet www.skog.is; Chair: Magnus

Jóhannesson; f 1930; publishing, education, planning of forests areas and planting trees.

Vinstrihreyfing-Grænt Frambod (Left–Green Alliance): c/o v/Austurvöll, 150 Reykjavík; tel 5630500; fax 5630520; founded by three dissident members of the People's Alliance—PA; political organization.

INDIA

Governmental Organizations

Ministry of Environment and Forests: Paryavaran Bhavan, CGO Complex Phase II, Lodi Rd, New Delhi 110 003; tel (11) 24361727; fax (11) 24362222; e-mail secy@menf.delhi.nic.in; internet www.envfor.nic.in; Minister: A Raja; environmental and forestry programmes; conservation and survey of flora, fauna, forests and wildlife; prevention and control of pollution; reforestation; regeneration of degraded areas; environmental impact assessment; research; education and training; international co-operation; awareness-raising.

Botanical Survey of India—BSI: P-8 Brabourne Rd, Calcutta 700 001; tel (33) 2424922; fax (33) 2429330; e-mail envis@cal2.vsnl.net.in; internet envfor.nic.in/bsi; Dir: Dr B D Sharma; botanical surveys and research.

Central Pollution Control Board—CPCB: CBD-cum-Office Complex, Parivesh Bhavan, East Arjun Nagar, New Delhi 110 032; e-mail cpcb@alpha.nic.in; internet envfor.nic.in/cpcb; Chair: Shri Dilip K. Biswas; f 1974; promotes cleanliness of streams and wells throughout India by the prevention, control and abatement of water pollution; improves the quality of air and prevents, controls or abates air pollution.

Central Zoo Authority—CZA: Annexe-IV, Bikaner House, Shahjahan Rd, New Delhi 110 011; tel (11) 3381585; fax (11) 3386012; e-mail cza@ndf.vsnl.net.in; internet envfor.nic.in/cza; f 1992; regulates zoos in India; seeks to ensure that zoo animals are provided with conditions congenial to their psychological and physical health and are able to breed, thus augmenting declining populations in the wild.

Environmental Information System—ENVIS: Paryavaran Bhavan, CGO Complex Phase II, Lodi Rd, New Delhi 110 003; tel (11) 4364687; fax (11) 4360678; e-mail harjit-s@nic.in; internet envfor.nic.in/envis; Sr Adviser: Dr Harjit Singh; f 1982; serves as the INFOTERRA national focal point; co-ordinates a network of 25 ENVIS Centres throughout India.

Wildlife Institute of India—WII: POB 18, Chandrabani, Dehra Dun 248 001, Uttar Pradesh; tel (135) 640111; fax (135) 640117; e-mail wii@wii.gov.in; internet www.wii.gov.in; training in wildlife biology and management; conservation research; comprises the Wildlife Biology, Wildlife Management and Wildlife Extension faculty divisions, the Ecodevelopment Cell and the Environmental Impact Assessment—EIA Cell.

Zoological Survey of India—ZSI: M Block, New Alipur, Calcutta 700 053; tel (33) 4786893; fax (33) 786893; e-mail zsi@envfor.delhi.nic.in; internet envfor.nic.in/zsi; Dir: Prani Vigyan Bhawan; f 1916; maintains the National Zoological Collections; conducts faunistic surveys and research on systematic ecology.

Ministry of Agriculture and of Consumer Affairs, Food and Public Distribution: Krishi Bhavan, Dr Rajendra Prasad Rd, New Delhi 110 001; tel (11) 23383370; fax (11) 23382756; internet www.fcamin.nic.in; Minister: Sharad Pawar.

Indian Council of Agricultural Research—ICAR: Krishi Bhavan, Dr Rajendra Prasad Rd, New Delhi 110 001; tel (11) 388991; fax (11) 387293; Pres: Bal Ram Jakhar; Dir-Gen: Dr V L Chopra; f 1929; part of the Department of Agricultural Research and Education of the Ministry of Agriculture.

Ministry of Chemicals and Fertilizers: Shastri Bhavan, New Delhi 110 001; tel (11) 23386519; fax (11) 23015477; Minister: Ram Vilas Paswan.

Ministry of Non-Conventional Energy Sources: Block 14, CGO Complex, New Delhi 110 003; tel (11) 24360744; fax (11) 24362554; e-mail secymnes@ren02.nic.in; internet www.mnes.nic.in; Minister of State: Vilas Muttemwar.

Ministry of Rural Development: Krishi Bhavan, New Delhi 110001; tel (11) 23383548; fax (11) 23385876; e-mail arunbhat@rural.delhi.nic.in; internet www.rural.nic.in; Minister: Raghubansh Prasad Singh.

Ministry of Water Resources: Shram Shakti Bhavan, Rafi Marg, New Delhi 110 001; tel (11) 23714663; fax (11) 23710804; e-mail webmaster@mowr.delhi.nic.in; internet www.wrmin.nic.in; Minister: Priyaranjan Dasmunsi.

Central Water and Power Research Station: PO Khadakwasla Research Station, Pune 411 024; tel (20) 4391801; fax (20) 4392004; e-mail wapis@mah.nic.in; internet www.mah.nic.in/cwprs; Dir: R Jeyaseelan; f 1916.

National Institute of Hydrology: Jal Vigyan Bhawan, Roorkee 247 667, Uttar Pradesh; tel (1332) 72106; fax (1332) 72123; Dir: Dr Satish Chandra; f 1979; research in all aspects of water resources.

National Institute of Oceanography—NIO: Miramar, Panaji 403 004, Goa; tel (832) 221322; fax (832) 223340; e-mail ocean@csnio.ren.nic.in; Dir: Dr E Desa; f 1966; a laboratory of the Council of Scientific and Industrial Research, New Delhi.

Non-Governmental Organizations

Applied Interdisciplinary Development Research Institute: 10 Nelson Manickam Rd, 2nd Floor, Choolaimedu, Chennai 600 094, Tamil Nadu; tel (44) 4726457; Exec Dir: Dr A Peter; f 1985; promotion of sustainable rural and urban development; management of urban wastes.

Atucon Research and Development Centre: A-7 Pushpanjali Enclave, New Delhi 110 034; tel (11) 7014590; fax (11) 7014572; e-mail atul.paliwal@sol.indiagate.com; Prin Officer: Dr Ratan C Paliwal; f 1980; part of the Atucon Group of Companies; research into social, educational and technical development, ecology and pollution control.

Central Marine Fisheries Research Institute, Calicut: West Hill, Calicut 673005, Kerala; tel and fax (495) 382011; Sr Scientist and Officer-in-Charge: Dr T M Yohanan; f 1947; affiliated to the Central Marine Fisheries Research Institute, Cochin, Kerala; studies the marine fish stocks of the Malabar area and develops methods to manage them; aquaculture.

Chemtech Foundation: 26, Maker Chambers VI, 2nd floor, Nariman Point, Mumbai - 400 021; tel (22) 22874758 / 59; fax

(22) 22870502; e-mail anil_mazumdar@jasubhai.com; internet www.chemtech-online.com; Chair: Jasu Shah; Sr Adviser: Anil Mazumdar; organization of trade fairs, conferences and workshops focusing on environmental issues confronting industry; provision of marketing base for pollution-control companies.

Commonwealth Human Ecology Council—CHEC India: Unnithan Farm, Todiramsanipura, Jagatpura, Jaipur Rajasthan 302 017; tel (141) 750047; fax (141) 516699; Pres: Prof T K N Unnithan; national office of the CHEC, London; education in the relationship between humans and nature.

Dasholi Gram Swarajya Mamdal (Chipko Movement): PO Gopeshwar District, Chamoli 246 401, Uttar Pradesh; works with rural communities to promote the sustainable use of forestry resources and to foster greater environmental awareness.

Environment and Ecology Communication Centre, Kalyani University: Fisheries Laboratory, Department of Zoology, Kalyani 741 235, West Bengal; tel (33) 5825548; e-mail skonar@klyniv.ernet.in; Dir: Prof S K Konar; research on all aspects of ecology and environmental biology; publications.

Environment Association of Bangalore: Antariksh Bhavan, New BEL Rd, Bangalore 560 094, Karnataka; tel (80) 3415275; Pres: Dr M G Chandrasekhar; f 1995; conducts seminars and workshops with the aim of finding technical solutions to environmental problems, with emphasis on the prevention of lake pollution.

Environment Society of India: Karuna Sadan, Sector 11-B, Chandigarh 160 011; tel (172) 546832; fax (172) 600531; Pres: S K Sharma; f 1976; promotion of art, heritage and environment; research and creation of public awareness; public-interest litigation in the High and Supreme Courts against environmental degradation; incorporates the Green Library for Environmental Sciences; publication of a journal; functions as the Regional Resource Agency of the Ministry of Environment and Forests; member of the World Conservation Union—IUCN.

Green Volunteers: Behind Telegraph Office, Mandir Marg, Gopeshwar, Chamoli, Uttrakhand 246 401; tel (1372) 52316; fax (1372) 52367; Prin Officer: Pradeep Chandra Dimri; f 1997; provision of advice and assistance to local organizations; dissemination of information on sustainable development to rural communities; affiliated with Dasholi Gram Swarajya Mamadal.

Greenpeace India: Lodi Rd, POB 3166, New Delhi 110 003; tel (11) 4313458; fax (11) 4310651; e-mail greenpeaceindia@vsnl.com; national office of Greenpeace International.

Indian Ecological Society: College of Agriculture, Punjab Agricultural University, Ludhiana 141 004, Punjab; tel (161) 401962; fax (161) 400945; Pres: Dr Avtar Singh Atwal; Gen Sec: Dr G S Dhaliwal; Editor-in-Chief: Dr Ramesh Arora; f 1974; development of appropriate technology for recycling of agricultural and industrial wastes; seminars, symposia and conferences; collaboration with organizations having related aims and activities.

Indian National Trust for Art and Cultural Heritage—INTACH: 71 Lodi Estate, New Delhi 110 003; tel (11) 4631818; fax (11) 4611290; Chair: Pupul Jayakar; Exec Dir and Sec: Ashis Banerjee; f 1984; preservation of the cultural and natural heritage of India, including places of archaeological, historical, artistic and scientific value; provision of specialist services through the architectural, material and natural heritage divisions; awareness-raising; workshops, seminars and publications.

IOI India: IC & SR Bldg, 1st Floor, Indian Institute of Technology, Chennai 600 036; tel (44) 2301138; fax (44) 2200559; e-mail ioi@vsnl.com; Dir: Prof R Rajagopalan; f 1993; Indian operational centre of the International Ocean Institute—IOI, Malta.

Mumbai Natural History Society: Hornbill House, Shaheed Bhagat Singh Rd, Mumbai 400 023; tel (22) 2821811; Pres: B G Deshmukh; f 1883; natural history; wildlife conservation; conservation education; environmental impact assessment.

National Environmental Engineering Research Institute: Nehru Marg, Nagpur 440 020, Maharashtra; tel (712) 226071; fax (712) 222725; e-mail dirneeri@nagpur.dot.net.in; Dir (acting): Dr N S Kaul; f 1958; independent, non-profit organization; research into water, sewage, industrial waste, air pollution, industrial hygiene and rural sanitation.

Pulp and Paper Research Institute: Rayagada Dist, Jaykapur 765 017, Orissa; tel (6856) 233456 /233550; fax (6856) 222238; e-mail directorpapri@myway.com or paprijkpur@sancharnet.in; Dir: Dr U K Deb; f 1971; research and development, consultancy, information services, training and documentation, testing services. Member of TAPPI, PAPTAC, APPITA.

Rhino Foundation for Nature in North East India: c/o Tollygunge Club, 120 DP Sasmal Rd, Calcutta 700 033, West Bengal; tel and fax (33) 4733306; Chair: Anne Wright; Chief Exec (Hon): Dr Anwaruddin Choudhury; f 1994; lobbying of the Government on conservation issues; fund-raising for conservation projects; public-interest litigation; liaison with other conservation NGOs; education.

Sri Aurobindo Society: Manukula Vinayagar Koil St, No 1, Rangapillai St, Pondicherry 605 001, Tamil Nadu; tel (413) 36396; fax (413) 34447; Gen Sec: Pradeep Narang; Mem, Exec Cttee and Int Sec: Gopal Bhattacharjee; f 1960; operation of Scientific Research Trust, which conducts research into natural energy sources, energy-efficient housing, and wind and solar energy technology; consultative status with the United Nations Economic and Social Council.

Tata Energy Research Institute—TERI: Darbari Seth Block, Habitat Pl, Lodi Rd, New Delhi 110 003; tel (11) 4682100; fax (11) 4682144; e-mail mailbox@teri.ernet.in; internet www.teriin.org; Dir: Dr R K Pachauri; f 1974; autonomous, non-profit organization engaged in research on energy policy, engineering and technology, industrial and rural energy, forestry, biodiversity, biosciences, and information and computer sciences; training programmes on energy and environment for senior officials in the government, private and public sectors; affiliated with Tata Energy and Resources Institute (Arlington, VA, USA); collaborating centre in the UNEP Global Environment Outlook—GEO Programme.

Thapar Centre for Industrial Research and Development: Thapar Technology Campus, POB 68, Patiala 147 001, Punjab; tel (175) 393501; fax (175) 212002; e-mail director@tcrdcpt.ren.nic.in; internet www.thapartech.org; Dir: Dr Mahesh P Kapoor; Deputy Dirs: Dr Pramod K Bajpai, Dr Amitabh Verma; f 1983; part of the Thapar Group; research and consultancy in environment, solid waste management and environmentally friendly technologies for the process industry; environmental monitoring; testing and analytical service; environmental laboratory accredited by the Central Pollution Control Board and Punjab Pollution Control Board.

TRAFFIC India—Regional Office: c/o WWF Secretariat, 172 B Lodi Estate, New Delhi 110 003; tel (11) 4698578; fax (11) 4626837; e-mail trfindia@del3.vsnl.net.in.

Tropical Botanic Garden and Research Institute: Pacha Palode, Trivandrum 695 562, Kerala; tel (472) 869226; fax (471) 869646; Dir: Dr G Sreekandan Nair; f 1979; autonomous organization financed by the State Government of Kerala through the Department of Science, Technology and

Environment, Trivandrum; maintains botanical gardens; development and marketing of products derived from tropical plants.

Wildlife, Environment and Tourism Institute: 3/16 Murai Tola, Ayodhya, Faizabad 224 123, Uttar Pradesh; tel (5278) 32214; Founder, Pres and Dir: Dr M R Yadav; f 1984; education, training, management, research and conservation; affiliated to the International Association for Turtle Research and Conservation.

World Wide Fund for Nature—WWF India: 172b Lodi Rd, POB 3058, New Delhi 110 003; tel (11) 4691760; fax (11) 4626837; e-mail root@wwfind.ernet.in; internet indev.nic.in/wwf; Pres: J N Godrej; Vice-Pres: A K Mahindra; national affiliate organization of the World Wide Fund for Nature—WWF International; fund-raising; research; education; support for conservation agencies; international co-operation; lobbying and other action in support of environmental causes.

INDONESIA

Governmental Organizations

Office of the Minister of State for the Environment: Jalan D I Panjaitan, Kebon Nanas Lt II, Jakarta 13110; tel (21) 8580103; fax (21) 8580101; internet www.bapedal.go.id; Minister of State: Rachmat Witoelar.

Ministry of Agriculture and Forestry: Jalan Harsono RM 3, Gedung D-Lantai 4, Ragunan Pasar Minggu, Jakarta Selatan 12550; tel (21) 7822638; fax (21) 7816385; e-mail eko@deptan.go.id; internet www.deptan.go.id; Minister of Agriculture: Anton Apriyantono; Minister of Forestry: M S Kaban.

Ministry of Energy and Mineral Resources: Jalan Merdeka Selatan 18, Jakarta 10110; tel (21) 3804242; fax (21) 3847461; e-mail pulahta@setien.dpe.go.id; internet www.dpe.go.id; Minister: Dr Ir Purnomo Yusgiantoro.

Centre for Scientific Documentation and Information—PDII-LIPI: Jalan Jenderal Gatot Subroto 10, Jakarta 12190; tel (21) 5733465; fax (21) 5733467; e-mail jusni@pdii.lipi.go.id; internet www.pdii.lipi.go.id; Head: Jusni Djatin; the Division for Dissemination of Scientific Information serves as the INFOTERRA national focal point; collecting of Indonesian scientific literature; provision of scientific services; provision of training and skills enhancement to librarian or information workers; conducting research of library and information science.

Pusat Penelitian dan Pengembangan Hutan dan Konservasi Alam (Forest and Nature Conservation Research and Development Centre): Jalan Gunung Batu 5, POB 165, Bogor 16001; tel and fax (251) 325111; e-mail slitbang@bogor.indo.net.id; internet www.forda.org; Head, Research Div: Ir Darmawan Budiantho; f 1913; affiliated to the Forestry and Estate Research and Development Agency, Ministry of Forestry; guidance, implementation, evaluation, documentation and publication of forest management and nature conservation research and development; affiliated with the Forest and Nature Conservation Research and Development Centre.

Pusat Penelitian Sumberdaya Manusia dan Lingkungan, Universitas Indonesia—PPSML-UI (Centre for Research on Human Resources and the Environment, University of Indonesia—CRHRE-UI): Jl. Salemba Raya No. 4 Ged. C Lt.V & VI, Jakarta 10430; tel (21) 31930318; fax (21) 31930266; e-mail ppsml-ui@indo.net.id; internet www.ppsml-ui.org; Dir: Dr Setyo S Moensidik; Advisers: Prof Emil Salim, Prof Retno soetarjono; Prof Corrie Wawolumaya; f 1979; affiliated to BKPSL Indonesia and the National Co-operative Network of Environmental Study Centres; education, training, research and community activities in the field of environmental management.

Non-Governmental Organizations

Bogor Agricultural University, Laboratory of Oceanography: Faculty of Fisheries, Jalan Rasamala, Campus IPB Darmaga, Bogor 16680; tel (251) 621761; fax (251) 621761; e-mail ikanmsp@indo.net.id; Prin Officers: Dr Mulia Purba, Dr John Pariwono; f 1970; courses, research and consultancy services on physical, chemical and biological oceanography, coastal and marine environments, marine and fisheries resource management and marine remote sensing.

Centre for Appropriate Technology, Gadjah Mada University: Bulaksumur, Yogyakarta 55281; Dir: Dr Ir Sunarto Ciptohadijoyo.

Centre for Environmental Studies, Gadjah Mada University: Bulaksumur, Yogyakarta 55281; Dir: Dr Ir Haryadi.

Federation for Indonesian Speleological Activities (FINSPAC): Jalan Ir H Juanda 30, POB 154, Bogor, West Java; tel (251) 257049; fax (251) 318160; e-mail speleoindonesia@mdo.net.id; Pres: Dr Robby K T Ko; Scientific Dir: Prof Dr Ir Go Ban Hong; Conservation Dir: Ir Rusmono; Technical Caiving Dir: Cahyo Alkantana; f 1983; activities include lobbying governmental organizations to protect the underground environment, with special emphasis on cave biota and karst springs; education; consultation; monitoring of water quality of karst springs.

Wahana Lingkungan Hidup Indonesia—WALHI (Indonesian Environmental Forum): Jalan Tegal Parang Utara 14, Jakarta 12790; tel (21) 7941672; fax (21) 7941673; e-mail walhi@walhi.or.id; internet walhi.or.id; Exec Dir: Emmy Hafild; network of environmental NGOs; liaises with government.

IRAN

Governmental Organizations

Department of the Environment: Ostad Nejotollahi Ave 187, POB 5181-15875, Tehran; tel (21) 8903720; fax (21) 8908230; e-mail info@ir-doe.org; internet www.ir-doe.org; Head: Ma'sumeh Ebtekar (Vice-Pres); Gen Dir, Public Relations and Int Affairs: Seyed Amir Ayafat; f 1971; scientific research on all environmental issues, including air, water, soil and noise pollution, and wildlife conservation; issue of regulations and setting of environmental standards; promotion of environmental education; monitoring, control and law enforcement by rangers; establishment and expansion of natural history museums and environmental exhibitions; the Public Relations and International Affairs dept serves as the INFOTERRA national focal point.

Ministry of Agricultural Jihad: 20 Malei Ave, Vali-e-Asr Sq, Tehran; tel (21) 8895354; fax (21) 8904357; e-mail webinfor@asid.moa.or.ir; internet www.moa.or.ir; Minister: Mahmud Hojjati.

Research Organization of Agriculture and Natural Resources: POB 13185-116, Tehran; tel (21) 6026574; fax (21) 6026575.

Ministry of Energy: North Palestine St, Tehran; tel (21) 890001; fax (21) 8801995; e-mail webmaster@moe.or.ir; internet www.moe.or.ir; Minister: Habibollah Bitaraf.

Research Institute of Forests and Rangelands—RIFR: Teheran Karaj, Freeway Km 15, POB 11385-116, Tehran; tel (21) 6026574; fax (21) 6026575; Dir: Dr M Jafari; f 1968; affiliated to the Ministry of Construction Jihad, Education and Research Division; areas of research interest include poplars and fast-growing trees, wood and paper science, botany, socio-economics, prevention of desertification, sand-dune fixation, medicinal plants and non-wood by-products, genetic and plant physiology, and mechanization of natural resource exploitation.

Soil and Water Research Institute: North Kargar Ave, Jajal Al Ahmad Rd, POB 14155-6185, Tehran; tel (21) 631065; fax (21) 634006; e-mail swri1357@rose.ipm.ag.ir; Dir-Gen: Dr M J Malakouti; Deputy Dir: Dr M H Banaei; f 1955; affiliated to the Agricultural Research, Education and Extension Organization and the Ministry of Agriculture; soil and land mapping; makes recommendations on the type and quantity of chemical fertilizers for use in Iran, on water use for different crops and locations, and on proper land use; routine chemical and physical analysis of soil and water samples.

Non-Governmental Organizations

Solar Energy Centre, Shiraz University: Department of Mechanical Engineering, School of Engineering, Shiraz; tel (71) 303051; fax (71) 52725; e-mail yaghoub@succ.shirazu.ac.ir; Dir: M Yaghoubi; f 1976; research on applications of solar energy, including water pumping, desalination and heating; design and construction of the first solar thermal power plant in Iran; measurement of solar radiation.

IRAQ

Governmental Organizations

Directorate of Environmental Protection and Improvement: POB 61294, Baghdad; tel (1) 415-8411; fax (1) 885-2345; Dir Gen: Khidhir E Putres; Tech Officer: Manal K Elias; serves as the INFOTERRA national focal point.

Ministry of Agriculture: Khulafa St, Khullani Sq, Baghdad; tel (1) 887-3251; e-mail min_of_agriculture@orha.centcom.mil; Minister: Dr Sawsan al-Majid ash-Sharifi.

Ministry of Oil: Baghdad; e-mail oil@uruklink.net; internet www.uruklink.net/oil; Minister: Thamir Abbas al-Ghadban.

General Directorate of Preventive Medicine and Environmental Protection: POB 10062, Baghdad; tel (1) 719-2033.

Ministry of the Environment: Baghdad; Minister: Prof Mishkat Moumiu.

Non-Governmental Organizations

Agriculture and Water Resources Research Centre: POB 2416, Karada Al-Sharkiya, Baghdad; tel (1) 751-2080; Dir-Gen: Dr Samir A H Al-Shakir; f 1980; research to improve and develop water and agricultural resources.

IRELAND

Governmental Organizations

Department of the Environment and Local Government: Custom House, Dublin 1; tel (1) 8882000; fax (1) 8742710; e-mail minister@environ.ie; Minister: Martin Cullen; institution with ministerial status.

Environmental Protection Agency—EPA: POB 3000, Johnstown Castle Estate, Co Wexford; tel (53) 60600; fax (53) 60699; e-mail info@epa.ie; internet www.epa.ie; Dir-Gen: Liam McComiskey; f 1993; promotes and implements the highest practicable standards of environmental protection and management that embrace the principles of sustainable and balanced development.

Department of Agriculture and Food: Kildare St, Dublin 2; tel (1) 6072000; fax (1) 66121165; e-mail infodaff@agriculture.gov.ie; internet www.agriculture.gov.ie; Minister: Joe Walsh; institution with ministerial status.

Department of Arts, Heritage, Gaeltacht and the Islands: Dún Aimirghin, 43–49 Mespil Rd, Dublin 4; tel (1) 6473000; fax (1) 6629271; e-mail eolas@ealga.irlgov.ie; internet www.irlgov.ie/ealga; Minister: Síle de Valera; responsible for Special Area of Conservation—SACs, terrestrial, freshwater and marine habitats protected under the European Union Habitats Directive.

Department of Marine and Natural Resources: Leeson Lane, Dublin 2; tel (1) 6785444; fax (1) 6618214; Minister: Frank Fahey; institution with ministerial status; responsible for regulation and development of the marine sector.

Agriculture and Food Development Authority—TEAGASC: 19 Sandymount Ave, Ballsbridge, Dublin 4; tel (1)

6376000; fax (1) 6688023; e-mail info@hq.teagasc.ie; internet www.teagasc.ie; Chair: Dr Tom O'Dwyer; Dir: Jim Flanagan; f 1988; autonomous body; responsible for providing integrated research and carrying out advisory and training functions for the agriculture and food industry.

An Chomhairle Oidhreachta (Heritage Council): Rothe House, Kilkenny; tel (56) 7770777; fax (56) 7770788; e-mail mail@heritagecouncil.ie; internet www.heritagecouncil.ie; Chief Exec: Michael Starrett; Wildlife Officer: Liam Lysaght; Architecture Officer: Mary Hanner; Planning Officer: Stephen Rhys Thomas; Archaeology Officer: Ian Doyle; Marine and Coastal Officer: Beatrice Kelly; f 1995; statutory body with advisory function; formulation of policies and priorities for the identification and protection of national heritage, including flora and fauna, wildlife habitats, inland and coastal areas of natural beauty and heritage gardens and parks.

Coillte Teoranta (Irish Forestry Board): Spruce House, Leeson Lane, Dublin 2; tel (1) 6615666; fax (1) 6789527; Chair: Patrick Cooney; Chief Exec: Martin D Lowery; f 1988.

Príomh-Bhord Iascaigh (Central Fisheries Board): Mobhi Boreen, Glasnevin, Dublin 9; tel (1) 8379206; fax (1) 8360060; Chief Exec: M S Breathnach; f 1980; co-ordinates the activities of the seven regional fisheries boards in the areas of protection, conservation, management and development of the country's inland fisheries and sea angling resources.

Non-Governmental Organizations

An Taisce/National Trust for Ireland: Tailors' Hall, Back Lane, Dublin 8; tel (1) 4541786; fax (1) 4533255; e-mail info@antaisce.org; internet www.antaisce.org; Pres: Prof David Jeffrey; Chair: Michael Smith; f 1948; independent, voluntary organization concerned with conservation of the natural and man-made environment; campaigning; lobbying; education; participation in planning.

Comhaontas Glas (Green Party): 5a Upper Fownes St, Dublin 2; tel (1) 6790012; fax (1) 6797168; e-mail greenpar@iol.ie; internet www.greenparty.ie.eu.org; Co-ordinator: Mary Bowers; co-operation with members in local government and the Dáil (House of Representatives) to ensure that environmental issues are a high political priority.

Conservation Volunteers Ireland: The Green, Griffith College, South Circular Rd, Dublin 8; tel (1) 4547185; fax (1) 4546935; e-mail info@cvi.ie; internet www.cvi.ie; Exec Dir: Melanie Hamilton; f 1990; provides opportunities for volunteers to participate in one-day, weekend and week-long conservation projects throughout Ireland; runs education and training courses; affiliated with the Tree Council of Ireland, Mountaineering Council of Ireland, Irish Georgian Society.

Dublin Naturalists' Field Club: c/o 35 Nutley Park, Donnybrook, Dublin 4; tel (1) 2697469; e-mail dnfc@eircom.net; internet www.dnfc.net; Pres: Mary Carson; Sec: Gerard Sharkey; study of all aspects of the natural history of the Greater Dublin area and butterflies of Ireland.

Earthwatch/Friends of the Earth—FoE Ireland: 20 Grove Rd, Rathmines, Dublin 6; tel (1) 4973773; fax (1) 4970412; e-mail foeeire@iol.ie; Co-ordinator: Jeremy Wates; national office of Friends of the Earth International—FoEI; campaigning membership organization working for environmental protection.

ECO—Irish National Youth Environmental Organisation: 26 Clare St, Dublin 2; tel (1) 6625491; fax (1) 6625493; e-mail eco@connect.ie; internet www.connect.ie/users/eco; Dir: Enda Connolly; Office Man: Marie Hamilton; f 1986; operates a national network of environmental youth clubs; urban forestry development service for local communities; education and training; publication of newsletters, magazines and environmental resource packs; affiliated with the European Environment Bureau, National Youth Council of Ireland, Tree Council of Ireland, UNESCO, and Youth and Environment Europe.

Environmental Institute, University College Dublin: Richview, Clonskeagh, Dublin 14; tel (1) 2697988; fax (1) 2837009; e-mail fconvery@ucd.ie; Dir: Prof Frank J Convery; Assoc Dir: Prof Vincent Dodd; f 1990; formation of multi-disciplinary research teams for environmental projects; teaching.

Greenpeace Ireland: 44 Upper Mount St, Dublin 2; tel (1) 6619836; fax (1) 6605258; e-mail greenpeace.ireland@green2.greenpeace.org; Campaigns Dir: John Bowler; Development Dir: Margaret Ward; national office of Greenpeace International.

Institute of Fisheries Management: c/o Marine Institute, Fisheries Research Centre, Abbotstown, Dublin 15; tel (1) 8210111; fax (1) 8205078; e-mail moriarty@frc.ie; Sec (Hon): Dr Christopher Moriarty; Chair: Vincent Roche; f 1987; aims to promote effective fisheries management and to advance the standing of fisheries management as a profession; provision of technical knowledge; awards a professional qualification; forum for information exchange; seminars.

Institute of Technology, Department of Environmental Science: Ballinode, Sligo; tel (71) 55284; fax (71) 46802; e-mail fitzgerald.billy@itsligo.ie; internet www.itsligo.ie; Head: Dr Billy Fitzgerald; undergraduate and postgraduate teaching, and research; environmental science and technology; pollution assessment and control; occupational safety and health; environmental protection by distance learning.

Irish Business and Employers' Confederation, Environment Policy Committee: Confederation House, 84–86 Lower Baggot St, Dublin 2; tel (1) 6601011; fax (1) 6601717; internet www.ibec.ie; Dir-Gen: John Dunne; Environment Exec: Dr Mary Kelly; f 1993; represents about 7,000 companies and organizations.

Irish Energy Centre: Glasnevin, Dublin 9; tel (1) 8369080; fax (1) 8372848; e-mail info@irish-energy.ie; internet www.irish-energy.ie; Dir: David Taylor; f 1994; affiliated to Enterprise Ireland; delivery and implementation of Irish national policy for energy efficiency and renewable energy.

Irish Peatland Conservation Council: Capel Chambers, 119 Capel St, Dublin 1; tel and fax (1) 8722397; e-mail ipcc@indigo.ie; internet indigo.ie/~ipcc; Chair: Dr Peter Foss; Prin Officer: Dr Catherine O'Connell; f 1982; charitable organization entirely funded by voluntary contributions; conservation projects include purchase of bogland nature reserves and repair of damaged bogs; provision of resources and training for teachers and educational groups; public awareness-raising; discouragement of behaviour harmful to the environment.

Organic Trust Ltd: Vernon House, 2 Vernon Ave, Clontarf, Dublin 3; tel and fax (1) 853 0271; e-mail organic@iol.ie; internet www.organic-trust.org; Nat Co-ordinator: Helen Scully; f 1991; European Union-approved organic inspection and certification body; training and education; international co-operation on organic standards; publications; member of the International Federation of Organic Agriculture Movements—IFOAM. EU (IRL-OIB3-EU) and UK (UK9) approved organic certification body.

Royal Zoological Society of Ireland: Phoenix Park, Dublin 8; tel (1) 6771425; fax (1) 6771425; Dir: P Wilson; Pres: Joe McCullogh; Education Officer: Michele Griffin; involved in the maintenance, exhibition, breeding, conservation and study for educational purposes of all species of living animals at Dublin Zoo; promotion of study and knowledge of zoology and cultivation of interest in the conservation of animals.

Tree Council of Ireland—Comhairle Crann na h Éireann: Cabinteely House, The Park, Cabinteely, Dublin 18; tel (1) 2849211; fax (1) 2849197; e-mail trees@treecouncil.ie; internet www.treecouncil.ie; Pres: John McLoughlin; Vice-Pres: Mary Keenan; Public Relations Officer: Miriam Rollins; f 1985; promotes planting, propagation, conservation and maintenance of trees in urban and rural areas; dissemination of knowledge on trees and tree-care; advisory functions; organization of annual National Tree Week.

ISRAEL

Governmental Organizations

Ministry of the Environment: POB 34033, 5 Kanfei Nesharim St, Givat Shaul, Jerusalem 95464; tel 2-6553777; fax 2-6553752; e-mail sergio@environment.gov.il; internet www.environment.gov.il; Minister: Shalom Simhon; environmental policy; strategies, standards and priorities for environmental protection; air, marine and water quality control; monitoring of solid waste, hazardous substances, radiation and noise; pest control; agro-ecology; land-use planning; legislation; international co-operation; education and information.

Ministry of Agriculture and Rural Development: POB 30, Beit Dagan, Tel-Aviv 50250; tel 3-9485571; fax 3-9485870; e-mail regeva@moag.gov.il; internet www.moag.gov.il; Minister: Yisrael Katz.

Ministry of the Interior: 2 Rehov Kaplan, Kiryat Ben-Gurion, POB 6158, Jerusalem 91008; tel 2-6701411; fax 2-6701628; e-mail pniot@moin.gov.il; internet www.moin.gov.il; Minister: Ophir Pines-Paz; incorporates the Environmental Protection Service.

Earth Sciences Research Administration: POB 1442, Jerusalem 91130; tel 2-5316130; fax 2-5373470; e-mail beyth@netvision.net.il; Dir: M Beyth; part of the Ministry of National Infrastructure; defines research and development needs in relation to natural resources.

Geological Survey of Israel: 30 Malkhei Yisrael St, Jerusalem 95501; tel 2-5314221; fax 2-5380688; Dir: Dr G Steinitz; f 1949; part of the Ministry of National Infrastructure.

Nature Reserves Authority: 78 Yirmeyahu St, Jerusalem 94467; tel 2-387471; fax 2-374887; Dir-Gen: Shaike Erez; Chair: Uri Baidatz; creation, restoration and preservation of nature reserves; protection of wildlife and other natural resources through scientific monitoring, education and law enforcement.

Soreq Nuclear Research Centre: Yavne 81800; tel 8-9434290; Dir: Uri Halavee; f 1958; affiliated to the Israeli Atomic Energy Commission.

Non-Governmental Organizations

Council for a Beautiful Israel: Yehoshua Gardens, 80 Rokach Blvd, POB 53250, Tel-Aviv 61532; tel 3-6423111; fax 3-6422839; Dir-Gen: Michael Oren; Int Pres: Aura Herzog; f 1970; implementation of projects to help preserve landscape and natural resources; prevention of pollution; education programmes.

Environmental and Water Resources Research Centre, Technion-Israel Institute of Technology: Technion City, Haifa 32000; tel 4-829359; fax 4-8228898; e-mail vardit@tx.technica.ac.il; internet www.technica.ac.il; Head: Assoc Prof Yancov Mamanc; research into environmental and water resource engineering.

Greenpeace Israel: POB 14423, Tel-Aviv 61143; tel 3-5102079; fax 3-5163301; e-mail gpmedisr@diala.greenpeace.org; internet www.greenpeacemed.org.mt; Energy Campaigner: Liad Ortar; Sea Pollution Campaigner: Mia Elasar; Fundraising Campaigner: Rick Gentry; f 1994; national office of Greenpeace International and of Greenpeace Mediterranean.

Institute for Nature Conservation Research—INCR, Tel-Aviv University: George S Wise Faculty of Life Sciences, Ramat Aviv 69978; tel 3-6409813; fax 3-6407304; e-mail incr@post.tau.ac.il; Dir: Prof Prof Baruch Sneh; ecological, biological and toxicological studies of the environment in Israel and their application to nature conservation; areas of research interest include: conservation of marine and freshwater ecosystems, ecosystem rehabilitation, biological pest control, ecotoxicology, chemical ecology, animal behaviour and population genetics, air pollution, insecticides and pollination.

Israel Society for Ecology and Environmental Quality Sciences: Department of Environmental Sciences, Hebrew University, Givat Ram, Jerusalem 91904; tel 2-6584192; fax 2-6585559; e-mail shimshon@vms.huji.ac.il; Pres: Prof Shimshon Belkin; Sec: Dr Asher Brenner; Treas: Dr Stilian Gelber; f 1956; promotion of environmental education and environmental best practice.

Keren Kayeneth LeIsrael—KKL (Jewish National Fund—JNF): Head Office, 1 KKL St, POB 283, Jerusalem 91002; tel 2-6707411; fax 2-6256941; e-mail info@kkl.org.il; internet www.kkl.org.il; World Chair: Moshe Rivlin; Dir-Gen: Yitzhak Elyashive; f 1901; land reclamation; infrastructure; afforestation; ecological conservation; development of forest recreation; drainage; water resources; river-bed rehabilitation.

Life and Environment: 4/7 Katzenelson St, Kfar-Saba 44405; tel 9-7678442; fax 9-7678451; e-mail karassin@netvision.net.il; Chair: Dr Alon; Adv Dir: Orr Karassin; f 1975; supports the Israeli environmental community by means of professional advice, references to volunteer experts, dissemination of information, and facilitation of dialogue and co-operation; strengthens the community by increasing public awareness, research and working with other groups; represents the environmental community and lobbies the Government; affiliated with 65 member organizations; member of ELCI.

Mitrani Center for Desert Ecology—MCDE, Ben-Gurion University of the Negev: Jacob Blaustein Institute for Desert Research, Sede Boqer Campus 84990; tel 7-6596771; fax 7-6596772; e-mail mitrani@bgumail.bgu.ac.il; Head: Prof Berry Pinshow; f 1980; researchers have two primary aims: to study deserts as model ecosystems in order to advance ecological knowledge at different levels of integration; and to understand the ecological properties of desert in Israel in order to provide data to government agencies, industry and the scientific community at large; information gained is used for the conservation, restoration and prudent, sustainable development of the Negev and other desert regions.

Palestinian–Israeli Environmental Secretariat—PIES: Faidy Al-Alami St Beit Hanina, POB 66839, Jerusalem 95908; tel and fax 2-5830712; e-mail pies@netvision.net.il; internet www.piesme.org; Co-Dirs: Dr Imad Khatib, Paul Amil; develops joint activities in the fields of environmental protection, education and public awareness.

Society for the Protection of Nature in Israel—SPNI: 4 Hashfela St, Tel-Aviv 66183; tel 3-6388666; fax 3-5374561; e-mail international@spni.org.il; Exec Dir: Eitan Golalizan; Chair: Dr Shimshan Shoshani; f 1953; non-profit organization; conservation of landscape and ancient relics; protection of plant and animal life; environmental protection; research; education.

ISRAEL, EMERGING PALESTINIAN AUTONOMOUS AREAS

Governmental Organizations

Ministry of Environmental Affairs: POB 3841, Ramallah; tel (2) 2403495; fax (2) 2403494; e-mail menawb@gov.ps; internet www.mena.gov.ps; Minister of State: Dr Yousuf Abu Saffieh; serves as the INFOTERRA national focal point.

Ministry of Agriculture: POB 197, Ramallah; tel (2) 2961080; fax (2) 2961212; e-mail moa@planet.edu; Minister: Dr Ibrahim Abu an-Naja.

Non-Governmental Organizations

Palestinian–Israeli Environmental Secretariat—PIES: See chapter on Israel.

ITALY

Governmental Organizations

Ministry of the Environment: Piazza Venezia 11, 00187 Rome; tel (06) 5722-55-02; fax (06) 5728-85-13; internet www.scn.minambiente.it; Minister: Altero Matteoli.

Ministry of Agriculture and Forestry: Via XX Settembre, 00187 Rome; tel (06) 46651; fax (06) 4742314; e-mail ministro.staff@politcheagricole.it; internet www.politcheagricole.it; Minister: Giovanni Alemanno; incorporates the Department for Mountain Economy and Forests.

Laboratorio Centrale di Idrobiologia (Central Laboratory of Hydrobiology): Via del Carvaggio 107 00147 Rome; tel (06) 51600178; fax (06) 5140296; e-mail hull@inea.it; Dir: Dr Sergio Panella; f 1921; ecology of aquatic systems; fisheries; aquaculture.

Centro di Documentazione Internazionale Parchi della Provincia di Firenze—CEDIP (International Park Documentation Centre, Province of Florence): Villa Demidoff, Via Fiorentina 6, Pratolino, 50036 Florence; tel (055) 409051; fax (055) 409155; e-mail cedip.direzione@provincia.firenze.it; internet www.provincia.firenze.it/cedip; Prin Officer: Prof Bonato Romano; f 1986; information on protected area issues; participation in development of protected areas system in Italy; maintains an extensive library (catalogue available on-line); organizes workshops and lectures; member of IUCN and EUROPARC.

Consiglio Nazionale delle Ricerche—CNR (National Research Council): Piazzale Aldo Moro 7, 00185 Rome; tel (06) 49931; fax (06) 4957241; Pres: Lucio Bianco; Dir-Gen: Dr Nunzio de Rensis; f 1923; activities include environmental research.

Istituto Nazionale per la Fauna Selvatica—INFS (National Wildlife Institute): Via Cà Fornacetta 9, Ozzano Emilia, 40064 Bologna; tel (051) 6512111; fax (051) 796628; e-mail infsammi@iperbole.bologna.it; internet www.infs.it; Chief Exec: Prof Mario Spagnesi; f 1993; national research and advisory agency for issues concerning wildlife conservation; affilated with EURING, the Bonn Convention, Ornis Commission and World Conservation Union—IUCN.

Non-Governmental Organizations

Animal and Nature Conservation Fund—ANCF: Viale Cassala 5, 20143 Milan; tel (02) 58113313; fax (02) 89428918; e-mail ancf@it.net; internet www.ancf.it; Dir: Giuli Cordara; f 1997; fmrly Bellerive Foundation, Italy.

Associazione Primatologica Italiana (Italian Primatological Association): Istituto di Antropologia, Via del Proconsolo 12, 50122 Florence; tel (055) 2338065; fax (055) 283358; e-mail entropos@unifi.it; Prin Officer: Giuseppe Ardito; affiliated to the International Primatological Society.

Federazione dei Verdi (Green Party): Via Antonio Salandra 6, 00187 Rome; tel (06) 4203061; e-mail federazione@verdi.it; Leader: Grazia Frances-Cato; f 1986; environmental political party; promotion of environmentally responsible and anti-nuclear policies; branch of the European Green movement.

Fondo Mondiale per la Natura (World Wide Fund for Nature—WWF Italy): Via Po 25c, 00198 Rome; tel (06) 844971; fax (06) 8554410; e-mail posto@wwf.it; internet www.wwf.it; Pres: Fulco Pratosi; Vice-Pres: Carlo Galli, Maurizio Santoloci; Sec-Gen: Michele Candotti; f 1966; national affiliate organization of the World Wide Fund for Nature—WWF International; conservation; education; management of wildlife areas (103 natural reserves); proposals for environmental policy in the fields of energy, transport, consumer affairs and land management; 300 local offices.

Friends of the Earth Italy/Amici della Terra: Via di Torre Argentina 18, 00186 Rome; tel (06) 6868289; fax (06) 68308610; e-mail amiterra@amicidellaterrra.it; internet www.amicidellaterra.it; f 1977; national branch of Friends of the

Earth International—FoEI; approx 26,000 mems and 100 local groups.

Greenpeace Italy: Viale Manlio Gelsomini 28, 00153 Rome; tel (06) 5729991; fax (06) 5783531; e-mail staff@greenpeace.it; national office of Greenpeace International.

Lega per l'Abolizione della Caccia—LAC (League for the Abolition of Hunting): Viale Bligny 22, Ufficio Pt Isola, CP 10489, 20100 Milan; tel and fax (02) 58306583; e-mail e.f.a.h@agora.stm.it; internet www.agora.stm.it/efah/lac.htm; Pres: Prof Carlo Consiglio; f 1978; campaigns on wildlife, abolition of hunting, animal rights, trapping, poaching and endangered species.

Legambiente (Environmental League): Via Salaria 280, 00199 Rome; tel (06) 8841552; fax (06) 8552976; Contact: Giovanna Melandri; environmental education in developing countries, in co-operation with local NGOs; active throughout Africa and South America.

Rete Fenologica Italiana (Italian Phenological Network): c/o Dipartimento di Biologia, Università di Padova, Viale Giuseppe Colombo 3, 35121 Padova; tel (049) 8276239; fax (049) 8276230; e-mail geobot.@civ.bio.unipd.it; botany; biology; agricultural research; phenology; ecology.

Società Italiana di Ecologia—SITE (Italian Ecological Society): c/o Dipartimento Scienze Ambientali, Viale delle Scienze, 43100 Parma; tel (081) 2538501; fax (081) 450165; e-mail site@eagle.bio.unipr.it; internet eagle.bio.unipr.it/site; Pres: Prof Amalia Viazo de Santo; Gen Sec: Prof Marino Gatto; f 1976; promotion of theoretical and applied ecological research; dissemination of ecological knowledge; development of exchange between researchers and facilitation of national and international co-operation through conferences, symposia and publications.

JAMAICA

Governmental Organizations

Ministry of Land and the Environment: 16a Halfway Tree Rd, Kingston 5; tel 920-3273; fax 929-7349; e-mail mehsys@hotmail.com; Minister: Dean Peart.

Natural Resources Conservation Authority: PO 53½ Molynes Rd, POB 305, Kingston 10; tel 923-5155; fax 923-5070; e-mail nrca@infochan.com; Exec Dir: Franklin McDonald; f 1991; principal government agency responsible for management, conservation and protection of natural resources.

Ministry of Agriculture: Hope Gardens, POB 480, Kingston 6; tel 927-1731; fax 927-1904; e-mail psoffice@moa.gov.jm; internet www.moa.gov.jm; Minister: Roger Clarke.

Department of Forestry and Soil Conservation: 173 Constant Spring Rd, Kingston 8; tel 924-2612; fax 924-2626.

Fisheries Division: POB 470, Kingston 13; tel and fax 923-8811; e-mail fisheries@cwjamaica.com; Dir, Fisheries Div: Gilbert Andre Kong; Dir, Marine Branch: Stephen Smikle; Dir, Aquaculture Branch: Avery Galbraith; Acting Sr Fisheries Officers: Ian Jones, Richard Kelly, June Masters; Fisheries Officer: Tenile Grant; f 1949; management, conservation and development of the capture and culture fisheries resources of Jamaica.

Ministry of Health: Oceana Hotel Complex, 2 King St, Kingston; tel 967-1092; fax 967-7293; e-mail junorj@moh.gov.jm; internet www.moh.gov.jm; Minister: John Junor; incorporates the Environmental Control Division; areas of activity include monitoring of pollution and environmental health.

Non-Governmental Organizations

Centre for Environment and Development, University of the West Indies: 3 Gibraltar Camp Rd, Mona Campus, Kingston 7; tel 977-5530; fax 977-1658; Prin Officers: Prof Albert Binger; f 1992; research; project implementation.

Electron Microscopy Unit, University of the West Indies: Mona Campus, Kingston 7; tel 977-1076; fax 927-1640; e-mail wreid@uwimona.edu.jm; Prin Officers: W A Reid, K Aiken, Prof V B Meyer-Rochow, I A N Stringer; f 1946; understanding and protecting cave environments and their inhabitants; research and consultancy.

Gosse Bird Club: 93 Old Hope Rd, Kingston 6; tel and fax 978-5881; e-mail gosse@infochan.com; internet www.jatoday.com/gossebird.html; Pres: Peter Vogel; Sec: Annika Lewinson; study of birds; lobbying for conservation of birds and their habitats; educational programmes; affiliated with the National Environmental Societies Trust and BirdLife International.

Inter-American Institute for Co-operation on Agriculture—IICA: 11 Fairway Ave, Kingston 6; tel 927-6462; fax 927-6933; e-mail iicajam@uwimona.edu.jm; f 1975; promotion of the sustainable development of the agri-food sector and the modernization and improvement of production.

National Environmental Societies Trust—NEST: 95 Dumbarton Ave, Kingston 10; tel 960-3316; fax 968-5872; e-mail nest@infochan.com; Exec Dir: Maureen Rowe; Programme Dir: Sybil Douglas-Ricketts; association of environmental NGOs; provision of networking services and training.

Natural History Society of Jamaica: c/o Dept of Life Sciences, University of the West Indies, Mona, Kingston 7; tel 977-6938; fax 977-1075; Pres: Dr Trevor Yee; Vice-Pres: Jillian Byles; Treas: Grace Smith; f 1940; study of Jamaican flora and fauna; organizes field trips and nature-walks; lectures and exhibitions; preservation and conservation of species and the natural environment; affiliated with BirdLife Jamaica.

Portland Environment Protection Association—PEPA: 6 Allan Ave, Port Antonio, Portland; tel 993-9632; fax 715-3705; e-mail ttpepa@cwjamaica.com; Pres and Founder: Marguerite Gauron; Chief Exec: Harvey Webb, III; f 1988; coalition of local organizations; creation of public awareness of environmental problems; lobbying of government, industry, food producers and the private sector to take affirmative action with regard to environmental protection; the Port Antonio Marine Park and Conservation Corridor project aims to protect the area's marine and terrestrial resources and maintain a continuous natural forest corridor from the coast to the summit of Jamaica's highest peak; affiliated with the Jamaica Conservation Development Trust and the Nature Conservancy.

St Ann Environment Protection Association: POB 212, Runaway Bay; tel and fax 973-4305; e-mail ee@cwjamaica.com; Pres: Wendy A Lee; Sec: Barbara Zampelli; Treas: Frank Lawrence; f 1989; conservation of natural resouces; environmental education; advocacy; wildlife rescue and rehabilitation for protected species, particularly Amazon parrots; facilitation of community involvement in environmental policy and management issues, such as local development; teacher training and outreach activities.

United Nations Association of Jamaica—UNAJ: 33 Anthurium Drive, Kingston 6; tel 962-9771; fax 978-6046; 977-2893; e-mail 946-2163; Pres: Dr Lucille Buchanan; Gen-Sec: Monica Blair; f 1949; activities include advocacy of environmental management; training and education of students at high schools island-wide in the UN model assembly; environmental advocacy—participating in environmental studies and conferences; members and youth members participate in environmental management; participating in international conferences, modern mass media, media communication.

JAPAN

Governmental Organizations

Environment Agency of Japan: 1-2-2, Kasumigaseki, Chiyoda-ku, Tokyo 100-8975; tel (3) 3581-3351; fax (3) 3502-0308; e-mail web@eanet.go.jp; Minister of State and Dir-Gen: Kayoko Shimizu; environmental management; establishment of environmental policy; protection of wildlife; sustainable development; research; statistics; environmental impact assessment; protection of marine, freshwater and terrestrial environments; pollution control.

Japan Environment Co-operation: Nitochi Bldg, 1-4-1, Kasumigaseki, Chiyoda-ku, Tokyo 100; tel (3) 5251-1014; provision of assistance to environmental NGOs.

National Institute for Environmental Studies—NIES: Centre for Global Environmental Research, 16-2, Onogawa, Tsukuba, Ibaraki 305-0053; tel (298) 51-6111; fax (298) 58-2645; e-mail nfp@nies.go.jp; internet www.nies.go.jp; Contact, INFOTERRA national focal point: Takashi Oshima; Tech Officer: Yasushi Nakajima; collaborating centre in the UNEP Global Environment Outlook—GEO programme; the Environmental Information Centre serves as the INFOTERRA national focal point.

Ministry of Agriculture, Forestry and Fisheries: 1-2-1, Kasumigaseki, Chiyoda-ku, Tokyo 100-8950; tel (3) 3502-8111; fax (3) 3592-7697; internet www.maa.go.jp; Minister: Yoshinobu Shimamura.

Forestry and Forest Products Research Institute—FFPRI: POB 16, Tsukuba Norin Kenkyu Danchi-nai, Ibaraki 305; Dir: Itsu-Hito Ohnuki; f 1878; part of the Forestry Agency of the Ministry of Agriculture, Forestry and Fisheries.

Ministry of Economy, Trade and Industry: 1-3-1, Kasumigaseki, Chiyoda-ku, Tokyo 100-8901; tel (3) 3501-1511; fax (3) 3501-6942; e-mail webmail@meti.go.jp; internet www.meti.go.jp; Minister: Shoichi Nakagawa; incorporates the Environmental Policy Division.

Ministry of Foreign Affairs: 2-11-1, Shiba-Koen, Minato-ku, Tokyo 105-8519; tel (3) 3580-3311; fax (3) 3581-2667; e-mail webmaster@mofa.go.jp; internet www.mofa.go.jp; Minister: Nobutaka Machimura; incorporates the Office of Global Environment.

Japan Atomic Energy Research Institute: Fukoku-Seimei Bldg, 2-2-2 Uchisaiwai-cho, Chiyoda-ku, Tokyo 100; Pres: Masaji Yoshikawa; f 1956; managed by the Science and Technology Agency (organization with ministerial status); publications.

Non-Governmental Organizations

Centre for Ecological Research, Kyoto University: 509-3 Otsuka, Kamitanakami-Hirano, Otsu City, Shiga 520-2113; Dir: Prof E Wada; f 1991.

Chikyu, Ningen Kankyo Forum (Global Environmental Forum): Iikura Bldg, 1-9-7, Azabudai, Minato-ku, Tokyo 106; tel (3) 5561-9735; fax (3) 5561-9737; e-mail idw00423@nifty.ne.jp; internet www.shonan.ne.jp/~gef20/gef; Chair: Jiro Kondo; Pres: Hiroshi Shimizu; f 1990; scientific research on domestic and international environmental issues; environmental policy research; dissemination of environmental information and research results by means of symposia, conferences and publications.

Ecological Society of Japan: c/o Centre for Ecological Research, University of Kyoto, 509-3 Otsuka, Kamitanakami-Hirano, Otsu City, Shiga 520-2113; tel (77) 549-8228; fax (77) 549-8229; e-mail esj-cer@ecology.kyoto-u.ac.jp; Sec: Dr Masahide Yuma; research; conferences and meetings; publications.

Elsa Nature Conservancy: Tsukauba-Gakuen, POB 2, Ibaraki 305; tel (298) 51-1637; fax (298) 51-1637; Pres: Eiji Fujiwara; conservation of fauna and flora; education.

Greenpeace Japan: Yoyogikaikan 4f, 1-35-1, Yoyogi, Shibuya-ku, Tokyo 151; tel (3) 5351-5400; fax (3) 5351-5417; e-mail greenpeace.japan@dialb.greenpeace.org; Exec Dir: Naoki Ohara; Admin Asst: Sanae Shida; national office of Greenpeace International.

IAIA Japan: c/o A Tree Co, Horiuchi 804-4-204, Hyama, Kanagawa 240-0112; tel and fax (4) 6876-3845; e-mail akiratanaka@attglobal.net; internet www.seiryo.ac.jp/iaia-japan/index.html; Sec-Gen: Dr Akira Tanaka; regional office of the International Association for Impact Assessment—IAIA; holds regular seminars in Tokyo six times a year; research, including environmental impact assessment; publications.

Institute for Himalayan Conservation: 3-5-7-207, Yoyogi, Shibuya-ku, Tokyo 151-0053; tel (3) 5350-8458; fax (3) 5350-8459; e-mail ihcjapan@par.odn.ne.jp; internet www.ngo.gr.jp/ihc; Chair: Masami Mizuno; Exec Sec: Hiroshi Tanaka; environmental conservation and development in the mountainous regions of Nepal; IHC-Nepal and IHC-Japan.

Institute of Cetacean Research: Tokyo Suisan Bldg, 4-18, Toyomi-cho, Chuo-ku, Tokyo 104; tel (3) 3536-6521; fax (3) 3536-6522; e-mail kujita@mxc.meshnet.or.jp; internet www.whalesci.org; Dir-Gen: Seiji Ohsumi; f 1987; research on cetaceans and other marine mammals; monitoring of the world-wide population of marine mammals; collection and dissemination of information.

Japan Centre for Human Environmental Problems, Tokyo Metropolitan University: Faculty of Economics, Meisei University 1-1 Hodokubo 2 chome Hino-shi, Tokyo 191-8506; tel (425) 91-5927; fax (425) 91-5927; e-mail inoueh@econ.meisei-u.ac.jp; internet www.wwe.jp/i.html; Pres: Dr Ichiro Kato; Prin Officers Prof Yoshihiro Nomura, Prof Fukashi Utsunomiya, Prof Hidenori Inoue; the JCHEP was established in 1973 as a new type of academic society; while its membership is kept to 50 or less, it cordially welcomes the exchange of information and views with specialists interested in environmental problems. The JCHEP is a member of IUCN.

Japan Committee for IUCN: c/o Nature Conservation Society of Japan Yamaji Sanbancho Bldg 3f, 5-24, Sanbancho, Chiyoda-ku, Tokyo 102; tel (3) 3265-0524; Chair: Prof Dr Makoto Numata; national office of the World Conservation Union—IUCN.

Japanese Association of Zoological Gardens and Aquariums—JAZGA: Vella Heights, Okachimachi 402, 4-23-10, Taito, Taito-ku, Tokyo 110; tel (3) 3837-0211; fax (3) 3837-1231; Chair: Takamasa Ikeda; f 1965; nature conservation activities; research and surveys on zoos and aquariums; conferences and technical study group meetings; training programmes; qualification tests for keepers of zoos and aquariums; publications.

Kokuritsu Kohen Kyokai (National Parks Association of Japan): c/o Toranomon Denki Bldg, 2-8-1, Toranomon, Minato-ku, Tokyo 105; tel (3) 3502-0488; fax (3) 3502-1377; internet www.npaj.or.jp; Pres: Nobuya Seta; Editor: Yumiko Wakabayashi; conservation of nature, areas of natural beauty and recreational areas, including national parks; environmental protection; education.

Nature Conservation Society of Japan—NACS-J: Yamaji Sanbancho Bldg 3f, 5-24, Sanbancho, Chiyoda-ku, Tokyo 102; tel (3) 3265-0524; Chair: Takao Fukiwara; conservation of primeval beech forest, coral reefs, rivers and wildlife.

Nettairin Kodo Network (Japan Tropical Forest Action Network—JATAN): 6-5, Uguisudani-cho 1f, Shibuya-ku, Tokyo 150; tel (3) 3770-6308; fax (3) 3770-0727; e-mail jatan@jca.apc.org; internet www.jca.apc.org/jatan; Co-ordinator: Yoichi Kuroda; campaigns against Japanese corporate and government involvement in the destruction of tropical, temperate and boreal forests around the world; concerned with Japan's contributions to development institutions.

Primate Society of Japan: Primate Research Institute, Kyoto Univ, Inuyama, Aichi 484; tel (568) 63-0567; fax (568) 63-0085; internet www.soc.nacsis.jp/psj2; Pres: Dr Yukimaru Sugiyama; f 1985; affiliated to the International Primatological Society.

Shizen Kankyo Kenkyo Senta (Japan Wildlife Research Centre—JWRC): 2-29-3, Yushima, Bunkyo-ku, Tokyo 113; tel (3) 3813-8897; fax (3) 3813-8898; e-mail komoda@jwrc.or.jp; Pres: Prof Dr Taisitiroo Satoo; Exec Dir: Kazuhiro Yamase; f 1978; formerly Nihon Yaseiseibutsu Kenkyu Senta; research on wildlife management, conservation and ecology; compilation and management of a data base on the natural environment of Japan.

Tokyo University of Fisheries: 5-7 Konan 4, Minato-ku, Tokyo 108-8477; tel (3) 5463-0400; fax (3) 5463-0359; internet www.tokyo-u-fish.ac.jp; Pres: Dr C Koizumi; f 1888.

Toyama-ken Shizen Hogo Kyoukai (Toyama Prefectural Association for Nature Conservation): Toyama-ken 930; Pres: K Wakabayashi; Vice-Pres: M Fujihira, M Nagai; research on nature in the region; education through publications and lectures; co-ordination with other organizations.

TRAFFIC East Asia, Japan: Nihonseimei Akabanebashhi Bldg, 6th Floor, 3-1-14, Shiba, Minato-ku, Tokyo 105; tel (3) 3769-1716; fax (3) 3969-1304; e-mail trafficj@twics.com; internet www.twics.com/~trafficj; national office of TRAFFIC International.

World Wide Fund for Nature—WWF Japan: 6f Nihonseimei Akabanebashi Bldg, 3-1-14, Shiba, Minato-ku, Tokyo 105-0014; tel (3) 3769-1711; fax (3) 3769-1717; e-mail communi@wwf.or.jp; internet www.wwf.or.jp; Chair: Teruyuki Ohuchi; Vice-Chairs: Hisako Hatakeyama, Hisanaga Shimazu; f 1971; research, communications and education projects at the WWF Coral Reef Conservation and Research Centre in Shiraho, Ishigaki Island of Nansei Shoto Islands; policy works and publicity for the conservation of dugongs in Okinawa and Yambaru forests in Nansei Shoto Islands; policy works, research and publicity for the wetlands such as Sanbanze and Isahaya Bay Tidal Flat; promoting an independent forest certification system with the Forest Stewardship Council (FSC); promoting energy saving and the use of renewable energy with industry to mitigate climate change; activities of TRAFFIC East Asia Japan to monitor trade in wildlife.

JORDAN

Governmental Organizations

Ministry of Municipal and Rural Affairs: POB 1799, Amman; tel (6) 4641393; fax (6) 4640404; Minister: Dr Amal Farhan.

General Corporation for Environment Protection: POB 1408, Amman 11941; tel (6) 830149; fax (6) 830084; Prin Officer: Faris Al-Juniadi; Contact, INFOTERRA national focal point: Sameh Tubaishat; f 1995; environmental legislation; monitoring of air pollution; solid waste and water treatment; preservation of wetland ecosystems.

Ministry of Agriculture: POB 2099, Amman; tel (6) 5686151; fax (6) 5686310; e-mail falah-a@moa.gov.jo; internet www.moa.gov.jo; Minister: Sharari Shakhanbeh.

Ministry of Water and Irrigation: POB 2412, Amman; tel (6) 5680100; fax (6) 5680075; e-mail infor@mwi.gov.jo; internet www.mwi.gov.jo; Minister: Dr Hazem an-Nasser.

Non-Governmental Organizations

Royal Scientific Society—RSS: POB 925819, Amman; tel (6) 844701; fax (6) 844806; comprises the Environmental Research Centre, the Renewable Energy Research Centre and the Industrial Chemistry Centre, which has an Environmental Division.

Environmental Research Centre—ERC: POB 1438, Jubeiha 11941 Amman; Dir: Ayman A Al-Hassan; studies and applied research; provision of specialized technical services, advice and training in the field of the environment for the public and private sectors in Jordan and neighbouring Arab countries; comprised of Water and Soil Division, Air and Hazardous Chemicals Division and Ecology Division; collaborating centre in the UNEP Global Environment Outlook—GEO programme.

Royal Society for the Conservation of Nature—RSCN: POB 6354, Amman; tel (6) 837931; fax (6) 647411; e-mail rscn@nets.com.jo; Pres: HE Anis Mouasher; f 1966; protection of wild flora and fauna and their habitats; captive breeding of endangered animals; establishment and management of protected areas; integrated conservation and development; co-operation with government and international agencies; public awareness-raising and education.

KAZAKHSTAN

Governmental Organizations

Ministry of Environmental Protection: 473000 Astana, Pobeda 31; tel (3172) 59-19-44; fax (3172) 59-19-73; internet www.nature.kz; Minister: Aitkul Samakova; f 1988; conservation and wise use of natural resources; ecological and hydrometeorological research; provision of scientific and technical expertise; comprises the Committee of Environmental Protection, the Committee of Forestry, Fish and Hunting Economy, the Committee on Water Resources and the Committee of Geological Protection; it includes local departments in each of the 16 regions and the two main cities.

International Aral Sea Rehabilitation Fund—IFAS: See International Environmental Organizations section.

Ministry of Agriculture: 473000 Astana, pr Abaya 49; tel (3172) 32-37-63; fax (3172) 32-62-99; e-mail mailbox@minagri.kz; internet www.minagri.kz; Minister: Serik Umbetov.

Forestry Committee: Almaty, Zheltoksan 112; tel (3272) 62-63-20; fax (3272) 62-20-26; Chair: Nikolai I Bayev; f 1946.

Kazekologiya Information Centre: Almaty; Gen Dir: Amangeldy A Skakov.

Non-Governmental Organizations

Aral-Asia-Kazakhstan International Public Committee: Almaty, Lenina 7; tel (3272) 33-14-94; Pres: Muhtar Shahanov.

Ecofund of Kazakhstan: 400037 Almaty, Toslonko 31; tel (3272) 29-55-59; fax (3272) 67-21-24; Co-Chair: Lev Ivanovich Kurlapov, Viktor Zonav; f 1988; formerly one of the largest environmental groups in Kazakhstan; subsequently split several times.

Eikos Company Ltd: Almaty, Nusupbekova 32; tel (3272) 30-49-90; fax (3272) 30-68-03; e-mail eikos@nursat.kz; Gen Dir: Tatyana Pilat; f 1990; provides equipment and technology for water purification and sea water desalination, disinfection, water pre-treatment, sewage and industrial waste treatment.

Ekologiya i Obshchestvennoye Mneniye—EKOM (Ecology and Public Opinion): 637046 Pavlodar, Suvarova 12/131; tel (3182) 72-67-75; Chair: Nikolai Stepanovich Savukhin; Sec: Valeri Pavlovich Galenko; f 1987; oldest registered environmental NGO in Kazakhstan.

Fund in Support of Ecological Education: 480005 Almaty, S Kovalevskaya 63, kv 13; tel (3272) 41-29-91; fax (3272) 63-66-34; Chair: Zharas Abu-uly Takenov; Prin Officers: Gulmira Djamanova, Raushan Kryldakova; f 1991; environmental education; library.

Green Movement Socio-Ecological Centre—GMC: 484006 Taraz, Lunacharsky 42-2; tel and fax (32622) 3-27-93; e-mail alex@zagribelny.jambyl.kz; Chair: Aleksandr Zagribelny; Sec: Luba Raupova; f 1990; opposition to manufacture of phosphorus fertilizer; publications and documentary films; participated in UNCED (Earth Summit, Brazil) in 1992.

Green Party: 480012 Almaty, Vinogradova 85, Rm 302; Chair: Mels Hamsayevich Elusizov; f 1991; environmental political party; political wing of the Tabigat EcoUnion, which campaigns for environmental health and against water pollution, particularly concerning the Sor-Bulak sewage lake near Almaty.

Institut Geografii, Natsionalnoi Akademii Nauk RK (Institute of Geography, Kazakstan Academy of Sciences): 480100 Almaty, Kabanbai Batyr 69[a]; tel (3272) 61-56-08; fax (3272) 61-53-14; Dir: Bespaev Khalei Abdulkhakovich; f 1940; research and development in the fields of mineral ore and petroleum geology; collection of data on mineral raw material resources.

International Ecology Centre—Biosphere Club: 493910 Leninogorsk, Microraion 3, d 19, kv 10; Chair: Vladimir Pavlovich Karamanov.

Kazakh Community for Nature Protection: Central Council, 480044 Almaty, Zhibek zholy 15; tel (3272) 61-65-16; Chair: Kamza B Zhumabekov.

Kazakhstan Academy of Sciences: 480021 Almaty, Shevchenko 28; tel (3272) 69-51-50; fax (3272) 69-61-16; Pres: V S Shkolnik; Sec-Gen: M K Suleimenov; several attached institutes are involved in environmental research.

Nevada–Semipalatinsk International Antinuclear Movement—IAM: House of Democracy, Almaty, 1 Karasai Batyr 85; tel (3272) 63-49-02; fax (3272) 50-71-87; Pres: S O Omarovich; Head, Exec Cttee: Erimbetov Myrzahan; f 1989; aims to ban the production and storage of nuclear and other types of weapons and to foster responsible attitudes to the environment; developed and implemented the Programme for the Ecological, Economic and Spiritual Regeneration of regions where nuclear testing took place; co-ordination of international anti-nuclear efforts through exchange of information and congresses; activism and charitable activities; divisions throughout Kazakhstan and offices world-wide.

Lop Nor Semipalatinsk Ecological Committee: Almaty; tel (3272) 63-04-64; fax (3272) 63-12-07; Chair: Azat M Akimbek; Deputy Chair: Irkiz Ileva; f 1992; semi-autonomous department; campaigns against nuclear testing in neighbouring parts of the People's Republic of China.

Zelonoye Spaseniye (Green Salvation Ecological Society): 480059 Almaty, Shagabutdinova 133, kv 66; tel (3272) 67-96-85; e-mail ecoalmati@glas.apc.org; Prin Officers: Sergei G Kuratov, Sergei Solyanik, Semen Svitelman; f 1990; maintains an ecological data base; media campaigns; direct action; ecological education.

KENYA

Governmental Organizations

Ministry of the Environment and Natural Resources: Maji House, Ngong Rd, POB 49720, Nairobi; tel (2) 716103; e-mail mec@nbnet.co.ke; internet www.environment.go.ke; Minister: Kalonzo Musyoka.

Forest Department: POB 30513, Nairobi; tel (2) 229261; fax (2) 340260.

Ministry of Agriculture: Kilimo House, Cathedral Rd, POB 30028, Nairobi; tel (2) 718870; fax (2) 720586; internet www.agriculture.go.ke; Minister: Kipruto Rono arap Kirwa.

Ministry of Livestock and Fisheries Development: Kilimo House, Cathedral Rd, POB 30028, Nairobi; tel (20) 718870; fax (20) 2711149; Minister: Joseph Konzolo Mungao.

Ministry of Tourism and Wildlife: Utalii House, 5th Floor, off Uhuru Highway, POB 30027, Nairobi; tel (2) 333555; fax (20) 318045; internet www.tourism.go.ke; Minister: Karisa Maitha; incorporates the Wildlife Conservation and Management Department.

Ministry of Water: Maji House, Ngong Rd, POB 49720, Nairobi; tel (2) 716103; internet www.kenya.go.ke/water; Minister: Wangari Karua.

Appropriate Technology Centre, Kenyatta University: POB 43844, Nairobi; Dir: Dr Thomas Nyaki Theruwa; f 1979; teaching and research in renewable energy technology, water technology and technology in the development of agriculture and entrepreneurship.

Kenya Industrial Research and Development Institute—KIRDI: POB 30650, Nairobi; tel (2) 556362 /535966; fax (2) 555738; e-mail kirdi@onlineKenya.com; internet www.kirdi.go.ke; Dir: Dr Patrick M Muturi; Programme Coordinators: Dr M C Z Moturi, Dr M Makayoto, Sam Wambugu; f 1942; affiliated to the Ministry of Trade and Industry. The Institute has been mandated to conduct research and development for the development of industrial technologies and for the provision of technology management services to the industrial sector in Kenya. It promotes cleaner production technologies and their implementation within industry.

Permanent Presidential Commission on Soil Conservation and Afforestation: Office of the President,Harambee House, Harambee Ave, POB 30510, Nairobi; tel (2) 227411.

Non-Governmental Organizations

ICRISAT Nairobi: POB 39063, Nairobi; tel (2) 521450; fax (2) 521001; e-mail icrisat-kenya@cgiar.org; national office of the International Crops Research Institute for the Semi-Arid Tropics—ICRISAT, India.

IUCN Kenya: the office of the World Conservation Union—IUCN in Kenya serves as the regional office for Eastern Africa (see International Environmental Organizations section).

Nature Kenya: POB 44486, Nairobi; tel (20) 3749957; fax (20) 3741049; e-mail office@naturekenya.org; internet www.naturekenya.org; Chair: Dr Ian Gordon; Vice-Chair: Theresa Aloo; Exec Dir: Paul Matiku; f 1909; study and conservation of the natural environment; action groups for forests, birds and mammals; annual water bird censuses; meetings and field workshops; publication of a bulletin and journal; affiliated with BirdLife International; environmental policy and legislation working group (ELPWG); Heipetofauna working group; Habitat Restoration initiative of East Africa (HARI); Friends of Nairobi Arboretum (FONA); Friends of Nairobi City Park (FoCP); conservation programme—Action for Biodiversity in Kenya.

TRAFFIC East/Southern Africa, Kenya: POB 68200, Mukoma Rd, Langata, Nairobi; tel (2) 506839; fax (2) 600543; e-mail traffic@iconnect.co.ke; national office of TRAFFIC International.

Wildlife Clubs of Kenya: POB 20184, Nairobi; tel (2) 891904; fax (2) 891906; e-mail iali@arcc.or.ke; f 1968; nature and wildlife conservation; preservation of natural resources; promotion of environmental education for young people through publications, teachers' workshops, student seminars and awareness-raising campaigns; member of the Kenya Wildlife Service, the World Conservation Union—IUCN and the World Wide Fund for Nature—WWF.

World Wide Fund for Nature—WWF Kenya: the office of WWF International in Kenya serves as the Eastern Africa Regional Programme Office (see International Environmental Organizations section).

KIRIBATI

Governmental Organizations

Ministry of the Environment, Lands and Agricultural Development: POB 234, Bikenibeu, Tarawa; tel 28211; fax 28334; e-mail ps@melad.gov.ki; Minister: Martin Tofinga.

Ministry of Internal Affairs and Social Development: POB 75, Bairiki, Tarawa; tel 21092; fax 21133; e-mail homeafffairs@tskl.net.ki; Minister: Amberoti Nikora.

Ministry of Natural Resources Development: POB 64, Bairiki, Tarawa; tel 21099; fax 21120; Minister: Tetabo Nakara.

Non-Governmental Organizations

Atoll Research and Development Unit, University of the South Pacific: POB 206, Bikenibeu, Tarawa; Programme Man: Temakei Terano; research areas include environmental and ecological aspects of atoll life.

Solar Energy Company: Tarawa; co-operative administering and implementing solar-generated electricity projects in North Tarawa and the outer islands.

KOREA, DEMOCRATIC PEOPLE'S REPUBLIC

Governmental Organizations

Ministry of Land and Environmental Protection: Gwangbok Dong, Mangyongdae Dist, Pyongyang; tel (2) 7214852; fax (2) 3814410; Minister: Jang Il Son.

Ministry of Agriculture: Pyongyang; Minister: Ri Kyong Sik.

Ministry of Fisheries: Pyongyang; Minister: Ri Song Un.

Ministry of Forestry: Pyongyang; Minister: Ri Sang Mu.

National Co-ordinating Committee for the Environment—NCCE: POB 44, Pyongyang; tel (2) 3827222; fax (2) 3814660; e-mail pang.yong.gun@undp.org; Contact: Pak Chun Il; serves as the INFOTERRA national focal point.

Non-Governmental Organizations

Institute of Environmental Protection, Academy of Sciences: Ryusong-dong, Central Dist, Pyongyang; tel (2) 51956; Dir: Kim Yong Chan.

Natural Conservation Union: 220-93-7-24, Dongsong St, Chung Dist, Pyongyang; tel (2) 814410; Vice-Chair: Han Sok-Yuk; Sec Gen: Pak Jae-Su; f 1959; conservation of wild flora and fauna; afforestation and forestry; conservation of protected areas, public lands and national parks; scientific research; management and planning of protected areas; environmental management and planning; flood control; training for management of protected areas; education in resource management.

KOREA, REPUBLIC

Governmental Organizations

Ministry of the Environment: 1 Jungang-dong, Gwachon City, Kyonggi Prov; tel (2) 2110-6546; fax (2) 504-9206; e-mail shinae@me.go.kr; internet www.me.go.kr; Minister: Kwak Kyul-Ho; formulation of environmental policies; enforcement of legislation.

Ministry of Agriculture and Forestry: 1, Jungang-dong, Gwachon City, Kyonggi Prov; tel (2) 503-7200; fax (2) 503-7238; e-mail wmaster@maf.go.kr; internet www.maf.go.kr; Minister: Huh Sang-Man.

Ministry of Government Administration and Home Affairs: 77-6, Sejong-no, Chongno-ku, Seoul; tel (2) 3703-2114; fax (2) 3703-5502; internet www.mogaha.go.kr; Minister: Huh Sung-Kwan.

National Parks Authority—NPA: Taeyong Bldg, 252-5, Kongduk-dong, Mapo-gu, Seoul; tel (2) 3272-7931; fax (2) 3272-8973; e-mail npa@www.npa.or.kr; internet www.npa.or.kr; Chair: Nam Kim; f 1987; administration of 18 of Korea's 20 national parks; conservation of natural resources; education of visitors.

Ministry of Maritime Affairs and Fisheries: 139, Chungjeong-no 3, Seo-daemun-gu, Seoul 120-715; tel (2) 3148-6040; fax (2) 3148-6044; internet www.momaf.go.kr; Minister: Chang Seung-Woo.

Korea Atomic Energy Research Institute: POB 105, Yu-Seong, Taejon 305-600; tel (42) 868-2000; fax (42) 868-2702; Pres: Seong-Yun Kim; f 1995; affiliated to the Ministry of Science and Technology.

Korea Institute of Energy and Research: POB 5, Taedok Science Town, Teajon 305-343; tel (42) 861-9700; fax (42) 861-6224; Pres: P Chung Moo Auh; f 1977; affiliated to the Ministry of Science and Technology.

Non-Governmental Organizations

Korea Environmental Preservation Association—KEPA: Korea Chamber of Commerce and Industry Bldg, 6th Floor, 497-66, Tapsimni-5-dong, Dongdaemun-ku, Seoul; tel (2) 216-3882; fax (2) 249-5267; e-mail kepa@hitel.kol.co.kr; internet www.green.co.kr; f 1978; environmental preservation; research and technical development work; surveying; education; public information; affiliated with the Ministry of the Environment.

National Parks Association of Korea: Sahak Bldg, 1-1, Jungreung-4-dong, Sungbuk-ku, Seoul 136-104; tel (2) 942-2420; fax (2) 912-4243; Pres: Lee Hyun Jae; Vice-Pres: Lee Hwa Young, Oh Whee Young; development of public parks and tourist sites.

KUWAIT

Governmental Organizations

Environmental Public Authority—EPA: POB 24395, 13104 Safat, Kuwait City; tel 4820580; fax 4820570; e-mail sarawi@epa.org.kw; internet www.epa.org.kw; Man: Dr Mohamed A Al-Sarawi; Tech Officer: Huda Munayes; formerly the Environment Protection Council; serves as the INFOTERRA national focal point.

Ministry of Energy: POB 12, 13001 Safat, Kuwait City; tel 4896000; fax 4897484; Minister: Sheikh Ahmad al-Fahd al-Ahmad as-Sabah.

Ministry of Health: POB 5, 13001 Safat, Arabian Gulf St, Kuwait City; tel 4877422; fax 4865414; Minister: Muhammad Ahmad al-Jarallah.

Ministry of Transport and Planning: POB 15, 13001 Safat, Kuwait City; tel 2428100; fax 2414734; e-mail info@mop.gov.kw; internet www.mop.gov.kw; Minister: Sheikh Ahmad Abdullah al-Ahmad as-Sabah; incorporates the General Department of Technical Co-operation (Environmental Affairs).

Non-Governmental Organizations

Kuwait Institute for Scientific Research—KISR: POB 24885, 13109 Safat; tel 4816988; fax 4846891; Dir-Gen: Prof Adnan H Al-Aqel; f 1967; autonomous national research organization; promotes and conducts research in the fields of food and water resources, petroleum sector support, environmental studies, infrastructure services and urban development.

KYRGYZSTAN

Governmental Organizations

Ministry of Environmental Protection: 720033 Bishkek, Isanova 131; tel (3312) 21-97-37; fax (3312) 21-36-05; e-mail min-eco@elcat.kg; Minister: Tynybek Alykulov; First Deputy Minister: Tilekbai Kyshtobaev.

State Hydrometeorology Enterprise: 720403 Bishkek, Karasuiskaya 1; tel and fax (3312) 214422; e-mail kgmeteo@kyrgyzmeteo.elcat.kg; Gen Dir: Mouratbek Bakanov; f 1926.

Ministry of Agriculture, Water Resources and Processing Industry: 720040 Bishkek, ul Kievskaya 96a; tel (312) 62-36-33; fax (312) 62-36-32; e-mail mawr@bishkek.gov.kg; Minister: Aleksandr Kostyuk.

Ministry of Ecology and Emergency Situations: 720055 Bishkek, ul Toktonaliyeva 2/1; tel (312) 54-79-56; fax (312) 54-11-79; e-mail mecd@bishkek.gov.kg; Minister: Temirbek Akmataliyev.

State Forestry Committee: 720033 Bishkek, Abdymomounov 276; tel (3312) 21-36-79; e-mail mail@forestagency.bishkek.gov.kg; Chair: Dzhanysh Rustenbekov.

Non-Governmental Organizations

'AGAT' Club: 720008 Bishkek, Intergelpo 2; tel (3312) 21-82-81; e-mail root@agat.freenet.bishkek.su; Head: Vyacheslav Charski.

Aleyne Ecological Movement of Kyrgyzstan: 720071 Bishkek, POB 50; tel (3312) 68-04-11; e-mail root@emil.cango.kg; Co-Chair: Emil Shukhurov, Ganadii Vorobjev; branch offices in Talas Oblast, Naryn Oblast, Issyk-Kul Oblast and Jalal-Abad Oblast.

Alpha-C Socio-Ecological Centre: 720038 Bishkek, Djal Microraion 7-76; tel (3312) 48-06-73; e-mail a-1_safonov@au.rus; Head: Andrei Safnov.

Altan-Tamyr: 715530 Uzgen dist, Mirza Aki 42; tel (3233) 498; Dir: Abdrali Sultanov.

Asian Ecological Group: 720033 Bishkek, Isanov 131; tel (3312) 61-04-11; e-mail azamat@sdnp.kyrnet.kg; Head: Azamat Hudaibergenov.

Beagle Environmental Association: 720001 Bishkek, Bokombaeva 153; tel (3312) 21-98-78; e-mail beagle@cango.net.kg; Pres: Ekaterina Kriatova.

BIOM Youth Ecological Movement: 720024 Bishkek, Abdymomunova 328, Rm 327; tel (3312) 25-18-78; e-mail biom@infotel.kg; Head: Natalia Kravzova; branch offices in Bishkek, Tchui Oblast, Issyk-Kul Oblast and Jalal-Abad Oblast.

Bishkek-ECO: 720020 Bishkek, Ahunbaeva 119[a]; tel (3312) 42-25-00; Head: Bakyt Duishembaev.

Booruker: 720001 Bishkek, Frunze 547; tel (3312) 66-01-54; e-mail gulia@kaf-i.kg; Pres: Gulija Tajieva; environmental charitable organization.

Botany Society of Kyrgyzstan: Bishkek; tel (3312) 24-25-19; Head: Rostislav Ionov.

Centre for Independent Economic and Social Research: 720075 Bishkek, 8 Micro-rayon 34, kv 2; tel (3312) 47-13-62; fax (3312) 22-44-14; Dir: Viktor Aleksandrovich Bobrov; f 1991; part of the Foundation for Social Initiatives; research into sustainable development; ecological education.

Committee for the Defence of Lake Issyk-Kul: 720023 Bishkek, 10 Micro-rayon 32-31; tel (3312) 22-19-68; Pres: Omor Sultanov; f 1990.

Consortium of Ecological NGOs: 720033 Bishkek, Isanov 131, Rm 403; tel (3312) 21-26-28; e-mail ecocons@cango.net.kg; Co-ordinator: Chinara Sydykova.

Democracy and Environment: 720001 Bishkek, Kievskaya 95-53; tel (3312) 22-00-38; fax (3312) 66-01-46; e-mail dos@cango.net.kg; Head: Madjit Husainov; Chief, Sociology and Rights Section: N Samlitov; Chief, Democracy Section: R Saitov; Chief, Environmental Protection Section: I Vasilev; Chief, Foreign Co-operation and Information Section: N Vasilenko; f 1997; promotion of democracy among the population, including increasing public awareness of civic rights and duties; environmental education.

Diamond Association: 720040 Bishkek, Panfilov 200-39; tel and fax (3312) 66-02-35; Pres: Bermet Tugelbaeva.

ECOBILIM Centre: 720040 Bishkek, Erkindik 38, Rm 133; tel (3312) 66-12-54; Chair: Mira Botbaeva; branch offices in Talas Oblast and Jalal-Abad Oblast.

ECODEM: 720021 Bishkek, Almaty 95; tel (3312) 43-13-37; Chair: Jypara Djanysakova.

ECOINFORM: 720033 Bishkek, Isanov 131; tel (3312) 21-06-82; fax (3312) 21-36-05; e-mail slava@sdnp.kyrnet.kg; Head: Vyacheslav Zuanov.

ECOLOG Club: 720040 Bishkek, Erkindik 27-9, POB 1041; tel (312) 621487; fax (312) 666334; e-mail igorho@mail.ru; Vice-Pres: Isak Jumaev.

Ecological Information Centre: Bishkek; tel (3312) 25-53-70; Man: Aleksandr Vorobjev.

Ecological Institute on Water Resources and Desert Problems: 720071 Bishkek, Tchui pr 265; tel (3312) 21-79-73; Dir: Lidia Orolbaeva.

Ecological Movement of Kyrgyzstan: Bishkek; tel (3312) 26-55-28; Chair: T Chodurayev; f 1994.

Ecology and Culture Centre: 714000 Osh, Alebastrovaya 31; tel (3222) 33249; Chair: Rashid Abdulmjanov.

Entomological Society of Kyrgyzstan: Bishkek; tel (3312) 24-39-98; e-mail kirgwwf@tarbin.bishkek.su; Head: Yurii Tarbiski.

Environmental Protection Fund of Kyrgyzstan: 720071 Bishkek, Tchui pr 265[a], Rm 13; tel (3312) 24-36-61; fax (3312) 24-34-07; e-mail eco@kyrnet.kg; Chair: Kazimir Karimov.

Genetics Society of Kyrgyzstan: Bishkek; tel (3312) 25-53-66; Head: Eugenia Lushihina.

Enviroteach: POB 1592, Swakopmund; tel (64) 461604; fax (64) 461300; Contact: T Sqazzin; environmental education.

Integrated Rural Development and Nature Conservation—IRDNC: POB 9681, Windhoek; tel (61) 228506; fax (61) 228530; Prin Officers: Garth Owen-Smith, Dr Margaret Jacobsohn; community-based natural resource management programmes; integration of conservation and development.

Lions Club: POB 691, Windhoek; tel (61) 224666; fax (61) 233914; Contact: H Lith.

Namibia Nature Foundation: POB 245, Windhoek; tel (61) 248345; fax (61) 248344; e-mail nnf@nnf.org.na; Exec Dir: Dr Christopher J Brown; Deputy Dir: Judy K Storm; f 1987; nature conservation; financial management; education and training; fundraising for conservation and environmental initiatives; plans, develops, implements and manages projects; establishes and maintains communications with national and international organizations; encourages community-based natural resource management; affiliated with the World Conservation Union—IUCN.

Rössing Foundation: Private Bag 13214, Windhoek; tel (61) 211721; fax (61) 211668; e-mail ecotree@rf.org.na; internet www.rf.org.na; Dir: Len le Roux; f 1978; co-founded and largely financed by Rössing Uranium Ltd; social, agricultural and community development; environmental education; craft development; adult education; natural resource management programme; enterprise development.

Save the Rhino Trust: POB 2159, Swakopmund; tel and fax (64) 403829; e-mail srt@rhino-trust.org.na; internet www.desertrhino.com; Dir: Blythe D Loutit; Acting Dir: Michael E Hearn: Deputy Dir: Simon Uri-Khob; Liaison Off: Bernd Brell; f 1982; rhinoceros monitoring.

Society for Environmental Protection, Education and Development: POB 1183, Tsumeb; tel (64) 220470; Contact: R Fable.

WWF Namibia: POB 9681, Windhoek; tel (61) 239945; fax (61) 239799; Contact: Amelia Crider; national office of the World Wide Fund for Nature—WWF.

NAURU

Governmental Organizations

Ministry of Foreign Affairs: Yaren; tel 444-3330; fax 444-3331; Minister: David Adeang; serves as the UNEP national focal point.

Ministry of Health: Yaren; tel 444-3166; fax 444-3136; Minister: Kieran Keke.

Nauru Fisheries and Marine Resources Authority: POB 449; tel 444-3733; fax 4444-3812; e-mail nrvms@cenpac.net.nr; f 1997.

Nauru Rehabilitation Corporation—NRC: Chair: (vacant); f 1999; devises and manages programmes for the rehabilitation of those parts of the island damaged by the over-mining of phosphate.

NEPAL

Governmental Organizations

Ministry of Population and Environment: c/o Central Secretariat, Singha Durbar, Kathmandu; tel 4245367; fax 4242138; e-mail info@mope.gov.npo; Minister: Bachaspati Devkota.

Department of National Parks and Wildlife Conservation: Baneswor, POB 860, Babar Mahal, Kathmandu; tel (1) 220912; fax (1) 227675; Dir-Gen: Dr Tirtha Maskey; Deputy Dir-Gen: Dr Bijaya Kattel; Ecologist: Narayan Poudel; f 1980; management of eight national parks, three wildlife reserves and one hunting reserve.

Ministry of Agriculture and Co-operatives: c/o Central Secretariat, Singh Durbar, Kathmandu; tel 4225108; fax 4225825; e-mail moa@fert.mos.com; Minister: Homnath Dohal.

Ministry of Forest and Soil Conservation: c/o Central Secretariat, Singh Durbar, Kathmandu; tel 4224892; fax 4223868; internet www.biodiv-nepal.goiv.np; Minister: Badri Prasad Mandal.

Ministry of Land Reform and Management: c/o Central Secretariat, Singha Durbar, Kathmandu; tel 4221660; fax 4220108; Minister: Jog Mehar Shrestha.

Ministry of Water Resources: Singha Durbar, Kathmandu; Minister of State: Thakur Prasad Sharma.

Non-Governmental Organizations

Community Welfare and Development Society—CWDS: POB 7923, Balaju-Kathmandu; tel (1) 4350196; fax (1) 4350038; e-mail cwds@wlink.com.np; internet cwds.domainvalet.com; Pres and Exec Dir: S P Yadav; B L Pandey; Programme Officer: Dr Rajesh Yadav; f 1990; promotion of sustainable agriculture and rural development, with specific emphasis on farming practices based on the use of local resources; conservation of natural resources, soil improvement and waste management; community mobilization; training; publications; member of the International Federation of Organic Agriculture Movements—IFOAM.

Environment Camps for Conservation Awareness—ECCA: POB 9210, Kathmandu; tel (1) 425867; fax (1) 224627; Programme Dir: Prachet K Shrestha; Programme Co-ordinator: Hari Banjara; Programme Officer: Narendra M Joshi; f 1987; awareness-raising activities for children on conservation and resource management; camps for children organized with a network of over 500 counsellors in Nepal and Bhutan; training camp for counsellors; field trips; liaison with local, national and international bodies; member of the World Conservation Union—IUCN.

Environmental Management and Research Centre—e-marc: POB 6117, Kathmandu; tel (1) 371708; fax (1) 227068; Exec Dir: Sushil Bhattarai; f 1993; promotion of sustainable development in accordance with the principles established at UNCED (Earth Summit, Brazil) in 1992, with special emphasis on the implementation of Agenda 21; promotion of protected areas; research; training; public awareness-raising activities.

IUCN Nepal Country Office: POB 3923, Kathmandu; tel (1) 528781; fax (1) 536786; e-mail iucn@wlink.com.np; Rep: (vacant); national office of the World Conservation Union—IUCN.

Kathmandu Environmental Education Project–KEEP: P O Box 9178, Jyatha, Thamel, Kathmandu; tel (1) 4412-944; fax (1) 4413-018; e-mail keep@info.com.np; internet www.keepnepal.org; Chair: Mangal Man Shakya; Exec Dir: P T Sherpa; aims to minimize the negative impact of tourism on the environment and culture of Nepal.

Nepal Nature Conservation Society—NENACOS: POB 3283, Kathmandu; Prin Officers: Sushil Shumsher, Sushil Bhattarai, Dr Madhusudan Sharma; f 1974; creation of public awareness of the need to conserve natural resources and preserve the environment; advice to government on policy matters; awarding of honours and prizes.

NETHERLANDS

Governmental Organizations

Ministry of Housing, Spatial Planning and the Environment (VROM): Rijnstraat 8, POB 20951, 2500 EZ The Hague; tel (70) 3393939; fax (70) 3391306; e-mail gilly.hendriks@minvrom.nl; internet www.minvrom.nl; Minister: Sybilla Maria Dekker; incorporates the Directorate-General for Environmental Protection—DGM.

Ministry of Agriculture, Nature and Food Quality: Bezuidenhoutseweg 73, POB 20401, 2500 EK The Hague; tel (70) 3786868; fax (70) 3786100; internet www.minlnv.nl; Minister: Dr Cornelis Pieter (Cees); incorporates the Directorate for Nature Protection, Environment and Fauna Management; administration of the Nature Conservation Act, designating protected nature monuments and animal and plant species; administration of the Birds Act, the Landscape Protection Act, the Endangered Exotic Animal Species Act and the Game Act; protection of natural areas, nature reserves, forests and national parks.

Ministry of Transport, Public Works and Water Management: Plesmanweg 1, POB 20901, 2500 EX The Hague; tel (70) 3516171; fax (70) 3517895; e-mail postduo@cend.minvenw.nl; internet www.minvenw.nl; Minister: Karla Hennëtte Peijs; responsibilities include land reclamation.

Afval Overleg Orgaan—AOO (Waste Management Council): POB 19015, 3501 DA Utrecht; tel (30) 2348800; fax (30) 2342260; e-mail secretariat@aoo.nl; internet www.aoo.nl; Chair: G ter Hurst; Dir: H Huisman; organization of national waste management; change of name, management and address pending in 2005.

Commission for Environmental Impact Assessment: Arthur van Schendelstraat 800, POB 2345, 3500 GH Utrecht; tel (30) 2347666; fax (30) 2331295; e-mail info@eia.nl; Chair: N G Ketting; Sec-Gen: J J Scholten; independent body advising competent authorities on the required content of environmental statements; assessment of the scientific quality and completeness of environmental statements.

DLO Winand Staring Centre for Integrated Land, Soil and Water Research—SC-DLO: Marijkeweg 11, POB 125, 6700 AC Wageningen; tel (317) 474200; fax (317) 424812; Dir: A N van der Zande; f 1989 by merger of four institutes; focuses on the problems of physical planning, water and soil management, and environmental protection in rural areas; research into the efficient, sustainable and socially acceptable use of rural resources nationally and internationally; affiliated with the Agricultural Research Department of the Ministry of Agriculture, Nature and Food Quality; publications.

Koninklijk Nederlands Meteorologisch Instituut—KNMI (Royal Netherlands Meteorological Institute): Wilhelminalaan 10, POB 201, 3730 AE De Bilt; tel (30) 2206911; fax (30) 2210407; e-mail komen@knmi.nl; internet www.knmi.nl; Dir: Dr J de Jong; Head, Research: Prof Dr G J Komen; f 1854; operational meteorology; climatology; climate research; seismology.

National Institute for Public Health and the Environment—RIVM: Antonie van Leeuwenhoeklaan 9, POB 1, 3720 BA Bilthoven; tel (30) 2743729; fax (30) 274435; e-mail fred.langeweg@rivm.nl; internet www.rivm.nl; Deputy Dir Netherlands Environmental Assessment Agency (RIVM): Fred Langeweg; collaborating centre in the UNEP Global Environment Outlook—GEO programme; Chair UNU-Forum on Global Integrated Envionmental Assessment Modelling focusing on sustainability and vulnerability.

Royal Netherlands Meteorological Institute, Research Department: Postbus 201, 3730 AE De Bilt; tel (30) 2206911; fax (30) 2210407; e-mail prv@knmi.nl; internet www.knmi.nl; Dir-in-Chief: Dr Ir F J J Brouwer; Head, Climatological Research: Prof Dr G J J Konen; f 1854; part of the Royal Netherlands Meteorological Institute—parent organization is the Ministry of Transport and Public Works; general weather forecasts and warnings in case of severe weather, climate monitoring, gathering meteorological data and aviation meteorology; meteorological and seismological infrastructure, research and development.

Staatsbosbeheer (State Forest Service of the Netherlands): POB 1300, 3970 BH Driebergen; tel (3404) 26111; fax (3404) 22978.

Non-Governmental Organizations

Centre for Clean Technology and Environmental Policy—CSTM, Twente University of Technology: POB 217, 7500 AE Enschede; tel (53) 4893203; fax (53) 4894850; e-mail secr@cstm.utwente.nl; internet www.utwente.nl/cstm; Scientific Dir: Prof Dr Hans Bressers; f 1988; research and consultancy; environmental and energy management and technology; environmental policy instruments and analysis; capacity-building, with emphasis on environmentally responsible planning and the local application of Agenda 21 of the UNCED (Earth Summit, Brazil); postgraduate teaching.

Centrum voor Energiebesparing (Centre for Energy Conservation and Environmental Technology): Oude Delft 180, 2611 HH Delft; tel (15) 2150150; fax (15) 2150151; e-mail ce@antenna.nl; internet www.cedelft.nl; Man Dir: Jan Paul van

Soest; research and provision of advice to governments and industry on environmental policy and energy conservation.

Centrum voor Landbouw en Milieu—CLM (Centre for Agriculture and Environment): POB 10015, 3505 AA Utrecht; tel (30) 2441301; fax (30) 2441318; e-mail clm@clm.nl; internet www.clm.nl; Dir-Gen: R V B Kroon; Dir, Research and Devt: W J van der Weijden; f 1981; independent association of farmers and environmentalists; promotion of sustainable agriculture which meets the needs of consumers; promotion of co-operation through research and policy proposals; information dissemination.

Centrum voor Milieu- en Klimaatstudies, Landbouw Universiteit Wageningen (Centre for Environment and Climate Studies, Wageningen Agricultural University): POB 9101, 6700 HB Wageningen; tel (317) 484812; fax (317) 484839; Dir: Prof L Hordijk; Environment Co-ordinator: Dr H de Jager; Climate Co-ordinator: Dr R S de Groot; f 1990; interdisciplinary research groups on regional environmental issues, global change, acidification, environmental impacts of agriculture and integrated control of the production chain; undergraduate courses on environmental protection and management.

Centrum voor Milieukunde—CML, Rijksuniversiteit Leiden (Centre of Environmental Science, Leiden University): Einsteinweg 2, 2333 CC Leiden; tel (71) 5277461; fax (71) 5275587; e-mail jstaats@eml.leidenuniv.nl; internet www.leidenuniv.nl/interface/eml; Prin Officers: Prof Dr H A Udo de Haes, Prof Dr W T de Groot, Prof G Huppes; f 1977; interdisciplinary research and education in landscape ecology, ecotoxicology, sustainable development, life-cycle assessment of products, substance flow analysis and market-oriented instruments for environmental policy; contract research for government departments, industry, the European Commission and other organizations; chain analysis in society and environment programme; environment and development programmes; affiliated with Dschang Univ (Cameroon) and Isabela State Univ (Philippines).

Communicatie en Adviesbureau over Energie en Milieu—CEA (Office for Information and Advice on Energy and the Environment): Westblaak 226, POB 21421, 3001 AK Rotterdam; tel (10) 2805666; fax (10) 2805654; e-mail advies@cea.nl; Man Dir: R W Boerée; consultancy and research; energy conservation; transport demand management; management of waste and emissions; consumer behaviour; communication planning.

De Groenen (Green Party): POB 6192, 2001 HD Haarlem; tel (23) 5427370; fax (23) 5144176; e-mail info@degroenen.nl; internet www.degroenen.nl; Chair: Ron van Wonderen; f 1983; environmental political party.

De Kleine Aarde (Small Earth): POB 151, 5280 AD Boxtel; tel (411) 684921; fax (411) 683409; e-mail info@dekleineaarde.nl; internet www.dekleineaarde.nl; Dir: Frank Zanderinh; f 1972; environmental education in co-operation with other organizations and local governments; maintains a public centre, ecological garden, sustainably-built visitors' centre and greenhouse; publications.

Expert Centre for Taxonomic Identification—ETI: Universiteit van Amsterdam, Mauritskade 61, 1092 AD Amsterdam; tel (20) 5257239; fax (20) 5257238; e-mail info@eti.uva.nl; internet www.eti.uva.nl; Dir: Dr Peter H Schalk; Chair, Int Board: Dr Wouter Los; f 1990; international networking of specialists and knowledge centres on biodiversity, taxonomy and systematics; development of information technology, data management and research tools for scientific and educational applications; CD-ROM and on-line publication of biodiversity information systems and taxonomic reference works; affiliated with UNESCO.

Fietsersbond enfb (Cyclists Federation): POB 2828, 3500 GV Utrecht; tel (30) 2918171; fax (30) 2918188; e-mail fbenfb@p2.nl; internet www.p2.nl/fbenfb; Co-ordinator: Jim A Schouten; f 1975; promotion of the use of bicycles; development of bicycle policy; information for cyclists.

Greenpeace Netherlands: Keizersgracht 174, 1016 DW Amsterdam; tel (20) 4223344; fax (20) 6221272; e-mail sdesk@ams.greenpeace.org; national office of Greenpeace International.

GroenLinks (The Green Left): POB 8008, 3503 RA Utrecht; tel (30) 2399900; fax (30) 2300342; e-mail partijbureau@groenlinks.nl; internet www.groenlinks.nl; Chair: Mirjam de Rijk; Parliamentary Leader: Paul Rosenmuller; f 1991 by the merger of the Communistische Partij van Nederland, Evangelische Volkspartij, Pacifistisch Socialistische Partij and Politieke Partij Radikalen; political alliance; programme includes environmental issues.

Hugo de Vries Laboratory, University of Amsterdam: Dept of Palynology, Kruislaan 318, 1098 SM Amsterdam; tel (20) 5257844; fax (20) 5257662; e-mail hooghiemstra@bio.uva.nl; Prin Officers: Prof Dr H Hooghiemstra, Prof Dr A M Cleef, Dr B van Geel, Dr F Bouman; f 1964; climate change studies; human impact on environment; paleoecology; paleoclimatology; ecology of tropical vegetation; sustainable land use; vegetation mapping; interactions between plants and animals.

Institute for Marine and Atmospheric Research, Utrecht University: Princetonplein 5, 3508 TA Utrecht; tel (30) 2533725; fax (30) 2543163; e-mail imau@phys.uu.nl; Prin Officer: Prof J Oerlemans; research and research training in climatology, meteorology, oceanography and marine sciences.

Instituut voor Milieuschade, Erasmus Universiteit Rotterdam (Institute of Environmental Damage, Erasmus University of Rotterdam): Faculty of Law, POB 1738, 3000 DR Rotterdam; tel (10) 4081630; fax (10) 4522878; e-mail despiegeleer@vbr@frg.eur.nl; Dir: Prof Dr Jan M van Dunné; Sec-Gen: Dr Babette de Spiegeleer; f 1985; research into liability for environmental damage according to civil, European and international public law; national and international comparative legal research; congresses, symposia, courses and publications; environmental law library.

International Institute for Geo-Information Science: Hengelosestraat 99, POB 6, 7500 AA Enschede; tel (53) 4874444; fax (53) 4874400; e-mail pr@itc.nl; internet www.itc.nl; Rector: Prof Dr Ir M Molenaar; Dirs External Affairs: Drs I J J Beerens, Drs G W van Dorp; f 1950; collection and interpretation of data in support of resource management and policy development.

International Institute for Land Reclamation and Improvement—ILRI: Lawickse Allee 11, Postbus 45, 6700 AA Wageningen; tel (317) 490967; fax (317) 417187; e-mail ilri@ilri.nl; Dir: J H P Pinkers; f 1955; member of Association Européenne des Institutions d'Aménagement Rural—AEIAR (European Association for Country Planning Institutions); collects and disseminates information on land reclamation and improvement and undertakes supplementary research work.

International Institute for the Urban Environment: Nickersteeg 5, 2611 EK Delft; tel (15) 2623279; fax (15) 2624873; e-mail iiue@urban.nl; internet www.urban.nl; Dir: Tjeerd Deelstra; f 1989; promotes the sustainable development of cities and towns, combining public administration and policy development with an interdisciplinary approach to science and technology; assists in the planning and realization of sustainable development projects; research, training and conferences.

International Tree Fund: Markweg 1[a], Postbus 160, 6870 AD Renkum; tel (317) 313616; fax (317) 318040; Pres: Wouter van Dam; preservation and restoration of forests, in particular tropical forests, in order to protect the economic, social and cultural development of their inhabitants and those who live nearby.

ISRIC— World Soil Information: POB 353, 6700 AJ Wageningen; tel (317) 471711; fax (317) 471700; e-mail soil .isric@wur.nl; internet www-isric.org; Dir: Dr David Dent; Deputy Dir: Dr Sjef Kauffman; Head World Data Centre: Dr Ottot Spaargaren; Research Team Leader: Dr Vincent van Engelen; Head World Soil Museum: Dr Alfred Hartemink; f 1966; documentation centre on land and resources; aims to improve methods of soil analysis; training and research on soils of the world.

It Fryske Gea/Het Friese Landschap (Frisian Landscape): POB 3, 9244 ZN Beetsterszwaag; tel (512) 381448; fax (512) 382973; Prin Officers: Henk Kroes, Dr Ultsje G Hosper; f 1930; nature conservation.

Koninklijk Zoologisch Genootschap (Royal Zoological Society): Plantage Kerklaan 38–40, 1018 CZ Amsterdam; tel (20) 5233400; fax (20) 5233481; internet www.artis.nl; Dir: Dr M T Frankenhuis.

Koninklijke Nederlandse Natuurhistorische Vereniging—KNNV (Royal Dutch Society for Natural History): Oude Gracht 237, 3511 NK Utrecht; tel (30) 2314797; fax (30) 2368907; e-mail bureau@knnv.nl; internet www.knnv.nl; Chair: M A Wildschut; f 1901; study of nature and conservation of wildlife; ecology; conservation strategies.

Landelijk Platform tegen Kernenergie (National Anti-Nuclear Platform): Ketelhuisplein 43, 1054 RD Amsterdam; tel (20) 6168294; e-mail akbstrategie@hetnet.nl; Prin Officers: Wim Kersten, Peer de Rijk; co-ordination of national activities against nuclear power.

Milieukontakt Oost-Europa (Foundation for Environmental Contacts with Eastern Europe and the Newly Independent States): POB 18185, 1001 ZB Amsterdam; tel (20) 6392716; fax (20) 6391379; e-mail info@milieukontakt.nl; Dir: Jerphaas Donner; strengthening and support of the environmental movement in Central and Eastern Europe through liaison, training, information exchange and advice.

MilieuTelefoon (Environment Telephone): Damrak 26, 1012 LJ Amsterdam; tel (20) 6262620; fax (20) 6275287; e-mail milieutelefoon@foenl.antenna.nl; Prin Officer: Simone de Jong; f 1987; dissemination of information by telephone and mail; environmental publications for consumers in the Netherlands and Belgium; affiliated with Vereniging Milieudefensie.

Nationaal Lucht- en Ruimtevaartlaboratorium—NLR (National Aerospace Laboratory): POB 90502, 1006 BM Amsterdam; tel (20) 5113113; fax (20) 5113210; Tech Dir: Prof F J Abbnik; research areas include air pollution caused by aircraft, and renewable energy sources, with special emphasis on wind energy.

Nederlands Instituut voor Volksontwikkeling en Natuurvriendenwerk—NIVON (Netherlands Institute for Friends of Nature Education and Work): Nieuwe Herengracht 119, 1011 SB Amsterdam; tel (20) 6269661; fax (20) 6388511; e-mail nivon@xs4all.nl; Man Dir: J E M Huige; Chair: M Ernsting; f 1924; documentation and signposting of footpaths; ecotourism; conferences, seminars and publications on nature and the environment.

Nederlandse Vereniging van Dierentuinen (Dutch Zoo Federation): Plantage Kerklaan 38–40, 1018 CZ Amsterdam; tel (20) 5233400; fax (20) 5233419; e-mail nvdzoos@nvdzoos .nl; internet www.nvdzoos.nl; Dir: Drs Koen Brouwer; f 1966; awareness-raising activities; implementation and monitoring of national and international agreements on zoos and wildlife protection.

Open Universiteit, Leerstofgebied Natuurwetenschappen (Open University of the Netherlands, Natural Sciences Department): Valkenburgerweg 167, POB 2960, 6401 DL Heerlen; tel (45) 5762359; fax (45) 5711486; open and distance teaching; M SC programmes in Environmental Science and Environmental Management.

Otterstation Nederland (Netherlands Otter Station): De Groene Ster 2, 8926 XE Leeuwarden; tel (511) 431214; fax (511) 431260; f 1985; public education and dissemination of information about otters, and freshwater plants and animals in general; operation of Otterpark Aqualutra.

Prince Bernhard Centre for International Nature Conservation: University of Utrecht, POB 80084, 3508 TB Utrecht; tel (30) 2536835; fax (30) 2518366; e-mail pbc@bio.uu .nl; internet www.bio.uu.nl/pbc; Prin Officer: Dr W Dijkman; f 1995; co-ordinates, initiates and executes research, training and extension programmes, focusing on conservation problems in developing countries; projects currently emphasize forests and tree resources in America, south-east Asia and southern Africa.

Raad voor het Milieu- en Natuuronderzoek—RMNO (Advisory Council for Research on Nature and Environment): POB 5306, 2280 HH Rijswijk; tel (70) 3364300; fax (70) 3364310; e-mail rmno@xs4all.nl; internet www.xs4all.nl/ ~rmno; publication of reports on major programmes of research into nature and the environment; co-ordination of research; monitoring of areas of neglect and duplication of effort in research.

Seas at Risk: Drieharingstraat 25, 3511 BH Utrecht; tel (30) 6701291; fax (30) 6701292; e-mail jmaggs@seas-at-risk.org; internet www.seas-at-risk.org; Chair: Hans Revier; Co-ordinator: John J Maggs; Policy Officer for Fisheries, Species and Habitats: Monica Verbech; Policy Officer for Hazardous Substances: Ute Meyer; f 1989; international lobbying for protection of European seas, especially the North Sea; publications; affiliated with Bond Beter Leefmilieu, Norges Naturvernforbund, Danmarks Naturfredningsforening, Liga para a Proteçcão da Natureza, Aktionskonferenz Nordsee, Irish Offshore Coalition, Voice of Irish Concern for the Environment, Grupo de Estudos de Ordenamento do Território e Ambiente, Marine Conservation Society, Wadden Vereniging, Stichting de Noordzee, European Environmental Bureau—EEB, Friends of the Earth International—FoEI and Youth and Environment Europe.

Stichte Milieufederatie (Environment Federation Foundation): Emmalaan 41, 3581 HP Utrecht; tel (30) 2519948.

Stichting Behoud Waterland (Waterland Preservation Foundation): POB 8, 1150 AA Broek in Waterland; tel (299) 622303; fax (299) 622414; e-mail wilrooij@wxs.nl; Chair: Rijk van den Hoek; Sec: Leendert van Pelt; Treas: Willem van Rooijen; f 1973; works to preserve the natural climate of the Waterland region.

Stichting de Faunabescherming (Wildlife Management Foundation): Amsteldijk Noord 135, 1183 TJ Amstelveen; tel (20) 6410798; fax (20) 6473700; e-mail faunabescherming@ wanadoo.nl; Sec: A P de Jong; f 1975; promotion of scientific and ethical wildlife management and minimal intervention.

Stichting de Noordzee (North Sea Foundation): Drieharingstraat 25, 3511 BH Utrecht; tel (30) 2340016; fax (30) 2302830; e-mail info@noordzee.nl; internet www.noordzee.nl; Dirs: F C Groenendijk, N A M Langendijk; f 1980; areas of interest include protection of the North Sea from pollution caused by shipping; spatial planning; fishing; hazardous substances; affiliated with Seas at Risk, Wadden Vereniging and Waterpakt.

Stichting Het Drentse Landschap (Drentse Landscape Foundation): Kloosterstraat 5, 9401 KD Assen; tel (5923) 13552; fax (5923) 18089; Dir: E W G van der Bilt; management of nature reserves and heritage buildings; seeks to influence national governmental and provincial policy on local nature conservation.

Stichting Het Noordhollands Landschap (Northern Holland Landscape Foundation): POB 257, 1900 AG Castricum; tel (251) 659750; fax (251) 652422; protection of nature reserves in Northern Holland province.

Stichting het Wereld Natuur Fonds—WNF Nederland (World Wide Fund for Nature—WWF Netherlands): Blvd 12, POB 7, 3707 BM Zeist; tel (3609) 37333; fax (3609) 12064; Chair: Dr Ed H T M Nijpels; Chief Exec and Exec Dir: Dr Siegfried Woldhek; national affiliate organization of the World Wide Fund for Nature—WWF International; conservation of marine, freshwater and terrestrial environments and wildlife; forestry; protected areas; land use; climatology; international co-operation; education and training.

Stichting IUCN-Ledencontact (Netherlands Committee for IUCN): Keizersgracht 384, 1016 GB Amsterdam; tel (20) 6261732; fax (20) 6279349; Prin Officer: Dr Wim Bergmans; national office of the World Conservation Union—IUCN.

Stichting Milieu Educatie—SME (Environmental Education Foundation): Australiëlaan 14, POB 13030, 3507 LA Utrecht; tel (30) 802444; fax (30) 801345; Man Dir: Frits Hasselink; Head, Operational Services: Kees Kil; non-profit consultancy on environmental communication and education; promotion of a sustainable society; information; support for dialogue between government and other interested parties; advice; courses; educational materials.

Stichting Natuur en Milieu (Society for Nature and Environment): Donkerstraat 17, 3511 KB Utrecht; tel (30) 2331328; fax (30) 2331311; e-mail snm@antenna.nl; Dir: A J M van den Biggelaar; f 1972; lobbying and other activities for nature conservation and environmental protection, primarily at national and European Union level; activities organized around themes of economy and environment, industry and environment, green areas, transport and urban environment.

Stichting Reinwater (Clean Water Foundation): Vossiusstraat 20-II, 1071 AD Amsterdam; tel (20) 6719322; fax (20) 6753806; e-mail rainwater@xsyall.nl; Dir: José Houweling; Water Quality and Research Officer: Tinco Lycklama à Nijeholt; Education Officer: Robert Ennema; f 1979; affiliated to Stichting Waterpakt; water quality research; collection of water samples at points of discharge for industrial waste water; legal action over unlawful pollution of surface waters, particularly the Rhine, Meuse and Scheldt rivers; environmental education in water issues.

Stichting Veldwerk Nederland (Netherlands Field Work Foundation): Het Woldhuis 11, 7325 WN Apeldoorn; tel (55) 3667199; fax (55) 3600615; e-mail info@veldnet.nl; internet www.weldwerk.nederland.nl; Man Dir: J W H van Ginkel; nature study courses for pupils from elementary, secondary and vocational schools; teacher training; study materials.

Studiecentrum voor Milieukunde, Erasmus Universiteit Rotterdam (Centre for Environmental Studies, Erasmus University of Rotterdam): Rm M7-22, POB 1738, 3000 DR Rotterdam; tel (10) 4082050; fax (10) 4089104; e-mail ulcemars@fsw.eur.nl; internet www.eur.nl/fsw; Prin Officers: Prof Dr W A Hafkamp, Prof Dr D Huisingh, Prof Dr H L F Saeijs, Prof Dr N Y Roome; f 1984; environmental research and studies; environmental management; research (non-technical) on cleaner production; affiliated with the European Asscn for Environmental Management Education.

TRAFFIC Europe, Netherlands: POB 7, 3700 AA Zeist; tel (30) 6937307; fax (30) 6912064; e-mail jjonkman@wwfnet.org; national office of TRAFFIC International.

Vakgroep Milieutechnologie, Landbouw Universiteit Wageningen (Sub-Department of Environmental Technology, Agricultural University of Wageningen): POB 8129, 6700 EV Wageningen; tel (317) 483339; fax (317) 382108; e-mail wim.rulkens@algemeen.mt.wau.nl; internet www.spb.wageningen-ur.nl/mt; Prin Officers: Prof W H Rulkens, Prof G Lettinga; f 1918; research centres on waste water treatment, soil and sediment remediation, treatment and reuse of solid waste and slurry, and treatment of contaminated air and gases.

Van Tienhoven Stichting—Stichting tot Internationale Natuurbescherming (Van Tienhoven Foundation for International Nature Protection): c/o Rijksherbarium, POB 9514, 2300 RA Leiden; tel (71) 5274727; fax (71) 5273511; Chair: Dr A C van Bruggen; Sec: Dr H P Nooteboom; promotion of international nature conservation projects, particularly those carried out by Dutch citizens.

Vereniging Milieudefensie (Friends of the Earth—FoE Netherlands): Damrak 26, POB 19199, 1000 GD Amsterdam; tel (20) 5507300; fax (20) 5507310; e-mail service@milieudefensie.nl; internet www.milieudefensie.nl; Chair: Chris Zydeveld; f 1972; national office of Friends of the Earth International—FoEI; campaigns on infrastructure, food and agriculture, transport and sustainable development.

Vereniging tot Behoud van Natuurmonumenten in Nederland (Dutch Society for the Preservation of Nature Monuments): Schaep en Burgh, Noordereinde 60, 1243 JJ 's-Graveland; tel (35) 6559933; fax (35) 6563174; internet www.natuurmonumenten.nl; conservation of buildings, monuments, archaeological sites, landscape, national parks and protected areas; land use.

Vereniging van Ondernemingen in de Milieudienstverlening tbv de Scheepvaart—VOMS: Kerkplein 3, 4209 AC Schelluinen; tel and fax (183) 623741; Prin Officers: T Lursen, A C P Nijdam; f 1991; association of companies providing environmental services to the shipping industry.

Vereniging voor Natuur- en Milieueducatie—IVN (Association for Natural and Environmental Education): Plantage Middenlaan 2c, POB 20123, 1000 HC Amsterdam; tel (20) 6228115; fax (20) 6266091; Dir: Gerard Jutten; Deputy Dir: Hugo Bunte; environmental education in primary schools; conservation of the environment by volunteers.

Vlinderstichting (Dutch Butterfly Conservation): POB 506, 6700 AM Wageningen; tel (317) 467346; fax (317) 420296; e-mail info@vlinderstichting.nl; internet www.vlinderstichting.nl; Man Dir: T J Verstrael; f 1983; research, and conservation and management of butterflies; compilation of inventories of butterfly and day-flying moth distributions in Europe, especially the Netherlands; butterfly monitoring; ecology and distribution of butterfly species; public education.

Wadden Vereniging (Dutch Society for the Preservation of the Waddensea): Voorstraat 16–18, POB 90, 8860 AB Harlingen; tel (517) 493693; fax (517) 493601; e-mail harlingen@waddenvereniging.nl; internet www.waddenvereniging.nl; Pres: D Eisma; Man Dir: J M Revier; f 1965; litigation; awareness-raising activities; environmental education; publications.

Netherlands Dependencies

ARUBA

Governmental Organizations

Department of Agriculture, Husbandry and Fisheries: Piedra Plat 114[a]; tel (8) 58102; fax (8) 55639; e-mail dirlvvm@toaruba.com; internet www.arubagricultureandfish.com; Head of Dept: Sylvester M Vrolijk; f 1976 as Government Experimental Station 'Santa Rosa'; part of the Ministry of Economic Affairs and Tourism; promotes more-efficient use of finances and natural resources; provides a centre for experimental work; aims to diversify the Aruban economy by encouraging and developing agriculture; major sections are Rain-fed Agriculture, Horticulture, Husbandry, Fisheries and Nature Management (responsible for Arikok National Park).

Ministry of Public Health and the Environment: L G Smith Blvd 76, Oranjestad; tel 5834966; fax 5835082; Minister: Candelario A S D Wever.

Non-Governmental Organizations

Aruban Foundation for Nature and Parks—FANAPA: POB 4014, Stimaruba, Spannslagoenweg 33b, Oranjestad.

NETHERLANDS ANTILLES

Governmental Organizations

Ministry of Public Health and Social Development: Santa Rosaweg 122, Willemstad, Curaçao; tel (9) 736-3530; fax (9) 736-3531; e-mail vornil@cura.net; Minister: Joan Theodora-Brewster.

Environment and Nature Section: APNA Bldg, 4th Floor, Schouwburgweg 24, Willemstad, Curaçao; tel (9) 465-5300; fax (9) 461-2154; e-mail info@mina.vomil.an; internet mina.vomil.an; formulation of environmental policy; development of environmental standards; maintains the Environmental Inspectorate, which monitors and enforces compliance with environmental laws and regulations.

Non-Governmental Organizations

Stichting Nationale Parken Nederlandse Antillen—STINAPA (National Parks Foundation of the Netherlands Antilles): POB 368, Bonaire; tel and fax (7) 8444; Chair: R R Hensen; Sec: V G Croes.

Stichting Nationale Parken Bonaire (National Parks Foundation of Bonaire): POB 368, Bonaire; tel 717-8444; fax 717-7318; e-mail stinapa@bonairelive.com; Pres (acting): J Chalk; Treas: H I Piar; protection and conservation of the natural environment of Bonaire; management of Washington National Park and Bonaire Marine Park.

NEW ZEALAND

Governmental Organizations

Ministry for the Environment: POB 10-362, Wellington; tel (4) 917-7400; fax (4) 917-7523; e-mail library@mfe.govt.nz; internet www.mfe.govt.nz; Minister for the Environment and for Disarmament and Arms Control: Marian Hobbs; Sec: Denise Church; f 1986; advice to government on environmental administration; environmental research and information.

Department of Conservation (Te Papa Atawbai): 59 Boulcott St, POB 10-420, Wellington; tel (4) 471-0726; fax (4) 471-1082; e-mail tsmith@doc.govt.nz; internet www.doc.govt.nz; Minister: Hon. Chris Carter; Dir Gen: Hugh Logan; f 1987; conservation of 14 national parks, three maritime parks, three World Heritage Sites, 20 conservation parks, reserves, forest parks and wildlife; production of educational material.

New Zealand Conservation Authority: 59 Boulcott St, POB 10420, Wellington; tel (4) 471 3211; fax (4) 471 3049; e-mail studhope@doc.govt.nz; internet www.conservationauthority.org.nz; Chair: Kerry Marshall; Man: Catherine Tudhope; Senior Issues Off: Gavin Rodley; approval agency for conservation management strategies and national park management plans; consulted by the Department for Conservation in the formulation of policies and plans and during the strategic planning phase of its annual business planning cycle.

Ministry of Agriculture and Forestry: POB 2526, Wellington; tel (4) 474-4100; fax (4) 474-4111; internet www.maf.govt.nz; Minister: Jim Sutton.

Biosecurity Authority: POB 2526, Wellington; tel (4) 474-4100; fax (4) 474-4111; internet www.maf.govt.nz/biosecurity.

Ministry of Fisheries: POB 1020, Wellington; tel (4) 470-2600; fax (4) 470-2601; internet www.fish.govt.nz; Minister: David Benson-Pope; Assoc Minister: Parekura Horomia; Assoc Minister: Dover Samuels.

Australian and New Zealand Environment and Conservation Council—ANZECC: See chapter on Australia, under the Department of the Environment and Heritage.

Environmental Risk Management Authority: POB 131, Wellington; Contact: Kevin Currie.

National Radiation Laboratory: POB 25099, Christchurch; tel (3) 366-5059; fax (3) 366-1156; Dir: Dr A C McEwan; Head, Environmental Section: Dr K M Matthews; administration of New Zealand radiation protection legislation for the Ministry of Health; monitoring of environmental radioactivity in New Zealand and the South Pacific region.

Natural Resource Management Ministerial Council—NRMMC: See chapter on Australia, under the Department of Agriculture, Fisheries and Forestry.

New Zealand Association of Crown Research Institutes—ACRI: POB 1578, Wellington; tel (4) 472-9979; fax (4) 472-4025; internet www.morst.govt.nz/nzsandt/cris.htm; Minister for Crown Research Institutes: Pete Hodgson; formerly Department of Scientific and Industrial Research—DSIR; association of government-owned research institutes registered as companies.

Forest Research: Private Bag 3020, Rotorua; tel (7) 343-5899; fax (7) 348-0952; e-mail info@forestresearch.co.nz; internet www.forestresearch.co.nz; Chief Exec: Dr Bryce Heard; Chief Operating Officer, Science: Tom Richardson; f 1947; a major provider of technology solutions and research services and consultancy to the forest and wood products sectors in New Zealand and internationally; one of the few forest research organizations in the world whose expertise spans the value chain, from seed to market; has extended its focus beyond wood to meet the growing consumer demand for renewable materials and products from plants.

Institute for Crop and Food Research Ltd: Private Bag 4704, Christchurch; tel (3) 325-6400; fax (3) 325-2074; e-mail info@crop.cri.nz; internet www.crop.cri.nz; CEO: Dr M W Dunbier; f 1992; multidisciplinary research for production and processing of arable crops, vegetables, seed crops, flowers, ornamental plants, essential oils, medicinal crops and seafood; plant breeding; soil science; aquaculture; post-harvest physiology; molecular biology; health and nutrition; personalized food; nutrigenomics.

Institute of Environmental Science and Research—ESR: Corporate Office, Kenepum Science Centre, Kenepum Dr, Porirua, POB 12444, Wellington; tel (4) 914-0700; fax (4) 914-0769; internet www.esr.cri.nz; CEO: Dr John Hay; Chief Financial Officer: Neil Wanden; Science and Research Man: Dr Val Orchard; research, analysis and consultancy in public health, environmental health and forensic sciences for the public and private sectors.

Institute of Geological and Nuclear Sciences Ltd: 69 Gracefield Rd, POB 30-368, Lower Hutt, Wellington; tel (4) 570-1444; fax (4) 569-0600; e-mail d.sheppard@gns.cri.nz; internet www.gns.cri.nz; CEO: Dr David Ross; Environmental Business Man: Dr Doug Sheppard; f 1992; research and consultancy services in earth sciences and applied nuclear sciences, particularly in the fields of natural hazards, resources, isotopic measurements, nuclear techniques, contaminated sites and ground-waters.

Landcare Research New Zealand Ltd (Manaaki Whenua): Canterbury Agriculture & Science Centre, POB 40, Lincoln 8152; tel (3) 325-6700; fax (3) 325-2127; e-mail GrindellJJ@landcareresearch.co.nz; internet www.LandcareResearch.co.nz; Chief Exec: Dr Andrew Pearce; Biosecurity and Pest Management: Dr Phil Cowan; Biodiversity and Ecosystem Processes: Dr David Choquenot; Urban Environmental Management: Dr Charles Eason; Rural Land Use: Dr Margaret Lawton; f 1992; research into management of national land environments through sustainable land use and prediction of the impacts of major resource utilization proposals; key areas are conservation of biodiversity, biosecurity, pest management, carbon storage and climate change, resource use efficiency and waste reduction, restoration and rehabilitation of land, land use effects and sustaining the quality of land, water and air.

National Institute of Water and Atmospheric Research: Private Bag 99940, Newmarket, Auckland; tel (9) 375-2090; fax (9) 375-2091; e-mail c.pridmore@niwa.cri.nz; internet www.niwa.co.nz; Chief Exec: Dr Rick Pridmore; Communications Man: Geoff Baird; f 1992; environmental research and services for the sustainable management and development of atmospheric, marine and freshwater resources, including atmosphere and climate, coasts and oceans, biodiversity and biosecurity, fisheries and aquaculture, bioactivities and biotechnology, database management and development, freshwater resources and biota, environmental forecasting.

Parliamentary Commissioner for the Environment: 2 The Terrace, Level 11, POB 10-421, Wellington; tel (4) 471-1669; fax (4) 495-8350; e-mail pce@pce.govt.nz; internet www.pce.govt.nz; Commr: Dr Morgan Williams; Asst Commr: Helen Beaumont; advice to local and central government on issues regarding environmental management.

Soil and Health Association of New Zealand, Inc: POB 46076, Auckland; tel and fax (9) 480-4440; e-mail soil@health.pl.net; internet www.soilhealth.org.nz; Pres: Brendan Hoare; Vice-Pres: Merial Watts; f 1942; promotion of organic cultivation; campaigning against pesticide use and genetic engineering; publications; conferences.

Non-Governmental Organizations

Australian and New Zealand Solar Energy Society Ltd—ANZSES: POB 1140, Maroubra, NSW 2035; See chapter on Australia.

Bio-Dynamic Farming and Gardening Association in New Zealand, Inc: POB 306, Napier; tel and fax (6) 835-2428; Exec Sec: David Wright; Chair: Gareth Bodle; f 1945; education in bio-dynamic agriculture; certification of produce according to the principles of the Demeter system; lobbying of government for the promotion of organic and bio-dynamic agriculture; publications; affiliated with Demeter International and the International Federation of Organic Agriculture Movements.

Cawthron Institute: Private Bag, Nelson; tel (3) 548-2319; fax (3) 546-9464; e-mail info@cawthron.org.nz; internet www.cawthron.org.nz; Chief Exec: Graeme Robertson; f 1917; offers expertise in aquaculture, coastal and river resource management, biological risk assessment and management of introduced marine organisms; research, development, consulting and laboratory services.

Centre for Resource Management, Lincoln University: POB 94, Canterbury; tel (3) 325-2811; fax (3) 325-2965; Prin Officer: Dr I Spellerberg.

Christchurch-Otautahi Agenda 21 Forum: POB 2657, Christchurch; e-mail anneandrhys@compuserve.com; Contact: Rhys Taylor.

Commonwealth Human Ecology Council—CHEC, New Zealand: POB 12-369, Hamilton; tel (7) 855-6578; fax (7) 855-2106; e-mail mqs@wave.co.nz; Convenor: Norman Stannard.

Ecologic Foundation: POB 756, Nelson; Contact: Guy Salmon.

Environment and Conservation Organisations of New Zealand—ECO: POB 11057, Wellington; tel (4) 385-7545; fax (4) 384-6971; e-mail eco@reddfish.org.nz; internet www.converge.org.nz/eco; Chair: Stephen Blyth; f 1971; alliance of over 70 environmental organizations; lobbying of government and officials; information dissemination; international networking.

Federated Mountain Clubs of New Zealand: POB 1604, Wellington; tel (4) 233-8244; fax (4) 233-8244; e-mail fmcsec@xtra.co.nz; internet www.fmc.org.nz; Pres: John Wilson; Vice-Pres: Owen Cox; Sec: Barabara Marshall; the national association of over 100 tramping, mountaineering, skiing, mountain-biking and deerstalking clubs; 14,000 members; advocate for recreation and wilderness areas.

Friends of the Earth—FoE New Zealand: 117 Houghton Bay Road, Wellington; tel and fax (9) 303-4319; e-mail foenz@kcbbs.gen.nz; Mike Ennis; f 1975; national office of Friends of the Earth International—FoEI; research on conservation and environmental issues; campaigning on local, national and international environmental issues; lobbying of government; research library.

Green Party of Aotearoa—New Zealand: POB 11652, Wellington; tel (4) 938-8622; fax (4) 938-6251; e-mail greenparty@greens.org.nz; internet www.greens.org.nz; Co-Leaders: Rod Donald, Jeanette Fitzsimons; formerly the Values Party; ecologist socialist party.

Greenpeace New Zealand: Private Bag 92507, 113 Valley Rd, Mount Eden, Auckland; tel (9) 630-6317; fax (9) 630-7121; e-mail greenpeace.new.zealand@dialb.greenpeace.org; Exec Dir: Di Paton; national office of Greenpeace International; environmental campaigns on climate change, energy, forestry, ocean ecology and nuclear issues.

ICLEI Australia-New Zealand (national office of the International Council for Local Environmental Initiatives; see chapter on Australia.

Maruia Society: POB 756, Nelson; tel (3) 548-3336; fax (3) 548-7525; e-mail maruia@nelson.planet.org.nz; Pres: Richard Thompson; Chief Exec: Guy Salmon; f 1975; development of policy on sustainable agriculture, climate change and transport; development of technical standards for ecological certification of food products; environmental litigation; conservation and management projects in tropical forests.

New Zealand Association for Environmental Education: POB 38341, Te Puni Mail Centre, Wellington; Contact: Pamela Williams.

New Zealand Association for Impact Assessment: Taylor Baines & Assocs, 75 Oxford Rd, Rangiora, North Canterbury; tel and fax (3) 313-8458; e-mail n.taylor@tba.co.nz; internet www.nziaia.org.nz; Co-convenors: Nick Taylor, Julie Warren.

New Zealand Biological Producers and Consumers Council, Inc: Marion Sq, POB 9693, Wellington; tel (4) 801-9741; fax (4) 801-9742; e-mail info@bio-gro.co.nz; internet www.bio-gro.co.nz; CEO: Dr Kaye McAulay; Tech Dir: Seager Mason; Project Dir: Michelle Glagau; certification of biological food producers and processors; affiliated with the International Federation of Organic Agriculture Movements.

New Zealand Committee for IUCN: c/o Dept of Conservation, POB 10420, Wellington; tel (4) 471-0726; fax (4) 471-1082; internet www.doc.govt.nz; Prin Officer: Andrew Bignell; national office of the World Conservation Union—IUCN.

New Zealand Council of Outdoor Recreation Associations: POB 1876, Wellington; e-mail hugh@infosmart.org; Contact: Hugh Barr.

New Zealand Fish and Game Council: POB 13-141, Johnsonville, Wellington; Contact: Bryce Johnston.

New Zealand Institute of Forestry: POB 19840, Christchurch; tel (3) 960-2432; fax (3) 960-2432; e-mail nzife@paradise.net.nz; internet www.forestry.org.nz; Pres: Ket Bradshaw; Sec: Peter Allan; independent advocate for forestry.

New Zealand Wind Energy Association: POB 553, Wellington; e-mail info@windenergy.org.nz; internet www.windenergy.org.nz; CEO: James Glennie; Chairman: Alistair Wilson; f 1997; association of members supporting the development of the wind energy industry in New Zealand.

Pacific Institute of Resource Management: POB 12-125, Wellington; e-mail pirmoffice@clear.net.nz; Contact: Kay Weir; education and awareness-raising activities on global and national social, environmental and economic policy issues; lobbying; seminars; international networking; publications.

Sustainable Energy Forum: POB 11-152, Wellington; tel (4) 5862003; fax (4) 5862004; e-mail info@sef.org.nz; internet www.sef.org.nz; Convenor: John Blakeley; Office Man: Ian Shearer; Sec: Kerry Wood; supporting and facilitating the use of energy for economic, environmental and social sustainability in New Zealand.

New Zealand's Dependent and Associated Territories

COOK ISLANDS

Governmental Organizations

Environment Services: Government of the Cook Islands, Avarua, Rarotonga; Minister of Environment Services and Cook Islands Natural Heritage Project: Norman George (Deputy Prime Minister).

Ministry of Agriculture: POB 96, Rarotonga; tel 28711; fax 21881; e-mail agricult@cookislands.gov.ck; Minister: Robert Wigmore.

Ministry of Energy: POB 72, Rarotonga; tel 24484; fax 24485; Minister: Tapi Taio.

Ministry of Health: POB 109, Rarotonga; tel 22664; fax 23109; e-mail aremaki@oyster.net.ck; Minister: Peri Vaevae Pare.

Ministry of Marine Resources: POB 85, Rarotonga; tel 28721; fax 29721; e-mail rar@mmr.gov.ck; Minister: Robert Wigmore.

NIUE

Governmental Organizations

Ministry for the Environment: c/o Office of the Secretary to Government, POB 40, Alofi; tel 4200; fax 4206; e-mail secgov.premier@mail.gov.nu; Minister for the Environment and Biodiversity: Toke Talagi.

Ministry for Agriculture, Forestry and Fisheries: c/o Office of the Secretary to Government, POB 40, Alofi; tel 4200; fax 4232; e-mail secgov.premier@mail.gov.nu; Minister: Bill Motufoou.

NICARAGUA

Governmental Organizations

Ministry of Agriculture and Forestry: Km 8½, Carretera a Masaya, Managua; tel (2) 76-0235; internet www.magfor.gob.ni; Minister: José Augusto Navarro Flores.

Ministry of Environment and Natural Resources: Km 12½, Carretera Norte, Apdo 5123, Managua; tel (2) 33-1111; fax (2) 63-1274; internet www.marena.gob.ni; Minister: Arturo Harding Lacayo.

Non-Governmental Organizations

Centro Intereclesial de Estudios Teológicos y Sociales—CIEETS (InterChurch Centre for Theological and Social Studies): Plaza El Sol 1 c al Sur, 1 c al Este, Residencial Los Robles, Managua; tel (2) 66-3033; fax (2) 67110; e-mail cieets@nicarao.apc.org.ni; f 1986; promotion of integrated rural development through local organizations; training in sustainable agriculture.

Solidaridad Unión Cooperación—SUCO (Solidarity, Unity and Co-operation, Nicaragua Office): Apdo Postal 1735, Edif Popol-Na, Plaza España, 3½ cuadras abajo, Managua; tel and fax (2) 66-4683; e-mail suco@nicarao.org.ni; Dir: Linda Gagnon; f 1961; active in the fields of local development, gender and the environment.

NIGER

Governmental Organizations

Ministry of Water, the Environment and the Fight against Desertification: BP 257, Niamey; tel 73-47-22; fax 72-40-15; Minister: Adamou Namata.

Ministry of Agricultural Development: BP 12091, Niamey; tel 73-35-41; fax 73-20-08; Minister: Abary Maï Moussa.

Ministry of Animal Resources: BP 12091, Niamey; tel 73-79-59; fax 73-31-86; Minister: Koroney Maoudé.

Ministry of Town Planning, the Environment and Public Property: Niamey; Minister: Gaya Mahaman Lawan.

Centre Régional de Formation et d'Application en Agrométéorologie et Hydrologie Opérationnelle—AGRHYMET (Regional Centre for Training and Application in Agrometeorology and Operational Hydrology): BP 11011, Niamey; tel 73-31-16; fax 73-24-35; Gen Dir: Jorge Santos Oliveira; f 1974; collection and dissemination of information about agriculture, control of desertification and management of natural resources in the Sahel region; design and implementation of food security and natural resource management projects; training in agrometeorology, hydrology and plant protection; application of remote sensing to natural resources; affiliated with Comité Permanent Inter-Etats de Lutte contre la Sécheresse dans le Sahel—CILSS.

Centre Régionale d'Enseignement Spécialisé en Agriculture—CRESA (Regional Centre for Specialized Training in Agriculture): Faculté d'Agronomie, Université Abdou Moumouni, BP 10960, Niamey; tel 73-39-42; fax 73-39-43; e-mail cresa@intnet.ne; Co-ordinator: Dr Adam Toudou; Dean: Dr Amoukou A Ibrahim; Asst Co-ordinator: Dr Ali Mahamadou; f 1992; selection of students for agricultural training courses; research programmes; joint collaborative projects with interested parties.

Office de l'Energie Solaire—ONERSOL (Solar Energy Office): BP 621, Niamey; tel 73-45-05; Dir: Albert Wright; f 1965; research and development of renewable sources of energy; training.

Non-Governmental Organizations

Institut de Recherche pour le Développement—IRD, Niger (Development Research Institute, Niger): BP 11416, Niamey; tel 75-38-27; fax 75-20-54; Dir: J P Guengant; formerly Centre ORSTOM à Niamey; a Centre of the Institute, which is based in Paris; research areas include medical entomology, hydrology, genetics, ecology, soil sciences, botany, agronomy.

Institut National de Recherches Agronomiques au Niger—INRAN (National Institute of Agricultural Research in Niger): BP 149, Niamey; Dir: J Nabos; soil science; stations at Tarna and Kolo.

IUCN Niger Country Office: BP 10933, Niamey; tel 72-40-28; fax 72-40-05; e-mail iucn@intnet.ne; Contact: Mamadou Mamane; national office of the World Conservation Union—IUCN; wildlife management; conservation and management of natural resources; environmental education; technical assistance; workshops.

Niger Green Alliance: c/o DFPP, BP 721, Niamey; tel 73-40-69; fax 73-27-84; e-mail faune@intnet.ne; Sec: Seyni Seydou; Asst Sec: Seydon Mamane; Treas: Ali Harouna; f 1998; affiliated to Direction Faune, Pêche et Pisciculture; conservation and use of biodiversity; environmental education; creation of public awareness; environmental impact assessment; involvement of local people in natural resource management.

NIGERIA

Governmental Organizations

Ministry of the Environment: New Federal Secretariat Complex, Shehu Shagari Way, 9th Floor, PMB 468, Abuja; tel (9) 5234014; Minister: Col Bala Mande.

Federal Environmental Protection Agency: PMB 265, Abuja; fax (1) 5235510; activities include monitoring of air and water quality.

Minister of Agriculture and Rural Development: Area 1, Secretariat Complex, Garki, Abuja; tel (9) 2341931; Minister:

Alhaju Sani Zango Daura; incorporates the Federal Department of Fisheries.

Federal Department of Forestry and Agricultural Land Resources: PMB 12613, Ikoye, Lagos; tel (1) 684178; areas of responsibilty include forestry and wildlife conservation.

Ministry of Water Resources: Area 1, Secretariat Complex, PMB 159, Garki, Abuja; tel (9) 2342376; Minister: Mallam Mukhtar Shagari.

National Centre for Energy Research and Development: University of Nigeria, Nsukka, Enugu State; tel (42) 771853; fax (42) 771855; e-mail misunn@aol.com; Dir: Dr O V Ekechukwu; f 1982; research and development of technologies for the utilization of all forms of energy resources, particularly solar and other renewable energy sources.

Natural Resources Conservation Council: c/o Office of the Secretary to the Federal Government, Abuja.

Sokoto Energy Research Centre: Usmanu Danfodiyo University, PMB 2346, Sokoto; tel (60) 237568; fax (60) 235120; Dir: Dr A A Tambuwal; f 1982; affiliated to the Energy Commission of Nigeria and the federal Ministry of Science and Technology; research and development in the field of renewable energy sources, particularly solar and biomass energy; establishment of pilot demonstration projects in solar, biomass and wind-energy technologies; dissemination of information through participation in trade fairs and exhibitions.

Non-Governmental Organizations

Chad Basin National Park: Ibrahim Abacha Way, PMB 1026, Maiduguri; tel (76) 234120; fax (76) 342184; f 1991; conservation of indigenous flora and fauna; promotion of ecotourism; anti-poaching patrols; education campaigns; development programmes to promote sustainable use of natural resources.

ICRISAT Kano: IITA Office, Sabo Bakin Zuwo Rd, PMB 3491, Kano; tel (64) 662050; fax (64) 669051; e-mail icrisat-w-nigeria@cgiar.org; national office of the International Crops Research Institute for the Semi-Arid Tropics—ICRISAT, India.

IUCN Nigeria: Hadejia-Nguru Wetlands Conservation Project, 36 Niger St, POB 32, Kano; tel (76) 740272; fax (64) 645779; e-mail iucn-nigeria@solkand.com; Project Adviser: Gert Polet; f 1987; national office of the World Conservation Union—IUCN.

Lake Chad Research Institute: Malamfatori, PMB 1293, Maiduguri, Borno State; tel (76) 232106; Dir: Dr B K Kagama; f 1975; research into the hydrological behaviour and characteristics of Lake Chad and the limnology of the associated surface and ground-waters.

Nigerian Affiliate of the IAIA: 13 Ekere St, POB 5166, Trans-Amadi, Port Harcourt; tel and fax (84) 334622; e-mail rust@alpha.linkserve.cm; Chair: Emuobonuvie Amy Akpofure; national office of the International Association for Impact Assessment—IAIA.

Nigerian Conservation Foundation: Plot 5, Moseley Rd, Ikoyi, POB 74638, Victoria Island, Lagos; tel (1) 2694020; fax (1) 2694021; Pres: Chief S L Edu; Chair, Cncl: Izoma P C Asiodu; f 1982; national associate organization of the World Wide Fund for Nature—WWF International; campaigns against pollution and wasteful use of renewable and non-renewable resources; conservation of nature; improvement of the quality of human life; public awareness-raising activities; environmental education.

Nigerian Environmental Study/Action Team—NEST: 1 Oluokun St, off Awolowo Ave, Bodija, Ibadan, Oyo State; tel (2) 8105167; fax (2) 8102644; e-mail nestnig@nest.org.ng; internet www.nest.org.ng; Programme Director: Damian Ihedioha; Programme Officer: Omolara Hassan; policy research and advocacy; local-level livelihood development; consultancy and training; dissemination of information; publications; promotes the objectives of the World Conservation Union—IUCN in Africa; affiliated with organizations including the World Resources Institute—WRI, UNDP and UNICEF.

NORWAY

Governmental Organizations

Ministry of the Environment: Myntgt 2, POB 8013 Dep, 0030 Oslo; tel 22-24-90-90; fax 22-24-95-60; e-mail postmottak@md.dep.no; Minister: Knut Arild Hareide.

Directorate for Cultural Heritage: Dronningensgt 13, POB 8196 Dep, 0034 Oslo; tel 22-94-04-00; fax 22-94-04-04; e-mail postmottak@ra.no; internet www.riksantikvaren.no/direct.html; responsible for the practical implementation of objectives laid down by the Storting (parliament) and the Ministry of the Environment; facilitates sound and efficient management throughout the country; ensures that a representative selection of monuments and sites is conserved; involved in environmental management, for which it is reponsible to the Department for Nature Conservation and Cultural Heritage; collaborates with other directorates in the environmental sector wherever appropriate.

Directorate for Nature Management: Tungasletta 2, 7485 Trondheim; tel 73-58-05-00; fax 73-58-05-01; e-mail service@dirnat.no; internet www.naturforvaltning.no; Dir: Bjørn Vindenes.

Norsk Polarinstitutt (Norwegian Polar Institute): Storgt 25[a], POB 399, 9008 Tromsø; tel 77-75-05-00; fax 77-75-05-01; Dir: Prof Olav Orheim; f 1928; mapping; scientific and environmental investigations of Norwegian polar regions in the Arctic and Antarctic.

SFT—Norwegian Pollution Control Authority: Strømsveien 96, POB 8100 Dep, 0032 Oslo; tel 22-57-34-00; fax 22-67-67-06; e-mail postmotbak@sft.telemax.no; internet www.sft.no; Dir: Hårard Holm; f 1974; combats pollution, noise and waste; regulates the use of environmentally hazardous substances and products.

Ministry of Agriculture: Akersgt 42, POB 8007 Dep, 0030 Oslo; tel 22-24-90-90; fax 22-24-95-15; e-mail postmottak@bfd.dep.no; internet odin.dep.no/bfd; Minister: Lars Spondheim.

Ministry of Fisheries: Grubbegt 1, POB 8118 Dep, 0032 Oslo; tel 22-24-90-90; fax 22-24-95-85; e-mail postmottak@fid.dep.no; internet odin.dep.no/fid; Minister: Svein Ludvigsen.

Havforskningsinstitutt (Institute of Marine Research): POB 1870, Nordnes, 5024 Bergen; tel 55-23-85-00; fax 55-23-85-31; Dir: Roald Vaage; f 1900; applied research related to fisheries; comprises three divisions: marine environment, marine living research and aquaculture.

Institutt for Energiteknikk (Institute for Energy Technology): POB 40, 2007 Kjeller; tel 63-80-60-00; fax 63-81-63-56; Man Dir: Kjell H Bendiksen; f 1948; part of the Ministry of Petroleum and Energy; national energy research establishment; main research areas are nuclear energy, petroleum technology, new energy technology, energy conservation, industrial process and materials technology, isotope production and irradiation services.

Norske Meteorologiske Institutt (Norwegian Meteorological Institute): POB 43, Blindern, 0313 Oslo 3; Dir: Dr A Grammeltvedt; f 1866.

Statens Stralevern (Norwegian Radiation Protection Authority—NRPA): Grini Naeringspark 13, 1361 Österaas; tel 67-16-25-00; fax 67-14-74-07; e-mail nrpa@nrpa.no; internet www.nrpa.no; Dir: Ole Harbitz; f 1993; the NRPA is the competent national authority in the area of radiation protection and nuclear safety in Norway. It was created through the consolidation of the former Nuclear Energy Safety Authority with the National Institute for Radiation Hygiene. It is organized under the Ministry of Health and provides assistance to all ministries on matters dealing with radiation, radiation protection and nuclear safety. It is responsible for: overseeing the use of radioactive substances and fissile material; co-ordinating contingency plans against nuclear accidents and radioactive fallout; monitoring natural and artificial radiation in the environment and at the workplace; increasing knowledge of the occurrence, risk and effects of radiation.

Non-Governmental Organizations

Agricultural University of Norway, Department of Biology and Nature Conservation: POB 14, 1432 Ås NLH; tel 64-94-81-22; fax 64-94-13-10; Head, Dept: Prof Sigmund Hagvar; education and research in nature conservation, wildlife and management of inland fisheries.

Friends of the Earth Norway/Norges Naturvernforbund: POB 342 Sentrum, 0101 Oslo; tel 22-40-24-00; fax 22-40-24-10; e-mail naturvern@naturvern.no; internet www.naturvern.no; Sec-Gen: Tore Killingland; Int Co-ordinator and Head of Information: Kare Olerud; f 1914; nation-wide membership organization active in all aspects of the environment; national office of Friends of the Earth International—FoEI; affiliated with Natur og Ungdom (Nature and Youth) and The Environmental Detectives (children's club).

Greenpeace Nordic, Norway: POB 6803, St Olavs Pass, 0130 Oslo 1; tel 22-20-51-01; fax 22-20-51-14; e-mail info@nordic.greenpeace.org; national office of Greenpeace International and Greenpeace Nordic.

Miljøpartiet de Grønne (Green Environmental Party): POB 2169, 7001 Trondheim; tel 73-53-09-11; fax 73-53-05-15; Chief Officer: Tore Bergum; f 1988; environmental political party.

Ministry of Petroleum and Energy: Einar Gerhardsens pl 3, POB 8019 Dep, 0030 Oslo; tel 22-24-90-90; fax 22-24-95-65; e-mail postmottak@oed.dep.no; internet odin.dep.no/oed; Minister: Thorhild Widveg.

Nansen Environmental and Remote Sensing Centre: Edvard Griegsvei 3[a], 5037 Solheimsviken; Dir: Ola M Johannessen; independent, non-profit institute affiliated with the University of Bergen.

Norges Geotekniske Institutt—NGI (Norwegian Geotechnical Institute): POB 3930, Ullevaal Stadion, 0806 Oslo; tel 22-02-30-00; fax 22-02-04-48; e-mail ngi@ngi.no; internet www.ngi.no; Man Dir: Suzanne Lucasse; f 1953; centre for geotechnical research and consulting services; undertakes remedial action against earthquakes and rockslides, snow avalanches, pollution of soil and ground-water, erosion and earthquake damage. Soil and rock mechanics and foundation engineering onshore and offshore, site investigations, instrumentation and performance monitoring.

Norges Naturvernforbund (Norwegian Society for the Conservation of Nature/Friends of the Earth—FoE Norway): POB 342 Sentrum, 0101 Oslo; tel 22-99-33-00; fax 22-99-33-10; e-mail naturvern@sn.no; internet www.naturvern.no; Sec-Gen: Jørund Ubø Soma; Int Co-ordinator and Head of Information: Kåre Olerud; f 1914; national office of Friends of the Earth International—FoEI; nation-wide membership organization active in all aspects of the environment; lobbying of government; information dissemination; publications; affiliated with Natur og Ungdom (Nature and Youth) and Blekkfulf.

Norsk Institutt for Luftforskning—NILU (Norwegian Institute for Air Research): POB 100, 2027 Kjeller; tel 63-89-80-00; fax 63-89-80-50; e-mail nilu@nilu.no; internet www.nilu.no; Dir: Øystein Hov; Information Sec: Sonja Berit Hauger; f 1969; private foundation; air pollution research; operates urban pollution monitoring programme for SFT, the Norwegian Pollution Control Authority; monitoring of global stratospheric air pollution, greenhouse gases, ozone depletion and the emission of hazardous substances for the Ministry of the Environment; environmental impact analyses; Chemical Co-ordinating Centre for the European Monitoring and Evaluation Programme—EMEP under the United Nations Economic Commission for Europe's Long-Range Transboundary Air Pollution—LRTAP Convention; Secretariat for the Nordic Council of Ministers Working Group on Air Pollution and its climate research programme; central data base for the Second European Stratospheric Arctic and Mid-latitude Experiment—SESAME and other European Union projects; major participant in various projects initiated by the European Commission, particularly in the field of ultraviolet radiation and climate change; wholly owned subsidiary, NILU Products AS, sells instruments for the monitoring of air pollution.

Norsk Institutt for Vannforskning—NIVA (Norwegian Institute for Water Research): POB 173, Kjelsas, 0411 Oslo; tel 22-18-51-00; fax 22-18-52-00; Dir-Gen: Haakon Thaulow; f 1958; research and contract projects on technical, economic and sanitary problems in connection with water supply, waste water and pollution in rivers, lakes or fjords.

Norwegian Forum for Environment and Development: POB 3894, 0805 Oslo; tel and fax 22-95-10-22; Exec Dir: Vegard Bye; fmrly Norwegian Campaign for the Environment and Development; co-ordinates Norwegian NGOs with regard to activities arising from UNCED (the Earth Summit).

Senter for Utviklingsstudier—SFU, Universitetet i Bergen (Centre for Development Studies, Bergen University): Stromgt 54, 5007 Bergen; tel 55-58-93-00; fax 55-58-98-92; Dir: Gunnar M Sørbø; research and training in development and environmental issues.

WWF Verdens Naturfond (World Wide Fund for Nature—WWF Norway): Kristian Augustsgt 7[a], St Olavs Plass, POB 6784, 0130 Oslo; tel 22-03-65-00; fax 22-20-06-66; e-mail info@wwf.no; internet www.wwf.no; Chair, Gen Ass: Christian N Sibbern; Chair, Bd: Jorgen Randers; Sec-Gen: Rasmus Hansson; f 1971; national affiliate organization of the World Wide Fund for Nature—WWF International; national and international projects for the conservation of biodiversity including conservation works in Norway and several African countries.

Norwegian External Territories

SVALBARD

Governmental Organizations

Norsk Polarinstitutt, Svalbardkontor (Norwegian Polar Institute, Svalbard Office): Næringsbygget, POB 505, 9170 Longyearbyen; tel 79-02-26-00; fax 79-02-26-04; e-mail haugland@npolar.no; Dir: Jan Erling Haugland; f 1928 (Norwegian Polar Institute, Oslo) as Norges Svalbard- og Ishavsundersøkelser; adopted present name 1948; mapping and research institute; responsible for advising the Government on matters concerning Svalbard, Jan Mayen and Norway's 'Antarctic' dependencies; monitors and investigates the environment of the territories; organizes regular Antarctic research expeditions; establishes and maintains aids to navigation in Svalbard waters.

Norsk Polarinstitutts Forskningsstasjon (Norwegian Polar Institute Research Station): Ny-Ålesund; tel 79-02-71-15; fax 79-02-70-02; permanent research base.

OMAN

Governmental Organizations

Ministry of Regional Municipalities and of the Environment: POB 323, Muscat 113; tel 696444; fax 602320; Minister: Abdullah bin Salem bin Amer ar-Rawas.

Ministry of Agriculture and Fisheries: POB 467, Ruwi 113; tel 694182; fax 6959091; internet www.maf.gov.om; Minister: Salim bin Hilal al-Khalili.

Ministry of Housing, Electricity and Water: POB 1491, Ruwi 112; tel 603906; fax 699180; internet www.mhew.gov.om; Minister: Dr Khamis bin Mubarak bin Isa al-Alawi.

Ministry of National Heritage and Culture: POB 668, Muscat 113; tel 602555; fax 602735; internet www.mnhc.gov.om; Minister: Sayyid Faisal bin Ali as-Said.

Office of the Adviser for Conservation of the Environment, Diwan of the Royal Court: POB 246, Muscat 113; tel 736482; fax 740550.

PAKISTAN

Governmental Organizations

Ministry of the Environment: CDA Block 4, Civic Centre, G-6, Islamabad; tel (51) 9202574; fax (51) 9223760; e-mail jawedalikhan@hotmail.com; internet www.environment.gov.pk; Minister: Maj (Retd) Tahir Iqbal; Dir-Gen: Jawed Ali Khan; national policy, plans and programmes for: environmental planning, pollution and ecology; physical planning and human settlements including urban water supply, sewerage and drainage. Dealing and agreements with other countries and international organizations in the fields of environment and human settlements: economic planning and policy making in respect of forestry and wildlife; administrative control of National Council for Conservation of Wildlife in Pakistan, Pakistan Forest Institute, Zoological Survey of Pakistan and National Energy Conservation Center (ENERCON).

National Council for Conservation of Wildlife—NCCW: 79-E Al-Rehman Chamber, 3rd Floor, Blue Area, Islamabad; tel (51) 9204515 / 9205058; fax (51) 9202211 / 9222472; e-mail nccw@isb.paknet.com.pk; Conservator (Wildlife): Umeed Khalid; Deputy Conservator: M Hafeez-ur-Rehman; Asst Sec: Mian Mohammad Shafiq; formulation of conservation policy; co-ordination of provincial activities; liaison with NGOs; wildlife management. Liaison with conventions such as CITES, CMS and RAMSAR.

Ministry of Food, Agriculture, Co-operatives and Livestock: Block B, Pakistan Secretariat, Islamabad; tel (51) 9210088; fax (51) 9221246; Minister: Saikandar Hayat Khan Bosan.

Ministry of Local Government and Rural Development: Block 4, Old Naval Headquarters, Civic Centre, G-6, Islamabad; tel (51) 9224291; fax (51) 9202211; Minister: Justice (retd) Abdul Razzaq Thahim.

Ministry of Petroleum and Natural Resources: Block A, 3rd Floor, Pakistan Secretariat, Islamabad; tel (51) 9209343; fax (51) 9206416; e-mail info@mpnr.gov.pk; internet www.mpnr.gov.pk; Minister: Amanullah Khan Jadoon.

Hydrocarbon Development Institute of Pakistan: 230 Nazimuddin Rd, F-7/4, POB 1308, Islamabad; tel (51) 9203958; fax (51) 9204902; e-mail hdip@apollo.net.pk; Operations Gen Mans: S N Sarwar, Ahmad Wasim; f 1975; national research and development organization for the petroleum sector; research areas include environmental protection and energy efficiency.

Ministry of Water and Power: Block A, 15th Floor, Shaheed-e-Millat, Pakistan Secretariat, Islamabad; tel (51) 9212442; fax (51) 9203187; Minister: Liaquat Ali Jatoi.

Fuel Research Centre, Pakistan Council of Scientific and Industrial Research: University Rd, Karachi 75280; tel (21) 4969761; fax (21) 4966671; e-mail pcsir@frc.khi.sdnpk.undp.org; Dir: Dr S M Abdul Hai; Sr Engineer: Ismat Ali; f 1953; affiliated to the Ministry of Science and Technology; research and analytical services for various industries;

monitoring of environmental pollution caused by liquid and gaseous effluents.

National Energy Conservation Centre—ENERCON: ENERCON Bldg, G-5/2, Islamabad; tel (51) 9206005; fax (51) 9206003; Man Dir: Arif Alaudin.

Pakistan Council of Research in Water Resources: H 3 and 5, St No 17, F-6/2, Islamabad; tel (51) 9218980; Chair: Dr Bashir Ahmad Chandio; Sec: Makhmoor-e-Ahmad Goheer; f 1964; research unit of the Ministry of Science and Technology; promotes research in the fields of hydraulics, irrigation, drainage, reclamation, tube wells and flood control; publications.

Pakistan Council of Scientific and Industrial Research—PCSIR: Press Centre, Shahrah-e-Kamal Ataturk, Karachi 74200; tel (21) 2628763; fax (21) 2636704; Chair: A Q Ansari; Sec: Dr R B Qadri; f 1953; part of the Ministry of Science and Technology; promotes scientific and industrial research and its application to the development of national industries and the use of the country's natural resources.

Pakistan Meteorological Department: Sector H-8, PO Box 1214, Islamabad; tel (51) 9257314; fax (51) 4432588; e-mail pmdcomp@pakmet-3.ptc.pk; internet www.pakmet.com; Dir-Gen: Dr Qamar-uz-Zaman Chaudhry; Chief Meteorologist: Arif Mahmood; f 1947; research; weather forecasting; agrometeorology; flood forecasting; seismology; drought studies; meteorology; climatology.

Non-Governmental Organizations

CABI Bioscience Pakistan Centre: POB 8, Rawalpindi; tel (51) 9290032; fax (51) 451147; e-mail bioscience.pakistan@cabi.org; a centre of CABI Bioscience, United Kingdom.

Commonwealth Human Ecology Council—CHEC, Pakistan: House 29[a], St 18, F-6/2, Islamabad; tel (51) 864077; fax (51) 563091; Chair: Dr Saadia Chishdi.

Energy and Environment Society of Pakistan: 123-J Model Town, Lahore 54700; tel (42) 5867642; fax (42) 5867697; e-mail nazim49@attglobal.net; Pres: Mohammad Nazim; Vice-Pres: Mohammad Sair Ali; f 1987; encourages and assists organizations campaigning for an improved environment; training, awareness-raising activities.

Environmental Protection Society of Pakistan: c/o Hassan & Hassan, PAAF Bldg, 2nd Floor, 7d Kashmir Egerton Rd, Lahore; tel (42) 6360800; fax (42) 6360811; e-mail phassan@brain.net.pk; Pres: Dr Parvez Hassan; f 1989; pollution control; environmental protection; research into environmental degradation; awareness-raising activities.

Institute of Environmental Studies, University of Karachi: University Campus, Karachi 75270; tel and fax (21) 4969054; e-mail ieku@paknet3.ptc.pk; Dir: Prof Dr M Altaf Khan; training and research; member of the International Water Association.

IUCN Pakistan Country Office: 1 Bath Island Rd, Karachi 75530; tel (21) 5861540; fax (21) 5870287; e-mail rafiq@iucn.khi.sdnpk.undp.org; Country Rep: Mohammad Rafiq; national office of the World Conservation Union—IUCN; formulation of policy; training; field programmes; creation of public awareness; environmental education.

Leadership for Environment and Development—LEAD Pakistan: 28 Margalla Rd, F-8/3, Islamabad; tel (51) 264742; fax (51) 264743; e-mail szaman@lead.org.pk; Chair, Bd of Governors: Dr Parvez Hassan; Nat Programme Dir: Ali Tauqeer Sheikh; f 1996; initiation and implementation of programmes for the development of a multidisciplinary cadre of mid-career professionals with a common perception of the local, national and global challenges integral to the pursuit of environmentally sensitive development; centre for the promotion of co-operative endeavour between scholars and institutions, both within Pakistan and in other countries.

Pakistan Forest Institute: Peshawar; tel (91) 9216123; fax (91) 9216203; e-mail pfilib@hotmail.com; Dir-Gen: B A Wani; f 1947; provides researchers and students with information about forestry, the environment and allied disciplines.

Pakistan Institute for Environment-Development Action Research—PIEDAR: 1st Floor, 64 e Masco Plaza, Blue Area, Islamabad; tel (51) 2820359 / 69; fax (51) 2820379; Vice-Pres: Faisal Imam; Environmental Engineer: Nadeam Afzal; f 1992; promotion of community organizations for sustainable development; sustainable development policy advice to governments and business; research; environmental education through participatory learning projects in low-income areas: WASH, water sanitation, hygiene in Pakistan; HID, human institutional development for CBO and local organizations.

Scientific and Cultural Society of Pakistan: B-7 St 25, Model Colony, Karachi 75100; tel and fax (21) 409336; Pres: Muhammad Zaheer Khan; f 1977; organization of seminars, conferences, training courses and workshops in environmental science, wildlife conservation and management, natural history and organic agriculture; member of the World Conservation Union—IUCN; affiliated with the International Federation of Organic Agriculture Movements—IFOAM.

Sub-Office: Scientific and Cultural Society of Pakistan, Zoology Dept (Wildlife and Fisheries), University of Karachi, Karachi; tel (21) 409336.

Shirkat Gah (Women's Resource Centre): F-25/A, Block 9, Clifton, Karachi; tel (21) 5832754; Prin Officers: Meher M Noshirwani, Hilda Saeed; f 1975; community development and environmental protection projects; education and training; workshops and conferences; publications.

Society for Conservation and Protection of Environment—SCOPE: D-141, Block 2, PECHS, Karachi 75400; tel (21) 4559448; fax (21) 4557009; e-mail scope@khi.compol.pk; internet www.scope.org.pk; Pres: Tanveer Arif; water purification; control of desertification; environmental action; rural development; control of industrial pollution; action for environmental and health awareness; consultancy; affiliated to the Scientific Committee on Problems of the Environment—SCOPE, France.

Solar Energy Centre: POB 356, GPO Super Highway, Hyderabad; tel (221) 650823; Officer-in-Charge: Pandhiani; product development, dissemination of solar energy technology, solar refrigeration, photovoltaic applications, solar water desalination, solar air-conditioning, solar architecture, solar energy storage and power generation, wind energy converters, development of tidal wave and geo-thermal energy.

Sungi Development Foundation: Latif Plaza, No 21, Block 13w, 2nd Floor, F7 Markaz, Islamabad; tel (51) 220878; Chair: Omar Asghar Khan; prevention of environmental degradation through awareness-raising activities and co-operation with other NGOs; community-based programmes.

Teachers Resource Centre—TRC: 67b Garden Rd, Karachi; tel (21) 7217967; fax (21) 7233578; Chair: Adeeba Kamal; Dir: Seema Malik; f 1986; training of teachers in ecology and the environment; development of educational materials; workshops and special events; library.

World Wide Fund for Nature—WWF Pakistan: Ferozepur Rd, POB 5180, Lahore 54600; tel (42) 5862360; fax (42) 5862358; Dir-Gen: Ali Hassan Habib; Pres: Brig Mukhtar Ahmad Chaudhry; f 1970; national affiliate organization of the World Wide Fund for Nature—WWF International; conservation; environmental education; wildlife management.

Karachi Office: WWF Pakistan, c/o Crescent Group of Industries, 606–607 Fortune Centre, 6th Floor, Block 6, PECHS Shahra-e-Faisal, Karachi 74500; tel (21) 4544791; fax (21) 4544790; e-mail wwfkhi@khi.compol.com.

PALAU

Governmental Organizations

Ministry of Resources and Development: c/o Office of the President, POB 100, Koror, PW 96940; tel 488-2701; fax 488-3380; Minister: Fritz Koshiba.

Palau Environmental Quality Protection Board: POB 100, Koror, PW 96940; tel 488-1639; fax 488-2963; e-mail eqpb@belau.com; Exec Officer: Lucio Abraham; Chair: Paula Holm; f 1981; promulgation and enforcement of environmental regulations; establishment of criteria and standards for the classification of air, soil and water quality; administration of appropriation grants and loans for the National Environmental Protection Program; publication of technical manuals.

PANAMA

Governmental Organizations

Autoridad Nacional del Ambiente—ANAM (National Environment Authority): Apdo 2016, Paraíso, Ancón; tel 232-5939; fax 232-6612; e-mail gmfranco@ns.inrenare.stri.si.edu; Admin-Gen: Ricardo Anguizola.

Ministry of Agricultural Development: Edif 576, Calle Manuel E Melo, Altos de Curundú, Apdo 5390, Panamá 5; tel 232-6254; fax 232-5044; e-mail infomida@mida.gob.pa; internet www.mida.gob.pa; Minister: Laurentio Cortizo.

Ministry of Public Health: Apdo 2048, Panamá 1; tel 225-6080; fax 212-9202; e-mail minsitro@minsa.gob.pa; internet www.minsa.gob.pa; Minister: Camila Alleyne.

Comisión Nacional del Medio Ambiente—CONAMA (National Environment Commission): Apdo 10120, Zona 4, Panamá; tel 264-3373; fax 264-3373; e-mail jcaldero@vasco.usma.ac.pa; internet www.usma.ac.pa/eco/conama.htm; Contact: Marta Loyda Robinson Adames; f 1983; advises the Government on all matters relating to protection of the environment and natural heritage of Panama; serves as the INFOTERRA national focal point; includes a Technical Secretariat.

Instituto de Recursos Naturales Renovables—INRENARE (Institute for Renewable Natural Resources): Apdo 2016, Paraíso, Ancón; tel 232-6612; fax 232-6601; e-mail inrenare@ns.inrenare.stri.si.edu; internet www2.usma.ac.pa/~eco1; Dir-Gen: Mirei Endara; Deputy Dir-Gen: Dimas Arcia; Sec-Gen: Luzmila Rodríguez; f 1986; planning and implementation of policy for the use, conservation and development of renewable natural resources.

Non-Governmental Organizations

Asociación Conservacionista Caribaró—ACCA (Caribaró Conservation Association): Apdo 17, Bocas del Toro, Isla Panamá; tel and fax 757-9488; Exec Dir: Isabel Alvendas; Exec Sub-Dir: Luis L Moul; Deputy Dir: Mario Garrido; f 1990; support for development of protected areas in Bocas del Toro and the Panamanian Caribbean; promotion of environmental education, applied research and exchange of related expertise in the conservation of natural resources; eco-development; preservation of the cultural heritage of the Panamanian Caribbean.

Centro de Estudios y Acción Social Panameña—CEASPA (Panamanian Centre for Research and Social Action): Vía Cincuentenario 84, Coco del Mar, San Francisco; tel 226-6602; fax 226-5320; e-mail ceaspa@sinfo.net; Dir: Mariela Arce; Direction Team: Raúl Leis, Olympia Díaz; f 1977; activities include the promotion of development alternatives, research into sustainable development, provision of support for environmental organizations and public education; consultant status with UNDP, UNESCO and UNICEF.

Comité Ambiental de Alanje—CAA (Alanje Environment Committee): Barqueta Beach, Alanje, Chiriquí; tel 774-0078; Pres: Teófilo Díaz Muñoz; Vice-Pres: Alcíbiades Rivera; f 1985; collection and hatching of turtle eggs; conservation of flora and fauna in coastal zones; education; affiliated with the Federación Chiricana de Medio Ambiente.

Fundación Dobbo Yala (Dobbo Yala Foundation): Vía España, Edif CROMOS, 1°, Of L5, Apdo 83-0308, Zona 3, Panamá; tel 265-4932; fax 265-1061; e-mail dobbo@pty.com; internet www.geocities.com/rainforest/4043; Exec Pres: Eligio Alvarado P; Exec Sec: José Mendoza Acosta; f 1990; affiliated to SAPIBE foundation, Spain; sustainable development for indigenous peoples; environmental conservation; research and consultation in mining, ecotourism, legal assistance and territorial demarcation on behalf of indigenous communities; affiliated with the World Conservation Union—IUCN.

Instituto de Ciencias Ambientales y Desarrollo Sostenible—ICADES, Universidad Autónoma de Chiriquí (Institute for Environmental Sciences and Sustainable Development, Autonomous University of Chiriquí): David, Chiriquí; tel 775-1114; fax 774-5329; Admin: Itzel Jaramillo; Dir: Prof Demetrio Miranda; f 1996; research and postgraduate teaching in the fields of environmental sciences, biodiversity and sustainable development; conferences and seminars; publications.

Programa de Ecología y Manejo de Areas Silvestres de Kuna Yala—PEMASKY (Ecological Programme and Management of the Wildlands of Kuna Yala): c/o Asociación de Empleados Kunas—AEK, Apdo 2012, Paraíso, Ancón; tel and fax 225-7603; e-mail gubi@pty.com; Exec Dir: Geodisio

Castillo; Pres: Jesús Smith, Jr; f 1983; affiliated to Asociación de Empleados Kunas—AEK; formerly Proyecto de Estudio para el Manejo de Areas Silvestres de Kuna Yala; human ecology; resource management; development planning; environmental management and planning; support for community conservation projects; development of environmental research and education projects in conservation, agroforestry and national park and forest management.

Sociedad Audubon de Panamá (Panama Audubon Society): POB 2026, Balboa; tel 224-9371; fax 224-4740; e-mail audubon@pananet.com; internet www.pananet.com/audubon; Pres: Loyda E Sánchez; Vice-Pres: Judith Anguizola; Treas: Hildegar Mendoza; Sec: Minerva Tapia; f 1967; conservation of birds and their habitats; environmental education; publications.

PAPUA NEW GUINEA

Governmental Organizations

Department of Environment and Conservation: Boroko 111, Kamul Ave, POB 6601, Waigani, NCD; tel 3011607; fax 3011691; Minister: William Duma; institution with ministerial status; environmental planning and monitoring; regulation of pesticides, hazardous substances, ozone-depleting substances and contaminants; conservation of water resources; establishment and management of protected areas; conservation of crocodiles and butterflies; design of management strategies for endangered species; marine and coastal resource management.

Department of Agriculture and Livestock: Spring Garden Rd, POB 417, Konedobu 125, NCD; tel 3231848; fax 3230563; internet www.agriculture.gov.pg; Minister: Matthew Siune; institution with ministerial status.

Department of Fisheries and Marine Resources: POB 2016, Port Moresby; tel 3271799; fax 3202074; internet www.fisheries.gov.pg; Minister: Ben Simri; institution with ministerial status.

Department of Mineral Resources: PMB, Konedobu, Port Moresby; tel 3227600; fax 3213701; internet www.mineral.gov.pg; Minister for Petroleum and Energy: Sir Moi Avei; institution with ministerial status.

National Forest Authority: Boroko, Frangipani St, Hohola, POB 5055, Port Moresby; tel 3277800; fax 3254433; internet www.datec.com.pg/government/forest/default.htm; Minister for Forestry: Patrick Pruaitch; institution with ministerial status.

Australian and New Zealand Environment and Conservation Council—ANZECC: See chapter on Australia, under the Department of the Environment and Heritage.

Non-Governmental Organizations

Appropriate Technology Development Institute, Papua New Guinea University of Technology: Private Mail Bag, Lae; tel 4734776; fax 4734303; Dir (acting): Charles Nakau; research and development of technologies suitable for application in Papua New Guinea communities.

Bulolo University College: POB 92, Bulolo, Morobe Prov; tel 4745236; fax 4745311; Prin Officer: T Nahuet; affiliated to Papua New Guinea Univ of Technology; specializes in forestry subjects; three-year training programme for foresters and park rangers.

Commonwealth Human Ecology Council—CHEC, Papau New Guinea: POB 6422, Boroko, Port Moresby; tel 257113; fax 3254743; Contact: Elizabeth Johnson.

Greenpeace Papua New Guinea: POB 136, Gerehu, NCD; tel and fax 3260560; e-mail greenpeace.pacific@dialb.greenpeace.org; national office of Greenpeace International and Greenpeace Pacific.

PARAGUAY

Governmental Organizations

Ministry of Agriculture and Livestock: Edif Aifra, 1°, Presidente Franco 472, Asunción; tel (21) 44-9614; fax (21) 49-7965; Minister: Antonio Ibañez Aquino; incorporates the Department of Agriculture, Forestry and Conservation of Natural Resources.

Subsecretaría de Estado de Recursos Naturales y Medio Ambiente—SSERNMA (Sub-Department of Natural Resources and the Environment): Presidente Franco, entre Alberdi y 14 de mayo, Asunción; tel (21) 44-3971; fax (21) 44-0306; Contact: D G Vega; Tech Officer, Office of Environmental Planning: Alfredo Molinas; serves as the INFOTERRA national focal point.

Ministry of Public Health and Social Welfare: Avda Pettirossi y Brasil, Asunción; tel (21) 20-7328; fax (21) 20-6700; internet www.mspbs.gov.py; Minister: Julio César Velázquez Tillería; incorporates the National Environmental Health Service.

PERU

Governmental Organizations

Consejo Nacional del Ambiente—CONAM: Avda San Borja Norte 226, San Borja, Lima 41; tel (1) 2255370; fax (1) 2255369; e-mail postmaster@conam.gob.pe; internet www.conam.gob.pe; Pres, Governing Bd: Luis Campos Baca; Exec Sec: Paul Remy F Vargas; promotes sustainable development in Peru by encouraging a balance between socioeconomic development and conservation of natural resources.

Ministry of Agriculture: Avda Salaverry s/n, Jesús María, Lima 11; tel (1) 4310424; fax (1) 4310109; e-mail postmast@minag.gob.pe; internet www.minag.gob.pe; Minister: Alvaro Quijandría Salmón; incorporates the Department of Forests and Wildlife.

Instituto Nacional de Recursos Naturales—INRENA (National Institute for Natural Resources): Calle 17 355, Urbanización El Palomar, Apdo 4452, San Isidro, Lima; tel (1) 2243298; fax (1) 2243218; e-mail inrena@correo.dnet.com.pe; Head: Dr Josefina Takahashi Sato; conservation of natural resources; research into viable renewable resources; co-ordination with the public and private sectors; policy proposals; evaluation of the environmental impact of agricultural projects.

Ministry of Health: Avda Salaverry Cuadra 8, Jesús María, Lima 11; tel (1) 4310408; fax (1) 4310093; e-mail webmaster@minsa.gob.pe; internet www.minsa.gob.pe; Minister: Pilar Mazzetti Soler.

Non-Governmental Organizations

Asociación Peruana para la Conservación de la Naturaleza—APECO (Peruvian Association for Conservation of Nature): Parque José de Acosta 187, Magdalena, Lima 17; fax (1) 2643027; e-mail apeco@amanta.rcp.net.pe; Exec Pres: Silvia Sánchez; Founder and Vice-Pres: Mariella Leo; f 1982; education and training for teachers in urban and rural areas to develop the awareness and skills for environmental protection; management of natural resources in the Abiseo River National Park, the Manu Biosphere Reserve and the cloud forest of the San Martín region; development of APECO RED, a network of autonomous organizations linked by common objectives.

Centro de Investigación y Capacitación Campesina—CICCA (Rural Research and Training Centre): Avda Mariño 101, Abancay, Apdo 58, Apurímac; tel (84) 321503; fax (84) 323884; e-mail postmast@cicca.org.pe; f 1979; soil conservation; productivity; occupational health; training and education; affiliated with Cáritas Neerlándica, the Catholic Relief Services, Desarrollo y Paz and Fondo Perú–Canadá.

Lima Office: Centro de Investigación y Capacitación Campesina, Avda Mariscal Miller 2214, Calle Lince, Apdo 11-0175, Lima 11; tel (1) 4700836.

Centro de Investigación, Educación y Desarrollo—CIED (Research, Education and Development Centre): Calle Buen Retiro 231, Urbanización Monterrico Chico, Surco, Lima 33; tel (1) 4342535; fax (1) 4378327; e-mail postmast@cied.org.pe; Pres and Dir: Juan Sánchez Barba; f 1973; rural development; local and regional development; sustainable development; agroecology; affiliated with Consorcio Latinoamericano sobre Agroecología y Desarrollo—CLADES.

Centro Eori de Investigación y Promoción Regional (Eori Regional Research and Promotion Centre): Jr Loreto 101, Puerto Maldonado, Madre de Dios; tel and fax (84) 571537; e-mail eori@sifocom.org.pe; Prin Officer: Thomas Moore; advice on development issues to organizations of indigenous people in the Madre de Dios region; project organization and management; legal defence of the land rights of indigenous people; advice to farmers' organizations; agroforestry and conservation; affiliated with the Forests, Tree and People Programme of the Environment Liaison Centre International.

Centro Ideas (Ideas Centre): Avda Arenales 645, Lima 1; tel (1) 4335060; fax (1) 4331073; e-mail postmast@ideas.org.pe; Prin Officers: Alfredo Stecher, Marina Irigoyen, Fernando Alvarado; f 1978; rural and urban development projects, with emphasis on agroecology, irrigation and local sanitation; co-operation with community organizations and local government agencies; participatory technology development.

Centro Panamericana de Ingeniería Sanitaria y Ciencias del Ambiente—CEPIS (Pan American Centre for Sanitary Engineering and Environmental Sciences): Calle Los Pinos 259, Urbanización Camacho, Lima 12; tel (1) 4371077; fax (1) 4378289; e-mail cepis@cepis.org.pe; internet www.cepis.org.pe; Dir: Sergio Augusto Caporali; environmental conservation and improvement; improvement of drinking water; reuse of waste water; information; technical training.

Centro para el Desarrollo de los Pueblos Ayllu—CEDEP-AYLLU (Centre for the Development of the Ayllu Peoples): Calle Vigil 246, Pisac, Cuzco; tel and fax (84) 203006/7; e-mail postmast@ayllu.org.pe; cedepayllu@wayna.rcp.net.pe; internet www.cedepayllu.org; Exec Dir: Alexander Chávez; rural development with Andean peasant communities in Southern Peru; agricultural production, including crop rotation, seeds and warehousing; natural resource management; irrigation; soil conservation; reafforestation; production of native and exotic plants; social development.

Centro Privado para el Desarrollo del Campesinado y del Poblador Urbano Marginal—CEDECUM (Private Centre for the Development of the Rural and Urban Poor): Avda Ejército 303, Apdo 542, Puno; tel and fax (54) 353461; e-mail credisol@geocities.com; Prin Officer: Dr Arturo Vásquez Salazar; f 1993; rural organization and management; environmental management; production methods; infrastructure; credit; marketing; affiliated with the Centro Regional de Crédito y Promoción Social—CREDISOL.

Comercio Alternativo de Productos No Tradicionales y Desarrollo en Latinoamérica—CANDELA Perú (Alternative Trade in Non-Traditional Products and Development in Latin America, Peru): Comandante Gustavo Jiménez 153, Lima 17; tel and fax (1) 4614556; Dir: Gaston Vizcarra; Gen Co-ordinator: Guadalupe Lanao; processing and marketing of Brazil nuts in Peruvian Amazonia; processing and marketing of agricultural products; research and development of new sustainable products in Amazonia; marketing of Peruvian crafts.

Comité Peruana para Cambio Global—COPECAM (Peruvian Committee for Global Change): Apdo 14-0363, Lima; tel and fax (1) 4336750; e-mail giescere@inictel.gob.pe; Chair: Alberto Giesecke M; f 1988; responsible to the National Cttee for the International Geosphere and Biosphere Programme; advice to policy-makers; seminars; forum for discussion; affiliated with the Peruvian Association for the Advancement of Science.

Escuela de Pintura Amazónica Usko-Ayar (Usko-Ayar Amazonian School of Painting): Jr Sánchez Cerro 465–467, Pucallpa; tel and fax (64) 573088; Prin Officers: Pablo César Amaringo Shuña, Luis Eduardo Luna Porres; f 1988; conservation of ethnobotanical gardens in the Amazon area; encouragement of respect for the Amazonian environment through artistic education for young people; artistic documentation of Amazon flora, fauna and ways of living; information about Amazonian nature, art and culture.

Intermediate Technology Development Group—ITDG (Grupo de Tecnología Intermedia para el Desarrollo): Avda Jorge Chávez 275, Miraflores, Apdo 18-0620, Lima 18; tel (1) 4475127; fax (1) 4466621; e-mail postmaster@itdg.org.pe; internet ekeko.rcp.net.pe/itdg/itdgint.htm; Dir: Alfonso Carrasco Valencia; f 1965; research and development of appropriate technologies for low-income rural and urban populations; programmes on agroprocessing, energy, disasters, irrigation and shelter.

PHILIPPINES

Governmental Organizations

Department of the Environment and Natural Resources: DENR Bldg, Visayas Ave, Diliman, Quezon City, 1100 Metro Manila; tel (2) 9296626; fax (2) 9204352; e-mail sechta@denr.gov.ph; internet www.denr.gov.ph; Sec: Michael Defensor; institution with ministerial status.

Ecosystems Research and Development Bureau: University of the Philippines, Los Baños, Laguna; tel (49) 5363628; fax (49) 5362850; e-mail erdb@laguna.net; internet www.laguna.net/~erdb/erdb.html; Dir: Celso P Diaz; formulates and promotes an integrated research and development programme relating to Philippine ecosystems and natural resources; provides technical assistance in the regional implementation and monitoring of research programmes; resource management.

Environmental Management Bureau: DENR Bldg, Visayas Ave, Diliman, Quezon City, Metro Manila; tel (2) 9247540; internet www.psdn.org.ph/emb; f 1987; management of air quality, chemicals, hazardous waste, solid waste and water quality; environmental education; environmental impact assessment; climate change; research and development.

Protected Areas and Wildlife Bureau: Quezon Ave, Diliman, Quezon City, 1100 Metro Manila; tel (2) 9246031/5; fax (2) 9240109; e-mail recreatoin@pawb.gov.ph; internet www.pawb.gov.ph; Dir: Theresa Mundita S Lim; Asst Dir: Asst Dir: Lorenzo C Agaloos; Nature Recreation and Extension: Angelita P Meniado; Wildlife Management: Carlo C Custodio; Biodiversity Management: Norma M Molinyawe; Protected Area Community Management: Marlynn M Mendoza; f 1987; conservation of protected areas, national parks, wildlife and biodiversity; monitors and co-ordinates the planning and implementation of the country's biodiversity programmes; provides technical assistance.

Department of Agrarian Reform: DAR Bldg, Elliptical Rd, Diliman, Quezon City, Metro Manila; tel (2) 9287031; fax (2) 9292527; e-mail nani@dar.gov.ph; internet www.dar.gov.ph; Sec: Rene Villa; institution with ministerial status.

Department of Agriculture: DA Bldg, 4th Floor, Elliptical Rd, Diliman, Quezon City, Metro Manila; tel (2) 9288741; fax (2) 9285140; e-mail dnotes@da.gov.ph; internet www.da.gov.ph; Sec: Arthur Yap; institution with ministerial status.

Philippine Atmospheric, Geophysical and Astronomical Services Administration: 1424 Quezon Ave, Quezon City, 1104 Metro Manila; tel and fax (2) 9228401; e-mail lamadore@mnl.sequel.net; Prin Officer: Leoncio A Amadore; f 1865; affiliated to the Department of Science and Technology; weather and flood forecasting; climatology; natural disaster management; agrometeorological services.

Philippine Council for Agriculture, Forestry and Natural Resources Research and Development—PCARRD: Paseo de Valmayor, Los Baños, 4030 Laguna; tel (49) 5360014; fax (49) 5360016; e-mail pcarrd@pcarrd.dost.gov.ph; internet www.pcarrd.dost.gov.ph; Exec Dir: Patricio S Faylon; Deputy Exec Dir, Research and Devt: Beatriz P del Rosario; Deputy Exec Dir, Institutional Devt and Resource Man: Aida R Ubreru; f 1972; formulates strategies, policies, plans and projects for the development of science and technology in agriculture, forestry and natural resources (AFNR); allocates government and external funds for research and development; monitors and evaluates research and development projects in the AFNR sectors; generates external funds for the same purposes.

Non-Governmental Organizations

Environmental Initiatives, Inc: Fersal II Condominium, Unit 2, 130 Kalayaan Ave, Diliman, Quezon City, Metro Manila; tel and fax (2) 9242333; Chair: Wilbur G Dee; Pres: Donato S de la Cruz; f 1996; environmental planning; watershed management; forest and natural resources management; environmental impact and risk assessment; project development, monitoring and evaluation; information dissemination and education.

Greenpeace Philippines: POB 2698, QC Central Post Office, Quezon City; e-mail athenar@dialb.greenpeace.org; national office of Greenpeace International.

Haribon Foundation for the Conservation of Natural Resources—HARIBON: 340 Villamor St, San Juan, 1500 Metro Manila; tel (2) 784179; fax (2) 704316; Science Programme Officer: Danilo S Balete; conservation of natural resources; management of protected areas; community-based resource management training; legal assistance; research; environmental information and education campaigns.

Marine Science Institute—MSI, University of the Philippines: UP-POB 1, Diliman, 1101 Quezon City; tel (2) 922-3959; fax (2) 924-7678; e-mail admin@upmsi.ph; internet www.upmsi.ph; Dir: Gil S Jacinto; Deputy Dir, Research: Maria Lourdes San Diego-McGlone; Deputy Dir, Education: Laura T David: Deputy Dir, Bolinao Marine Laboratory: Edna G Fortes; f 1974; affiliated to the College of Science, University of the Philippines; teaching and research in marine biology, marine chemistry, physical oceanography, marine geology and related disciplines.

Water Resources Center—WRC, University of San Carlos—USC: Talamban Campus, Nasipit, Talamban, 6000 Cebu City; tel (32) 3440523; fax (32) 3464615; e-mail wrc@mangga.usc.edu.ph; Dir: Herman van Engelen; Deputy Dir: F B Walag; f 1975; hydrometeorology; environmental and ground-water hydrology; water supply; watershed management; urban drainage; environmental impact assessment.

World Wide Fund for Nature—WWF Philippines: 23[a] Maalindog St, UP Village, Diliman, 1101 Quezon City; tel (2) 4333220; fax (2) 4263927; e-mail kkp@wwf-phil.org.oh; Rep: Dick Edwards; national affiliate organization of the World Wide Fund for Nature—WWF International.

POLAND

Governmental Organizations

Ministry of the Environment: 00-922 Warsaw, ul Wawelska 52/54; tel (22) 5792400; fax (22) 5792224; e-mail Minister@mos.gov.pl; internet www.mos.gov.pl; Minister: Jerzy Swatoń.

Biuro Urządzania Lasu i Geodezji Leśnej (Forest Management and Geodesy Bureau): 00-922 Warsaw, ul

Wawelska 52/54; tel (22) 8253423; fax (22) 8258399; e-mail buligl@buligl.pl; Dir: Adam Szempliński; f 1957; large-scale forest inventory; forest management inventory and preparation of forest management plans; forest monitoring; forest mapping.

Inspekcja Ochrony Środowiska—IOŚ (State Inspectorate for Environmental Protection): 00-922 Warsaw, Wawelska 52/54; internet www.pios.gov.pl.

Instytut Meteorologii i Gospodarki Wodnej (Institute of Meteorology and Water Management): 01-673 Warsaw, Podleśna 61; tel (22) 8341651; fax (22) 7342996; e-mail sekretariat@imgw.pl; internet www.imgw.pl; Dir Gen: Prof Dr Jan Zieliński; Deputy Dirs: Dr Jerzy Kloze (Water and Engineering Man), Dr Ryszard Klejinowski (Meteorological Forecasting), Roman Skapski (Hydrological and Meteorological Service), Prof Dr Henryk Slota (Research), Witold Drozdowiez (Technical and Economic Affairs); f 1919; research and expert appraisals in hydrology, meteorology, oceanology and water management; conducts meteorological and hydrological monitoring and measurements, and quantitative and qualitative monitoring of surface water; collects and stores the results of measurements and observations; prepares and disseminates long- and short-term meteorological forecasts on a regional and local scale, and designated aviation, marine, hydrological, agrometeorological, biometeorological, water and air pollution forecasts.

Państwowa Rada Ochrony Przyrody (National Council for Protection of Nature in Poland): 00-922 Warsaw, ul Wawelska 52–54; tel (22) 251114; fax (22) 254705; e-mail ekolog@warman.com.pl; Sec: Dr Ewa Symonides; f 1921; advice to government on environmental issues; conservation of forests, highlands, birds, endangered flora and fauna, landscape and national parks; design, planning and management of protected areas.

Ministry of Agriculture and Rural Development: 00-930 Warsaw, ul Wspólna 30; tel (22) 6231000; fax (22) 6232750; e-mail kanceleria@minrol.gov.pl; internet www.minrol.gov .pl; Minister: Wojciech Olejniczak.

Instytut Medycyny Pracy i Zdrowia Środowiskowego (Institute of Occupational Medicine and Environmental Health): 41-200 Sosnowiec, ul Kościelna 13; tel (32) 660885; fax (32) 661124; Dir: Jerzy A Sokal; f 1950; part of the Ministry of Health.

Państwowy Instytut Geologiczny (Polish Geological Institute): 00-975 Warsaw, ul Rakowiecka 4; tel (22) 495351; fax (22) 495342; Dir: Prof Stanisław Speczik; f 1919; adopted current name 1987; under the general supervision of the Ministry of the Environment; geological, geophysical, hydrogeological and environmental research.

Państwowy Zakład Higieny (National Institute of Hygiene): 00-791 Warsaw, ul Chocimska 24; tel (22) 494051; fax (22) 497484; Dir-Gen: Prof Dr Janusz Jeljaszewicz; part of the Ministry of Health; epidemiology, bacteriology, virology, parasitology, vaccines and sera control, medical statistics, radiological control and radiobiology, immunopathology, communal hygiene, foodstuffs, environmental toxicology, school hygiene, health education and biological contamination control.

Senate Committee on the Environment: 00-902 Warsaw, Wiejska 6–8; tel and fax (22) 6941639; Chair: Ryszard Ochwat; responsibilities include the formulation of legislation concerned with environmental protection.

Zakład Badania Ssaków, Polska Akademia Nauk—PAN (Mammals Research Institute, Polish Academy of Sciences): 17-230 Białowieza, woj Podlaskie, ul Gen Waszkiewicza 1; tel (85) 6812278; fax (85) 6812289; e-mail mripas@bison.zbs.bialowieza.pl; internet www.zbs.bialowieza.pl; Dir: Prof Dr Jan Marek Wójcik; Vice-Dir: Prof Dr Włodzimierz Jędrzejewski; f 1954; research into population biology, genetics and ecology of mammals; ecology of carnivore communities; mammal conservation; mammal phylogeny; journal.

Non-Governmental Organizations

Akademicki Klub Przyrody 'Terra' (Terra Association): 00-927 Warsaw, ul Krakowskie Przedmiescie 24 p 24; tel (22) 6200381; e-mail terra@free.polbox.pl; Contact: Zbigniew Kieras.

Biuro Obsługi Ruchu Ekologicznego—BORE (Service Office for the Environmental Movement—SOME): 03-727 Warsaw, Al Zieleniecka 6/8; Prin Officer: Jolanta Pawlak; organization of public meetings with the Ministry of the Environment; publication of information bulletin; organization of courses; support for environmental campaigns.

Centrum Informacji o Srodowisku UNEP/GRID, Warszawa (UNEP/GRID Environmental Information Centre, Warsaw): 02-511 Warsaw, ul Merliniego 9; tel (22) 6274623; fax (22) 488561; e-mail gridw@plearn.edu.pl; Contact: Teresa Galczynska.

Ecological Library Foundation: 61-715 Poznań, ul Kościuszki 79; tel (61) 521325; fax (61) 528276; Pres: Jacek Purat; Dir: Jarek Fiszer; protection of endangered species in Poland; ecological library; publications; book distribution; film presentations.

Ekologiczny Klub Obywatelski 'Czuwanie' w Darlowie (VIGIL Citizens' Environmental Club): 76-150 Darlowo, ul Zeromskiego 16; tel (94) 142582; Contact: Jadwiga Czarnoleska-Gosiewska.

Ekologiczny Klub UNESCO—Pracownia na Rzecz Bioróznorodnosci (UNESCO Ecological Club—Workshop for Biodiversity): 21-050 Piaski, ul Partyzantów 19; tel (81) 5821001; e-mail bel@dtm.dtm.lublin.pl; Contact: Janusz Kusmierczyk.

Europejskie Stowarzyszenie Ekonomistów Środowiska i Zasobów Naturalnych, Oddzial Polski (European Association of Environmental and Resource Economists, Polish Division): 30-067 Kraków, ul Gramatyka 10; tel and fax (12) 6373529; e-mail preisner@wzn4.2arz.agh.edu.pl; internet galaxy.uci.agh.pl/~esesizn/e; Chair: Dr Leszek Preisner; f 1990; promotes and develops environmental economics and management; initiates and implements research on environmental aspects of economic development; works towards an environmental policy in Poland.

Franciszkanski Ruch Ekologiczny (Franciscan Environmental Movement): 80-286 Gdańsk, ul Walowa 28; tel (601) 614982; fax (58) 479490; f 1989; areas of interest include the impact of the environment on humans, law of health and environmental protection, environmental medicine; international congresses; publications.

Fundacja Agrolot (Agrolot Foundation): 00-971 Warsaw, POB 122; tel (22) 8460031; fax (22) 8462701; Contact: Andrzej Slociński.

Fundacja Aktywna Ochrona Ptaków Ginacych (Active Conservation of Threatened Birds Foundation): 87-800 Włocławek, ul 3 Maja 22; tel and fax (54) 324927.

Fundacja Centrum Edukacji Ekologicznej Wsi (Rural Environmental Education Centre): 38-400 Krosno, ul Lewakowskiego 14, POB 159; tel and fax (13) 4368406; e-mail fceew@fceew.most.org.pl; Contact: Adam Kowalak; f 1991; environmental education, training and publications.

Fundacja 'Centrum Edukacji Zarzadzania Ochrona Środowiska' (Environmental Management Training Centre): 02-078 Warsaw, ul Krzywickiego 9; tel (22) 253852; fax (22) 6254305; Contact: Michal Chlebowski.

Fundacja 'Czysta Woda' (Clean Water Foundation): 00-175 Warsaw, Al Jana Pawla II 70; tel (22) 6360370; fax (22) 6360371; Contact: Janusz Strzyzewski.

Fundacja Czyste Powietrze i Czysta Woda (Clean Air and Clean Water Foundation): 05-074 Halinów, ul Hipolitowska 9; tel (22) 7836247; fax (22) 7836595; Contact: Piotr Grudziński.

Fundacja Ekologia i Zdrowie (Ecology and Health Foundation): 02-904 Warsaw, ul Bernardyńska 5/73[a]; tel (22) 6427337; fax (22) 318443; e-mail ecoheal@ikp.atm.com.pl; Prin Officers: Dr Marek Siemiński, Dr Stefan Bogusławski, Janusz Olton; information on environmental threats to health; education; information dissemination through bulletins, brochures and books; advice and research for local communities.

Fundacja Ekologiczna 'Zielony Górny Slask' (Green Upper Silesia Ecological Foundation): 40-012 Katowice, ul sw Jana 10; Contact: Krzysztof Dziadak.

Fundacja Ekonomistów Środowiska i Zasobów Naturalnych (Foundation of Environmental and Resource Economists): 15-875 Białystok, ul Krakowska 9; tel (85) 446096; Contact: Andrzej Demianowicz.

Fundacja Kultury Ekologicznej (Foundation for Ecological Culture): 58-500 Jelenia Góra 1, POB 145; tel and fax (75) 7526887; Contact: Jacek Jakubiec.

Fundacja Ochrony Środowiska 'Ekologika' (Ekologika Environmental Foundation): 90-965 Łódź, POB 5; tel (43) 773433; Contact: Henryk Romanowski.

Fundacja Ochrony Środowiska i Sluzby Zdrowia 'RENER' im Tadeusza Ratajczyka (RENER Foundation for Environmental Protection and Health Service): 30-036 Kraków, ul Mazowiecka 16–18; tel and fax (12) 2661625; Contact: Tadeusz Ratajczyk.

Fundacja Poszanowania Energii (Energy Conservation Foundation): 00-611 Warsaw, ul Filtrowa 1; tel (22) 255285; fax (22) 258670; e-mail fpeecf@it.com.pl; Contact: Ludomir Duda.

Fundacja Rolnicza (Agricultural Foundation): 01-015 Warsaw, Skwer Kard S Wyszynskiego 6; tel (22) 388569; fax (22) 389505; Contacts: Andrzej Kuliszewski, Krzysztof Gasiewicz.

Fundacja 'Terapia Homa' (Agnihotra Foundation): 34-240 Jordanów, Wysoka 151; tel (18) 447-5773; e-mail terapiahoma@idea.net.pl; internet www.terapiahoma.com.pl; Pres: J Bizberg; Vice-Pres: Dr Ulrich Berk; Sec: Maria Kalisz; f 1992; workshops, classes and lectures in Homa organic farming and gardening; there is a demonstration model farming community in south-east Poland between Kraków and Zalopane in the Tatra Mountain region. Publications: Homa farming for the new age, J Bizberg; Homa therapy—our last chance, Vasant Paranjpe.

Fundacja Wspierania Inicjatyw Ekologicznych (Foundation for the Support of Ecological Initiatives): 31-014 Kraków, ul Sławkowska 12; tel and fax (12) 4222264; e-mail fwie@free.polbox.pl; Contacts: Wojciech Krawczuk, Andrzej Zwawa.

Wydawnictwo 'Zielone Brygady' ('Green Brigades' Publishing House): 31-014 Kraków, ul Sławkowska 12/24; tel and fax (12) 4222147; e-mail zb@zb.most.org.pl; internet www.most.org.pl/zb; Prin Officer: Andrzej Zwawa; f 1989; publishing house; books and magazines on environmental subjects.

Fundacja Wspomagajaca Zaopatrzenie Wsi w Wode (Water Supply Foundation): 01-015 Warsaw, Skwer Kard S Wyszynskiego 6; tel (22) 386713; fax (22) 384683; e-mail wsfound@pol.pl; Contact: Piotr Szczepanski.

Instytut Badawczy Leśnictwa (Forest Research Institute): 00-973 Warsaw, ul Bitwy Warszawskiej 1920, r 3; tel (22) 8222457; fax (22) 8224035; e-mail ibl@ibles.waw.pl; internet www.ibles.waw.pl; Dir: Prof A Klocek; f 1930; comprises 23 scientific sections covering all aspects of forestry; wildlife and forest management; forest protection; receives 20% of its funding from the Ministry of the Environment.

Instytut Ekologii, Polska Akademia Nauk—PAN (Institute of Ecology, Polish Academy of Sciences): 05-092 Łomianki, Dziekanów Lesny, ul M Konopnickiej 1; tel (22) 7513046; fax (22) 7513100; Dir: Prof Dr Leszek Grüm; f 1952; popultation and community studies, landscape ecology, ecological bio-energetics, biogeochemistry, agroecology, human ecology, polar research, hydrobiology, plant ecology, ecological processes.

Mikołajki Hydrobiological Station: Instytut Ekologii, 11-730 Mikołajki, Leśna 13; tel (87) 216133; fax (87) 216051; Prin Officers: Jolanta Ejsmont-Karabin, Andrzej Karabin, Joanna Królikowska, Irena Kufel, Lech Kufel; research into aquatic ecology; monitoring of freshwater ecosystems; analysis of watersheds and surface waters.

Instytut Gospodarki Przestrzennej i Komunalnej, Polska Akademia Nauk—PAN (Institute of Physical Planning and Municipal Economy, Polish Academy of Sciences): 02-078 Warsaw, ul Krzywickiego 9; tel and fax (22) 8250937; Dir: Dr Jacek Malasek; f 1986; research on physical planning, municipal economy and architecture.

Instytut na rzecz Ekorozwoju—INE (Institute for Sustainable Development—ISD): 02-502 Warsaw, ul Lowicka 31; tel (22) 6460510; fax (22) 6460174; e-mail ine@ine-isd.org.pl; internet www.ine-isd.org.pl; Prin Officer: Dr Andrzej Kassenberg; f 1990; publishes reports and organizes seminars, conferences and training courses with the aim of promoting sustainable development in Poland.

Instytut Ochrony Przyrody, Polska Akademia Nauk—PAN (Institute of Nature Conservation, Polish Academy of Sciences): 31-120 Kraków, Mickiewicza 33; tel and fax (12) 4210348; e-mail okarma@ib-pan.krakow.pl; internet botan.ib-pan.krakow.pl/przyroda/index.htm; Dir: Prof Dr Zygmunt Denisiuk; Deputy Dir: Dr Henryk Okarma; f 1952; scientific research into natural biological processes; preservation of rare and threatened plant and animal species; research into the degradation of natural resources, changing ecosystems, water resources, landscape and geoconservation; member of the World Conservation Union—IUCN.

Instytut Podstaw Inżynierii Środowiska, Polska Akademia Nauk—PAN (Institute of Environmental Engineering, Polish Academy of Sciences): 41-800 Zabrze, ul M Skłodowskiej-Curie 34; tel (37) 2716481; fax (37) 2717470; e-mail ipis@ipis.zabrze.pl; internet www.ipis.zabrze.pl; Dir: Assoc Prof Jan Kapała; f 1961; air and water pollution control.

'Jastrzab' Fundacja Ochrony Środowiska Naturalnego (Goshawk Foundation for Environmental Protection): 31-010 Kraków, Rynek Glówny 34; tel and fax (12) 4221983; Pres: Leslaw Chmiela.

Klub Publicystów Ochrony Środowiska (Club of Environmental Journalists): 00-922 Warsaw, ul Wawelska 52/54; tel (22) 259056; fax (22) 258556; Contact: Slawomir Trzaskowski.

Komitet Ochrony Przyrody Polskiej Akademii Nauk (Nature Conservation Committee of the Polish Academy of Sciences): 50-335 Wrocław, ul Sienkiewicza 21; tel (71) 225041; fax (71) 222817; Contact: Prof Ludwik Tomialojc.

Liga Ochrony Przyrody, Zarzad Glówny (League for the Conservation of Nature in Poland, Central Board): 00-355

Warsaw, ul Tamka 37/2; tel and fax (22) 6358171; e-mail ecekd@tg.com.pl; internet www.tg.com.pl/ecekd; Pres: Władyśsław Skalny; Vice-Pres, European and Overseas Co-operation: Tomasz Cieślik; f 1928; nature conservation; campaigning; ecological education, especially in schools; co-operation with other organizations in Poland and abroad; publications; affiliated with Europa Nostra, European Youth Forest Action, the World Conservation Union—IUCN, the World Society for the Protection of Animals and Youth and Environment Europe.

Europejskie Centrum Ekologiczne Ligi Ochrony Przyrody—ECEKOL (European Environmental Centre of the League for the Conservation of Nature): 02-384 Warsaw, ul Wlodarzewska 13–15; tel (22) 6580164; fax (22) 6580135; e-mail ecekol@ecekol.edu.pl; internet www.ecekol.edu.pl; Dir: Tomas Cieślik; environmental education; promotes environmental protection and sustainable development; operates the Polish–British Co-operation Centre for the Conservation of Nature and National Heritage, Green Towns of Poland programme and Green Business Club.

Lubelska Fundacja Ochrony Środowiska—Regionalne Centrum Edukacji Ekologicznej (Lublin Foundation for Environmental Protection): 20-346 Lublin, ul Dluga 13[a]; tel (81) 7444825; fax (81) 7444657; Dir: Andrzej Karas.

Narodowa Fundacja Ochrony Środowiska (National Foundation for Environmental Protection): 02-078 Warsaw, ul Krzywickiego 9; tel (22) 252127; fax (22) 251018; e-mail nfos@warman.com.pl; Sec: Jolanta Passini.

Ogólnopolskie Towarzystwo Ochrony Ptaków (Polish National Bird Protection Society): 80-958 Gdańsk, POB 335; tel and fax (58) 412693; Contact: Bogumila Blaszkowska.

Ogólnopolskie Towarzystwo Zagospodarowania Odpadów '3R' (3R Waste Prevention Association): 31-014 Kraków, ul Sławkowska 12; tel (12) 4222264; fax (12) 4212107; e-mail office@otzo.most.org.pl; internet www.rec.org/poland/wpa; Contacts: Pawel Gluszynki, Piotr Rymarowicz, Slawomir Pietrasik.

Osrodek Edukacji Ekologicznej (Eco-Pedagogical Centre): 05-530 Czersk, Plac Tysiaclecia 5; tel and fax (22) 7273495; Contact: Dr Maria Janina Dietrich.

Polska Asocjacja Ekologii Krajobrazu (Polish Association for Landscape Ecology): 61-701 Poznań, ul Fredry 10; tel (61) 8520694; Contact: Prof Daniela Solowiej; promotes development of landscape ecology.

Polska Fundacja Lesna 'Pol-Forest' (Polish Forest Foundation—Pol-Forest): 50-357 Wrocław, ul Grunwaldzka 90; tel (71) 223011; fax (71) 211401; Contact: Bolesław Spring.

Polska Fundacja Partnerstwo dla Środowiska (Polish Environmental Partnership Foundation): 31-005 Kraków, ul Bracka 6/6; tel and fax (12) 225088; e-mail epce@kki.krakow.pl; Country Rep: Rafal Serafin; f 1991; provides small grants and technical and organizational assistance to NGOs and local authorities in Poland and other Central European countries; part of a network of similar organizations in the Czech Republic, Hungary and Slovakia.

Polski Klub Ekologiczny—PKE (Polish Ecological Club): 31-014 Kraków, ul Sławkowska 26[a]; tel (12) 4232047; fax (12) 4232098; e-mail biuro@zgpke.most.org.pl; internet www.most.org.pl/pke-zg; Mem, Bd: Prof Stanisław Juchnowicz; f early 1980s as part of the Solidarity trade union movement; federation of regional groups; promotion of sustainable development; human ecology; environmental education; landscape protection; energy conservation; waste recycling; biological pest control; organic farming; conservation strategies; campaigns on urban planning and human health; affiliated with Friends of the Earth International—FoEI.

West Pomeranian Branch: 71-550 Szczecin, Kaz Królewicza 4h; tel (91) 231061; fax (91) 231347; Prin Officers: Piotr Gruszka, Juliusz C Chojnacki; f 1991.

Polskie Towarzystwo Ekologiczne (Polish Ecological Society): 00-330 Warsaw, ul Nowy Swiat 72; tel (22) 6200381; fax (22) 8267991; e-mail kozak@plearn.edu.pl; Contact: Michal Kozakiewicz.

Polskie Towarzystwo Ochrony Przyrody 'Salamandra' (Polish Society for Nature Protection—Salamandra): 60-514 Poznań, ul Szamarzewskiego 11/6; tel and fax (61) 8432160; e-mail biuro@salamandra.org.pl; internet www.salamandra.org.pl; Contact: Ewa Olejnik; f 1993; education and lobbying in the field of nature protection.

Polskie Towarzystwo Przyjaciól Przyrody 'Pro Natura' (Pro Natura Society of Wildlife Friends): 50-449 Wrocław, ul Podwale 75; tel (71) 445055; fax (71) 446135; Contact: Roman Guziak; conservation of species and river habitats; promotes environmental awareness.

Polskie Towarzystwo Rolnictwa Ekologicznego (Polish Ecological Agriculture Society): 20-718 Lublin, Al Krasnicka 85; Contact: Jerzy Szymona.

Sandomierskie Towarzystwo Ekologiczne 'Terra Vita' (Terra Vita Sandomierz Ecological Association): 27-600 Sandomierz, ul Retmanska 1; tel (15) 8322098; fax (15) 8322225; e-mail witold@terravita.most.org.pl; Contact: Zenon Sobowiec.

Siec Biur Informacji Ekologicznej—Polska Zielona Siec (Network of Environmental Information Offices—Polish Green Net): 31-014 Kraków, ul Sławkowska 12; tel (12) 4222264; fax (12) 4222147; e-mail stroes@fwie.most.org.pl; Contact: Ernst Jan Stroes.

Spolecsny Instytut Ekologicsny—SIE (Social Institute for Ecology): 03-127 Warsaw Zieleniecka 6/8; tel (22) 6183781; fax (22) 6182884; e-mail bore@plearn.edu.pl; Pres: Justyna Krynacka; promotes research and projects in environmental protection; collects and disseminates information on the state of the environment in Poland; provides services and advice to the public and to NGOs in all matters concerning environmental protection.

Stowarzyszenie Agro Bio Test (Agro Bio Test Association): 02-766 Warsaw, ul Nowoursynowska 166; tel and fax (22) 437904; Contact: Dr Urszula Soltysiak.

Stowarzyszenie 'Centrum Informacji Ekologicznej' (Environmental Information Centre): 03-729 Warsaw, ul Targowa 59/1; tel (601) 207530; e-mail office@cie.most.org.pl; Contacts: Artur Siedlarek, Jaroslaw Dubiel, Piotr Flader.

Stowarzyszenie Ekologiczno-Kulturalne 'Klub GAJA' (Ecological and Cultural Association—Club GAIA): 43-301 Bielsko-Biała 1, POB 361; tel and fax (33) 8123694; e-mail klubgaja@klubgaja.pl; internet www.klubgaja.pl; Pres: Jacek Božek; Sec: Wojciech Owczarz; f 1988; independent organization; organizes the Vistula Now campaign, promoting ecological and social use of the River Vistula, and An Animal is not a Thing, promoting animal rights in Poland; environmental education; publications; afilliated with the Royal Society for the Prevention of Cruelty to Animals—RSPCA (United Kingdom) and the World Wide Fund for Nature—WWF.

Stowarzyszenie Grupa Reakcji A Eko-Front (Eko-Front Reaction Group): 58-310 Szczawno-Zdrój, ul Wojska Polskiego 2/1; tel (604) 974329; Chair: Rafał Jakubowski; f 1991; autonomous branch of Pracownia na rzecz Wszystkich Istot (Workshop for All Beings); dedicated to the protection of local nature; organizes lectures on ecology, health and spiritual traditions, and annual ecological and cultural workshops (Ekofrontacje); library.

Stowarzyszenie na rzecz Ekologicznego i Rolniczego Rozwoju (Society for Ecological and Rural Progress): 02-954 Warsaw, ul Królowej Marysienki 33-105; tel (22) 6420887; Contact: Dr Jacek J Nowak.

Stowarzyszenie Producentów Zywnosci Metodami Ekologicznymi 'Ekoland' w Przysieku k/Torunia (Ekoland Association of Organic Producers): 87-134 Zlawies Wielka, miejsc Przysiek k/Torunia; tel and fax (56) 781915; e-mail biobabalscy@poczta.onet.pl; Chair: Mieczysław Babalski; Prin Officer: Ewa Koreleska; f 1989; promotion of agriculutural diversification in co-operation with local administration and non-governmental organizations.

Stowarzyszenie Zdrowych Miast Polskich (Polish Healthy Cities Association): 90-113 Łódź, ul Sienkiewicza 5; tel and fax (42) 331344; Contact: Elzbieta Roczek.

Stowarzyszenie Zieloni (Greens Association): 40-012 Katowice, ul sw Jana 10; Contact: Andrzej Sochanski.

Towarzystwo Ekologiczne 'Ziemia przede Wszystkim' (Earth Above All Environmental Society): 61-966 Poznań 31, POB 63; tel (61) 8517801; fax (61) 8530923; Contact: Robert Rybicki.

Towarzystwo Ekologicznego Transportu (Ecological Transportation Association): 31-014 Kraków, ul Sławkowska 12; tel and fax (12) 4222264; e-mail fwie@free.polbox.pl; Contact: Olaf Swolkien.

Towarzystwo Naukowe Prawa Ochrony Środowiska (Polish Environmental Law Association): 50-138 Wrocław, ul Kuźnicza 46/46; tel (71) 444747; fax (71) 3410234; Contact: Tomasz Tatomir.

Towarzystwo Opieki nad Zwierzetami (Society for the Protection of Animals): 00-666 Warsaw, ul Noakowskiego 4; tel (22) 257535; fax (22) 256049; Contact: Wojciech Muza.

Zakład Badan Środowiska Rolniczego i Leśnego, Polska Akademia Nauk—PAN (Research Centre for Agricultural and Forest Environmental Studies, Polish Academy of Sciences): 60-809 Poznań, ul Bukowska 19; tel (61) 475601; fax (61) 473668; Dir: Prof Dr Lech Ryszkowski; f 1979; study of energy flows and the cycling of matter; evaluation of ecological guidelines for landscape management and strategy for nature conservation.

Zespół Prawnych Ochrony Środowiska, Polska Akademia Nauk—PAN (Research Group on Environmental Law, Polish Academy of Sciences): 50-168 Wrocław, ul Kuźnicza 46/47; tel and fax (71) 3444747; e-mail sommer@iuris.prawo.uni.wroc.pl; Dir: Prof Jerzy Sommer; f 1975; analysis and preparation of environmental legislation; education and consultancy.

PORTUGAL

Governmental Organizations

Ministry of the Environment and Territorial Planning: Rua de O Século 51, 1200-433 Lisbon; tel (21) 3232500; fax (21) 3232531; e-mail sg.ambienta@sg.mcota.gov.pt; internet www.ambiente.gov.pt; Minister: Luís Nobre Guedes.

Comissão do Programa o Homem e a Biosfera (Man and Biosphere Programme Commission): Rua de O Século 51, 4°, 1200-433 Lisbon; tel (21) 3231500; fax (21) 3231515; Pres: Prof Dr Fernando Henriques.

Conselho Nacional da Água (National Water Council): Rua de O Século 51, 3°, 1200-433 Lisbon; tel (21) 3231626; fax (21) 3231530; Sec-Gen: António Raúl Eita Leitão.

Conselho Nacional do Ambiente e Desenvolvimento Sustentável—CNADS (National Council for the Environment and Sustainable Development): Rua de S Domingos à Lapa 26, 1200-835 Lisbon; tel (21) 3929926; fax (21) 3929929; e-mail cnads.aleitao@ipamb.pt; Pres: Prof Dr Mário João O Ruivo.

Inspecção-Geral do Ambiente (Environment Inspectorate-General): Rua da Murgueira-Zambujal, Apdo 6154 Alfragide, 2720-601 Amadora; tel (21) 4728324; fax (21) 4728389; Prin Officer: Dr António Leones Dantas.

Instituto da Água—INAG (Water Institute): Avda Almirante Gago Coutinho 30, 1049-066 Lisbon; tel (21) 8430000; fax (21) 8473571; e-mail inforag@inag.pt; internet www.inag.pt; Pres: Carlos Alberto Mineiro Aires.

Instituto da Conservação da Natureza—ICN (Nature Conservation Institute): Rua da Lapa 73, 1200-701 Lisbon; tel (21) 3938900; fax (21) 3901048; e-mail icn@icn.pt; internet www.icn.pt; Pres: Carlos Alberto Guerra; nature conservation; management of natural parks, reserves, protected areas and coastal zones; information dissemination.

Instituto de Navegabilidade do Douro—IND (Douro Navigability Institute): Rua dos Camilos 90, 2°, 5050-272 Peso da Régua; tel (254) 320020; fax (254) 320023; e-mail ind@mail.telepac.pt; internet www.utad.pt/ind; Dir: Mário Manuel Fernandes.

Instituto de Promoção Ambiental—IPAMB (Environmental Promotion Institute): Rua de O Século 63, 1249-033 Lisbon; tel (21) 3215500; fax (21) 3432777; e-mail ipamb@mail.telepac; internet www.ipamb.pt; Pres: Dr José Manuel Pereira Alho.

Instituto do Ambiente (Environment Institute): Rua da Murgueira-Zambujal, 2721-865 Amadora; tel (21) 4728200; fax (21) 4719075; internet www.iambiente.pt; Pres: Eugenio Goncalves; f 2002.

Instituto dos Residuos (Wastes Institute): Avda Almirante Gago Coutinho 30, 5°, 1049-017 Lisbon; tel (21) 8424051; fax (21) 8424059; e-mail inr@inresiduos.pt; Pres: Prof Dr António Sarmento Lobato de Faria.

Instituto Regulador de Águas e Residuos—IRAR (Institute of Water and Waste Regulation): Rua Francisco Metrass 107, 1350-141 Lisbon; tel (21) 3841500; fax (21) 3841515; e-mail iraresiduos@instrar.pt; Pres: Pedro Cunha Serra; regulation of the drinking water, waste water and solid waste sectors.

Intervenção Operacional do Ambiente (Operational Intervention in the Environment): Rua de O Século 51, 3°, 1200-433 Lisbon; tel (21) 3231609; fax (21) 3231619; Man: Dr Luísa Maria Leitão do Vale.

Secretário de Estado do Ambiente (State Secretariat for the Environment): Rua de O Século 51, 1200-433 Lisbon; tel (21) 3231430; fax (21) 3232561; e-mail sea@maot.gov.pt; Sec: Rui Nobre Gonçalves.

Secretário do Estado do Ordenamento do Território e Conservação da Natureza (State Secretariat for Land

Management and Nature Conservation): Rua de O Século 51, 1200-433 Lisbon; tel (21) 3231456; fax (21) 3232588; e-mail seotcn@maot.gov.pt; Sec: Dr Manuel Pedro Cunha da Silva Pereira.

Ministry of Agriculture, Fisheries and Forests: Praça do Comércio, 1149-010 Lisbon; tel (21) 3234652; fax (21) 3234604; e-mail geral@min-agricultura.pt; internet www.min-agricultura.pt; Minister: Carlos da costa Neves.

Instituto Nacional de Engenharia e Tecnologia Industrial—INETI (National Institute for Engineering and Industrial Technology): Estrada do Paço do Lumiar, 1649-038 Lisbon Codex; tel (21) 7165141; fax (21) 7163688; internet www.ineti.pt; Pres: Prof Carlos Campos Morais; f 1992; materials and production technologies; biotechnology; chemistry; food technologies; environmental technologies; energy technologies (non-nuclear); information technologies; engineering and training management; technology and innovation management.

Non-Governmental Organizations

Amigos da Terra/Associação Portuguesa de Ecologistas (Friends of the Earth—FoE/Portuguese Ecologists' Association): Travessa Marquês de Sampaio 42 r/c esq, 1200 Lisbon; tel (1) 3479599; fax (1) 3473586; Prin Officers: Mario Alves, Nicole Mueller; national office of Friends of the Earth International—FoEI; projects on tourism, biotechnology, packaging, tropical rainforests, oceans and the coastal environment; environmental education.

Amigos dos Açores/Associação Ecológica (Friends of the Azores/Ecological Association): Rua Capitão Manuel Cordeiro s/n, Pico da Pedra, 9600 Ribera Grande; tel (296) 498474.

Associação Cívica para a Defesa do Mar—Amigos do Mar (Civic Association for the Protection of the Sea—Friends of the Sea): Marina no Edif Complexo Turístico, Apdo 533, 4900 Viana do Castelo; tel (258) 827427.

Associação para a Conservação do Lobo e do seu Ecosistema—Grupo Lobo (Association for the Preservation of the Wolf and its Ecosystem): Departamento Zoologia e Antropologia, Faculdade Ciencias de Lisboa, Campo Grande, Bloco C2, 1700 Lisbon; tel (21) 7500000; fax (21) 7500028; e-mail globo@fc.ul.pt; internet lobo.fc.ul.pt; Pres: Francisco Petrucci-Fonseca; f 1985; affiliated to Grupo Lobo; development of projects that involve the preservation of the wolf, its habitats and its prey; educational programme in schools and other conservation organizations.

Associação Portuguesa de Agricultura Biológica (Portuguese Association for Organic Agriculture): Calçada da Tapada 30, r/c Dto, 1300 Lisbon; tel (21) 3641554.

Associação Portuguesa de Educação Ambiental—ASPEA (Portuguese Association for Environmental Education): Apdo 4021, 1501 Lisbon Codex; tel (21) 7788371.

Associação Portuguesa de Empresa de Tecnologias Ambientais—APEMETA (Portuguese Association of Environmental Technology Businesses): Campo Grande 294, 3° Dto, 1700 Lisbon; tel (21) 7576174.

Associação Portuguesa de Engenheiros do Ambiente (Portuguese Association of Environmental Engineers): Rua Cidade da Horta 14b, Sala 22, 1000 Lisbon; tel (21) 3520305; fax (21) 3157636; e-mail apeambiente@mail.telepac.pt; internet www.apea.pt; Pres: Graça Martinho; Vice-Pres: João Pedro Rodrigues; f 1985; promoting awareness and development of environmental engineering and co-operation among environmental engineers.

Associação Portuguesa de Guardas e Vigilantes da Natureza (Portuguese Association of Nature Wardens): Olho de Boi, Apdo 131, Cova de Piedade, 2806 Almada Codex.

Associação Portuguesa dos Recursos Hídricos (Portuguese Water Resources Association): a/c do LENEC, Avda do Brasil 101, 1799 Lisbon Codex; tel (21) 8440100.

Associação Portuguesa para o Direito do Ambiente (Portuguese Environmental Law Association): Rua S Marçal 77, 1200 Lisbon; tel (21) 3479466.

Centro de Estudios da Avifauna Ibérica (Study Centre for Iberian Bird Life): Prolongamento da Avda Infante D Henrique Talhão 7, r/c, 7000 Evora; tel (266) 746102.

Centro de Estudos em Economia da Energia dos Transportes e Ambiente—CEEETA (Study Centre for Energy Efficiency in Transport and the Environment): Rua Dr António Cándido 10, 4°, 1050 Lisbon; tel (21) 3172052.

Comunidade Ecológica Europeia do Ambiente—CEEA (European Ecological Community for the Environment): Costa–Outeiro, 4900 Viana do Castelo; tel (258) 842026.

Grupo de Estudos de Ordenamento do Território e Ambiente—GEOTA (Research Group for Land and Environmental Management): Travessa Moinho de Vento 17 cv Direita, 1200 Lisbon; tel (21) 3956120; fax (21) 3955316; e-mail geota@mail.telepac.pt; internet www.despodata.pt/geota; Prin Officers: Conceição Martins, Isabel Moura, Filipa Ramalhota, Teresa Leonardo, Rute Eurto; f 1981; formulation and assessment of environmental policy; drafting and analysis of regulations and legislation on environmental impact assessment, land management, water resource management, forests, nature conservation, natural and cultural heritage, and energy planning; lobbying; environmental conservation; co-ordination of Coast Watch Portugal; environmental education.

Liga para a Protecção da Natureza (Nature Protection League): Estrada do Calhariz de Benfica 187, 1500 Lisbon; tel (21) 7780097; fax (21) 7783208; Vice-Pres: Luis Miguel Vieira; conservation of marine, freshwater and terrestrial environments, endangered wildlife and plants, protected areas, national parks and forests; environmental policy; conservation strategies; training courses for teachers; publications.

Núcleo Português de Estudo e Protecção da Vida Selvagem—NPEPVS (Portuguese Centre for the Study and Protection of Wildlife): Bairro Fundo de Fomento de Habitação, Bloco D r/c, 5300 Bragança; tel (27) 3324632; fax (27) 326514; Pres: Armando José Pereira; study of flora and fauna; awareness-raising activities; advice on conservation issues.

Partido Ecologista Os Verdes—PEV (The Greens): Calçada Salvador Correia de Sá 4, 1° Direito, 1200-399 Lisbon; tel (1) 3432763; fax (1) 3432764; e-mail osverdes@mail.telepac.pt; internet www.osverdes.pt; Leader: Maria Santos; f 1982; ecological political party.

Planeta Verde—Associação para a Protecção e Defesa da Floresta (Green Planet—Association for Forest Protection and Defence): Estrada do Calhariz de Benfica 187, 1500 Lisbon; tel (21) 7743542.

Quercus—Associação Nacional de Conservação da Natureza (Quercus National Association for Nature Conservation): Rua António Macedo, Apdo 4333, 1500 Lisbon; tel (21) 7787749; fax (21) 7788474; e-mail quercus@mail.telepac.pt; Chair: Prof Dr Viriato Soromenho Marques; Nat Sec: Dr José Paulo Martins; study of Portuguese flora, fauna and ecosystems; nature conservation; environmental education; promotion of sustainable development in political and economic decision-making.

QATAR

Governmental Organizations

Ministry of Municipal Affairs and Agriculture: POB 820, Doha; tel 4336336; fax 4430239; e-mail mmaa@mmaa.gov.qa; internet www.mmaa.gov.qa; Minister: Sultan bin Hassan adh-Dhabit ad-Dousary.

Environment Department: Bldg No 318, Rd No 230, Area No 24, 'C' Ring Rd, POB 7634, Doha; tel 320825; fax 415246; internet www.mmaa.gov.qa/english/org-chart/environment.html; Man: Khalid Ghanim Al Ali; fmrly the Permanent Committee for the Protection of Environment; formulation of environmental policy; serves as the INFOTERRA national focal point.

Ministry of Energy, Industry, Electricity and Water: POB 2599, Doha; tel 4832121; fax 4832024; internet www.kahramaa.com; Minister: Abdullah bin Hamad al-Attiya.

Ministry of Public Health: POB 42, Doha; tel 4441555; fax 4446294; internet www.hmc.gov.qa; Minister: Dr Hajar bin Ahmad Hajar.

Non-Governmental Organizations

Scientific and Applied Research Centre, Qatar University: POB 2713, Doha; Dir: Himead Al-Midffa; develops experience in scientific, industrial and agricultural fields with special reference to industries, natural resources, agriculture and animal resources.

ROMANIA

Governmental Organizations

Ministry of Environment and Water Management: 040129 Bucharest, Bd Libertăţii 12, Sector 5; tel (1) 4100243; fax (1) 3124227; e-mail Ianculesco@mappm.ro; internet www.mappm.ro; Minister: Speranţa Maria Ianculescu; establishment of guide-lines for environmental policy; monitoring of activities affecting waters, forests and the environment; incorporates the National Commission for Control of Nuclear Activity.

Ministry of Agriculture, Forestry and Rural Development: 70312 Bucharest, Bd Carol I 24; tel (1) 6144020; fax (1) 3124410; internet www.maap.ro; Minister: Petre Daea.

Administration of Biosphere Reservation of the Danube Delta: 8800 Tulcea, Taberei 32; tel (405) 50950.

Agency of Environmental Protection and Surveillance: Bucharest, Bd Mihail Kogălniceanu 27; tel (1) 6135535.

Institutul National de Cercetăre Dezvoltare Delta Dunării (Danube Delta National Institute for Research and Development): 8800 Tulcea, Babadag 165; tel (40) 531520; fax (40) 533547; e-mail office@indd.tim.ro; internet indd.timm.ro; Exec Dir: Romulus Stiuca; Science Dir: Mircea Staras; f 1970; research supporting the conservation and rehabilitation of the ecology of the Danube delta, especially the Danube Delta Biosphere Reserve; monitoring of water and soil quality; promotion of sustainable resource use; hunting; grazing; agriculture, fisheries and forestry; tourism research.

Non-Governmental Organizations

Action XXI Romania: 78136 Bucharest, Dr Iacob Felix 41; tel (1) 6507472; e-mail radu@hansi.deperm.pub.ro; Contact: Radu Gruescu.

Actiunea Ecologica Romana (Romanian Ecological Action): 1100 Craiova, Nicolae Titulescu 38; tel (51) 196686; Contact: Codresi Teclu.

Asociatia Eco Conseil (Eco-Council Association): 2200 Brasov, Barbu Lautaru 11/27/C/5; tel (68) 189587; fax (68) 315436; e-mail ecoenvironment@hotmail.com; Contact: Rusu Constantin Marius.

Asociatia Ecologista 'Scutierii Naturii' ('Nature's Shield-Bearer' Ecological Association): Bacau, Vasile Porvan 22; tel (32) 124071; Contact: Popa Mircea.

Asociatia Hidrogeologilor din Romania (Romanian Association of Hydrogeologists): 70139 Bucharest, Traian Vuia 6; tel and fax (1) 2123385; e-mail ahr@hidro.sbnet.ro; Contact: Florian Zamfirescu.

Asociatia pentru Politici Energetice din Romania (Romanian Energy Policy Association): 76117 Bucharest, Calea 13 Septembrie 13; tel and fax (1) 4104100; e-mail aper@pcnet.pcnet.ro; Contact: Violeta Kogalniceanu.

Asociatia pentru Promovarea Agriculturii Biodinamice (Biodynamic Agriculture Promotion Association): 78416 Bucharest, Pajurei 26, Sc B, Apt 22, Sector 1; tel (1) 6673700; Contact: Papacostea Petre.

Asociatia pentru Protectia Mediului si a Naturii (Nature and Environmental Protection Association): 76256 Bucharest, Dr Leonte 1/3; tel (1) 6384010; fax (1) 6374202; Contact: Prof Manole Cucu.

Asociatia Romana pentru Management Ecologic si Dezvoltare Durabila (Romanian Ecological Management for Sustainable Development Association): Bucharest, Calea Victoriei 125, Et 1, Cam 31; tel (1) 6504889; fax (1) 3125342.

Asociatia 'Terra Nostra' (Terra Nostra Association): Bucharest, B-dul Mihai Bravu 288; tel (1) 3222678; fax (1) 3222615; Contact: Strambu Bogdan.

Asociatia Tinerii Ecologisti–Natura (Young Ecologists–Nature Association): 71441 Bucharest, Giuseppe Garibaldi 3; tel (1) 2305276; Contact: Gavria Cecilia.

Centrul de Consultanta Ecologica (Eco-Counselling Centre): 6200 Galati, Basarabiei 2; tel (36) 435521; fax (36) 460827; e-mail eco@sisnet.ro; internet 212.93.435.99; Pres: Petruta Moisi; creation of environmental awareness; education and information.

Centrul de Educare Informare şi Resurse Pentru Marea Neagră Cier (Information, Education and Resource Centre for the Black Sea): 8700 Constanţa, Bd Mamaia 296; tel (41) 831099; e-mail cier@impromex.ro; internet www.sitex.ro/cier; Man: Ionică Bucur; Information Asst: Ivaşcu Carmen; promotes new environmental initiatives for the

Black Sea; supports public participation in policy formulation; provides co-ordination, training and resources for NGOs; environmental education; affiliated with the Black Sea Environmental NGOs Network.

Centrul de Informare Documentară pentru Energetică (Energy Information and Documentation Centre): 74568 Bucharest 3, Bd Energeticienilor 8; tel and fax (1) 3239552; fax (1) 3211010; e-mail cide@mail.gsci.vsat.ro; Dir: Constantin Tuică; f 1966; information relating to power stations, heat distribution and utilization, environmental protection, electricity generation and power system technology.

Centrul de Instruire, Informare si Mediere pentru Eco Dezvoltare (Training, Information and Mediation Centre for Eco-Development): 2000 Ploiesti, B-dul Bucuresti 39; tel (44) 118457; fax (44) 116549; e-mail timced@csd.univ.ploiesti.ro; Contact: Georgescu Adrian.

Centrul Mondial de Protectie a Mediului (World Environmental Protection Centre): 73546 Bucharest, Şos Pantelimon 309, Block 8, Apt 158; tel (1) 6272004; fax (1) 3128063; Contact: Liviu Ionescu.

Centrul National de Informare Ecologica si Turistica (National Centre for Ecological and Tourist Information): 6600 Iasi, Aleea Rozelor 38, Block A10, Sc A, Apt 9; tel (232) 258324; fax (232) 258326; e-mail office@chiven.ro; Contact: Podoleanu Dan.

'Cer Senin' la Baia Mare (Clear Sky in Baia Mare): 4800 Baia Mare, Avram Iancu 17\7; tel (62) 416362; fax (62) 417198; e-mail cer_senin@arhimedes.ro; internet www.arhimedes.ro/cer_senin; Contact: Ion M Mihai; f 1992; action against pollution and for the preservation, recovery and protection of the environment in Baia Mare; ecological education.

Clubul Montan Roman (Romanian Mountain Club): 5600 Piatra Neamt 3, POB 83; tel (94) 913941; fax (33) 233561; e-mail rom.mountainclub@decebal.ro; internet www.mountain.ro; Pres: Constantin Lacatusu; Vice-Pres: Irina Pacurariu, Dan Calinescu; Exec Dir: Florin Mihai; ecotourism and mountain camps in the Romanian Carpathian mountains; video documentaries; management of protected areas; affiliated with Lynx Alpine Club.

Clubul Naturalistilor ARCER (ARCER Nature Club): 4800 Baia Mare, I L Caragiale 9/33; tel (62) 423216; fax (62) 415769; Contact: Ubelhart Walter.

Earthkind Romania: 76201 Bucharest, Splaiul Independentei 91–95; tel and fax (1) 3122310; e-mail anvadi@bio.bio.unibuc.ro; Contact: Angheluta Vadineanu.

ECO Black Sea: Constanţa, Bd Mamaia 294; tel (41) 664392; environmental protection of the Black Sea.

Eco-Civica: 73337 Bucharest, B-dul Ferdinand 141; tel and fax (1) 6426478; Pres: Niculae Rădulescu-Dobrogea; f 1993; environmental inspections and provision of information.

Eco Club Independent: 4150 Odorheiu Secuiesc, Piaţa Libertăţii 22; tel and fax (66) 212783; e-mail office@ecihr.sbnet.ro; Contact: Szamo Iosif.

Eco Eden: Bucharest, W A Mozart 29; tel (1) 6799411; Contact: Curt Ghizela.

Eco Rural: 79656 Bucharest, Şos Oltenitei 35–37; tel (1) 6347070; fax (1) 3121154; Contact: Codreanu Mateiu.

ECODELTA: Tulcea, Piaţa Republicii 2; tel (405) 514660.

Ecology Society: Sighetu Marmatiei, Str T Vladimirescu 55; tel (62) 512660.

Federaţia Ecologistă din România (Romanian Ecological Federation—REF): 73226 Bucharest, Str Matei Voievod 102; tel (1) 6352743; fax (1) 6104858; Leader: Gugui Edward; federation of environmental political parties; part of the Democratic Convention of Romania—DCR alliance.

Fundatia Ecologica 'Floare de Colt' (Edelweiss Ecological Foundation): Bucharest, Aleea Fizicienilor 11; Contact: Stoinea Petre.

Fundatia Ecologica Romana (Romanian Ecological Foundation): 72442 Bucharest, Şos Colentina 56, Block 100, Apt 69; tel and fax (1) 6553355; Contact: Puiu Ilie Izvoranu.

Fundatia Eurofocus (Eurofocus Foundation): Bucharest, Rocadei 9; tel (1) 6333430; Contact: Vasile Costea.

Fundatia Europeana de Educatie şi Cultura Ecologica (European Foundation for Ecological Education and Culture): Bucharest, Şos Kiseleff 31; tel (1) 2225916; Contact: Toma George Maiorescu.

Fundatia pentru Eco Dezvoltarea Societatii (Eco-Development Foundation): 74538 Bucharest, Papiu Ilarian 4; Contact: Carmela Floroiu.

Fundatia Prietenii Deltei Dunarii (Friends of the Danube Delta Foundation): 8829 Sulina, Str Intai 202; tel and fax (40) 543542; e-mail azls@tim.ro; Pres: Nicolae Raducu; Vice-Pres: Dumitru Fatulescu; Councillor: Mihai Andrei; f 1992; environmental projects; social integration; co-ordination of regional and transborder cultural and scientific projects; affiliated with Earth Action, Inforce and Clean Up the World.

Institutul de Cercetări pentru Ingineria Mediului, Academia de ştiinţe Agricole şi Silvice 'Gheorghe Ionescu-Şişeşti' (Environmental Engineering and Research Institute, 'Gheorghe Ionescu-Şişeşti' Academy of Agricultural and Forest Sciences): 71552 Bucharest, Splaiul Independenţei 294; tel (1) 6373035; fax (1) 2229139; Dir: Dr V Rojanski.

Institutul de Cercetări şi Amenajări Silvice—ICAS, Academia de Ştiinţe Agricole şi Silvice 'Gheorghe Ionescu-Şişeşti' (Forest Research and Management Plans Institute, 'Gheorghe Ionescu-Şişeşti' Academy of Agricultural and Forest Sciences): Bucharest 2, Şos Stefăneşti 128; tel and fax (1) 6556845; Gen Man: Dr M Ianculescu; f 1933; silviculture; genetics; ecology; game management; trout farming; protection against soil erosion.

Institutul de Igienă şi Sănătate Publică Bucureşti (Institute of Hygiene and Public Health): 76256 Bucharest, Str Dr Leonte 1–3; tel (1) 6383970; fax (1) 3123426; Dir: Prof Manole Cucu; f 1927.

Institutul Naţional de Meteorologie şi Hidrologie, Academia de Ştiinţe Agricole şi Silvice 'Gheorghe Ionescu-Şişeşti' (National Institute of Meteorology and Hydrology, 'Gheorghe Ionescu-Şişeşti' Academy of Agricultural and Forest Sciences): 71552 Bucharest, Şos Bucureşti-Ploieşti 97; tel (1) 3129842; fax (1) 2229139; Dir: M Ioana; f 1884; affiliated with the Romanian Water Authority of the Ministry of Waters, Forestry and Environmental Protection.

Institutul Român de Cercetări Marine, Academia de Ştiinţe Agricole şi Silvice 'Gheorghe Ionescu-Şişeşti' (Romanian Marine Research Institute, 'Gheorghe Ionescu-Şişeşti' Academy of Agricultural and Forest Sciences): 70259 Bucharest, Alexandru Philippide 11; tel (1) 6412943; fax (1) 6104858; e-mail rmri@alpha.rmri.ro; Dir: Dr Toma George Maiorescu; f 1970; affiliated to the Ministry of Waters, Forestry and Environmental Protection; physical and chemical oceanography; ecology; pollution research; ecological rehabilitation; research into natural resources; biochemistry; technological research.

Mişcarea Ecologistă din România (Romanian Ecological Movement): Bucharest; tel (1) 6412943; fax (1) 6104858; Chair: Toma George Maiorescu; f 1990; environmental political party; part of the Romanian Ecological Federation—REF.

Oamenii şi Mediul Înconjurator (People and Environment): 2000 Ploiesti, Soarelui 10; tel and fax (44) 116537; e-mail peag@csd.univ.ploiesti.ro; Contact: Adrian Georgescu.

Organizatia Ecologica 'Ecomond' Timiş (Ecomond Ecological Organization, Timiş): 1900 Timişoara, Giuseppe Verdi 21; tel (56) 153374; Contact: Socaciu Radu Stefan.

Organizatia Ecologica 'Eden' (Eden Ecological Organization): 5500 Bacau, Carpati 18; tel (34) 172428; Contact: Nanu Eugenia Catalina.

Organizatia Neguvernamentala de Mediu 'Eco Alpex 024' (Eco Alpex 024 Non-Governmental Environmental Organization): 6100 Brăila, Rahova 259; tel (39) 638643; fax (39) 627434; e-mail ecoalpex@braila.net; internet members .xoom.com/ecoalpex; Pres: Cornel Apostol; Project Man: Daniela Platon; Scientific Consultant: Gabriel Ioan; environmental education; promotion of local and regional partnerships for environmental projects.

Organizatia pentru Protectia Mediului 'Rhododendron' (Rhododendron Environmental Protection Organization): 4300 Târgu Mureş, Victor Babes 11\221; tel (65) 216446; fax (65) 210545; e-mail office@rhodo.sbnet.ro; Contact: Sido Istvan.

Organizatia Regionala pentru Sanatatea Mediului (Regional Organization for Environmental Health): 3400 Cluj-Napoca, Brincusi 18/A; tel and fax (64) 414250; e-mail hero@mail.soroscj.ro; internet www.soros.ro/cctc/others/hero/ index.html; Contact: Ildiko Mocsy.

Partidul Ecologist Român (Romanian Ecological Party): Bucharest; tel (1) 6158285; Chair: Otto Weber; environmental political party; part of the Democratic Convention of Romania—DCR alliance.

Prietenii Pamantului (Earth Friends): 6200 Galaţi, Siderurgistilor SD4 A/12; tel and fax (36) 462564; e-mail zamfir@ sisnet.ro; Co-ordinator: Ion Constantin Zamfir; Exec Dir: Camelia Zamfir; environmental campaigning, consultancy and education; training for the public and members of other organizations.

Progresul Silvic (Silvan Progress): 3400 Cluj-Napoca, Padurii 22; tel (64) 128762; fax (64) 132373; Contact: Ciuca Dorin.

Romanian National Society of Soil Science: 71331 Bucharest, B-dul Marasesti 61; tel (1) 2227620; fax (1) 2225979; Contact: Lacatusu Radu.

Romanian Plant Protection Society: 71592 Bucharest, B-dul Ion Ionescu de la Brad 8; tel (1) 6335850; fax (1) 6335361; Contact: Valerian Severin.

Romanian Solar Energy Society: 77206 Bucharest, Splaiul Independenţei 313; tel (1) 4104585; e-mail anesr@nare .renerg.pub.ro; internet www.renerg.pub.ro; Contact: Ileana Creanga; non-profit organization, which acts under the aegis of the Romanian Academy; facilitates the construction of an independent infrastructure for the use of renewable energy sources (RES); promotes solar energy (solar thermal and photovoltaic), wind energy, biomass energy, small-scale hydro energy and geothermal energy; encourages international renewable energy technologies (RET), technology transfer, products manufactured by companies specialized in RET, RES applications software, educational programmes on RES utilization and environmental protection; provides consulting and engineering services for RES projects, market studies on utilization of RES in Romania, feasibility studies for solar and other renewable energy projects, data bases and testing of specialized equipment; co-operates with universities, research and design institutes, Romanian and foreign companies; organizes scientific and technical workshops, seminars, conferences and exhibitions.

Romanian Wildlife Society: 73232 Bucharest, Intr Plut Luicu Vasile 2; tel and fax (1) 6351761; Contact: Horia Almasan.

Societatea Carpatina Ardeleana (Transylvanian Carpathian Society): 1/41 Cluj-Napoca, Iuliu Maniu 7/5; e-mail gyopar@ekecj.sbnet.ro; Contact: Magdalena Behabetz.

Societatea de Protectie a Mediului Geologic (Geological Heritage Protection Society): 70111 Bucharest, Nicolae Balcescu 1; tel (1) 6143508; fax (1) 2118120; Contact: Dan Grigorescu.

Societatea Ecologista 'Sighetu Marmatiei' (Sighetu Marmatiei Ecological Society): 4925 Sighetu Marmatiei, Tudor Vladimirescu 55; tel (62) 512660; fax (62) 514067; Contact: Pop Alexandru Zoltan.

Societatea pentru Optimizarea Consumurilor Energetice din România—SOCER (Romanian Association for Optimization of Energy Consumption): 1100 Craiova, Ion Maiorescu 10; tel (51) 194118; fax (51) 417015; e-mail socer@ cisnet.ro; internet www.cisnet.ro/socer; Pres: Alexandru Marinescu; Exec Dir: Daniela Duta; Programmes Man: Ovidiu Marinescu; gathers and disseminates information concerning the efficient use of energy sources; supports implementation of energy efficiency projects; promotes renewable energy sources; assists local authorities in formulating regional energy strategy and policy; publications.

Societatea 'Salvati Terra' (Save the Earth Society): 5500 Bacau, Cornisa Bistritei, Block 6, Sc A, Apt 7; tel (34) 131151; fax (34) 171056; Contact: Barabas Octavian.

Staţiunea Centrală de Cercetări pentru Combaterea Eroziunii Solului—Perieni, Academia de Ştiinţe Agricole şi Silvice 'Gheorghe Ionescu-Şişeşti' (Central Research Station for Soil Erosion Control—Perieni, 'Gheorghe Ionescu-Şişeşti' Academy of Agricultural and Forestry Sciences): Barlad, jud Vaslui; tel (35) 413770; fax (35) 412837; e-mail perieni@axel.ro; internet www.spectral .ro/perieni; Dir: Dumitru Nistor; Scientific Sec: Ion Ionita; Technical Dir: Valentin Raseanu; Accounting Man: Ion Lita; f 1954; study of land degradation processes in agricultural land in small watersheds; improvement of land management methods; study of run-off and erosion of agricultural land in small watersheds; study of land degradation by gullying and landslides; assessment of reservoir sedimentation in small- to medium-sized basins; improvement of land planning and land treatment methods; conservation practices and tillage.

Tineretul Ecologist din România—TER (Ecologist Youth of Romania): 70108 Bucharest, Academiei 27, Suite 2, Apt 5, Sector 1; tel (1) 3126639; fax (1) 3124263; e-mail ter@ ter.sbnet.ro; Pres: Bogdan Paranici; Exec Dir: Eduard Petrescu; f 1991; institutional development of environmental NGOs; encouragement of NGO and public participation in environmental decision-making; nature conservation; organic agriculture programme; education, training and public information.

Tinerii Prieteni ai Naturii (Romanian Nature Friends): 1900 Timişoara, Calea Sagului 55, Et 8, Apt 35; tel and fax (56) 165408; Contact: Albu Tiberiu Calin.

RUSSIAN FEDERATION

Governmental Organizations

Ministry of Natural Resources: 123812 Moscow, ul Bolshaya Gruzinskaya 4/6; tel (095) 254-48-00; fax (095) 254-43-10; e-mail admin@mnr.gov.ru; internet www.mnr.gov.ru; Minister: Yurii P Trutnev; assumed the functions of the abolished State Committee for Environmental Protection in May 2000; the Ministry was instructed by President Vladimir V Putin to introduce a proposal for creating an independent ecological impact commission in July 2000.

Ministry of Agriculture: 107139 Moscow, Orlikov per 1/11; tel (095) 207-83-86; fax (095) 207-95-80; e-mail info@mcx.ru; internet www.mcx.ru; Minister: Aleksei V Gordeyev.

Ministry of Civil Defence, Emergencies and Clean-up Operations: 104012 Moscow, Teatralnyi proyezd 3; tel (095) 926-39-01; fax (095) 923-57-45; e-mail info@mchs.gov.ru; Minister: Col-Gen Sergei K Shoigu.

Ministry of Health and Social Development: 101431 Moscow, ul Nelinnaya 25; tel (095) 927-28-48; fax (095) 928-58-15; e-mail press-center@minzdrav-rf.ru; internet www.minzdrav-rf.ru; Minister: Mikhail Yu Zurabov.

Ministry of Industry and Energy: 125889 Moscow, pl Miusskaya 3, POB 47; tel (095) 972-70-51; fax (095) 229-55-49; e-mail pr@mte.gov.ru; internet www.mte.gov.ru; Minister: Viktor B Khristenko.

State Committee for Social Protection of Citizens and the Rehabilitation of Territories Affected by Chornobyl and Other Radiation Accidents: 103132 Moscow, Staraya pl 8/2, pod 3; tel (095) 206-48-81; Chair: Vasilii Y Voznyak.

Arctic and Antarctic Research Unit: 199397 St Petersburg, ul Beringa 38; tel (812) 352-00-96; fax (812) 352-26-88; Dir: I Ye Frolov; f 1920; research into Arctic and Antarctic ecology; responsible for the Russian Antarctic Expedition.

Federal Nuclear Radiation and Safety Authority: 109147 Moscow, Taganskaya ul 34; Chair: Yurii Vishnevskii; fmrly the State Committee for the Supervision of Nuclear and Radiation Safety.

Institute for Water and Environmental Problems (IWEP), Siberian Branch of the Russian Academy of Sciences: 1 Molodyozhnaya St., Barnaul 656038; tel (3852) 66-64-60; fax (3852) 24-03-96; e-mail iwep@iwep.ab.ru; internet www.iwep.ab.ru; Dir: Prof Yurii I Vinokurov; Deputy Dir: Dr Alexander V Puzanov; f 1987; research into use of water resources, land reclamation and environmental protection in Siberia; experimental and mathematical methods for analysis of hydrophysical, hydrochemical and other natural processes in the aquatic environment; environmental assessment of large-scale engineering projects; development of information and modelling systems for specific research projects and management resources; decision support systems; library of 30,000 vols; affiliated to the Novosibirsk Department of the Institute for Water and Environmental Problems.

Interdepartmental Commission for Ecological Security, Security Council: Moscow; Chair: Aleksei Yablokov; f 1993.

Losiny Ostrov National Park: 107113 Moscow, Poperechny pr 1[a]; tel (095) 268-60-45; fax (095) 268-46-26; e-mail nplosing@asvt.ru; Gen Dir: Aleksandr Jangutov; Chief Forester: Vladimir Seleznev; Heads, Dept of Science: Galina Mihalchenko, Mark Shapochkin; f 1983; affiliated to the Federal Forestry Service; environmental monitoring and protection; information exchange; education; affiliated with Moscow municipal government and the Federation of Nature and National Parks of Europe—FNNPE.

Perm Clinical Research Institute on Ecological Pathology in Children: 614601 Perm, Ordjonikidze 82; tel (3422) 33-50-25; fax (3422) 33-50-25; Prin Officer: Prof Dr Zaitseva Nina Vladimirovna; f 1995; monitoring of environmental pollution and its effects on children's health; clinical diagnostics; affiliated with the Perm Regional Health Care Administration Board.

Scientific Research Centre for the Ecological Safety of the Population: 614600 Perm, Pushkin 3/81; tel and fax (3422) 34-86-76; Prin Officer: Prof Dr Zaitseva Nina Vladimirovna; f 1992; development of methods for the evaluation of changes in environmental health conditions; formulation of criteria for the assessment of sustainable development levels; monitoring of environmental pollution.

Non-Governmental Organizations

A N Sysin Institute of Human Ecology and Environmental Health, Russian Academy of Medical Sciences: 119833 Moscow, Pogodinskaya 10; tel (095) 247-04-28; fax (095) 246-04-28; Dir: Yu A Rakhmanin; Deputy Dir: N V Rusakov; f 1931; assessment of the effects on human health of exposure to environmental pollution; monitoring and analysis of water, air and soil pollution; assessment of the environmental health risks of residential and public buildings; development of uses and methods of treatment for industrial waste.

ASEKO Association for Environmental Education: 249020 Kaluzhskaya obl, Obninsk 9, POB 9081; fax (095) 497-88-42; e-mail web@online.ru; internet www.ecoline.ru/aseko; Co-ordinator: Vadim Kalinin; environmental education in the CIS; initiation of environmental education at all levels; support for teachers; development of a network for information exchange; improvement of teaching skills; projects and workshops; publications.

Association of Ecological Centres: 107078 Moscow, ul Novobasmannaya 10.

Chita Institute of Natural Resources, Russian Academy of Sciences: 672014 Chita, ul Nedorezova 16, POB 147; tel and fax (302) 221-25-82; e-mail root@cinr.chita.ru; internet www.chita.ru/public_htm/cinr/cinr.htm; Dir: A B Ptitsyn; f 1981; scientific research into the ecosystems of the Siberian region.

Ecological Centre Dront: 603163 Nizhnii Novgorod, POB 34; tel (8312) 30-28-81; fax (8312) 39-11-91; e-mail dront@glas.apc.org; Exec Dir: Angela I Bakka; Treas: Askhat A Kayumov; f 1989; nature conservation.

Ekologicheskaya Partiya Rossii 'Kedr' (Kedr Ecological Party of Russia): Moscow; Leader: Anatolii Panfilov; environmental political party.

Greenpeace Russia: 101428 Moscow, Novaya Bashilovka 6, GSP 4; tel (095) 257-41-16; fax (095) 257-41-10; e-mail greenpeace.russia@diala.greenpeace.org; internet www.greenpeace.ru; Exec Dir: Sergei Tsiplenkov; Campaign Co-ordinator: Ivan Blokov; f 1992; national office of Greenpeace International; activities include campaigns on the protection of Lake Baikal and the Russian Federation's forests; projects on climate change, marine bio-resources and world natural heritage in the Russian Federation.

Institut Evolyutsionnoi Morfologii i Ekologii Zhivotnykh imeni A N Severtsova, Rossiiskaya Akademiya Nauk (A N Severtsov Institute of Evolutionary Morphology and Animal Ecology, Russian Academy of Sciences): Section of Chemical, Technological and Biological Sciences, 117071 Moscow, Leninskii pr 33; tel (095) 954-64-76; fax (095) 954-55-34; e-mail sevin@glas.apc.org; Dir: D S Pavlov; f 1936; research into ecology, morphology and ethology of animals, animal evolution and problems of biodiversity and nature conservation.

Institute of Atmospheric Physics, Russian Academy of Sciences: 109017 Moscow, per Pyzhevskii 3; tel (095) 951-55-65; fax (095) 953-16-52; e-mail and@omega.ifaran.ru; Dir: Prof G S Golitsyn; Deputy Dirs: Dr I P Malkov, Dr Y A Volkov; f 1956; monitoring of urban air pollution; modelling of the increase of air pollution in the atmosphere; modelling and forecasting of global change in climate and ecosystems.

Institute of Biological Problems of the North, Russian Academy of Sciences: 685000 Magadan, ul K Marksa 24; tel (41322) 2-47-30; fax (41322) 2-01-66; e-mail ibpn@ibpn.magadan.su; Dir: Prof F B Chernyavskii; f 1972.

Institute of Global Climate and Ecology—IGCE, Russian Academy of Sciences: 107258 Moscow, Glebovskaya 20b; tel (095) 169-24-30; fax (095) 160-08-31; e-mail yu.izrael@g23.relcom.ru; Dir: Yurii Izrael; f 1990; research on climate change and ecosystem responses to climate change, anthropogenic influences on the state of the environment, anthropogenic ecology of oceans, critical levels/loads of anthropogenic impact on the environment and climate, monitoring and assessment of anthropogenic changes in the environment and climate.

Institute of Industrial Ecology, Russian Academy of Sciences: 620219 Yekaterinburg, ul Sophy Kovalevskoi 20[a]; tel and fax (3432) 74-37-71; e-mail chukanov@ecko.uran.ru; Dir: Prof V N Chukanov; f 1992; environmental research; research into health, socio-economics, demographic consequences of environmental contamination, risk assessment and radioecology.

Institute of Limnology, Russian Academy of Sciences: 664033 Irkutsk, ul Ulan-Batorskaya 3; tel (3952) 46-05-04; fax (3952) 46-04-05; e-mail info@lin.irk.ru; internet www.lin.irk.ru; Dir: M A Grachev; f 1961; research into lake ecology; particularly concerned with the conservation of Lake Baikal.

Institute of Plant and Animal Ecology, Russian Academy of Sciences: 620008 Yekaterinburg, ul 8-go Marta 202; tel (3432) 22-85-70; fax (3432) 29-41-61; e-mail common@ipae.uran.ru; internet www.ipae.uran.ru; Dir: V N Bolshakov; f 1944; formation and development of ecological systems; theoretical bases of monitoring and conservation of biological and natural diversity; radioecology; natural resource management; ecological prediction and environmental impact assessment; rehabilitation of biological resources in degraded areas; affiliated with Labytnangi Research Station and the Biophysical Station in Zarechny.

Institute of Soil Science and Photosynthesis, Russian Academy of Sciences: 142292 Moskovskaya obl, Serpukhovskii raion, Pushchino; tel (095) 923-35-58; fax (0967) 79-05-32; Dir: V I Kefelya; research areas include soil conservation and land reclamation.

Institute of State and Law, Sector on Environmental Law, Russian Academy of Sciences: Dept of Philosophy and Law, Section of Social Sciences, 119841 Moscow, ul Znamenka 10; tel (095) 291-33-81; Dir: Boris N Topornin; f 1925; research into Soviet and Russian environmental law.

Institute of the Biology of Inland Waters, Russian Academy of Sciences: Dept of General Biology, Section of Chemical, Technological and Biological Sciences, 152742 Yaroslavl obl, Nekuzskii raion, P/O Borok; tel and fax (852) 25-38-45; e-mail mail@ibiw.yaroslavl.ru; internet www.ibiw.yaroslavl.ru; Dir: Dr S I Genkal; Deputy Dirs: Dr A I Kopylov, Dr D F Pavlov; incorporates the Commission on the Conservation of Natural Waters; studies the structure and function of freshwater ecosystems.

Institute of the Ecology of the Volga River Basin, Russian Academy of Sciences: Dept of General Biology, Section of Chemical, Technological and Biological Sciences, 445003 Toglyatti, ul Komzina 10; tel (8469) 23-54-78; fax (8469) 48-95-04; Dir: G S Rozenburg; f 1983; monitoring of the environment of the lower Volga River.

Institute of Water and Ecological Problems, Far Eastern Division of the Russian Academy of Sciences: 680063 Khabarovsk, ul Kim Yu Chena 65; tel (4212) 22-75-73; fax (4212) 22-70-85; e-mail dmitry@ivep.khv.ru; Dir: Boris A Voronov; Deputy Dir, Science: Aleksei N Makhinov; f 1968; research into characteristics of surface and ground-water formation; monitoring of water resources; formulation of basic principles for the sustainable use and management of water resources; study of the ecological impact of anthropogenic stress on terrestrial and aquatic ecosystems of the Far East.

Institute of Water Problems, Russian Academy of Sciences: Dept of Oceanology, Atmospheric Physics and Geography, Section of Earth Sciences, 107078 Moscow, ul Novo Basmannaya 10, POB 231; tel (095) 265-97-57; fax (095) 265-18-87; e-mail iwapr@iwapr.msk.su; Dir: M G Khublaryan; f 1968; complex evaluation of water resources; development of scientific substantiation for their rational use and protection.

International Foundation for the Survival and Development of Humanity—IF: 121002 Moscow, ul Vesnina 9/5; tel (095) 241-82-43; fax (095) 230-26-08; Exec Dir: Rustem Khairov; f 1988; identification of opportunities for global change and promotion of solutions to global problems; active in the fields of the environment, international economy, education and culture.

Krasnoyarsk Ecological Movement: 121596 Moscow, ul Tobolchina 4/2/21; tel (095) 316-75-43.

Krasnoyarskii Zelenii Svet (Krasnoyarsk Green World): Krasnoyarsk; local environmental group; campaigns against nuclear power and weapons.

Laboratory for Radiation Control: c/o Kemerovo State University, 650070 Kemerovo, ul Tukhachevskaya 33, g 113; tel and fax (3842) 31-14-98; e-mail nl@irk.da.su; Head, Lab: Nadezhda Aluker; f 1994; control of radiation pollution; environmental monitoring.

Laboratory for the Monitoring of the Environment and Climate, Russian Academy of Sciences: c/o Dept of Oceanology, Atmospheric Physics and Geography, Section of Earth Sciences, 117901 Moscow, Leninskii pr 14; tel (095) 234-14-24; Dir: (vacant).

Moscow Ecological Federation: 121019 Moscow, POB 211; tel and fax (095) 298-30-87; e-mail lun@glas.apc.org; Co-Chair: Lyubov Rubinchik, Nikolai Shalimov; concerned with Moscow's ecological problems; dissemination of information; assistance with urban development plans.

Moscow Society of Naturalists: 103009 Moscow, ul Bolshaya Nikitskaya 6; tel (095) 203-67-04; Chair: V A Sadovnichii; f 1805; library of 500,000 vols.

Movement for a Nuclear-Free North: Murmansk; advocates demilitarization of the Kola peninsula and an end to nuclear testing on Novaya Zemlya.

Pacific Oceanological Institute, Russian Academy of Sciences: 690041 Vladivostok, ul Baltiiskaya 43; tel (4232) 31-14-00; fax (4232) 31-25-73; Dir: Prof V A Akulichev.

Research Institute for Building Physics—NIISF: 127238 Moscow, Locomotivni pr 21; tel (095) 482-40-76; fax (095) 482-40-60; e-mail niisf@ipc.ru; Dir: George L Osipov; Deputy Dir: I L Schubin; f 1956; affiliated to the Russian Academy of Architecture and Building Sciences; heat engineering for buildings; energy conservation; climatology; research into thermal transmission, acoustics, noise control, and natural and artificial lighting; publications.

Rostovskii Oblastnoi Ekologicheskii Tsentr (Rostov Regional Ecological Centre): 344007 Rostov, ul Stanislavskogo 114, kv 1; tel (863) 32-33-70; Pres: Dr Valerii Privalenko; f 1988; urban ecology and geochemistry; research into atmospheric, soil, surface- and ground-water pollution; environmental rehabilitation programmes.

Russian Association for the Indigenous Peoples of the North—RAIPON: Moscow 117415, Vernandskogo pr 37, Suite 527, Korp 2; tel (7095) 930-44-68; fax (7095) 930-71-97; e-mail udege@glasnet.ru.

Russian Green Party: Moscow; Chair: Dmitri Likhachev.

Partiya Zelyenykh Prikamya (Green Party of Prikamya): 614006 Perm, Lenin 51, Rm 816; tel (3422) 12-57-69; fax (3422) 12-69-33; e-mail pn@plccl.ru; internet www.perm.ru/pn; Chair: Ivan Yezhikov; Deputy Chair: G Vereyvkin, V Sretenskii; f 1991; environmental political party.

Socio-Ecological Union—SEU: 125319 Moscow, ul Krasnoarmeyskaya 25, kv 85; e-mail soceco@glas.apc.org; internet www.cci.glasnet.ru; f 1987; co-ordinates 250 environmental committees, clubs and societies in all countries of the CIS, as well as Estonia, Norway and the USA; international co-operation; campaigns on issues of the environment, human rights, biodiversity protection, energy efficiency, nuclear energy and radioactive pollution; environmental education.

Socio-Ecological Union—SEU, Perm Department: 614081 Perm, POB 5786; tel (3422) 33-48-58; internet art.perm.ru; Chair: Yuly Schipakin; Vice-Chair: Aleksandr Nikonov; f 1987; local office of the SEU; formulation of sustainable development policy; pollution monitoring; ecological simulations; litigation; information; debates; scientific research on biodiversity and protected territories; affiliated with the UN Commission on Sustainable Development.

Tambov Green Party: 392032 Tambov, bul Entuziastov 32, kv 47; tel (752) 35-01-33; Chair: Ludmila Spiridonova; environmental political party.

TRAFFIC Europe, Russia: c/o WWF Russia Programme Office, 125319 Moscow, POB 55; tel (095) 264-99-48; fax (095) 264-99-27; e-mail vaisman@deol.ru; internet www.deol.ru/nature/protect; Head: Aleksei L Vaisman; Programme Admin Asst: Natalia A Dronova; national office of TRAFFIC International.

Vserossiiskii Nauchno-issledovatel'skii Institut Elektrifikatsii Sel'skogo Khoztstva (All-Russian Research Institute for the Electrification of Agriculture): 109456 Moscow, 1-st Veschnjakovskii pr. 2; tel (095) 1711920; fax (095) 1705101; e-mail viesh@dot.ru; Dir: Prof Dmitrii Strebkov; Deputy Dir: Anatolii Tichomikov; Information and Library: Dr Nikolai Mologno; f 1930; affiliated to the Russian Academy of Agricultural Sciences (Mechanization and Electrification Section); renewable energy for rural development; solar cell and solar module production line; agricultural waste utilization; development of electric transport; renewable energy for the power grid; equipment for dairy farms and milk production; solar and wind atlas of the Russian Federation.

Vserossiiskoye Obshchestvo Okhrany Prirody (All-Russian Society for Nature Conservation): 103012 Moscow L-12, Kuiybyshevskii pr 3; tel (095) 924-77-65; Chair: I F Bazishpol; f 1924; civilian association focusing on environmental conservation and education.

Vsesoyuznyi Nauchno-issledovatelskii Institut Okhrany Prirody i Zapovednovo Dela, Rossiiskaya Akademiya Nauk (All-Russia Research Institute for Nature Conservation, Russian Academy of Sciences): 113628 Moscow, Znamenskoye-Sadki, VNII Priroda; tel (095) 423-03-22; fax (095) 423-23-22; Dir: Prof V A Krasilov; f 1981; research, general methodology, environmental protection strategy, and domestic and international co-ordination; comprises five departments; major repository of research material.

Zelenyi Mir (Green World Environmental Association): 603047 Nizhnii Novgorod, ul Krasniye Zory 15, kv 409; tel (831) 224-39-41; fax (831) 244-02-85; Gen Co-ordinator: Valentina Malakhova; nature conservation; environmental education; conferences.

RWANDA

Governmental Organizations

Ministry of Lands, Resettlements and Environment: Kigali; Minister: Drocella Mugorewera; Minister of State for Lands and the Environment: Patricia Hajabakiga.

Ministry of Agriculture and Livestock: BP 621, Kigali; tel 85008; fax 85057; Minister: Dr Patrick Habamenshi; incorporates the Department of Rural Engineering and Soil Conservation.

Institut des Sciences Agronomiques du Rwanda—ISAR (Rwanda Institute of Agronomic Sciences): BP 138, Butare; Dir-Gen: Prof Bikoro Munyanganizi; f 1962; library of 2,500 vols; publications.

Ministry of Energy, Water and Natural Resources: Kigali; internet www.minerena.gov.rw; Minister of State of Water and Natural Resources: Bikoro Munyanganiai.

Ministry of Health: BP 84, Kigali; tel 573481; e-mail sgsante@rwandatel.com; Minister: Dr Jean-Demascène Ntawukuriryayo; incorporates the Department of Public Hygiene and Environment.

Office Rwandais du Tourisme et des Parcs Nationaux—ORTPN (Rwandan Office of Tourism and National Parks): BP 905, Kigali; tel 76515; fax 76512; Dir: Juvénal Uwilingiyimana; f 1973; incorporates the Department of National Parks; management and conservation of nature and national parks.

Non-Governmental Organizations

Parti Ecologiste—Peco (Ecologist Party): Kigali; f 1992; environmental political party.

SAINT CHRISTOPHER AND NEVIS

Governmental Organizations

Ministry of Finance, Technology and Sustainable Development, Tourism, Sports and Culture: Church St, POB 186, Basseterre; tel 465-2521; fax 465-1001; e-mail sknpmoffice@caribsurf.com; internet www.fsd.gov.kn; Minister: Dr Denzil Llewellyn Douglas.

Ministry of Health and the Environment: Church St, POB 186, Basseterre; tel 465-2521; fax 465-1316; e-mail minhwa@caribsurf.com; Minister: Dr Earl Asim Martin.

Department of Environment: Pelican Mall, Basseterre; tel 465-4040; fax 466-3915; e-mail sknmtcce@caribsurf.com; Dir: Raymond Solomon.

Ministry of Housing, Agriculture, Fisheries and Consumer Affairs: Church St, POB 186, Basseterre; tel 465-2521; fax 465-2635; e-mail minafclh@caribsurf.com; Minister: Cedric Roy Liburd.

Department of Agriculture and Fisheries: La Guerite, Basseterre; tel 465-2335; fax 465-5202; e-mail doastk@caribsurf.com; Dir: Dr Jerome Thomas.

Multi-Purpose Laboratory and Bureau of Standards: La Guerite, Basseterre; tel 465-5279; fax 465-3852.

Physical Planning Unit: Ministry of Finance, Govt Headquarters, Basseterre; tel 465-2521; fax 466-7398; e-mail planningstk@caribsurf.com; Dir: Oliver Knight.

Public Health Department: Cornell St, Basseterre; tel 465-2521; fax 466-2296; Chief Environmental Health Officer: Oliver Lawrence.

Non-Governmental Organizations

Admiral's Enterprises Ltd: Old Rd Bay, St Kitts; tel 465-3677; Man Dir: Harrington Bristol.

National Conservation Commission: Pelican Mall, Basseterre; tel 465-4040; fax 466-3915; e-mail raycsolomon@hotmail.com; Chair: Raymond Solomon.

Nevis Environmental Education Committee: POB 563, Charlestown, Nevis; tel 469-5786; fax 469-0274; Vice-Chair: Miriam Knorr; Sec and Treas: joan Robinson.

Nevis Historical and Conservation Society: POB 563, Charlestown, Nevis; tel 469-5786; fax 469-0274; e-mail nhcs@caribsurf.com; internet www.nevis-nhcs.org; Pres: John Gilbert; f 1980; environmental conservation; preservation of historic monuments; operation of the Museum of Nevis History and the Horatio Nelson Museum; library; archives; environmental education; affiliated with the Museums Association of the Caribbean and with the Caribbean Conservation Association.

Saint Christopher Heritage Society: POB 888, Basseterre; tel and fax 465-5584; e-mail schs@caribsurf.com; internet www.islandimage.com/schs/home.htm; Exec Dir: Jacqueline Armony; f 1989; protection and promotion of the natural, historic and cultural heritage.

Soho Environment Committee: Soho Village, Basseterre; tel 465-6786; Sec: Sharon Drew.

Solid Waste Management Corporation: New St, Basseterre; tel 465-9507; fax 465-5483; e-mail scanswmc@caribsurf.com; Chair: Alphonso Bridgewater.

SAINT LUCIA

Governmental Organizations

Ministry of Physical Development, Environment and Housing: Greaham Louisy Administrative Bldg, Waterfront, Castries; tel 468-4402; fax 452-2506; e-mail sde@planning.gov.lc; Minister: Theophilus Ferguson John; formulation of economic policy; monitoring development; formulation of environmental policy.

Ministry of Agriculture, Forestry and Fisheries: NIS Bldg, 5th Floor, Waterfront, Castries; tel 468-4210; fax 453-6314; e-mail admin@candw.lc; internet www.slumaffe.org; Minister: Ignatius Jean.

Water and Sewerage Co Inc: L'Anse Rd, POB 1481, Castries; tel 452-5344; fax 452-6844; e-mail wasco@candw.lc; f 1999.

Non-Governmental Organizations

Folk Research Centre: POB 514, Castries; tel 452-2279; fax 451-9365; environmental and cultural activities.

National Research and Development Foundation of Saint Lucia: POB 3067, Castries; tel 452-4253; fax 453-6389; activities include co-operation with international environmental organizations.

Saint Lucia National Trust: POB 595, Castries; tel 452-5005; fax 453-2791; e-mail naturst@candw.lc; Exec Dir: Giles Romulous; f 1975; protection of the natural and cultural heritage.

Saint Lucia Naturalists Society: POB 783, Castries; tel 451-6957; fax 451-6958; nature conservation; environmental education.

SAINT VINCENT AND THE GRENADINES

Governmental Organizations

Ministry of Health and the Environment: Government Bldgs, Kingstown; tel 457-2586; fax 457-2684; Minister: Dr Douglas Slater.

Ministry of Agriculture, Land and Fisheries: Government Bldgs, Kingstown; tel 457-1380; fax 457-1688; Minister: Selmon Walters.

State Committee for Hydrometeorology: c/o Cabinet of Ministers, 252008 Kiev, vul M Hrushevskoho 22/2; Chair: Viacheslav Lipinskiy.

State Committee for Land Resources: c/o Cabinet of Ministers, 252008 Kiev, vul M Hrushevskoho 22/2; Chair: Leonid Novakovskiy.

State Committee for Water Economy: c/o Cabinet of Ministers, 252008 Kiev, vul M Hrushevskoho 22/2; Chair: Victor Khoryev.

G N Vysotsky Ukrainian Scientific Research Institute of Forestry and Forest Melioration: 310024 Kharkov, vul Pushkinska 86; tel (57) 243-15-49; fax (57) 243-25-20; e-mail root@u-fri.kharkov.ua; Dir: Dr Igor Patlaj; Deputy Dir: Dr Victor Tkach; f 1929; fundamental and applied research, and development projects in the areas of forestry and environmental conservation.

P S Pasternak Ukrainian Mountain Forestry Research Institute: 284000 Ivano-Frankovsk, vul Hrushevsky 31; tel and fax (3422) 2-52-16; e-mail lis@il.if.ua; Dir: Dr Vasil Parpan; Deputy Dirs: Dr Roman Brodovych, Dr Volodymyr Korzhov; Head, Selection Lab: Dr Roman Yatsyk; Head, Forest Protection Lab: Dr Yaroslav Sloboda; Head, Ecology Lab: Dr Yuriy Shparyk; f 1964; genetics and selection research; development of forestry technologies; improvement of techniques for multi-purpose forest utilization and sustainable use of forest resources; monitoring; forest and nature preservation; hunting.

Non-Governmental Organizations

Agrarian Party of Ukraine: Kiev; Chair: Kateryna Vashchuk; f 1996; political party; advocates revival of the Ukrainian countryside.

EcoPravo Kharkiv: 61202 Kharkiv, POB 2050; tel and fax (57) 219-10-21; e-mail eco@online.kharkiv.ua; internet www.ecopravo.kharkiv.ua; Dir: Prof Aleksei M Shumilo; f 1993; concerned with environmental law; provision of legal assistance to citizens and NGOs; development and promotion of environmental legislation in Ukraine; clinical education.

Greenpeace Ukraine: 252030 Kiev, vul Pirogova 5, Apt 6; tel (44) 244-38-34; fax (44) 244-38-47; e-mail greenpeace.ukraine@green2.greenpeace.org; Exec Dir: Olga Savran; Deputy Exec Dir: Andrey Pleskonos; national office of Greenpeace International.

IOI Affiliate Operational Centre, Ukraine: Institute of Biology of the Southern Seas, 335011 Sevastopil, pr Nakhimov 3/2; tel (692) 52-52-49; fax (692) 59-28-13; e-mail radalpin@ibss.iuf.net; Dir: Dr Victoria N Radchenko; national office of the International Ocean Institute—IOI, Malta.

Ukrainian Chornobyl Union: c/o 254655 Kiev, pl Lvivska 8; Pres: Yuriy Andreyev; f 1991; represents victims of the Chornobyl disaster; 420,000 mems.

Ukrainian Institute of Energy-Saving Problems: 254070 Kiev, ul Pokrovskaya 11; tel (44) 412-20-44; fax (44) 417-07-37.

Ukrainian State Steppe Reservation, Ukrainian Academy of Sciences: Donetsk obl, Khomutoho, Novoazov raion; Dir: A P Genov; affiliated to the Department of General Biology of the Academy; has some environmental responsibilities.

Zeleniy Svit (Friends of the Earth Ukraine/Green World—Ukrainian Environmental Association): 320051 Dnipropetrovsk, POB 49; tel (56) 778-13-01; fax (56) 727-34-25; e-mail zsfoe@zsfoe.melp.dp.ua; Exec Dir: Yuriy Scherbak; f 1988; affiliated to Friends of the Earth International—FoEI; federation of various Ukrainian ecological groups; monitoring of acid rain and the consequences of the Chornobyl disaster; involved in discussions on the future of the energy industry; planting of trees for river protection; promotion of recycling; protection of nature reserves; affiliated with Rukh (People's Movement of Ukraine) and the Ukrainian Peace Council's campaign against nuclear power.

Partiya Zelenykh Ukrainy (Green Party of Ukraine): 252024 Kiev, vul Luteranska 24; tel (44) 293-69-09; fax (44) 293-52-36; internet www.green.ukrpack.net; Pres: Vitaliy Kononov; f 1990; political wing of Zeleniy Svit; democratic nationalist party; party branches in 18 administrative regions of Ukraine.

UNITED ARAB EMIRATES

Governmental Organizations

Federal Environmental Agency: POB 5951, Abu Dhabi; tel (2) 6777363; fax (2) 6770501; e-mail uaefea@emirates.net.ae; Man: Tarek Mahmoud Ibrahim; serves as the INFOTERRA national focal point.

Ministry of Agriculture and Fisheries: POB 213, Abu Dhabi; tel (2) 6662781; fax (2) 6654787; e-mail maf@uae.gov.ae; internet www.uae.gov.ae/maf; Minister: Said Muhammad ar-Ragabani.

Ministry of Energy: POB 629, Abu Dhabi; tel (2) 6274222; fax (2) 6269738; e-mail moew@uae.gov.ae; internet www.uae.gov.ae/moew; Minister: Muhammad bin Dhaen al-Hamili.

Ministry of Health: POB 848, Abu Dhabi; tel (2) 6330000; fax (2)6726000; e-mail moh@uae.gov.ae; internet www.uae.gov.ae/moh; Minister: Hamad Abd ar-Rahman al-Madfa.

Non-Governmental Organizations

Abu Dhabi Environmental Group: POB 25566, Abu Dhabi.

Arabian Leopard Trust: POB 24444, Sharjah.

Commission of Environmental Research, Emirates Heritage Club—CEREHC: POB 42959, Abu Dhabi; tel (2) 5584440; fax (2) 5582224; e-mail cerehc@emirates.net.ae; internet www.cerehc.org.ae; Head: Abdulmonem Darwish; research into the country's ecosystems and development.

Friends of the Environment Society: POB 4940, Dubai.

UNITED KINGDOM

Governmental Organizations

Department for Environment, Food and Rural Affairs (DEFRA): Nobel House, 17 Smith Square, London SW1A 3JP; tel (20) 7238-6034; fax (20) 7276-8861; e-mail environment@defra.gsi.gov.uk; internet www.defra.gov.uk; Sec of State: Margaret Beckett; Ministers: Elliot Morley, Rt Hon Alun Michael; planning and land use; housing and construction; inner city areas; new towns; environmental protection; conservation areas and countryside affairs; health and safety at work; energy efficiency.

Global Wildlife Division: c/o DEFRA, Tollgate House, Rm 814, Houlton St, Bristol BS2 9DJ; tel (117) 987-8032; fax (117) 987-8373; e-mail cites.ukma@defra.gsi.gov.uk; internet www.defra.gov.uk/wildlife-countryside/wacd/index.htm; conservation of globally threatened species through the United Kingdom's membership of the Convention on International Trade in Endangered Species of Wild Flora and Fauna—CITES.

Environment Agency: Rio House, Waterside Drive, Aztec West, Almondsbury, Bristol BS12 4UO; tel (1454) 624400; fax (1454) 624409; e-mail enquiries@environment-agency.gov.uk; internet www.environment-agency.gov.uk; Chair: Sir John Harman; Vice-Chair: Sir Richard George; Chief Exec: Barbara Young; f 1996; non-departmental public body; incorporates the former National Rivers Authority—NRA, HM Inspectorate of Pollution and county council waste regulation authorities; protection and improvement of the air, land and water environment of England and Wales through education, prevention of pollution and enforcement of regulations; responsibilities include protection and management of natural water resources, water quality, flood warning and flood defence, inland fisheries and navigation, conservation, recreation, minimization of waste and regulation of industrial processes.

Department for Environment, Food and Rural Affairs (DEFRA): Lower Ground Floor, Ergon House, c/o Nobel House, 17 Smith Sq, London SW1P 3JR; tel (20) 7238-3000; fax (20) 7238-6609; e-mail helpline@defra.gsi.gov.uk; internet www.defra.gov.uk; Sec of State: Margaret Beckett; Minister of State for the Environment: Elliot Morley; Minister of State for Rural Affairs: Alun Michael.

Advisory Committee on Dangerous Substances—ACDS: Health and Safety Executive, Policy Group, Nuclear and Hazardous Installations Division, Rose Court, 5th Floor North Wing, 2 Southwark Bridge, London SE1 9HS; e-mail Colin.Potter@hse.gsi.gov.uk; internet www.hse.gov.uk/aboutus/hsc/iacs/acds; Sec: Colin Potter; advises the Health and Safety Commission on methods of ensuring safety by controlling risks to workers and others connected with the manufacture, import, storage, conveyance and use of dangerous substances.

Agricultural and Environmental Science Division: Newforge Lane, Belfast BT9 5PX; tel (28) 9025-0666; fax (28) 9025-3486; e-mail fraserm@dani.gov.uk; Prin Officers: Dr J Stevens, Prof F Gordon, Dr I Heaney; f 1994; semi-governmental organization affiliated to both the Department of Agriculture and Rural Development for Northern Ireland and Queen's University of Belfast; research and education relevant to agriculture, fisheries and agri-environment.

British Library Environmental Information Service—EIS: 96 Euston Rd, London NW1 2DB; tel (20) 7412-7955; fax (20) 7412-7954; e-mail eis@bl.uk; internet www.bl.uk/collections/environment/environment.html; Head of Section: Dr Paula S Owen; f 1989; provides enquiry, consultancy and research services in all areas of environmental information; publications and training courses on environmental information.

Countryside Agency: John Dower House, Crescent Pl, Cheltenham GL50 3RA; tel (1242) 521381; fax (1242) 584270; e-mail info@countryside.gov.uk; internet www.countryside.gov.uk; Chair: Pam Warhurst; Chief Exec: Richard Wakeford; f 1999; affiliated to DEFRA (Department for Environment, Food and Rural Affairs); adviser to government on countryside issues; works to conserve English countryside; promotes social equity and economic opportunity for the rural population; designates national parks, areas of outstanding natural beauty and 'heritage coasts'.

Department for Environment, Planning and Countryside, National Assembly for Wales: Cathays Park Cardiff CF10 3NQ; tel (29) 2082-5111; fax (29) 2082-5180; internet www.wales.gov.uk/cabinet/members; Minister: Carwyn Jones; Dir: Gareth Jones; responsible for agriculture, environment, fisheries, food, forestry, planning and rural affairs in Wales.

Department of Agriculture and Rural Development for Northern Ireland: Dundonald House, Upper Newtownards Rd, Belfast BT4 3SF; tel (28) 9052-4999; fax (28) 9052-5546; e-mail dardhelpline@dardni.gov.uk; internet www.dardni.gov.uk; Minister: Ian Pearson.

Department of the Environment for Northern Ireland: Clarence Court, 10–18 Adelaide St, Belfast BT2 8GB; tel (28) 9054-0540; fax (28) 9054-0029; e-mail press.office@doeni.gov.uk; internet www.doeni.gov.uk; Minister: Angela Smith; Permanent Sec: Stephen Peover.

English Heritage: PO Box 569, Swindon, Wilts SN2 2YP; tel 0870 333 1181; fax (1793) 414926; internet www.english-heritage.org.uk; Chair: Sir Neil Cossons; Chief Exec: Dr Simon Thurley; f 1984; statutory body which awards grants to thousands of listed cathedrals, churches, other buildings, archaeological sites, historic parks, gardens and ancient monuments across the country; promotes regeneration of city centres, towns and villages; maintains nine regional offices and the Centre of Archaeological Excellence at Fort Cumberland in Portsmouth; conducts survey work and archive services; manages over 400 of the nation's historic houses and monuments, attracting 12m. visitors a year; publications.

English Nature—Nature Conservancy Council for England: Northminster House, Peterborough PE1 1UA; tel (1733) 455100; fax (1733) 568834; e-mail enquiries@english-nature.org.uk; internet www.english-nature.org.uk; Chair: Sir Martin Doughty; Chief Exec: Dr Andrew Brown; f 1991; English Nature is the government-funded body the purpose of which is to promote conservation of the wildlife and natural features of England; this is achieved by taking action and by working through and enabling others.

Environment, Food and Rural Affairs Select Committee: 7 Millbank, London SW1P 3JA; tel (20) 7219-5774; fax (20) 7219-2094; e-mail efracom@parliament.uk; internet http://www.parliament.uk/parliamentary_committees/environment_food_and_rural_affairs.cfm; Chair: Rt Hon Michael Jack; f 1979; parliamentary committee of MPs responsible for monitoring the policy, expenditure and administration of the Department of the Environment, Food and Rural Affairs.

Environment, Planning and Countryside Committee of the National Assembly for Wales: Cardiff Bay, Cardiff

CF99 1NA; tel (29) 2089-8151; fax (29) 2089-8021; e-mail environment.plan@wales.gsi.gov.uk; internet www.wales.gov.uk/keypubassemenvplancou; Chair: Alun Ffred Jones; the Committee has a very broad remit based on the portfolio of the Minister for Environment, Planning and Countryside; the Minister's portfolio embraces the environment and sustainable development; town and country planning; countryside and conservation issues; and agriculture and rural development including forestry and food production.

ESRC Global Environmental Change Programme: Mantell Bldg, University of Sussex, Falmer, Brighton BN1 9RF; tel (1223) 678935; fax (1223) 604483; e-mail gec@sussex.ac.uk; internet www.gecko.ac.uk; Dirs: Dr Frans Berkhout, Dr Ian Scoones, Dr Melissa Leach; Asst Dir: Alister Scott; f 1991; formerly based at Wye College (University of London); research programme of the Economic and Social Research Council—ESRC; social science element of the national Global Environmental Research—GER programme; aims to understand the societal and economic forces behind global environmental change and to determine suitable policy objectives to address environmental problems.

Fisheries Conservancy Board for Northern Ireland: 1 Mahon Rd, Portadown BT62 3EE; tel (1762) 334666; fax (1762) 338912; e-mail fiona@fcbni.org; internet www.fcbni.com; Chief Exec: Karen Simpson; f 1966; affiliated with the Department of Agriculture and Rural Development for Northern Ireland, Fisheries Division; conservation and protection of salmon and inland fisheries in Northern Ireland, with the exception of the Foyle area, Co Londonderry.

Forestry Commission: Dept of Forestry, Silvan House, 231 Corstorphine Rd, Edinburgh EH12 7AT; tel (131) 334-0303; fax (131) 334-3047; e-mail info@forestry.gov.uk; internet www.forestry.gov.uk; Chair: Rt Hon Lord Clark of Windermere; Dir-Gen: Tim Rollinson; responsible for protection and expansion of Britain's forest and woodlands, and increasing their value to society and the environment; the Commission implements the Government's forestry policy within the framework of the Forestry Act, administers the Woodland Grant Scheme, controls tree-felling through the issue of licences and administers plant health regulations to protect woodlands against tree-pests and diseases; its two agencies, Forest Enterprise and Forest Research, manage the Commission's estates and undertake research, respectively.

Hadley Centre for Climate Prediction and Research: Hadley Centre, FitzRoy Rd, Exeter, Devon EX1 3PB; tel (1344) 856653; fax (1344) 854898; e-mail hadley@metoffice.gov.uk; internet www.metoffice.com/research/hadleycentre; Dir: Dr David Griggs; climate monitoring; detection and prediction of climate change; development of climate change modelling methodologies.

LGC: Queen's Rd, Teddington TW11 0LY; tel (20) 8943-7000; fax (20) 8943-2767; e-mail info@lgc.co.uk; internet www.lgc.co.uk; CEO: Dr R Worswick; Dir, Business Devt: Dr R Ah-Sun; f 1996; fmrly Laboratory of the Government Chemist; fully integrated analysis, research and consultancy addressing environmental issues such as solid waste management, environmental management and waste minimization, contaminated land risk assessment and management, water and effluent management, and chemical disposal; affiliated with Pipeline Developments Ltd, LGC France, LGC Nordic and LGC (Northwest).

Met Office: Fitzroy Road, Exeter EX1 3PB; tel (1392) 885680; fax (1392) 885681; e-mail enquiries@metoffice.com; internet www.metoffice.com; Chief Executive: Peter Ewins; f 1854; provider of environmental and weather related services; its solutions and services meet the needs of many communities of interest from the general public, government and schools, through broadcasters and online media, to civil aviation and almost every other industry sector both in the UK and around the world.

National Radiological Protection Board: Chilton, Didcot OX11 0RQ; tel (1235) 831600; fax (1235) 833891; Chairman: Sir William Stewart; Dir: Prof Roger Cox; Board Sec: Dr J R Cooper; independent statutory body; information, advice, technical services, monitoring and field and laboratory research on radiological hazards.

Natural Environment Research Council—NERC: Polaris House, North Star Ave, Swindon SN2 1EU; tel (1793) 411500; fax (1793) 411501; e-mail requests@nerc.ac.uk; internet www.nerc.ac.uk; Chair: Rob Margetts; Chief Exec: Prof John Lawton; research and consultancy in environmental aspects of the terrestrial, marine, freshwater, atmospheric and biological sciences.

British Antarctic Survey—BAS: High Cross, Madingley Rd, Cambridge CB3 0ET; tel (1223) 221400; fax (1223) 362616; e-mail basinfo@bas.ac.uk; internet www.antarctica.ac.uk; Dir: Prof C J Rapley; Deputy Dir: Dr J R Dudeny; f 1962; government-funded research body responsible for study of the natural environment of Antarctica, South Georgia and the South Sandwich Islands; BAS discharges the United Kingdom's responsibilities in compliance with the Antarctic Treaty and assists in the administration of the British Antarctic Territory; played an important role in the discovery of ozone layer depletion; research in glaciology, meteorology, geology, biology, oceanography and atmospheric physics; operates three stations in the Antarctic, five aircraft and two research ships.

Royal Commission on Environmental Pollution: Third Floor, The Sanctuary, Westminster, London SW1P 3JS; tel (20) 7799 8970; fax (20) 7799 8971; e-mail enquiries@rcep.org.uk; internet www.rcep.org.uk; Chair: Sir Tom Blundell; f 1970; provides advice on national and international matters concerning pollution of the environment, the adequacy of research in that field and the possibility of future threats to the environment.

Scottish Environment Protection Agency—SEPA: SEPA Corporate Office, Erskine Court, Castle Business Park, Stirling FK9 4TR; tel (1786) 457700; fax (1786) 446885; e-mail info@sepa.org.uk; internet www.sepa.org.uk; Chair: Sir Ken Collins; Chief Exec: Campbell Gemmell; f 1996; non-departmental public body affiliated with the Scottish Office; superseded the Scottish River Purification Boards' Association—SRPBA and Her Majesty's Industrial Pollution Inspectorate—HMIPI, and assumed the waste regulation and air pollution powers of the former Scottish councils; responsible for the control of pollution to land, air and water in Scotland; aims to provide an integrated environmental protection system for Scotland in order to improve the environment and contribute to the Government's goal of sustainable development.

Scottish Executive Development Department—SEDD: SEDD Secretariat, Area 3-H, Victoria Quay, Edinburgh EH6 6QQ; tel (131) 244-0763; fax (131) 244-0785; e-mail ceu@scotland.gov.uk; internet www.scotland.gov.uk/About/Departments/DD; Head: Nicola Munro.

Scottish Executive Environment and Rural Affairs Department: Room 440, Pentland House, 47 Robb's Loan, Edinburgh EH14 1TY; tel (131) 556-8400; fax (131) 244-6116; e-mail ceu@scotland.gsi.gov.uk; internet www.scotland.gov.uk/About/Departments/ERAD; Minister for Environment and Rural Affairs: Ross Finnie; responsible for environment and natural heritage, land reform, water, sustainable development, agriculture, fisheries, rural development including aquaculture and forestry.

Scottish Natural Heritage: 12 Hope Terrace, Edinburgh EH9 2AS; tel (131) 447-4784; fax (131) 446-2277; e-mail enquiries@snh.gov.uk; internet www.snh.org.uk; Chair: Dr John Markland; Chief Exec: Ian Jardine; f 1992; statutory body responsible to the Scottish Executive; works to conserve and enhance the wildlife, habitats and landscapes that constitute Scotland's natural heritage, to promote awareness, understanding and responsible enjoyment of it and its sustainable use.

Welsh Development Agency—WDA: Plas Glyndŵr, Kingsway, Cardiff CF10 3AH; tel (1443) 845500; fax (1443) 845589; e-mail enquiries@wda.co.uk; internet www.wda.co.uk; Chair: Roger S Jones; Chief Exec: Gareth Hall; f 1976; stimulation of economic development and environmental improvement.

Non-Governmental Organizations

Advisory Committee on Packaging: ACP Secretariat, Zone 7/F8, Ashdown House, Defra, 123 Victoria Street, London SW1E 6DE; tel (20) 7082 875; internet www.defra.gov.uk/environment; Chair: John Turner; f 1996; advises UK Government on the drafting of Regulations implementing parts of the EC Directive on Packaging and Packaging Waste 94/62/EC; designed to give balanced representation to the four main industry sectors in the packaging chain (i.e. packaging raw material manufacturers, convertors, packer/fillers and sellers) and the main packaging materials.

Association for the Conservation of Energy: Westgate House, Prebend St, London N1 8PT; tel (20) 7359-8000; fax (20) 7359-0863; e-mail info@ukace.org; internet www.ukace.org; Dir: Andrew Warren; lobby group; conducts policy research into energy conservation; membership limited to 24 United Kingdom-based companies with substantial interests in energy conservation equipment and services.

Bat Conservation Trust—BCT: Unit 2, 15 Cloisters House, 8 Battersea Park Rd, London SW8 4BG; tel (20) 7627-2629; fax (20) 7627-2628; e-mail enquiries@bats.org.uk; internet www.bats.org.uk; Pres: Prof David Bellamy; conservation of bats and their habitats; comprises over 90 organizations involved in bat conservation.

BIAZA–British & Irish Association of Zoos and Aquariums: Regent's Park, London NW1 4RY; tel (20) 7586-0230; fax (20) 7722-4427; internet www.zoofederation.org.uk; Dir: Miranda Stevenson; Admin: Mimi Frame; Conservation Coordinator: Nicola Charlton; f 1966; fmrly Federation of Zoological Gardens of Great Britian and Ireland; represents the zoological community in the United Kingdom; conservation, education and scientific charity dedicated to the maintainance of biodiversity, welfare of animals in zoos and the advancement of scientific knowledge; non-invasive research and coordinated, *ex situ* breeding programmes for endangered species; provision of financial and technical support for habitat and species conservation in the wild, both in the United Kingdom and abroad; collaboration with English Nature and the Joint Nature Conservation Committee—JNCC; affiliated to the European Association of Zoos and Aquaria, World Zoo Organization, World Conservation Union—IUCN and Conservation Breeding Specialist Group.

Bio-Dynamic Agricultural Association: Painswick Inn Project, Gloucester St, Stroud GL5 1QS; tel and fax (1453) 759501; e-mail office@biodynamic.org.uk; internet www.biodynamic.org.uk; Chair: Nick Raeside; Exec Dir: Bernard Jarman; Sec: Jessica Standing; f 1928; promotion of the bio-dynamic approach to farming, gardening and forestry, based on the principles of Rudolf Steiner. Demeter (Organic Certification UK6) certification in UK, offers training and disseminates information (books), membership.

Blake Shield BNA Trust: POB 5681, Rushden NN10 8ZF; tel and fax (1933) 314672; Hon Admin: J F Pearton; Chair of Trustees: Rev T W Gladwin; Hon Treas: T R Burbidge; f 1998; promotes and organizes the Blake Shield conservation competition for young people.

Born Free Foundation: 3 Grove House, Foundry Lane, Horsham, West Sussex RH13 5PL; tel (1403) 240170; fax (1403) 327833; e-mail info@bornfree.org.uk; internet www.bornfree.org.uk; CEO: William Travers; Founders: Virgina McKenna (Bill Travers (deceased)); f 1984; co-ordinates and develops effective campaigns to prevent animal suffering, protect endangered species and their habitats, and keep wildlife in the wild; programmes include: Zoo Check, Elefriends, the Big Cats, Wolves, Bears, Primate, Marine and Education.

British Association of Nature Conservationists,: BANC Membership Services, Lings House, Billings Lings, Northampton NN3 8BE; internet www.banc.org.uk; Pres: Derek Ratcliffe; f 1979.

British Ecological Society: 26 Blades Court, Deodar Rd, Putney, London SW15 2NU; tel (20) 8871-9797; fax (20) 8871-9779; e-mail info@BritishEcologicalSociety.org; internet www.britishecologicalsociety.org; Pres: Prof Alastair Fitter; Exec Sec: Dr H J Norman; f 1913; promotes the science of ecology through publications and scientific meetings; encourages research and ecological education in schools.

British Herpetological Society: c/o Zoological Society of London, Regent's Park, London NW1 4RY; tel (20) 8452-9578; internet www.thebhs.org; Pres: Dr H Robert Bustard; Sec: Monica Green; conservation and captive breeding of amphibians and reptiles.

British Naturalists' Association: 17 Beatty Gardens, Lodge Park, Corby NN17 2RT; f 1905; natural history and wildlife conservation society.

British Society of Dowsers: Sycamore Barn, Hastingleigh, Ashford TN25 5HW; tel (1684) 576969; fax (1684) 576969; e-mail secretary@britishdowsers.org; internet www.britishdowsers.org; Administrator: Ian Clements; f 1933; lectures, short courses, advisory service, conferences and production of a quarterly journal on dowsing.

British Trust for Ornithology—BTO: The Nunnery, Thetford IP24 2PU; tel (1842) 750050; fax (1842) 750030; e-mail info@bto.org; internet www.bto.org; Dir: Dr Jeremy Greenwood; Patron: HRH The Duke of Edinburgh; f 1933; provision of objective scientific information and advice to numerous groups; areas of interest include ornithological surveys, habitat research, population studies and site and impact assessments.

British Waterways: Willow Grange, Church Road, Watford, Hertfordshire, WD17 4QA; tel 01923 226422; fax 01923 201400; e-mail enquiries.hq@britishwaterways.co.uk; internet www.britishwaterways.co.uk; Chair: Dr G P Greener; Chief Exec: R Evans.

Butterfly Conservation: Manor Yard, East Lulworth, Wareham, Dorset BH20 5QP; tel 0870 7744309; fax 0870 7706150; e-mail info@butterfly-conservation.org; internet www.butterfly-conservation.org; Pres: Sir David Attenborough; Chair: Dudley Cheesman; f 1968; creates public awareness, funds research and establishes butterfly reserves.

Campaign for Nuclear Disarmament—CND: 162 Holloway Rd, London N7 8DQ; tel (20) 7700-2393; fax (20) 7700-2357; e-mail enquiries@cnduk.org; internet www.cnduk.org; Chair: Kate Hudson; Sec: Linda Hugl; campaigns against the use of nuclear power to process weapons-grade plutonium and the development, production, transport and deployment of all weapons of mass destruction, on environmental, political, economic and moral grounds.

Campaign for the Protection of Rural Wales/Ymgyrch Diogelu Cymru Wledig: Tŷ Gwyn, 31 High St, Welshpool

SY21 7YD; tel (1938) 552525; fax (1938) 552741; e-mail info@cprw.org.uk; internet www.cprw.org.uk; Dir: Peter Ogden; works to protect the coast and countryside of Wales; encourages sustainable development.

Campaign to Protect Rural England—CPRE: CPRE National Office, 128 Southwark Street, London, SE1 0SW; tel (20) 7981-2800; fax (20) 7981-2889; e-mail info@cpre.org.uk; internet www.cpre.org.uk; Pres: Sir Max Hastings; CEO: Shaun Spiers; f 1926; charitable organization which campaigns for the protection of the countryside and sustainable use of land and other resources; 43 county branches, 200 local groups and more than 45,000 members.

Centre for Air Transport and the Environment, Manchester Metropolitan University: Faculty of Science and Engineering, Chester St, Manchester M1 5GD; tel (161) 247-3658; fax (161) 247-6332; e-mail aric@mmu.ac.uk; internet www.cate.mmu.ac.uk; Dir: Prof D Raper; f 1984; to facilitate the integrated social, economic and environmental sustainability of the aviation industry through critical research and analysis, and through knowledge transfer between the academic, industry, regulatory and NGO sectors.

Centre for Alternative Technology—CAT: Machynlleth SY20 9AZ; tel (1654) 705950; fax (1654) 702782; e-mail help@cat.org.uk; internet www.cat.org.uk; Devt Dir: Paul Allen; f 1975; education and display centre demonstrating renewable energy, organic horticulture and environmental building; provides residential courses, information and consultancy services; publishes periodicals and operates mail-order service for environmental books and products.

Centre for Environmental Initiatives—CEI: The Old School House, Mill Lane, Carshalton SM5 2JY; tel (20) 8770-6611; fax (20) 8647-0719; e-mail info@thecei.org.uk; internet www.thecei.org.uk; Dir: Lisa Loughlin; f 1987; formerly Centre for Environmental Information; charitable trust; acts as adviser to local government; produces reports and feasibility studies; encourages public environmental action.

Centre for Human Ecology: 12 Roseneath Pl, Edinburgh EH9 1JB; tel (131) 624-1972; fax (131) 228-9630; e-mail info@che.ac.uk; internet www.che.ac.uk; Exec Dir: Osbert Lancaster; f 1972; research and M SC course on human ecology and international PH D in ecological economics; sustainable development courses and consultancy; seminars, conferences, workplace learning and public information.

Centre for Population Studies, London School of Hygiene & Tropical Medicine: 49-51 Bedford Square, London WC1B 3DP; tel (20) 7299-4614; fax (20) 7299-4637; internet www.lshtm.ac.uk/cps; Head: Ian Timaeus; Sr Staff: Basia Zaba, Prof John Clelland, Prof Emily Grundy; research and advanced teaching on measurement of population change, effective family planning provision, reproductive health and interactions between population and environment.

Centre for Social and Economic Research on the Global Environment—CSERGE, Norwich Office: University of East Anglia, School of Environmental Sciences, Norwich NR4 7TJ; tel (1603) 593738; fax (1603) 593739; e-mail d.turnbull@uea.ac.uk; internet www.uea.ac.uk/env/cserge; Dir: Prof R K Turner; f 1991; located at twin sites; funded by the Global Environmental Change programme of the Economic and Social Research Council; central interdisciplinary research programme has three main focuses: global warming, including the cost of damage caused by climate change, the economic costs of controlling greenhouse gas emissions, the incentives for international agreements and the instruments for securing them; conservation of biological diversity, concentrating on various economic, ethical and political aspects of species and habitat loss; and institutional adaptation to global environmental change, including an evaluation of the policy-making process, the role of science within this and the communication of scientific findings to society; other areas of research include waste management and recycling, resource valuation and accounting, sustainable development indicators and coastal zone management.

Centre for Social and Environmental Accounting Research—CSEAR, University of Glasgow: c/o Dept of Accountancy and Finance, 73 Southpark Ave, Glasgow G12 8LE; tel (141) 330-6315; fax (141) 330-4442; e-mail csear@accfin.gla.ac.uk; internet www.gla.ac.uk/departments/accounting/csear; Dir: Prof R H Gray; f 1991; focuses on social and environmental accounting, auditing and reporting, with a special interest in the sustainability of organizations; organizes an annual conference and publishes a journal twice a year; gathers and disseminates information; enables academics and practitioners to exchange ideas and experience; associate organizations in over 30 countries.

Chartered Institution of Wastes Management—IWM: 9 Saxon Court, St Peter's Gardens, Northampton NN1 1SX; tel (1604) 620426; fax (1604) 621339; e-mail technical@ciwm.co.uk; internet www.ciwm.co.uk; Chair: Peter Ager; Pres: Alistair Lamont; Chief Exec: Michael Philpott; Chair, IWM Business Services Ltd: Roger Hewitt; professional association of waste managers; affilliated to the International Solid Waste Association—ISWA.

Chartered Institution of Water and Environmental Management—CIWEM: 15 John St, London WC1N 2EB; tel (20) 7831-3110; fax (20) 7405-4967; e-mail admin@ciwem.org; internet www.ciwem.com; Pres: Geoff Bateman; Exec Dir: Nick Reeves; f 1987; multidisciplinary professional and examining body for engineers, scientists and others engaged in water and environmental management; conferences and seminars; publications on all aspects of water and waste water treatment.

Christian Ecology Link: 20 Carlton Rd, Harrogate HG2 8DD; tel (1524) 33858; e-mail info@christian-ecology.org.uk; internet www.christian-ecology.org.uk; Chair: Tim Cooper; Correspondence Off: Laura Deacon; Sec: Barbara Echlin; Vice-Chair: Paul Bodenham; f 1981; provides resources to churches in the UK and individuals to enable them to care for the environment; membership network, topic leaflets, thrice-yearly magazine and regular news bulletins.

Civic Trust: Winchester House, 259–269 Old Marylebone Rd, London NW1 5RA; tel (20) 7170-4299; fax (20) 7170-4298; e-mail info@civictrust.org.uk; internet www.civictrust.org.uk; Patron: HRH The Prince of Wales; Chair: Nigel Burton; f 1957; charitable federation of nearly 950 local groups; aims to improve living and working environments through promotion of high standards of planning and conservation; administers the Environmental Action Fund on behalf of the Department of the Environment, Transport and the Regions; grants annual awards for good development; operates the Regeneration Unit; organizes heritage open days, green flag park awards and civic trust awards; regional office in Liverpool.

Combined Heat and Power Association: 35–37 Grosvenor Gardens, 4th Floor, London SW1W 0BS; tel (20) 7828-4077; fax (20) 7828-0310; e-mail info@chpa.co.uk; internet www.chpa.co.uk; Dir: David Green; Office Man: Jane Marsh; promotion of combined heat and power (CHP) and community heating.

Common Ground: Gold Hill House, 21 High Street, Shaftesbury, Dorset SP7 8JE; tel (1747) 850820; fax (1747) 850821; internet www.commonground.org.uk; Dirs: Sue Clifford, Angela King, Kate O'Farrell.

Commons, Open Spaces and Footpaths Preservation Society: 25A Bell St, Henley-on-Thames, Oxfordshire RG9

2BA; e-mail hq@oss.org.uk; internet www.oss.org.uk; Chair: Rodney Legg; Gen Sec: Kate Ashbrook.

Compassion in World Farming Ltd—CIWF: Charles House, 5a Charles St, Petersfield GU32 3EH; tel (1730) 264208; fax (1730) 260791; e-mail compassion@ciwf.co.uk; internet www.ciwf.org.uk; Dir: Joyce D'Silva; f 1967; organizes campaigns banning transportation of live animals, factory farming, genetic engineering of farm animals, battery cages and farrowing crates for sows.

Confederation of British Industry—CBI, Environment, Health and Safety Group: Centre Point, 103 New Oxford St, London WC1A 1DU; tel (20) 7379-7400; fax (20) 7497-2597; e-mail web.editor@cbi.org.uk; internet www.cbi.org.uk; Head of Health and Safety Policy: Dr Janet Asherson; Head of Environment Policy: Matthew Farrow; f 1965; represents business interests on environmental issues and provides information to help companies improve their environmental performance; CBI Environment Business Forum promotes voluntary action on environment.

Conservation Foundation: 1 Kensington Gore, London SW7 2AR; tel (20) 7591-3111; fax (20) 7591-3110; e-mail info@conservationfoundation.co.uk; internet www.conservationfoundation.co.uk; Prin Officers: Prof David Bellamy, David Shreeve; f 1982; organization and management of a variety of programmes and projects providing funding and publicity for a wide range of environmental initiatives; conferences and publications; promotion of positive environmental news, awareness and action.

CORDaH Environmental Management Consultants, the European Environmental Management Institute: Scotstown Road, Bridge of Don, Aberdeen AB23 8NG; tel (1224) 414200; fax (1224) 414250; e-mail main@astp.cordah.co.uk; Man Dir: Dr Owen Harrop, Prof Brian D Clark, Dr Gordon Picken; f 1996; superseded AURIS Environmental and Altra Safety and Environment Ltd; independent environmental management consultancy; environmental training and information; consultancy to the petroleum and gas industries; aquatic scientific services; safety management; incorporates the Centre for Environmental Management and Planning—CEMP and the Oil Pollution Research Unit—OPRU.

Corus Steel Packaging Recycling: Port Talbot Works, SA3 2NG; tel (1639) 872516; fax (1639) 872693; internet www.cspr.co.uk; Communications Man: Anna Richards; Commercial Man: David Williams; works with local authorities to establish recycling schemes for all metal cans; provides assistance, advice and support material.

Council for Environmental Education (Youth Unit): 94 London Street, Reading RG1 4SJ; tel (118) 950-1550; fax (118) 959-1955; e-mail info@cee.org.uk; internet www.cee.org.uk; national membership body for organizations and individuals in England committed to environmental education and education for sustainable development.

Council for Environmental Education—CEE: University of Reading, London Rd, Reading RG1 5AQ; tel (118) 950-2550; fax (118) 959-1955; e-mail info@cee.org.uk; internet www.cee.org.uk; Dir: Libby Grundy; Information Officer: Helen Rose; f 1968; co-ordinating and policy development body for England, Wales and Northern Ireland; represents and promotes members' interests; information, advice, research and publications.

Council for National Parks: 246 Lavender Hill, London SW11 1LJ; tel (20) 7924-4077; fax (20) 7924-5761; e-mail info@cnp.org.uk; internet www.cnp.org.uk; Pres: Brian Blessed; Chief Executive: Kathy Moore; f 1936; independent charity; campaigns to protect the 11 national parks of England and Wales; more than 40 affiliated conservation and recreation associations.

Countryside Council for Wales: c/o Joint Nature Conservation Committee, Monkstone House, City Rd, Peterborough PE1 1JY.

CSV Environment: 3rd Floor, Parker Bldg, St Peter's Urban Village Trust, College Road, Saltley, Birmingham B8 3TE; tel (121) 322-2008; e-mail mike@csvenvironment.org.uk; internet www.csvenvironment.org.uk; Prin Officers: Mike Williams, Guy Dixon; creation of opportunities for members of the public to participate fully in their community and to improve their environment.

Ecology Building Society: Belton Rd, Silsden, Keighley, West Yorkshire BD20 0EE; tel (0845) 674-5566; fax (1535) 650780; e-mail info@ecology.co.uk; internet www.ecology.co.uk; Chief Exec: Paul Ellis; f 1981; mutual society providing ethical savings facilities; provides mortgage finance for ecologically sound purchases, energy-efficient housing, renovation of derelict inner-city and rural properties, organic smallholdings and other small businesses; opposes the wasteful use of land and resources.

ENCAMS: Elizabeth House, The Pier, Wigan WN3 4EX; tel (1942) 612639; fax (1942) 824778; internet www.encams.org; Chair: John Markland; runs the Keep Britain Tidy campaign.

Energy Action Scotland: Suite 4a, Ingram House, 227 Ingram Street, Glasgow, G1 1DA; tel (141) 226-3064; fax (141) 221-2788; e-mail eas@eas.org.uk; internet www.eas.org.uk; promotes energy efficiency and conservation; provides training in energy awareness and home energy auditing.

Energy and Environment Research Unit, Open University: Faculty of Technology, Walton Hall, Milton Keynes MK7 6AA; tel (1908) 653335; fax (1908) 858407; e-mail d.a.elliott@open.ac.uk; internet www-tec.open.ac.uk/eeru; Dir: Dr David Elliott; university group concerned with the development and assessment of sustainable energy technologies and policies; publishes a bimonthly newsletter.

Energy Institute: 61 New Cavendish Street, London W1G 7AR; tel (20) 7467-7100; fax (20) 7255-1472; e-mail info@energyinst.org.uk; internet www.energyinst.org.uk; Chief Exec: Louise Kingham; f 2003; merger of the former Institute of Energy and the Institute of Petroleum; brings together individuals and organisations across the energy sectors of industry, academia and government; topics: upstream and downstream oil, gas and other primary fuels and renewables, through power generation, transmission and distiribution to sustainable development, demand side management and energy efficiency; conferences and courses; extensive library and information service.

Environment Council: 212 High Holborn, London WC1V 7BF; tel (20) 7836-2626; fax (20) 7242-1180; e-mail info@envcouncil.org.uk; internet www.the-environment-council.org.uk; Chair: Dr Malcolm Aickin; CEO: Mike King; f 1988; independent charity; aims to protect and enhance the United Kingdom's environment by increasing public awareness and finding effective solutions to environmental problems.

Environmental Change Institute, University of Oxford: 1a Mansfield Rd, Oxford OX1 3SZ; tel (1865) 281180; fax (1865) 281202; e-mail enquiries@eci.ox.ac.uk; internet www.eci.ox.ac.uk; Dir: Prof Diana Liverman; f 1991; centre for promotion of research and teaching on the environment; principal areas of research are: climate impacts and responses under the Core Programme Office for Climate Impacts, funded by the Department of the Environment, Transport and the Regions; energy and environment, covering demand-side management of personal energy use, particularly domestic appliances and lighting, and rural transport; land degradation and rehabilitation, with emphasis on soil erosion by water; mountain regions; solar energy; and asthma and the environment; areas of staff expertise include anthropology, botany, earth sciences, energy policy, geography, geographic information systems, operational research, statistical modelling, technical design and transport; MSC in Environmental Change and Management.

Environmental Impact Assessment—EIA Centre, University of Manchester: Dept of Planning and Landscape, Oxford Rd, Manchester M13 9PL; tel (161) 275-6873; fax (161) 275-6893; e-mail eia.centre@man.ac.uk; internet www.art.man.ac.uk/eia/eiac.htm; Head: Dr Carys E Jones; f 1988; research and other studies in the field of environmental impact assessment (EIA) and strategic environmental assessment; provision and facilitation of EIA educational and training programmes; provision of advice to those involved in the implementation of EIA in developed, transitional and developing countries; publications.

Environmental Investigation Agency: 62/63 Upper Street, London N1 0NY; tel (20) 7354-7960; fax (20) 7354-7961; e-mail ukinfo@eia-international.org; internet www.eia-international.org; Chair: Allan Thornton; Exec Dir: Dave Currey; f 1984; investigation of environmental abuses worldwide and campaigns for environmental protection legislation.

Environmental Services Association: 154 Buckingham Palace Rd, London SW1W 9TR; tel (20) 7824-8882; fax (20) 7824-8753; e-mail info@esauk.org; internet www.esauk.org; Chair: Stephen Jenkinson; Chief Exec: Dirk Hazell; Dir of Policy: Stuart McLanaghan; f 1969; The Environmental Services Association (ESA) is the trade association for the UK's waste management industry. The regulated industry contributes approximately £5,000m. annually to the UK's economy. ESA's Members provide essential waste and secondary resources management services to the public and private sectors across the UK. From small, locally-based concerns to large international organizations, ESA Members span the full spectrum of operations (including collection, treatment, disposal, recovery, recycling and re-use of waste), specialist equipment manufacturers and environmental consultancies. ESA wants to build our industry's future not on managing rising quantities of waste but on managing more waste as a resource by recovering secondary materials and energy. Given the right opportunity, our Members will achieve high levels of recycling and recovery which they are providing in other European countries. Working closely with industry, Government and regulators, ESA wants to ensure economically and environmentally sustainable waste and secondary resource management across the UK.

Ethical Consumer: Unit 21, 41 Old Birley Street, Manchester, M15 5RF; tel (161) 226 2929; e-mail mail@ethicalconsumer.org; internet www.ethicalconsumer.org; Editor: Rob Harrison; f 1987; publishes bi-monthly magazine on corporate responsibility and runs database/consultancy service screening of companies.

Fauna and Flora International—FFI: Great Eastern House, Tenison Rd, Cambridge CB1 2DT; tel (1223) 571000; fax (1223) 461481; e-mail info@fauna-flora.org; internet www.fauna-flora.org; Exec Dir: Mark Rose; f 1903; formerly Fauna and Flora Preservation Society; charitable membership organization working to protect endangered plant and animal species world-wide.

Federation of City Farms and Community Gardens: The Greenhouse, Hereford St, Bristol BS3 4NA; tel (117) 923-1800; fax (117) 923-1900; e-mail admin@farmgarden.org.uk; internet www.farmgarden.org.uk; Dir: Jeremy Iles; f 1980; supports and advises city farms and community gardens; provides technical, managerial, financial and legal assistance and training programmes; prepares feasibility studies.

Field Studies Council—FSC: Head Office, Preston Montford, Montford Bridge, Shrewsbury SY4 1HW; tel (1743) 852100; fax (1743) 01743 852101; e-mail enquiries@field-studies-council.org; internet www.field-studies-council.org; Chief Exec: Tony Thomas; f 1943; environmental courses, publications, consultancy and research; 13 study centres in the United Kingdom and overseas.

Filtration Society: 5 Henry Dane Way, Newbold Coleorton, Nr Coalville, Leicestershire LE67 8P; tel (1530) 223124; e-mail filtech.exhibitions@btinternet.com; internet www.lboro.ac.uk/departments/cg/research/filtration/; Chair: R Lydon; Sec: Prof Richard Wakeman; f 1964.

Findhorn Foundation, Eco-Village Project: The Park, Findhorn Bay, Forres IV36 3TZ; tel (1309) 690154; fax (1309) 691387; e-mail info@ecovillagefindhorn.com; internet www.ecovillagefindhorn.com; Project Dir: John Talbott; Int Liaison: May East; aims to create an ecologically, economically and socially sustainable human settlement of 400–500 people.

Fish Conservation Centre: Gladshot, Haddington, East Lothian EH41 4NR; tel and fax (1620) 823691; e-mail SavingFish@maitland60.freeserve.co.uk; Prin Officer: Prof Peter S Maitland; f 1985; research on fish conservation and ecology; consultancy work in freshwater ecology and impact assessment.

Forest Stewardship Council, UK Working Group: Station Buildings, Unit K, Llanidloes SY18 6EB; tel (1686) 413916; fax (1686) 412176; e-mail fsc@fsc-uk.org; internet www.fsc-uk.org; Dir: Anna Jenkins; Advisory Service Man: Amy Mulkern; Construction Industry Advisor: Beck Woodrow; f 1995; promotion of the protection and conservation of the world's forests through introduction of a forest certification and timber labelling scheme, and development of consensus on standards for forest certification in the United Kingdom; public awareness-raising activities.

Forth Naturalist and Historian, University of Stirling: 30 Dunmar Drive, Alloa, Clackmannanshire; tel (1259) 215091; fax (1786) 464994; e-mail lc2@stir.ac.uk; internet www.fnh.stir.ac.uk; Chair: John Proctor; Hon Editor and Sec: Lindsay Corbett; f 1975; charitable body; promotion of the environment and heritage of Central Scotland through publications and annual symposia.

Forward Scotland Ltd: Portcullis House, 6th Floor, 21 India St, Glasgow G2 4PZ; tel (141) 222-5600; fax (141) 222-5601; e-mail enquiries@forward-scotland.org.uk; internet www.forward-scotland.org.uk; Convener: Anne Mearns; Partnership Projects Man: Frazer Scott; f 1996; formerly UK2000 Scotland; sustainable development initiatives through partnership in economic, environmental and community affairs.

Foundation for International Environmental Law and Development—FIELD: 3 Endsleigh St, London WC1H 0DD; tel (20) 7388-2117; fax (20) 7388-2826; e-mail field@field.org.uk; internet www.field.org.uk; Co-Dir: Alice Palmer; formerly the Centre for International Environmental Law—CIEL; independent, non-profit organization working towards the progressive development of international law for the protection of the environment and the attainment of sustainable development; international legal advice and assistance; promotes long-term research on climate change and energy, trade, investment and sustainable development, and biodiversity and marine resources; works in collaboration with the Centre on International Co-operation—CIC at New York University; research, publishing and teaching.

Friends of Conservation: 16-18 Denbigh Street, London SW1V 2ER; tel (20) 7592-0110; fax (20) 7828-4856; e-mail focinfo@aol.com; internet www.foc-uk.com; using invaluable experience gained in East Africa, FOC now participates in and helps fund community conservation projects around the world. Countries where FOC is currently involved in projects include: Belize, Brazil, Costa Rica, the Caribbean, India, Kenya, Namibia, Patagonia, Sri Lanka, Seychelles, Spain, Tanzania, Uganda and Zimbabwe.

Friends of the Earth Scotland: Lamb's House, Burgess St, Edinburgh EH6 6RD; tel (131) 554-9977; fax (131) 554-8656; e-mail info@foe-scotland.org.uk; internet www.foe-scotland .org.uk; Dir: Kevin Dunion; Head of Research: Dr Richard Dixon; Information Officer: Lang Banks; f 1978; regional office of Friends of the Earth International—FoEI; head office for nine local groups.

Safe Energy Unit: Bonnington Mill Business Centre, 70–72 Newhaven Rd, Edinburgh EH6 5QG; Contact: Helen Snodin; fmrly the Scottish Campaign to Resist the Atomic Menace—SCRAM; areas of interest include renewable and alternative energy sources, energy policy, energy efficiency, the petroleum and natural gas industries, and climate change; journal.

Friends of the Earth—FoE England, Wales and Northern Ireland: 26–28 Underwood St, London N1 7JQ; tel (20) 7490-1555; fax (20) 7490-0881; e-mail info@foe.co.uk; internet www.foe.co.uk; Exec Dir: Tony Juniper; f 1971; national office of Friends of the Earth International—FoEI; co-ordination of 250 local groups.

Friends of the Earth Birmingham Ltd: The Warehouse, 54-57 Allison St, Digbeth, Birmingham B5 5TH; tel (121) 632-6909; fax (121) 643-3122; e-mail info@birminghamfoe .org.uk; internet www.birminghamfoe.org.uk; Chair: Tracey Fletcher; Campaigns Co-ordinators: Brett Rehling, Andy Pryke; Bldg and Information Man: Dave Clare; f 1972; campaigns on energy, transport, pollution, climate and biodiversity; information service; sells recycled paper.

Friends of the Earth Cymru: 33 The Balcony, Castle Arcade, Cardiff CF10 1BY; tel (29) 2022-9577; fax (29) 2022-8775; e-mail cymru@foe.co.uk; internet www .foecymru.co.uk; Dir: Julian Rosser; Local Groups Devt Officer: Bleddyn Lake; Assembly Campaigner: Gordon James; Campaigner: Neil Crumpton; f 1984; Welsh office of Friends of the Earth—FoE; campaigns on Welsh environmental and social issues; priorities are energy, waste, agriculture, food and transport, with particular emphasis on sustainable development.

Friends of the Earth Northern Ireland: 7 Donegall St Pl, Belfast, BT1 2FN; tel (28) 9023-3488; fax (28) 9024-7556; e-mail foe-ni@foe.co.uk; internet www.foe.co.uk/northern_ ireland.

Sheffield Friends of the Earth: Voluntary Action, 69 Division St, Sheffield S1 4GE; tel (114) 267-0508; e-mail steveg@doctors.org.uk; internet www.sheffieldfoe.co.uk/; Co-ordinator: Steve Goodacre; f 1974; campaigns on transport, land- use and energy issues.

Game Conservancy Trust: Burgate Manor, Fordingbridge SP6 1EF; tel (1425) 652381; fax (1425) 655848; e-mail admin@ gct.org.uk; internet www.gct.org.uk; Chair: Andrew Christie-Miller; Chief Exec: Teresa Dent; Dir for Policy and Public Affairs: Stephen Tapper; researches methods for conservation of game species and associated wildlife; encourages less-intensive rural land use and offers management advice to landowners and farmers.

Global Commons Institute: 60 Lordship Park, London N16 5OA; tel and fax (20) 8800-6075; e-mail aubrey@gci.org .uk; internet www.gci.org.uk; Dir: Aubrey Meyer, Anandi Sharan; f 1990; conducts socio-economic analysis of global climate change and acts as adviser to UN intergovernmental negotiations.

Global Ideas Bank: 6 Blackstock Mews, Blackstock Rd, London N4 2BT; tel (20) 7359-8391; fax (20) 7354-3831; e-mail ideas@globalideasbank.org; internet www.globalideasbank .org; Co-ordinator and Editor: Nick Temple; fmrly Institute for Social Inventions; aims to promote and disseminate good creative ideas to improve society.

Global Islands Network: 3 Douglas Row, Portree, Isle of Skye IV51 9DD; tel (1478) 611213; fax (1478) 612519; e-mail graeme@globalislands.net; internet www.globalislands.net; Contact: Graeme Robertson; f 2002; connects and co-ordinates efforts to help ensure a healthy and productive future for islanders.

Grass Roots Conservation Group: c/o 24 Deramore Park, Belfast BT9 5JU; tel (28) 9020-1809; e-mail dha.brown7@ ntworld.com; Prin Officers: Dr A Hardcastle, D Brown; practical conservation activities such as tree planting and dry stone wall building.

Green Alliance: 40 Buckingham Palace Road, London, SW1W 0RE; tel (20) 7233-7433; fax (20) 7233-9033; e-mail ga@ green-alliance.org.uk; internet www.green-alliance.org.uk; Dir: Guy Thompson; Chair: Dorothy MacKenzie; aims to increase the prominence of the environment among the concerns of the major policy-making institutions in the United Kingdom.

Green Liberal Democrats: 19 Village Road, London N3 1TL; e-mail info@greenlibdems.org.uk; internet www .greenlibdems.org.uk; f 1977 as Liberal Ecology Group; interest group of the Liberal Democrat Party; promotion of environmental protection, resource conservation, population reduction and economic stabilization for a sustainable future; establishment of links with other pressure groups in similar fields.

Green Light Trust: Lawshall Green, Bury St Edmunds, Suffolk IP29 4QJ; tel (1284) 828754; fax (1284) 827273; e-mail info@greenlighttrust.org; internet www.greenlighttrust.org; Operations Dir: Rick Edmunds; Strategy Dir: Nigel Hughes; undertakes environmental education through creative action.

Green Party: 1a Waterlow Rd, London N19 5NJ; tel (20) 7272-4474; fax (20) 7272-6653; e-mail office@greenparty .org.uk; internet www.greenparty.org.uk; Exec Chair: Hugo Charlton; Principal Speakers: Dr Caroline Lucas, Dr Mike Woodin; f 1973; formerly the Ecology Party; contests local, national and European Parliament elections, campaigns on environmental, peace and economic issues; aims to increase public awareness of green politics and the need for a sustainable economy.

GreenNet: Islington High Street, London N1 9LH; tel (20) 7713-1941; fax (20) 7837-5551; e-mail info@gn.apc.org; internet www.gn.apc.org; Contact: Viv Kendon; global computer communications network for environmental, peace, human rights and development groups; member of the Association for Progressive Communications—APC.

Greenpeace UK: Canonbury Villas, London N1 2PN; tel (20) 7865-8100; fax (20) 7865-8200; e-mail info@uk.greenpeace .org; internet www.greenpeace.org.uk; Exec Dir: Stephen Tindale; f 1977; independent non-profit global campaigning organization that uses non-violent, creative confrontation to expose global environmental problems and their causes.

Habitat Scotland: Hazelmount, Heron Pl, Portree, Isle of Skye IV51 9EU; tel (1478) 612898; fax (1478) 613254; Dir: Graeme Robertson; f 1977; independent charitable research organization established to investigate conservation issues arising from varied use of land and sea in the Scottish highlands and islands; provides a range of consultancy services for voluntary and community groups working in the field of sustainable resource management.

Industry Nature Conservation Association—INCA: 1 Belasis Court, Belasis Hall Technical Park, Greenwood Rd, Billingham TS23 4AZ; tel (1642) 370319; fax (1642) 370288; e-mail plover@inca.uk.com; internet www.inca.uk.com; Prin Officers: D H Robinson, D P Muir; f 1987; association of industrial companies, local authorities and nature conservation organizations; encourages industry to consider the environment and undertake habitat improvement schemes; offers advice on practical conservation schemes.

Institute for Earth Education: P.O. Box 91, Tring, Hertfordshire HP23 4RS; tel (144) 289-0875; e-mail office@earthed.org.uk; internet www.earthed.org.uk; develops and disseminates model Earth Education (the process of helping people build an understanding of, appreciation for, and a harmony with the earth and its life) programmes; conducts workshops for teachers, leaders and others involved in educating people about environmental problems; hosts regional and international gatherings; publishes a mail order catalogue of books and programme materials, and various journals and newsletter; supports a network of major and local branches.

Institute for Social Inventions: 6 Blackstock Mews, Blackstone Road, London N4 2BT; tel (20) 7359-8391; fax (20) 7354-3831; e-mail ideas@alberyfoundation.org; internet www.globalideasbank.org; Director: Nick Temple; f 1985; runs the Global Ideas Bank website; publishes an annual compendium of new ideas; runs social invention workshops.

Institute of Biology: 20–22 Queensberry Pl, London SW7 2DZ; tel (20) 7581-8333; fax (20) 7823-9409; e-mail info@iob.org; internet www.iob.org; Pres: Dr Nancy Lane; Chief Exec: Alan Malcolm; f 1950; professional body; activities include formulation and submission to government and other bodies of policy documents; publications and symposia.

Institute of Chartered Foresters—ICF: 7[a] St Colme St, Edinburgh EH3 6AA; tel (131) 225-2705; fax (131) 220-6128; e-mail icf@charteredforesters.org; internet www.charteredforesters.org; Pres: Wilma C G Harper; Vice-Pres: Brian Salter; Exec Dir: Peter H Wilson; f 1925; safeguards the public interest in forestry matters; advises the Government on matters of policy; maintains standards regulating entry into the forestry profession.

Institute of Ecology and Environmental Management: 45 Southgate Street, Winchester, Hants SO23 9EH; tel (1962) 868626; fax (1962) 868625; e-mail enquiries@ieem.demon.co.uk; internet www.ieem.org.uk; Pres: Dr Chris Spray; Exec Dir: Dr Jim Thompson; professional membership organization; aims to publicize the profession of ecology and environmental management, maintain and enhance professional standards and promote an ethic of care for the environment among members and their clients and employers.

Institution of Environmental Sciences: 1 Ebury St, London SW1W 0LU; tel (20) 7730-5516; e-mail enquires@ies-uk.org.uk; internet www.ies-uk.org.uk; Chair: J Baines; Hon Sec: Dr R A Fuller; Hon Treas: J Motture; f 1971; representative body for professional environmental scientists; meetings, seminars and conferences; responses to government papers; accreditation of environmental courses; publications and information dissemination.

Intermediate Technology Development Group—ITDG: Shumacher Centre for Technology and Development, Bourton Hall, Bourton on Dunmore, Rugby CV23 9QZ; tel (1926) 634400; fax (1926) 634401; e-mail itdg@itdg.org.uk; internet www.itdg.org; Chief Exec: Cowan Coventry; f 1966; aims to enable people with little or no income to develop and use productive technologies and methods which contribute to the sustainable, long-term development of their communities; offices in Bangladesh, Kenya, Nepal, Peru, Sri Lanka, Sudan and Zimbabwe; member of the Business Advisory Service and Information Network—BASIN.

International Centre for Conservation Education—ICCE: Brocklebank, Butts Lane, Woodmancote, Cheltenham GL52 9QH; tel and fax (1242) 674839; e-mail enquiries@icce.org.uk; internet www.icce.org.uk; Dir: Mark Boulton; environmental education, especially in Africa, training and consultancy; design and production of resource materials, especially audio-visual and CD-ROM; specialized desk-top publishing (DTP) and print department for conservation publications and reports.

International Centre for Protected Landscapes, University College of Wales: Unit 8e, Science Park, Aberystwyth SY23 3AH; tel (1970) 622620; fax (1970) 622619; e-mail icpl@protected-landscapes.org; internet www.protected-landscapes.org; Exec Dir: Dr Liz Hughes; resource centre, research, consultancy, training and conferences on protected landscape planning and management; aims to promote the concept of protected landscapes and sustainable development world-wide.

International Tree Foundation: Sandy Lane, Crawley Down RH10 4HS; tel (1342) 712536; fax (1342) 718282; e-mail info@internationaltreefoundation.org; internet www.internationaltreefoundation.org; Chair: Spencer G Keys; Deputy Chair: Peter Rapson; f 1922; formerly Men of the Trees; plants trees world-wide to increase land fertility; particularly active in desert areas and tropical rainforests; membership of approx 3,000.

International Tree Planting Committee: International Tree Foundation, Sandy Lane, Crawley Down RH10 4HS; tel (1342) 712536; fax (1342) 718282; Chair: Peter Rapson; assists in tree planting and related research, particularly in developing countries, in order to combat desertification and land erosion, sustain tropical rainforests, secure water-table improvement and provide timber, fuelwood, edible produce and shade.

Invertebrate Link–Joint Committee for the Conservation of British Invertebrates: c/o Royal Entomological Society, 41 Queen's Gate, London SW7 5HR; tel (20) 7584-8361; Prin Officers: Prof M Morris, S J Brooks; f 1968; monitoring of threatened species and promotion of conservation measures.

Irish Sea Forum: Department of Earth and Ocean Sciences, University of Liverpool, Liverpool L69 3BX; tel (151) 794-4089; fax (151) 794-5196; e-mail h.davies@liverpool.ac.uk; internet www.liv.ac.uk/~isf1/isfhome.html; Dir and Chair: Dr D F Shaw; Admin: H Davies; f 1992; formerly Irish Sea Study Group; organizes seminars on issues related to the environmental health of the Irish Sea and the sustainable use of its resources.

Jane Goodall Institute UK: 15 Clarendon Park, Lymington SO41 8AX; tel (23) 8033-5660; fax (23) 8033-5661; e-mail info@janegoodall.org.uk; internet www.janegoodall.org.uk; Founding Trustee: Dr Jane Goodall; Chair: David Gregson; Office Administrator: Claire Quarendon; f 1989; wildlife research, education and conservation; research on chimpanzees, an endangered species; provision of sanctuaries for orphaned chimpanzees illegally taken from the wild in the Republic of the Congo, Kenya, Tanzania and Uganda; concerned with conservation of habitats and endangered species; conservation education, particularly through the 'Roots and Shoots' global environmental and humanitarian programme for young people; part of a network of institutes in Austria, Canada, the Republic of the Congo, Germany, Italy, the Netherlands, South Africa, Tanzania, Uganda and the USA.

Joint Nature Conservation Committee—JNCC: Monkstone House, City Rd, Peterborough PE1 1JY, United

Kingdom; tel (1733) 562626; fax (1733) 555948; e-mail communications@jncc.gov.uk; internet www.jncc.gov.uk; Man Dir: Deryck Steer; Communications Man: Alex Geairns; f 1991; represents the UK nature conservation agencies in international forums and is itself a forum through which English Nature, Scottish Natural Heritage and the Countryside Council for Wales deliver their statutory responsibilities for Great Britain and internationally; also includes representatives from the Countryside Agency and the Environment and Heritage Service (Northern Ireland).

Jupiter Environmental Research Unit: PO Box 300, West Malling ME19 4YY; tel (20) 7412-0703; fax (20) 7412-0705; e-mail green.dept@jupiter-group.co.uk; internet jupiteronline.co.uk; Dir: Emma Howard Boyd; f 1988; affiliated to Jupiter Asset Management, a member of the Commerzbank Group; research into corporate environmental performance in support of green investment funds.

Just World Partners: 45 York Pl, Edinburgh EH1 3HP; tel (131) 557-0705; fax (131) 558-2501; e-mail jw@justworld.org.uk; internet www.justworld.org.uk; Chair: Canon Rex Davis; Exec Dir: Dorothy McIntosh; f 1981; formerly UK Foundation for the Peoples of the South Pacific; promotes sustainable forestry projects; assists in the marketing of timber from certified sources and encourages the promotion of fair trade non-timber products; supports the sustainable use of marine resources and the protection of coral reefs; works with a number of partners in the Pacific Islands and South-east Asia.

Land Heritage: Summerhill Farm, Hittisleigh, Exeter EX6 6LP; tel (1647) 24511; fax (1647) 24588; e-mail enquiries@landheritage.org.uk; internet www.landheritage.org.uk; Dir: Robert Brighton; Business and Office Man: Ruth Curtis; f 1984; managed by a board of seven trustees; rental of land to organic farmers; concerned with the protection and sustainability of small family farms, committed to organic husbandry, provides conversion advice, concerned with conservation of our native flora and fauna.

Landscape Institute: 33 Great Portland St, London W1W 8QG; tel (20) 7299-4500; fax (20) 7299-4501; e-mail mail@l-i.org.uk; internet www.l-i.org.uk; Pres: Rod Edwards; Exec Dir: Christine Jackson (acting); professional body for landscape architects, managers and scientists.

London Ecology Unit,: Befdford House 125 Camden High Street London NW1 7JR; tel (20) 7267 7944; fax (20) 7267 9334.

London Green Belt Council: 13 Oakleigh Park Ave, Chislehurst BR7 5PB; tel (20) 8467-5346; e-mail greenbelt@brookmans.com; internet website.lineone.net/~greenbelt; Chair: R W G Smith; Hon Sec: L G Holt; f 1954; monitoring of the preservation of the London Metropolitan Green Belt, areas of open land surrounding Greater London; advice and assistance to affiliated members on planning applications, local plans and public inquiries; representations to the Department of the Environment, Transport and the Regions; approx 130 affiliated amenity organizations.

London Wildlife Trust: Skyline House, 200 Union St, London SE1 0LW; tel (20) 7261-0447; fax (20) 7633-0811; e-mail enquiries@wildlondon.org.uk; internet www.wildlondon.org.uk; Chief Exec: Carlo Laurenzi; protection and improvement of London's green spaces for the benefit of people and wildlife; management of 57 wildlife sites; organization of over 400 free events annually.

Lothian and Edinburgh Environmental Partnership: 36 Newhaven Rd, Edinburgh EH6 5PY; tel (131) 555-4010; fax (131) 555-2768; e-mail DGreenhill@LEEP.org.uk; internet www.leep.org.uk; Dir: Simon Lee; supports initiatives to supply or promote green products and services, and environmental technology; particularly active in the areas of recycling, energy conservation and transport.

LSE Environment: Centre for Environmental Policy and Governance: Department of Geography and Environment, Houghton St, London WC2A 2AE; tel (20) 7955-7903; fax (20) 7955-7412; e-mail Environ.Policy@lse.ac.uk; internet www.lse.ac.uk/collections/geographyAndEnvironment/CEPG; Dir: Yvonne Rydin; acts as a base for premier social science research on the environment and hosts a variety of seminars, workshops and public events.

Mammal Society: 2b Inworth St, London SW11 3EP; tel (20) 7350-2200; fax (20) 7350-2211; e-mail enquiries@mammal.org.uk; internet www.abdn.ac.uk/mammal; Chair: Prof Stephen Harris; Pres: Dr D W Yalden; Hon Sec: G Bemmeny; f 1954; promotes study and conservation of mammals; publications and meetings.

Marine Biological Association of the United Kingdom: The Laboratory, Citadel Hill, Plymouth PL1 2PB; tel (1752) 633207; fax (1752) 633102; e-mail sec@mba.ac.uk; Pres: Sir Neil Chalmers; Dir and Sec: Prof Steve Hawkins; research, publications, scientific meetings and an educational programme in co-operation with Plymouth Aquarium.

Marine Conservation Society—MCS: Unit 3' Wolf Business Park, Alton Rd, Ross-on-Wye, Hereford HR9 5NB; tel (1989) 566017; fax (1989) 567815; e-mail info@mcsuk.org; internet www.mcsuk.org; Pres: HRH The Prince of Wales; Vice-Pres: Prof David Bellamy; Dir, Conservation: Sam Fanshawe; f 1983; protection of the marine environment through research promotion, information dissemination and lobbying; organizes volunteer projects, diving surveys and beach cleaning operations.

Maritime Foundation: 202 Lambeth Road, London SE1 7JW1; tel (20) 7928 8100; fax (20) 7401 2537; e-mail secretary@bmcf.org.uk; internet www.bmcf.org.uk; Hon Pres: The Countess Mountbatten of Burma.

Mid Career College: POB 20, Cambridge CB1 5DG; tel (1223) 880016; fax (1223) 881604; e-mail courses@mid-career-college.ac.uk; internet www.mid-career-college.ac.uk; Dir: Dr Alan Sherratt; Head of Admin: Ann Chapman; courses and conferences on energy and environmental issues, building services, management and good practice design.

National Council for the Conservation of Plants and Gardens—NCCPG: The Pines, RHS Garden, Wisley, Woking GU23 6QP; tel (1483) 211465; fax (1483) 212404; e-mail info@nccpg.org.uk; internet www.nccpg.com; Gen Admin: Genevieve Melbourne Webb; Plant Conservation Officer: Ros Johnson; f 1978; conservation of British garden plants using the skills of amateur and professional gardeners; 42 county groups with over 6,000 members.

National Council for Voluntary Organisations—NCVO: Regent's Wharf, 8 All Saints St, London N1 9RL; tel (20) 7713-6161; fax (20) 7713-6300; e-mail HelpDesk@ncvo-vol.org.uk; internet www.ncvo-vol.org.uk; Chair: Norman Warner; Chief Exec: Stuart Etherington; Environment Support Team provides information and assistance for environmental voluntary action by working with networks, producing publications and operating a referrals service.

National Society for Clean Air and Environmental Protection: 136 North St, Brighton BN1 1RG; tel (1273) 326313; fax (1273) 735802; e-mail info@nsca.org.uk; internet www.nsca.org.uk; Sec-Gen: (vacant); f 1899; conferences, training events and campaigns on air pollution, noise and environmental protection issues.

National Trust for Scotland: Wemyss House, 28 Charlotte Sq, Edinburgh EH2 4ET; tel (131) 243-9300; fax (131) 243-9301; e-mail information@nts.org.uk; internet www.nts.org.uk; Pres: The Earl of Dalkeith; Chair: Prof Roger

Wheater; Chief Exec: Robin Pellew; promotes the preservation of places of historic and architectural interest or natural beauty.

National Trust—National Trust for Places of Historic Interest or Natural Beauty: 36 Queen Anne's Gate, London SW1H 9AS; tel (20) 7222-9251; fax (20) 7222-5097; e-mail enquiries@thenationaltrust.org.uk; internet www.nationaltrust.org.uk; Chair: William Proby; Dir-Gen: Fiona Reynolds; f 1895; charitable membership organization; promotion of environmental protection through the development and application of innovative, sustainable solutions to land and building conservation, for the benefit of the natural and cultural environment.

Northern Ireland Regional Office: National Trust, Rowallane House, Saintfield, Ballynahinch BT24 7LH; tel (1238) 510721; fax (1238) 511242; internet www.nationaltrust.org.uk; Dir: Ruth Laird; The National Trust in Northern Ireland holds and protects some of the most important landscape features, countryside and historic houses and gardens in the country.

Natural Death Centre: 6 Blackstock Mews, Blackstone Road, London N4 2BT; tel (0871) 288-2098; fax (20) 7354-3831; e-mail ndc@alberyfoundation.org; internet www.naturaldeath.org.uk; Dir: Stephanie Wienrich (acting); concerns include promotion of environmentally friendly burial in nature reserves or on the deceased's own land, using cardboard or recycled coffins.

Natural History Museum: Cromwell Rd, London SW7 5BD; tel (20) 7938-9123; e-mail direct@nh.ac.uk; internet www.nhm.ac.uk; Chair, Bd of Trustees: Lord Oxburgh; Dir: Dr M Dixon; formerly the Natural History Depts of the British Museum, and a branch comprising the Zoological Museum at Tring, Hertfordshire; became a separate institution 1963; incorporates the Geological Museum.

NEA: St Andrew's House, 90–92 Pilgrim St, Newcastle upon Tyne NE1 6SG; tel (191) 261-5677; fax (191) 261-6496; e-mail info@nea.org.uk; internet www.nea.org.uk; Chief Exec: William Gillis; f 1981; formerly Neighbourhood Energy Action; promotion of efficient use of energy to relieve poverty and protect health; dissemination of information to government and the private sector.

Network for Alternative Technology and Technology Assessment—NATTA: c/o Energy and Environment Research Unit, Open University, Faculty of Technology, Walton Hall, Milton Keynes MK7 6AA; tel (1908) 654638; e-mail s.j.dougan@open.ac.uk; internet technology.open.ac.uk/eeru/natta; Prin Officers: Prof David Elliott, Tam Dougan; f 1976; promotes renewable energy technologies and offers information, advice and a careers service.

New Economics Foundation: 3 Jonathan Street, London SE11 5NH; tel (20) 7820-6300; fax (20) 7820-6301; e-mail info@neweconomics.org; internet www.neweconomics.org; Deputy Dir: Andrea Westfall; disseminates information to members; supports research into viable alternatives to unrestrained growth; promotes the formulation of alternative economic indicators and the conservation of energy.

Northern Ireland Environment Link: 77 Botanic Ave, Belfast BT7 1JL; tel (28) 9031-4944; fax (28) 9031-1558; e-mail info@nienvironmentlink.org; internet www.nienvironmentlink.org; Dir: Dr Susan Christie; Chair: Isabel Hood; f 1990; forum for voluntary conservation organizations; liaison between the voluntary sector and government; provision of environmental information; promotion of environmental interest and activity.

Open Spaces Society: 25a Bell St, Henley-on-Thames RG9 2AB; tel (1491) 573535; fax (1491) 573051; e-mail hq@oss.org.uk; internet www.oss.org.uk; Chair: Rodney Legg; Gen Sec: Kate Ashbrook; f 1865; campaigns to create and conserve common land, town and village greens, open spaces and rights of public access, in town and country, in England and Wales.

Open University, Environmental Education and Training Programme: Walton Hall, Milton Keynes MK7 6AA; tel (1908) 652590; fax (1908) 858407; internet www.open.ac.uk; Liaison Officer: Mark Yoxon; courses offered: Environmental Decision Making: a systems approach is one of a number of individual 6-month courses which can be put together to build a diploma or an MSc in Environmental Ethics; Enterprise and the Environment; Integrated Safety Health and Environmental Management; courses are offered twice yearly in May and November.

Organic Organisation/HDRA: Ryton Organic Gardens, Ryton-on-Dunsmore, Coventry CV8 3LG; tel (24) 7630-3517; fax (24) 7663-9229; e-mail enquiry@hdra.org.uk; internet www.hdra.org.uk; Chair: Bill Blyth Chief Exec: Susan Kay-Williams; Exec Dir: Jackie Gear; f 1958; educational charity, formerly the Henry Doubleday Research Association; researches, demonstrates and promotes environmentally friendly organic gardening, farming and food in the United Kingdom and small-scale agriculture in developing countries; operates three organic display gardens, a waste management consultancy, a garden landscape and design consultancy and the Heritage Seed Library to conserve old and unusual vegetable varieties; 70 affiliated local groups.

Oxfam: Oxfam House, 274 Banbury Rd, Summertown, Oxford OX2 7DZ; tel (1865) 311311; fax (1865) 313770; e-mail oxfam@oxfam.org.uk; internet www.oxfam.org.uk; Chair: David Bryer; Dir: Barbara Stocking; Exec Dir, Oxfam International: Jeremy Hobbs; f 1942; involved in sustainable development and emergency relief projects in over 70 countries.

Panos London: 9 White Lion St, London N1 9PD; tel (20) 7278-1111; fax (20) 7278-0345; e-mail info@panoslondon.org.uk; internet www.panos.org.uk; Exec Dir: James Deane; independent information organization promoting socially, environmentally and economically sustainable development world-wide; conducts research into major global issues; numerous publications; photographic library, Panos Pictures; offices in France, Nepal, India, Zambia, Uganda, Ethiopia.

Peak District National Park Authority: Aldern House, Baslow Rd, Bakewell DE45 1AE; tel (1629) 816200; fax (1629) 816310; e-mail aldern@peakdistrict-npa.gov.uk; internet www.peakdistrict.org; Nat Park Officer: C J Harrison; Dir of Admin and Sec: K M Francis; f 1951; formerly Peak Park Joint Planning Board; protection of the natural beauty of the Peak National Park; affiliated with the Countryside Commission.

Pedestrians Association: 31–33 Bondway, 3rd Floor, London SW8 1SJ; tel (20) 7820-1010; fax (20) 7820-8208; e-mail info@pedestrians.org.uk; internet www.pedestrians.org.uk; Pres: Terence Bendixson; Dir: Tom Franklin; f 1929; encourages provision of safe and convenient footways; promotes road safety; provides advice and information for pedestrians; affiliated with the European Federation of Pedestrian Associations and the International Federation of Pedestrian Associations.

Permaculture Association UK: BCM Permaculture Association, London WC1N 3XX; tel and fax ((0845) 4581805; e-mail office@permaculture.org.uk; internet www.permaculture.org.uk; Co-ordinator: Andrew Goldring; f 1984; educational charity concerned with permaculture design for sustainable land use and regional and community development; provides training courses; member of international network.

Pesticides Action Network UK: 56-64 Leonard Street, London EC2A 4JX; tel (20) 7065-0905; fax (20) 7065-0907;

e-mail admin@pan-uk.org; internet www.pan-uk.org; Dir: Barbara Dinham; f 1987; independent charity; addresses health and environmental problems caused by pesticides; advocates sustainable agricultural practices; affiliated with the Pesticides Action Network; publications.

Pesticides Trust: Eurolink Centre, 49 Effra Road, London SW2 1BZ; tel (20) 274-8895; fax (20) 274-9084; e-mail pesttrust@gn.apc.org; internet www.ful.ac.be/hotes/sandrine/pestrust.htm; f 1987; work nationally and internationally with like-minded organizations and individuals to minimize and ultimately to eliminate the hazards of pesticides.

Plantlife: 14 Rollestone St, Salisbury, Wilts SP1 1DX; tel (1722) 342730; fax (1722) 329035; e-mail enquiries@plantlife.org.uk; internet www.plantlife.org.uk; Chief Exec: Dr Jane Smart; Chair: Adrian Darby; Pres: Prof David Bellamy; Patron: HRH The Prince of Wales; f 1990; conservation of wild plants and their habitats; conferences; research and publishing; member of the European Environmental Bureau—EEB, Wildlife and Countryside Link and the World Conservation Union—IUCN.

Policy Studies Institute: 100 Park Village East, London NW1 3SR; tel (20) 7468-0468; fax (20) 7388-0914; e-mail postmaster@psi.org; internet www.psi.org.uk; Dir: Prof Jim Skea; independent research organization aiming to inform public policy; undertakes studies of economic, industrial and social policy and of matters relating to the environment.

Population Concern: 3–5 Highgate Studios, 53–79 Highgate Rd, London NW5 1TL; tel (20) 7421-8500; fax (20) 7267-6788; e-mail info@populationconcern.org.uk; internet www.populationconcern.org.uk; Chief Exec: Wendy Thomas; Information Officer: Kathy Siddle; seeks to create awareness of the related issues of population, development and the environment through the UK Advocacy Programme; operates family planning programmes in 18 developing countries.

Primate Society of Great Britain: Dept of Psychology, University of Reading, Bldg 3, Earley Gate, Whiteknights, Reading RG6 2AL; e-mail info@psgb.org; internet www.psgb.org; Prin Officer: Dr Hilary Box; affiliated to the International Primatological Society.

Railway Development Society—RDS: 15 Clapham Rd, Lowestoft NR32 1RQ; tel (1584) 890807; fax (1584) 891300; Chair: Steve Wilkinson; Gen Sec: Trevor Garrod; f 1978; campaigns for the retention, expansion and greater use of rail transport for passengers and freight; approx 70 local rail users' groups are affiliated to RDS.

Ramblers' Association: Camelford House, 2nd Floor, 87–90 Albert Embankment, London SE1 7TW; tel (20) 7339-8500; fax (20) 7339-8501; e-mail ramblers@london.ramblers.org.uk; internet www.ramblers.org.uk; Chief Exec: Nick Barrett; Dir of Campaigns: Jacquetta Fewster; promotes walking for pleasure; protects rights of way; campaigns for access to open country; defends the beauty of the countryside.

Recovered Paper Sector Confederation of Paper Industries: 1 Rivenhall Road, Swindon, Wiltshire SN5 7BD; tel (1793) 889638; fax (1793) 878700; e-mail pseggie@paper.org.uk; internet www.recycledpaper.org.uk; Pres: Richard Bourdon; Man: Peter Seggie; trade association representing the waste paper recovery and processing industry; acts as a centre for national and international exchange of ideas and knowledge.

Reforesting Scotland: 62–66 Newhaven Rd, Edinburgh EH6 5QB; tel (131) 554-4321; fax (131) 554-0088; e-mail info@reforestingscotland.org; internet reforestingscotland.gn.apc.org; Chair: Les Bates; Projects Co-ordinator: Nick Marshall; creates awareness of the deforestation of Scotland and its implications; promotes ecological restoration and development through reforestation and the concepts of sustainable forestry and integrated land use; collaborates with similar organizations world-wide.

Robens Centre for Public and Environmental Health: University of Surrey, Building AW 02, Guildford, Surrey, GU2 7XH; tel (1483) 689209; fax (1483) 689971; internet www.robenscentres.com/rcpeh; Dir: Prof Gareth Rees; aims to advance human and environmental health and well-being through the provision of independent scientific research, investigation, advice and training world-wide; activities cover environmental health, human factors, ergonomics and chemical safety.

Royal Botanic Gardens: Kew, Richmond TW9 3AB; tel (20) 8332-5000; fax (20) 8332-5197; e-mail info@kew.org.uk; internet www.kew.org; Chair: Lord Selborne; Dir: Prof Sir Peter Crane; f 1759; scientific research institute and tourist attraction; maintains a second garden at Wakehurst in Sussex.

Royal Entomological Society: 41 Queen's Gate, London SW7 5HR; tel (20) 7584-8361; fax (20) 7581-8505; e-mail reg@royensoc.co.uk; internet www.royensoc.co.uk; Pres: Dr Hugh Loxdale; Registrar: W H F Blakemore; f 1833; improvement and diffusion of entomological science; holds meetings, seminars and symposia; publications.

Royal Environmental Health Institute of Scotland: 3 Manor Place, Edinburgh, EH3 7DH; tel (131) 225-6999; fax (131) 225-3993; e-mail rehis@rehis.org.uk; internet www.rehis.com; Pres: David Cameron.

Royal Geographical Society/Institute of British Geographers: 1 Kensington Gore, London SW7 2AR; tel (20) 7591-3000; fax (20) 7591-3001; e-mail info@rgs.org; internet www.rgs.org; Pres: Sir Neil Cossons; Dir and Sec: Dr Rita Gardner; f 1830; promotion and diffusion of geographical knowledge.

Royal Scottish Forestry Society—RSFS: Hagg-on-Esk, Canonbie DG14 0XE; tel (1387) 371518; fax (1387) 371418; e-mail rsfs@ednet.co.uk; internet www.rsfs.org; Pres: Sir Michael Strang-Steel; Dir: Andrew G Little; f 1854; membership organization; involved in forest education and training, field excursions, formulation of forest policy, research, forest industry development and dissemination of information.

Royal Society: 6-9 Carlton House Terrace, London SW1Y 5AG; tel (20) 7451-2500; fax (20) 7930-2170; e-mail info@royalsoc.ac.uk; internet www.royalsoc.ac.uk; Exec Sec: S. J. Cox; f 1660; national academy of science; promotes and advances all fields of physical, biological, medical and agricultural science, mathematics and engineering, and their applications.

Royal Society for the Protection of Birds—RSPB: The Lodge, Sandy SG19 2DL; tel (1767) 680551; fax (1767) 692365; e-mail enquiries@rspb.org.uk; internet www.rspb.org.uk; Chair: Prof Ian Newton; Chief Exec: Graham Wynne; f 1889; conservation of wild birds and their environment; creation of nature reserves.

Royal Society of Wildlife Trusts: The Kiln, Waterside, Mather Road, Newark, Nottinghamshire, NG24 1WT; tel (0870) 036 1000; fax (0870) 036 0101; e-mail grants@rswt.org; internet www.rswt.org; Chief Exec: Tim Cordy; Dir of Conservation: Tim Sands; voluntary organization concerned with all aspects of wildlife protection; partnership of 47 wildlife trusts, 47 urban wildlife groups and a junior wing; protects almost 2,000 sites nationally; formerly the Society for the Promotion of Nature Reserves (SPNR) then the Society for the Promotion of Nature Conservation (SPNC) and most recently the Royal Society for Nature Conservation (RSNC).

Wildlife Trust for Bedfordshire, Cambridgeshire, Northamptonshire and Peterborough: The Manor House, Broad Street, Cambourne CB3 6DH; tel (1954)

713500; fax (1954) 710051; e-mail cambridgeshire@wildlifebcnp.org; internet www.wildlifebcnp.org; Dir: Nick Hammond; Conservation Dir: Brian Eversham; f 1957; formerly Bedfordshire and Cambridgeshire Wildlife Trust; acquisition and management of land; creation of awareness of biodiversity and conservation issues among policy-makers and the public; member of The Wildlife Trusts.

***rural*Scotland**: Gladstone's Land, 3rd Floor, 483 Lawnmarket, Edinburgh EH1 2NT; tel (131) 225-7012; fax (131) 225-6592; e-mail aprs@aprs.org.uk; internet www.aprs.org.uk; Pres: Jimmie Macgregor; f 1926; fmrly Association for the Protection of Rural Scotland; protection of Scotland's rural environment through research and involvement in the planning process; presents annual awards for design excellence in the countryside.

Salmon & Trout Association: Fishmongers' Hall, London Bridge, London EC4R 9EL; tel (20) 7283-5838; fax (20) 7626-5137; e-mail hq@salmon-trout.org; internet www.salmon-trout.org; Chair: Tony Bird; Exec Dir: Paul Knight; f 1903; campaigns to safeguard the game fishing environment and represents the views and concerns of members.

Science and Technology Policy Research—SPRU, Environment and Energy Programme: Mantell Bldg, University of Sussex, Falmer, Brighton BN1 9RF; tel (1273) 686758; fax (1273) 685865; e-mail a.r.wilson@sussex.ac.uk; internet www.sussex.ac.uk/spru; Programme Administrator: Andy Wilson; f 1965; policy-oriented academic research in four main fields: technological innovation and environmental performance, science and appraisal in environmental decision-making, energy and environment.

Scott Polar Research Institute, University of Cambridge: Lensfield Rd, Cambridge CB2 1ER; tel (1223) 336540; fax (1223) 336549; e-mail enquiries@spri.cam.ac.uk; internet www.spri.cam.ac.uk; Dir: Prof Julian Dowdeswell; study of the Arctic, Antarctica and glaciology, with particular emphasis on environmental change; offers one-year postgraduate course in polar studies.

Scottish Conservation Bureau: Longmore House, Salisbury Place, Edinburgh, EH9 1SH; tel (131) 668 8668; fax (131) 668 8669; e-mail hs.conservation.bureau@scotland.gsi.gov.uk; internet www.historic-scotland.gov.uk.

Scottish Council for National Parks: 15 Park Terrace, Stirling FK8 2JT; tel (1786) 465714; fax (1786) 473843; Chair and Hon Sec: Brian K Parnell; Vice-Chair: The Countess of Glasgow; organizes meetings and produces publications with the aim of securing legislation for the establishment of national parks in Scotland.

Scottish Field Studies Association: Kindrogan Field Centre, Enochdhu, Blairgowrie PH10 7PG; tel (1250) 881286; fax (1250) 881433; Chair: Dr Alan Pyke; Dir: Alison Gimingham; f 1943; provides courses on aspects of natural history and the environment for educational establishments, groups of adults and individuals with the aim of instilling understanding of the Scottish countryside; also operates professional training courses.

Scottish Solar Energy Group: School of Engineering, Napier University, 10 Colinton Rd, Edinburgh EH10 5DT; tel (131) 455-2660; fax (131) 455-2264; e-mail k.macgregor@napier.ac.uk; internet www.sseg.org.uk; Chair: Kerr MacGregor; Sec: Dr Colin Porteous; promotion of research, development and use of solar energy in Scotland through meetings, visits and conferences.

Scottish Wildlife and Countryside Link—SWCL: 2 Grosvenor House, Shore Rd, Perth PH2 8BD; tel (1738) 630804; fax (1738) 643290; e-mail enquiries@scotlink.org; internet www.scotlink.org; Co-ordinator: Jennifer Anderson; f 1987; liaison body for voluntary organizations concerned with the conservation of the wildlife and countryside of Scotland; member of the World Conservation Union—IUCN.

Council for Scottish Archaeology—CSA: c/o National Museums of Scotland, Chambers St, Edinburgh EH1 1JF; tel (131) 247-4119; fax (131) 247-4126; e-mail info@scottisharchaeology.org.uk; internet www.britarch.ac.uk/csa; Dir: Eila MacQueen; f 1945; encourages study and conservation of Scotland's archaeological heritage; promotion of good conservation policy and practice; co-ordination of Young Archaeologists' Club Scottish Network; extensive educational outreach programme; publications.

Sea Watch Foundation: 11 Jersey Road, Oxford OX4 4RT; tel and fax (1865) 717276; e-mail info@seawatchfoundation.org.uk; internet www.seawatchfoundation.org.uk; Head of Research: Dr P G W Evans; f 1991; cetacean monitoring unit; maintenance of an observer network throughout the United Kingdom to record the status and distribution of whales, dolphins and porpoises; monitoring of conservation threats, including pollution incidents, accidental capture in fishing nets and noise disturbance; dissemination of information to environmental bodies, government, industry and the public; affiliated with the Mammal Society.

Socialist Environment and Resources Association—SERA: 11 Goodwin St, London N4 3HQ; tel (20) 7263-7389; fax (20) 7263-7424; e-mail seraoffice@btconnect.com; internet www.serauk.org.uk; Chair: Bill Eyres; Nat Co-ordinator: Alex Morrell; f 1973; environmental group affiliated to the Labour Party; aims to ensure that environmental considerations are included in all aspects of policy- and decision-making.

Society for Environmental Exploration—Frontier: 50–52 Rivington Street, London EC2A 3QP; tel (20) 7613-2422; fax (20) 7613-2992; e-mail enquiries@frontierprojects.ac.uk; internet www.frontierprojects.ac.uk; Man Dir: Eibleis Fanning; f 1989; non-profit company; aims to provide a channel through which individuals can make a significant contribution to urgent research and conservation projects; projects are developed in collaboration with institutions in co-operating countries and operate under the name Frontier; teams of self-financing volunteers assembled in the United Kingdom implement specific programmes of field research and practical conservation work in Madagascar, Tanzania and Viet Nam.

Society for Responsible Use of Resources in Agriculture and on the Land—RURAL: Chester House, Hillbury Rd, Alderholt, Fordingbridge SP6 3BQ; internet www.rural.org.uk; Dir: Brig H J Hickman; policy studies for food, farming and countryside.

Society for the Protection of Animals Abroad—SPANA: 14 John St, London WC1N 2EB; tel (20) 7831-3999; fax (20) 7831-5999; e-mail hq@spana.org; internet www.spana.org; Pres: Lord Newall; Chair: Ian W Frazer; Chief Exec: Jeremy F Hulme; protection of working animals and other animals in North Africa and the Middle East; education programme introduces concepts of animal welfare and care for the environment.

Society of Chemical Industry—SCI: 14–15 Belgrave Sq, London SW1X 8PS; tel (20) 7598-1500; fax (20) 7598-1545; e-mail secretariat@soci.org; internet www.soci.org; World Pres: Sir Tom McKillop; Chair, Cncl: Dr John Beacham; Gen Sec and Chief Exec: Richard Denyer; learned society aiming to advance the application of chemistry and related sciences for public benefit; produces publications and holds meetings on diverse subjects, including environmental issues.

Soil Association: Bristol House, 40–56 Victoria St, Bristol BS1 6BY; tel (117) 929-0661; fax (117) 925-2504; e-mail info@soilassociation.org; internet www.soilassociation.org;

Dir: Patrick Holden; Chair: Helen Browning; f 1946; educates the public, offers advice, produces publications, lobbies the authorities and conducts research on organic agriculture and food; sets independent standards for organic farming based on international guide-lines; subsidiary company inspects and licenses growers, food processors and manufacturers.

Solar Energy Society: c/o School of Technology, Oxford Brookes University, Headington Campus, Gypsy Lane, Oxford OX3 0BP; tel (1865) 484367; fax (1865) 484263; e-mail ukises@brookes.ac.uk; internet www.thesolarline.com; Chair: Prof M Hutchins; Sec: Fred Treble; Contact: Christiane Buckle; national section of the International Solar Energy Society; learned society; conferences and publications.

Solar Trade Association Ltd: The National Energy Centre, Davy Ave, Knowlhill, Milton Keynes MK5 8NG; tel (1908) 442290; fax (0870) 0529194; e-mail enquiries@solartradeassociation.org.uk; internet www.solartradeassociation.org.uk; Chair: John Blower; Chief Exec: Gareth Ellis; f 1978; acts as a national enquiry centre on solar energy and a forum for organizations involved with the solar industry; regulates the activities and standards of member organizations.

Southampton Oceanography Centre—SOC: University of Southampton Waterfront Campus, European Way, Southampton SO14 3ZH; tel (2380) 596666; fax (2380) 596032; e-mail noi@soc.soton.ac.uk; internet www.soc.soton.ac.uk; Dir: Dr Prof Howard S Roe; Dir (designate): Dr Edward Hill; f 1995; superseded the Institute of Oceanographic Sciences, Deacon Laboratory—IOSDL (formerly the National Institute of Oceanography—NIO) and the James Rennell Centre for Ocean Circulation; joint venture between the Natural Environment Research Council—NERC and the University of Southampton; centre for all aspects of marine research and earth sciences; deep ocean research covers climate change and ocean circulation, ocean biology, hydrography, computer modelling, satellite remote sensing, including oil-slick analysis; training, and undergraduate and postgraduate teaching; maintains the Marine Information and Advisory Service—MIAS and the National Oceanographic Library.

Sunseed Trust: c/o 18a Kirkhill Gardens, Penicuik, Midlothian EH26 8JE; tel (1274) 869061; internet www.sunseed.org.uk; Chair: Adrian Windisch; Project Man: Carol Biggs; f 1984; research and development and education on ways of containing or reclaiming deserts and on technologies, such as tree-planting and solar ovens, of benefit to those living on desert margins; activities conducted principally by paying volunteers; research base in south-eastern Spain.

Sustain: 94 White Lion Street, London N1 9PF; tel (20) 7837-8980; fax (20) 7837-1141; e-mail sustain@sustainweb.org; internet www.sustainweb.org; Chair: Tim Lang.

SustainAbility Ltd: 20-22 Bedford Row, London, WC1R 4EB; tel (20) 7269-6900; fax (20) 7269-6901; e-mail info@sustainability.co.uk; internet www.sustainability.co.uk; Exec Dir: Peter Zollinger; f 1987; sustainable development campaigns; consultancy; environmental strategy, management, reporting and verification; environmental communication.

Sustainable Urban Neighbourhood Initiative: 5th Floor, 10 Little Lever St, Manchester M1 1HR; tel (161) 200-5500; fax (161) 200-5500; e-mail sun@urbed.co.uk; internet www.urbed.com; Prin Officer: David Rudlin; f 1996; research, networking and advocacy in the field of sustainable urban development; publications; affiliated with the Urban and Economic Development Group—URBED.

SUSTRANS: National Cycle Network Centre, 2 Cathedral Square, College Green, Bristol BS1 5DD; tel (845) 113-0065; fax (117) 915-0124; e-mail info@sustrans.org.uk; internet www.sustrans.org.uk; Dir and Chief Engineer: John Grimshaw; Devt Dir: Carol Freeman; Information Officer: Nic Antonini; f 1984; sustainable transport charity; works on practical projects to encourage people to walk, cycle and use public transport in order to reduce motor traffic and its adverse effects; 5000 miles of their flagship project, the National Cycle Network, opened in 2000 with 5,000 miles, will increase to 10,000 miles by 2005; other projects include Safe Routes to Schools, Home Zones and other practical responses to transport and environmental challenges; 40,000 Supporters; also receives support from charitable trusts, companies, the National Lottery and local authority programmes.

The Land is Ours: 16b Cherwell St, Oxford OX4 1BG; tel (1865) 460171; e-mail office@tlio.demon.co.uk; internet www.tlio.org.uk; f 1994; campaigns for access to land and to decision-making processes concerning land use, employing both conventional approaches and non-violent direct action; advocates the provision of land for homes, including low-cost and self-designed housing in cities and low-impact settlements in the countryside, farming practices that take into account the environment and rural employment, protection of common space, reform of planning and public consultation procedures, and mandatory land registration.

Timber Research and Development Association—TRADA: Stocking Lane, Hughenden Valley, High Wycombe HP14 4ND; tel (1494) 563091; fax (1494) 565487; internet www.trada.co.uk; f 1934; independent association offering research, advisory and information services in order to promote the efficient use of timber and wood products.

Tourism Concern: Stapleton House, 277–281 Holloway Rd, London N7 8HN; tel (20) 7133-3320; fax (20) 7133-3331; e-mail info@tourismconcern.org.uk; internet www.tourismconcern.org.uk; Dir: Tricia Barnett; f 1989; membership organization; promotes awareness of the impact of tourism on host communities and environments; advocates tourism that is just, sustainable and participatory.

Town and Country Planning Association: 17 Carlton House Terrace, London SW1Y 5AS; tel (20) 7930-8903; fax (20) 7930-3280; e-mail tcpa@tcpa.org.uk; internet www.tcpa.org.uk; Pres: Sir Peter Hall; Dir: Gideon Amos; f 1898; policy and research work on planning and environmental matters; reports, conferences and lobbying.

Trade Plus Aid: 17 Paxton Close, Kew Rd, Richmond TW9 2AW; tel (20) 8948-0533; fax (20) 8255-3881; e-mail charlotte@tradeplusaid.com; internet www.tradeplusaid.com; Dir: Charlotte di Vita; f 1992; initiates environmentally sound self-sufficiency programmes, ensures fair payments, decent working conditions, interest-free credit, effective quality training and access to reliable markets for some of the world's poorest communities in Brazil, the People's Republic of China, Ethiopia, Ghana, India, Kenya, Namibia, Peru, the Philippines, South Africa, Thailand and Zimbabwe.

Transport 2000: The Impact Centre, 12–18 Hoxton Street, London N1 6NG; tel (20) 7613-0743; fax (20) 7613-5280; e-mail info@transport2000.org.uk; internet www.transport2000.org.uk; Pres: Michael Palin; Exec Dir: Stephen Joseph; campaigns for environmentally sensitive transport policies.

Tree Council: 71 Newcomen Street, London SE1 1YT; tel (20) 7407-9992; fax (20) 7407-9908; e-mail info@treecouncil.org.uk; internet www.treecouncil.org.uk; Dir-Gen: Pauline Buchanan Black; f 1974; independent registered charity; aims to improve the urban and rural environment in the United Kingdom by promoting the planting and conservation of trees, to disseminate information about tree management and to act as a forum for similar organizations.

Trees for Life: The Park, Findhorn Bay, Forres IV36 0TZ; tel (1309) 691292; fax (1309) 691155; e-mail trees@findhorn.org; internet www.treesforlife.org.uk; Exec Dir: Alan Watson Featherstone; Field Officer: Adam Powell; f 1987; conducts practical conservation work for the regeneration of the Caledonian Forest in Scotland; operates a volunteer programme; produces audio-visual material and publications; works in partnership with the Forestry Commission, the Millennium Forest for Scotland Trust, the National Trust for Scotland and landowners.

Tusk Trust: 5 Townbridge House, High Street, Gillingham, Dorset SP8 4AA; tel (1747) 831005; fax (1747) 831006; e-mail tim@tusk.org; internet www.tusk.org; Contacts: Tim Jackson, Sarah Watson.

UK Centre for Economic and Environmental Development—UK CEED: Priestgate House, Suite 1, 317 Priestgate, Peterborough DE1 1JN; tel (1733) 311644; fax (1733) 312782; e-mail info@ukceed.org; internet www.ukceed.org; Exec Dir: Jonathan Selwyn; f 1984; independent charitable research institute specializing in economic analysis of environmental issues.

UK Overseas Territories Conservation Forum: 15 Insall Rd, Chipping Norton OX7 5LF; tel (1608) 644425; e-mail fmarks@ukotcf.org; internet www.ukotcf.org; Chair: Dr Mike Pienkowski; Co-ordinator: Frances Marks; promotion of the conservation of biodiversity in United Kingdom Overseas Territories through liaison with local conservation agencies, NGOs in the United Kingdom, experts and government departments; provides information and co-ordinates project activities.

University of Dundee, Department of Environmental Management: School of Town and Regional Planning, Perth Rd, Dundee DD1 4HT; tel (1382) 345239; fax (1382) 204234; e-mail a.a.jackson@dundee.ac.uk; internet www.trp.dundee.ac.uk; Course Dir: A A Jackson; f 1991; provision of four-year honours degree in Environmental Management; programme recognized by Institute of Environmental Management.

University of East Anglia, School of Environmental Sciences: Norwich NR4 7TJ; tel (1603) 592534; fax (1603) 591452; e-mail Jean.Bryant@uea.ac.uk; internet www.uea.ac.uk/env; Dean: Prof C E Vincent; Research Admin: Dr J P Darch; teaching and research in environmental sciences, including atmospheric chemistry, hydrology, hydrogeology, earth sciences, ecology, soil science, environmental economics, risk assessment, health studies, natural hazards, oceanography and geophysics.

University of Hertfordshire, Department of Geography and Environmental Sciences: College Lane, Hatfield AL10 9AB; tel (1707) 284516; fax (1707) 284514; Prin Lecturers: Dr Les Allen-Williams, Dr Avice Hall, Keith Pinn; undergraduate and postgraduate interdisciplinary courses; postgraduate research; contract research and consultancy; short courses.

University of Wales, Cardiff, School of Earth, Ocean and Planetary Science: Main Bldg, Park Pl, Cardiff CF10 3YE; tel (29) 2087-4830; fax (29) 2087-4326; e-mail earth@cf.ac.uk; internet www.earth.cardiff.ac.uk; Head: Prof Dianne Edwards; research addresses some of the most significant themes in world science, including global change, the origin and evolution of life, environmental science, natural resource exploration and the evolution of Earth and the planets.

Volunteering England: Regents Wharf, 8 All Saints Street, London, N1 9RL; tel (0845) 305-6979; fax (20) 7520 8910; e-mail information@volunteeringengland.org; internet www.volunteeringengland.org; Chief Exec: Christopher Spence; provides information and advice to volunteer organizers and campaigns for policy changes to make volunteering more accessible.

Wadebridge Ecological Conference and Study Centre: Worthyvale Manor Farm, Camelford PL32 9TT; tel (1840) 212375; fax (1840) 212808; Prin Officers: Edward Goldsmith, Peter Bunyard; conferences on Gaia; environmental study courses.

Wales Green Party/Plaid Werdd Cymru: POB 10, Mountain Ash CF45 4YZ; tel (0845) 458-1026; e-mail info@walesgreenparty.org.uk; internet walesgreenparty.org.uk; Gen Sec: Jonathan Spink; environmental political party; campaigns on social and environmental issues; contests local, general and European Parliament elections; research and promotion of relevant educational material.

Waste Watch: 56–64 Leonard St, London EC2A 4JX; tel (20) 7549-0300; fax (20) 7549-0301; e-mail info@wastewatch.org.uk; internet www.wastewatch.org.uk; Pres: Baroness Gloria Hooper; Chair: Jane Stephenson; Exec Dir: Barbara Herridge; national agency working with local and national government, industry and other groups to promote waste reduction and recycling; operates Wasteline telephone information service; publications.

West Wales Eco Centre: The Old School Business Centre, Lower St Mary St, Newport SA42 0TS; tel (1239) 820235; fax (1239) 820801; e-mail westwales@ecocentre.org.uk; internet www.ecocentre.org.uk; Chair: Colin Luker; Dir: Jake Hollyfield; f 1980; charitable organization concerned with energy conservation, promotion of renewable energy, the environmental impact of various energy technologies, planning issues and fuel poverty; free advice service and publications; consultancy and provision of energy conservation services to local authorities.

Whale and Dolphin Conservation Society—WDCS: Brookfield House, 38 St Paul St, Chippenham, Wiltshire SN15 1LJ; tel (1249) 449500; fax (1249) 449501; e-mail info@wdcs.org; internet www.wdcs.org; Conservation Dir: Alison Smith; CEO: Chris Stroud; Marketing and Communication Dir: Chris Vick; f 1987; conservation of whales, dolphins and porpoises; campaigns against commercial whaling, marine pollution and the keeping of killer whales and other species in captivity; active field projects in 30 countries.

Wildfowl and Wetlands Trust—WWT: Slimbridge GL2 7BT; tel (1453) 891900; fax (1453) 890827; e-mail enquiries@wwt.org.uk; internet www.wwt.org.uk; Man Dir: Martin Spray; f 1946; charitable organization; maintains nine visitor centres; concerned with wetland habitats, wildfowl biology and conservation; conducts research on wetland ecology; public education.

Wildlife and Countryside Link: 89 Albert Embankment, London, SE1 7TP; tel (20) 7820 8600; fax (20) 7820 8620; e-mail enquiry@wcl.org.uk; internet www.wcl.org.uk; Dir: Pippa Langford; brings together environmental voluntary organizations in the UK united by their common interest in the conservation and enjoyment of the natural and historic environment.

Willing Workers on Organic Farms—WWOOF: POB 2675, Lewes BN7 1RB; tel (1273) 476286; internet www.wwoof.org; Prin Officer: Fran Whittle; f 1971; membership organization for voluntary workers in the organic agriculture movement in the United Kingdom and world-wide.

Women's Environmental Network—WEN: POB 30626, London E1 1TZ; tel (20) 7481-9004; fax (20) 7481-9144; e-mail info@wen.org.uk; internet www.wen.org.uk; Contact: Liz Sutton; information and campaigns to educate and empower women and promote care for the environment.

Woodland Trust: Autumn Park, Grantham, Lincolnshire, NG31 6LL; tel (1476) 581135; fax (1476) 590808; e-mail enquiries@woodland-trust.org.uk; internet www.woodland-trust.org.uk; Chief Exec: Julian Purvis (acting);

f 1972; charitable organization concerned with the conservation and creation of native and broad-leaved woodland as places of recreation and as wildlife habitats; manages over 1,050 woods of varying size covering a total of 43,000 acres (17,402 ha).

Woodland Trust Northern Ireland: 1 Dufferin Court, Dufferin Avenue, Bangor, County Down BT20 3BX; tel (28) 9127-5787; fax (28) 9127-5942.

Woodland Trust Scotland: Glenruthven Mill, Abbey Road, Auchterarder, Perthshire, PH3 1DP; tel (1764) 662554; fax (1764) 662553; internet www.the-woodland-trust.org.uk.

Woodland Trust Wales (Coed Cadw): Yr Hen Orsaf, Llanidloes, Powys, SY18 6EB; tel (1686) 412508; fax (1686) 412176; internet www.woodland-trust.org.uk.

World Land Trust: Blyth House, Bridge St, Halesworth IP19 8AB; tel (1986) 874422; fax (1986) 874425; e-mail info@worldlandtrust.org; internet www.worldlandtrust.org; Chief Exec: John A Burton; f 1989; World Land Trust is the operating name of the World Wide Land Conservation Trust; registered charity involved in conservation initiatives, including land purchase and management of threatened areas; emphasis on local community involvement; operates the Philippine Reef and Rainforest Project, the Programme for Belize, Rainforest Action Costa Rica and Wyld Court Rainforest Conservation Centre (Newbury).

Programme for Belize—PFB: World Land Trust, POB 99, Saxmundham IP17 2LB; principally involved in fundraising and public awareness campaigns for the head office in Belize.

World Resource Foundation: 1st Floor, The British School, Otley, Skipton, North Yorkshire BD23 1EP; tel (1756) 709800; fax (1756) 709801; e-mail admin@residua.com; internet www.residua.com; Dir: Kit Strange; formerly Warmer Campaign; collects and disseminates information on the minimization, reuse and recycling of waste and the generation of energy from waste.

WRc PLC: Frankland Rd, Blagrove, Swindon, Wiltshire SN5 8YF; tel (1793) 865000; fax (1793) 865001; e-mail solutions@wrcplc.co.uk; internet www.wrcplc.co.uk; CEO: Ron Chapman; Man Dir: Dr John Moss; f 1927; independent research and development and consultancy organization specializing in water, waste water and total environmental management; WRc is recognized as the United Kingdom National Centre for Ecotoxicology and is the principal organization for the European Topic Centre on Inland Waters in support of the European Environment Agency; offices in Brussels (Belgium), Rome (Italy), Philadelphia and Washington, DC (USA).

WWF United Kingdom—World Wide Fund for Nature: Panda House, Weyside Park, Godalming GU7 1XR; tel (1483) 426444; fax (1483) 426409; internet www.wwf.org.uk; Pres: HRH Princess Alexandra: Chair: Christopher Ward; Chief Exec. Robert Napier; f 1961; the mission of WWF is to stop the degradation of the planet's natural environment and to build a future in which humans live in harmony with nature, by: conserving the world's biological diversity; ensuring that the use of renewable resources is sustainable, and promoting the reduction of pollution and wasteful consumption.

Yorkshire Naturalists' Union: Stonecroft, 3 Brookmead Cl, Sutton Poyntz, Weymouth, Dorset DT3 6RS; tel (1305) 837384; e-mail webmaster@ynu.org.uk; internet www.ynu.org.uk; Treasurer: J A Newbould; dedicated to the study of natural history in England, particularly Yorkshire; issues publications.

Young People's Trust for the Environment and Nature Conservation—YPTENC: 8 Leapale Rd, Guildford GU1 4JX; tel (1483) 539600; fax (1483) 301992; e-mail info@yptenc.org.uk; internet www.yptenc.org.uk; Dir: Peter Littlewood; Chair, Bd of Trustees: Ray Dafter; f 1981; independent charity dedicated to educating young people about conservation of the world's natural resources and the environment; lectures, publications and projects on most environmental topics.

Zoological Society of London: Regent's Park, London NW1 4RY; tel (20) 7722-3333; fax (20) 7586-5743; internet www.zsl.org; Pres: Sir Patrick Bateson; Sec: Prof Paul H Harvey; Treas: Paul Rutteman; Dir-Gen: Ralph Armond; Zoological Dir: Prof Chris West; f 1828; field conservation; consultancy; conservation science; education; owns London Zoo, Whipsnade Wild Animal Park and the Institute of Zoology.

United Kingdom Overseas Territories

ANGUILLA

Governmental Organizations

Ministry of Home Affairs, Tourism, Agriculture, Fisheries and Environment: c/o Office of the Chief Minister, The Secretariat, The Valley; tel 497-2518; fax 497-3389; e-mail chief-minister@gov.ai; internet www.gov.ai; Chief Minister and Minister: Osbourne Fleming.

Ministry of Education, Health, Social Development and Lands: c/o Office of the Chief Minister, The Secretariat, The Valley; tel 497-2518; fax 497-3389; internet www.gov.ai; Minister: Eric Reid.

Department of Physical Planning: The Valley; Prin Officer: Orris Proctor.

Office of the Parliamentary Secretary for the Environment: POB 60, The Valley; tel 497-3089.

Non-Governmental Organizations

Anguilla Archaeological and Historical Society: POB 252, The Valley; tel 497-2651; fax 497-5571; Pres: Kenn Banks; Sec: Aileen Smith; f 1982; preservation of all aspects of Anguillan history and archaeology; dissemination of information for educational purposes; affiliated with Anguilla National Trust.

Anguilla Beautification Club: POB 934, The Valley.

Anguilla National Trust: POB 1234, The Valley; tel 497-4440; fax 497-4067; Exec Dir: Ijahnya Christian.

Concerned Anguillan Youth for the Environment: POB 20, The Valley.

BERMUDA

Governmental Organizations

Ministry of the Environment: Government Administration Bldg, 30 Parliament St, Hamilton HM 12; tel 297-7590; fax 292-2349; e-mail browlinson@gov.bm; Minister: Neletha I Butterfield.

Division of Agriculture, Fisheries and Parks: POB HM 834, Hamilton HM CX; tel 236-4201.

Non-Governmental Organizations

Bermuda Biological Station for Research, Inc: 17 Biological Lane, Ferry Reach GE 01; tel 297-1880; fax 297-8143; e-mail biostation@bbsr.edu; internet www.bbsr.edu; Dir: Dr A H Knap; f 1929; marine science research; oceanography; study of the local environment; educational programmes; undergraduate and postgraduate courses.

Bermuda National Trust: POB HM 61, Hamilton HM AX; tel 236-6483; fax 236-0617.

Keep Bermuda Beautiful: POB HM 2227, Hamilton HM JX; tel 295-5142; fax 292-2977.

BRITISH VIRGIN ISLANDS

Governmental Organizations

Ministry of Natural Resources and Labour: Road Town, Tortola; tel 494-3614; e-mail tsm@bvigovernment.org; internet www.bvigovernmnet.org; Minister: C Alvin Christopher.

National Parks Trust: POB 860, Road Town, Tortola; tel 494-3904; fax 494-6383; e-mail bvinpt@bvinationalparkstrust.org; internet www.bvinationalparkstrust.org; Chair: Janice George-Creque; Dir: Joseph Smith-Abbott; f 1961; manages 19 national parks, 18 of which are terrestrial and one marine; environmental research; runs Darwin Initiative biodiversity project at Gorda Peak National Park and on the island of Anegada; reef protection and mooring programmes; species restoration and rehabilitation; reforestation; environmental education; affiliated with the Caribbean Conservation Association and the World Conservation Union—IUCN.

CAYMAN ISLANDS

Governmental Organizations

Ministry of Health Services, Agriculture, Aviation and Works: Government Administration Bldg, Elgin Ave, George Town, Grand Cayman; tel 949-7900; fax 949-7544; e-mail tgo.gov.ky; internet www.gov.ky; Minister: Gilbert McLean.

Department of the Environment: Tower Bldg, George Town, Grand Cayman; tel 949-7999; fax 949-8487.

Marine Survey Office: Tower Bldg, George Town, Grand Cayman; tel 949-7999; fax 949-8487.

Ministry of Tourism, Environment, Development and Commerce: Government Administration Bldg, Elgin Ave, George Town, Grand Cayman; internet www.gov.ky; Minister: W McKeeva Bush.

Non-Governmental Organizations

National Trust for the Cayman Islands: POB 31116, Seven Mile Beach, Grand Cayman; tel 949-0121; fax 949-7494; e-mail fjburton@candw.ky.

GIBRALTAR

Governmental Organizations

Ministry for the Environment, Roads and Utilities: Joshua Hassan House, Secretary's Lane; tel 59833; fax 71143; e-mail environ@gibnynex.gi; Minister: Fabian Vinet.

Environmental Agency: 37 Town Range; tel 70620; fax 74119; e-mail envag@gibnet.gi; Chief Environmental Health Officer: J L Delgado; environmental health; control of pollution.

Gibraltar Botanic Gardens: The Alameda, Red Sands Rd, POB 843; tel 72639; fax 74022; e-mail wildlife_gib@compuserve.com; Dir: Dr John E Cortés; Curator: Brian M Lamb; Horticulturalist: Andrew Abrines; f 1991; conservation of plants; environmental education; member of Botanic Garden Conservation International.

Nature Conservancy Council: Government House.

Non-Governmental Organizations

Gibraltar Heritage Trust: Wellington Front, POB 683; Chair: A Almeda.

Gibraltar Ornithological and Natural History Society: Gibraltar Natural History Field Centre, Jews' Gate, Upper Rock Nature Reserve, POB 843; tel 72639; fax 74022; e-mail gonhs@gibnet.gi; internet www.gibnet.gi/~gonhs; Gen Sec: Dr John E Cortés; f 1976; conservation; research on ornithology, botany, mammalogy, entomology, malacology, herpetology and marine biology; partner-designate of BirdLife International; member of the Iberian Council for the Defence of Nature and of the World Conservation Union—IUCN.

MONTSERRAT

Governmental Organizations

Ministry of Agriculture and the Environment: Government Headquarters, POB 292, Brades; tel 491-2075; fax 491-9275; Minister: Margaret Dyer-Howe; development and implementation of policy on management of land and natural resources to ensure sustainable development of agriculture, fisheries and the environment; promotion of affordable, comfortable housing.

Ministry of Communications and Works, Land, Housing and Disaster Mitigation: Government Headquarters, POB 292, Brades; tel 491-2521; fax 491-4534; Minister: John Wilson.

Non-Governmental Organizations

Montserrat National Trust: tel 491-3086; fax 491-3046; previously located at Parliament St, POB 393, Plymouth, but relocated to temporary accommodation owing to volcanic activity; conservation of areas of natural beauty; preservation of historic sites and buildings; acquisition of property for the benefit of the island; compilation of records; public education on the island's heritage; fund-raising.

ST HELENA

Governmental Organizations

Ministry of Agriculture and Natural Resources: c/o Office of the Chief Secretary, The Castle, Jamestown; tel 2555; fax 2598; e-mail ocs@helanta.sh; Minister: Mervyn Robert Yon.

Council Committee on Public Health: c/o Office of the Governor, The Castle, Jamestown; tel 2555; fax 2598; e-mail ocs@atlantic.co.ac; Chair: Eric W George.

Non-Governmental Organizations

St Helena Heritage Society: Broadway House, Jamestown; tel 4525; fax 4924; Chair: Basil George; f 1980; liaison with government departments on heritage matters; lobbying for the restoration of heritage sites; advisers to the Building Authority; museum; walks; lectures.

TURKS AND CAICOS ISLANDS

Governmental Organizations

Ministry of Health, Social Services and Gender Affairs: c/o Office of the Chief Minister, Government Compound, Grand Turk; tel 946-2801; fax 946-2777; e-mail unfptc@tciway.tc; Minister: Karen Lorene Delaney; protection of the marine and terrestrial environment; management of 33 marine and terrestrial protected areas, including national parks, nature reserves, sanctuaries and historical sites; close co-operation with the National Trust of the Turks and Caicos Islands.

Ministry of Communications, Works and Utilities: c/o Office of the Chief Minister, Government Compound, Grand Turk; tel 946-2801; fax 946-2777; e-mail unfptc@tciway.tc; Minister: McAllister Eugene Hanchell.

Coastal Resources Management Project: Sam's Bldg, Providenciales; tel 941-5122; fax 941-4793; e-mail crmp.garland@tciway.tc; Project Man: Judith Garland-Campbell.

Department of Environmental and Coastal Resources: South Base, Grand Turk; tel 946-2970; fax 946-1895; e-mail decr@tciway.tc; Dir: Michelle Fulford.

Non-Governmental Organizations

National Trust of the Turks and Caicos Islands: Sam's Bldg, Providenciales; tel and fax 941-5710; e-mail tc.nattrust@tciway.tc; Dir: Ethlyn Gibbs-Williams; preservation of sites of historic or cultural significance; conservation of lands, natural features and underwater areas of beauty or environmental importance; protection of native flora and fauna.

Turks and Caicos National Museum: Guinep House, Front St, Grand Turk; tel 946-2160; fax 946-2161; Chair: Grethe Seim.

UNITED STATES OF AMERICA

Governmental Organizations

Environmental Protection Agency—EPA: Ariel Rios Bldg, 1200 Pennsylvania Ave, NW, Washington, DC 20460; tel (202) 272-0167; internet www.epa.gov; Admin (acting): Stephen L Johnson; f 1970; independent federal agency; the Administrator of the EPA has Cabinet rank; responsible for control and abatement of pollution in air and water, and for solid waste; also concerned with noise, radiation and toxic substances; co-operation with state and local governments.

Environmental Research Laboratory—Gulf Ecology Division: 1 Sabine Island Drive, Gulf Breeze, FL 32561-5299; tel (904) 934-9200; fax (904) 934-9201; internet www.epa.gov/ged; Dir: William H Benson, PhD; Asst Dir for

Science: J Kevin Summers, PhD; Asst Dir for Program Operations: Connie Shoemaker; f 1970; the mission of the National Health and Environmental Effects Research Laboratory's Gulf Ecology Division is to conduct research to understand the physical, chemical and biological dynamics of coastal systems in order to: assess the ecological condition of coastal resources in the Gulf of Mexico and southeastern US; determine cause(s) of affected and declining systems; predict future ecological risk to populations, communities and ecosystems from multiple aquatic stressors; support development of criteria to protect coastal environments; and transfer scientific technology to federal and state agencies, private sector and the public.

National Air and Radiation Environmental Laboratory: 540 South Morris Ave, Montgomery, AL 36115-2601; tel (702) 270-3402; fax (702) 270-3454; Dir: Ed Sensintaffar; environmental radiation monitoring, radiochemical analysis, radon studies, electromagnetic field measurement, mixed waste analysis, emergency response to nuclear accidents, and site characterization and assessment.

Western Ecology Division—WED: Environmental Protection Agency, 200 SW 35th St, Corvallis, OR 97333; tel (541) 754-4600; fax (541) 754-4799; e-mail webmaster@mail.cor.epa.gov; internet www.epa.gov/wed; Dir: Thomas D Fontaine; f 1962; formerly Environmental Research Laboratory—Corvallis; EPA national centre for research on plant and terrestrial ecology and western US inland ecosystems.

Council on Environmental Quality: 722 Jackson Pl, NW, Washington, DC 20503; tel (202) 395-5750; fax (202) 456-6546; internet www.whitehouse.gov/ceq; Chair. James Laurence Connaughton; affiliated to the Executive Office of the President.

Department of Agriculture: 1400 Independence Ave, SW, Washington, DC 20250; tel (202) 720-2791; fax (202) 720-6314; internet www.usda.gov; Sec: Mike Johanns; f 1889; federal institution with ministerial status.

Office of the Under-Secretary for Natural Resources and the Environment: 14th St and Independence Ave, SW, Washington, DC 20250; tel (202) 720-7173; fax (202) 720-4732; Under-Sec: James R Lyons; Deputy Under-Sec, Conservation: Tom Hebert; Deputy Under-Sec, Forestry: Brian Burke.

Agricultural Research Service—ARS: 1400 Independence Ave, SW, Whittenburg, Washington, DC 20250 302[a] 14th St and Independence Ave, SW, Washington, DC 20250; tel (202) 720-3656; fax (202) 720-5427; e-mail eknipling@ars.usda.gov; internet www.ars.usda.gov; Dir Edward B Knipling; ensures high-quality, safe food and other agricultural products, assesses the nutritional needs of Americans, sustains a competitive agricultural economy, enhances the natural resource base and the environment and provides economic opportunities for rural citizens, communities and society as a whole.

Natural Resources Conservation Service: POB 2890, Washington, DC 20013; tel (202) 720-4525; fax (202) 720-7690; internet www.ncg.nrcs.usda.gov; Head: Paul W Johnson.

United States Forest Service: 1400 Independence Ave, SW, Washington, DC 20250-0003; tel (202) 205-1661; e-mail mailroom_wo@fs.fed.us; internet www.fs.fed.us; Chief: Dale Bosworth.

USDA Council on Sustainable Development: Room 112[a], Jamie E Whitten Bldg, Washington, DC 20250-3810; tel (202) 720-2456; fax (202) 690-4915; e-mail abackiel@oce.usda.gov; internet www.usda.gov/sustainable; Dir: Adela Backiel.

Department of Commerce: 14th St and Constitution Ave, NW, Washington, DC 20230-0001 DC 20230-0001; tel (202) 219-3605; fax (202) 219-4247; internet www.doc.gov; Sec: Carlos Gutierrez; f 1919; federal institution with ministerial status.

National Oceanic and Atmospheric Administration—NOAA: Silver Spring Metro Centre 3, 1315 East-West Highway, Silver Spring, MD 20910-3282; tel (202) 482-3436; fax (202) 482-6203; e-mail jbaker@rdc.noaa.gov; internet www.noaa.gov; Under Sec, Oceans and Atmosphere and NOAA Admin: Dr D James Baker; Deputy Under Sec: Scott B Gudes; f 1970; weather forecasting, oceanic and atmospheric research, management of marine fisheries, protection of endangered species, sea and sky charting.

National Environmental Satellite, Data and Information Service: 14th St and Independence Ave, SW, Washington, DC 20250; tel (301) 713-3578; fax (301) 713-1249; internet www.nesdis.noaa.gov; Asst Admin: Gregory W Withee; Deputy Asst Admin: Mary Glackin.

National Marine Fisheries Service: 1315 East–West Highway, Silver Spring, MD 20910; tel (301) 713-2239; internet www.nmfs.noaa.gov; Asst Admin: Penelope D Dalton; Deputy Asst Admin: Andrew A Rosenburg.

National Ocean Service: 14th St and Constitution Ave, NW, Washington, DC 20230; tel (301) 713-3074; fax (301) 713-4269; internet www.nos.noaa.gov; Asst Admin: Nancy Foster; Deputy Asst Admin: Capt Ted I Littlestone.

National Weather Service: 14th St and Constitution Ave, NW, Washington, DC 20230; tel (301) 713-0689; fax (301) 713-0610; internet www.nws.noaa.gov; Asst Admin: John R Kelly, Jr; Deputy Asst Admin: John Jones, Jr.

NOAA Coastal Services Center—CSC: 2234 South Hobson Ave, Charleston, SC 29405-2413; tel (843) 740-1200; fax (843) 740-1224; e-mail csc@csc.noaa.gov; internet www.csc.noaa.gov; Dir: Margaret Davidson; Deputy Dir: Jeff Payne; f 1994; works with various branches of NOAA and other federal agencies to bring information, services and technology to the nation's coastal resource managers.

Oceanic and Atmospheric Research: 1315 East–West Highway, Silver Spring, MD 20910; tel (301) 713-2458; fax (301) 713-0613; internet www.oar.noaa.gov; Asst Admin: Elbert W Friday, Jr; Deputy Asst Admin: Alan R Thomas.

Department of Energy: Forrestal Bldg, 1000 Independence Ave, SW, Washington, DC 20585 1000 Independence Ave, SW, Washington, DC 20585; tel (202) 586-5000; fax (202) 586-4403; internet www.energy.gov; Sec: Samuel W Bodman; f 1977; federal institution with ministerial status.

Office of Scientific and Technical Information—OSTI: POB 62, Oak Ridge, TN 37831; tel (865) 576-1188; fax (865) 576-2865; e-mail ostiwebmaster@osti.gov; internet www.osti.gov; Dir: Walter L Warnick; affiliated to the Department of Energy; responsibilities include maintenance of data bases on energy-related environmental, safety and health issues, and management of the Department of Energy's files on world-wide energy research and development.

Department of State: 2201 C St, NW, Washington, DC 20520 Washington, DC 20520; tel (202) 647-6575; fax (202) 647-6783; internet www.state.gov; Sec: Dr Condoleezza Rice; f 1789; federal institution with ministerial status.

Bureau of Oceans and International Environmental and Scientific Affairs—OES: 2201 C St, NW, Rm 7831, Washington, DC 20520; tel (202) 647-4000; fax (202)

647-0217; Exec Dir, OES: Carol S Fuller; central contact point with other US government departments and agencies for liaison on environmental issues.

Department of the Interior: 1849 C St, NW, Washington, DC 20240 Washington, DC 20240; tel (202) 208-3171; fax (202) 208-5048; internet www.doi.gov; Sec: Gale Norton; f 1849; federal institution with ministerial status.

Office of Environmental Policy and Compliance: 1849 C St, NW, Washington, DC 20240; tel (202) 208-3891; Dir: Willie R Taylor; part of the Office of the Assistant Secretary for Policy, Management and Budget.

Office of the Assistant Secretary for Fish and Wildlife and Parks: 1849 C St, NW, Washington, DC 20240; tel (202) 208-4416; fax (202) 208-5048; e-mail contact@fws.gov; internet www.fws.gov; Asst Sec: Craig Manson.

National Park Service: 1849 C St, NW, Washington, DC 20240; tel (202) 208-6843; fax (202) 219-0910; internet www.nps.gov; Dir: Fran Mainella.

United States Fish and Wildlife Service: 1849 C St, NW, Washington, DC 20240; e-mail contact@fws.gov; internet www.fws.gov; Dir: Steven A Williams; conservation of biodiversity for the continuing benefit of mankind; concerned with issues such as endangered species, acquisition and management of refuge land, fish hatcheries, mitigation of the adverse impacts of federal activities, enforcement of conservation law, conservation and management of migratory birds and anadromous fish, and research related to fish and wildlife, and their habitats.

United States Geological Survey—USGS: USGS National Center, 12201 Sunrise Valley Drive, Reston, VA 20192; tel (703) 648-4000; fax (703) 648-4454; e-mail cgroat@usgs.gov; internet www.usgs.gov; Dir: Charles G Groat; Deputy Dir: Kathryn R Clement; Asst Dir for Operations: Barbara J Ryan; incorporates the Biological Resources Division and the Water Resources Division.

Agency for Toxic Substances and Disease Registry—ATSDR: Office of Public Health and Science, 1600 Clifton Rd, Atlanta, GA 30333; tel (404) 498-0070; fax (404) 498-0093; e-mail atsdric@cdc.gov; internet www.atsdr.cdc.gov/atsdrhome.html; Dir Henry Falk; agency of the federal Department of Health and Human Services; aims to prevent exposure to hazardous substances from waste sites, unplanned releases and other sources of pollution present in the environment, and to prevent adverse human health effects and diminished quality of life associated with such exposure.

Alabama Department of Conservation and Natural Resources: POB 301450, Montgomery, AL 36130-1450; tel (334) 242-3486; fax (334) 242-1880; e-mail jperry@dcnr.state.al.us; internet www.dcnr.state.al.us; Commr: Barnett Lawley; state executive agency; incorporates the Division of Marine Resources.

Alabama Department of Environmental Management: 1400 Coliseum Blvd, POB 301463, Montgomery, AL 36130-1463; tel (334) 271-7700; fax (334) 271-7950; e-mail public.affairs@adem.state.al.us; internet www.adem.state.al.us; Dir: James W Warr; Deputy Dir: Marilyn Elliott; state executive agency.

Alaska Department of Environmental Conservation: 410 Willoughby Ave, Suite 303, Juneau, AK 99801-1795; tel (907) 465-5010; fax (907) 465-5097; e-mail website@dec.state.ak.us; internet www.state.ak.us/dec; Commr: Ernesta Ballard; state executive agency; formulation of environmental policy; enforcement of regulations; regional offices in Anchorage and Fairbanks; library.

Alaska Department of Natural Resources: 550 West Seventh Ave, Suite 1260, Anchorage, AK 99501-3557; tel (907) 269-8400; fax (907) 269-8901; internet www.dnr.state.ak.us; Commr: Tom Irwin; Deputy Commrs: Marty Rutherford, Dick Lefebvre; state executive agency.

Arizona Department of Environmental Quality: 1110 West Washington St, Phoenix, AZ 85007; tel (602) 771-2215; fax (602) 771-2270; internet www.azdeq.state.az.us; Dir: Stephen A Owens; state executive agency; environmental monitoring and regulation, covering air and water quality and waste management.

Arizona Department of Land: 1616 W Adams St, Phoenix, AZ 85007; tel (602) 542-4631; fax (602) 542-2590; e-mail inquiry@land.state.az.us; internet www.land.state.az.us/asld/htmls/natural_99.htm; Commr: Mark Winkleman; Dir, Div of Natural Resources: Bill Dowdle; state executive agency.

Arizona Department of Water Resources: Park Pl, 4th Floor, 500 N 3rd St, Phoenix, AZ 85004; tel (602) 417-2400; fax (602) 417-2401; e-mail jrlavelle@adwr.state.az.us; internet www.adwr.state.az.us; Dir: Rita P Pearson; state executive agency.

Arkansas Department of Heritage: 1500 Tower Bldg, 323 Center St, Little Rock, AR 72201; tel (501) 324-9150; fax (501) 324-9154; e-mail info@arkansasheritage.com; internet www.heritage.state.ar.us; Dir: Cathie Matthews; state executive agency.

Arkansas Environmental Quality Department: POB 8913, Little Rock, AR 72219-8913; tel (501) 682-0959; fax (501) 682-0798; e-mail help-custsvs@adeq.state.ar.us; internet www.adeq.state.ar.us; Dir: Randall Mathis; Deputy Dirs: Becky Keogh, Jim Shirell, Larry Wilson; state executive agency.

California Energy Commission: 1516 9th St, MS-29, Sacramento, CA 95814-5512; tel (916) 654-4287; e-mail mediaoffice@energy.ca.gov; internet www.energy.ca.gov; Exec Dir: Bob Therkelsen.

California Environmental Protection Agency: 1001 I St, POB 2815, Sacramento, CA 95814; tel (916) 445-3846; fax (916) 445-6401; e-mail cepacomm@calepa.ca.gov; internet www.calepa.ca.gov; Sec: Terrence Tamminem.

California Resources Agency: Resources Bldg, Rm 1311, 1416 9th St, Sacramento, CA 95814 Resources Bldg, Rm 1311, Sacramento, CA 95814; tel (916) 653-5656; fax (916) 653-8102; internet resources.ca.gov; Sec: Mike Chrisman.

Carbon Dioxide Information Analysis Center: Oak Ridge National Laboratory, POB 2008, Oak Ridge, TN 37831-6335; tel (865) 574-0390; fax (865) 574-2232; e-mail cdiac@ornl.gov; internet cdiac.esd.ornl.gov; Dir: Robert M Cushman; provides data on the greenhouse effect and global climate change; data holdings include records of the concentrations of carbon dioxide and other radiatively active gases in the atmosphere, the role of the terrestrial biosphere and the oceans in the biogeochemical cycles of greenhouse gases, emissions of carbon dioxide to the atmosphere, long-term climate trends, the effects of elevated carbon dioxide on vegetation and the vulnerability of coastal areas to rising sea-level.

Colorado Department of Natural Resources: 1313 Sherman St, Rm 718, Denver, CO 80203; tel (303) 866-3311; fax (303) 866-2115; e-mail feedback.dnr@state.co.us; internet www.dnr.state.co.us; Exec Dir: Russell George; state executive agency.

Division of Wildlife: 6060 Broadway, Denver, CO 80216; tel (303) 297-1192; internet wildlife.state.co.us; management of state wildlife resources.

Colorado Department of Public Health and Environment: 4300 Cherry Creek Dr. S, Denver, CO 80246-1530; tel (303) 692-2000; e-mail cdphe.information@state.co.us; internet www.cdphe.state.co.us/cdphehom.asp; Exec Dir: Douglas H Benevento; state executive agency.

Connecticut Council on Environmental Quality: 79 Elm St, Hartford, CT 06106; tel (860) 424-4000; fax (860) 424-4070; e-mail karl.wagener@po.state.ct.us; internet dep.state.ct.us/ceq.htm; Chair: Donal C O'Brien, Jr; Exec Dir: Karl J Wagener; f 1971; investigation of citizens' complaints; review of environmental assessments of state construction activities; preparation of annual reports to the Governor on the status of Connecticut's environment.

Connecticut Department of Environmental Protection: 79 Elm St, Hartford, CT 06106-5127; tel (860) 424-3000; fax (860) 424-4053; e-mail dep.webmaster@po.state.ct.us; internet dep.state.ct.us; Commr: Arthur J Rocque; state executive agency.

Delaware Department of Natural Resources and Environmental Control: 89 Kings Highway, POB 1401, Dover, DE 19903-1401; tel (302) 739-4403; fax (302) 739-6242; e-mail mpolo@state.de.us; internet www.dnrec.state.de.us; Sec: Nicholas A DiPasquale; Deputy Sec: Mary L McKenzie; state executive agency.

Exxon Valdez Oil Spill Trustee Council, Restoration Project Office: 645 G St, Suite 401, Anchorage, AK 99501; tel (907) 278-8012; fax (907) 276-7178; Exec Dir: Molly McCammon; institution of the state of Alaska.

Florida Bureau of Natural Resources: 3900 Commonwealth Blvd, Mail Station 530, Tallahassee, FL 32399-3000; tel (850) 245-3104; fax (850) 245-3114; internet www.dep.state.fl.us/parks/bncr; Bureau Chief: Dana C Bryan.

Florida Department of Environmental Protection: 3900 Commonwealth Blvd, MS-49, Tallahassee, FL 32399-3000; tel (850) 245-2118; fax (850) 245-2128; internet www.dep.state.fl.us; Sec: Colleen M Castille; state executive agency.

Florida Game and Fresh Water Fish Commission: 620 South Meridian St, Tallahassee, FL 32399-1600; tel (850) 487-3796; e-mail webmaster@fs.fed.us; internet www.state.fl.us/fwc; Exec Dir: Allan Egbert; Asst Exec Dir: Victor J Heller; law enforcement, management and research concerning wildlife.

Georgia Department of Natural Resources: 2 Martin Luther King Jr Dr., SE, Suite 1252, East Tower, Atlanta, GA 30334; tel (404) 656-3500; fax (404) 656-0770; internet www.gadnr.org; Commr: Noel Holcomb; state executive agency.

Environmental Protection Division: 205 Butler St, SE, Suite 1252, Atlanta, GA 30334; tel (404) 656-4713; internet www.ganet.org/dnr/environ; Dir: Harold Reheis; Asst Dirs: David Word, Bruce Osborn.

Hawaii Department of Land and Natural Resources: Kalanimoku Bldg, 1151 Punchbowl St, Honolulu, HI 96813; tel (808) 587-0400; fax (808) 587-0390; e-mail dlnr@hawaii.gov; internet www.hawaii.gov/dlnr; Chair: Peter Young; state executive agency.

Idaho Department of Health and Welfare, Division of Environmental Quality: 1410 North Hilton, Boise, ID 83720-0036; tel (208) 373-0502; fax (208) 373-0417; Admin, Div of Environmental Quality: Wally Cory; state executive agency.

Idaho Department of Lands: 954 West Jefferson, POB 83720, Boise, ID 83720-0050; tel (208) 334-0200; fax (208) 334-2339; e-mail sterry@idl.state.id.us; internet www2.state.id.us/lands; Dir: Winston Wiggins; state executive agency; manages Idaho's endowment trust lands for maximum long-term benefit; involved in forest, fire and minerals management, rangeland and surface leasing, and soil conservation.

Idaho Department of Water Resources: 1301 North Orchard St, Boise 83706; tel (208) 327-7900; fax (208) 327-7866; e-mail idwrinfo@idwr.id.us; internet www.idwr.state.id.us/idwr/idwrhome.htm; Dir: Karl J Dreher; state executive agency.

Illinois Department of Natural Resources: 1 Natural Resources Way, Springfield, IL 62702-1271; tel (217) 785-3602; internet www.dnr.state.il.us; Dir: Joel Brunsvold; state executive agency.

Illinois Environmental Protection Agency—IEPA: 1021 North Grand Ave East, POB 19276, Springfield, IL 62794-9726; tel (217) 782-3397; fax (217) 782-9039; e-mail dennis.mcmurray@epa.state.il.us; internet www.epa.state.il.us; Dir: Renee Cipriano; regulatory agency responsible for implementation of the Illinois Environmental Protection Act and of federal environmental laws at state level.

Indiana Department of Environmental Management: Indiana Government Center North, 100 North Senate Ave, POB 6015, Indianapolis, IN 46206-6015; tel (317) 232-8603; fax (317) 232-6647; internet www.in.gov/idem; Commr: Lori F Kaplan; state executive agency.

Indiana Department of Natural Resources: Indiana Government Center South, Rm W255B, 402 West Washington St, Indianapolis, IN 46204; tel (317) 232-4200; fax (317) 232-8036; internet www.in.gov/dnr; Dir: John Goss; state executive agency.

Iowa Department of Natural Resources: Henry A. Wallace Bldg, 502 East 9th St, Des Moines, IA 50319-0034; tel (515) 281-5918; fax (515) 281-8895; e-mail webmaster@dnr.state.ia.us; internet www.iowadnr.com; Dir: Jeffrey R. Vonk; incorporates the Division of Environmental Protection.

Kansas Department of Health and Environment: Curtis State Office Bldg, 100 SW Jackson St, Topeka, KS 66012-1290; tel (785) 296-1500; fax (785) 368-6368; e-mail info@kdhe.state.ks.us; internet kdhe.state.ks.us; Sec: Roderick Bremby; state executive agency.

Kentucky Environment and Public Protection Cabinet: Capitol Plaza Tower, 500 Mero St, 5th Floor, Frankfort, KY 40601; tel (502) 564-3350; fax (502) 564-3354; internet www.environment.ky.gov; Dir: LaJuana S Wilcher.

Kentucky Natural Resources: 663 Teton Trail, Frankfort, KY 40601; tel (502) 564-2184; fax (502) 564-6193; e-mail naturalresources@ky.gov; internet www.naturalresources.ky.gov; Sec: Susan C Bush; state executive agency.

Louisiana Department of Agriculture and Forestry: POB 631, Baton Rouge, LA 70821-0631; tel (225) 922-1234; fax (225) 922-1253; e-mail info@ldaf.state.la.us; internet www.ldaf.state.la.us; Commr: Bob Odom; state executive agency.

Louisiana Department of Environmental Quality: 602 North 5th St, Baton Rouge, LA 70802; tel (225) 219-3953; fax (225) 219-3971; e-mail mike.mcdaniel@la.gov; internet www.deq.state.la.us; Sec: Dr Mike McDaniel; state executive agency.

Louisiana Department of Natural Resources: 617 North Third St, POB 94396, Baton Rouge, LA 70802-5428; tel (225) 342-4500; fax (225) 342-5861; e-mail info@dnr.state.la.us; internet www.dnr.state.la.us; Sec: Scott Angelle; state executive agency.

Maine Department of Conservation: 22 State House Station, Augusta, ME 04333-0022; tel (207) 287-2211; fax

(207) 287-2400; internet www.state.me.us/doc; Commr: Patrick K McGowan; state executive agency.

Maine Department of Environmental Protection: 17 State House Station, Augusta, ME 04333-0017; tel (207) 287-2812; fax (207) 287-2814; e-mail dawn.r.gallagher@maine.gov; internet www.maine.gov/dep; Commr: Dawn R Gallagher; state executive agency.

Maine Department of Inland Fisheries and Wildlife: 41 State House Station, 284 State St, Augusta, ME 04333-0041; tel (207) 287-5202; fax (207) 287-6395; internet www.state.me.us/ifw/homepage.htm; Commr: Lee E Perry; Deputy Commr: Frederick Hurley; state executive agency.

Maine Department of Marine Resources: 21 State House Station, Augusta, ME 04333; tel (207) 624-6553; fax (207) 624-6024; internet www.state.me.us/dmr; Commr: George Lapointe; Deputy Commr: E Penn Estabrook; state executive agency.

Marine Mammal Commission: 4340 East–West Highway, Rm 905, Bethesda, MD 20814; tel (301) 504-0087; fax (301) 504-0099; Exec Dir: John R Twiss, Jr; Scientific Programme Dir: Dr Robert J Hofman; f 1974; furthers the purposes and policies of the US Marine Mammal Protection Act 1972.

Maryland Department of Natural Resources: Tawes State Office Bldg, 580 Taylor Ave, Annapolis, MD 21401-2397; tel (410) 260-8367; fax (410) 260-8111; e-mail customerservice@dnr.state.md.us; internet www.dnr.state.md.us; Sec: C Ronald Franks; state executive agency.

Tree-mendous Maryland: Tawes State Office Bldg, 580 Taylor Ave, Annapolis, MD 21401; tel (410) 260-8531; fax (410) 260-8595; e-mail tgalloway@dnr.state.md.us; internet www.gacc.com/dnr; Dir and Volunteer Co-ordinator: Terry Galloway; f 1989; fund-raising and organization of volunteer groups to plant almost 1m. trees and seedlings each year, focusing on riparian forest buffers; educational programmes on tree planting and maintenance.

Maryland Department of the Environment: 1800 Washington Blvd, Baltimore, MD 21230; tel (410) 537-3000; fax (410) 631-3888; internet www.mde.state.md.us; Sec: Kendl P Philbrick; state executive agency; incorporates the Air and Radiation Management Administration; activities include the monitoring of air quality, development of plans and regulations, and establishment of emission standards to prevent air pollution, implementation of state and federal laws, review of permits for air pollution sources and regulation of all sources of ionizing radiation; responsible for asbestos.

Massachusetts Department of Environmental Protection: 1 Winter St, Boston, MA 02108-4746 Boston, MA 02202; tel (617) 292-5500; e-mail edmund.coletta@state.ma.us; internet www.mass.gov/dep; Sec: Robert W Golledge; state executive agency.

Michigan Department of Environmental Quality: POB 30473, Lansing, MI 48909-7973; tel (517) 373-7917; fax (517) 241-7401; internet www.michigan.gov/deq; Dir: Steven Chester; state executive agency.

Michigan Department of Natural Resources: Mason Bldg, 6th Floor, POB 30028, Lansing, MI 48909-7528; tel (517) 373-2329; fax (517) 335-4242; internet www.michigan.gov/dnr; Dir: Rebecca A Humphries; state executive agency.

Minnesota Department of Natural Resources: 500 Lafayette Rd, St Paul, MN 55155-4040; tel (612) 296-6157; fax (612) 296-5484; e-mail info@dnr.state.mn.us; internet www.dnr.state.mn.us; Commr: Gene Merriam; state executive agency.

Minnesota County Biological Survey—MCBS: 500 Lafayette Rd, POB 25, St Paul, MN 55155; tel (651) 296-9782; fax (612) 297-4961; e-mail carmen.converse@dnr.state.mn.us; internet www.dnr.state.mn.us/ecological_services/mcbs/index.html; Co-ordinator: Carmen K Converse; systematic identification of the locations of Minnesota's rare natural ecosystems, their component natural communities and rare species.

Natural Heritage Program: 500 Lafayette Rd, St Paul, MN 5155-4001; tel (651) 297-2276; fax (651) 296-1811; e-mail info@dnr.state.mn.us; internet www.dnr.state.mn.us/fish_and_wildlife/natural_heritage.html; Prin Officer: Bonita Eliason; identification and protection of Minnesota's rare native plants and endangered natural communities.

Mississippi Department of Environmental Quality: POB 20305, Jackson, MS 39289-1305; tel (601) 961-5171; fax (601) 961-5349; internet www.deq.state.ms.us; Exec Dir: Charles Chisolm; state executive agency.

Missouri Department of Natural Resources: POB 176, Jefferson City, MO 65102; tel (573) 751-3443; fax (573) 751-7627; e-mail oac@dnr.mo.gov; internet www.dnr.mo.gov; Dir: Steve Mahfood; state executive agency; comprises Division of Environmental Quality, Division of State Parks, Division of Energy and Division of Geology and Land Survey.

Montana Department of Environmental Quality: POB 200901, Helena, MT 59620-0901; tel (406) 444-2544; fax (406) 444-4386; internet www.deq.mt.gov; Dir: Mark Simonich; Deputy Dir: Curt Chisholm; state executive agency.

Montana Department of Natural Resources and Conservation: 1625 11th Ave, POB 201601, Helena, MT 59620-1601; tel (406) 444-2074; fax (406) 444-2684; e-mail dabushnell@state.mt.us; internet www.dnrc.state.mt.us; Dir: Arthur 'Bud' Clinch; f 1995; state executive agency.

Montana Environmental Quality Council: State Capitol, POB 201704, Helena, MT 59620-1704; tel (406) 444-3742; fax (406) 444-3971; e-mail teverts@state.mt.us; Head of Staff: Todd Everts; part of the Montana Legislative Services Division; advisory service for the development and co-ordination of environmental policy, particularly in relation to natural resource management; discussion and consultation forum for government, industry, conservation groups and the general public; statutory responsibility to provide citizens with prompt, authoritative and unbiased information on environmental issues.

Nebraska Department of Environmental Quality: 1200 North St, Suite 400, POB 98922, Lincoln, NE 68509-8922; tel (402) 471-0275; fax (402) 471-2909; e-mail moreinfo@ndeq.state.ne.us; internet www.deq.state.ne.us; Dir: Mike Linder; state executive agency.

Nebraska Department of Water Resources: 301 Centennial Mall South, Lincoln, NE 68509-4676; tel (402) 471-2363; fax (402) 471-2900; e-mail webmaster@dnr.state.ne.us; internet www.dnr.state.ne.us; Dir: Roger K Patterson; state executive agency.

Nevada Department of Conservation and Natural Resources: 123 West Nye Lane, Rm 230, Carson City, NV 89706-0818; tel (702) 687-4360; fax (702) 687-6122; Dir: Peter G Morros; state executive agency; incorporates the Division of Environmental Protection.

New Hampshire Department of Environmental Services: 29 Hazen Drive, POB 95, Concord, NH 03302-0095; tel (603) 271-3503; fax (603) 271-2867; internet www.des.state.nh.us; Commr: Michael Nolin; state executive agency.

New Hampshire Department of Resources and Economic Development: 172 Pembroke Rd, POB 1856, Concord, NH 03302-1856; tel (603) 271-2411; fax (603) 271-2629; e-mail sboucher@dred.state.nh.us; internet www.dred.state.nh.us; Commr: Sean O'Kane; state executive agency; incorporates the Division of Forests and Lands.

National Council for Environment Protection in Yemen: 202 Zubeiry St, San'a; responsible for the formulation of environmental policy.

Non-Governmental Organizations

Friends of the Environment Society—FES: POB 16898, San'a.

ZAMBIA

Governmental Organizations

Ministry of Tourism, Environment and Natural Resources: Electra House, Cairo Rd, POB 30575, Lusaka; tel (1) 227645; fax (1) 222189; internet www.tourism.gov.zm; Minister: Patrick Kalifungwa; formulation of environmental policy; co-ordination of the implementation of environmental legislation; conservation and management of natural resources; protection of the environment; control of pollution.

Ministry of Agriculture, Food and Fisheries: Mulungushi House, Independence Ave, Nationalist Rd, POB RW50291, Lusaka; tel (1) 213551; Minister: Mundia Sikatana.

Ministry of Energy and Water Development: Mulungushi House, Independence Ave, Nationalist Rd, POB 36079, Lusaka; tel and fax (1) 252589; Minister: George Mpombo.

Ministry of Mines and Mineral Development: POB 31969, 10101 Lusaka; tel (1) 251402; fax (1) 252095; internet www.mines.gov.zm; Minister: Kaunda Lembalemba.

Anti-Corruption Commission, Species Protection Department: POB 50486, Lusaka; tel (1) 251328; fax (1) 251397; e-mail accspd@zamnet.zm; conservation of wildlife; investigation and prosecution of poaching; community education.

Luangwa Integrated Resource Development Project: POB 510249, Chipata; tel (62) 21126; fax (62) 21092; e-mail lirdp@zamnet.zm; Dir: F E C Munyenyembe; Asst Dir, Research: G C Kalyocha; f 1987; community-based wildlife and natural resource management programmes in Luangwa National Park and the basin of the Luangwa River; research; affiliated with the National Parks and Wildlife Service and the Ministry of Tourism.

National Council for Scientific Research: POB 310158, Chelston, Lusaka 15302; tel (1) 281081; fax (1) 283502; Chair: Prof M N Siamwiza; f 1967; part of the Ministry of Science, Technology and Vocational Training; state body to advise the government on scientific research policy, to promote and co-ordinate research and to collect and disseminate scientific information; incorporates Livestock and Pest Research Centre, Tree Improvement Research Centre, Radioisotopes Research Centre, Food Technology Research Unit, Water Resources Research Unit, Building and Industrial Minerals Research Unit.

National Heritage Conservation Commission: POB 60124, Livingstone; tel (3) 324239; fax (3) 324509; Exec Dir: Donald C Chikumbi; Information and Public Relations: Isaac Kanguya; Editor: Maxwell Zulu; statutory body; conservation and management of Zambia's natural and cultural heritage.

National Parks and Wildlife Service: Private Bag 1, Chilanga; tel (1) 278526.

Non-Governmental Organizations

IUCN Zambia Country Office: Lotti House, 7th Floor, North end of Cairo Rd, Private Bag W 356, Lusaka; tel (1) 231866; fax (1) 231867; Country Rep: (vacant); national office of the World Conservation Union—IUCN.

World Wide Fund for Nature—Zambia Co-ordination Office: POB 50551, Ridgeway, Lusaka; tel and fax (1) 253749; e-mail wwfzam@zamnet.zm; Nat Programme Co-ordinator: Monica Chundama; Project Man: Nyambe Nalumino; Team Leader: George Muwowd; Programme Admin: Happy Mbuld; f 1991; national affiliate organization of the World Wide Fund for Nature—WWF International.

Zambian Ornithological Society: POB 33944, Lusaka; Chair: L O'Keeffe; Vice-Chair: A Bowden; f 1971; field meetings; publications; affiliated with BirdLife International.

ZIMBABWE

Governmental Organizations

Ministry of Environment and Tourism: Karigamombe Centre, Private Bag 7753, Causeway, Harare; tel (4) 751720; fax (4) 734075; Minister: Francis Nhema; Permanent Sec: M Sangarwe; review, formulate and co-ordinate policy and communications in support of environmentally sustainable development; contribute to the development of action plans for the national conservation strategy; advise and oversee the implementation of international environmental convention and protocols to which Zimbabwe is part; integrate socio-economic valuations of environmental costs and benefits in development, review and decision making; integrate environmental research and analysis in support of environmentally sustainable development.

Department of Natural Resources: Makombe Complex, Block 1, Harare St and Herbert Chitepo Ave, POB 8070, Causeway, Harare; tel (4) 705671; fax (4) 793123; e-mail zpn143@mureb.co.zw; Dir: M D Chasi; ensures that the natural resources of Zimbabwe are used on a sustainable basis; environmental: monitoring, education and extension, planning and management, law enforcement, impact assessment policy implementation; resource inventorying.

Ministry of Agriculture and Rural Resettlement: Ngungunyana Bldg, Private Bag 7701, Causeway, Harare; tel (4) 792223; fax (4) 734646; Minister: Joseph Made.

Institute of Agricultural Engineering: POB BW 330, Borrowdale, Harare; tel (4) 860019; fax (4) 860136; part of the Department of Agricultural, Technical and Extension

Services—AGRITEX; agricultural engineering research, development, training and extension services.

Ministry of Water Resources, Infrastructural Development, and Small and Medium Enterprises: Makombe Complex, Private Bag 7701, Causeway, Harare; tel (4) 706081; Minister: Joyce Mujuru.

Non-Governmental Organizations

ICRISAT Bulawayo: Matopos Research Station, POB 776, Bulawayo; tel 838311; fax 838253; e-mail icrisatzw@cgiar.org; office of the International Crops Research Institute for the Semi-Arid Tropics—ICRISAT, India.

Ornithological Association of Zimbabwe: POB CY 161, Causeway, Harare; tel (4) 794611; e-mail birds@harare.iafrica.com; Pres: J R Paxton; f 1962; promotion of interest, conservation and research in relation to the birds of Zimbabwe; magazine; affiliated with BirdLife International.

Zambezi Society: POB HG774, Highlands, Harare; tel (4) 747002; e-mail zambezi@samara.ca.zw; internet www.zambezi_society.com; Dir: Richard G Pitman; f 1982; protection of biodiversity and the ecological, aesthetic and other natural resources of the Zambezi Basin; affiliated with Biodiversity Foundation for Africa and Fauna and Flora International.

Zimbabwe National Conservation Trust—ZIMNET: POB CY 358, Causeway, Harare; tel and fax (4) 496105; Exec Dir: Joseph C Tasosa; f 1974; promotion of the sustainable use of natural resources in order to maintain balanced ecosystems, genetic resources and species diversity; public education programmes; target groups are rural communities, schools and youth organizations.

Zimbabwe Trust—Zimtrust: 4 Lanark Rd, POB 4027, Belgravia, Harare; tel (4) 730543; fax (4) 795150; e-mail zimtrust@samara.co.zw; Gen Sec: R H T Monro; f 1980; dedicated to the relief of poverty and improvement of the quality of life; operates the Communal Areas Management Programme for Indigenous Resources—CAMPFIRE, which provides training support to rural communities in natural resource management; affiliated to the Africa Resources Trust—ART.

BIBLIOGRAPHY
of
ENVIRONMENTAL PERIODICALS

BIBLIOGRAPHY of ENVIRONMENTAL PERIODICALS

3 Umwelt: a3 Zeitschriftenverlagges mbH, 2372 Giesshübl, Hagenauertalstr 40, Austria; tel (2236) 42-5-28; fax (2236) 26-3-11; e-mail a3@a3verlag.co.at; Editor: Erich Rotomer; 8 a year; specialized business magazine for those responsible for environmental technology and environment policy; with supplement, *a3 ECO*; circ 16,000.

Abfallwirtschaftlicher Informationsdienst: Rhombos Verlag, 10785 Berlin, Kurfürstenstr 17, Germany; tel (30) 2616854; fax (30) 2619461; Editor: Bernhard Reiser; f 1994; 8 a year; newsletter containing items on the concepts, policies and technologies of recycling and waste prevention and minimization; circ 1,000.

Acid News: Swedish NGO Secretariat on Acid Rain, POB 7005, 402 31 Göteborg, Sweden; tel (31) 711-45-15; fax (31) 711-46-30; e-mail info@acidrain.org; internet www.acidrain.org; Editor: Christer Ågren; f 1982; 4–5 a year; newsletter of the Swedish NGO Secretariat on Acid Rain; covers air pollution, acidification of the environment, and climate change; in English; available online; circ 4,000.

Acqua Aria: Editrice Arti Poligrafiche Europee, Via Casella 16, 20156 Milan, Italy; tel (02) 392281; fax (02) 39214341; e-mail acquaaria@ape.apenet.it; internet www.ape.apenet.it; Editor: Stefano Meinardi; f 1948; 9 a year; magazine dealing with ecology, pollution and the environment; circ 5,500.

Acta Agriculturae Nucleatae Sinica: Institute for Application of Atomic Energy, Chinese Academy of Sciences, POB 5109, Beijing 100094, People's Republic of China; tel (10) 62581177; Editor: Ji Xiaobing; f 1987; quarterly; joint publication of the Chinese Society of Nuclear Agricultural Sciences; concerned with agricultural applications of atomic energy and isotopes; in Chinese and English; electronic formats available; circ 2,000. Publishing house: Agricultural Publishing House, 2 Nongzhanguan North Rd, Chaoyang Dist, Beijing 100026, People's Republic of China.

Acta Hydrochimica et Hydrobiologica—Zeitschrift für Wasser- und Abwasser-Forschung (Journal for Water and Wastewater Research): Wiley-VCH, 69451 Weinheim, Pappelallee 3, Postfach 101161, Germany; tel (6201) 6060; fax (6201) 606328; e-mail acta@hebi-wasser.uni-karlsruhe.de; internet www.wiley-vch.de; Editor: F H Frimmel; f 1968; every 2 months; Water Chemical Society, a division of the German Chemical Society; journal dedicated to research in the areas of water and waste water; intended for scientists, engineers and ecologists; in English and German; circ 1,000.

Acta Theriologica: c/o Mammal Research Institute, Polish Academy of Sciences, 17-230 Bialowieza, Poland; tel (85) 6812278; fax (85) 6812289; e-mail acta@bison.zbs.bialowieza.pl; internet bison.zbs.bialowieza.pl; Editor-in-Chief: Zdzislaw Pucek; Assoc Editors: Leszek Rychlik, Joseph F Merritt, Andrzej Zalewski, Krzysztof Schmidt; f 1958; quarterly; international journal of mammalogy and related subjects including mammalian ecology; contains reports of research in over 40 countries, short articles and book reviews; many papers deal with the biology and ecology of mammals in the pristine lowland forest of Bialowieza National Park; in English; circ 700.

Action Alert: Rainforest Action Network—RAN, 221 Pine St, 5th Floor, San Francisco, CA 94104, USA; tel (415) 398-4404; fax (415) 398-2732; e-mail rainforest@ran.org; internet www.ran.org; Editor: Mark Westland; every 2 months; members' bulletin dealing with environmental issues requiring urgent public action; circ 20,000.

Adirondack Journal of Environmental Studies—AJES: Paul Smith's College, Paul Smiths, NY 12970, USA; tel (518) 327-6377; fax (518) 327-6369; e-mail gcchilson@aol.com; Editor: Gary Chilson; f 1994; 2 a year; deals with sustainable development within the Champlain-Adirondack biosphere reserve.

Advances in Ecological Research: Academic Press, Inc, 525 B St, San Diego, CA 92101, USA; tel (619) 231-0926; fax (619) 699-6715; Editors: A MacFadyen, E D Ford; f 1962.

Advances in Environmental Science: Research Centre for Eco-Environmental Sciences, Chinese Academy of Sciences—CAS, POB 2871, Beijing 100085, People's Republic of China; tel (10) 62925511; fax (10) 62923563; Editor: Yu Zhendong; f 1993; every 2 months; deals with environmental chemistry, chemical hazards, health and safety, pollution and pollution control, and water treatment; circ 2,000.

African Journal of Ecology: Blackwell Science Ltd, Osney Mead, Oxford OX2 0EL, United Kingdom; tel (1865) 206206; fax (1865) 721205; Editor: F I B Kayanja; f 1962; quarterly; the East African Wildlife Society; former title *East African Wildlife Journal*; includes indexes and book reviews; circ 380.

African Primates: Zoo Atlanta, 800 Cherokee Ave, SE, Atlanta, GA 30315-1440, USA; tel (404) 624-5808; fax (404) 627-7514; e-mail forthmand@mindspring.com; internet www.primate.wisc.edu/pin; Editors: T M Butynski, D Forthman; f 2 a year; conservation of African primates; in English and French; circ 1,300.

African Wildlife: Wildlife and Environment Society of Southern Africa, POB 44189, Linden 2104, South Africa; tel (21) 5351818; fax (21) 5351937; e-mail wildmag@yebo.co.za; internet www.wildlifesociety.org.za; Editor: Sandie Anderson; Asst Editor: Pat Coles; Chair, Editorial Cttee: Jan Giliomee; f 1946; every 2 months; conservation magazine; includes indexes and book reviews; in English; available on microform; circ 20,000.

African Wildlife News: African Wildlife Foundation, 1400 16th St, NW, Suite 120, Washington, DC 20036, USA; tel (202) 939-3333; fax (202) 265-2361; internet www.awf.com; Editor: Rebecca Villareal; circ 40,000.

African Wildlife Update: African Wildlife News Service, POB 546, Olympia, WA 98507-0546, USA; tel (360) 459-8862; fax (360) 459-8771; e-mail awnews@aol.com; internet www.africanwildlife.org; Editor: Ellis Paguirigan; f 1990; every 2 months; newsletter on issues relating to African wildlife protection; available online.

Agazen: Ethiopian Wildlife and Natural History Society, POB 13303, Addis Ababa, Ethiopia; tel (1) 183520; Editors: Tilaye Negissie, Zewditu Tessema; f 1970; 2 a year; deals with environmental and conservation issues; in English and Amharic; circ 15,000.

Agrifack: POB 2062, Lilla Nygt 14, 103 12 Stockholm, Sweden; tel (8) 613-49-10; fax (8) 20-20-81; e-mail agrifack@saco.se; Editor: Lars-Erik Liljebaeck; f 1942; 11 a year; trade union issues and politics in the areas of agriculture, horticulture, forestry, food and nutrition, and the environment; circ 5,000.

Agroforestry Today: ICRAF, United Nations Ave, POB 30677, Nairobi, Kenya; tel (2) 521450; fax (2) 521001; e-mail icraf@cgnet.com; internet www.cgiar.org/icraf; quarterly; in English.

Air Currents: Bay Area Air Quality Management District, 939 Ellis St, San Francisco, CA 94109, USA; tel (415) 749-4900; fax (415) 749-5101; internet www.baaqmd.gov/aircurrents; Sr Editor: Teresa Galvin Lee; Jr Editor: Aaron Richardson; f 1959; every 2 months; aims to increase public awareness of issues related to air quality, new regulations and other local environmental issues; circ 4,500.

The Air Pollution Consultant: Elsevier Science, Inc, POB 945, New York, NY 10159-0945, USA; tel (212) 633-3730; fax (212) 633-3680; internet www.elsevier.nl; Editor: Vicki A Dean; f 1991; 7 a year; information on issues related to air pollution.

Air Pollution Control: Bureau of National Affairs, Inc, 1231 25th St, NW, Washington, DC 20037, USA; tel (202) 452-4200; fax (202) 452-4150; e-mail reline@bna.com; internet www.bna.com; Editor: Regina P Klein; f 1980; monthly, with a newsletter supplement every 2 weeks; provides information and advice on the control of air pollution; intended for those responsible for compliance with relevant Environmental Protection Agency—EPA and state regulations; available online and on CD-ROM.

Air Pollution Control—Annual Report: Department of Mines, Mining Commissioner, Private Bag 0049, Gaborone, Botswana; tel 352641; fax 352141; Editor: Choma J Matale; f 1978; annual; provides information on air quality in specific areas of Botswana and deals with the environmental effects of pollution; circ 200.

Air Pollution Management: McIlvaine Co, 2970 Maria Ave, Northbrook, IL 60062, USA; tel (847) 272-0010; fax (847) 272-9673; e-mail editor@mcilvainecompany.com; internet www.mcilvainecompany.com; Editor: Marilyn McIlvaine; Publr: Robert McIlvaine; quarterly, with monthly supplements; former title *Gold Dust*; international journal providing commercial information, management advice and assistance for the air pollution control industry.

Air Quality Management: Information for Industry Ltd, 4 Valentine Pl, London SE1 8RB, United Kingdom; tel (20) 7654-7100; fax (20) 7654-7171; e-mail aqm@ifi.co.uk; internet www.ifi.co.uk; Editor: Jack Pease; Publr: James Benn; f 1996; monthly; monitoring and management of local air quality for those active in local government, public health, business or industry; includes details of policy announcements, research findings and local initiatives; circ 600.

Air Quality Monographs: Elsevier Science BV, Books Division, POB 211, 1000 AE Amsterdam, Netherlands; tel (20) 4853911; fax (20) 4853705; e-mail nlinfo-f@elsevier.nl; internet www.elsevier.nl; f 1994; includes studies on atmospheric pollution, particularly relating to environmental chemistry and biochemistry.

Airone: L'Airone di Giorgio Mondadori & Associati SpA, Corso Magenta 55, 20123 Milan, Italy; tel (02) 43313375; fax (02) 43313574; e-mail edgmona@tin.it; f 1981; monthly; environmental issues.

Air-Water Pollution Report: Business Publishers, Inc, 8737 Colesville Rd, Suite 1100, Silver Spring, MD 20910-3928, USA; tel (301) 587-6300; fax (301) 587-1081; e-mail bpinews@bpinews.com; internet www.bpinews.com; Editor: Stephen Goldstein; f 1963; weekly; former title *Environment Week*; reports on environmental protection, covering legislation, regulation, policy, relevant research and commercial information; includes statistics and book reviews.

Alabama Wildlife: POB 1109, Montgomery, AL 36102, USA; fax (334) 832-9454; Editor: Dan Dumont; quarterly.

Alaska Center for the Environment: Center News: 519 W 8th Ave, Suite 201, Anchorage, AK 99501-3549, USA; tel (907) 274-3621; fax (907) 274-8733; Editor: Cliff Eames; circ 9,000.

Albania: Partia Ekologjike Shqiptare, Rruga Aleksander Moissi 26, POB 135, Tirana, Albania; tel (42) 22503; fax (42) 34413; Editor: Dr Namik Vehbi Fadile Hoti; f 1991; twice a week; journal of the Albanian Ecological Party; environmental issues; circ 13,000.

Albany Report: Environmental Advocates, 353 Hamilton St, Albany, NY 12210, USA; tel (518) 462-5526; fax (518) 427-0381; e-mail webmistress@envadvocates.org; internet www.envadvocates.org; every 2 months; available online.

Alberta Naturalist: Federation of Alberta Naturalists, POB 1472, Edmonton, AB T5J 2N5, Canada; tel (403) 453-8639; fax (403) 453-8553; e-mail fan@connect.ab.ca; internet www.connect.ab.ca/~fan; Editor: Brian Parker; f 1970; quarterly; cumulative index available; circ 500.

Alternative Energy Retailer: Zackin Publications, Inc, 70 Edwin Ave, POB 2180, Waterbury, CT 06722-2238, USA; tel (203) 755-0158; Editor: Sybil Blau; monthly; circ 9,500.

Alternatives Journal: University of Waterloo, Faculty of Environmental Studies, Waterloo, ON N2L 3G1, Canada; tel (519) 885-1211; fax (519) 746-0292; e-mail subscriptions@alternativesjournal.ca; internet www.alternativesjournal.ca; Editor: Dr Robert Gibson; f 1971; bi-monthly; the Environmental Studies Association of Canada; environmental thought, policy and action; in English; circ 5,000.

Ambiente Risorse Salute (Environment, Resources and Health): Centro Studi 'l'Uomo e l'Ambiente', Via delle Palme 13, 35137 Padua, Italy; tel (049) 36435; fax (049) 8761945; e-mail ars@csinternational.com; internet www.csinternational.com/ars; Editor: Franco Spelzini; f 1982; monthly; deals with pollution, use of resources and the development of clean technologies, with particular emphasis on public health and law; in Italian, with summaries in English; circ 6,000.

Aménagement et Nature: Association pour les Espaces Naturels, 21 rue du Conseiller-Collignon, 75116 Paris, France; tel 1-45-20-15-00; fax 1-45-20-45-36; e-mail ibispress@iscpubs.com; internet www.iscpubs.com; Editor: Bernard Cesari; f 1966; quarterly; publication dealing with planning and the natural world; includes bibliographies, indexes and book reviews; circ 2,400. Publishing house: IBIS Press, Paris.

American Forests: American Forestry Association, POB 2000, Dept SPD, Washington, DC 20013, USA; tel (202) 667 7751; f 1894; quarterly.

American Shore and Beach Preservation Association Newsletter: 410 O'Brien Hall, University of California at Berkeley, Berkeley, CA 94720-1718, USA; tel (510) 642-2666; fax (510) 339-6710; e-mail lvida@library.berkeley.edu; internet www2.ncsu.edu/ncsu/CIL/ncsu_kenan/shore_beach; Editor: Linda Vida; f 1954; quarterly; circ 900.

ANJEC Report: Association of New Jersey Environmental Commissions—ANJEC, 300 Mendham Rd, POB 157, Mendham, NJ 07945, USA; tel (973) 539-7547; fax (973) 539-7713;

e-mail anjec@aol.com; internet www.anjec.org; Editor: Sandy Batty; f 1969; quarterly; reports on environmental education and training for municipal officials; circ 2,300.

Annual Activities Report of Maryland Department of Natural Resources: Tawes State Office Bldg, D-4, 580 Taylor Ave, Annapolis, MD 21401, USA; tel (410) 974-3990; fax (410) 974-5206; Editor: R L Gould; circ 1,500.

Annual Editions: Environment: Dushkin-McGraw-Hill, Sluice Dock, Guildford, CT 06437-9989, USA; tel (203) 453-4351; fax (203) 453-6000; Editor: John L Allen; f 1979; annual.

Annual Report of Chesapeake Bay Foundation: 162 Prince George St, Annapolis, MD 21401, USA; tel (410) 268-8816; fax (410) 268-6687; e-mail chesapeake@cbf.org; internet savethebay.cbf.org; Man Editor: Flannery Davis; includes interviews, profiles, accounts of ongoing programmes, lists of contributors to the Foundation and financial statements; in English.

Annual Report of Citizens Advisory Council to the Department of Environmental Protection: Citizens Advisory Council, Rachel Carson State Office Bldg, 5th Floor, POB 8459, Harrisburg, PA 17105-8459, USA; tel (717) 787-4527; fax (717) 772-2291; e-mail wilson.susan@a1.dep.state.pa.us; Editor: Susan M Wilson; annual; circ 900.

Annual Report of Marine Mammal Protection Act of 1972: US National Marine Fisheries Service, National Oceanic and Atmospheric Administration, Office of Protected Resources, 1335 East–West Highway, Silver Spring, MD 20910, USA; tel (301) 713-2332; fax (301) 731-0376; internet www.nmfs.noaa.gov/prot_res; Office Dir: Donald Knowles; Editor: Nicole Leboeuf; f 1972; annual; NMFS implementation of the Marine Mammal Protection Act of 1972; in English; available as a PDF file; circ 5,000.

Annual Report of New Jersey Conservation Foundation: Bamboo Brook, 170 Longview Rd, Far Hills, NJ 07931, USA; tel (908) 234-1225; fax (908) 234-1189; e-mail info@njconservation.org; Editor: David Yaskulka.

Annual Report of the Australian Conservation Foundation: 340 Gore St, Fitzroy, Vic 3065, Australia; tel (3) 9416-1166; fax (3) 9416-0767; f 1968; annual.

Annual Report of the Commission for the Preservation of Natural and Historical Monuments and Relics: POB 60124, Livingstone, Zambia; tel (3) 3204841; fax (3) 324509; e-mail nhccsowe@zamnet.zm; Editor: Maxwell Zulu; Advisory Editors: Donald C Chukumbr, Isaac Kanguya; f 1948; Archaeological research findings; reports. Publishing house: NHCC.

Annual Report of the Executive Director: United Nations Environment Programme, POB 30552, Nairobi, Kenya; tel (2) 621234; fax (2) 226890.

Annual Report of the International Institute for Applied Systems Analysis (IIASA): 2361 Laxenburg, Austria; tel (2236) 80-70; fax (2236) 7-31-49; internet www.iiasa.ac.at; Editors: Christoph Schneider, Mary Ann Williams; available as a PDF file; circ 7,000.

Annual Report of the International Union of Forest Research Organizations: Mariabrunn (BFW), Hauptstrasse 7, A-1140 Vienna, Austria; tel (1) 877-0151-0; fax (1) 877-0151-50; e-mail office@iufro.org; internet www.iufro.org; Pres: Risto Seppala, Exec Sec: Dr Peter Mayer; f 1892; English, French, Spanish, German.

Annual Report of the International Whaling Commission—IWC: The Red House, 135 Station Rd, Impington, Cambridge CB4 4NP, United Kingdom; tel (1223) 233791; fax (1223) 232876; e-mail iwcoffice@compuserve.com; internet ourworld.compuserve.com.homepages/iwcoffice.

Annual Report of the Ministry of the Environment: Ministry of the Environment, Environment Bldg, 40 Scotts Rd, Singapore 228231, Singapore; tel 7327733; fax 7319456; e-mail tanls@cs.gov.sg; internet www.gov.sg/env; f 1972; concerned with the environmental infrastructure; in English; available online; circ 650.

Annual Report of the National Parks Board: POB 787, Pretoria 0001, South Africa; tel (12) 3430905; fax (12) 3431991; includes statistics.

Annual Report of the Natural Resources Department: Government Printing Department, POB 30136, Lusaka, Zambia; f 1964, relaunched 1976; former title *Annual Report of the Office of the Conservateur of Natural Resources.*

Annual Report of the Western Australia Environmental Protection Authority: Westralia Sq, 141 St George's Terrace, Perth, WA 6000, Australia; tel (8) 9222-7000; fax (8) 9322-1598; f 1972; annual; circ 1,000.

Annual Review of Ecology and Systematics: Annual Reviews, Inc, 4139 El Camino Way, POB 10139, Palo Alto, CA 94303-0139, USA; tel (650) 493-4400; fax (650) 424-0910; e-mail service@annurev.org; internet ecolsys.annualreviews.org; Editor: Samuel Gubins; f 1970; annual.

Annual Review of Energy and the Environment: Annual Reviews, Inc, 4139 El Camino Way, POB 10139, Palo Alto, CA 94303-0139, USA; tel (415) 493-4400; fax (415) 855-9815; e-mail service@annurev.org; internet energy.annualreviews.org; Editor: Robert H Socolow; Assoc Editors: Dennis Anderson, John Harte; f 1932; contains approx 20 updated chapters dealing with issues in the field of energy and the environment.

Annual Summary: National Air Pollution Surveillance: Environment Canada, EP Publications, Ottawa, ON K1A 0H3, Canada; fax (819) 953-5921; f 1970; produced by Environment Canada Conservation and Protection Service; monitors and assesses air quality in urban regions of Canada; available in English and French editions; circ 600.

ANS Digest: Freshwater Foundation, Gray Freshwater Center, 2500 Shadywood Rd, Excelsior, MN 55331, USA; tel (612) 471-9773; fax (612) 471-7685; e-mail freshwater@freshwater.org; internet www.freshwater.org; Editor: Nils C Halker; f 1996; quarterly; published in co-operation with the US Fish and Wildlife Service and the National Aquatic Nuisance Species Task Force; current information for specialists and the general public on study, monitoring and control of the spread of harmful, non-indigenous aquatic species; circ 12,000.

Antarctica Project: Antarctic and Southern Ocean Coalition, POB 76920, Washington, DC 20013, USA; tel (202) 544-0236; fax (202) 544-8483; e-mail antarctica@igc.org; internet www.asoc.org; Editor: Beth Marks; f 1982; quarterly; events in Antarctica and the international campaign for its protection; circ 1,000.

Antinquinamento: Tecniche Nuove SpA, Via Ciro Menotti 14, 20129 Milan, Italy; tel (02) 75701; fax (02) 7610351; e-mail inquin@tecnet.it; internet www.tecnet.it; Editor: Luciano Lucchetti; f 1986; quarterly; concerned with the environmental protection technology sector.

Anuario de Mercado Ambiental: SPA, Lagasca 21, 1°E, 28001 Madrid, Spain; tel (91) 5759350; fax (91) 5759962; f 1992; annual; environmental directory of 2,000 industrial and service corporations, as well as national, regional and local government institutions, NGOs and others.

Appalachian Alternatives: Appalachia—Science in the Public Interest, 50 Lair St, Mount Vernon, KY 40456, USA; tel (606) 256-0077; fax (606) 256-2779; e-mail aspi@kiuh.net; internet www.kih.net/aspi; Editor: Albert Fritsch; f 1977; quarterly.

Applied Environmental Education and Communication: Taylor & Francis, Inc, 325 Chestnut St, Suite 800, Philadelphia, PA 19106, USA; tel (215) 625-8900; fax (215) 625-2940; e-mail info@aeec.org; internet www.tandf.co.uk/journals/titles/1533015X.asp; Editor: Brian A Day; f 2002; quarterly; presents the latest developments in the fields of environmental social marketing, environmental journalism, environmental education, sustainability education, environmental interpretation and environmental health communication.

Applied Geography: Elsevier Science Ltd, The Boulevard, Langford Lane, Kidlington OX5 1GB, United Kingdom; tel (1865) 843000; fax (1865) 843010; Editor: Dr James Hansom; f 1981; quarterly; covers research into the evaluation, exploration and management of natural resources.

Aquatic Conservation: Marine and Freshwater Ecosystems: John Wiley and Sons Ltd, The Atrium Southern Gate, Terminus Rd, Chichester, West Sussex PO19 8SQ, United Kingdom; tel (1243) 779777; fax (1243) 775878; e-mail cs-journals@wiley.cp.uk; internet www.wiley.co.uk/journals-sci; Editors-in-Chief: Dr P J Boon, Dr R M Mitchell; Publ Editor: Sally Wilkinson; f 1991; quarterly; international scientific journal; circ 500.

Aquatic Sciences and Fisheries Abstracts (ASFA) Series, Part 1—Biological Sciences and Living Resources: Cambridge Scientific Abstracts, 7200 Wisconsin Ave, Bethesda, MD 20814-4823, USA; tel (301) 961-6750; fax (301) 961-6720; e-mail market@csa.com; internet www.csa.com; Editor: R Pepei; Publr: Ted Caris; f 1971; monthly; electronic formats available.

Aquatic Sciences and Fisheries Abstracts (ASFA) Series, Part 3—Aquatic Pollution and Environmental Quality: Cambridge Scientific Abstracts, 7200 Wisconsin Ave, Bethesda, MD 20814-4823, USA; tel (301) 961-6750; fax (301) 961-6720; e-mail market@csa.com; internet www.csa.com; Editor: R Pepei; Publr: Ted Caris; f 1990; every 2 months; abstracts journal devoted to contamination problems in seas, lakes, rivers and estuaries; electronic formats available.

Aquatic Toxicology: Elsevier Science Publishers BV, POB 181, 1000 AD Amsterdam, Netherlands; tel (20) 4853417; fax (20) 4853325; e-mail nlinfo-f@elsevier.nl; internet www.elsevier.com/locate/aquatox; Editors-in-Chief: P. Pärt, J Stegeman; f 1981; 20 a year in 5 vols; intended for environmental and biochemical toxicologists, marine biologists, ecotoxicologists, and conservationists; in English; available online.

Arab Water World: Chatila Publishing House, POB 13-5121, Chouran, Beirut, Lebanon; tel (1) 352413; fax (1) 352419; e-mail info@chatilapublishing.com; internet www.chatilapublishing.com; Publr and Editor-in-Chief: Fathi Chatila; Exec Editor: Einas Abu-Hatab; Manager: Mona Chatila; f 1977; every 2 months; trade journal of the water and waste-water industries in the Middle East and Africa; intended for public- and private-sector firms; in English and Arabic; circ 9,262.

Arbor Age: Gold Trade Publishing, Inc, 68-860 Perez Rd, Cathedral City, CA 92234, USA; tel (619) 770-4370; Editor: Bruce Shank; Advertising Manager: Denne Goldstein; f 1981; monthly; publication concerned with tree care; intended for professionals responsible for trees in urban environments; circ 15,500.

Arbor Day: National Arbor Day Foundation, 100 Arbor Ave, Nebraska City, NE 68410, USA; tel (402) 474-5655; fax (402) 474-0820; Publr: John E Rosenow; every 2 months; promotes tree-planting projects in the USA.

Arboricultural Journal: The International Journal of Urban Forestry: A B Academic Publishers, POB 42, Bicester OX6 7NW, United Kingdom; tel and fax (1869) 320949; e-mail jrnls@abapubl.demon.co.uk; Editor: Gerald Dawe; quarterly; published for the Arboricultural Association; covers ecology and other branches of science relevant to arboriculture; includes original research papers, news and reviews; in English, with summaries in French; circ 2,200.

Archives of Complex Environmental Studies—ACES: ACES Publishing Ltd, POB 114, 33101 Tampere, Finland; tel and fax (3) 2608807; Editor-in-Chief: Dr Olavi Manninen; f 1989; quarterly; published for the International Society of Complex Environmental Studies; evaluates the significance of complex environmental conditions; identifies the combined effects of environmental factors and the methodological and technical applications needed in related research and projects; in English.

Archives of Environmental Contamination and Toxicology: Springer-Verlag New York, Inc, 175 Fifth Ave, New York, NY 10010, USA; tel (212) 460-1500; fax (212) 473-6272; e-mail orders@springer-ny.com; internet www.springer-ny.com; Editor: H M Bolt; f 1972; 8 a year; describes important developments and discoveries in the fields of air, soil and water pollution; available online; circ 1,100.

Archives of Environmental Health: (Subscriptions) Allen Press, Inc, POB 1897, Lawrence, KS 66044-8897, USA; tel (202) 296-6267; fax (202) 296-5149; e-mail aeh@heldref.org; internet www.heldref.org; Editor: Patricia McCreedy; f 1950; every 2 months; the Helen Dwight Reid Educational Foundation; circ 2,000. Publishing house: Heldref Publications, 1319 18th St, NW, Washington, DC 20036-1802, USA; tel (202) 296-6267; fax (202) 296-5149; e-mail tkelly@caheldref.org.

Archiwum Ochrony Srodowiska (Archives of Environmental Protection): Zaklad Narodowy im Ossolinskich-Wydawnictwo, 50 062 Wroclaw, pl Solny 14^a, Poland; e-mail suchecki@ipis.zarbze.pl; Editor: S Godzik; f 1975; quarterly; the Polish Academy of Sciences; environmental protection archives; deals with natural resources and their conservation; atmospheric air, water, earth surface protection against pollution; exhaust gas, sewage and soil purification technology; transformations and transport of pollutants in the environment; measurement techniques used in environmental engineering and monitoring; land reclamation and waste management; protection of human and natural environment; other problems directly related to environmental engineering and environment protection; in Polish and English with summaries in both languages; circ 450.

Arctic: Arctic Institute of North America, University of Calgary, Library Tower, 2500 University Drive, NW, Calgary, AB T2N 1N4, Canada; tel (403) 220-7515; fax (403) 282-4609; Editor: Dr Karen McCullough; quarterly; multidisciplinary journal of circumpolar research; includes papers on the life, physical and social sciences, the humanities, engineering and technology; also includes book reviews, letters and profiles of northern people and places.

Arhiv za Higijenu Rada i Toksikologiju (Archives of Industrial Hygiene and Toxicology): Institute for Medical Research and Occupational Health, 10001 Zagreb, Ksaverska c 2, POB 291, Croatia; tel (1) 4673188; fax (1) 4673303; e-mail archiv@imi.hr; internet mimi.imi.hr/arhiv/oarhivu.html; Editor: R Pleština; Deputy Editor-in-Chief: M Piasek; Sec: D Cakalo; f 1950; quarterly; includes abstracts, bibliographies, statistics, indexes and book reviews on the subjects of occupational health, toxicology, health physics, radiation protection and environmental health from analytical and/or epidemiological perspectives; in Croatian and English; circ 1,000.

Arid Land Research and Management: Taylor & Francis, Inc, 325 Chestnut St, 8th Floor, Philadelphia, PA 19106, USA; tel (215) 625-8900; fax (215) 625-2940; Editor-in-chief: J Skujinš; f 1986; quarterly; formerly Arid Soil Research and Rehabilitation; native and agricultural environments; available online.

Arid Lands Newsletter: University of Arizona, Office of Arid Lands Studies, 1955 East Sixth St, Tucson, AZ 85719, USA; tel (520) 621-8584; fax (520) 621-3816; e-mail kwaser@ag.arizona.edu; internet ag.arizona.edu/oals/ALN/ALNHome.html; Editor: Katherine Waser; 2 a year; conservation problems and sustainable development in arid and semi-arid lands; available online; summary version ISBN: 0277-9455; full text as e-mail ISBN: 1092-5481; circ 3,500.

Arizona Wildlife Views: AZ Game and Fish Dept, 2221 West Greenway Rd, Phoenix, AZ 85023, USA; tel (602) 942-3000; internet www.azgfd.com; Editor: Bob Miles; f 1953; bimonthly; government publication on wildlife in Arizona; circ 20,000.

Arkansas Wildlife: Game and Fish Commission, 2 Natural Resources Drive, Little Rock, AR 72205, USA; tel (501) 223-6331; fax (501) 223-6447; Editor: Keith Sutton; circ 72,000.

Ársrit Skógræktarfélag Islands (The Journal of the Icelandic Forestry Association): Ránargata 18, 150 Reykjavík, Iceland; tel 5518150; fax 5627131; e-mail skog@skog.is; internet www.skog.is; Dir: Brynjólfur Jónsson; f 1931; 2 a year; forestry year book; Icelandic; circ 4,300. Publishing house: The Icelandic Forestry Association.

Asia Environmental Business Journal: Environmental Business International, Inc, 4452 Park Blvd, Suite 306, San Diego, CA 92116-4039, USA; tel (619) 295-7685; fax (619) 295-5743; e-mail vyvyan.tenorio@together.org; Editor: Grant Ferrier; f 1995; every 2 months; covers environmental markets in the Asia-Pacific region.

Asia Environmental Review: Asia Environmental Trading Ltd, 55 Exhibition Rd, London SW7 2PG, United Kingdom; tel (20) 7581-5277; fax (20) 7691-7019; e-mail aet@asianenviro.com; internet www.asianenviro.com; Editor: Murray Griffin; f 1995; monthly (10 a year); business publication concerning environment and sustainability issues and their impact for companies operating in the region; available by e-mail.

Asian Environment: Asian Environment Journals, POB 90 MCC, Makati City, Metro Manila, Philippines; Editor: E A R Ouano; f 1978; quarterly; journal of environmental science and technology; published for the Pollution Control Association of the Philippines and the Philippine and Indonesian Societies of Sanitary Engineers; in English; circ 2,000.

Asian Environmental Technology: The Old Station, London Rd, St Albans AL1 1JD, United Kingdom; tel (1727) 858840; fax (1727) 840310; e-mail enviro.tech@pub.btinternet.com; Publr: M A I Pattison; f 1997; quarterly; circ approx 26,000.

Asian Water, Asia's journal of environmental technology: MTP (S) Pte Ltd, Chuan Bldg, 112 Lavender St, 02-00, Singapore 338728, Singapore; fax 2958801; e-mail yayna@mtp.com.sg; Editors: Kevin Chang, Evelyn Tang; f 1984; monthly; circ 6,850.

Asia-Pacific Journal of Environmental Law: Kluwer Law International, POB 85889, 2508 CN The Hague, Netherlands; tel (70) 3081500; fax (70) 3081515; e-mail services@wkap.nl; internet www.wkap.nl/journalhome.htm/1385-2140; Editor-in-Chief: Donald R Rothwell; Publisher: Lukas Claerhout; f 1996; quarterly; regional and comparative orientation with a special focus on the development of regional and subregional law; in English.

Atmosphere Crisis: Social Issues Resources Series, POB 272348, Boca Raton, FL 33427-2348, USA; tel (561) 994-0079; fax (561) 994-4704; e-mail custserv@sirs.com; internet www.sirs.com; Editor: Trudy Collins; f 1989.

Atmospheric Environment: Pergamon Press, Elsevier Science Ltd, POB 800, Kidlington, Oxford OX5 1DX, United Kingdom; tel (1865) 843000; fax (1865) 843010; e-mail nlinfo-f@elsevier.nl; internet www.elsevier.nl/inca/publications/store/2/4/6; Sr Exec Editor: P Brimblecombe; f 1967; 30 a year; covers all aspects of the interaction of people and ecosystems with their atmospheric environment; in English, French and German; electronic formats available; circ 2,600.

Audubon: National Audubon Society, 700 Broadway, 5th Floor, New York, NY 10003-9562, USA; tel (212) 979-3126; fax (212) 477-9069; internet www.audubon.org; Editor: Lisa Gosselin; f 1899; every 2 months; environment, wildlife, preservation of natural habitats; electronic formats available.

Australian Forest Grower: POB E18, Kingston, ACT 2604, Australia; fax (2) 6285-3855; Editor: Jo Vondra; quarterly; circ 3,000.

Australian Geographic Journal: Australian Geographic Pty Ltd, 321 Mona Vale Rd, POB 321, Terrey Hills, NSW 2084, Australia; tel (2) 9450-2344; fax (2) 9450-2990; e-mail order@ausgeo.com.au; internet www.australiangeographic.com.au; Editor: Terry Cowley; Chief Exec: Rory Scott; f 1986; quarterly; promotes respect for Australia's environment; covers: animals, endangered species, flora and fauna; topics affecting Australian life — drought, salinity; education and Australian led scientific projects; excerpts available online; circ 200,000. Publishing house: John Fairfax Holdings Ltd, Sydney, NSW, Australia.

Australian Journal of Ecology: Blackwell Science Pty Ltd, 54 University St, Carlton, Vic 3053, Australia; tel (3) 9347-0300; fax (3) 9347-5001; internet www.blackwell-science.com; Editors: P Fairweather, J Ludwig, S McIntyre; f 1976; quarterly; published for the Ecological Society of Australia; includes details of basic and applied ecological research conducted in Australasia, in addition to articles, reviews and news items relating to marine, freshwater and terrestrial ecosystems; circ 1,350.

Australian Journal of Environmental Management: Environment Institute of Australia, GPOB 211 D, Melbourne, Vic 3001, Australia; tel (3) 9654-7473; fax (3) 9650-1242; e-mail adoff@eia.asn.au; internet www.vicnet.au/~eia; Editor: Eric Anderson; f 1994; quarterly.

Australian Waste Disposal Catalogue: Editorial and Publishing Consultants Pty Ltd, 29 First Ave, Klemzig, SA 5087, Australia; tel (8) 8261-5837; fax (8) 8261-2697; Editor: Frank H Schmidt; f 1979; annual; describes equipment, products and services for solid and liquid waste management; circ 2,500.

Australian Wildlife Newsletter: Wild Life Preservation Society of Australia, 36 Diamond Rd, Pearl Beach, NSW 2256, Australia; tel and fax (2) 4344-4708; Editor: P Medway; f 1966; quarterly; members' periodical; circ 1,000.

Australians and the Environment: Australian Bureau of Statistics, POB 10, Belconnen, ACT 2616, Australia; tel (2) 6252-5249; fax (2) 6252-6778; internet www.abs.gov.au; f 1995; superseded *Australia's Environment: Issues and Facts*; includes statistics relating to human activities and the environment.

AWT—Abwassertechnik: Bauverlag GmbH, Redaktion Abwassertechnik, 65004 Wiesbaden, Postfach 1460, Germany; tel (6123) 700236; fax (6123) 700122; e-mail bauverlag.zeitschr.journals@t-online.de; Editor: Hans Wendt; f 1950; every 2 months; professional journal for the waste-water treatment industry; covers sewage treatment, recycling and water pollution; circ 5,400.

Bankcheck Quarterly: International Rivers Network, 1847 Berkeley Way, Berkeley, CA 94703, USA; tel (510) 848-1155; fax (510) 848-1008; Editor: Juliette Maiot.

BATE's ISO 14000 Update: Cutter Information Corpn, 37 Broadway, Suite 1, Arlington, MA 02174-5539, USA; tel (617) 641-5125; fax (617) 648-1950; e-mail crowley@cutter.com; internet www.cutter.com; Editor: Kathleen Victory; f 1995; monthly; deals with the ISO 14000 procedures for environmental management formulated by the International Organization for Standardization—ISO; electronic formats available.

BBC Wildlife Magazine: BBC Worldwide, Broadcasting House, Whiteladies Rd, Bristol BS8 2LR, United Kingdom; tel (117) 973-8402; fax (117) 946-7075; e-mail wildlifemag@gn.apc.org; Editor: Rosamund Kidman Cox; f 1983; monthly; deals with wildlife, conservation and environmental issues; circ 61,350.

BBR—Wasser und Rohrbau: Verlagsgesellschaft Rudolf Müller GmbH & Co KG, 50933 Köln, Stolberger Str 84, Postfach 410949, Germany; tel (221) 54970; fax (221) 5497326; monthly; scientific and technical journal concerned with water supply, pipeline construction, structural engineering and environmental technology; circ 3,400.

BBU Info-Dienst—Meldungen und Meinungen aus der Umwelt- und Friedensbewegung: Bundesverband Bürgerinitiativen Umweltschutz—BBU, 53113 Bonn, Prinz-Albert-Str 43, Germany; tel (228) 214032; fax (228) 214033; f 1987; every 2 months; newsletter of the Federal Association for Citizens' Environmental Protection Initiatives.

The Bear Deluxe: Orlo, POB 10342, Portland, OR 97210, USA; tel (503) 242-1047; fax (503) 243-2642; e-mail bear@teleport.com; internet www.orlo.org; Editor-in-Chief Thomas L Webb; f 1994; 3 a year; explores environmental issues through the creative arts; circ 17,000.

Behavioral Ecology and Sociobiology: Springer-Verlag, 14197 Berlin, Heidelberger Pl 3, Germany; tel (30) 82070; fax (30) 8214091; e-mail subscriptions@springer.de; internet link.springer.de/link/service/journals/00265/index.htm; Chief Editor: Tatiana Czeschlik; f 1976; monthly; quantitative, empirical and theoretical research into animal behaviour; in English; available online and on microform.

Beifang Huanjing (Northern Environment): Heilongjiang Sheng Huanjing Baohuju, Keji Chu, 72 Dazhi Jie, Nangang-qu, Harbin, Heilongjiang 150001, People's Republic of China; tel (451) 249272; Editor: Liu Hongnian; quarterly; in Chinese.

Beitrage der Arbeitsgruppe Systemforschung der Universität Osnabrück/Contributions of the Systems Research Working Group of the University of Osnabrück: Institute of Systems Research, University of Osnabrück, 49069 Osnabrück, Germany; tel (541) 9692547; fax (541) 9692570; e-mail helmut.lieth@usf.uni-osnabrueck.de; Editor: H Lieth; f 1988; reports on environmental modelling and international environmental activities; in German and English; circ 500.

Bereginya (Guardian): c/o Socio-Ecological Union—SEU, 125319 Moscow, ul Krasnoarmeyskaya 25, kv 85, Russian Federation; tel and fax (8312) 30-28-90; e-mail ber@dront.ru; internet www.gubernia.mov.ru/bereginya; Editor-in-Chief: Tatyana Selyvanovskaya; Journalists: Irina Fufaeva, Natalia Pchelina, Albina Blizhenskaya; f 1990; monthly; independent ecological newspaper published in Nizhnii Novgorod; receives financial aid from the US Agency for International Development—AID and the Environmental Protection Dept of Nizhnii Novgorod, and is assisted by volunteers from ISAR, formerly the Institute for Soviet-American Relations (USA); in Russian; circ 10,000.

Berlin-Brandenburger Naturmagazin: Natur und Text in Brandenburg GmbH, 15834 Rangsdorf, Friedensallee 21, Germany; tel (708) 120432; fax (708) 120433; Editor: Severin Zillich; Publr: Rheinhard Baier; f 1987; every 2 months; former title *Ökowerkmagazin*; concerned with nature conservation in Berlin and Brandenburg; circ 7,000.

Bezpecnost a Hygiena Prace (Occupational Safety and Health): Vydavatelstvi a Nakladatelstvi Prace, SRO, Vaclavske nám 17, 112 58 Prague 1, Czech Republic; tel (2) 2356008; fax (2) 2356043; Editor: Jan Drha; f 1951; monthly; publication dedicated to working environments and conditions; covers developments in safety engineering, legislation, health risks and protection; in Czech and Slovak; circ 48,000.

Bio-Cleanup Report: King Communications Group, 627 National Press Bldg, Washington, DC 20045, USA; tel (202) 662-9748; Editor: Hunter McCleary; f 1991; 2 a month.

Biocycle, The Journal of Composting and Recycling: JG Press, Inc, 419 State Ave, Emmaus, PA 18049, USA; tel (610) 967-4135; e-mail biocycle@igress.com; internet www.igress.com; Editor: Jerome Goldstein; f 1960; monthly; former title *Biocycle, The Journal of Waste Recycling*; contains reports and technical papers on recycling and composting of urban waste; available online and on microform; circ 14,000.

Biodiversity and Conservation: Kluwer Academic Publishers, Spuiboulevard 50, Postbus 17, 1016 DW Amsterdam, Netherlands; tel (78) 6576000; fax (78) 6576254; e-mail services@wkap.nl; internet www.kluweronline.com/issn/0960-3115; Editor-in-Chief: Alan T Bull; f 1991; 14 issues a year; publication of articles on all aspects of biological diversity, its description, analysis, and controlled and rational use by mankind; available online.

Biollania (Biology of the Plains): Museo de Zoología, Universidad de los Llanos Ezequiel Zamora—UNELLEZ, Edif Portuguesa, Guanare 3310, Venezuela; tel (57) 68007; fax (55) 68130; e-mail taphorn@cantv.net; Editor: Dr Crispulo Marrero; f 1984; 2 a year; biodiversity, systematics, biology, ecology, conservation and flora and fauna in the neotropics; in Spanish and English. Publishing house: BioCentro, UNELLEZ, Barinas, Venezuela.

Biologia: Institute of Zoology, Slovak Academy of Sciences, Dúbravská cesta 9, 84206 Bratislava, Slovkia; tel 7-373507; fax 7-374494; e-mail bilogia@savba.sk; internet www.zoo.sav.sk/biologia_editorial.htm; Editor-in-Chief: František Hindák; Exec Editor, Section Zoology: M Kazmirová; an international journal covering all fields of botany, zoology and cellular and molecular biology. Publishing house: Editor: Iveta Pekárová.

Biological Agriculture and Horticulture: An International Journal for Sustainable Production Systems: A B Academic Publishers, POB 42, Bicester OX6 7NW, United Kingdom; tel and fax (1869) 320949; e-mail jrnls@abapubl.demon.co.uk; Editor: P J C Harris; f 1981; quarterly; original scientific and economic research reports.

Biological Conservation: Elsevier Science Ltd, The Boulevard, Langford Lane, Kidlington OX5 1GB, United Kingdom; tel (1865) 843000; fax (1865) 843010; Editor-in-Chief: Dr Eric Duffey; Editors: B N K Davis, C M Schonewald-Cox, T R New; f 1969; 12 a year in 4 vols; contains original papers on wildlife conservation and the wise management of biological and other natural resources.

Biologie in unserer Zeit (Biology in Our Time): Wiley-VCH, 69451 Weinheim, Postfach 101161, Germany; tel (6201) 606147; fax (6201) 606117; e-mail kkuehl@bio-redaktion.de; internet www.wiley-vch.de/vch/journals/2008; Editor: K Kühl; every 2 months; in German.

Biology and Fertility of Soils: Springer Verlag, 69042 Heidelberg, Postfach 105280, Germany; tel (6221) 4870; fax (6221) 4878699; e-mail czeschlik@springer.de; internet springeronline.com/journal/00374; Editor-in-Chief: P Nannipieri; Editorial Manager: L Falchini; f 1985; 6 a year; deals with pure and applied soil biology and productivity studies, including techniques for assessing ecological stresses on soils; in English; available online; circ 600.

Biomass & Bioenergy: Elsevier Science Ltd, The Boulevard, Langford Lane, Kidlington OX5 1GB, United Kingdom; tel (1865) 843000; fax (1865) 843010; e-mail nlinfo-f@elsevier.com; internet www.elsevier.com; Editors: C Mitchell, R P Overend; f 1991; monthly; concerned with the use of biomass as a renewable energy resource; formerly *Biomass*, incorporates *Energy in Agriculture*; available online via www.sciencedirect.com.

Biosphère: Canadian Wildlife Federation/Fédération Canadienne de la Faune, 2740 Queensview Drive, Ottawa, ON K2B 1A2, Canada; tel (613) 721-2286; fax (613) 721-2902; Editor: Martin Silverstone; f 1985; every 2 months; magazine concerned with the environment, wildlife and the conservation of natural resources; in French; circ 20,000.

Bird Red Data Book: Wellbrook Court, Girton Road, Cambridge, CB3 0NA, United Kingdom; tel (1223) 277318; fax (1223) 277200. Publishing house: BirdLife International.

Blue Goose Flyer: National Wildlife Refuge Association, 1776 Massachusetts Ave, NW, Suite 200, Washington, DC 20036-1916, USA; tel (202) 298-8095; fax (202) 298-8155; e-mail nwra@refugenet.org; internet www.refugenet.org; f 1975; quarterly; circ 5,000.

BNA's Environmental Compliance Bulletin: Bureau of National Affairs, Inc, 1231 25th St, NW, Washington, DC 20037, USA; tel (202) 452-4200; (800) 372-1033 (Subscriptions); fax (202) 822-8092; e-mail kfetherston@bna.com; internet www.bna.com; Man Editor: Kevin Fetherson; f 1986; every 2 weeks; contains news of recent pollution legislation and regulations, and of trends in government and industry; available online.

BNA's Federal Environment and Safety Regulatory Monitoring Report: Bureau of National Affairs, Inc, 1231 25th St NW, Washington, DC 20037, USA; tel (202) 452-4200; fax (202) 822-8092; internet www.bna.com; Editor: Susan Korn; f 1993; every 2 weeks; summarizes all regulatory activities of the Environmental Protection Agency—EPA and Occupational Safety & Health Administration—OSHA as well as Department of Transportation hazardous materials transportation issues.

Bogong: Canberra and South-East Region Environment Centre, POB 1875, Canberra, ACT 2601, Australia; tel (2) 6248-0885; fax (2) 6247-3064; e-mail caserec@peg.apc.org; internet www.spirit.net.au/envoz/eccser/core.html; Editorial Working Group: Ian Fraser, Roger Green, Michael Hissink, David Pfanner, Ann Rawson, Jane Rawson, Kerrie Tucker; f 1980; quarterly; articles emphasize regional environmental matters, but also cover national and international issues; circ 500.

Boletín ARA-Noticias (ARA News Bulletin): Red ARA, Apdo 63.109, Caracas 1067-A, Venezuela; tel (2) 284-8189; fax (2) 284-7489; e-mail econatu@viptel.com; internet www.fpolar.org.ve; Editors: Isabel Novo, Anita Reyna, Deborah Bigio, Franklin Rojas, Soliria Menegatti; f 1991; 2 a month; published in association with EcoNatura, Fundación Tierra Viva, Fudena, Provita and SCAV; news, events, courses in conservation; in Spanish; available by e-mail.

Borealis: Canadian Parks and Wilderness Society—CPAWS, 401 Richmond St W, Suite 380, Toronto, ON M5V 3A8, Canada; tel (416) 979-2720; fax (416) 979-3155; Editor: David G Dodge; f 1965; quarterly; former title *Park News*; concerned with Canadian wilderness, ecosystems and environmental issues; circ 15,000.

Boston College Environmental Affairs Law Review: Boston College, School of Law, 885 Centre St, Newton, MA 02159, USA; tel (617) 552-4354; Editorial Bd; quarterly.

Brach Flächen Recycling (Recycling Derelict Land): Verlag Glückauf GmbH, 45206 Essen, Postfach 185620, Germany; tel (2054) 924120; fax (2054) 924129; e-mail vge-verlag@t-online.de; Editor in Chief: B Litke; Deputy Editor: Kirsten Lackmann; f 1994; quarterly; in English and German; circ 3,000.

Brownfields Report: King Publishing, Inc, 627 National Press Bldg, NW, Washington, DC 20045, USA; tel (202) 638-4260; fax (202) 662-9744; e-mail proberson@home.com; internet www.kingpublishing.com; Editor: Peggy Roberson; f 1996; every 2 weeks; provides information on the restoration and redevelopment of contaminated industrial and commercial sites, including financing insurance and remediation technology; state, local and federal programmes, risk assessment and site use standards; case histories and company profiles; in English; electronic formats available.

Bugle: Rocky Mountain Elk Foundation, POB 8249, Missoula, MT 59807-8249, USA; tel (406) 523-4570; fax (406) 523-4550; e-mail bugle@rmef.org; internet www.rmef.org; Editor: Dan Crockett; f 1984; every 2 months; conservation of elk and other animal species and their habitat; circ 195,000.

Bulletin of Environmental Contamination and Toxicology: Springer-Verlag New York, Inc, 175 Fifth Ave, New York, NY 10010, USA; tel (212) 460-1500; fax (212) 473-6272; e-mail orders@springer-ny.com; internet www.springer-ny.com; Editor: H N Nigg; f 1966; monthly; deals with issues in the fields of environmental pollution, toxicology and biochemistry; available online; circ 1,700.

Bulletin of International Institute for Land Reclamation and Improvement: Lawickse Allee 11, POB 45, 6700 AA Wageningen, Netherlands; tel (317) 495549; fax (317) 495590; e-mail ilri@ilri.nl; internet www.ilri.nl.

Bulletin of the Niger Basin Authority: BP 729, Niamey, Niger; tel 72-31-02; fax 72-42-08; e-mail abnsec@intnet.ne; internet www.abn.ne.

Bulletin of the Strategy of the Centre Naturopa: Council of Europe, 67075 Strasbourg Cédex, France; tel 3-88-41-20-00; fax 3-88-41-27-15; f 1969; every 2 months; former title *Newsletter of the Centre Naturopa*; provides information on implementation of the Pan-European Biological and Landscape Diversity Strategy; available in English, French, German, Italian and Russian editions; circ 20,000.

Business and the Environment: Aspen Publishers, 7201 McKinney Circle, Frederick, MD 21704, USA; tel (301) 644-3599; fax (301) 698-7100; internet www.aspenpublishers.com/environment.asp; f 1989; monthly; available online.

Business Strategy and the Environment: John Wiley and Sons Ltd, The Atrium Southern Gate, Terminus Rd, Chichester, West Sussex PO19 8SQ, United Kingdom; tel (1243) 779777; fax (1243) 775878; e-mail cs-journals@wiley.co.uk; internet www.wiley.co.uk/journals-sci; Editor: Prof Richard Welford; Publishing Editor: Lesley Valentine; f 1992; every 2 months; environmental business journal; circ 450.

Butterworths Environmental Regulation: Butterworths Ltd, Halsbury House, 35 Chancery Lane, London WC2A 1EL, United Kingdom; tel (20) 2400-2500; fax (20) 2400-2842; 2 a year, with a supplementary bulletin every 2 months.

Bygd och Natur—tidskrift før hembygdsvaard: Sveriges Hembygdsførbund—SHF, POB 6167, 102 33 Stockholm, Sweden; tel (8) 34-55-11; fax (8) 34-74-74; Editors: Henrik Axioe, Peter Johansson, Gunilla Lindberg; f 1920; every 2 months; published by the National Association for the Preservation of Swedish Culture and Nature; circ 12,000.

CA Selects Plus—Environmental Pollution: Chemical Abstracts Service, 2540 Olentangy River Rd, POB 3012, Columbus, OH 43210-0012, USA; tel (614) 447-3600; fax (614) 447-3751; e-mail help@cas.org; internet www.cas.org; Editor: Dr David Weisgerber; every 2 weeks.

CA Selects Plus—Pollution Monitoring: Chemical Abstracts Service, 2540 Olentangy River Rd, POB 3012, Columbus, OH 43210-0012, USA; tel (614) 447-3600; fax (614) 447-3751; e-mail help@cas.org; internet www.cas.org.

CA Selects—Air Pollution: Chemical Abstracts Service, 2540 Olentangy River Rd, POB 3012, Columbus, OH 43210-3012, USA; tel (614) 447-3600; fax (614) 447-3751; e-mail help@cas.org; internet www.cas.org; Editor: Dr David Weisgerber; every 2 weeks.

Cahiers de la Réunion et de l'Océan Indien: 24 blvd des Cocotiers, 97434 Saint Gilles les Bains, Réunion; monthly; environmental publication covering the island of Réunion and the Indian Ocean.

California Coast and Ocean: California State Coastal Conservancy, 1330 Broadway, Suite 1100, Oakland, CA 94612-2530, USA; tel (510) 286-0934; fax (510) 286-0470; e-mail calcoast@igc.org; internet www.coastalconservancy.ca.gov; Editor: Rasa Gustaitis; f 1985; quarterly; environmental issues affecting the Californian coast; circ 10,000.

California Environmental Law and Land Use: Matthew Bender & Co, Inc, 2 Park Ave, New York, NY 10016, USA; tel and fax (212) 448-2000; e-mail international@bender.com; internet www.bender.com; Editor: Daniel Selmi; f 1989; irreg.

CALMScience: Western Australian Journal of Conservation and Land Management: Dept of Conservation and Land Management, Locked Bag 104, Bentley Delivery Centre, WA 6983, Australia; tel (8) 9334-0333; fax (8) 9334-0498; e-mail cathyb@calm.wa.gov.au; internet www.calm.wa.gov.au; Editor: Marianne Lewis; f 1993; 2 a year; published by the Dept of Nature Conservation and Land Management; scientific journal; in English; circ 600.

Caminhos da Terra: Editora Abril, Av Nações Unidas 7211, 05425-902 São Paulo, SP, Brazil; tel (11) 3037-2000; fax (11) 3037-5638; e-mail terra.atleitor@email.abril.com.br; Editor: Jorge Parquini; f 1992; monthly; consumer publication dealing with ecology, nature and travel; electronic formats available; circ 110,000.

Canadian Environmental Compliance Manual: CCH Canadian Ltd, 90 Sheppard Ave E, Suite 300, North York, ON M2N 3A1, Canada; tel (416) 224-2224; fax (416) 224-0729; internet www.ca.cch.com; Editor: Joseph Weisberg; quarterly; covers corporate management policies, environmental management systems, environmental auditing, site assessments, data management, emergency programmes, training, etc; in English.

Canadian Environmental Law Reports: Carswell Publications, Corporate Plaza, 2075 Kennedy Rd, Scarborough, ON M1T 3V4, Canada; tel (416) 609-8000; fax (416) 298-5094; Editors: Rick Lindgren, Paul Muldoon, Marcia A Valiante, Ramani Nadarajah; f 1972, resumed 1987; monthly; former title *Canadian Environmental Law News*; published for the Canadian Institute for Environmental Law and Policy; includes indexes, articles and annotations; in English, with some items in French; electronic formats available.

Canadian Environmental Mediation Newsletter: Conflict Management Resources, c/o Osgoode Hall Law School, York University, 4700 Keele St, North York, ON M3J 1P3, Canada; Editors: Doreen C Henley, D Paul Emond; f 1986; 3 a year; circ 600.

Canadian Forest Industries: Southam Business Communications, 1 rue Pacifique, Ste-Anne-de-Bellevue, PQ H9X 1C5, Canada; tel (514) 457-2211; fax (514) 457-2558; e-mail jcft@aei.ca; Scott Jamieson; f 1880; 8 a year; circ 13,000.

Canadian Wildlife: Canadian Wildlife Federation/Fédération Canadienne de la Faune, 2740 Queensview Drive, Ottawa, ON K2B 1A2, Canada; tel (613) 721-2286; fax (613) 721-2902; e-mail stone@odysee.net; Editor: Martin Silverstone; f 1965; 5 a year; former title *International Wildlife*; members' magazine; concerned with wildlife and the environment; in English; circ 75,000.

Capitalism Nature Socialism: Taylor & Francis Group (Journals), 4 Park Sq, Milton Park, Abingdon OX14 4RN, United Kingdom; tel (20) 7017-6000; fax (20) 7117-6336; internet www.tandf.co.uk/journals/titles/10455752.asp; Editor: Joel Kovel; quarterly; international red-green journal of theory and politics; dialectics of human and natural history, labor and land, workplace and community struggles, economics and ecology, and the politics of ecology and ecology of politics. Publishing house: Routledge.

CASOLink: South Pacific Regional Environment Programme—SPREP, POB 240, Apia, Samoa; tel 21929; fax 20231; quarterly.

C&D Recycling: GIE, Inc, 4012 Bridge Ave, Cleveland, OH 44113-3320, USA; tel (216) 961-4130; fax (216) 961-0364;

e-mail btaylor@recyclingtoday.com; internet www.recyclingtoday.com; Editor: Brian Taylor; f 1999; covers the construction and demolition materials recycling industry; circ 5,000.

CEASE News: Concerned Educators Allied for a Safer Environment, 17 Gerry St, Cambridge, MA 02138, USA; tel and fax (617) 864-0999; Editor: Peggy Schirmer; f 1980; 3 a year; circ 900.

Center View: Peninsula Conservation Center, 3921 East Bayshore Rd, Palo Alto, CA 94303-4303, USA; tel (650) 962-9876; fax (650) 962-8234; e-mail info@pccf.org; internet www.pccf.org; Editor: Laurie Mueller; f 1976; 3 a year; articles on local environmental issues, Peninsula Conservation Center Foundation programmes and lists of volunteer projects; circ 1,400.

CEP News: Caribbean Environment Programme—CEP, UNEP Regional Co-ordinating Unit, 14–20 Port Royal St, Kingston, Jamaica; tel 922-9267; fax 922-9292; e-mail uneprcuja@cwjamaica.com; internet www.cep.unep.org/pubs/cepnews/cepnews.html; Editors: Matt Soderstrom, Lesma Levy; Programme Officers: Tim Kasten, Alessandra Vanzella-Khouri; f 1987; quarterly; environmental newsletter; in English; available online; circ 5,000.

ÇEVRE (Environment): Türkiye Çevre VakfiI—TÇV/Environment Foundation of Turkey—EFT, Tunali Hilmi 50/20, 06660 Ankara, Turkey; tel (312) 4255508; fax (312) 4185118; Contact: Engin Ural; f 1978; quarterly; provides information on the activities and publications of the EFT; circ 6,500.

CHEC Points: Commonwealth Human Ecology Council, Church House, Newton Rd, London W2 5LS, United Kingdom; tel (20) 7792-5934; fax (20) 7792-5948; e-mail chec@dialin.net; internet www.tcol.co.uk/comorg/chec.htm; Editor: Zena Daysh; f 1978; irreg.

Chemie-Umwelt-Technik: Hüthig GmbH, 69121 Heidelberg, Im Weiher 10, Germany; tel (6221) 489403; fax (6221) 489479; Editor-in-Chief: Dr Sieghard Neufeldt; Publishing Dir: Hans-Jörn Hoffmann; f 1972; annual; annual supplement to *Chemie Technik*; intended for process engineers, chemists, chemical engineers and plant managers in the chemical, pharmaceutical and food processing industries; circ 20,000.

Chemistry and Ecology: Taylor & Francis Group (Journals), 4 Park Sq, Milton Park, Abingdon, OX14 4RN, United Kingdom; tel (20) 7017-6000; fax (20) 7017-6336; e-mail danovaro@unian.it; internet www.tandf.co.uk/journals/titles/02757540.asp; Editor: Prof Roberto Danovaro; every 2 months; relationship between chemistry and ecological processes; helps to promote the ecological assessment of changing chemical environment and in the development of a better understanding of ecological functions.

Chemistry and Industry: Society of Chemical Industry, 14 Belgrave Sq, London SW1X 8PS, United Kingdom; fax (20) 7235-9410; e-mail enquiries@chemind.demon.co.uk; internet mci.mond.org; Editor: Maria Burke; every 2 months; circ 9,161. Publishing house: (Subscriptions) Distribution Centre, Blackhorse Rd, Letchworth SG6 1HN, United Kingdom.

Chile Forestal: Avda Bulnes 259, Of 706, Santiago, Chile; tel (2) 671-1850; fax (2) 696-6724; Dir: Mariela Espejo Suazo; f 1974; monthly; technical information and features on forestry sector; circ 4,000.

China Environment News: 15a Xiaoxinglongjie, Chongwen Dist, Beijing 100062, People's Republic of China; tel (10) 67122478; fax (10) 67013772; e-mail cenv@public3.bta.net.cn; Pres: Zhang Lijun; Co-ordinator: Zhang Xiaoqing; f 1984; publicizes China's basic national policies, guide-lines, laws and regulations governing environmental protection; also deals with environmental protection world-wide; available in Chinese and English editions; circ 250,000 (Chinese edition), 20,000 (English edition).

China Environmental Protection Industry: 9 Sanlihe Rd, Haidian Dist, Beijing 100835, People's Republic of China; tel (10) 68393896; fax (10) 68393748; Editor-in-Chief: Jiang Xiaoyu; f 1995; every 2 months; covers aspects of the environmental protection industry, including policies, regulations and rules, scientific and technological development, and foreign and local news; in Chinese, with English abstracts; circ 5,000.

Chinese Geographical Science: Changchun Institute of Geography, Chinese Academy of Sciences, 16 Gongnong Rd, Changchun 130021, People's Republic of China; tel (431) 5658067; fax (431) 5652931; Editor: Huang Xichou; f 1991; quarterly; concerned with physical geography and its subdisciplines, cartography, remote sensing, geographical information systems and issues such as population, resources, development and the environment; in English. Publishing house: Science Press China, 16 Donghuangchenggen North St, Beijing 100717, People's Republic of China; fax (10) 64034205.

Chinese Journal of Applied and Environmental Biology: Chengdu Institute of Biology, Chinese Academy of Sciences, POB 416, Chengdu 610041, People's Republic of China; tel (28) 85229903; fax (28) 85237341; e-mail biojaeb@cib.ac.cn; internet www.cibj.com; Editor-in-chief: Tang Ya; f 1995; bimonthly; sponsored by and based at Chengdu Inst of Biology, Chinese Acad of Sciences; applied biology; environmental biology; promotes utilization of biological information and data; protection of environment; rehabilitation of degenerated ecosystem and sustainable development of bioresources; in Chinese and English; available as PDF; circ 1,000. Publishing house: Science Press, 16 Donghuangchenggen Beijie St, Beijing 100717, People's Republic of China.

Chinese Journal of Applied Ecology: Institute of Applied Ecology, Academia Sinica, 72 Wenhua Lu, POB 417, Shenyang, Liaoning 110015, People's Republic of China; tel (24) 3916250; fax (24) 3843313; Editors: Shen Shanmin, Sun Shunjiang; f 1990; every 2 months; contains comprehensive articles, original papers and short communications on aspects of applied ecology, including forest ecology, agricultural ecology, grassland ecology, fisheries ecology, natural resource ecology, landscape ecology, global ecology, urban ecology, pollution and ecological engineering; in Chinese; circ 1,200. Publishing house: Science Press, 16 Donghuangchenggen Beijie St, Beijing 100717, People's Republic of China.

Chinese Journal of Arid Land Research: Allerton Press, Inc, 150 Fifth Ave, New York, NY 10011, USA; tel (212) 924-3950; fax (212) 463-9684; Editor: Xia Xuncheng; f 1988; quarterly; translations of papers from Chinese journals published by the Institute of Geography of the Chinese Academy of Sciences (People's Republic of China); concerned with the natural resources of arid territories, anthropological impacts and managed development of desert environments, desertification and desert landscape ecology, the economics of desert lands, population centres, population-carrying capacity, desert soils, historical evaluation of deserts, weather patterns and climatic factors; in English.

Chinese Journal of Atmospheric Sciences: Allerton Press, Inc, 150 Fifth Ave, New York, NY 10011, USA; tel (212) 924-3950; fax (212) 463-9684; Editor: Huang Ronghui;

f 1987; quarterly; translation of *Daqi Kexue*, a journal of the Institute of Atmospheric Physics, Chinese Academy of Sciences (People's Republic of China); covers environmental studies and pollution monitoring, remote sensing and monitoring techniques, atmospheric and oceanic circulation, sea–air interaction, atmospherical boundary layer physics and turbulence analysis, weather modification research, geophysical fluid dynamics, weather and climate pattern simulations, and weather forecasting; in English.

Chongqing Huanjing Kexue (Chongqing Environmental Sciences): Chongqing Huanjing Kexue Xuehui, 12 Renmin Lu, Chongqing, Sichuan 630015, People's Republic of China; fax (811) 3850021; Editor: Yongshu Qiu; f 1979; every 2 months; publication of Chongqing Society of Environmental Sciences; co-published by Chongqing Bureau of Environmental Protection; in Chinese; circ 5,000.

Chronmy Przyrode Ojczysta (Let us Protect the Nature of Our Homeland): Instytut Ochrony Przyrody, Polska Akademia Nauk, 31-512 Kraków, ul Lubicz 46, Poland; tel (12) 4215144; fax (12) 4210348; e-mail nonkorzen@cyf-kr.edu.pl; Editor-in-Chief: Zygmunt Denisiuk; Sec: Irena Szczygiel; f 1945; every 2 months; publication of the Institute of Nature Conservation of the Polish Academy of Sciences and the State Council of Nature Protection; contains results of original research on protected and endangered plant and animal species and natural areas; deals with contemporary issues in environmental conservation, trends in relevant research in Poland and other countries, and the activities of the State and of the public within the field of conservation; includes bibliographies, indexes and book reviews; in Polish, with summaries in English; circ 2,000.

Citizens' Report on Sri Lankan Environment and Development: National Environmental Information Centre, 434/4, Sri Jayawardenapura, Sri Lanka; tel (1) 2873131 or (1) 2813568; fax (1) 2883187; e-mail slejf_na@sri.lanka.net; internet www.environmentaljournalists.lk; Editorial Dir: Dr Dharman Wickremaratne; f 1992; annual; superseded *SLEJF Newsletter*; in English; circ 3,000. Publishing house: Sri Lanka Environmental Journalists Forum.

City Sierran: Sierra Club, New York City Group, 116 John St, Rm 3100, New York, NY 10038-3401, USA; Editor: Henry Pita; quarterly; preservation of the environment in New York City, news items and calendar of events.

Civic Focus: Civic Trust, 17 Carlton House Terrace, London SW1Y 5AW, United Kingdom; tel (20) 7930-0914; fax (20) 7321-0180; e-mail focus@civictrust.org.uk; internet www.civictrust.org.uk; Editor: Jane Smythson; quarterly; former title *Urban Focus*; regeneration of towns, cities and villages; available online; circ 3,000.

Clean Air: Clean Air Society of Australia and New Zealand—CASANZ, 12 Pall Mall, Mount Waverley, Vic 3149, Australia; tel (3) 9807-1942; fax (3) 9888-3545; internet www.casanz.org.au/journal.htm; Editor: J N O'Heare; f 1967; quarterly; includes book reviews, bibliographies and indexes; circ 4,000.

Clean Air & Environmental Protection: National Society for Clean Air and Environmental Protection, 136 North St, Brighton BN1 1RG, United Kingdom; tel (1273) 326313; fax (1273) 735802; e-mail twilliamson@nsca.org.uk; internet www.greenchannel.com/nsca; Editor: Tim Williamson; f 1929; every 2 months; articles and news items on all aspects of pollution control, air quality and environmental management; circ 2,000.

Clean Air Report: Inside Washington Publishers, POB 7167, Benjamin Franklin Station, Washington, DC 20044-7167, USA; tel (703) 416-8500; fax (703) 416-8543; e-mail service@iwp.news.com; every 2 weeks.

Clean Water Report: Business Publishers, Inc, 9737 Colesville Rd, Suite 1100, Silver Spring, MD 20910-3928, USA; tel (301) 587-6300; fax (301) 587-1081; e-mail bpinews@bpinews.com; internet www.bpinews.com; Editor: Elaine R Eiserer; Publisher: Leonard A Eiserer; every 2 weeks (25 a year); newsletter.

Clearwater Navigator: Hudson River Sloop Clearwater, Inc, 112 Market St, Poughkeepsie, NY 12601, USA; tel (914) 454-7673; fax (914) 454-7953; e-mail office@mail.clearwater.org; internet www.clearwater.org; Editor: Mary Evans; f 1969; every 2 months; circ 11,100.

Climate Alert: Climate Institute, 120 Maryland Ave, NE, Washington, DC 20002-5616, USA; tel (202) 547-0104; fax (202) 547-0111; e-mail ncwilson@climate.org; internet www.climate.org; Editor: Nancy C Wilson; f 1988; quarterly; deals with international climate change, covering sea-level rise, drought and stratospheric ozone depletion; focuses on developing countries and alternative energy; circ 1,500.

Coastal Management: Taylor & Francis, Inc, 325 Chestnut St, Suite 800, Philadelphia, PA 19106, USA; tel (215) 625-8900; fax (215) 625-2940; e-mail coastjnl@u.washington.edu; internet www.tandf.co.uk/journals/titles/08920753.asp; Editor: Marc J Hershman; quarterly; peer-reviewed, applied research journal dedicated to exploring the technical, legal, political, social and policy issues relating to the use of coastal resources and environments on a global scale; in English; available online.

Coastal Society Annual Conference Proceedings: Coastal Society, POB 25408, Alexandria, VA 22313-5408, USA; tel (703) 768-1599; fax (703) 768-1598; f 1975; annual; circ 400.

Coastal Zone Management: Nautilus Press, Inc, 1054 National Press Bldg, Washington, DC 20045, USA; tel (202) 347-6643; Editor: John R Botzum, Sr.

Coastal Zone Topics: Joint Nature Conservation Committee—JNCC, Monkstone House, City Rd, Peterborough PE1 1JY, United Kingdom; tel (1733) 62626; fax (1733) 555948; internet www.jncc.gov.uk; Editor: N V Jones; f 1995; irreg; co-published with the Estuarine and Coastal Science Association; monographic series.

Code Permanent Environnement et Nuisances: Editions Législatives, 80 ave de la Marne, 92546 Montrouge Cédex, France; tel 1-40-92-68-68; fax 1-46-56-00-15; e-mail mtudez@editions-legislatives.fr; internet www.editions-legislatives.fr; Deputy Editor: Martine Tudez; Man Editor: Emmanuel de Baiillen; f 1973; quarterly, with monthly update bulletins; French and EC regulations affecting pollution, nature protection and other aspects of the environment; French.

Colorado Wildlife: Colorado Wildlife Federation, 445 Union Blvd, Suite 302, Lakewood, CO 80228-1243, USA; tel (303) 987-0400; fax (303) 987-0200; e-mail cwfed@aol.com; internet www.coloradowildlife.com; Editor: Diane Gansauer; f 1981; every 2 months; circ 5,500.

Combat Nature: BP 3046, 24003 Périgueux Cédex, France; tel 5-53-08-29-01; fax 5-53-09-52-52; Dir: Alain de Swarte; f 1971; quarterly; former title *Revue des Associations Ecologiques et de Défense de l'Environnement*; circ 10,000.

Commonwealth Forestry Review: Commonwealth Forestry Association, Oxford Forestry Institute, South Parks Rd, Oxford OX1 3RB, United Kingdom; tel (1865) 271037; fax (1865) 275074; e-mail cfa_oxford@hotmail.com; Editor: A J Grayson; f 1921; quarterly; reports of international forestry conferences, debates on governmental and private forest management policy, original academic and practical papers on all aspects of forestry, reports of research in progress and book reviews; circ 1,200.

Compendium Newsletter: Educational Communications, Inc, POB 351419, Los Angeles, CA 90035-9119, USA; tel and fax (310) 559-9160; e-mail ecnp@aol.com; internet www.ecoprojects.org; Editor: Nancy Pearlman; f 1971; bi-monthly; Educational Communications, Inc; conservation, sustainability, wildlife, land use, natural resources, parks; circ 800.

Connect: UNESCO/Education Sector, 7 pl de Fontenoy, 75352 Paris 07 SP, France; tel 1-45-68-08-09; fax 1-45-68-56-28; e-mail d.bhagwut@unesco.org; internet www.unesco.org/education/ste; Editor: Diileep Bhagwut; f 1976; quarterly; in 1997 *Connect* was merged with the International Network of Institutions of Science and Technology Educations (INISTE) bulletin to become the International Science, Technology and Environmental Education (STEE) newsletter of UNESCO; the objective of the newsletter, especially since the Budapest World Conference on Science of 1999, is to promote UNESCO's action in STEE world-wide, as well as to provide a platform for the exchange of information among individuals and institutions throughout the world; available in English, French, available in Spanish, Arabic, and Russian editions; English edition available online; circ 28,600.

Conscious Choice: 920 North Franklin, Suite 202, Chicago, IL 60610-3121, USA; tel (312) 440-4373; e-mail cc@consciouschoice.com; internet www.consciouschoice.com; Editor: Sheri Reda; Rights Officer: James Faber; f 1988; every 2 months; reports on environmental issues and natural alternatives in health care, food and nutrition; available online.

Conserva: Ministry of Environmental Affairs and Tourism, Private Bag X447, Pretoria 0001, South Africa; tel (12) 3103457; fax (12) 3222476; Editor: Maryke Arnoldi; f 1974; quarterly; former title *Environment RSA*; education and information about current environmental issues; in Afrikaans and English; circ 30,000.

Conservation Biology: c/o Gary K Meffe, Wildlife Ecology and Conservation, Newins-Ziegler 303, POB 110430, University of Florida, Gainesville, FL 32661-0430, USA; tel (781) 388-8250; fax (781) 388-8255; e-mail journals@blacksci.com; internet www.blackwell-science.com; Editor: Gary K Meffe; f 1987; quarterly; scientific journal published for the Society for Conservation Biology; in English, with abstracts in Spanish; available online; circ 5,000.

Conservation Directory: National Wildlife Federation, 8925 Leesburg Pike, Vienna, VA 22184-0001, USA; tel (703) 790-4000; fax (703) 790-4468; e-mail gordonr@nwf.org; internet www.nwf.org; Editor: Rue Gordon; f 1955; annual; circ 6,000.

Conservation Ecology: Ecological Society of America, 2010 Massachusetts Ave, NW, Suite 400, Washington, DC 20036, USA; tel (202) 833-8773; fax (202) 833-8775; e-mail info@consecol.org; internet www.comsecol.org; Editor: C S Holling; irreg; available only online.

Conservation Education: Young People's Trust for the Environment and Nature Conservation—YPTENC, 8 Leapale Rd, Guildford GU1 4JX, United Kingdom; tel (1483) 539600; fax (1483) 301992; e-mail info@yptnec.org.uk; internet www.yptnec.org.uk; Editor: Cyril Littlewood; f 1981; 2 a year; newsletter for teachers and youth leaders; circ 40,000.

Conservation Matters: Cleveland Wildlife Trust, Bellamy House, 2a Brighouse Business Village, Riverside Park, Middlesbrough, Cleveland TS2 1RT, United Kingdom; tel (1642) 253716; fax (1642) 251072; informs members of Cleveland Wildlife Trust of local and national wildlife issues.

Conservation Science Western Australia: Department of Conservation and Land Management – Science Division, Locked Bag 104, Bentley D.C., WA 6983, Australia; tel (08) 9334-0324; fax (08) 9334-0327; e-mail Brownyn.Mathiesen@calm.wa.gov.au; internet science.calm.wa.gov.au/cswajournal; Editor: Bronwyn Mathiesen; f 2002; irregular; scientific journal covering biology, entomology, construction, conservation, environmental studies and forestry; continues *CALMScience: Western Australian Journal of Conservation and Land Management* vol1(1) 1993 – vol3(4) 2001; previous journal: *Research Bulletin* (Western Australia Dept. of Conservation and Land Management) no1 (1986) – no6 (1992); available as PDF; circ circ 1,500.

Conservation Sciences: Manomet Center for Conservation Sciences, POB 1770, Manomet, MA 02345, USA; tel (508) 224-6521; fax (508) 224-9220; e-mail webresponse@manomet.org; internet www.manomet.org/quarterly/csqrt.htm; quarterly; available online.

Conserve: Western Pennsylvania Conservancy, 209 Fourth Ave, Pittsburgh, PA 15222-1707, USA; tel (412) 288-2777; fax (412) 281-1792; Editor: Bill Randour; 2 a year; members' magazine reporting on the Conservancy's efforts to preserve the water, land and wildlife of Western Pennsylvania; circ 20,000.

Conserve Wildlife: Dept of Environmental Protection, Division of Fish, Game and Wildlife, POB 400, Trenton, NJ 08625-0400, USA; tel (609) 292-9400; fax (609) 984-1414; Editor: Kate McGuire; f 1978; quarterly; state government publication, concerning the protection and management of wildlife, principally in New Jersey; circ 50,000.

The Conserver: British Trust for Conservation Volunteers—BTCV, 36 St Mary's St, Wallingford OX10 0EU, United Kingdom; tel (1491) 821600; fax (1491) 839646; e-mail information@btcv.org.uk; internet www.btcv.org; Editor: Katherine Taylor; f 1972; 3 a year; practical conservation news, intended particularly for members of the Trust; circ 16,000.

Container Recycling Report: Resource Recycling, Inc, POB 42270, Portland, OR 97242-0270, USA; tel (503) 233-1305; fax (503) 233-1356; e-mail info@resource-recycling.com; internet www.resource-recycling.com; Editor: Jerry Powell; f 1989; monthly; covers the recycling of glass, plastic, aluminium, steel and composite containers.

Corporate Environmental Strategy: PRI Publishing, 441 Main St, 2nd Floor, Metuchen, NJ 08840-1836, USA; tel (908) 548-5827; fax (908) 548-2268; Editor: Bruce Plasecki; Rights Officer: Richard Mansfield; f 1993; quarterly.

Corsonat: Corbett Society of Naturalists, 342 Shivaji Rd, Meerut 250 001, India; Editor: Y M Rai; f 1979; quarterly; concerned with the conservation of Indian flora and fauna; in English; circ 300.

Countryside: Countryside Commision, John Dower House, Crescent Pl, Cheltenham GL50 3RA, United Kingdom; tel (1242) 521381; fax (1242) 584270; Editor: Calvin Pugsley; f 1983; quarterly; circ 12,000.

Country-Side: c/o Mrs J Pearton, 48 Russell Way, Higham Ferrers NN10 8EJ, United Kingdom; fax (1933) 314672; Editor: David Applin; 2 a year; the British Naturalists' Association; circ 14,000.

Critical Reviews in Environmental Science and Technology: CRC Press, Inc, 2000 Corporate Blvd, NW, Boca Raton, FL 33431, USA; tel (561) 994-0555; fax (561) 998-9784; internet www.crcpress.com; Editor: Terry J Logan; f 1970; quarterly; former title *Critical Reviews in Environmental Control*; circ 600.

Croner's Environmental Management: Croner.CCH Group Ltd, 145 London Rd, Kingston upon Thames KT2 6SR, United Kingdom; tel (20) 8547-3333; fax (20) 8547-2638; e-mail info@croner.co.uk; internet www.croner.co.uk; Editor: K Hunter; f 1991; annual, with quarterly updates.

Croner's Waste Management: Croner.CCH Group Ltd, 145 London Rd, Kingston upon Thames KT2 6SR, United Kingdom; tel (20) 8547-3333; fax (20) 8547-2638; e-mail info@croner.co.uk; internet www.croner.co.uk; Editor: A Trunder; quarterly updates; information on waste management for professionals; available online and on CD-ROM.

CSEB Newsletter/Bulletin de la SCBE: Canadian Society of Environmental Biologists—CSEB, POB 962, Station F, Toronto, ON M4Y 2N9, Canada; quarterly; informs members of the CSEB's activities, current events and progress in the field of environmental sciences; encourages an integrated approach to environmental problems.

Current Bibliography on Science and Technology: Environmental Pollution—Kagaku Gijutsu Bunken Sokuho, Kankyo Kogai-hen: Kagaku Gijutsu Shinko Gigyodan, 5-3 Yonbancho, Chiyoda-ku, Tokyo 100, Japan; tel (3) 5214-8413; fax (3) 5214-8410; f 1975; monthly; published by Japan Science and Technology Corporation, Information Centre for Science and Technology; includes abstracts; circ 600.

Daily Environment Report: (Publr) Bureau of National Affairs, Inc, 1231 25th St, NW, Washington, DC 20037, USA; tel (202) 452-4200 (Publr); (800) 372-1033 (Subscriptions); fax (202) 452-7891; internet www.bna.com; Managing Editor: Larry E Evans; f 1992; daily; US and international environmental news; available online.

Daziran Tansuo (Exploration of Nature): Sichuan Kexue Jishu Chubanshe, 3 Yandao Jie, Chengdu, Sichuan 610016, People's Republic of China; tel (28) 664688; Editor: Guangwe Li; quarterly; in Chinese; circ 4,000.

Découvertes/Discoveries: Centro Nacional de Cultura, Rua António Maria Cardoso 68, 1200 Lisbon, Portugal; tel (1) 3466722; fax (1) 3428250; e-mail info@cnc.pt; internet www.cnc.pt; Editor: Helena Vaz da Silva; annual abridged edition of *Descobertas*; dedicated to the promotion of Portugal's cultural and natural heritage; in English and French.

The Deer Farmer: Wham Deer Ltd, Manners St, POB 11-092, Wellington, New Zealand; tel (4) 473-9243; fax (4) 473-4530; e-mail info@wham.co.nz; internet www.deerfarmer.co.nz; Editor: Diane Barton; Publishers: Angela Fussell, Trevor Walton; f 1979; 11 a year; deer health, farm management, product marketing and general interest; associated publication: *New Zealand Deer Farming Annual*; in English; circ 4,000.

Defenders: Defenders of Wildlife, 1101 14th St, NW, Suite 1400 Washington, DC 20005-5601, USA; tel (202) 682-9400; fax (202) 682-1331; e-mail info@defenders.org; internet www.defenders.org; Editor: James G Deane; f 1930; quarterly; circ 130,000.

Development Education: 1 St Paul's Close, Clitheroe BB7 2NB, United Kingdom; Editor: Colin Scott; 3 a year.

Developments in Environmental Economics: Elsevier Science BV, Books Division, POB 211, 1000 AE Amsterdam, Netherlands; tel (20) 4853911; fax (20) 4853705; e-mail niinfo-f@elsevier.nl; internet www.elsevier.nl; f 1991; irreg; in English.

Die Naturwissenschaften (Natural Science): Springer-Verlag GmbH & Co KG, 14197 Berlin, Heidelberger Pl 3, Germany; tel (30) 8207730; fax (30) 82787448; e-mail subscriptions@springer.de; internet link.springer.de/link/service/journals/00114/index.htm; monthly; Max-Planck-Gesellschaft zur Förderung der Wissenschaften (Max Planck Society for the Advancement of Science), Munich; nature magazine intended for the general public; circ 3,000.

Die Naturwissenschaften: Springer Verlag AG, 14197 Berlin, Heidelberger Pl 3, Germany; tel (30) 827870; fax (30) 8214091; e-mail service@spinger.de; Editor: Dr D Czeschlik; Advertising Manager: Edda Lückermann; monthly; circ 1,700.

Die Rote Mappe (The Red Map): Niedersächsischer Heimatbund eV, 30159 Hannover, Landschaftstr 6a; Germany; tel (511) 3681251; fax (511) 3632780; e-mail NHBev@t-online.de; internet www.niedersaechsischer-heimatbund.de; Editor: Dr Wolfgang Rüther; f 1960; annual; critical reports relating to conservation of the nature and heritage of Lower Saxony; in German; circ 7,000. Publishing house: Federation for Lower Saxony.

Different Drummer Magazine: Thoreau Institute, 1425 Beach Loop Rd, Bandon, OR 97411-8801, USA; tel (503) 652-7049; e-mail rot@ti.org; internet www.ti.org/ddrummer.html; Editor: Randal O'Toole; f 1980; quarterly; circ 1,500.

Directory of Environmental Websites on the Internet: US Environmental Directories, Inc, POB 65156, St Paul, MN 55165, USA; tel (612) 331-6050; e-mail john.brainard@worldnet.att.net; internet www.geocities.com/usenvironmentaldirectories; Editor: John C Brainard; f 1997; annual; also available online and as CD-ROM.

Diyadama (Nature): POB 20, Nugegoda, Sri Lanka; tel and fax (1) 827810; e-mail afej@sri.lanka.net; monthly; concerned with communication and awareness-raising in the areas of environment and development; in Sinhala.

DNR-Kurier: Deutscher Naturschutzring—DNR, 53117 Bonn, Am Michaelshof 8, Germany; tel (228) 359005; fax (228) 359096; f 1980; every 2 months; bulletin of the German Association for the Conservation of Nature; circ 5,000.

Documentos Taller Multidisciplinario del Medio Ambiente: Universidad Católica de Valparaíso, Vicerrectoría Académica, Dirección General de Investigación, Casilla 4059, Valparaíso, Chile; f 1978; annual; abstracts from multidisciplinary environmental research papers; circ 1,000.

Dokumentation Natur und Landschaft: Bundesamt für Naturschutz, 53179 Bonn, Konstantinstr 110, Germany; tel (228) 8491132; fax (228) 8491200; internet www.king.de/bfn; Editor: Dr Rainer Flück; f 1961; quarterly; publication of the Federal Agency for Nature Conservation; general ecological issues; includes abstracts, indexes and book reviews; circ 800. Publishing house: Verlag W Kohlhammer GmbH, 70549 Stuttgart, Hessbrühlstr 69, Postfach 800403, Germany.

Ducks Unlimited: Ducks Unlimited, Inc, 1 Waterfowl Way, Memphis, TN 38120-2351, USA; tel (901) 758-3825; fax (901) 758-3850; internet www.ducks.org; Editor: Tom Fulgham; Pres: Julius Wall; Exec Vice-Pres: Don Young; f 1938; every 2 months; concerned with water-fowl conservation; illustrated; includes book reviews and statistical information; circ 610,000.

DUMAC: Ducks Unlimited de Mexico, AC—DUMAC, Apdo 776, 64000 Monterrey, NL, Mexico; tel (83) 78-6757; fax (83) 78-6439; Editor: Sergio Robles Puga; f 1978; every 2 months; members' magazine; circ 11,000.

Eagle: Eagle Foundation, POB 1767, Madison, WI 53701-1767, USA; Editor: James C Clark; f 1972; quarterly; circ 1,700.

Earth First!: Daily Planet Publishing, POB 1415, Eugene, OR 97440-1415, USA; tel (541) 741-9191; fax (541) 741-9192; e-mail earthfirst@igc.apc.org; internet www.envirolink.org/orgs/ef; Publrs: Jim Flynn, Lacey Phillabaum, Theresa Kintz; f 1979; 8 a year; reports on direct action in defence of the environment; circ 15,000.

Earth Island Journal: Earth Island Institute—EEI, 300 Broadway, Suite 28, San Francisco, CA 94133-3312, USA; tel (415) 788-3666; fax (415) 788-7324; e-mail audreywebb@earthisland.org; internet www.earthisland.org/ei; Editor: Chris Clarke; Man Editor: Audrey Webb; f 1986; quarterly; international environmental news; in English; circ 20,000.

Earth Surface Processes and Landforms: John Wiley and Sons Ltd, The Atrium Southern Gate, Terminus Rd, Chichester, West Sussex PO19 8SQ, United Kingdom; fax (1243) 775878; e-mail cs-journals@wiley.co.uk; internet www.wiley.co.uk; Editor: Michael J Kirkby; Rights and Permissions Officer: Diane Southern; 13 a year; the British Geomorphological Research Group; circ 1,050.

Earthmatters: Friends of the Earth, 26–28 Underwood St, London N1 7JQ, United Kingdom; tel (20) 7490-1555; fax (20) 7490-0881; e-mail info@foe.co.uk; internet www.foe.co.uk; Editor: Lesley Sneardon; Deputy Editor: Nicola Baird; f 1988; quarterly; news and features about current key environmental issues; in English; circ 73,000.

East African Agricultural and Forestry Journal: c/o Narc-Muguga, POB 30148, Nairobi, Kenya; tel (154) 32880; fax (154) 32090; Editor: Dr J O Muga; f 1935; quarterly; findings of original research in agriculture, animal husbandry, forestry and related fields; circ 1,000.

Ecoalert: Conservation Council of New Brunswick, 180 St John St, Fredericton, NB E3B 4A9, Canada; tel (506) 458 8747; fax (506) 458-1047; e-mail ccnb@web.apc.org; Editor: Lori Sinclair; every 2 months; former title *Conservation*; deals with environmental issues, particularly those affecting the province of New Brunswick; circ 1,300.

Ecocycle: Environment Canada, Ottawa, ON K1A 0H3, Canada; tel (819) 953-0459; fax (819) 953-7970; e-mail ecocycle@ec.gc.ca; internet www.ec.gc.ca/ecocycle; Editor: Duncan Bury; f 1995; 2 a year; newsletter on product policy and management issues; available in English and French editions; circ 3,000.

Ecodesign: Ecological Design Association, The British School, Slad Rd, Stroud GL5 1QW, United Kingdom; tel (1453) 765575; fax (1453) 759211; e-mail ecological@designassociation.freeserve.co.uk; Editor: David Pearson; Deputy Editor: Charlie Ryrie; f 1991; 3 a year; ecological issues relating to all design disciplines, including architecture, interior design, textiles, furniture, product and industrial design, graphics, packaging and landscape design; circ 6,000.

Ecoforum: Environment Liaison Centre International—ELCI, POB 72461, Nairobi, Kenya; tel (2) 562015; fax (2) 562175; e-mail ecoforum@africaonline.co.ke; Editor: John Scully; f 1974; quarterly; environment and development issues for NGOs; covers campaigns and environmental struggles; links local and global issues; in English; circ 3,000.

Ecography: Oikos Nordic Ecological Society, Lund University, Dept of Ecology, Ecology Bldg, 223 62 Lund, Sweden; tel (46) 222-37-91; fax (46) 222-37-90; e-mail ecography@ekol.lu.se; internet www.oikos.ekol.lu.se; Editors: Esa Ranta, Mikko Monkkönen, Martin Zobel; Man Editor: Linus Svensson; every 2 months; former title *Holarctic Ecology*; studies in biodiversity, biogeography, ecological conservation and natural history of species; in English; circ 600. Publishing house: Munksgaard International Publishers Ltd, 35 Nørre Søgade, POB 2148, 1016 Copenhagen K, Denmark; tel 77-33-33-33; fax 77-33-33-77.

Ecohydrology and Hydrobiology: International Centre for Ecology PAS, Dziekanów Lesny, 05-092 Lomianki, Poland; tel and fax: (22) 7514116; e-mail mce-pan@mail.unicom.pl; Editor-in-Chief: Maciej Zalewski; Editors: Prof David M Harper, Prof Richard D Robarts; Technical Editor: Dr Pawel Bijok; f 2001; quarterly; Polish Academy of Sciences; Ecohydrology & Hydrobiology publishes: original research papers, invited or submitted review papers, short communications in every field of ecohydrology and hydrobiology, especially invited are contributions which provide an integrative (also interdisciplinary) approach to aquatic sciences, explaining ecological and hydrological processes at a river-basin scale, or propose practical applications of this knowledge; the journal is a continuation of two previous scientific periodicals *Polish Archives of Hydrobiology* and *Acta Hydrobiologica*; in English; circ 650.

EcoLog Canadian Pollution Legislation: Southam Information Products Ltd, 1450 Don Mills Rd, Don Mills, ON M3B 2X7, Canada; tel (416) 442-2041; fax (416) 442-2200; e-mail llubka@corporate.southam.ca; internet www.ecolog.com; Assoc Publisher: Lidia Lubka; f 1983; every 2 months; environmental legislation in Canada; in English; available on CD-ROM.

EcoLog Week: Southam Information Products Ltd, 1450 Don Mills Rd, Don Mills, ON M3B 2X7, Canada; tel (416) 445-6641; fax (416) 442-2020; Editor: Deborah Orchard; Publr: Margaret Nearing; Vice-Pres: Ian Rhind; f 1972; weekly; reports on events, legislation, policy, research, technological developments, private- and public-sector programmes and initiatives in the environmental field; focuses on waste management and control of industrial pollution; intended for readers in Canadian business, industry and government.

Ecológica: ECOE Ediciones, Calle 26, No 5-52, Torre B, Apdo Aéreo 58308, Santafé de Bogotá, DC, Colombia; tel (1) 282-8974; fax (1) 235-2750; f 1989; quarterly; deals with global environmental issues, with particular attention to South and Central America; includes features, interviews and ecological news.

Ecological Abstracts: Elsevier/Geo Abstracts, The Old Bakery, 111 Queens Rd, Norwich NR1 3PL, United Kingdom; tel (1603) 626327; fax (1603) 667934; e-mail geoabs@elsevier.co.uk; internet www.elsevier.nl/locate/spol; f 1974; monthly; contains summaries of international scientific papers and conferences on all aspects of aquatic and terrestrial ecology; areas covered include global, marine, tidal, estuarine, freshwater and microbial ecology, weeds, pests and diseases, pollution, conservation, agriculture, forestry, fisheries, evolution and palaeoecology; available online and on CD-ROM.

Ecological Applications: Ecological Society of America, 2010 Massachusetts Ave, NW, Suite 400, Washington, DC 20036, USA; tel (202) 833-8773; fax (202) 833-8775; e-mail esahq@esa.org; internet www.esa.org/publications; Editor: Louis F Pitelka; f 1991; quarterly; circ 4,200.

Ecological Complexity: Cologicla Complexity and Modelling Laboratory, Dept of Botany and Plant Sciences, University of California, Riverside, CA 92521-0124, USA; tel bai-lian.li@ucr.edu; internet www.sciencedirect.com/science/journal/1476945X; Editor-in-Chief: Bai-Lian (Larry) Li; f 2004; 4 issues per year; an international journal devoted to the publication of high quality, peer-reviewed articles on all aspects of biocomplexity in the environment, theoretical ecology, and special issues on topics of current interest; the scope of the journal is wide and interdisciplinary with an integrated and quantitative approach; the journal particularly encourages submission of papers that integrate natural and social processes at appropriately broad spatio-temporal scales; available online. Publishing house: Elsevier.

Ecological Economics: Elsevier Science Publishers BV, POB 211, 1000 AE Amsterdam, Netherlands; tel (20) 4853911; fax (20) 4853598; e-mail nlinfo-f@elsevier.nl; internet www.elsevier.nl; Editor-in-Chief: Robert Costanza; f 1989; 12 a year in 4 vols; scientific journal published for the International Society for Ecological Economics; includes ecological impact studies and promotes investigation of the connections between ecological and economic systems.

Ecological Engineering: Elsevier Science Publishers BV, POB 211, 1000 AE Amsterdam, Netherlands; tel (20) 4853911; fax (20) 4853598; e-mail nlinfo-f@elsevier.nl; internet www.elsevier.nl; Editor-in-Chief: William J Mitsch; f 1992; 8 a year in 2 vols; scientific journal containing papers on ecotechnology, including synthetic ecology, bioengineering, pollution control and sustainable agriculture.

Ecological Modelling: Elsevier Science Publishers BV, POB 211, 1000 AE Amsterdam, Netherlands; tel (20) 4853911; fax (20) 4853325; e-mail nlinfo-f@elsevier.nl; internet www.elsevier.nl; Editor-in-Chief: Sven E Jørgensen; f 1975; 36 a year in 10 vols; international scientific journal concerned with ecological modelling and engineering, and systems ecology; includes book reviews, bibliographies and indexes; in English.

Ecological Monographs: Ecological Society of America, 2010 Massachusetts Ave, NW, Suite 400, Washington, DC 20036, USA; tel (202) 833-8773; fax (202) 833-8775; e-mail esahq@esa.org; internet www.esa.org/publications; Editor: Robert K Peet; f 1931; quarterly; circ 3,400.

Ecological Research: Blackwell Science Asia, 54 University St, Carlton, Vic 3053, Australia; tel (3) 9347-0300; fax (3) 9347-5001; e-mail info@blacksci-asia.com.au; internet www.blackwell-science.com/australi; Editor-in-Chief: Prof Takuya Abe; f 1986; 3 a year; members' journal published for the Ecological Society of Japan; original research papers in all fields of ecology; in English; circ 3,200.

Ecological Restoration: Society for Ecological Restoration, University of Wisconsin Press, Journals Division, 2537 Daniels St, Madison, WI 53718-6772, USA; tel (608) 263-7889; fax (608) 262-5209; e-mail rmn@macc.wisc.edu; internet www.wisc.edu/wisconsinpress; Editor: William R Jordan III; f 1981; 2 a year; formerly *Restoration and Management Notes*; available on microform; circ 3,000.

Ecological Society of America Bulletin: Ecological Society of America, 2010 Massachusetts Ave, NW, Suite 400, Washington, DC 20036, USA; tel (202) 833-8773; fax (202) 833-8775; e-mail esahq@esa.org; internet www.esa.org/publications; Editor: Allen Solomon; f 1917; quarterly; circ 7,300.

The Ecologist: Unit 18, Chelsea Wharf, 15 Lots Rd, London SW10 0QJ, United Kingdom; tel (20) 7351-3578; fax (20) 7351-3617; e-mail editorial@theecologist.org; internet www.theecologist.org; Editors: Zac Goldsmith; f 1970; monthly; journal containing reports, commentary and investigative journalism on social, political and environmental matters; English; circ 20,000. Publishing house: Tyler Moorehead.

Ecology: Ecological Society of America, 2010 Massachusetts Ave, NW, Suite 400, Washington, DC 20036, USA; tel (202) 833-8773; fax (202) 833-8775; e-mail esahq@esa.org; internet www.esa.org/publications; Editor: Robert K Peet; f 1920; 8 a year; circ 6,400.

Ecology Abstracts: Cambridge Scientific Abstracts, 7200 Wisconsin Ave, Bethesda, MD 20814-4823, USA; tel (301) 961-6750; fax (301) 961-6720; e-mail market@csa.com; internet www.csa.com; Editor: Robert Hilton; Publr: Ted Caris; f 1975; monthly; interdisciplinary journal of ecological research; focuses on the ways in which living organisms of all kinds interact with their environment and with other organisms; electronic formats available.

Ecology and Farming: International Federation of Organic Agriculture Movements—IFOAM, 66636 Tholey-Theley, Ökozentrum Imsbach, Germany; tel (6853) 5190; fax (6853) 30110; e-mail ifoam@t-online.de; internet www.ifoam.org; 3 a year; provides information on the activities of organic agriculture world-wide, relevant research and new agricultural publications; in English; circ 2,000.

Ecology Law Quarterly: University of California Press, 2120 Berkeley Way, Suite 5812, Berkeley, CA 94720, USA; tel (510) 643-7154; fax (510) 642-9917; e-mail journals@ucop.edu; internet www.ucpress.edu/journals; f 1973; quarterly; journal of environmental law and policy; intended for legal professionals, academics, scientists, engineers, planners and government officials; circ 1,150.

Ecology USA: Business Publishers, Inc, 8737 Colesville Rd, Suite 1100, Silver Spring, MD 20910-3928, USA; tel (301) 587-6300; fax (301) 587-1081; e-mail bpnews@bpnews.com; internet www.bpnews.com; Editor: Elaine Eiserer; f 1972; every 2 weeks (25 a year); newsletter.

Eco-Management and Auditing: John Wiley and Sons Ltd, The Atrium Southern Gate, Terminus Rd, Chichester, West Sussex PO19 8SQ, United Kingdom; tel (1243) 779777; fax (1243) 775878; internet www.wiley.co.uk/journals.sci; Editor: Prof Richard Welford; Publishing Editor: Lesley Valentine; f 1994; 3 a year; business journal concerning environmental auditing; circ 400.

ECONEWS: Northcoast Environmental Center, 879 9th St, Arcata, CA 95521, USA; tel (707) 822-6918; fax (707) 822-0827; e-mail nec@igc.org; internet www.necandeconews.to; Editor: Sidney Dominitz; f 1971; monthly (11 a year); concerned with the environment and ecology; circ 7,500.

Ecos: Commonwealth Scientific and Industrial Research Organization—CSIRO, POB 1139, Collingwood, Vic 3066, Australia; tel (3) 9662-7500; fax (3) 9662-7555; internet www.publish.csiro.au; Editor: Bryony Bennett; Assoc Editor: Wendy Pyper; f 1974; quarterly; reports on CSIRO environmental research findings for the non-specialist reader; in English; circ 8,000.

Ecos: 119021 Moscow, Zubovsky bul 4, Russian Federation; tel (095) 201-25-86; fax (095) 230-21-70; Editor: V B Rudenko; f 1990; quarterly; publication of the International Peace Fund

Association, the Socio-Ecological Union—SEU and RIA—Novosti Press Agency; covers nature conservation issues and the global ecological situation; includes articles by non-Russian authors; available in Russian and English editions; circ 100,000.

Ecosystems: Springer-Verlag, 175 Fifth Ave, New York, NY 10010, USA; tel (212) 460-1500; fax (212) 473-6272; e-mail ecosys@macc.wisc.edu; internet www.springer-verlag.com; Editors: Stephen Carpenter, Monica Turner; f 1998; every 2 months.

Ecotoxicology & Environmental Safety: Academic Press, Inc, Journal Division, 525 B St, Suite 1900, San Diego, CA 92101-4495, USA; tel (619) 250-1840; fax (619) 699-6800; e-mail apsubs@acad.com; internet www.academicpress.com/ees; Editors: Frederick Coulson, Friedheim Korte; f 1977; 9 a year; available online.

Ecovillages: Global Village Institute, POB 90, Summertown, TN 38483, USA; tel (931) 964-3992; e-mail dx@gaia.org; internet www.gaia.org/dx; Editor: Albert Bates; f 1978; quarterly; circ 1,500.

Ecumene: A Journal of Cultural Geographies: Arnold, 388 Euston Rd, London NW1 3BH, United Kingdom; tel (20) 7873-6000; fax (20) 7873-6325; e-mail arnoldjournals@hodder.co.uk; internet www.arnoldpublishers.com/journals/journpages/09674608.htm; Editors: Denis Cosgrove, Prof Don Mitchell; f 1994; quarterly; concerned with the cultural appropriation of nature, landscape and the environment; circ 9,100.

EDF Letter: Environmental Defense Fund—EDF, 257 Park Ave S, New York, NY 10010, USA; tel (212) 505-2100; fax (212) 505-2375; e-mail members@edf.org; internet www.edf.org/pubs/edf-letter; Editor: Norma H Watson; f 1970; every 2 months; documents the activities of the EDF, which aims to safeguard human health, reduce pollution of air, land and water, and conserve wildlife; circ 300,000.

Eesti Loodus (Estonian Nature): Perioodika Publishing House, Veski 4, Tartu 5002, Estonia; tel (7) 421-186; fax (7) 427-432; e-mail toimetus@el.loodus.ee; internet www.loodus.ee/el; Editor-in-Chief: Uno Siitan; f 1933; monthly; popular science and nature; circ 5,200.

EHP Supplements: Dept of Health and Human Services, National Institute of Environmental Health Sciences, Environmental Health Information Service, POB 12233, Research Triangle Park, NC 27709, USA; tel (919) 541-3406; fax (919) 541-0273; e-mail ehponline@niehs.nih.gov; internet ehp.niehs.nih.gov; Editor: Thomas Goehl; f 1972; monthly; Chinese – quarterly; online, open access; circ 2,900.

EI Digest: Environmental Information Ltd, POB 390266, Minneapolis, MN 55439-0266, USA; tel (612) 831-2473; fax (612) 831-6550; e-mail ei@mr.net; internet www.envirobiz.com; Editor: Cary Perket; monthly (10 a year); national journal on industrial and hazardous waste management; includes analysis of new or proposed regulations and technologies, and documentation of specific waste services.

EIC Envirotech News: 45 Weymouth Street, London W1G 8ND, United Kingdom; tel (20) 7935-1675; fax (20) 7486-3455; e-mail info@eic-uk.co.uk; internet www.eic-uk.co.uk; Editor: H Hyman; f 1991; quarterly; covers developments in UK and EU environmental policy.

EIS Cumulative: Cambridge Scientific Abstracts, 7200 Wisconsin Ave, 6th Floor, Bethesda, MD 20814, USA; tel (301) 961-6750; fax (301) 961-6720; e-mail market@csa.com; internet www.csa.com; Editor: Edward J Reid; Publr: Ted Caris; f 1970; every 2 months; available on CD-ROM.

EIS: Digests of Environmental Impact Statements: Cambridge Information Group, 7200 Wisconsin Ave, Bethesda, MD 20814, USA; tel (301) 961-6750; fax (301) 961-6720; e-mail market@csa.com; internet www.csa.com; Editor: Edward J Reid; f 1970; every 2 months.

Ekistics: Str Syndesmou 24, 106 73 Athens, Greece; tel 2103623216; 2103623373; fax 2103629337; e-mail ekistics@otenet.gr; Editor: Panayotis C Psomopoulos; Circulation Man: Myto Connou; f 1955; every 2 months; the Athens Technological Organization; the problems and science of human settlements at the global, regional, national, internation and local scales; circ 2,000. Publishing house: Athens Technological Organization, Athens Center of Ekistics.

El Niño Southern Oscillation Diagnostic Advisory: National Center for Environmental Protection, W-NP52, 5200 Auth Rd, Suite 65, Camp Springs, MD 20746-4304, USA; e-mail wd52vk@hp31.wwb.noaa.gov; internet nic.fb4.noaa.gov/products/analysis_monitoring/enso-advisory; weekly; observation, detection and prediction of short-term fluctuations in climatic conditions; available only online.

Elainmaailma: Sanoma Corporation/Sanomaprint Magazines, POB 100, 00040 Helsinki, Finland; tel (9) 1201; fax (9) 1205456; Editor-in-Chief: Jyrki Leskinen; Sales Manager: Kirsi-Marja Skogster; f 1979; monthly; dedicated to aquatic and terrestrial wildlife; circ 44,000.

Electromagnetics and VDU News: POB 25, Liphook GU30 7SE, United Kingdom; tel (1428) 751430; fax (1428) 658802; Editor and Publr: Simon Best; f 1989; quarterly; newsletter on the health hazards of electromagnetic fields from power lines, visual display units (VDUs) and microwave systems and the use of such fields in orthodox and alternative medicine.

Electronic Green Journal: University of Idaho, Library, Moscow, ID 83843-4198, USA; tel (208) 885-6631; fax (208) 885-6817; e-mail majanko@uidaho.edu; internet egj.lib.uidaho.edu; Gen Editor: Maria A Jankowska; Man Editor: William T Johnson; f 1994; 2 a year; former title *Green Library Journal*; professional journal on international environmental topics such as assessment, conservation, development, education, information resources and technology; in English; available only online.

Elpa: Jana seta 5, Riga 1050, Latvia; tel 721-1776; fax 722-3280; Editor-in-Chief: Mairita Solima; f 1990; weekly; circ 3,900.

EM: Air and Waste Management Association, 1 Gateway Center, 3rd Floor, Pittsburgh, PA 15222, USA; tel (412) 232-3444; fax (412) 232-3450; internet www.awma.org; Editor: William Tony; f 1907; monthly; magazine for environmental managers; available online; circ 16,000.

EMAP Monitor: EPA Office of Research and Development, 401 M St, SW, Washington, DC 20460, USA; tel (202) 655-4000; monthly.

Endangered Species Update: University of Michigan, School of Natural Resources, Dana Bldg, 430 East University, Ann Arbor, MI 48109-1115, USA; tel (313) 763-3243; fax (313) 936-2195; e-mail esupdate@umich.edu; internet www.umich.edu/~esupdate; Editor: Misty McPhee; f 1983; every 2 months; incorporates the US Fish and Wildlife Service *Endangered Species Bulletin*; current news and information on endangered species protection; includes articles on research, management and policy issues, book reviews and technical notes; in English.

Endangered Wildlife: Endangered Wildlife Trust, Private Bag X11, Parkview 2122, South Africa; tel (11) 4861102; fax

(11) 4861506; e-mail ewts@ewt.org.za; internet www.ewt.org.za; Editor: David Holt-Biddle; Advertising Manager: Lynn Ras; f 1983; 4 a year; former title *Quagga*; journal devoted to endangered wildlife in southern Africa; includes features, news items, reports from southern African nations and book reviews; in English; circ 6,000. Publishing house: Future Publishing.

ENDS Environment Daily: Environmental Data Services Ltd, Finsbury Business Centre, 40 Bowling Green Lane, London EC1R 0NE, United Kingdom; tel (20) 7814-5320; fax (20) 7415-0106; e-mail envdaily@ends.co.uk; internet www.ends.co.uk; Editor: Nick Rowcliffe; f 1997; daily.

ENDS Report: Environmental Data Services Ltd, Finsbury Business Centre, 40 Bowling Green Lane, London EC1R 0NE, United Kingdom; tel (20) 7814-5320; fax (20) 7415-0106; e-mail subs@ends.co.uk; internet www.ends.co.uk/report; Editor: Julian Rose; Advertising Manager: Maria Sztayer; Subscriptions: Vivienne Plax; f 1978; monthly; environmental information for business; intended for environmental professionals in industry, consultancy and central and local government; circ 25,000.

ENEA Report—Energia, Ambiente e Innovazione: Comitato Nazionale per la Ricerca e per lo Sviluppo dell'Energia Nucleare e delle Energie Alternative—ENEA, Lungotevere Thaon di Revel 76, 00196 Rome, Italy; tel (06) 36272401; fax (06) 36272299; e-mail alida.laococe@caseeia.enea.it; f 1951; every 2 months; bulletin of the National Committee for Research and Development of Nuclear and Alternative Energy; includes information on new technology, energy and the environment; circ 3,500.

Energetica: Icemenerg-Oide, 74568 Bucharest, Bd Energeticienilor 8, Romania; tel and fax (1) 3239552; e-mail oide@icemenerg.vsat.ro; internet www.icemenerg.vsat.ro/energetica; Editor: Vasile Rugina; f 1953; every 2 months; covers sources of energy, means of generation and specific problems; practical review for researchers, designers, investors, teachers and technical experts in the power sector; in Romanian, with abstracts in English, French, German and Russian; circ 2,500.

Energi & Planlægning (Energy and Planning): Foreningen Sun Media, Ryesgade 19, 2200 Copenhagen N, Denmark; tel 31-35-22-11; fax 31-35-65-45; e-mail sunmedia@dk-online.dk; internet www2.dk-online.dk/users/sun_media; Editor: Jette Johnsen; f 1985; every 2 months; covers energy and environmental issues; in Danish; available online; circ 5,000.

Energia és Atomtechnika: Scientific Society for Energy Economy, 1055 Budapest, Kossuth Lajos tér 6-8, Hungary; tel (1) 153-2751; fax (1) 156-1215; Editor-in-Chief: Dr G Boki; f 1947; every 2 months.

Energiagazdálkodás (Energy Economy): Scientific Society for Energy Economy, 1055 Budapest, Kossuth Lajos tér 6-8, Hungary; tel (1) 153-2751; fax (1) 153-3894; Editor: Dr Andor Anesini.

Energie en Europe/Energy in Europe: Euroffice, Office for Official Publications of the EC, rue Mercier 2, BP 1003, 2985 Luxembourg, Luxembourg; 3 a year; in English, French, German and Spanish.

Energie Extra (Energy Extra): Bundesamt für Energie/Office Fédéral de l'Energie/Ufficio Federale dell'Energia, 3003 Bern, Switzerland; tel (31) 3225664; fax (31) 3232510; e-mail olivier.grandjean@bfe.admin.ch; internet www.admin.sh/bfe; f 1977; every 2 months; former title *Courrier de l'Antigaspillage/Bollettino sul Risparmio d'Energia/Energie-Spar-Nachrichten*; government publication dealing with energy research, energy efficiency and renewable energy; covers the Swiss action programme, Energy 2000; includes statistical tables; in French and German; circ 70,000.

Energie Plus: Association Technique pour l'Efficacité Energétique, 47 ave Laplace, 94117 Arcueil Cédex, France; tel 1-43-07-08-00; fax 1-49-85-06-27; Editor: Michel Hoez; monthly; technical journal dedicated to efficient energy use and environmental protection; circ 5,890.

Energie Spektrum—Magazin für Energiemanagement: Henrich Publikationen GmbH, 60528 Frankfurt am Main, Schwanheimerstr 110, Germany; fax (69) 96777111; e-mail du@verlag.henrich.de; internet www.henrich.de; Editor: Ralf Dunker; f 1986; monthly (10 a year); professional energy management publication; deals with environmental protection, solar energy, emission measurement, marketing and the ethics of the energy industry; includes statistics, indexes, book reviews, notification of events and reviews of new products; circ 9,573.

Energiemonitor: Centraal Bureau voor de Statistiek, Prinses Beatrixlaan 428, Voorburg, Netherlands; tel (70) 3375364; fax (70) 3377429; e-mail rlnk@cbs.nl; internet www.cbs.nl; f 1998; quarterly; statistics on quantities and prices of energy and water; in Dutch; circ 200.

Energy Conversion and Management: Elsevier Science Ltd, POB 800, Kidlington, Oxford OX5 1DX, United Kingdom; fax (1865) 843010; e-mail nlinfo-f@elsevier.com; internet www.elsevier.com; Editor: Jesse Denton; f 1961; 20 a year; provides an interdisciplinary forum through which energy conversion, fuel conservation and energy management can be treated; in English, French and German; available online via www.sciencedirect.com.

Energy Daily: King Communications Group, Inc, 627 National Press Bldg, Washington, DC 20045, USA; tel (202) 638-4260; fax (202) 662-9744; Exec Editor: Dennis Wamstead; provides daily news of the international energy industry, covering conventional energy sources, energy technology, financial and legislative developments, alternative sources of energy and environmental issues; available online.

Energy Design Update: Cutter Information Corpn, 37 Broadway, Suite 1, Arlington, MA 02174-5539, USA; tel (617) 648-8700; fax (617) 648-1950; e-mail crowley@cutter.com; internet www.cutter.com; Editors: Don Best, Charlie Wardell; monthly; covers issues related to energy-efficient building design and residential load management; electronic formats available.

Energy Economist: Financial Times Energy Publishing, Pearson Professional Ltd, Maple House, 149 Tottenham Court Rd, London W1P 9LL, United Kingdom; tel (20) 7896-2241; fax (20) 7896-2275; e-mail eninfo@pearson-pro.com; internet www.ftenergy.com; Editor: Lucy Plaskett; f 1983; monthly; news and analysis of energy issues world-wide.

Energy in Europe: The Stationery Office, Publications Centre, 51 Nine Elms Lane, London SW8 5DR, United Kingdom; tel (20) 7873-0011; fax (20) 7873-8463; e-mail book.enquiries@theso.co.uk; internet www.the-stationery-office.co.uk; 3 a year; published for the European Commission; concerned with management of the energy sector in Europe.

Energy Policy: Elsevier Science Ltd, The Boulevard, Langford Lane, Kidlington OX5 1GB, United Kingdom; tel (1865) 843000; fax (1865) 843010; e-mail nlinfo-f@elsevier.com; internet www.elsevier.com; Editor: Nicky France; Man

Editor: Lyndon Driscoll; f 1973; journal containing articles on various energy issues, including energy-efficiency potential in the domestic sector, environmental impacts of fossil fuel use and energy demand management in developing countries; intended for managers, planners, politicians and researchers; available online via www.sciencedirect.com; circ 18 a year.

Energy World: Institute of Energy, 8 Devonshire St, London W1N 2AU, United Kingdom; tel (20) 7580-7124; fax (20) 7580-4420; e-mail info@instenergy.org.uk; internet www.instenergy.org.uk; Editor: Steve Hodgson; 10 a year; all aspects of energy use, from exploration and mining to energy consumption and efficiency; circ 6,000.

Enfo: Florida Conservation Foundation, Environmental Information Center, 1251 Miller Ave, Suite B, Winter Park, FL 32789-4827, USA; Editor: Linda Lord; f 1971; quarterly; circ 1,200.

ENFO Newsletter: Asian Institute of Technology, POB 4, Klong Luang, Pathumthani 12120, Thailand; tel (2) 524-5863; fax (2) 524-5870; e-mail enterrc@ait.ac.th; internet www.ait.ac.th/clair/center/ensic; quarterly; in English.

ENTA Newsletter (Environmental Technology Assessment): United Nations Environment Programme—UNEP, Industry and Environment Programme Centre—IE, 39–43 quai André Citroen, 75539 Paris Cédex 15, France; tel 1-44-37-14-50; fax 1-44-37-14-74; e-mail unepie@unep.fr; f 1994; 2 a year; supplement to *Industry and Environment* (qv).

Entsorga-Magazin/Entsorgungswirtschaft: Deutscher Fachverlag, 60326 Frankfurt am Main, Mainzer Landstr 251, Germany; tel (69) 759501; fax (69) 759515; e-mail entsorga@dfv.de; Editors: B Wassmann, H Wilkus; f 1982; 10 a year; waste handling, waste water treatment, air measurement, pollution prevention, soil decontamination, environmental management; circ 15,800.

Entsorgungs Praxis: Bertelsmann Fachzeitschriften GmbH, 33335 Gütersloh, Avenwedder Str 55, Postfach 120, Germany; tel (5241) 802165; fax (5241) 8060660; e-mail entsorgungs.praxis@bertelsmann.de; Editor: Ulrick Knorra; Publishing Manager: R Brummel; f 1982; 10 a year; trade journal concerned with developments in environmental technology affecting the refuse industry, air purification and waste-water treatment; circ 16,200.

Enviroaction: National Wildlife Federation, 8925 Leesburg Pike, Vienna, VA 22184-0001, USA; tel (703) 790-4000; fax (703) 772-7332; e-mail mclean@nwf.org; internet www.nwf.org; Editor: Tim McLean; f 1980; monthly; available on microform; circ 32,000.

EnviroKids: Wildlife and Environment Society of Southern Africa, POB 44189, Linden 2104, South Africa; tel (21) 6718344; e-mail rgriffs@iafrira.com; internet www.wildlifesociety.org.za; Editor: Dr Roberta Griffiths; f 1980; every 2 months; former titles *Toktokkie*; conservation, natural history and environmental magazine for young people; in English; circ 11,000.

Environ: A Magazine for Ecological Living and Health: Wary Canary Press, POB 2204, Fort Collins, CO 80522, USA; tel (303) 224-0083; Editor: Suzanne Randegger; f 1987; quarterly; covers various environmental health issues; circ 5,000.

Environment: Heldref Publications, Inc, 1319 18th St, NW, Washington, DC 20036-1802, USA; tel (202) 296-6267; fax (202) 296-5149; e-mail env@heldref.org; internet www.heldref.org; Editor: Barbara T Richman; f 1958; 10 a year; covers environmental science and policy issues; circ 12,000. Publishing house: (Subscriptions) Allen Press, Inc, POB 1897, Lawrence, KS 66044-8897, USA.

Environment: Environment Centre of Western Australia, Inc, 587 Wellington St, Perth, WA 6000, Australia; fax (8) 9322-3045; Editor: Jeff Bryant; f 1975; quarterly; exploration of the social and physical environments of Western Australia; circ 800.

Environment Abstracts: Congressional Information Service, 4520 East–West Highway, Suite 800, Bethesda, MD 20814-3389, USA; tel (301) 654-1550; fax (301) 652-1273; e-mail cainet@cispubs.com; internet www.cispubs.com; Man Editor: Larry Sheridan; Asst Man Editor: Todd Malfa; f 1971; monthly in 2 vols; contains approx 1,400 abstracts of articles from periodicals, conference proceedings and special reports from universities, NGOs and government agencies in the USA and abroad; available online and quarterly on CD-ROM.

Environment Abstracts Annual: Congressional Information Service, 4520 East–West Highway, Suite 800, Bethesda, MD 20814-3389, USA; tel (301) 654-1550; fax (301) 652-1273; e-mail cainet@cispubs.com; internet www.cispubs.com; Editor: Larry Sheridan; Dir, Environmental and Statistical Services: Darlene Montgomery; f 1971; annual; cumulative annual edition and index of *Environment Abstracts* (qv); in English; available online (monthly) and on CD-ROM (quarterly).

Environment and Ecology: Environment and Ecology Communication Centre—EECC, Kalyani University, Fisheries Laboratory, Dept of Zoology, Kalyani 741 235, West Bengal, India; tel (3162) 98220; e-mail laser@cal2.vsnl.net.in; Editor and Exec Dir: Prof S K Konar; f 1983; quarterly; publishes results of research in all branches of environmental science and ecology, including agriculture, forestry, pest control and air, water and soil pollution control; in English; circ 50,000. Publishing house: MKK Publications, 91A Ananda Palit Rd, Calcutta 700 014, West Bengal, India.

Environment and Planning A: Pion Ltd, 207 Brondesbury Park, London NW2 5JN, United Kingdom; tel (20) 8459-0066; fax (20) 8451-6454; e-mail sales@pion.co.uk; internet www.pion.co.uk/ep; Hon Editor: A G Wilson; Editor: N J Thrift; Publishing Dir: J Schubert; f 1969; monthly; concerned with the planning of cities and regions; issues covered include inner-city regeneration and damage to the environment; contains papers, reviews of research policy and book reviews.

Environment and Urbanization: Human Settlements Programme, International Institute for Environment and Development—IIED, 3 Endsleigh St, London WC1H 0DD, United Kingdom; tel (20) 7388-2117; fax (20) 7388-2826; e-mail eandu@iied.org; internet www.iied.org/human/eandu/; Editor: Dr David Satterthwaite; f 1989; 2 a year; dedicated to urban environmental problems, primarily in Latin America, Asia and Africa; intended for specialist and general readers; each issue focuses on an issue selected by subscribers; includes papers by experts, reviews, news, profiles of relevant NGOs and a bibliography; co-published by the Instituto Internacional de Medio Ambiente y Desarrollo—IIED, América Latina; on-line at www.ingentaselect.com/titles/09562478.htm; in English, with summaries in English, French and Spanish; available online; circ 2,600.

Environment Bulletin: CERAM Research Ltd, Queens Rd, Penkhull, Stoke-on-Trent ST4 7LQ, United Kingdom; tel (1782) 764444; fax (1782) 412331; e-mail enquiries@ceram.co.uk; internet www.ceram.com; Information Officer: Joanne Dawson; f 1997; quarterly; news on environmental law, policy, best practice and health and safety; in English; available online; circ 400. Publishing house: CERAM.

Environment Business Magazine: Information for Industry Ltd, 4 Valentine Pl, London SE1 8RB, United Kingdom; tel (20) 7654-7100; fax (20) 7654-7172; e-mail ebm@ifi.co.uk; internet www.ifi.co.uk; Editor: Karen Harries-Rees; Publr: James Benn; f 1990; 9 a year; reports on pollution control, environmental protection and sustainable management of resources; intended for senior industrial managers; incorporates *Environment Today*; circ 9,400.

Environment Business News Briefing: Information for Industry Ltd, 4 Valentine Pl, London SE1 8RB, United Kingdom; tel (20) 7654-7100; fax (20) 7654-7171; e-mail ebnb@ifi.co.uk; internet www.ifi.co.uk; Editor: Ian Grant; Publr: James Benn; f 1990; 24 a year; circ 1,200.

Environment Canada: Selected Publications and Websites: EPS Publications, Ottawa, ON K1A 0H3, Canada; tel (819) 953-5750; fax (819) 994-5629; e-mail epspubs@ec.gc.ca; internet www.ec.gc.ca/etad/infopr_e.html; f 1993; annual; Environment Canada, Technology Outreach; former title *Environment Canada. Environmental Protection Series Reports*; lists more than 350 publications and websites from Environment Canada; in English and French; available online; circ 2,000.

Environment Information Bulletin–EIB: Halsbury House, 35 Chancery Lane, London WC2A 1EL, United Kingdom; tel (20) 8662-2000 (Customer Services); fax (20) 7400-2570; e-mail zira.skeats@lexisnexis.co.uk; internet www.lexisnexis.co.uk; Editors: Louis Wustermann, Paul Suff; 10 issues per year; guide to forthcoming regulation and environmental best practice; information about the environmental challenges facing organizations in the public and private sectors; subjects covered: enforcement activities; court cases; EU developments; environmental standards; office guidance; cleaner technology; industry sector initiatives; international trends; government policy and company practice.

Environment Management Abstracts: Referativnyi Zhurnal, Vsesoyuznyi Institut Nauchno-Tekhnicheskoi Informatsii—VINITI, 125315 Moscow, Usievicha ul 20a, Russian Federation; tel (095) 152-64-41; f 1987; monthly; in English.

Environment New Jersey: Parker Publishing, Inc, 254 Mercer St, Stirling, NJ 07980, USA; tel (908) 766-6960; Editor: Elizabeth K Parker; f 1995; quarterly; concerned with the preservation of New Jersey's natural and historic resources, recreation facilities and wildlife; also covers Green Acres funding, solid waste management, pollution control and air and water quality; circ 50,000.

Environment News: International Mass Retail Association, 1700 North Moore St, Suite 2250, Arlington, VA 22209-1903, USA; tel (202) 861-0774; fax (202) 785-4588; Editor: Robin Lanier; f 1991; monthly; circ 270.

Environment News Digest: National Environmental Health Association, South Tower, Suite 970, 720 South Colorado Blvd, Denver, CO 80246, USA; tel (303) 756-9090; fax (303) 691-9490; e-mail neha.org@juno.com; internet www.neha.org; Editor: Julie Collins; f 1933; quarterly; circ 6,000.

Environment Newsletter: South Pacific Regional Environment Programme—SPREP, POB 240, Apia, Samoa; tel 21929; fax 20231; f 1971; quarterly; available in English and French editions; circ 1,000.

Environment NSW: Nature Conservation Council of New South Wales, Level 5, 362 Kent St, Sydney, NSW 2000, Australia; tel (2) 9279-2466; fax (2) 9279-2499; e-mail ncc@nccnsw.org.au; internet www.nccnsw.org.au; Chair: Robert Pallin; Exec Off: Brooke Flanagan; f 1955; quarterly; available online; circ 2,500.

Environment Protection Engineering: Oficyna Wydawnicza Politechniki Wroclawskiej, 50-370 Wroclaw, Wybrzeze Wyspianskiego 27, Poland; fax (71) 3282940; Editors: Tomasz Winnicki, Lucjan Pawlowski; f 1975; quarterly; contains papers on water purification, waste-water treatment, and disposal and sterilization of solid waste; in English, with summaries in Polish and Russian; circ 600.

Environment Report: Trends Publishing, Inc, 1079 National Press Bldg, Washington, DC 20045, USA; tel (202) 393-0031; fax (202) 393-1732; Pres and Publr: Arthur Kranish; environmental newsletter.

Environment Reporter: Bureau of National Affairs, Inc, 1231 25th St, NW, Washington, DC 20037, USA; tel (202) 452-4200; fax (202) 452-7891; internet www.bna.com; Editor: Larry E Evans; f 1970; weekly; US and North American environmental news; available online.

Environment, Science and Society: Kluwer Academic Publishers, POB 17, 3300 AA Dordrecht, Netherlands; tel (78) 6392392; fax (78) 6392254; e-mail services@wkap.nl; internet www.wkap.nl; f 1994; irreg; monographic series.

Environment South Australia: Australia; tel (8) 8223-5155; fax (8) 8232-4782; e-mail general@ccsa.asn.au; internet www.ccsa.asn.au; Editor: Margaret Bolster; Co-ordinator: Christine Nankivell; f 1991; quarterly; covers environmental events, news and opinions in South Australia; circ 6,000.

Environment Statistics: Central Statistics Office, c/o Government Statistician, Private Bag 0024, Gaborone, Botswana; tel (31) 352200; fax (31) 352201; e-mail csbots@gov.bw; Editor: G M Charumbira; f 1999; annual.

Environment Victoria News: Environment Victoria, Inc, 19 O'Connell St, North Melbourne, Vic 3051, Australia; tel (3) 9348-9044; fax (3) 9348-9055; e-mail envict@metspace.net.av; Editor: Geoffrey Heard; f 1973; monthly; deals with urban environmental issues, pollution of land, forests and water, public transport and air pollution, sewerage and the alleviation of the greenhouse effect; circ 2,000.

Environment Watch: Europe: Agra Informa Ltd, 80 Calverley Rd, Tunbridge Wells, Kent TN1 2UN, United Kingdom; tel (1892) 533813; fax (1892) 544895; e-mail marketing@agra-net.com; internet www.cutter.com; f 1992; every 2 weeks; environmental policy news and analysis; covers EU environmental policy, environmental regulations, waste management, environmental liability and environmental labelling; electronic formats available.

Environmental & Waste Management: EPP Publications, 52 Kings Rd, Richmond TW10 6EP, United Kingdom; tel (20) 8948-7165; fax (20) 8747-9663; e-mail epppublications@btinernet.com; internet www.epppublications.com; Editor: Prof P C Coggins; Publr: M G Pomel; f 1997; quarterly; formed by the merger of *Environmental Policy and Practice* and *Journal of Waste Management and Resource Recovery*; articles on environmental policy and law, with particular emphasis on issues relating to waste management; in English; circ 550.

Environmental and Resource Economics: Kluwer Academic Publishers, Spuiblvd 50, POB 17, 3300 AA Dordrecht, Netherlands; tel (78) 6576000; fax (78) 6576254; e-mail services@wkap.nl; internet www.kluweronline.com/issn/0924-6460; Editor-in-Chief: R K Turner; Exec Editor: Ian J Bateman; f 1991; 12 a year; published for the European Association of Environmental and Resource Economists—EAERE; deals with the application of economic theory and methods to environmental problems; in English.

Environmental Approvals in Canada: Butterworths Canada Ltd, 75 Clegg Rd, Markham, ON L6G 1A1, Canada; tel (905) 479-2665; fax (905) 479-2826; Editor: Michael I Jeffery; f 1989; 3 a year; covers the entire environmental approvals procedure, including public consultation and hearings; circ 500.

Environmental Awareness: International Society of Naturalists—INSONA, Oza Bldg, Salatwada, Baroda 390 001, India; tel and fax (265) 421009; Editor: Dr Gunavant M Oza; f 1977; quarterly; members' magazine on conservation of wildlife and the global environment; devoted to environmental conservation for human welfare; in English; circ 1,000.

Environmental Business Journal: Environmental Business International, Inc, 4452 Park Blvd, Suite 306, San Diego, CA 92116-4039, USA; tel (619) 295-7685; fax (619) 295-5743; Editor: Grant Ferrier; Vice-Pres, Publs: Lynette Thwaites; monthly; available online.

Environmental Communicator: North American Association for Environmental Education, 410 Tarvin Rd, Rock Spring, GA 30739-2502, USA; tel and fax (937) 676-2514; internet www.naaee.org; Editor: Andrea Shotkin; f 1972; every 2 months; circ 2,500.

Environmental Compliance: Environmental Compliance Group, 1402 Glenhill Lane, Lewisville, TX 75067-2832, USA; tel (214) 436-8797; f 1990; annual.

Environmental Compliance in Florida: Lexis Law Publishing, POB 7587, Charlottesville, VA 22906-7587, USA; tel (804) 972-7600; fax (804) 972-7666; e-mail llp.customer.support@lexisnexis.com; internet www.lexislawpublishing.com; f 1993; 2 a year, with irreg updates.

Environmental Compliance in New Jersey: Business & Legal Reports, Inc, c/o Tom Picinich, 39 Academy St, Madison, CT 06443, USA; tel (203) 318-0000; fax (203) 245-2559; Editor: Claire Condon; f 1990; 12 a year.

Environmental Conservation: Dept of Marine Science and Coastal Management, University of Newcastle, Newcastle upon Tyne NE1 7RU, United Kingdom; tel (191) 222-5868; fax (191) 222-7891; e-mail envcons@ncl.ac.uk; internet www.cup.cam.ac.uk/journals/jnlscat/enc/enc.html; Editor: Dr Nicholas V C Polunin; Publishing Dir: Conrad Guettler; f 1974; quarterly; published for the Foundation for Environmental Conservation; international journal dedicated to exposing and opposing environmental degradation resulting from human over-population and unsound technology; advocates urgent action to protect and improve the global environment; in English; circ 800.

Environmental Control Law: Illinois State Bar Association, Section for Environmental Law, Illinois Bar Center, Springfield, IL 62701, USA; tel (217) 525-1760; fax (217) 525-0712; Editors: Johnie Brown, Ellen Partridge; f 1970; quarterly; circ 725.

Environmental Education: National Association of Environmental Education—NAEE, Wolverhampton University, Walsall Campus, Gorway Rd, Walsall WS1 3BD, United Kingdom; tel and fax (1922) 31200; Editor: Philip Neal; Sec-Gen: G Back; Pres: D Bellamy; f 1960; 3 a year; circ 1,500.

Environmental Education: China Environmental Science Press, 8 Beigangzi St, Chongwen Dist, Beijing 100062, People's Republic of China; tel (10) 67145444; fax (10) 67144442; Editor-in-Chief: Ding Mei; f 1995; quarterly; published for the China National Environmental Protection Agency; concerned with environmental education at various levels in the People's Republic of China; in Chinese.

Environmental Education and Information: School of Environment and Life Sciences, University of Salford, Allerton Bldg, Frederick Rd, Salford M6 6PU, United Kingdom; tel (161) 295-2140; fax (161) 295-2138; Editor-in-Chief: Prof Graham Ashworth; Man Editor: Dr Philip James; f 1981; quarterly; aims to promote protection of the global environment by provision of accurate information and sound education; directed at practitioners, policy-makers and academics in the field of the environment; published in association with the Tidy Britain Group (United Kingdom); circ 300.

Environmental Education Research: Taylor & Francis Group (Journals), 4 Park Sq, Milton Park, Abingdon, OX14 4RN, United Kingdom; tel (20) 7017-6000; fax (20) 7017-6336; e-mail eer@bath.ac.uk; internet www.tandf.co.uk/journals/titles/13504622.asp; Editor: Prof William Scott; 5 a year; help advance understanding of environmental and sustainability education through a focus on papers reporting research and development activities. Publishing house: Routledge.

Environmental Engineering: Springer-Verlag, 14197 Berlin, Heidelberger Pl 3, Germany; tel (30) 82070; fax (30) 8214091; e-mail subscriptions@springer.de; internet www.springer.de; f 1994; irreg; concerned with the development of technologies for the treatment and prevention of environmental pollution; in English.

Environmental Engineering News: School of Civil Engineering, Purdue University, West Lafayette, IN 47907, USA; tel (317) 474-4211; fax (317) 496-1107; Editor: John M Bell; f 1994; every 2 months; newsletter; includes short news and statistical items relating to the environment; circ 3,000.

Environmental Engineering Selection Guide: American Academy of Environmental Engineers, 130 Holiday Court, Suite 100, Annapolis, MD 21401, USA; tel (410) 266-3311; fax (410) 266-7653; Editor: William C Anderson; f 1985; annual; circ 10,000.

Environmental Forensics: Taylor & Francis, Inc, 325 Chestnut St, Suite 800, Philadelphia, PA 19106, USA; tel (215) 625-8900; fax (215) 625-2940; e-mail bob@rmorrison.com; internet www.tandf.co.uk/journals/titles/15275922.asp; Editor: Robert Morrison; quarterly; provides a forum for scientific investigations that address environmental contamination which is subject to law court, arbitration, public debate, or formal argumentation; encompasses all aspects of contamination within the environmental media of air, water, soil and biota.

Environmental Forum: Environmental Law Institute, 1616 P St, SW, Suite 200, Washington, DC 20036, USA; tel (202) 939-3800; fax (202) 939-3868; Editor: Stephen Dujack; f 1988; every 2 months.

Environmental Geosciences: Blackwell Science, Inc, Commerce Pl, 350 Main St, Malden, MA 02148-5018, USA; tel (781) 388-8250; fax (781) 388-8255; e-mail journals@blacksci.com; internet www.blackwell-science.com; Editor: Stephen M Testa; f 1996; quarterly; in English.

Environmental Health: Chadwick House Group, Chadwick Court, 15 Hatfields, London SE1 8DJ, United Kingdom; fax (20) 7827-5866; e-mail m.randall@chgl.com; internet www.cieh.org.uk; Editor: Claire Brown; Rights Officer: Nicola Rose; the Chartered Institute of Environmental Health; circ 11,725.

Environmental Health and Pollution Control (Section 46): Excerpta Medica, Elsevier Science Publishers, Molenwerf 1, 1014 AG Amsterdam, Netherlands; tel (20) 4853507;

fax (20) 4853222; e-mail embase-europe@elsevier.nl; internet www.elsevier.nl; f 1971; monthly; deals with air, soil and water pollution, noise prevention, radioactivity and general environmental matters; in English; available online; circ approx 100.

Environmental Health and Safety CFRs: Government Institutes, 4 Research Pl, Suite 200, Rockville, MD 20850, USA; tel (301) 921-2355; fax (301) 921-0373; e-mail giinfo@govinst.com; internet www.govinst.com; annual; compilation of monthly *Code of Federal Regulations* (CFR) updates: *Environmental Protection Agency, Occupational Safety and Health Administration* and *Department of Transportation: Hazardous Materials*; covers environmental risk assessment, new regulatory requirements and the Environmental Protection Agency's enforcement priorities; available as quarterly compilations on CD-ROM.

Environmental Health Criteria: World Health Organization—WHO, Marketing and Dissemination Dept, ave Appia, 1211 Geneva 27, Switzerland; tel (22) 7912111; fax (22) 7913111; e-mail publications@who.ch; internet www.who.int; f 1976; approx 12 a year; discusses the dangers that certain chemicals and pollutants pose to human health and the environment; in English, with summaries in French and Spanish; circ 5,000.

Environmental Health News: Chadwick House Group Ltd, Chadwick Court, 15 Hatfields, London SE1 8DJ, United Kingdom; tel (20) 7827-9929; fax (20) 7827-9930; Publr and Editor: Bill Randall; Advertising Manager: Paul Prior; f 1986; weekly; journal published for the environmental health profession; circ 11,800.

Environmental Health Perspectives: US Dept of Health and Human Services, National Institute of Environmental Health Sciences, POB 12233, Research Triangle Park, NC 27709, USA; tel (919) 541-3406; fax (919) 541-0273; e-mail ehponline@niehs.nih.gov; internet ehp.niehs.nih.gov; Editor: Thomas Goehl; f 1993; monthly; Chinese edn quarterly; available online, open access; circ 6,500.

Environmental History: Forest History Society, 701 Vickers Ave, Durham, NC 27701, USA; tel (919) 682-9319; fax (919) 682-2349; Editor: Adam Rome; Man Editor: Eve Munson; f 1957; quarterly; documents the history of human interaction with the environment, particularly forests, and of the conservation of natural resources; includes bibliographies, book reviews and news of current research by the Forest History Society; circ 2,000. Publishing house: Duke University Press, College Station, POB 90660, Durham, NC 27708, USA; tel (919) 687-3600; fax (919) 688-2615.

Environmental Impact Assessment Review: Elsevier Science, Inc, POB 945, New York, NY 10159-0945, USA; tel (212) 633-3730; fax (212) 633-3680; e-mail usinfo-f@elsevier.com; internet www.elsevier.nl; Editor: Eric Johsani; f 1985; every 2 months; Massachusetts Institute of Technology; describes developments in environmental impact assessment, environmental decision-making and the settlement of environmental disputes.

Environmental Impact Review in New York: Matthew Bender & Co, Inc, 2 Park Ave, New York, NY 10016, USA; tel (212) 448-2000; e-mail international@bender.com; internet www.bender.com; f 1990; monthly; available on CD-ROM.

Environmental Industry Yearbook: Environmental Economics, POB 6852, Lawrenceville, NJ 08648-0852, USA; tel (215) 925-7168; Editor: Michael Silverstein; f 1988; annual.

Environmental Judicial Review Bulletin: John Wiley and Sons Ltd, The Atrium Southern Gate, Terminus Rd, Chichester, West Sussex PO19 8SQ, United Kingdom; tel (1243) 779777; fax (1243) 843232; e-mail cs-journals@wiley.co.uk; Editor: Michael Fordham; f 1994; 3 a year.

Environmental Key Contacts and Information Sources: ABS Group, Inc, Institutes Division, 4 Research Pl, Suite 200, Rockville, MD 20850-3226, USA; tel (301) 921-2300; fax (301) 821-0373; e-mail ginfo@govinst.com; internet www.govinst.com; Editors: Charles Ikonomou, Diane Pacchione; f 1998; irreg.

Environmental Law: Northwestern School of Law, Lewis and Clark College, 10015 South West Terwilliger Blvd, Portland, OR 97219, USA; tel (503) 768-6700; fax (503) 768-6671; e-mail envtl@lclark.edu; internet www.lclark.edu/envtl; Editor: Katherine Kellner; Law Review Program Asst: Theresa Edwards; f 1975; quarterly; deals with environmental and natural resource issues; electronic formats available; circ 900.

Environmental Law Monthly: Agra Informa Ltd, 80 Calverley Rd, Tunbridge Wells, Kent TN1 2UN, United Kingdom; tel (1892) 533813; fax (1892) 544895; e-mail marketing@agra-net.com; internet www.agra-net.com; monthly; UK environmental law and regulations; legislation, statutes, EU Directives, court rulings, industry news.

Environmental Law Reporter: Environmental Law Institute, 1616 P St, NW, Suite 200, Washington, DC 20036, USA; tel (202) 939-3800; fax (202) 939-3868; e-mail law@eli.org; internet www.eli.org; Editor-in-Chief: John Turner; Man Editor: Linda Johnson; f 1970; monthly, with weekly updates; in English.

Environmental Law Update: Data Research, Inc, 4635 Nicols Rd, Suite 100, Eagan, MN 55122-3337, USA; tel (612) 452-8267; fax (612) 452-8694; f 1994; monthly; contains case reports, updates on legislation and law review articles.

Environmental Management: Springer-Verlag New York, Inc, 175 Fifth Ave, New York, NY 10010, USA; tel (212) 460-1500; fax (212) 473-6272; e-mail orders@springer-ny.com; internet www.springer-ny.com; Editor: David Alexander; 8 a year; intended for environmental scientists, engineers, legal professionals and sociologists; serves as a forum for the assessment of techniques, ideas and research results that may be applied to specific environmental management problems; available online; circ 1,200.

Environmental Management and Audit Manual: C C H Australia Ltd, POB 230, North Ryde, NSW 2133, Australia; tel (2) 9857-1300; fax (2) 9857-1600; f 1995; 2 a year; includes guide-lines on the implementation of environmental practices and planning, and the integration of environmental considerations into overall management.

Environmental Management and Health: MCB University Press Ltd, 62 Toller Lane, Bradford BD8 9BY, United Kingdom; tel (1274) 777700; fax (1274) 785200; Editor: Dr John Peters; f 1990; quarterly; concerned with environmental factors and their effects on human health.

Environmental Manager: Agra Informa Ltd, 80 Calverley Rd, Tunbridge Wells, Kent TN1 2UN, United Kingdom; tel (1892) 533813; fax (1892) 544895; e-mail marketing@agra-net.com; internet www.agra-net.com; Editor: Murray Griffin; f 1994; monthly; environmental controls, regulatory requirements, European Directives and UK legislation in pollution prevention and control, land contamination, environmental audits, transport, energy efficiency, waste management, recycling, environmental management systems, eco-labelling.

Environmental Monitoring and Assessment: Kluwer Academic Publishers, POB 17, 3300 AA Dordrecht, Netherlands; tel (78) 6334911; fax (78) 6334254; e-mail services@wkap.nl; internet www.kluweronline.com/issn/0167-6369; Editor: G Bruce Wiersma; f 1981; 18 a year; available online.

Environmental News: Federation of Pakistan Chambers of Commerce and Industry, 454/9 F B Area, POB 13631, Karachi 75950, Pakistan; tel (21) 4987127; fax (21) 6349806; monthly; ISO standards, national environment quality standards, requirements for environmental impact assessment, environmental audit, eco-labelling, present environmental regulations and other environmental problems.

Environmental News Briefing: Environmental Book Publishing Programme, POB 26, 434/3 Sri Jayawardenapura, Sri Lanka; tel (1) 2873131 or (1) 2813568; fax (1) 2883187; e-mail slejf@sltnet.lk; internet www.environmentaljournalists.lk; Dr Dharman Wickremaratne; f 1987; monthly; IUCN–World Conservation Union; concerned with communication and awareness-raising in the areas of environment and development; in English; circ 5,000. Publishing house: Sri Lanka Environmental Journalists Forum.

Environmental News Digest: Sahabat Alam Malaysia (Friends of the Earth), 19 Kelawai Rd, 10250 Pulau Pinang, Malaysia; tel (4) 376930; fax (4) 375705; Editor: David Heath; f 1983; every 2 months; includes statistics, diagrams and book reviews; in English; circ 500.

Environmental News Service: 733 Bishop St, No 170–218, Honolulu, HI 96813, USA; tel (800) 446-1211; weekly.

Environmental Opportunities: 103 Roxbury St, No 5, Keene, NH 03431, USA; tel and fax (603) 357-5940; Publr and Editor: Sanford Berry; f 1982; monthly; national newsletter listing primarily entry- and mid-level environmental employment opportunities in the private, non-profit sector; circ 1,500.

Environmental Periodicals Bibliography: Environmental Studies Institute, International Academy at Santa Barbara, 5385 Hollister Ave, Suite 210, Santa Barbara, CA 93111-2305, USA; tel and fax (805) 965-0790; Editor: Jana Carey; f 1972; monthly; guide to recent articles on environmental issues; covers human ecology and planning, air pollution, energy, land resources, water resources, nutrition and health; available online (monthly) and on CD-ROM (quarterly); circ 400.

Environmental Planning: 5 Gazit St, Tel-Aviv 69417, Israel; tel 3-6473432; fax 6-6423456; Editor: Prof Dan Soën; f 1966; 3 a year; published for the Israeli Association for Environmental Planning; concerned with urban, rural and regional environmental planning, urban and regional economy, urban and regional sociology, and urban geography; in English and Hebrew; circ 500.

Environmental Policy Alert: Inside Washington Publishers, POB 7167, Benjamin Franklin Station, Washington, DC 20044, USA; tel (703) 416-8500; fax (703) 416-8543; e-mail service@wpnews.com.

Environmental Policy and Law: IOS Press, Van Diemenstr 94, 1013 CN Amsterdam, Netherlands; tel (20) 6382189; fax (20) 6203419; e-mail market@iospress.nl; internet www.iospress.nl; Editor-in-Chief: Wolfgang E Burhenne; Editor: Marlene Jahnke; f 1975; every 2 months; published for the International Council of Environmental Law (Germany); information on policy, legal and administrative matters relating to the natural environment and to sustainable development; in English and French.

Environmental Policy Review: Marjorie Mayrock Center for Russian, Eurasian and East European Research, Hebrew University of Jerusalem, Mount Scopus, Jerusalem 91905, Israel; tel 2-5883180; fax 2-5322545; e-mail msrissia@pluto.mscc.hujiac.il; Editor: Dr Zeev Wolfson; f 1987; 2 a year; environmental problems and policy in the CIS and Eastern Europe, with particular attention to the military-industrial complex; in English; circ 200–300.

Environmental Pollution: Elsevier Earth Sciences Department, Molenwerf 1, 1014 Amsterdam, Netherlands; tel (20) 485-3467; fax (20) 485-2696; e-mail nlinfo-f@elsevier.com; internet www.elsevier.com/locate/envpol; Editor-in-Chief: J W Manning; Publishing Ed: N van Dijk; f 1970; 18 a year in 6 vols; addresses issues relevant to the nature, distribution and ecological effects of all types and forms of chemical pollutants in air, soil and water; available online.

Environmental Problems and Remediation: Merton Allen Associates, Infoteam, Inc, POB 15640, Plantation, FL 33318-5640, USA; tel (954) 473-9560; fax (954) 473-0544; Editor: Merton Allen; f 1994; monthly; concerned with the effects of environmental problems and with methods and technologies for their remediation and mitigation; available online.

The Environmental Professional: National Association of Environmental Professionals, 6524 Ramoth Drive, Jacksonville, FL 32226-3202, USA; tel (904) 251-9900; fax (904) 251-9901; internet www.naep.org; Editors: John Perkins, Cathy French; f 1979; irreg; published for the National Association of Environmental Professionals—NAEP; forum for interdisciplinary discussion and analysis of environmental issues; available on microform; circ 4,000. Publishing house: Blackwell Science, Inc, 238 Main St, Cambridge, MA 02142, USA; tel (617) 876-7022; fax (617) 492-5263.

Environmental Progress (New York): American Institute of Chemical Engineers, 3 Park Ave, New York, NY 10016-5991, USA; tel (212) 591-8100; fax (212) 591-8888; e-mail xpress@aiche.org; internet www.aiche.org; Editor: Gary F Bennett; f 1982; quarterly; covers current technological advances with an emphasis on practical applications of chemical engineering to pollution abatement; circ 5,000.

Environmental Protection: China Environmental Science Press, 8 Beigangzi St, Chongwen Dist, Beijing 100062, People's Republic of China; tel (10) 67145444; fax (10) 67144442; Editor-in-Chief: Tang Dawei; f 1976; monthly; published for the National Environmental Protection Agency—NEPA; covers issues relating to environmental protection, including policy, management and technology; also contains news and reviews.

Environmental Protection Bulletin: Institution of Chemical Engineers, Davis Bldg, 165–189 Railway Terrace, Rugby CV21 3HQ, United Kingdom; tel (1778) 578214; fax (1778) 560833; e-mail delyth@icheme.org.uk; internet www.icheme.org; Editor: Delyth Forsdyke; f 1989; every 2 months; includes technical case studies on environmental topics, such as waste minimization, clean technology, land remediation, environmental management systems and water treatment; intended for chemical and process engineers and those in related disciplines; electronic formats available; circ 400.

Environmental Quality in Connecticut (Annual Report of Council of Environmental Quality): 79 Elm St, Hartford, CT 06106, USA; tel (860) 424-4000; fax (860) 424-4070; e-mail karl.wagener@po.state.ct.us; internet dep.state.ct.us/ceq.htm; Exec Dir: Karl J Wagener; annual; status and trends of Connecticut's environment; in English; available online.

Environmental Quality Management: John Wiley & Sons, Inc, Journals, 605 Third Ave, New York, NY 10158-0012, USA; tel (212) 850-6645; fax (212) 850-6021; e-mail subinfo@jwiley.com; internet www.wiley.co.uk; Editor: John T Willig; f 1991; quarterly.

Environmental Radioactivity in New Zealand and Rarotonga: Annual Report: National Radiation Laboratory, POB 25099, Christchurch, New Zealand; tel (3) 366-5059; fax (3) 366-1156; e-mail kmm@nrl.moh.govt.nz; Editor: Dr K M Matthews; f 1960; former title *Environmental Radioactivity Annual Report*; contains summaries of the results of environmental monitoring operations in New Zealand and Rarotonga; circ 150.

Environmental Regulation and Permitting: John Wiley & Sons, Inc, Journals, 605 Third Ave, New York, NY 10158-0012, USA; tel (212) 850-6645; fax (212) 850-6021; e-mail subinfo@jwiley.com; internet www.wiley.co.uk; Editor: Joseph R Guida; f 1991; quarterly; formed by the merger of *Journal of Environmental Permitting* and *Journal of Environmental Regulation*; provides detailed accounts of major environmental legislation, explanations of existing legislation and advice on planning for and implementing new regulations.

Environmental Regulation—State Capitals: Wakeman / Walworth, Inc, POB 7376, Alexandria, VA 22307, USA; tel (703) 768-9600; fax (703) 768-9690; e-mail newsletters@statecapitals.com; internet www.statecapitals.com; Pres: Keyes Walworth; Vice-Pres: Christy Walworth; Marketing Dir: Christine Ryan; f 1945; weekly; covers all states in the US; includes laws, court decisions, regulations, governors' proposals, solid wastes, resource recovery, recycling, hazardous waste, air and water pollution, nuclear waste storage, sewage disposal, pesticide policies, ground-water protection, fish kills, beach renourishment, shoreland protection, genetically modified crops, green space, fish, game and wildlife protection, methyl tert-butyl ether (MTBE), urban sprawl, endangered species, wetland protection, livestock regulation and brownfields; in English; available by e-mail.

Environmental Research: Academic Press, Inc, Journal Division, 525 B St, Suite 1900, San Diego, CA 92101-4495, USA; tel (619) 230-1840; fax (619) 699-6800; e-mail apsubs@acad.com; internet www.academicpress.com/er; Editor: Ellen K Silbergeld; f 1967; 8 a year; available online.

Environmental Research Forum: Transtec Publications Ltd, Brandrain 6, 8707 Zurich, Switzerland; tel (1) 9221020; fax (1) 9221033; e-mail ttp@ttp.net; internet www.ttp.net; Editors: A Gerhardt, S Incecik, J Staudenmann, J Niemczynowicz; f 1996; irreg; biomonitoring of polluted water; air quality management; resource recycling; integrated water management; in English.

Environmental Research in Japan: Environment Agency of Japan, 1-2-2, Kasumigaseki, Chiyoda-ku, Tokyo 100, Japan; fax (3) 3580-3542; f 1975; annual; in English; circ 320.

Environmental Research News: Oak Ridge National Laboratory, Environmental Sciences Division, POB 2008, Oak Ridge, TN 37831-6038, USA; tel (423) 374-7374; fax (423) 574-7287; e-mail webmaster@esd.ornl.gov; internet www.esd.ornl.gov/ern; irreg; available online.

Environmental Review: International Air Transport Association, 800 pl Victoria, POB 113, Montréal, PQ H4Z 1M1, Canada; tel (514) 874-0202; fax (514) 874-9632; e-mail sales@iata.org; internet www.iata.org; f 1995; annual; concerned with environmental regulatory requirements and issues, and environmental performance indicators for the civil aviation industry.

Environmental Reviews: National Research Council of Canada, Ottawa, ON K1A 0R6, Canada; tel (613) 991-5475; fax (613) 952-7656; e-mail pubs@nrc-cnrc.gc.ca; internet pubs.nrc-cnrc.gc.ca; Editor: Dr John P Smol; f 1993; quarterly; covers environmental sciences, focusing in particular on the relationship between anthropogenic stress and ecosystems; in English; available online.

Environmental Risk Watch: Ebasco Environmental, 2 World Trade Center, New York, NY 10048, USA; tel (206) 451-4619; Editor: Elaine Doward King; f 1991; every 2 weeks.

Environmental Science: Springer-Verlag, 14197 Berlin, Heidelberger Pl 3, Germany; tel (30) 82070; fax (30) 8214091; e-mail orders@springer.de; internet www.springer.de; f 1995; irreg; monographic series; in English.

Environmental Science and Policy: Elsevier Science, Inc, POB 945, New York, NY 10159–0945, USA; tel (212) 633-3730; fax (212) 633-3680; e-mail usinfo@elsevier.com; internet www.elsevier.nl; f 1998; every 2 months.

Environmental Science and Pollution Research: Ecomed Verlagsgesellschaft AG & Co KG, 86899 Landsberg, Rudolf-Diesel Str 3, Germany; tel (8191) 125500; fax (8191) 125600; e-mail a.heinrich@ecomed.de; internet www.ecomed.de/journals.htm; Editor: Otto Hutzinger; f 1994; quarterly; in English; circ 2,000.

Environmental Science and Technology: Centcom Ltd, 676 East Swedesford Rd, Suite 202, Wayne, PA 19087, USA; tel (610) 964-8061; fax (610) 964-8071; f 1967; monthly; the American Chemical Society; pollution control and environmental management; circ 12,400.

Environmental Science and Technology Library: Kluwer Academic Publishers, POB 17, 3300 AA Dordrecht, Netherlands; tel (78) 6576000; fax (78) 6576254; e-mail services@wkap.nl; internet www.wkap.nl; f 1995; irreg; monographic series; in English.

Environmental Science Research: Plenum Publishing Corpn, 233 Spring St, New York, NY 10013-1578, USA; tel (212) 620-8000; fax (212) 463-0742; e-mail info@plenum.com; internet www.plenum.com; Editor: Herbert S Rosenkranz; f 1972; irreg.

Environmental Sciences: Journal of Integrative Environmental Research: Taylor & Francis Group (Journals), 4 Park Sq, Milton Park, Abingdon, OX14 4RN, United Kingdom; tel (20) 7017-6000; fax (20) 7017-6336; internet www.tandf.co.uk/journals/titles/15693430.asp; Editors-in-Chief: Jan J Boersema, Andrew Blowers, Adrian Martin; f 2004; quarterly; international journal providing an interdisciplinary forum for the publication of research into contemporary environmental issues; especially concerned with the relationships between science, society and policy and a key aim is to advance understanding of the theory and practice of sustainable development.

Environmental Technology: Adams Business Media, 2101 South Arlington Heights Rd, Suite 150, Arlington Heights, IL 60005, USA; tel (847) 427-9512; fax (847) 427-2097; internet www.abm.net; Editor: Michael Hilts; f 1991; 7 a year; circ 65,000.

Environmental Technology: Selper Ltd, Unit 66, 2 Old Brompton Rd, London SW7 3DQ, United Kingdom; fax (20) 8995-4160; internet www.environtechnol.co.uk; Editors: J N Lester, R M Harrison; Subscription and Circulation Manager: J Roberts; f 1980; monthly; former title *Environmental Technology Letters*; includes indexes and book reviews; in English and French.

Environmental Technology Monitor: UNIDO Publications Sales Office, 1400 Vienna, POB 300, Austria; tel (1) 213-46-5031; fax (1) 260-26-5031; e-mail publications@unido.org; quarterly.

Environmental Times: Environmental Assessment Association, 1224 North Nokomis NE, Alexandria, MN 56308-5072, USA; tel (320) 763-4320; fax (320) 763-9290; Editor: Robert Johnson; monthly (except December).

Environmental Toxicology & Chemistry: Society of Environmental Toxicology and Chemistry, 1010 North 12th St, Pensacola, FL 32501-3370, USA; tel (850) 469-1500; fax (85) 469-9778; e-mail setac@setac.org; internet www.setac.org; Editor: Calvin H Ward; f 1982; monthly.

Environmental Toxicology and Pharmacology: Dept of Toxicology, POB 8000, 6700 EA Wageningen, Netherlands; tel (317) 482137; fax (317) 484931; Editors: J H Koeman, N P E Vermeulen; f 1996; 8 a year; former title *European Journal of Pharmacology—Environmental Toxicology and Pharmacology Section*; contains details of research into the toxic effects on man and other vertebrate animals of drugs and environmental contaminants; in English; circ approx 300. Publishing house: Elsevier Science Publishers BV, POB 181, 1000 AD Amsterdam, Netherlands; tel (20) 5803911; fax (20) 5803249; internet www.elsevier.nl/inca/publications/.

Environmental Viewpoints: Gale Group, 27500 Drake Rd, Farmington Hills, MI 48331-3535, USA; tel (248) 699-4253; fax (248) 699-8061; e-mail glaeord@gale.com; internet www.gale.com; Editor: Janet Witalec; f 1992; annual.

Environments in Transition: European Bank for Reconstruction and Development—EBRD, 1 Exchange Sq, London EC2A 2JN, United Kingdom; tel (20) 7338-7553; fax (20) 7338-6102; e-mail pubsdesk@ebrd.com; 2 a year.

Environments—a journal of interdisciplinary studies: University of Waterloo, Faculty of Environmental Studies, Waterloo, ON N2L 3G1, Canada; tel (519) 885-4567; fax (519) 746-2031; e-mail environ@fes.uwaterloo.ca; Editor: J Gordon Nelson; f 1982; 3 a year; intended for non-specialist readers; covers geography, architecture, the environment, development and urban and regional planning; includes articles and reviews; in English and French; circ 700.

Environmetrics: John Wiley and Sons Ltd, The Atrium Southern Gate, Terminus Rd, Chichester, West Sussex PO19 8SQ, United Kingdom; tel (1243) 779777; fax (1243) 775878; internet www.wiley.co.uk/journals-sci; Editors: Dr A H El-Shaarawi, Helen Ramsey; f 1990; 6 a year; deals with the development of statistical methodology and its application to the environmental sciences.

Environnement Africain/African Environment: ENDA Publications, Environmental Development Action in the Third World—ENDA, 4–5 rue Kleber, BP 3370, Dakar, Senegal; tel 23-63-91; fax 23-51-57; e-mail editions@enda.sn; Editor: Jacques Bugnicourt; f 1974; quarterly, supplemented by occasional papers (10 a year) dealing with single issues; environmental studies and regional planning bulletin; contains original papers on environment and development problems in Africa and developing countries in general, documents, book and magazine reviews, articles on technologies for ecological development and a calendar of events; available in French and English editions, with a small number of issues in Arabic; circ 2,200.

Environnement et Technique: Societe Alpine de Publications, 7 chemin de Gordes, 38100 Grenoble, France; fax 4-76-56-94-09; Editor: Roger Pellegrini; f 1980; monthly (10 a year); formerly Information Déchets; economic and technical strategies for resolving environmental problems; circ 8,000.

Environs: Dept of Environmental Protection, Office of Public Affairs, 59–17 Junction Blvd, 3rd Floor, Flushing, NY 11373-5107, USA; tel (212) 669-8200; fax (212) 669-2522; Editor: Andrew McCarthy; f 1982; quarterly; circ 6,000.

EnviroReport: Swedish Environmental Protection Agency, Blekholmsterrassen 36, 106 68 Stockholm, Sweden; tel (8) 698-10-00; fax (8) 698-14-90; internet www.environ.se/enviroreport; Editor-in-Chief: Katrin Hallman; f 1999; quarterly; facts and Swedish views on important environmental issues; in English; available online and as a PDF file; circ 4,000.

Envirorisk: Reactions Group, 39–41 North Rd, London N7 9DP, United Kingdom; monthly; newsletter for the Risk Management of Pollution and Protection of the Environment. Publishing house: Corporate Cover Ltd.

Era Solar: SAPT Publicaciones Técnicas, SL, Costa Rica 13, 4°, Apdo 19295, 28016 Madrid, Spain; tel (91) 3505885; fax (91) 3459313; every 2 months; technical journal concerned with renewable energy, the environment and energy conservation.

Eretz Magazine: Eretz Ha-tzvi, Inc, 5 Ma'avar Yabok St, Tel-Aviv 67440, Israel; tel 3-6912211; fax 3-6091890; Editor: Yadin Roman; f 1985; every 2 months; available in English and Hebrew editions; circ 40,000.

Eretz Ve'Teva: Nature Reserves Authority, Rehov Yermeyahn 78, Jerusalem 94467, Israel; tel 2-536271; Editor: Y Roman; f 1975; every 2 months; former title *Yidion*; information on nature reserves in Israel; includes book reviews; circ 45,000.

Erosion Control: Forester Communications, Inc, 5638 Hollister Ave, Suite 301, Goleta, CA 93117-3474, USA; tel (805) 681-1300; fax (805) 681-1312; e-mail ecinfo@ieca.org; internet www.ieca.org; Editor: John Trott; Publr: Daniel Waldman; f 1973; every 2 months; circ 20,000.

ESCAP/UNEP Asia-Pacific Environment: Environment Section, United Nations Economic and Social Commission for Asia and the Pacific—UN/ESCAP, United Nations Bldg, Rajadamnern Nok Ave, Bangkok 10200, Thailand; tel (2) 288-1234; fax (2) 288-1000; f 1995; quarterly; published jointly for ESCAP and UNEP; concerned with environmental issues and activities in the Asian and Pacific region; electronic formats available; circ 3,500.

ESE Notes: University of NC at Chapel Hill, School of Public Health, Dept of Environmental Sciences and Engineering, CB 7400, Chapel Hill, NC 27599-7400, USA; tel (919) 966-1024; fax (919) 966-2581; e-mail laura_marcial@unc.edu; internet www.spl.unc.edu/env/esenotes.htm; Editor: Laura Marcial; f 1964; quarterly; circ 1,750.

Estudios en Ecología Social (Studies in Social Ecology): Centro Latinoamericano de Ecología Social—CLAES, Amezaga 1891, Casilla 13000, 11700 Montevideo, Uruguay; tel (2) 922362; fax (2) 201908; Editor: Alain Santandreu; f 1989; monographic studies and reviews on social ecology; circ 500.

E/The Environmental Magazine: Earth Action Network, 28 Knight St, Norwalk, CT 06851, USA; tel (203) 854-5559; fax (203) 866-0602; e-mail info@emagazine.com; internet www.emagazine.com; Editor: Jim Motavalli; Publisher: Doug Moss; f 1990; every 2 months; available online; circ 70,000.

EU Energy Policy: Financial Times Energy Publishing, Pearson Professional Ltd, Maple House, 149 Tottenham Court Rd, London W1P 9LL, United Kingdom; tel (20)

7896-2450; fax (20) 7896-2620; e-mail andreas.arvanitakis@ft.com; internet www.ftenergy.com; Editor: Andreas Arvanitakis; f 1989; monthly; EU energy legislation and its impact on EU member states; in English.

Europe Environnement—Europe Environment: Europe Information Service SA, 66 ave Adolph Lacomble, 1030 Brussels, Belgium; fax (2) 732-67-57; e-mail eis@eis.be; internet www.eis.be; Editor: E Van Puyvelde; Principal Officer: Anne Van Goethem; f 1973; every 2 weeks (22 a year); in English and French; available online and on CD-ROM.

European Bulletin: Nature and National Parks: Federation of Nature and National Parks of Europe—FNNPE, 94481 Grafenau, Kröllstr 5, Postfach 1153, Germany; tel (8552) 96100; fax (8552) 961019; internet www.greenchannel.com/fnnpe; f 1973; quarterly; information concerning the management of protected areas throughout Europe, and the programmes and activities of the FNNPE; in German, English and French; circ 1,400.

European Directory of Sustainable Energy Efficient Building: James & James (Science Publishers) Ltd, 35–37 William Rd, London NW1 3ER, United Kingdom; tel (20) 7387-8558; fax (20) 7387-8998; e-mail orders@jxj.com; Editorial Manager: Edward Milford; annual; former title *European Directory of Energy Efficient Building*; circ 10,000.

European Energy Report: Financial Times Energy Publishing, Pearson Professional Ltd, Maple House, 149 Tottenham Court Rd, London W1P 9LL, United Kingdom; fax (20) 7896-2275; Editor: Paul Whitehead; every 2 weeks (25 a year).

European Environment: John Wiley and Sons Ltd, The Atrium Southern Gate, Terminus Rd, Chichester, West Sussex PO19 8SQ, United Kingdom; tel (1243) 779777; fax (1243) 775878; internet www.wiley.co.uk/journals-sci; Editors: Andrew Gouldson, Prof Peter Roberts; f 1991; every 2 months; dedicated to European, national and local environmental policy and practice; intended for universities, local authorities, financial institutions and environmental pressure groups; circ 500.

European Environmental Law: Kluwer Law International, Postbus 85889, 2508 CN The Hague, Netherlands; fax (70) 3081515; internet www.kluwerlaw.com; Editor: John R Salter; f 1994; 3 supplements a year; concerned with law pertaining to the natural, human and man-made environments of Europe.

European Environmental Law for Industry: Agra Informa Ltd, 80 Calverley Rd, Tunbridge Wells, Kent TN1 2UN, United Kingdom; tel (1892) 533813; fax (1892) 544895; e-mail marketing@agra-net.com; internet www.agra-net.com; f 1994; 2 a year; complete and consolidated texts of EU environment legislation on air, waste, water, IPPC, SEVESO and pollution from biotechnology.

European Environmental Law Review: Kluwer Law International, Postbus 85889, 2508 CN The Hague, Netherlands; fax (70) 3081515; e-mail enquiries@kluwerlaw.co.uk; internet www.kluwerlaw.com; Editor: Pauline M Callow; f 1991; monthly (11 a year); includes developments in European environmental law, investigations of major issues and a digest of the latest developments in the EU.

Evaluation Report of the United Nations Environment Programme: UNEP, POB 30552, Nairobi, Kenya; tel (2) 621234; fax (2) 226890; annual.

Everwild: Everwild International Creations, Inc, 4041 East Hastings St, Burnaby, BC V5C 2J1, Canada; tel (604) 298-0766; fax (604) 298-0765; Editor: Lauren Mulholland; Publr: Sharon Richlark; f 1989; every 2 months; environmental newspaper; circ 50,000.

Everyone's Backyard: Center for Health, Environment and Justice, 150 South Washington St, POB 6806, Falls Church, VA 22046-6806, USA; tel (703) 237-2249; fax (703) 237-8389; e-mail cchw@essential.org; internet www.essential.org/cchw; Editor: Patty Lovera; f 1982; quarterly; circ 6,000.

Expressdienst Umweltrecht: Erich Schmidt Verlag GmbH & Co, 10785 Berlin, Gentiner Str 30g, Germany; tel (30) 2500850; fax (30) 25008521; e-mail vertrieb@esvmedien.de; internet www.erich-schmidt-verlag.de; f 1995; 10 a year; environmental law bulletin.

Facets of Freshwater: Freshwater Foundation, Gray Freshwater Center, 2500 Shadywood Rd, Navarre, MN 55331, USA; tel (612) 471-9773; fax (612) 471-7685; e-mail freshwater@freshwater.org; internet www.freshwater.org; Editor: Kemp W Powers; f 1971; every 2 months; newsletter for members of the Freshwater Foundation; concerned with the sustainable protection and use of freshwater resources; circ 3,000.

Filtration and Separation: Elsevier Science Ltd, The Boulevard, Langford Lane, POB 150, Kidlington OX5 1GB, United Kingdom; tel (1865) 843726; fax (1865) 843971; Editor: S Barrett; Publr: Robert Feather; f 1964; 10 a year; technical magazine covering all aspects of industrial filtration and separation, including waste-water, effluent and chemical processing; details of the activities of the Filtration Society (Horsham); circ 7,500.

Florida Environments: Florida Environments Publishing, Inc, 4010 Newberry Rd, Suite 8, Gainesville, FL 32607, USA; tel (904) 373-1401; fax (904) 373-1405; Editor: David Newport; f 1986; monthly; circ 12,000.

Florida Specifier: National Technical Communications Co, Inc, POB 2027, Winter Park, FL 32790-2027, USA; tel (407) 671-7777; fax (407) 671-7757; e-mail mreast@enviro-net.com; internet www.enviro-net.com; Editor: Melora Gratten; f 1979; monthly; circ 13,000.

Florida Wildlife: Florida Game and Fresh Water Fish Commission, 620 South Meridian St, Tallahassee, FL 32399-1600, USA; tel (904) 488-5563; fax (904) 488-1961; e-mail subletd@mail.state.fl.us; internet www.state.fl.us/gfc; Editor: Dick Sublette; f 1947; every 2 months; concerned with fishing, hunting and natural areas in Florida; includes information on outdoor pursuits, Florida's natural history and wildlife management areas; circ 25,000.

Focus: World Wildlife Fund, 1250 24th St, NW, Washington, DC 20037, USA; tel (202) 293-4800; fax (202) 293-9211; Editor: Ann Ferber; f 1977; every 2 months; members' newsletter; reports World Wildlife Fund activities; circ 1m.

Focus on British Environmental Sciences: British Library, Document Supply Centre, Boston Spa, Wetherby LS23 7BQ, United Kingdom; tel (1937) 546080; fax (1937) 546286; f 1994; monthly; bibliographic series.

Focus on Renewable Natural Resources: FWR Publications, College of Forestry, Wildlife and Range Sciences, University of Idaho, Moscow, ID 83844-1130, USA; tel (208) 885-6673; fax (208) 885-6226; e-mail ortiz@uidaho.edu; internet www.uidaho.edu/~forpub; Editor: Denise Ortiz; f 1974; annual; includes features describing research conducted at the College and an appendix listing scientists, research projects, educational courses, the year's publications and fiscal information.

FoE Link: Friends of the Earth International—FoEI, POB 19199, 1000 GD Amsterdam, Netherlands; tel (20) 6221369; fax (20) 6392181; e-mail foei@foei.org; internet www.foei.org; Information Officer: Ann Doherty; quarterly; covers various environmental subjects, including genetically modified organisms (GMOs) and agriculture; in English, French and Spanish; online in French and Spanish; circ 800.

Fondeko Svijet: 71000 Sarajevo, Branilaca Sarajeva 47, Bosnia and Herzegovina; Editor-in-Chief: Nijaz Abadzic; f 1997; popular science magazine; concerned with nature, mankind and ecology.

Food, Cosmetics and Drug Packaging: P J Barnes and Assocs, POB 200, Bridgwater TA7 0YZ, United Kingdom; tel (1823) 698973; fax (1823) 698971; e-mail editor@pbarnes .demon.co.uk; Editor: Simon Atkinson; f 1978; monthly; aims to enable managers to make informed, responsible decisions in relation to packaging; covers news, technology, patents, markets and environmental considerations affecting the packaging industry world-wide; circ 2,000.

Food Management News: Yaffa Publishing Group Pty Ltd, 17–21 Bellevue St, Surry Hills, NSW 2010, Australia; tel (2) 9281-2333; fax (2) 9281-2750; e-mail kenmcgregor@yaffa .com.au; internet www.yaffa.com.au; Editor: Ken McGregor; f 1952; monthly (11 a year); deals with all aspects of packaging for food and beverages, including relevant legislation, environmental considerations, recycling and new materials; circ 5,533.

Forest and Bird: Royal Forest and Bird Protection Society, POB 631, Wellington, New Zealand; tel (4) 385-7374; fax (4) 385-7373; e-mail general.manager@forestandbird.org.nz; internet www.forestandbird.org.nz; Editor: Gordon Ell; f 1923; quarterly; includes indexes and book reviews, conservation briefs, feature articles on nature, environmental conservation and updates on branch activities; in English; circ 20,500. Publishing house: Bush Press Communications Ltd, POB 33-029, Takapuna, Auckland 9, New Zealand.

Forest Science: Society of American Foresters, 5400 Grosvenor Lane, Bethesda, MD 20814-2198, USA; tel (301) 897-8720; fax (301) 897-3690; e-mail safweb@safnet.org; internet www.safnet.org; Editor: Edwin J Green; f 1955; bimonthly; technical journal covering recent forestry research around the world; in English; circ 1,300.

Forest Voice: Native Forest Council, POB 2190, Eugene, OR 97402, USA; tel (541) 688-2600; fax (541) 689-9835; e-mail zerocut1@aol.com; internet www.forestcouncil.org; Editor: Ed Dorsch; f 1989; quarterly; circ 15,000.

Forestry: Journals Marketing Dept, Oxford University Press, Great Clarendon St, Oxford OX2 6DP, United Kingdom; tel (1865) 556767; fax (1865) 267835; internet www.oup.co.uk/jnis; Editors: P S Savill, H L Wright, H G Miller; f 1927; quarterly; journal of the Institute of Chartered Foresters—ICF; contains original international research papers and book reviews; particular emphasis on European forestry; circ 1,000.

Forestry Abstracts: CABI Publishing, Wallingford OX10 8DE, United Kingdom; tel (1491) 832111; fax (1491) 833508; e-mail publishing@cabi.org; internet www.cabi.org; f 1939; monthly; CAB International; concerned with literature published world-wide on all aspects of forestry including conservation and land use; English; available in electronic formats, CD-ROM and web access. Publishing house: CABI Publishing.

Forestry Chronicle: Canadian Institute of Forestry/Institut Forestier du Canada—CIF/IFC, 151 Slate St, Suite 606, Ottawa, ON K1P 5H3, Canada; tel (613) 234-2242; fax (613) 234-6181; e-mail cif@cif-ifc.org; internet www.cif-ifc.org; Editors: V Nordin, B Haddon; f 1925; every 2 months; official journal of the CIF/IFC; professional and technical periodical published in order to encourage a wider understanding of forestry; in English and French; circ 3,000.

The Forestry Source: Society of American Foresters, 5400 Grosvenor Lane, Bethesda, MD 20814-2198, USA; tel (301) 897-8720; fax (301) 897-3690; internet www.safnet.org; Editor: Joseph Smith; f 1996; 11 a year; concerned with forestry and environmental policy; in English; circ 17,000.

The Forth Naturalist And Historian: University of Stirling, Stirling FK9 4LA, United Kingdom; tel (1259) 215091; fax (1786) 464994; e-mail lindsay.corbett@stir.ac.uk; internet www.stir.ac.uk/departments/forthnat; Editors: Lindsay Corbett, Neville Dix; Chair: John Proctor; f 1975; annual; vol 26 (2003) covers Scotland's weather and climate change, phenology and Bird Reports; circ 400.

Franc-Vert: Union Québécoise pour la Conservation de la Nature—UQCN, 690 Grande Allée Est, 4ème étage, Québec, PQ G1R 2K5, Canada; tel (418) 648-2104; fax (418) 648-0991; Editor: Karen Grislis; f 1984; quarterly; former title *Franc-Nord*; conservation publication; circ 6,000.

Freshwater Biology: Blackwell Scientific Publishing Ltd, Osney Mead, Oxford OX2 0EL, United Kingdom; tel (1865) 206206; fax (1865) 721205; e-mail journals.cs@blacksci.co.uk; internet www.blackwell-science.com; Editors: C R Townsend, A G Hildrew; Rights Officer: Sophie Savage; 8 a year; circ 840.

From Duck Country: Ducks Unlimited Canada, POB 1160, Stonewall, MB R0C 2Z0, Canada; tel (204) 467-3000; fax (204) 467-9028; Editor: T G Neraasen; f 1939; every 2 months; members' magazine; circ 48,000.

From the Ground Up: Ecology Center of Ann Arbor, Inc, 117 North Division St, Ann Arbor, MI 48104-1528, USA; tel (734) 761-3186; e-mail news@ecocenter.org; internet www .ecocenter.org; Editor: Mike Garfield; f 1970; every 2 months; circ 2,000.

Frontier News: World Resources Institute, 10 G St, NE, Suite 800, Washington, DC 20002, USA; tel (202) 729-7600; fax (202) 729-7610; e-mail lauralee@wri.org; internet www .wri.org/ffi/news; news reports concerning natural forests.

Fujian Huanjing (Fujian Environment): Fujiansheng Huanjing Kexue Xuehui, Hualin Lu, Fuzhou, Fujian 350003, People's Republic of China; tel (591) 570146; Editor: Gan Jinggao; every 2 months; Fujian Association of Environmental Science; deals with environmental management and education, and the prevention of industrial pollution; in Chinese.

Functional Ecology: Blackwell Science Ltd, Osney Mead, Oxford OX2 0EL, United Kingdom; tel (1865) 206206; fax (1865) 721205; e-mail fecol@blacksci.co.uk; internet www.blackwell-science.com/fec; Editors: Prof David Robinson (Botany), Dr Kevin Gaston (Zoology); Publr: Amanda McLean-Inglis; f 1987; every 2 months; published in association with the British Ecological Society; specializes in physiological, biophysical and evolutionary ecology; available online and on microform.

Fuoristrada: G Cassini Editore Srl, Viale Matteotti 14, 13051 Biella, Italy; tel and fax (015) 352766; 2 a year; general environmental publication.

Future Survey: World Future Society, 7910 Woodmont Ave, Suite 450, Bethesda, MD 20814, USA; tel (301) 656-8274; fax (301) 951-0394; internet www.wfs.org/fsurv.htm; Editor and Founder: Michael Marien; f 1979; monthly; includes abstracts of books, reports and articles on trends, forecasts and proposals relating to alternative futures; supplemented by *Future Survey Annual*, which comprises categorized abstracts and indexes; circ 2,300.

Game Bird Breeders and Conservationists Gazette: Allen Publishing LLC, POB 171227, Salt Lake City, UT 84117, USA; tel (801) 575-1111; fax (801) 277-7832; e-mail gamebirdgazette@gamebird.com; internet www.gamebird.com; Editor: George Allen; f 1952; every 2 months; all aspects of conserving, breeding and keeping game birds; circ 20,000.

Gansu Environmental Study and Monitoring: Gansu Environmental Monitoring Centre, 2 Office Bldg, Dongfanghong Sq, Lanzhou, Gansu, People's Republic of China; tel (931) 8412162; Editor: Zhao Heping; f 1982; quarterly; published for the Gansu Provincial Bureau of Environmental Protection and the Gansu Provincial Institute of Environmental Protection; covers environmental management, assessment, study and monitoring, and pollution prevention and control; also includes reviews, discussion and research findings; in Chinese; circ 5,000.

Garner's Environmental Law: Butterworths Tolley, Halsbury House, 35 Chancery Lane, London WC2A 1EL, United Kingdom; tel (20) 7400-2500; fax (20) 7400-2842; internet www.butterworths.com; Editor: Prof D J Harris; Scottish Editor: Charles Smith; Practice Editor: Ian Doolittle; f 1994; 3 a year, with a supplementary bulletin every 2 months and service issues; superseded *Control of Pollution Encyclopedia*; collection of annotated United Kingdom and European Union source materials on environmental law arranged by subject; covers legislation, EC Directives and government circulars; in English; available online and on CD-ROM.

GECKO: ESRC Global Environmental Change Programme, Mantell Bldg, University of Sussex, Falmer, Brighton BN1 9RF, United Kingdom; tel (1223) 678935; fax (1223) 604483; internet www.gecko.ac.uk; Editor: Alister Scott; f 1992; 3 a year; newsletter; news of the most recent developments in the Global Environmental Change Programme, including information about new phases of research, events and items by researchers; circ 4,250.

Gefahrstoffe—Reinhaltung der Luft: Springer-VDI-Verlag GmbH & Co KG, 40239 Düsseldorf, Heinrichstr 24, Germany; tel (211) 61030; fax (211) 6103148; e-mail helpdesk@link.springer.de; internet link.springer.de/link/service/journals/94190/index.htm; Editors: K Grefen, R Michaelis; f 1920; monthly (11 a year); former title *Staub—Reinhaltung der Luft*; professional journal dedicated to air purification, with particular emphasis on dust particles; in English and German; available online; circ 2,000.

Generator's Journal: US Pollution Control, Inc, 515 West Greens Rd, Suite 500, Houston, TX 77067, USA; tel (713) 775-7910; Editor: Priscilla Proctor; f 1988; quarterly; issues of concern to hazardous waste generators; circ 18,000.

Geo Australasia: Geo Productions Pty Ltd, 2a Blakesley St, Chatswood, NSW 2067, Australia; tel (2) 9411-1766; fax (2) 9413-2689; e-mail geomag@ozemail.com.au; Editor: Michael Hohensee; Publr: Grant Young; f 1978; every 2 months; former title *Geo—Australasia's Geographical Magazine*; subjects covered include wildlife, flora and fauna, history, anthropology, travel and traditional crafts; circ 35,000.

Geodes: Europress Srl, Via del Salviantino 1, Fiesole, 50016 Florence, Italy; fax (02) 2948031; Editor: Riccardo Venchiarutti; monthly; circ 44,000.

Geomicrobiology Journal: Taylor & Francis, Inc, 325 Chestnut St, 8th Floor, Philadelphia, PA 19106, USA; tel (215) 625-8900; fax (215) 625-2940; e-mail journals@tandf.co.uk; internet www.tandf.co.uk/journals; Editors: Henry L Ehrlich, William C Ghiovse; f 1983; quarterly; deals with microbial transformations of materials composing the earth's crust, including oceans, seas, lakes, bottom sediments, soils, mineral deposits and rocks, and the geological impact these transformations have, or have had over geological time; available online.

Geomundo: Editorial America SA, 6355 North-west 36th St, Virginia Gardens, FL 33166-7099, USA; tel (305) 871-6400; fax (305) 871-8769; e-mail subscriptions@editorialtelevisa.com; Editor: Gabriel González; monthly; in Spanish, with summaries in English and Spanish; circ 116,000.

Georgetown International Environmental Law Review: Georgetown University Law Center, 600 New Jersey Ave, NW, Washington, DC 20009, USA; tel (202) 662-9468; f 1991; 3 a year; available online.

Geospatial Solutions: Advanstar Communications, 859 Willamette St, POB 10460, Eugene, OR 97440, USA; tel (541) 984-5237; fax (541) 984-5333; e-mail sbarnes@advanstar.com; Editor: Scottie Barnes; f 1990; monthly; former title *Geo Info Systems*; benefits of GIS and related geospatial technologies; in English; circ 29,000.

The Gildea Review: Community Environmental Council—CEC, 930 Miramonte Drive, Santa Barbara, CA 93109, USA; tel (805) 963-0583; fax (805) 962-9080; Editor: Michael Colin; f 1986; 3 a year; newsletter reporting on events and programmes at the CEC's Gildea Resource Center; circ 700.

Global and Planetary Change: Institute for Earth Sciences, Vrije Universiteit, Post Box 7161, 1007 MC Amsterdam, Netherlands; fax (20) 4852696; e-mail f.wallien@elsevier.com; internet www.elsevier.com/locate/gloplacha; Editors: Sierd Cloetingh, Thomas M Cronin, Ann Henderson-Sellers, Paolo A Pirazolli; f 1988; 5 vols per year; aims to achieve a multidisciplinary view of the causes, processes and limits of variability in planetary change; focuses on the record of change in earth history and the analysis and prediction of recent and future changes; topics include, but are not limited to, changes in the chemical composition of the oceans and atmosphere, climate change, sea level variations, human geography, global geophysics and tectonics, global ecology and biogeography; available as PDF and HTML files. Publishing house: Elsevier.

Global Environmental Change: Elsevier Science Ltd, The Boulevard, Langford Lane, Kidlington OX5 1GB, United Kingdom; tel (1865) 843000; fax (1865) 843010; Editors: Prof Martin Parry, M Williams, A Goudie; Group Editor: Barbara Clarke; f 1990; quarterly; international journal; examines environmental issues with particular emphasis on human ecology and public policy.

Global Environmental Change Report: Cutter Information Corpn, 37 Broadway, Suite 1, Arlington, MA 02174-5539, USA; tel (212) 242-3887; e-mail gecr@cutter.com; internet www.cutter.com; Editor: Mike Dipaola; f 1989; every 2 weeks; concerned with climate change, ozone depletion, global warming and climate policy; electronic formats available.

Global Report on Human Settlements: UNCHS (Habitat), POB 30030, Nairobi, Kenya; tel (2) 623091; fax (2)

624060; e-mail ramadhan.indiya@unchs.org; internet www.unchs.org; Ramadhan M Indiya.

The Globe: Global Environmental Research—GER Office, Polaris House, North Star Ave, Swindon SN2 1EU, United Kingdom; tel (1793) 411734; fax (1793) 444513; e-mail ukgeroff@nerc.ac.uk; internet www.nerc.ac.uk/ukgeroff/ welcome.htm; Editor: Dr I Simpson; every 2 months; national and international environmental science and policy developments; available online; circ 3,000.

Globus—Begleithefte: Bund für Umwelt- und Naturschutz Deutschland, 70569 Stuttgart, Landesverband Baden-Württemberg eV, Rotebühlstr 86-1, Germany; tel (711) 6197011; fax (711) 6197070; Editor: Erich Kimmich; f 1982; monthly; published by the German Association for Environment and Nature Conservation; supplement dealing with conservation issues; circ 15,000.

Golob's Oil Pollution Bulletin: World Information Systems, POB 535, Harvard Sq Station, Cambridge, MA 02238, USA; fax (617) 491-5100; Editor: Eric Brus; Publr: Richard Golob; f 1989; every 2 weeks; former title *Oil Pollution Bulletin*; available online.

Government & Policy: Environment and Planning C: Pion Ltd, 207 Brondesbury Park, London NW2 5JN, United Kingdom; tel (20) 8459-0066; fax (20) 8451-6454; e-mail sales@pion.co.uk; internet www.pion.co.uk/ep; Editor: R Bennett; Publishing Dir: J Schubert.

Grasduinen: Angel Business Communications Ltd, 34 Warwick Rd, Kenilworth CV8 1HE, United Kingdom; Editor: Anne-Marie Tassier; f 1979; monthly; nature magazine; covers wildlife, landscapes, travelling, art, gardens and nature photography; in Dutch; circ 55,000.

Great Lakes United: Buffalo State College, Cassety Hall, 1300 Elmwood Ave, Buffalo, NY 14222, USA; tel (716) 886-0142; fax (716) 886-0303; e-mail glu@igc.org; internet www.great-lakes.net/glu; Editor: Reg Gilbert; f 1986; quarterly; available online; circ 3,000.

The Green Book: Environmental Resource Directory: Green Book, Inc, 54 Burroughs St, Boston, MA 02130-4017, USA; tel (508) 474-5000; fax (508) 474-5054; f 1991; annual; circ 120,000.

Green Brigades: 31-014 Kraków, ul Stawkowska 12/24, Poland; tel and fax (12) 4222147; e-mail zb@zb.most.org.pl; internet www.most.org.pl/zb; Editors: Piotr Rymarowicz, Katy Wimble, Andrzej Zwawa; f 1989; quarterly; English-language edition of *Zielone Brygady* (qv); published in order to inform the international community about the activities of the environmental movement in Poland; in English; available online.

Green Chemistry: Royal Society of Chemistry, Thomas Graham House, Science Park, Milton Rd, Cambridge CB4 0WF, United Kingdom; tel (1223) 420066; fax (1223) 423429; e-mail sales@rsc.org; internet www.rsc.org/greenchem; Man Editor: Harpal Minhas; f 1999; monthly; deals exculsively with green chemistry and related issues, is at the frontiers of the science and encompasses all aspects of research that lead to a reduction in the environmental impact of chemicals, whether from their use or manufacture; available online; circ 500.

Green City Calendar: Planet Drum Foundation, POB 31251, San Francisco, CA 94131, USA; tel (415) 285-6556; fax (415) 285-6563; e-mail planetdrum@gc.org; internet www.planetdrum.org; f 1992; every 2 months; circ 5,000.

The Green Disk: Ecology Media, 151 Hallowell Rd, Jefferson, ME 04348-3832, USA; fax (888) 476-3475; e-mail greendisk@gc.apc.org; internet www.igc.org/greendisk; Editor: William C Sugg, III; f 1992; every 2 months; circ 1,000.

Green Drum: Greenspur Enterprise, 18 Cofton Lake Rd, Rednal, Birmingham B45 8PL, United Kingdom; tel (121) 445-2576; Hon Editor: Ron Andrews; f 1974; quarterly; issues covered include conservation, animal welfare, green politics, economics and human rights; circ 2,000.

The Green Guide: Mothers & Others for a Liveable Planet, Inc, 40 West 20th St, New York, NY 10011-4211, USA; tel (212) 242-0545; fax (212) 242-0010; e-mail greenguide@ mothers.org; internet www.mothers.org; Editor: Mindy Pennybacker; f 1994; 12 a year.

Green Guide for Everyday Life: Mothers & Others for a Liveable Planet, Inc, 49 West 20th St, New York, NY 10011, USA; tel (212) 242-0010; Editor: Annie Berthold-Bond; f 1994; 15 a year; newsletter.

Green Lines: Publishing Centre, Polar House, Bowburn North Industrial Estate, Bowburn, Durham DH6 5FP, United Kingdom; tel (191) 377-9805; fax (191) 377-3131; e-mail publishingcentre@btinternet.com; Editor: David Pike; f 1996; every 2 months; environmental issues relating to business, industry and local government; circ 25,000.

GREEN Lines: Grassroots Environmental Effectiveness Network, 1101 14th St, NW, Suite 1400, Washington, DC 20005-5601, USA; tel (202) 682-9400; fax (202) 682-1331; e-mail rfeather@defenders.org; internet www.defenders.org/ gline-h.html; f 1995; daily; available online.

Green Sheet: Environmental Advocates of New York, 353 Hamilton St, Albany, NY 12210, USA; tel (518) 462-5526; fax (518) 462-0381; e-mail info@eany.org; internet www.eany.org; Managing Editor: Jeff Jones (Communications Director); Editor: Laura DiBetta (Communications Associate); monthly; available online.

Green World: Green Party, 49 York Rd, Aldershot GU11 3JQ, United Kingdom; tel (1252) 330506; fax (1252) 330506; e-mail greenworld@ntlworld.com; internet www .greenparty.org.uk; Editor: Peter Barnett; f 1978; quarterly; former title *Econews*; concerned with politics, animal rights, civil liberties, the arms trade and energy production, with particular emphasis on environmental aspects of those issues; circ 12,000.

Greenbits: Cleveland Wildlife Trust, Bellamy Pavilion, Kirkleatham Old Hall, Kirkleatham, Redcar TS10 5NW, United Kingdom; tel (1642) 480033; fax (1642) 480401; Editor: John Hardy; 3 a year; informs members of Cleveland Wildlife Trust of local and national wildlife issues; circ 90,000. Publishing house: *The Evening Gazette*, Middlesbrough, United Kingdom.

GreeNotes: American Library Association, Social Responsibilities Round Table, Task Force on the Environment, 50 East Huron St, Chicago, IL 60611-2795, USA; tel (312) 944-6780; fax (312) 440-9374; e-mail fstoss@acsu.buffalo .edu; internet www.ala.org/alaorg/table/srrt/greenotes/ greenotes.html; Editor: Frederick W Stoss; f 1998; quarterly; available online.

Greenpeace: Greenpeace Switzerland, Heinrichstr 147, 8005 Zürich, Switzerland; tel (1) 4474141; fax (1) 4474199; quarterly; members' magazine; circ 23,000.

Greenpeace Magazin: Greenpeace Umweltschutzverlag GmbH, 20459 Hamburg, Deichstr 17, Germany; tel (40)

31186198; fax (40) 31186212; e-mail gpm.germany@greenpeace.de; internet www.greenpeace.de; Editor: Jochen Schildt; f 1982; quarterly; former title *Greenpeace Nachrichten*; circ 130,000.

Greenpeace Quarterly: Greenpeace USA National Office, 1436 U St, NW, Washington, DC 20009, USA; tel (202) 462-1177; fax (202) 462-4507; internet www.greenpeace.org/~usa; Editor: Jay Townsend; f 1981; quarterly; provides information on the campaign group's activities in the USA and worldwide; circ 900,000.

GréngeSpoun: GréngeSpoun Société Coopérative, 51 ave de la Liberté, BP 684, 2016 Luxembourg, Luxembourg; tel 29-79-99-0; fax 29-79-79; e-mail grespoun@pt.lu; Editors: Robert Garcia, Richard Graf, Renée Wagener; f 1988; weekly; newspaper presenting environmental and global ecological news and suggesting social and ecological alternatives; special sections on individual ecological issues, including air quality, climate change, species conservation, energy and transport; circ 3,000.

Groundwork: Tides Center, POB 14141, San Fransisco, CA 94114, USA; tel (415) 255-7623; f 1992; irreg; circ 6,500.

Grünstift—das Umweltmagazin für Berlin und Brandenburg: Stiftung Naturschutz Berlin, 10785 Berlin, Potsdamerstr 68, Germany; tel (30) 2626001; fax (30) 2615277; e-mail snb-gst@mail.blinx.de; internet www.snb.blinx.de; Editor: Jürgen Herrmann; Publr: Holger Wonneberg; f 1983; every 2 months; encourages conservation of nature and environmental protection in Berlin and Brandenburg; circ 10,000.

Grünstift—Umweltmagazin für Düsseldorf: Bund-NW eV, 40878 Rattingen, Graf-Adolph-Str 9, Germany; f 1989; quarterly; environmental periodical for the Düsseldorf area; circ 3,000.

GSF Mensch und Umwelt: GSF—Forschungszentrum für Umwelt und Gesundheit GmbH, 85764 Neuherberg, Ingolstädter Landstr 1, Germany; tel (89) 31872711; fax (89) 31873324; e-mail oea@gsf.de; internet www.gsf.de; Editors: Cordula Klemm, Heinz-Jörg Haury; f 1984; annual; provides public information on environmental and health research issues; published by the National Research Centre for Environment and Health; circ 20,000.

Guide de l'Eau: Editions Johanet, 30 rue René Boulanger, 75010 Paris, France; tel 1-44-84-78-78; fax 1-42-40-26-46; e-mail info@editions-johanet.com; internet www.editions-johanet.com; Editor: Vincent Johanet; Tech Deputy Editor: Jean-Louis Mathieu; f 1970; annual; directory of European, national and local water information and management bodies; covers European and national legislation relating to water quality and water management policies; lists specialists in the field and suppliers of water products and services; in French; available on CD-ROM; circ 4,800.

Guidebook on Environmental Courses and Conferences: World Information Systems, POB 535, Harvard Sq Station, Cambridge, MA 02238, USA; tel (617) 491-5100; fax (617) 492-3312; Editor: George Stubbs; circ 7,500.

Habitat Australia: Australian Conservation Foundation, 340 Gore St, Fitzroy, Vic 3065, Australia; tel (3) 9416-1166; fax (3) 9416-0767; e-mail acflib@peg.apc.org; internet www.peg.apc.org/~acfenv; Editor: Louise Ray; f 1973; every 2 months; covers environmental, conservation and energy issues; circ 15,000.

Habitat Debate: UNCHS (Habitat), POB 30030, Nairobi, Kenya; tel (2) 623988; fax (2) 623080; e-mail rasna.warah@unchs.org; internet www.unchs.org; Editor: Rasna Warah; f 1995; quarterly; newsletter of shelter and urban issues; in English, Arabic and Russian, some issues available in French and Spanish; available online.

Habitat (incorporating NEWS): The Environment Council, 212 High Holborn, London WC1V 7VW, United Kingdom; tel (20) 7836-2626; fax (20) 7242-1180; e-mail info@envcouncil.org.uk; internet www.the-environment-council.org.uk/habarch/habindex.htm; Editor: Susan Joy; f 1964; 10 a year; formerly published by the Council for Nature; environmental news digest, incorporating members' newsletter; includes local, national and international news, reports on environmental affairs concerning government and industry, and book reviews; available online; circ 4,000.

Haiyang Huanjing Kexue: Guojia Haiyang Ju, Haiyang Huanjing Baohu Yanjiusuo, POB 303, Dalian, Liaoning 116023, People's Republic of China; tel (411) 471429; Editor: Qian Wanying; quarterly; in Chinese.

Hakusan (Hakusan Nature Conservation Centre News): Ishikawa-ken Hakusan Shizen Hogo Senta, Kinameri, Yoshinodanimura, Ishikawa-gun, Ishikawa-ken 920-23, Japan; tel (7619) 55321; fax (7619) 55323; f 1973; quarterly; in Japanese.

Haramata (Bulletin of the Drylands): International Institute for Environment and Development—IIED, 3 Endsleigh St, London WC1H 0DD, United Kingdom; tel (20) 7388-2117; fax (20) 7388-2826; e-mail drylands@iied.org; internet www.iied.org; Editors: Ced Hesse, Lorenzo Catula, Nicole Kenton, Su Fei Yan, Baz Gueye, Christele Rion, Marie Jaecky; f 1988; 2–3 a year; bulletin of IIED Drylands Programme; covers issues of sustainable development in arid and semi-arid regions; each issue is accompanied by three Issue Papers dealing with specific subjects in greater depth; contains reviews of publications and reports on research activities; in English and French; available online; circ approx 3,000. Publishing house: International Institute for Environment and Development.

Hazard Alert: C C H Australia Ltd, POB 230, North Ryde, NSW 2113, Australia; tel (2) 9857-1300; fax (2) 9857-1600; internet www.cch.com.au; f 1995; 2 a year; includes information and advice on working with hazardous substances safely and in compliance with Australian law; in English.

Hazardous Cargo Bulletin: Intapress Publishing Ltd, 8th Floor, 29 Bressendon Pl, London SW1E 5DR, United Kingdom; tel (20) 7976-4021; fax (20) 7931-0516; e-mail mail@hazardouscargo.com; internet www.hazardouscargo.com; Editor: Rachel Jones; Publr: Stuart Fryer; f 1980; monthly; international magazine on storage, transport and handling of oils, gases and chemicals world-wide; circ 7,000.

Hazardous Cargo Bulletin Tank Guide: Intapress Publishing Ltd, 8th Floor, 29 Bressendon Pl, London SW1E 5DR, United Kingdom; tel (20) 7976-4021; fax (20) 7931-0516; e-mail mail@hazardouscargo.com; internet www.hazardouscargo.com; Publr: Barry Perry; supplement to Hazardous Cargo Bulletin.

Hazardous Materials and Waste Management: Hazardous Materials and Waste Management Association, 243 West Main St, Kutztown, PA 19530, USA; tel (215) 683-5098; fax (215) 593-3171; 6 a year.

Hazardous Materials Management: CHMM, Inc, 951 Denison St, Unit 4, Markham, ON L3R 3W9, Canada; tel (905) 305-6155; fax (905) 305-6255; e-mail sales@hazmatmag.com; internet www.hazmatmag.com; Pres and Editor-in-Chief: Guy Crittenden; Publr: Todd Latham; Sales Dir: Arnie Gess; f 1989; every 2 months; explains new environmental regulations, waste treatment and pollution prevention technologies; intended for readers in industry and local government; with supplement, *Compliance Monitor*; circ 20,000.

Hazardous Materials Transportation: Washington Business Information, Inc, 300 North Washington St, Suite 200, Falls Church, VA 22046, USA; tel (703) 247-3424; fax (703) 247-3421; Editor: Robert Billings; 2 a week.

Hazardous Substances and Public Health: Agency for Toxic Substances and Disease Registry, Department of Health and Human Services, 1600 Clifton Rd, NE, Mailstop E-33, Atlanta, GA 30333, USA; tel (404) 639-5040; fax (404) 639-0560; e-mail gbm7@cdc.gov; internet atsdr1.atsdr.cdc.gov:8080/atsdrhome.html; Editor: Georgia Moore; f 1990; quarterly; concerned with environmental health, hazardous waste and other hazardous substances, chemicals, public health, health education and communication; circ 9,650.

Hazardous Waste News: Business Publishers, Inc, 8737 Coleville Rd, Silver Spring, MD 20910-3928, USA; tel (301) 587-6300; fax (301) 585-9075; e-mail bpinews@bpinews.com; internet www.bpinews.com; Editor: Erich Shea; Publr: Leonard A Eiserer; f 1979; weekly; covers the testing, classification, transportation, storage, treatment, recycling and disposal of industrial hazardous waste; anticipates new regulations and changes in the status of waste substances; available online.

Haznews: Profitastral Ltd, Park House, 140 Battersea Park Rd, London SW11 4NB, United Kingdom; tel (20) 7498-2511; fax (20) 7498-2343; e-mail editor@haznews.com; Editor and Publr: David Coleman; f 1988; monthly; international business newsletter dealing with hazardous waste management and the rehabilitation of contaminated sites.

Health and Safety Science Abstracts: Cambridge Scientific Abstracts, 7200 Wisconsin Ave, Bethesda, MD 20814-4823, USA; tel (301) 961-6750; fax (301) 961-6720; e-mail market@csa.com; internet ww.csa.com; Editor: Evelyn Beck; quarterly; available online.

Heimatschutz/Sauvegarde: Schweizer Heimatschutz/Ligue Suisse du Patrimoine National, Geschäftsstelle, Merkurstr 45, Postfach, 8032 Zürich, Switzerland; tel (1) 2522660; fax (1) 2522870; Editor: Hans Gattiker; Publr: Marco Badilatti; f 1906; quarterly; published by the Swiss League for the Preservation of the National Heritage; in German and French; circ 20,000.

The Highlands Voice: West Virginia Highlands Conservancy, Inc, POB 306, Charleston, WV 25321-0306, USA; tel (304) 284-9548; Editor: William E Reed; f 1967; 12 a year; concerned with the conservation and sound management of the natural resources of the highlands of West Virginia; circ 3,700.

HIMAL South Asia: POB 7251, Kathmandu, Nepal; tel 523845; fax 521013; e-mail himal@himpc.mos.com.np; internet www.himalmag.com; Editor: Kanak Mani Dixit; f 1987; every 2 months; contains features on South Asia, including many items on environmental issues; in English; circ 10,000.

Hokkaido No Shizen (Nature in Hokkaido): Hokkaido Shizen Hogo Kyokai, 064 Kamori Bldg, Nishi 1, Kita 3, Chuoh-ku, Sapporo, Japan; annual; publication of the Nature Conservation Society of Hokkaido.

Holistic Management in Practice: Center for Holistic Management, POB 7128, Albuquerque, NM 87194, USA; tel (505) 842-5252; fax (505) 843-7900; e-mail jody@holisticmanagement.org; internet www.holisticmanagement.org; Editor: Jody Butterfield; f 1983; every 2 months; agricultural, economic and environmental issues, discussed in a Holistic Resource Management perspective; circ 2,000.

The Holocene: Arnold, 338 Euston Rd, London NW1 3BH, United Kingdom; tel (20) 7873-6336; fax (20) 7873-6325; e-mail arnoldjournals@hodder.co.uk; internet www.arnoldpublishers.com/journals/journpages/09596836.htm; Editor: Prof John Matthews; Journals Dir: Mary Altree; f 1991; quarterly; describes scientific research into long-term natural and anthropogenic environmental processes; concerned with environmental change during the Holocene, the current geological period, and with long-term environmental forecasting; available online.

Honeyguide: Ornithological Association of Zimbabwe, POB CY 161, Causeway, Harare, Zimbabwe; tel (4) 794611; e-mail birds@zol.co.zw; Editor: Michael P S Irwin; f 1962; 2 a year; members' magazine; concerned with ornithology, particularly the ecology, distribution, behaviour, systematics and conservation of birds; includes research papers, recent observations and items of general interest on birds in Zimbabwe; circ 1,000.

Huanjing Daobao (Environment Herald): Jiangsu Sheng Huanbao Ju, 241 Fenghuang Xijie, Nanjing 210002, People's Republic of China; fax (25) 6513105; Editor: Shi Zhenhua; f 1984; every 2 months; published by Jiangsu Environmental Protection Bureau; disseminates information on the theory and technology of practical environmental protection; in Chinese; circ 120,000.

Huanjing Huaxue (Environmental Chemistry): Science Press, 16 Donghuangchenggen North St, Beijing 100717, People's Republic of China; fax (10) 64034205; f 1982; every 2 months; Chinese Academy of Sciences; original papers and reviews on the theory and practice of all aspects of environmental chemistry, including atmospheric, aquatic and soil chemistry; in Chinese with summaries in English; circ 21,000.

Huanjing Kexue (Chinese Journal of Environmental Science): Research Centre for Eco-Environmental Sciences—RCEES, Chinese Academy of Sciences, POB 2871, Beijing 100085, People's Republic of China; tel (10) 62925511; fax (10) 62923563; Editor-in-Chief: Ouyang Ziyuan; f 1976; every 2 months; interdisciplinary journal concerned with significant advances in studies of environmental protection and ecology; published by RCEES in association with the Chinese Research Academy of Environmental Science, Tsinghua University Department of Environmental Engineering and Beijing Municipal Research Academy of Environmental Protection; available on CD-ROM; circ 3,500.

Huanjing Kexue Xuebao—Acta Scientiae Circumstantiae: Science Press, Marketing and Sales Dept, 16 Donghuangchenggen Beijie, Beijing 100707, People's Republic of China; fax (10) 64034205; f 1981; quarterly; published for the Ecology and Environment Research Centre of the Chinese Academy of Sciences; contains results of original research on the theory and practice of environmental science, including details of new technology and methods; in Chinese, with summaries in English; circ 11,000.

Huanjing Wuran Yu Fangzhi (Environmental Pollution and Control): 43 Tianmushan Lu, Hangzhou, Zhejiang 310007, People's Republic of China; tel (571) 7961293; fax (571) 7051838; e-mail zjshksjn@mail.hz.zj.cn; Editorial Bd: Jin Jun, Wang Wei, Liu Jian, Teng Huamei; f 1979; every 2 months; published for Zhejiang Environmental Science Research and Design Institute and Zhejiang Province Environmental Protection Bureau; includes articles on environmental protection policy and law, environmental economics and management, research papers on environmental protection and the technology of pollution control and environmental monitoring in China; in Chinese; circ 5,000.

Huanjing yu Jiankang Zazhi (Journal of Environment and Health): Tianjin Shi Weisheng Fangbin Zhongxin, 76 Tianshanshe Dajie, Hedong-qu, Tianjin 300011, People's Republic of China; tel (22) 24414403; fax (22) 24329940; e-mail tjfbsz@public.tpt.tj.cn; Editor: Dong Shanheng; f 1984; every 2 months; published by Tianjin Municipal Centre for Health and Disease Prevention; includes findings of scientific research into pollution of natural and man-made environments and its effects on the human body, and results of monitoring of the quality of the human environment; in Chinese with a summary in English; available on CD-ROM.

Hudson Valley GREEN Times: Hudson Valley Grass Roots Energy and Environmental Network—GREEN, POB 208, Red Hook, NY 12571, USA; tel (914) 758-4484; Editor: Dale Cotton; f 1981; quarterly; focuses on sustainable development, safe energy options, recycling, environmental legislation and environmental activism; circ 10,000.

Human Dimensions of Wildlife: Taylor & Francis, Inc, 325 Chestnut St, Suite 800, Philadelphia, PA 19106, USA; tel (215) 625-8900; fax (215) 625-2940; e-mail jerryv@cnr.colostate.edu; internet www.tandf.co.uk/journals/titles/10871209.asp; Co-Editors: Jerry J Vaske, Mike Manfredo; f 1996; 6 a year; deals with the human aspect of wildlife management issues; circ 300.

The Human Ecologist: Human Ecology Action League, POB 29629, Atlanta, GA 30359-0629, USA; tel (404) 248-1898; fax (404) 248-0162; e-mail healatnl@aol.com; internet members.aol.com/healatnl/index.html; Editor: Diane Thomas; f 1977; quarterly; concerned with human health and environmental exposures.

Human Ecology: CHEC Journal: Commonwealth Human Ecology Council, Church House, Newton Rd, London W2 5LS, United Kingdom; tel (20) 7792-5934; fax (20) 7792-5948; e-mail chec@dialin.net; internet www.tcol.co.uk/comorg/chec.htm; Editor: Eleanor Morris; Co-Editor: Apurva Narain; Consulting Editor: Zena Daysh; Editorial Team: David Hall, Julie Kingsland, Arif Naqvi; monthly.

Hydrogeology Journal: International Association of Hydrogeologists (IAH), IAH Secretariat, POB 9, Kenilworth CV8 1JG, United Kingdom; tel (1926) 56561; fax (1926) 56561;
e-mail cwatson@iah.org; internet www.iah.org/journal.htm; Exec Editor: C I Voss; f 1992; 6 issues a year; International Association of Hydrogeologists; English; abstracts available in French, Spanish; available online; circ 4,000. Publishing house: Springer Verlag.

Hydrological Processes: John Wiley and Sons Ltd, The Atrium Southern Gate, Terminus Rd, Chichester, West Sussex PO19 8SQ, United Kingdom; fax (1243) 775878; e-mail cs-journals@wiley.co.uk; internet www.wiley.co.uk; Editor: M G Anderson; Rights Officer: Diane Southern; 18 a year; circ 600.

Hydrological Sciences Journal: International Association of Hydrological Sciences—IAHS, Centre for Ecology and Hydrology, Wallingford OX10 8BB, United Kingdom; tel (1491) 692405; fax (1491) 692448; e-mail frances@iahs.demon.co.uk; internet www.iahs.info; Editor: Zbigniew W Kundzewicz; f 1956; every 2 months; the official journal of IAHS providing a forum for original papers and for the exchange of information and views on significant developments in hydrology world-wide; IAHS also publishes a series of books which are mostly collections of selected, peer-reviewed papers and discussion selected from IAHS and other conferences; in English and French; available online (from 2005); circ 700. Publishing house: IAHS Press.

Hydrological Summary for Great Britain: Institute of Hydrology/British Geological Survey, Maclean Bldg, Crowmarsh Gifford, Wallingford OX10 8BB, United Kingdom; tel (1491) 692468; fax (1491) 692424; e-mail nwamail@ioh.ac.uk; internet www.nwl.ac.uk:80/~nrfadata/nwa.html; f 1989; monthly; concerned with rainfall, river flow and groundwater.

Hydrological Yearbook: Danube Commission, 1068 Budapest, Benczúr u 25, Hungary; tel (1) 352-1835; fax (1) 352-1839.

Hydroplus International Water Review: 38 rue de Villias, 92300 Levallois, France; tel 1-41-49-00-00; fax 1-41-49-00-09; Publishing Dir: Jean-Jacques Lemoigne; Editorial Manager: France Lemoigne; f 1989; monthly; international magazine intended for water industry professionals, local authorities and international organizations; in French and English; circ 7,000.

IAIA Newsletter: International Association for Impact Assessment—IAIA, 1330 23rd St S, Suite C, Fargo, ND 58103, USA; tel (701) 297-7908; fax (701) 297-7917; e-mail info@iaia.org; internet www.iaia.org; f 1982; quarterly; newsletter concerned primarily with environmental impact assessment and matters pertaining to members of the Association; in English; circ 2,000.

IBSRAM Highlights: International Board of Soil Resources and Management—IBSRAM, POB 9-109, Bangkhen, Bangkok 10900, Thailand; tel (2) 579-7590; fax (2) 561-1230; e-mail oibsram@nontri.ku.ac.th; f 1983; annual report of IBSRAM's activities; in English; circ 2,500.

ICASALS Newsletter: International Center for Arid and Semi-Arid Land Studies, Texas Technical University, POB 41036, Lubbock, TX 79409-1036, USA; tel (806) 742-2218; fax (806) 742-1954; Editor: Idris R Traylor, Jr; f 1968; 2 a year; circ 3,000.

ICCET Annual Report: Imperial College Centre for Environmental Technology—ICCET, Imperial College of Science, Technology and Medicine, 48 Prince's Gardens, London SW7 2PE, United Kingdom; tel (20) 7594-9283; fax (20) 7581-0245; e-mail iccet@ic.ac.uk; internet www.et.ic.ac.uk; Editor: V James; f 1978; annual; circ 300.

ICE—Inland, Coastal, Estuarine Water: UK Centre for Economic and Environmental Development—UK CEED, Priestgate House, Suite 1, 3–7 Priestgate, Peterborough PE1 1JN, United Kingdom; tel (1733) 311644; fax (1733) 312782; e-mail ice@ukceed.org; internet www.ukceed.org; Editor: Hazel Church; f 1999; every 2 months; sustainable use of the United Kingdom's water resources; circ 3,000.

ICEL References: International Council of Environmental Law—ICEL, 53113 Bonn, Adenauerallee 214, Germany; tel (228) 2692240; fax (228) 2692250; Editor: Torsten Wäsch; f 1970; quarterly; covers international environmental law and policies; includes bibliographies; in English; circ 300.

ICES Techniques in Marine Environmental Sciences: International Council for the Exploration of the Sea—ICES, H.C. Andersens Boulevard 44-46, 1553 Copenhagen V, Denmark; tel 33-38-67-00; fax 33-93-42-15; e-mail info@ices.dk; internet www.ices.dk; Editor: J Pawlak; f 1987; irreg; in English; available online at www.ices.dk/products/techniques.asp; circ 300.

ICF News: Institute of Chartered Foresters—ICF, 7a St Colme St, Edinburgh EH3 6AA, United Kingdom; tel (131) 225-2705; fax (131) 220-6128; e-mail icfor@ibm.net; Editor: E H Kensington; quarterly bulletin; circ 1,500.

ICRISAT Annual Report: International Crops Research Institute for the Semi-Arid Tropics—ICRISAT, Patancheru, Andhra Pradesh 502 324, India; tel (40) 596161; fax (40) 241239.

ICSU World Data Centre for Soils—ISRIC, Bi-Annual Report: POB 353, 6700 AJ Wageningen, Netherlands; tel (317) 471711; fax (317) 471700; e-mail soil.isric@wur.nl; internet www-isric.org; Dir: Dr D L Dent; Deputy Dir: J H V Kauffman; f 1966; every 2 years; in English; available as a PDF file; circ more than 500. Publishing house: ISRIC.

IEQ Strategies: Cutter Information Corpn, 37 Broadway, Suite 1, Arlington, MA 02174-5539, USA; tel (781) 648-8700; fax (617) 648-1950; e-mail crowley@cutter.com; internet www.cutter.com; Editor: Carlton Vogt; f 1988; monthly; concerned with the indoor environment and indoor air problems; electronic formats available.

IES Newsletter: Institute of Ecosystem Studies, Education Program, POB R, Millbrook, NY 12545, USA; tel (914) 677-5359; fax (914) 677-6455; e-mail cadwalladerj@ecostudies.org; internet www.ecostudies.org; Editor: Jill Cadwallader; f 1983; every 2 months; newsletter of the research and education programmes of the Institute of Ecosystem Studies—IES; local ecology and natural history; circ 1,000.

IIMI Review: International Irrigation Management Institute (IIMI), POB 2075, Colombo, Sri Lanka; tel (1) 567404; fax (1) 566854.

Illinois Annual Air Quality Report: Illinois Environmental Protection Agency, Bureau of Air, 1001n Grand Ave E, Springfield, IL 62794, USA; tel (217) 782-5811; fax (217) 524-4710; f 1970; contains data summaries of ambient air quality and air emissions; circ 1,000.

Illinois Audubon: Illinois Audubon Society, POB 2418, Danville, IL 61834-2418, USA; tel (217) 446-5085; fax (217) 446-6375; internet www.illinoisaudubon.org; Editor: Debbie Scott Newman; f 1897; quarterly; circ 3,000.

Illinois Wildlife: Illinois Wildlife Federation, 112 East Attica St, Rossville, IL 60963-1106, USA; tel (217) 748-6365; fax (217) 748-6304; e-mail wildlife@htb.net; internet www.htb.net/~wildlife; Editor and Publr: Tom Mills; circ 12,500.

Ilmamsuojelu-uutiset (Finnish Air Pollution Prevention News): Ilmamsuojeluyhdistys ry, POB 335, 00131 Helsinki, Finland; tel (9) 631442; fax (9) 631442; Editor-in-Chief: Mika Railo; every 2 months; information bulletin of the Finnish Air Pollution Prevention Society—FAPPS; in Finnish, with English abstracts.

Indian Forester: Forest Research Institute, PO New Forest, Dehra Dun 248 006, India; tel (135) 627021; fax (135) 626865; e-mail amrish@jpl.vsnl.net.in; Editor: R S Mathur; Publr and Advertising Manager: Dr N S Bisht; f 1875; monthly; forestry science journal; intended for foresters, botanists, environmentalists, ecologists, industries dependent on forestry products and research institutes; in English; circ 1,700.

Indian Journal of Ecology: Punjab Agricultural University, Dept of Entomology, Ludhiana 141 004, Punjab, India; tel (161) 401962; fax (161) 400945; Editor: Dr Ramesh Arora; f 1974; 2 a year; the Indian Ecological Society; all aspects of basic and applied ecology; in English; circ 700.

Indian Journal of Environmental Protection: Kalpana Corpn, POB 5, Varanasi 221 010, India; tel (542) 350809; Editor: Dr Surendra Kumar; f 1981; monthly; deals with environmental pollution and its control; includes book reviews and indexes; in English; circ 5,000.

Indian Journal of Power and River Valley Development: Books & Journals (Pvt) Ltd, 6–2 Madan St, Kolkata 700 072, West Bengal, India; tel (33) 271711; fax (33) 2155867; Editor: P K Menon; f 1950; monthly; journal dedicated to all aspects of thermal and hydroelectric power generation, irrigation and river valley projects; in English; circ 3,000.

Indian Literature in Environmental Engineering: National Environmental Engineering Research Institute, Documentation and Library Services, Nehru Marg, Nagpur 440 020, India; f 1971; annual; bibliographic review; indexed according to author and place of origin; in English; circ 2,662.

Indoor and Built Environment: International Society of the Built Environment, POB 73, Buckden PE18 9SS, United Kingdom; tel (1480) 810687; fax (1480) 810768; Editor: J A Hoskins; f 1989; every 2 months; concerned with indoor air quality and the built environment; circ 400. Publishing house: S Karger AG, Allschwillerstr 10, Postfach, 4009 Basel, Switzerland.

Industria e Ambiente: Publindústria, Lda, Praça da Corujeira, 38, Apdo 3825, 4300-144 Porto, Portugal; tel (22) 5899620; fax (22) 5899629; e-mail industriaeambiente@publindustria.pt; internet www.publindustria.pt; f 1993; quarterly; environmental science and technology; Portuguese; circ 5,000.

Industrial Waste Management: Faversham House Group Ltd, Faversham House, 111 St James's Rd, Croydon, Surrey CR9 2TH, United Kingdom.

Industry and Environment: UNEP Industry and Environment Programme Centre, Tour Mirabeau, 39–43 quai André Citroen, 75539 Paris Cédex 15, France; tel 1-44-37-14-50; fax 1-44-37-14-74; e-mail unepie@unep.fr; internet www.unepie.org/home.htm; Editor: Jacqueline Aloisi de Larderel; f 1978; quarterly; thematic publication; includes abstracts and book reviews; in English, French and Spanish; circ 40,000. Publishing house: (Subscriptions) UN Bookshop, Sales Unit, Palais des Nations, 1211 Geneva 10, Switzerland.

Infectious Waste News: Environmental Industry Associations, 4301 Connecticut Ave, Suite 300, Washington, DC 20008, USA; tel (202) 244-4700; fax (202) 966-4868; e-mail michella@envasns.org; internet www.envasns.org; Editor: Mike Malloy; every 2 weeks.

Info-Nature: Société Réunionnaise pour l'Etude et la Protection de l'Environnement, BP 1109, 97482 Saint-Denis, Réunion; f 1974; nature and conservation publication of the Réunion Society for the Study and Protection of the Environment.

INFORM Reports: INFORM, Inc, 120 Wall St, New York, NY 10005, USA; tel (212) 361-2400; fax (212) 361-2412; e-mail inform@igc.apc.org; Editor: Gina Goldstein; f 1974; quarterly; former title *INFORM News*; members' newsletter; details of current INFORM research programmes and their findings; subjects include municipal solid waste management, chemical risk reduction, air quality and alternative fuels for transport; circ 4,000.

Informar: Instituto de Promoção Ambiental, Rua de O Século 63, 1200 Lisbon, Portugal; tel (1) 3211360; fax (1) 3432777; e-mail np67cd@mail.telepac.pt; internet www.ipamb.pt; Editor: Mário Cartaxo; f 1996; monthly; covers environmental issues, training and education; in Portuguese; circ 7,500.

Information on Environmental Pollution in Foreign Countries: Tokyo Metropolitan Research Institute for Environmental Protection, 7-5, Shinsuna 1-chome, Koto-ku, Tokyo 136, Japan; f 1968; annual.

Informazione Innovativa (Inf-Inn): Centro Studi 'l'Uomo e l'Ambiente', Via delle Palme 13, 35137 Padua, Italy; tel (049) 36435; fax (049) 8761945; Editor: Franco Spelzini; f 1985; every 2 weeks; Centro Studi 'l'Uomo e l'Ambiente' ('Man and the Environment' Studies Centre); examines the relationship between technological development and environmental quality, with particular attention to clean technology, energy sources and industrial health and safety; in Italian, with summaries in English; circ 1,000.

INFOTERRA Bulletin: INFOTERRA Programme Activity Centre, UNEP, POB 30552, Nairobi, Kenya; tel (2) 624299; fax (2) 624269; every 2 weeks; newsletter of UNEP's international environmental information system; available in English, French, Russian and Spanish editions.

In-Pak: Teknisk Forlag A/S, Skelbaekgade 4, 1780 Copenhagen V, Denmark; tel 53-48-28-00; fax 53-48-22-05; Publr: Kjeld Krogh; Editor: Gitte Søndergaard; f 1980; monthly (11 a year); concerned with all aspects of packaging, including recycling, disposal and environmental considerations; intended for the packaging industry in Denmark and its external territories, and Iceland; circ 7,000.

Inquinamento: Gruppo Editoriale Jackson SpA, Via Gorki 69, 20082 Cinisello Balsamo, Italy; tel (02) 66034327; fax (02) 66034333; internet www.jackson.it; f 1959; monthly; technical journal dealing with water, air and soil quality, waste treatment, noise reduction, recycling and clean energy; circ 8,500.

Insectes (Insects): Office pour l'Information Eco-Entomologique, BP 30, 78041 Guyancourt Cédex, France; tel 1-30-44-13-43; fax 1-30-83-36-58; e-mail guilbot@jouy.inra.fr; internet www.inra.fr/internet/hebergement/opie.insectes/pa.htm; Editor-in-Chief: Remi Coutin; Dir: Robert Guilbot; f 1967; quarterly; insects, their relationship with the environment and environmental protection.

Inside FERC: McGraw-Hill, Inc, 1221 Ave of the Americas, New York, NY 10020, USA; tel (202) 383-2141; fax (202) 383-2125; Editor-in-Chief: Larry Foster; Man Editor: Chris Newkumet; f 1979; weekly; concerned with the US Federal Energy Regulatory Commission—FERC and the natural gas industry; available online.

Inside Waste: Sturtevant and Cupps Assocs, Capitol Reports, 1500 West El Camino Ave, Suite 216, Sacramento, CA 95833-1945, USA; tel (916) 920-9094; fax (916) 920-9096; Editor: John Cupps; f 1990; monthly.

Institute of Environmental Sciences Tutorial Series: Institute of Environmental Sciences, 940 E Northwest Hwy, Mt Prospect IL 60056, USA; tel and fax (847) 255 1561.

Intec-Urbe: Tecnipublicaciones España, SL, Paseo de la Castellana 232, 28046 Madrid, Spain; tel (91) 7321010; fax (91) 7321014; e-mail urbe@correo.tpesp.es; internet www.tpesp.es; Editor: Eduardo Lizarraga; f 1993; 6 a year; industrial environment and municipal suppliers; intended for government and for business sectors associated with environmental problems.

International Directory of Solid Waste Management: James & James (Science Publishers) Ltd, 35–37 William Rd, London NW1 3ER, United Kingdom; tel (20) 7387-8558; fax (20) 7387-8998; e-mail orders@jxj.com; Editorial Manager: Edward Milford; annual; published for the International Solid Waste Association—ISWA (Denmark); circ 10,000.

International Environment and Safety News: The Old Station, London Rd, St Albans AL1 1JD, United Kingdom; tel (1727) 858840; fax (1727) 840310; Editor: Anne Satchell; Publr: M A I Pattison; f 1978; quarterly; newsletter for the environmental health industry; circ 10,500.

International Environmental Law: Kluwer Law International, Postbus 85889, 2508 CN The Hague, Netherlands; fax (70) 3081515; internet www.kluwerlaw.com; Editor: W E Burhenne; 2 supplements a year; includes texts of multilateral conventions and agreements; lists treaties in chronological order; in English, French and German.

International Environmental Technology: The Old Station, London Rd, St Albans AL1 1JD, United Kingdom; tel (1727) 858840; fax (1727) 840310; e-mail enviro.tech@pub.btinternet.com; Editor: Simon Latter; Publr: M A I Pattison; f 1990; every 2 months; concerned with pollution monitoring and control; circ approx 51,000.

International Ground Water Technology: National Trade Publications, Inc, 13 Century Hill Drive, Latham, NY 12110-2197, USA; tel (518) 783-1281; fax (518) 783-1386; e-mail igwt@waternet.com; internet waternet.com; Editor: Susan Wheeler; f 1995; monthly; superseded *Ground Water Age*; concerned with the detection, monitoring and treatment of contaminants in ground-water and the exploration of ground-water sources; circ 10,000.

International Hydrographic Bulletin: International Hydrographic Organization—IHO, International Hydrographic Bureau—IHB, 4 quai Antoine 1er, BP 455, Monte Carlo 98011 Cédex, Monaco; tel 93-10-81-00; fax 93-25-20-03; e-mail ihb@unice.fr; monthly; in English, French and Spanish; circ 600.

International Hydrographic Review: International Hydrographic Organization—IHO, International Hydrographic Bureau—IHB, 4 quai Antoine 1er, BP 455, Monte Carlo 98011 Cédex, Monaco; tel 93-10-81-00; fax 93-25-20-03; e-mail ihb@unice.fr; 2 a year; available in English and French editions; circ 1,050.

International Journal of Ambient Energy: Ambient Press Ltd, POB 25, Lutterworth LE17 4FF, United Kingdom; tel (1455) 202281; Co-ordinating Editor: A F C Sherratt; Editor: J C McVeigh; Dir: David Bennett; Head of Admin: Margaret Wrathall; f 1980; quarterly; applied and theoretical aspects of renewable energy sources, their importance in energy policy and their impact on the environment; includes research papers, case studies and book reviews.

International Journal of Biometeorology: International Society of Biometeorology, Secretariat, c/o Dr Scott Greene, Department of Geography, University of Oklahoma, Norman, OK 73071, Canada; tel (405) 325-4319; fax (405) 447-8455; e-mail jgreene@ou.edu; internet www.biometeorolgy.org; Editor-in-Chief: Prof Masaaki Shibita; quarterly; original research papers, reviews of articles and short communications on studies concerning the interactions between living organisms and factors of the natural and artificial physical environment; the environment includes climate and weather, electromagnetic fields, radiations of all frequencies and physical and biological pollutants; scope embraces basic and applied research and practical aspects such as living conditions, health and disease; available online at www.springerlink.com. Publishing house: Springer-Verlag Heidelberg, 69121 Heidelberg, Tiergartenstr 17, Germany; tel (6221) 487444; internet www.springer.de.

International Journal of Ecology and Environmental Sciences: International Scientific Publications, 50b Pocket C, Sidhartha Extension, New Delhi 110 014, India; e-mail ispdel3@hotmal.com; internet www.members.tripod.com/nicindia/index.htm; Editor: Brij Gopal; f 1974; quarterly; journal of the Scientific and Environmental Educational Society; in English; circ 600.

International Journal of Environment and Pollution: Inderscience Enterprises Ltd, World Trade Centre Bldg, 110 ave Louis Casai, CP 306, 1215 Geneva-Aéroport, Switzerland; fax (22) 7910885; e-mail info@inderscience.com; internet www.inderscience.com; Editor: Dr M A Dorgham; f 1991; 8 a year; initiated by UNESCO; covers sustainable development, environmental policy, pollution control and waste management; in English; circ 10,000.

International Journal of Environmental Analytical Chemistry: Taylor & Francis Group (Journals), 4 Park Sq, Milton Park, Abingdon OX14 4RN, United Kingdom; tel (20) 7017-6000; fax (20) 7017-6336; e-mail albgam@cid.csic.es; internet www.tandf.co.uk/journals/titles/03067319.asp; 15 a year; research papers on pollutants and harmful substances.

International Journal of Environmental Health Research: Taylor & Francis Group (Journals), 4 Park Sq, Milton Park, Abingdon, OX14 4RN, United Kingdom; tel (20) 7017-6000; fax (20) 7017-6336; e-mail smit-ce0@wpmail.paisley.ac.uk; internet www.tandf.co.uk/journals/titles/09603123.asp; Editors-in-Chief: Prof Paul G Smith; f 1991; 6 a year; devoted to rapid publication of research in environmental health, acting as a link between the diverse research communities and practitioners in environmental health; publishes articles on all aspects of the interaction of the environment with human health, divided into three main areas: the natural environment and health, the built environment and health, and communicable diseases; available online.

International Journal of Environmental Studies: Taylor & Francis Group (Journals), 4 Park Sq, Milton Park, Abingdon, OX14 4RN, United Kingdom; tel (20) 7017-6000; fax (20) 7017-6336; internet www.tandf.co.uk/journals/titles/00207233.asp; Editor: J Rose; 6 a year; devoted to the examination and solution of potential and existing environmental problems in both the industrialized and developing countries.

International Journal of Global Energy Issues: Inderscience Enterprises Ltd, World Trade Centre Bldg, 29 route de Pré-Bois, CP 896, 1215 Geneva 15, Switzerland; fax (22) 7910885; e-mail info@inderscience.com; internet www.inderscience.com; Editor: Dr M A Dorgham; f 1988; 8 a year in 2 vols; initiated by UNESCO and published with the support of the UN Economic Commission for Europe (Energy Division) and the International Institute for Applied Systems and Research; covers all aspects of energy policy, technology, management and conservation; in English; circ 10,000.

International Journal of Land Management: John Wiley and Sons Ltd, The Atrium Southern Gate, Terminus Rd, Chichester, West Sussex PO19 8SQ, United Kingdom; tel (1243) 779777; fax (1243) 843232; e-mail cs-journals@wiley.co.uk; internet www.wiley.co.uk; Editor: Richard Bullard; f 1997; quarterly.

International Journal of Plant Sciences: University of Chicago Press, Journals Division, POB 37005, Chicago, IL 60637, USA; tel (312) 702-3347; fax (312) 753-0811; e-mail subscriptions@journals.uchicago.edu; internet www.journals.uchicago.edu/IJPS/home.html; Editors: Edward D Garber, Manfred Ruddat; f 1875; every 2 months; former title *Botanical Gazette*; presents results of original research in all areas of plant biology, including ecology; in English; circ 1,500.

International Journal of Radioactive Materials Transport: Nuclear Technology Publishing, POB 7, Ashford TN23 1YW, United Kingdom; tel (1233) 641683; fax (1223) 610021; e-mail subscriptions@ntp.org.uk; internet www.ntp.org.uk; Editor: M S T Price; f 1990; quarterly; circ 450.

International Journal of Sustainable Development: Inderscience Enterprises Ltd, World Trade Centre Bldg, 29 route de Pré-Bois, CP 896, 1215 Geneva 15, Switzerland; fax (22) 7910885; e-mail info@inderscience.com; internet www.inderscience.com; Editor: Dr M A Dorgham; f 1997; quarterly; covers sustainable development and all associated socio-economic issues, environmental management, education and training, economic instruments and incentives for environmental protection, environmental regulations and standards, population control, urbanization, food technology, deforestation and water management; in English.

International Journal of Sustainable Development and World Ecology: Parthenon Publishing Group, Casterton Hall, Carnforth LA6 2LA, United Kingdom; tel (15242) 72084; fax (15242) 71587; e-mail mail@parthpub.com; internet www.parthpub.com/susdev/home.html; Editor-in-Chief: Prof John Jeffers; f 1994; quarterly; concerned with the ecological, economic and sociological aspects of sustainable development and the use of world resources.

International Journal of Water Resources Development: Taylor & Francis Group (Journals), 4 Park Sq, Milton Park, Abingdon OX14 4RN, United Kingdom; tel (20) 7017-6000; fax (20) 7017-6336; e-mail akbiswas@att.net.mx; internet www.tandf.co.uk/journals/titles/07900627.asp; Editor: Prof Asit K Biswas; quarterly; covers all aspects of water development and management in both industrialized and Third World countries. Publishing house: Routledge.

International Journal of Wilderness: FWR Publications, Wilderness Research Center, College of Forestry, Wildlife and Range Sciences, University of Idaho, Moscow, ID 83844-1130, USA; tel (208) 885-2267; fax (208) 885-2268; e-mail ortiz@uidaho.edu; internet www.uidaho.edu/~forpub; Man Editor: John Hendee; f 1995; quarterly; includes articles, book reviews and chronologies of major events; concerned with world-wide strategies and research for the stewardship of wilderness areas.

International Magazine of African Timber: African Timber Organization, BP 1077, Libreville, Gabon; tel 73-29-28; fax 73-40-30; 2 a year.

The International Tree Crops Journal: the journal of agroforestry: A B Academic Publishers, POB 42, Bicester OX6 7NW, United Kingdom; tel and fax (1869) 320949; e-mail jrnls@abapubl.demon.co.uk; Editors: Michelle A Pinard, Martin G Barker; f 1980; quarterly; journal for the communication of research results and the exchange of practical experience in the fields of multipurpose tree crops, agroforestry, environmental management, tree crop processing, economics and forestry for local community development.

International Wildlife: National Wildlife Federation, 8925 Leesburg Pike, Vienna, VA 22184-0001, USA; tel (703) 790-4000; fax (703) 790-4544; e-mail pubs@nwf.org; internet www.nwf.org; Editor: Bob Strohm; available online and on microform; circ 250,000.

International Wolf: International Wolf Center, 5930 Brooklyn Blvd, Minneapolis, MN 55429, USA; tel (612) 560-7374; fax (612) 569-7368; e-mail mplspack@wolf.org; internet www.wolf.org; Editor: Mary Ortiz; f 1990; quarterly; circ 10,000.

International Zoo Yearbook: Zoological Society of London, Regent's Park, London NW1 4RY, United Kingdom; tel (20) 7449-6282; fax (20) 7449-6411; e-mail yearbook@zsl.org; internet www.londonzoo.co.uk; Editors: F A Fisken, P J S Olney; f 1960; annual; serves as an international forum for the exchange of information between zoos concerning the conservation of biodiversity and increasing public awareness of the need for conservation of species and habitats; circ approx 1,000.

Intervenor: Canadian Environmental Law Association—CELA, 517 College St, Suite 401, Toronto, ON M6G 4A2, Canada; tel (416) 960-2284; fax (416) 960-9392; e-mail cela@web.net; internet www.web.net/cela; Editor: David McLaren; f 1976; every 2 months; former title *CELA Newsletter*; reports on environmental issues and legal cases; news of conferences and other events; circ 2,888.

Inzynieria Srodowiska: Akademia Górniczo-Hutnicza im Stanislawa Staszica, Wydawnictwo A G H, 30-059 Kraków, Al Mickiewicza 30, Poland; tel and fax (12) 3634038; Editor: A Wichur; f 1996; environmental engineering publication.

IPPL News: International Primate Protection League—IPPL, POB 766, Summerville, SC 29484, USA; tel (803) 871-2280; fax (803) 871-7988; e-mail ippl@awod.com; internet www.ippl.org; Editor: Shirley McGreal; f 1972; 3 a year; deals with primate issues; circ 20,000.

Irish Forestry: Irish Forestry Society, Coillite Teoranta, Sidmonton Pl, Bray, Co Dublin, Ireland; fax (1) 2868126; Editor: Alistair Pfeifer; 2 a year; circ 750.

Irish Naturalists' Journal: Irish Naturalists' Journal Ltd, c/o Ulster Museum, Botanic Gardens, Belfast BT9 5AB, United Kingdom; tel (1232) 383152; fax (1232) 383103; e-mail catherine.tyrie.um@nics.gov.uk; Editor: Dr Robin Govier; Treas: Catherine Tyrie; Sec: Dr Valerie Hall; f 1925; quarterly; papers, articles, notes, book reviews and special supplements on the natural history of the Republic of Ireland and of Northern Ireland; covers botany, zoology, ecology and geology; circ 500.

Irrigation Journal: Gold Trade Publishing, Inc, 68-860 Perez Rd, Cathedral City, CA 92234, USA; tel (619) 770-4370; Editor: Bruce Shank; Publr: Denne Goldstein; Exec Editor: Anne Goldstein; f 1951; every 2 months; technical journal dedicated to agricultural irrigation systems; circ 15,000.

Irrigazione e Drenaggio: Edagricole SpA, Via Emilia Levante 31, 40139 Bologna, Italy; fax (051) 493660; Editor: Roberto Genovesi; quarterly; Centro Internazionale Studi sull'Irrigazione (International Centre for Irrigation Studies); in Italian with summaries in English.

Ishikawa-ken Hakusan Shizen Hogo Senta Kenkyu Hokoku (Annual Report of Hakusan Nature Conservation Centre): Ishikawa-ken Hakusan Shizen Hogo Senta, Kinameri, Yoshinodanimura, Ishikawa-gun, Ishikawa-ken 920-23, Japan; tel (7619) 55321; fax (7619) 55323; f 1974; in Japanese, with summaries in English and Japanese.

Israel Environment Bulletin: Ministry of the Environment, POB 34033, Jerusalem 95464, Israel; tel 2-6553777; fax 2-6535934; internet www.environment.gov.il; Editor: Shoshana Gabbay; f 1974; quarterly; covers environmental policy, legislation, initiatives and events in Israel; in English; circ 3,500.

Israel Journal of Earth Sciences: Laser Pages Publishing Ltd, POB 35409, Merkaz Sapir 6/37, Givat Shaul, Jerusalem 91352, Israel; tel 2-6522226; fax 2-6522277; Editors: Dr Y Bartov, Y Kolodny; f 1951; quarterly.

Israel Journal of Plant Sciences: Laser Pages Publishing Ltd, POB 35409, Merkaz Sapir 6/37, Givat Shaul, Jerusalem 91352, Israel; tel 2-6522226; fax 2-6522277; Editor: Prof A M Mayer; f 1994; quarterly.

Israel Journal of Zoology: Laser Pages Publishing Ltd, POB 35409, Merkaz Sapir 6/37, Givat Shaul, Jerusalem 91352, Israel; tel 2-6522226; fax 2-6522277; Editor: Prof D Grauer; f 1951; quarterly.

ISRI Digest: Institute of Scrap Recycling Industries, 1325 G St, NW, Washington, DC 20005-3104, USA; tel (202) 737-1770; fax (202) 626-0900; Editor: Jason Gindele; f 1995; 2 a month; circ 1,800.

Issues in Environmental Science and Technology: Royal Society of Chemistry, Thomas Graham House, Cambridge CB4 0WF, United Kingdom; tel (1223) 420066; fax (1223) 423429; e-mail sales@rsc.org; internet www.rsc.org/issues; Editors: R E Hester, R M Harrison; f 1994; 2 a year; the series presents a multidisciplinary approach to pollution and environmental science and, in addition to covering the chemistry of environmental processes, focuses on broader issues; notably economic, legal and political considerations; available online.

ISWA Times: International Solid Wastes Association—ISWA, Overgarden Oven Vandet 48e, 1415 Copenhagen K, Denmark; tel 32-96-15-88; fax 32-96-15-84; e-mail iswa@inet.uni2.dk; internet www.iswa.org.

IUCN Red List of Threatened Speicies: IUCN—World Conservation Union, rue Mauverney 28, 1196 Gland, Switzerland; tel (22) 9990001; fax (22) 9990002; internet www.iucnredlist.org; 3–4 a year; supercedes the Redlist of Threatened Animals and is available as a searchable on-line database via the website; available online.

IUFRO News: International Union of Forestry Research Organizations—IUFRO, 1131 Vienna-Schönbrunn, Austria; tel (1) 877-01-51; fax (1) 877-93-55; e-mail iufro@forvie.ac.at; internet www.iufro.org; newsletter of International Union of Forestry Research Organizations—IUFRO; available online.

IVL—Nytt: Institutet för Vatten- och Luftvårdsforskning—IVL, POB 21060, 100 31 Stockholm, Sweden; tel (8) 729-15-00; fax (8) 31-85-16; e-mail carina.bergqvist@ivl.se; internet www.ivl.se; Editor: Carina Bergqvist; f 1966; quarterly; the Swedish Environmental Research Institute; contains articles on research conducted at the Institute; in Swedish.

IVL—Referat: Institutet för Vatten- och Luftvårdsforskning—IVL, POB 21060, 100 31 Stockholm, Sweden; tel (8) 729-15-00; fax (8) 31-85-16; e-mail carina.bergqvist@ivl.se; internet www.ivl.se; Editor: Carina Bergqvist; f 1985; quarterly; the Swedish Environmental Research Institute; summaries of research reports from the Institute; in English and Swedish.

Jaegaren: Jaegarnas Centralorganisation, Oestersundom, 01100 Helsinki, Finland; tel (9) 8777677; fax (9) 8777617; e-mail klaus.ekman@mkj-jco.fi; internet www.mkj-jco.fi; Editor: Klaus Ekman; f 1951; every 2 months; published by the Central Hunters' Organization; Swedish-language edition of *Metsastaja* (qv); in Swedish.

Jahrbuch des Vereins zum Schutz der Bergwelt: Verein zum Schutz der Bergwelt eV, 80538 München, Praterinsel 5, Germany; tel and fax (89) 479053; Editor: Hans Smettan; f 1900; annual; published by the Society for the Conservation of Mountain Areas; concerned with the protection of alpine plants and wildlife.

Jamaica Naturalist: Natural History Society of Jamaica, c/o Peter Vogel, University of the West Indies, Dept of Zoology, Kingston 7, Jamaica; tel 927-1202; fax 927-1640; Editor: Peter Vogel; f 1991; 2 a year; conservation, biodiversity, sustainable development and natural heritage in Jamaica; circ 2,000.

Jersey Sierran: Sierra Club, New Jersey Chapter, 57 Mountain Ave, Princeton, NJ 08540, USA; tel (609) 924-3141; Editor: Mary Penney; every 2 months; circ 15,000.

Jiaotong Huanbao (Environmental Protection in Transportation): Information Service for Environmental Protection in Transportation, Ministry of Communications, 37 Xingang Ergao Rd, Tianjin 300456, People's Republic of China; tel (22) 25708587; fax (22) 25795125; Editor: Li Shuhua; f 1980; every 2 months; concerned with pollution caused by ships and motor vehicles, environmental protection in the vicinity of ports and roads, environmental impact assessment in relation to traffic and the sustainable development of transportation; in Chinese.

Journal of Agriculture and Water Resources Research: Agriculture and Water Resources Research Centre, Jadiriyah, POB 2441, Baghdad, Iraq; Editor: Semir A Al-Shaker; in Arabic and English; circ 500.

Journal of Animal Ecology: Blackwell Science Ltd, Osney Mead, Oxford OX2 0EL, United Kingdom; tel (1865) 206206; fax (1865) 721205; Editors: L R Taylor, J M Elliott; Dir, Journal Publishing: Allen Stevens; f 1932; every 2 months; published for the British Ecological Society; original research papers on most aspects of animal ecology; reviews; available on CD-ROM; circ approx 2,900.

Journal of Applied Ecology: Blackwell Science Ltd, Osney Mead, Oxford OX2 0EL, United Kingdom; tel (1865) 206206; fax (1865) 721205; Editors: J Miles, W Block; quarterly; f 1964; results of research into the application of ecological ideas to the use of natural resources; circ 2,720.

Journal of Arid Environments: Academic Press Ltd, 24–28 Oval Rd, London NW1 7DX, United Kingdom; tel (20) 7424-4200; fax (20) 7482-2293; internet www.academicpress.com/ae; Editor: Prof John L Cloudsley-Thompson; f 1978; 8 a year; covers a broad range of subjects, including physiological, ecological, anthropological, geological and geographical studies relating to arid environments; circ 400.

Journal of Biogeography: Blackwell Science Ltd, Osney Mead, Oxford OX2 OEL, United Kingdom; tel (1865) 206206; fax (1865) 200918; e-mail jbiog@blacksci.co.uk; internet www.blackwell-science.com/products/journals/jbiog.htm; Editor: Prof Philip Stott; f 1973; every 2 months; incorporates *Global Ecology and Biogeography* and *Diversity and Distributions*; covers all aspects of spatial, ecological and historical biogeography; available online; circ 750.

Journal of Chemical Ecology: Plenum Publishing Corporation, 233 Spring St, New York, NY 10013-1578, USA; tel (212) 620-8000; fax (212) 463-0742; e-mail info@plenum.com; internet www.plenum.com; f 1975; monthly; available online.

Journal of Cleaner Production: Elsevier Science Ltd, The Boulevard, Langford Lane, Kidlington OX5 1GB, United Kingdom; tel (1865) 843000; fax (1865) 843010; Gen Editor: Prof Donald Huisingh; f 1993; quarterly; international journal devoted to technologies, concepts and policies for pollution prevention, waste minimization and cleaner products.

Journal of Ecology: Blackwell Science Ltd, Osney Mead, Oxford OX2 0EL, United Kingdom; tel (1865) 206206; fax (1865) 721205; Editor: Jonathan Silvertown; f 1912; every 2 months; official journal of the British Ecological Society; circ 3,345.

Journal of Energy and Natural Resources Law: International Bar Association, 271 Regent St, London W1R 7PA, United Kingdom; tel (20) 7629-1206; fax (20) 7409-0456; e-mail editor@int-bar.org; internet www.ibanet.org; Editors: Prof Thomas Wälde, Prof Alan Page; f 1982; quarterly; concerned with law applicable to energy and natural resources; covers issues such as mining, petroleum and natural gas production, privatization of utilities and the international energy trade; circ 1,800. Publishing house: Kluwer Law International, Sterling House, 66 Wilton Rd, London SW1V 1DE, United Kingdom.

Journal of Environmental Biology: 1/206 Vikasnagar, Lucknow 226 022, India; tel (522) 2769181; fax (522) 2768752; e-mail rcdalela@sancharnet.in; internet www.geocities.com/j_environ_biol/; Editor-in-Chief: Dr R C Dalela; Man Editor: Smt Kiran Dalela; f 1980; quarterly; international, multidisciplinary, peer-reviewed research journal dealing with environmental pollution, toxicology and all areas of the environmental sciences; in English; circ 1,000. Publishing house: Smt Kiran Dalela (Proprietor), Triveni Enterprises (Educational Publishers).

Journal of Environmental Economics and Management: Elsevier Inc, 525 B St, Suite 1900, San Diego, CA 92101-4495, USA; tel (407) 345-4020; fax (407) 363-1354; e-mail usjcs@elsevier.com; internet www.sciencedirect.com; Editor: Joseph Herriges; f 1975; 6 a year; theoretical and empirical papers devoted to specific natural resources and environmental issues; available online.

Journal of Environmental Education: Heldref Publications, 1319 18th St, NW, Washington, DC 20036-1802, USA; tel (202) 296-6267; fax (202) 296-5149; e-mail jee@heldref.org; internet www.heldref.org; Editor: Catherine Simon; f 1969; quarterly; the Helen Dwight Reid Educational Foundation; available on CD-ROM; circ 1,200. Publishing house: (Subscriptions) Allen Press, Inc, POB 1897, Lawrence, KS 66044-8897, USA.

Journal of Environmental Engineering: American Society of Civil Engineers, 1801 Alexander Graham Bell Drive, Reston, VA 20191, USA; tel (703) 548-2723; fax (703) 295-6222; e-mail marketing@asce.org; internet www.pubs.asce.org; Editor: Steve C McCutcheon; monthly; concerned with research in environmental engineering science, systems engineering and sanitation; also includes material on design, development of engineering methods, governmental management policies, the societal impacts of waste-water collection and treatment; available on CD-ROM; circ 6,100.

Journal of Environmental Engineering and Science: National Research Council of Canada, Ottawa, ON K1A 0R6, Canada; tel (613) 991-5475; fax (613) 952-7656; e-mail pubs@nrc-cnrc.gc.ca; internet pubs.nrc-cnrc.gc.ca; Editors: Prof D W Smith, Prof D S Mavinic; f 2002; monthly; addresses all aspects of environmental engineering and environmental science. Publishing house: NRC Research Press.

Journal of Environmental Health: National Environmental Health Association, 720 South Colorado Blvd, Suite 970, South Tower, Denver, CO 80246-1925, USA; tel (303) 756-9090; fax (303) 691-9490; e-mail staff@neha.org; internet www.neha.org; Editor: Nelson E Fabian; f 1938; 10 a year; deals with all aspects of environmental health protection and sanitation; areas of interest include air, food, water, waste, shelters, recreational facilities, institutional facilities, pools and spas, workplace environments, hazardous materials and other related topics; available online; circ 6,000.

Journal of Environmental Law: Oxford University Press, Oxford Journals, Pinkhill House, Southfield Rd, Eynsham, Oxford OX8 1JJ, United Kingdom; tel (1865) 882283; fax (1865) 882890; Editor: Richard Macrory; f 1989; 2 a year; deals with legislation concerning pollution control, waste management, protection of habitats, biotechnology and public natural resources; circ 850.

Journal of Environmental Management: Academic Press Ltd, 24–28 Oval Rd, London NW1 7DX, United Kingdom; tel (20) 7424-4200; fax (20) 7482-2293; internet www.hbuk.co.uk/ap/jem; Editor: R H Young; f 1973; 12 a year in 2 vols; contains articles on all aspects of the management and use of the natural and built environments.

Journal of Environmental Monitoring: Royal Society of Chemistry, Thomas Graham House, Science Park, Milton Rd, Cambridge CB4 0WF, United Kingdom; tel (1223) 420066; fax (1223) 423429; e-mail sales@rsc.org; internet www.rsc.org/jem; Man Editor: Harpal Minhas; f 1999; monthly; presents a mixture of news and views on current and future issues in environmental monitoring from state-of-the-art research in the latest analytical breakthroughs to perpsectives and reviews on the pertinent research fields; available online; circ 500.

Journal of Environmental Planning and Management: Taylor & Francis Group (Journals), 4 Park Sq, Milton Park, Abingdon OX14 4RN, United Kingdom; tel (20) 7017-6000; fax (20) 7017-6336; e-mail info@tandf.co.uk; internet www.tandf.co.uk/journals/titles/09640568.asp; Man Editor: Kenneth G Willis; f 1958; every 2 months; international research on environmental planning and management; available online. Publishing house: Routledge.

Journal of Environmental Policy & Planning: Taylor & Francis Group (Journals), 4 Park Sq, Milton Park, Abingdon OX14 4RN, United Kingdom; tel (20) 7017-6000; fax (20) 7017-6336; e-mail kevin.bishop@wlga.gov.uk; internet www.tandf.co.uk/journals/titles/1523908x.asp; Editors: Editors: Dr Kevin Bishop, Dr Andrew Flynn, Prof Terry Marsden; quarterly; critical analysis of environmental policy and planning; explores the environmental dimensions of common policies such as transport, agriculture and fisheries, urban and rural policy.

Journal of Environmental Quality: American Society of Agronomy, Inc, 677 South Segoe Rd, Madison, WI 53711, USA; tel (608) 273-8080; fax (608) 273-2021; e-mail sernst@agronomy.org; internet www.agronomy.org; Editor: G M Pierzynski; f 1972; every 2 months; comprises original research dealing with environmental quality in relation to natural and agricultural ecosystems; includes sections on heavy metals, plant and environmental interactions, organic chemicals, water quality, land reclamation, soil processes, chemical transport, waste management, atmospheric pollutants and trace gases, ecosystem processes, wetlands and aquatic processes, biodegradation and bioremediation; circ 3,000.

Journal of Environmental Radioactivity: Elsevier Science Ltd, The Boulevard, Langford Lane, Kidlington OX5 1GB, United Kingdom; tel (1865) 843000; fax (1865) 843010; e-mail nlinfo@elsevier.com; internet www.sciencedirect.com/science/journal/0265931X; Editor-in-Chief: S Sheppard; f 1984; monthly; publication of ECOMatters Inc., Suite 105, W.B. Lewis Business Centre, 24 Aberdeen Avenue, PO Box 430, PINAWA, Manitoba, R0E 1L0, Canada; tel (204) 753 2747; fax (204) 753 2170; e-mail SheppardS@ecomatters.com; provides a coherent international forum for publication of original research or review papers on any aspect of the occurrence of radioactivity in natural systems; available online. Publishing house: Elsevier Earth Sciences Department, Molenwerf 1, 1014 Amsterdam, The Netherlands; tel (20) 485-2746, fax (20) 485-2696.

Journal of Environmental Science and Health, Part A—Toxic/Hazardous Substances and Environmental Engineering: Taylor & Francis, Inc, 325 Chestnut St, Suite 800, Philadelphia, PA 19106, USA; tel (215) 625-8900; fax (215) 625-2940; e-mail Skhan6@gmu.edu; internet www.tandf.co.uk/journals; Editor: Shahamat U Khan; f 1965; 10 a year; serves as an international forum for information pertinent to environmental problems affecting air, water or soil; includes engineering innovations, effects of pollutants on health, control systems, laws and projections; circ 300.

Journal of Environmental Science and Health, Part B—Pesticides, Food Contaminants and Agricultural Wastes: Taylor & Francis, Inc, 325 Chestnut St, Suite 800, Philadelphia, PA 19106, USA; tel (215) 625-8900; fax (215) 625-2940; e-mail Skhan6@gmu.edu; internet www.tandf.co.uk/journals; Editor: Shahamat U Khan; f 1965; every 2 months; original research reports on advances relating to residues of pesticides, natural food contaminants and additives, other contaminants and their metabolites in the ecosphere; covers the persistence and biodegradation of chemicals, their toxicological consequences and methods of detoxification; circ 325.

Journal of Environmental Science and Health, Part C—Environmental Carcinogenesis & Ecotoxicology Reviews: Taylor & Francis, Inc, 325 Chestnut St, Suite 800, Philadelphia, PA 19106, USA; tel (215) 625-8900; fax (215) 625-2940; e-mail Woo.YinTak@epamail.epa.gov; internet www.tandf.co.uk/journals; Editor: Yin-Tak Woo; f 1982; 2 a year; multidisciplinary journal reviewing various aspects of environmental carcinogenesis; subjects covered include synergism and antagonism, theoretical models, inhibition of carcinogenesis and interaction of physical, chemical and biological factors; circ 250.

Journal of Environmental Sciences (China): IOS Press, Van Diemenstr 94, 1013 CN Amsterdam, Netherlands; e-mail market@iospress.nl; internet www.iospress.nl; Editor: Tie Xiaoshan; f 1989; quarterly; specialized academic journal; English-language edition of *Huanjing Kexue* (qv); in English; available online.

Journal of Environmental Systems: Baywood Publishing Co, Inc, 26 Austin Ave, POB 337, Amityville, NY 11701, USA; tel (516) 691-1270; fax (516) 691-1770; e-mail baywood@baywood.com; internet www.baywood.com; Editor: Sheldon Reavon; f 1971; quarterly.

Journal of Forestry: Society of American Foresters, 5400 Grosvenor Lane, Bethesda, MD 20814-2198, USA; tel (301) 897-8720; fax (301) 897-3690; internet www.safnet.org; Managing Editor: Fran Pflieger; f 1902; monthly; members' journal; concerned with forestry, forest products, management of wildlife and natural resources, water quality and environmental practices and policy; in English; circ 17,000.

Journal of Health Science: Pharmaceutical Society of Japan, Nihon Yakugakkai, 2-12-15, Shibuya 2-chome, Shibuya-ku, Tokyo 150-0002, Japan; tel (3) 3406-3325; fax (3) 3498-1835; e-mail naganuma@mail.pharm.tohoku.ac.jp; internet jhs.pharm.or.jp; Editor: Akira Naganuma; f 1955; every 2 months; former title *Eisei Kagaku / Japanese Journal of Toxicology and Environmental Health*; published by the Pharmaceutical Society of Japan; includes statistics and indexes; in English; available online; circ 1,700.

Journal of Himalayan Studies and Regional Development: Institute of Himalayan Studies and Regional Development, Garhwal University, POB 12, Srinagar, Garhwal 246 174, India; Editor: Dr Shankar Kala; f 1977; annual; former title *Himalaya*; in English; circ 1,000.

Journal of Human Ecology: Kamla-Raj Enterprises, 2273 Gali Bari Paharwali, Chawri Bazar, Delhi 110 006, India; tel (11) 23284126; fax (11) 27666473; e-mail kre@vsnl.com; internet www.krepublishers.com; Man Editor: Dr Xeena; f 1990; 8 issues in 2 vols per year; published in association with the Indian Society for Human Ecology—ISHE; contains articles, technical reports and short items on environmental issues in the fields of human ecology, biology, genetics, social and biological anthropology, life sciences, geography, medical sciences, public health, demography, botany and zoology; in English; available online; circ 300.

Journal of Industrial Ecology: MIT Press, Journals Dept, 5 Cambridge Center, Cambridge, MA 02142-1493, USA; tel (617) 253-2889; fax (617) 577-1545; e-mail indecol@yale.edu; internet mitpress.mit.edu/jiec; Editor: Reid Lifset; f 1997; quarterly; International Society for Industrial Ecology; in English; circ 1,100.

Journal of Industrial Pollution Control: Enviro Media, 2nd Floor, Rohan Heights, POB 90, Karad 415 110, India; tel (2164) 44369; fax (2164) 71645; Editor: R K Trivedy; 2 a year; circ 2,000.

Journal of International Wildlife Law & Policy: Taylor & Francis, Inc, 325 Chestnut St, Suite 800, Philadelphia, PA 19106, USA; tel (215) 625-8900; fax (215) 625-2940; e-mail JIWLP@internationalwildlifelaw.org; internet www.tandf.co.uk/journals/titles/13880292.asp; Editor-in-Chief: William C G Burns; quarterly; to address legal and political issues concerning the human race's inter-relationship with and management of wildlife species, their habitats and the biosphere.

Journal of Land Use and Environmental Law: College of Law, Florida State University, Tallahassee, FL 32306-1601, USA; tel (850) 644-4240; fax (850) 644-7282; e-mail landuse@law.fsu.edu; internet www.law.fsu.edu/journals/landuse; Editor-in-Chief: elected each year; f 1984; 2 a year; covers land use, environmental law and zoning; available online; circ 500. Publishing house: Western Publishing, Indianapolis, Indiana.

Journal of Plant Resources and Environment: Institute of Botany, Chinese Academy of Sciences, Zhongshanmenwai, Nanjing, Jiangsu 210014, People's Republic of China; tel (25) 4432128; fax (25) 4432074; e-mail jsszzzzz@public1.ptt.js.cn; internet www.chinajournal.net/zwzy.html; Editor-in-Chief: He Shuan-an; Editorial Staff: Hui Hong, Xu Ding-fa, Zong Shi-Xian, Deng Ying; f 1992; quarterly; published jointly with the Jiangsu Society of Botany and the Botanical Garden Conservation branch of the Chinese Society of Environmental Sciences; contains original articles, research notes and reviews dealing with basic and applied studies in plant resources and the environment; concerned with the investigation, evaluation, rational exploitation and sustained utilization of plant resources in the People's Republic of China, plant conservation, the construction and management of botanical gardens and nature reserves, the role of plants in protecting and beautifying the environment and environmental influences on plants; in Chinese, with English abstracts, tables and figures; electronic formats available; circ 2,000.

Journal of Soil and Water Conservation: Soil and Water Conservation Society—SWCS, 7515 North-east Ankeny Rd, Ankeny, IA 50021-9764, USA; tel (515) 289-2331; fax (515) 289-1227; e-mail swcs@swcs.org; internet www.swcs.org; Editor: Deb Happe-von Arb; f 1946; quarterly; members' journal; intended for professional conservationists, educators, planners, landscape architects and other interested parties; circ 5,000.

Journal of Soil and Water Conservation in India: All India Soil and Land Use Survey, IARI Bldgs, New Delhi 110 012, India; tel (11) 5743811; e-mail aislus@vsnl.com; Editor: T K Sarkar; f 1952; 2 a year; in English; circ 1,500.

Journal of Solar Energy Research: Solar Energy Research Council, Scientific Research Centre, Jadiriyah, POB 13026, Baghdad, Iraq; Editor: Nidhal I Al-Hamdani; f 1983; 2 a year; in Arabic and English; circ 500.

Journal of Sustainable Agriculture: Biology Department, Fairfield University, Fairfield, CT 06430-7524, USA; tel (203) 254-4000; fax (203) 254-4253; e-mail rpoincelot@fair1.fairfield.edu; Sr Editor: Raymond P Poincelot; f 1989; quarterly; international journal comprising peer-reviewed papers on sustainable agriculture; circ 750. Publishing house: Haworth Press, Inc, 10 Alice St, Binghamton, NY 13904-1580, USA.

Journal of the Air and Waste Management Association: 1 Gateway Center, 3rd Floor, Pittsburgh, PA 15222, USA; tel (412) 232-3444; fax (412) 232-3450; internet www.awma.org; Editor: William Tony; f 1907; monthly; covers environmental management, air pollution control, waste management and soil remediation; fosters the exchange of current technological information and encourages the development of new technology for a cleaner environment; circ 16,000.

Journal of the American Water Works Association: American Water Works Association, 6666 West Quincy Ave, Denver, CO 80235, USA; tel (303) 794-7711; fax (303) 794-7310; internet www.awwa.org; Editor: Nancy Zeilig; f 1908; monthly; concerned with water quality and management of water utilities; circ 45,000. Publishing house: Banta Publishing, 100 Banta Rd, Long Prairie, MN 56347, USA.

Journal of the Atomic Energy Society of Japan: Atomic Energy Society of Japan—AESJ, 1-1-13 Shimbashi, Minato-ku, Tokyo 105-0004, Japan; tel (3) 3508-1261; fax (3) 3581-6128; e-mail atom@aesj.or.jp; internet wwwsoc.nacsis.ac.jp/aesj; Editor-in-Chief: C Kinoshita; f 1959; monthly; nuclear science and technology; in Japanese; circ 8,950.

Journal of Tropical Ecology: Cambridge University Press, The Edinburgh Bldg, Shaftesbury Rd, Cambridge CB2 2RU, United Kingdom; tel (1223) 325757; fax (1223) 315052; Chief Exec: Anthony K Wilson; Man Dir: Jeremy Mynott: Editor: Adrian G Marshall; Publishing Dir: Conrad Guettler; quarterly; includes articles on the ecology of tropical regions, original research papers and reviews; circ 700.

Journal of Water and Environmental Management: Chartered Institution of Water and Environmental Management—CIWEM, 15 John St, London WC1N 2EB, United Kingdom; tel and fax (20) 7831-3110; Editor: Malcolm Haigh; every 2 months; deals with world-wide aspects of the water cycle and other environmental matters, including the management of solid and hazardous wastes, control of air pollution and environmental conservation.

Journal of Wildlife Management: Wildlife Society, 5410 Grosvenor Lane, Bethesda, MD 20814, USA; tel (301) 897-9770; fax (301) 530-2417; e-mail tws@wildlife.org; internet wildlife.org/index.html; Editor: Guy A Baldassare; f 1937; quarterly; circ 7,000.

Journal of World Forest Resource Management: the journal of forest policy: A B Academic Publishers, POB 42,

Bicester OX6 7NW, United Kingdom; tel (1869) 320949; e-mail jrnls@abapubl.demon.co.uk; Editor: Alan Grainger; f 1984; 2 per volume; focuses on ecology, resource assessment and all aspects of forests as a global resource; intended for those concerned with the future of the world's forest resources and with the impact of forest exploitation on the global environment; contains research papers, news items, book reviews and details of conferences.

Jumin Katsudo: Shinseikatsu Undo Kyokai, 1-3, Hibiya Koen, Chiyoda-ku, Tokyo, Japan; f 1972; quarterly.

KA–Abwasser, Abfall (KA–Wastewater, Waste): Gesellschaft zur Förderung der Abwassertechnik, 53758 Hennef, Postfach 1165, Germany; tel (2242) 872190; fax (2242) 872151; e-mail bringewski@atv.de; internet www.atv.de; Editor: Dr Frank Bringewski; f 1954; monthly; ATV-DVWK–Die Deutsche Vereinigung für Wasserwirtschaft, Abwasser und Abfall (German Association for Water Management, Wastewater and Waste); published by the Association for Promotion of Wastewater Technology; concerned with water, sewage and solid waste; in German, with summaries in German, and English; available online; circ 16,000.

Kanagawa-kenritsu Shizen Hogo Senta Hokoku: Kanagawa-kenritsu Shizen Hogo Senta, 657 Nanasawa, Atsugi-shi, Kanagawa-ken 243-01, Japan; tel (462) 480323; fax (462) 482560; f 1984; annual; reports on research conducted at Kanagawa Prefectural Nature Conservation Centre; in Japanese.

Kankyo Gijutsu (Environmental Conservation Engineering): Kankyo Gijutsu Kenkyu Kyokai, Matsumo Bldg, 2-1-20, Tenma, Kita-ku, Osaka 530, Japan; tel (6) 357-7611; fax (6) 357-7612; e-mail riet@osk.3web.ne.jp; internet www3.osk.3web.ne.jp/~riet; Editor: Y Shimazu; f 1972; monthly; publication of Kankyo Gijutsu Kenkyu Kyokai (Environmental Technology Research Institute); concerned with basic technical issues relating to global environmental problems; in Japanese, with summaries in English; circ 8,000.

Kansas Wildlife and Parks: Dept of Wildlife and Parks, Education-Information, 512 South-east 25th Ave, Pratt, KS 67124-8174, USA; tel (316) 672-5911; fax (316) 672-6020; Editor: Mike Miller; f 1938; every 2 months; circ 45,000.

Keep Tahoe Blue: League to Save Lake Tahoe, 955 Emerald Bay Rd, South Lake Tahoe, CA 96150, USA; tel (916) 541-5388; Editor: Jeff Cutler; f 1965; quarterly; circ 5,500.

Kobus: Wildlife Conservation Society of Zambia, POB 30255, Lusaka, Zambia; tel (1) 254226; fax (1) 222906; Editor: Mwape Sichilongo; f 1987; every 2 months; conservation magazine intended for the general public.

Koedoe: South African National Parks/Suid Afrikanse Nasionale Parkse, POB 787, Pretoria 0001, South Africa; tel (12) 3439770; fax (12) 3432832; e-mail kobier@parks-sa.co.sa; internet www.parks-sa.co.za; Editor: J C Rautenbach; f 1958; 2 a year; journal of research relevant to the national parks of South Africa; contains original papers relating to the preservation of biodiversity and the conservation of natural resources; includes abstracts, bibliographies and cumulative indexes; in English and Afrikaans; circ 1,200.

Komba (Bushbaby): Wildlife Clubs of Kenya Association, Langata Rd, POB 20184, Nairobi, Kenya; tel (2) 891904; fax (2) 891906; Editor: Rupi Mangat; f 1969; 3 a year; wildlife and environmental conservation; in English; circ 6,000.

Környezetvedelem: Magyar Mediprint Szakkiado Kft, 1055 Budapest, ul Balassi Ba'lint 7, Hungary; tel (1) 269-2033; fax (1) 266-1375; e-mail medinfo@mail.lang.hu; internet www.lang.hu/mediprint/korny; Editor: Dr Erika Zador; Publisher: Dr Tamas Kolosi; f 1993; 10 a year; environmental protection and technology, ecology; circ 8,000.

Környezetvedelmi Szakirodalmi Tajekoztato (Environmental Control Abstracts): (Publr) Orszagos Muszaki Informacios Kozpont es Konyvtar—OMIKK, 1428 Budapest, Museum u 17, POB 12, Hungary; Editor: Eszter Molnar; f 1983; every 2 months; published by the National Technical Information Centre and Library; circ 380.

Kosmos: Deutscher Verlags-Anstalt, 70190 Stuttgart, Neckarstr 121, Germany; tel (711) 26310; fax (711) 2631292; Editor: Dr R Köthe; Advertising Manager: Barbara Behn; f 1904; monthly; the discovery, understanding and protection of nature; circ 63,000.

La Chaîne/De Keten: La Chaîne Bleue Mondiale, 39 ave de Visé, 1170 Brussels, Belgium; tel (2) 673-52-30; fax (2) 672-09-47; e-mail bwk@vt4.net; Man Dir: Marleen Elsen-Verlodt; f 1962; quarterly; publication of the World Blue Chain for the Protection of Animals, dedicated to the prevention of cruelty to domestic and farm animals; in French and Dutch; circ 70,000.

La Forêt (The Forest): Economie Forestière Suisse, Rosenweg 14, 4501 Solothurn, Switzerland; tel (32) 6258800; fax (32) 6258899; e-mail info@wvs.ch; internet www.wvs.ch; Head Public Relations: Marcel Güntensperger; f 1947; 11 a year; aimed at the owners of forests, forest engineers, foresters, loggers, workmen, forest apprentices, contractors, various associations, federal, cantonal and communal administrations; French; circ 2,045.

La Houille Blanche: Revue Internationale de l'Eau (Water Power: International Water Review): Revue Générale de l'Electricité SA, 48 rue de la Procession, 75724 Paris Cédex 15, France; tel 1-44-49-60-00; fax 1-44-49-60-35; Editor: Hélène Girardet; f 1946; 6 a year, including 2 double issues; publication dedicated to water resources, fluid mechanics and hydroelectric power; in English and French with summaries in English, French, German and Spanish; circ 4,000.

Lake Line: North American Lake Management Society, POB 5443, Madison, WI 53705-5443, USA; Editor: Jeffrey Thornton; f 1981; quarterly; environmental issues concerning lake management; circ 2,300.

Land and Water: Land and Water, Inc, Route 3, POB 1197, Fort Dodge, IA 50501, USA; tel (515) 576-3191; fax (515) 576-5675; e-mail landandwater@dodgenet.com; internet www.landandwater.com; Editor: Teresa C Doyle; Pres: Kenneth M Rasch; f 1974; 8 a year; magazine of natural resource management and restoration, covering erosion control and water management; circ 20,000.

Land Contamination and Reclamation: EPP Publications Ltd, 6 Eastbourne Road, Chiswick, London W4 3EB, United Kingdom; tel (20) 8400-1601; fax (20) 8747-9663; e-mail enquiries@epppublications.com; internet www.epppublications.com; Man Editor: M G Pomel; f 1993; quarterly; covers soil decontamination, land reclamation, ground-water and restoration of habitats; in English; circ 700.

Land Degradation and Rehabilitation: John Wiley and Sons Ltd, The Atrium Southern Gate, Terminus Rd, Chichester, West Sussex PO19 8SQ, United Kingdom; tel (1243) 779777; fax (1243) 775878; internet wiley.co.uk/journals-sci; Publishing Editor: Lesley Valentine; f 1989; every 2 months; promotes the study of the recognition, monitoring, control and rehabilitation of degradation in terrestrial environments; circ 500.

Land Letter: Environmental and Energy Publishing, LLC, 112 C St, NW, Suite 700, Washington, DC, 20001, USA; tel (202) 628-6500; fax (202) 737-5299; internet www.eenews.net; Editor: Kevin Braun.

Landscape and Urban Planning: Elsevier Science Publishers BV, POB 211, 1000 AD Amsterdam, Netherlands; tel (20) 4853911; fax (20) 4853325; e-mail nlinfo-f@elsevier.nl; internet www.elsevier.nl; Editor-in-Chief: Jon E Rodiek; Editor: M M McCarthy; f 1974; 16 a year in 4 vols; international scientific journal covering landscape design, reclamation and conservation, and urban planning and ecology; includes bibliographies, book reviews and indexes; in English.

Landscape Architecture Europe: Landscape Design Trust, 13a West St, Reigate RH2 9BL, United Kingdom; tel (1737) 221116; fax (1737) 224206; e-mail ldt@landscape.co.uk; internet www.landscape.co.uk; Editor: Ken Fieldhouse; f 1992; annual; in English and French.

Landscape Ecology: Kluwer Academic Publishers, Postbus 17, 3300 AA Dordrecht, Netherlands; tel (78) 6576000; fax (78) 6576254; e-mail services@wkap.nl; internet www.kluweronline.com/issn/0921-2973; Editor: David J Mladenoff; f 1987; 8 issues a year; comprises research papers on the structure and functioning of landscapes and their interaction with humans, and relevant papers on land use, nature management and environmental conservation; in English.

Landscope: Western Australia–Department of Conservation and Land Management, Locked Bag 104, Bentley Delivery Centre, WA 6983, Australia; tel (8) 9334-0333; fax (8) 9334-0498; internet www.naturebase.net; Editor: Ron Kawalilak; f 1985; quarterly; former title *Forest Focus* (published by the Forest Department); general publication dealing with conservation, parks and wildlife; index online; circ 7,000.

L'Eau, L'Industrie, Les Nuisances: Editions Johanet, 30 rue René Boulanger, 75010 Paris, France; tel 1-42-40-00-08; fax 1-42-40-26-46; internet www.editions-johanet.com; Editor: Vincent Johanet; f 1975; monthly; journal covering water treatment, industry and water pollution; includes technical articles by experts, news items, reviews of new products, a bibliography and a calendar of relevant dates; circ 6,500.

Le Courrier de la Nature: Société Nationale de Protection de la Nature—SNPN, 9 rue Cels, 75014 Paris, France; tel 1-43-20-15-39; fax 1-43-20-15-71; e-mail snpn@wanadoo.fr; internet www.snpn.com; Editor: Marc Gallois; Dir: Christian Jouanin; 7 a year; journal of the National Nature Protection Society; concerned with flora and fauna, natural areas, national and international conservation activities; includes news items and a bibliography; French; circ 6,000.

Leben und Umwelt—unabhängige Zeitschrift für Biologie, Umwelt und Lebensschutz: Biologie-Verlag, 65004 Wiesbaden, Postfach 1449, Germany; Editor: Herbert Bruns; f 1964; quarterly; independent journal dealing with biology, the environment and preservation of biodiversity; includes indexes and book reviews; principally in German.

Les (Forest): Redakcia LES, Ministry of Agriculture, Dobrovicova 12, 812 66 Bratislava, Slovakia; tel (7) 306-6520; fax (7) 306-6517; Editor-in-Chief: Ján Fillo; f 1947; concerned with problems relating to forestry and the natural environment.

Lesnaya Promyshlennost (Forest Industry): Roslesprom, 101934 Moscow, Arkhangelskii per 1-330, Russian Federation; tel (095) 207-91-53; Editor: V G Zayedinov; f 1926; 3 a week; formerly organ of the USSR State Committee for Forestry and the Central Committee of the Timber, Paper and Wood Workers' Union of the USSR; circ 250,000.

L'Età Verde (Green Age): Associazione l'Età Verde, Via S Quintino 5, Palazzo C, CP 443, 00100 Rome, Italy; tel (06) 70453308; fax (06) 77206257; e-mail etaverde@tin.it; Editors: Eleonora Matini, Maria Letizia Parrett, Stefania Zuccati; Dir: Prof Augusta Busico; Deputy Dir: Prof Donato Tambli; f 1974; every 2 months; in Italian and English; circ 5,000.

Lettre de l'Environnement: Victoires Editions, 38 rue Croix des Petits Champs, 75001 Paris, France; tel 1-42-96-67-22; fax 1-40-20-07-75; Editor-in-Chief: Cécile Clicquot de Hentque; 20 a year; newsletter specializing in environmental technology and industrial practices.

Lifewatch: London Zoo, Regent's Park, London NW1 4RY, United Kingdom; tel (20) 7449-6228; fax (20) 7586-6177; e-mail marketing@zsl.org; internet www.zsl.org; Editor: Gina Guarnieri; f 1989; quarterly; includes news of the Zoological Society of London and items on animal conservation, environmental issues and animal-related topics; circ 40,000.

Links: International Society of Biometeorology, Secretariat, c/o Dr Scott Greene, Department of Geography, University of Oklahoma, Norman, OK 73071, Canada; tel (405) 325-4319; fax (405) 447-8455; e-mail jgreene@ou.edu; internet www.biometeorolgy.org; Editor: Dr Scott Greene.

Linnut: Lintutieteellisten Yhdistysten Liitto (BirdLife Finland), POB 1285, 00101 Helsinki, Finland; tel (9) 6854700; fax (9) 6854722; e-mail birdlife@iki.fi; Editor: Janne Taskinen; f 1965; every 2 months; former title *Lintumies*; devoted to the conservation and study of birds in Finland; in Finnish.

Living Earth: Soil Association, Bristol House, 40–56 Victoria St, Bristol BS1 6BY, United Kingdom; tel (117) 929-0661; fax (117) 925-2504; e-mail info@soilassociation.org; internet www.soilassociation.org; Editor: Charlotte Russell; f 1946; quarterly; concerned with organic farming, sustainable agriculture and responsible forestry.

Local Authority Waste and Environment: Faversham House Group Ltd, Faversham House, 232a Addington Rd, South Croydon CR2 8LE, United Kingdom; tel (20) 8651-7100; fax (20) 8651-7117; e-mail fhg.env@dial.pipex.com; Editor: Alex Catto; Publr: Amanda Barnes; f 1993; monthly; environmental protection and waste management issues that affect local authorities; circ 5,800.

Local Environment: the International Journal of Justice and Sustainability: Taylor & Francis Group (Journals), 4 Park Sq, Milton Park, Abingdon, OX14 4RN, United Kingdom; tel (20) 7017-6000; fax (20) 7017-6336; e-mail julian.agyeman@tufts.edu; internet www.tandf.co.uk/journals/titles/13549839.asp; Editors: Julian Agyeman, Bob Evans; 6 a year; focuses on local environmental, justice and sustainability policy, politics and action. Publishing house: Routledge.

London Energy News: Ambient Press Ltd, POB 25, Lutterworth LE17 4FF, United Kingdom; tel (1455) 202281; Editor: Jonathan David; f 1985; 4 a year; official newsletter of the London Energy and Environment Group; promotes efficient use of energy in the built environment.

Loris: Wildlife and Nature Protection Society of Sri Lanka, 5 19th Lane, Colombo 3, Sri Lanka; tel (1) 325248; fax (1) 580721; e-mail metalixc@mail.slt.lk; Editor: Sirancee Gunawardena; f 1936; 2 a year; journal of wildlife in Sri Lanka; in English; circ 2,000.

Louisiana Conservationist: Dept of Wildlife and Fisheries, POB 98000, Baton Rouge, LA 70898-9000, USA; tel (504) 765-2918; fax (504) 763-3568; Editor: Bob Dennie; f 1923; every 2 months; circ 45,000.

Lucht: Sansom HD Tjeenk Willink BV, Postbus 316, 2400 AH Alphen aan den Rijn, Netherlands; tel (1720) 66822; fax (1720) 66639; f 1984; quarterly; formerly *Lucht en Omgeving*; air pollution control.

Lufthygienischer Monatsbericht: Hessisches Landesamt für Umwelt und Geologie, Postfach 32 09 65022 Wiesbaden, Germany; tel (611) 69390; fax (611) 6939555; e-mail info@hlug.de; internet www.hlug.de; Editor: Klaus Hanewald; f 1973; monthly; published by Hesse Regional Institution for the Environment; air quality reports; in German; circ 400.

Luftverunreinigung: Deutscher Kommunal Verlag GmbH, 40470 Düsseldorf, Roseggerstr 5a, Germany; tel (211) 624417; fax (211) 622998; Editor: H J Schumacher; f 1973; annual; concerned with air pollution.

Madoqua: Ministry of Environment and Tourism, PMB 13346, Windhoek, Namibia; tel (61) 248345; fax (61) 248344; e-mail chrisbrown@nnf.org.na; Editor: C J Brown; f 1969; 2 a year; namibia Nature Foundation; journal containing original papers on applied research relating to arid zone biology in the Namib Desert and the conservation of nature in Namibia and neighbouring countries; in English; circ 750.

Magazine Recycling Benelux: Elsevier Business Information BV, Postbus 4, 7000 BA Doetinchem, Netherlands; fax (314) 363638; e-mail m.beck@misset.nl; f 1966; 8 a year; publication intended for recycling and reclamation professionals; marketing information and technical reports on recycling equipment; in Dutch and French; circ 2,470.

Maine Environment: Natural Resources Council of Maine, 3 Wade St, Augusta, ME 04330-6351, USA; tel (207) 622-3101; fax (207) 622-4343; Editor: Leslie Hahn; circ 8,000.

Maine Fish and Wildlife: Maine Department of Inland Fisheries and Wildlife, 284 State St, 41 State House Station, Augusta, ME 04333-0041, USA; tel (207) 287-8000; fax (207) 287-6395; e-mail mag.mfw@state.me.us; internet www.state.me.us/ifw; Editor: V Paul Reynolds; f 1959; quarterly.

Malaysian Forester: c/o Forestry Department Headquarters, Jalan Sultan Salahuddin, 50660 Kuala Lumpur, Malaysia; tel (3) 2988244; fax (3) 2925657; Editors: Thai See Kiam, Dr Ahmad Said bin Sajap; quarterly; publication dedicated to technical aspects of tropical forestry and forestry products; in English; circ 400.

Manitoba Naturalists Society Bulletin: 401 63 Albert St, Winnipeg, MB R3B 1G4, Canada; tel and fax (204) 943-9029; e-mail mns@escape.ca; Editor: Jenny Gates; f 1975; 10 a year; members' bulletin; deals with issues related to natural history, environmental protection and outdoor pursuits; circ 1,500.

Marine Bulletin: National Coalition for Marine Conservation, 3 West Market St, Leesburg, VA 20176-2901, USA; tel (703) 777-0037; Editor: Ken Hinman; circ 1,000.

Marine Conservation: Marine Conservation Society—MCS, 9 Gloucester Rd, Ross-on-Wye HR9 5BU, United Kingdom; tel (1989) 566017; fax (1989) 567815; e-mail info@mcsuk.org; internet www.mcsuk.org; Editor: Richard Harrington; f 1983; 4 a year; members' magazine; circ 5,000. Publishing house: Nature Conservation Bureau, Newbury, United Kingdom.

Marine Conservation News: Center for Marine Conservation, 1725 DeSales St, NW, Washington, DC 20036, USA; tel (202) 429-5609; fax (202) 872-0619; e-mail cmc@dccmc.org; internet www.cmc-ocean.org; Editor: Rose Bierce; f 1977; quarterly; conservation of marine species and habitats; circ 120,000.

Marine Environmental Research: Elsevier Applied Science, The Boulevard, Langford Lane, Kidlington OX5 1GB, United Kingdom; tel (1865) 843000; fax (1865) 843010; e-mail nlinfo-f@elsevier.nl; internet www.elsevier.nl; Editors: G Roesijadi, R B Spies; f 1978; 8 a year in 2 vols; contains results of research on biological, chemical and physical interactions in marine waters; concerned with toxicology; incorporates *Oil and Chemical Pollution.*

Marine Fish Management: Nautilus Press, Inc, 1201 National Press Bldg, Washington, DC 20045, USA; tel (202) 347-6643; Editor: John R Botzum, Sr; monthly.

Marine Georesources and Geotechnology: Taylor & Francis, Inc, 325 Chestnut St, 8th Floor, Philadelphia, PA 19106, USA; tel (215) 625-8900; fax (215) 625-2940; e-mail journals@tandf.co.uk; internet www.tandf.co.uk/journals/tf/1064119X.html; Editors-in-Chief: Ronald C Chaney, Michael J Cruickshank; quarterly; publishes research devoted to scientific, engineering and georesource aspects of the management and utilization of sea-floor sediments and rocks; available online.

Marine Mammal News: Nautilus Press, Inc, 1201 National Press Bldg, Washington, DC 20045, USA; tel (202) 347-6643; internet www.mediafinder.com/newsletter/~1560001.htm; Editor: John R Botzum, Sr; monthly.

Marine Pollution Bulletin: Elsevier Science Ltd, The Boulevard, Langford Lane, Kidlington OX5 1GB, United Kingdom; tel (1865) 843000; fax (1865) 843010; e-mail nlinfo-f@elsevier.nl; internet www.elsevier.nl; Editor: Charles Sheppard; f 1970; 24 a year; use of marine resources, measurement, analysis and control of pollution.

Marine Pollution Research Titles: Plymouth Marine Laboratory, Citadel Hill, Plymouth PL1 2PB, United Kingdom; tel (1752) 633266; fax (1752) 633102; e-mail d.moulder@pml.ac.uk; Editor: D S Moulder; f 1974; monthly; bibliography and book reviews; available on CD-ROM; circ 200.

Maryland Air Quality Data Report: Maryland Department of the Environment, Air and Radiation Management Administration, 2500 Broening Highway, Baltimore, MD 21224, USA; tel (410) 631-3280; f 1956; annual; contains ambient air quality data; electronic formats available; circ 400.

Massachusetts Wildlife: Massachusetts Division of Fisheries and Wildlife, Field Headquarters, 1 Rabbit Hill Rd, Westborough, MA 01581-3337, USA; tel (508) 792-7270; fax (508) 792-7275; e-mail pmirick@state.ma.us; Exec Editor: Ellie Horwitz; Editor: Peter Mirick; f 1952; quarterly; information on wildlife and related issues in the state of Massachusetts; circ 20,000.

Materials Reclamation Handbook (Year): EMAP Maclaren, Leon House, 19th–20th Floors, 233 High St, Croydon CR0 9XT, United Kingdom; tel (20) 8277-5000; fax (20) 8277-5550; internet www.emap.com; lists companies in the materials reclamation and recycling industries; includes details of legislation, testing procedures and identification tables.

Materials Recycling Week: EMAP Maclaren, Leon House, 19th Floor, 233 High St, Croydon CR0 9XT, United Kingdom; tel (20) 8277-5540; fax (20) 8277-5560; e-mail recycling@maclaren.emap.co.uk; Editor: Steve Eminton; Advertising Manager: Steve Dickinson; f 1912; weekly (50 a year); former title *Materials Reclamation Weekly*; business publication intended for the recycling industry and associated bodies such as local authorities; deals with the reclamation of all types of materials, including metals, waste paper, glass, plastics, textiles, chemicals and oils; circ 4,500.

Medio Ambiente: Universidad Austral de Chile, Facultad de Ciencias, Casilla 567, Valdivia, Chile; fax (63) 22-1344; Editor: Carlos A Moreno; f 1975; 2 a year; general environmental publication; in Spanish and English with summaries in English; available online; circ 500.

Meguro-ku No Kankyo (Environmental Pollution in Meguro Ward): Meguro Ward Office, Environmental Pollution Dept, 4-5, 2-chome, Chuo-Chiyou, Meguro-ku, Tokyo 152, Japan; fax (3) 5722-9338; f 1971; annual; in Japanese.

Meio Ambiente: Thesaurus Editora, SIG, Quadra 8, Lote 2356, 70610 Brasília, DF, Brazil; tel (61) 225-3011; fax (61) 321-0401; f 1978; quarterly; general environmental publication.

Merigal: Australian Native Dog Conservation Society Ltd, 590 Arina Rd, POB 91, Bargo, NSW 2574, Australia; tel (2) 4684-1156; fax (2) 4684-1830; e-mail merigal@zip.com.au; internet www.zip.comau/~merigal; Editors: David Steward, Berenice Walters; f 1976; quarterly; dedicated to the conservation of the Australian dingo and other canids; also concerned with animal behaviour, animal welfare, ecology, pets and wildlife; circ 300.

The Messenger: Citizens Energy Council, POB U, Hewitt, NJ 07421, USA; tel (201) 208-1620; fax (201) 728-0915; quarterly; environmental and health journal.

Messergebnisse des Zentralen Immissionsmessnetzes—Monatsbericht: Landesamt für Umweltschutz und Gewerbeaufsicht Rheinland-Pfalz, 55118 Mainz, Rheinallee 97, Abteilung 3, Germany; tel (6131) 608306; Editor: Horst Borchert; f 1978; monthly; published by the Rhineland-Palatinate Regional Office for Environmental Protection and Industrial Health and Safety; compilation of statistics relating to the environmental effects of industry; circ 500.

Metsälehti (Journal of Forestry): Forestry Development Centre Tapio, Saidinkuja 4, 00700 Helsinki, Finland; tel (9) 1562333; fax (9) 1562335; e-mail paavo.seppanen@metsalehti.mailnet.fi; internet www.metsalehti.fi; Editor: Paavo Seppänen; Advertising Manager: Tomi Silvonen; f 1933; every 2 weeks; forestry magazine; in Finnish; circ 41,355.

Metsäliiton Viesti: Osuuskunta Metsäliitto, Revontulentie 6, 02100 Espoo, Finland; tel (9) 46941; fax (9) 4550025; concerned with forest economy.

Metsastaja (Hunter): Hunters' Central Organization, Fantsintie 13–14, 01100 Itäsaimi, Finland; tel (9) 8777677; fax (9) 8777617; e-mail klaus.ekman@riista.fi; internet www.riista.fi; Editor: Klaus Ekman; f 1951; every 2 months; in Finnish; available in a Swedish-language edition, *Jaegaren* (qv); available online; circ 314,000 (including Swedish edition).

Michigan Natural Resources: Kolka & Robb, Inc, 30600 Telegraph Rd, Suite 1255, Bingham Farms, MI 48025-4531, USA; tel (810) 642-9580; fax (810) 642-5290; internet www.dnr.state.mi.us/www/mp/magazine.html; Editor: Richard Morschek; Publr: Vicki Robb; f 1931; every 2 months; circ 90,000.

Microbial Ecology: Springer-Verlag New York, Inc, 175 Fifth Ave, New York, NY 10010, USA; tel (212) 460-1627; fax (212) 473-6272; e-mail orders@springer-ny.com; internet link.springer.de/link/service/journals/00248/index.htm; Editor: James K Fredrickson; 8 a year; original papers on all aspects of ecology involving micro-organisms; available online; circ 1,200.

Mie-ken Kogai Senta Nenpo (Annual Report of Mie-ken Environmental Science Institute): Kogai Senta, 8-12, Shinjo 4-chome, Yokkaichi, Mie-ken, Japan; f 1973; annual; in Japanese.

Milieudefensie: Vereniging Milieudefensie, Damrak 26, 1012 LJ Amsterdam, Netherlands; tel (20) 6221366; fax (20) 6275287; e-mail redactie@milieudefensie.nl; Editors: Koen Vink, Han van de Wiel; f 1972; 10 a year; environmental protection magazine; covers sustainable development, climate change, conservation, pollution, energy and environmental education; in Dutch; circ 20,000.

Miljø Danmark (Danish Environment): Miljøstyrelsen, Strandgade 29, 1401 Copenhagen K, Denmark; tel 32-66-01-00; fax 32-66-04-79; internet www.mst.dk; Editor: Søren Jensen; f 1974; 10 a year; published by the Danish National Agency of Environmental Protection; in Danish; circ 17,000.

Minnesota Conservation Volunteer: Department of Natural Resources, DNR Bldg, 500 Lafayette Rd, St Paul, MN 55155-4046, USA; tel (612) 296-0888; fax (612) 296-0902; e-mail sue.ryan@dnr.state.mn.us; internet www.dnr.state.mn.us/magazine; Editor: Kathleen Weflen; f 1940; every 2 months; natural resources conservation, outdoor recreational opportunities in Minnesota, USA; available online; circ 170,000.

Misset's Milieu Magazine: Misset uitgeverij bv, POB 9000, 6800 DA Arnheim, Netherlands; tel (26) 3209911; fax (26) 3230223; Editor: D van Lent; Manager: D D F Wiedemeijer; 10 a year; environmental business magazine covering environmental policy, management and technology; circ 15,000.

Mississippi Outdoors: Department of Wildlife, Fisheries and Parks, POB 451, Jackson, MS 39025, USA; tel (601) 364-2123; Editor: David L Watts; f 1935; every 2 months; circ 40,000.

Missouri Conservationist: Department of Conservation, Outreach and Education, POB 180, Jefferson City, MO 65102, USA; tel (573) 751-4115; fax (573) 751-2660; e-mail internet@mail.conservation.state.mo.us; internet www.conservation.mo.us/conmag; Editor: Tom Cwynar; f 1938; monthly; available online; circ approx 400,000.

Missouri Wildlife: Conservation Federation of Missouri, 728 West Main St, Jefferson City, MO 65101-1534, USA; tel (314) 634-2322; fax (314) 634-8205; Editor: Charles F Davidson; f 1939; every 2 months; circ 40,000.

Mondo Sommerso: Editoriale Olimpia SpA, Viale Milton 7, 50129 Florence, Italy; fax (055) 5016281; e-mail mondo.sommerso@edolimpia.it; internet www.edolimpia.it; Editor: Sabina Cupi; f 1959; monthly; includes biological reports on marine and freshwater environments and their wildlife; also covers issues related to diving; in Italian and English; circ 38,000.

Monitor: Angel Business Communications Ltd, Kingsland House, 361 City Rd, London EC1V 1LR, United Kingdom; tel (20) 7417-7400; fax (20) 7417-7500; e-mail monitor@eurosemi.demon.co.uk; Editor: Bryan Rimmer; f 1991; monthly; environmental news bulletin for the packaging industry.

Monti e Boschi: Edagricole SpA, Via Emilia Levante 31/2, 40139 Bologna, Italy; tel (051) 492211; fax (051) 493660; internet www.agriline.it/edagri; Editor: Prof Umberto Bagnaresi; Advertising Manager: Giuliano Avoni; f 1949; 2 a month; ecology and forestry; circ 16,700.

Müllmagazin: Rhombos Verlag, 10785 Berlin, Kurfürstenstr 17, Germany; tel (30) 261-6854; fax (30) 261-6300; e-mail verlag@rhombos.de; internet www.rhombos.de; Editor: Bernhard Reiser; Marketing: Steffi Miller; f 1988; quarterly; specialist journal dealing with waste prevention and minimization, ecological waste management and recycling; circ 3,300.

Municipal and Industrial Water and Pollution Control: Zanny Publications Ltd, 11966 Woodbine Ave, Gormley, ON L0H 1G0, Canada; tel (905) 887-5048; fax (905) 887-0764; Editor: Amy Margaret; Publr: Janet Gardiner; f 1893; every 2 months; available online and on microform.

Musu gamta (Our Nature): Rudens 33b, Vilnius 2600, Lithuania; tel (2) 696-964; Editor-in-Chief: Vytautas Klovas; f 1964; every 2 months; popular science, nature preservation and tourism; circ 2,500.

Nabat (Alarm Bell): Chornobyl Socio-Ecological Union, 220002 Minsk, vul Krapotkina 44, Belarus; tel (17) 234-22-41; fax (17) 210-18-80; Editor-in-Chief: Vasily Yakovenka; f 1990; monthly; ecological survival and the socio-economic prosperity of peoples and nations; in Belarusian and Russian; circ 61,000.

NAEP News: National Association of Environmental Professionals, 6524 Ramoth Drive, Jacksonville, FL 32226-3202, USA; tel (904) 251-9900; fax (904) 251-9901; internet www.naep.org; Editor: Monica Prado; f 1975; every 2 months; circ 4,000.

Naga, The ICLARM Quarterly: International Center for Living Aquatic Resources Management—ICLARM, POB 500, GPO, 10670 Penang, Malaysia; e-mail m.v.gupta@cgiar.org; internet www.cgiar.org/iclarm/naga/nagainfo.htm; comprises papers containing interpretative, applied and descriptive scientific and technical information of interest to researchers, scientists and policy-makers in the field of fisheries and living aquatic resources; includes news on research projects, summaries, notices of new publications, meetings, workshops, courses and symposia, and news of ICLARM personnel; available online; circ 5,500.

National Association of Conservation Districts, Tuesday Letter: National Association of Conservation Districts, POB 855, League City, TX 77574-0855, USA; tel (713) 332-3402; fax (713) 332-5259; Editor: Russell Slaton; f 1952; quarterly; circ 25,000.

National Geographic Magazine: National Geographic Society, 1145 17th St, NW, Washington, DC 20036-4688, USA; tel (202) 857-7000; fax (202) 775-6141; Editor: William Allen; f 1888; monthly; popular periodical dedicated to geography, natural history, the environment and cultural anthropology; circ 9,013,000.

National Laws on the Marketing and Use of Dangerous Substances and Preparations: Agra Informa Ltd, 80 Calverley Rd, Tunbridge Wells, Kent TN1 2UN, United Kingdom; tel (1892) 533813; fax (1892) 544895; f 1995; base vol with updates; implementation of EU directives into UK national legislation.

National Parks Journal: National Parks Association of New South Wales, Inc, POB A96, Sydney S, NSW 1235, Australia; tel (2) 9299-0000; fax (2) 9290-2525; e-mail editors@npansw.org.au; internet www.npansw.org.au; Editor: Rosemary Pryor; f 1957; every 2 months; concerned with environmental issues relating to national parks and nature conservation; excerpts available online; circ 4,000.

National Trust Magazine: National Trust, 36 Queen Anne's Gate, London SW1H 9AS, United Kingdom; tel (20) 7222-9251; fax (20) 7222-5097; internet www.nationaltrust.org.uk; Editor: Gaynor Aaltonen; f 1969; 3 a year; contains news of the National Trust and items on history, heritage and the countryside; circ 1,300,000.

National Trust Quarterly: National Trust of Australia, Observatory Hill, Watson Rd, The Rocks, POB 518, Sydney, NSW 2000, Australia; tel (2) 9258-0123; fax (2) 9258-1264; e-mail clevins@nsw.nationaltrust.org; internet www.nsw.nationaltrust.org.au; Editor: Julian Faigan; includes bibliographies and book reviews; circ 25,000.

National Wildlife: National Wildlife Federation, 8925 Leesburg Pike, Vienna, VA 22184-0111, USA; tel (703) 790-4000; fax (703) 790-4544; e-mail pubs@nwf.org; internet www.nwf.org; Editor and Publr: Bob Strohm; f 1962; every 2 months; intended for a general readership interested in nature and wildlife; available online and on microform; circ 900,000.

Natour: European Centre for Professional Training in Environment and Tourism—CEFAT, Viriato 20, 3°, 28010 Madrid, Spain; tel (91) 5930831; fax (91) 5930980; e-mail eol@eol.cestel.es; internet www.eol.es/natour; Editor: Arturo Crosby; f 1987; concerned with aspects of tourism and development including ecotourism, rural tourism, sustainable tourism development and tourism management in protected areas.

Natur + Kosmos: Natur Media GmbH, Belfortstr 8, 81667 München, Germany; tel (89) 4587270; fax (89) 45872724; e-mail redaktion@natur.de; internet www.natur.de; Editor-in-Chief: Ilona Jerger; CEOs: Katja Kohlhammer, Peter Dilger; f 1981; monthly; presents information on the environment and nature preservation; intended for a European readership; German; circ 129,400.

Natur & Miljø (Nature and Environment): Norges Naturvernforbund—NNV, POB 342 Sentrum, 0130 Oslo, Norway; tel 22-40-24-00; fax 22-40-24-10; e-mail naturvern@naturvern.no; internet www.naturvern.no; f 1914; adopted current title 1989; every 2 months; former title *Miljømagasinet*; publication of the Norwegian Society for the Conservation of Nature; concerned with biodiversity, transport, energy and climate, sustainable consumption and production, and persistent toxic substances.

Natur und Landschaft: Verlag W Kohlhammer GmbH, 70565 Stuttgart, Hessbruhlstr 69, Postfach 800430; Germany; tel (711) 78631; fax (711) 7863393; Editors: Dr W Mrass, Marlies Petzoldt; f 1925; monthly; published for the Bundesanstalt für Naturschutz (Federal Agency for Nature Protection); concerned with nature conservation, landscape management and environmental protection; includes indexes and book reviews; circ 5,700.

Natural Areas Journal: Natural Areas Association, POB 1504, Bend, OR 97709-1504, USA; tel (541) 317-0199; e-mail naa@natareas.org; internet www.allenpress.com; Editor: Don Leopold; f 1981; quarterly; circ 2,300.

Natural History: American Museum of Natural History, Central Park, West 79th St, New York, NY 10024, USA; tel (212) 769-5500; fax (212) 769-5511; Editor: Bruce Stutz; f 1900; 10 a year; magazine dedicated to nature, culture and environmental issues; intended for a general readership; available online and on CD-ROM; circ 400,000. Publishing house: (Subscriptions) Natural History, POB 5000, Harlan, IA 51537-5000, USA.

Natural Resources and Environmental Issues: Utah State University, College of Natural Resources, Logan, UT 84322-5200, USA; Editor: Amy Lyn De Zwart; f 1993; irreg.

Natural Resources Forum: Elsevier Science Ltd, The Boulevard, Langford Lane, Kidlington OX5 1GB, United Kingdom; tel (1865) 843000; fax (1865) 843010; Editor-in-Chief: Marcia Brewster; f 1977; quarterly; international, multidisciplinary journal; focuses on the management of energy, mineral and water resources in developing countries; examines economic, financial, legal and environmental aspects of natural resource development, in order to identify factors which influence and promote sustainable development.

Natural Resources Journal: Natural Resources Center, School of Law MSC11 6070, 1 University of New Mexico, Albuquerque, NM 87131-0001, USA; tel (505) 277-4910; fax (505) 277-8342; e-mail nrj@law.unm.edu; internet lawschool.unm.edu/Nrj/index.htm; Editor-in-Chief: G Emlen Hall; f 1961; quarterly; environmental policy and law; natural resources; conservation; ecology; in English; available on microform and online; circ 2,000. Publishing house: Univ of New Mexico School of Law.

Natural Resources Research: Kluwer Academic/Plenum Publishers, 233 Spring St, New York, NY 10013, USA; tel (919) 677-0977; fax (919) 677-1303; Editor-in-Chief: Daniel F Merriam; Deputy Editor: Douglas C Peters; Founding Editor: Richard B McCammon; quarterly; journal of the International Association for Mathematical Geology; articles cover natural resource economics, energy and mineral management, and exploration geology; sponsored by the Energy Minerals division of the American Association of Petroleum Geology.

Natural World: River Publishing Ltd, Victory House, Leicester Sq, London WC2H 7QH, United Kingdom; tel (20) 7306-0304; fax (20) 7413-9409; Editor: Sarah-Jane Forder; f 1998; concerned with wildlife, nature conservation and the environment, principally within the United Kingdom; circ 175,000.

Naturalist: SM Publications, 20 Collens Rd, Maraval, Port of Spain, Trinidad and Tobago; tel 622-3428; Publr and Editor-in-Chief: Stephen Mohammed; f 1975; every 2 months; the Field Naturalists Club of Trinidad and Tobago; natural heritage and conservation in the Caribbean; circ 20,000.

Naturalist: Yorkshire Naturalists' Union, c/o University of Bradford, Bradford BD7 1DP, United Kingdom; tel (1274) 384212; fax (1274) 384231; e-mail m.r.d.seaward@bradford.ac.uk; Hon Editor: Prof M R D Seaward; f 1875; natural history in England, particularly Yorkshire; circ 2,800.

Nature: Nature Publishing Group, Porters South, Crinan St, London N1 9XW, United Kingdom; tel (20) 7833-4000; fax (20) 7843-4596; e-mail subscriptions@nature.com; internet www.nature.com; Editor: Dr Philip Campbell; f 1869; weekly; Macmillan Publishers; multi-disciplinary science; Naturejobs (career and recruitment listing and information); in English; Japanese translation; available online; circ 65,000.

Nature and Resources: UNESCO, 7 pl de Fontenoy, 75352 Paris, France; tel 1-45-68-43-00; fax 1-45-68-57-41; internet www.unesco.org/publications; Editor: Malcolm Hadley; f 1958; quarterly.

Nature Canada: Canadian Nature Federation—CNF, 1 Nicholas St, Suite 606, Ottawa, ON K1N 7B7, Canada; tel (613) 562-3447; fax (613) 562-3371; e-mail cnf@cnf.ca; internet www.cnf.ca; Editor: Barbara Stevenson; f 1971; quarterly; former title *Canadian Audubon*; magazine focusing on Canadian natural history and conservation issues; includes statistics, bibliographies and book reviews; available on microform; circ 15,000.

The Nature Conservancy Magazine: Nature Conservancy, 4245n Fairfax Drive, No 100, Arlington, VA 22203-1606, USA; tel (703) 841-5300; internet www.tnc.org; Editor: Ron Geatz; f 1951; every 2 months; available online; circ 800,000.

Naturen: Universitetsforlaget, POB 2959, Tøyen, 0608 Oslo 6, Norway; tel 22-57-53-00; e-mail journals@scup.no; internet www.scup.no/naturen; Editor: Prof Per Min Jørgensen; f 1877; every 2 months; comprehensive natural sciences journal; intended for the general public; circ 2,200.

Naturfreund/Ami de la Nature: Naturfreunde Schweiz, Pavillonweg 3, POB 7364, 3001 Bern, Switzerland; every 2 months; in French and German.

Natürlich: AZ Fachverlag AG, Neumattstr 1, 5001 Aarau, Switzerland; tel (62) 8366565; fax (62) 8366566; internet www.az-verlag.ch; Editor-in-Chief: Walter Hess; Deputy Editor: Heinz Knieriemen; Publr: Riccarda Heddenburg; f 1984; monthly; information on nature and health; areas of interest include ecology, nature preservation, education, building with environmentally friendly natural materials and enjoyment of the countryside; in German; circ 90,000.

Naturmiljön i Siffor (The Natural Environment in Figures): Statistics Sweden, POB 24300, 104 51 Stockholm, Sweden; tel (8) 50694000; fax (8) 50694763; e-mail scb@scb.se; internet www.scb.se; Statistical Editor: Solveig Danell; Environmental Editor: Eva Ahuland; f 1977; every 4 years; published jointly by Statistics Sweden and the Swedish Environmental Protection Agency.

Naturopa: Council of Europe, 67075 Strasbourg, Cedex, France; tel 3-88-41-23-98; fax 3-88-41-37-51; e-mail maguelonn.dejeant-pons@coe.int; internet www.nature.coe.int/english/cadres/issue.htm; Head, Spatial Planning and Landscape Division: Maguelonne Dejeant-Pons; f 1968; 2 per year; national and cultural heritage and landscape in a perspective of sustainable developent and enhancement of the quality of life; English and French; PDF online.

Naturschutz Heute (Nature Conservation Today): NABU, 53225 Bonn, Postfach 301054, Germany; tel (228) 9756141; fax (228) 9756194; e-mail 106025.2414@compuserve.com; internet www.umwelt.de/nabu; Editor-in-Chief: J Michael Schroeren; Editor: Helge R May; f 1969; 5 a year; published for Naturschutzbund Deutschland eV—NABU (Nature Preservation Society); concerned with nature conservation, conservation of species and habitats, biodiversity, protected areas, land-use policy and landscape planning; also includes items on resource management, soil conservation, water and air pollution, environmental law and consumer politics; circ 170,000.

Naturschutz und Naturparke (Nature Protection and Nature Parks): Verband Deutscher Naturparke, 29646 Bispingen, Niederhaverbeck, Germany; tel (5198) 987033; fax (5198) 987039; e-mail vdn@naturpark.de; internet www.naturpark.de; Manager: Maria Stadtler; f 1926; quarterly; concerned with Luneburg Heath, nature parks, protected areas and natural habitats; circ 5,000.

Naturwissenschaftliche Rundschau: 70191 Stuttgart, Birkenwaldstr 44, Germany; tel (711) 25820; fax (711) 2582390; e-mail nr@wissenschaftliche-verlagsgesellschaft.de; internet www.naturwissenschaftliche-rundschau.de; Editors:

Dr Klaus Rehfeld; f 1948; monthly; publication of Gesellschaft Deutscher Naturforscher und Ärzte eV (GDNÄ), www.gdnae.de; natural sciences; German. Publishing house: Wissenschaftliche Verlagsgesellschaft mbH.

Natuur en Milieu: Stichting Natuur en Milieu—SNM, Donkerstraat 17, 3511 KB Utrecht, Netherlands; tel (30) 2331328; fax (30) 2331311; e-mail snm@snm.nl; internet www.snm.nl; Editor: M Jehae; f 1977; monthly; published by the Society for Nature and Environment; includes indexes, diagrams and book reviews; circ 10,000.

Nebraska Resources: Natural Resources Commission, 301 Centennial Mall S, POB 94876, Lincoln, NE 68509, USA; tel (402) 471-2081.

Nebraskaland: Game and Parks Commission, POB 30370, Lincoln, NE 68503, USA; tel (402) 471-0641; fax (402) 471-5528; Editor: Donald Cunningham; f 1926; 10 a year; government publication covering wildlife, conservation and outdoor recreation; circ 50,000.

NERC News: Natural Environment Research Council—NERC, Polaris House, North Star Ave, Swindon SN2 1EU, United Kingdom; tel (1793) 411500; fax (1793) 411501; internet www.nerc.ac.uk; Editor: Leslie Jones; f 1970; quarterly; former title *NERC News Journal*; includes research papers, news of recent advances in all fields of environmental science, notification of forthcoming conferences and book reviews; circ 6,000.

NETA News: National Environment Training Asscn, 2930 East Camelback Rd, No 185, Phoenix, AZ 85016-4412, USA; tel (602) 956-6099; fax (602) 956-6399; internet ehs-training.org; Editor: Charles L Richardson; f 1977; quarterly; circ 1,500.

New Forests News: New Forests Project—NFP, International Centre for Development Policy, 731 Eighth St, SE, Washington, DC 20003, USA; tel (202) 547-3800; fax (202) 546-4784; internet www.newforestsproject.com; Dir, New Forests Project: Thomas Minney; f 1996; quarterly; quarterly newsletter concerned with reforestation in developing countries, watershed protection and renewable energy; in English, French and Spanish; circ 2,000.

New Hampshire Audubon: Audubon Society of New Hampshire, 3 Silk Farm Rd, POB 528-B, Concord, NH 03301-8200, USA; tel (603) 224-9909; fax (603) 226-0902; e-mail nhaudubon@igc.apc.org; internet www.nhaudubon.org; Editor: Miranda Levin; f 1921; every 2 months; circ 7,000.

New Jersey Conservation: New Jersey Conservation Foundation, 170 Longview Rd, Far Hills, NJ 07931, USA; e-mail info@njconservation.org; internet www.njconservation.org; Editor: David Yaskulka; quarterly; protection and appreciation of the environment in New Jersey; circ 8,000.

New Jersey Outdoors: Dept of Environmental Protection, Office of Communications and Public Education, POB 402, Trenton, NJ 08625, USA; tel (609) 984-0364; Editor: Denise Damiano Mikics; f 1974; quarterly; circ 20,000.

New Mexico Wildlife: Department of Game and Fish, POB 25112, Santa Fe, NM 87504, USA; tel (505) 827-7911; fax (505) 827-7915; f 1961; every 2 months; circ 10,000.

New World Journal: New World Journal Corpn, 330 West 56th St, Suite 3g, New York, NY 10019-4241, USA; tel (212) 265-7970; fax (212) 265-8052; Editor: Rick Bard; f 1992; quarterly; magazine concerned with science, business, the arts and popular culture as they relate to the emerging new world and its environment; circ 100,000.

New York City Environmental Bulletin: Council on the Environment of NYC, 51 Chambers St, Rm 228, New York, NY 10007, USA; tel (212) 788-7909; fax (212) 788-7913; e-mail cenyc@cenyc.org; internet www.cenyc.org; Admin Manager: Julie Walsh; f 1970; every 2 months; environmental events and publications in New York; circ 200.

The New York State Conservationist: Department of Environmental Conservation, 50 Wolf Rd, Rm 548, Albany, NY 12233-4502, USA; tel (518) 457-5547; fax (518) 457-0858; internet www.dec.state.ny.us/website/about/pubconl4.html; Editor: R W Groneman; f 1946; every 2 months; circ 100,000.

New Zealand Environment: Environmental Publications Trust, 34 Norana Ave, Remuera, Auckland 1105, New Zealand; tel (9) 524-2949; Editor: Robert Mann; f 1971; quarterly; contains book reviews, statistics and a cumulative index; circ 1,000.

New Zealand Forest Industries: POB 5544, Auckland, New Zealand; tel (9) 630-8940; fax (9) 630-1046; e-mail info@nzforest.co.nz; Editor: Vicki Jayne; Publr: Reg Birchfield; Advertising Manager: Andrew Moses; f 1960; monthly; circ 2,500.

New Zealand Journal of Forestry: New Zealand Institute of Forestry—NZIF, POB 19840, Christchurch, New Zealand; tel (3) 960-2432; fax (3) 960-2430; e-mail nzjf@paradise.net.nz; internet www.nzjf.org.nz; Editor: Dr Bruce Manley; Pres: Ket Bradshaw; Sec: Peter Allan; f 1927; quarterly; provides professional umbrella for those involved in forest management; circ 1,200.

News Sweep: National Solid Wastes Management Asscn, Contract Sweepers Institute, 5 Commonwealth Rd, Natick, MA 01760-1526, USA; tel (908) 704-9646; Editor: Denise Naidu; quarterly; circ 6,000.

Newsletter of the Environmental Resources Research Institute: Pennsylvania State University, Land and Water Research Bldg, University Park, PA 16802, USA; tel (814) 863-0291; fax (814) 865-3378; e-mail jrr131@psu.edu; internet www.erri.psu.edu; f 1970; quarterly; reports on the Institute's research and on environmental issues; intended for researchers in the USA and world-wide, legislators, civil servants and the general public; available online; circ 2,500.

Newsletter of the International Association of Fish and Wildlife Agencies: 444 North Capitol St, NW, Suite 544, Washington, DC 20001, USA; tel (202) 624-7890; fax (202) 624-7891; e-mail iafwa@sso.org.

Newsletter of the International Union of Societies of Foresters: c/o Society of American Foresters, 5400 Grosvenor Lane, Bethesda, MD 20814, USA; tel (301) 897-8720; fax (301) 897-3690; e-mail istfisuf@igc.apc.org; Editor: Bill Banzhaf; f 1965; 2–4 a year; concerned with all subjects relating to forestry and natural resource management.

Newsletter of the World of Birds Wildlife Sanctuary CC: Valley Rd, Hout Bay 7800, South Africa; tel (21) 7902730; fax (21) 7904839; Editor: Walter Mangold; f 1974; every 2 months; in English; circ 1,000.

Nieuwsblad Stromen (Energy and Environment Newspaper): Ten Hagen & Stam B, POB 34, 2501 AG The Hague, Netherlands; tel (70) 3045730; fax (70) 3045803; e-mail stromen@wkths.nl; Editor-in-Chief: Rolf de Vos; f 1999; 2 a week; superseded *Energie & Milieuspectrum* 1999; sustainable development in energy supply, industry, transport and the built environment; circ 7,000.

NIS Nuclear Profiles Database: Center for Non-Proliferation Studies, Monterey Institute of International Studies, 425

Van Buren St, Monterey, CA 93940, USA; tel (831) 647-4154; fax (831) 647-3519; e-mail cns@miis.edu; internet cns.miis.edu; Editor: John Lepingwell; f 1991; monthly; concerned with environmental issues in the Newly Independent States (NIS—the former Soviet republics), such as radioactive waste, spent nuclear fuel, civilian nuclear power facilities, uranium mining and milling, and submarine decommissioning and dismantlement; electronic formats available.

NJ Audubon: New Jersey Audubon Society, POB 126, Beardsville, NJ 07924-0126, USA; Editor: Peter Dunne; f 1940; quarterly; available on microform; circ 9,000.

NOAA Coastal Ocean Program Update: National Coastal and Atmospheric Administration, Coastal Ocean Program, 1315 East–West Highway, SSMC3, Rm 9608, Silver Spring, MD 20910, USA; tel (301) 713-3338; e-mail coastalocean@cop.noaa.gov; internet www.cop.noaa.gov/pubs/newsletters.html; irreg; available only online.

Nongye Huanjing Baohu (Agro-Environmental Protection): Chinese Society of Agro-Environmental Protection, 31 Fukang Lu, Nankai Qu, Tianjin 300191, People's Republic of China; tel and fax (22) 23361247; e-mail caep@public.tpt.tj.cn; Editors: Fu Kewen; f 1982; every 2 months; journal containing reports of progress, new technology and new methods of scientific research into agro-environmental pollution and its control; in Chinese; electronic formats available; circ 2,000.

Nongye Huanjing Yu Fazhan (Agro-Environment and Development): Chinese Society of Agro-Environmental Protection, 31 Fukang Lu, Nankai Qu, Tianjin 300191, People's Republic of China; tel (22) 23674336; fax (22) 23369542; e-mail caed@public.tpt.tj.cn; Editors: Wang Xiwu, Tao Zhan; f 1984; quarterly; papers and essays on impacts of governmental decisions on the agro-environment, management of the agro-environment and related policies, monitoring and quality assessment of the agro-environment, handling of accidents resulting from severe pollution of the agro-environment, promotion of eco-agriculture, technologies for the control and treatment of the agricultural eco-environment, and conservation of biodiversity; electronic formats available; circ 2,000.

North American Wildlife and Natural Resources Conference, Transactions: Wildlife Management Institute, 1101 14th St, NW, Suite 801, Washington, DC 20005, USA; tel (202) 371-1808; fax (202) 408-5059; e-mail wmihq@aol.com; internet www.wildlifemgt.org/wmi; Editor: Richard E McCabe; annual conference proceedings; circ 1,000.

North Dakota Outdoors: Game and Fish Department, 100 North Bismarck Expressway, Bismarck, ND 58501, USA; tel (701) 328-6300; fax (701) 328-6352; Editor: Harold Umber; f 1933; 10 a year; circ 20,000.

North Woods Call magazine: North Woods Call, Inc, Route 1, 00509 Turkey Run, Charlevoix, MI 49720, USA; tel (616) 547-9797; fax (616) 547-0367; Editor: Glen Sheppard; f 1953; every 2 weeks; circ 10,000.

Northern Journal of Applied Forestry: Society of American Foresters, 5400 Grosvenor Lane, Bethesda, MD 20814-2198, USA; tel (301) 897-8720; fax (301) 897-3690; internet www.safnet.org; Editor: James W Hornbeck; f 1984; quarterly; scientific journal concerned with forestry practice in the northern USA; in English; circ 900.

Northern Line: Northern Alaska Environmental Center—NAEC, 218 Driveway, Fairbanks, AK 99701, USA; tel (903) 452-5021; fax (903) 452-3100; e-mail northern@northern.org; internet www.northern.org; Communications Dir: Beth Caissie; f 1960; quarterly; articles related to the NAEC; in English; circ 1,400.

Northern Perspectives: Canadian Arctic Resources Committee (CARC), 1276 Wellington St, Ottawa, ON K1Y 3A7, Canada; tel (613) 236-7379; fax (613) 232-4665; internet www.carc.org; Editor: Alan Saunders; f 1973; quarterly; policy journal; includes Members' Update, supplement covering CARC activities and issues; circ circ 14,000.

Nouvelles de PRONAT (PRONAT News): ENDA Publications, Environmental Development Action in the Third World—ENDA, 54 rue Carnot, BP 3370, Dakar, Senegal; tel 22-55-65; fax 23-51-57; e-mail pronat@endadak.gn.apc.org; f 1982; quarterly; collaborative newsletter; concerned with research and training in natural crop protection and agroecology; circ 700.

NSCA Pollution Handbook: National Society for Clean Air and Environmental Protection—NSCA, 136 North St, Brighton BN1 1RG, United Kingdom; tel (1273) 326313; fax (1273) 735802; e-mail admin@nsca.co.uk; Editor: Loveday Murley; f 1978; annual; former titles *Clean Air Year Book, NSCA Yearbook* and *NSCA Reference Book*; guide to all aspects of United Kingdom pollution control and to European and international developments in pollution control.

NTIS Alerts: Environmental Pollution and Control: US National Technical Information Service, 5285 Port Royal Rd, Springfield, VA 22161, USA; tel (703) 605-6060; fax (703) 605-6880; internet www.ntis.gov/alerts.htm; summaries of research titles and articles received by NTIS; available on microform.

The Nuclear Engineer: Institute of Nuclear Engineers, 1 Penerley Rd, London SE6 2LQ, United Kingdom; tel (20) 8698-1500; fax (20) 8695-6409; Editor: K J Simm; f 1959; every 2 months; available on microfiche; circ 1,800. Publishing house: S & K Publishing, 179 Brighton Rd, Purley CR8 4HF, United Kingdom.

Nuclear News: American Nuclear Society, 555 North Kensington Ave, La Grange Park, IL 60526, USA; tel (708) 352-6611; fax (708) 352-6464; e-mail nucnews@ans.org; internet www.ans.org; Editor-in-Chief: Gregg M Taylor; f 1959; 13 a year; concerned with nuclear power plant operations and maintenance, relevant training, radioactive waste management and nuclear applications other than power generation; electronic formats available; circ 14,000.

Nuclear Waste Bulletin/Bulletin sur les Déchets Nucléaires: OECD Nuclear Energy Agency, 12 blvd des Iles, 92130 Issy-les-Moulineaux, France; tel 1-45-24-10-15; fax 1-45-24-11-10; e-mail compte.pubsinq@oecd.org; internet www.nea.fr; f 1979; irreg; former title *Newsletter on Radionuclide Migration in the Geosphere*; available on microfiche; circ 1,000.

Nuclear Waste News: Business Publishers, Inc, 951 Pershing Drive, Silver Spring, MD 20910-4464, USA; tel (301) 587-6300; fax (301) 587-1081; e-mail bpinews@bpinews.com; internet www.bpinews.com; Editor: Thecia Fabian; Publr: Leonard A Eiserer; f 1981; weekly; available online.

NUYTSIA: Western Australia Department of Conservation and Land Management, Locked Bag 104, Bentley Delivery Centre, WA 6983, Australia; tel (8) 9334-0333; fax (8) 9334-0498; f 1970; irreg; journal of taxonomic botany in Western Australia; circ 600.

Nyala: Wildlife Society of Malawi, POB 30360, Blantyre 3, Malawi; tel 643428; fax 643765; Editor: Cornell Dudley; f 1975; 2 a year; in English; circ 625.

Oasis: Industrie Grafiche Editoriali Musumeci SpA, Località Amérique 99, 11020 Quart, Italy; tel (0165) 765853; fax (0165) 765106; Editor: Pietro Giglio; f 1985; monthly (10 a year); nature, ecology, travel and photography; circ 65,000.

Oasis: Green Movement Socio-Ecological Centre—GMC, 484006 Taraz, Lunacharsky 42-2, Kazakhstan; tel and fax (32622) 3-27-93; e-mail alex@zagribelny.jambyl.kz; Editor: Aleksandr Zagribelny; f 1991; monthly; independent ecological publication intended for NGOs, particularly the Socio-Ecological Union (Russian Federation); contains articles on social and environmental issues; in Russian; circ 1,000–5,000.

Occupational Hazards: Penton Publishing, Inc, 1100 Superior Ave, Cleveland, OH 44114, USA; tel (216) 696-7000; fax (216) 696-8765; e-mail corpcomm@penton.com; internet www.penton.com; Editor: Stephen G Minter; Publr: David Gursky; f 1938; monthly; publication concerned with health, safety and environmental management; available online; circ 60,000.

Ocean & Coastal Management: Elsevier Applied Science, The Boulevard, Langford Lane, Kidlington OX5 1GB, United Kingdom; tel (1865) 843000; fax (1865) 843010; Editors-in-Chief: B Cicin-Sain, R W Knecht; f 1988; monthly; reports on the environment and resources of oceans and coastal regions; formerly *Journal of Shoreline Management and Ocean Management*.

Ocean Development & International Law: University of Victoria, Faculty of Law, POB 2400, Victoria, BC V8W 3H7, Canada; tel (604) 721-8181; fax (250) 721-8146; e-mail tlmcdorm@uvic.ca; internet www.tandf.co.uk/journals/tf/00908320.html; Editor: Ted L McDorman; f 1973; quarterly; concerned with international law, oceans law and marine resources law; available online; circ 600. Publishing house: Taylor & Francis, Inc, 325 Chestnut St, 8th Floor, Philadelphia, PA 19106, USA.

Ocean Realm: Friends of the Sea, Inc, 4067 Broadway, San Antonio, TX 78209, USA; tel (210) 824-8099; fax (210) 820-3522; e-mail oceanica@eden.com; Editor: Charlene deJori; Publr: Cheryl Schorp; f 1988; quarterly; protection of marine ecosystems; circ 30,000.

Ocean Science News: Nautilus Press, Inc, 1201 National Press Bldg, Washington, DC 20045, USA; tel (202) 347-3043; Editor: John R Botzum, Sr; 3 a month.

Ocean Yearbook: University of Chicago Press, Journals Division, 1427 E 60th St, Chicago, IL 60637, USA; tel (773) 702-7600; fax (773) 702-0172; e-mail subscriptions@journals.uchicago.edu; internet www.press.uchicago.edu; Editors: Aldo Chircop and Moira L McConnell; f 1978; annual; International Ocean Institute and Dalhousie University Law School; published for the International Ocean Institute (Malta); devoted to the assessment of the resources, ecology and strategic importance of the world's oceans; includes reports from relevant organizations, selected documents and proceedings, and tabular data on marine activities and resources; in English.

Ochrona Przyrody (Nature Conservation): Instytut Ochrony Przyrody, Polska Akademia Nauk, 31-512 Kraków, ul Lubicz 46, Poland; tel and fax (12) 4210348; e-mail nojuchie@cyf-kr.edu.pl; internet botan.ib-pan.krakow.pl/przyroda/index_htm; Editor-in-Chief: Prof Dr Zofia Alexandrowicz; Assoc Editors Halina Piekos-Mirkowa, Zbigniew Witkowski, Zbigniew Dzwonk, Henryk Okarma; f 1920; annual; publication of the Institute of Nature Conservation of the Polish Academy of Sciences; includes research on wildlife conservation and documentation of protected species and sites in Poland; monitors changes in the natural environment, identifies dangers and suggests appropriate courses of action for conservationists; in Polish and English; circ 400.

Oecologia: Springer-Verlag, 14197 Berlin, Heidelberger Pl 3, Germany; tel (30) 8207730; fax (30) 8207300; e-mail subscriptions@springer.de; internet link.springer.de/link/service/journals/00442/index.htm; f 1968; 20 a year; the International Association for Ecology—INTECOL; describes advances in research into the interaction between plant and animal species and their environment; in English; circ 1,000.

OET al Día (OTS Up to Date): Organization for Tropical Studies—OTS, Apdo 676, 2050 San Pedro, Costa Rica; tel 240-6696; fax 240-6783; e-mail comunica@ots.ac.cr; internet www.ots.ac.cr; Editors: Silvia Alvaredo, José María Rodríguez, Rodney Vargas; f 1983; 2 a year; contains news items on OTS courses, field stations and current research; in Spanish; circ 2,000.

Oikos, A Journal of Ecology: Oikos Nordic Ecological Society, Lund University, Dept of Ecology, Ecology Bldg, 223 62 Lund, Sweden; tel (46) 222-37-91; fax (46) 222-37-90; e-mail oikos@ekol.lu.se; internet www.oikos.ekol.lu.se; Editor: Nils Malmer; Man Editor: P H Enckell; f 1949; 12 a year; in English; circ 1,350. Publishing house: Munksgaard International Publishers Ltd, 35 Nørre Søgade, POB 2148, 1016 Copenhagen K, Denmark; tel 77-33-33-33; fax 77-33-33-77.

Oil Spill Intelligence Report: Cutter Information Corpn, 37 Broadway, Suite 1, Arlington, MA 02174-5539, USA; tel (781) 648-8700; fax (781) 648-1950; e-mail crowley@cutter.com; internet www.cutter.com; Editor: Jim Polson; f 1978; weekly; documents petroleum spills; concerned with related contingency planning and legislation; electronic formats available.

Okayama no Shizen (Nature in Okayama): Okayama no Shizen o Mamoru Kai, 1-21, Tsushima Minami 1-chome, Okayama-shi, Okayama-ken 700, Japan; every 2 months; published by the Study Group for Nature Protection in Okayama; in Japanese.

ÖkologiePolitik: Ökologisch-Demokratische Partei, Bundesgeschäftsstelle, 97070 Würzburg, Bohnesmühlgasse 5, Germany; tel (931) 40486-0; fax (931) 40486-25; e-mail florence.bodisco@oedp.de; internet www.oedp.de; Editor: Raphael Mankau; f 1984; every 2 months; former title *Ökologie und Politik*; publication of the Ecological-Democratic Party; concerned with political ecology; provides information for party members and all those with an interest in environmental politics; in German; available as PDF; circ 7,500. Publishing house: Ecological-Democratic Party (Germany).

Öko-Mitteilungen—Informationen aus dem Institut für angewandte Ökologie eV: Institut für angewandte Ökologie—Öko-Institut eV, 79038 Freiburg, Postfach 6226, Germany; tel (761) 452950; fax (761) 475437; Editor: Jörn Ehlers; f 1977; quarterly; bulletin of Institute for Applied Ecology; circ 7,000.

Öko-Test-Magazin für Gesundheit und Umwelt (Eco-Test Magazine for Health and the Environment): Öko-Test-Verlag, Öko-Haus, 60486 Frankfurt am Main, Germany; tel (69) 977770; fax (69) 97777139; e-mail oet.verlag@oekotest.de; internet www.oekotest.de; Editor: Jürgen Stellpflug; f 1985; monthly; ecological and health consumers' magazine; product tests; in German; circ 150,000.

On the Edge: Wildlife Preservation Trust International, 1520 Locust St, Suite 704, Philadelphia, PA 19102, USA; tel

(215) 731-9770; fax (215) 731-9766; e-mail homeoffice@wpti.org; internet www.columbia.edu/cu/cerc/wpti.html; Editor: Mary C Pearl; f 1976; 3 a year; circ 3,000.

On Track: the Environmental Magazine: POB 1290, Kengray 2100, South Africa; f 1994; every 2 months; in English.

One Country: Bahá'í International Community, Office of Public Information, 866 United Nations Plaza, Suite 120, New York, NY 10017, USA; tel (212) 803-2543; fax (212) 803-2566; e-mail 1country@bic.org; internet www.onecountry.org; Editor: Brad Pokorny; f 1989; quarterly; newsletter of the Bahá'í International Community; religious publication with items on environmental conservation and sustainable development; available in English and Russian editions; available online; circ 30,000.

OnEarth: Natural Resources Defense Council—NRDC, 40 West 20th St, New York, NY 10011, USA; tel (212) 727-2700; fax (212) 727-1773; e-mail onearth@nrdc.org; internet www.nrdc.org/onearth; Editor-in-Chief: Douglas S Barasch; f 1979; quarterly; intended for members of the NRDC and other interested parties; includes indexes and book reviews; circ 175,000.

Open Space: Open Spaces Society, 25a Bell St, Henley-on-Thames RG9 2BA, United Kingdom; tel (1491) 537535; e-mail osshq@aol.com; internet www.oss.org.uk; Editor: Kate Ashbrook; access to public open spaces; circ 2,300.

Options: International Institute for Applied Systems Analysis—IIASA, 2361 Laxenburg, Schlosspl 1, Austria; tel (2236) 80-70; fax (2236) 73-149; e-mail options@iiasa.ac.at; Editor: Mary Ann Williams; f 1978; quarterly; publishes findings of IIASA's research; available online; circ 8,000.

Organic Gardening Magazine: Rodale Press, Inc, 33 East Minor St, Emmaus, PA 18098, USA; tel (610) 967-5171; fax (610) 967-7846; Editor-in-Chief: Mike McGrath; Pres, Magazine Div: John Griffin; Publr: Fritz Craiger; f 1942; 9 a year; circ 721,000.

Organic Grower: Organic Growers' Association of Western Australia, POB 213, Wembley, Perth, WA 6014, Australia; fax (8) 9397-6005; Editor: Julie Woodman; f 1977; quarterly; aims to demonstrate that practical alternatives exist for gardeners and farmers concerned by the increasing use of harmful chemicals; encourages soil restoration; circ 245.

Orion Afield: Orion Society, 195 Main St, Great Barrington, MA 01230, USA; tel (413) 528-4422; fax (413) 528-9667; e-mail orion@orionsociety.org; internet www.orionsociety.org; Editor: Jennifer Sahn; f 1995; quarterly; newsletter; formerly *Orion Society Notebook*; includes *Orion*.

Oryx: Fauna and Flora Preservation Society, c/o Zoological Society of London, Regent's Park, London NW1 4RY, United Kingdom; tel (20) 7586-0872; Editor: Jacqui Morris; f 1950; quarterly; journal of the Flora and Fauna Preservation Society, covering all aspects of wildlife conservation. Publishing house: Blackwell Science Ltd, Osney Mead, Oxford OX2 0EL.

Österreichische Wasser- und Abfallwirtschaft: Springer-Verlag Komm Gesellschaft, 1200 Vienna, Sachsenpl 4–6, POB 89, Austria; tel (1) 330-24-15; fax (1) 330-24-26; Editors: W Stalfer, F Oberleitner, H Werner; f 1949; monthly; former title *Österreichische Wasserwirtschaft*; publication dedicated to water supply and management; circ 2,300.

OTS Liana: Organization for Tropical Studies—OTS, POB 90630, Durham, NC 27708-0630, USA; tel (919) 684-5774; fax (919) 684-5661; e-mail nao@acpub.duke.edu; internet www.ots.duke.edu; Editors: C Cheatham, J Giles, C Mozell; f 1963; 2–3 a year; includes updates on the activities of the OTS biological field stations, information on research funding and travel opportunities, details of the OTS education programme, general interest articles on tropical biology and tropical studies; circ 3,500.

Our Planet/Notre Planète/Nuestro Planeta: UNEP Information Service, POB 30552, Nairobi, Kenya; fax (2) 226831; e-mail anthony@smibooks.com; internet www.unchs.union.org; Editor: Shane Cave; f 1974; quarterly; encourages global awareness of all aspects of the contemporary environmental crisis; available in English, French and Spanish editions; available online; circ 17,000.

Outdoor Alabama: Conservation Department, 64 North Union St, Montgomery, AL 36130, USA; tel (334) 242-3151; fax (334) 242-1880; e-mail magazine@outdooralabama.com; internet www.outdooralabama.com; Editor: Kim G Nix; f 1929; 5 issues a year; natural resources, hunting, fishing, Alabama State parks, conservation issues, endangered species; focus on Alabama's vast natural resources; circ 10,000. Publishing house: Alabama Department of Conservation and Natural Resources.

Outdoor America: Izaak Walton League of America, 707 Conservation Lane, Gaithersburg, MD 20878-2983, USA; tel (301) 548-0150; fax (301) 548-0146; internet www.iwla.org; Editor: Zachary Hoskins; f 1924; quarterly; current events in the USA concerning conservation, environmental and recreation; circ 50,000.

Outdoor Delaware: Dept of Natural Resources and Environmental Control, 89 Kings Highway, POB 1401, Dover, DE 19903, USA; tel (302) 739-4506; fax (302) 739-6242; e-mail dnrec@state.de.us; Editor and Publr: Kathleen M Jamison; f 1956; quarterly; formerly *Delaware Conservationist*; circ 15,000.

Outdoor Illinois: Department of Natural Resources, 524 South Second St, Rm 510, Springfield, IL 62701-1787, USA; tel (217) 782-7454; fax (217) 782-9552; e-mail editor@dntmail.state.il.us; internet www.dnr.state.il.us; Editor: Gary Thomas; f 1972; monthly; circ 25,000.

Outdoor Indiana: Department of Natural Resources, 402 West Washington St, W 2558, Indianapolis, IN 46204-2748, USA; tel (317) 232-4004; fax (317) 232-8036; Editor: Steve Sellers; f 1934; every 2 months; circ 30,000.

Outdoor News: Oklahoma Wildlife Federation, POB 60126, Oklahoma City, OK 73146-0126, USA; tel (405) 524-7009; fax (405) 521-9270; e-mail owf@nstar.net; Editor: Lance Meek; f 1951; 9 a year; circ 10,000.

Outdoor News Bulletin: Wildlife Management Institute, 1101 14 St, NW, Suite 801, Washington, DC 20005, USA; tel (202) 371-1808; fax (202) 408-5059; e-mail wmihq@aol.com; internet www.wildlifemgt.org; Editor: Richard E McCabe; f 1947; monthly; circ 4,000.

Outdoor Oklahoma: Department of Wildlife Conservation, 1801 North Lincoln, Oklahoma City, OK 73105, USA; tel (405) 521-3855; fax (405) 521-6898; Editor: Nels Rodefeld; f 1945; every 2 months; circ 21,000.

Ozone News: International Ozone Association—IOA, Editorial Office, 7534 Squirrel Creek Drive, Cincinnati, OH 45247-3611, USA; tel and fax (513) 385-3906; e-mail blloeb@fuse.net; internet www.int-ozone-assoc.org; Editor: Barry L Loeb; f 1973; every 2 months; circ 1,500.

P2—Pollution Prevention Review: John Wiley & Sons, Inc, Journals, 605 Third Ave, New York, NY 10158, USA; tel

(212) 850-6645; fax (212) 850-6021; e-mail subinfo@wiley.com; internet www.wiley.co.uk; Editor: Ginger Griffin; f 1990; quarterly; former title *Pollution Prevention Review*; incorporates *Environmental Manager*; comprises articles intended for pollution prevention specialists, managers in industry, consultants, researchers and government agency representatives.

Pacific Conservation Biology: Surrey Beatty and Sons, 43 Rickard St, Chipping Norton, NSW 2170, Australia; tel (2) 9602-3888; fax (2) 9821-1253; e-mail surreybeatty@iform.com.au; internet wwwscience.murdoch.edu.au/centres/others/pcb; Editor: Harry Recher; Publr: Ivor Beatty; f 1993; quarterly; articles and research on conservation and land management in the Pacific; circ 500.

Pakistan Journal of Forestry: Pakistan Forest Institute, Peshawar, Pakistan; tel (91) 9216123; fax (91) 9216203; e-mail pfilib@hotmail.com; Editor: K M Siddiqui; f 1951; 2 a year; provides researchers and students with information about forestry, the environment and allied disciplines; in English; circ 400.

Pan European Biological and Diversity Strategy, Bulletin: Council of Europe Publishing, Council of Europe, 67075 Strasbourg Cédex, France; tel 3-88-41-25-81; fax 3-88-41-39-10; e-mail publishing@coe.fr; internet book.coe.fr; Editor: Jean-Pierre Ribaut; f 1997; every 2 months; available in English, French, available in German, Italian and Russian editions; circ 20,000.

Panos Features: Panos Institute, 9 White Lion St, London N1 9PD, United Kingdom; tel (20) 7278-1111; fax (20) 7278-0345; e-mail info@panoslondon.org.uk; internet www.panos.org.uk; Editor-in-Chief: Dipankar De Sarkar; Asst Editor: Alex Whiting; monthly; environment and development news.

Panos Media Briefings: Panos Institute, 9 White Lion St, London N1 9PD, United Kingdom; tel (20) 7278-1111; fax (20) 7278-0345; e-mail info@panoslondon.org.uk; internet www.panos.org.uk; 6 a year; environment, development and health news.

Panoscope—PS: Panos Institute, 9 White Lion St, London N1 9PD, United Kingdom; tel (20) 7278-1111; fax (20) 7278-0345; e-mail markc@panoslondon.org.uk; internet www.panos.org.uk; Editor: Kelly Haggart; Editorial Dir: Joe Hanlon; Publications Dir: Liz Carlile; quarterly; environment and development magazine; includes *Focus* section containing several articles on a single, specific topic.

Paper Recycling Markets Directory: GIE, Inc, 4012 Bridge Ave, Cleveland, OH 44113-3320, USA; tel (216) 961-4130; fax (216) 961-0364; internet www.recyclingtoday.com; f 1963.

Paperbark: North Queensland Conservation Council, Inc, 340 Flinders Mall, POB 364, Townsville, Qld 4810, Australia; tel (7) 4771-6226; fax (7) 4721-1713; e-mail nqcc@byte-tsv.net.au; internet www.nqcc.org.au; Editor: Carolyn Pike; f 1975; every 2 months; formerly *Conservation North Queensland*; concerned with world heritage sites, the Great Barrier Reef Marine Park, mining, tree clearance, rainforests, aquaculture, trawling, endangered species, coastal development and biodiversity; in English; circ 300.

Parks: Nature Conservation Bureau Ltd, 36 Kingfisher Court, Hambridge Rd, Newbury RG14 5SJ, United Kingdom; tel (1635) 550380; fax (1635) 550230; e-mail parks@naturebureau.co.uk; Editor: Paul Goriup; f 1990; 3 a year; published in association with the World Commission on Protected Areas and the World Conservation Union—IUCN; international publication concerned with the world's protected areas; in English, summaries in English, French and Spanish; circ 1,700.

Parks: US National Park Service, Interior Bldg, Washington, DC 20240, USA; tel (202) 343-1100; f 1976; quarterly; international journal for staff of national parks, historic sites and protected areas.

Peak and Prairie: Sierra Club, Rocky Mountain Chapter, 1410 Grant St, Suite B-205, Denver, CO 80203-1846, USA; tel (303) 861-8819; fax (303) 861-2436; e-mail kfmacke@bewellnet.com; internet www.rmc.sierraclub.org; Editors: Kevin Mackessy, Lou Snyder; Publr: Diane Benjamin; every 2 months; circ 14,000.

Pedosphere: Institute of Soil Science, Chinese Academy of Sciences, POB 821, 71 East Beijing Rd, Nanjing 210008, People's Republic of China; tel (25) 8688-1256; (25) 8688-1235; fax (25) 8688-1256; e-mail rmdu@issas.ac.cn; pedo@issas.ac.cn; internet www.pedosphere.issas.ac.en; www.periodicals.net.cn; www.cnki.net; Editor-in-Chief: Prof Jian-Min Zhou; Deputy Editors-in-Chief: Prof Du Rongmin, Prof B A Stewart; f 1991; quarterly –2004, bimonthly 2005–; sponsored jointly by the Soil Science Society of China, the Institute of Soil Science (Chinese Academy of Sciences); and the State Key Laboratory of Soil and Sustainable Agriculture (China); in co-operation with several important universities and research institutions in China; timely high quality original research findings, especially up-to-date achievements and advances in all fundamental and applied aspects of soil science, including original papers, reviews and short communications; in English; circ 2,500. Publishing house: Science Press, 16 Donghuangchenggen N St, Beijing 100717, People's Republic of China.

People and the Planet: Planet 21, Suite 112, Spitfire Studios, 63–71 Collier St, London N1 9BE, United Kingdom; tel (20) 7713-8108; fax (20) 7713-8109; e-mail planet21@netcomuk.co.uk; internet www.peopleandplanet.net; Editor: John Rowley; Asst Editor: Maya Pastakia; f 1992; quarterly; sponsored by the World Wide Fund for Nature—WWF International (Switzerland), the UN Population Fund (USA), the International Planned Parenthood Federation (United Kingdom), the Hewlett Foundation and the Swedish Development Co-operation Agency; concerned with the links between population, reproductive health, development and the environment; in English; available only online; circ 15,000.

Peregrine Fund Newsletter: The Peregrine Fund, Inc, 5668 West Flying Hawk Lane, Boise, ID 83709, USA; tel (208) 363-3716; fax (208) 362-2376; e-mail tpf@peregrinefund.org; internet www.peregrinefund.org; Editor: William Burnham; f 1970; 2 a year; reports on activities of The Peregrine Fund; available online; circ 6,000.

Perspectives in Energy: Moscow International Energy Club—MIEC, 127412 Moscow, Izhorskaya 13–19, Russian Federation; tel (095) 485-96-63; fax (095) 485-79-36; e-mail shpilrn@termo.msk.su; internet turpion.ioc.ac.ru; Editor-in-Chief: A Sheindlin; Assoc Editors: G Aslanian, S Malyshenko; f 1991; irreg; intends to inform scientists, politicians, business people, journalists and the general public of developments and problems in the field of energy; area of interest includes the environmental impact of energy development; in English; available online; circ 150. Publishing house: Aleksandr Sheindlen & Anatolii Diakov, Editors and Publrs, Moscow, Russian Federation.

Pesticide Outlook: Royal Society of Chemistry, Thomas Graham House, Science Park, Milton Rd, Cambridge CB4 4WF, United Kingdom; tel (1223) 432360; fax (1223) 423429; e-mail cockheadg@rsc.org; internet chemistry.rsc.org/rsc;

Editor: Hamish Kidd; f 1989; every 2 months; journal of current developments in all aspects of pesticides and their use; circ 400.

Pesticides News: Eurolink Centre, 49 Effra Rd, London SW2 1BZ, United Kingdom; tel (20) 7274-8895; fax (20) 7274-9084; internet www.pan-uk.org; Editor: David Duffin; quarterly; UK, European and international pesticides issues; in English; articles available on website; circ 2,000.

Phoenix: Voice of the Scrap Recycling Industries: Institute of Scrap Recycling Industries, Inc, 1325 G St, SE, Suite 1000, Washington, DC 20005, USA; tel (202) 737-1770; fax (202) 626-0900; Editor: David Krohne; f 1969; 2 a year; circ 40,000.

Physiological Zoology: University of Chicago Press, Journals Division, POB 37005, Chicago, IL 60637, USA; tel (312) 702-3347; fax (312) 753-0811; e-mail subscriptions@journals.uchicago.edu; internet www.journals.uchicago.edu/PZ/home.html; Editor: Gregor K Snyder; f 1928; every 2 months; presents recent, original research in environmental, adaptational and comparative physiology and biochemistry; in English; circ 1,400.

Physiology and Ecology Japan: Kyoto Daigaku Rigakubu Dobutsugaku Kyoshitsu, Kita-Shirakawa, Sakyo-ku, Kyoto 606, Japan; fax (75) 751-6149; Editor: Hiroya Kawanabe; f 1947; 2 a year; published by Kyoto University Department of Zoology; deals with all aspects of physiology and ecology; in English; circ 700.

The Pilot: UNEP, POB 30552, Nairobi, Kenya; tel (2) 621234; fax (2) 226890; f 1987; quarterly; newsletter of the Marine Mammal Action Plan; circ 5,000.

The Pines Post: Pinelands Preservation Alliance, 114 Hanover St, Pemberton, NJ 08068, USA; tel (609) 894-8000; Editor: Lisa Thibault; f 1996; 2 a year; contains news items and information on activities, events, funding and legislative issues relating to the work of the Pinelands Preservation Alliance.

Planning and Design: Environment and Planning B: Pion Ltd, 207 Brondesbury Park, London NW2 5JN, United Kingdom; tel (20) 8459-0066; fax (20) 8451-6454; e-mail sales@pion.co.uk; internet www.pion.co.uk/ep; Editor: Michael Batty; Co-Editors: Michael Breheny, Helen Couclelis; Reviews Editor: Paul Longley; Publishing Dir: J Schubert; f 1974; every 2 months; international journal of the theory and practice of urban and regional planning and design.

Plant Genetic Resources Newsletter: International Board for Plant Genetic Resources Institute, Via dei Tre Denari 472/a, 00057 Maccarese, Rome, Italy; tel (06) 6118233; fax (06) 61979661; e-mail p.neate@cgiar.org; internet www.ipgri.cgiar.org; Editor: Paul Neate; f 1957; quarterly; contains research papers, articles and reports on work relating to plant genetic resources; includes reports of conferences, news items, abstracts, bibliographies and book reviews; available in English, French and Spanish editions; circ 5,000. Publishing house: IPGRI/FAO.

Plastics Recycling Update: Resource Recycling, Inc, POB 42270, Portland, OR 97242-0270, USA; tel (503) 233-1305; fax (503) 233-1356; e-mail info@resource-recycling.com; internet www.resource-recycling.com; Editor: Katie Sparks; Publr: Judy Roumpf; f 1987; monthly; covers the recycling of all types of plastics, market pricing and trends, technology, corporate developments, legislation, etc; circ 2,000.

Polish Journal of Ecology: Centre for Ecological Research, Polish Academy of Sciences, 05-092 Lomianki, Dziekanów Lesny, Poland; tel (22) 7513046; fax (22) 7513100; e-mail cbe@cbe-pan.pl; internet www.cbe-pan.pl; Editor-in-Chief: Anna Hillbricht-Ilkowska; Assoc Editors: Krassimira Ilieva-Makulec, Pawel Prus; f 1952; quarterly; the Institute of Ecology of the Polish Academy of Sciences; former title *Ekologia Polska*; publishes original scientific papers dealing with all aspects of ecology, both fundamental and applied, physiological ecology, ecology of population, community, landscape as well as global ecology; there is no bias regards taxon, environment or geographical area; English; circ 500.

Pollution Abstracts: Cambridge Scientific Abstracts, 7200 Wisconsin Ave, Bethesda, MD 20814-4823, USA; tel (301) 961-6750; fax (301) 961-6720; e-mail market@csa.com; internet www.csa.com; Editor: Evelyn Beck; Publr: Ted Caris; f 1970; monthly; contains summaries of articles on scientific research and government policy relating to all aspects of pollution and pollution control; electronic formats available.

Pollution Engineering: Cahners Business Information, 2000 Clearwater Drive, Oak Brook, IL 60523, USA; tel (630) 320-7154; fax (630) 320-7150; e-mail polengineering@cahners.com; internet www.pollutionengineering.com; Editor: John Krukowski; Publr: Diane Pirocanac; f 1969; 13 a year; technical journal; annual supplement concerning relevant equipment; circ 70,000.

Pollution Equipment News: Rimbach Publishing, Inc, 8650 Babcock Blvd, Pittsburgh, PA 15237-5821, USA; tel (412) 364-5366; fax (412) 369-9720; e-mail rimbach@sgi.net; internet www.rimbach.com; Editor: David C Lavender; Pres: Norberta Rimbach; f 1968; every 2 months; available online; circ 91,000.

Pollution Research: Enviro Media, 2nd Floor, Rohan Heights, POB 90, Karad 415 110, India; tel (2164) 44369; fax (2164) 71645; Editor: R K Trivedy; f 1986; quarterly; international journal published for the Indian Association for Pollution Chemists and Biologists; in English; circ 2,000.

Polymer Recycling: RAPRA Technology Ltd, Shawbury, Shrewsbury SY4 4NR, United Kingdom; tel (1939) 250383; fax (1939) 251118; e-mail kmevans@rapra.net; internet www.rapra.net; Editor: Kate Evans; f 1995; quarterly; concerned with recent scientific and technological advances in plastics and rubber recovery and recycling.

Power in Asia: Financial Times Energy, Maple House, 149 Tottenham Court Rd, London W1P 9LL, United Kingdom; fax (20) 7896-2275; internet www.ftenergy.com; Editor: Frank Gray; every 2 weeks (25 a year); available online.

Power in Europe: Financial Times Energy, Maple House, 149 Tottenham Court Rd, London W1P 9LL, United Kingdom; fax (20) 7896-2275; internet www.ftenergy.com; Editor: Niamh Kenny; every 2 weeks; available online.

Practical Alternatives: Victoria House, Bridge St, Rhayader LD6 5AG, United Kingdom; tel (1597) 810929; Editor and Publr: David Stephens; f 1981; irreg; reports on the construction of a village of passive solar houses designed to facilitate significant reductions in energy consumption; circ 2,000.

Practice Periodical of Hazardous, Toxic and Radioactive Waste Management: American Society of Civil Engineers, Alexander Graham Bell Drive, Reston, VA 20191, USA; tel (703) 295-6300; e-mail marketing@asce.org; internet www.asce.org; Editor: Yee Cho; f 1997; quarterly; articles concerning the environmental aspects of hazardous and radioactive waste management; electronic formats available.

Prairie Forum: Canadian Plains Research Center, University of Regina, Regina, SK S4S 0A2, Canada; tel (306) 585-4795; fax (306) 585-4699; e-mail patrick.douaud@uregina.ca;

internet www.cprc.uregina.ca; Editor: Patrick Douaud; f 1976; 2 a year; journal of interdisciplinary research relating to the interaction of humans and nature in the Canadian plains region; in English and French; circ 300.

Priroda (Nature): 117069 Moscow, Maronovskii per 26, Russian Federation; tel (095) 238-23-33; fax (095) 238-26-33; Editor: A F Andreyev; f 1912; monthly; the Presidium of the Academy of Sciences; popular natural science publication; in Russian; circ 2,500. Publishing house: Nauka (Science) Publishing House, 117864 Moscow, ul Profsoyuznaya 90, Russian Federation; tel (095) 336-02-66; fax (095) 420-22-20.

Priroda, casopis za popularizaciju prirodoslovlja i ekologije: 10000 Zagreb, Rooseveltov trg 6, Croatia; tel (1) 442804; fax (1) 4552645; e-mail oskar.springer@public.srce.hr; Contact: Dr Oskar Springer; nature magazine.

Pro Natura Genova: Pro Natura Genova, c/o Museo di Storia Naturale, Via Brigata Liguria 9, 16121 Genoa, Italy; Dir: Pierluigi Oneto; f 1976; quarterly; members' bulletin; contents include book reviews and statistics; circ 1,000.

Pro Natura Magazin: Pro Natura—Schweizerischer Bund für Naturschutz, Postfach, 4020 Basel, Switzerland; tel (61) 3179191; fax (61) 3179166; f 1935; every 2 months; former title *Schweizer Naturschutz/Protection de la Nature*; ensures communication between Pro Natura (the Swiss League for Nature Protection), its sections and members; special issues on specific subjects; in German and French.

Problems of Desert Development: Allerton Press, Inc, 150 Fifth Ave, New York, NY 10011, USA; tel (212) 924-3950; fax (212) 463-9684; internet www.allerton.com; Editor: A G Babayev; f 1980; every 2 months; translation of *Problemy Osvoyeniya Pustyn*, a Russian-language publication of the Desert Research Institute, Turkmen Academy of Sciences (Turkmenistan); concerned with arid zone geography and exploration, monitoring of desert ecosystems, phytomelioration of desert pastures, desert soil degradation and its prevention, anthropogenic factors in desert environments, drift sands, afforestation, evolution and biological diversity, and desert flora and fauna; in English.

Problemy Bol'shikh Gorodov (Problems of Large Metropolitan Areas): Moskovskii Gorodskoi Territorial'nyi Tsentr Nauchno-Tekhnicheskoi Informatsii i Propagandy, 101958 Moscow, pr Serova 5, Russian Federation; tel (095) 921-67-05; Editor: Irina P Mikhailova; f 1971; 2 a month; general survey dealing with the development of urban environmental protection, urban traffic improvement, problems related to energy and water supply, and recreation; circ 800.

Process Safety and Environmental Protection: Institution of Chemical Engineers—IChemE, Davis Bldg, 165–189 Railway Terrace, Rugby CV21 3HQ, United Kingdom; tel (1788) 578214; fax (1788) 560833; e-mail jcheshire@icheme.org.uk; internet www.icheme.org; Exec Editor: Richard Wakeman; Hon Editor: John Garside; f 1990; every 2 months; deals with clean-process and minimum-waste technology, solid waste management, water and waste-water treatment, air pollution, atmospheric dispersion, safety, industrial hygiene, dust explosions, hazardous reactions, hazard and risk analysis, and noise; available in electronic formats; circ 940.

Progress in Environmental Science: Arnold, 338 Euston Rd, London NW1 3BH, United Kingdom; tel (20) 7873-6000; fax (20) 7873-6325; e-mail arnoldjournals@hodder.co.uk; internet www.arnoldpublishers.com/journals/journpages/14604094.htm; Man Editors: Donald A Davidson, Michael F Thomas; f 1999; quarterly; available online.

Project Appraisal: Beech Tree Publishing, 10 Watford Close, Guildford GU1 2EP, United Kingdom; tel and fax (1483) 567497; Editors: Dr Vary Coates, Dr John Weiss; Publr: William Page; f 1986; quarterly; international journal concerned with environmental impact assessment and related fields such as cost-benefit analysis and technology assessment; contains articles, book reviews and details of forthcoming conferences.

PROTERRA: Boletín Jurídico Ambiental: Calle Madrid 166, Miraflores, Lima 18, Peru; tel (1) 4466363; fax (1) 2420238; e-mail postmaster@proter.org.pe; f 1993; quarterly; environmental law; circ 1,000.

PROVITA Bulletin: PROVITA, Edif Catuche, 1°, Of 105–106, Apdo 47552, Caracas 1041-A, Venezuela; tel (2) 576-2828; fax (2) 576-1579; e-mail provita@telcel.net.ve; f 1987; quarterly; conservation of wildlife and protected areas; environmental education; management of protected areas; in Spanish; circ 600.

Public Lands News: Resources Publishing Co, 1010 Vermont Ave, NW, Suite 708, Washington, DC 20005, USA; tel (202) 638-7529; fax (202) 393-2075; e-mail coffinj@clark.net; internet www.plnfpr.com; Editor: James B Coffin; f 1976; every 2 weeks.

Pulse of the Planet: Orgone Biophysical Research Laboratory, POB 1148, Ashland, OR 97520, USA; tel and fax (541) 552-0118; e-mail demeo@mind.net; Editor: James DeMeo; f 1989; quarterly; research report and journal containing original and reprinted articles on social, environmental and geophysical subjects.

Quarterly Journal of Forestry: Royal Forestry Society of England, Wales and N Ireland, 102 High St, Tring HP23 4AH, United Kingdom; tel and fax (1939) 261722; e-mail qjf@ancesorsuk.com; Editor: Lesley Trotter; f 1906; quarterly; scientific and general articles on forestry and woodland management in England Wales and Northern Ireland; circ 5,000.

Rachel's Environment and Health Weekly: Environmental Research Foundation, POB 5036, Annapolis, MD 21403-7036, USA; tel (410) 263-1584; fax (410) 263-8944; e-mail erf@rachel.org; internet www.rachel.org; Editor: Peter Montague; f 1986; weekly; in English and Spanish; available online.

Radiation Protection Dosimetry: Nuclear Technology Publishing, POB 7, Ashford TN23 1YW, United Kingdom; tel (1233) 641683; fax (1233) 610021; e-mail subscriptions@ntp.org.uk; internet www.ntp.org.uk; Editor: Dr J C McDonald; f 1981; every 2 weeks; all aspects of personal and environmental radiation dosimetry; circ 700.

Radiological Protection Bulletin: National Radiological Protection Board, Chilton, Didcot OX11 0RQ, United Kingdom; tel (1235) 831600; fax (1235) 833891; e-mail publications@nrpb.org.uk; internet www.nrpb.org.uk; Editor: Michael C O'Riordan; Rights Officer: A Sharma; f 1972; monthly; concerned with radiological protection, non-ionizing radiation, radon, medical uses of radiation, biological studies and environmental assessments; includes news, reviews, articles and reports; circ 1,200.

Radwaste Magazine: American Nuclear Society, 555 North Kensington Ave, La Grange Park, IL 60525, USA; tel (708) 352-6611; fax (708) 352-6464; e-mail dschabes@ans.org; Editor-in-Chief: Gregg M Taylor; Publr: Jon Payne; f 1994; quarterly; contains reports and proceedings relevant to nuclear waste management, industry news, technical notes and information on products and services; circ 1,500.

Rain: POB 30097, Eugene, OR 97403-1097, USA; Editors: Greg Bryant, Danielle Janes; f 1974; quarterly; ceased publication 1987, resumed 1991; contains articles which aim to encourage environmental improvement; circ 8,000.

Ramsar Newsletter: Ramsar Convention Bureau, c/o World Conservation Union—IUCN, 28 rue Mauverney, 1196 Gland, Switzerland; tel (22) 9990170; fax (22) 9990169; e-mail ramsar@hq.iucn.org; internet iucn.org/themes/ramsar; f 1971; 2–3 a year; publication of the Ramsar Convention on Wetlands of International Importance, Especially as Waterfowl Habitat; concerned with the conservation and sustainable use of wetlands; in English, French and Spanish; circ 5,000.

RCO Update: Recycling Council of Ontario, 489 College St, Suite 504, Toronto, ON M6G 1A5, Canada; tel (416) 960-1025; fax (416) 960-8053; e-mail rco@web.net; internet www.web.net/rco; Editor: Katharine Partridge; Publr: Brenda Berry; f 1981; monthly; former title *Ontario Recycling Update*; members' bulletin; concerned with reduction, reuse and recycling of waste; circ 800.

Reclamation Newsletter: Canadian Land Reclamation Association, POB 61047, Kensington PO, Calgary, AB T2N 4S6, Canada; tel (403) 427-5553; fax (403) 422-8233; Editor: Chris Powter; f 1977; 3 a year; co-published by the American Society for Surface Mining and Reclamation; includes statistical tables and book reviews; in English and French; circ 1,200.

Recover: Recover Enterprises, 114 Dollery Court, North York, ON M2R 3P1, Canada; Editors: Nancy Phillips, Heather Sangster; f 1990; quarterly; environmental magazine.

Recyclage Magazine: Les Editions Montmartre, 142 rue Montmartre, 75080 Paris, France; tel 1-40-26-83-21; fax 1-40-39-97-52; Editor: Claude Platier; f 1995; 9 a year; concerned with the recovery, recycling and pricing of all solid wastes; in French; circ 6,000.

Recyclage-Récupération (Recycling-Reclamation): Les Editions Montmartre, 142 rue Montmartre, 75080 Paris, France; tel 1-40-26-83-21; fax 1-40-39-97-52; Editor: Claude Platier; Publr: Simone Leportier; f 1909; weekly (47 a year); waste reclamation bulletin, including listings of all solid waste prices; in French; circ 5,000.

Recycling: Verlagsgruppe Handelsblatt GmbH, 40213 Düsseldorf, Kasernenstr 67, Germany; tel (211) 8870; Editor: Rolf Willeke; 3 a year; former title *Schrottbetrieb*; published for Bundesverband der Deutschen Schrottwirtschaft (National Association of German Scrap Metal Traders); includes statistics and trade information; circ 1,000.

Recycling and Resources Management: Taylor Marketing Ltd, Oakwood House, 3 Moulton Park Office Village, Northampton, NN3 1AP, United Kingdom.

Recycling Today: GIE, Inc, 4012 Bridge Ave, Cleveland, OH 44113-3320, USA; tel (216) 961-4130; fax (216) 961-0364; e-mail btaylor@recyclingtoday.com; internet www.recyclingtoday.com; Editor: Brian Taylor; Publr: James R Keefe; f 1963; monthly; concerned with the secondary commodity processing and recycling markets, from traditional scrap metal and paper to post-consumer municipal waste and construction and demolition debris; available online; circ 18,000.

Recycling World Magazine: The Scrap Market Ltd, Hilltop, off Church Rd, Webheath, Redditch B97 5PQ, United Kingdom; tel (1527) 404550; fax (1527) 404644; e-mail recycling@tecweb.com; internet www.tecweb.com/recycle/eurorec.htm; Editor: Jackie Oliver; Publr: G Shaw; f 1987; every 2 weeks; trade magazine for the commercial recycling industry; available online; circ approx 6,150.

Références INERIS: Institut National de l'Environnement Industriel et des Risques—INERIS, Parc Technologique ALATA, BP 2, 60550 Verneuil-en-Halatte, France; tel 3-44-55-66-77; fax 3-44-55-66-99; e-mail ineris@ineris.fr; internet www.ineris.fr; Editor: Marie-Pierre Bigot; f 1991; 2 a year; concerned with air pollution, ecotoxicology, geotechnics, industrial hazards and health and the environment; circ 12,000.

Reflets Sahéliens: Comité Permanent Inter-Etats de Lutte contre la Sécheresse dans le Sahel—CILSS, 03 BP 7049, Ouagadougou 03, Burkina Faso; tel 30-67-58; fax 30-67-57; quarterly.

Reinwater Kwartaal-Tijdschrift: Stichting Reinwater, Vossiusstraat 20, 1071 AD Amsterdam, Netherlands; tel (20) 6719322; fax (20) 6753806; e-mail reinwater@xs4all.nl; Editor: Herman de Kievith; f 1980; quarterly; publication of the Clean Water Foundation; in Dutch; circ 1,000.

Remediation Review: Dept of Environmental Conservation, Division of Hazardous Waste Remediation, Bureau of Program Management, 50 Wolf Rd, Albany, NY 12233-7010, USA; tel (518) 457-1684; Editor: D M Ritter; f 1988; 4 a year; reports on the Hazardous Waste Site Remediation Program operated in New York; circ 13,000.

Remote Sensing of Environment: Elsevier Science, Inc, POB 945, New York, NY 10159-0945, USA; tel (212) 633-3730; fax (212) 633-3680; e-mail usinfo-f@elsevier.com; internet www.elsevier.nl; Editor-in-Chief: Marvin E Bauer; Advertising Manager: Arthur J Carlucci; f 1969; monthly; concerned with remote sensing applications in the fields of ecology, the environment and management of natural resources; circ 12,000.

Renew: Network for Alternative Technology and Technology Assessment—NATTA, c/o Energy and Environment Research Unit, The Open University, Faculty of Technology, Walton Hall, Milton Keynes MK7 6AA, United Kingdom; tel (1908) 654638; fax (1908) 654052; e-mail s.j.dougan@open.ac.uk; internet eeru.open.ac.uk/; Editor: Prof David Elliott; f 1976; every 2 months; newsletter concerned with renewable energy issues; digest available online; circ 550.

ReNew: Technology for a Sustainable Future: Alternative Technology Association, POB 2001, Lygon St North, Brunswick East, Vic 3057, Australia; tel (3) 9388-9311; fax (3) 9388-9322; e-mail ata@ata.org.au; internet www.ata.org.au/renew.htm; Editor: Michael Linke; f 1979; quarterly; former title *Soft Technology*; concerned with renewable energy and appropriate technologies, solar-, wind- and water-power generation, recycling and alternative transport; circ 14,000.

Renewable Energy World: James & James (Science Publishers) Ltd, 35–37 William Rd, London NW1 3ER, United Kingdom; tel (20) 7387-8558; fax (20) 7387-8998; e-mail james@jxj.com; news on developments in renewable energy.

Renewable Resources Journal: Renewable Natural Resources Foundation—RNRF, 5430 Grosvenor Lane, Bethesda, MD 20814, USA; tel (301) 493-9101; fax (301) 493-6148; e-mail info@rnrf.org; internet www.rnrf.org; Editor:

and Prediction, Meteorological Office. *Present Position*: Dirr, World Climate Research Programme Department, World Meteorological Organization. *Address*: 7 bis ave de la Paix, CP 2300, 1211 Geneva 2, Switzerland. *Tel*: (22) 7308246. *Fax*: (22) 7308036. *E-mail*: carson_d@gateway.wmo.ch.

CARTER, Jimmy (James Earl, Jr), BS; American farmer and politician; b 1 Oct 1924, Plains, GA; s of James Earl Carter, Sr and Lillian Gordy; m Rosalynn Smith 1946; 3s 1d. *Education*: Georgia Southwestern Coll, Georgia Inst of Tech, US Naval Acad, Annapolis, MD. *Career*: US Navy 1946–53; peanut farmer and warehouseman, Carter Farms and Carter Warehouses, GA 1953–77; mem, School Bd, Sumter Co, GA 1955–62, Chair 1960–62; mem, Americus and Sumter Co Hosp Authority 1956–70, Sumter Co Library Bd 1961; Dir, GA Crop Improvement Asscn 1957–63, Pres 1961; Senator (Democrat), GA 1962–66, Gov 1971–74; Pres, Plains Devt Corpn 1963; Chair, W Cen GA Area Planning and Devt Comm 1964; Georgia Planning Asscn 1968; Chair, Congressional Campaign Cttee and Democratic Nat Cttee 1974; Pres, USA 1977–81; Prof, Emory Univ 1982–; Founder, Carter Presidential Center 1982; Chair, Carter-Menil Human Rights Foundation 1986–, Global 2000 Inc 1986–; leader, int observer team, Panama 1989, Nicaragua, Dominican Repub and Haiti 1990; host, peace negotiations, Ethiopia 1989. *Present Position*: Chair, Bd of Trustees, Carter Center, Inc 1986–. *Awards and Honours*: World Methodist Peace Award 1984; Albert Schweitzer Prize for Humanitarianism 1987; Matsunaga Medal of Peace 1993; Houphouët Boigny Peace Prize, UNESCO (jtly) 1995. *Publications*: Why Not the Best? 1975; A Government as Good as its People 1977; Keeping Faith: Memoirs of a President 1982; The Blood of Abraham: Insights into the Middle East 1985; Everything to Gain: Making the Most of the Rest of Your Life 1987; An Outdoor Journal 1988; Turning Point: A Candidate, a State and a Nation Come of Age 1992. *Address*: The Carter Centre, Inc, 1 Copenhill Ave, NE Atlanta, GA 30307, USA.

CATANIA, Prof Dr Peter Joseph, M SC, PH D; Canadian professor of engineering; b 6 Jan 1942; m; 1s 1d. *Present Position*: Prof, Faculty of Engineering, Univ of Regina 1973–; Chair, International Energy Foundation—IEF 1989–. *Address*: Faculty of Engineering, University of Regina, Regina, SK S4A 0A2, Canada. *Tel*: (306) 585-4364. *Fax*: (306) 585-4855. *E-mail*: peter.catania@uregina.ca.

CATLEY-CARLSON, Margaret, BA; Canadian international organization official; b 6 Oct 1942, Regina, SK; d of George Lorne and Helen Margaret Catley; m Stanley Frederick Carlson 1970. *Education*: Univ of British Columbia. *Career*: Officer, Dept of External Affairs, seconded to Sri Lanka 1968, UK 1975, Asst Under-Sec 1981; Vice-Pres, Canadian Int Devt Agency—CIDA 1978, Sr Vice-Pres and Pres (acting) 1979–80, Pres 1983–89; Asst Sec-Gen, UN and Deputy Exec-Dir of Operations, UNICEF 1981; Deputy Minister of Health and Welfare 1989–92; Pres, Population Council 1993–99. *Present Position*: Chair, Bd of Govs, CABI; Comm, World Bank Comm on Water for the 21st Century, Int Advisory Cttee, International Food Policy Research; mem, Advisory Bd, Global Public Policy Project. *Address*: Population Council, 1 Dag Hammarskjöld Plaza, New York, NY 10017, USA. *Tel*: (212) 339-0500. *Fax*: (212) 755-6052.

CEROVSKÝ, Dr Jan, D SC; Czech conservationist; b 2 Feb 1930, Prague; m Jarmila Misikova 1992; 2c. *Education*: Science Faculty, Charles Univ. *Career*: Ed-in-Chief, ABC of Young Technologists and Naturalists 1956–59; Sen Officer, Nature Conservation Dept, State Inst for Protection of Monuments and Nature 1959–68, 1973–90; Education Exec Officer, IUCN HQ 1969–73; Chief Scientist, Czech Agency for Nature and Landscape Protection 1990–2002. *Present Position*: Environmental consultant. *Awards and Honours*: J P Lenne Gold medal 1993; hon mem IUCN 1996; J P Galland Prize 2001. *Publications*: 8 conservation textbooks in Czech, some with Co-Authors; Endangered Plants 1995; Transboundary Biodiversity conservation in Europe (ed) 1996; Czech and Slovak Red Data Book, vol. 5 1999. *Address*: Pernerova 50, CZ-18600 Praha 8, Czech Republic. *Tel and fax*: (2) 22321121. *E-mail*: jan@cerovsky.net.

CHADWICK, Prof Michael John, PH D, FIBIOL; British environmental scientist; b 13 Sept 1934, Leicester; s of the late Henry Chadwick and Hilda Corman; m Josephine Worrall; 1s 2d. *Education*: Univ Coll of N Wales, Downing Coll, Cambridge. *Career*: Lecturer, Dept of Botany, Khartoum Univ, Sudan 1959–62; Univ Demonstrator, Univ of Cambridge 1962–66; Lecturer, Sr Lecturer, Reader then Prof of Biology, Univ of York 1966–91; Dir, Stockholm Environment Inst, Sweden 1991–96; Dir, LEAD-Europe, Geneva, Switzerland 1996–98; Sec, British Ecological Soc. *Present Position*: environmental consultant 1998–. *Publications*: Restoration of Land (Co-Author) 1980; The Relative Sensitivity of Ecosystems in Europe to Acidic Depositions (Co-Author) 1990; Environmental Management of Low-grade Fuels (Co-Author) 1996. *Address*: 3 Skipwith Rd, Escrick, York YO19 6JT, United Kingdom. *Tel and fax*: (1904) 728025. *E-mail*: Cmjchadwick@aol.com.

CHAMBERLAIN, Kenneth L, BA; British retired consultant; b 5 April 1928, London; m S E Chamberlain. *Education*: Univ of Oxford, UK; Stanford Univ, USA. *Career*: employee, P & O Shipping Co 1952–73 (fmr Vice-Pres, P & O N America, Inc); Exec, Pacific and Asia Travel Asscn—PATA 1974–90, Chief Exec 1979–90, involved in devt of heritage and conservation in the Asia-Pacific area; Exec Dir, PATA Foundation 1990–96, involved in mission to protect area's environment and cultural heritage; adviser, World Tourism Organization—WTO 1990–2000. *Present Position*: adviser, UNEP, Himalaya Environmental Foundation; mem, World Tourism Asscn, WTO Leadership Forum of Advisers; Trustee, PATA Foundation. *Awards and Honours*: UNEP Global 500. *Publications*: numerous papers and reports on tourism development and conservation, with emphasis on sustainable development and strategic planning. *Address*: Orchard Stables, Higher Hapsden, Castle Cary BA7 7LX, United Kingdom. *Tel*: (1963) 350616. *Fax*: (1963) 351810.

CHARLIER, Albert; Belgian biologist. *Present Position*: Consultant and Int Del, Amis de la Terre (FoE Belgium). *Address*: rue du Piroy 88, 5200 Malonne, Belgium. *Tel*: (81) 44-15-58.

CHARTERS, Prof William W S (Bill), B SC; Australian/British solar energy researcher and university professor; b 10 Feb 1935, Shanghai, People's Repub of China; s of William and Elizabeth (Bailey) Charters; m Sheila Maria Charters 1961; 1d. *Education*: Univ of Leeds, UK, Princeton Univ, USA. *Career*: Dir, University-Melbourne Ltd Energy Research and Devt Corpn; Consultant, UNDP, NY, USA, UNIDO, Vienna, Austria, UNESCO, Paris, France, UNEP, Nairobi, Kenya, Commonwealth Science Council, UK 1974–; Chair, Vic Solar Energy Council 1980–88; Dean, Faculty of Eng, Univ of Melbourne 1988-97; Bd Dir, ACRE Ltd, APPL Ltd, Engineer Research and Devt Center—ERDC 1990–99, CVC (REIF). *Present Position*: Prof of Mechanical Eng and Chair, Dept of Mechanical and Manufacturing Eng, Univ of Melbourne. *Awards and Honours*: Hon M ENG; CSIRO Research Award 1988; Achievement through Action Award, Int Solar Energy Soc 1991; Farrington Daniels Award, Int Solar Energy Soc 1993. *Publications*: Solar Energy—Principles and Applications (Co-Author) 1982; Solar Systems—Installation and Design (Co-Author) 1983; Theory and Design of Solar Thermal Systems (Co-Author) 1984. *Addresses*: Univ of Melbourne, Parkville, Vic 3052, Australia; Nyoka, 5

Andromeda Way, Lower Templestowe, Vic 3107, Australia. *Tel*: (3) 8344-6751. *Fax*: (3) 8347-8784. *E-mail*: b.charters@eng.unimelb.edu.au. *Internet*: www.unimelb.edu.au.

CHATURVEDI, Amar Nath, M SC; Indian energy researcher; b 5 Feb 1930, Agra; s of Hira Lal and A Chaturvedi; m Manorama Chaturvedi 1953; 1s 2d. *Education*: Agra Univ. *Career*: Uttar Pradesh Forest Dept 1955–67, seconded to Forest Research Inst and Coll, Dehradun 1967–73, seconded to Bhutanese govt 1976–79; Consultant, FAO, Philippines 1987; fmr Sr Fellow, Tata Energy Research Inst. *Awards and Honours*: Hari Om Ashram Award, Indian Council of Agricultural Research 1986. *Publications*: Forest Mensuration; Eucalyptus for Farming; Bamboos for Farming; Popular Farming in Uttar Pradesh. *Address*: c/o Tata Energy Research Institute, Darbari Seth Block, Habitat Pl, Lodi Rd, New Delhi 110 003, India. *Tel*: (11) 4622246. *Fax*: (11) 4621770. *E-mail*: mailbox@teri.ernet.in.

CHEN HUAI-MAN, Dr; Chinese environmental chemist; b 9 Dec 1939, Jiangsu; m Zheng Chun-Rong 1965; 2s. *Education*: Univ of Manila, Philippines; Nanjing Univ; Inst of Soil Science, Chinese Acad of Sciences—CAS. *Career*: researcher and teacher in the field of soil chemistry; specialist, environmental analysis and heavy metal pollution of various ecosystems, combined pollution and interaction of inorganic and organic pollutants, remediation of contaminated soil and metal tailings, land remediation. *Present Position*: Prof of Environmental Chemistry, Inst of Soil Science, Chinese Acad of Sciences 1993–. *Awards and Honours*: Outstanding Contributor of Chinese Doctor's Degree 1991; Second Award for Natural Science, CAS 1998; Second Award for Advances in Science and Technology, CAS 1999. *Publications*: Heavy Metal Pollution in Soil–Plant Systems 1996. *Address*: 71 E Beijing Rd, POB 821, Nanjing, Jiangsu 210008, People's Republic of China. *Tel*: (25) 3600672. *Fax*: (25) 3353590. *E-mail*: hmchen@issas.ac.cn.

CHEN PANQIN, Prof; Chinese atmospheric scientist; b 8 Oct 1944, Sichuan; 1s 1d. *Education*: Chengdu Meteorological Inst. *Career*: Research in the fields of air pollution, meteorology, climatology and global change. *Present Position*: Sec-Gen, Chinese Nat Cttee, Int Geosphere–Biosphere Programme—IGBP. *Publications*: Air Pollution 1980; Global Change Study in China 1991. *Address*: 52 Sanlehe Rd, Beijing 100864, People's Republic of China. *Tel*: (10) 68597546. *Fax*: (10) 68511095. *E-mail*: chenpq@sun.ihep.ac.cn.

CHEN PEIXUN, Prof; Chinese scientist; b 5 July 1927; m 1956; 1d. *Education*: Hua Zhong Univ, Wuhan. *Career*: Asst, Agricultural Coll, Wuhan 1951–52; Lecturer and Assoc Prof, Inst of Hydrobiology, Chinese Acad of Sciences 1953–78, Prof 1978–; mem, IUCN 1987–. *Present Position*: Head, Dept of River Dolphin Research, Inst of Hydrobiology, Chinese Acad of Sciences 1978–. *Publications*: Fishes of the Yangtze River 1976; Baiji, A Rare Treasure 1992. *Address*: Institute of Hydrobiology, Chinese Academy of Sciences, Luojiashan, Wuhan, Hubei 430072, People's Republic of China. *Tel*: (27) 7801331. *Fax*: (27) 7875132. *E-mail*: dolphin@wip.whcnc.ac.cn.

CHEN SHUPENG, Prof; Chinese scientist; b 28 Feb 1920, Jiangxi; s of Chen Yuoyuan and Lee Manlian; m Zhang Diehua 1944; 1s 1d. *Education*: Zhejiang Univ. *Career*: Pres, Geographical Soc of China 1991–94; Hon Dir, Inst of Remote Sensing Applications 1988–, Co-Chair, Space Science Application Cttee 1990–; Fellow, Third World Acad of Sciences 1992; Academician, Intl Eurasia Adac of Sciences (IWAS) 1992; mem, China Council for Intl Cooperation on Environment and Development (CCIED) 2002; Chief Engineer of Application Subsystems of CBEA-S Sattelite 1989–92; Adviser, National Remote Sensing Centre, SSTCC 1985–; Prof on Life, Univ of Science and Technology CAS 2003–. *Present Position*: research prof, Geography Inst of Academia Sinica 1978–, mem, Dept of Earth Sciences, Academia Sinica 1980–. *Awards and Honours*: Hon State Prize of Science of China 1988, State Gold Prize for Environmental Science 1993; O. Miller Cartographic Award, American Geographical Soc 1998; Award of Natural Science CAS 1st Class 2000; Highest Honour Award of Int Cartographical Assoc 2001; First Award of Intern Karst Assoc 2001. *Publications*: Selected Works in Geo-Sciences, Vols I–VI 1990–2003; Dictionary of Remote Sensing (Ed-in-Chief) 1990; Atlas of Multidisciplinary Analysis of Meteorological Satellite Imagery in China 1992; The Start of Remote Sensing and Geo-information Systems in China 1993; Geo-Analysis of Remote Sensing 1985; An Introduction on Geo-Information Systems 1998; Geo-Analysis on Landsat Imagery in China 1985; Machenisium of Remote Sensing 1999; Graphic Analysis of Geo-Information Science 2001; Geo-Information Science and Digital Earth 2004. *Addresses*: Institute of Geographic Science and Natural Resources–CAS, Datun Rd, Beijing 100101, People's Republic of China; No 1301, Bldg A-801, Apt of CAS, Zhongguansun, 190090, Beijing. *Tel*: (10) 6488-9318. *Fax*: (10) 6488-9530. *E-mail*: chensp@ires.ac.cn.

CHERKASOVA, Dr Maria V; Russian ecologist; b 13 March 1938, Moscow; m Dr Aleksandr A Dulov; 1s 1d. *Education*: Moscow State Univ, Moscow Inst of Nature Prot. *Career*: Staff mem, Zoological Museum of Moscow Univ 1962–68; Scientific Researcher, Research Inst for the Protection of the Environment 1973–88; mem, Russian Cttee, UNESCO Man and Biosphere programme, Women's Int Policy Action Cttee; Environmental Adviser, ISAR. *Present Position*: Founder and Dir, Centre for Ind Ecological Programmes and Dir, Socio-Ecological Union 1988–. *Awards and Honours*: UNEP Global 500 1992; Ford European Conservation Award 1993. *Publications*: They Must Live 1987; more than 150 books and articles. *Address*: Centre for Independent Ecological Programmes, 103104 Moscow, Malaya Bronnaya 12, Apt 12, Russian Federation. *Tel and fax*: (095) 118-86-86. *E-mail*: cnep@glas.apc.org.

CHOU CHANG-HUNG, PH D; Taiwanese researcher and academic; b 5 Sept 1942, Tainan; s of F K Chou and C Y Shih Chou; m Ruth L H Yang Chou 1970; 1s 1d. *Education*: Nat Taiwan Univ, Taipei, Univ of California, Santa Barbara, USA, Univ of Toronto, Canada. *Career*: Assoc Research Fellow, Inst of Botany, Academia Sinica, Taipei 1972–76; Prof, Dept of Botany, Nat Taiwan Univ 1976–; mem, various nat cttees for Int Council of Scientific Unions (ICSU) 1974–; Sec for Int Affairs Academia Sinica 1988–, Dir, Inst Botany 1989–96, mem, Council, Academia Sinica 1989–; mem, Council, Pacific Science Asscn 1989; mem, Cttee for Science Educ, Ministry of Educ 1986–, Cttee for Environmental Educ 1991–, Cttee for Cultural and Natural Preservation, Council of Agric 1990, Council, Taiwan Livestock Research Inst 1976–, Taiwan Forestry Research Inst 1989–, Council Nat Sustainable Devt 1997–; Dir, Life Science Research Promotion Centre, Nat Science Council 1989–; Visiting Scholar, Oklahoma Univ, Univ of Texas, Washington State Univ 1979–80; Pres, Botanical Soc of the Repub of China (Taiwan) 1983–84, Biological Soc of Repub of China (Taiwan) 1987–88; Chair, Nat Cttee, Int Union of Biological Sciences 1990–; mem, Exec Cttee of IUBS; Chair, SCOPE Nat Cttee; Vice-Pres, Int Union of Biological Sciences 1997–; Ed, *Botanical Bulletin*, Academia Sinica 1989–; Fellow, Third World Acad of Sciences; mem, Academia Sinica, Taipei. *Present Position*: Research Fellow, Inst of Botany, Academia Sinica 1976–, Dir 1989–. *Awards and Honours*: awards from Ministry of Educ and Science Council of Taiwan. *Publications*: over 200 scientific papers, one univ textbook. *Address*: Institute of Botany, Academia Sinica, 280 Yean Jiou Yuan Rd, Sec 2,

Taipei 115, China (Taiwan). *Tel*: (2) 27899590. *Fax*: (2) 27827954.

CIFUENTES-ARIAS, Miguel, M SC; Ecuadorean biologist; b 29 May 1951, Baños; s of Ruperto Cifuentes and Clara Arias; m Rosa V Jara; 2s. *Education*: Centro Agronómico Tropical de Investigación y Enseñanza—CATIE, Costa Rica. *Career*: Asst to Dir, Charles Darwin Research Station, Galápagos Islands 1974–76; Dir, Galápagos Nat Park Service 1976–86. *Present Position*: WWF Rep for Cen America 1986–; Prof, Faculty member, Centro Agronómico Tropical de Investigación y Enseñanza—CATIE, Costa Rica 1987–; Pres, Charles Darwin Foundation for the Galápagos Islands 1998–. *Awards and Honours*: WWF Conservation Merit Award 1985; Thomas Alflants Merit Award. *Publications*: Algunos aspectos sobre la ecología de la reproducción de la tortuga negra de Galápagos (Chelonia mydas agassizi) 1975; Evaluación de la implementación del Plan Maestro del Parque Nacional Galápagos 1978; El turismo en el Parque Nacional Galápagos y el desarollo de la provincia insular 1978; Esquema histórico de suceso relacionados con la conservación y el desarollo de la provincia insular de Galápagos 1978; Manejo y ordenación de cuencas hidrográficas en Costa Rica 1982; Producción de leña en sistema agroforestales 1982; Plan de manejo y desarollo. Reserva Biológica Carara 1982; Plan de interpretación del sendero natural Los Espaveles 1982; Reservas de biosfera: Clarificación de su marco conceptual y aplicación de una metodología para la planificación estratégica de un subsistema nacional 1983; Strategic Planning of National or Regional Systems of Biosphere Reserves: a methodology and case study from Costa Rica 1983; Parque Nacional Galápagos: plan de manejo y desarollo, II fase 1984; Guía a los sitios de visita del Parque Nacional Galápagos (Co-Author) 1987; Metodología para la planificación de sistemas de areas protegidas 1988; Sistema Regional de Areas Silvestres Protegidas en América Central, Plan de Acción 1989–2000 (Co-Ed) 1989; Determinación de capacidad de carga turistica en areas protegidas 1992. *Address*: Centro Agronómico Tropical de Investigación y Enseñanza, Apdo 7.170, Turrialba, Costa Rica. *Tel*: 556-1383. *E-mail*: mcifuent@catie.ac.cr.

CINGAL, Georges René Charles, BA; French teacher and environmentalist; b 13 Feb 1945, Douvres; s of Louis and Charlotte (Lizoret) Cingal; m Marie-Hélène Bédrède 1970; 1s 1d. *Career*: Gen Sec, Sepanso 40 1970–93; Editor, *Ecolandes*. *Present Position*: Pres, Sepanso 40 1993–; Officer, France Nature Environnement 1986–; mem, Bd, European Environmental Bureau 1996–, Vice-Pres 1999–. *Address*: Money Box, 1581 route de Cazordite, 40300, France. *Tel*: 5-58-73-68-11. *Fax*: 5-58-73-14-53. *E-mail*: georges.cingal@wanadoo.fr.

CLAYTON, Dame Barbara, PH D, FRCP, FRCPE; British professor of medicine; b 1 Sept 1922, Liverpool; d of William and Constance Clayton; m William Klyne 1949; 1s 1d. *Education*: Edinburgh Univ. *Career*: Prof, Inst of Child Health, Univ of London 1970–78; Pres, Asscn of Clinical Biochemists 1977–78, Soc for the Study of Inborn Errors of Metabolism 1981–82, Royal Coll of Pathologists 1984–87; Chair, MRC Cttee on Toxic Hazards in the Workplace and Environment, Standing Cttee on Postgraduate Medical and Dental Education, Medical and Scientific Panel, Leukaemia Research Fund; mem, Royal Comm on Environmental Pollution 1981–96, Gen Medical Council 1983–87. *Present Position*: Pres, Nat Soc for Clean Air and Environmental Protection 1995–. *Publications*: Clinical Biochemistry and the Sick Child (Co-Author) 1994. *Address*: c/o National Society for Clean Air and Environmental Protection, 136 North St, Brighton BN1 1RG, United Kingdom. *Tel*: (1273) 326313. *Fax*: (1273) 735802.

CLOUDSLEY-THOMPSON, John Leonard, PH D, FRES, C BIOL, FIBIOL, FZS, FWAAS; British zoologist; b 23 May 1921, Murree, India; s of A G G and Muriel Elaine (Griffiths) Thompson; m J Anne Cloudsley 1944; 3s. *Education*: Marlborough Coll, Pembroke Coll, Cambridge. *Career*: war service 1940–44; Lecturer in Zoology, King's Coll, London 1950–60; Prof of Zoology, Univ of Khartoum and Keeper, Sudan Natural History Museum 1960–71; Nat Science Senior Research Fellow, Univ of New Mexico, Albuquerque, USA 1969; Prof of Zoology, Birkbeck Coll, London 1972–86; Leverhulme Emer Fellowship, Univ Coll London 1987–89; Visiting Prof, Univ of Kuwait 1978, 1983, Univ of Nigeria, Nsukka 1981, Univ of Qatar 1986; Chair, British Naturalists' Asscn 1974–83; Chair, Biological Council 1977–82 (Medal 1985); Pres, British Arachnological Soc 1982–85, British Soc for Chronobiology 1985–87; Vice-Pres, Linnean Soc 1975–76, 1977–78; Ed-in-Chief, *Journal of Arid Environments* 1978–97. *Present Position*: Vice-Pres, British Naturalists' Asscn 1985–; Prof Emer, Birkbeck Coll, London 1986–. *Awards and Honours*: hon mem, Royal African Soc 1969 (Medal 1969), British Herpetological Soc 1983 (Pres 1991–96), Centre Int de Documentation Arachnologique, Paris 1995; Liveryman, Worshipful Co of Skinners 1952; Hon FLS 1997; Hon D SC (Khartoum Univ) and Silver Jubilee Gold Medal 1981; Inst of Biology Charter Award 1981; J H Grundy Memorial Medal, Royal Army Medical Coll 1987; Peter Scott Memorial Award 1993. *Publications*: Spiders, Scorpions, Centipedes and Mites 1958; Animal Behaviour 1960; Rhythmic Activity in Animal Physiology and Behaviour 1961; Animal Conflict and Adaptation 1965; Animal Twilight 1967; Zoology of Tropical Africa 1969; The Temperature and Water Relations of Reptiles 1971; Desert Life 1974; Terrestrial Environments 1975; Insects and History 1976; Man and the Biology of Arid Zones 1977; The Desert 1977; Animal Migration 1978; Biological Clocks 1980; Tooth and Claw 1980; Evolution and Adaptation of Terrestrial Arthropods 1988; Ecophysiology of Desert Arthropods and Reptiles 1991; Nile Quest 1994; Predation and Defence Amongst Reptiles 1994; Biotic Interactions in Arid Lands 1996; Teach Yourself Ecology 1998; The Diversity of Amphibians and Reptiles 1999. *Address*: 10 Battishill St, London N1 1TE, United Kingdom.

COCKLIN, Prof Christopher Reid, B SCMA, PH D; New Zealand professor of geography and environmental science; b 19 April 1958, New Zealand; s of Ralph and Mollie Cocklin; m Marjorie Cutting 1990. *Education*: Univ of Waikato, Univ of Guelph, Canada, McMaster Univ, Canada, Massey Univ. *Career*: faculty mem, Dept of Geography, Univ of Auckland 1985–97; Sr Scientist, Forest Research Inst, Rotorua 1995–96; mem, Scientific Steering Cttee, Global Environmental Change and Human Security project of INDP, Int Geographical Union—IGU Comm on Sustainable Rural Systems;mem Bd, Greening Australia; Australian Academy of Science, National Sustainability Cttee; Victoria, Environmental Protection Authority Sustainability Network. *Present Position*: Logan Prof of Geography and Environmental Science, Monash Univ, Australia 1997–, Dir, Monash Environmental Inst 2000–. *Publications*: approx 160 papers, monographs and reports. *Address*: School of Geography and Environmental Science, Monash University, POB 11a, Vic 3800, Australia. *Tel*: (3) 9905-2926. *Fax*: (3) 9905-2948. *E-mail*: chris.cocklin@arts.monash.edu.au. *Internet*: www .arts.monash.edu.au/ges/who/cocklin.html.

COLLIE, John Stuart; Seychellois/British environmental manager; b 6 Oct 1958, Seychelles; s of Dr B and Dr E Collie; m Thérèse Delpech; 2d. *Education*: Glasgow Univ, UK; Adelaide Univ, Australia. *Career*: officer, Nat Trust for Scotland 1980–86; Man, Aldabra World Heritage Site 1983–87. *Present Position*: Man Dir, Marine Parks Authority 1996–. *Address*: Ministry of Foreign Affairs and

Environment, POB 445, Seychelles. *Tel*: 322891. *Fax*: 324570. *E-mail*: mpa@seychelles.net.

COLLIN, Dr Gérard; French heritage curator; b 22 Dec 1945, Le Raincy. *Education*: Univ of Paris I. *Career*: Head, Dept of Human Sciences, Cevennes Nat Park 1973–94; Consultant, UNESCO Training Centre in Africa, Nigeria 1975–76, UNESCO Man and the Biosphere Programme 1985–2000; mem, IUCN Steering Cttee for European Protected Areas; leader, missions to protected areas in Algeria, Bulgaria, Guinea, India, Italy, Nigeria, South Africa and Spain. *Present Position*: Head, Heritage Dept, Regional Council for Languedoc-Roussillon 1995–. *Publications*: La muséologie selon G H Riviere 1989; Rural Society and Protected Areas: Which Dialogue in Landscape and Urban Planning 1990; Cultural Landscapes in France 1995. *Addresses*: 3 ave du Mont St-Clair, Apt 19, 34280 Carnon, France; Hôtel de la Région, 201 ave de la Pompigane, 34000 Montpellier, France. *Tel*: 4-67-22-80-83. *Fax*: 4-67-22-90-98. *E-mail*: collin@cr_languedocroussillon.fr.

COLLINS, Dr (Nicholas) Mark; British zoologist; b 23 April 1952, Cheltenham; m Melanie Margaret Collins; 1s 1d. *Education*: Univ of Oxford, Imperial Coll, London, Open Univ. *Career*: Officer, Centre for Overseas Pest Research, London 1977–80; Officer, Int Centre for Insect Physiology and Ecology, Nairobi, Kenya 1980–82; Officer, World Conservation Monitoring Centre—WCMC 1982–. *Present Position*: Dir, WCMC 1994–. *Publications*: IUCN Invertebrate Red Data Book (Co-Author) 1983; Threatened Swallowtail Butterflies of the World (Co-Author) 1985; The Last Rain Forests 1990; The Conservation Atlas of Tropical Forests (Co-Author) 1991. *Address*: World Conservation Monitoring Centre, 219 Huntingdon Rd, Cambridge CB3 0DL, United Kingdom. *Tel*: (1223) 277314. *Fax*: (1223) 277136. *E-mail*: info@wcmc.org.uk.

COMBIER, Elizabeth, BA; American journalist and theatrical and video producer; b 7 Nov 1949, New York, NY; d of Hon P Hodges and Julia Strauss (Taschereau) Combier; m David Kapel 1984; 4d. *Education*: Johns Hopkins School for Advanced Int Studies; Northwestern Univ, IL; New York Univ. *Career*: Producer, ABC News 1976–80; initiator, Solar-powered Television and Video Scheme, Egypt 1978–83; Speaker, UN conferences on Appropriate Technology in Rural Development, Popular Participation for Successful Project Implementation, Environmental and Solar Communication for the Third World. *Present Position*: Pres, Artslink, Inc, Theatre Kids, Inc; creator, Arts Together community partnership. *Awards and Honours*: AECT Memorial Scholarship Award; Hon Mention, Rolex Awards; Nominee, Marconi Int Award, Berta Mertz Award; Fellow, Salyburg Seminar. *Publications*: numerous articles in *Environmental Nutrition, American Health, VITA News, Working Woman*. *Address*: Ecomedia, 315 East 65th St, New York, NY 10021, USA.

CONIGLIARO MICHELINI, Anna Olga; Italian sociologist; b 5 Sept 1954; m Livio Michelini 1983; 6c. *Education*: Univ of Trento. *Career*: Exec, Confederazione Italiana Sindacati Lavoratori—CISL 1980–84; Visiting Prof, Pontifical Catholic Univ of Minas Gerais, Brazil 1984–90; Project Man for Urban Environmental Rehabilitation, Associacção Voluntários para o Serviço Internacional—AVSI, Salvador, Brazil 1984–87. *Present Position*: Co-ordinator for Environmental Interventions, AVSI 1992–. *Awards and Honours*: Order of Merit of the Italian Republic 1993; UNEP Global 500 1995 (jtly). *Publications*: Methodologies and Experience in Upgrading Informal Urban Areas: Research Project on Seven Brazilian State Capitals 1996. *Address*: Praça da Sé 1, Centro, 40020-210 Salvador, BA, Brazil. *Tel*: (71) 322-1366. *Fax*: (71) 322-1367. *E-mail*: avsiba@ajax.e-net.com.br.

CONTRÉRAS-MANFREDI, Hernán; Chilean agricultural engineer and environmentalist consultant; b 14 Sept 1931, Curicó; s of Hernán Contréras Barrenechea and Yolanda Manfredi; m 1st Margarita Fernández Papic 1955; m 2nd América Graciela Cordero Velásquez 1978; 4d. *Education*: Univ of California, Davis, USA. *Career*: Pres, Sociedad de Amigos del Arbol de ñuñoa; Dir, Cursos Nacionales I–III de Conservación de la Naturaleza y sus Recursos Renovables 1972, 1974 and 1992; Dir, América Latina hacia una Cultura Ambiental para el Año 2000 programme 1975–78; Dir, Curso Nacional de Conservación de los Recursos Naturales y Equilibrio Ecológico en Venezuela 1976–79. *Present Position*: Consultant to int environmental organizations. *Awards and Honours*: UNESCO Environmental Education award 1976; UNEP Global 500 1987. *Publications*: Conservación de la Naturaleza y sus Recursos Renovables; Educación Ambiental para un Desarrollo Sustentable 1993; Ambiente, Desarrollo Sustentable y Calidad de Vida 1994. *Address*: Apdo 81, Correo Carmelitas, Caracas, Venezuela. *Tel and fax*: (2) 573-8596.

COONEY, Paul Antony, M SC, DMS, MCIEH; British environmental health officer; b 19 Sept 1942, Paddock Wood; s of the late Ronald Albert and of Winifred Cooney; m Barbara Hoiles 1966; 1s 2d. *Education*: Thames Polytechnic. *Career*: Officer, Blood Lead/Immune Status Survey, Dept of the Environment/Dept of Health and Social Security 1984–87; Chair, Nat Soc for Clean Air and Environmental Protection 1996–97; mem, Chartered Inst of Environmental Health—CIEH 1966–, Environmental Protection Cttee, Steering Group on Emission Monitoring, London Air Quality Steering Group, Working Group on the Implementation of EC Directive on Municipal Waste Incineration. *Present Position*: Principal Environmental Health Officer, Pollution Control Service, London Borough of Greenwich 1974–. *Address*: c/o National Society for Clean Air and Environmental Protection, 136 North St, Brighton BN1 1RG, United Kingdom. *Tel*: (1273) 326313. *Fax*: (1273) 735802.

CORELLA, Cosme Casals; Cuban environmental engineer; b 1956, Holguín. *Career*: Studies of coastal dynamics in the provinces of Granma, Guantánamo, Holguín and Santiago de Cuba; organizer, conferences, education and training in the field of environmental protection, with emphasis on coastal zones; Prin Officer, Parque Nacional Bariay project; fmr Dir, Flora and Fauna Unit, Bahía de Naranjo Nat Park; Sec, Nat Council, Brigadas Técnicas Juveniles de la Unión de Jóvenes Comunista de Cuba 1986–88; mem, Comisión Municipal y Provincial de la Flora, la Fauna y el Medio Ambiente de Holguín 1979–, Sociedad de Geología de Cuba, Sociedad Pro-Naturaleza en Holguín, Tourism Research Group. *Present Position*: Head and nature tourism specialist, Cristóbal Colón Nat Park. *Awards and Honours*: Julio A Mella Order of Merit1986; Forjadores del Futuro award (six times). *Publications*: numerous articles. *Address*: Apdo 246, Holguín 80100, Cuba. *Tel*: (24) 30238. *Fax*: (24) 30926. *E-mail*: inverotet@mares.solmelia.cma.net.

CORVALÁN, Osvaldo Bernabé; Argentinian environmentalist; b 6 April 1943, Santiago del Estero; s of Bernabé Corvalán and Eloísa Saganías; 2c. *Education*: Santiago del Estero Nat Univ. *Present Position*: Prin Officer, Amigos de la Tierra (FoE) Argentina 1992–. *Address*: Casilla 3560, 1000 Buenos Aires, Argentina. *Tel*: (11) 773-5947. *Fax*: (11) 777-9837. *E-mail*: postmaster@amigos.org.ar.

COVACEVICH, Jeanette Adelaide, BA, M SC; Australian museum curator and herpetological taxonomist; b 26 March 1945, Innisfail, Qld; d of Anthony Thomas Covacevich and Gladys Rose (Bryant) Covacevich. *Education*: Univ of Qld, Griffith Univ, Brisbane. *Career*: projects in the taxonomy of reptiles and in conservation. *Present Position*: Sr Curator (vertebrates), Qld Museum, Brisbane. *Awards and Honours*:

Australian Medal for services to science, particularly herpetology and conservation. *Publications*: Toxic Plants and Animals: A Guide for Australia 1987; Venoms and Victims 1988; History, Heritage and Health 1996; approx 120 papers in herpetological and historical journals. *Address*: Queensland Museum, POB 3300, South Brisbane, Qld 4101, Australia. *Tel*: (7) 3840-7708. *Fax*: (7) 3846-1226.

CRANBROOK, Gathorne Gathorne-Hardy, Earl, PH D, FLS, FZS, FRGS; British environmental administrator; b 20 June 1933; s of Earl and Countess Cranbrook; m Caroline Jarvis 1967; 2s 1d. *Education*: Corpus Christi Coll, Cambridge, Univ of Birmingham. *Career*: Asst, Sarawak Museum 1956–58; Fellow, Yayasan Siswa Lokantara, Indonesia 1960–61; Sr Lecturer in Zoology, Malaya Univ 1961–70; Ed, *Ibis* 1973–80; Non-Exec Dir, Anglian Water 1987–; Vice-Pres, Nat Soc for Clean Air and Environmental Protection; Chair, Foundation for European Environmental Policy 1990–95; mem, Suffolk Coastal Dist Council 1974–83, Environment Sub-Cttee, House of Lords Select Cttee on EC 1979–85, 1987–90, Chair 1980–83, 1987–90, Chair, English Nature 1990–98; Royal Comm on Environmental Pollution 1981–92, Natural Environment Research Council 1982–88, Broads Authority 1988–, Harwich Haven Authority 1989–; Pres, Suffolk Trust for Nature Conservation 1979–. *Present Position*: Chair, ENTRUST, Regulator of Environmental Bodies under Landfill Tax Regulations 1996–. *Publications*: Birds of the Malay Peninsula 1976; Mammals of Borneo 1977; Mammals of Malaya (Co-Author) 1978; Riches of the Wild: Mammals of South-East Asia 1987; Key Environments: Malaysia (Ed) 1988; Belalong: a Tropical Rainforest (Co-Author) 1994; Wonders of Nature in South-East Asia (Ed) 1997. *Address*: ENTRUST, Acre House, 2 Market Square, Sale M33 7WZ, United Kingdom. *Tel*: (161) 972-0044. *Fax*: (161) 972-0055.

CRICKHOWELL, The Rt Hon Lord, PC, MA, LLD (HON); British politician and administrator; b 25 Feb 1934, London; s of Ralph Edwards and Marjorie Ingham Brooke; m Ankaret Healing 1963; 1s 2d. *Education*: Westminster School, Trinity Coll, Cambridge. *Career*: MP (Cons) for Pembroke 1970–87; Sec of State for Wales 1979–87; Dir, Associated British Ports Holdings PLC, HTV Group PLC; Pres, Univ of Wales, Cardiff; mem, Cttee, Automobile Asscn; Chair, Nat Rivers Authority 1989–96. *Publications*: Opera House Lottery; Westminster, Wales and Water. *Address*: 4 Henning St, London SW11 3DR, United Kingdom. *E-mail*: Ncrickhowell@aol.com.

CROFT, Trevor Anthony, B SC; British town planner; b 9 June 1948, Bradford; s of Kenneth Edward and Gladys (Bartle) Croft; m Janet Frances Halley 1980; 2d. *Education*: Univ of Hull, Univ of Sheffield. *Career*: Sr Asst Govt Planning Officer, N Ireland 1971–72; Asst Planning Officer, Countryside Comm for Scotland 1972–75; Park Planning Officer, Dept of Nat Parks and Wildlife, Malawian Govt 1976–82, responsible for establishment of Lake Malawi Nat Park 1982–97; Planning Officer, National Trust for Scotland 1982–84, Head of Policy Research 1984–88, Regional Dir 1988–95, Deputy Dir 1995–97; mem, Royal Town Planning Inst. *Present Position*: Dir, Nat Trust for Scotland 1997–. *Publications*: Lake Malawi National Park—A Case Study in Conservation 1991. *Address*: National Trust for Scotland, 28 Charlotte Sq, Edinburgh EH2 4ET, United Kingdom. *Tel*: (131) 243-9519. *Fax*: (131) 243-9592. *E-mail*: tcroft@nts.org.uk. *Internet*: www.nts.org.uk.

CRONSTRÖM, Ulrica; Finnish biologist; b 25 March 1941; 1c. *Education*: Univ of Helsinki, Univ of Copenhagen, Denmark. *Present Position*: Sec-Gen, Natur och Miljo 1973–. *Address*: Natur och Miljo, Bulevarden 30, 00120 Helsinki, Finland. *Tel*: (9) 644731. *Fax*: (9) 605850. *E-mail*: ulrica.cronstrom@sll.fi.

CROSS, Nigel; *Career*: Dir, SOS Sahel UK 1985–95; Exec Dir, Panos Inst 1995–99; Chair, Accord. *Present Position*: Exec Dir, Int Inst for Environment and Devt—IIED 1999–. *Address*: International Institute for Environment and Development—IIED, 3 Endsleigh St, London WC1H 0DD, United Kingdom.

CROSSETT, Robert Nelson (Tom), D PHIL; British environmentalist; b 27 May 1938; s of Robert and Mary (Nelson) Crossett; m Susan Marjorie Legg 1966; 2s. *Education*: Queen's Univ, Belfast, Lincoln Coll, Oxford, Univ of E Anglia. *Career*: Sr Scientific Officer, ARC Letcombe Lab 1969; Devt Officer (Crops), Scottish Agricultural Devt Council 1972; Prin Scientific Officer, Dept of Agric and Fisheries for Scotland 1975; Scientific Liaison Officer (Horticulture and Soils), Ministry of Agric, Fisheries and Food 1978, Head of Food Science Div 1984, Chief Scientist (Fisheries and Food) 1985–89; Head of Environmental Policy, National Power 1989–90, Environment Dir 1990–91; Sec-Gen, Nat Soc for Clean Air and Environmental Protection 1992–96; mem, Natural Environment Research Council 1985–89, Agricultural and Food Research Council 1985–89, Advisory Group to Sec of State for Environment on Eco-Management and Audit 1994–, UK Round Table on Sustainable Devt 1995–. *Present Position*: Dir-Gen, Int Union of Air Pollution Prevention and Environmental Protection Asscns 1996– and Chair, Southern Regional Environmental Protection Advisory Cttee, Environment Agency 1996–. *Publications*: Numerous scientific papers. *Address*: International Union of Air Pollution Prevention and Environmental Protection Associations, 136 North St, Brighton BN1 1RG, United Kingdom. *Tel*: (1273) 326313. *Fax*: (1273) 735802. *E-mail*: admin@nsca.org.uk.

CRUTZEN, Prof Paul J, PH D; Dutch meteorologist; b 3 Dec 1933, Amsterdam; s of Josef C Crutzen and Anna Gurk; m Terttu Soininen 1958; 2d. *Education*: Stockholm Univ. *Career*: Asst, Stockholm Univ 1973–74; Scientific Dir, NCAR, Boulder, CO, USA 1974–77, Research Dir 1977–80; Dir, Max Planck Inst for Chemistry 1980–2000. *Present Position*: Dir (Emeritus), Max Planck Inst for Chemistry 2000–. *Awards and Honours*: Scientist of the Year 1984; Tyler Prize 1989; Nobel Prize for Chemistry (jtly) 1995. *Publications*: Environmental Consequences of Nuclear War 1985; Schwarzer Himmel 1986; Atmospheric Change: An Earth System Perspective 1993. *Address*: Max-Planck Institute for Chemistry, 55020 Mainz, Johann-Joachim-Becher-Weg 27, POB 3060, Germany. *Tel*: (6131) 305333. *Fax*: (6131) 305577. *E-mail*: air@mpch-mainz.mpg.de.

D

DADE, Drita; Albanian interpreter; b 4 Aug 1968; d of Drita Zefloka and Prena Loka; m Shyqyri Dade 1989; 1d. *Education*: Tirana Univ. *Career*: Head of Information and Public Awareness, Cttee of Environmental Protection. *Present Position*: National Environment Agency. *Address*: Blvd Zhan D'Ark 1, Tirana, Albania. *Tel and fax*: (42) 65229. *E-mail*: cep@cep.tirana.al.

DAHL, (Rut) Birgitta, BA; Swedish politician and international environmental organization executive; b 20 Sept 1937,

Gothenburg; m Enn Kokk; 1s 2d. *Education*: Univ of Uppsala. *Career*: Minister with special responsibility for Energy Issues 1982–86, Ministry of Environment and Energy 1986–90, Minister of Environment 1990–91; Chair, Socialist Int Environment Comm 1986–93; Vice-Chair, UN Secr-Gen High-Level Advisory Bd on Sustainable Development 1993–96, Chair 1996–98. *Present Position*: Talman (Speaker), Riksdagen (Swedish Parl) 1994–; Sr Adviser, World Bank Global Environment Facility—GEF 1998–. *Address*: Riksdagen, 100 12 Stockholm, Sweden. *Tel*: (8) 786-40-00. *Fax*: (8) 786-61-43.

DAMMANN, Erik; Norwegian writer and environmental activist; b 9 May 1931, Oslo; s of Erik and Gerd (Müller) Dammann; m Ragnhild Østby; 4s 3d. *Education*: Univ of Oslo. *Career*: founder and leader, The Future in Our Hands popular movement 1974–80; founder, Alternative Future (later renamed Prosus) research programme 1982, Forum for System Debate 1998. *Present Position*: Adviser, The Future in Our Hands and Forum for System Debate. *Awards and Honours*: Right Livelihood Award 1982; Oso Environment Prize 1998; Frittord Honour Prize 1998. *Publications*: The Future in Our Hands 1972; Revolution in the Affluent Society 1979; Beyond Time and Space 1987; Your Money or Your Life 1989. *Address*: Loftuveien 48, 1450 Nesoddtangen, Norway. *Tel*: 66-91-61-16.

DANIELS, Dr R J Ranjit, M SC, PH D; Indian naturalist; b 5 June 1959, Nagercoil; s of David D W and Thilagavathy Daniels; m; 1d. *Education*: Tamil Nadu Agricultural Univ, Indian Inst of Science, Bangalore. *Career*: researcher, distribution of biodiversity in tropical rainforests and mangroves (with emphasis on human impact on biodiversity) 1983–; worker, tribal communities. *Present Position*: freelance naturalist. *Awards and Honours*: Young Scientist Lectureship Award, NAM Centre for Science and Technology 1998. *Publications*: A Field Guide to the Birds of Southwestern India 1997; more than 60 papers on biodiversity. *Address*: Chennai Snake Park, Raj Bhavan PO, Chennai 600 022, India. *Tel*: (44) 2353623. *E-mail*: rjrdaniels@usa.net.

DANILOV-DANILIAN, Prof Dr Victor I; Russian ecologist; b 1938, Moscow; 3s 1d. *Education*: Moscow State Univ. *Career*: Chair of Ecology, Acad for Nat Economy, USSR Council of Ministers; Minister of Environmental Protection 1991–96; Chair, State Cttee for Environmental Protection 1997–2000. *Present Position*: mem, Council of Managers, UNEP. *Publications*: More than 200 scientific publications. *Address*: c/o Ministry of Natural Resources, 123812 Moscow, Bolshaya Gruzinskaya 4/6, Russian Federation.

DASMANN, Prof Raymond Frederic, PH D; American professor of ecology; b 27 May 1919, San Francisco; m Elizabeth Sheldon 1944; 3d. *Education*: Univ of California, Berkeley. *Career*: Chair, Div of Natural Resources, Humboldt State Univ 1962–65; Int Dir, Conservation Foundation 1966–70; Pres, Wildlife Soc 1970; Sr Ecologist, IUCN, Morges, Switzerland 1970–77; Chair, Environmental Studies, Univ of California, Santa Cruz 1977–89; mem, Bd of Dirs, World Wildlife Fund 1978–84. *Present Position*: Prof Emer, Univ of California, Santa Cruz 1989– and Pres, Golden Gate Biosphere Reserve Asscn 1996–. *Awards and Honours*: Browning Medal, Smithsonian Institution 1974; Leopold Medal, Wildlife Society 1978; Order of the Golden Ark, Netherlands 1978. *Publications*: Environmental Conservation 1959; Wildlife Biology 1964; Planet in Peril? 1972; Ecological Principles for Economic Development (Co-Author) 1973; The Conservation Alternative 1975. *Address*: Environmental Studies, University of California, 116 Meadow Rd, Santa Cruz, CA 95060, USA. *Tel*: (408) 426-5261.

DAVAAJAMTS, Prof Tsevegyn; Mongolian botanist; b 1 July 1924; 1s 1d. *Education*: Irkutsk Univ, Russian Fed. *Present Position*: Sr Scientist and Principal Consultant, Inst of Biological Sciences, Mongolian Acad of Sciences. *Awards and Honours*: UNEP Global 500 1993. *Publications*: Map of the Mongolian Ecosystems (Co-Author) 1995. *Address*: Institute of Biological Sciences, Mongolian Academy of Sciences, Ulan Bator 51, Mongolia. *Tel*: (1) 358851. *Fax*: (1) 321638. *E-mail*: ibiotech@magicnet.mn.

DAVIS, Prof Dr Bruce W, B ECONS, DIP ENG, PH D; Australian retired academic and civil servant; b 30 Jan 1931, Hobart, Tas; m Rosalie Jean Davis. *Education*: Univ of Tasmania. *Career*: sr exec in the fields of natural resources policy and environmental management; consultant and adviser, Australian federal and Tasmanian state govts; consultant, IUCN. *Present Position*: Commr, Resource Planning and Devt Comm, Tas. *Awards and Honours*: mem, Order of Australia; various university prizes. *Publications*: Oceans Law and Policy in the Post-UNCED Era (Co-Ed); numerous publications in learned journals and books. *Addresses*: Antarctic Cooperative Research Centre, University of Tasmania, POB 252-77, Hobart, Tas 7001, Australia; 30 Wandella Ave, Hobart, Tas 7053, Australia. *Tel*: (3) 6226-2972. *Fax*: (3) 6226-2973. *E-mail*: b.w.davies@utas.edu.au.

DAYSH, Zena; British/New Zealand ecologist. *Career*: Founder mem and Chief Admin, Commonwealth Human Ecology Council—CHEC 1953–93. *Present Position*: Exec Vice-Chair, CHEC. *Awards and Honours*: Most Distinguished Fellowship, Delhi School of Non-Violence, India 1992; Scholarly Achievement Award, Institute of Oriental Philosophy 1992. *Publications*: Human Ecology—An Indian Perspective (Co-Ed) 1985; Human Ecology, Environmental Education and Sustainable Development (Co-Ed) 1991. *Address*: Church House, Newton Rd, London W2 5LS, United Kingdom. *Tel*: (20) 7792-5934. *Fax*: (20) 7792-5948.

DE CARVALHO, Prof Dr G Soares; Portuguese geologist and consultant; b 26 March 1920, Oliveira de Azeméis; 1s 1d. *Education*: Univ of Minho, Braga. *Career*: Asst Prof, Univ of Coímbra 1944–54; Head Geologist, Geological Map Div, Angola Survey 1954–58; Research Fellow, Council of Overseas Research, Lisbon 1958–61, with geological missions to Guinea-Bissau 1959, Goa 1960 and Angola 1960; Asst Prof and Aggregate Prof, Univ of Oporto 1961–70; Head, Dept of Earth Sciences, Inst for Scientific Research, Mozambique 1970–75, Dir 1975–76; Head, Dept of Earth Sciences, Univ of Minho 1976–88, Prof 1980–90. *Present Position*: Prof Emer, Univ of Minho 1990–. *Address*: Rua Elísio do Moura 62 r/c, 4710-422 Braga, Portugal. *Tel*: (253) 253081. *Fax*: (253) 240256.

DE RIJK, Peer; Dutch anti-nuclear campaigner; b 13 June 1966, Brokopondo, Suriname; 1s 1d. *Career*: Campaigner, FoE Netherlands 1985–96. *Present Position*: Campaigner, World Information Service on Energy—WISE 1997–. *Publications*: numerous articles on nuclear energy. *Address*: WISE International, POB 59636, 1040 LC Amsterdam, Netherlands. *Tel*: (20) 6126368. *Fax*: (20) 6892179. *E-mail*: wiseamster@antenna.nl.

DECAMPS, Henri, D SC; French research director. *Education*: Centre Nat de la Recherche Scientifique—CNRS, Toulouse. *Career*: Dir, Centre d'Ecologie des Ressources Renouvelables, CNRS 1980–95; Pres, Int Asscn for Landscape Ecology 1991–95. *Present Position*: Dir, Centre d'Ecologie des Systèmes Aquatiques Continentaux, CNRS, Toulouse. *Awards and Honours*: Corresp Mem, Acad des Sciences and Acad d'Agric de France; Hon Mem, Ecological Soc of America. *Publications*: Ecology and the Management of Aquatic-Terrestrial Ecotones (Co-Author) 1990. *Address*: Centre d'Ecologie des Systèmes Aquatiques Continentaux, 29 rue Jeanne-Marvig, BP 4009, 31055 Toulouse Cédex 04, France. *Tel*: 5-62-26-99-69. *Fax*: 5-62-26-99-99. *E-mail*: decamps@cesac.cemes.fr.

DECROLY, Vincent; Belgian politician and environmentalist; b 12 March 1963, Charleroi; 1s 1d. *Education*: Liège Univ. *Career*: Asst to mem of EP 1989–92; Fed Sec, Ecolo (Ecologist Party) 1992–94. *Present Position*: mem of Parl (Ecolo) 1994–. *Publications*: Si c'était à refaire 1999. *Address*: Maison des Parlamentaires, 1008 Brussels, Belgium. *Tel*: (2) 549-89-09. *Fax*: (2) 549-87-98. *E-mail*: vincent.decroly@lachambre.be. *Internet*: www.ecolo.be/vip/decroly/defaut.html.

DELKOV, Alexander; Bulgarian landscape architect; b 23 April 1959, Sofia; s of Nedyalko Delkov and Maria Delkova. *Education*: Forestry Univ, Sofia. *Career*: Designer, Agrolesproekt Inst for Surveys and Planning 1984–90. *Present Position*: Researcher and Dendrologist, Forest Research Inst, Bulgarian Acad of Sciences 1990–. *Publications*: Results from the Introduction of Cedrus Atlantica Manetti in South-West Bulgaria 1993; The Arboretum of the Forest Research Institute 1994; Analysis of the Geographical Provenance of the Dendroflora in Some Parks in Sofia 1998; Black Poplars in Bulgaria 2000. *Address*: Forest Research Institute, Bulgarian Academy of Sciences, 1756 Sofia, St Kliment Ohridski Blvd 132, Bulgaria. *Tel*: (2) 62-29-61. *Fax*: (2) 62-29-65. *E-mail*: forestin@bulnet.bg.

DELL'ALBA, Gianfranco; Italian politician and environmentalist; b 24 May 1955, Livorno. *Career*: Deputy Sec-Gen, Green Group, EP 1989–94. *Present Position*: MEP 1989–. *Address*: rue Wiertz, 1047 Brussels, Belgium. *Tel*: (2) 284-21-11. *Fax*: (2) 284-78-37. *E-mail*: g.dellalba@agora.stm.it.

DEMYANENKO, Valerii; Ukrainian lecturer and ecologist; b 13 Nov 1940, Romny, Sumsky region; s of Ivan Demyanenko and Anna Timchenko; m 1969; 2s. *Education*: Pedagogical Inst, Cherkassy. *Career*: teacher of biology, Nikolayevsky and Cherkassky regions 1968–82; founder, Ecological Education NGO 1994. *Present Position*: Docent (Assoc Prof), English and Ecology Dept, Eng and Tech Inst, Cherkassy 1995–. *Awards and Honours*: UNEP Global 500 1998. *Publications*: Creating the system of students permanent environmental eduaction in the Cherkassy region 1997; Ecology Reader, collected texts 1997; Environmental English (Ukrainian Dictionary for students) 1997; Methodological Approaches to the Stage Formation of Ecological Culture Value of a Personlaity 1998; Qualitative Estimation of Stage Engineering Education in a Framework of Sustainability 1999; Quality Assessment for Good Engineering Education 2003. *Addresses*: Ecological Education, 20700-719 Cherkasska oblast.Smila, Sverdlova St 72a, 80, Ukraine; Cherkassy 18006, Shevchenko St 460, Ukraine. *Tel*: (472) 43-36-80. *Fax*: (472) 31-00-89. *E-mail*: demyanen@chiti.uch.net. *Internet*: www.ecofond.com.au.

DI VITA, Charlotte, MA, MBE; British business executive; b 5 Sept 1966, London; d of John di Vita OBE and Esme (Chapman) di Vita (now Cope). *Education*: Edinburgh Univ; Lancing Coll. *Career*: Co-ordinator, first Anglo–Brazilian Conference on the Environment (on behalf of Dept for Int Devt) 1990–91. *Present Position*: Founder and Dir, Trade Plus Aid 1992–. *Awards and Honours*: Finalist, Cosmopolitan Women of Achievement Award 1993, 1999, Finalist and Special Mention 1998; Jt Runner Up, Exporter of the Year Japan Cup 1996; Midland Bank Sage Award for Business of the Year 1997; Certificate of Excellence for Outstanding Commitment to Quality Business Man, Parcel Force Worldwide Business Awards 1997; Surrey Business Creative Marketing Award 1997; Tate & Lyle Award for Sustainable Devt 1997; First Runner-Up, World Vision Award for Devt Initiative 1997, 1998; Women into Business Special Achievement Award 1998; Worldaware Award for Small Business 1998; Barclays Bank Customer of the Year 1999. *Address*: Trade Plus Aid, 17 Paxton Close, Kew Rd, Richmond TW9 2AW, United Kingdom. *Tel*: (20) 8948-0533. *Fax*: (20) 8255-3881. *E-mail*: charlotte@tradeplusaid.com. *Internet*: www.tradeplusaid.com.

DIMAS, Stavros, LLM; Greek politician and lawyer; b 30 April 1941, Athens. *Education*: Univ of Athens, New York Univ. *Career*: lawyer, Sullivan and Cromwell 1969–70; World Bank 1970–75; Deputy Gov, Hellenic Industrial Bank 1975–77; elected mem of Parl 1977; Deputy Minister of Econ Coordination 1977–80; Minister of Trade 1980–81; parl spokesman for New Democracy party 1985–89, Sec-Gen 1995–2000, Sr Mem, Political Analysis Steering Cttee 2000–03; Minister for Agriculture 1989–90, for Industry, Energy and Tech 1990–91; Head, Del to Council of Europe 2000–04; EU Commr for Employment and Social Affairs 2004. *Present Position*: EU Commr for Environment 2004–. *Address*: European Commission, Rue de la Loi 200, 1049 Brussels, Belgium. *Tel*: (2) 299-11-11. *Fax*: (2) 295-01-38.

DIOUF, Jacques; Senegalese agronomist and international civil servant; b 1 Aug 1938, Saint-Louis; m Aissatou Seye 1963; 1s 4d. *Education*: Nat School of Agric, Paris/Grignon, Nat School of Tropical Agronomy Applications, Paris/Nogent, Sorbonne Univ, Paris, France. *Career*: Exec Sec, African Groundnut Council, Lagos, Nigeria 1965–71, W African Rice Devt Asscn, Monrovia, Liberia 1971–77; Sec of State for Science and Tech, Govt of Senegal 1978–83; mem, Nat Assembly, Chair, then Sec, Foreign Relations Comm 1983–84; Sec-Gen, Banque Centrale des Etats de l'Afrique de l'Ouest 1985–90; Perm Rep of Senegal, UN 1991; leader (Chair of 1st Comm), Senegalese del, UN Confs on Science and Tech, Austria 1979, Industrial Devt, India 1980, New and Renewable Energy Sources, Kenya (Vice-Chair) 1981, Peaceful Use of Space, Austria 1982; African Rep, Consultative Group on Int Agricultural Research, USA. *Present Position*: Dir-Gen, FAO 1994–. *Awards and Honours*: Officier, Légion d'honneur, des Palmes Académiques, France; Grand Commdr, Order of the Star of Africa, Liberia. *Publications*: La détérioration du pouvoir d'achat de l'arachide 1972; Les fondements du dialogue scientifique entre les civilisations euro-occidentale et négro-africaine 1979; The Challenge of Agricultural Development in Africa 1989. *Address*: Food and Agriculture Organization—FAO, Viale delle Terme di Caracalla, 00100 Rome, Italy. *Tel*: (06) 57051. *Fax*: (06) 57053152. *E-mail*: fao-hq@fao.org.

DJATENG, Flaubert; Cameroonian environmentalist; b 25 Aug 1964, Nkongsamba; m Madeleine Yankoué 1986; 3d. *Education*: Univ of Dschang. *Career*: Founder, Cercle International pour la Promotion de la Création—CIPCRE 1993; Perm Sec, Fédération des Organisations Non-Gouvernementales pour l'Environnement au Cameroun—FONGEC 1994–. *Present Position*: Nat Dir, CIPCRE 1995–. *Address*: Cercle International pour la Promotion de la Création, BP 1256, Bafoussam, Cameroon. *Tel*: 44-62-67. *Fax*: 44-66-69. *E-mail*: cipcre@geod.geonet.de.

DJENBAEV, Prof Dr Bekmamat, D SC; Biogeochemist; b 29 Dec 1960. *Education*: Kyrgyz State National Univ. *Career*: Biogeochemist, Env Inst Geochemistry and Analytical Chemistry 1988–90. *Present Position*: Prof of Biology, Biology Soil Institute 2001–. *Publications*: over 70 scientific works including: The Geochemical Ecology of Land-Water Organisms 1999; The Biogenic Chemistry Elements and Selenium States 1999. *Address*: Biology Soil Institute, Kyrgyz National Academy of Science, 265 Ave. Chui 720071 Bishkek, Kyrgyzstan. *Tel*: (312) 243991. *Fax*: (312) 243607. *E-mail*: bekmamat2002@yandex.ru.

DOLMAN, Paul, B SC, PH D; British lecturer in ecology; b 19 Oct 1962, Kenya; 2c. *Education*: Univ of E Anglia—UEA. *Career*: involved in nature conservation –1988; researcher, habitat man for conservation; founder, Landscape Ecology Research Group, Centre for Ecology Evolution and Conservation, Univ of E Anglia. *Present Position*: Lecturer in Ecology, UEA 1995–. *Address*: School of Environmental Sciences, University of East Anglia, Norwich NR4 7TJ, United Kingdom. *E-mail*: p.dolman@uea.ac.uk. *Internet*: www.uae.ac.uk/env.

DOMOTO, Akiko, BA; Japanese politician and business executive; b 31 July 1932, Oakland, CA, USA; d of Takayuki and Ryuo Domoto. *Education*: Tokyo Women's Christian Coll. *Career*: Reporter, Camera Operator, Newscaster and Dir, Tokyo Broadcasting System 1959–89; Bd Mem, Family Planning Fed; Int Vice-Pres, Global Legislators Org for a Balanced Environment—GLOBE International, Japan 1990–99, Chair, Biodiversity Working Group 1991–; Co-Founder, Japan Women's Global Environment Network Int—GENKI 1992–. *Present Position*: Mem, House of Councillors, Japanese Diet 1989–, Research Cttee on Soc of Co-operative Way of Life, Nat Welfare Cttee, Special Cttee on Okinawa and Northern Problems; Vice-Pres and Regional Councillor, IUCN 1997–; Int Pres, GLOBE Int 1999–. *Awards and Honours*: Japanese Citizens' Broadcasting League Award 1991; Conference of Japanese Journalists Award; Bronze Medal, New York Film Festival 1985. *TV Documentaries*: Baby Hotel 1981; Tibetan New Year 1984; The Phantom Kingdom 1984; Monitoring Japan's ODA 1987; The Age of Child Slavery 1989; A Threat to Life: The Impact of Climate Change on Japan's Biodiversity (Co-Ed) 2000. *Address*: House of Councillors, 1-7-1, Nagata-cho, Chiyoda-ku, Tokyo 100, Japan. *Tel*: (3) 3508-8422. *Fax*: (3) 3506-8085. *E-mail*: ab04-ad@t3.rim.or.jp. *Internet*: www.domotoakiko.to.

DONOSO, Prof Claudio, M SC; Chilean forest engineer; b 27 June 1933, Santiago; s of Jorge Donoso and Elsa Zegers; m Sonia Hiriart (died 1993); 2s 2d. *Education*: Inst Nacional, Univ de Chile, Santiago. *Career*: first Pres and Vice-Pres, Ingenieros Forestales por el Bosque Nativo; Prof of Botany and Forest Ecology, Univ de Chile. *Present Position*: Prof (semi-retd) of Forest Ecology, Forest Dynamics and Silviculture of Native Forests, Univ Austral de Chile. *Awards and Honours*: Luis Oyarzún Prize. *Publications*: Forest Ecology—The forest and its environment; The Temperate Forests of Chile and Argentina; Silviculture of Chilean Native Forests; Forest Typology of Chile. *Addresses*: Casilla 567, Universidad Austral de Chile, Valdivia, Chile; Eleodoro Yóñez 1688, Valdivia, Chile. *Tel*: (63) 22-1228. *Fax*: (63) 22-1230. *E-mail*: silvicul@uach.cl.

DONOVAN, Richard Zell, MS; American forestry specialist; b 6 Aug 1952; s of Lawrence and Berniece Donovan; m Karen Ann Alfonsi 1982; 1s 1d. *Education*: S Florida Univ. *Career*: Peace Corps volunteer, Paraguay 1975–77; logger, photographer and training consultant 1978–81; natural resource consultant 1982–86; Sr Fellow and Project Dir, WWF Costa Rica 1987–91; Visiting Lecturer, Univ of Vermont 1992–. *Present Position*: Dir, Timber Project and Smart Wood, Rainforest Alliance 1992–. *Address*: 2 Kettle Creek Rd, Jericho, VT 05465, USA. *Tel*: (802) 899-1383. *Fax*: (802) 899-2018. *E-mail*: rzd@smartwood.org.

DOTTO, Lydia Carol; Canadian science writer and editor; b 29 May 1949; d of August and Assunta (Paron) Dotto. *Education*: Carleton Univ, Ottawa. *Career*: Science writer, *Toronto Globe and Mail* 1972–78; Pres, Canadian Science Writers Asscn 1979–80; Exec Ed, Canadian Science News Service 1982–92; mem, Science and Technology Advisory Bd, Environment Canada. *Present Position*: Freelance science writer and ed. *Publications*: Ozone War (Co-Author) 1978; Planet Earth in Jeopardy 1986; Canada in Space 1987; Thinking the Unthinkable: Civilization and Rapid Climate Change 1988; Asleep in the Fast Lane 1990; Blue Planet 1991; Ethics and Climate Change 1993; The Astronauts: Canada's Voyagers in Space 1993; Yucca Mountain Socio-economic Study 1993; Storm Warning 1999. *Address*: c/o Doubleday Canada, Random House of Canada Ltd, 2775 Matheson Blvd E, Mississauga, ON L4W 4P7, Canada.

DOWDESWELL, Elizabeth, MS; Canadian international civil servant. *Education*: Univ of Saskatchewan, Utah State Univ, USA. *Career*: Minister of Culture and Youth, Prov Govt of Saskatchewan; Man Consultant, Govt of Canada; Perm Rep of Canada, WMO, mem, Exec Council; Asst Deputy Minister, Environment Canada, Head, Atmospheric Environment Service, Deputy Minister of the Environment 1992; Exec Dir, UNEP 1993–98; Jt Chair, Great Lakes Water Quality Bd, Canada–USA Int Jt Comm; Canadian Prin Del, Intergovernmental Panel on Climate Change, UNCED, Rio de Janeiro, Brazil 1992, Jt Chair, Working Group on the Framework Convention on Climate Change; mem, Advisory Bd, Action by Canadians 1999–. *Present Position*: Pres, Nuclear Waste Man Org (NWMO) 2002–. *Address*: Nuclear Waste Management Organization, 49 Jackes Ave, First Floor, Toronto, ON M4T 1E2, Canada. *Tel*: (416) 934-9814. *Fax*: (416) 934-9526. *Internet*: www.nwmo.ca.

DRAMMEH, Halifa; international civil servant. *Career*: Sr Environmental Affairs Officer, Regional Office for Africa, UNEP. *Present Position*: Sr Programme Officer, Programme Management Div, UNEP. *Address*: POB 47074, Nairobi, Kenya. *Tel*: (2) 624278. *Fax*: (2) 624249.

DRDOŠ, Ján, PROF, D SC; Slovakian Geographer and Landscape Ecologist; b 6 Jan 1934, Viglas; s of Jozef and Amalia Drdos; m Oldriska Drdoš. *Education*: Comenius Univ, Bratislava. *Career*: Institute of Nature Protection 1958–63; Institute of Landscape Ecology 1963–76; Institute of Geography 1976–96; Dept of Landscape Ecology, Comenius Univ 1996–97. *Present Position*: Professor of Lanscape Ecology and Environmental Planning, Dept of Geography and Geoecology, Univ of Presov 1998–. *Awards and Honours*: Slovak Academy of Sciences, Silver Medal; Univ of Presov, Silver Medal. *Publications*: Landscape Synthesis (Editor) 1984; Landscape-Ecological Conditions of Sustainable Development (Co-Author) 1996; Geoecology and Environment I 1999; Environmental Planning 2001. *Address*: Pupavova 26, 84104 Bratislava, Slovak Republic.

DREGNE, Harold; American soil scientist; b 25 Sept 1916, WI; s of Carl John and Clementine Ellen (Wagner) Dregne; m Mary Mihevc 1943; 4d. *Education*: Univ of Wisconsin, Madison, Oregon State Univ, Corvallis. *Career*: Prof, Texas Tech Univ; Chair, Cttee on Arid Lands, American Soc for the Advancement of Science 1970–76; Contributor, UNEP Environmental Data Report 1992. *Present Position*: Prof, Int Center for Arid and Semiarid Land Studies, Texas Tech Univ 1972–. *Publications*: Soils of Arid Regions 1976; Managing Saline Water for Irrigation 1977; Desertification of Arid Lands 1983; numerous books and journal articles. *Address*: International Center for Arid and Semiarid Land Studies—ICASALS, Texas Technical University, POB 41036, Lubbock, TX 79409-1036, USA. *Tel*: (806) 742-2218. *Fax*: (806) 742-1954. *E-mail*: aihed@ttacs.ttu.edu.

DRYEF, Dr M Muhammad; Moroccan urban planner and administrator; b 30 June 1948, Taounat; 4c. *Education*: Univ of Social Sciences, Grenoble, France. *Career*: Gov, cities of Fez and Benslimane; Dir, Urban Agency, Casablanca; Dir-Gen, City Planning and Environmental Protection; Chief of Cabinet, Ministry of Town Planning, Habitat, Tourism and Environmental Protection. *Present Position*: Dir-Gen of

Interior Affairs, Ministry of the Interior. *Awards and Honours*: Commandeur de l'Ordre du Trône. *Publications*: Urban Planning of Casablanca City 1986; Urbanization and Urban Law 1993. *Address*: Ministry of the Interior, Rabat, Morocco.

DU TOIT, Raoul Frederic, M SC; Zimbabwean ecologist; b 6 June 1957, Zimbabwe; s of Friedrich P and Phyllis E du Toit; m Deleen Lourens 1989; 1d. *Education*: Univ of Rhodesia (now Univ of Zimbabwe); Univ of Cape Town, S Africa. *Career*: researcher, ecological impacts of Lake Kariba's discharge regime 1981–85; co-ordinator, environmental impact assessment (EIAs) of proposed hydroelectric schemes, Zambezi River 1981–83, consultant, similar EIAs 1992; co-ordinator, survey for Zimbabwe Rural Afforestation Project 1984–85, survey of woodfuel use, Harare 1987–88; researcher, environmental problems in communal farming areas 1983–85; scientific officer, rhinoceros conservation in Africa 1987–; planner, commercial wildlife projects 1989–; adviser to Namibian and Angolan Govts on EIA of proposed hydroelectric scheme, Cunene River 1994–98; developer, sectoral guidelines for EIAs in Zimbabwe 1997. *Present Position*: Project Executant, WWF Zimbabwe 1987–. *Awards and Honours*: Best B SC Student Award, Univ of Rhodesia 1977, 1978, 1979; Three Feathers Scholarship 1978–79; S African Breweries Scholarship, Univ of Cape Town 1980. *Publications*: Preliminary Assessment of the Environmental Implications of the Proposed Mupata and Batoka Hydro-Electric Schemes (Zambezi River) 1982; The Save Study: Zambezia (Supplement—Co-Author) 1989; African Elephants and Rhinos, IUCN Action Plan (Co-Author) 1990. *Addresses*: Rhino Conservancy Project, POB 1409, Causeway, Zimbabwe; POB BW 164, Borrowdale, Zimbabwe. *Tel*: (4) 252533. *Fax*: (4) 252534. *E-mail*: rdutoit@wwf.org.zw.

DUDLEY, Barbara; American environmentalist; 1d. *Education*: Stanford Univ, Univ of California at Berkeley. *Career*: Sr Counsel and Counsel to Bd, Agricultural Labor Relations Bd, CA 1979–83; Pres and Exec Dir, Nat Lawyers Guild 1983–87; Program Officer, then Exec Dir, Veatch Program 1987–92. *Present Position*: Exec Dir, Greenpeace USA 1992–. *Address*: Greenpeace USA, 1436 U St, NW, Washington, DC 20009, USA. *Tel*: (202) 462-1177. *Fax*: (202) 462-4507. *E-mail*: barbara.dudley@g2.bos.ns.gl3.

DUNCAN, Dr Patrick, BA, PH D; British research scientist; b 14 Feb 1948, Maseru, Lesotho; s of P B Duncan and Cynthia Ashley Cooper; m Alison M Phillips; 1s. *Education*: Univ of Oxford, Nairobi Univ, Kenya, Serengeti Research Inst, Tanzania. *Career*: research scientist, int consultancies, Station Biologique de la Tour du Valat, France 1975–85; Scientific Dir 1985–90; research scientist, Centre d'Etudes Biologiques de Chizé—CEBC, Centre Nat de la Recherche Scientifique—CNRS, France 1990–98. *Present Position*: Dir, CEBC 1999–. *Awards and Honours*: OBE. *Publications*: Horses and Grasses: the nutritional ecology of equids and their impact on the Camargue 1992; The European Roe Deer: the biology of success (Co-Ed) 1998; over 100 other scientific publications. *Address*: Centre d'Etudes Biologiques de Chizé—CEBC, CNRS, 79360 Beauvoir-sur-Niort, France. *Tel*: 5-49-06-61-11. *Fax*: 5-49-09-65-26. *E-mail*: duncan@cebc.cnrs.fr. *Internet*: www.cebc.cnrs.fr.

DUNLOP, Nicholas James, B A; Irish/New Zealand social entrepreneur; b 6 Jan 1956, New Zealand; s of Dennis and Paddy Dunlop; 1s 1d. *Career*: Sec-Gen, Parliamentarians for Global Action; Co-ordinator, Six Nations Peace Initiative 1984; Founder and Dir Earth Action Network; Founder and Sec Gen, e-Parliament. *Present Position*: Exec Dir, Earth Action 1992–. *Awards and Honours*: Indira Gandhi Peace Prize 1987. *Address*: e-Parliament, 35 Church Street, Wye, Kent, TN25 5BN, United Kingdom. *Tel*: (1233) 813796. *Fax*: (1233) 813795. *E-mail*: info@e-parl.net. *Internet*: www.e-parl.net.

DYANKOVA, Elena Detecheva, MS; Bulgarian forester; b 30 June 1957, Isperih; d of Detcho Dyankov and Ivanka Dyankova; m Ivo Simeonov Nedyalkov; 1d. *Education*: Higher Institute of Forestry and Forest Industries—HIFFI, Sofia. *Career*: Staff mem, *New Ludogorie* newspaper 1980–85. *Present Position*: Researcher, Forest Research Institute, Bulgarian Academy of Sciences 1985–. *Publications*: Biological Cycling of Minerals in Temperate Deciduous Forest Ecosystems in Northeastern Bulgaria; Throughfall and Stem-Flow Chemistry in a Spruce Forest on Rila Mountain. *Address*: Forest Research Institute, Bulgarian Academy of Sciences, 1756 Sofia, St Kliment Ohridski Blvd 132, Bulgaria. *Tel*: (2) 62-20-52. *Fax*: (2) 62-29-65.

E

EATON, Prof Dr Peter Price; British university professor; b 25 July 1935, Nuneaton; s of Allan and Margaret Eaton; m; 1s 2d. *Education*: London School of Econs, Univ of Hull. *Career*: Dir, Land Studies Centre, Univ of Papua New Guinea 1973–88; Visiting Scholar, Univ of Cambridge; Assoc, Inst of Pacific Studies, Univ of the S Pacific; mem, IUCN World Comm on Protected Areas; environmental man, conservation, land policy and environmental assessment consultant, int and govtal orgs, Brunei Shell and S Pacific Regional Environment Programme—SPREP. *Present Position*: Prof of Environmental Man, Dept of Public Policy and Admin, Univ of Brunei Darussalam 1993–. *Awards and Honours*: Vivian Stewart Award, Univ of Cambridge 1984; Pingat Indah Kerjah Baik—PIKB, Govt of Brunei 1999. *Publications*: Land Tenure and Conservation: Protected Areas in the South Pacific 1985; Borneo: Change and Development (Co-Author) 1992; Tradition and Reform: Land Tenure and Rural Development in SE Asia (Co-Author) 1996; Environment and Conservation in Borneo (Ed) 1999. *Addresses*: University of Brunei Darussalam, Tungku Link 1410, Brunei; 20 Wye Bank, Bakewell DE45 1BH, United Kingdom. *Tel*: 2249001. *Fax*: 2249003. *E-mail*: ppeaton@ubd.edu.bn.

EDINBURGH, HRH The Prince Philip, Duke of, KG, KT, OM, GBE; British royal and environmentalist; b 10 June 1921, Corfu, Greece; s of HRH Prince Andrew of Greece and Denmark and HRH Princess (Victoria) Alice Elizabeth Julia Marie; m HRH Princess Elizabeth (now HM Queen Elizabeth II); 3s 1d. *Education*: Royal Naval Coll, Dartmouth. *Career*: Mediterranean Fleet and British Pacific Fleet, SE Asia and Pacific 1939–45; renounced right of succession to thrones of Greece and Denmark and became naturalized British subject, adopting surname Mountbatten 1947; Personal ADC to HM King George VI 1948–52; Chancellor, Univ of Wales 1948–76, Univ of Edinburgh 1952–, Univ of Salford 1967–91, Univ of Cambridge 1977–; PC 1951–; Pres, English-Speaking Union of Commonwealth and RSA 1952–, Commonwealth Games Fed 1955–90, Royal Agricultural Soc of Commonwealth 1958–, BMA 1961–81, Council for Nat Academic Awards 1965–75, Scottish Icelandic Asscn 1965–, Maritime Trust

1969–, Nat Council of Social Service 1970–73, Australian Conservation Foundation 1971–76, David Davies Memorial Inst of Int Studies 1979–83, WWF Int 1981–96; Admiral of Fleet, Field Marshal, RAF 1953–, RNZAF and NZ Army 1978; Patron and Chair of Trustees, Duke of Edinburgh's Award Scheme 1956–; PC, Canada 1957–; Visitor, RCA 1965–. *Present Position*: Pres Emer, WWF Int 1997–. *Publications*: Birds from Britannia 1962; Wildlife Crisis (Co-Author) 1970; The Environmental Revolution: Speeches on Conservation 1962–77 1978; Men, Machines and Sacred Cows 1984; Down to Earth 1988; Living off the Land 1989. *Address*: Buckingham Palace, London SW1A 1AA, United Kingdom.

EDWARDS, Gordon Douglas, MS, MA, PH D; Canadian academic and consultant; b 18 July 1940, Landsdowne, ON; s of Alan Edwards and Mary Coghlan; m Karen Rogers 1966; 3s. *Education*: Toronto Univ, ON; Chicago Univ, IL, USA; Queen's Univ, Kingston, ON. *Career*: Ed, *Survival* int ecology magazine 1970–74; Dir, Green Energy Conf, Montréal 1989. *Present Position*: Prof of Mathematics, Vanier Coll, Montréal 1974–; Founder and Pres, Canadian Coalition for Nuclear Responsibility 1975–. *Awards and Honours*: Woodrow Wilson Fellow 1971; White Owl Conservation Award 1976; Gold Medal in Math and Physics 1971. *Publications*: Nuclear Waste: What, Me Worry? 1978; The Myth of the Peaceful Atom: Canada's Nuclear Dilemma 1983. *Address*: CP 236, Snowdon Station, Montréal, PQ H3X 3T4, Canada. *Tel and fax*: (514) 489-5118. *E-mail*: ccnr@web.net. *Internet*: www.ccnr.org.

EHRENFELD, David, MD, PH D; American biologist and writer; b 15 Jan 1938, New York, NY; s of the late Dr Irving and Anne Ehrenfeld; m Joan Gardner 1970; 2s 2d. *Education*: Harvard Univ, Harvard Medical School, Univ of Florida. *Career*: Founder Ed, *Conservation Biology*; columnist, *Orion Magazine*; mem, Bd of Dirs, Caribbean Conservation Corpn, E F Schumacher Soc, Educational Foundation of America. *Present Position*: Prof of Biology, Rutgers Univ 1974–. *Awards and Honours*: Fellow, American Asscn for the Advancement of Science 1986; Distinguished Achievement Award, Soc for Conservation Biology 1993. *Publications*: Biological Conservation 1970; Conserving Life on Earth 1972; The Arrogance of Humanism 1981; Beginning Again: People and Nature in the New Millennium 1993; Swimming Lessons: Keeping Afloat in the Age of Technology 2002. *Address*: Cook Coll, Rutgers Univ, New Brunswick, NJ 08901-8551, USA. *Tel*: (732) 932-9553. *Fax*: (732) 932-8746.

EHRLICH, Paul Ralph, PH D; American population biologist; b 29 May 1932, Philadelphia, PA; s of William and Ruth (Rosenberg) Ehrlich; m Anne Fitzhugh Howland 1954; 1d. *Education*: Univ of Pennsylvania, Univ of Kansas. *Career*: Assoc Investigator, USAF Research Project, Univs of Alaska and of Kansas 1956–57; Research Assoc, Chicago Acad of Sciences and Entomology Dept, Univ of Kansas 1957–59; Faculty mem, Stanford Univ 1959–; Correspondent, NBC News 1989–; Fellow, American Acad of Arts and Sciences and NAS. *Present Position*: Prof of Biology, Stanford Univ 1966–, Prof of Population Studies 1976–; Pres, Centre for Conservation Biology 1988–. *Awards and Honours*: Gold Medal, WWF Int 1987; UNEP Global 500 1989; Crafoord Prize, Swedish Acad of Sciences 1993, UNEP Sasakawa Prize 1994; Heinz Prize 1996; Tyler Prize (jtly) 1998; Blue Planet Prize (jtly) 1999. *Publications*: How to Know the Butterflies 1961; Population Resources; The Population Bomb 1971; How to be a Survivor (Co-Author) 1971; Man and the Ecosphere (Co-Ed) 1971; Global Ecology (Co-Author) 1971; Environment (Co-Author) 1972; Human Ecology (Co-Author) 1973; Ark II (Co-Author) 1974; The Process of Evolution (Co-Author) 1974; The End of Affluence (Co-Author) 1974; Biology and Society (Co-Author) 1976; The Race Bomb (Co-Author) 1977; Ecoscience: Population, Resources, Environment (Co-Author) 1977; Introduction to Insect Biology and Diversity (Co-Author) 1978; The Golden Door: International Migration, Mexico and the US (Co-Author) 1979; Extinction: The Causes and Consequences of the Disappearance of Species (Co-Author) 1981; Machinery of Nature 1986; Earth (Co-Author) 1987; The Birder's Handbook (Co-Author) 1988; New World, New Mind (Co-Author) 1989; The Population Explosion (Co-Author) 1990; Healing the Planet (Co-Author) 1991; Birds in Jeopardy 1992; The Stork and the Plow (Co-Author) 1995; Betrayal of Science and Reason (Co-Author) 1996; Human Natures 2000; Butterflies (Co-Author) 2003; On the Wings of Checkerspots (Co-Author) 2004; One with Nineveh (Co-Author) 2004. *Address*: Stanford University, Stanford, CA 94305-2493, USA. *Tel*: (415) 723-3171. *Fax*: (415) 723-5920. *Internet*: www.stanford.edu/group/CCB/Staff/paul.htm.

EKINS, Paul, M SC, M PHIL, PH D; British environmental economist; b 24 July 1950, Jakarta, Indonesia; s of John Robert and Lydia Mary Ekins; m Susan Anne Lofthouse 1979; 1s. *Education*: Imperial Coll and Birkbeck Coll, Univ of London. *Career*: General Sec, Ecology (later Green) Party 1979–82; Dir, The Other Economic Summit, New Economics Foundation 1985–87; Exec Dir, Right Livelihood Award 1987–90, Special Advisor 1990–94, Trustee 1994–; Programme Dir, Forum for the Future 1996–; mem, National Consumer Council 1996–2002; mem, Royal Comm on Environmental Pollution 2002–. *Present Position*: Head of Environment Group, PSI 2002–; Prof of Sustainable Dev, Univ of Keele 2000–. *Awards and Honours*: UNEP Global 500, 1994. *Publications*: The Living Economy: a New Economics in the Making (Ed) 1986; A New World Order: Grassroots Movements for Global Change 1992; Wealth Beyond Measure: an Atlas of New Economics (Co-Author) 1992; Real-Life Economics: Understanding Wealth Creation (Co-Author) 1992; Global Warming and Energy Demand (Co-Author) 1995; Economic Growth and Environmental Sustainability: the Prospects for Green Growth 2000. *Addresses*: c/o PSI–Policy Studies Institute, 100 Park Village East, London NW1 3SR, United Kingdom; Spire, Keele University, ST5 5BE. *Tel*: (20) 7468-0468. *Fax*: (20) 7388-0914. *E-mail*: p.ekins@psi.org.uk. *Internet*: www.psi.org.uk.

EL-ASHRY, Mohamed, PH D; American geologist and environmental scientist; b 21 Jan 1940, Cairo, Egypt; s of Taha El-Ashry and Faika Fadda; m Patricia R Murphy 1962; 2d. *Education*: Cairo Univ, Egypt, Illinois Univ. *Career*: Asst Prof of Geology, Cairo Univ, Egypt 1966–69, Prof of Environmental Sciences 1969–79; Sr Scientist, Environmental Defense Fund 1975–79; Dir of Environment, TN Valley Authority 1979–83; Sr Vice-Pres, World Resources Inst 1983–91; Chief Environmental Adviser to Pres and Dir of Environment, IBRD (World Bank) 1991–94. *Present Position*: Chair and CEO, Global Environment Facility 1994–. *Awards and Honours*: Air Conservationist of the Year 1983; UNCED Honour for Outstanding Contribution to Earth Summit 1992. *Publications*: Air Photography and Coastal Problems 1977; Water and Arid Lands of the Western United States 1988; Air Pollution's Toll on Forests and Crops 1989. *Address*: GEF Secretariat, 1818 H Street, NW, Sixth Floor, Washington, DC 20433, USA. *Tel*: (202) 473-0508. *Fax*: (202) 522-3240. *E-mail*: melashry@worldbank.org.

ELIAS, Dr Pavel, M SC, RNDR, PH D; ecologist and environmental botanist; b 19 Oct 1949, Velcice; m 1972; 3s. *Education*: Comenius Univ, Bratislava; Slovak Academy of Sciences; Slovak Technical Univ. *Career*: senior scientific worker, Comenius Univ 1979-90; Scientific Sec of the Scientific Council, Czech Research Prog on Structure and Function of Terrestrial Ecosystems 1981–86; Senior Research

Scientist SAS Inst of Botany 1990–94, Chief Research Scientist 1994–; Scientific Sec, Cttee for Environment at Presidium of the SAS 1990–96; Head, Dept of Ecology and mem, Faculty Council, Slovak Agricultural Univ 1997–2004. *Present Position*: Head, Dept of Ecology and mem, Faculty Council, Faculty of European Studies and Regional Development, Slovak Agricultural Univ 2004–. *Awards and Honours*: Agrokomplex Nitra Award 1984; Thai Biological Soc Award 1995; Slovak Botanical Soc Award 1999. *Publications*: numerous titles on plant biology and ecology; over 150 scientific papers. *Address*: Department of Ecology, Faculty of European Studies and Regional Development, Slovak Agricultural University, Marianska 10, 949 76 Nitra, Slovakia. *Tel*: (37) 6524-004. *Fax*: (37) 6526-637. *E-mail*: pavol.elias@uniag.sk.

ELIÁŠ, Pavol, PH D; Slovak ecologist and environmental botanist; b 19 Oct 1949, Velcice; m 1972; 3s. *Education*: Comenius Univ of Bratislava. *Career*: Fellow, Inst of Botany, Slovak Acad of Sciences 1972–74, Sr Research Scientist 1990–; Fellow, Inst of Experimental Biology and Ecology, Slovak Acad of Sciences 1975–79; presented *Ecology* exhibition 1984; Scientific Sec, Cttee for Environment, Presidium, Slovak Acad of Sciences. *Present Position*: Chair, Nat Cttee, ICSU Scientific Cttee on Problems of the Environment—SCOPE 1993–, Head, Dept of Ecology, Slovak Univ of Agric, Nitra 1995–, Programme Sec, European Ecological Fed 1995– and Pres, Slovak Ecological Soc—SEKOS 1997–. *Awards and Honours*: Agrokomplex Nitra Award 1984. *Publications*: A Dictionary of Ecology and the Environment (Co-Author) 1993; Principles of Ecology (Co-Author) 1994; Plant Population Biology III and IV (Ed) 1994, 1996; Monitoring of Biota in the Slovak Republic (Ed) 1996. *Address*: Mariánska 10, 949 01 Nitra, Slovakia. *Tel*: (87) 414-748. *Fax*: (87) 414-987. *E-mail*: elias@uniag.sk.

ELKINGTON, John Brett, M PHIL, FID, FRSA; British environmental consultant and writer; b 23 June 1949; 2d. *Education*: Univ of Essex, Univ Coll London. *Career*: Regular contributor on environmental, energy and devt issues, *New Scientist* magazine 1975–78; columnist, *The Guardian* 1989–92; Sr Planner and Assoc, Transport and Environment Studies; Co-Founder, Environmental Data Services Ltd—ENDS 1978, Man Dir 1981–83, Founder Ed, *ENDS Report*; Ed, *Biotechnology Bulletin* 1981–96; independent environmental consultant 1983–88; mem, EU Gen Consultative Forum on the Environment 1993–; Chair, Environment Management Faculty, Herning Inst, Denmark 1995–98; adviser, Business in the Environment; Non-Exec Dir, Asscn of Environmental Consultancies; mem, Council, Inst of Environment Management, Royal Town Planning Inst, Strategic Planning Soc, Council, Environmental Law Foundation; Patron, New Economics Foundation. *Present Position*: Chair, Environment Foundation 1986–, Co-Founder and Chair, SustainAbility 1987–. *Awards and Honours*: Hon Fellow, Centre for Social and Environmental Accounting Research—CSEAR, Univ of Dundee; UNEP Global 500 1989. *Publications*: The Green Consumer Guide; The Green Capitalists 1987; The Green Business Guide 1988; The Corporate Environmentalists 1992; Holidays That Don't Cost the Earth 1992; Coming Clean 1993; Cannibals with Forks 1997; Manual 2000: Life Choices for the Future you Want 1998. *Address*: 20-22 Bedford Row, London WC1R 4EB, United Kingdom. *Tel*: (20) 7269-6900. *Fax*: (20) 7269-6901. *E-mail*: elkington@sustainability.co.uk. *Internet*: www.sustainability.com.

ENCKELL, P H, PH D; Swedish academic and writer. *Education*: Lund Univ. *Present Position*: Man Ed, *Oikos, A Journal of Ecology* 1978–. *Address*: Dept of Ecology, Ecology Bldg, 223 62 Lund, Sweden. *Tel*: (96) 222-37-91. *Fax*: (96) 222-37-90. *E-mail*: oikos@ekol.lu.se. *Internet*: www.oikos.ekol.lu.se.

F

FAIRCLOUGH, Anthony John, CMG, MA; British environmental consultant; b 30 Aug 1924; s of the late Wilfrid and Lillian Anne (Townsend) Fairclough; m Patricia Monks 1957; 2s. *Education*: St Catharine's Coll, Cambridge. *Career*: Head of New Towns Div, Dept of the Environment 1970–72, Under-Sec 1973–74, Dir, Cen Unit on Environmental Pollution 1974–78; Chair, Environment Cttee, OECD 1976–79; Del, Governing Council, UNEP and Cttee on Challenges of Modern Soc, NATO; mem, Nat Cttee on Problems of Environment, Royal Soc; Dir of Int Transport, Dept of Transport 1978–81; Dir for Environment, Directorate-Gen XI (Environment, Consumer Protection and Nuclear Safety), Comm of the EC 1981–85, Acting Dir-Gen 1985–86, Deputy Dir-Gen, Directorate-Gen VIII (Devt) 1986–89, Special Adviser 1989–95; Dir, Groundwork Foundation 1989–95. *Present Position*: Sr Adviser, Environmental Resources Man 1989–. *Addresses*: 6 Cumberland Rd, Kew TW9 3HQ, United Kingdom; 32 quai aux Briques, Appt 12, 1000 Brussels, Belgium. *Tel*: (20) 8940-6999. *Fax*: (20) 8940-3758. *E-mail*: tony.fairclough@virgin.net.

FAITONDJIEV, Lyudmil, PH D; Bulgarian soil chemist and environmentalist; b 11 Aug 1943, Pernik; s of Petar Faitondjiev and Nadya Faitondjieva; m Kichka Faitondjieva 1972; 1s 1d. *Education*: Sofia Higher Inst of Chemistry and Tech. *Career*: Scientist, N Poushkarov Inst of Soil Science and Agroecology 1975–85. *Present Position*: Sr Scientist and Head, Dept of Soil Chemistry, N Poushkarov Inst of Soil Science and Agroecology 1985–. *Publications*: Problems of Soil Pollution 1984; Problems of Land Use and the Control of Heavy Metal Pollution of Soils 1987; Air Pollution and its Effect on Soils and Vegetation 1989; Essence and Problems of the Modern Agroecology 1997. *Address*: 1080 Sofia, Bankya 7, Bulgaria. *Tel*: (2) 911-71. *Fax*: (2) 24-89-37.

FALKENMARK, Prof Malin, PH D; Swedish academic. *Career*: Consultant on water-scarcity issues, FAO; Chair, Cttee on Water Strategies for the 21st Century, Int Water Resources Asscn, Working Group on the Int Geosphere-Biosphere Programme, Int Asscn for Hydrological Sciences; Prof of Int Hydrology, Nat Science Research Council 1986–92, Prof Emer 1992–. *Present Position*: Chair, Scientific Programme Cttee, Stockholm Int Water Inst—SIWI. *Awards and Honours*: UNEP Global 500 1991; KTH Prize 1995; Int Hydrology Prize 1998. *Publications*: more than 150 water science publications. *Address*: Stockholm International Water Institute—SIWI, Sveavägen 59, 113 59 Stockholm, Sweden. *Tel*: (8) 522-139-77. *Fax*: (8) 522-139-61. *E-mail*: malin.falkenmark@siwi.org. *Internet*: www.siwi.org/siwi/siwi.html.

FARMAN, Dr Joe; British scientist. *Career*: Discoverer of hole in ozone layer over Antarctic. *Present Position*: mem, British Antarctic Survey; consultant, European Ozone Research Co-ordinating Unit. *Address*: High Cross,

Madingley Rd, Cambridge CB3 0ET, United Kingdom. *Tel*: (1223) 251400. *Fax*: (1223) 362616.

FEARNSIDE, Philip Martin, PH D; American academic; b 25 May 1947, Berkeley; s of William Ward and Margaret Ellen (Martin) Fearnside; 2d. *Education*: Colorado Coll, Univ of Michigan. *Career*: Researcher, human carrying capacity studies, Osa Peninsula, Costa Rica 1972, Trans-Amazonian Highway, Brazil 1974–; participant, creation of Ouro Preto do Oeste Ecological Research Reserve, Brazil; researcher, sustainability studies, Amazonia, Brazil 1978–, tropical deforestation and global warming studies, Brazil 1983–, environmental impact studies, IBRD projects 1988–89. *Present Position*: Research Prof, Dept of Ecology, Nat Inst for Research in the Amazon—INPA, Brazil 1978–. *Awards and Honours*: Nat Ecology Prize, Brazil 1988/89; UNEP Global 500 1991. *Publications*: Human Carrying Capacity of the Brazilian Rainforest 1986; Human Occupation of Rondonia 1989. *Address*: CP 478, 69011-970 Manaus, AM, Brazil. *Tel*: (92) 642-3300. *Fax*: (92) 236-3822. *E-mail*: pmfearn@cr-am.rnp.br.

FERNÁNDEZ-GALIANO, Eladio, PH D; Spanish international civil servant and environmentalist; b 27 June 1953; s of Dimas Fernández-Galiano and Maria Victoria Ruiz Pérez; m Leticia Asensio 1983; 2s. *Education*: Univ Complutense de Madrid. *Career*: Researcher and Asst Prof, Ecology Dept, Univ Autónoma de Madrid 1977–85. *Present Position*: Civil servant, Council of Europe, Strasbourg, France 1985– and Sec, European Wildlife Convention (Berne Convention). *Address*: BP 431, 67075 Strasbourg Cédex, France. *Tel*: 3-88-41-22-59. *Fax*: 3-88-41-37-51. *E-mail*: eladio.galiano@coe.fr.

FERNANDO, Dr Ranjen Lalith; Sri Lankan medical consultant and naturalist; b 7 April 1939, Colombo; m Preethi Fernando 1971; 1d. *Education*: Univ of Ceylon, Royal Coll of Anaesthetists, UK. *Career*: Asst Sec then Sec, Wildlife and Nature Protection Society of Sri Lanka 1965–84; mem, Environmental Council of Sri Lanka 1984–87 and 1989–93, Fauna and Flora Advisory Cttee of Sri Lanka 1984–, Ministerial Cttee for Formulation of Nat Policy for Protection of Fauna and Flora in Sri Lanka, Nat Steering Cttee for the Biodiversity Convention in Sri Lanka, Coordinating Group for Man and Biosphere Areas in Sri Lanka, Int Forestry Advisory Group—Global 2000, Jimmy Carter Center, USA, UNDP–IUCN Forest Conservation Review Cttee, World Bank Forestry Devt Project; Hon Consultant, Species Survival Comm, IUCN, Switzerland, Fauna and Flora Preservation Soc, UK; Sri Lankan Rep, Convention on Int Trade in Endangered Species of Wild Flora and Fauna—CITES, Japan 1992, IUCN Regional Meeting 1993, IUCN General Assemblies in Costa Rica, Perth, Australia and Buenos Aires, Argentina. *Present Position*: Pres, Wildlife and Nature Protection Soc of Sri Lanka 1984–. *Awards and Honours*: UNEP Global 500 1992. *Publications*: Forestry Master Plan for Sri Lanka (Co-Ed); Role of the NGO in Wildlife Conservation; Development and Conservation; New Organizational Structure for Management of Wildlife in Sri Lanka; National Parks and the People; Changing Role of National Parks in Sri Lanka. *Address*: Wildlife and Nature Protection Society of Sri Lanka, Chaitiya Rd, Colombo 1, Sri Lanka. *Tel*: (1) 325248. *E-mail*: metalixc@mail.slt.lk.

FÉRON, Eric Maurice, M SC; French wildlife manager; b 11 April 1963, Fria-Kimbo, Guinea. *Education*: Univ of Edinburgh, UK, Univ of Lyons. *Career*: Co-Founder and Dir, Vetaid, UK and Mozambique 1988–90; Asst Project Man, Communal Areas Goat Research, Zimbabwe 1989–90; Consultant, Campfire and French Ministry of Foreign Affairs, Zimbabwe 1990–91; Country Rep, IUCN Zambia 1991–92. *Present Position*: Wildlife and Livestock Project Man, Centre de Coopération Internationale en Recherche Agronomique pour le Développement—CIRAD, Zimbabwe 1992–. *Awards and Honours*: Ordre des Vétérinaires 1987, 1988; Fondation pour la Vocation-Promotion, Guinea 1989. *Publications*: New Food Sources and Sustainable Development 1994. *Address*: Centre de Coopération Internationale en Recherche Agronomique pour le Développement—CIRAD, POB 1378, Harare, Zimbabwe. *Tel and fax*: (4) 722850.

FERRAZ, Bernardo Pedro; Mozambican environmental planner and politician; b 13 Nov 1938, Quelimane; s of Pedro Ferraz and Isabel Lino; m Arlinda António Das Monjane; 1s. *Education*: New York Univ, Univ of Pittsburgh, Harvard Univ, USA. *Career*: Promotion of community participation in decision-making in rural areas 1975–82; Dir, State Marketing Bd 1977–82; Officer, Nat Inst of Physical Planning 1982–94; Founder, Government Agency for Environmental Issues. *Present Position*: Minister for Environmental Coordination 1994–. *Awards and Honours*: UNEP Global 500 1995. *Address*: Ministry for Environmental Co-ordination, Avda Acordos de Lusaka 2115, CP 2020, Maputo, Mozambique. *Tel*: (1) 465843. *Fax*: (1) 465849. *E-mail*: micoa@ambinet.uem.mz.

FERRETTI, Janine Helene, BA; Canadian environmentalist; b 26 June 1958, Washington, DC, USA; d of James Joseph and Irma (Priebe) Ferretti; m Gary T Gallon 1983; 2d. *Education*: Univ of California, Santa Cruz, USA, York Univ, Downsview, ON. *Career*: Research Asst, IUCN 1977; Researcher, Environment Liaison Centre, Nairobi, Kenya 1980–81; Researcher, Pollution Probe Foundation 1984–87, Dir of Int Programmes 1987–89, Exec Dir 1989–94; mem, Int Trade Advisory Cttee, New Directions Group, Bd of Dirs, Canadian Council for Int Co-operation; mem then Vice-Chair, ON Round Table on Environment and Economy. *Present Position*: Dir, North American Comm for Environmental Cooperation 1994–. *Awards and Honours*: Environmentalist of the Year, City TV 1991. *Address*: 393 St Jacques St W, Suite 200, Montréal, PQ H2Y 1N9, Canada.

FETZ, Anita; Swiss historian, activist in women's movement and environmentalist; b 19 March 1957, Basel. *Career*: Councillor, Basel Canton 1984–88; mem, Nat Council (parl) 1985–89; Founder and Dir, FEMMEDIA Büro für Frauenförderung 1985–; mem, Advisory Council, Alternative Bank Schweiz—ABS 1990–. *Present Position*: Mem, Steering Cttee, Greenpeace Switzerland 1992–. *Publications*: Strukturwandel der Gesellschaft und Veränderung der Frauenrolle (Co-Author) 1988; Hoffnungspotential Frauen in der Wirtschaft 1989; Mut zur Karriere, Laufbahnplanung für Frauen (Co-Author) 1992; Rahmenkonzept für die Aus- und Weiterbildung in den öffentlichen Verwaltungen des Kantons Zürich, der Stadt Winterthur und der Stadt Zürich (Co-Author) 1992. *Address*: Oberer Rheinweg 37, 4058 Basel, Switzerland. *Tel*: (61) 6810972.

FISCHER, Joschka; German politician; b 12 April 1948, Gerabronn. *Career*: mem, Green Party 1982–; mem, German Bundestag (Fed Ass) 1983–85; Minister for Environment and Energy, Hesse state 1985–87, for Environment, Energy and Fed Affairs 1991–94; mem, Bundesrat (Fed Council) 1985–; Chair, Green Parl Group, Hessian Parl 1987–91; Deputy Minister-Pres of Hesse 1991–98; Speaker, Bündnis 90/Die Grünen (Alliance 90/The Greens) parl group, Bundestag 1994–98. *Present Position*: Vice-Chancellor and Minister of Foreign Affairs, Fed Govt 1998–. *Address*: Ministry of Foreign Affairs, 10117 Berlin, Werderscher Markt 1, Germany.

FISK, David John, PH D; British civil servant; b 9 Jan 1947; s of the late John Howard and Rebecca Elizabeth (Haynes) Fisk; m Anne Thoday 1972; 1s 1d. *Education*: St John's Coll, Cambridge, Univ of Manchester. *Career*: Higher Scientific Officer, Bldg Research Establishment 1972–73, Sr Scientific

Officer (Energy Conservation Research) 1973–75, Prin Scientific Officer 1975–78, Sr Prin Scientific Officer, Mechanical and Electrical Eng Div 1978–84; Asst Sec, Cen Directorate of Environmental Protection, Dept of the Environment 1984–87; Under-Sec, Dept of Environment (renamed Dept of the Environment, Transport and the Regions 1997) 1987–, Chief Scientist 1988–97, Dir, Air, Climate and Toxic Substances 1990–95, Environmental Protection and Int Directorate 1995–98; Visiting Prof, Liverpool Univ 1988–. *Present Position*: Dir, Central Strategy Directorate, Dept of the Environment, Transport and the Regions 1999–. *Publications*: Thermal Control of Buildings 1981; numerous papers on bldg science, systems theory and econs. *Address*: c/o Department of the Environment, Transport and the Regions, Ashdown House, 123 Victoria St, London SW1E 6DE, United Kingdom.

FITTER, Perin Savakshaw, B SC; British teacher and conservationist; b 24 Aug 1948, Eldoret, Kenya; d of Savakshaw and Narges Fitter. *Education*: Univ of Poona, India. *Career*: Teacher and Patron, Wildlife Clubs of Kenya 1969–79, Western Kenya Regional Co-ordinator 1980–86; Founder, Environment Conservation Volunteers Project 1987–. *Present Position*: Project Dir, Kenya NEEM Foundation 1997–. *Awards and Honours*: UNEP Global 500 1992. *Address*: POB 1268, Kisumu, Kenya. *Tel*: (35) 44360.

FOGG, Gordon Elliott, CBE, PH D, SC D, LL D, FIBIOL, FRS; British marine biologist; b 26 April 1919, Langar; s of Rev Leslie Charles and Doris Mary (Elliott) Fogg; m Elizabeth Beryl Llechid Jones 1945 (died 1997); 1s 1d. *Education*: Queen Mary Coll, Univ of London; St John's Coll, Cambridge. *Career*: Asst, Seaweed Survey of Britain, Marine Biological Asscn 1942; Plant Physiologist, Pest Control Ltd 1943–45; Asst Lecturer, Dept of Botany, Univ Coll London 1945–47, Lecturer 1947–53, Reader 1953–60; Prof of Botany, Univ of London at Westfield Coll 1960–71; Prof of Marine Biology, Univ Coll of N Wales 1971–85; Rockefeller Fellow 1954; Hon Sec, Inst of Biology 1953–56, Vice-Pres 1961–62, Pres 1976–77; Botanical Sec, Soc for Experimental Biology 1957–60; Pres, British Phycological Soc 1962–63, Int Phycological Soc 1964; Jt Hon Sec, 10th Int Botanical Congress, Edinburgh 1964; Visiting Research Worker, British Antarctic Survey 1966, 1974, 1979; Gen Sec, British Asscn 1967–72; Royal Soc/Leverhulme Visiting Prof, Univ of Kerala 1970; Chair, Council, Freshwater Biology Asscn 1974–85; Trustee, British Museum (Natural History) 1976–85, Royal Botanic Gardens, Kew 1983–89; mem, Royal Comm on Environmental Pollution 1979–85, Natural Environment Research Council 1981–82. *Present Position*: Prof Emer, Univ of Wales 1985–. *Awards and Honours*: Leverhulme Emer Fellowship 1986–88. *Publications*: The Metabolism of Algae 1953; The Growth of Plants 1963; Algal Cultures and Phytoplankton Ecology (Co-Author) 1967; Photosynthesis 1968; The Blue-Green Algae (Co-Author) 1973; The Explorations of Antarctica (Co-Author) 1990; A History of Antarctic Science 1992; The Biology of Polar Habitats 1998. *Address*: c/o Marine Science Laboratories, Menai Bridge, Anglesey LL59 5EY, United Kingdom. *Tel*: (1248) 351151. *Fax*: (1248) 716367.

FONSECA, Ivan Claret Marques, MD; Brazilian doctor and environmental educator; b 11 July 1938, Santo Estevau; s of Jose and ZuPetra Fonseca; m Leonidia Freire Fonseca; 1s 2d. *Education*: Escola Baiana da Medicina. *Career*: Gen Surgeon, Nanuque 1964–; mem, Geographic and Historic Inst of Minas Gerais state; mem, Brazilian Surgeons' Coll; mem, Brazilian Soc of Doctors and Writers. *Present Position*: Lecturer on the environment, secondary schools and univs 1973–; Traumatologist, Govt Hospital, Nanuque 1995–. *Awards and Honours*: Lyons Int Prize, Nanuque 1975; UNEP Global 500 1990; Rotary Int Prize 1995; Grande Oriente do Brasil Prize 1995. *Publications*: 72 books, numerous articles in professional journals. *Address*: Rua Pocas de Caldas 165, Nanuque, MG, Brazil.

FONTANA, Walter Atilio; Argentinian journalist and teacher; b 1968, Córdoba. *Career*: Founder, Juvenile Action and Ecological Fight Centre—CALEJ 1984; Pres, Organizing Cttee, 'World Year of Children and Environment' 1990; radio writer and dir. *Present Position*: Pres, Fundación de Acción Ecológica 1989– and Dir, Foro Ecologista. *Awards and Honours*: UNEP Global 500 1991. *Address*: Foro Ecologista, Casilla 1296, Correo Central, 5000 Córdoba, Argentina. *Tel*: (351) 512631. *Fax*: (351) 894353.

FORAND, Liseanne; Canadian environmental administrator; b 13 Sept 1957, Montréal. *Education*: Concordia Univ, Montréal. *Career*: Int Fisheries Officer, Dept of Fisheries and Oceans—DFO 1987–89, Int Dir 1990–93; Exec Dir, Canadian Asscn of Prawn Producers 1993–94; mem, Advisory Cttee on Environmental Man Systems, Standards Council of Canada 1996–99; Dir-Gen, Canadian Council of Ministers of the Environment—CCME 1994–99. *Present Position*: Assistant Deputy Minister, Policy, Fisheries and Ocans Canada 1999–. *Awards and Honours*: Canada 125 Medal 1994; Merit Award, DFO 1994. *Address*: c/o Fisheries and Oceans Canada, 200 Kent St, Ottawa ON K1A 0E6, Canada. *Tel*: (613) 993-1808. *Fax*: (613) 993-6958.

FORBES, Lynn Mackenzie, M SC; Canadian public land administrator (retired); b 22 Aug 1932, Berkeley, CA, USA; s of Robin S M and Helen E (Laundy) Forbes; m Dorothy Lee 1957; 1s 1d. *Education*: Univ of Alberta, Utah State Univ, USA. *Career*: Officer, Public Lands Div, Alberta Forestry, Lands and Wildlife 1955–88, Asst Deputy Minister 1981–87; mem, Soc for Range Man. *Address*: 647 Eagle Tree Close, Parksville, BC V9P 2N9, Canada. *Tel and fax*: (250) 954-0774.

FOX, Allan Maitland, BA; Australian environmental consultant; b 16 March 1931, Coonabarabran; 3s. *Career*: school teacher; Education Officer then Chief Wildlife Officer, NSW Parks and Wildlife Service; Officer-in-Charge, Education and Training, Fed Nat Parks and Wildlife Service; mem, Comm on Education and Communication, IUCN. *Present Position*: Man Dir, Allan Fox & Assocs environmental consultancy 1982–. *Awards and Honours*: Whitley Award (twice). *Publications*: Social Studies for Junior School 1962; Of Birds and Billabongs 1984; Australia's Wilderness Experience 1985; Kakadu Man 1985; Field Guide to South Australia's National Parks 1992; National Parks of Australia 2000. *Address*: RMB, 35 Beaumont Crescent, The Ridgeway, Queanbevan, NSW 2620, Australia. *Tel*: (2) 6297-3434. *Fax*: (2) 6284-4200. *E-mail*: afox@beatentrack.com.au.

FRASER, John Allen, LL B; Canadian politician; b 15 Dec 1931, Yokohama, Japan; m Catherine Findlay; 3d. *Education*: Univ of British Columbia. *Career*: barrister 1955–; mem, House of Commons (Parl) 1972–, Speaker 1986–93; Minister of the Environment and Postmaster-Gen 1979; Minister of Fisheries and Oceans 1984–85; fmr Caucus Spokesman for the Post Office, Labour and Environment; fmr mem, Parl Special Cttee on Acid Rain; fmr Chair, Environmental Law Subsection, Canadian Bar Asscn. *Present Position*: Canadian Amb for the Environment 1994–. *Address*: Lester B Pearson Bldg, 125 Sussex Drive, Ottawa, ON K1A 0G2, Canada. *Tel*: (613) 944-0886. *Fax*: (613) 944-0892.

FULLER, Dr Kathryn Scott, MS; American international organization executive; b 8 July 1946, New York, NY; d of Delbert Orison and Carol Scott Fuller; m Stephen Doyle

1977; 2s 1d. *Education*: Brown Univ, RI, Univ of Texas, Univ of Maryland. *Career*: Research Asst, Yale Univ 1968–69, American Cetacean Soc 1970–71, Harvard Univ 1971–73; Law Clerk, New York and Texas 1974–76, to Chief Justice John V Singleton, Jr, US Dist Court, Southern Dist of TX 1976–77; Attorney and adviser, Office of Legal Counsel, US Dept of Justice, Washington, DC 1977–79, Attorney, Wildlife and Marine Resources Section 1979–80, Chief 1981–82; Exec Vice-Pres and Dir, TRAFFIC USA 1982–89; mem, Council on Foreign Relations, Int Council of Environmental Law, Overseas Devt Council. *Present Position*: Pres and CEO, World Wildlife Fund—WWF USA 1989–. *Awards and Honours*: UNEP Global 500 1990. *Publications*: numerous articles in journals. *Address*: World Wildlife Fund—WWF USA, 1250 24th St, NW, Suite 500, Washington, DC 20037-1175, USA. *Tel*: (202) 293-4800. *Fax*: (202) 775-8287.

FUTRELL, J William, BA, LL B; American lawyer; b 6 July 1935; s of J W and Sarah Ruth (Hitesman) Futrell; m Iva MacDonald 1966; 1s 1d. *Education*: Columbia Univ, NY, Tulane Univ of Louisiana, Free Univ of Berlin, Germany. *Career*: Prof of Law, Univ of Alabama 1971–74, Univ of Georgia 1974–80; mem, bd of trustees, Sierra Club 1977–78. *Present Position*: Pres, Environmental Law Inst 1980–. *Publications*: Sustainable Environmental Law 1993. *Address*: Environmental Law Institute, 1616 P St, NW, Suite 200, Washington, DC 20036, USA. *Tel*: (202) 939-3800. *Fax*: (202) 939-3868.

G

GAKAHU, Christopher Gatama, PH D; Kenyan conservation biologist; b 15 Sept 1951, Kenya; m Alice Nyambura 1977; 2s 2d. *Education*: Nairobi Univ. *Career*: Observer, feeding habits and dispersal of large mammals, Amboseli Nat Park 1977–80; teacher and researcher, Nairobi Univ 1981–84; Founder, Sr Lecturer and Head, Wildlife Programme, Moi Univ 1985–88; mem, IUCN Species Survival Comm, East African Wildlife Soc, Nat Council for Science and Tech, Ecotourism Soc of Kenya. *Present Position*: Conservation Biologist and Consultant, Wildlife Conservation Int 1988–. *Publications*: Research Needs for Sustainable Wildlife Use 1991; African Rhinos: Current Numbers and Distribution; Visitor Dispersal Strategies in Ecotourism Management 1992; Tourist Attitudes and Use Impacts in Maasai Mara National Reserve 1992. *Address*: Wildlife Conservation International, POB 62844, Nairobi, Kenya. *Tel*: (2) 221699. *Fax*: (2) 215969.

GALAZY, Dr Gregory Ivan; Ukrainian botanist, ecologist and limnologist; b 5 March 1922, Mechebilovo, Kharkovskaya; s of Jogan Semenov and Praskovia Laurentia Galazy; m Alevtina Vasilievna Galazy 1945; 2s. *Education*: Irkutsk Univ. *Career*: Dir, Limnological Inst, USSR Acad of Sciences 1954–87, Baikal Ecology Museum 1988–93, Consultant 1993–; mem, Societas Internationalis Limnologiae—SIL, Hydrobiological Soc of Russia, Vice-Pres 1975–. *Present Position*: Co-Dir, Tacho-Baikal Int Limnological Inst 1991–and Head, Pribaikalia Dept of Ecological Research, Russian Acad of Sciences 1993–. *Awards and Honours*: UNEP Global 500 1987. *Publications*: Baikal and the Problem of Pure Water in Siberia 1968; Problems of Baikal 1973; The History of Botanic Research on Baikal 1980. *Address*: Dept of Ecological Research, Institute of Limnology, Russian Academy of Sciences, 666063 Irkutsk, ul Lermontova 281, Russian Federation. *Tel*: (3952) 46-05-04. *Fax*: (3952) 46-69-33. *E-mail*: galazy@bem.irkutsk.su.

GALLAGHER, Edward, C ENG, FIEE, FCIWEM; British environmental administrator. *Education*: Univ of Sheffield. *Career*: Staff mem, various industrial cos including Vauxhall Motors, Sandoz Products, Switzerland and Robinson Willey; Dir, Business Analysis and Dir, Market and Product Devt, Black and Decker 1971–86; Dir of Corporate Devt then Div Chief Exec and Bd Dir responsible for world-wide manufacturing, Amersham Int 1986–92; Chief Exec, Nat Rivers Authority 1992–96; Visiting Prof, Faculty of Technology and Business School, Middlesex Univ; mem, Council of Bristol Univ, Royal Inst of GB; Vice-Pres, Council for Environmental Educ. *Present Position*: Chief Exec, Environment Agency 1996–. *Address*: Environment Agency, Rio House, Waterside Drive, Aztec West, Almondsbury BS12 4UD, United Kingdom. *Tel*: (1454) 624400. *Fax*: (1454) 624409.

GALLOPIN, Dr Gilberto C; Canadian ecologist. *Career*: Int Inst for Sustainable Devt; Stockholm Environment Institute. *Present Position*: Regional Adviser on Envtl Policies, Economic Comm for Latin America and the Caribbean, Envt and Human Settlements Division. *Address*: POB 179D, Santiago, Chile. *Tel*: (2) 210-2000. *Fax*: (2) 208-1946. *E-mail*: ggallopin@eclac.cl.

GÁPER, Ján, PH D; Slovak forestry specialist and academic; b 18 Sept 1957, Fackov; m Svetlana Gáperová 1991; 1c. *Education*: Comenius Univ of Bratislava. *Career*: Scientist, Inst of Dendrobiology, Slovak Acad of Sciences 1981–90. *Present Position*: Deputy Dir, Inst of Forest Ecology, Slovak Acad of Sciences 1991– and Teacher, Ecology Faculty, Zvolen Univ 1993–. *Address*: Institute of Forest Ecology, Slovak Academy of Sciences, Štúrova 2, 960 53 Zvolen, Slovakia. *Tel*: (855) 223-12. *Fax*: (855) 274-85.

GARBA, Adamou; Niger civil servant and environmentalist; b 1954, Bara; m Fati Ibrahim 1988. *Education*: Lomé Univ, Togo, Poitiers Univ, France, Univ of Paris, France. *Career*: Dir of Communications, Ministry of Communications 1993–95. *Present Position*: Pres, Rassemblement des Verts 1994–, Sec-Gen, Federation of Ecologist Parties of Africa 1998–. *Publications*: La Télévision au Niger 1983. *Address*: BP 12515, Niamey, Niger. *Tel and fax*: 74-11-25. *E-mail*: adamou@intnet.ne.

GARCÍA-DURÁN, Germán, MS; Colombian environmental engineer and diplomat; b 30 Dec 1938, Cúcuta; s of Hermes and Francisca de García-Baldó; m Margarita González 1965; 3s. *Education*: Los Andes Univ, Santafé de Bogotá, DC, Univ of Notre Dame, IN, USA. *Career*: Prof of Environmental Sciences, Los Andes Univ 1970–76; Consultant in Environmental Engineering 1972–86; fmr Pres, Colombian Soc of Ecology; Gen Dir, Instituto Nacional de los Recursos Naturales Renovables y del Ambiente—INDERENA 1986–90; Vice-Pres, UNEP Gov Council, Kenya 1989–91, Vienna Convention for the Protection of the Ozone Layer 1991–93; Chair, G77 1996, 1999. *Present Position*: Colombian Amb to Kenya, Ethiopia and Tanzania, and Perm Rep, UNEP and UN Centre for Human Settlements—UNCHS (Habitat) 1990–93, 1995–, Chair, UN Comm on Human Settlements 1999–2001; Chair, Preparatory Comm for the 'Istanbul + 5' Conf 1999–2001. *Awards and Honours*: UNEP Global 500 1989. *Publications*: Rapid Filtration through Granular Media 1967; The Environmental Philosophy 1971; The Rio Conference 1992. *Address*: Embassy of Colombia, Muthiaga Rd,

POB 48494, Nairobi, Kenya. *Tel*: (2) 246770. *Fax*: (2) 246772. *E-mail*: emkenia@form-net.com.

GARRETT, Peter, LL B; Australian singer and environmentalist; b 16 April 1953, Wahroonga, NSW; s of Peter and Betty Garrett. *Education*: Australian Nat Univ, Univ of New South Wales. *Career*: Mem, Midnight Oil (pop group) 1977–2002, toured USA and Europe 1989, performed protest concert outside New York offices of Exxon petroleum co 1990; Parl Cand, Nuclear Disarmament Party 1984, 1987; campaigner for Koori (Australian Aborigines) rights; bd mem, Greenpeace 1993–94. *Present Position*: Pres, Australian Conservation Foundation 1989–93 and 1997–. *Awards and Honours*: D Litt, Univ of NSW 1999; Australian Humanitarian Foundation Envt Award 2000; Australia's Living Treasures Award, Nat Trust of Australia 1999. *Recordings/albums*: Midnight Oil 1978; Head Injuries 1979; Place without a Postcard 1980; 10,9,8,7,6,5,4,3,2,1 1982; Red Sails in the Sunset 1985; Diesel and Dust 1987; Blue Sky Mining 1990; Scream in Blue–Live 1992; Earth and Sun and Moon 1993; Breathe 1996; 20,000 Watt RSL 1997; Redneck Wonderland 1998; The Real Thing 2000. *Address*: Australian Conservation Foundation, 340 Gore St, Fitzroy, Vic 3065, Australia. *Tel*: (3) 9416-1166. *Fax*: (3) 9416-0767. *Internet*: www.petergarrett.com.au.

GARRIDO, Francisco; Spanish politician and environmentalist. *Present Position*: Spokesperson, Los Verdes (Green Party). *Address*: Apdo 565, 38400 Puerto de la Cruz, Tenerife, Spain.

GASKIN, Molly R; Trinidadian environmentalist; b 3 Dec 1941, Trinidad; d of Mildred and Henry Gaskin. *Education*: University of Wales, Cardiff, UK. *Career*: drafted programme for environmental education 1977; Chair, Council of Presidents of Environmental NGOs—COPE 1994–96; mem, Nat Wetland Cttee 1995–, Bd, Environmental Man Authority 1995–, Nat Selection Cttee, UNDP Global Environment Facility—GEF Small Grants Programme 1995–, Bd, Chaguaramas Devt Authority 1997–. *Present Position*: Pres, Pointe-a-Pierre Wildfowl Trust 1978–. *Awards and Honours*: Humming Bird Gold Medal, Trinidad and Tobago 1987; UNEP Global 500 1989; Woman of the Year, *Trinidad Guardian* 1989; one of 25 Women Leaders in the Environment, UNEP 1997. *Publications*: Sea Turtles and their Habitats, Trinidad and Tobago and the Caribbean. *Address*: 38 La Reine Townhouse, Flagstaff Hill, Long Circular Rd, St James, Trinidad and Tobago. *Tel and fax*: 628-4145. *E-mail*: molly.gaskin@petrotrin.com.

GAŠPARÍKOVÁ, Otília, PH D; Slovak plant physiologist; b 1935, Tvrdošín; d of Jozef Kubáni and Žofia Kubáni-Zmrazek; m 1963; 1d. *Education*: Comenius Univ of Bratislava. *Career*: Fellow, Inst of Botany 1961–64, Scientist 1965–, Head, Plant Physiology Dept 1977–90, appointed Dir of Inst 1990; Deputy Dir, Inst of Experimental Biology and Ecology 1982–90; mem and Chair, Scientific Bd for Biological and Ecological Sciences 1985–; Lecturer, Comenius Univ of Bratislava 1988–; Vice-Pres, Int Soc for Root Research. *Present Position*: mem, Plant Physiology Dept, Inst of Botany. *Awards and Honours*: Slovak Acad of Sciences Gold Medal 1989. *Publications*: Structural and Functional Aspects of Transport in Roots 1989; Root Metabolism 1992. *Address*: Institute of Botany, Slovak Academy of Sciences, Dúbravská cesta 14, 842 23 Bratislava, Slovakia. *Tel*: (7) 5941-2845. *Fax*: (7) 5941-2861. *E-mail*: botugasp@savba.savba.sk.

GAYOOM, Maumoon Abdul, MA; Maldivian politician; b 29 Dec 1937, Malé; s of the late Abdul Gayoom Ibrahim and Khadheeja Moosa; m Nasreena Ibrahim 1969; 2s 2d. *Career*: Head, Maldivian Del, major int environmental conventions and protocols; speaker, nat and int forums on global climate change and its implications for low-lying small island states; major participant, establishment of Commonwealth Group of Experts on Climate Change 1987; speaker, UN Gen Ass Special Debate on Issues of Environment and Devt, New York, USA 1987; host, Small States Conference on Sea Level Rise leading to adoption of the Malé Declaration on Global Warming and Sea Level Rise 1989; mem, Group of Eminent Persons on Sustainable Devt of Small Island States 1994–; prin speaker, World Municipal Leaders' Summit on Climate Change, Berlin, Germany 1995. *Present Position*: Pres, Maldives 1978–. *Awards and Honours*: UNEP Global 500 1988; Man of the Sea Award, Lega Navale Italiana 1991. *Address*: The President's Office, Mulee-agee Medhuziyaaraiy Magu, Malé 20-05, Maldives. *Tel*: 323701. *Fax*: 325500. *Internet*: www.presidencymaldives.gov.mv.

GEBRE EGZAIBHER, Dr Tewolde Berhan, B SC, PH D; Ethiopian plant ecologist, educator and environmental administrator; b 19 Feb 1940, Adwa; s of Gebre Egzaibher Gebre Yohannes and Maiza Tewolde Medhin; m; 4d. *Education*: Haile Selassie I Univ, Univ Coll of N Wales, Bangor, UK. *Career*: Graduate Asst, Dept of Biology, Haile Selassie I Univ 1963–64, Asst Lecturer 1964–66, Asst Prof 1969–74; Dean, Faculty of Science, Addis Ababa Univ 1974–78, Assoc Prof of Biology 1978–83; Leader, Research and Development in Rural Settings project (sponsored by Int Devt Research Centre—IDRC, Canada, and UN Univ, Japan), Ethiopian Science and Tech Comm 1978–82; Project Leader, Ethiopian Flora Project 1980–96; Pres, Asmara Univ 1983–91; Dir, Ethiopian Conservation Strategy Secr, Addis Ababa 1991–94; Negotiator Ethiopia and Africa Group in Cartagena Protocol for Biosafety; Leader, Ethiopian Delegation to WSSD, Johannesburg 2002. *Present Position*: Dir-Gen, Environmental Protection Authority, Ethiopia 1995–. *Awards and Honours*: Right Livelihood Award 2001. *Publications*: Soil Conservation in Ethiopia I: Conservation in its ecological perspective 1972; Soil Conservation in Ethiopia II: Some problems of afforestation and forest management 1974; preliminary ecological study. *Address*: POB 30231, Addis Ababa, Ethiopia. *Tel*: (9) 211274. *Fax*: (1) 669466. *E-mail*: sustain@telecom.net.et.

GELL-MANN, Murray, B SC, PH D; American theoretical physicist; b 15 Sept 1929, New York, NY; s of Arthur Gell-Mann and Pauline Reichstein; m 1st Josephine Margaret Dow 1955 (died 1988); m 2nd Marcia Ann Southwick 1992; 1s 1d. *Education*: Yale Univ, MIT. *Career*: Dir, J D and C T MacArthur Foundation 1979–, Chair, World Environment and Resources Cttee 1982–97; Co-Founder, Santa Fe Inst 1984; Dir, Aero Vironment Inc 1971–; Chair, Bd of Trustees, Aspen Center for Physics 1973–79; mem, Bd of Dirs, California Nature Conservancy 1984–93; mem, Scientific Advisory Cttee, Conservation Int 1993–; Trustee, Wildlife Conservation Soc 1994–; mem, US Pres's Council of Advisors on Science and Tech 1994–. *Present Position*: Prof and Distinguished Fellow, Santa Fe Inst 1993–; R A Millikan Prof Emer, California Inst of Technology 1993–. *Awards and Honours*: Hon D SC, Yale Univ 1959, Univ of Chicago 1967, Univ of Illinois 1968, Wesleyan Univ 1968; Hon Doctorate, Univ of Turin, Italy 1969, Univ of Utah 1970, Columbia Univ 1977, Univ of Cambridge 1980, Univ of Oxford 1992, S Illinois Univ 1993; Doctorate of Natural Resources, Univ of Florida 1994, S Methodist Univ 1999; Dannie Heineman Prize of the American Physical Soc 1959; Ernest O Lawrence Award 1966; Franklin Medal, Franklin Inst of Philadelphia 1967; John J Carty Medal, National Acad of Sciences 1968; Research Corpn Award 1969; Nobel Prize in Physics 1969; UNEP Global 500 1988; Erice Prize 1990. *Publications*: The Quark and the Jaguar: Adventures in the Simple and the Complex 1994. *Address*: California Institute of Technology, Pasadena, CA 91125, USA. *Tel*: (505) 984-8800. *Fax*: (505)

982-0565. *E-mail*: mgm@santafe.edu. *Internet*: www.santafe.edu/sfi/people/mgm.

GHABBOUR, Prof Dr Samir Ibrahim, M SC, PH D; Egyptian ecologist; b 30 Oct 1933; m; 2c. *Education*: Cairo University. *Career*: Prof of Ecology, Cairo University; collaborator, IUCN, UNESCO, UN ECOSOC and numerous int environmental orgs. *Present Position*: Prof Emer, Univ of Cairo; Pres, Egyptian Soc for Environmental Applications; Chair, Egyptian National MAB Cttee. *Awards and Honours*: Kuwait Prize for Environment 1986; Charles Sauvage Prize, French Ecological Soc 1990. *Publications*: The Status of the Rural Environment in Developing Countries 1990; Identification of Natural Heritage Sites in Arab Countries 1998; Environmental Values in Arab Countries 2000; The Arab Biodiversity Strategy: an Analytical Study 2001; Natural Resources and Their Conservation in Egypt and Africa 2004. *Address*: Dept of Natural Resources, Institute of African Research and Studies, Cairo University, 12613 Giza, Egypt. *Tel*: (2) 3924804. *Fax*: (2) 5780979. *E-mail*: ghabbour@aucegypt.edu.

GHEBRAY, Tekeste, M SC, PH D; Eritrean agricultural engineer; b 1949. *Education*: Alemaya Coll of Agric, Haile Selassie Univ (now Addis Ababa Univ), Ethiopia; Cranfield Inst of Tech, Silsoe Coll, UK; Centre of Remote Sensing, Imperial Coll of Science, Tech and Medicine, London, UK. *Career*: Land Devt Supervisor, HVA, Netherlands and Metahara Sugar Estate, Ethiopia 1972; Supervisor, Ethiopia–Swedish Extension and Project Implementation Dept—EPID, Ethiopia 1973–74, Project Co-ordinator 1974–80; Lecturer in soil and water conservation, Addis Ababa Univ 1980–81; soil and water conservation expert 1980–81; Head, Soil and Water Conservation and Forestry Dept, Ministry of Agriculture 1981–83, Co-ordinator, Project Preparation, Monitoring and Education Service 1994–96; Consultant in Soil and Water Conservation, Oxfam UK, Tigray 1989; environmental consultant and remote sensing specialist, Hunting Tech Services (UK), Algeria and Kenya 1989–91; consultant in environment, remote sensing and agriculture 1991–94. *Present Position*: Exec Sec, Intergovtal Authority on Devt—IGAD, Djibouti 1996–. *Address*: IGAD, POB 2653, Djibouti, Djibouti. *Tel*: 356452. *Fax*: 353520. *E-mail*: igad@intnet.dj.

GIBBS, Lois Marie. *Career*: founder, Citizens Clearinghouse for Hazardous Waste (later renamed the Center for Health, Environment and Justice) 1981–. *Present Position*: Exec Dir, Center for Health, Environment and Justice. *Address*: Center for Health, Environment and Justice, POB 6806, Falls Church, VA 22040, USA. *Tel*: (703) 237-2249. *E-mail*: chej@chej.org. *Internet*: www.chej.org.

GIRARDET, Herbert, B SC; German writer and film-maker; b 28 May 1943; s of Dr Herbert Girardet and Ingrid Hoffman; m Barbara Hallifax 1967; 2s. *Education*: Eberhard-Karls-Univ Tübingen, Free Univ of Berlin, LSE, UK. *Career*: Environmentalist, writer and film-maker 1974–; co-founder, Sustainable London Trust. *Present Position*: Patron, Soil Assoc, UK 1992–, Trustee, Earth Love Fund, UK 1993–, Visiting Prof, Middlesex Univ, UK 1994–, Visiting Prof, Sustainable Urban Devt, Univ of Northumbria 2003–, Visiting Prof. of Cities and the Environment, Univ. of the W of England 2004–; Chair, Schumacher Soc, UK 1995–; Dir, Under the Sky Urban Regeneration Co, Bristol 2004–; Dir of Research, World Future Council 2004–. *Awards and Honours*: UNEP Global 500 1992. *Publications*: Land for the People (Ed) 1976; Far From Paradise (Co-Author) 1986; Blueprint For A Green Planet (Co-Author) 1987; Earthrise 1992; The Gaia Atlas of Cities 1992; Making Cities Work (Co-Author) 1996; Getting London in Shape for 2000 1997; Creating a Sustainable London (Co-Author) 1998; Creating Sustainable Cities 1999; Tall Buildings and Sustainable Development (Co-Author) 2001; Creating a Sustainable Adelaide 2003; Cities, People, Planet 2004; *Films*: Jungle Pharmacy 1989; Halting The Fires 1990; Metropolis 1994; Countdown 2000 1997; Creating Sustainable Cities 1999. *Address*: 93 Cambridge Gardens, London W10 6JE, United Kingdom. *Tel*: (20) 8969-6375. *Fax*: (20) 8960-2202. *E-mail*: herbie@easynet.co.uk.

GIVEN, Assoc Prof David R, B SC, PH D; New Zealand botanist, conservation biologist and administrator; b 8 Nov 1943, Nelson; s of Bruce B and Brenda V G Given; m Karina C (Jansen) Given 1968; 3c. *Education*: Nelson Coll, Univ of Canterbury, Christchurch. *Career*: faculty mem, Dept of Botany, Univ of Canterbury 1965–90; field researcher, approx 60 countries world-wide; adviser and cttee mem, numerous nat and int environmental orgs. *Present Position*: Man, Int Centre for Nature Conservation, and Visiting Lecturer, Lincoln Univ; Head, David Given & Assocs 1992–; Consultant, Sarawak Biodiversity Council, Malaysia. *Awards and Honours*: Fellow, Linnean Soc, London, UK 1973; Canadian NRC Post-Doctoral Fellow 1973–74; Christchurch Int Photographic Exhibition, Bronze Medal 1976; NZ Int Photographic Exhibition, Bronze Medal 1977; Commonwealth Science Council Travelling Fellowship 1980, 1983; Finalist, A H & A W Reed Book of the Year 1981; Sir Joseph Banks Lecturer, Royal NZ Inst of Horticulture 1986; Tennant Lecturer, Univ of Otago 1991; Artiste of the Fédération d'Art Photographique (AFIAP) 1992; Assoc of Honour, Royal NZ Inst of Horticulture 1993; Guest Lecturer, NZ Photographic Soc Convention 1994, 1997; Life Member, WWF New Zealand 1995; Loder Cup for services to NZ and int conservation 1995; McCaskill Memorial Lecturer, Royal NZ Forest and Bird Soc 1998. *Publications*: The Arctic-Alpine Element of the Vascular Flora at Lake Superior (Co-Author) 1981; Rare and Endangered Plants of New Zealand 1981; Techniques and Methods of Ethnobotany: A Training Manual (Co-Author) 1994; Principals of Plant Conservation 1994. *Addresses*: International Centre for Nature Conservation, POB 84, Lincoln University, Canterbury, New Zealand; David Given & Assocs, 101 Jeffreys Rd, Christchurch 5, New Zealand. *Tel*: (3) 325-2811. *Fax*: (3) 325-3844. *E-mail*: givend@lincoln.ac.nz.

GLANTZ, Michael Howard, PH D; American scientist; b 14 Dec 1939, Providence, RI; m A Karen Lynch 1981; 1d. *Education*: Univ of Pennsylvania. *Career*: Researcher, climate-related environmental impact studies, with particular reference to the Aral and Caspian Seas. *Present Position*: Researcher then Sr Scientist, Environmental and Societal Impacts Group, Nat Center for Atmospheric Research 1974–. *Awards and Honours*: World Hunger Year Award 1988; UNEP Global 500 1990; Mitchell Prize 1992. *Publications*: Currents of Change: El Niño's Impact on Climate and Society; Drought Follows the Plough; Climate, Climate Variability and Fisheries; Societal Aspects of Regional Climate Change. *Address*: Environmental and Societal Impacts Group, National Center for Atmospheric Research, POB 3000, Boulder, CO 80307-3000, USA. *Tel*: (303) 497-8119. *Fax*: (303) 497-8125. *E-mail*: glantz@ucar.edu.

GLOWACINASKI, Prof Dr Zbigniew Andrzej; Polish biologist; b 5 Feb 1943, Dukla; s of Andrzej and Franciszka Glowacinski; m Maria Glacinski 1974; 2s. *Education*: Jagiellonian Univ. *Career*: Scientist, Polish Acad of Sciences 1966–. *Present Position*: Prof and Head, Species Protection Unit, Inst of Nature Conservation, Polish Acad of Sciences; Vice-Pres, State Council for the Conservation of Nature. *Awards and Honours*: European Prize for Nature Protection 1993; Gold Medal, Polish Ministry of the Environment 1995. *Publications*: Polish Red Data Book of Animals (Ed) 1992. *Address*: Institute of Nature Conservation, Polish Academy of

Sciences, 31-120 Kraków, Mickiewicza 33, Poland. *Tel*: (12) 4219701. *Fax*: (12) 4210348. *E-mail*: noglowac@cyf-kr.edu.pl.

GOLDBERG, Edward David, PH D; American marine chemist; b 2 Aug 1921, Sacramento, CA; s of Edward Davidow and Lillian (Rothholz) Goldberg; 1s 3d. *Education*: Univ of California, Berkeley, and Univ of Chicago. *Career*: practising marine chemist specializing in marine pollution and wastes; served as consultant to many nat and int bodies including UNESCO and FAO; mem, NAS. *Present Position*: Prof of Chem, Scripps Inst of Oceanography 1960–. *Awards and Honours*: Guggenheim Fellow 1960; NATO Fellow 1970; B H Ketchum Award 1984, Tyler Prize for Environmental Achievement 1989, John H Martin Medal of Excellence in Marine Science 1996. *Publications*: The Health of the Oceans 1976; Black Carbon in the Environment 1985; Coastal Zone Space 1994; more than 200 scientific papers. *Addresses*: Scripps Institution of Oceanography, La Jolla, CA 92093, USA (Office); 750 Val Sereno Drive, Encinitas, CA 92024, USA (Home). *Tel*: (619) 534-2407.

GOLDSMITH, Edward René David, MA; British environmentalist and writer; b 8 Nov 1928; s of Frank B H and Marcelle (Mouiller) Goldsmith; m 1st Gillian Marion Pretty 1953; m 2nd Katherine Victoria James 1981; 3s 2d. *Education*: Magdalen Coll, Oxford. *Career*: Founder, *The Ecologist* magazine, Ed 1969–89; Parl Cand, Ecology Party, Eye 1974, Cornwall and Plymouth 1979; Adjunct Assoc Prof, Univ of Michigan, USA 1975; Visiting Prof, Sangamon State Univ, USA 1984. *Present Position*: Publisher, *The Ecologist* 1970–. *Awards and Honours*: Chevalier de la Légion d'honneur, France 1991; Right Livelihood Award (Alternative Nobel Prize) 1991. *Publications*: Can Britain Survive? (Ed) 1971; A Blueprint for Survival (Co-Author) 1972; The Future of an Affluent Society: The Case of Canada 1976; The Stable Society 1977; La Médecine à la Question (Co-Ed) 1981; The Social and Environmental Effects of Large Dams (Co-Author) 1986; Green Britain or Industrial Wasteland? (Co-Ed) 1986; The Earth Report (Co-Ed) 1988; The Great U-Turn 1988; 5,000 Days to Save the Planet (Co-Author) 1990; The Way: An Ecological World View 1991; The Case Against the Global Economy and For a Turn Towards the Local (Co-Author and Ed) 1996. *Address*: 46 The Vineyard, Richmond TW10 6AN, United Kingdom.

GOLITSYN, Prof Georgii Sergeyevich, D SC; Russian atmospheric scientist; b 23 Jan 1935, Moscow; s of Sergei Golitsyn and Klaudia Golitsyna; m Ludmila Lissitskaya 1956; 2d. *Education*: Moscow State Univ. *Career*: Staff mem, Inst of Atmospheric Physics, USSR (now Russian) Acad of Sciences 1958–, mem, Presidium 1992–; Deputy Chair, Cttee of Scientists for Global Security; bd mem, Soc for Ecology and Peace. *Present Position*: Dir, Inst of Atmospheric Physics, Russian Acad of Sciences 1990–. *Awards and Honours*: Acad of Sciences Friedmann Prize in Meteorology 1990. *Publications*: An Introduction to the Dynamics of Planetary Atmospheres 1973; Study of Conventions with Geophysical Applications and Analogies 1980; Climatic Catastrophes (Co-Author) 1987. *Address*: 109017 Moscow, per Pyzhevskii 3, Russia. *Tel*: (095) 231-55-65. *Fax*: (095) 233-16-52. *E-mail*: mail_adm@omega.ifaran.ru.

GOODALL, Jane, PH D; British primatologist, environmentalist and writer; b 3 April 1934, London; d of Mortimer Herbert and Vanne (Joseph) Morris-Goodall; m 1st Hugo Van Lawick 1964 (divorced 1974); m 2nd M Derek Bryceson 1975 (died 1980); 1s. *Education*: Univ of Cambridge. *Career*: Sec, Univ of Oxford; Asst Ed, Documentary Film Studio; waitress; Asst Sec, Olduvai Gorge then Gombe Stream Game Reserve, Tanzania 1964; Founder, Cttee for Conservation and Care of Chimpanzees 1986; Visiting Lecturer, numerous univs including Yale Univ, USA; speaker on conservation issues, TV appearances include: 20/20, Nightline, Good Morning America; contrib to *New York Times*. *Present Position*: Founder, Jane Goodall Inst for Wildlife Research, Educ and Conservation, USA. *Awards and Honours*: Conservation Award, New York Zoological Soc; Franklin Burr Award (twice), Nat Geographic Soc; Nat Geographic Soc Centennial Award; Hubbard Medal 1995. *Publications*: Shadow of Man, Chimpanzees of Goombe 1986; Through a Window 1990; The Chimpanzee: The Living Link between 'Man' and 'Beast' 1992. *Address*: Jane Goodall Institute for Wildlife Research, Education and Conservation, POB 41720, Tucson, AZ 85717, USA.

GOODMAN, Gordon, M SC, D SC, FI BIOL; British environmental scientist; b 19 May 1926, Bridgend, Glamorgan, UK; s of F C and E M Goodman; m Margaret Rowe Laver 1951; 3d. *Education*: University of Wales. *Career*: Gen Sec, British Ecological Soc 1961–66; Co-Founder, Industrial Ecology Group; Prof of Applied Biology, Chelsea College, Univ of London 1972–77; Founding Dir, Monitoring and Assessment Research Centre, Univ of London; Founding Dir, Int Inst for Energy, Resources and the Human Environment, Royal Swedish Acad of Sciences 1977–88; Exec Dir, Stockholm Environmental Research Inst, UN, Sweden 1988–90, Chair 1990–92; Co-Founder and Pres, Int Service for Acquisition of Agri-Biotech Applications 1990–93; Co-Founder and mem of bd, Leadership for Environment and Devt. *Present Position*: Environmental researcher. *Awards and Honours*: Winston S Churchill Fellow; mem, Royal Swedish Acad of Sciences, 'Older' Linnean Gold Medal Award; Royal Order of Swedish Polar Star, 1st Class; D Sc hc Durham Univ, York. *Publications*: Sub-maritime Ecology of South Wales 1954; Ecology and the Industrial Society (Ed) 1965; Quantitative Methods for Heavy Metal Air Pollution 1972; The Ecology of Resource Degradation and Renewal (Ed) 1975; Reclamation of Derelict and Disturbed Land 1975; Energy Risk Management (Ed) 1979; European Oil Transition (Ed) 1981; Agenda for Science for Environment and Development into the 21st Century (Ed) 1992. *Address*: 20 Dunstal Field, Cottenham, Cambridge CB4 8UH, United Kingdom. *Tel*: (1954) 200077. *Fax*: (1954) 200078. *E-mail*: gordon.goodman@networld.com.

GOODWIN, Richard Hale, PH D; American botanist; b 14 Dec 1910, Brookline, MA; s of Harry M and Mary B (Linder) Goodwin; m Esther Bemis 1936; 1s 1d. *Education*: Harvard Univ. *Career*: Prof of Botany, Connecticut Coll, New London 1944–76, Dir, Interdepartmental Major in Human Ecology 1969–76; Dir, Connecticut Arboretum 1944–65, 1968; Commr, Connecticut Geological and Nat History Survey 1945–71; Pres, Conservation and Research Foundation, Inc 1953–94; Pres, Nature Conservancy 1956–58, 1964–66; Treas, Inst of Ecology 1975–77. *Present Position*: Prof Emer of Botany, Connecticut Coll 1976– and Trustee, Conservation and Research Foundation 1994–. *Awards and Honours*: George B Fell Award, Natural Areas Association 1993; Francis K Hutchinson Medal, Garden Club of America 1994. *Publications*: Inland Wetlands of the United States (Co-Author) 1975. *Addresses*: Conservation and Research Foundation, Connecticut College, POB 5261, New London, CT 06320, USA; POB 2040, Salem, CT 06420-2040, USA. *Tel and fax*: (860) 873-8514.

GORBACHEV, Mikhail Sergeyevich; Russian politician and environmentalist; b 2 March 1931, Privolnoye, Krasnogvardeisky Dist; s of Sergei Andreevich and Maria Panteleimonovna (Gopcalo) Gorbachev; m Raisa Titirenko 1953 (died 1999); 1d. *Education*: Moscow State Univ, Stavropol Agricultural Inst. *Career*: Machine operator 1946; joined CPSU 1952; 1st Sec, Stavropol Komsomol City Cttee 1955–58, then Deputy Head, Propaganda Dept, then 2nd then 1st Sec, Komsol Territorial Cttee; Party Organizer, Stavropol

Territorial Production Bd of Collective and State Farms 1962–66; Head, Party Bodies Dept, Territorial Cttee, CPSU 1962–66; 1st Sec, Stavropol City Party Cttee 1966–68; 2nd Sec, Stavropol Territorial Cttee, CPSU 1968–70, 1st Sec 1968–70; mem, Cen Cttee, CPSU 1971, Agric Sec 1978–85, alt mem, Political Bureau 1979–80, mem 1980–91, Gen Sec 1985–91; Deputy, Supreme Soviet 1970–91, Chair, Foreign Affairs Comm, Soviet of Union 1984–85, mem, Presidium 1985–88, Chair 1988–90; Chair, Supreme Soviet, RSFSR 1979–90; mem, Congress of People's Deputies 1989, Chair 1989–91; Pres, USSR 1990–91. *Present Position*: Founder and Head, Int Foundation for Socioeconomic and Political Studies—Gorbachev Foundation 1992–; Pres, Green Cross Int 1992–. *Awards and Honours*: Nobel Peace Prize 1990; Albert Schweitzer Leadership Award (jtly); Ronald Reagan Freedom Award 1992. *Publications*: A Time for Peace 1985; The Coming Century of Peace 1986; Speeches and Writings 1986–90; Peace Has No Alternative 1986; Moratorium 1986; Perestroika: New Thinking for Our Country and the World 1987; The August Coup (Its Cause and Results) 1991; My Stand 1992; The Years of Hard Decisions 1993; Mikhail Gorbachev: Life and Reforms 1995; Reflections on the Past and the Future 1998; On My Country and the World 1999. *Address*: Int Foundation for Socioeconomic and Political Studies, 125168 Moscow, Leningradsky pr 39, Bldg 14, Russian Federation. *Tel*: (095) 945-59-99. *Fax*: (095) 945-78-99. *E-mail*: info@gorbachev.net. *Internet*: www.gorbachev.net.

GORCHAKOVSKY, Pavel Leonidovich; Russian biologist; b 3 Jan 1920; m; 1s. *Education*: Siberian Inst of Wood Tech. *Career*: Chair, Urals Inst of Wood Tech 1945–58; Head of Lab, Inst of Ecology of Plants and Animals, Urals br, USSR (now Russian) Acad of Sciences 1958–88; corresp mem, USSR Acad of Sciences 1990, mem 1994; research in ecology and geography of plants, genesis of flora, protection of environment. *Awards and Honours*: Merited Worker of Science. *Publications*: Main Problems of Historical Phytogeography of the Urals 1969; Flora of the High-Mountain Urals 1975; numerous articles in scientific journals. *Address*: c/o Institute of Ecology of Plants and Animals, Urals Branch of the Russian Academy of Sciences, 620219 Yekaterinburg, 8 March Str 202, Russia. *Tel*: (3432) 29-40-92.

GORDON, John Keith, MA; British environmentalist; b 6 July 1940; s of Prof James and Theodora (Sinker) Gordon; m Elizabeth Shanks 1965; 2s. *Education*: Univ of Cambridge, Yale Univ, USA and LSE. *Career*: Diplomatic Service 1966–90, Perm Del to UNESCO 1984–85; Head, Nuclear Energy Dept, Foreign and Commonwealth Office 1986–88; Visitor, Centre for Environmental Tech, Imperial Coll, London 1988–90; Deputy and Policy Dir, Global Environmental Research Centre, Imperial Coll, London 1990–94; mem, Royal Soc of Arts, Gov Bd, Int Security Information Services—ISIS; Special Adviser, UNED-UK; mem, Nat Comm for UNESCO. *Present Position*: Ind consultant and author 1994–. *Publications*: Institutions and Sustainable Development: Meeting the Challenge 1991; 2020 Vision: Britain, Germany and a New Environmental Agenda 1994; Canadian Round Tables 1994. *Address*: 68 Hornsey Lane, London N6 5LU, United Kingdom. *Tel*: (20) 7263-3725. *Fax*: (20) 7263-3898.

GORE, Al (Albert, Jr); American politician; b 31 March 1948; s of Albert and Pauline (LaFon) Gore; m Mary E Aitcheson 1970; 1s 3d. *Education*: Harvard Univ, Vanderbilt Univ. *Career*: Investigative reporter and editorial writer, *The Tennessean* 1971–76; home-builder and land developer, Tanglewood Home Builders Co 1971–76; livestock and tobacco farmer 1973–; mem, House of Reps 1977–79; Senator for Tennessee 1985–93; Vice-Pres of USA 1993–2000, Presidential Candidate (Democrat) 2000; Head, Community Enterprise Bd 1993–. *Publications*: Earth in the Balance: Ecology and the Human Spirit 1992. *Address*: Office of the Vice-President, The White House, Old Executive Office Bldg, NW, Washington, DC 20503, USA.

GOULANDRIS, Niki, PH D; Greek ecologist and museum manager; b 9 Jan 1925, Athens; d of Minas and Eleni Kefala; m Angelos Goulandris (died). *Education*: Univ of Athens, Goethe Univ, Frankfurt am Main, Germany. *Career*: Co-Founder, Goulandris Natural History Museum 1964; Deputy Minister for Social Services 1974–75; Deputy Pres, Hellenic Radio and TV 1974–81, Nat Tourism Organization of Greece 1989–91; Pres, Save the Children Asscn in Greece 1980–; exec, women's org; mem, Linnean Soc of London, UK, Bd of Govs, Int Devt Research Centre of Canada for the Third World, Int Comm for Culture and Devt, Int Centre for Conservation Educ. *Present Position*: Pres, Goulandris Natural History Museum. *Awards and Honours*: mem The Dinnaean Acad; Woman of Europe 1991; Dr hc Univ Uppsala; mem French Academy of Moral and Political Sciences; mem Academia Europa; Legion of Honour (Officer). *Publications*: Wild Flowers of Greece 1968; Peonies of Greece. *Address*: Goulandris Natural History Museum, 13 Levidou St, 145 62 Kifissa, Greece. *Tel*: 2108014813. *Fax*: 2108080674. *E-mail*: goul@gnhm.gr. *Internet*: www.gnhm.gr.

GRANDTNER, Prof (Emer) Dr Miroslav Marian, D SC, M SC, B SC; Canadian educator; b 1928, Liptovska-Teplicka, Czechoslovakia (now Slovakia); s of Michael Grandtner and Sophia Sucha; m divorced; 1d. *Education*: Louvain Univ, Belgium; Laval Univ, Québec. *Career*: Prof of Forest Ecology, Laval Univ 1958–93; researcher, ecology of Senegalese savannah 1982–84, ecology of Malian-Sudanese-Saharan savannah 1987–92, classification and cartography of the vegetation of Belgium, Canada, Italy, Japan, Senegal, Spain, the USA and the West Indies 1985–94, study of the potential of B-chromosomes in plants as bioindicators (in collaboration), world diversity of trees (many collaborators); ed, co-ed, mem of ed cttee of several research or professional periodicals. *Present Position*: Adjunct Prof in Int Forestry, Laval Univ 1993–. *Awards and Honours*: foreign mem, Slovak Acad of Agricultural Sciences; hon mem, Québec Biologists Asscn; life mem, Québec Geographical Soc; hon researcher, CRBF. *Publications*: La Végétation Forestière du Québec Méridional 1966; Vegetation in Eastern North America 1994; World Dictionary of Trees: vol 1, North America 2005. *Address*: Department of Wood and Forest Sciences, Laval University, Ste-Foy, PQ G1K 7P4, Canada. *Tel*: (418) 656-2838. *Fax*: (418) 656-5262. *E-mail*: miroslav.grandtner@sbf.ulaval.ca. *Internet*: www.wdt.qc.ca.

GRASSL, Prof Dr Hartmut; German climatologist; b 18 March 1940, Salzburg, Austria; m Renate Grassl; 1d. *Education*: Univ of München, Univ of Hamburg. *Career*: Jr Scientist, Univ of Mainz 1971–76; Sr Scientist, Max-Planck Inst for Meteorology, Hamburg 1976–81; Prof of Theoretical Meteorology, Univ of Kiel 1981–84; Dir, Inst of Physics, Gesellschaft für Kernenenergieverwertung in Schiffbau und Schiffahrt 1984–88; Prof of Meteorology and Dir, Univ of Hamburg 1988–94; Chair, Global Change Advisory Bd, Govt of Germany 1992–94. *Present Position*: Dir, World Climate Research Programme, WMO 1994–. *Awards and Honours*: Max-Planck Prize, Humboldt Foundation 1991. *Publications*: Remote Sensing and Global Climate Change: Water Cycle and Energy Budget 1991; Environmental Aspects of Orbital Transport 1993. *Address*: World Meteorological Organization—WMO CP 2300, 41 ave Giuseppe Motta, 1211 Geneva 2, Switzerland. *Tel*: (22) 7308111. *Fax*: (22) 7342326.

GRAU VILARRUBIAS, Dr Juan, MD; Chilean/Spanish physician and ecologist; b 6 July 1917, Santiago, Chile; s of José Grau Coch and Dora Vilarrubias Vilalta; m Mitsuyo

Tanabe 1975; 2s 4d. *Education*: Univ of Barcelona, Spain, Univ of Chile, Santiago. *Career*: Founder and Officer, Instituto de Ecología de Chile 1974–, Nat Comm of Ecology 1984; Prof, Ecology Police School 1974–97; Chilean Del, UNEP, Nairobi, Kenya 1984–89. *Present Position*: Gen Sec, Instituto de Ecología de Chile. *Awards and Honours*: Ecologist Award, Sociedad Científica de Chile 1985; Distinguished Ecologist, Municipality of Santiago 1986; UNEP Global 500 1987; Premio Nacional del Medio Ambiente 2000 (Chile). *Publications*: Ecología del Pequeño José (4 vols) 1993–97; Ecología y Ecologismo 1996; Palmeras 2000; Voces Indigenas de Uso Comun en Chile 2002; Palmeras de Chile 2004. *Address*: Agustinas 641, Oficina 11, Santiago, Chile. *Tel*: (2) 633-1904. *Fax*: (2) 633-0963. *E-mail*: grau-eco@entelchile.net. *Internet*: www.doctorjuangrau.cl.

GREEN, Reg, B SC; British health, safety and environment officer; b 12 March 1952, Liverpool; s of James Reginald and Lilian (Tulloch) Green; m Marianne Wenning 1987; 1d. *Education*: LSE. *Career*: Asst, Legal Dept, Nat Union of Agricultural and Allied Workers 1980–85, Legal and Health and Safety Depts 1985–87; Health, Safety and Environment Officer, ICFTU 1987–92; Trade Union Co-ordinator, UNCED 1992, Contributor, *Earth Summit '92* report. *Present Position*: Health, Safety and Environment Officer, Int Fed of Chemical, Energy and Gen Workers' Unions—ICEF (now ICEM) 1992–. *Awards and Honours*: Fellow, Collegium Ramazzini. *Publications*: Priorities for the Future 1992. *Address*: ICEM, 109 ave Emile de Béco, 1050 Brussels, Belgium. *Tel*: (2) 626-20-20. *Fax*: (2) 648-43-16. *E-mail*: reg.green@icem.org.

GREGUS, Ctibor, D SC; Slovak forestry engineer and ecologist; b 6 Oct 1923, Kremnica; s of Jan Gregus and Julia Dobrá; m Mária Pajerská 1951; 1s 2d. *Education*: Slovak Tech Univ in Bratislava; Agricultural Coll, Brno (now in Czech Repub). *Career*: Head, Forestry Man Inst, Zilina 1951–64; Co-ordinator, Forestry Research Inst, Zvolen 1964–85; Head, Working Party, Int Union of Forestry Research Orgs 1967–75; mem, Forestry Man Union, Fed Repub of Germany 1972–80. *Present Position*: Scientific Consultant, Inst of Forestry Ecology, Slovak Acad of Sciences 1985–. *Awards and Honours*: Slovak Environmental Union Prize 1984; Slovak Acad of Sciences Gold Medal 1992. *Publications*: Management of the Shelterwood Forest 1976; Conception of Forestry Ecology Research 1985; Forestry Strategy in the Slovak Republic 1987. *Address*: Institute of Forest Ecology, Slovak Academy of Sciences, Štúrova 2, 960 53 Zvolen, Slovakia. *Tel*: (855) 320-313. *Fax*: (855) 274-85. *E-mail*: gregus@sav.savzv.sk.

GRINEVALD, Dr Jacques M L; French ecologist and writer; b 14 Jan 1946, Strasbourg; s of Théo Grinevald and Madeleine Promenet. *Education*: Univ of Paris; Univ of Geneva, Switzerland. *Career*: ecological writer and activist 1970–; founder mem, ECOROPA, Int Soc for Ecological Econs; promoter, int awareness of global ecology and the biosphere. *Present Position*: dir of ecological studies, Ecole Polytechnique Fédérale de Lausanne 1981–, IUED 1982–, Univ of Geneva, Switzerland 1987–. *Awards and Honours*: First Prize in Philosophy, Facultés Catholiques de Lyon 1966; Humbert Prize in Philosophy, Univ of Geneva 1968. *Publications*: numerous scientific papers. *Address*: 6 chemin Rieu, 1208 Geneva, Switzerland. *Tel*: (22) 7357015. *Fax*: (22) 6095947.

GROZEV, Ognyan, PH D; Bulgarian forester; b 6 April 1955, Plovdiv; s of Grozu Grozev and Elena Simeonova; m Mariana Penkova Grozeva 1979; 1s 1d. *Education*: Higher Inst of Forestry and Forestry Eng, Sofia. *Career*: Agrolesproject 1980–85; Forest Research Inst, Bulgarian Acad of Sciences 1985–. *Present Position*: Researcher, Ecology and Environmental Conservation Dept, Forest Research Inst, Bulgarian Acad of Sciences 1987–. *Publications*: Genesis of Growth Variety of Pinus Sylvestris L in the Region of the West Rhodopes 1989; Results of the Introduction of Cedrus Atlantica Manetti in South-West Bulgaria 1993. *Address*: Forest Research Institute, Bulgarian Academy of Sciences, 1756 Sofia, St Kliment Ohridski Blvd 132, Bulgaria. *Tel*: (2) 62-29-61. *Fax*: (2) 62-29-65.

GRUBB, Dr Michael; British environmentalist and energy researcher. *Present Position*: Adviser, UNEP; mem, Energy and Environmental Programme, Royal Inst of Int Affairs. *Publications*: Emerging Energy Technologies: Impacts and Policy Implications 1992; The Earth Summit Agreements: A Guide and Assessment 1993; Renewable Energy Strategies for Europe Vol I 1995, Vol II 1997; The Kyoto Protocol: A Guide and Assessment 1999. *Address*: Chatham House, St James Sq, London SW1Y 4LE, United Kingdom. *Tel*: (20) 7957-5711. *Fax*: (20) 7957-5710. *E-mail*: mjgrubb@compuserve.com.

GUDYNAS, Eduardo; Uruguayan social ecologist; b 4 Jan 1960, Montevideo; m Rosario Acosta 1983; 1d. *Education*: Uruguay Univ, Franciscan Multiversity, Montevideo. *Career*: Co-ordinator for Environment and Devt, CIPFE 1984–, 1st Latin American Congress on Ecology 1989; Duggan Fellow, Natural Resources Defense Council 1990; Faculty Assoc, Atlantic Coll, USA 1991–97, Inst of Ecology, Univ of San Andrés, Bolivia 1995–97; mem, Centro Latinoamericana de Ecología Social—CLAES 1991. *Present Position*: Co-ordinator for Environment and Devt, then Dir, CLAES 1989–. *Publications*: Ethics, Environment and Development in Latin America 1989; Praxis for Life: Social Ecology Methodologies 1991; Selling Nature: Trade, Integration and Ecology 1996; Democracy and Ecology 1997. *Address*: Centro Latinoamericana de Ecología Social—CLAES, Amezaga 1891, Casilla 13125, 11700 Montevideo, Uruguay. *Tel*: (2) 922362. *Fax*: (2) 201908. *E-mail*: claes@adinet.com.uy.

H

HAFIDI, My El Mehdi, M SC; Moroccan environmental engineer; b 28 Oct 1956; s of Hafidi My Ali and Hassani Zahra; m Benlafqih Hafida 1985; 1s 1d. *Education*: Mediterranean Agronomic Inst, Ecole Nationale Forestière d'Ingénieurs, Sale. *Career*: Head, Desertification Control Office, Service Forestier d'Errachidia 1983–91, Dir, Service Forestier d'Errachidia 1991–94, project for creation of Parc National du Haut Atlas Oriental. *Present Position*: Head of Protected Areas, Div of Water and Forests, Ministry of Agric and Agricultural Investments, Errachidia 1996–. *Awards and Honours*: UNEP Global 500 1993. *Publications*: Food Habits and Preferences of Barbary Sheep in the Eastern High Atlas National Park, Morocco 1996. *Address*: Div of Water and Forests, Errachidia, Morocco. *Tel*: (5) 573572. *Fax*: (5) 572252.

HAIGH, Nigel, OBE, MA; British director of environmental policy institute; b 23 Feb 1938, Tokyo, Japan; s of Anthony and Pippa (Dodd) Haigh; m Carola Pickering 1971; 2d. *Education*: King's Coll, Cambridge. *Career*: Civic Trust 1973–80; Vice-Pres, European Environmental Bureau 1975–79; Dir, London Office, Inst for European Environmental Policy—IEEP 1980–89, fmr Dir, IEEP. *Present Position*: Dir, Foundation for European Environmental Policy and Chair then Dir, Green Alliance 1989–; Visiting Research Fellow, Imperial Coll Centre for Environmental Technology. *Publications*: EEC Environmental Policy and Britain 1987; A Manual of Environmental Policy: The EC and Britain 1992. *Address*: c/o IEEP, 52 Horseferry Rd, London SW1P 2AG, United Kingdom. *Tel*: (20) 7799-2244. *Fax*: (20) 7799-2600.

HAILES, Julia Persephone, MBE; British environmental consultant and author; b 23 Sept 1961, Templecombe, Somerset; m 1991; 3s. *Career*: Consultant, Cargill Pow Polymers, Procter & Gamble, Marks & Spencer, Novo Nordisk; Vice-Chair, ACCPE (Advisory Committee on Consumer Products and the Environment); Jt Founder, Sustain-Ability Ltd, Dir 1987–95; Dir, Creative Consumer Cooperative Ltd 1994–. *Present Position*: Dir, Jupiter Global Green Investment Trust. *Awards and Honours*: UNEP Global 500 1989. *Publications*: Green Consumer Guide (Co-Author) 1988; Green Pages: The Business of Saving the World 1988; The Green Consumer's Supermarket Shopping Guide 1989; The Young Green Consumer Guide 1990; The Green Business Guide 1991; Holidays That Don't Cost the Earth 1992; The Life-Cycle Assessment Sourcebook 1993; Manual 2000, Life Choices for the Future you Want 1998; The New Foods Guide, what's here, what's coming, what it means for us 1999; Mamal 2000 1998; The New Foods Guide 1999. *Address*: Tintinhull House, Tintinhull Yeovil BA22 8PZ, United Kingdom. *Tel*: (1935) 823972. *Fax*: (1935) 826176. *E-mail*: julia@juliahailes.com. *Internet*: www.juliahailes.com.

HAIR, Jay D, PH D; American environmental consultant; b 30 Nov 1945, Miami, FL. *Education*: Clemson Univ, Univ of Alberta. *Career*: Lecturer; writer, weekly syndicated column; mem, Bd of Dirs, Clean Sites Inc, The Windstar Foundation, Kids for Saving Earth, Earth Day 1990; Pres and CEO, Nat Wildlife Fed 1981–94; Chair of Bd and Pres, IUCN USA; Pres, IUCN 1994–97; mem, Investment Policy Advisory Cttee—INPAC to US Trade Rep, National Wetlands Policy Forum 1987–88, US Environmental Protection Agency Biotechnology Science Advisory Cttee 1987–89. *Present Position*: organizational devt and public policy consultant. *Awards and Honours*: Environmental Educator of the Year, Ball State Univ 1988; Edward J Cleary Award, Nat Acad of Environmental Engineers 1989; Theodore and Conrad Wirth Environmental Award, National Park Foundation 1990. *Publications*: Numerous articles on the environment for periodicals. *Address*: c/o National Wildlife Federation, 1400 16th St, NW, Washington, DC 20036-2266, USA. *Tel*: (202) 797-6800. *Fax*: (202) 797-6646.

HALIM, Dr Youssef, D ès SC; Egyptian marine biologist; b 27 Jan 1925, Cairo; m Amal Yassa 1958; 1s 1d. *Education*: Cairo Univ, Alexandria Univ, Univ of Paris, France. *Career*: Researcher and teacher in marine biology, specializing in dinoflagellates, red tides, estuarine ecology and heavy metal aquatic pollution; Principal Investigator, Aquatic Environmental Pollution Project, UNDP/UNESCO/Alexandria Univ 1980–86. *Present Position*: Prof of Marine Biology, Alexandria Univ 1973–. *Awards and Honours*: UNEP Global 500 1992. *Publications*: Ocean Margin Processes in Global Change (Co-Author) 1991; The Impact of the Nile and the Suez Canal on the Living Marine Resources of the Egyptian Mediterranean (1958–1986) 1995. *Addresses*: Department of Oceanography, Univ of Alexandria, 21511 Alexandria, Egypt; (home) 97 Abdel Salam Aref St, 21411 Alexandria, Egypt. *Tel*: (3) 5869661. *Fax*: (3) 5457611. *E-mail*: asclub@soficom.com.eg.

HALL, Prof David Oakley, PH D; British biologist; b 14 Nov 1935; s of Charles Hall and Ethel Oakley; m Peta Smyth 1981; 2d. *Education*: Univ of Natal, S Africa, Univ of California, Berkeley, USA. *Career*: Fellow, Johns Hopkins Medical School, Baltimore, MD 1963–64; Lecturer, King's Coll, London 1964–68, Reader 1968–74; Visiting Sr Research Scientist, Princeton Univ, USA 1990–91; Contrib, *Environmental Data Report,* UNEP 1992. *Present Position*: Prof of Biology, King's Coll, London 1974–. *Publications*: Photosynthesis 1972; Plants as Solar Collectors 1983; Biomass 1987. *Address*: Campden Hill Rd, London W8 7AH, United Kingdom. *Tel*: (20) 7333-4317. *Fax*: (20) 7333-4500.

HALLIDAY, Prof Timothy Richard, D PHIL; British biologist; b 11 Sept 1945, Marlborough; s of J H and E Halliday; m Carolyn Wheeler 1970; 1s 2d. *Education*: Univ of Oxford. *Career*: Lecturer then Sr Lecturer and Reader in Biology, Open Univ 1977–91. *Present Position*: Prof of Biology, Open Univ 1991– and Int Dir, Declining Amphibian Populations Task Force 1995–. *Publications*: Vanishing Birds 1978; Sexual Strategy 1980. *Address*: Department of Biology, Open University, Milton Keynes MK7 6AA, United Kingdom. *Tel*: (1908) 653831. *Fax*: (1908) 654167.

HAMANN, Prof Dr Ole Jørgen; Danish botanist; b 4 Feb 1944, Frederiksberg; s of Børge and Katrine Hamann; m Michelle Worning; 1d. *Education*: Univ of Copenhagen. *Career*: UNESCO Assoc Expert in Plant Ecology applied to Wildlife Conservation 1971–72; IUCN Plants Programme Officer 1984–87; Vice-Pres (Europe), Charles Darwin Foundation for the Galapagos Islands, Ecuador 1984–97; founding mem and Trustee, Botanic Gardens Conservation Int 1987; mem bd, WWF Denmark 1990–; mem, Danish Nat Nature Council 1998–2000. *Present Position*: Prof of Botany and Dir, Botanic Garden, Univ of Copenhagen 1989–; Vice-Chair, WWF Denmark 1998–. *Awards and Honours*: Rasch's Legat 1981; Top-Danmark Prize 1989; Merit Award, Charles Darwin Foundation 1999. *Publications*: Plant Communities of the Galapagos Islands 1981; Botanic Gardens and the World Conservation Strategy (Co-Ed) 1987; Botanical Research and Management in Galapagos (Co-Ed) 1990; Ex Situ Conservation in Botanical Gardens (Ed) 1992. *Address*: Botanic Garden, University of Copenhagen, Øster Farimasgade 2b, 1353 Copenhagen K, Denmark. *Tel*: 35-32-22-22. *Fax*: 35-32-22-21. *E-mail*: oleh@bot.ku.dk. *Internet*: www .botanic-garden.ku.dk.

HANBALI, Suleiman, B SC; Jordanian geographer; b 9 Sept 1938, Nablus; m Basima Baker Hilmi 1970; 2s 1d. *Education*: Beirut Arab Univ, Lebanon. *Career*: Head of Projects, Municipality of Kuwait City, Kuwait 1970–81; Head, Nature Protection Section, Dept of Environment 1981–92, Jordan Environmental Strategy 1989–92; Ed, *National Environmental Strategy* 1989–91. *Present Position*: Exec Dir, Jordan Environment Soc 1992–. *Address*: Jordan Environment Society, POB 922821, Amman 11192, Jordan. *Tel*: (6) 699844. *Fax*: (6) 695857. *E-mail*: jes@go.com.jo.

HANBURY-TENISON, Dr Robin, OBE; British environmentalist; b 7 May 1936, London; s of Maj Gerald and Ruth Tenison; m Louella Hanbury-Tenison; 2s 1d. *Education*: Magdalen Coll, Oxford. *Career*: Self-employed farmer, writer, journalist and photographer; leader, numerous expeditions to investigate natural resource management issues and human rights abuses; Co-Founder and Chair, Survival Int 1969–82; mem, Royal Geographical Soc (fmrly Councillor and Vice-Pres), Explorers' Club, Ecology Foundation; Pres, Cornwall

Wildlife Trust. *Present Position*: Pres, Survival Int (now Survival) 1982–. *Awards and Honours*: Dr hc, Mons-Hainaut 1992; Gold Medal, Royal Geographical Soc 1979; Krug Award for Excellence 1980. *Publications*: Mulu: The Rain Forest 1980; Worlds Apart 1984; The Oxford Book of Exploration 1993. *Address*: Cabilla Manor, Cardinham, Bodmin PL30 4DW, United Kingdom. *Tel*: (1208) 821224. *Fax*: (1208) 821267. *E-mail*: robin@cabilla.co.uk.

HARRISON, Paul Anthony, PH D; British writer and development consultant; b 10 July 1945, Oldham; s of George and Phyllis Harrison; m 1973; 2s. *Education*: Univ of Cambridge, LSE. *Career*: Social Policy Ed, *New Society* 1972–75; Assoc Ed, *People* 1982–84; Contributing Ed, *People and Planet* 1992–; Ed-in-Chief, *Independent Commission on Population and Quality of Life*; Consultant, IBRD, FAO, UNFPA, UN and nat devt agencies. *Present Position*: Self-employed writer and devt consultant; Pres, World Pantheist Movement. *Awards and Honours*: UNEP Global 500 1988; Global Media Award, Population Inst 1992. *Publications*: Inside The Third World 1979; The Third World Tomorrow 1980; Inside The Inner City 1983; The Greening of Africa 1987; The Third Revolution 1993; Caring for the Future (Ed) 1994; Elements of Pantheism 1999. *Address*: 29a Nassington Rd, London NW3 2TX, United Kingdom. *Tel and fax*: (20) 7794-6921. *E-mail*: harrison@dircon.co.uk.

HARROP, Sir Peter John, KCB, MA; British civil servant; b 18 March 1926, Morecambe; s of Gilbert and Frances May Harrop; m Margaret Elliott-Binns 1975; 2s. *Education*: Peterhouse, Cambridge. *Career*: New Towns Dept, Ministry of Town and Country Planning 1949–51; Ministry of Housing and Local Govt 1951–70, Head, Clean Air Branch 1956–60, SE Regional Planning Dept 1966–69; Regional Dir for Yorks and Humberside, Dept of Environment 1971–73, Head, Inner Cities Div 1977, Water and Environmental Protection Directorate 1977–79, Cabinet Office 1979–80, 2nd Perm Sec for Environment, Water and Housing 1981–86; Treasury 1973–76; Chair, UK Cttee for European Year of Environment 1986–88, UK Cttee for UNEP and UNED 1988–93, Advisory Cttee, Waste Watch 1988–91; Trustee, British Museum; Dir, Thames Water Authority and Thames Water PLC 1986–95, Chair, Bd Environmental Audit Cttee; Trustee, River Thames Boat Project 1988–. *Present Position*: Chair, River Thames Boat Project 2001–. *Address*: 19 Berwyn Rd, Richmond TW10 5BP, United Kingdom. *Tel and fax*: (20) 8876-0752. *E-mail*: harrop@globalnet.co.uk.

HASSAN, Parvez, LL M, SJD; Pakistani lawyer; b 30 Sept 1941; s of S Ahmad Hassan; m 1967 (divorced 1992); 1s 1d. *Education*: Harvard Univ, Yale Univ, USA, Punjabi Univ. *Career*: Mem, UNEP Mission to Bangladesh 1978, Pakistan 1981; Consultant, Workshop on Environmental Co-ordination of Devt Agencies, IBRD 1979, ESCAP 1985–86; Chair, Scientific Cttee, WWF Pakistan 1987–93; Chair, Comm of Environmental Law, IUCN 1991–96; mem, WWF Pakistan 1979–, Int Council on Environmental Law 1979–, Pakistan Environmental Protection Council 1989–. *Present Position*: Pres, Environmental Protection Soc of Pakistan 1989– and Chair, Leadership for Environment and Devt 1995–. *Awards and Honours*: UNEP Global 500 1991. *Address*: Environmental Protection Society of Pakistan, c/o Hassan & Hassan, PAAF Bldg, 2nd Floor, 7d Kashmir Egerton Rd, Lahore, Pakistan. *Tel*: (42) 6360800. *Fax*: (42) 6360811. *E-mail*: phassan@brain.net.pk.

HASSELL, Prof Michael Patrick, D PHIL, D SC, FRS; British biologist; b 2 Aug 1942, Tel-Aviv, Israel; s of Albert Marmaduke Hassell and Gertrude Loeser; m 1st Glynis Mary Everett 1966; m 2nd Victoria Anne Hassell 1982; 3s 1d. *Education*: Clare Coll, Cambridge, Oriel Coll, Oxford. *Career*: Research Fellow, Entomology Dept, Univ of Oxford 1968–70; Lecturer, Biology Dept, Imperial Coll, London 1970–75, Reader 1975–79, Prof 1979–, Deputy Head 1984–93, Head 1993–2001. *Present Position*: Prin, Faculty of Life Sciences, Imperial Coll, London 2001–; Trustee, Natural History Museum, London 1999–. *Awards and Honours*: Scientific Medal, Zoological Soc of London 1981; British Ecological Soc Gold Medal 1994; Weldon Memorial Prize, Univ of Oxford 1995. *Publications*: Insect Population Ecology (Co-Author) 1973; The Dynamics of Competition and Predation 1975; The Dynamics of Arthropod Predator–Prey Systems 1978; The Spatial and Temporal Dynamics of Host-Parasitoid Interactions 2000. *Address*: Silwood Lodge, Silwood Park, Ascot SL5 7PZ, United Kingdom. *Tel*: (1344) 294207. *Fax*: (1344) 874957. *E-mail*: m.hassell@ic.ac.uk.

HASSI, Satu, MP, M SC; Finnish politician and writer; b 1951, Helsinki; d of Osmo and Maija-Liisa Hassi; m Dr Jukka Valijakka; 2d. *Education*: Helsinki Univ of Tech, Tampere Univ of Tech. *Career*: Design and Research Engineer, Oy Tampella Ab 1979–81; teacher, Tampere Univ of Tech 1981–85; freelance writer 1985–91; mem Tampere City Council 1985–2000; Chair, Green League 1997–2001; Min of Environment and Devp Cooperation 1999-2003; candidate for EP 2003. *Present Position*: Mem, Tampere City Council 1985–, mem, Eduskunta (parl) 1991– and Chair, Parl Group, Vihreä Liitto (Green League) 1991–93, 1997, 2003–. *Awards and Honours*: Priyadarshni Acad Award, Mumbai 2002. *Publications*: Magdalene, Ashamed No More (poems) 1984; A Stationary Model of an MHD Generator as a Part of an Electrical Network 1985; Female Screw (novel) 1986; The Serpent and the Tree of Knowledge (report on women and tech) 1987; A Wall Which Does Not Exist 1988; The New Clothes of King Adam (TV play) 1989; Physics for High School (8 vols—Co-Author) 1994–2000; My Hair on a Hat Rack: a Minister's Diary 2003. *Address*: Mäikikatu 16, 33250 Tampere, Finland. *Tel*: (9) 4321. *Fax*: (9) 4322722. *E-mail*: satu.hassi@eduskunta.fi. *Internet*: www.satuhassi.net.

HAYNES-SUTTON, Dr Ann Marilyn, B SC, PH D; British conservation ecologist; b 7 Nov 1951, London; d of Ferdinand William and Marta (Piepzic) Haynes; m Robert Sutton 1987. *Education*: Dundee Univ; Univ of West Indies, Kingston, Jamaica. *Career*: ecologist, Govt of Jamaica, Natural Resources Conservation Dept 1979; Pres, Natural History Soc of Jamaica 1986–88; Founder, Jamaica Conservation and Devt Trust 1987. *Present Position*: Chair, Jamaica Jr Naturalists 1990–; ecological consultant 1987–. *Publications*: Human Exploitation of Seabirds in Jamaica 1987; Plan for System of Protected Areas for Jamaica 1992; Management Plan for Black River Managed Resource Protected Area (Co-Author) 1999. *Address*: Marshall's Pen, POB 48, Mandeville, Jamaica. *Tel*: 904-5454. *Fax*: 964-6383. *E-mail*: asutton@uwimona.edu.jm.

HAZARE, Kisan (Anna); Indian social scientist; b 15 Jan 1940, Bhingar. *Career*: Researcher and admin in the field of watershed devt and man, including soil and water conservation, afforestation through gully plugs, contour trenches, earthen and live bunds, percolation tanks, gabion structures, check dams, plantation and fodder devt. *Present Position*: Chair, State Govt Cttee for the Devt of 300 Model Villages in Maharashtra State. *Awards and Honours*: Vriksha Mitra Award 1988; Padma Shree and Padma Bhushan, Govt of India 1990, 1992; UNEP Global 500 1993; Vivekanand Seva Puraskar 1995. *Publications*: Ralegan Siddhi: A Veritable Transformation 1994. *Address*: Ralegan Siddhi, Parner Tehsil, Ahmednagar Dist, Maharashtra, India. *Tel*: (2488) 40227. *Fax*: (2488) 40224.

HEIDE, Ola Mikal, M SC, DR AGR; Norwegian botanist; b 26 April 1931, Trondenes; s of Hans Kr and Marit Heide; m Gerd Lillebakk 1955; 3s 2d. *Education*: Agric Univ of Norway; Univ of Wisconsin, USA. *Career*: Research Fellow, Agric Univ of Norway 1961–70; Prof of Plant Sciences, Makerere Univ of Kampala, Uganda 1970–72; Prof of Plant Physiology, Univ of Tromsø 1972–76; Prof of Botany, Agric Univ of Norway 1976–, Head, Dept of Biology and Nature Conservation 1990–95, Rector 1978–83; Vice-Chair, Agric Research Council of Norway 1979–84; mem, various advisory cttees, etc; mem, Norwegian Acad of Sciences, Royal Soc of Sciences of Uppsala 1991, Finnish Acad of Science and Letters 1994; Pres, Scandinavian Soc of Plant Physiology 1976–82, 1988–94, Fed of European Socs of Plant Physiology 1988–90. *Awards and Honours*: Kellogg Foundation Fellowship 1965; Norsk Varekrigsforsikrings Fund Science Prize 1968. *Publications*: more than 90 primary scientific publs in the field of plant physiology and ecophysiology. *Addresses*: Agricultural University of Norway, Dept of Biology and Nature Conservation, 1432 Ås-NLH, Norway (Office); Skogvegen 34, 1430 Ås, Norway (Home). *Tel*: 64-94-84-86. *Fax*: 64-94-85-02.

HENS, Prof Dr Luc; Belgian professor of human ecology; b 5 Dec 1951, Mechelen; s of Gaston Hens and Louisa Bruyninckx; m Erna De Cock 1982; 2s. *Education*: Free Univ of Brussels. *Career*: Asst, Dept of Human Ecology, Free Univ of Brussels 1974–80, Assoc Prof 1980–90; mem, Fed Council for Sustainable Devt; Editor, *International Journal of Environment, Development and Sustainabiltiy*. *Present Position*: Prof and Head, Dept of Human Ecology, Free Univ of Brussels 1990–. *Awards and Honours*: Laureate, Belgian Acad of Sciences. *Publications*: Numerous contribs to Belgian and int journals. *Address*: Laarbeeklaan 103, 1090 Brussels, Belgium. *Tel*: (2) 477-42-81. *Fax*: (2) 477-49-64. *E-mail*: gronsse@meko.vub.ac.be. *Internet*: www.vub.ac.be/MEKO.

HESSELINK, Frederik Joost; Dutch environmentalist; b 25 Aug 1945, Amersfoort; s of Sander Hesselink and Sofia Cornelia Klaertje Leusvelt; m Namgyal Lhamo 1979. *Education*: Utrecht State Univ. *Career*: Fellow, Inst of Int Law, Utrecht State Univ 1968–72; Co-Founder, Inst for Environmental Communication—SME 1975; Vice-Chair, IUCN Comm on Educ and Communication—CEC 1987–93; mem, Dutch Nat Comm on Environmental Educ. *Present Position*: Man Dir, SME 1983– and Chair, IUCN CEC 1994–. *Address*: Institute for Environmental Communication—SME, POB 13030, 3507 LA Utrecht, Netherlands. *Tel*: (30) 802444. *Fax*: (30) 801345.

HILLARY, Sir Edmund Percival; New Zealand explorer and diplomatist; b 20 July 1919, Auckland; s of Percival Augustus and Gertrude Hillary; m 1st Louise Mary Rose 1953 (died 1975); m 2nd June Mulgrew 1989; 1s 2d (1 deceased). *Education*: Auckland Univ. *Career*: RNZAF, Pacific 1944–45; NZ Garwhal Expedition to Himalayas 1951; British Expedition to Cho Oyu 1952; British Mt Everest Expedition 1953; Leader, NZ Alpine Club Expedition to Barun Valley 1954; NZ Antarctic Expedition 1956–58, S Pole 1957; Leader, Himalayan Expeditions 1961, 1963, 1964; Pres, Volunteer Service Abroad 1963–64; built hosp for Sherpa tribesmen, Nepal 1966; Leader, expedition to Mt Herschel, Antarctica 1967; River Ganges Expedition 1977; High Commr to India 1984; Consultant, Sears Roebuck & Co, Chicago. *Present Position*: bee farmer. *Publications*: High Adventure 1955; The Crossing of Antarctica (Co-Author) 1958; No Latitude for Error 1961; High in the Thin Cold Air (Co-Author) 1963; Schoolhouse in the Clouds 1965; Nothing Venture, Nothing Win 1975; From the Ocean to the Sky: Jet-Boating up the Ganges 1979; Two Generations (Co-Author) 1983. *Address*: 278a Remuera Rd, Auckland SE2, New Zealand.

HILLER, Wolfgang; Austrian economic and environmental consultant; b 8 Nov 1960; s of Ernst and Edeltrude Hiller; m; 1s. *Education*: Vienna Univ of Econs. *Career*: Project Man, Voest-Alpine AG, Linz 1985–86; Econ Co-ordinator, Fed Chancellery 1986–90; Foreign Dept, Girozentrale and Bank of Austrian Loan and Savings Asscns 1990–91;Admin, Cttee on Environment, Public Health and Consumer Protection. *Present Position*: Co-ordination of parliamentary work and sessions, Socialist Group, EP. *Address*: European Parliament, 97–113 rue Belliard, 1040 Brussels, Belgium. *Tel*: (2) 284-17-93. *Fax*: (2) 230-66-64. *E-mail*: whiller@europarl.eu.int.

HINDÁK, František, D SC, RNDR, PH D; Slovak biologist; b 25 March 1937, Trnava; s of Jozef Hindák and Vilma Bocková; m Dr Magdaléna Hindáková 1961; 1s 1d. *Education*: Charles Univ, Prague. *Career*: Inst of Microbiology, Czechoslovak Acad of Sciences, Trebon 1962–68, Inst of Botany, Bratislava 1968–, Sciences Faculty, Comenius Univ of Bratislava 1968–; Construction Faculty, Slovak Tech Univ, Bratislava 1987–; Pres, Slovak Botanical Soc 1990–93; Ed, *Preslia*, *Biologia*, *Algol Studies*, *Annales de Limnologie* and *Algologia*; Chair, Cttee for Environment, Presidium, Slovak Acad of Sciences. *Present Position*: Head, Dept of Lower Plants, Inst of Botany, Slovak Acad of Sciences and Prof, Sciences Faculty, Comenius Univ of Bratislava. *Awards and Honours*: Holuby Prize of the Slovak Botanical Soc 1985; Prize of the Slovak Academy of Sciences 1987, 1997; Prize of the Comenius Univ of Bratislava 2001, 2003; Prize of the Slovak Agriculture Uni of Hitra 2003. *Publications*: The Key to the Determination of Algae 1975; Freshwater Algae 1978; Studies on the Chlorococcal Algae 1990; Atlas of Cyanophytes. *Addresses*: Dúbravská cesta 14, 845 23 Bratislava, Slovakia; Astrová 16, 821 01 Bratislava, Slovakia. *Tel*: (2) 5942-6104. *Fax*: (2) 5477-1948. *E-mail*: frantisek.hindak@savba.sk.

HOFFMANN, Luc, D PHIL; Swiss biologist; b 23 Jan 1923, Basel; s of Emanuel and Maja (Stehlin) Hoffmann; m Daria Razumovsky 1952 (died 2002); 1s 3d. *Education*: Univ of Basel. *Career*: Founder and Dir Station Biologique de la Tour du Valat, Camarge 1954—1974, Chair 1974—2003, Hon Chair 2004–; Dir IWRB (now Wetlands International) 1962—68, now Dir Emeritus; Organizer project MAR joint wetland conservation programme by IUCN, IWRB nad ICBP (now Birdlife Int) resulting in the RAMSAR convention for wetlands; C0founder WWR, Pres 1961—88, now Vice-Pres emeritus; Vice-Pres IUCN 1966-69; Co-initiator Donana Nat Park, Spain and Prespa Nat Park, Greece; Chair, Soc for the preservation of Prespa 1990—2002; mem, exec cttee ICBP; Vice-Pres Wildfowl and Westlands Trust 1979, now hon life Fellow; Founder FIBA, Chair –2002; Chair WWF France 1996–2000, later Hon Pres; Co-founder WWF Greece 1994; Vice-Pres Hoffmann-LaRoche & Co 1990–96. *Present Position*: Pres, MAVA for conservation of nature; Hon Pres Station Biologique de la Tour du Valat. *Awards and Honours*: hon mem WWF Int 1986; Chevalier Legion of Honour 1989; Commander Order of the Golden Ark, Netherlands 1989; Dr hc Univ Basle 1990; Dr hc Univ of Thessaloniki 1992; Pro Natura 1993; Kai-Curry-Lindahl Award Colonial Waterbird Soc 1994; Order of Merit Mauritania 1998; Off Order of Honour, Greece 1998; Duke of Edinburgh Conservation Medal (WWF Int) 1998; Konrad-Lorenz-Medal 2003. *Publications*: Over 60 publications in the fields of ornithology, wetland ecology and conservation. *Address*: Station Biologique de la Tour du Valat, Le Sambuc, 13200 Arles, France. *Tel*: (4) 90-97-29-60. *Fax*: (4) 90-97-20-19. *E-mail*: l.hoffmann@tourduvalat.org. *Internet*: www.tourduvalat.org.

HOLCÍK, Vladimír, ENG; Slovak dam specialist, water engineer and researcher; b 8 Oct 1944, Bratislava; m Dr Marta Holcíková 1975. *Education*: Slovak Tech Univ in Bratislava, City Univ, Bratislava. *Career*: Hydroprojekt, Bratislava 1967–69; Slovak Hydrometeorological Inst 1969–72; Danube Basin Admin 1972–91; URBACO, Constantine, Algeria 1980–83, Eastern Hydroproject 1986–87; Danube Basin Admin– Gabcikovo WPP 1988–90; Dir, Water Research Inst 1990–95; Water Management Development Auth 1995–2000; Danube Commission 2001–03 (Novi Sad Proj Dept Dir). *Present Position*: WMDA, Bratislava 2004–. *Publications*: List of Danube Bridges. *Addresses*: Karloveská 2, 842 04 Bratislava, Slovakia; Hanulova 11m 841 01 Bratislava. *Tel*: (2) 6029-2508. *Fax*: (2) 6254-0839. *E-mail*: vladimir.holcik@vvb.sk.

HOLLO, Prof Dr Erkki Johannes; Finnish professor of environmental law; b 28 Nov 1940, Janakkala; s of Chancellor J A Hollo and Iris Walden. *Education*: Helsinki Univ; Tübingen Univ, Germany. *Career*: Asst Prof, Land and Water Law, Helsinki Univ 1974–77; Prof, Economic Law, Helsinki Univ of Tech 1977–93; Justice, Supreme Admin Court of Finland 1983–86, 1993–97. *Present Position*: Prof of Environmental Law, Helsinki Univ 1997–. *Awards and Honours*: mem, Finnish Acad of Sciences 1989. *Publications*: Water Pollution Control (doctoral thesis—in Finnish) 1976; Land Use Law (in Finnish) 1984; Environmental Law (in Finnish) 1989; Environmental Damages Law (in Finnish) 1995. *Address*: Kiskontie 16 B 18, 00280 Helsinki, Finland. *Tel*: (9) 2711992. *Fax*: (9) 4775004. *E-mail*: erkki.hollo@helsinki.fi.

HOLOWESKO, Lynn Pyfrom, CBE, JP; Bahamian attorney at law; b 11 Oct 1935, Meriden, Conn, USA; m William Paul Holowesko 1955; 7c. *Education*: St Francis Xavier Acad, Nassau; Rosarian Acad, W Palm Beach, FL, USA; Albertus Magnus Coll, New Haven, CT, USA; Catholic Univ of America, Washington, DC, USA. *Career*: Hon Sec, Bahamas Nat Trust 1974–76, Pres 1976–82, Deputy Pres 1982–84, Pres 1984–91, Hon Life Mem, Hon Vice-Pres; Bahamas Rep to UN Comm on Sustainable Devt 1995–98, Head of Del, Conf of the Parties to the Convention on Biological Diversity 1995–98, Head of Del, Conf of the Parties to the UN Framework Convention on Climate Change—UNFCCC 1997–98, mem, council, Ad Hoc Cttee for the Protocol on Biosafety 1998–99; Bahamas rep and Head of Del, 2nd Consultative Meeting on Devt and the Environment, Inter-American Devt Bank 1989, 3rd Consultative Meeting on Devt and the Environment 1991, Conf on Environmental Law 1993; Deputy Chair, IUCN Comm on Nat Parks and Protected Areas 1992–; Chair, Panel of Experts on Environmental Law for the Montevideo Programme, UNEP 1999–; Chair, Legal Drafting Cttee for the Protocol on Biosafety 1999–; Bahaman Ambassador for the Environment 1995–2000; Senator 2002. *Present Position*: Council mem, IUCN–The World Conservation Union; Chair, IUCN Task Force on Governance. *Awards and Honours*: Outstanding Achievement as Pres Award, Bahamas National Trust 1982; Outstanding Business Woman of the Year Award, Business and Professional Women's Asscn of New Providence 1984; 25th Anniversary Award for Outstanding Contribution to Conservation, Bahamas National Trust 1984; Resolution of Council for Devoted Service and Inspired Leadership, Bahamas Nat Trust 1991; Commonwealth of the Bahamas Silver Jubilee Award 1998; Women Living Legend Award, Zonta Club of New Providence 1998; Contribution to the Growth of Catholicism in the Bahamas Award, Catholic Diocese of Bahamas 1998. *Address*: Holowesko & Co, POB 7776-348, Nassau, Bahamas. *Tel*: 362-6251. *Fax*: 362-8571. *E-mail*: lholowesko@bahamas.net.bs.

HORVAT, Manfred; Austrian academic and environmentalist. *Present Position*: Dir, Univ Extension Centre, Vienna Tech Univ, Austrian Bureau for International Research and Technology Co-operation. *Address*: 1040 Vienna, Wiedner Hauptstrasse 76, Austria. *Tel*: (1) 581-16-16-114. *Fax*: (1) 581-16-16-16. *E-mail*: horvat@bit.ac.at.

HOTI, Dr Namik Vehdi Fadile, PH D; Albanian politician, writer, journalist and environmentalist; b 13 July 1927, Vushtri; 1s 2d. *Career*: Member, Central Commission for the Albanian Constitution, Co-Founder; adviser to President of the Republic of Albania. *Present Position*: Chair, Partia Ekologjike Shqiptare (Albanian Ecological Party). *Address*: Rruga Aleksander Moisiu 26, POB 135, Tirana, Albania. *Tel*: (42) 22503. *Fax*: (42) 34413.

HOUGHTON, Sir John, CBE, D PHIL, FRS; British physicist; b 30 Dec 1931, Dyserth; s of Sidney and Miriam (Yarwood) Houghton; m 1st Margaret E Broughton 1962 (died 1986); m 2nd Sheila Thompson 1988; 1s 1d. *Education*: Jesus Coll, Oxford. *Career*: Research Fellow, Royal Aircraft Establishment, Farnborough 1954–57; Lecturer in Atmospheric Physics, Univ of Oxford 1958–62, Fellow, Jesus Coll 1960–, Reader 1962–76, Prof 1976–83; Pres, Royal Meteorological Soc 1976–78; Dir, Rutherford Appleton Lab, Science and Eng Research Council 1979–83; Chair, Earth Observation Advisory Cttee, European Space Agency 1980–, Jt Scientific Cttee, World Climate Research Programme 1981–83; Dir-Gen, Meteorological Office 1983–90, Chief Exec 1990–91; mem, Exec Cttee, WMO 1983–91, Vice-Pres 1987–91; Chair, Royal Comm on Environmental Pollution 1992–98. *Present Position*: Scientific co-Chair, Scientific Assessment Working Group, UN Intergovernmental Panel on Climate Change—IPCC, WMO 1988–. *Awards and Honours*: Symons Gold Medal, Royal Meteorological Soc 1991; Gold Medal, Royal Astronomical Soc 1995; Int Meteorological Org Prize 1998. *Publications*: Infra Red Physics (Co-Author) 1966; Remote Sounding of Atmospheres (Co-Author) 1984; The Physics of Atmospheres 1986; Does God Play Dice? 1988; Global Warming: A Complete Briefing 1994, 1997; The Search For God: Can Science Help? 1995. *Addresses*: Hadley Centre for Climate Prediction and Research, Meteorological Office, London Rd, Bracknell RG12 2SY, United Kingdom; Space Science Dept, Rutherford Appleton Lab, Chilton, Didcot OX11 0QX, United Kingdom. *Tel*: (1344) 856888. *Fax*: (1344) 856912.

HOWE OF ABERAVON, Elspeth Rosamund Morton, Lady, JP, B SC; British organization executive; b 8 Feb 1932; d of Philip Morton Shand and Sybil Mary Sissons; m Rt Hon Sir Geoffrey Howe 1953; 1s 2d. *Education*: LSE. *Career*: JP, Inner London Juvenile Court Panel 1964–; mem, Lord Chancellor's Advisory Cttee on Appointment of Magistrates for Inner London Area 1965–75, on Legal Aid 1971–75; Vice-Chair, Cons London Area Women's Advisory Cttee 1966–67, Pres, Contact Group 1973–77, mem, Cons Women's Nat Advisory Cttee 1966–71; Gov, Cumberlow Lodge Remand Home 1967–70; mem, Inner London Educ Authority 1967–70, Briggs Cttee on Nursing Profession 1970–72; Gov, Froebel Educ Inst 1968–75, LSE 1985–, James Allen's Girls School 1988–; Chair, Inner London Juvenile Court 1970–90; Parole Bd 1972–75; Broadcasting Standards Council 1993– 98; Deputy Chair, Equal Opportunities Comm 1975–79; Pres, Peckham Settlement 1976–, Women's Gas Fed 1979–93, Fed of Recruitment and Employment Services 1980–94; Vice-Pres, Pre-School Playgroups Asscn 1979–83; Dir, United Biscuits (Holdings) plc 1988–94; Chair, Hansard Soc Comm on Women at Top 1989–90, Women's Econ Devt Target Team and Business in Community; Trustee, Westminster Foundation for Democracy 1992–96. *Present*

Position: Dir, Kingfisher plc 1986–, Legal & Gen Group 1989–, Chair, BOC Foundation for Environment and Community 1990–, and Pres, UNICEF UK. *Publications*: Under Five 1966. *Address*: c/o Broadcasting Standards Council, 5 The Sanctuary, London SW1, United Kingdom. *Tel*: (20) 7233-0544. *Fax*: (20) 7233-0397.

HROMAS, Dr Jaroslav; Czech nature conservationist; b 2 June 1943; m Eva Ruprechtová 1962; 1s 2d. *Education*: Charles Univ, Prague. *Career*: Admin, Koneprusy Cave 1962–65; Head, Geological Section, Beroun Dist Museum 1965–67; State Inst for Care of Historical Monuments and Nature Conservation 1967–90; Founder, Czech Asscn for Nature Conservation 1981–; Dir, Czech Agency for Nature Conservation and Landscape Protection 1990–95. *Present Position*: Head, Dept for Care of Caves, Czech Agency for Nature Conservation and Landscape Protection 1995– and Pres, Czech Speleological Soc. *Address*: Kališnická 4, 130 00 Prague 3, Czech Republic. *Tel*: (2) 6974928. *Fax*: (2) 6970012.

HTUN, Nay; Thai international civil servant. *Career*: Deputy Exec Dir, and Dir, Asia and the Pacific Office, UNEP 1974–90; Dir of Programme, UNCED 1990–92. *Present Position*: Asst Admin and Regional Dir with rank of Asst Sec-Gen, UNDP 1994–. *Address*: UN Bldg, 10th Floor, Rajadamnern Ave, Bangkok 10200, Thailand. *Tel*: (2) 288-1234. *Fax*: (2) 288-1000.

HUBA, Mikuláš, PH D; Slovak geographer and landscape ecologist; b 24 March 1954, Bratislava; s of Alexander Huba and Olga Hubová; m Olga Martinceková 1985; 2s 1d. *Education*: Comenius Univ of Bratislava. *Career*: Sr Researcher, Inst of Geography, Slovak Acad of Sciences 1987–; Chair, Slovak Union of Nature and Landscape Conservation—SZOPK 1989–93, FoE Slovakia 1990–93; mem, Nat Council (parl) and Chair, Parl Environment Cttee 1990–92; mem, IUCN, ICSU Scientific Cttee on Problems of the Environment—SCOPE, Slovak Asscn, Club of Rome, Slovak Cttee, UNESCO, Core Expert Group for Indicators, UNCSD 1995–, Transparency Int—Slovakia 1998–, Governmental Bd for Sustainable Devt 1999–, Slovak Parl Expert Group for the Environment 1999–. *Present Position*: Chair, Sustainable Living Soc 1993–; Team Leader, Sustainability Indicators Project, Inst of Geography, Slovak Acad of Sciences. *Awards and Honours*: Slovak Acad of Sciences Medal 1987; Slovak Nature and Landscape Union Medal 1989; Tatras National Park Medal 1990; Lanza Foundation Award 1996; Josef Vavroušek Award 1997. *Publications*: Bratislava (Co-Author) 1987; Slovak Forests SOS 1990; The Danube Story 1990; Slovak Agriculture SOS 1991; Slovakia's Message for Rio 1992; Slovakia's Message for Lucerne 1993; Prosperity Shock—Towards Sustainable Slovakia 1995; Endangered Planet 1998; The World Perceived by the Heart of Europe (Ed) 2000; *Film*: Settlement and Landscape 1987. *Address*: Staroturský chodník 1, 811 01 Bratislava, Slovakia. *Tel*: (7) 5249-2751. *Fax*: (7) 5441-0647. *E-mail*: huba@savba.sk.

HUETING, Dr Roefie, D ECON; Dutch economic consultant and researcher; b 16 Dec 1929, The Hague; s of Bernardus Hueting and Elisabeth Hueting-Steinvoorte; m Erna Jans Postuma 1957; 2d. *Education*: Econs Univ of Amsterdam. *Career*: Ind consultant, lecturer and writer on environmental and resource econs 1965–; Founder and Head, Dept of Environmental Statistics, Statistics Netherlands 1969–94. *Present Position*: Adviser on Environmental Statistics, Statistics Netherlands. *Awards and Honours*: Officer in the Orde van Oranje Nassau, Netherlands Govt 1991; UNEP Global 500 1994. *Publications*: New Scarcity and Economic Growth 1974; Methodology for the Calculation of Sustainable National Income 1992; Environmental Valuation and Sustainable National Income According to Hueting 2000. *Address*: Roelofsstraat 6, 2596 VN The Hague, Netherlands. *Tel*: (70) 3249744. *Fax*: (70) 3877429.

HUGHES, Dr George Ritchie, BSC; PHD; South African zoologist and administrator; b 13 May 1939, Aberdeen, UK; s of Mitchell and Constance Mary Hughes; m Alethea Joy Cooper; 1s 1d. *Education*: Univ of Natal, Pietermaritzburg. *Career*: Game Ranger, Natal Parks Bd 1964–65, Professional Officer 1974–75, Chief Conservator 1975–80, Asst Dir, Admin 1980–82, Asst Dir, Conservation 1982–87, Chief Exec 1988–98; Professional Officer, Oceanographic Research Inst, Durban 1969–74; Chief Exec (acting), KwaZulu Natal Nature Conservation Service 1998–2002. *Present Position*: CEO, The Conservation Trust 2002–. *Awards and Honours*: Dr Edgar Brooke Award for Human Freedom and Endeavour; Green Trust Environmental Award; Chairman's Award, SATOUR. *Publications*: The Sea Turtles of South East Africa 1974. *Addresses*: The Conservation Trust, POB 13053, Cascades 3202, South Africa; 4 Thorngate Road, Pietermaritzburg 3201. *Tel*: (33) 3432660. *Fax*: (33) 3432666. *E-mail*: ghughes@kznnwildlife.com. *Internet*: www.kznwildlife.com.

HUNT, Prof Julian Charles Roland, PH D, FRS; British meteorologist and environmentalist; b 5 Sept 1941; s of Roland Charles Colin Hunt; m Marylla Ellen Shephard 1965; 1s 2d. *Education*: Trinity Coll, Cambridge, Univ of Warwick. *Career*: Fellow, Trinity Coll, Cambridge 1966–; Researcher, Cornell Univ 1967, Cen Electricity Research Labs 1968–70; Lecturer in Applied Mathematics and Eng, Cambridge Univ 1970–78, Reader in Fluid Mechanics 1978–90, Prof 1990–; mem, Cambridge Co Council 1971–74, Leader, Lab Group 1972; Visiting Prof, Colorado State Univ 1975, N Carolina State Univ and Environmental Protection Agency 1977, Colorado Univ 1980, Nat Center for Atmospheric Research, Boulder, CO 1983; Sec, Int Meteorological Asscn 1984–89, Vice-Pres 1989–93, Pres 1993–95; Founder Dir, Cambridge Environmental Research Consultants Ltd 1986–91; Gen Sec, European Research Community for Flow Turbulence and Combustions 1988–95; mem, Exec Council, WMO 1992–, Natural Environment Research Council 1994–. *Present Position*: Chief Exec, Meteorological Office 1992–. *Address*: Meteorological Office, Bracknell RG12 2SZ, United Kingdom. *Tel*: (1344) 854463.

HURNI, Hans, PH D; Swiss geographer and university teacher; b 21 Dec 1950, Erlenbach; m Marlies Alt 1978; 2s. *Education*: Univ of Berne. *Career*: Park Warden, WWF Int, Ethiopia 1975–77; Founder and Dir, Ethiopia Soil Conservation Research Project, Univ of Berne 1981–87, Lecturer 1989–91, Privat-docent 1991–; Asst Prof, Addis Ababa Univ 1986; Pres, World Asscn of Soil and Water Conservation, Ankeny, IA, USA 1991–96; Prof 1997–; Pres, European Forum on Agricultural Research for Development 2004–; Pres, Swiss Commission for Research Partnerships with Developing and Transition Countries 1999–2002. *Present Position*: Co-Dir, Centre for Devt and Environment, Geography Inst, Univ of Bern Univ 1988– and Chair, Programme Review Cttee, Int Bd for Soil Research and Man; Ed-in-Chief, *Mountain Research and Development* 1999–. *Publications*: Climatic and Geomorphic Studies in the Simen High Mountains 1982; Soil Conservation in Ethiopia 1986; Simen Mountains Management Plan 1986; Soil Erosion in Agricultural Environments 1990; Soil Conservation for Survival 1992; Soil Conservation and Small-scale Farming 1992; Precious Earth 1996; Research Partnerships with Developing Countries 2001; Research for Mitigating. *Address*: Hallerstr

12, 3012 Berne, Switzerland. *Tel*: (31) 6318822. *Fax*: (31) 6318544. *E-mail*: hans.hurni@cde.unibe.ch.

HURTADO, Mario; Ecuadorean maritime and coastal conservationist; b 13 Dec 1953; s of Segundo Hurtado and María Gualán; m Laura Domínguez 1978; 2s 1d. *Education*: Guayaquil Univ. *Career*: Research on sea turtles, Galapagos Islands 1978–82, co-ordinator, biological research on fish 1983–84, Deputy Dir, Charles Darwin Scientific Station 1984–86, Research Dir, Charles Darwin Foundation for Galapagos Islands 1987, Administrative and Technical Officer. *Present Position*: Scientific and Technical Co-ordinator, Charles Darwin Foundation for the Galapagos Islands. *Publications*: Tortugas Marinas en Galápagos 1984; Aves Marinas en el Ecuador 1986; Mamíferos Marinos en el Ecuador 1990. *Address*: Avda Colón y 6 de diciembre, 6°, Casilla 17-01-3891, Quito, Ecuador. *Tel*: (2) 244-803. *Fax*: (2) 443-935.

HUSBAND, Victoria (Vicky), BA; Canadian environmentalist; b 1940, Victoria, BC; m 1973. *Education*: Univ of British Columbia, Vancouver. *Career*: Film dir and producer 1980–82; public speaker; Pres, Friends of Ecological Reserves, BC 1985–89; campaign for S Moresby/Gwaii Haanas Nat Park Reserve 1987; Conservation Chair, Sierra Club of Western Canada 1989–90; campaign for Khutzeymateen, first Canadian grizzly bear sanctuaries 1992; campaign for Clayoquot Sound temperate rainforest 1993. *Present Position*: Chair, Sierra Club of British Columbia 1990–. *Awards and Honours*: Fred M Packard Award, IUCN 1987; UNEP Global 500 1988; Order of British Columbia 2000. *Publications*: Ancient Rainforests at Risk 1993. *Address*: 1525 Amelia St, Victoria, BC V8X 2K1, Canada. *Tel*: (604) 386-5255. *Fax*: (604) 386-4453.

HU TAO, Tom, PH D; Chinese environmental researcher; b 23 Oct 1962; s of Hu Qiwen and Hao Guilan; m Xie Hong 1990; 1d. *Education*: Chinese Acad of Sciences. *Career*: Project Team Leader, Devt Policy for Sustainable Agric Project, IBRD (World Bank) 1992–; Consultant, Urban Environment Improvement Project, Asian Devt Bank 1994–95. *Present Position*: Sr Fellow and Head, Div of Environmental Econs, Policy Research Centre, Nat Environmental Protection Agency 1994–. *Awards and Honours*: SciTech Prize for Chinese Youth 1992. *Publications*: China's Sustainable Development: From Concepts to Action 1995; China's Environmental Economics: From Theory to Practice 1996. *Address*: 115 Xizhimennei Nanxiaojie, Beijing 100035, People's Republic of China. *Tel*: (10) 64924241. *Fax*: (10) 62200399. *E-mail*: hutao@public.bta.net.ch.

HUTTUNEN, Prof Dr Satu Orvokki, PH D; Finnish botanist; b 27 Sept 1945, Sotkamo; d of Sulo O Pulkkinen and Sirkka O Achren-Pulkkinen; m Jouko Huttunen 1965; 1d. *Education*: Univ of Oulu. *Career*: Jr Fellow then Sr Fellow, Finland Acad 1976–88; Lecturer, Environmental Hygiene, Univ of Kuopio 1977; Lecturer, Botany, Univ of Oulu 1982–89. *Present Position*: Prof of Botany, Univ of Oulu 1989–. *Awards and Honours*: Finnish Woman of the Year 1986; UNEP Global 500 Laureate 1987. *Publications*: Pelastakaa metsät 1988; numerous works relating to air pollution, climate change and plant conservation. *Address*: University of Oulu, POB 3000, 90014 Oulu, Finland. *Tel*: (8) 5531527. *Fax*: (8) 5531500. *E-mail*: satu.huttunen@oulu.fi.

I

IGLESIAS, Enrique V; Uruguayan international official; b 26 July 1931, Asturias, Spain; s of Isabel García de Iglesias. *Education*: Univ de la República, Montevideo. *Career*: held several positions including Prof Agregado, Faculty of Political Economy, Prof of Econ Policy and Dir, Inst of Econs, Univ de la República, Montevideo 1952–67; Man Dir, Unión de Bancos del Uruguay 1954; Technical Dir, Nat Planning Office of Uruguay 1962–66; Pres (Gov), Banco Central del Uruguay 1966–68; Chair, Council, Latin American Inst for Econ and Social Planning—ILPES, UN 1967–72, Interim Dir-Gen 1977–78; Head, Advisory Mission on Planning, Govt of Venezuela 1970; Adviser, UN Conf on Human Environment 1971–72; Exec Sec, Econ Comm for Latin America and the Caribbean—ECLAC 1972–85; Minister of External Affairs 1985–88; Acting Dir-Gen, Latin American Inst for Econ and Social Planning 1973–78; Pres, Third World Forum 1973–76; mem, Steering Cttee, Soc for Int Devt 1973–92, Pres 1989, Selection Cttee, Third World Prize 1979–82; Sec-Gen, UN Conf on New and Renewable Sources of Energy 1981; Chair, UN Inter-Agency Group on Devt of Renewable Sources of Energy; mem, North–South Round Table on Energy; Chair, Energy Advisory Panel, Brundtland Comm 1984–86. *Present Position*: Pres, Inter-American Devt Bank—IDB 1988–. *Awards and Honours*: Hon LL D (Liverpool) 1987; Prince of Asturias Award 1982. *Address*: Inter-American Development Bank, 1300 New York Ave, NW, Washington, DC 20577, USA.

IRA, Dr Vladimír, C SC; Czech geographer and researcher; b 25 Oct 1952, Písek; m Lubica Rapošová 1975; 1s 1d. *Education*: Comenius Univ of Bratislava, Slovakia. *Career*: Asst Researcher, Czechoslovak Environment Centre 1980–81; Asst Researcher, Inst of Geography, Slovak Acad of Sciences 1981–85, Sr Researcher 1985–90, Prin Researcher 1990–; mem, Nat Cttee, Man and Biosphere Programme 1990–; Dir, Postgraduate Environmental Studies Programme, Acad of Sciences 1991–; Lecturer in Geography and Environment, Comenius Univ of Bratislava 1991–. *Present Position*: Vice-Pres, Soc for Sustainable Living in Slovakia—SSL/SR 1993–. *Publications*: Bratislava (Co-Author) 1987. *Address*: Staroturskу chodnik 1, 811 01 Bratislava, Slovakia. *Tel*: (7) 392-751. *Fax*: (7) 5441-3968. *E-mail*: ira@savba.savba.sk.

ISAEV, Prof Dr Alexander Sergeevich; Russian biologist and forester; b 26 Oct 1931, Moscow; m Lidia Pokrovskaya 1953; 2d. *Education*: Leningrad (now St Petersburg) Acad of Forestry. *Career*: Engineer, Moscow Forest Inventory Expedition 1954–60; Scientific Researcher, Head of Lab, Deputy Dir, then Dir, Krasnoyarsk Inst for Forest and Wood, USSR 1960–88; Chair, State Cttee on Forests of the USSR 1988–91; Chair, Higher Ecological Council, Russian Supreme Soviet 1991–93. *Present Position*: Chair, Scientific Council on Forests, Russian Acad of Sciences—RAS 1978–; Editor-in-Chief, *Lesovedenye* 1978–; Dir, Centre for Ecological Problems and Forest Productivity, RAS 1991–. *Awards and Honours*: Gold Medal, INFRO; UNEP Global 500 1989; Gold Medal, RAS 1992, Academician; Hon Mem, American Soc of Foresters. *Publications*: 230 articles and 7 books, including Dynamics of Insect Number 1984; Airspace Monitoring of Forests 1991; Forestry at the Threshold of the 21st Century 1991; Carbon in the Forests of Northern Eurasia 1999. *Address*: Forestry Cttee, 117418 Moscow, Novocheremushkinskaya 69, Russian Federation. *Tel*: (095) 332-86-52. *Fax*: (095) 332-29-17. *E-mail*: postmaster@spepb.msk.su.

ISHI, Hiroyuki, BA; Japanese environmental scientist; b 28 May 1940, Tokyo; m Tokiko Ishi 1968; 3d. *Education*: Univ of Tokyo. *Career*: Journalist, *Asahi Shimbun* newspaper 1965–82, Science Ed 1982–94; Special Adviser to Sec-Gen, UNEP, Kenya 1985–87, to Pres, Japanese Int Co-operation Agency 1994–2000; Chair, Japan Council of Sustainable Devt 1997–2000, mem, bd, Regional Environmental Centre for Cen and E Europe, Budapest, Hungary 1997–2000. *Present Position*: Prof of Advanced Environmental Science, Univ of Tokyo 1996–. *Awards and Honours*: FAO Boerma Award 1987; UNEP Global 500 1989. *Publications*: Report on Global Environment I 1988; Degradation of Global Ecosystem 1997; Report on Global Environment II 1999; Unveiled Destruction of Earth 2000. *Addresses*: 2-20-9, Nozawa, Setagaya-ku, Tokyo 154-0003, Japan; Univ of Tokyo, 7-3-1, Hongo Bunkyoku, Tokyo 113-8656, Japan. *Tel and fax*: (3) 3795-2277. *E-mail*: ishi@k.u-tokyo.ac.jp.

IZRAEL, Prof Yurii Antonievich, DR SC; Russian geophysicist and ecologist; b 15 May 1930, Tashkent, Uzbek SSR (now Uzbekistan); s of Antony I Izrael and Antonina S Shatalina; m Elena Sidorova 1958; 1s 1d. *Education*: Tashkent State Univ. *Career*: engineer, research assoc, Geophysics Inst, USSR Acad of Sciences 1953–63; Deputy Dir, Dir, Inst of Applied Geophysics 1963–70; Head, Main Admin, Hydrometeorological Service of USSR 1974–78; First Deputy Head 1970–74, Prof 1973; Corresp mem, USSR (now Russian) Acad of Sciences 1974, mem 1994, Acad-Sec, Dept of Oceanography, Atmospheric Physics and Geography 1996–; mem, Russian Acad of Ecology 1994; Chair, USSR State Cttee for Hydrometeorology and Environmental Control 1978–88; Chair, USSR State Cttee for Hydrometeorology 1988–91; Deputy, Supreme Soviet 1979–89; Sec and First Vice-Pres, World Meteorological Org 1975–87; Vice-Chair, Intergovernmental Panel on Climate Change; mem, Int Acad of Astronautics 1990. *Present Position*: Dir, Inst of Global Climate and Ecology 1990–. *Awards and Honours*: State Prize in the field of Environment 1981; Gold Medal, USSR Acad of Sciences in the field of Ecology 1983; Gold Medal (for *Chornobyl*), 'Ettore Majorana' Int Centre, Italy 1990; Gold Medal of Soviet State Exhbn 1991; Sasakawa Environmental Prize (jtly) 1992; Gold Medal and Prize of Int Meteorological Org 1992; state orders. *Publications*: Peaceful Nuclear Explosions and Environment 1974; Ecology and Control of Environment 1983; Global Climatic Catastrophes 1986; Anthropogenic Climate Change 1987; Anthropogenic Ecology of the Ocean 1989; Chornobyl: Radioactive Contamination of the Environment 1990; Earth's Ozone Shield and its Changes (Co-Author) 1992; Radioactive Fallout after Nuclear Explosions and Accidents 1996; numerous other scientific books and articles. *Addresses*: c/o Institute of Global Climate and Ecology, 107258 Moscow, Glebovskogo str 20b, Russian Federation; Moscow, Romanov per 3, Apt 84, Russian Federation. *Tel*: (095) 169-24-30.

J

JACOBS, Prof Peter Daniel Alexander, ML ARCH; Canadian architect and landscape architect; b 12 March 1939, Montréal; m Prof Ellen Jacobs. *Education*: Antioch College; Harvard Graduate School of Design. *Career*: invited scholar, Univ of British Columbia and Harvard Univ, MA, USA; Chair, Environmental Planning Comm, World Conservation Union—IUCN 1978–90; mem, Québec Round Table on Environment and the Economy 1988–93; Chair, Public Advisory Cttee on Canada's State Environmental Reporting 1989–92; mem, editorial cttee, *Eco-Decision* 1990–97. *Present Position*: Prof of Landscape Architecture, School of Landscape Architecture, Faculty of Environmental Planning, Univ of Montréal; adviser to the City of Montréal on urban development and urban open space design; Chair, Kativik Environmental Quality Comm 1979–. *Awards and Honours*: Tammsaare Medal, World Conservation Union—IUCN; Gov-Gen's Medal, 125th Anniversary of Canadian Fed; President's Medal, Canadian Soc of Landscape Architects. *Publications*: Conservation with Equity: Strategies for Sustainable Development (Co-Ed) 1987; Sustainable Development and Environmental Assessment: Perspectives on Planning for a Common Future (Co-Ed) 1990; Territoire du Nouveau Québec: savoirs, pratiques, politiques (Ed). *Addresses*: Université de Montréal, Faculté de l'Aménagement, Ecole d'Architecture de Paysage, CP 6128 succursale Centre-ville, Montréal, PQ H3C 3J7, Canada; 644 Belmont Ave, Westmount, PQ H3Y 2W2, Canada. *Tel*: (514) 343-7119. *Fax*: (514) 343-6104. *E-mail*: peter.jacobs@umontreal.ca.

JACOBY, Abbes; Luxembourgeois politician and environmentalist. *Career*: mem, then Sec, Déi Gréng. *Present Position*: Treas, Déi Gréng 2000–. *Address*: BP 454, 2014 Luxembourg, Luxembourg. *Tel*: 46-37-40. *Fax*: 46-37-41.

JAIN, Dr Sudhanshu Kumar, PH D; Indian botanist and retired scientist; b 30 June 1926, Amroha; s of Sridevi and P C Jain; m Satya Jain 1948; 3s. *Education*: Univ of Pune and Allahabad. *Career*: Dir, Botanical Survey of India 1978–84; Nat Environment Fellow, Govt of India 1984–86; Scientist, Dept of Science and Tech and Council of Science and Industries, Govt of India 1986–94; Consultant, UNDP projects 1994–95; Investigator, Wanner-Gren Foundation 1997–98. *Present Position*: Hon Dir, Inst of Ethnobiology 1995–2002. *Awards and Honours*: Distinguished Econ Botanist, Soc of Econ Botany, USA 1999. *Publications*: 36 books and over 300 papers. *Addresses*: A-26 Mall Ave Colony, Lucknow 226 001, Uttar Pradesh, India; c/o National Botanical Research Institute, Lucknow 226 001, India. *Tel and fax*: (522) 224556.

JAKIMAVICIUS, Dr Algimantas, D SC; Lithuanian biologist and entomologist; b 23 June 1939, Rokiškis; m Jurate Tomonyte 1968; 2s. *Education*: Lithuanian Univ of Agric. *Career*: Biological and entomological research in protected areas of Lithuania, the Crimea, Kazakhstan, the Carpathians and the northern Caucasus. *Present Position*: Sr Research Fellow, Inst of Ecology, Lithuanian Acad of Sciences 1970–, Scientific Sec 1987–. *Awards and Honours*: Lithuanian Nature Protection Cttee awards 1981, 1984. *Publications*: Entomoparasites of Insects—Orchard Pests in Lithuania 1979; Key to the Insects of the European Region of the USSR (3 vols—Co-Author) 1981–86; Lithuanian Zoologists in the 18th–20th centuries 1997. *Address*: Institute of Ecology, Lithuanian Academy of Sciences, Akademijos 2, Vilnius 2600, Lithuania. *Tel*: (2) 729-244. *Fax*: (2) 729-257. *E-mail*: ekoi@ekoi.lt.

JAKOWSKA, Prof Dr Sophie, M SC, PH D; American biologist; b 12 Feb 1922, Poland; d of Józef Jakowski and Maria Swiergocka; m C L Jeannopoulos 1941 (died 1980); 2s 1d. *Education*: Univ of Perugia, Univ of Rome, Italy; Fordham Univ, USA. *Career*: founding mem, Comisión Nacional de Bioética, Dominican Repub. *Present Position*: Prof Emer, Dept of Biological Sciences, Coll of Staten Island, City Univ of NY 1977–; mem, Bd of Trustees, Parque Nacional Mirador del Norte, Dominican Repub 1995–; mem, IUCN Comm on Educ—CEC. *Awards and Honours*: Tree of Learning Award; IUCN Comm on Educ—CEC Gold Medal; Liga Ochrony Przyrody—LOP, Poland. *Address*: Arz Merino 154, Santo Domingo, Dominican Republic. *Tel and fax*: 687-3984. *E-mail*: jakowska@hotmail.com.

JANZEN, Daniel Hunt, PH D; American professor of biology; b 18 Jan 1939, Millwaukee, WI; s of Daniel Hugo and Floyd (Foster) Janzen; 1s 1d. *Education*: Univ of Minnesota, Univ of California, Berkeley. *Career*: Asst and Assoc Prof, Univ of Kansas 1965–68; Assoc Prof, Univ of Chicago 1969–72; Assoc Prof and Prof of Ecology and Evolutionary Biology, Univ of Michigan 1972–76; field research in tropical ecology, supported mainly by grants from Nat Science Foundation, USA 1963. *Present Position*: teacher, Org for Tropical Studies in Costa Rica 1965–; Prof of Biology, Univ of Pennsylvania 1976–. *Awards and Honours*: MacArthur Fellow 1989; Gleason Award, American Botanical Soc 1975; Crafoord Prize, Coevolutionary Ecology, Swedish Royal Acad of Sciences 1984. *Publications*: Herbivores (Co-Ed) 1979; Costa Rican Natural History (Ed) 1983; over 250 papers in scientific journals. *Addresses*: Dept of Biology, University of Pennsylvania, Philadelphia, PA 19104, USA; Parque Nacional Santa Rosa, Apdo 169, Liberia, Guanacaste Province, Costa Rica. *Tel*: (215) 898-5636 (USA).

JARIWALA, Prof Dr C M, B COM, LL M, PH D; Indian professor of law; b 19 Oct 1937, Barhanpur; s of Shri M P Jariwala; m Prof Savitri Jariwala; 1s 1d. *Education*: Banaras Hindu Univ, Varanasi; Univ of London, UK. *Career*: founder, first environmental law course in India, Banaras Hindu Univ 1977; mem, Environmental Law Comm, Switzerland, Int Council of Environmental Law, Germany. *Present Position*: Prof of Law, Banaras Hindu Univ 1985–. *Publications*: five books and numerous articles. *Address*: G/12 Arvindo Colony, Banaras Hindu University, Varanasi 221005, Uttar Pradesh, India. *Tel*: (542) 316294.

JARZEBSKI, Prof Dr Stefan; Polish environmental engineer and politician; b 16 Sept 1917, Bedzin; m Wanda Jarzebski 1945; 2s. *Education*: Warsaw Univ of Tech, Kraków Univ of Tech. *Career*: Prof of Environmental Engineering, Kraków Univ of Tech 1946–58, Univ of Silesia 1978–83; Head, Inst of Environmental Engineering, Polish Acad of Sciences 1962–83; Co-Founder and Head, Ministry of Environmental Protection and Natural Resources (later renamed Ministry of Environmental Protection, Natural Resources and Forestry) 1983–97. *Publications*: 17 books and more than 200 scientific papers. *Address*: 02-954 Warsaw, ul Wiktorii Wiedenskiej 1 m 2, Poland. *Tel*: (22) 423317.

JAYAL, Nalni Dhar, BA; Indian retired civil servant and environmentalist; b 11 Feb 1927, Almora; s of Chandra Dhar and Shambhavi Jayal; m Amena Jayal 1960; 1s. *Education*: Delhi Univ. *Career*: Devt Admin, tribal Himalayan areas of the (then) North-East Frontier Agency and Kinnaur Dist of Himachal Pradesh 1956–67; Joint Sec, Ministry for Forests and Wildlife 1976–80; Adviser, Planning Comm for Oceanic Islands of Andaman and Nicobar 1983–95; mem, Island Devt Authority 1987–90; Regional Councillor for E Asia, IUCN 1978–84; Dir, Nat Heritage, Indian Nat Trust for Art and Cultural Heritage—INTACH 1985–96. *Present Position*: Sec, Himalaya Trust 1993–. *Publications*: Indian Environment: Crises and Responses 1985; Eliminating Poverty: An Ecological Response 1986; Tropical Deforestation: An Unseen Crisis 1990; Himalaya: Our Fragile Heritage 1991; Forest Policy: An Overview 1991; Ecology and Human Rights 1993. *Address*: Himalaya Trust, 274/II Vasant Vihar, Dehra Dun 248 006, Uttar Pradesh, India. *Tel*: (135) 773081. *Fax*: (135) 620334. *E-mail*: ubcentre@del2.vsnl.net.in.

JEFTIC, Ljubomir, PH D; Croatian United Nations official; b 25 Feb 1936, Novi Becej; s of Mile and Slavica (Elblinger) Jeftic; m Ivanka Karacic 1961; 2s. *Education*: Zagreb Univ, Warsaw Univ, Poland. *Career*: Marine Scientist, Rudjer Boskovic Inst, Centre for Marine Research, Zagreb 1961–81; Research Assoc, Kansas Univ, USA 1967–69, Brookhaven Nat Lab, NY, USA 1969–70; Prof of Environmental Protection, Zagreb Univ 1979–85; Dir, Planning and Environmental Protection Dept, Croatian Cttee for Bldg, Housing and Environmental Protection 1981–85; Deputy Co-ordinator and Sr Marine Scientist, Mediterranean Action Plan Co-ordinating Unit, UNEP, Athens, Greece 1985–96. *Present Position*: Dir of Programmes and Chair, Cttee on Pollution Control and Prevention, Advisory Cttee on Protection of the Sea. *Publications*: Electrochemical Oxidation Pathways of Benzo(a)pyrene (Co-Author) 1970; Chromium (II) Catalyzed Aquation of Hexacyanochromate (III) to Pentacyanomonohydroxychromate (III) (Co-Author) 1971; Climatic Change and the Mediterranean (Co-Ed) 1992. *Address*: 11 Dartmouth St, London SW1H 9BN, United Kingdom. *Tel*: (20) 7799-3033. *Fax*: (20) 7799-2933. *E-mail*: jeftic@hol.gr.

JENDROSKA, Jerzy, LL D; Polish lawyer; b 4 Jan 1956, Wroclaw; s of Jan and Krystyna (Jandy) Jendroska; m Ewa Jendroska 1979; 2s. *Education*: Wroclaw Univ. *Career*: Asst Prof, Research Group on Environmental Law, Polish Acad of Sciences 1978; mem, Polish Environmental Law Reform Cttee 1989–92; Visiting Fellow, Univ Coll, London 1990–91;Adjunct Prof, Research Group on Environmental Law, Polish Acad of Sciences 1986–; Sec, Bd of Dirs, Polish Environmental Law Asscn 1987–; legal expert of the Polish Chamger of Ecology 2002–; vice-Chair, GMO Commission of the Gov of Poland; Vice-Chair, Aarhus Convention Bureau 2002-; Arbitrator, Permanent Court of Arbitration, The Hague 2002–(08). *Present Position*: Managing Partner. *Publications*: editor or author of over 170 articles and 20 books in Polish, English, Russian and German, mostly relating to the organizational structure of environmental administration and governance. *Address*: Polish Environmental Law Asscn, ul. Uniwersytecka 1 50-951 Wroclaw, Poland. *Tel*: (71) 3410234. *Fax*: (71) 3410197. *E-mail*: jerzy.jendroska@kjjb .com.pl. *Internet*: www.jjb.com.pl.

JIANG, Xiaoyu; Chinese hydrologist and editor; b 14 Dec 1940, Shanghai; m 1965; 2s 1d. *Education*: Nanjing Hohai Univ. *Career*: Formulation of measurement and management procedures for water resource protection and water quality monitoring, Nat Environmental Protection Agency 1972–79; Vice-Pres and Dir, Editorial Office, China Environmental Newspaper House 1983–92. *Present Position*: Deputy Sec-Gen, China Asscn of the Environmental Protection Industry 1993– and Chief Ed, *China Environmental Protection Industry* 1995–. *Awards and Honours*: China State Science Conf award 1973; UNEP Global 500 1985. *Address*: China Association of the Environmental Protection Industry, 9 Sanlihe Rd, Haidian Dist, Beijing 100835, People's Republic of China. *Tel*: (10) 68393892. *Fax*: (10) 68393748. *E-mail*: caepi@public3.bta.net.cn.

JIM, Prof Dr Chi Yung, BA, PH D; British university professor and environmental assessment specialist; b 23 Jan 1953; m. *Education*: Univ of Hong Kong; Reading Univ, UK. *Career*: researcher, Univ of Hong Kong 1981–; mem, editorial team, *Environmental Conservation*, *Arboricultural Journal*, *Pedosphere*, *Asian Geographer* and *Environmental Awareness*; Guest Prof, Zhongshan Univ and Guangdong Normal Univ, Guangzhou, People's Repub of China 1999–2000; Chair, Country Parks Man Cttee, Hong Kong 1993–, Country and Marine Parks Bd, Hong Kong 1995–; mem, Town Planning Bd, Metro Development Bd, Hong Kong 1994–; Deputy Project Leader, Urban Forestry Project, Int Union of Forestry Research Orgs 1995–; mem, Univ Grants Cttee, IUCN Task Force on Tourism and Protected Areas, Liaison Meeting with Green Groups, Green Hong Kong Campaign, Education Panel, Steering Cttee, US–Asia Environmental Partnership 1996–, Special Cttee on Cleaning Up Black Spots in New Territories, IUCN World Comm on Protected Areas 1997–, *Ad Hoc* Cttee on Green Matters; Vice-Chair, Hong Kong Cttee, Int Geographical Union 1998–; mem, Exec Cttee, Friends of Country Parks, Chair 1999–; mem, Advisory Council, Royal Geographical Soc, Hong Kong; mem, Exec Cttee, Green Fun Cttee 1999–, Chair, Subcttee on Publicity and Public Educ 2000–; Chair, Organizing Cttee, IUCN World Comm on Protected Areas Conference 2000–. *Present Position*: Head, Dept of Geography, Univ of Hong Kong 1998–. *Awards and Honours*: Commonwealth Scholarship 1977–80; Best Teacher Award, Univ of Hong Kong 1986; Croucher Foundation Fellowship 1987; Better Environment Award, Shell Green Fund 1990; Green Project Award, Caltex Green Fund and South China Morning Post 1994; Badge of Honour, Queen's New Year Honours List 1995; Justice of the Peace, Hong Kong Govt 1997; Oustanding Researcher Award, Hong Kong Univ 1999–2000. *Publications*: numerous articles in scientific journals. *Address*: Departments of Geography and Geology, University of Hong Kong, Pokfulam Rd, Hong Kong Special Administrative Region, People's Republic of China. *Tel*: 28597020. *Fax*: 25598994. *E-mail*: hragjcy@hkucc.hku.hk. *Internet*: geog.hku.hk.

JIMÉNEZ-BELTRÁN, Domingo; Spanish environmental executive; b 2 April 1944, Zaragoza; s of Mariano Jiménez and María Gloria Beltrán; m Elin Solem; 1c. *Education*: High Tech School of Industrial Engineers, Polytechnic Univ, Madrid. *Career*: Lecturer, Polytechnic Univ, Madrid 1978–86; Exec Adviser to Min for Public Works and Planning 1983–85; Deputy Dir-Gen for Int and EC Relations, Ministry of Public Works and Urban Planning 1985–86; Attaché for Environment and Public Works, Perm Mission to EC, Belgium 1986–87; Head, Div of Health, Physical Safety and Quality, Consumer Policy Service, Comm of the EC (European Comm), Belgium 1987–91; Dir-Gen for Environmental Policy, Sec of State for Environment and Housing, Ministry of Public Works, Transport and Environment 1991–94; Exec Dir, European Environment Agency—EEA 1994–2002. *Address*: c/o European Environment Agency, Kongens Nytorv 6, 1050 Copenhagen K, Denmark.

JONAITIS, Dr Vytautas; Lithuanian entomologist; b 10 Nov 1938; m Vladislava Jonaitiene 1960; 1s. *Education*: Lithuanian Agric Univ, Kaunas; Inst of Forest Eng, Moscow, Zoological Inst, St Petersburg, Russian Fed. *Career*: Forester, Širvintos 1962–65; Researcher, Inst of Ecology, Lithuanian Acad of Sciences 1965–. *Present Position*: Head, Lab of Entomology, Inst of Ecology, Lithuanian Acad of Sciences 1991–. *Publications*: Key to the Insects of the European Region of the USSR (3 vols, Co-Author) 1981–86; Ichneumonids of Lithuania 1983; Resources, Formation and Functioning of Host–Parasite Entomocomplexes in Ecosystems 1990; The Inter-Ecosystematic Relationships of Insects and Their Dynamics 2000. *Address*: Institute of Ecology, Lithuanian Academy of Sciences, Akademijos 2, Vilnius 2600, Lithuania. *Tel*: (2) 729-280. *Fax*: (2) 729-257. *E-mail*: jpl@ekoi.lt.

JÜSSI, Fred; Estonian biologist, zoologist, writer and photographer; b 29 Jan 1935, Aruba; s of Johannes and Elise-Anita Jüssi; m Helju Jüssi 1960; 2s 1d. *Education*: Univ of Tartu. *Career*: Teacher, Hiiumaa 1958–60; Insp of Nature Protection 1961–75; radio broadcaster 1961–91; freelance writer and nature campaigner 1976–93. *Present Position*: Pres, Eestimaa Looduse Fond (Estonian Fund for Nature) 1992–. *Awards and Honours*: Estonian Renaissance Award 1997. *Publications*: Numerous books and audio recordings related to nature in Estonia. *Address*: Eestimaa Looduse Fond, Riia 185a, POB 245, Tartu 2400, Estonia. *Tel*: (7) 428-443. *Fax*: (7) 428-166. *E-mail*: elf@elfond.tartu.ee.

K

KAASIK, Arne; Estonian nature conservationist; b 2 Oct 1947, Rutja; s of Valdek and Signe Kaasik; m Mare Kaasik 1968; 1s 1d. *Education*: Agric Univ, Tartu. *Career*: Ranger then Deputy Dir, Lahemaa Nat Park and Section Chair, Estonian Nature Conservation Union 1972–; mem, IUCN World Comm on Protected Areas (formerly Comm on Nat Parks and Protected Areas) 1991–; mem, Council, Asscn of Baltic Nat Parks 1991–98; Dist Gov, Lions Clubs Int 1993–94; mem, Council, Fed of Nature and Nat Parks of Europe 1996–, Council, Estonian State Forest Man Centre, Nat Contact Point, European Parks for Life Programme. *Present Position*: Dir, Lahemaa Nat Park 1988–, Chair, Laheman Nat Park Foundation 1998– and Chair, Eesti Kaitsealade Liit (Union of Protected Areas of Estonia) 1991–. *Awards and Honours*: Fred M Packard Parks Merit Award, IUCN 1993; Int Leadership Award, Lions Clubs Int 1999. *Publications*: Lahemaa National Park 1983. *Address*: Lääne-Virumaa, Viitna 45202, Estonia. *Tel*: (50) 38-577. *Fax*: (32) 93-939. *E-mail*: arne.kaasik@mail.ee.

KAKABADSE, Yolanda; Ecuadorean international environmental organization executive; b 15 Sept 1948, Quito; m divorced; 2s. *Education*: Pontifical Catholic Univ of Ecuador. *Career*: Exec Dir, Fundación Natura 1979–80; NGO Liaison Officer, UNCED (Earth Summit) 1990–92; Pres, Fundación Natura 1990–; Exec Pres, Fundación Futuro Latinoamericano 1993–; mem, WWF Int Bd, Advisory Bd to the Vice-Pres for Sustainable Devt of the IBRD (World Bank), Soc for Int Devt, Int Advisory Council of the WorldWIDE Network. *Present Position*: Pres, World Conservation Union—IUCN. *Awards and Honours*: UNEP Global 500 1991. *Publications*: Movers and Shapers—NGOs in International Development 1994. *Address*: World Conservation Union—IUCN, 28 rue Mauverney, 1196 Gland, Switzerland. *Tel*: (22) 9990001. *Fax*: (22) 9990002.

KAMARUZAMAN, Prof Jusoff, M SC, PH D; Malaysian professor of forestry; b 28 March 1958, K Bharu; s of Jusoff Che Wok; m Rohaita 1985; 4d. *Education*: Cranfield Univ, UK. *Career*: expert, sustainable forest operation and devt using geospatial information tech. *Present Position*: Head, Faculty of Forestry and Centre for Precision Agric and Bioresource Remote Sensing, Universiti Pertanian Malaysia Serdang. *Awards and Honours*: UPM Excellence Service Award; Comstech Fellowship; Man of the Year 1998–99; 20th Century Award; Commonwealth Book Prize. *Publications*: more than 150 papers, reports and articles for learned journals. *Addresses*: Faculty of Forestry, Universiti Pertanian Malaysia Serdang, 43400 Selangor, Malaysia; Centre for Precision Agricultural and Bioresource Remote Sensing, Inst of Bioscience, Universiti Pertanian Malaysia Serdang, 43400 Selangor, Malaysia. *Tel*: (3) 98468101. *Fax*: (3) 9166003. *E-mail*: kamaruz@forr.upm.edu.my.

KANWAR, Prof Dr Jaswant Singh; Indian agricultural scientist; b 10 Dec 1922; s of Thakur Fateh Singh and Risal Devi; m Vidya Kanwar 1950; 2s 3d. *Education*: Univ of Adelaide, Australia, Univ of Ohio, USA. *Career*: Prof of Soil Science, Punjab Agricultural Univ 1954–62, Dir of Research 1962–65; Deputy Dir-Gen, Indian Council of Agricultural Research 1966–73, Int Crops Research Inst for Semi-Arid Tropics—ICRISAT 1973–88. *Present Position*: Ind consultant and Hon Deputy Dir-Gen, ICRISAT 1988–. *Publications*: Management of Soil and Water Resources and Environment 1994; Nutrient Management and Sustainability of Intensive Agriculture 1997. *Address*: 17 Krishni Nagar, ICRISAT Colony Phase II, Hashmetpet Rd, Secunderabad 500 009, Andhra Pradesh, India. *Tel*: (40) 7755104.

KARACA, Hayrettin; Turkish environmentalist; b 1922; m 1949; 2s 2d. *Education*: Bogaziccçi Lise, Istanbul. *Career*: Founder, first private arboretum in Turkey 1980; Founder and Publr, *Karaca Arboretum* magazine 1991–; mem, Int Dendrology Soc. *Present Position*: Founder and Pres, TEMA Foundation 1992–. *Awards and Honours*: UNEP Global 500 1992; Environmentalist of the Year, *Nokta* magazine 1992; Environmental Prize, Int Olympic Cttee 1993; Melvin Jones Fellowship Award, International Lions Club 1994; Tolerancy Award, Journalist and Author Foundation 1996. *Address*: PK 213, 80620 Levent, Istanbul, Turkey. *Tel*: (212) 2811027. *Fax*: (212) 2811132. *E-mail*: tema@tema.org.tr. *Internet*: www .tema.org.tr.

KAROLES, Dr Kalle, PH D; Estonian forestry expert and nature conservationist; b 1 June 1953, Tartu; s of Aleksei and Ilme Karoles; m Malle Karoles 1978; 1s 2d. *Education*: Agric Univ, Tartu, Forest Acad, St Petersburg, Russian Fed. *Career*: Kaarepere Forest Research Station 1971–76; Estonian Research Inst of Forestry and Nature Conservation 1980–; mem, Int Council, Int Union of Forestry Research Orgs 1996–2000. *Present Position*: Dir, Estonian Research Inst of Forestry and Nature Conservation. *Address*: Estonian Research Institute of Forestry and Nature Conservation, Rõõmu tee 2, Tartu 2400, Estonia. *Tel*: (7) 436-375. *Fax*: (7) 436-381.

KASSAS, Prof Mohamed, PH D; Egyptian professor of botany; b 6 July 1921, Kafr-el-Sheikh, Egypt; m Freda K Hosny 1957; 1s 1d. *Education*: Cairo Univ, Univ of Cambridge, UK. *Career*: Science Faculty, Cairo Univ 1950–; Ecological Cartographer, Mediterranean Basin, UNESCO Arid Lands Research Programme 1950–60; Sr Adviser, UNEP 1973–92; UN Conf on Desertification 1977; Pres, IUCN 1978–84; mem, Shoura Council 1981–, Egyptian Acad of Science, Egypt Inst, World Acad of Art and Science, Club of Rome. *Present Position*: Emer Prof of Botany, Cairo Univ 1981–. *Awards and Honours*: First Order of Science and Arts 1959; UN Prize for Environment 1978; ALECSO Gold Medal 1978; State Science Prize 1982. *Publications*: The World Environment 1972–1982 (Co-Ed) 1982; Desertification (Co-Ed) 1987; Financial Support for the Biosphere (Co-Ed) 1987. *Addresses*: 12613 Giza, Egypt; 41 Dokki St., Dokki, Giza. *Tel*: (2) 5676890 or (2) 7485742. *Fax*: (2) 5688884.

KAVALIAUSKAS, Prof Dr Paulius; Lithuanian geographer; b 1945, Anykšciai; s of Povilas Kavaliauskas and Monika Kavaliauskiene; 1s 1d. *Education*: Vilnius Univ. *Career*: Regional Planning Man, Nat Parks Protected Areas, Inst of Architecture and Bldg 1967–76; Founder, Nature Frame (Lithuanian ecological network). *Present Position*: Prof and Head, Land Management Section, Nature Faculty, Vilnius Univ 1976–. *Publications*: Perspectives of Protected Areas in Lithuania 1988; The Integrated Environmental Protection Scheme of Lithuania for the Period to the Year 2000 1988; Nature Framework Theory Questions 1990; Methodological Fundamentals of Land Management 1992. *Address*: Ciurlionio 21–27, Vilnius 2031, Lithuania. *Tel*: (2) 631-417.

KAWANABE, Prof Hiroya, D SC; Japanese ecological researcher; b 10 May 1932, Kyoto; s of Osamu and Tsuya (Fujii) Kawanabe; m Aya Minoguchi 1963; 1d. *Education*: Kyoto Univ. *Career*: Instructor, Kyoto Univ 1960–61, Lecturer 1961–67, Assoc Prof 1967–77, Prof, Zoology Dept, Science Faculty 1977–; Dir, Centre for Ecological Research, Kyoto Univ 1991–96; fmr Pres, Ecological Soc of Japan; mem, WWF Japan. *Present Position*: Dir-Gen, Lake Biwa Museum, Shiga 1996–. *Awards and Honours*: Asahi Subsidy Prize 1970; Japan Academy Duke of Edinburgh Prize 1996; Foreign hon mem American Acad of Arts and Sciences 1999; Award of Ecological Soc of Japan 2003. *Publications*: Ecology of Rivers and Lakes 1985; Fishes in Rivers and Lakes (2 vols, Co-Author) 1989–90; Symbiotic Biosphere (6 vols, Ed) 1992–93; Mutualism and Community Organization 1993; Global Biodiversity Assessment (Co-Author) 1995; Fish Communities in Lake Tanganyika 1999; Ancient lakes: Their Cultural and Biological Diversity 1999; Ancient Lakes: BIODIVERSITY, Ecology and Evolution 2000; An Integrated Study on Biodiversity Conservation under Global Change and Bioinventory Management Systems (Ed) 2002. *Address*: Lake Biwa Museum, 1091, Oroshimo, Kusatsu, Shiga 525-0001, Japan. *Tel*: (77) 568-4800. *Fax*: (77) 568-4848. *E-mail*: kawanabe@lbm.go.jp. *Internet*: www.lbm.go.jp.

KAZALIEV, George; Bulgarian teacher and ecologist; b 5 Sept 1927, Smoljan, Rajkovo; m Elena Kazalieva 1952; 2d. *Education*: Sofia Univ. *Career*: Teacher, Smoljan 1952–63, instructor 1963–. *Present Position*: Principal Officer, Smolyan br, Nat Ecological Club. *Address*: Nat Ecological Club, 4700 Smolyan, Rajkovo Dist, ZhK Izvor, etr B, Flat 4, Bulgaria. *Tel*: (301) 228-80.

KEAR, Dr David, CMG, FRSNZ; British/New Zealand scientific administrator and geologist; b 29 Oct 1923, London; s of Harold and Constance (Betteridge) Kear; m Joan Kathleen Bridges 1948; 2s 1d. *Education*: Imperial Coll, London. *Career*: Geologist, NZ Geological Survey 1948–67, Dir 1967–74; UN Ground-water Consultant, Western Samoa 1969–78; Asst, NZ Scientific and Industrial Research Dept 1974–80, Dir-Gen 1980–83; Geothermal Energy Consultant, Kenya and Indonesia 1985–87; mem, Commonwealth Science Council, Ross Dependency Research Cttee and NZ Medical Research Council 1980–83, UN Advisory Cttee on Science and Tech for Devt 1988–90; Consultant, South Pacific Geoscience Comm 1991–. *Present Position*: Private consultant geology, minerals and environment. *Publications*: Geology and Hydrology of Western Samoa (Co- Author) 1959; Jurassic Sequence at Kawhia Harbour (Co-Author) 1960; Geology of Northland: Accretion, Allochtons and Arcs (Co-Author) 1989; New Zealand's Volcanoes—Their Story 1999; Reassessment of Neogene Tectonism and Volcanism in North Island, New Zealand 2004. *Address*: 34 W End, Ohope, Whakatane, New Zealand. *Tel*: (7) 312-4635.

KEC, Jirí; Czech forestry engineer; b 1956, Brno; m Maria Kec 1980; 3s. *Education*: Forestry Tech Univ, Brno. *Career*: Army Forest Man 1980–91. *Present Position*: Dir, Šumava Nat Park and Landscape Protected Area Admin 1991–. *Address*: Zámek, 385 01 Vimperk, Czech Republic. *Tel*: (339) 21334. *Fax*: (339) 22947.

KELANY, Prof Dr Ibrahim, PH D; Egyptian environmentalist; b 1944, Minia; s of Mohamed and Niefesa Kelany; m Amany Abd el-Maksoud 1976. *Education*: Assuit Univ, Zagazig Univ, Justus-Liebig-Univ, Giessen, Fed Repub of Germany. *Career*: Lecturer 1979–84, Assoc Prof 1984–88, Prof 1988–; Bd mem, Entomological Soc of Egypt 1992–, Soc of Biological Pest Control 1994–, Egyptian Soc of Environmental Protection 1995–, Environmental NGOs Steering Cttee 1996–; adviser, Environment and Community Devt, Zagazig Univ 1996–; Head, Plant Protecton Dept; Consultant, environment and natural resources. *Present Position*: Prof of Environmental Protection and Applied Zoology, Zagazig Univ 1988–, Sec-Gen, Tree Lovers Asscn 1993– and Vice-Pres, Egyptian Green Party 1993–. *Awards and Honours*: International Cultural Diploma of Honour, American Biographical Inst 1996. *Publications*: numerous scientific papers in the fields of medical and veterinary entomology, plant extracts as alternative means of pesticides, arthropod parasites in birds, pheromones 1980–. *Address*: POB 262, El-Maadi, 11728 Cairo, Egypt. *Tel and fax*: (2) 5197009 or (2) 5196949. *E-mail*: iakelany@thewayout.net.

KELLY, Michael; British environmental executive; b 13 Sept 1955, London. *Career*: Mem, CBI Environment Forum, Environment Issues Advisory Panel, British Bankers Asscn; Liaison Del, World Business Council. *Present Position*: Head of Environmental Man, NatWest Group PLC 1996–. *Address*: NatWest Group plc, Mezzanine, 41 Lothbury, London EC2P 2BP, United Kingdom. *Tel*: (20) 7726-1166. *Fax*: (20) 7726-1183. *E-mail*: mike.kelly@natwestuk.co.uk.

KENMOGNE, Rev Jean-Blaise; Cameroonian pastor and environmentalist; b 10 June 1952, Bahouan; m Suzanne Djouinndje 1977; 1s. *Education*: Montpellier Univ, Univ of Paris. *Career*: Founder mem, Fédération des Organisations Non-Gouvernementales pour l'Environnement au Cameroun—FONGEC 1992–; Del-Gen, North–South Movt for Environment and Devt—MONSED. *Present Position*: Dir-Gen, Cercle International pour la Promotion de la Création—CIPCRE 1990–, Man Ed, *ECOVOX* magazine and Pastor, Cameroon Evangelical Church. *Publications*: Pour la vie en abondance, 2002; Commuiquer en Afrique, 2002; Le message du VIH-SIDA a l'Afrique, 2002; Pour vaincre le tribalisme, 2002. *Address*: Cercle International pour la Promotion de la Création, BP 1256, Bafoussam, Cameroon. *Tel*: 44-62-67. *Fax*: 44-66-69. *E-mail*: cipcre.dg@cipcre.org.

KENT, Bruce, LL B; British campaigner for nuclear disarmament, justice and peace; b 22 June 1929, London; s of Kenneth and Rosemary (Marion) Kent; m Valerie Flessati 1988. *Education*: Univ of Oxford. *Career*: Curate, Kensington N and S 1958–63; Sec, Archbishop's House, Westminster 1963–64; Chair, Diocesan Schools Comm 1964–66; Catholic Chaplain, London Univ 1966–74; Chaplain, Pax Christi 1974–77; Parish Priest, Somers Town, London 1977–80; Gen Sec, Campaign for Nuclear Disarmament—CND 1980–85, Vice-Chair 1985–87, Vice-Pres 1985, Chair 1987–90; retired from ministry 1987; Lab Party Parl Cand, Oxford W and Abingdon 1992. *Present Position*: Consultant, Int Peace Bureau, Geneva, Switzerland 1985–; Chair, Movement for the Abolition of War 2002–. *Awards and Honours*: DLIT Manchester; DLIT Middlesex. *Publications*: Building the Global Village 1991; Undiscovered Ends 1992. *Address*: 11 Venetia Road, London N4 1EJ, United Kingdom. *E-mail*: (20) 8347-6162. *Internet*: www.psi.org.uk.

KERN, Eha; Swedish environmentalist; b 1947, Hultsfred; m Bernd Kern; 2d. *Career*: Primary school teacher 1977–; Founder mem, Int Children's Rainforests Network 1987–; active in the promotion and protection of rainforest, particularly in Cen and S America. *Present Position*: Co-Founder and Chair, Barnens Regnskog (Children's Rain Forest) 1987–. *Awards and Honours*: Goldman Environmental Foundation Prize 1991. *Address*: Barnens Regnskog, POB 4471, 137 94 Västerhaninge, Sweden. *Tel and fax*: (8) 53-02-31-73.

KETCHUM, Robert Glenn; American photographer, artist and writer; b 12 Jan 1947, Los Angeles, CA; m Amy Holm 1996; 1s. *Education*: Univ of California, Los Angeles, California Inst of the Arts. *Career*: Faculty mem, California Inst of the Arts; Exec Dir and Chair, Bd of Dirs, Los Angeles Center for Photographic Studies; Curator of Photography, Nat Park Foundation; UNCED (Earth Summit) exhibition, Nat Museum of Fine Arts, Rio de Janeiro, Brazil 1992. *Present Position*: Trustee, Alaska Conservation Foundation, Pres, Advocacy Arts Foundation and Councillor, American Land Conservancy. *Awards and Honours*: Ansel Adams Award for Conservation Photography 1989; UNEP Global 500 1991; United Nations Outstanding Environmental Achievement Award 1991; Chevron–Times Mirror Magazines Conservation Awards 1995. *Publications*: American Photographers and the National Parks 1981; The Hudson River and the Highlands 1985; The Tongass: Alaska's Vanishing Rain Forest 1987; Overlooked in America: The Success and Failure of Federal Land Management 1991; CLEARCUT: The Tragedy of Industrial Forestry 1993; The Legacy of Wilderness: The Photographs of Robert Glenn Ketchum 1993; Presidio Gateways 1993; Northwest Passage 1996. *Address*: 696 Stone Canyon Rd, Los Angeles, CA 90077, USA. *Tel*: (310) 472-3681. *Fax*: (310) 440-2654. *E-mail*: peace2rth@aol.com.

KHAMDAN, Shaker A A, PH D; Bahraini marine biologist; b 8 June 1960, Manama; m Zainab R Al-Aali 1989; 2d. *Education*: Univ of Wales, Bangor, UK. *Career*: Marine biology research, including population genetics, habitat protection, sustainable devt and environmental law 1983–; Founder, Bahrain Soc for Pearls and Oysters 1997. *Present Position*: Sr Marine Biologist, Ministry of Housing, Municipalities and Environment 1995–. *Address*: Environmental Affairs Div, Ministry of Housing, Municipalities and Environment, POB 26909, Adliya, Bahrain. *Tel*: 293693. *Fax*: 293694. *E-mail*: skhamdan@batelco.com.bh.

KHAN, Anisuzzaman, M SC; Bangladeshi wildlife ecologist; b 30 Sept 1954, Dhaka; s of K M Joaher Ali and Aeysha Khatun; m Sakeba Khatun 1986; 1s. *Education*: Rajshahi Univ. *Career*: Research Fellow, WWF/IUCN Elephant Project 1983–84; Nat Co-ordinator, Wetlands Int 1991–97, BirdLife Int 1993–97; Fellow, Ashoka: Innovators for the Public, USA 1992–95; Nat Consultant, Canadian Int Devt Agency, Norwegian Agency for Devt Corpn, IBRD (World Bank) 1993–96; Nat Rep, Oriental Bird Club, UK 1996–97. *Present Position*: Exec Dir, Nature Conservation Movement—NACOM 1995– and Nat Consultant, Nat Conservation Strat. *Publications*: Biodiversity in Animals 1994; CITES: A Handbook for Bangladesh 1996; Ecology of Floodplains 1997. *Address*: Nature Conservation Movement, Mohammadia Supermarket, Rm 125–127, 4 Sobhanbag, POB 3413, Dhaka 1207, Bangladesh. *Tel*: (2) 819237. *Fax*: (2) 816442.

KHAN, Niaz Ahmad, B SC; Pakistani environmental engineer; b 24 April 1954; s of Fazal A Khan and Alam Bibi; m Pervaan Akhtar; 1s 2d. *Education*: Univ of Eng and Tech, Lahore; Univ of Wisconsin, Madison, USA. *Career*: Environmental Specialist, ERT, Jubail, Saudi Arabia 1982–87; Environmental Engineer, fertilizer co, Jubail 1987–89; Head, Environmental Affairs Section, Health Services Dept, Royal Comm of Jubail 1989–95. *Present Position*: Sr Environmental

Officer and Environmental Adviser, Jebel Ali Free Zone Authority/Dubai Ports Authority 1995–. *Awards and Honours*: Man of the Year 1997 Gold Medal, ABI, NC, USA 1998. *Publications*: articles in nat and int journals. *Addresses*: Jebel Ali Free Zone Authority/Dubai Ports Authority, POB 17000, Dubai, United Arab Emirates; Villa 41, St C, Jebel Ali Village, Dubai, United Arab Emirates. *Tel*: (4) 8040292. *Fax*: (4) 8818857. *E-mail*: niaz.k@dpa.co.ae. *Internet*: www.dpa.co.ae.

KHATIB, Hisham, PH D; Jordanian energy and environmental consultant; b 5 Jan 1936, Acre; s of Mohamed and Fahima Khatib; m Maha Khatib 1968; 2s 1d. *Education*: Univ of Cairo, Egypt; Univ of Birmingham, Univ of London, UK. *Career*: Chief Engineer, Jerusalem Electricity Co 1965–73; Deputy Dir-Gen, Jordanian Electricity Authority 1974–76, Dir-Gen 1980–84; Sr Energy Expert, Arab Fund, Kuwait 1976–80; Minister of Energy 1984–90; Vice-Chair, World Energy Council 1989–92; int energy consultant 1990–93; Minister of Water and Planning 1993–95; Chair, Cttee for Developing Countries, World Energy Council 1992–95. *Present Position*: Hon Vice-Chair, World Energy Council; int consultant 1995–. *Awards and Honours*: Achievement Medal, Inst of Electrical Engineers, UK 1998; decorations from Jordan, Sweden, Italy, Indonesia and Austria. *Publications*: Economics of Reliability 1978; Financial or Economic Evaluation of Projects 1997; Palestine and Egypt under the Ottomans 2003; Economics of Electricity 2003; numerous articles in professional journals. *Address*: POB 925387, Amman, Jordan. *Tel*: (6) 5621532. *Fax*: (6) 5698556. *E-mail*: khatib@nets.com.jo.

KHOSHOO, Triloki Nath, PH D; Indian scientist and biodiversity researcher; b 7 April 1927, Srinagar, Kashmir; s of S C and Vanamala Khoshoo; m Mohini Khoshoo 1945; 2s. *Education*: Punjab Univ, Lahore (now Pakistan); Punjab Univ, Chandigar. *Career*: Sr Lecturer, Punjab Univ 1948–62; Head, Postgraduate Botany Dept, Srinagar Univ 1962–64; Asst Dir, Nat Botanic Garden, Lucknow 1964–74, Deputy Dir 1974–76; Dir, Nat Botanical Research Inst, Lucknow 1976–82; Vice-Chair, Governing Council, UNEP 1982; Sec, Dept of Environment, Govt of India 1982–85; Distinguished Scientist, Council of Scientific and Industrial Research 1985–90; mem, Plant Advisory Group, IUCN 1988–, Advisory Cttee to Dir-Gen, UNESCO 1993–. *Present Position*: Fellow, Tata Energy Research Inst 1991–. *Publications*: Plant Systematics and Evolution (Co-Ed) 1978; Silvae Genetica (Co-Ed) 1984; Mahatma Gandhi: An Apostle of Applied Human Ecology 1995. *Address*: Tata Energy Research Institute, Darbari Seth Block, Habitat Pl, Lodi Rd, New Delhi 110 003, India. *Tel*: (11) 4622246. *Fax*: (11) 4621770.

KIERNAN, Ian Bruce Carrick, AO; Australian master builder and environmental campaigner; b 4 Oct 1940, Sydney; s of George Arthur and Leslie K M Kiernan; m Judith Anne Kiernan; 2d. *Education*: Armidale School, NSW. *Career*: Founder, Clean Up Sydney Harbour 1989, Clean Up Australia 1990, Clean Up The World 1991; Presenter, *Living Earth* series and two documentaries, *Global Heroes* 1993 and *Spaceship Earth* 1997. *Present Position*: Chair, Clean Up Australia 1990–; Exec Chair, Clean Up The World 1991–; Chair, CRC Waste Control and EB CRC (Environmental Biotechnology); Chair, Enviro-Finance Ass of Austrlia, Dir. *Awards and Honours*: Order of Australia Medal (OAM); UNEP Global 500 1993; Officer of the Order of Australia (AO); UN Sasakawa Award; Australian of the Year 1994; Centenary Medal ACOPS; National Living Treasure. *Publications*: Coming Clean (autobiography) 1995. *Addresses*: 37 Willoughby Rd, Kirribilli, Sydney, NSW 2061, Australia; 18 Bridge Rd, Glebe, NSW 2037, Australia. *Tel*: (2) 9552-6177. *Fax*: (2) 9552-4468. *E-mail*: cleanup@cleanup.com.au. *Internet*: www.cleanup.com.au.

KINNERSLEY, David John, MA; British water consultant; b 28 May 1926, Essex; s of Arthur Kinnersley and Doris Angier; m Barbara Fair 1950; 1s 2d. *Education*: Corpus Christi Coll, Cambridge. *Career*: Energy industry 1951–64; water industry 1964–83; Fellow, Nuffield Coll, Oxford then Mansfield Coll, Oxford 1983–86; Consultant, IBRD (World Bank), Washington, DC, USA 1984–94; Special Consultant to Sec of State for Environment 1987–88; mem, Bd, Nat Rivers Authority 1989–91; mem, Council, Monopolies and Mergers Comm. *Present Position*: Mem, Council, WATERAID 1983–. *Awards and Honours*: Churchill Fellowship 1973. *Publications*: Troubled Water: Rivers, Politics and Pollution 1988; Coming Clean 1994. *Address*: 111 Church St, Chesham HP5 1JD, United Kingdom. *Tel*: (1494) 778118. *Fax*: (1494) 791444.

KINTANAR, Roman, MA, PH D; Filipino government official and university professor; b 13 June 1929, Cebu City; s of Augustin Y Kintanar and Pureza Lucero; m Generosa Pérez-Kintanar 1959; 2s 1d. *Education*: Univ of the Philippines; Univ of Texas, USA. *Career*: Prof of Physics, Univ of the Philippines 1955–56, Feati Univ 1958–65; Professorial lecturer, Ateneo de Manila Univ 1966–68; Chief Geophysicist, Philippine Weather Bureau 1953–58, Dir 1958–72; Admin, Philippine Atmospheric, Geophysical and Astronomical Services Admin—PAGASA 1972–77, Dir-Gen 1977–; Chair, Scientific and Tech Cttee, UN Int Decade for Natural Disaster Reduction 1994; Pres, ROMAROSA Realty Devt Corpn Inc 1996–; Co-ordinator, WMO/ESCAP Typhoon Cttee Secr 1967–79; Perm Rep to WMO 1958; Vice-Pres, Regional Asscn V for S-W Pacific (WMO-RA V) 1966–74, Pres 1974–78; Vice-Pres, WMO 1978–79, Pres 1979; mem, Philippine Asscn for the Advancement of Science, Nat Research Council of the Philippines, Philippine Meteorological Soc, Int Asscn of Seismology and Physics of the Earth's Interior, UNESCO/UNDRO Int Advisory Cttee on Earthquake Damage and Mitigation. *Present Position*: Vice-Chair, Manila Observatory Inc 1996–; Chair of Bd, Typhoon Cttee Foundation 1996–. *Awards and Honours*: Int Meteorological Org Prize 1995; Fulbright Smidthmundt Scholarship, US Educational Foundation; Office of the Pres Ecology Award; Budiras Award for Outstanding Performance, Bureau Dirs Asscn; Parangal ng PAGASA Award; Lingkod Bayan Award 1982; Padre Faura Astronomy Medal 1982. *Publications*: A Study of Typhoon Microseisms 1958; many articles in scientific journals. *Address*: 100 Don Primitivo St, Don Antonio Heights, Quezon City, Philippines. *Tel*: (2) 9317069. *Fax*: (2) 9318484.

KIRA, Tatuo, PH D; Japanese environmental scientist; b 1919, Osaka; s of Tetumyo and Hukue Kira; m Yasuko Kametani 1948. *Education*: Kyoto Univ. *Career*: Asst Prof, Kyoto Univ 1948–49; Prof, Osaka City Univ 1949–81; Pres, Ecological Soc of Japan 1980–83; Dir, Lake Biwa Research Inst 1982–94; Chair, Scientific Cttee, Int Lake Environment Cttee Foundation 1986–95, Pres, Japan Soc of Tropical Ecology 1990–97. *Present Position*: Prof Emer, Osaka City Univ. *Awards and Honours*: hon mem, British Ecological Soc 1985, Ecological Soc of Japan 1995; Purple Ribbon Medal 1984; Order of the Sacred Treasure (Second Class) 1990; Int Cosmos Prize 1995; Duke of Edinburgh Prize of Japan Acad 1998. *Publications*: Nature and Life in SE Asia, Vols I–VII (Ed) 1961–76; Primary Production of Japanese Forests (Ed) 1977; Biological Production in a Warm-temperate Evergreen Oak Forest (Ed) 1978; Data Book of World Lake Environments I–III (Ed) 1989–91; ten textbooks and compiled essays, 100 scientific papers. *Addresses*: Lake Biwa Research Institute, 1-10 Uchidehama, Otsu, Shiga 520, Japan (Office); Nango 2-21-9, Otsu, Shiga 520, Japan (Home). *Tel*: (775) 264800.

KISBY, Steve; Canadian environmentalist. *Career*: Fmr Exec Sec, Green Party of Canada/Canadian Greens. *Present*

Position: membership Chair, Green Party of Canada. *Address*: 320 East 37th Avenue, Vancouver, BC, V5W 1E7, Canada. *Tel*: (604) 323-0204. *Fax*: (604) 323-0224. *E-mail*: skisby@web.net.

KNILL, Prof Sir John Lawrence, D SC, FICE, F ENG; British engineering geologist; b 22 Nov 1934, Wolverhampton; s of William Cuthbert and Mary (Dempsey) Knill; m Diane Constance Judge 1957; 1s 1d. *Education*: Imperial Coll, London. *Career*: Geologist, Sir Alexander Gibb & Partners 1957; Asst Lecturer, Lecturer, Reader then Prof of Eng Geology, Imperial Coll, London 1957–93; Chair, Natural Environment Research Council 1988–93. *Present Position*: Emer Prof of Eng Geology, Imperial Coll, London 1993–. *Awards and Honours*: British Geotechnical Soc Prize 1966; Whitaker Medal (IWEM) 1969; Hon FCGI 1988; Aberconway Medal (IGEOL) 1989, Hon D SC (Kingston) 1992, (Exeter) 1995; William Smith Medal (GEOLSOC) 1995; Hon D TECH (Nottingham Trent) 1996. *Publications*: Industrial Geology 1977; Geological and Landscape Conservation 1994. *Address*: Highwood Farm, Shaw-cum-Donnington, Newbury RG14 2TB, United Kingdom. *Tel*: (1635) 40542. *Fax*: (1635) 36826.

KNOX, George Alexander, MBE, M SC, FRSNZ; New Zealand zoologist, consultant and writer; b 1919, Pleasant Point. *Education*: Univ of New Zealand. *Career*: school teacher 1937–41, 1946–48; served with NZ armed forces, Middle East 1942–43; Asst Lecturer, Canterbury Univ Coll 1949–51, Lecturer, then Sr Lecturer 1952–59, Prof of Zoology and Head of Dept, Univ of Canterbury 1958–78, Prof of Zoology 1978–83; Leader, Chatham Island Expedition 1954, mem, Royal Soc of London Chile Expedition 1958–59, Dir, Univ of Canterbury Antarctic Research Unit 1962–83, 12 field expeditions to McMurdo Sound region; mem, Freshwater Fisheries Advisory Council 1966–70, NZ Nat Cttee for UNESCO 1970–74, Nat Research Advisory Council, Environment and Energy Cttee 1973–77, NZ Nat Cttee for the Man and Biosphere Programme 1972–80; mem, Scientific Cttee on Antarctic Research—SCAR 1969–, Sec 1974–78, Pres 1978–84; mem, Special Cttee on Oceanic Research—SCOR 1971–75, Special Cttee for Int Biological Programme—SCIBP 1965–72; Bd mem, Int Asscn of Ecology—INTECOL 1965–89, Sec Gen 1973–78, Pres 1978–82; Foundation mem, Council mem, NZ Ecological Soc 1952–54, NZ Marine Sciences Soc 1961–64 (Pres 1961–62); mem, NZ Limnological Soc, NZ Antarctic Soc, Ecological Soc of America, AAAS, Australian Marine Sciences Soc; Fellows Councillor, Royal Soc of NZ 1970–76, Int Sec 1970–73. *Present Position*: Prof Emer, Univ of Canterbury 1983–. *Awards and Honours*: Hutton Medal, Royal Soc of NZ 1979; Conservation Trophy, NZ Antarctic Soc 1980; NZ Marine Sciences Soc Award for Outstanding Contribution to Marine Science in NZ 1985; Sir Ernest Marsden Medal for Service to Science, NZ Asscn for the Advancement of Science 1985. *Publications*: six books, 125 papers. *Address*: Department of Zoology, University of Canterbury, Private Bag 4800, Christchurch, New Zealand.

KODRÍK, Milan, PH D; Slovak research scientist; b 16 May 1957, Zvolen; m Emília Kodríková 1987; 2s. *Education*: Tech Univ in Zvolen. *Career*: Chair, Dept of Soil Ecology, Inst of Forest Ecology, Slovak Acad of Sciences 1989–90. *Present Position*: Chair, Dept of Forest Ecology, Inst of Forest Ecology, Slovak Acad of Sciences 1990–. *Publications*: Forest Ecosystems in Slovakia (Ed) 1991; Distribution of Root Biomass and Length in Picea Abies Ecosystem 1994. *Address*: Institute of Forest Ecology, Slovak Academy of Sciences, Štúrova 2, 960 53 Zvolen, Slovakia. *Tel*: (855) 209-14. *Fax*: (855) 274-85. *E-mail*: kodrik@sav.savzv.sk.

KOESTER, Prof Veit; Danish lawyer and conservationist; b 30 May 1934, Berlin, Germany; m Winnie Kjoelby 1993; 2c. *Education*: Univ of Copenhagen. *Career*: Mem, Bar 1962–; civil servant, responsible for formulation and implementation of environmental legislation and EU and other int legal instruments; Head, Danish dels to Bonn Convention of the Conservation of Migratory Species of Wild Animals, Ramsar Convention on Wetlands of Int Importance, the Convention on Int Trade in Endangered Species of Wild Fauna and Flora—CITES and the Convention on Biodiversity; Chair and Vice-Chair, numerous int confs, including the Ramsar Convention, the Convention on Biodiversity, the Bern Convention, the Council of Pan-European Biodiversity and Landscape Strategy, the Biodiversity Protocol and CITES; Councillor, IUCN 1988–93. *Present Position*: Dir, Int Div, Nat Forest and Nature Agency, Ministry of Environment and Energy. *Awards and Honours*: Danish Animal Welfare Soc Annual Prize 1981; UNEP Global 500 1996. *Publications*: numerous articles and other publications on nature conservation and int environmental law. *Address*: National Forest and Nature Agency, Haraldsgade 53, 2100 Copenhagen Ø, Denmark. *Tel*: 39-47-20-00. *Fax*: 39-27-98-99. *E-mail*: vko@sns.dk.

KOHANOFF, Roberto; Argentinian politician and environmentalist. *Career*: Mem, Los Verdes (Green Party). *Present Position*: Mem then Sec of Ecology, Partido Humanista 1994–. *Address*: Callao 152, 1022 Buenos Aires, Argentina. *Tel*: (11) 4371-3122. *Fax*: (11) 4371-3130. *E-mail*: kohanoff@cvtci.com.ar.

KOITJÄRV, Teet; Estonian environmentalist; b 17 March 1963, Tapa; s of Helle and Heino Koitjärv; m Lea Koitjärv 1990; 1s. *Education*: History and Archives Inst, Moscow, Russian Fed. *Career*: Archivist, Tallinn City Archives 1986–87; Historian and Guide, Lahemaa Nat Park 1987–88; Sec, Asscn of Baltic Nat Parks—ABNP 1991–98; project leader, Kadrina Commune Devt Plan 1998–99. *Present Position*: Deputy Dir, Lahemaa Nat Park 1988–; Sec, Union of Protected Areas of Estonia 1991–. *Publications*: Estonia Biodiversity Strategy and Action Plan (Co-Author) 1999; *TV Documentary Script*: Even One Vargamäe (Co-Author) 1989. *Address*: Lahemaa 202, Viitna 45202, Estonia. *Tel*: 3293932. *Fax*: 3293931. *E-mail*: teet.koitjarv@neti.ee. *Internet*: www.hot.ee/teetkoitjarv.

KOLLÁR, Miroslav; Slovak chemical engineer, water technology specialist; b 21 Nov 1961; s of Emil Kollár and Helena Kollárová; m Eleonóra Kollárová 1987; 1c. *Education*: Slovak Tech Univ, Bratislava, City Univ, Bratislava. *Career*: National Focal Point, Europarc Expertise Exchange 1997–2001. *Present Position*: Scientific Sec, Váyskumnáy ústav vodného hospodárstva (Water Research Inst) 1992–; project leader, Kadrina Commune Devt plan and spatial planning 2003–04. *Publications*: Avast Eestimaa uut Moodi (Ed) 2001; Review of the Current Protected Areas Management System (Co-Author) 2000. *Address*: nábr L Svobodu 5, 812 49 Bratislava, Slovakia. *Tel*: (7) 343-111. *Fax*: (7) 315-743.

KONDO, Jiro, D ENG; Japanese engineer; b 23 Jan 1917; s of Shuzo and Fujie Kondo; m Kei Mizuki 1944; 1s 1d. *Education*: Kyoto Univ, Tokyo Univ. *Career*: Prof of Applied Mathematics and Gas Dynamics, Tokyo Univ 1958–77, Dean, Eng Faculty 1975; Dir-Gen, Nat Inst of Environmental Studies 1977–85; Pres, Science Council of Japan 1985–94. *Present Position*: Pres, Central Council for Environmental Pollution Control 1988–. *Awards and Honours*: Deming Prize 1967; Purple Ribbon Medal 1982; Grand Cordon of the Order of the Sacred Treasure 1990; Person of Cultural Merit 1995. *Publications*: Integral Equations 1991; Supercomputing (Ed) 1991. *Address*: 9-2, Kizugawadai, Kizu-cho, Soraku-gun, Kyoto 619-02, Japan.

KONDRATYEV, Prof Dr Kirill Yakovlevich; Russian environmental scientist; b 14 June 1920, Rybinsk; s of Ya I

Kondratyev and S I Kondratyeva; m; 2s 1d. *Education*: Leningrad State Univ. *Career*: researching atmospheric and environmental physics, remote sensing, climate dynamics and global change; Ed-in-Chief, *Earth Observation and Remote Sensing*. *Present Position*: Counsellor, Research Centre for Ecological Safety, Russian Acad of Sciences; Nansen Scholar, Nansen Int Fund Environmental and Remote Sensing, St Petersburg. *Awards and Honours*: Academician, Russian Acad of Sciences; Hon PH D, Univ of Athens, Greece, Univ of Budapest, Hungary, and Univ of Lille, France; Gold Medal World Meteorological Org; Symons Gold medal UK; hon mem: International Astronautical Academy; Academia Scientiarium, Austria; Akademia Leopoldina, Germany; American Meteorological Soc; Royal Meteorological Soc, UK. *Publications*: About 100 scientific papers in refereed journals and over 100 monographs most recently: Key Issues of Global Change 1996; Volcanic Activity and Climate Change 1996; Multidimensional Global Change 1998. *Addresses*: Research Centre for Ecological Safety, St Petersburg Centre of the Russian Academy of Sciences, 197110 St Petersburg, Korpusnaya 18, Russia; 199106 St Petersburg, Nalichnaya Str 3, Apt 79. *Tel*: (812) 230-78-37. *Fax*: (812) 230-79-94. *E-mail*: kondratyev@kk10221.spb.edu.

KOSTENKO, Dr Yuriy I; Ukrainian engineer and politician; b 12 June 1951, Vinnitsia Region; s of Ivan Kostenko and the late Lyudmila Kostenko; m Alina Kostenko; 1s. *Education*: Machinery Inst, Zaporizjia, Electric Welding Inst, Ukrainian Acad of Sciences. *Career*: Engineer, Electric Welding Inst, Ukrainian Acad of Sciences 1973–82, student 1982–85, staff mem 1985–90; elected to Supreme Soviet (parl) 1990, Deputy Chair, Parl Ecological Comm 1990–92; Minister for Environmental Protection (later Minister of Environmental Protection and Nuclear Safety) 1992–98; Head, Parl Working Group on Nuclear Disarmament 1993; presidential candidate 1999; mem, perm del of Supreme Council of Ukraine to PACE; mem European People's Party group in PACE 2003–; Chair, Ukrainian People's Party –1999; First Deputy Head, Our Ukraine parliamentary faction 2002–; fellow, World Academy of Arts and Sciences. *Present Position*: MP, Supreme Council (parl) 1990–; mem permanent delegation of European People's Party Group in PACE. *Awards and Honours*: Honourable Diploma (for participation in rescue expedition in Armenia) 1989. *Publications*: Kiev and the Bomb: Ukrainians Reply 1993; Ukraine's Nuclear Weapons: A Political, Legal and Economic Analysis of Disarmament 1993; Chornobyl 'Sarcofag' and the Problems of Nuclear Policy 1994. *Address*: c/o Supreme Council, 252019 Kiev, vul M Hrushevskoho 5, Ukraine.

KOTHARI, Ashish, MA; Indian researcher, trainer and environmental activist; b 15 Aug 1961, Vadodara; s of Dr Rajni and Hansa Kothari; m Sunita Rao 1992. *Education*: Delhi School of Econs. *Career*: Founder mem, Kalpavriksh environmental action group 1979–; Research Assoc, Indian Inst of Public Admin—IIPA 1985–92; Project Dir, Status Report on Biodiversity in India 1992–94; mem, IUCN Comms on Protected Areas and on Environmental Sustainability 1995–. *Present Position*: Lecturer, IIPA 1992–. *Publications*: Narmada Valley Project: A Critique 1988; Management of National Parks and Sanctuaries in India 1989; State Directories of National Parks and Sanctuaries in India (Ed) 1990–; Conservation of Biological Diversity in India 1993; Birds in India: Status and Conservation 1993; Conserving Life: Implications of the Biodiversity Convention for India 1995; People and Protected Areas (Ed) 1996; Understanding Biodiversity: Life, Equity and Sustainability 1997. *Address*: c/o Rajni Kothari, A-19, Sector 56, Noida 201 301, Uttar Pradesh, India. *Tel*: (11) 3317309. *Fax*: (11) 3319954. *E-mail*: ashish@giasdloi.vsnl.net.in.

KOUDA, Dr Michel; Burkinabé environmentalist; b 1949; s of N Kouda and K Dianda; m Marie Bonafos 1981; 1s 2d. *Education*: Paul Sabatier Univ, Toulouse, France. *Career*: Ministry of Rural Devt 1979; Deputy Dir, CRTO 1980; mem, UN Group of Experts on Remote Sensing 1983; Dir-Gen of Agriculture 1984; Dir of Research and Planning, Ministry of Water 1985, Minister of Water 1986; Head, Canada–Burkina Faso Devt Project 1988. *Present Position*: Head, IUCN Mission to Burkina Faso 1993–. *Publications*: Analyse synchronique et diachronique de l'évolution de la végétation par télédétection multispectrale en zone semi-aride de Haute Volta 1981. *Address*: IUCN Burkina Faso, 515 rue Agostino Neto, 01 BP 3133, Ouagadougou 01, Burkina Faso. *Tel*: 31-31-54. *Fax*: 30-13-51. *E-mail*: kouda@fasonet.bf.

KOVÁCS-LÁNG, Prof Edit, PH D; Hungarian botanist; b 1938, Budapest. *Education*: Loránd Eötvös Univ. *Career*: Assoc Prof, Dept of Plant Taxonomy and Ecology, Univ of Budapest 1976–90; Dir, Ecological and Botanical Research Inst, Hungarian Acad of Sciences 1991–96; Team Leader, CORINE Biotopes Project 1993–96; Project Leader, PHARE Biodiversity Monitoring Project 1995–96. *Present Position*: Head, Dept of Ecology, Ecological and Botanical Research Inst, Hungarian Acad of Sciences 1997–. *Publications*: Numerous scientific papers and textbooks. *Address*: Ecological and Botanical Research Institute, Hungarian Academy of Sciences, 2163 Vácrátót, Alkotmány u 2–4, Hungary. *Tel*: (28) 360-122. *Fax*: (28) 360-110. *E-mail*: h6867lan@ella.hu.

KOZHOVA, Prof Dr Olga Mikhailovna; Russian hydrobiologist; b 3 Feb 1931, Irkutsk; d of Prof Dr M M Kozhov and R V Kozhova; m 1st Ravil Gasin (deceased); m 2nd Boris Pavlov; 1d. *Education*: Irkutsk State Univ. *Career*: Scientist, Limnological Inst, Acad of Sciences 1956–69; Chair of Invertebrate Zoology and Dir, Biological Inst, Irkutsk State Univ 1969–82; Head of Lab, Ecological Toxicology Inst 1982–90. *Present Position*: Dir, Biology Research Inst, Irkutsk State Univ 1990–. *Publications*: Phytoplankton of Lake Baikal: Structural and Functional Characteristics 1987. *Address*: 664003 Irkutsk, ul Lenin 3, POB 24, Russian Federation. *Tel*: (3952) 34-42-77. *Fax*: (3952) 43-52-07. *E-mail*: root@bio.isu.runnet.ru.

KOZLOV, Dr Mikhail Vasilievich, PH D; Finnish biological researcher; b 31 Aug 1962, Leningrad (now St Petersburg), Russian Fed; s of Vasilii Ivanovich and Mira Arsenieva; m Elena Zvereva 1989; 2s. *Education*: Leningrad State Univ 1984. *Career*: researcher, impact of pollution on insect biodiversity and insect–plant relationships 1981–, ecosystems responses to environmental contamination 1995–, interactions between environmental contamination and global change 1998–; sr researcher, Sect Ecology, Univ of Turku, Finland 1991–. *Present Position*: Docent (Asst Prof), Univ of Turku 2001sr researcher, Sect Ecology, Univ of Turku, Finland 1991–. *Publications*: Human Impact on Population of Terrestrial Insects (Ed) 1990; Aerial Pollution in the Kola Peninsula 1993; ca 190 publications in professional journals and book chapters. *Address*: Section of Ecology, University of Turku, 20014 Turku, Finland. *Tel*: (2) 3335716. *Fax*: (2) 3336550. *E-mail*: mikoz@utu.fi. *Internet*: www.sci.utu.fi/biologia/ekologia/index_eng.html.

KOZOVÁ, Dr Mária, PH D; Slovak landscape ecologist and environmentalist; b 6 Dec 1948, Cierny Brod; m Peter Koza 1971; 1d. *Education*: Comenius Univ of Bratislava. *Career*: Asst Researcher, Inst of Landscape Biology 1972–74; Researcher, Inst of Experimental Biology and Ecology 1975–89; Sr Researcher and Deputy Dir, Inst of Landscape Ecology and mem, Council of Experts, Federal Ministry of Environment 1990–91; Asst Prof, Faculty of Natural Sciences, Comenius Univ of Bratislava 1992–96, Assoc Prof 1997–. *Present Position*: Founder and Head, Landscape

Ecology Dept, Faculty of Natural Sciences, Comenius Univ of Bratislava and Dir, Centre for Environmental Impact Assessment. *Publications*: Use of Measures of Network Connectivity in the Evaluation of Landscape Stability (Co-Author) 1986; Umweltbericht Wien–Bratislava (Co-author) 1993; The EIA Process: The Present State of Methodology, Legislation, Implementation and Education in the Slovak Republic 1993; Carrying Capacity and Environmental Impact Assessment (Co-Author) 1995. *Address*: Faculty of Natural Sciences, Comenius University, Mlynská Dolina B-2, 842 15 Bratislava, Slovakia. *Tel*: (7) 796-579. *Fax*: (7) 728-938. *E-mail*: kozova@fns.uniba.sk.

KRAJCOVIC, Dr Roman Štefan; Slovak environmental and regional development executive; b 25 April 1958, Bratislava; s of Dr Štefan Krajcovic and Adriena Krajcovicová; m Jana Trencianska; 2d. *Education*: Comenius Univ of Bratislava; Univ Coll London, UK. *Career*: Scientist, State Nature Conservation 1982–88; Sec-Gen, Regional Org of the Slovak Union for Nature and Landscape Protection 1984–88; Environmental Affairs Consultant, Slovak Youth Union 1984–89; Sr Researcher, Inst of Landscape Ecology, Slovak Acad of Sciences 1988–94; Dir, Slovak Environmental Agency 1991–94; Man Dir, Technum Slovakia consultancy 1995–97. *Present Position*: Dir, CPC Consultancy Ltd 1999–. *Awards and Honours*: Bronze Medal, Slovak Union for Nature and Landscape Protection 1986, Dept of Culture and Nature Protection, Regional Govt 1988. *Publications*: Environmental Manual for Local Authorities in Central and Eastern Europe—A Tool for Modern Physical Planning 1994. *Address*: Stodolova 6, 949 01 Nitra, Slovakia. *Tel and fax*: (87) 567-037. *E-mail*: cpc@nextra.sk.

KRASNOV, Prof Dr Eugene V, D SC; Russian geo-ecologist; b 16 Aug 1933, Urzhum; s of Maria and Vasilii Krasnov; m Galina Krasnova 1957; 1d. *Education*: Kishinev State Univ, Kishinev (Chisinau), Moldova. *Career*: Head, Dept of Ecology, and Head, Palaeo-ecological Lab, Inst of Marine Biology, USSR Acad of Sciences, Far East Br, Vladivostok 1967–78; Head, Dept of Palaeontology, Far East Geological Inst, Vladivostok 1978–87; mem, European Union of Coastal Conservation 1992; mem, Editorial Bd, *International Coastal Conservation* 1995; Chair, Baltic Ecological Parl 1996–98. *Present Position*: Head, Geo-ecological Dept, Kaliningrad State Univ 1987–. *Awards and Honours*: USSR State Medals 1970, 1976; Hon Academician, Russian Fed 1996; Prizewinner, Baltic Sea Foundation, Åland, Finland 1997. *Publications*: Interaction between Water and Living Matter (Ed) 1979; Corals in Reefal Facies of USSR Mesozoic 1983; The Calcium–Magnesium Method in Marine Biology (Co-Author) 1982. *Address*: 236016 Kaliningrad, Pionerskaya 70-26, Russian Federation. *Tel*: (0112) 43-62-54. *Fax*: (0112) 46-58-13.

KRASSILOV, Prof Valentin A; Russian ecologist and palaeontologist; b 1 Dec 1937; m 1970; 1d. *Education*: Kharkiv State Univ, Ukraine. *Career*: Head, Evolutionary Dept, Far Eastern Br of the USSR Acad of Sciences 1961–89; Dir, All-Russian Research Inst on Nature Conservation and Reserves, Moscow 1990–92. *Present Position*: Chief Scientist, Inst of Geography, Russian Acad of Sciences, Moscow 1992–. *Awards and Honours*: Int Org of Palaeontology Medal 1992. *Publications*: Palaeontology of Terrestrial Plants 1975; Nature Conservation: Principles, Problems, Priorities 1993. *Address*: 109017 Moscow, Staromonetnii per 29, Russian Federation. *Tel*: (095) 238-82-77. *Fax*: (095) 230-20-29.

KRAUŠKOPF, Konrad Bates, PH D; American geologist; b 30 Nov 1910, Madison, WI; s of Francis Craig and Maude Luvan Bates Krauskopf; m Kathryn Isabel McCune 1936; 1s 3d. *Education*: Wisconsin Univ, Univ of California, Berkeley, Stanford Univ. *Career*: Chemistry Instructor, Univ of California 1934–35; Physical Science Instructor, Stanford Univ 1935–39, Asst Prof, Assoc Prof then Prof 1939–76; Geologist, US Geological Survey 1943–; Head, G2 Geographic Section, US Army, Tokyo, Japan 1947–49; Pres, American Geological Inst 1964, Geological Soc of America 1967, Geochemical Soc 1970. *Present Position*: Emer Prof of Geochemistry, Stanford Univ 1976–. *Awards and Honours*: Day Medal, Geological Soc of America 1961; Goldschmidt Medal, Geochemical Soc 1982; Ian Campbell Medal, American Geological Inst 1984. *Publications*: Introduction to Geochemistry 1967; The Third Planet 1974; Radioactive Waste Disposal and Geology 1988; The Physical Universe (Co-Author) 1992. *Address*: Geology Department, Stanford University, CA 94305, USA. *Tel*: (415) 723-3325. *Fax*: (415) 725-2199. *E-mail*: konrad@pangea.stanford.edu.

KREBS, Prof John Richard, D PHIL, FRS; British research professor and science administrator; b 11 April 1945, Sheffield; s of Sir Hans Adolf Krebs and Margaret Cicely Fieldhouse; m Katherine Anne Fullerton 1968; 2d. *Education*: Pembroke Coll, Oxford. *Career*: Zoology Lecturer, Pembroke Coll, Oxford 1969–70, Zoology Fellow 1981–88; Asst Prof, Animal Resource Ecology Inst, British Columbia Univ, Vancouver, Canada 1970–73; Zoology Lecturer, N Wales Univ Coll, Bangor 1973–74; Researcher, Animal Behaviour Research Group, Zoology Dept, Univ of Oxford 1975–76; Zoology Lecturer, Edward Grey Field Ornithology Inst 1976–88; Fellow, Royal Soc 1984; External Scientific Mem, Max Planck Soc 1985; Pres, Int Soc of Behavioural Ecology 1988–90; mem, Agricultural and Food Research Council—AFRC 1988–94, Dir, Ecology and Behaviour Unit 1989–94, Chair, Animals Research Cttee 1990–94, Sr Scientific Consultant 1991–94; Dir, Behavioural Ecology Unit, Natural Environment Research Council—NERC 1989–94, Chair, NERC Review of Evolution, Taxonomy and Biodiversity 1991–92; mem, Council, Zoological Soc of London and Chair, BBC Science Consultative Group 1992–93; Pres, Asscn for Study of Animal Behaviour 1993–; mem, Academia Europaea 1995. *Present Position*: Research Prof, Univ of Oxford, Fellow, Pembroke Coll and Chief Exec, NERC 1994–. *Awards and Honours*: Bicentenary Medal, Linnean Society 1983; Frink Medal, Zoological Soc of London 1997. *Publications*: Foraging Theory 1987; Foraging Behaviour 1987; Behavioural and Neural Aspects of Learning and Memory (Co-Ed) 1991; Behavioural Ecology: An Evolutionary Approach (Co-Ed) 1991; Behavioural Ecology (Co-Author) 1997. *Address*: Polaris House, North Star Ave, Swindon SN2 1EU, United Kingdom. *Tel*: (1793) 411653. *Fax*: (1793) 411780. *E-mail*: hqpo@wpo.nerc.ac.uk.

KREISEL, Wilfried, PH D; German physical chemist and international civil servant; b 20 July 1942, Mährisch-Ostrau; m 1971; 1s 1d. *Education*: Univ of Heidelberg. *Career*: Asst Prof, Inst for Environmental Research, Univ of Dortmund 1973–77; WHO Air Quality Man Adviser, WHO, Seoul, Repub of Korea then Kuala Lumpur, Malaysia 1977–85, Regional Adviser, Manila, Philippines 1985–86, Dir, Div of Environmental Health, Geneva, Switzerland 1986–93. *Present Position*: Exec Dir, Health and Environment, WHO 1993–. *Publications*: Representation of the Environmental Quality Profile of a Metropolitan Area 1984; Water Quality and Health 1991; Health of People, Health of Planet (Co-Author) 1993; International Programme on the Health Effects of the Chernobyl Accident 1995. *Address*: World Health Organization—WHO, ave Appia, 1211 Geneva 27, Switzerland. *Tel*: (22) 7913582. *Fax*: (22) 7914849. *E-mail*: kreiselw@who.ch.

KRIKIS, Dr Andris, PH D; Latvian forester and chemical engineer; b 16 Feb 1936, Livani; s of Julijs and Nora-Emma Krikis; m Dagnija Snikere 1964; 1d. *Education*: Agric Acad,

Jelgava, Riga Polytechnic Inst. *Career*: Sr Engineer 1959–68; Head, Research Group, Inst of Organic Synthesis, Latvian Acad of Sciences 1968–78, Deputy Head of Lab 1980–90; Polymer Researcher, Science Inst, Jelgava and Head, Nature Conservation Lab 1978–80; Head, Environmental Protection Co-ordination Dept, Environmental Protection Cttee 1990–92; Head, Helsinki Comm Bureau of Latvia 1992–95. *Present Position*: Sr Officer, Chemicals Inspectorate, Environmental State Inspectorate 1995–. *Publications*: Structure and Functions of Low Molecular Peptides 1980; Modified Amino Acids 1987. *Address*: Environmental State Inspectorate, Rupniecibas iela 25, Riga 1877, Latvia. *Tel*: (7) 325-195. *Fax*: (7) 243-077. *E-mail*: krikis@main.vvi.gov.lv.

KRIŠTÍN, Dr Anton; Slovak zoologist and ecologist; b 2 Aug 1956, Bratislava; s of Anton and Eva Krištín; m Maria Krištín 1977; 1s 1d. *Education*: Comenius Univ of Bratislava, Bayreuth Univ, München, Fed Repub of Germany. *Career*: Sr Scientist, Inst of Forest Ecology, Slovak Acad of Sciences 1993–. *Present Position*: Head, Dept of Animal Ecology, Inst of Forest Ecology, Slovak Acad of Sciences 1993–. *Awards and Honours*: INCHEBA Gold Medal; Prof Ferinac Bonze Medal for Ornithology. *Publications*: Trophic Ecology of the Tree Sparrow 1984; Stictocephala bisonia 1988; Trophic Relationships of Birds and Invertebrates 1992; Foraging Ecology and Habitat Preferences in the Lesser Grey Shrike 1995; Biology of Hoopoes 2001; Mating Strategies in the Lesser Grey Shrike 2003. *Addresses*: Institute of Forest Ecology, Slovak Academy of Sciences, Štúrova 2, 960 53 Zvolen, Slovakia; Kremnická 13, 960 06 Zvolen, Slovakia. *Tel*: (855) 5330914. *Fax*: (855) 5479845. *E-mail*: kristin@sav.savzv.sk. *Internet*: www .savzv.sk/english/people/Kristin.htm.

KÖSTER, (Kajus) Friedemann, DR RER NAT; German zoologist, film-maker and environmentalist; b 14 Jan 1944, Breslau; m Heide Köster-Stoewesand; 2s. *Education*: Rheinische Friedrich Wilhelms Univ, Bonn, Hamburg Univ. *Career*: Zoologist and Deputy Dir, Inst Columbo-Alemán de Investigaciones Científicas 'Punta de Betin', Santa Marta, Colombia 1974–80; Expert, German Agency for Tech Co-operation—GTZ 1976–80; Dir, Charles Darwin Research Station, Galapagos Islands, Ecuador 1980–84; Scientific Adviser, Zoologist and Wildlife Photographer, Anglia TV/ Survival Anglia Ltd 1984–93; ind wildlife film-maker in Ecuador 1994–95; mem, Bd of Dirs, Charles Darwin Foundation for the Galapagos Islands, Ecuador. *Present Position*: Dir, Yasuni Research Station, Pontifical Catholic Univ of Ecuador, Quito and Scientific Dir, 'Condor Huasi' Andean Condor Recovery Station, Cayambe, Ecuador. *Awards and Honours*: Silver Edelweiss Award for Best Ecological Film, 11th Int Festival of Mountain Films, Torello, Spain. *TV Documentaries*: Vampires on Wolf Island 1986; How They Got There 1986; Cold on the Equator 1986; The Ocean Travellers 1986; Man Came To Eden 1986; Perspectives of Paradise 1986; The Year of El Niño 1986; An Ocean Nursery 1986; Quest in the Flooded Forest 1989; The Green Leaf Code 1990; The Last Flight of the Condor 1993. *Address*: Casilla 17-16-1014, Quito, Ecuador. *Tel*: (2) 372-860. *E-mail*: koester@ecnet.ec.

KUCERA, Dr Bohumil; Czech scientist and nature conservationist; b 15 Oct 1939, Prague; s of Bohumil Kucera and Bozena Kucerová; m Alena Šittnerová 1969; 2s 1d. *Education*: Charles Univ, Prague. *Career*: Alpine research expeditions 1961, 1962, 1963, 1968; Specialist, State Inst for Care of Historical Monuments and Nature Conservation 1963; geological researcher, Mongolia 1970–71; study expeditions, India 1974, Cen Asia 1988, Lake Baikal 1989 and Vladivostok region 1990 (Russian Fed); geomorphological mapping, Zagros Mountains, Iraq 1976. *Present Position*: Nat Co-ordinator, CORINE biotopes and Emerald projects; Ed-in-Chief, *Magazina Ochrana Prírody*. *Publications*: Development of State Conservancy 1986; Principles of Karsology and Speleology (Co-Author) 1992. *Address*: Kališnická 4, 130 00 Prague 3, Czech Republic. *Tel*: (2) 6974928. *Fax*: (2) 6970012.

KUKHAR, Prof Valeriy P, D CHEM SCI; Ukrainian chemist; b 26 Jan 1942, Kiev; s of Pavel Kukhar and Valentina Loseva; m Natalia Mirayn 1964; 2s 1d. *Education*: Dnepropetrovsk Chemical Tech Inst. *Career*: Head, Nat Comm, UNESCO Man and Biosphere Programme 1988–93, Council on Biosphere Problems, Nat Acad of Sciences of Ukraine 1988–93; Head, Presidential Comm on Nuclear Policy and Environmental Safety 1995–; Co-Chair, Ukraine–US Programme to Promote Sustainable Devt 1995–99; mem, Nat Comm on Sustainable Devt, Ukraine 1998–. *Present Position*: Dir, Inst of Bio-organic Chemistry and Petrochemistry, Ukrainian Acad of Sciences 1987–. *Awards and Honours*: Academician, Nat Acad of Sciences of Ukraine; UNEP Global 500 1993; San Valentino Prize 1999; Nat Award, Ukraine 1999. *Publications*: Ecotechnology, The Optimization of Industrial Technology 1989; Chemistry of Bioregulator Processes 1992; Fluorine-containing Amino Acids 1994; Comprehensive Risk Assessment of the Chornobyl Accident 1998. *Address*: Institute of Bio-organic Chemistry and Petrochemistry, Ukrainian Academy of Sciences, 02094 Kiev, Murmanska 1, Ukraine. *Tel*: (44) 558-53-88. *Fax*: (44) 573-25-52. *E-mail*: kukhar@bpci.kiev.ua.

KULFAN, Ján, RNDR, PH D; Slovak entomologist and ecologist; b 10 June 1957, Myjava, Slovakia; s of Viliam Kulfan and Olga Kulfanová; m Katarína Kulfanová 1984; 2s. *Education*: Comenius Univ of Bratislava. *Career*: Researcher, Inst of Forest Ecology, Slovak Acad of Sciences 1984–; Chair, Scientific Bd and Co-ordinator, UNESCO Zoological Investigations in Forest Ecosystems in Polana Biosphere Reserve 1991–93; mem, Cttee, Slovak Entomological Soc, Slovak Ecological Soc, Slovak Zoological Soc. *Present Position*: Mem, Zoology Dept, Inst of Forest Ecology, Slovak Acad of Sciences 1984–. *Awards and Honours*: Slovak Literary Fund Prize 2000. *Publications*: Structure of Heliophilous Butterflies' Taxocoenoses in Several Biotopes of Western Slovakia 1990; Changes in the Distribution of Thermophilous Butterflies in Slovakia (Co-Author) 1992; The Structure and Seasonal Dynamics of Caterpillar Communities on Oak Trees 1992; Ecology and Seasonal Dynamics of Moth Larvae on Beech Trees (Co-Author) 1995; Assemblages and Diurnal Activity of Butterflies in the Sambucetum Ebuli Plant Community 1997; Structure of Moth Larvae Assemblages on Fir and Spruce 1998; Oak Pests and their Natural Enemies (Co-Author and Co-Ed) 1999; Beetle Assemblages on Willow Trees: do Phenolic Glycosides Matter? (Co-Author) 2002; The Oak Pests and their Natural Enemies (Co-Author and Co-Ed) 1999; The Seasonal Dynamics of Moth Larvae Assemblages on Norway Spruce 1994; Changes in the Distribution of Thermphilous Butterflies in Slovakia (Co-Author) 1992. *Address*: Institute of Forest Ecology, Slovak Academy of Sciences, Štúrova 2, 960 53 Zvolen, Slovakia. *Tel*: (45) 532-0313. *Fax*: (45) 547-9485. *E-mail*: kulfan@sav .savzv.sk.

KULICH, Jirí; Czech environmental educator; b 26 May 1961, Sušice; s of Víteslav and Marie (Zapleralová) Kulich; m Hana Donbková 1987; 2d. *Education*: Charles Univ. *Career*: mem, Brontosaurus Movement 1980–; organizer, TOUCH conference on Environmental Educ through Experience 1990, 1997. *Present Position*: Head, SEVER environmental educ centre 1990–; mem, IUCN Comm on Educ and Communication 1997–; Vice-Pres, Soc for Sustainable Living 1999–. *Awards and Honours*: Ford Conservation Award 1997; Sasakawa Peace Foundation Award 1998. *Address*: SEVER

Environmental Education Centre, 542 26 Horní Marsov, Czech Republic. *Tel and fax*: (439) 874181. *E-mail*: jiri.kulich@ecn.cz. *Internet*: www.ecn.cz/sever.

KULL, Prof Kalevi, PH D; Estonian biologist; b 12 Aug 1952, Tartu; m Tiiu Veski 1978; 2s 2d. *Education*: Tartu Univ. *Career*: Research Asst, Zoology and Botany Inst, Tartu 1977–83, Head, Dept of Ecophysiology 1984–98, Prof of Ecophysiology 1992–98; Pres, Estonian Naturalists' Soc 1991–94. *Present Position*: Head, Int Cttee, Jakob von Uexküll Centre 1993–; Prof of Biosemiotics, Zoology and Botany Inst, Tartu 1998–. *Awards and Honours*: Estonian Life Science and Culture Award 1984. *Publications*: Lectures in Theoretical Biology (Co-Ed) 1993; Wooded Meadows (Co-Author) 1997; Jakob von Uexküll: A Paradigm for Biology and Semiotics (Ed) 2000. *Address*: Riia 181, Tartu 51014, Estonia. *Tel*: (7) 428-619. *Fax*: (7) 383-013. *E-mail*: kalevi@zbi.ee. *Internet*: www.zbi.ee/@~kalevi.

KÜLVIK, Mart, M SC, PH D; Estonian ecologist and conservationist; b 20 Dec 1960, Tallinn; 3s 1d. *Education*: Tartu Univ. *Career*: Research Fellow, Tallinn Botanical Gardens, Acad of Sciences 1984–85; Ecological Adviser, Tartu City Council 1985–91; mem, Congress 1990–92, Tartu City Council and Bd, Estonian Fund for Nature 1990–, Advisory Group, E European Programme, IUCN 1992–; Head, Nature Conservation Dept, Ministry of Environment 1991–92; Dir-Admin, Nature Conservation Research Centre, Ministry of the Environment 1993–96; Vice-Chair, Estonian Greek Mov't 1990–92; Head Dept Nature Conservation of the Ministry of the Env; Head, Nature Conservation Research Centre 1993–2001; Chair, cttee of Experts European Ecological Network under the Pan European Biological and Landscape Strategy 1997–99, mem 1995–. *Present Position*: Head, Nature Conservation Research Centre, Inst of Environmental Protection, Estonian Agric Univ 1996–. *Awards and Honours*: Int Man of the Year Award 1993; mem IUCN-WCPA; mem IUCN SSC ESUGS. *Publications*: IUCN Environmental Status Report: The Baltic Countries—Estonia (Ed) 1993; Implementation of the Convention on Biological Diversity in Estonia: Country Report (Ed) 1996; Biodiversity Management Strategy for Commercial Forests in Estonia (Ed) 1998; Ecological Networks in Estonia: Concepts and Appications 2002. *Address*: POB 222, Tartu 50002, Estonia. *Tel*: (7) 427-435. *Fax*: (7) 427-432. *E-mail*: mkulvik@envinst.ee. *Internet*: www.envinst.ee.

KUSSAKIN, Oleg Grigoryevich; Russian zoologist and hydrobiologist; b 12 July 1930, Leningrad (now St Petersburg); s of Gregor P Kussakin and Eugenia A Kussakina; m Alla P Kassatkina 1960; 1s. *Education*: Leningrad State Univ, jr researcher Zool Inst. *Career*: asst of Chair, Leningrad State Univ, Prof, Far E State Univ in Vladivostok; Head of Lab, Inst of Marine Biology, Far E Br, USSR (now Russian) Acad of Sciences; corresp mem, USSR Acad of Sciences 1990, mem 1994; mem, American Crustacean Soc. *Publications*: books and numerous articles on isopod taxonomy, intertidal ecology, megataxonomy and biogeography of ocean in scientific journals. *Address*: Institute of Marine Biology, Far East Branch, Russian Academy of Sciences, 690041 Vladivostok, ul Palchevskogo 17, Russian Federation. *Tel*: (4232) 31-09-21. *Fax*: (4232) 31-09-00.

KVASNICKOVA, Danuse, PH D; Czech teacher and researcher (retired); b 15 Nov 1935, Mladá Boleslav; d of Stanislav Prochazka and Vera Prochazkova; m 1st Václav Kvasnicka 1957 (died 1980); m 2nd Vladimír Kalina 1988; 2s. *Education*: Charles Univ, Prague. *Career*: teacher 1958–67; environmental education researcher 1967–84; leader, UNESCO projects, Agric Univ 1984–90; researcher, Ministry of Agriculture 1990–93; leader, Environmental Educ Club, Dir, Ekogymnazium 1993–97; leader, Netherlands–Czech Project on Environmental Education for basic schools, UNDP project for secondary schools and teacher population 1999–2001; leader, UNESCO project, Education for Sustainability 2002–04. *Present Position*: mem, Czech UNESCO Cttee; consultant, Ministries of the Environment and of Education; mem of bd, Czech Soc for the Environment; lecturer, of ESMA of Barcelona in Prague; leader of EEC professional NNO of teachers and schools. *Awards and Honours*: Krkose Natl Park Prize 1972, 1985; Prize of the Minister of the Environment; UNEP Global 500 1996. *Publications*: numerous methodological books on environmental education in schools 1972–98; approx 20 textbooks 1980–98; TV programs. *Address*: Na Strzi 47, 140 00 Prague 4, Czech Republic. *Tel*: (2) 6922650. *E-mail*: d.kvasnickova@volny.cz.

L

LAFFERTY, William M, PH D; Norwegian/Irish political scientist; b 2 May 1939, Cleveland, OH, USA; m Gro Helgesen 1978; 2d. *Education*: King's Point Acad, NY, USA, Florida Univ, Gainesville, USA. *Career*: Research Dir, Social Research Inst, Oslo, Norway 1970–77; Prof of Political Science, Oslo Univ, Norway 1977–. *Present Position*: Prof, Political Science and Dir Programme for Research and Documentation for a Sustainable Society ProSus 1992, Centre for Dev and the Env (SUM), Univ of Oslo; Prof of Sustainable Strategy, Centre for Clean Technology and Environment Policy (CSTM) Univ of Twente; Research Assoc at the Inst for the Study of Social Change (ISC), Univ Coll Dublin. *Publications*: Economic Development and the Response of Labour in Scandinavia 1971; Industrialization, Community Structure and Socialism 1973; Participation and Democracy in Norway 1981; Democracy and the Environment: Problems and Prospects (Co-Author) 1996; From Earth Summit to Local Forum: Studies of Local Agenda 21 in Europe (Co-Author) 1997; Towards Sustainable Development: The Goals of Development and the Conditions of Sustainability (Co-Author) 1998; Realizing Rio in Norway (Co-Author) 2002; Governance for Sustainable Development 2004. *Address*: ProSus/SUM, Post Box 1116 Blindern, 0317 Oslo, Norway. *Tel*: 22-85-87-88. *Fax*: 22-85-87-90. *E-mail*: william.lafferty@prosus.uio.no. *Internet*: www.prosus.uio.no.

LAL, Prof Devendra, PH D; Indian nuclear geophysicist; b 14 Feb 1929, Varanasi; m Aruna Lal 1955 (deceased). *Education*: Banaras Hindu Univ, Bombay Univ. *Career*: Research student, Tata Inst of Fundamental Research 1949–50, Fellow, Assoc Prof, Prof, Sr Prof 1950–72; Visiting Prof, Univ of California, Los Angeles, USA 1966–67, 1983–84; Dir, Physical Research Lab, Scripps Inst of Oceanography, Univ of California, San Diego, USA 1972–83, Sr Prof 1987–89; Pres, Int Asscn for Physical Sciences of the Ocean 1979–83; Int Union of Geodesy and Geophysics 1983–87; mem, Advisory Cttee on Environment—ACE, Exec Cttee, Int Council of Scientific Unions. *Present Position*: Prof, Scripps Inst of Oceanography 1967–. *Awards and Honours*: Outstanding Scientist, Federation of Indian Chambers of Commerce and Industry Award in Science and Technology 1974; NASA Group Achievement Award (Skylab III) 1986; Pandit

Jawaharlal Nehru Award for Sciences 1986; Sir C V Raman Birth Centenary Award 1996–97; V M Goldschmidt Medal 1997. *Address*: Scripps Institute of Oceanography, GRD 0220, University of California, San Diego, 9500 Gilman Drive, La Jolla, CA 92093-0220, USA. *Tel*: (619) 534-2134. *Fax*: (619) 534-0784. *E-mail*: dlal@ucsd.edu. *Internet*: orpheus.ucsd.edu/earth/faculty/dlal.html.

LALONDE, (Olivier) Brice Achille; French politician and environmentalist; b 10 Feb 1946, Neuilly-sur-Seine; s of Alain Lalonde and Fiona Forbes; m Patricia Raynaud 1986; 2s (1 deceased) 2d; 1s 1d from previous marriage. *Education*: Univ of Paris. *Career*: Man, textile co 1969–70; journalist, *Sauvage* 1971–77; Chair, Friends of the Earth 1972, French br 1978; Presidential Cand, Green Party 1981; mem, Nat Planning Comm, Ministry of Planning and Land Man; Admin, European Environment Bureau 1982–85; Dir, Paris Office, Inst for European Environment Policy 1986–88; Sec of State for the Environment 1988–89, Sec of State for the Environment and the Prevention of Tech and Natural Disasters 1989–90, Minister-Del 1990–91, Minister of the Environment 1991–92; Chair, Cttee to Free Aleksandr Nikitin (qv) 1996–2000; leader, missions to Kosovo, Yugoslavia, and Afghanistan 1999. *Present Position*: Leader, Génération Ecologie 1990–. *Address*: Génération Ecologie, 7 Villa Virginie, 75014 Paris, France. *Tel*: 1-56-53-53-73. *Fax*: 1-56-53-53-70. *E-mail*: generation.ecologie@wanadoo.fr. *Internet*: www.generation-ecologie.com.

LANGE, Otto Ludwig, DR RER NAT; German botanist; b 21 Aug 1927, Dortmund; s of Otto and Marie (Pralle) Lange; m Rose Wilhelm 1959; 2d. *Education*: Georg-August-Univ Göttingen, Albert-Ludwigs-Univ Freiburg. *Career*: Asst Prof, Georg-August-Univ Göttingen 1953–61; Dozent, Technische Hochschule Darmstadt 1961–63; Prof of Forest Botany, Georg-August-Univ Göttingen 1963–67; Prof of Botany, Bayerische-Julius-Maximilians-Univ Würzburg 1967–92; Visiting Scientist, Utah State Univ, USA 1973, 1985, Australian Nat Univ, Canberra, ACT 1978–79; Dir, Botanical Garden, Bayerische-Julius-Maximilians-Univ Würzburg 1967–92; Co-Ed, scientific journals, *Flora*, *Trees*; mem, Deutsche Akad der Naturforscher Leopoldina, Bayerische Akad der Wissenschaften, Acad Europaea, Acad Scientiarum et Artium Europaea; Corresp mem, Akad der Wissenschaften Göttingen. *Present Position*: Prof Emer, Bayerische-Julius-Maximilians-Univ Würzburg 1992–. *Awards and Honours*: Foreign Hon mem, American Acad of Arts and Sciences 1994; DR HC, Bayreuth 1995, Tech Univ, Lisbon, Portugal 1996; Antarctic Service Medal, US Govt 1974; Gottfried-Wilhem-Leibniz Prize, Deutsche Forschungsgemeinschaft 1986; Balzan Prize for 'Applied botany including ecology' 1988; Adalbert Seifriz Prize for transfer of technology 1990; Bayerische Maximiliansorden for Science and Art 1991; Acharius Medal, Int Asscn of Lichenology 1992; Bundesverdienstkreuz (1st Class) 1985. *Publications*: Physiological Plant Ecology in Encyclopedia of Plant Physiology (4 vols—Co-Ed) 1981–83; books on water and plant life, plant responses to stress, forest decline and air pollution; Ecological Studies book series; 320 scientific papers. *Address*: Julius-von-Sachs-Institut für Biowissenschaften der Universität Würzburg, 97082 Würzburg, Julius-von-Sachs-Pl 3, Germany. *Tel*: (931) 8886205.

LANGRAND, Olivier, M SC; French environmentalist; b 19 Sept 1958, Angers. *Education*: Louis Pasteur Univ of Strasbourg, Univ of Natal, S Africa. *Career*: Madagascar Rep, WWF. *Present Position*: Regional Rep for Central Africa, WWF 1997. *Awards and Honours*: Ordre National de la République de Madagascar 1997. *Address*: BP 9144, Libreville, Gabon. *Tel*: 73-00-28. *Fax*: 73-80-56. *E-mail*: 73674.745@compuserve.com.

LANGSLOW, Dr Derek Robert, PH D; British environmentalist; b 7 Feb 1945, London; s of Alec and Beatrice (Wright) Langslow; m Helen Katherine Langslow 1969; 1s 1d. *Education*: Queens' Coll, Cambridge. *Career*: Research Fellow, Cambridge Univ 1969–72; Lecturer in Biochemistry, Edinburgh Univ 1972–78; Sr Ornithologist, Nature Conservancy Council 1978–84, Asst Chief Scientist 1984–87, Dir, Policy and Planning 1987–90. *Present Position*: Chief Exec, English Nature—Nature Conservancy Council for England 1990– and Chair, Asia Wetland Bureau 1994–. *Address*: English Nature, Northminster House, Peterborough PE1 1UA, United Kingdom. *Tel*: (1733) 455100. *Fax*: (1733) 455103.

LANTOS, Tamás; Hungarian agronomist and sociologist; b 23 May 1956; s of Ferenc and Mária (Vecsey) Lantos. *Education*: Univ of Agric, Keszthely; Loránd Eötvös Univ, Budapest; Johns Hopkins Univ, MD, USA. *Career*: Limnologist, Biological Research Inst, Tihany 1982–84; Sociologist, Apáczai Csere János Educational Centre 1986–88, Nat Inst for Health Promotion 1989–90, Min of Welfare 1990–92. *Present Position*: Pres, Ormánság Foundation 1989– and Man, Ormánság Open School of Sustainable Devt. *Awards and Honours*: UNEP Global 500 1995. *Publications*: Angaben zur Kenntnis der Sedimentoberfläche im Balaton 1987; A Nonprofit Szervezetek Tervezési Sajátosságai 1995; Sustainable Rural Development in Ormánság 1998. *Address*: Ormánság Foundation, 7625 Pécs, István u 52, Hungary. *Tel and fax*: (73) 352-333. *E-mail*: ormansag@mail.matav.hu.

LAPTEV, Ivan Dmitrievich, D PHIL; Russian editor and journalist; b 15 Oct 1934, Sladkoye; m Tatyana Kareva 1966; 1d. *Education*: Siberian Rd Transport Inst, Acad of Social Sciences. *Career*: employee, Cen Cttee, Communist Party of the Soviet Union—CPSU, mem, CPSU 1960–91; employee, Omsk River Port 1952–60; teacher 1960–61; instructor, Soviet Army Sports Club 1961–64, literary collaborator and special corresp, *Sovietskaya Rossiya* 1964–67; Consultant, *Kommunist* 1967–73; worked with CPSU Cen Cttee 1973–78; Section Ed, *Pravda* 1978–82, Deputy Ed 1982–84; Chief Ed, *Izvestiya* 1984–90; mem, USSR Supreme Soviet 1989–91; Chair, Council of Union 1990–91; Gen Man, Isvestiya Publrs 1991–94; Deputy Chair, Fed Press Cttee 1994–95; mem, Int Acad of Information 1993; Pres, Asscn of Chief Eds and Publrs 1993–95. *Present Position*: Chair, Fed Press Cttee 1995–; Head of Service, Professional Acad of State Service to Russian Presidency 1995–. *Publications*: Ecological Problems 1978; The World of People in the World of Nature 1986; more than 100 scientific articles on ecological problems. *Address*: Goskompetchat, 101409 Moscow, Strastnoy blvd 5, Russian Federation. *Tel*: (095) 209-63-52. *Fax*: (095) 200-22-81.

LARSEN, Kai, D SC (HON), M SC, BA; Danish botanist; b 15 Nov 1926, Hillerød; s of Axel G and Elisabeth (Hansen) Larsen; m Supee Saksuwan 1971; 1s 3d. *Education*: Copenhagen Univ. *Career*: Scientific Asst, Botanical Lab, Copenhagen Univ 1952–55; Asst Prof, Royal Danish School of Pharmacy 1955–62; Leader, botanical expeditions to Thailand 1958–59, 1961–62, 1963, 1966, 1968, 1970, 1972, 1974, 1990, 1991, 1992, 1993; Prof of Botany, Univ of Aarhus 1963–96; Visiting Prof, Royal Forestry Dept and Kasetsart Univ, Bangkok, Thailand 1974; mem, Royal Danish Acad of Sciences and Letters, Royal Norwegian Acad of Sciences, Linnean Soc, London, Presidium, WWF Denmark and Exec Bd, Flora Malesiana Foundation, Leiden, Netherlands; corresp mem, Musée Nat d'Histoire Naturelle, Paris, France; hon mem, Botanical Garden Asscn, Thailand; Pres, Int Asscn of Botanical Gardens 1981–87; Ed-in-Chief, *Nordic Journal of Botany* and *Opera Botanica;* Danish Ed, *Flora Nordica;* Ed, *Flora of Thailand;* Adviser, *Flora of China*; Consultant, Royal Forest Dept, Thailand. *Awards and Honours*: Dr Sc hc. *Publications*: more than 250 books and articles on plant taxonomy and plant geography in international journals. *Address*: Botanical Institute, Nordlandsvej 68, 8240 Risskov, Denmark. *Tel*: 89-42-47-00 or 89-42-47-08. *Fax*: 89-42-47-47. *E-mail*: kai.larsen@biology.au.dk.

LAVALLE DE VIGNAROLI, Prof Vivian Beatriz; Argentine teacher and environmentalist; b 18 Nov 1948, New Orleans, LA, USA; d of Hector Lavalle and Lidia Castiglioni de Lavalle; m Ernesto O Vignaroli 1973; 3s 3d. *Career*: Coordinator and Dir, Educ Dept, Vida Silvestre Foundation 1977–87; tutor, environmental educ workshops 1983–; Asst, 1st Pan-American Conservation Congress 1990; Argentine Rep, Centre for Instruction, Staff Devt and Evaluation—CISDE, Carbondale, IL, USA; mem, North American Asscn for Environmental Educ—NAAEE. *Present Position*: Dir, Grupo de Educadores Ambientalistas (Environmentalist Teachers Group) 1991–; leader, environmental educ workshops. *Address*: Grupo de Educadores Ambientalistas, Laprida 1373, San Isidro, 1642 Buenos Aires, Argentina. *Tel*: (11) 4723-5467. *Fax*: (11) 756-2767. *E-mail*: vivvigna@ciudad.com.ar.

LAY, Vladimir; Croatian environmentalist. *Present Position*: Mem, Zelena Akcija Zagreb (Green Action Zagreb). *Address*: Zelena Akcija Zagreb, 10000 Zagreb, Radnicka 22, Croatia. *Tel and fax*: (1) 6110951. *E-mail*: vladimir.lay@zg.tel.hr.

LEAKEY, Richard Erskine Frere, FRAI; Kenyan palaeontologist; b 19 Dec 1944, Nairobi; s of the late Louis and Mary Leakey; m Dr Maeve Gillian Epps 1970; 3d. *Career*: Leader, expeditions to W Natron, Tanzania 1963, 1964, Baringo, Kenya 1966, Omo River, Ethiopia 1967 and E Rudolf, Kenya 1968–70; Researcher on origin of man, Lake Turkana 1971–76, Bulok 1980, Rusinga Island 1983–84, W Turkana 1984–85; Admin Dir, Nat Museums of Kenya 1968–74, Dir 1974–89, mem, Bd of Govs; Chair, Kenya Exploration Soc 1969–72, Wildlife Clubs of Kenya 1969–84, Foundation for Research into Origin of Man; Dir, Wildlife and Conservation Man Service 1989–90; Dir and Exec Chair, Kenya Wildlife Service 1990–94 and 1998–99; Trustee, Nat Fund for Disabled and Foundation for Social Habilitation; Co-Founder and Sec-Gen, Safina Party 1995–98. *Present Position*: Chair, E African Wildlife Soc 1985–; Perm Sec, Sec to the Cabinet and Head of the Public Service, Office of the Pres, Kenya 1999–. *Awards and Honours*: Franklin Burr Prize 1965, 1973; Golden Ark Medal for Conservation 1989; James Smithsonian Medal, Smithsonian Inst 1990; Gold Medal, Royal Geographic Soc, UK 1990; Chevalier de l'Ordre de Léopold II, Belgium 1990; Companion of the Order of the Gold Ark, Netherlands 1991; Order of the Burning Spear, Kenya 1993; Hubbard Medal, Nat Geographic Soc 1994; World Ecology Medal, Int Centre for Tropical Ecology, St Louis 1997. *Publications*: Origins (Co-Author); People of the Lake: Man, His Origins, Nature and Future (Co-Author) 1979; The Making of Mankind 1981; Human Origins 1982; One Life 1984; Origins Reconsidered (Co-Author) 1992; The Origin of Humankind 1994; The Sixth Extinction (Co-Author) 1995; *TV Series*: The Making of Mankind 1981. *Addresses*: East African Wildlife Society, POB 20110, Nairobi, Kenya; POB 24926, Nairobi, Kenya (home). *Tel*: (2) 240445. *Fax*: (2) 210150.

LEBRETON, Philippe Alain, D SC; French biochemist and ornithologist; b 25 Oct 1933, St Etienne; m Paulette Boffy 1956; 1s 1d. *Education*: Univ of Lyons; Inst Chem Ind, Lyon. *Career*: Founder, Centre Ornithologique Rhône-Alps—CORA; Founder, Fédération Rhône-Alps Protection Nature—FRAPNA; fmr Rhône-Alps Regional Councillor, Groupe Ecologique; mem, Scientific Cttee, Vanoise Nat Park, fmr Admin Council, WWF France; Admin Council SNPN. *Present Position*: fmr Prof Biology and Ecology, Lyon State Uni. *Awards and Honours*: Officier Mérite Agricole 2002; UNEP Global 500 1990. *Publications*: Nine books and more than 400 scientific papers. *Address*: 56 chemin du Lavoir, 01700 Beynost, France. *Tel and fax*: 4-78-55-36-66.

LEEFLANG, Seitz Arnold; Dutch scientist and environmentalist; b 29 May 1933; m Anna Maria de Jong 1960; 2d. *Education*: Univ of Amsterdam. *Career*: Royal Dutch Air Force 1951–52; Foreign Desk Ed, *Algemeen Handelsblad* 1952–58, Science Ed 1958–69; Founder, De Kleine Aarde (Dutch experimental environment centre) 1972. *Present Position*: Dir, Stichting De Twaalf Ambachten (Centre for Alternative Techniques) 1979–. *Awards and Honours*: UNEP Global 500 1990. *Publications*: Op Zoek naar Leefruimte (Co-Author) 1966; Vandaag Beginnen 1979. *Address*: Stichting De Twaalf Ambachten, Mezenlaan 2, 5282 HB Boxtel, Netherlands. *Tel*: (411) 672621. *Fax*: (411) 672854.

LEGGETT, Jeremy, PH D; British scientist and environmentalist. *Education*: Univ of Oxford. *Career*: Dir, Science, Atmosphere and Energy Div, Greenpeace Int 1990–94, Solar Initiative 1995–96; Visiting Fellow, Green Coll Centre for Environmental Policy and Understanding, Univ of Oxford 1996. *Present Position*: Chief Exec, The Solar Century 1996–; Charterhouse Fellow in Solar Energy, Climate Change Unit, Univ of Oxford. *Awards and Honours*: Global Warming (Author-Ed) 1990; Climate Change and the Financial Sector 1996; The Carbon War 1999. *Address*: Unit 5, Sandycombe Centre, 1–9 Sandycombe Road, Richmond TW9 2EP, United Kingdom.

LIANG CONGJIE; Chinese environmental activist and retired academic; b 1932. *Career*: teacher; fmr Prof of History, Acad for Chinese Culture, Beijing. *Present Position*: founder and Pres, Friends of Nature 1994–. *Awards and Honours*: Ramon Magsaysay Award for Public Service, Philippines 2000. *Address*: Beijing, People's Republic of China. *E-mail*: office@fon.org.cn. *Internet*: www.fon.org.cn.

LIKENS, Gene Elden, PH D; American ecologist; b 6 Jan 1935, Pierceton, IN; s of Colonel Benjamin and Josephine Estella Garner Likens; m Phyllis Irene Craig Likens 1983; 1s 2d. *Education*: Univ of Wisconsin, Madison. *Career*: Instructor, Asst Prof then Assoc Prof, Dartmouth Coll 1961–69; Assoc Prof, Prof then Chair, Ecology and Systematics Section, Cornell Univ 1969–83; mem, NAS; Co-Dir, Hubbard Brook Ecosystem Study. *Present Position*: Dir, Inst of Ecosystem Studies 1983–, Pres 1993–, Adjunct Prof, Ecology and Systematics Section, Cornell Univ, Prof, Grad Ecology and Evolution Program, Rutgers Univ and Prof of Biology, Yale Univ. *Awards and Honours*: USDA Forest Service 75th Anniversary Award 1980; Tyler World Prize for Environmental Achievement 1993; Australia Prize 1994; Naumann-Thienemann Medal, Societas Internationalis Limnologiae 1995; Eminent Ecologist, Ecological Soc of America 1995. *Publications*: Long-Term Trends in Precipitation Chemistry at Hubbard Brook, New Hampshire (Co-Author) 1984; An Ecosystem Approach to Aquatic Ecology: Mirror Lake and its Environment (Ed) 1985; Evidence for Sulfate-Controlled Phosphorus Release from Sediments of Aquatic Systems (Co-Author) 1989; Changes in the Chemistry of Surface Waters: 25-year Results at the Hubbard Brook Experimental Forest (Co-Author) 1989; Some Aspects of Air Pollution on Terrestrial Ecosystems and Prospects for the Future 1989; Dry Deposition of Sulfur: A 23-year Record for Hubbard Brook Forest Ecosystem (Co-Author) 1990; The Impact of Changing Regional Emissions on Precipitation Chemistry in the Eastern United States (Co-Author) 1991; The Ecosystem Approach: Its Use and Abuse 1992; Long-term Effects of Acid Rain: Response and Recovery of a Forest Ecosystem (Co-Author) 1996; Limnological Analyses (Co-Author) 2000. *Address*: Institute of Ecosystem Studies, Box AB, Millbrook, NY 12545-0129, USA. *Tel*: (914) 677-5343. *Fax*: (914) 677-5976. *E-mail*: likensg@ecostudies.org.

LINET, Christiane; Belgian environmentalist; b 29 April 1934; m Marc Linet 1957. *Education*: Univ of Brussels. *Career*: Publr, *Panda* and *Pandapress* magazines 1967–; mem, Species Survival Comm, IUCN; contrib, World Conservation Strategy, IUCN/UNEP/WWF 1980 and Caring for the Earth, IUCN/UNEP/WWF 1991–; Pres, WWF Belgium 1992–99. *Present Position*: mem of bd, WWF Belgium. *Awards and Honours*: Order of the Golden Ark, Netherlands 1975; Mem of Honour, WWF Int 1998. *Address*: 608 chaussée de Waterloo, 1050 Brussels, Belgium. *Tel*: (2) 340-09-99. *Fax*: (2) 340-09-33.

LISICKÝ, Mikuláš Juraj, PH D; Slovak ecologist and conservation scientist; b 9 May 1946, Trencín; s of Mikuláš Lisický and Wanda Kucielová; m Eva Jelínková 1973. *Education*: Comenius Univ of Bratislava; Charles Univ, Prague, Czech Repub. *Career*: Lecturer then Sr Lecturer in Zoology, Comenius Univ of Bratislava 1968–76, in Conservation Science 1976–85, 1990–93, Founder and Head, Conservation Science Dept 1990–93; Founder, Co-ordinator and Biomonitor, Danube in Slovakia 1989–91; Vice-Pres, Nature and Landscape Protection Union 1990–93; Co-ordinator, Morava River Restoration Project 1994–98. *Present Position*: Founder and Head, Dept of Conservation Science and Monitoring, Slovak Acad of Sciences 1989–91 and 1993–; mem, Scientific Advisory Bd, Donau Auen Nat Park, Austria 1998–. *Publications*: Living Symmetry 1980; Nature Conservation: An Introduction 1987; Molluscs of Slovakia 1991; Ekosozológia (Conservation Science) 1996. *Addresses*: Dúbravská cesta 9, 842 06 Bratislava, Slovakia; Dlhá 15, 900 31 Stupava, Slovakia. *Tel*: (7) 5930-2630. *Fax*: (7) 5930-2646. *E-mail*: lisicky@savba.savba.sk.

LITVIN, Prof Dr Vladimir M; Ukrainian professor of ocean geography; b 10 Feb 1932, Dnepropetrovsk; s of Michail Litvin and Anisia Streltsova; m Ludmila Semirova 1960; 2d. *Education*: Moscow State Univ. *Career*: Scientist, Polar Inst of Fishery and Oceanography 1955–66, Atlantic Dept, Inst of Oceanology 1966–86; specialist, geomorphology, geoecology and cartography; mem, EU Coast Conservation 1993–; mem, Working Group, HELCOM NATURE/CZM 2000. *Present Position*: Head, Ocean Geography Dept, Kaliningrad Univ 1987–; Chair, Kaliningrad Council, Russian Soc of Nature Conservation 1989–, Kaliningrad Dept, Russian Geographical Soc 1990–. *Awards and Honours*: Academician, Russian Ecological Acad; F P Litko Gold Medal, Russian Geographical Soc. *Publications*: 300 publications, including Morphostructure of the Ocean Bottom 1987; Dissection and the Ocean Bottom Morphometry 1990; The World of Submarine Landscapes 1994; Morphostructure of the Earth 1995; Islands in the Ocean 1999. *Address*: Ocean Geography Department, Kaliningrad University, 236041 Kaliningrad, ul A Nevskogo 14, Russian Federation. *Tel*: (0112) 43-24-34. *Fax*: (0112) 46-58-13.

LOVEJOY, Thomas E, PH D, FLS; American tropical and conservation biologist; b 22 Aug 1941, New York, NY; m Charlotte Seymour 1966 (divorced 1978); 3d. *Education*: Yale Univ. *Career*: Zoological Asst, Yale Peabody Museum 1963–64; Yale Carnegie Teaching Fellow 1964–65; Research Asst, Natural History Museum, Smithsonian Inst 1970, Exec Asst to Scientific Dir and Asst to Vice-Pres for Resource and Planning 1972–73, Asst Sec for External Affairs 1987–94, Counsellor to Sec for Biodiversity and Environmental Affairs 1994–; Program Dir, WWF USA 1973–78, Vice-Pres for Science 1978–85; Exec Vice-Pres, WWF 1985–87; mem, Pres's Council of Advisors in Science and Tech 1989–92, Advisory Bd, New York Botanical Garden, Acad of Natural Sciences, Philadelphia, Resources for the Future, World Resources Inst, Environmental Defense Fund, WWF, Center for Plant Conservation, Rainforest Alliance and Rachel Carson Council, Inc; Pres, American Inst of Biological Sciences; Hon Chair, Wildlife Preservation Trust Int, Advisory Council, Center for Applied Biodiversity; Chair, Advisory Bd, Earth Communications Office, Biosphere II Scientific Advisory Cttee and US Man and Biosphere Program. *Present Position*: Science Adviser to Sec of Interior 1993–; Chief Biodiversity Adviser, World Bank 1998–. *Awards and Honours*: UNEP Global 500 1992. *Publications*: Nearctic Avian Migrants in the Neotropics (Co-Author) 1983; Key Environments: Amazonia; Conservation of Tropical Forest Birds (Co-Ed) 1985; Global Warming and Biological Diversity (Co-Ed) 1992. *Address*: Castle Bldg 320, Washington, DC 20560, USA. *Tel*: (202) 786-2263. *Fax*: (202) 786-2304.

LOVELOCK, Prof James Ephraim, CBE, PH D, D SC, FRS; British scientist; b 26 July 1919; s of Tom Arthur and Nellie Ann Elizabeth (March) Lovelock; m 1st Helen Mary Hyslop 1942 (died 1989); m 2nd Sandra Jean Orchard 1991; 2s 2d. *Education*: Univ of Manchester, London School of Hygiene & Tropical Medicine, Univ of London. *Career*: Staff Scientist, Nat Inst for Medical Research 1941–61; Fellow, Harvard Univ, USA 1954–55, Yale Univ, USA 1958–59; Prof of Chemistry, Baylor Univ Coll of Medicine, USA 1961–64; Visiting Prof, Univ of Reading 1967–90; Pres, Marine Biology Asscn 1986–90. *Present Position*: Ind scientist 1964– and Hon Visiting Fellow, Green Coll, Oxford. *Awards and Honours*: Norbert Gerbier Prize, World Meteorological Asscn 1988; Amsterdam Prize for the Environment, Royal Netherlands Acad of Arts and Sciences; Volvo Environment Prize 1996; Nonino Prize 1996; Blue Planet Prize 1997. *Publications*: Gaia 1979; The Great Extinction (Co-Author) 1983; The Greening of Mars (Co-Author) 1984; The Ages of Gaia 1988; Gaia: The Practical Science of Planetary Medicine 1991. *Address*: Coombe Mill, St Giles on Heath, Launceston PL15 9RY, United Kingdom.

LOZA SAWIRIS, Yousriya, MA; Egyptian financial adviser, accountant and environmentalist; b 1936, Assuit; d of Nassif and Rosa Loza; m Onsi Naguib Sawiris 1953; 3s. *Education*: Cairo Univ. *Career*: Voluntary work 1973; Founder, Asscn for the Protection of the Environment 1984. *Present Position*: Pres, Bd of Dirs, Asscn for the Protection of the Environment 1984–; mem, Egyptian Parl Health and Environment Cttee 1995–; Treas, Gen Foundation of Social Agencies 1997–. *Awards and Honours*: Ministry of Agriculture Golden Prize 1987; Distinguished Alumni Volunteer Award 1998; certificate and award, Medical Syndicate for Community Health Services 2000. *Publications*: Solid Waste Management—A Pilot Project in Cairo 1997; Win Win Situation—Journal of the Arab Bankers' Asscn 1999; Integrated Solid Waste Management System 2000. *Addresses*: 160 26th July St, 4th Floor, Agouza, Giza, POB 1191, Cairo, Egypt; 3 El Nessim St, Zamalek, Egypt (home). *Tel*: (2) 3033028. *Fax*: (2) 3440201. *E-mail*: yousriya@intouch.com.

LUPASCU, Prof Dr Mihail F; Moldovan plant biologist; b 27 Aug 1928, Kuizovka Rezina; s of Feodosie and Eudokia Lupascu; m Zinaida Armash 1953; 1s 1d. *Education*: Agric Inst of Chisinau. *Career*: Dir-Gen and Dir of Research, Plant-Growing Inst, Beltsy 1962–78; Vice-Pres, Acad of Sciences of Moldova 1978–80 and 1990–94, Head of Lab 1986–90; Ministry of Agric 1980–85. *Present Position*: Pres, Man and Biosphere Programme Nat Cttee, UNESCO 1979– and Head of Lab, Acad of Sciences of Moldova 1994. *Publications*: Environmental Protection and Intensive Agricultural Production 1984; Ecology and Intensification of Forage Production 1989; Alfalfa 1992; The Agriculture of Moldova and Improvement of its Ecological Condition 1996; numerous scientific books and papers. *Address*: 2004 Chisinau, Serghei Lazo 26, apt 10, Moldova. *Tel*: (2) 23-76-05. *Fax*: (2) 23-76-78.

LUTZENBERGER, José Antônio; Brazilian agronomist and environmentalist; b 17 Dec 1926, Porto Alegre; s of Josef and Emma (Kroeff) Lutzenberger; m Annemarie Wilm (deceased); 2d. *Education*: Univ do Rio Grande do Sul; Louisiana State Univ, USA. *Career*: BASF 1955–70, founder, Asociacção Gaúcha de Proteção ao Ambiente Natural 1971, Vida Produtos Biológicos 1979, Tecnologia Convivial 1985; State Sec for the Environment 1990–92. *Present Position*: Founder and Pres, Gaia Foundation 1987–. *Awards and Honours*: D SC, Univ of São Francisco, Universität für Bodenkultur, Vienna, Austria 1995; Brazilian Agronomic Soc Award 1980; Bodo-Manstein Medal 1981; Right Livelihood Award 1988; Order of the Rio Branco 1990; Order of the Andes 1990; Order of Merit of the Italian Republic 1991. *Publications*: Fim do Futro? 1976; Pesadelo Atômico 1980; Ecologia—Do Jardim ao Poder 1985; Política e Meio Ambiente (Co-Author) 1986; Giftige Ernte 1988; Gaia—O Planeto Vivo 1991; Knowledge and Wisdom Must Come Back Together 1994. *Address*: Rua Jacinto Gomes 39, 90040-270 Porte Alegre, RS, Brazil. *Tel*: (51) 331-3105. *Fax*: (51) 330-3567. *E-mail*: fundgaia@zaz.com.br. *Internet*: www.fgaia.org.br.

LYNGSTAD, Anni-Frid (Frida); Swedish singer and environmentalist; b 15 Nov 1945, Ballangen, Norway; d of Alfred Haase and the late Synni Lyngstad; m 1st Ragnar Fredriksson; m 2nd Benny Andersson 1978 (divorced 1981); m 3rd Ruzzo Reuss 1992 (died 1999); 1s 1d (deceased). *Career*: Entered Eurovision Song Contest as solo artist 1971; mem, ABBA pop group 1973–82; fmr Chair, Det Naturliga Steget (The Natural Step), Stockholm; active mem, Artister för Miljö (Artists for the Environment), 1987–. *Present Position*: Solo singer 1982– and environmental campaigner. *Awards and Honours*: Winner, Eurovision Song Contest (with ABBA) 1974. *Address*: c/o Det Naturliga Steget, Wallingt 22, 111 24 Stockholm, Sweden.

M

MAATHAI, Wangari, PH D; Kenyan environmentalist; b 1 April 1940, Nyeri; d of Muta Njugi and Wanjiru Kibicho; m Mwangi Maathai 1969 (divorced 1980); 2s 1d. *Education*: Mount St Scholastica Coll, KS, USA, Univ of Pittsburgh, PA, USA, Univ of Nairobi. *Career*: Visiting Research Asst, Univ of Munich, Fed Repub of Germany 1967–69; lecturer, Univ of Nairobi 1969–71, Sr Lecturer 1973, Assoc Prof 1978, Chair, Dept of Anatomy 1973–81; Dir, Kenya Red Cross Soc 1973–77; mem then Chair, Bd, Environment Liaison Centre Int 1974–84; mem, Club of Rome 1987–; fmr mem, Forum for Restoration of Democracy; Presidential Canidate 1997. *Present Position*: Founder and Co-ordinator, Green Belt Movement, Nat Council of Women of Kenya 1977– and public speaker on human rights, women's issues, devt and the environment. *Awards and Honours*: Woman of the Year Award 1983; Better World Soc Award 1986; UNEP Global 500 (first woman laureate) 1987; Woman of the World Award 1989; Green Century Environmental Award for Courage 1990; Goldman Environmental Prize 1991; Edinburgh Medal 1993; Nobel Peace Prize 2004; Judge Sasakawa Environment Prize. *Publications*: The Green Belt Movement 1985; The Green Belt Movement: Sharing the Approach and the Experience 1988; numerous scientific papers in journals. *Address*: Green Belt Movement, National Council of Women of Kenya, POB 67545, Nairobi, Kenya. *Tel and fax*: (2) 504264.

MCCALLA, Dr Winston, LL B, PH D; Jamaican environmental lawyer. *Education*: Univ of London, UK, Univ of Melbourne, Australia, London School of Econs. *Career*: drafter, legislation establishing environmental agencies in Belize, Guyana and Jamaica; Chair, Nat Environmental Socs Trust. *Present Position*: Consultant on environmental policy and legislation, various int orgs active in the Caribbean region. *Publications*: Guide for Developers, Environmental Law in Belize. *Address*: 3 Markway, Kingston 8, Jamaica. *Tel*: 946-1978. *Fax*: 924-7459. *E-mail*: wmcc@cwjamaica.com.

MCDOWELL, David Keith, MA; New Zealand diplomatist and conservationist; b 30 April 1937, Palmerston North; s of Keith and Gwen McDowell; m Jan Ingram 1960; 1s 3d. *Education*: Victoria Univ of Wellington. *Career*: Ministry of Foreign Affairs 1959–, Head, UN and African and Middle East Divs 1973, Dir of External Aid 1973–76, Head, Economic Div 1980–81; First Sec, Perm Mission to UN 1964–68, Perm Rep 1985–88; Special Asst to Sec-Gen, Commonwealth Secr, London, UK 1969–72; High Commr in Fiji 1977–80, India, Nepal and Bangladesh 1983–85; Asst Sec of Foreign Affairs for Asia, Australia and the Americas 1981–85; Dir-Gen, Dept of Conservation 1988–89; CEO, Dept of the Prime Minister and Cabinet 1989–91; Amb to Japan 1992–94; Dir-Gen, World Conservation Union—IUCN, Switzerland 1994–99. *Present Position*: environmental consultant 1999–. *Addresses*: 86 Waerenga Rd, Otaki, New Zealand; Flat 5, 268 Oriental Parade, Wellington, New Zealand. *Tel*: (6) 364-6296. *Fax*: (6) 364-6209. *E-mail*: jan.david.mcdowell@xtra.co.nz.

MCEWAN, Angus David, PH D, FAA; British/Australian oceanographer; b 20 July 1937, Alloa, UK; s of David N R and Annie Marion (Rowan) McEwan; m Juliana R Britten 1962 (divorced 1982); 2d. *Education*: Melbourne Univ, Cambridge Univ. *Career*: Research Scientist, Australian Aeronautical Research Labs 1961–62, 1966–69; Queen Elizabeth Fellow 1969–71; Research Scientist then Programme Leader and Chief Research Scientist, Div of Atmospheric Research, CSIRO, Vic, Australia 1972–81, Foundation Chief, Div of Oceanography, Hobart, Australia 1981–95; Fellow, Woods Hole Oceanographic Inst 1975; Chair, Cttee on Climatic Changes and Ocean 1987–90; Hon Research Prof, Univ of Tasmania 1988–; Chief Australian Nat Del, Intergovernmental Oceanographic Comm—IOC, ASEAN 1995–, Nat Del, Scientific Cttee for Oceanic Research—SCOR, IOC; mem, Scientific Cttees for the Global Climate Observing System 1990– and Global Ocean Observing System. *Present Position*: Sr Scientific Adviser to CSIRO and Australian Bureau of Meteorology 1995–. *Address*: Bureau of Meteorology, POB 727G, Hobart, Tas 7001, Australia. *Tel*: (3) 6221-2000.

MCGLADE, Prof Jacqueline, B SC, PH D; British/Canadian. *Career*: Prof, Biological Sciences, Univ of Warwick; Dir, Theoretical Ecology Group, Jülich; Dir, Natural Environment Research Council (NERC) Centre for Coastal and Marine Sciences –2000; NERC Professorial Fellow in Environmental Informatics and Mathematics, University College London 2000–03. *Present Position*: Exec Dir, European Environment Agency 2003–. *Address*: Kongens Nytorv 6, 1050 Copenhagen K, Denmark. *Tel*: 33-36-71-00. *Fax*: 33-36-71-99. *E-mail*: Jacqueline.Mcglade@eea.eu.int. *Internet*: org.eea.eu.int.

MACINNIS, Joseph Beverley, CM, MD; Canadian marine research scientist; b 2 March 1937, Barrie, ON; s of Allistair MacInnis and Beverly Saunders; m Deborah J Ferris 1971; 1s 3d. *Education*: Univ of Toronto; Univ of Pennsylvania, USA. *Career*: founder, SUBLIMNOS, Canada's first underwater manned station programme 1969; leader, 14 scientific expeditions into Arctic 1970–79, established world's first polar dive station under ice during third expedition, SUB-IGLOO; co-ordinator, diving programme for ICE Station LOREX 1979; presenter and consultant, numerous films and TV programmes 1974–79; consultant, Titanic Project 1985; Co-leader IMAX-Titanic Expedition 1991; mem, Canadian Environmental Advisory Council, Canadian Council of Fitness and Health; Fellow, Royal Canadian Geographical Soc. *Present Position*: Pres, Undersea Research Ltd. *Awards and Honours*: Hon LL D; Hon FRCP, Dr hc (Queen's) 1990. *Publications*: Underwater Images 1971; Underwater Man 1974; Coastline Canada 1982; Shipwreck Shores 1982; The Land that Devours Ships 1984; Titanic: In a New Light 1992; Saving the Oceans 1992 (Gen Ed); more than 30 scientific papers and articles in *Scientific American*, *National Geographic Magazine*, etc. *Address*: 14 Dale Ave, Toronto, ON M4W 1K4, Canada.

MCKENNA, Virginia; British actress and conservationist; b 1931, London; d of Terence McKenna and Anne Oakeley; m Bill Travers 1957 (deceased); 3s 1d. *Education*: Herons Ghyll, Horsham; Herschel, S.A.;. *Career*: Film, theatre and TV actress 1949–including: Born Free, Ring of Bright Water. *Present Position*: Founder, Born Free Foundation 1984–. *Awards and Honours*: British Film Acad Award 1954; Variety Club Award 1965; Soc of W End Producers Award 1979; OBE 2004. *Publications*: On Playing with Lions (co-author Bill Travers) 1966; Some of My Friends Have Tails 1971; Beyond the Bars 1987; Into the Blue 1992; Back to the Blue 1997; *Films*: A Town like Alice 1954; Carve Her Name with Pride 1957; Born Free 1964; Ring of Bright Water 1968. *Address*: Coldharbour RH5 4LW, United Kingdom. *Tel*: (1306) 712091. *Fax*: (1306) 713350.

MCQUAID, James, PH D, D SC, CB; British engineer; b 5 Nov 1939; s of James and Brigid McQuaid; m Catherine Anne Hargan 1968; 2s 1d. *Education*: Univ Coll, Dublin, Ireland, Jesus Coll, Cambridge. *Career*: Grad eng apprentice, British Nylon Spinners 1961–63; Sr Research Fellow, Safety in Mines Research Establishment 1966–68, Sr Scientific Officer 1968–72, Prin Scientific Officer 1972–78; Safety Adviser, Petrochemicals Div, ICI 1976–77; Deputy Dir, Safety Eng Lab 1978–80, Dir 1980–85; Research Dir, Health and Safety Exec 1985–92, Dir, Strategy and Gen Div, and Chief Scientist 1992–96; Chair, Electrical Equipment Certification Management Bd; mem, Safety in Mines Research Advisory Bd and Court, Sheffield Univ 1985–, Council, Midland Inst of Mining Engineers 1987–97; Dir, Science and Tech, and Chief Scientist, Health and Safety Exec 1996–99; Pres, Sheffield Trades History Soc and mem, Council, S Yorks Trades History Trust 1989–. *Present Position*: Visiting Prof, Univ of Sheffield and Univ of Ulster. *Awards and Honours*: Hon D ENG (Univ of Sheffield). *Address*: 61 Pingle Rd, Sheffield S7 2LL, United Kingdom. *Tel*: (114) 236-5349. *Fax*: (114) 249-0866. *E-mail*: jim@mcquaid.demon.co.uk.

MAHER ALI, Prof Dr Abd (El Moneim); Egyptian academic and environmental administrator; b 9 March 1922, Dammanhour; s of Ali Elsayed Shehata and Nagia M Manna; m Fardos Abbas Abd-el Aal 1948; 2s. *Education*: Univ Coll London, UK; Natural Resources USA. *Career*: Founder then Consultant, Wady El Assiuty protected area 1980–95; Founder and Gen Sec, Egyptian Soc of Youth Hostels 1955–71; Gen Sec, Egyptian Asscn for Conservation of Nature and Natural Resources, Egyptian Asscn for Medicinal Plants; Man, Cen Agricultural Pesticides Lab, UNDP Project 1963–69; Dir, Plant Protection Dept, Univ of Assiut 1969–82. *Present Position*: Prof Emer, Assiut Univ; Pres, Aradis Co 1982–. *Awards and Honours*: Conservation Merit Award, World Wildlife Fund; Order of the Republic; Science and Arts First Class Order. *Publications*: more than 180 scientific papers. *Addresses*: Aradis Co, POB 318, Dokki, Giza, Egypt; 50 Wizaret Zeraa Street, Dokki, Gizah. *Tel*: (2) 3360846. *Fax*: (2) 3462029. *E-mail*: webmail.aradis-co.com. *Internet*: www.aradis-co.com.

MAHOUD, Omar Osman, B SC; Sudanese agronomist and plant conservationist; b 1 Jan 1955, Taiba; 2s 2d. *Education*: Alexandria Univ, Egypt. *Career*: Plant Protection Dept, Ministry of Agric 1979–81; Head of Agric Devt, Islamic Relief 1981–86; Founder and Dir, NAFIR environmental NGO 1986–88; NGO consultant 1988–89; Deputy Exec Dir, Council of NGOs 1989–91. *Present Position*: Sec-Gen, Sudanese Red Crescent 1992–. *Publications*: The Role of NGOs and the International Community in Crisis Management in Sudan (Co-Author) 1990. *Address*: POB 235, Khartoum, Sudan. *Tel*: (11) 72011. *Fax*: (11) 72877. *E-mail*: srcs@sudanmail.net.

MAKÝŠ, Oto, PH D; Slovak architect and environmental administrator; b 27 June 1962, Bratislava. *Education*: Slovak Tech Univ, Bratislava. *Career*: Field conservation activities 1985–; Consultant, PHARE Environmental Educ and Public Awareness in Slovakia Project 1995–96. *Present Position*: Chair and Dir, Strom Života (Tree of Life) 1990–. *Address*: Strom Života, Pražská 11, 816 36 Bratislava, Slovakia. *Tel*: (7) 498-703. *Fax*: (7) 398-703.

MALMBERG, Prof Torsten, PH D; Swedish human ecologist; b 5 July 1923, Helsingborg; s of Sigfrid Vilhelm Malmberg and Lydia Henrietta Bergsten; m Aimée Birgit Magnussen 1948; 1s 1d. *Education*: Lund Univ. *Career*: Asst, Museum Keeper and Librarian, Zoology Dept, Lund Univ 1946–55, Asst Prof, Geography Dept 1978–83. *Present Position*: Head and Prof, Human Ecology Div, Lund Univ 1984–. *Awards and Honours*: Swedish Ornithological Soc Bronze Medal; Swedish Nature Protection Soc Silver Medal; UNEP Global 500 1992. *Publications*: Population Fluctuations and the Influence of Pesticide in the Rook, Corvus frugilegus L, in Scandinavia 1970; Human Territoriality 1980. *Address*: Human Ecology Division, Lund University, POB 2015, 220 02 Lund 2, Sweden. *Tel*: (46) 10-96-85. *Fax*: (46) 10-42-07.

MALUKA, Nevruz; Albanian politician and environmentalist; b 23 March 1931, Erseke; s of Rustem and Shefikat Maluka; m Violeta Shkembi 1961; 2s 1d. *Education*: Tirana Univ. *Career*: Hydrometeorology Inst 1955–60; Prof, Tech School, Tirana 1960–80; staff mem, plastics and rubber factory 1980–88. *Present Position*: Adviser, Industrial Science Inst, Tirana 1988– and Chair, Partia e Blertë Shqiptare (Albanian Green Party) 1991–. *Publications*: Inorganic Chemistry; Chemical Technology. *Address*: Partia e Blertë Shqiptare, POB 749, Tirana, Albania. *Tel and fax*: (42) 33309.

MANSER, Bruno; German environmental campaigner; b 25 Aug 1954, Basel, Switzerland. *Career*: Alpine herdsman, Switzerland 1973–84; ethnological fieldwork in Graübunden canton, Switzerland and in Sarawak state, Malaysia 1984–90. *Present Position*: Founder and Pres, Bruno Manser Foundation, Switzerland 1990– and campaigner for Amazonian rainforests. *Awards and Honours*: Binding Prize for Environmental Protection 1994. *Publications*: Stimmen aus dem Regenwald 1992. *Address*: Bruno Manser Foundation, Heuberg 25, 4051 Basel, Switzerland. *Tel*: (61) 2619474. *Fax*: (61) 2619473.

MARGUBA, Lawan Bukar, M SC; Nigerian ecologist and environmentalist; b 4 July 1951, Marguba; s of Akil Ali and Mrs Fanna Bure; m Saadatu Kilishi B Marguba 1980; 4s 1d. *Education*: Govt College Keffi, Nigeria, New Mexico State Univ, USA, Coll of African Wildlife Man, Mweka, Tanzania, Salford Univ, UK. *Career*: Range man asst 1970–72; livestock devt asst 1973–74; wildlife ecologist 1979–85; Chief Conservator of Forests, Borno state 1985–89, Dir of Forestry 1989–90; Head, Afforestation Programme Co-ordinating Unit 1990–91; Dir, National Parks of Nigeria (Pioneer Chief Exec)

1991–99. *Present Position*: Conservator-Gen and Chief Exec, Nigeria National Park Service 1999–. *Awards and Honours*: Fellow of the Forestry Association of Nigeria (FFAN); Fellow Institute of Corporate Administration of Nigeria (FICAN); Winner Africa International Award of Merit (AFIAM) 2000; Pioneer Role Model Award by Tribune Newspapers 2001; Merit Award for Conservation by National Asscn of Geography Students (ABU) 2001. *Publications*: Desertification and its Impact on the Sahel Region of West Africa 1984; National Parks and Their Benefits to Local Communities in Nigeria 2001; Social and Economic Considerations in Rural Tourism Development in Nigeria, A Paper Presented to a Three-Day National Workshop on Integrated Community Based Tourism and Rural Economic Endowment 2002; Ecotourism in National Parks, the Business Perspective 2003; Biodiversity Loss, the Anthropogenic Factor in Nigeria 2004. *Addresses*: Nigeria National Park Service, Nnamdi Azikiwe Int. Airport Expressway, PMB 0258, Garki-Abuja, Nigeria; Chad Basin national Park, PMB 1026, Maiduguri, Borneo State. *Tel*: (9) 5230429. *Fax*: (9) 5230409.

MARK, Alan Francis, CBE, PH D, FRSNZ, DCNZM; New Zealand professor of botany; b 19 June 1932, Dunedin; s of Cyril L Mark and Frances E Marshall; m Patricia K Davie 1957; 2s 2d. *Education*: Univ of Otago; Duke Univ, NC, USA. *Career*: Otago Catchment Bd, Dunedin 1959–61; Sr Research Fellow, Hellaby Indigenous Grasslands Research Trust 1961–65, Adviser in Research 1965–; Lecturer, Univ of Otago 1960, Sr Lecturer 1966, Assoc Prof 1969, Prof of Botany 1975–98; Visiting Asst Prof, Duke Univ 1966. *Present Position*: Prof Emer, Univ of Otago 1998–. *Awards and Honours*: Fulbright Travel Award 1955; James B Duke Fellowship 1957; Loder Cup 1975, NZ 1990 Commemoration Medal, The Awards of NZ (Conservation/Environment) 1994; Hutton Medal (Botanical/Conservation Research), Royal Soc of NZ 1997; FRSNZ 1978. *Publications*: New Zealand Alpine Plants (Co-Author) 1973; about 160 scientific papers. *Addresses*: Department of Botany, University of Otago, POB 56, Dunedin, New Zealand (Office); 205 Wakari Rd, Helensburgh, Dunedin, New Zealand (Home). *Tel*: (3) 4797573. *Fax*: (3) 4797583. *E-mail*: amark@otago.ac.nz.

MARTIN, Dr Claude, PH D; Swiss conservationist; b 20 July 1945, Zürich; m Judy Martin; 2s 2d. *Education*: Univ of Zürich. *Career*: Sr Game Warden, Dept of Game and Wildlife, Ghana 1975–78; Dir, WWF Switzerland 1979–90; Deputy Dir-Gen, WWF Int 1990–93. *Present Position*: Dir-Gen, WWF Int 1993–. *Publications*: The Rainforests of West Africa 1991. *Address*: World Wide Fund for Nature—WWF International, ave de Mont Blanc, 1196 Gland, Switzerland. *Tel*: (22) 3649111. *Fax*: (22) 3645468. *Internet*: wwf.panda.org.

MARTÍNEZ DE PISÓN, Dr Eduardo; Spanish physical geographer; b 1 Jan 1937; s of Eduardo and Julia Martínez de Pisón; m Margarita Fernández-Ahuja 1976. *Education*: Complutense Univ of Madrid. *Career*: Co-Founder, Spanish Association for Environmental Regulation—AEORMA 1972; Prof of Geography and Physical Geography, Univ of La Laguna, Canary Islands 1977; Researcher in Physical Geography, Autonomous Univ of Madrid; geographer, Spanish expedition to North Pole 1999. *Present Position*: Prof of Physical Geography, Autonomous Univ of Madrid 1981–. *Awards and Honours*: Premio J M Quadrado (Research); Premio Nacional de Medio Ambiente 1991. *Publications*: El Teide, Estudio Geográfico 1981; Geomorfología del Mount Everest, Tibet 1988; La Antártica y las Regiones Glaciares 1992; Cuadernos de Montaña 1999. *Address*: Department of Geography, Autonomous Univ of Madrid, Carretera de Colmenar, Campus de Canto Blanco, 28049 Madrid, Spain. *Tel*: (91) 3974577. *Fax*: (91) 3974042.

MARUSIK, Dr Yurii M; Russian zoologist and biogeographer; b 13 May 1962, Ukraine; m divorced. *Education*: Leningrad (St Petersburg) State Univ, Russian Fed. *Career*: expert, arachnology. *Present Position*: Assoc Prof, Int Magadan Univ 1993–; Sr Scientific Fellow, Inst for Biological Problems of the North—IBPN 1997–. *Awards and Honours*: (Russian) Far East Award for Scientists 1993. *Publications*: Spiders of Tuva, South Siberia 2000. *Address*: Institute for Biological Problems of the North, Magadan 685000, Portovaya Str 18, Russian Federation. *Tel*: (41322) 345-57. *Fax*: (41322) 344-63. *E-mail*: ibpn@online.magadan.su.

MASTROVIC, Margita, M SC; Croatian environmental adviser; b 2 July 1939, Rijeka; d of Ivo and Andra Ivaniševic; m Jurica Mastrovic 1967; 2s. *Education*: Univ of Zagreb. *Career*: Ecology Team Leader, INA oil refinery, Rijeka 1962–79; Adviser, Physical Planning and Environmental Protection Inst, Ass of Communes, Rijeka 1979–90; Sr Adviser, Ministry of Civil Eng and Environmental Protection 1990–95; participant, Environment for Europe programme 1992–98. *Present Position*: Sr Adviser, Ministry of Environment and Physical Planning 2000–; Co-ordinator, MED-POL 1998–. *Publications*: Climate and the Mediterranean 1996; Rijeka–Istria Region Solid Waste Study 1996; Environmental Management Plan for the Cres-Lošinj Archipelago 1996; The State of the Environment in Croatia 1998. *Address*: Ministry of Environment and Physical Planning, Office for the Sea and Coastal Protection, 51000 Rijeka, Užarska 2/I, Croatia. *Tel*: (51) 213499. *Fax*: (51) 214324.

MATHEW, Thomas; environmentalist. *Present Position*: Mem, WWF India. *Address*: World Wide Fund for Nature—WWF India, 172b Lodi Rd, New Delhi 110 003, India. *Tel*: (11) 616532. *Fax*: (11) 4626837.

MATOLA, Sharon, BA; Belizean (b American) environmentalist and zoo director; b 3 June 1954, MD, USA; d of Edward and Janice Matola. *Education*: New Coll of the Univ of S Florida. *Career*: Circus exotic dancer and lion tamer, Mexico; natural history film co. *Present Position*: Founder and Dir, Belize Zoo and Tropical Educ Center 1983– and Chair, Tapir Specialist Group, IUCN. *Publications*: Hoodwink the Owl 1988; The Further Adventures of Hoodwink the Owl 1993. *Address*: Belize Zoo and Tropical Centre, POB 474, Belize City, Belize. *Tel*: (2) 45523. *Fax*: (2) 78808.

MATTHIESSEN, Peter, BA; American environmentalist and writer; b 22 May 1927; s of Erard and Elizabeth (Carey) Matthiessen; m 1st Patricia Southgate 1951 (divorced 1957); m 2nd Deborah Love 1963 (died 1972); m 3rd Maria Eckhart 1980; 3s 1d. *Education*: Univ of Paris (Sorbonne), France, Yale Univ. *Career*: Co-Founder, *Paris Review*; ordained a Zen monk 1981; Correspondent, *New Yorker* magazine; Trustee, New York Zoological Soc; mem, American Acad of Arts and Letters, American Acad of Arts and Sciences. *Awards and Honours*: National Book Award 1978; John Burroughs Medal 1982; African Wildlife Leadership Award 1982; Gold Medal for Distinction in Natural History, Academy of Natural Sciences 1985; UNEP Global 500 1991; Orion-John Hay Award 1999; Soc of Conservation Biologists Award 1999. *Publications*: Wildlife in America 1959; The Cloud Forest 1961; Under the Mountain Wall 1963; The Shore Birds of North America 1967; Oomingmak: The Expedition to the Musk Ox Island in the Bering Sea 1967; Blue Meridian 1971; The Snow Leopard 1978; Sand Rivers 1981; Nine-Headed Dragon River 1986; Men's Lives 1986; African Silences 1991; Baikal 1992; African Shadows 1992; Tigers in the Snow 1999. *Address*: 527 Bridge Lane, POB 392, Sagaponack, NY 11962, USA. *Tel*: (516) 537-0837. *Fax*: (516) 537-5372.

MEADOWS, Dennis Lynn, PH D; American professor, writer and consultant. *Education*: Massachusetts Inst of Tech—MIT.

Koltsov Institute of Biology of Development, 117334 Moscow, ul Vavilova 26, Russian Federation. *Tel*: (095) 135-75-83.

VOYNET, Dominique, D EN MED; French politician; b 4 Nov 1958, Montbéliard; d of Jean and Monique (Richard) Voynet; 2d. *Education*: Univ of Besancçon. *Career*: Co-Founder, Les Verts (French Green Party) 1984, Nat Spokesperson 1992–93, Cand, Presidential Election 1995; MEP 1991, Sec-Gen, Green Group 1989–91; mem, Dole Municipal Council 1989–97, Regional Councillor, Franche-Comté 1992–97. *Present Position*: Leader, Les Verts 1993–; Min of Town and Country Planning and the Env 1997–. *Address*: Min of Town and Country Planning and the Environment, 20 ave de Ségur, 75302 Paris Cédex, France. *Tel*: 1-42-19-20-21. *E-mail*: dvoynet@verts.imaginet.fr.

W

WADAEE, Abdul-Elah Al-, PH D; Bahraini industrial hygienist; b 27 Dec 1955, Manama; s of Jawad Al-Wadaee and Batool Al-Asfour; m Naaem Al-Asfour 1984; 2s 1d. *Education*: Baghdad Univ, Iraq, Paris Univ and Aix-Marseilles Univ, France. *Career*: Occupational hygienist, Ministry of Health 1979–90; chemist, Environmental Protection Cttee, Ministry of Housing, Municipalities and Environment 1990–97. *Present Position*: Regional Network Co-ordinator, Ozone Depletion Substances, UNEP Regional Office for West Asia 1997–. *Address*: POB 10880, Manama, Bahrain. *Tel*: 297072. *Fax*: 276075.

WAECHTER, Dr Antoine; French ecologist and engineer; b 11 Feb 1949, Mulhouse; m Martine Charbonnel 1990; 2c. *Education*: Strasbourg Univ. *Career*: Co-Founder, Ecologie et Survie 1973; environmental consultant 1978–; Regional Fed of Nature and Environment Protection Asscns Rep, Social and Econ Cttee, Alsace 1980, Vice-Pres 1983; Regional Councillor, Alsace 1986–98, Vice-Pres, Regional Council; Co-Founder and mem, Les Verts (French Green Party) 1984–94, Spokesperson 1986–89, Nat Spokesperson 1991–94, Cand, Presidential Election and Cantonal Election, Mulhouse E 1988, Municipal Councillor, Mulhouse and Cand, European Election 1989; MEP, Pres, EP Cttee on Regional Policy and Land Man, Substitute, Cttee on External Econ Relations 1989–91; Founder, Mouvement Ecologiste Indépendant 1994–, Pres 1994–98. *Present Position*: Pres, Mouvement Ecologiste Indépendant 1999–. *Publications*: Dessine-Moi une Planète 1990. *Address*: c/o Mouvement Ecologiste Indépendant, 7 rue de Verbois, 75003 Paris, France.

WALKER, Brian Wilson, MA; British environment and development consultant; b 31 Oct 1930, Chipping Norton; s of Arthur Harrison and Eleanor Charlotte Mary Walker; m Nancy Margaret Gawith 1954; 1s 5d. *Education*: Manchester Univ, Leicester Coll of Tech. *Career*: Business 1952–74; Founder and Chair, New Ulster Movt 1969–74; Dir-Gen, Oxfam 1974–83; Dir, Independent Comm on Int Humanitarian Issues, Geneva, Switzerland 1983–84; Pres, Int Inst for Environment and Devt 1985–89; Exec Dir, Earthwatch Europe 1989–95; Int Comm on Food and Peace 1989–94. *Present Position*: Ind consultant. *Awards and Honours*: Annual Brian Walker Lecture on Environment and Devt, Green Coll, Oxford. *Address*: 'Biskets', Church Hill, Arnside LA5 0DW, United Kingdom. *Tel*: (1524) 761949. *Fax*: (1524) 761933. *E-mail*: brian@bisketsw.fsnet.co.uk.

WALLACE, Catherine Ceinwen, BA; New Zealand economist and environmentalist; b 17 Jan 1952, Hamilton; d of Dr Lindsay and Dr Charlotte Wallace. *Education*: Victoria Univ, Wellington. *Career*: Fisheries Economist, Ministry of Agric and Fisheries 1980–81, 1983–85; Chair, Environment and Conservation Orgs of NZ 1985–86, 1989–93, Vice-Chair 1987–88; Convenor, Antarctic and Southern Ocean Coalition New Zealand 1985–92; Regional Councillor for Oceania, IUCN 1991–96; mem, IUCN Bureau and Chair, Membership Cttee 1994–96. *Present Position*: Sr Lecturer in Econs and Public Policy, Victoria Univ, Wellington 1986–. *Awards and Honours*: New Zealand Medal 1990; Goldman Environmental Foundation Prize 1991. *Address*: 3 Finnimore Terrace, Vogeltown, Wellington 6002, New Zealand. *Tel*: (4) 389-1696. *Fax*: (4) 463-5804. *E-mail*: cath.wallace@vuw.ac.nz.

WALLÉN, Carl Christian, PH D; Swedish meteorologist and climatologist; b 17 Aug 1917; s of Axel and Ingerd (Hessle) Wallén; m Betty Hyltén-Cavallius; 1d. *Education*: Stockholm Univ. *Career*: Deputy Dir, Swedish Meteorological and Hydrological Inst 1955–68; Chief of Div, WMO 1968–76, Organizer, Stockholm Conf 1972, Consultant 1994–98; Deputy Dir, Global Environment Monitoring System—GEMS, UNEP 1976–80, Chair, Scientific Advisory Cttee 1994–98. *Addresses*: World Meteorological Organization—WMO, 41 ave Giuseppe Motta, CP 5, 1211 Geneva, Switzerland; Oxenstiernsgt 35, Stockholm, Sweden. *Tel*: (22) 7308111. *Fax*: (22) 7342366.

WARREN, Andrew; British environmental organization administrator. *Present Position*: Dir, Asscn for the Conservation of Energy. *Address*: Association for the Conservation of Energy, Westgate House, Prebend St, London N1 8PT, United Kingdom. *Tel*: (20) 7359-8000. *Fax*: (20) 7359-0863.

WASAWO, Prof Dr David Peter Simon; Kenyan zoologist; b 17 May 1923, Gem; s of Petro Onyango Osare and Ana Omondi; m Ruth Peace Lusinde; 2s 2d. *Education*: Makerere Univ, Uganda; Univ of Oxford, Univ of London, UK. *Career*: Lecturer, then Head of Dept of Zoology and Vice-Prin, Makerere Univ 1953–65; Deputy Prin, Univ of Nairobi 1965–68, Prof, Head of Dept of Zoology and Dean, Faculty of Science 1970–71; mem, Advisory Cttee on Natural Resources Research, UNESCO 1965–71; Chair, Working Party on Research Priorities and Research Admin, E African Community 1969; Trustee, WWF 1970–74; UNESCO Scientific Adviser, Tanzania 1971–73; Vice-Pres, IUCN 1972–74; Chief, Natural Resources Div, UN Econ Comm for Africa—ECA 1973–79; Man Dir, Lake Basin Devt Authority 1979–82; Devt Adviser, Ministry of Energy and Regional Devt 1982–86. *Present Position*: devt consultant 1987–; Chair, Nairobi Univ Council 1998–. *Awards and Honours*: Hon D SC, Kenyatta Univ, Univ of Nairobi. *Publications*: A Dry-Season Burrow of Protopterus Aethipicus 1959; The Taxonomic Position of Alma Worthingtoni 1962; Quality of Life in Different Cultures 1973; Strategic Resources in Uganda (Ed). *Address*: POB 41024, Nairobi, Kenya. *Tel and fax*: (2) 715582. *E-mail*: wasawo@insightkenya.com.

WATERS, Joe; American ozone researcher. *Present Position*: Chief Researcher on the Ozone Layer in the Northern Hemisphere, NASA. *Address*: NASA Jet Propulsion Laboratory, 4800 Oak Grove Dr, Pasadena, CA 91109, USA. *Tel*: (818) 354-4321. *E-mail*: joe@mls.jpl.nasa.gov.

WATKINS, Ron G; Australian farmer and environmentalist; b 15 March 1950, Kosonup, WA; m Suzanne Watkins 1973; 1s

2d. *Education*: Curtin Univ, Northam. *Present Position*: Environmental management consultant 1973– and Pres, Land Man Soc. *Awards and Honours*: Churchill Fellowship 1987; UNEP Global 500 1995; UNEP Effective Action Against Desertification Award 1995; Australian Department of the Environment Award 1995. *Publications*: Practical Application (Applied Science) 1982. *Address*: 'Payneham', Haynesdale Rd, Frankland, WA 6396, Australia. *Tel and fax*: (8) 9826-7013.

WAYBURN, Edgar, BA, MD; American physician, environmentalist and conservationist; b 17 Sept 1906, Macon, GA; s of Emanuel and Marian Wayburn; m Cornelia Elliot 1947; 1s 3d. *Education*: Univ of Georgia, Harvard Medical School. *Career*: leader, campaigns to establish Redwood Nat Park 1968, 1978, Golden Gate Nat Recreation Area 1972, Alaska Nat Interest Lands Act 1980. *Present Position*: Hon Pres, Sierra Club 1993–, Sierra Club Foundation 1998–. *Awards and Honours*: President's Medal of Freedom 1999; Albert Schweitzer Prize for Humanitarianism 1995. *Address*: 314 30th Ave, San Francisco, CA 94121, USA. *Tel*: (415) 386-4241. *E-mail*: edgarwayburn@aol.com.

WEBER, Otto; Romanian politician and environmentalist. *Present Position*: Chair, Partidul Ecologist Român (Romanian Ecological Party). *Address*: Basarab Matei Nr 106, Bucharest, Romania. *Tel*: (1) 3201168.

WEEKS, John; British administrator and environmentalist. *Present Position*: Prof and Dir, Centre for Devt and Policy Research, School of Oriental and African Studies—SOAS, Univ of London, UK. *Address*: Centre for Development and policy Research, School of Oriental and African Studies—SOAS, Thornhaugh St, Russell Sq, London WC1H 0XG, United Kingdom. *Tel*: (20) 7323-6241. *Fax*: (20) 7323-6605. *E-mail*: jw10@soas.ac.uk.

WEREKO-BROBBY, Charles Y; Ghanaian energy consultant. *Career*: Mem, Commonwealth Science Council; Energy Policy Adviser, Ministry of Fuel and Power, Nat Energy; Exec Dir, Centre for Energy, Environment and Development. *Present Position*: founder and leader, United Ghana Movement, Presidential Candidate 2000. *Address*: POB TU 39 Accra, Ghana. *Tel*: (21) 225581. *Fax*: (21) 231390.

WERGER, Marinus Johannes Antonius, PH D; Dutch botanist and ecologist; b 3 May 1944, Enschede; s of Johannes G Werger and Gezina M Zwerink; m Karin E Klein 1968; 1d. *Education*: Jacobus Coll, Enschede, Univ Utrecht, Rijksuniv Groningen, Katholieke Univ Nijmegen. *Career*: professional research officer, Botanical Research Inst, Pretoria, S Africa 1968–73; Asst Prof, Assoc Prof, Katholieke Univ Nijmegen 1974–79; Dean, Faculty of Biology, Univ Utrecht 1990–93; Consultant Prof, SW China Univ Beibei 1984–; Visiting Prof, Univ of Tokyo, Japan 1985. *Present Position*: Prof of Plant Ecology and Vegetation Science, Univ Utrecht 1979–. *Awards and Honours*: mem, Royal Netherlands Acad; Mid-America State Univs Asscn Award 1986. *Publications*: Biogeography and Ecology of Southern Africa (2 vols) 1978; The Study of Vegetation 1979; Man's Impact on Vegetation 1983; Plant Form and Vegetation Structure 1988; Tropical Rain Forest Ecosystems. *Addresses*: Department of Ecology and Evolutionary Biology, Heidelberglaan 8, POB 80125, 3508 TC Utrecht, Netherlands; Nieuwe Gracht 145, 3512 LL Utrecht, Netherlands. *Tel*: (30) 536700.

WERIKHE, Michael Mayeku; Kenyan conservationist; b 25 May 1956, Mombasa; m Hellen Werikhe 1987; 2d. *Career*: Wildlife Conservation and Man Dept 1976–77; walked from Mombasa to Nairobi (490 km) to raise awareness and funds for rhinoceros projects in Africa 1982, across E Africa (2300 km) 1985, Europe (2200 km) 1988, N America 1991; Advisory Dir, World Soc for Protection of Animals 1987; Chair, E African Wildlife Soc 1988. *Present Position*: Founder, Michael Werikhe Conservation Centre 1993–. *Awards and Honours*: Guinness Stout Award 1983; E African Wildlife Soc Conservation Medal 1987; UNEP Global 500 1989; Goldman Environment Prize 1990; Eddie Bauer Heroes of the Earth Prize 1991; Conservation Medal, San Diego Zoological Soc 1991. *Address*: POB 80310, Mombasa, Kenya. *Tel*: (11) 433311. *Fax*: (11) 434461.

WESTING, Arthur H, MF, PH D; American ecologist and environmental consultant; b 18 July 1928, USA; s of S W and Paula R Westing; m Carol E Westing 1956; 1d 1s. *Education*: Columbia Univ, Yale Univ. *Career*: Sr Researcher, Stockholm Int Peace Research Inst, Sweden 1976–78, 1983–87; Prof of Ecology and Dean of Natural Science, Hampshire Coll, Amherst 1978–83; Sr Researcher, Int Peace Research Inst, Oslo, Norway 1988–90, Consultant in Environoment Security 1990–. *Present Position*: CEO, Westing Assocs in Environment, Security and Educ 1990–. *Awards and Honours*: Hon D SC (Windham Coll) 1973; UNEP Global 500 1990. *Publications*: Ecological Consequences of the Second Indochina War 1976; Warfare in a Fragile World: The Military Impact on the Human Environment 1980; Cultural Norms, War and the Environment 1988; Transfrontier Reserves for Peace and Nature 1993; Hazards of War: releasing dangerous forces in an industrialized world 1990. *Address*: Westing Associates, 134 Fred Houghton Rd, Putney, VT 05346, USA. *Tel*: (802) 387-2152. *E-mail*: westing@sover.net.

WHITAKER, Romulus Earl, B SC; Indian herpetologist and film-maker; b 23 July 1943, New York, USA; s of Dr R and Doris (Norden) Whitaker; m divorced; 2s. *Education*: Pacific Western Univ, USA. *Present Position*: Founder and Dir, Centre for Herpetology and Madras Crocodile Bank; Dir, Irula Snake-catchers Co-operative; producer, environmental films. *Awards and Honours*: Rolex Award; WWF Gold Medal; Peter Scott Award; Order of the Golden Ark; Asoka Fellow. *Publications*: Common Indian Snakes 1978; Crocodile Fever 1999; 130 scientific publications, 140 popular publications; 12 documentary films. *Addresses*: Centre for Herpetology/Madras Crocodile Bank, Post Bag 4, Mammallapuram 603 104, Tamil Nadu, India; POB 21, Chingleput 603 001, Tamil Nadu, India. *Tel*: (98) 41023543. *Fax*: (44) 4918747. *E-mail*: draco@vsnl.com.

WHITE, Robert Mayer, SC D; American meteorologist; b 13 Feb 1923, Boston, MA; s of David and Mary (Winkeller) White; m Mavis Seagle 1948; 1s 1d. *Education*: Harvard Univ and Massachusetts Inst of Tech—MIT. *Career*: war service with USAF, exec at Atmospheric Analysis Lab, Geophysics Research Directorate, Air Force Cambridge Research Center 1952–58, Chief, Meteorological Devt Lab 1958; Research Assoc, MIT 1959; Travelers Insurance Cos 1959–60, Pres, Travelers Research Center, Hartford 1960–63; Chief, Weather Bureau, US Dept of Commerce 1963–65; Admin, Environmental Science Services Admin, US Dept of Commerce 1965–70; Perm Rep and mem, Exec Cttee, World Meteorological Org 1963–77; Admin, Nat Oceanic and Atmospheric Admin 1971–77; Chair, Jt Oceanographic Inst, Inc 1977–79; Chair, Climate Research Bd, NAS 1977–79; Admin, Nat Research Council, Exec Officer 1979–80; Pres, Univ Corpn for Atmospheric Research 1979–83; Karl T Compton Lecturer, MIT 1995–96; mem, Exec Cttee, American Geophysical Union, Council, Nat Acad of Eng (Pres 1983–), Marine Tech Soc, Royal Meteorological Soc, Nat Advisory Cttee on Oceans and Atmosphere 1979–84, Nat Advisory Cttee on Govt and Public Affairs, Univ of Ill 1987–; Bd of Overseers, Harvard Univ 1977–79; mem, numerous weather research cttees, Commr, Int Whaling Comm 1973–77. *Present Position*: Sr Fellow, Univ Corpn for Atmospheric Research 1995–;

Pres, Washington Advisory Group 1996–. *Awards and Honours*: Cleveland Abbe Award, American Meteorological Soc 1969; Rockefeller Public Service Award 1974; David B Stone Award, New England Aquarium 1975; Matthew Fontaine Maury Award, Smithsonian Inst 1976; Int Conservation Award, Nat Wildlife Fed 1976; Neptune Award, American Oceanic Org 1977; Charles Franklin Brooks Award 1978; Int Meteorological Asscn Prize 1980; Tyler Prize, Univ of California 1992; Vannevar Bush Award 1998. *Addresses*: 1200 New York Ave, NW, Suite 410, Washington, DC 20005, USA (Office); Somerset House II, 5610 Wisconsin Ave, Apt 1506, Bethesda, MD 20815, USA (Home).

WIGLEY, Prof Tom M L; climatologist; b 18 Jan 1940, Adelaide. *Education*: Univ of Adelaide. *Career*: Meteorologist, Research Section, Commonwealth Bureau of Meteorology, Melbourne, Australia 1962–63; Asst Prof, Dept of Mechanical Engineering, Univ of Waterloo, Canada 1968-72, Assoc Prof 1972–75; Sr Research Assoc, Climatic Research Unit, Univ of E Anglia, UK 1975–78, Dir 1978–93, Prof of Climatology, School of Environmental Sciences 1987–93; Dir, Office for Interdisciplinary Earth Studies, Univ Corpn for Atmospheric Research 1993–94. *Present Position*: Sr Scientist, Nat Center for Atmospheric Research, Boulder, CO 1994–. *Address*: National Centre for Atmospheric Research, 1800 Table Mesa Drive, Boulder, CO 80307-3000, USA. *Tel*: (303) 497-2690. *E-mail*: wigley@ucar.edu.

WILLIAMS, Philip; American environmentalist. *Present Position*: Pres, Int Rivers Network. *Address*: International Rivers Network, 1847 Berkeley Way, Berkeley, CA 94703, USA. *Tel*: (510) 848-1155. *Fax*: (510) 848-1008. *E-mail*: irn@irn.org.

WINFIELD, Mark, PH D; Canadian researcher and environmentalist; b 1 Nov 1963, Toronto. *Education*: Toronto Univ. *Career*: Instructor, Toronto Univ 1991–. *Present Position*: Dir of Research, Canadian Inst for Environmental Law and Policy—CIELAP and mem, Exec Cttee, Conservation Council, ON 1992–. *Publications*: The Ultimate Horizontal Issue: Environmental Policy and Politics in Ontario and Alberta 1971–1992 1992. *Address*: Canadian Institute for Environmental Law and Policy—CIELAP, 517 College St, Suite 400, Toronto, ON M6G 4A2, Canada. *Tel*: (416) 923-3529. *Fax*: (416) 923-5949.

WITT, Ronald George, MA; American environmental administrator; b 12 Nov 1954; m A Witt 1987; 2d. *Education*: Univ of Utah. *Career*: Officer, UNEP, Nairobi, Kenya 1986–87, Geneva, Switzerland 1987–. *Present Position*: Regional Co-ordinator, DEIA/Global Resource Information Database—GRID, UNEP 1996–. *Address*: United Nations Environment Programme—UNEP, DEIA/GRID, 11 chemin des Anémones, 1219 Châtelaine/Geneva, Switzerland. *Tel*: (22) 9799295. *Fax*: (22) 9799020. *E-mail*: rgwitt@gridi.unep.ch.

WOOLF, Timothy, MBA; American environmental researcher; Boston, MA. *Career*: Research Man, Asscn for Conservation of Energy, London, UK. *Present Position*: Sr Scientist, Tellus Inst, Boston, MA 1993–. *Address*: Tellus Institute, 11 Arlington St, Boston, MA 02116, USA. *Tel*: (617) 266-5400.

WRIGHT, Wayne, B SC, MBA; Canadian petroleum industry executive and naturalist; b 10 June 1943, Toronto, ON; s of Archie and Ethel (Majury) Wright; m Nancy Wright 1966; 1s 1d. *Education*: McGill Univ, Harvard Business School, MA, USA. *Career*: Vice-Pres, Canadian Petroleum Products Inst 1990–95; mem, Bd of Dirs, Long Point Bird Observatory. *Present Position*: Dir of Supply Chain Man and Pres, Waterfront Shipping Co, Methanex Corpn 1996– and Chair, Bd of Trustees, Nature Conservancy of Canada. *Address*: Methanex Corpn, 200 Burrard St, Suite 1800, Vancouver, BC V6C 3M1, Canada.

Y

YABLOKOV, Prof Dr Aleksei Vladimirovich, D BIOL SC; Russian ecologist and politician; b 3 Oct 1933, Moscow; s of Vladimir S Yablokov and Tatiana S Sarchyeva; m 1st Eleonora Bakulina 1955 (died 1987); m 2nd Dil'bar N Klado 1990; 1s. *Education*: Moscow State Univ. *Career*: Chair, Youth Section, All-Russian Soc for the Protection of Nature 1951–53; Head, Post-Natal Onthogenesis Lab, Devt Biology Inst, USSR Acad of Sciences 1969–91; Chair, Fish Studies Cttee, Ministry of Fisheries 1988–91; People's Deputy, Supreme Soviet USSR and Deputy Chair, Ecology Cttee 1989–91; Chair, Presidential Ecological Policy Council 1992–93; Presidential Adviser on Ecology and Health 1991–93, Chair, Interdepartmental Comm on Ecological Security, Nat Security Council 1993–97. *Present Position*: Pres, Centre for Russian Environmental Policy 1993–; Adviser, Russian Acad of Science 2004–; Vice-Pres, IUCN, the World Conservation Union 2002–. *Awards and Honours*: IUCN Roll of Honour 1991; Green House Int Award 1993; WWF Gold Medal 2000; WASA Env Prize 1994; BASK Medal Roy Geog Soc 1995; Nuclear Free Future Award 2002. *Publications*: over 450 scientific papers and 24 books including: Whales and Dolphins 1972; Variability of Mammals 1976; Levels of Protection of Living Nature 1985; Conservation of Living Nature and Resources: Problems, Trends and Prospects 1987; Nuclear Mythology 1997; Pesticides: Chemical Weapons that Kill Life 2004; Non-invasive Study of Mammalian Populations 2004. *Addresses*: Centre for Russian Environmental Policy, 119991 Moscow, Vavilov 26, Russian Federation; 119296 Moscow, Vavilov street 56-1-36. *Tel*: (095) 952-80-19. *Fax*: (095) 952-30-07. *E-mail*: yablokov@ecopolicy.ru. *Internet*: www.ecopolicy.ru.

YAKAVENKA, Vasily Timofeevich; Belarusian geologist, journalist and writer; b 5 May 1936, Gomelskaya Region; s of Timofey and Vera Yakavenka; m Larisa Furmanova 1972; 2d. *Education*: Belarus State Univ, Minsk. *Career*: Engineer and geologist 1966–72; Ed, Essay Dept, *Polyma* literary magazine 1972–77; writer 1977–; Chief Ed, *Nabat* newspaper 1990–. *Present Position*: Pres, Chornobyl Socio-Ecological Union 1990–. *Publications*: The Earth Opened to Us 1971; Test Stone 1974; While a Sun is at the Zenith 1978; Getting to a Harmony 1985; Rural Public Debates 1987; The Trace of the Black Wind 1995; The Black Rose 1996; Chornobyl: 10 Years After 1996. *Address*: 220116 Minsk, Izvestiya pr, Bldg 17, Flat 72, Belarus. *Tel*: (17) 234-22-41. *Fax*: (17) 210-18-80.

YAMIN, Farhana Naz, MA, LL M; British/Pakistani environmental lawyer; b 22 Feb 1965. *Education*: Somerville Coll, Oxford, King's Coll, London. *Present Position*: Lawyer, Foundation for Int Environmental Law and Devt—FIELD, SOAS 1991–, Dir 1996–, Legal Adviser, UN Framework Convention on Climate Change 1992– and Lecturer in Int Environmental Law, School of Oriental and African Studies—SOAS, Univ of London, UK 1993–. *Publications*: Numerous articles and reports. *Address*: Foundation for

International Environmental Law and Development—FIELD, School of Oriental and African Studies—SOAS, University of London, 47 Russell Sq, London WC1B 4JP, United Kingdom. *Tel*: (20) 7637-7950. *Fax*: (20) 7637-7951. *E-mail*: fy1@soas.ac.uk. *Internet*: www.field.org.uk.

YEROULANOS, Marinos; Greek environmental and fisheries consultant; b 13 May 1930, Athens; s of Ioannis and Despina Yeroulanos; m Aimilia Kalliga 1955; 1s 3d. *Education*: Swiss Fed Inst of Tech, Zürich, Switzerland. *Career*: Dir-Gen, Environment and Physical Planning 1975–81; mem, Nat Water Bd, Nat Coastal Planning Council, Nat Council for Scientific Research 1977–81; Consultant, UNEP 1981–86; ind fisheries consultant 1981–96. *Present Position*: Pres, Cephalonian Fisheries Ltd 1992– and Benaki Museum 1995–. *Awards and Honours*: UNEP Global 500 1987. *Address*: Lykiou St 10, 106 74 Athens, Greece. *Tel*: 2107216600. *Fax*: 2107221618. *E-mail*: yeroulano@netor.gr.

YOUMBI, Augustin; Cameroonian environmentalist; b 9 June 1952, Mbalmayo; s of Marcel Youmbi and Marie Motoum; m Matagne Henliatte 1980; 3s 1d. *Education*: Yaoundé Univ, Free Univ of Brussels, Belgium. *Career*: Researcher, Nat Cttee for Man and the Biosphere—MAB 1980–81, Tech Adviser 1981–86, Head of Research 1986–89; environmental consultant 1990–. *Present Position*: Sec-Gen, Int Asscn for the Protection of the Environment in Africa—ENVIRO-PROTECT 1990–. *Publications*: Dégradation de la Végétation et des Sols au Cameroun 1979; Les Problèmes de l'Environnement au Cameroun 1990. *Address*: International Association for the Protection of the Environment in Africa—ENVIRO-PROTECT, BP 13623, Yaoundé, Cameroon. *Tel*: 21-46-09. *Fax*: 22-18-73.

YOUNES, Talal, D SC; French biologist; b 10 Jan 1944, Lebanon; m Jeanine Gebrane 1984; 3s. *Education*: Claude Bernard Univ, Lyon. *Career*: Prof, Faculty of Sciences, Lebanese Univ, Beirut 1970–75; Sr Consultant, Science and Tech Educ, UNESCO 1976–79. *Present Position*: Exec Dir, Int Union of Biological Sciences 1980–; Ed, *Biology International* 1980–. *Publications*: Environmental Education in the Arab Region 1978; The State of Biology in the Arab States 1984; The Significance of Species Diversity in the Tropical Ecosystem 1984; New Challenges for Biological Education 1986; Biology Nomenclature Today 1986; Ecosystem Function of Biodiversity: Biodiversity, Science and Development 1996; Biological and Cultural Diversities 1999. *Addresses*: 51 blvd de Montmorency, 75016 Paris, France; 5–7 allée Henri Matisse, 92130 Issy-Les-Moulineaux, France. *Tel*: 1-45-25-00-09. *Fax*: 1-44-25-20-29. *E-mail*: tyounes@iubs.org. *Internet*: www.iubs.org.

Z

ZABELINA, Dr Natalia M, PH D; Russian geographer and ecologist; b 9 March 1937, Moscow; d of Mihail Idelson and Tatiana Dementieva; m Vyacheslav Zabelin; 1s. *Education*: Moscow State Univ. *Career*: Scientist, All-Russia Inst for Nature Protection 1961–70, Chief, Scientific Information Dept 1970–80, Sr Scientist 1980–90. *Present Position*: Chief Scientist, Reserves Dept, All-Russia Inst for Nature Protection. *Publications*: more than 100 scientific and review papers on problems of biodiversity and protected areas, national parks planning and design, nature reserves systems, ecological education, marine reserves and so forth, in particular: National Park "Mysl"1987; Zapovedniks and National Parks of Russia "Logata" (Co-Author) 1998. *Addresses*: 113628 Moscow, Znamenskoye Sadki, Russian Federation; 117208 Moscow, Sumskoy Proezd St 21-1-19. *Tel*: (095) 423-19-55. *Fax*: (095) 423-23-22. *E-mail*: nzabelina@mail.ru.

ZACH, Peter, C SC; Slovak forest entomologist; b 15 Sept 1962, Zvolen; s of Dr Oto Zach and Veronika Zachová; m Gabriela Zachová 1990; 2s. *Education*: Zvolen Tech Univ. *Career*: Ornithologist and waterfowl conservationist 1980–85. *Present Position*: Forest Entomologist, Inst of Forest Ecology, Slovak Acad of Sciences 1986–, Scientific Sec 1994–. *Address*: Institute of Forest Ecology, Slovak Academy of Sciences, Štúrova 2, 960 53 Zvolen, Slovakia. *Tel*: (855) 223-12. *Fax*: (855) 274-85.

ŽALAKEVICIUS, Dr Mecislovas, PH D; Lithuanian zoologist and ornithologist; b 19 May 1949, Dotnuva; s of Mecys and Ona Žalakevicius; m Irena Žalakeviciene 1972; 1s 1d. *Education*: Vilnius Univ. *Career*: staff mem, Inst of Ecology, Lithuanian Acad of Sciences 1972–, Head, Lab of Avian Ecology 1984, Head, Dept of Terrestrial Ecosystems 1992; corresponding member of Lithuanian Academy of Science (2000); Lithuanian Co-ordinator of BSCE; President of Lithuanian Ornithological Union; member of Scientific Advisory Council of the Ministry of Environment; Task Force of National Environmental Strategy (PHARE, 1994–85); Task Force of National Climate Change Strategy (CC: TRAIN-UNITAR, 1985–86); Task Force in prepartaion of National Concept of Envionmental Education, Public Awareness and Information; UNDP Lithuanian office adviser. *Present Position*: Deputy Dir, Inst of Ecology, Lithuanian Acad of Sciences 1992–; Dir, Inst of Ecology, Vilnius Univ, Head of the Lab and Dept. *Awards and Honours*: Lithuanian Science Award. *Publications*: Bird Migration and Wintering in Lithuania 1995; Nocturnal migration of thrushes in Eastern Baltic region, 2002; Global climate change impact on birds; Birds of Western Palearctic, 1997. *Address*: Institute of Ecology, Lithuanian Academy of Sciences, Akademijos str 2, Vilnius 2600, Lithuania. *Tel*: (2) 729-257. *Fax*: (2) 729-257. *E-mail*: mza@ekoi.lt. *Internet*: www.ekoi.lt.

ZÁVODSKÝ, Dušan, PH D; Slovak physicist and environmentalist; b 8 June 1941, Bratislava; s of Andrej and Emma Závodský; m Eva Závodský 1965; 1s 1d. *Education*: Comenius Univ of Bratislava. *Career*: Asst Prof, Nitra Agric Univ 1963–64; Head, Deputy Dir then Dir, Radioactivity Lab, Air Pollution Dept, Slovak Hydrometeorological Inst, Bratislava 1964–; Lecturer in Air Pollution Meteorology, Comenius Univ of Bratislava 1972–; Pres, Slovak Meteorological Soc 1992–. *Present Position*: Sr Scientist, Slovak Hydrometeorological Inst 1993–. *Publications*: Air Pollution Analysis (Co-Author) 1981; Monitoring Trans-Boundary Movement of Air Pollutants (Co-Author) 1987; Atmospheric Ozone 1992. *Address*: Slovak Hydrometeorological Institute, Jeséniova 17, 833 15 Bratislava, Slovakia. *Tel*: (7) 378-5377. *Fax*: (7) 375-670.

ZHANG SHEN, Prof, PH D; Chinese environmental scientist; b 24 Oct 1933, Jiangsu Prov; m Kuang Tingyun 1962; 1s. *Education*: Nanjing Inst of Agric, Moscow Nat Univ, Russian Fed. *Career*: mem, Chinese Nat Cttee, Scientific Cttee on Problems of the Environment—SCOPE, ICSU 1988–, Int Geosphere–Biosphere Programme—IGBP 1988–, Collegium Ramazzini 1990–, European Acad of Arts and Sciences, Chinese Acad of Sciences. *Present Position*: Professor of Environmental Geochemistry and Biogeochemistry 1985– and Vice-Pres, Environment Science Council 1991– and

Chinese Soc of Environmental Sciences 1996–. *Address*: Bldg 917, Datun Rd, Beijing 100101, People's Republic of China. *Tel*: (10) 64916364. *Fax*: (10) 64911844. *E-mail*: szhang@dls.iog.ac.cn.

ZHANG YAOMIN, Prof, BS; Chinese environmental biologist; b 1 April 1938, Henan Prov; s of Zhang Zhigui and Yangxian Fan; m Huansi Kang 1957; 1s 2d. *Education*: Nankai Univ, Tianjin. *Career*: Researcher on agro-environmental quality, pollution ecology, greenhouse gas emissions 1979–; Del, FAO Conf on Agric and Environment, Netherlands 1991. *Present Position*: Deputy Ed-in-Chief, *Agro-Environmental Protection* 1995– and Dir, Key Lab, Ministry of Agric 1996–. *Awards and Honours*: Scientific and Tech Advancement, Environmental Protection Bureau of China 1989; Excellent Achievement Award, Ministry of Agric 1989; Scientific and Tech Advancement, Ministry of Agric 1994. *Publications*: Agro-Environmental Quality of China 1983; Acid Rain 1989; China: Greenhouse Gas Control in the Agricultural Sector 1994. *Address*: Agro-Environment Protection Institute—AEPI, 31 Fukang Rd, Tianjin 300191, People's Republic of China. *Tel*: (22) 23361247. *Fax*: (22) 23369542. *E-mail*: aepi@v7610.tisti.ac.cn.

ZHANG YOUFU; Chinese ecologist; b 21 May 1940, Jiangsu prov; s of Zhang Changsheng and Zhong Yinlan; m Xu Chunlan 1968; 2s. *Education*: Nanking Univ, Chengdu Inst of Geography, Chinese Acad of Sciences. *Career*: engaged in long-term research work on biological prevention of mudflow damage, soil and water conservation and forest ecology. *Publications*: An Observational Study on Mudflow at Jiangjia Gully, Three Papers 1982; A Brief Introduction of the Programme Against Mudflow By Means of Biology 1983; Mudflow and Forest Vegetation in Jiangjia Gully Basin 1987; Mudflow and its Comprehensive Control 1989; Process of the Erosion, Transportation and Deposition of Debris Flow, Application of Information from Remote Sensing for Preventing Disasters caused by Debris Flow 1990; Application of Information from Remote Sensing for Preventing Disasters caused by Landslides 1990; papers on measures against debris flow and on remote sensing information. *Address*: POB 417, Chengdu 610041, Sichuan, People's Republic of China. *Tel*: (28) 5581260. *Fax*: (28) 5552258.

ZHOU KAIYA, Prof, BS; Chinese biologist; b 8 Oct 1932, Fenghua, Zhejiang Prov; s of Zhou Jingting and Wang Cuilian; m Xu Luoshan 1959; 1s 1d. *Education*: Jiangsu Normal Coll, Suzhou. *Career*: Teaching Asst, Jiangsu Coll 1953–55, Nanjing Coll 1955–60, Lecturer 1960–78, Assoc Prof and Dir, Cetacean Research Lab 1973–83; Dean, Biology Dept, Nanjing Univ 1983–90. *Present Position*: Prof, Nanjing Univ 1983–. *Awards and Honours*: Certificate of Merit, IUCN Marine Turtle Specialist Group 1989; First Prize, Nat Educ Cttee Science and Tech Advancement 1996. *Publications*: On the Conservation of the Baiji 1982; A Project to Translocate the Baiji, Lipotes vexillifer, from the Mainstream of the Yangtze River to Tongling Baiji Semi-Nature Reserve 1986; The Baiji and its Preservation 1989; Baiji: The Yangtze River Dolphin and Other Endangered Animals 1991. *Address*: Nanjing Normal University, 122 Ninghai Rd, Nanjing 210097, People's Republic of China. *Tel and fax*: (25) 3598328. *E-mail*: kyzhounj@jlonline.com.

ZHVANIA, Zurab B; Georgian biologist, politician and environmentalist; b 9 Dec 1963, Tbilisi; 2c. *Education*: Tbilisi State Univ. *Career*: Jr Researcher, Chair of Physiology of Man and Animals, Tbilisi State Univ 1985–92; Chair, Cen Exec Bd, Green Party of Georgia 1988–93, Speaker 1990–93; Chair, European Union of Green Parties 1992–93. *Present Position*: Co-Founder and Sec-Gen, Union of Citizens of Georgia 1993–, mem, Sakartvelos Parlamenti (parl) 1992–, Leader, Green Party of Georgia 1993– and Chair, parl Green Faction 1995–. *Address*: Green Party of Georgia, c/o Sakartvelos Mtsvaneta Modzraoba, 380012 Tbilisi, Davit Aghmashenebeli Ave 182, Green House, Mushthaid Park, Georgia.

ZOUHA, Sekkou; Moroccan forestry engineer; b 1947, Aït Yacoub; m Dich Fadma 1970; 2s 4d. *Education*: Royal Tech Forestry Inst, Salé. *Career*: Desertification control, Draa Valley, Zagora and Ziz Valley, Errachidia 1979–. *Present Position*: Dist Chief, Div of Water and Forests, Errachidia 1987–. *Awards and Honours*: UNEP Global 500 1987. *Publications*: Manuel de Lutte contre l'Ensablement Continental 1993. *Address*: Division of Water and Forests, Errachidia, Morocco. *Tel*: (5) 571543. *Fax*: (5) 572147.

ZWAHLEN, Dr Robert; Swiss environmental consultant; b 19 May 1947, Saanen; m; 1s 1d. *Education*: Univ of Berne. *Career*: environmental consultant 1977–, specializing in water resources devt projects and environmental impact assessments (EIA); lecturer on EIA, Swiss Inst of Tech, Lausanne. *Present Position*: Sr Environment and Social Devt Specialist, Electrowatt Eng Ltd. *Address*: Elektrowatt Ingenieurunternehmung, Hardtumstr 161, POB, 8037 Zürich, Switzerland. *Tel*: (76) 3562113. *Fax*: (1) 3555512. *E-mail*: robert.zwahlen@ewe.ch.

ZYLICZ, Tomasz, PH D; Polish environmental economist; b 22 Sept 1951, Lublin; s of Marek Zylicz and Teresa Zmigrodzka; m Barbara Stankiewicz 1976; 2d. *Education*: Univ of Warsaw, Univ of Wisconsin—Madison, USA. *Career*: Teacher, Econs Dept and Mathematics Dept, Univ of Warsaw 1974–, Assoc Prof 1989–96; environmental researcher and project co-ordinator 1981–; Researcher, Environment and Behaviour Project, Colorado Univ, USA 1988–89; Head, Econs Dept, Ministry of Environment 1989–91, Adviser 1991–; mem, Scientific, Man and Editorial Bd, *Ecological Economics* 1989– and *Environmental and Resource Economics,* and Beijer Int Inst of Ecological Ergonomics, Royal Swedish Acad of Sciences, Sweden 1991–. *Present Position*: Chair, Microeconomics, Univ of Warsaw 1996–. *Awards and Honours*: Pew Scholar in Conservation and Environment, USA 1993. *Publications*: Principles of Environmental and Resource Economics 1995. *Address*: 00-241 Warsaw, 44 ul Dluga, Poland. *Tel*: (22) 8313201. *Fax*: (22) 8312846. *E-mail*: tzylicz@wne.uw.edu.pl.

INDEX
of
ENVIRONMENTAL ORGANIZATIONS
by
FIELD OF ACTIVITY

Index of Environmental Organizations by Field of Activity

INDEX OF ENVIRONMENTAL ORGANIZATION BY FIELD OF ACTIVITY

AIR QUALITY

CLIMATE CHANGE

CONSERVATION OF PLANT AND ANIMAL SPECIES

PROTECTED AREAS AND NATURAL HABITATS

SOILS AND DESERTIFICATION

TRANSPORT

WASTE MANAGEMENT

WATER RESOURCES AND WATER QUALITY

Contents

Home Scar

THE SEA WAS WHAT his father called a cowshitty sea – a brownish, algae green, that meant it would be good fishing, even though it sounded like it would be bad fishing. But when he said something was bullshit, like the landlord raising the rent, or not fixing the oven, or mentioning that he might put the flat up for sale, then that was definitely something bad. Except when he was in the pub, in a group, and then it could be said and the laughter would be low and raucous as seagulls. To Ivor, it was all in the same murky category as words like restive – Ivor is a very restive boy, his teacher would say into the phone, is everything alright? Apparently that didn't mean that he was calm and easy.

The beach had been scraped and dragged by the winter storms. It was almost March now and where there had been sand there were stones, and where there had been stones there were channels that kept their water long after the tide had gone back out.

Crystal and Gull Gilbert were throwing stones at a limpet on a rock. The rock was covered in a rind of barnacles and there were anemones deep in the cracks; dark red and glistening like sweets.

Crystal picked up a handful of stones and threw them. One of them hit the limpet but it didn't move. She went up and pressed her hand against it. The limpet grated a few millimetres across the rock. 'That one up there looks empty,' she said. She was pushing the limpet, but staring at a house on the cliff.

'Let's do something else,' Ivor said. The week billowed and sagged around them, like a tent that might stay up, or might at any moment collapse. It was a school holiday. They'd already wrecked Crystal's TV and been forced out of Gull Gilbert's house by his brother, who had a girl hidden in his sour, dim bedroom. Ivor had seen her feet sticking out from under the bed.

He put his hand in one of the pools. Sea gooseberries rolled in the wind, scattering like a smashed chandelier. The ripples in the pool were dark and bright. Crystal's hair was the same dark, dry colour as charcoal – you could rub your hand over it and get an electric shock. Sometimes it got tangled and clumps had to be cut off with scissors. She was the biggest person in their class, bigger even than Gull Gilbert, and could put a safety pin through the skin on her elbow. Last year she'd pushed over a teacher.

'We've been in there already,' Gull Gilbert said. He picked up a stone with two hands and swung it through the air. There were blotchy freckles on his wrists and neck. He never wore a coat. He picked up another stone. He was frowning like he always did when he was concentrating. He would throw for hours until he hit his target in exactly the way he wanted. When he was concentrating, you knew exactly what he was doing. When he wasn't, anything could happen.

'We haven't,' Crystal said.

'Let's go back into town,' Ivor said. There was an indent in the rock, shallow and easy to miss at first, where the limpet had been before it moved. It was exactly the same size as the limpet's shell and it had the same rough curves, the same fluted edge.

'I want to go in that one.' Crystal pushed her foot in the sand and turned a fast, lumbering pirouette.

Gull Gilbert put his stone down slowly.

Ivor closed his eyes and leaned into the wind. If he did it right, it was like falling without ever hitting the ground. The cold found its way through his jumper, puckering the folds of skin. Goosebumps, goose barnacles, sea gooseberries. There weren't as many geese as there used to be, his father said. He had mended Ivor's jumper with lumpy stitches.

They'd already tried most of the other empty houses. There was the white one with the blue door, which had a porcelain doll on the windowsill that stared at them with its cracked face. The ceilings were streaked with yellow and the whole place smelled like a stale tin of biscuits. They would prise up slates and scratch their names underneath, pick at the bare walls until the plaster flaked off like confetti, and lie on the stiff, damp beds. But this time all the lights were on and someone was standing in the kitchen.

The big house with the red roof had people in it as well – there were bags and suitcases by the door, and the sound of voices and laughing. Sometimes they would go in there, sit on the leather chairs and read the tourist leaflets, then open all the cupboards to see if anyone had forgotten to pack anything – they'd found watches, cigarettes, a silky dressing gown that they'd taken turns wearing. Once, the oven had been left on by accident and Gull Gilbert had turned it off, then, after a moment, turned it back on again.

And then, just as they were testing the door of the stone house, the cleaning man had driven up and shouted something. Crystal had run first, then Gull Gilbert, Ivor struggling behind, his armpits streaming, that shivery almost-laughing feeling in his throat and bladder.

The wind dropped and Ivor tipped forward. The sand creaked under his knees like polystyrene. He opened his eyes. The others had already gone. Their footprints crossed the beach, sloppy as leftover cereal. Water rose up and filled each print, stretching them until they disappeared.

They were circling the house when he got up there. Gull Gilbert was trying the front windows, which faced out to sea and were rimed with salt.

Ivor lifted the doormat and looked under. There was nothing there. Sometimes the keys would be in locked boxes on the wall and you had to know the right combination to get in. There were a thousand different combinations, maybe a million. You could try all day and never get it right. Sometimes people left back doors open. Sometimes you could slide old windows down. Or, if you watched long enough, you might see someone hiding a spare key – behind flowerpots, underneath paving slabs, pushed into the thumb of a glove.

He let the doormat drop back down. What they should do was go into town and get some of those coins out of the wishing well. Then they could sit in a café and order drinks and talk about things, even though he couldn't imagine what drinks they would order, or what things they would talk about.

There was a grinding sound and Gull Gilbert swore. The grinding got louder, then, suddenly, Crystal was standing inside the house.

Ivor went round the back. One of the windows was open – the bottom pane had been forced up and there were splinters of paint and wood on the ground. A saucepan crashed onto the floor somewhere, something kept clicking, there was a drift of gas, then nothing.

He climbed in. The house was cold in that deep, quiet way that meant no one had been inside it for a long time. The window went into the bathroom, then the bathroom went out to a narrow hallway lined with pictures. The pictures showed the same three faces over and over – a man and a woman and a boy who was sometimes a baby, sometimes older.

Gull Gilbert was methodically checking each room. 'There's a load of crud in here,' he said. 'Shoes and shit.' He disappeared into the bathroom, then came back out. 'Where's all those small, wrapped-up soaps?'

'It's not one of those places,' Ivor told him.

'What?'

'Where different people come every week. It's not one of those places,' Ivor said. He kept looking at the pictures on the wall. The family were eating together round the table, they were walking outside on the cliffs, they were sitting on a rug on the beach.

'There's weird food they've left behind back here,' Crystal shouted.

Gull Gilbert jerked round, almost skidded. 'I'll pay you to eat it.'

'How much?'

Crystal was in the kitchen. The fridge was open and there was a pot of something on the table that smelled bitter and plasticky, like dentists' gloves. She was chewing on a strand of her hair. Whenever she got it cut, there would be a smooth, pale strip of skin on the back of her neck. In the lunch queue, she'd slipped her hand up Ivor's sleeve and held her palm against his shoulder blade. But that skin had been rough and almost scorching.

'How much?' Crystal said again.

Gull Gilbert went over and examined the food. 'Twelve,' he told her.

'When?'

'Tomorrow.'

'You won't have it tomorrow.'

Gull Gilbert was pacing the room with long strides. 'I'll pay you every week. Over the summer.'

Crystal reached out towards the pot, then stopped. 'I won't be here in the summer.'

The sea sounded like gunshots through the house. 'Again?' Ivor said, too loudly.

The last time Crystal had gone away, it had been to Cyprus, and it had been for a whole year. Before that, it was somewhere he'd forgotten the name of, for six months. One day she was here, the next she wasn't. It was because her parents worked at the Dishes. They had to go to places where there were other Dishes. She'd been born on an island called Ascension, which meant going up in the air and not coming back down.

'How long for?' he said, quieter this time.

'I'll pay you two pounds right now,' Gull Gilbert said.

Crystal's hand went back towards the pot. 'Three,' she said.

Ivor took a step backwards then turned and kept walking until he was out of the room and in the hallway. Something moved in the glass of one of the pictures and he glanced round quickly, then realised it was just himself. He went into the front room and stood by the window. The sea was rock-coloured and surging. There was the familiar feeling in his chest – tight and untethered at the same time, like a straining balloon. They said it was his asthma and gave him an inhaler to use. But asthma was what happened when you'd been running or fighting, it wasn't what happened when you were standing still.

Along the window there were yoghurt pots crammed with sand and shells and bits of smooth blue glass. The bits of glass were so small; it would have taken a long time to find them all.

Gull Gilbert might leave too. Then he wouldn't have to move up to the big school next year like everyone else. Any time he felt like it, he could wave goodbye to his grandparents and go and live with his father, who worked in a town with shops so huge you could walk around in them all day, and eat in them, and stay in them until it was night.

Ivor reached out and gently knocked each pot over, until the sand and the stones and the glass spilled down the wall and onto the floor.

He didn't know how long he'd been standing there before he saw Crystal and Gull Gilbert outside. They were running towards the path, shouting back at him that it was boring, it was a bit bollocks. They were going back into town. What they wanted was helium, cheap biscuits from the out-of-date food shop, that sticky hairspray that smelled like the bottle of drink they'd found washed up on the beach last summer.

His father was kneeling in the grass by the front door. The road out the front was full of parked cars, cats sleeping against wheels, a skip loaded with rubble and cracked sinks and flowerpots. At the bottom of the street there was a wedge of sea, strung between the houses like a wrinkled sheet.

Ivor opened the gate and his father looked up and then back down again. There was a bike strewn in pieces around him –

handlebars, wheels, a seat with a split like the skin of a tomato. His father picked up the chain and held it for a moment. There was a bottle of oil on the grass and oil on his hands.

Ivor pulled at an oily dandelion. 'Why are you doing that?' he said.

'Fixing it.' When his father was kneeling like that, the top of his head showed through his hair and there were bright blue veins, so thin they looked like they might break, behind his ears. But when he looked up there was the same face as ever – creased eyes from squinting into the sun, cheeks that scraped when they touched against Ivor's, the bent top tooth like a door off its hinges. There was a hole in his eyebrow with a ring in it, which he'd got when he was sixteen, just before Ivor was born.

'Why are you?' Ivor said.

His father ran his hand down the back of his neck. 'Dean's brother asked if I could. He's paying me.'

'Can you?'

'There's loads of bikes when you think about it,' his father said. 'Think how many bikes there are that need fixing.'

Ivor ripped at the dandelion. 'Can you?'

'Almost everyone has a bike. They always need fixing, don't they.'

The window in the flat above them opened and TV and laughing came out. A seagull lifted itself off the roof and circled the chimney, barking sharply.

Ivor leaned against the wall until the pebbles dug into his spine. His father was turning the bike wheel with his finger.

Ivor took his inhaler out of his bag and puffed it. He moved his arm into the shape of a gun and aimed at the seagull, bang bang. He would never hurt a seagull. Bang. If his father could fix the bike then there would be a lot more bikes he could fix, almost everyone had a bike. But if he stopped turning the wheel, got up and went inside without saying anything, then he couldn't fix the bike.

Then it would be like that time the hotel management changed and they could stick their longer shifts with no extra pay up their arses. And when the car park closed where his father gave out tickets and they played guess who would be fattest when they stepped out of their car. Or when everyone stopped coming on his walking tours because whenever he took people out onto the headland, where the cliff suddenly sloped and there was the beach for three miles and the rocks in horseshoes and waves galloping in and everything was silver, his father would just stand there shaking his head and say, fucking delectable, absolutely fucking delectable.

The bike wheel kept turning like it was a clock slowly being wound.

'Did you ring Mev yet?' Ivor said.

'Did I do what?'

'Did you ring Mev.'

Still the wheel kept turning, grating softly each time.

'She said she needed to know,' Ivor said.

'What?'

'About the restaurant. She said she needed to know.' Before Mev moved away, she used to stay over, and in the mornings

Ivor was allowed to get in their bed and keep sleeping. But that was last year, when he was a little kid.

'I know,' his father said. 'I told you that.'

'Why don't we?'

'What?'

'Go and live with Mev and work in her restaurant.'

The church bells near the beach tolled five times. 'That's a hundred miles away, Ivor.'

It was almost dark. If his father could fix the bike, there would be potatoes frying in oil and tomatoes sliced with sugar on them. 'So?' Ivor said.

And for dessert they would shake up cans of cream and spray them straight into their mouths.

The bike wheel went round and round.

His father got up, put the screwdriver down carefully on the grass and went into the house without saying a word.

Ivor pushed the window up until his wrists burned. The frame shuddered and jammed, then finally opened.

Below him, the cliff was slumped and worn, the rock underneath pale as a shinbone. Green waves crumbled onto the beach, then pulled back against the stones like a rasping intake of breath. A surfer drifted in the darker water.

He climbed inside, checked the window wouldn't fall shut behind him, then checked again. When he looked back out, the surfer had gone.

It was colder than before. The quiet was thick as dust. The floorboards creaked softly under his feet. That morning he'd put on his coat, found the shopping list and money his father had left next to the sink, and walked down the road into town. He'd got to the shop, picked up a basket, then put the basket down and kept walking until the road turned to the path along the cliffs, and then the house, and then the loose back window.

He moved slowly through each room, opening empty drawers and cupboards, running his fingers over a shelf of maps and books, a crackling bunch of dried flowers. There were patterned plates and glasses that looked like they'd hardly been used, and bowls that were too small for anything. There were leaflets heaped by the door and he picked some up, read something about window cleaning, something about gardening services, then he put them back down where he'd found them.

There were three pairs of sandals by the front door, three raincoats, three wetsuits folded over hangers. Ivor looked them over one by one. Nothing had sand on it, or mud, or crusts of salty rain. There was no torn and snapped umbrella, no piles of old newspaper, no takeaway pots flattened and ready for the outside bin. There were no tangled keys, no stacks of bills hidden behind the microwave. He looked under every bed but there were no cardboard boxes, reinforced with gaffer tape, waiting.

Nothing moved except Ivor. No clocks ticked. There were three yellow chairs round one of the windows and he sat in each one, then got up and watched the dents he'd made spring slowly back to smoothness. He opened and closed the curtains.

He turned on the lamp. His trainers left faint treads of sand. There were some clothes in the small bedroom – not many, just a few shirts and a jumper – and he unfolded each one, studied them carefully, then folded them back up, matching the creases exactly.

In the bathroom, he opened the cabinet above the sink and took out the bottles and jars. He opened the lids one by one and dipped his fingers into the creams, then scooped up talcum powder, leaving behind shallow indents and the half-moon shapes of his nails. He tipped up a bottle and white tablets fell onto his palm. When he tipped them back in, one tablet stuck to his skin. It was small and perfectly round. He thought about swallowing it, then shook his head and lifted his hand to drop it back in. But now that the thought had appeared, there was nothing else he could do. It was like locking and unlocking the door three times, or touching the wing mirrors of every red car.

His breath fogged up the mirror and he wiped it away with his sleeve, but it stayed on there for a long time after he'd left.

~

Every day his father would go fishing. His lines and nets were always by the door. He would leave early, depending on the tide, and there would be the sound of him in the kitchen, packing his kit, the thump of the car boot. He would hum that song he liked where the tune went so low it was as if his chest was vibrating.

When he came in to say goodbye he would put his hand on the top of Ivor's head and it would be warm and smell like bait. Ivor would pretend to be asleep. When he went downstairs, his breakfast would be on the table: milkshake, cereal that had soaked up everything, a plate of crackers to dip in. His father always said he'd only be gone a few hours, but he was never only gone for a few hours.

Ivor came down off the cliffs and glanced back once more in the direction of the house. There were bits of chipped paint on his hands from the window, and bits of talcum powder under his nails. He rubbed them off and crossed the beach towards the road. His father was down at the edge of the water. His silhouette was like a hawthorn bending. His line was arched over the sea and there were a couple of cans by his feet.

'Did you get the shopping?' his father said. Cold radiated off him, and he pulled the hood of his sweatshirt up against it.

Ivor stood as close as he could without knocking anything. The sky over the sea had turned dark yellow, like a very old piece of paper.

The line tensed and began to buckle, and his father gave his can to Ivor and put his hand on the reel.

'I forgot,' Ivor told him. He took his father's other hand and blew on each stiff knuckle.

His father played out the line. The bones in his fingers made popping noises under Ivor's mouth. 'Remember when your breath smelled like those onion crisps for a week?' his father said. 'I almost took you to the doctor.'

'Remember when you ate that whole sweetcorn and your beard smelled like butter?'

The line tensed some more, and it was important to watch it, and bring it in slowly. Now his father needed both hands.

The line went tighter and tighter, then slackened. His father took the can back and sipped it. 'I'll catch us something,' he said. He still held the record for catching the biggest fish in town.

The dark yellow turned to dark blue. A ship flashed on the horizon. Somewhere the oystercatchers whistled and scolded like boiling kettles.

'How about this then,' his father said.

Sometimes Ivor didn't think his father really even minded if he caught a fish or not, because then he could just stand out there all day, all night even, and sip his beer and listen to the sea, until the mist came in and rose up around his feet, and everyone else had gone home a long time ago, and their lights would be on along the streets, and their curtains would start to close, and cooking smells would come out, and it would just be him and Ivor left on the beach, waiting and watching the line.

~

Crystal ate chips like a seagull – she held one up in her mouth, then dropped it straight down her throat. She sat cross-legged by the swings, the beach sloping down in front of them. Ivor dug in the sandy grass with his fingers.

'We should be sitting on a rug,' he said.

‘A what?’

‘A rug. We should probably be sitting on one.’

The tide was just going out and the stones were still wet – they looked like they were splashed with blue paint. A dog ran up, soaked and quivering, holding a crushed barbecue as if it was a stick to throw. Behind them, Gull Gilbert swung standing up, the bent chains clanking.

‘Why?’ Crystal said.

Ivor dug his fingers in deeper. ‘I don’t know.’

Crystal held her chips against her chest until the dog went away. ‘You’d have to know you were going to sit on it, then carry it down especially.’

‘I suppose.’

‘How would you know?’

‘What?’

‘If you were definitely going to sit on it,’ Crystal said. Her weird lacy skirt was rucked and there was sand high up on her legs.

The swing behind them thumped as Gull Gilbert rode it like a bull at a rodeo.

‘I don’t know,’ Ivor said. His chest started to tighten. ‘Maybe you’re just supposed to know.’

Crystal ate another chip. Sometimes she would pass one to Ivor, sometimes she wouldn’t. This round he missed out.

‘They’ve probably got them at that house,’ he said.

‘What house?’

‘The one on the cliff.’ His fingers hit against a stone and he started digging around it, working it loose. ‘We could go there.’

'I'm not walking any more.'

'Tomorrow,' Ivor said. The stone was almost loose; he could nearly get his finger under it. 'All of us.' He thought about the lamps, the three yellow armchairs. He'd gone there again that morning and stood by the kitchen table in the strange, cool quiet, and thought something that wouldn't go away. 'We could stay there.'

Gull Gilbert jumped off the swing and staggered up behind them, his cheeks mottled almost purple. His tracksuit snapped like a flag in the wind. 'That dog's got itself a dead fish,' he said. He dipped his hand in the bag of chips, then skipped away from Crystal's fist. She was known for conjuring the blackest bruises. 'Stay where?' he said.

Ivor's heart raced under his coat. 'At that house.' A hot feeling pushed at the backs of his eyes. If anyone asked why, he didn't know what he would do.

Crystal finished eating, put her arms behind her head and lifted her hips until she was doing the bridge. 'Like, living?' she said. Her hair swung against the ground.

Gull Gilbert scanned the tideline, watching the dog's owner chasing it over the seaweed. 'Do you reckon that dog'll eat that fish?' he said. His eyes looked glassy and far away. Who knew what thoughts were teeming.

Ivor prised the stone out and clenched it in his muddy hand.

The dog started to eat the fish.

Gull Gilbert leaned forward, spat on his palm, said he was in, and shook on it, which was as binding as a triple-signed contract, amen.

~

When Ivor got home the light was on but his father's shoes weren't on the mat. That meant he was still wearing them, which meant he'd gone straight onto the kitchen sofa. Ivor went in quietly. His father was asleep under the scratchy blanket. Ivor had saved up for that blanket from the gift shop. It didn't seem right that people could sell a blanket that was scratchy, to tourists, or to anyone.

His father murmured something and his cheek twitched. There was a scar under there from when Ivor was three and had bit him. 'Is it right?' his father said. 'Is it right?' He sat up suddenly, opened his eyes and rubbed his hand over his face. 'Christ, Ivor, how long have you been standing there?'

He reached out and pulled him down onto the sofa. It was soft and dusty, and Ivor sneezed, then sneezed again.

The fridge hummed next to his ear. Ivor picked at the fraying cushion threads. 'Did you ring Mev yet?' he said.

His father moved the cushion away. 'You'll tear it.'

'Did you?' Ivor said again.

'These aren't our cushions. If you tear them I'll have to try and buy new ones exactly the same.'

The clock on the oven glowed red – you could see the shapes of all the other numbers behind the lit-up parts.

'Don't you want to be here?' his father said.

Ivor looked around. There was the kitchen, the dark outside the window. 'Here?' he said.

Once, in town, his father had passed someone he used to know from school. His father had recognised the man, Jody, straight away, but it had taken Jody a moment to come up with Ivor's father's name. Jody had been down visiting his parents and now he wanted to go – he kept looking towards his car and nodding in all the wrong places.

Ivor had pulled on his father's hand but his father had kept talking. About the state of the tides, what was biting, the blue shark, the development out the back of town. Remember that party out at the Jennings' place? he said. Remember the ambulance?

Ivor had pulled again at his father's hand, until his father let go. And still Jody kept glancing round and checking his watch, and nodding, until finally he said, I have to get back.

His father had run his hand down his neck and watched him walk away. 'Back,' he said. Then he'd shrugged and walked into the pub. A beer for him and a Coke for Ivor, and those chewy scratchings that were so tough and salty they made your teeth ache.

His father's eyes were closing again.

His phone started to ring in the front room. It rang and rang but he didn't get up to answer it.

'The warehouse might be hiring next week,' he said.

Over went the blanket with its smoky, ketchupy smells. Ivor leaned in and his teeth were against his father's cheek, and his father's hand came up and smoothed and smoothed, like he did with the fish he caught when they were thrashing and gleaming.

~

Ivor got to the house first. It was late afternoon and the sky was dark, the cliffs silhouetted like breaching whales. He'd told his father he was staying at Gull's and would be back in the morning. The town glinted in the distance, supermarket floodlights bright as haloes.

It was raining and he put his bags down and pushed at the window. It didn't move. He leaned forward and pushed harder. The frame was wet and heavy. It shook but didn't budge.

He ran round the side of the house, tried the other windows, then rattled the front door. The rain came down in sharp pieces. He looked towards the town, then back at the house. He shoved the door, then leaned all his weight against it. Something gave and he shoved again. A gap appeared and he forced it with his shoulder. The door jolted open. The wood around the lock was spongy and on his way in he pushed the screws of the metal plate until they nestled back in place.

When the others arrived he met them at the front. Crystal was carrying a rucksack. Gull Gilbert had brought nothing.

They stood inside, too close, Crystal's arm pressed against Ivor's. She smelled like apples and petrol and she was wearing lace-up boots that reached almost to her knees, and a pyjama jacket with clouds on it. Gull Gilbert had slicked down the sides of his hair.

Ivor's cheeks were hot. Everyone was just standing in the open doorway, waiting.

Gull Gilbert prodded Ivor's bags with his foot. 'What's in these?'

'Nothing,' Ivor said. It was just the food he'd brought. There was a packet of crackers, cheese he'd cut off a bigger piece, half a carton of orange juice, a tin of soup, eggs although he had no idea what to do with eggs. Also three cans of beer he'd found lurking at the back of the fridge.

When he'd packed it, he'd thought there was too much – he'd almost taken out the cheese – but now everything looked small and awful. Any moment now, Gull Gilbert's lip would twist and everything would crumble.

A handful of rain flung itself across the wall. Gull Gilbert reached out and closed the door. 'We should get all that in the fridge,' he said.

They went into the kitchen. Ivor put the eggs in the cupboard, then took them out and put them in the fridge. He thought the orange juice should go in the fridge door, the soup on a shelf. He spent a long time deciding, even though he knew he'd be getting it all out again in a minute.

Crystal went to the sink and ran the tap. She opened all the cupboards and looked inside, got out plates and slammed them down on the table. Then she picked them up and placed them gently. Then she pursed her lips, crossed her arms over her chest and pretended to smoke. Finally she slumped down over the table with her head in her hands. 'What are we supposed to do?' she said.

Gull Gilbert pulled out a chair, sat down, and got up again. The chair screeched against the floor and made everyone flinch.

He opened the fridge. 'We should have a drink,' he said.

'Now?' Ivor said.

'It's Friday, isn't it?'

The cans opened with a hiss. When Ivor drank, all he felt was very cold. He realised that the lights were off. He clicked them on and the kitchen turned orange. The room appeared in the black window, three faces staring back in.

He went out into the hall and looked at the pictures. In one of them, the table was laid with all the different types of cutlery, and the food was on mats in the middle. He went back into the kitchen and started laying out knives and forks and spoons, then he opened the soup and glooped it into a pan.

Gull Gilbert had one leg up on the table. His fingers drummed. 'We need to turn the lights off,' he said. He tipped his can and drained it to the dregs. His voice sounded huskier, as if his throat was very dry.

'I want the lights on,' Ivor said. He tipped his can up until the bubbles burned his throat. The taste was getting better, or maybe his mouth was going numb.

'Someone will see us,' Gull Gilbert said. He got up and clicked off the lights. The kitchen plunged into gloom. He sat back down and up went his leg. He stared at Crystal's beer. 'Are you going to finish that?' he said.

Ivor got up and drew the curtains, glanced at Gull Gilbert, then turned on the lamp in the corner. He took another long drink, then clicked the burner under the pan of soup. Nothing happened. He clicked it again.

'The gas is broken,' Crystal said. 'I tried it before.'

'Shitting frick,' Ivor said.

'You have to hit something when you say that,' Crystal told him. 'Then you have to go and lock yourself in the bathroom.'

Ivor drank some more beer, then spooned the soup into bowls. There were only a few cold spoonfuls in each one but he laid them out anyway, then the cheese. 'Someone else could help,' he said.

Gull Gilbert got up and brought over the crackers and spread them across a plate. He took out the eggs and looked at them, then put one down in front of each of them. 'I'm not hungry yet,' he said.

Ivor looked at his watch. 'I think this is the time we're supposed to eat.' He cut the cheese into slices and gave them out. The rain hit the windows with clinking sounds. 'We should have a conversation,' he said.

'Us?' Crystal said. She had taken more than her share of the crackers.

'Say something,' Ivor said.

Gull Gilbert was pushing his spoon around his bowl. 'Did you make this soup yourself, Ivor?'

Crystal snorted into her bowl. 'Why are you talking in that voice?'

Gull Gilbert's spoon clattered down. 'He said we had to have a conversation.' His leg wouldn't stop drumming.

Ivor poured out the orange juice, which looked too thick. He couldn't remember how long it had been open. 'Don't actually drink this,' he said as he passed it round.

Crystal took hers and started drinking.

'I said don't drink it.' Ivor tried to slow his breathing. There was sand everywhere. He should have got everyone to take their shoes off. He crouched down and started scooping it up into his hand.

'Where are we going to sleep?' Crystal said.

Gull Gilbert leaned back in his chair. 'Depends,' he said. 'Do you snore?'

'How would I know?'

Ivor crawled under the table. The sand was everywhere. The grains he'd already picked up kept scattering out of his hand. 'I think we should take our shoes off,' he said. He followed the gritty trail out into the middle of the kitchen.

Gull Gilbert was staring at Crystal. 'You'll have to sleep in my room.'

Crystal stared back, harder. 'Why?'

Gull Gilbert's eyes shifted away, he put his leg down from the table, got up and started pacing. He pointed to her beer. 'Are you going to finish that?'

'I already did.'

He reached out and shook it to check, then crushed the empty can in his fist.

Ivor tipped the sand in the bin then sat back down. No one had finished their soup. Gull Gilbert was circling the edges of the room, wall to wall to wall.

Ivor took a spoonful and raised it to his mouth, but he couldn't do it. He pushed his bowl away. His spoon had rust on the handle. His stomach made a thin, hollow noise. 'Soon we have to go and sit in the armchairs,' he said.

Crystal was moving her chair closer. Ivor sat very still. What he was probably meant to do was lean in to her and smell her hair, like his father used to do to Mev.

His breathing was so fast and shallow it was as if he couldn't catch up with it.

'You took too many crackers,' Ivor told her.

Crystal stopped moving for a moment, then tipped her chair back and swung on its spindly legs. She started humming something fast and looping.

Gull Gilbert turned on the TV. There was someone on there doing a magic trick with cards, but you could see where she'd tucked the spare ones in her pocket. He picked up the remote and changed the channel. A zebra was running through a wide river. He changed the channel again and there was a crowd of people. He flicked it again and again.

The room was cold and dark. The blue from the TV and the orange from the lamp cast a strange, underwater light. Crystal's chair was almost at the point where it would snap. Gull Gilbert was staring at the screen with unfocused eyes. His hair had sprung up slowly from under its layer of gel. He kept moving from channel to channel without stopping, one image blurred into the next; there was a voice, then music, then more voices. The zebra was still in the river, the crowd of people was getting bigger. The magician's hidden cards fell on the ground like leaves from a wilting plant.

Ivor pushed his plate off the table. It slid across the shiny wood and kept sliding, then seemed to pause for a moment before it hit the floor and shattered.

Crystal stopped tipping. Gull Gilbert blinked and looked around.

Ivor picked up his glass. It glinted in the TV's light. He held it out over the floor, then he dropped it.

Slowly, Gull Gilbert's elbow moved towards his plate. It teetered on the edge of the table, then broke with a hard clunking sound across his shoes.

Crystal picked up her plate, licked off the last crumbs, and dropped it. She got up and kicked her chair over behind her.

Then they all picked up their stupid eggs, raised them in the air, and smashed them into a million glorious pieces.

Ivor finally caught up with his own breath. His hand touched against Crystal's hand and he tried to make it mean that he would miss her when she wasn't there. Even though he didn't know if you could say that just with hands.

The sea paced with its heavy boots through the house. If you listened closely, you could tell how high the tide was, and what kind of waves were breaking. Ivor's father could walk out the front door and know that the waves were mushy, or that it was low tide and the waves were clean as a damn whistle.

Ivor picked up his can and rubbed the back of his neck. Later, but not now, he would clean up the house, and whoever came in next, whenever they came in next, would find, what? Not anything worth mentioning really: a scatter of crumbs, a few missing plates, a lamp that had been left on by mistake, sand in the floorboards, a smudge of breath on the bathroom mirror that could have been anyone's.

The Dishes

THE BABY WAS TEETERING on the edge of speech. Bru, she would say. Da Da Da. She had a way of looking at him as if she knew. Her forehead would furrow and her eyes would go dark as oil. Then he would pick her up and carouse around the room, giddy up, giddy up horsey, while the mist pressed against the windows from the sea, wet and dripping like bedding on a line.

They were there for three months. His wife, Lorna, had a temporary posting and they'd been given the use of a small, brick house in a terraced row. Theirs was on the end and it backed onto rough ground: tussocks, bracken, horned sheep sprayed blue and red, as if they were going into battle. Beyond that were fields, hedges tangled like wires, a few lonely farmhouses. The beaches were stony. The trees were not in leaf. In front of the house there was a road that hardly anyone drove along, then a barbed-wire fence with No Entry signs and cameras that pointed in all directions. Behind the fence were the dishes, where his wife went to work every morning and came back later and later into the evening. Sometimes she would have a shift in the middle of the night, and when Jay turned over in bed to hold her, she would be gone.

The dishes were on the edge of the cliff and could be seen for miles – hard white shapes that looked like a chess set waiting to be played. They were data gatherers, listening stations, bigger than the house and smooth and silent. Some were full spheres, some were hexagonal, others hollowed like the dip in an ear. At the centre of each tilted dish there was an antenna that reached upwards, and, sometimes, if Jay watched carefully, he would see them slowly turn, like a flower might, or someone following a voice that no one else could hear.

It was early morning and Lorna had already left. Jay was in the kitchen clearing away the breakfast things. It was cold outside. Rain blew across the road in thin lines. He turned the heating up higher.

The baby was strapped in her chair. He wiped her face with a warm cloth. Her skin was so soft, almost translucent, except for all the dried food stuck to it – it was on her cheeks and on the floor. Some was in her wispy hair. She laughed and squirmed while he wiped around her mouth, then puckered her lips and blew a bubble. Jay crouched down and tried to blow one too but it didn't work and he ended up drooling down one corner of his mouth. The baby laughed and blew another one.

'How are you doing that?' he said.

'Hamna fla,' the baby told him.

'Oh, OK,' Jay said. 'I thought you were doing it a different way.'

He picked up the plates and put them in the sink, then ran the hot water until the washing liquid foamed up. He plunged his hands in and his wrists went red.

'What do you want to do today?' he said.

The baby banged her hands against her tray.

'Do you want to go out anywhere?'

She banged again.

'Or we could play that xylophone game you seem to like so much.'

She kept banging.

'Bang your hands if you've got food in your hair.'

She kept banging.

'Bang your hands if you woke me up five times last night.'

She banged again.

'Bang your hands if you think I'm the best.'

She stopped banging.

Jay ran more hot water and swiped plate after plate with the cloth, until they were all stacked on the draining board. He liked washing up now – the hot water, the steam, how, when he rinsed out a tin of tomatoes, he pretended there'd been a shark attack. He liked the way the bubbles had bits of colour in them. He would blow them off his hands so that the baby could watch them floating. He hardly ever felt like smashing it all against the wall any more.

He dried his hands and lifted the baby out of the chair and onto her mat. There was an arched bar over it with bells hanging down. They made a dull, jangling noise when she grabbed at them. They sounded like a doorbell and he wished he'd packed her other mat – the one without any bells. They hadn't brought much from home – just a suitcase for him and Lorna and a few boxes of the baby's things. He liked it that this house was small and empty. He could walk around each room seeing nothing that reminded him; just a table, a couple of chairs, a sofa, a wilting pot plant on top of the fridge that he watered every day.

He sat down next to the baby, then got up again. If he sat down he would fall asleep. He had that heavy, dull feeling behind his eyes which pushed down towards his jaw. It had been five times last night; the night before he'd lost count after seven. He straightened the curtains, the chairs, then picked up the cloth and wiped at another weird stain on the floor.

'Was this you?' he said to the baby.

She looked at him, frowning, like it was inappropriate to even ask.

It wasn't even nine o' clock yet.

After a while he noticed the sound of low voices coming through the kitchen wall. He stopped wiping the floor. There it was again: a low murmur of voices.

The wall was thin and connected with next door, but he didn't think there was anyone living there. When they'd arrived there weren't any lights on, and there were no cars parked at the front. The curtains were half-drawn and there was a pile of rubble by the steps – bricks and plaster – that looked as if a room had recently been knocked through.

He couldn't hear what they were saying. He stayed kneeling on the floor. Water dripped off the cloth and pooled next to his leg. The voices rose and fell and then they stopped. The baby let out a cry and he turned to her quickly, thought he heard a door open and close somewhere. The baby cried out again and he picked her up and cupped her warm head with his wet hands.

The front door of the house next door opened then shut with a bang. Jay sat upright in the kitchen chair, where he'd been slumped over a cup of coffee, on the edge of sleep. It was mid-morning the next day. He glanced over at the window. There was a man crossing the road further up, heading towards the dishes. Jay glimpsed the back of his coat before he disappeared through the gates.

An hour later there were footsteps behind the wall, someone ran up the stairs and there was a strange rattling, which might have been curtains closing across their runners.

It was misty again, and too cold to go out. He brought the baby into the living room and turned on the electric fire. Soon the room was warm and fuggy and smelled like burned dust. He brought out a box of toys and emptied it onto the floor. He put the rattle and the fraying bear in front of the baby, then found the spinning top, spun it up, and let it go. It whirled and clinked out tinny music. He spun it up again.

When he got bored he styled the baby's hair into a Mohican.

At lunchtime, someone drove up near the house. The engine revved, idled for a moment, then finally stopped. Jay glanced out. There was a dark blue van parked by the side of the road, in the lay-by in front of the terrace.

He strapped the baby in her chair and put her food in a pan to warm up. 'Mashed peas and potato,' he told her. 'A classic choice.'

'Forofoo,' the baby said. She'd twisted her bib up into her mouth and she was chewing on it.

'It'll be ready in a minute,' Jay told her. 'I just want to make sure it's warm.'

He went over to the sink to wash his hands. He washed them twice, then scrubbed under his nails. He'd read something somewhere about how easy it was to contaminate a baby's food and since then he'd started washing his hands more and more every day. The skin around his nails was sore to the touch.

He dried his hands and filled the baby's bowl with food. He sat down next to her and blew on it to cool it down. 'I just heated this up, now we have to wait for it to cool down,' he said.

'Forofoo,' the baby said, trying to grab the spoon. She took a handful of food and aimed at her mouth, but most of it ran down her wrist and back into the bowl.

After a while the voices started up behind the wall. They were louder this time, closer, although he couldn't make out any actual words. One was deep, the other sounded like a woman's voice. There was a lot of low, drawn-out laughter.

Jay spooned the food into the baby's mouth. He wiped around her lips, then hooked his finger gently inside her cheek to make sure she wasn't storing any of it in there. She'd gone through a stage of doing that – he would find bits of food that she'd kept hidden all night.

She squirmed and sucked at his finger.

'I'm only checking,' he said. 'You have previous, remember?'

The voices came again through the wall. He got up and went over to the window. The van was still there. 'I'll be back in a second,' he said.

He went outside and knocked at next door. He waited, checking his hands for mashed-up peas. What would he say? He didn't know. All he wanted was to speak to someone and not have them say forofoo, or whatever the hell it was, back. But there was no sound from inside. Nothing moved. There were no lights on. Upstairs, the curtains were all drawn. Downstairs, there were net curtains that were frayed and yellowing. He would have to go right up and stare in to see past them. He turned round and looked at the road. The mist had almost covered the dishes. He could only see the one closest to the fence. The metal was dripping. The antenna was tilted towards the road. It almost looked like it was pointing at him. Was it pointing at him? He took a step towards it, then stopped and shook his head. It was pointing upwards, above the houses, like it always did.

He knocked once more, then turned and went back into his own house.

He sat down at the table, spooned up the last bit of the baby's food and put it in her mouth.

The voices started up again, and someone laughed.

He got up so quickly that his chair tipped over. He went back outside and stood there, looking around. There was no one. The van was still parked by the side of the road. It was dusty and there was sand on the tyres.

When he looked out again later, the van had gone.

At night, he watched his wife sleeping. She slept straight away, as soon as she'd checked the baby and got into bed. There were dark smudges under her eyes, as if soot had gathered in a fireplace.

Sometimes she murmured and rolled away from him to the other side of the bed. Sometimes she rolled onto his chest and buried her face in his ribs. She mumbled things he couldn't really hear. 'What?' he would ask her. 'What?' He smoothed back her hair and rubbed her shoulder blades to settle her back into sleep.

'What do you do over there all day?' he asked, but he knew she wasn't allowed to answer.

Often, the pillow would have creased the side of her cheek, and the creases would run into the fine lines that had started to gather around her eyes. When her nightdress rode up, there were lines across her stomach and the tops of her legs, the skin puckering like clay. He couldn't take his eyes off them.

Finally he would fall asleep, but after a few moments he would jolt awake and freeze, sure that he'd been muttering, talking. What had he been saying? What if Lorna had woken up and heard him saying something?

It was only once, it had only happened once. The doorbell had rung and he'd opened it and Lorna had been working, she was always working, and he'd been on his own for such a long time.

The baby had been in the other room. He'd put music on, and afterwards he'd checked and she was deep in sleep, her arms and legs flung outwards, her hand clutching her rabbit, and that warm, sour, milky smell clinging to her which reminded him of the corridors of school many years before; how he used to get lost in the twisting maze of them.

~

He pressed his ear closer to the kitchen wall. The van had arrived at midday, while Jay was changing the baby. There'd been no sound from next door all morning, and he'd started to think that the van was probably there to do repairs to one of the houses further along the row. Now and again, drilling and hammering would reverberate down the terrace like a heartbeat.

But then someone had run up the stairs. The banister had creaked. A door somewhere further back seemed to shut softly.

He turned away from the wall and back to the baby, who was tipping herself backwards in her chair, trying to get out.

She'd been restless all morning – crying whenever he went out of the room and throwing down toys, but if he picked her up she would go rigid and try to twist out of his arms. Her cheeks were hot and she kept scratching at her belly, and when he rubbed it for her, she just cried again. He offered up her favourite toys – the rabbit, the jangly ball – but she batted them away.

He looked around; saw only the road, the mist, the cliffs, the dishes.

He slumped down in a chair and rested his head on the table. It had not been possible, before, to know that this kind of tiredness existed. He could hardly even lift his head. When he did manage to look up, the baby had slumped down too, in her chair, and she was watching him with her head cocked sideways.

He sat up, then covered his eyes with his hands.

The baby did the same.

He waved his hands, and the baby waved her hands.

She watched him, without blinking, to see what he would do next.

Then someone said 'Ssshhhh' suddenly and loudly from behind the wall.

The baby opened her eyes wide. 'Ssshhh,' she said.

'Ssshhh,' the voice came again from behind the wall.

The baby looked around the room, then back at Jay. 'Ssshhhhh,' she said.

Jay shook his head. 'You don't need to do that,' he told her.

'Ssshhhh,' the baby said again.

Jay got up and went over to her. 'Don't do that.'

She looked at him with her wide, dark eyes.

The sound came again from the wall.

Jay went over and knocked on it, once, twice, loud and hard.

Above him, on the roof, a tile slipped and grated in the wind.

'Sshhhh,' the baby said, quieter this time.

~

There was a swing tied to a branch of a tree at the back of the house. It was small and sturdy, with high sides for a child. Jay had tested it, and tested again, pulling down with all his strength to see if anything gave.

He put the baby in her coat and opened the back door. The misty rain had finally stopped. It was good to feel the wind against his face.

He put the baby in the swing and pushed gently. The chains creaked as they moved against the tree. He pushed and pushed and it was cold and quiet and he thought of nothing except pushing the swing and the wet, salty smell of the fields behind him.

When he looked up at the house, there was someone standing in the window.

He fumbled with the swing, missed the middle of it, and ended up pushing the baby sideways. The swing lurched outwards, rocked, then righted itself.

Jay steadied the chains. It was just his wife, wearing her coat and carrying her bag ready to leave for work. He didn't know how long she'd been standing there; he thought she'd already gone. She was wearing the green scarf he'd bought for her just after they'd first met. He hadn't seen her wearing it for a long time. He raised his hand and waved. Lorna's mouth moved but he couldn't tell what she was saying.

He realised he'd been pushing the swing quite high, and probably harder than he should. The baby was laughing and kicking her legs with each push but now he slowed it down, keeping it low, feeling himself making a show of how careful he was being.

The baby screamed indignantly, but he kept pushing the swing very gently. The next time he looked up, the window was empty, except for the blurred reflection of the swing moving backwards and forwards slowly across the glass.

A phone rang next door. It rang, then cut out, then rang again. No one answered it.

Jay strapped the baby in the pram and pushed her hat further down over her head. She looked up at him and her face creased. Her eyes were exactly the same as Lorna's – sometimes it seemed like she was right there, staring out at him. When Lorna and the baby looked at each other, it was as if something secret passed between them, something that he wasn't allowed to know.

‘Ha fa ma?’ she asked. Her cheeks were already red in the cold.

‘We need to get out of the house,’ Jay told her.

‘Bada shlam.’

‘Yeah, I know. It’s bloody cold, but we need to get out of the house.’

He put another blanket over her. She stared out sternly from under all the layers. He tucked the blanket in, then started walking down the road. The pram’s wheels sent up spray from the wet tarmac. The road was steep and narrow, with high hedges on both sides. If a car came, there would be nowhere to go. They would have to turn and walk all the way back. But he needed to get out of the house. It had rained for three days in a row – heavy showers that didn’t stop. The gutters had spilled over and poured down the windows. They’d stayed in and turned the heaters up high. Small noises had come through the wall: murmurs, footsteps, low laughter. Sometimes he was sure it was just the pipes, or the rain.

There was a thin, raw mist, as if the ground couldn’t absorb any more water so the wetness had moved into the air itself. Soon his nose was numb and dripping and his fingers were stiff against the handle of the pram. The road sloped down and small trees twisted on either side, their trunks bright with moss.

It got colder the lower he went into the valley. He could hear the sea somewhere in the distance. Water ran down the road and splashed up his legs. It looked orange, like it was leaking through rusty iron.

The mist thickened into drizzle and he shivered. He crouched down and tucked the baby in tighter. She was making cooing sounds at the gorse, trying to reach out and grab it. He showed her the prickles but she grabbed at it anyway. There was gorse everywhere, like lamps in the hedges. It gave out a sweet, heavy smell.

The drizzle came in waves, sweeping across the tops of the trees, and hanging there like curtains. The road narrowed again. Something moved in the dead leaves under a tree. He walked slowly, checking every bend before carrying on. He came to the bottom of the road and it forked: one way turned into a track that followed a stream, the other seemed to bend inland. He took that one and kept going. There were no road signs, just hedges and fields and the valley below him: the trees huddled like a herd of animals escaping the weather.

'Sa?' the baby asked.

He stroked her damp cheek with his finger.

There was the sound of a motor in the distance, coming closer, and he walked forward to find a wider bit of road. Whatever it was, it was moving fast, the engine revving. He smelled the petrol before he saw it. There was no wider bit of road. He walked back quickly, away from the bend. He crammed the pram in sideways against the hedge, mounting the wheels up on the bank and pressing it in as far as it would go.

It was a dark blue van. It came careening round the corner of the lane and revved past him before he could see who was in it. The wing mirror brushed against him as it went.

Jay jumped out and shook his fist at the back of the van. 'You arsehole,' he shouted. 'You irresponsible son-of-a-bitch arsehole.'

He got the pram out of the hedge. The baby had a handful of dried leaves in each fist and was chewing on a stick. He took the stick out of her mouth and crouched down to check she was OK.

'Don't ever repeat what I just said,' he told her.

The baby looked at him, then back down at the leaves she was holding.

He stood in the middle of the road. No one else went past. He saw no one except a farmer, small and faint, walking through a field in the distance. The baby went to sleep. Her hand slackened and the leaves fell out. He turned and started walking back. Soon the dishes rose up in front of him. One of them was pointing down at the valley. It stayed like that all night.

His wife hummed low, monotonous tunes in the shower. She used to sing pop songs, ballads, those deep, soulful ones where she used the showerhead as a microphone, but now she just hummed the same thing over and over, quietly and without stopping, like static on an old radio.

While she was in the shower, music started up behind the wall. It was slow but with a heavy beat that thrummed through the floor. It was coming from somewhere near the kitchen, then it faded and seemed to move into the living room, then down the hall, as if it was in the pipes or the wires.

Jay's heart gave a strange lurch. He banged on the wall. 'Stop it,' he said. He banged again. 'Stop it.'

The music didn't stop. He followed it through the house. It was louder near the bathroom. When he went in, it sounded like it was in the room, low and slow and echoing off the tiles.

He could see Lorna through the steam. She was washing her hair and there was soap and bubbles all over her head. She was humming and her eyes were closed.

There was a thump near the door, and then the sound of breathing only a few inches from where Jay was standing. A cold draught came under the door. Any moment now Lorna would rinse off the soap and take her hands away from her ears and then she would hear.

The breathing got louder. The music surged. Lorna ducked her head under the water and shampoo ran down her neck and onto her shoulders.

He stood in the middle of the room, clenching his hands. His nails dug into his palms. He could tell, even behind the music, the particular way the body would be pressing against the wall.

Stop, he said silently. Stop it.

Lorna shook her wet hair and turned off the shower.

The music stopped.

She opened her eyes and when she saw Jay she let out a faint cry and put her hand on her chest, looking at him for a moment as if she didn't recognise him at all.

~

The phone rang from behind the wall. It rang and then it cut out, then it rang again. Still no one answered it.

~

It was lunchtime and Jay was cleaning up. The baby had woken him every few hours in the night and he kept knocking things onto the floor – cups, bits of food. The baby would lean down out of her chair and try to help him pick them up, then almost topple out, so he would straighten her, and then she would do it again, clapping her sticky hands.

Soon Lorna would be home and he would start cooking something for dinner.

He ran the sink full of hot water. It was cold in the house, his hands were cold and he was looking forward to dipping them in.

An engine revved suddenly and he looked up just in time to see the van speed away past the window. The tyres left a burning smell on the air.

He picked up a plate and put it in the sink. He washed it and stacked it on the draining board. Bubbles ran down and pooled in the grooves. He started on another plate.

A door slammed and someone shouted from behind the wall.

He fumbled with the plate, dropped it in the sink, and hot water splashed over his feet.

There was a bang, then voices. 'Why did you?' someone said. 'Why did you do it?' There was another bang, and a long silence.

Jay picked up the plate. It had cracked down the middle. He stroked the baby's cheeks. She seemed fine; she was pushing a bit of cracker around her tray, jabbing at it until it was wet and crumbly.

'Ham nu for,' she said, pointing to it.

'It's OK,' Jay told her. 'It's OK.'

He dried his hands, sat down, then got up and opened the door. He went outside and paced around the front of the houses. There were no cars; the house next door looked empty. In another house, further up the row, washing billowed on the line; trousers and shirts straining against their pegs as if they were trying to get away.

Something moved behind next door's window. Jay ran to the door and raised his hand to knock, his hand was in a fist, it was almost on the door, then he stopped and brought his hand down. He stood on the step for a long time.

The baby watched him. 'Wayha do int?' she said one morning. She looked at him carefully, as if she was waiting for an answer.

His wife got home late and they sat, almost asleep, on the sofa in front of the TV. Jay flicked through the channels – there were old programmes on that they used to watch, repeats that seemed half-familiar, the jokes coming in slightly different places than he remembered.

He put his arm round Lorna and she leaned her head back against him. He could see the freckle behind her ear. It was tiny, hardly more than a dot. He used to kiss her there.

She yawned and leaned in closer. Her hair was kinked from wearing headphones at work most of the day. Her eyes were dry and flecked with red.

The audience on the TV laughed raucously at something and he found the remote and turned it down.

He could hear her watch ticking. There was a phrase they used to say to each other when they'd first met – something about clocks or time, because she always used to be late, and he was about to say it to her, it used to make her laugh. But he couldn't remember it.

He'd seen her earlier on his phone and he'd grabbed it, almost yanked it out of her hands, but she was just checking a friend's number. His hand had been shaking and he'd gone upstairs so that she wouldn't notice.

He turned the volume up on the TV again and Lorna sighed and shifted her head so that it was against the cushion instead of his chest, and her hips moved, just slightly, away from his. His hand started to shake again, but it was nothing, he'd deleted everything, there had been no more phone calls. Any moment now she would turn back and lean against him again.

He was putting away the washing up – the cups and plates and glasses – in the cupboards and drawers. Everything was clean. Dinner was cooking. He was ahead for once. He lined the cups up carefully, and stacked the plates on top of each other. The glasses caught the light and gleamed.

A glass fell and smashed against the floor.

He reached up automatically to the shelf to stop any more falling but nothing had fallen, there was no broken glass anywhere.

There was another loud smash from behind the wall.

He put his hands over his ears and waited for it to stop.

The dishes were moving. If he hadn't been watching them every day, he might not have noticed, but he did watch them every day, and he saw them move. Soon they would be pointing straight in at the kitchen window.

The van was there again. He hadn't heard it drive up, or any doors opening and closing. But it was there. Jay watched it out of the window. He checked on the baby. He went back to the window, waited a moment, then went outside. He walked over to the van and looked in. There was an empty plastic bottle under the seat, and a newspaper on the dashboard from a week earlier.

He circled the van twice in the drizzle, then thought about the number plate. What he should do was write down the number plate. He ran inside and found a pen, then crouched down next to the van to write. The number plate was covered in mud and he rubbed at it, saw an X and a 7, then rubbed again but the mud was too thick and wouldn't come off.

When he looked up, there was a light in one of next door's windows. It flicked on, then off. The curtains upstairs moved.

He walked over to the house. He glanced back at the road, then went closer, right up to the window. The rooms downstairs were dark. He pressed his ear against the glass but couldn't hear anything. Something moved further back in the house – maybe it was an arm, or someone's back, he just glimpsed something crossing into another room.

He ran round the side of his house, down the alley and through the long grass on the bank. He scrabbled over the brambles, dropped the pen, and scratched his hand on a broken bit of fence. There was a low wall behind the house next door. He jumped down softly. The back door was padlocked. The windows were shut and dark.

He stayed crouched against the concrete. The net curtains swayed against the glass.

Something rustled in the bank above him. The rustling got louder, and then a blackbird ran out towards him, scolding loudly.

He moved closer to the windows. They were smeared and dusty but he was sure there was something back there, in the darkness. He went closer. A voice murmured and someone laughed.

There was a shout behind him. He turned quickly. It was the farmer he'd seen in the field. She was walking towards him, calling out, asking what he was doing. He looked at her, then back at the window. He realised his hand was on the latch. His fingers were rigid and scratched, the nails bitten right down. It didn't look like his hand. He turned and ran, disappearing into his own house.

He jumped at small noises. When the baby broke her bowl he brushed up every single piece with the dustpan. He picked out the tiny shards from the cracks between the tiles.

It turned very cold. He stayed up late into the night with his ear to the kitchen wall, just the blue light from the fridge, and the white security lights coming in through the thin curtains. He paced the kitchen. When the baby cried he went straight to her and lifted her out of her cot and held her while he paced. She shaped her mouth into a sound and then gave up and blew a sticky bubble instead, and sighed.

'It's OK,' he told her. 'It's OK.'

Then he went back to the wall and listened. He pressed so hard that bits of paint flaked off onto the floor.

He left Lorna sleeping in bed and came downstairs and listened all night.

He heard the music again, faintly this time, somewhere towards the back of the house.

Another time there was a hushed, crying sound, like someone had left a tap slowly running.

~

'I no, I no,' the baby said. She opened her eyes wide. 'Sshhh,' she said.

~

The phone next door rang, cut out, then rang again. Jay stopped turning his phone on. He put it under a box in the wardrobe, then in a drawer. After a few days he took it out and threw it into the brambles behind the back window.

~

Someone was leaving. He heard it clearly and distinctly.

The baby looked at him, her head to one side. 'Wha?' she said. She frowned.

A very cold feeling washed over Jay – it went from his neck down to his feet, almost rooting him to the floor.

The voice came again through the wall. It was a man's voice, but not as deep as the one he usually heard. 'Going,' it said. 'The only thing to …' A cupboard opened, then drawers opened, and something heavy was dragged across the floor. A zip crunched.

Jay picked up the baby and held her to his chest. He stood by the front door. Footsteps thudded through the wall, more cupboards creaked open.

He put the baby in her coat, then went outside. He crossed the front yard. The van wasn't there. It was cold and the dishes seemed poised, tensed. They were pointing straight at him.

His breathing was fast and shallow. He held the baby tight and she pressed into his neck. 'Da?' she said.

There was no sound except for a rook cawing from a wet branch.

The house next door was in front of him. The door was half-open. Jay walked over to it slowly. He went up the step.

The rubble was still there. It was wet and bits of plaster had spread over the ground like snow. He pushed the door slowly and it swung inwards. It was quiet in there. There were no shoes by the door, no coats on the hooks. The hallway was long and dark. He turned and looked back towards where Lorna would be working. He imagined her at a desk, by a computer, listening.

He thought of how he would tell her.

He suddenly remembered the phrase they used to say to each other.

The phone rang. He held the baby tight. He took a deep breath and stepped inside.

Dreckly

Tide: 7.5 metres

What I'm about to tell you is the stupidest thing I've ever done. There was this one time when I bet Jory he couldn't swim out to that rock with all the seagull shit on it and of course he went right ahead and did it and I lost my entire savings which weren't really anything in the first place but still. I should have made sure the bet was for swimming out and back again because I had to call the coastguard for him on the way back, but there's no time to go into that now. And this other time I wanted to pierce my ears but they were charging way too much at the hairdresser's so I did it myself using an ice cube and a needle and a bottle of gin, and look how that turned out. I won't even tell you about that time with Leon – remember him? I can hardly even think about it. If I even just glimpse a gold tooth now I get this deep-down shudder – less like I've walked over my own grave than I've fallen right down into it. But now there's this, and this is probably worse.

It was me and it was Freya and it was Jory. We'd picked up the metal detectors and we were driving in my van to the beach. I was living in the van at that time, but that's another story. I had my shoes in the footwells, my toothbrush in the glove compartment, and dresses hanging across the windows for curtains. I'd strung up all these air-freshener things that looked like baubles, but it turned out all they did was make everything smell like a toilet at a festival.

Freya was sitting on the mattress in the back and she slid every time we cornered. Her dog, Mercury, kept whining and pawing at my pillow, right in the dent where my face went. Mercury's a greyhound that Freya found sleeping in a wheelbarrow somewhere. Freya can't ever leave anything if she finds it. Once there was a pigeon with a broken leg and after that no one saw her for weeks. She made a splint for it out of a cocktail stick. The thing about Mercury, though, is that she's really uptight. She'll stand completely still for hours, with just a single muscle quivering in her jaw, then suddenly, for no reason at all, she'll bolt and disappear. One of her eyes is milky blue, like a planet or something, but one of those planets that's completely screwed itself up and imploded.

'Hey look,' Freya said.

'I'm driving,' I told her.

'Hey look,' Freya said.

I turned around. She'd got a pair of my pants from God knows where and put them on Mercury's head. Mercury was walking backwards over the mattress and trying to shake them off. I didn't even know dogs could walk backwards.

'Get them off her you stupid bint,' I said.

Freya didn't move. She just sat there watching Mercury with that dopey expression she gets – as if the dog was her kid performing Shakespeare. The van bounced over a pothole and she had to duck her head to stop it hitting the roof. She's been six foot since we were about nine and then just got wider, like one of those giant redwoods. Whenever she goes in the supermarket someone always asks if she'll get something down for them from a high shelf – if she's in the right mood she does it, if she isn't she'll pass them something like absinthe or itch-relief cream and then walk away.

Jory finally turned around from the front seat and looked at what was going on. He's always about five minutes behind everyone else. I think it's because he's thinking about things but who can tell. He once found a body that had washed in and got caught amongst all these rock pools and it took him about an hour to realise it wasn't just someone sleeping. If you ask him why he thought someone was sleeping all twisted up in a rock pool he'll just shrug and say it was quiet down there. He likes quiet places where no one else goes. He took me out in his boat once and we just drifted for a long time.

I turned at the crossroads and the sea spread out in front of us. The beach down there is long and wide and packed every summer. We never go there in the summer. But now the holidays were over, and it was empty again, and we were doing what we always did after everyone had gone back home: scouring around with our metal detectors for whatever had been left behind. Sometimes we found money, jewellery, watches, and unopened cans of beer. But mostly we found belt buckles, keys, forks, tins smothered in barnacles. One time I found a bra and another time I found a crutch – I can understand forgetting a bra but I sometimes wonder how the person with the crutch got off the beach without noticing they'd left something important behind.

We borrow the metal detectors off Mr Warner. His son, Buddy, used to hang around with us before he moved away. Buddy had one of those BB guns and he once shot Jory in the leg with it. Jory won't tell anyone what it was about. Freya heard Buddy had made it out to Alaska and was working on boats – he and Jory had always wanted to work on boats – but I heard he'd married some crazy woman and had five kids in a caravan somewhere. Mr Warner told us he was working in insurance upcountry, which I guess is about in between.

The thing is, there used to be a whole load of us that came down to the beach and went around with the metal detectors. But, one by one, they all moved away and now we were the only ones left, doing the same old things over and over. We'd be twenty-six soon, then we'd be almost thirty. I didn't want to do it this year, and I sure as hell didn't want to be back doing it again next year – digging up tins and rusty coins, the three of us stuck up to our knees in the damp sand.

Tide: 7.0 metres

I parked next to the steps and we got out and looked down across the beach. There was no one around. There were brambles and sloe thickets all over the cliffs and Freya picked a blackberry, ate it, then spat it back out. She did the same thing every year – the blackberries up there are always sour, she just forgets every time.

The tide was high, a lot higher than usual, just like they'd told me at the pub the night before. I'd been in there with Jake and Lyn and Ricky and the talk had got around to how there was going to be this huge tide. It would come right up the beach, and then, when it dropped out, it would be so low that it would be possible to get round to the cove. Usually no one can get round to the cove but this time all the rocks leading over to it would be uncovered. And according to Lyn, who'd heard it from Morrie, who'd heard it from someone who'd seen it from their boat, all the sand and shingle had been scraped away by the spring storms.

'So you know what that means,' Lyn had said, leaning back on her stool until the legs looked like they were about to snap.

'Exactly,' Jake said, nodding slowly.

'Exactly,' Ricky said.

Then they all looked at me.

I knew what they were trying to do. Everyone knows about the cache that's supposed to be buried in the cove and how impossible it is to even get round there, let alone find it. They were trying to rile me up. They love getting me riled up. They'd done it before with the lottery ticket, and when the hospital was opened by that prince, and I wasn't going to give them the satisfaction this time.

So I just shrugged and didn't say a word. After a while I asked Ricky about his mother – she has this thing where she woke up one morning speaking French, even though she'd never spoken it before in her life – and then, when I thought I'd left it long enough, I finished my drink, stretched and got up. 'I guess I better head off,' I said.

'Early start in the morning then?' Lyn said, and they all practically stopped breathing from laughing, the bastards. That's the thing – people get certain ideas in their heads about you, and they never let you forget them. After a while, you find yourself doing exactly what they expect because mostly it's just easier.

I leaned into the van and got out the metal detectors and passed them round. Buddy's dad has a whole collection of them – most of them are old and a bit knackered; the kind with rusty coils rather than digital screens. He collects clocks too – they all chime on the hour across the house, slightly out of sync, and all these cuckoos and other weird crap jump out. I stayed over there once and I swear I still sometimes hear the ticking of each passing second.

'I've got the buggered one,' Freya said. She turned her detector over and examined it. 'Remember last year, there was one that kept whining? It just whined the whole time without stopping. I think I've got that one.'

'It's not doing it now,' Jory said.

'That's because I haven't turned it on yet dumb-ass.'

I locked up the van and put the key on a chain around my neck. I wasn't about to lose it like I did that other time, which wasn't even my fault, it was that bloody scarecrow in that bloody field, but that's another story.

'The tide's really high today,' I said.

'Turn it on then,' Jory said to Freya.

'It's going to go out really far,' I said.

'I'm trying to turn it on,' Freya said.

'We'd be able to get round to the cove.'

Freya kicked her metal detector and it made a screeching noise. She kicked it again and it stopped. 'Why would we want to do that?' she said.

I told them what everyone had been saying in the pub. 'Ricky was there,' I said. 'And Jake. They said Morrie said he'd heard it from someone on a boat.'

Freya broke off a bit of stick from the hedge and threw it for Mercury. Mercury didn't move. She pulled a long bit of ivy out and draped it around Mercury's neck.

'Cache,' she said. 'What the hell is a cache?'

'A hoard,' I said.

'A hoard?'

'Treasure for fuck's sake.'

Jory did up his rucksack and put it on his back. 'It's supposed to be drugs,' he said. 'From South America.' He always carries a rucksack. No one knows what he keeps in it.

'Ricky was being a total moron the other night,' Freya said. 'Did you see him doing that thing with the snooker cue?'

'Or gold coins,' Jory said. 'Bullion.'

'We just need to wait for the tide to drop,' I said.

'He's going to get kicked out soon,' Freya said. 'And then where will he go every night?' She opened a bottle and started walking down the steps to the beach.

'They let him back in before,' Jory said. He followed Freya down the steps. Mercury watched them, then suddenly ran, skidding down the loose stones.

I waited at the top for a minute, watching the tide. A few rocks had already started to appear. The water looked thick and creased, like oil. There was a container ship on the horizon. The sand was wide and empty. Sometimes, when I see the sand and rocks all bare like that, it looks like a building site: all brown and heaped up like it's going to become something else, but it never does, does it.

Tide: 5.8 metres

Freya and Jory started going over the beach from right to left, their detectors making those low, steady beeps that remind me of monitors in a hospital.

After a while Freya's detector started to beep faster and she stopped and moved it around until it became a single, high-pitched note. Jory opened his bag, got out a plastic kids' spade and started digging.

'It sounds like something big,' Freya said.

Jory kept digging. His hair blew around in the wind. There are streaks in there that look almost red – you know when the light catches that sandstone in the cliff? Not the crappy, claggy bits, the other stuff. But I don't know why I'm telling you that anyway.

'It's going to be big,' Freya said.

Jory dug some more, then reached in and pulled something out of the sand. It was a screwed-up bit of foil.

'Treasure,' Freya said. 'We've struck frigging foil.' She slapped me on the back and shook my hand. I slapped her back, harder. She took the foil from Jory and threw it for Mercury to chase, but Mercury was staring down at the sand and didn't even notice. Jory picked it up and put it in his pocket. He hates litter.

'We never find anything good,' I said.

'We found that bracelet last year,' Freya said. 'We sold it to Lyn.'

'It made her wrist go green.'

'Yeah but that's not our problem,' Freya said. She took another drink and passed the bottle round.

I turned and watched the tide. I could almost see it falling back, millimetre by millimetre. One minute there was a wrinkle in the water, the next it was the top of a rock. The first line of rocks had already appeared, wet and dripping, in front of us.

'If we find it we could do something,' I said.

'We are doing something,' Freya said.

'Something else.'

I scuffed my foot in the sand. Something else. Don't ask me what, exactly – all I know is that for years now Freya has been scrabbling around for shifts at the restaurant, only getting a few a week because her boss can pay all the younger people less. Once it's winter, her shifts will halve again and she'll be back to living off rice – all she eats is rice through the winter even though I told her I read somewhere that it's laced with arsenic. And then there's Jory, doing whatever the hell he always does – going for day-long walks on his own, doing up his boat, sleeping on Jake's floor which is the floor of a shed at the bottom of his dad's garden. They've dragged an old gas stove in there, which is probably going to blow up. I can smell gas as soon as I walk in but apparently, like with the arsenic, I'm just being paranoid. And then there's me, working in the same old hotel which is about to go under at any minute, serving the same old breakfasts of one stingy piece of bacon and mushrooms swimming in their own grey liquor, seeing things go on that would make your eyes water. Did I ever tell you about the thing with the machete? Remind me later.

Jory's detector started making a noise. It got louder, then softer, then it stopped.

'I heard Letty finally made it into acting,' Freya said.

'She was wooden,' I said. 'Remember that play we did at school?'

'She's got in some advert.'

'Toothpaste,' Jory said.

'Furniture,' I said, but it wasn't even funny.

'She's living with Mylo.'

'Mylo,' Jory said. 'What's she doing that for?'

'I heard he's manager of some hotel chain.'

'Which one?'

'We need to start going round in a minute,' I told them.

'The one with those beds in them.'

'They've all got beds in them,' I practically shouted. 'We've got to start in a minute, OK?' We had to time it exactly right, so that we'd have long enough to get round and back again before the tide came in and cut us off.

'There's loads of stuff here,' Freya said. She swept her detector over my shoes and it started beeping. What was supposed to happen was that I would take off my shoes and throw them at her, then she'd get the spade and dig at my feet, then I would trip her up and throw sand, and then she'd chase me even though she knows she'll never be able to catch me. When she starts doing it I just kind of have to go along with it. It's like when we order Chinese food and we share crispy beef – I don't even like crispy beef any more, but I don't exactly know how to tell her.

I took a step back, and then another. 'We've got to start going round,' I said.

Jory's detector beeped again. Freya stood there for a moment, watching me walk towards the rocks. I thought she was going to chase me and try to take my shoes, but she just watched me, a small frown on her face, then she bent down and started to dig.

I kept going. When I glanced back they'd got whatever it was out of the sand. It looked like someone's bent and rusty retainer. I knew that any minute now they would do that thing where they pretend the detector is a microphone, and Freya would hold up the retainer, lean right in, and pretend to be the first person she thought of with big teeth.

I got down to the rocks and started climbing.

Freya belted out a bar of June Carter.

Tide: 4.0 metres

The rocks were dark and slick with seaweed. My thighs scraped against mussels, which shone like wet lumps of coal. A gull banged at a limpet with its beak while I slipped and scraped, and slipped all over again, trying to keep hold of the bloody metal detector with one hand, and using the other to grip with. The beach already looked very far away.

I tried humming that crazy tune that always gets stuck in my head – you know how things always get stuck in my head, don't you – but all I kept doing was going over and over all the times I've tried to get out of here. I've tried a lot of things, if you want to know. I had this idea once, for example, that I would do up an old bus and go around selling food at campsites and festivals – everyone knows that people want to eat about a tonne of food when they've had too much to drink. Freya is like a hog at a trough when she's had a few, I can tell you, and probably I am too.

No one thought I would even get the bus, let alone the right paperwork for Chrissake, but I did, even though trying to memorise different types of bacteria for the hygiene certificate was one of the low points of my life. Have you heard of listeria?

That one's a complete bastard because it can grow even if the food's in the fridge.

I got Freya and Jory to come and help me at the first place; I paid them actually, and we got everything set up the night before, ready for breakfast the next morning. Then Freya got us these drinks and I swear I don't know what was in them, but I don't remember anything after that except being on Freya's shoulders with this ukulele band playing next to us, and a lot of lights flashing, and Jory, Jory was definitely there, very close. We didn't make any food the whole weekend.

Then I thought I would train as a lifeguard, because those lifeguards go everywhere, don't they. I figured I'd better practise first though because it'd been a while since I'd been in the sea and I don't really like it that much to be honest – all that stuff brushing past your legs, and stones bruising your feet, and have you ever been in when you know the kid next to you is peeing? I hate that; their innocent face, the slightly strained look around their eyes. I don't know about you but I probably wouldn't save a kid if I knew it had just pissed through its wetsuit.

So I got Freya to paddle out and pretend to drown so that I could rescue her, only it turns out that the stupid bint can't actually swim. Somehow she forgot to tell me that piece of information. She started flailing around, and I thought she was just pretending really well until she panicked and clung onto my neck and dragged me under. We came up gasping then went under again. I don't actually remember how we got out, but somehow, finally, we were lying on the sand and Freya was coughing and I was coughing and we were pummelling each other's backs for a lot longer than we probably needed to.

After that I sort of lost interest in the whole idea.

There are about a million other things I've tried as well, but I don't feel like going into them now. I even started drawing this book for kids, about a man who forgets where he lives and just wanders around from door to door, knocking. Sometimes people let him in but mostly they don't. It took ages and then I showed it to Jory and he looked at it for a long time, just sitting there, turning the pages slowly without saying anything. He never said anything at all.

Tide: 2.5 metres

There was this clattering noise and Mercury ran straight past me over the rocks, almost ramming into my legs. The gull lifted upwards, screaming. Freya's voice came over on the wind, I couldn't hear what she said, but I turned round and they were both climbing over the rocks behind me.

I stopped and waited. The tide was very low now. All I could see in either direction were wide, flat rocks, like shelves, and rock pools in between them. There were rock pools everywhere – some were shallow, some were smaller than my hand, others were so narrow and deep that I couldn't see the bottom. They smelled leafy, sort of vegetable, and they were full of this bright red and green weed. I kept glimpsing things darting around, but whatever they were they always disappeared before I could properly look, leaving the water rippling. There's a programme

I saw on TV once about rock pools; how, every moment, something is trying to kill something else: limpets crushing barnacles, anemones rasping bits off other anemones, starfish cracking open mussels like walnuts. But on the surface they look so still.

Jory stepped quickly from rock to rock, like some kind of pro. He was carrying his metal detector in one hand, his other hand was in his pocket, and Freya's bottle clinked in his rucksack. Freya didn't have her metal detector any more. She was down on all fours, making low, grumbling sounds.

'I'm getting cramp,' she said. 'I can feel it in my leg.'

'We're almost there,' I told her, which was a lie because I couldn't even see the cove yet. Those rocks went on for ever; they were so bony and ridged it was like being on the moon or something.

'I'm hungry,' Freya said. 'God I am so hungry.'

'We need to get round quicker than this,' I said.

'I need salt. That's why my leg's cramping. I need salt.'

'Stop thinking about it,' I told her. Freya's cramps are the bane of my entire life. One time there was this guy at a party and let's just say things were progressing, and then Freya came up, almost bent over with pain. I asked her what the matter was, thinking it was something really bad, and she said it was cramp in her little toe. She couldn't walk or do anything. She took herself off into a corner like some animal nobly going off to die, and I ended up having to bend her toe back for over an hour until it eased, and the guy went off with someone else.

Freya stopped crawling and sat down.

'We don't have any salt,' I said. 'Get up.'

Jory kept walking, thank God, but a second later he stopped. He pointed at a clump of seaweed. 'You can eat that,' he said.

Freya looked at it. 'You serious?' she said.

'We don't have time for this,' I told her.

'That one's nice,' Jory said. 'Salty.'

'Jesus Christ,' I said.

Freya went over and picked up a handful of dark green strands. She raised them up to her mouth.

'Wait,' Jory said suddenly. 'Don't eat it.'

Freya dropped the seaweed and spat on the rocks. She kept spitting even though she hadn't eaten any of it. 'You're a frigging fiend,' she said. 'You told me I could eat it.'

Jory pointed at a trickle of yellow on the rock. 'I think Mercury got there first.'

I started clambering again. This time I didn't wait to see if they were following. If I'd waited, we'd have been stuck on those rocks until the tide covered us slowly over.

Tide: 1.2 metres

After a while Jory caught up with me. I could hear Freya somewhere far behind, crawling along next to Mercury, complaining to her, and Mercury making these small yippy replies, complaining right back.

I watched Jory as he moved. He never slipped. I was watching him quite a lot and then there was a trench between two rocks and my foot went down into it. He came over and pulled me back up. His hand was cool and slightly rough. I got this sudden memory of him from when we were younger. He was in the playground looking at the bark chips under the swing, and he reached down, picked one up, and ate it. He chewed on it for a long time with his eyes closed, as if there was no one else around him. I actually get that memory quite a lot. I wonder what he'd think if he knew I'd watched him, and that I remembered it almost every time I saw him.

I looked over towards the cove. 'What if we find it?' I said.

Jory was looking at the cliffs. 'There are supposed to be fossils round here.'

'It's meant to be right over there,' I said. 'Just waiting, under the sand.'

'They're like ferns,' Jory said. 'We could look for them sometime, if you want.'

'What will Lyn say if I find it?' I said. The edge of the cove was up ahead. I'd only ever seen it from the top of the cliff before. It looked wider from here and it was covered in bits of thin, slatey rock, as if a roof had slid down and smashed. A few more minutes and we'd be there.

My hand started to feel very hot. I looked down and realised I was still holding Jory's hand.

Jory looked down too but he didn't move. My hand was so hot it was almost burning.

'The tide's going to turn soon,' I said.

Jory nodded. His hand slipped away. He jumped over a rock and I followed.

‘I liked that bit in your drawings,’ he said, ‘where the man knows it’s about to get dark, but he decides to try one more door.’

I stopped. ‘You remember that?’ I said.

A crow glided down and landed on top of a rock. It watched us carefully.

Jory reached across and helped me over another gulley and suddenly we were there, on the gritty sand in the cove.

Tide: 0.5 metres

No one had been there for a long time. I know the tide comes in and washes everyone’s footprints away so it’s not like you can ever really tell, but it felt like no one had been on that beach for a really long time. There was a sort of hush to it. I could hear every loose stone that rolled.

I called to Freya to hurry up, then I switched my metal detector on and started going up the beach towards the cliffs.

‘Wait a second,’ Freya shouted. ‘Mercury’s seen something.’ I turned round and Mercury was staring down past the rocks towards the water. Even her tail was tense.

‘Come on,’ I said.

‘I think there’s a bird down there,’ Jory said.

‘She’s about to go,’ Freya said. ‘Look at her.’

‘It might be a plover,’ Jory said.

‘Come on,’ I said again, but I knew they weren’t going to. They were never going to. Bloody Jory with his seaweed and his birds and his boat. Bloody Freya, clinging on to my neck and dragging me under. I turned away and walked faster up the beach.

The slates crunched under my shoes. I went past heaps of rope and wood and plastic boxes that had been stranded by the tide. There were little flies jumping all over them. The seaweed looked baked and brittle.

Behind me, Mercury let out a low, strangled bark, Freya shouted something, and there was the sound of feet against stones. When I looked back, the three of them were no more than specks on the wet sand at the edge of the water.

I moved my detector in slow wide arcs. It made no sound, not even the smallest beep.

Tide: 1.1 metres

I started walking in zigzags to cover more ground, swinging the metal detector from one side to the other. I went up a stony slope and then across a series of slates that cracked under my feet. I went back down the slope. I climbed over a pile of wooden boards and pushed the detector right under them, but still there was nothing.

I moved across the back of the cove, so that I was right at the base of the cliffs. The stones and shingle had all been churned up, just like they'd told me. A whole layer of stones had been scraped off, and underneath the sand looked raw and pale.

I went right over that sand, every millimetre of it. At one point my detector let out a faint whine and I stopped, bent down and dug with my hand around the stones, but all I found was a sweet wrapper. I dropped it and kept going. I went back the way I'd come, working backwards and forwards over the same area. I must have missed something, I must have. I checked my metal detector, running it over the tops of my shoes. It let out a long, wavering beep.

I turned and looked down at the water. The lowest line of rocks had already been covered back over. I ran to the next bit of sand, and then the next.

Something moved behind me and I think I must have swung the metal detector round because there was a yelp and then I saw it was Mercury, standing right next to my legs.

'Shit, you stupid hound,' I said. 'What are you doing?'

She leaned into me and I stroked her bony head.

'Find it then,' I said. 'Find the cache.'

She looked at me like she didn't know what a cache was either.

'The treasure,' I said. 'Find the treasure, OK?'

She barked at me, then arched her back and sidled away, like a crab or something. Then she darted back and lunged at my shoes.

'Piss off,' I said.

She darted backwards, then lunged again. I skipped my feet out of the way.

'Piss off Mercury,' I said.

I pushed the detector under a stone and moved it around. Then I pushed it under another one. Mercury lunged again. Bits of grit clattered quietly down the cliff.

I let her gnaw on my shoes until she got bored and ran away. I sat down and let the metal detector fall on the ground behind me. My legs were aching. My throat was parched. Bits of shingle dug into my thighs and arse.

I closed my eyes. When I opened them again I could see Freya and Jory messing around down in the shallows. They were pushing each other and suddenly Freya reared up and tipped Jory right in. I heard the splash from up the beach. He got up, flailing, and ran at her. She sidestepped somehow. They were both laughing their asses off and I was laughing too and I got up and started making my way down to them. I was going to push Freya in and then I was going to, well I didn't know what I was going to do to Jory.

That's when my detector started to beep. I stopped and looked back at it. I thought I'd probably knocked it on my way past. The noise was louder than usual, more insistent. 'What the hell are you doing?' I said. I kicked it but it didn't stop.

I looked back at the sea. My throat got even dryer. I crouched down and dug into the cold sand. My fingers touched something hard. I kept digging until I found an edge. I scraped the sand off and there was the corner of a metal box.

I stopped digging and looked up. Freya was waving her arms at me and Jory was wringing out his wet hair. The water was glinting off them and the light was this weird haze – I can't really describe it, but you know when light comes down through trees, it was kind of like that, and I know there were no trees, I'm not an idiot, but the beach was deserted, it was all empty and it was ours, the whole place was ours.

They started coming back up the beach and Jory called something but I couldn't hear what he said.

My hand was on the edge of the box.

Freya waved again. 'Next year,' she shouted. 'Next year, OK, dickhead?'

Next year. I ran my hands right around the edge. It was big. It felt heavy and wedged in. I started to prise it out.

What had I said to them? We could do something else. I thought of being on Freya's shoulders, and heaving her out of the water. I thought of the way I had drifted in Jory's boat.

I moved my hand and pushed a heap of sand forwards. Then I did it again, pushing with both hands. I kept pushing the sand until the box was buried. I covered it right over and stamped down around it. When I looked up Mercury was back. She was just standing there, watching me. We stared at each other for a moment, then I got up and ran down to the rocks.

'I was calling you for ages,' Freya said.

I brushed the sand off my hands. It was so sharp and gritty it had left scratches in my nails.

'What's the matter with you?' Freya said. 'You look weird.'

Jory looked at me closely. He smelled like sweat and the sea. 'It was a good idea,' he said. 'Coming over here.'

'You almost made me eat piss,' Freya said.

We went back across the rocks. The sea came in and covered them over, millimetre by millimetre, and all the time it glittered like coins.

Mercury kept turning round to look at me and I had this feeling that she'd seen the box, that she knew, and would look at me differently now with her wild blue eye.

We got back to the beach. We got in the van. We drove away.

And that's it. That's all I really have to tell you. I suppose you'll want to know that I went back. It was a few days later and I waited until low tide and tried to get round but it wasn't low enough and halfway there I had to turn back, water lapping at my ankles.

I suppose you'll want to know that I didn't tell Lyn or Ricky or Jake anything about it. And that, after the autumn storms, in reference to another conversation entirely, they said that the sand and the stones in the cove had heaped up and shifted around and that it all looked completely different again.

I suppose you'll want to know that last night I ate crispy titting beef.

And I suppose you'll want to know why I did it. What am I supposed to tell you? Sometimes you feel like opening something up and sometimes you don't. And that's all I have to say about it, OK.

One Foot in Front of the Other

SHE WALKS DOWN THE track and climbs the first gate. Her legs ache. They are heavy as wet bales. She's been walking for a long time, although she can't remember how long exactly. Her jeans are soaked to the knees; there's a bramble hooked on the back of her shirt and another around her foot. Her grey hair is damp, brittle, and there's a moth caught in it. There's a scratch along the bottom of her jaw.

She climbs the first gate. She's been walking for a long time. She doesn't have anything with her unless you count the brambles or the moth. She walks over the field, which is bare and dewy. The barley has just been cut back to stubble. It's early and the air is wet – damp gusts blow in like smoke before the fire's got going properly. It will be hot later; the sun will break through and parch everything. She walks faster. A gunshot goes off in the distance. All she wants to do is get back. There is the constant sound of hammering from somewhere, and chainsaws, and the terrible screech of an angle grinder.

She crosses the field and comes to the next gate. There are cows standing on the other side of it. She stops for a moment in the churned, hoofy mud. The tree next to her is bent at the hips, staring at the ground. There's a line of ants down there, carrying a green dragonfly. She goes over to the gate and climbs the first rung. The cows huddle together and press against the bars. They are a dark brown mass. She claps her hands but they don't move. She rattles the gate but they don't move. She climbs down. The cows' skin twitches, as if something has run over it.

She crosses the field and goes into the next one. There is the constant sound of hammering from somewhere. The gate is in the far corner and she walks over to it. A gunshot goes off in the distance. There is a drinking trough in front of the gate. She's suddenly thirsty. It feels like a long time since she's eaten or had a drink of anything. She goes over to the trough and dips her hands in. The water is dark and cold. There are flies stuck on the creased surface. She dips her hands in and cups some water and splashes it over her face and down her throat. The water is so cold she almost can't feel it. She splashes some more. Her hands and throat are numb. She still feels thirsty.

When she looks up there's a herd in front of her, pressing against the other side of the gate. They are pressed silently, tensely, as if they are waiting for something. She doesn't know if they're the same cows or not. They are a dark brown mass. A cow leans its head over the top bar and rubs its jaw along the metal. One eye watches her while the other rolls.

'What are you doing?' she says. 'Get away.' She climbs the gate and bangs her hands on the bar. The cows don't move. Their breath comes in thick shapes on the air.

She waits a moment. All she wants to do is get back. She's been walking a long time and her legs ache. But the cows still don't move. Only their tails flick. She turns and looks back at the way she's come. There is a chainsaw somewhere, and the terrible screech of an angle grinder. The gates through the fields are the quickest way of crossing down, she remembers that much, even though she hasn't been back for a long time. Otherwise she'd have to loop right up to the main road, hike along for a few miles, then come down that way. She doesn't want to go up to the main road. What she needs to do is cross the fields, get onto the lane, aim for the slope, then cut across the trees from there.

She follows the edge of the field until she finds a gap in the hedge. She pushes through it. Brambles catch at her clothes. The sleeve of her shirt tears. She gets a scratch across the wrist. Finally she is out and in the lane. The lane is narrow and stony. The nettles on the banks are taller than her and there's cow parsley with stems as thick as fingers. She keeps going. Her legs ache and her hair is damp. The potholes are filled with oily water. A jackdaw is splayed on the ground.

There's a low noise ahead but she keeps going. The nettles thicken on either side until she's brushing past them with both shoulders. Flies knock into her. The nettles lean. There's a sort of clopping noise coming from somewhere. A gunshot goes off in the distance. The lane dips downwards. She turns a corner and the cows are coming up the lane in front of her, three abreast, walking slowly and looking straight at her.

She raises her arms. 'Get away,' she says. 'Get away.'

The cows come forwards slowly. They're pressed into each other, their flanks are rasping, and the cows at each edge push into the nettles, bending and trampling them.

She waves her arms. They don't stop. She stamps her feet and shouts but they keep coming. She turns and walks back to the hedge and goes along it, looking for the gap. The cows are closer now. She walks quickly along the hedge. There's no gap. The cows are right behind her. They're walking slowly and steadily. She pushes her hands into the hedge. It's too thick to get through. She pushes again. Something cuts her hand. A nettle loops over her foot. She pushes harder. There's the gap. She stumbles in and crouches on the ground. The cows walk carefully, pressing into the hedge. When they reach the gap they slow down and then stop. They smell of old grass and dry skin and the sticky mud around their feet. They stand in the lane and shift their weight from side to side. She stays crouched. Her legs ache. There is a small bone and some fur on the ground by her foot. Whenever she moves the cows' skin twitches.

She backs out of the hedge and into the field on the other side. She looks once more through the gap. The cows snort. One of them stamps. She walks back through the empty field. There's only one way left to go – over to the main road and across from there. All she wants to do is get back. She can't remember how long she's been walking. It must be a long time.

There is the constant sound of hammering from somewhere. She crosses the first field and enters the second. A gunshot goes off in the distance. This field has long wet grass that sticks to her legs. It tangles in clumps and trips her up. It's tough and doesn't snap across her boots. She keeps going. There isn't far to go. She can't remember exactly, but surely there isn't far to go.

The next field is wide and open and more land, more fields, stretch in front of her, strung with telegraph poles and bending trees. The road is in the distance – it thrums with tractors and brewery lorries and lorries delivering frozen food. They flash on the horizon – red, blue, red, blue. They seem very far away.

The gate she needs is ahead, in the opposite corner of the field. There is the sound of a chainsaw somewhere. She starts to cross the field. As soon as she starts crossing, she sees the cows. They are on the far side, walking towards the gate. She walks faster. The cows seem to quicken their pace. She keeps walking. The cows will reach the gate first, she knows it, they are closer than she is. She walks faster. The cows are in a line, now they are in a group, pushing against each other. Her legs ache. She doesn't want to run but she starts running. The cows start running.

All she sees is the gate. The cows' hooves strike at the ground. A gunshot goes off in the distance. The cows' bodies send out a wave of heat and it is behind her as she runs. She stumbles in the mud. Her foot sticks. She gets to the gate. Her foot is on the rung. She slips. A cow thuds against the gate and it shakes on its hinges. She slips again, then she is up and over and on the other side.

She doesn't turn around until she's almost across the field. All she wants to do is get back. Then she makes herself turn. The cows have gone. The fields are empty all around her. Below there is a dark line of trees. She hobbles down, mud on her legs, grass on her legs, brambles hooked to the back of her shirt. A crow circles.

Her foot catches on a stone and she stumbles and almost falls. Her nose is raw and the numb feeling has spread up her hands and into her arms. The mist is pushing in thicker now, dropping down so that everything is swathed up to knee height. She can't see the ground, or her feet. Her feet are very cold. There is the constant sound of hammering from somewhere, and chainsaws, and the terrible screech of an angle grinder.

She makes her way down the slope and towards the trees. Once she is on the other side of the trees she will almost be back. There is a sound ahead of her and at first she thinks it must be her boots hitting the stones in the grass. She keeps going. The crow is still circling. A gunshot goes off in the distance.

The first cow comes up the slope towards her. There are two more behind it. They come in ones and twos, slowly, with their heads down, pushing closer. The whole herd is there, coming up the slope about twenty yards ahead, wading through the mist, and spreading out in a semicircle around her.

She stops. The fields are dark and empty for miles in all directions.

The cows don't run, they don't stamp, they just press slowly forwards.

She stays where she is. Her legs ache. She's been walking for a long time. She can't remember how long exactly. All she wants to do is get back. There is the constant sound of hammering from somewhere.

She takes a step backwards, and then another, until she's up the slope and across the field, one slow step at a time. The cows don't move. She doesn't take her eyes off them. Their tails flick. Their breath comes out in thick shapes on the air.

She reaches the gate. She climbs back over it. What she needs to do is circle around another way. She needs to go back to the first field and start again. She needs to climb the gate, and then the next one, and then she will be back. She hasn't been back for a long time – she can't remember how long, exactly.

She crosses into the first field. The nettles along the edges are taller than she is. She doesn't have anything with her unless you count the brambles or the moth in her hair. The field is bare and dewy. The barley has just been cut back to stubble. It will be very hot later; the sun will break through and parch everything. She walks faster. She sees the gate. There is the terrible sound of an angle grinder somewhere. Damp gusts blow in like smoke before the fire's got going properly.

Way the Hell Out

'DID YOU HEAR ABOUT the Ellis house,' Fran says. She picks up her mug, holds it, but doesn't take a drink.

'It's coming up for sale again,' Morrie says. He leans back and looks around the café. It's empty apart from a table of three people in the corner. He nods to them. 'Lyn,' he says. 'Jake. Ricky.'

They all nod back then carry on eating.

The owner of the café stands behind the counter, drying glasses and wrapping cutlery.

'That's because it's happened again,' Fran says.

Morrie nods and takes a drink. The windows drip with steam from the inside, rain from the outside.

Fran puts her mug back down on the table but keeps hold of the handle. 'It didn't start straight away,' she says. 'When they first moved in, everything was fine.'

'It's a nice old house,' Morrie says.

'They bought it for their holidays.'

'They paid a lot for it.'

'They wanted the quiet.'

'It's definitely quiet out there.'

'No other houses.'

'No lights.'

'No traffic on the road.'

'You wouldn't see anyone else for days.'

'Shut the door, light the fire, close up all the curtains.'

Morrie finishes his drink. 'I like a fire.'

'You'd need it out there, the way the wind comes in.'

'It does come in.'

'The sea mists.'

'They do come in.'

They both look out of the window. A sheet of rain presses against the glass. The group in the corner sit in silence. There's just the sound of their forks against their plates.

'It's a shame,' Fran says. 'The way it starts happening.'

Morrie leans back further in his chair. 'Same as before?'

Fran leans forward. 'Same as before. They come back and find the door's unlocked. At first they think they just forgot.'

'It's easy to forget.'

'But the next day it's unlocked again,' Fran says. 'So they lock everything carefully, check it over, and go to bed. In the morning, all the windows have been flung wide open.'

The café owner starts stacking the glasses on a shelf. Each glass grates against the next as he stacks them.

'They close all the windows and put the latches down. They check the doors. And for a few days nothing happens. They almost forget about it. The windows are so loose that maybe they just opened by themselves in the wind.'

Morrie nods again, slowly.

'Then one night they hear someone walking down the road.'

'No one walks down that road.'

'But they hear someone. Boots on the gravel. A cough. Whoever it is they're dragging their feet. They're unhurried.'

'It doesn't seem right,' Morrie says.

'They look out and try to see who it is. The road's empty. There's no one there.'

'What about in the trees?'

'That's next.' Fran twists her mug around and holds it with her other hand. 'They start thinking there's something in the trees.'

'What?'

'They can't tell – they have those thick pines out there, and at first it just looks like a dark shape, maybe a gap between branches.'

'Pine trees can do that.'

'Then they realise what it is.'

A glass falls off the shelf and cracks against the floor. No one jumps.

'It's a figure, just standing there, looking across at the house. It's standing very still. There's some kind of hood, maybe a long coat.'

Morrie sighs and shakes his head.

'It's dark, so they try to find a torch, but by the time they've found it there's nothing there, just the gaps between the trees.' Fran glances at the group at the other table. They've all got their heads down, eating. 'They try to tell themselves it was nothing. Just the wind, just the trees moving.'

'The wind does move things out there.'

'It does.'

A chair scrapes against the floor but no one gets up.

'They tell themselves they didn't really see anything. And for a while they don't see anything else. Everything goes back to how it was, until they come back one day and, as they're getting out of their car, they happen to look across at their kitchen window.' Fran stops again and looks down at her tea. There's half left but she still doesn't drink it. 'There's hands pressed against it, from the inside.'

'The inside?'

'Two hands, just pressing.'

'I don't like that,' Morrie says. 'That's not right.'

'After that they stay up all night. They check each room. They sit on the sofa and put the TV on loud. They try not to think about the hands.'

The cutlery clinks as the café owner wraps some more and puts it carefully in a basket.

'But they must fall asleep, because when they wake up, someone's in the room.'

'I don't like that,' Morrie says.

'There's someone behind them, near the door. It's dark and they've just woken up, and the figure is moving carefully, keeping close to the walls. It walks through the house, down the corridor and out of the front door.'

Someone on the other table knocks their plate with their elbow and it rocks, then slowly stills.

Morrie shakes his head. 'They only bought it for their holidays.'

'Not much of a holiday.'

'What did they do?'

'Same as the people before. They phoned the police.'

'There's no signal out there.'

'There's no anything out there.'

'They probably had to go to the top of the road to ring.'

'In the wind.'

'In the rain.'

They look out of the window. The café owner folds and wraps. The cutlery basket is almost full.

'Of course, by the time the police get there the house is empty again.'

'What did they say?'

'What could they say.'

'The doors are all locked.'

'No marks, no traces. No other witnesses. What can they do.'

'It is quiet out there.' Morrie watches a drop of water as it moves slowly down the window. 'I suppose they don't really know anyone.'

'I suppose not.'

'Out there on their own.'

Someone walks past the door and slows down. Everyone watches. The person stops, looks in, then carries on walking.

'They stop coming down so often.'

'No one they can talk to,' Morrie says. 'That's not right.'

'And when they do come back, things have been moved.'

'The rugs?'

'The books. Furniture.'

'The rugs?'

'Probably the rugs.'

'I don't like that,' Morrie says.

The door of the café rattles in the wind. The café owner crosses the room, looks out, then pulls it tightly shut.

'Eventually they stop coming down at all,' Fran says.

'That's what the others did.'

'That's what they always do.' She watches the café owner as he walks back behind the counter. 'They stop coming down and then they sell it.'

'They'll want to sell it quick.'

'They will.'

'It'll probably be a bargain.'

'No doubt.'

'Like last time.'

'No doubt.'

Fran turns and gets her coat off the back of her chair. She puts it on but stays sitting. She picks up her mug again. The group on the other table hold their forks but don't eat.

'Someone'll snap it up of course,' Fran says.

'Buy it cheap, sell it expensive,' Morrie says.

'That's what she always does.'

Fran finishes her drink in one mouthful. The café owner wraps the last of the cutlery and switches on the radio. The group on the other table start to talk amongst themselves.

'She's always been a swine, that Jane Ellis,' Fran says. She gets up and zips her coat.

Salthouse

WINTERS ARE WHEN PEOPLE disappear. One minute you're elbow to elbow on the street, the next you walk along side-stepping nothing but the wind. Cafés put down their blinds. Houses are locked and dark. The car parks slowly empty and all that's left on the beaches are a few forgotten shoes. Waiters and waitresses go away to work the ski season, cleaning chalets in a bright glare of snow. Lifeguards pack their tents and dented surfboards and get on planes, following the sun like a flock of migrating birds.

I wait for Gina by the door, my coat and trainers on, and the old, dried-up Christmas tree leaning against the wall. It's not even four o'clock, but the sky is already dim – one of those days where it never really gets light except for a pale streak above the sea. Gina lives a few streets away in a bungalow that's almost identical to mine, except hers has a hole in the wall from where her mother once tried to decorate and then gave up halfway through. There's a TV in front of it now but you can still see the cracked edges. It takes two minutes to walk between our houses; one and a half if you take the alley with the mattress and the bin bags. Down the sides of each street there's clumps of sea beet, burdock, grass that knots into sandy bouquets. The grass is sharp and tough. We used to take turns ripping out handfuls and seeing who would get cuts across their fingers. There's sand everywhere around here. When you walk in the wind, grains crunch against your teeth. We're out on the edge of town, where the cliffs start to crumble and turn to sloping dunes. The dunes are heavy and soft, like flour in a bowl. They never stay still. They slip and shift around; sometimes growing, sometimes flattening out. When the gales come, loose sand blows down the road and heaps at our front doors.

Gina finally knocks and I go straight out, dragging the tree behind me. 'I thought you were coming earlier,' I say. I prop the tree up and lock the door, then start hauling it down the steps. We always bury the tree together first thing in the new year, but somehow it's already halfway through February. The tree's almost bare, except for a few brown needles clinging on.

'You look like you're moving a body,' Gina says.

'I thought you were coming earlier,' I tell her.

Gina turns and looks back up the road, as if she's seen something, but there's nothing there. 'How late are your parents working tonight?'

'Late,' I say. The care home they manage is full and low on staff. 'Mr Richards is sleepwalking again. He's started getting out and trying to hitch-hike at the side of the road. No one's stopped for him yet.'

Gina picks up her end of the tree but doesn't move off the step. 'Late,' she says.

I start walking backwards, then turn and hold the tree behind me so that I can walk facing the right way. We cross the street and cut across a few front gardens. A cat follows us, then yawns and sits by somebody's door. We pass the last of the bungalows, with their banging shutters and wind-cracked paint, and come out onto the road. The dunes spread out ahead of us, humped and dark. We start down the road towards them. Every year we take our tree down to Salthouse and bury it along with everyone else's, to try and stop the sand moving and the dunes disappearing. There are rows and rows of old trees. Gina and I always find the best place, and dig ours in the deepest. We do it with my tree because Gina's is plastic and has flashing lights and a singing snowman on top.

The wind slaps into us. I pull up my hood and button it under my chin. I wait for Gina to do the same, then realise she isn't wearing her coat. She's had the same one since we were about seven – a parka with a broken zip and sand in the pockets, which used to come down past her knees. Once, we both fitted into it at the same time. Instead of her coat, she's wearing a white jumper that looks too small for her, and instead of trainers, she's wearing long brown boots that must be her mother's. They're too big for her. She keeps stopping to wedge her feet in tighter.

'Where's your coat?' I ask. I have a loose tooth and I keep touching it with my tongue. I've lost all the others – some have grown back, some are halfway through – and this is the only original tooth I have left. It should have come out weeks ago. Usually I would bend it until it makes a gristly, crackling noise, and twist it at the root, but with this tooth I push downwards, into the gum, until it settles back in place.

Another gust of wind hits us, and Gina rakes her hair out of her mouth. Her hair is very pale and so is her skin and the tips of her eyelashes, like grass that has dried in a heatwave. Her mouth is small and dark and she smells sweet and sour, like a vinegary strawberry I once ate and spat back out.

'You need to pull that tooth,' she says. 'I'll do it for you if you want.' I can feel her shivering – it goes down through the tree and into my hands.

We come to the crossroads and I start following the path down to the dunes, but Gina stops and puts her end of the tree down. She picks at a dry needle in her finger. By now our palms are stippled. The lights of town are in the distance. A bus goes past, its insides lit up like an aquarium. There is the sound of the wind, and the sound of the sea, and something I can't place, a low beating that isn't the wind or the sea.

'The fair's back,' Gina says.

I can just make out a small, hazy glow past the hotels. 'Remember the dodgems?' I say.

'You rammed that old woman until she almost got concussion.'

'She did it to me first.'

'They still have them,' she says. 'And they have this pendulum thing now, where you get tipped upside down.'

I watch her pick at her finger. 'You've been,' I say.

She gets the needle out and drops it. 'It's been here all week.' Again, she turns and looks down the road, at nothing.

The beat of the fair surges in, louder for a moment, on the back of the wind. I start dragging the tree towards the path.

'Evie,' Gina says.

'We need to get going with this.'

'Evie,' Gina says.

'We're almost there.'

'We could do the tree after.'

'It'll be too dark after.'

'We'll get covered in sand if we do it first.'

'It'll be too dark after,' I say.

'Remember that time you broke your parents' window, kicking that stone, and I told them it was me?' Gina says. 'Remember when I taught you how to go really hot and dizzy, so you look too ill for school?'

I keep my eyes on the ground and push at my tooth. If I look up that's it; especially with her reminding me of all the things I owe her like that.

'Remember when we were five and we found those flies on top of each other and we pulled off their wings, and afterwards you couldn't bear it, so I stuck them back on with superglue?'

I look up. 'It didn't work,' I say.

Gina picks up her end of the tree. She knows we're going. 'What didn't?'

'The glue,' I say. 'You told me they flew away but I found them later, on the windowsill.'

The beat of the fair gets louder as the road slopes into town. Every time the wind hits, the tree loses more needles. They scatter along the pavement like hair on the floor of a hairdresser's.

'Look at Mrs Bradley's house,' Gina says. 'She got another ornament.' She points to a hare by the front door. There are moths attached to the wall, and a few hunched gnomes on the grass. Mrs Bradley taught us for years, before we moved to secondary school last September. She showed us how to dissect a frog. All I remember are the thin, splayed-out hands, how they looked like they were asking for something. Behind her desk she had a shelf of jars with frogs' legs and fish and something small and twisted, all floating around in salty water. They looked like the pickle jars in Gina's mother's kitchen – gherkins, silverskins, shredded beetroot – and once Gina made us eat some and pretend we were eating whatever it was in Mrs Bradley's jars. I threw up first. One time, Mrs Bradley brought in a pair of calf's lungs and inflated them by blowing with a straw and one kid fainted and smacked his mouth on the wall and his two front teeth slid right out next to my shoes.

As we go past, Gina glances round, then climbs over the wall and puts the hare in the middle of the driveway. Then she moves a weird baby gnome so that it's nestled behind the front wheel of the car. We don't like babies and swore a long time ago that having them was nothing to do with us.

As we get closer the music thrums up the road and into my stomach. The fair's in a car park that belongs to a hotel that's been boarded up for as long as I can remember. There's a pile of beer cans at the entrance, and ticket stubs scattered over the ground. Lights throb like something painful. Someone lets out a long low howl. It's still only the afternoon but suddenly it feels dark and late. A kid runs past wearing a crash helmet, chasing a balloon that's been wrenched away by the wind.

We go through the gate, then Gina stops and I almost trip. I can see that she's trying to smooth down her hair and her clothes without me noticing.

'What are you doing?' I say.

'Do I look OK?'

There's sand on her boots and her hair has gone curly in the salty wind. Her cheeks are pink along the top of the bone. I haven't noticed those bones before. They make her face look narrower, as if a new one is slowly being chiselled out. How haven't I noticed them?

I put my hand up to my own face and feel the same soft, puckering skin that is already flushed and burning under the hot lights.

'OK for what?' I say. I say it stiffly, my mouth suddenly dry. I start backing in with the tree, but Gina doesn't move.

'We have to put this somewhere,' she says.

'We'll just hold it.'

'Put it under this van.'

'I'll hold it.'

'Just slide it under here.'

There are a group of vans with the fair's logo on parked near the wall. Gina crouches down and slides the tree under one of them. 'We'll come and get it later,' she says. 'OK?'

She doesn't wait for me to answer. I turn back one more time to check the tree as she disappears into the crowd.

There are people everywhere: families moving too slowly, stopping at every stall for candyfloss, flashing bracelets, bags of bright yellow sweets. There's the smell of burnt sugar, sweat, fat, the tang of hot metal, like overheated brakes. A group of older girls almost walks into me, then veers off, laughing, towards the big wheel. Cigarette smoke blows in and out. Music blares at each ride – disco, jazz, and the fair boys whistle and call out that there are two minutes before their rides start, one minute, then gates clank, more music starts up and someone screams.

Gina threads her way through everything, looking from side to side, past the stalls, past the rides, out to where the generators and cables thump.

I finally catch up with her. 'Let's go on the waltzer,' I say.

She looks at me as if she's forgotten I'm there. 'The waltzer?'

'We always go on that first,' I say. I don't ask why she's gone all the way out to the back of the fair, what she's looking for. Instead I turn and make my way over to the ride, pushing through the crowds, hoping Gina is following behind. Just as I get in the queue I realise I don't have my purse with me. I'm almost at the front and I search my pockets but all I have is a crumpled tissue and a decoration I found that was still hanging on the tree. Just as the woman ahead of me finishes paying, I remember the five-pound note I keep in a tiny, inner pocket in my coat. It's meant to be for emergencies. I've never used it. It's my turn in the queue and the man on the ride asks how many I want. I take out the money and pay for two tickets.

I get on and sit back in the sticky seat. After a few minutes Gina climbs in next to me. Our seat is already tilting at almost a right angle and my stomach starts to feel very light, like it's rising above me. I lean forward and put both hands on the bar. The music starts and the ride jolts. Gina is still sitting back, not even holding the bar, but I can tell she's braced.

We circle around once, the cup spinning slowly on its axis as we go. Gina still isn't holding the bar. Her arms are crossed in front of her. I catch her eye and she gives me a small smile, the kind where only one corner of her mouth rises up. It's her pretend smile – the one she uses for parents or people she passes in the street and doesn't want to speak to – but there's hardly any time to take it in because suddenly the music surges, the lights flare, and we're tipping forwards, spinning backwards and being flung around the ride, our hair whipping across our cheeks. Gina's mouth snaps open but no sound comes out. I shut my eyes. Colours bleed and distort in front of me. The music pulses. Gina slides across the seat and bangs hard into my hip. She grabs at the bar and ends up gripping my wrist, her nails in my skin. We spin three times and I know I'm going to fall out, I'm lifting upwards, my hands are clutching the bar and I open my eyes and see a row of pale, blurred faces. I bang back down. The ride tips then slows. The music cuts out and we come to a shuddering halt. Someone unhooks our bar and we stumble out, the ground rolling under our feet.

'I need to eat something,' I say. I try to walk, but it comes off more like lurching. I follow the smell of onions and burnt coffee. If I don't eat something I'll throw up, I know it. There's a queue. A man at the front doesn't want sugar in his coffee but they've already put sugar in. I have to stop myself pushing him out of the way. Finally it's my turn and I buy a hot dog and a bag of doughnuts. The hot fat seeps through the paper bag, turning it translucent.

I offer the bag to Gina but she shakes her head. I offer again and she takes a doughnut and nibbles at the edges. She wipes her mouth with a napkin after every bite. I eat a doughnut, then half the hot dog, then another doughnut, making sure I don't chew anywhere near my loose tooth. I finish the rest of the hot dog, then lick my fingers one by one to get the last of the mustard and the sugar. Just as I'm about to start on my thumb, someone puts their hand on top of my head. I turn around. It's Mrs Bradley. Her hand is strong and stiff, like a tree root.

'Are my girls having fun?' she asks. There's a wisp of candy-floss trailing from her sleeve. 'Have you been on any rides?'

I finish licking. 'Just the waltzer,' I say.

'With your parents?'

Gina stops eating her doughnut. 'Why would we do that?'

Mrs Bradley smiles quickly. At school we always sat at the front, right by her desk. She gave us gold stars to stick on our books when we got the answers right. 'I saw the merry-go-round earlier,' she says, 'and it reminded me of my girls. Remember when all you wanted to do was be horses? You'd gallop into class and when I asked you a question you'd neigh the answer.'

I nod, but Gina shakes her head. 'I don't remember that,' she says. She looks around and then freezes. The top of her cheeks go even pinker, just for a moment, then fade back to how they were. 'I have to go,' she says. She thrusts the half-eaten doughnut at me. I catch a glimpse of her jumper as she vanishes back into the crowd.

I turn to follow then stop, but Mrs Bradley waves me on. 'You go,' she says. 'Have fun, OK?'

'I'll look at the merry-go-round later,' I call back, but I don't know if she hears or not.

I walk fast. The fair is even busier now. Queues spill into the walkways and I keep having to stop and go round them. I see a kid from school and then another, bigger group of people I half-recognise. They're sitting on the steps by the big dipper. I almost walk past them but then I see Gina sitting at the top, next to a boy who has his arm slung over her shoulder. I stop. Someone's shoe bangs into my heel. I don't cry out. I don't move. I'm about to back away when Gina sees me and beckons me over. I go slowly, not looking at anyone, and stand near her, at the edge of the steps. I stand very still, like a hare that's caught the scent of something. They're passing a can around and when it comes to me I shake my head, then can't stop watching while Gina drinks. The boy next to her takes the can and finishes it. He has a very thin nose and bloodshot eyes. His hair is so dark it has a blue sheen to it.

I keep licking my lips to make sure there's no sugar left on them, but they taste salty, not sweet, and they sting in the cold.

Gina looks over at me and smiles, properly this time.

I don't smile back.

The group watches everything and everyone that goes past. They watch and then someone says something, or does an imitation, and laughter ripples out, as if a stone has just hit water. No one talks to me and it gets colder and later, until one of the boys turns round suddenly and asks if Gina and I want a ride on anything, we can have a go on anything for free.

Everyone sniggers and the colour comes again into Gina's cheeks, darker this time, more like when she was ten and she grabbed a teacher's hand instead of mine and pulled him around the playground.

I lean in and say something, just to her. She doesn't hear. I say it again. 'Let's go on the dodgems.' I wait for her to get up and come with me, but then everyone is standing up and moving and the boy with his arm around Gina says, 'Good idea. Let's go on the dodgems.'

'Evie,' Gina says. She jerks her head for me to follow. There's nothing else to do except follow. Gina gets in a car next to the boy and someone presses a ticket into my hand and puts me in a car by myself. The music rolls. My car powers up. The man running the ride shoves me off the side. Sparks crackle on the ceiling and I drive.

I circle the edge slowly, looking for Gina. We used to pick someone and follow them mercilessly, ours hands overlapping on the wheel. This time I can pick for myself. I see Gina's car and start following. The boy isn't driving properly. What you're meant to do is go round in a circuit, but he's zigzagging across the floor, ramming into everyone, throwing Gina back in her seat, then reversing and going the other way. His friends are all doing the same. One of them smacks into me from behind and I jolt forward, then spin into the wall. I try to reverse but another car rams me back and my steering locks. I reach out and push off the wall and the wheel turns. I slam my foot on the pedal and aim for Gina's car. I'm almost on them when another car comes in and knocks me sideways and I spin again. I curse and turn the wheel hard. By now I'm leaning my whole weight against the pedal, circling the floor, sparks chipping off above me. I see Gina ahead and I lean forward. She hasn't seen me and I bear down on them, gritting my teeth. Her head is resting on the boy's shoulder. I turn my car so that I can hit them from the side. I'm almost there. I'm already imagining her face, the way she'll look at me when I send them slamming into the wall. I'm five metres away, four metres, and I pump my foot down on the pedal but it feels different, lighter, there's no traction. I pump again but my car is slowing, the music's stopped, the sparks are dying away and I'm stranded in the middle of the floor.

I sit there, my hands still on the wheel, until I'm told I have to get out unless I want to pay for another go. I get up and make my way off the ride. There's a white line across my hands and my knee almost locks from where I've been leaning so hard on the pedal. I go down the steps and stand on the concrete. Behind me, the dodgems start up all over again.

I walk back through the fair. By now it's completely dark and more lights have come on. I step over a crumpled balloon, candyfloss sticks and spilled drinks, which spread across the ground like the fair's tideline.

I can see the vans ahead and I speed up. All I want to do is leave. I try to remember which one we put the tree under. I think it's the middle one and I head straight over. There are two people ahead of me and I realise that it's Gina and the boy. They're walking towards the vans. Gina tilts her face up to look at the boy and for a moment I don't even recognise her. The ground tips and rolls under me. I watch as they disappear round the back of the vans.

A dog barks somewhere behind me – it sounds strange and far away.

Then Mrs Bradley walks past. She's walking towards the exit but the way she's going will take her right between the vans. I take a step forward, stop, then take another. Mrs Bradley keeps going, swinging her arms, stepping over all the sticks and bags. The candyfloss is still stuck to her sleeve. My skin burns under the hot lights. She's going to see Gina, any moment now she will see Gina, and somehow I know that, if Mrs Bradley sees Gina, she'll never call us 'my girls' again, and she'll never touch the tops of our heads or tell us that we used to pretend we were horses.

She's almost there. She's at the first van and her head is about to turn, and suddenly I'm rushing forward, my hand is at my mouth, my tooth is out and there's blood and I must have yelled because now everyone is turning and looking at me instead.

Mrs Bradley comes straight over, and then Gina is there adjusting her jumper. Someone hands me a glass of warm, salty water. Blood swirls back into the glass.

The pain comes suddenly, sharply, and I move my tongue around the gum, feeling the rough, fleshy socket. Someone passes me my tooth – I think it's the boy – and I put it in my pocket because I don't know what else to do with it. It's big and black with dried blood.

Gina puts her mouth to my ear – all beer and doughnuts and bubblegum shampoo – and says, 'Let's go get the tree.'

We slip away from the crowd, away from Mrs Bradley, who is explaining to the boy about the best way to remove teeth. We slide the tree out from under the van and go.

The wind has picked up. It's blowing the sea into white water that gleams in the dark. The fair music throbs behind us, and my mouth throbs in the cold, so I keep it shut.

We take the turning to the coast path, past the fence that marks the edge of the cliff. The path is narrower now, because some of the cliff has fallen away. It's too dark to really see anything but we keep going, following the fence and feeling for when the path turns from gravel to the sand at the start of the dunes. We know the path so well that we automatically step over the cracks and stones and then our feet hit packed sand and we're there.

The dunes have spread since last year. They're lower and flatter, and seem to go on further than I can see. We start walking down, our feet sinking with each step. The sand is very cold and the beach is very quiet. The marram grass rustles, like a group of people waiting for something.

Gina stumbles and puts out her hand. 'Shit, that's sharp,' she says. 'I think I've found the trees.'

We stop and look. I can just make them out – rows of small dark shapes standing stiffly against the wind.

We follow the line until we find a gap. The trees are bare, skeletal, no more than husks. Gina stops and lays our tree down. 'Shall we put it in here?' she says.

I watch the dark blowing sand. It blows over my face and my hair. Suddenly it doesn't seem to matter so much where we put it, or if we put it in at all. The sand will move and bury the trees, the dunes will spread and heap up and flatten again. The trees seem too small; there will never be enough of them.

'I guess so,' I say.

Gina crouches down and starts digging with her hands, scooping out handfuls of sand and piling it up next to her.

I can feel the dunes moving under my feet – big, shifting movements, like an iceberg creaking.

Gina digs hard. She kneels in the sand and leans right into the hole she's making. She works at it steadily, scooping and piling the sand. After a while I kneel down next to her and dig too. The damp grains clog under my nails. We find a rhythm – Gina scoops, then I scoop, one hand after the other. There's just the sound of the sea, and the grass, and our hands in the sand.

The hole is already deep enough but we keep digging anyway, just more slowly. Neither of us stops. But after a while I say, 'That's deep enough now,' and then there's nothing else to do except stop. We stand up and put the tree in the hole and I hold it upright while Gina pushes the sand back around it and stamps it down to keep it in place.

We stand there, not moving, watching the tree.

'Does it hurt?' Gina asks.

Usually I would say something about primary versus secondary lesions – I know all about them from the care home. I would talk about how Mr Samuels, who has been in the home since before I was born, has a gold earring so that, if he'd ever drowned at sea and washed up in another country, he would have had enough money to pay for his own burial. But I don't feel like saying any of that.

'Yes,' I say. 'It hurts.'

The sand blows around us. It's cold and dark. The lights of a fishing boat move past in the distance. I am the first to turn to go. I brush the sand off my hands and start making my way across the dunes. Gina follows me, then, just as we're almost back on the path, she darts away and runs up the side of the highest dune. She runs slowly, fighting against the toppling sand, the way it falls back under her feet. Then she's at the top. I wait on the path. The lights of our bungalows glint in the distance. They look as small and far away as the boat. I think of the sandy streets and the shiny clumps of sea beet. The shortcut down the alley. The cracks in the wall of Gina's front room. I think about those lungs – how they rose up, so full of air, that it was impossible to imagine they didn't really work any more.

Above me, Gina spreads her arms out wide. I look once more at the lights, then I run up the sand and stand next to her. The rows of trees are somewhere behind us. I spread my arms out just as Gina starts running down, and I run too, flying so fast that sand rises up everywhere and we are lost in it; it whirls and kicks up and it's in my eyes, my mouth, it's so loose under my feet that it seems as if there's nothing there.

Gina is ahead of me. She's running so fast that I lose sight of her – one moment she's there, the next she disappears in a scatter of sand. I see an arm, a leg, her pale hair streaming as she moves away, and I am right behind, I am almost right behind, as the dunes below us creak and shift and catch in the wind, like they always do, like they'll always keep doing.

Flotsam, Jetsam, Lagan, Derelict

MARY AND VINCENT LAYTON lived in a small house that overlooked an empty beach. The beach was wide and rocky – there were rocks the size of doors that had been thrown up by the tide, and smaller stones that banked up in drifts. Rows of low, dark rocks radiated out across the beach like the hands of a clock. These were the worn-down layers where the cliff used to be, before it had been whittled down to its bones.

Their house was painted white, with a porch at the side and a garage at the back. There was a road behind it. In front there was the cliff – still being worn down, still being whittled – but they'd been assured it would not affect them in their lifetime.

They'd moved in over the summer. Finally everything was sorted and in order: their work had reached its natural end point; finances were tied up; their children were married and settled. There were no loose ends. They'd been together for a very long time – they could hardly believe how long really – but now, finally, there were no loose ends.

Vincent had found a job to fill his weekday mornings, doing gardening work for people around the town. Sometimes Mary would go and help but more often she would walk along the cliffs or the beach, or just sit looking out. It was quiet and everything else seemed very far away. There was no TV, no mobile signal. They didn't have to think about anything at all.

One morning Mary was out walking when she saw something glinting further down the beach. She made her way towards it. Clouds hung low in the sky – they were pale, almost yellow, like eyes that were old or tired. The rocks were slippery and she walked carefully – if she fell and broke something then that could be it, and all the work, all the years of planning, would be for nothing. She avoided the places with wet mossy weed, and stepped instead on the fat brown ribbons, which creaked softly under her shoes. It was still early. She'd always woken early but now, instead of lying awake in bed, she got up and came down to the beach.

She crossed the rocks and stepped down onto the sand, which was coarse and flecked with colours. Sometimes it looked bronze. Sometimes it looked silver. It always felt cold, even in the sun, and she often wondered how deep it was.

The glinting thing was half-buried. It was a plastic bottle; one of those small water bottles with ridges all around it. The plastic was tinged blue and the top was sticking up amongst all the stones and shells. It didn't look right. It didn't look like it was supposed to be there. She crouched down, scraped up a handful of sand and pressed it over the top of the bottle. Then she dug up another handful and did the same, until it was completely covered. She stood back and looked. There. She couldn't even tell where it was any more. And later, the tide would take it away for good.

The next morning there were five bottles strewn across the rocks below the house.

Mary stood on the path looking down at them. It was mizzling. The beach seemed flatter and washed of colour, except for the blue of the bottles. She went down the path and over the rocks. They were rectangular five-litre bottles and the plastic was thick and shiny. She collected them one at a time and put them in a pile, then turned and looked back at the cliffs, across the sand, and out at the sea. There was nowhere for the bottles to go. They were too big to bury, and there were too many of them. The tide wouldn't reach that far for hours. She thought about putting them behind one of the big rocks, but she would probably still be able to see them from the house. And, even if she couldn't see them, she would know they were there.

She picked the bottles up awkwardly, holding one under each arm and the rest against her chest, and carried them over the rocks. There was a car park at the other end of the beach which had a bin in it. She crossed the beach and went over to the bin. It was overflowing and there were extra bags stuffed with rubbish on the ground underneath it. She put the bottles down by the bags and turned to leave. The wind knocked one of the bottles over and it fell with a hollow thud. Another one blew back towards the sand. Mary watched it moving. It made a scraping sound as it skidded against the gravel. She picked the bottles up again and carried them back across the beach. She walked up the path towards the house and unlocked the garage. There was a shelf against the back wall and she put them on there, lining them up neatly in a row.

She locked the garage and went inside. Vincent was in the kitchen, making lunch. She went up behind him and put her arms around his waist. He smelled of bonfires and paint. His waist had thickened over the years. So had hers. Sometimes their bones clicked. She leaned into his warm back. He reached his arm around and rubbed her hip.

'Our daughter phoned,' he said. He took four slices of bread out of the bag and put them on plates. The kitchen was small and white and clean. There were white plates in the cupboard, a few white mugs, two bowls and two glasses. They had got rid of almost everything.

'I didn't think we'd given her this number yet,' Mary said.

Vincent put cheese in the bread and cut each sandwich in half. He wiped the crumbs up carefully. 'Something's happening with Jack again.'

Mary watched as Vincent got up the last crumbs with the tip of his finger. He passed Mary her plate and picked up his own.

'Let's eat these in bed,' Mary said. They could do things like that now. They could close all the curtains and afterwards they could sleep until dinner if they wanted to. There was nothing to stop them.

The next morning she took her usual route out of the house and along the path down to the rocks. Before she opened the gate she stopped and scanned the beach. For a moment she thought she saw something glinting and her heart began to beat faster than usual. But it was nothing. The beach was clear and empty. She undid her hair and let it stream out. She started humming. There were shallow pools among the rocks and they rippled in the wind.

She took the long way round, past her favourite rock, which was covered in a dark sheet of mussels.

Her boots crunched on the stones. She passed heaps of seaweed that must have been pushed in by the tide. Some of it was orange, and some was blue, and there were hundreds, maybe thousands, of tiny blue and white shells.

She reached down and picked up some of the seaweed. She liked the way it popped under her hands. But it wasn't seaweed. This was stiff and tough and fraying at the edges. She dropped it and looked at the other piles. They weren't seaweed, none of them were – they were heaps of twisted nylon rope. She crouched down and picked up one of the shells. The edge of it dug into her finger. It was a fragment of plastic. All along the tideline, as far as she could see, the beach was covered in small, sharp fragments.

She turned quickly and went back to the house. The wind knotted her hair into clumps and she tied it up tightly away from her face. She looked through the cupboards for the bin bags. There weren't any left. She went into the garage and found the bucket, which they used to clean the car and the windows. She took the bucket down to the beach with her, knelt in the damp sand, and started picking everything up – rope, plastic, translucent strips of polythene. After a while she stopped doing it piece by piece and scraped up entire handfuls. When the bucket was full she stood up and stretched her legs. Her back was stiff and there was a faint, dull ache in the joints of her hands. She carried the bucket back to the house, opened the garage and emptied it onto the floor. Everything spread in a tangled heap. She locked the garage and went inside.

Vincent was pouring drinks. 'Where've you been?' he asked. He kissed the soft skin on the side of her neck.

'Just my usual walk,' Mary said. She took off her coat.

'You missed lunch,' he told her. He emptied crisps into a bowl and passed them to her.

Mary looked at the clock. Her stomach was empty. When she reached into the bowl the salt stung her fingers.

There was a letter on the table. The envelope was thick and cream-coloured and headed with Vincent's old company's logo. It was still sealed.

Vincent saw her looking and went over to the table, but neither of them opened the letter.

'Why are they writing to you?' Mary said.

'I don't know.'

'I didn't think they needed to write to you any more.'

Vincent took the letter and put it in the drawer underneath the phone. The white kitchen and the white lights made his skin look almost grey. It was Mary who'd first persuaded him to take that job, even though he hadn't wanted to. There were other things he'd wanted to do. She held his hand. Their fingers laced between each other's. Vincent reached over and picked something out of her hair – it was tangled in and it took him a moment to loosen it. It was a strip of blue plastic. That night, when she was getting undressed, she found another strip caught in the cuff of her shirt.

The next day she got up early, took the bucket and went straight to work on the beach. She picked and sifted until her knees throbbed and her hands felt like they were about to seize up. The more she picked up, the more she saw – there were ring pulls, tin lids, bottle caps, tags, rusty springs coiled under stones, watch batteries, translucent beads that she could only see if she squinted, hidden among the grains like clutches of eggs. There were bits of Styrofoam that were exactly the same colour as the sand, and bright specks of glass.

The sun slipped down lower. The tide came in. Finally she stopped and stood up. She'd only covered a few square metres.

When she went to bed there were bits stuck to her feet. When she brushed them off they scattered across the floor and fell down between the boards. She got up and tried to pick them out without waking Vincent. He murmured and reached for her. She got back into bed. Bits of plastic blew in on a draught under the door.

On the day before the rubbish collection Mary took their bin bag from the kitchen and put it by the side of the road. She'd bought a new roll of black bags, and she took them down to the garage, unlocked it, and went in. The room was full. A fetid smell rose up, like something in a ditch that hadn't drained away. The floor was a teeming mass of boxes and crates, ropes, plastic bottles, wet shoes, chipped and broken toys. There were reams of greasy netting with tins and plastic beads and pen lids caught in them; and a heap of oil cans and rubber gloves and mouldy bits of fabric. In the far corner there was a pile of sand and a sieve. Sometimes things looked like sand, but they weren't sand, really.

She stood in the middle of the garage and looked around. There was so much of it – it was piled halfway up the walls. She gripped the roll of bags. What she needed to do was fill each one and then leave them out for the collection. Then, by tomorrow, it would all be gone. She went over to the edge of the pile and started filling the first bag. She filled it quickly, tied the top and started on another, breaking up the boxes and crates, not stopping until everything in the garage was cleared. It took a long time. When she'd finished she dragged the bags outside one by one and put them by the road. A car drove past and slowed down, looking at the vast, toppling pile. Her cheeks burned. But they would be gone by the morning.

Vincent was waiting in the hall. 'We'd better go,' he said. He was buttoning his coat.

'Go?'

'To the Gleesons'. They invited us, remember?'

'I don't remember,' Mary said.

'They said we should go over.'

'Why?'

'I suppose to have a drink. Talk.'

'Talk,' Mary said. 'About what?'

Vincent leaned down and tied his shoes. 'I guess they want to get to know us. Where we lived before, what we're like, what we did.'

Mary leaned back against the wall. 'Before?' she said. She could still smell the rank saltiness on her hands. It was probably on her clothes. She took off her coat and her shoes. She slipped her hand up the back of Vincent's shirt. 'Let's stay in,' she said. His skin was creased and soft. She knew each bone of his spine.

Vincent phoned in for them and they spent the evening listening to music and eating leftovers from the fridge, with the radiators turned up high.

The wind picked up and surged all night. The tiles clattered like bits of stone falling off the cliff. Hail chipped at the windows. Mary lay awake listening to the waves hitting against the beach. She thought about the bags out on the road. The palm tree scratched against the wall. She sat up suddenly. Where would it all go, after it had been collected? It wouldn't really be gone, would it? It would just be somewhere else. It would be somewhere else, instead of here. Maybe, eventually, some of it would end up back on the beach. Her heart beat hard, almost painfully. She couldn't think about that. She'd done what she could. Eventually she lay back down and closed her eyes.

She slipped out early the next morning, while Vincent was still asleep. There were two messages from their daughter on the answerphone. The red light flashed slowly.

When she opened the front door it hit against something. She pushed harder but it still wouldn't open more than a few inches. There was something on the other side – she could almost see it through the letter box. She shoved harder and the door finally opened. There was a pile of wet netting slumped against it. She pushed it away with her foot and went out. The grass was strewn with rope and shoes and tins. Some of the bin bags had ripped open, some had tipped over and come untied, some had rolled down the path and burst. Plastic had been flung across the road. There were bottles and strips of cardboard caught in the hedge; packets flapped on the ground like injured birds. There was a rubber glove pressing against the downstairs window.

Mary stood in the middle of the garden for a long time. Then she turned, picked up her bucket, and walked slowly down to the beach.

The sand was churned; stones had been flung around into new trenches and drifts. Water trickled off the cliff in thin streams, as if a cloth were being wrung out. Mary looked across the rocks, deciding which way to go first. There was something bright further ahead, on the other side of the beach – a row of something that she couldn't quite make out. The rocks on that side were taller, more jagged. She didn't normally go that way. The early-morning sun flashed on whatever it was. They looked like discs. Mary closed her eyes but still saw the shapes on the backs of her eyelids.

There were no flat places to rest her feet so she just kept going – stepping quickly from rock to rock without giving herself time to lose balance. Finally she could see what they were – it was a mass of hub-caps, piles of them, like a stranding. Some had been thrown up on top of the rocks. Others were cracked in half. Her heart beat hard again. There were so many of them. More were washing in and rolling at the edge of the tide.

She left the bucket and picked up as many as she could, tucking them under her arms and holding a stack in both hands. Then she turned and made her way across the rocks. She would leave them by the path, then go back for more.

She was almost on the sand when she slipped. She reached out with her foot but found nothing. The hub-caps clattered down. She stretched out her arm but still there was nothing, then her wrist twisted against the ground and something rough grated against her cheek. The stones and the sand were very cold.

When Vincent found her she'd managed to drag herself so that she was almost on the path. He leaned her against him, taking her weight, and walked her slowly back to the house, lifting her with each step.

'What have you been doing?' he asked. He gently prised the hub-cap out of her hands.

She couldn't get out of bed. Vincent brought her breakfast on a tray in the morning, then, when he got back from work, he made lunch and they ate together, sitting propped up on the pillows. He brought in the radio and rubbed her swollen ankle while they listened.

The room was small and bare and white. Once, a piece of plastic, or maybe a wrapper, caught on the window and flapped in the wind. Mary closed her eyes. When she opened them again it had gone. The walls smelled like fresh paint and she lay there, breathing it in. This was how it was meant to be: the quiet, the sea somewhere outside the window. She slept deeply and for a long time.

One lunchtime she woke up from a nap and Vincent wasn't there. 'Vincent?' she called. 'Are you back?' She was hungry. She tried getting out of bed but as soon as she put any weight on her foot it wrenched and gave way. She sat back down. It grew slowly dark. Finally she heard the front door open and a few moments later Vincent came in. His hands were cold but the tops of his cheeks looked very hot.

'It's late,' Mary said.

'It was work,' he told her. 'I overran doing the Millers' garden.' He brought her tea and a sandwich and straightened the covers. He sat next to her and switched on the radio. He turned the volume up high.

After a while he said, 'Do you ever think about it?'

'What?' Mary asked.

'It was your parents' and I …'

Mary must have moved suddenly because a shot of pain went through her foot. 'Why are you talking about that?' she said.

'I just thought about it today.'

'We said we wouldn't,' she told him.

Vincent nodded. He turned and patted her pillow so that it was more comfortable. 'I should have checked it all out,' he said. 'I don't know why I didn't check.'

'We said we wouldn't go over it any more,' Mary said. All that was done with now. She hadn't thought about it for a long time.

Vincent was late back again the next day, and the day after that. He fell asleep straight after dinner but woke up through the night, his legs and arms moving restlessly.

The following morning he was gone before she was awake. Mary's breakfast was on the bedside table. She drank cold tea and ate cold toast. The phone rang. She got up and put her foot carefully on the floor. There was a dull ache but she could stand. She walked slowly through the house. The phone stopped ringing. The message light was flashing red. The kitchen was clean and quiet. There was another unopened letter from Vincent's company on the table.

She went out and turned towards the living room. The door was closed. It was never closed. The phone rang again and she went back into the kitchen. She watched it ringing for a moment, then picked it up. 'Hello?' she said. She nodded slowly, said a few words, then put the phone down. Vincent hadn't been showing up for work.

She opened the living-room door. The room was full. Every shelf, every inch of floor space, every chair, was covered with things from the beach. The hub-caps were stacked in tall piles, like coins. There were fruit crates, balls of rope, a bag overflowing with what looked like computer parts. There were fishing buoys – some orange, some green, some bleached to no colour at all. Sheets of plastic leaned against the window, casting a warped light. The room smelled stale, but also humid. Water droplets collected and rolled down the walls.

She went over to the window and looked out. Vincent was standing near the edge of the water, on the far side of the beach, staring at something. She closed the living-room door, put on her shoes, tied them carefully over her ankle, and went outside. The wind was picking up again. The tiles clacked. The palm tree was frayed. There were bin bags stuffed full in the porch and more along the side of the path.

She went down to Vincent and slipped her arms around his waist. He put his hand on her hip.

'It was just there,' he said. 'I came down and it was just there.'

Mary followed where he was looking, past the rocks, and over towards the water. At first she thought it was another rock. It towered up next to the cliff. Then she saw dark red metal. There was some kind of writing painted on the side. It was a shipping container, almost the size of the house, draped in seaweed and barnacles. It was padlocked. The metal was thick and corrugated. One side was bent inwards, like a chest when someone is holding their breath.

'I thought it was going to be different,' Mary said. She held Vincent tighter and leaned into his back.

'Maybe by the morning …' Vincent said.

But they both knew it would still be there in the morning. It was, perhaps, unmovable.

They stood there, together, watching it.

The Life of a Wave

All waves begin as nothing more than a wrinkle, called a cat's paw – a crease in the surface of the water caused by a gust of wind.

You're lying on a blanket on the hot sand and shadows move across your eyes. Above you, there is the sky and in it there's a hard, white crescent. You reach out and try to cram it in your mouth. Sand scratches your elbows. The blanket smells sweet and dusty and there are small stones under it. Something is drumming somewhere. It's deep and regular and it sounds close one moment and far away the next. You can't see it. You rock and kick your feet but you still can't see it. You roll onto your side. The drumming comes through the ground and into your ear.

A shadow moves and grows bigger and your father sits down next to you. Heat radiates off him in waves. There's a lot of bright orangey hair on his legs and neck and when you pull on it he sucks in his breath and says, 'He's trying to kill me.' His hand is hot and damp when it unclenches your fist. You try and grab again. You want him to come closer but he's holding something and now he's looking at that instead. It's another body. It's red and blue and he's pulling it over himself. The sun beats down. He sweats and breathes heavily. Sometimes he grunts. He pulls the rubbery legs up over his knees and his elbows fly out. Your mother says, 'Merl, mind the baby.' She puts down her book. Your father swoops towards you, except it's not your father – it's some kind of dark bird. There's no skin, there's no hair. It slips and creaks under your fingers. It smells hot and burning and you cry out because he's disappeared.

In the night he stumbles into your room to quieten you. His face looms, then turns away, like the moon's does. Sometimes he paces the house and your eyes follow him as he crosses from room to room.

He shows you the sea from your window. He lifts you up ('By God,' he says, 'your heart's going about a thousand times a second') and there is a glimpse of it in the distance. It glints and it's different to the sky and that is where he always goes. He holds you for a long time, until he lets out a shout because you've peed down his wrist.

The deep galumphing noises are the same as the ones in his chest.

Your mother goes away for the day and your father looks after you by himself. When you cry he gives you a chamois leather to hold, and you clutch it, gumming the soft corners. You fall asleep. When you wake up you're on the beach. You reach out and grab for him. There's a dent in the sand where his surfboard has been, and footprints disappearing down towards the sea.

This is the story that's told: an hour passed and then another. You slept and then woke up and the friend of your father's who was watching over you, knowing nothing about babies, bought you an ice cream which melted over your legs. You were taken to the lifeguard hut and someone put a jacket over you. You clutched the chamois leather. Eventually your father came back in, wet and dripping, elated: the waves were glassy, he caught a tube at the end. When the lifeguard handed you over your father blinked, laughed, then slapped everyone on the back. He'd completely forgotten you existed.

The wave grows as it comes into conflict with the surface tension of the water. This is called the capillary wave.

You grow up with a language you think everyone knows: lumpy, crumbly, clean, hollow, walling up. You know reef breaks and shore breaks. You know of faraway, gigantic waves: Waimea, Mavericks, Pipeline. To you they are as strange and magical as Lapland. You know that to drop in on someone, to steal their wave, is the worst crime. In town, when your mother says that the two of you could drop in on her friend, you scream and won't go in the door.

When you stand at the edge of the water the waves tug and hiss at your feet. They break suddenly; rearing up then smashing down, like the jars you throw as hard as you can at the recycling centre. Your father wades in ahead of you and, after a moment, you follow. Seaweed wraps around your toes. Something sharp hits your ankle. You're in over your waist. Cold water flushes down your trunks. You turn and look back. There's your mother on the sand, holding your baby sister. They look small and far away. You take another step forwards, then turn back again. Everything onshore looks different. You hardly recognise it. You wave, and someone waves back, distantly.

You find an injured crab and bring it home in your pocket. You put it under your bed and bring it crisps and grapes. After a few days you forget to check. When you look again, the crab has died. The smell stays in the room for weeks.

You lose your chamois leather. You look everywhere for it, then crawl and hide under the table. Your mother asks what's wrong. A few minutes later your father comes in from the garden. His voice is louder than usual. His eyes don't quite meet yours. He leans down to you – his face is huge, it crowds out everything; it's flecked with golden stubble and bits of soap and shaving nicks – and then he presents the chamois with a flourish. You take it slowly. It's damp and it smells strong and different and there's wax around the edges.

Your mother reads you a book about the moon. You can't stop thinking about it. You get up at night and look at it out of your window – sometimes you can see the face clearly, sometimes it's sly and shadowy and you can't make it out at all.

When he waxes his surfboard you stand behind and watch. He leans down and scrapes the old wax off. The flakes curl onto the ground like bits of grey snow. Then he wipes the board down with a rag and a bottle of something called turpentine. When he's finished you're allowed to rub on the new layer. The block of wax is shaped like a lady's you-know-what.

When he puts on his wetsuit it's your job to pass him his zip. The zip is very long and he can't reach it by himself. You do it carefully and solemnly. It is your task alone.

Finally, you're given a lesson. It's so windy that you can hardly carry your end of the board across the beach. The waves don't look that big from the car park but as soon as you step into the water they seem to tower over you. Spray stings your face. The board knocks against your shoulder. You can't even climb onto it. You finally get one leg on it and then you slip off the side. A wave flips the board and knocks it against your neck. 'Move forward,' your father shouts. You can hardly hear him over the wind. 'Move forward.' His main teaching technique is to say the same thing, just louder. 'Where?' you ask. Another wave hits you and you roll under, swallowing what feels like lungfuls. Your throat and stomach burn. You come up retching and try to climb on again. The board seems huge and awkward now that you're on it. The waves are relentless. Your father is talking to someone who's just about to paddle out. They shout and laugh. They talk about yesterday's waves. He nods to someone else. He knows everyone. The waves don't knock him off balance; they seem to pass through him, glistening. You fall off again. The board thwacks you across the arse and you rip off your leash, trip, half-running, half-wading back into shore. Your father picks up the board and starts to follow you. Behind him, a perfect set develops. He turns to watch it. You know what he wants – he wants to go back out and catch them. He looks at you. You fold your arms and demand to be taken home.

In the summer he hoses off the salt and leaves his wetsuit dripping on the line all night. By the morning it's dry and stiff and ready to put on again.

Water particles in a wave don't move forward: the wave moves through the water particles, which stay in exactly the same place – these are called rolling particles.

In the winter, you watch him break ice off his wetsuit before he puts it on. You're in the steamy kitchen, eating breakfast and playing slaps with your sister. He chips away the ice from his hat, his gloves, his boots. His leash is an icicle. You shake your head and turn away.

You catch him phoning in sick to work. You should have left for school already but you've forgotten your PE kit and have to go back. He isn't sick but he's been working ten-hour shifts at the warehouse, coming in late smelling of sweat and dust. He puts the phone down and unbuttons the top of his shirt. He sees you standing there and something passes across his eyes – not fear exactly, not exactly distaste. You see yourself suddenly as an interloper: skinny, buck-toothed, always hungry, a rash of coppery freckles on your forehead. 'I forgot my bag,' you tell him, holding it up for evidence. He nods, watching you carefully, then his bear paw comes down and ruffles your hair, boom, boom, kneading his fingers into the back of your neck.

You decide you will be an astronaut. The moon seems very still. Nothing up there would drum and knock you off your feet; you would just float, silently.

You find a mouse behind the fridge. The cat must have brought it in and then lost it. The mouse has made a dusty nest and had five babies. You put them all in a shoebox and keep them in your room. The first few days you dig up worms but the mice don't seem interested. You switch to cheese but they don't eat that either. After a while you forget about them, and when you remember and open the box, they're all dead.

In school you learn that to be an astronaut you must not only be top of the class, but also fit, strong and mentally resilient. You start doing sit-ups every morning. You sit in your tiny wardrobe, very hunched and still, for half an hour every day.

One Saturday morning your father takes your sister down to the beach for her first lesson. She has your old wetsuit and your old polystyrene board. You wait in the house. You can't settle to anything. You walk down to the library, where your mother works. She's laughing with the other librarian about some graffiti a kid has left in a *Where's Wally?* book. She shows you but you don't crack a smile. You grab the first book you see and walk home. You pace the house. When you hear the car you run upstairs and lie on the bed with the book. The doors thump shut. There's no other sound – no talking, no laughing. You turn the pages without reading them. Then your sister laughs and your father squirts water at her with the hose. You close the book slowly.

You can do twenty sit-ups, and you can sit in the wardrobe for an hour – any longer than that and your hips cramp up.

He misses the school play. You've been cast as a servant and your sister is Juliet. Your sister is secretly in love with Romeo – a boy called Jackson who wears henna tattoos. The seat next to your mother is empty. She puts her coat and bag on it. Halfway through, there's a noise at the back of the auditorium and your father comes in with dripping hair. 'Did I miss much?' he asks.

He has a black eye. 'Someone dropped in on me,' he whispers loudly. Everyone has to stand up as he moves down the row. His hair drips onto the floor. Your sister flushes, almost forgets her lines. 'Is that your dad?' Romeo mouths.

Your sister stops her surfing lessons. The wetsuit and the polystyrene board are sold.

You can spend two hours in the wardrobe. The time seems to fly by.

His black eye takes two months to disappear. First it's purple, then green, then yellow. Your mother says he deserves it, then gently holds a bag of peas on it for a few minutes every day.

Things always seem to happen when your father is out in the water. For example, your sister falls through the downstairs window. You were chasing her around the house because she used your telescope without asking and now, somehow, it's broken. She wasn't even looking at stars – she was trying to see if she could spy on her friend on the other side of town. There's a lot of glass but not as much blood as you might expect. When the ambulance comes everything goes very quiet. You're too old to do it, but you hide under the kitchen table. The ambulance doors shut and everyone leaves. The house is dark. When your father comes home he opens the door and calls out, then he switches on the lights, hums, cleans off his wetsuit and hangs it on the line. He turns on the radio to a station you don't recognise. He sings along. He opens the fridge, heaps food onto a plate and eats with his hands: tearing bread, wiping sauce with his fingers. You crawl out. He jumps and bellows, almost chokes on a slice of cold chicken. You explain where everyone is. It takes him a while to understand. He hasn't noticed the broken window. The calm glassiness in his eyes slowly fades. He sits down and rubs his jaw, then drives you both to the hospital.

The next time he goes in your rabbit, Millicent, is run over by a motorbike.

Another time a swarm of bees comes down the chimney and gets trapped in the living room. A bee catches in your sleeve and stings you on the wrist, even though you were trying to help it. The sting swells to the size of a planet.

'How?' your father asks each time. 'How has it happened?' You don't know it yet but what he means is: how can life, this other life, have carried on so drastically without him, while he was just drifting on his board, a basking shark slipping under him like a submarine?

Each time he paces the house. No one speaks to him. Everything has been taken care of without him – he has not-helped, he is not-needed. He frowns and mutters to himself. He looks around but no one looks at him. He mutters again. Then, suddenly, he runs out. He takes the box of bees, which your mother stunned with smoke while wearing a snorkel, and leaves it at the bottom of the mean neighbour's garden. He draws rude, elaborate pictures over your sister's cast. He digs a grave for the rabbit under the buddleia, puts on a suit, and conducts a service: prayers, hymns, a soaring eulogy that speaks of her kindness, her penchant for cereal and how she stared at herself, almost smiling, in the glass of the oven.

When you walk past the beach on your way into town, your eyes scan the water. There are many small, dark dots drifting out the back. It's hard to tell which one he is. There he is, no there, no there, no there.

The drag of the water against the seabed begins to slow the wave down.

You bring home a girlfriend. She has waves in her hair and a nose that broke and healed crookedly. Over dinner, your father is loud and animated. The light shines on his orange hair; his skin looks darker, more golden. He stands up to speak and moves his arms in wide, expansive gestures. He presses more food on everyone and makes them all laugh. There's roast chicken and he pulls the skin off in sheets and rolls them delicately – usually you both do it. He looks over at you. You don't do it. You eat quickly and ask to be excused. Just as you're getting up he launches into a long, complicated anecdote about you – the time that you locked yourself out of the house and tried to get in through the roof hatch. The mean neighbour saw you and phoned the police. Your girlfriend laughs in all the right places, but this actually happened to your sister, not to you. No one mentions it. You roll your eyes and say, 'Yeah, good one, Dad.' Then you go upstairs with your girlfriend and into the wardrobe. It's good that you got so much practice in that cramped space. You hear your father calling to you, and you both sit there, silently, the light striping you through the slats, your underwear slung over your old bears and books and space rockets.

You get up late and come home late. By the time you're up, he's already left for work. By the time you get back, he's asleep, sometimes on the sofa, his head tipping up against the cushions, his mouth slightly open as if he's about to speak.

His wetsuit drips on the line. It has a rip down one side and the gloves are fraying.

You are applying for courses in astrophysics. They are competitive and far away. Your girlfriend tells you to stop worrying. She takes off her shirt. She closes your books. When the exams come you race through the papers, writing essays, ticking multiple choices. You don't look up. You reach the final section. It doesn't seem familiar. The questions are strange and unfamiliar. You haven't revised for this – there is a whole section you haven't revised for. You put down your pen. You look up.

Your father is having some trouble at work. He has to go in at weekends, and in the evenings. There's something wrong with the accounts, there's talk of cutting back, redundancies, but he doesn't want to talk about it. What he does want to talk about is the rumour of the fifteen, maybe twenty-foot wave that's going to hit the coast in the next few weeks. He's going to try and ride it. He asks if anyone will be coming to watch. Your mother will be away, visiting friends. Your sister shakes her head. You're about to do the same but then you stop yourself. Exams were a long time ago, the summer is dragging. You have nothing to do except wait. You say that you'll go. Your father glowers – his whole body darkens and seems to fill the room – he glowers like he does when he wants to disguise being pleased.

On the day of the wave you can hear the sea from the front door. He packs the car and puts the board across the front seat. You sit in the back, behind him, like a little kid. Your father talks non-stop, almost babbling, about angles of approach, velocity, the height of the drop. When you get there, the car park is full. There are surfers everywhere and lots of people with cameras. He gets out, unpacks and gets changed. He's still talking. He puts on his boots and gloves. You do up your coat. He locks the car and you follow him down the beach. The waves are huge already – they're dark green and they break like thunder rolling. Your father turns back and looks at the car. Something in his face reminds you of the time he thought there was an intruder in the house and he came out carrying a rolling pin. Or the time he was cornered by that weird dog down the road and it was crawling towards him. 'It's big,' you say. He nods, staring out at the water. 'You don't have to do it,' you tell him. He turns and looks at you, then back at the car. Then someone waves at him and calls him over and the next moment he's gone – he's gone so quickly you don't have time to pass him his zip. You see him laughing, slapping someone on the back. Someone else does up his zip.

There are hands over your eyes and your girlfriend says, 'Guess who?' You say you have no idea: then, when she takes her hands away, you pretend you still don't recognise her. Her smile falters. For some reason you think of those mice in the dark box. You smile and kiss her fingers. You slip your hand into the back pocket of her jeans. You watch the waves. The results came in this morning and you've failed an exam. You haven't told anyone. You're not sure exactly what you'll do, but you guess you'll take up another course instead – there's ones in business or IT that look OK, at the same place your girlfriend is going, which is probably better anyway, it's probably much better this way.

It's impossible to tell how big the wave is until you see your father begin to paddle into it. It rises higher and he rises with it. Then he's up on his board and dropping. For a moment you lose sight of him in the speed and tumult. The roar is almost deafening. Spray hits against your face. The moon is thin as a bone. You see him wipe out for the first time.

While the base of the wave slows, the top rushes on, becoming steeper and more unstable. In a green wave, this top is glassy and smooth, like a mirror.

In your office at your work placement, there's a lot of paperwork to get through. It seems so much, it's like an impossible wave. Sometimes, when you work late, you look out of the window and see the moon and the stars in the yellow-black sky. All the telescopes at observatories around the world will be looking up at this exact moment: Arecibo, Coonabarabran, Jodrell Bank. They are as magical and far away as Lapland. You unpack the food your new girlfriend has made you – cold chicken, tomatoes, bread – it's delicious. You've only been with her for two months. You count down the minutes until you can see her.

One night you're driving in the dark on your way to her parents' house. She's already there. You haven't met her parents yet. You're spending the whole weekend. It's a long drive and you're running late. There's a meteor shower scheduled which is going to start at any moment. You keep driving. You glimpse something bright falling in the rear-view mirror. You don't stop. You're not even halfway yet and the roads are busy. The meteor shower is about to peak. You're very late now. You pull over, stop the car, and get out.

When you visit other people's houses – friends from university, from work, your girlfriend's parents – what you notice is that their fathers are usually there. They are in the kitchen or in the garden or in the garage with drills and paint. They sit and try to talk to you, and you become strange and tongue-tied. 'Uh,' you say. 'Pardon?' They are not rushing around because the tide and the swell and the wind have all come together to make the best possible conditions at an inconvenient time. They are not out chasing anything.

Sometimes, when you're half-asleep, you think you can hear the sea. Sometimes it's the wind in the roof. Sometimes it's your heart beating. Sometimes it's the sound of the motorway in the distance, that restless, relentless thrumming.

Your sister is having a baby. You go and see her and touch her stomach. You can feel the foot sticking out in there. She's grown her hair long and has a dog that lies across her feet. At first you play with the dog a lot, but it always does the same things and soon you get bored and ignore it. The dog seems to understand and shuffles away. There's a scar on your sister's arm from where she fell through the window – it feels like a long time ago, but also not very long ago at all. She craves meat even though she's a vegetarian and you break her in, you roast her a chicken and show her how to peel the skin off in sheets and roll them up. 'You look just like him when you do that,' she tells you.

The wave collapses under its own weight. It topples and begins to break.

He has an injury. You phone up and he answers and speaks in a subdued voice. 'Pardon?' he says all the time. 'What?' He's always had terrible surfers' ear, where the bone, after long immersion in cold water, begins to grow to try and stop the water getting in. He was once told his ears were about fifty per cent covered, but now it's more like sixty-five. 'Your injury,' you shout. 'What is it?' It turns out that he hurt his back hanging out the washing in cold weather and he hasn't been in the sea for months. 'Months?' you say. He passes the phone to your mother.

You pull in out the front and go inside. The house is tidy, spotless actually. There's no surf kit on the line. There's no wax or turpentine anywhere. Over the years he's had a few accidents and injuries: the black eye, a cut leg, a broken wrist, concussion, but nothing stopped him going back in before. You pace the house, crossing from room to room. Finally you find him upstairs, in his bedroom, where there's a new TV. He's sitting up watching it. 'Hanging out the washing?' you say. He shows you his new favourite programme. It's about people who buy old crappy cars and then do them up so that they look new. 'They made that one turn out alright,' he keeps saying. 'They made that turn out alright, didn't they?'

The moon is almost full. It's very pale, very yellow. He says that when he hears the waves now they sound like windows smashing. 'OK, Dad,' you tell him. 'OK.'

You catch him bending down to clip the lawn, his back strong and flexible, his arms stretching out. He reaches up to the top shelf in the kitchen and twists when he backs out of the driveway in the car.

Your girlfriend rings you. Her voice is different – higher, she's talking faster, she's trying to tell you something important. Your hands start to shake. You think you know what she's saying. 'It's definite,' she says. 'I did the test twice.' She cries and then starts to laugh. You do the same. You definitely know what she's saying. The room suddenly feels very small. You get up and look out of your old bedroom window. By God, your heart is going about a thousand times a second. Maybe this is what going into space feels like.

The wind direction is good, the tide is right, the swell is small but likely to pick up. You pack the wetsuits and the boards in the car, get your father and drive him down to the sea. You park up and look out. He buries his chin into his chest. The skin around his jaw seems looser, there's more weight across his stomach and face. He's still a big man but the extra weight seems to diminish him somehow, as if he's slowly disappearing. He watches the sea, leaning back in the seat, holding one hand in the other. You tell him your news. He sits there, staring out. 'Was I …' he says. Then he stops. 'I'll have to find a new chamois leather,' he says. There's a shaving nick on his face, golden stubble, bits of soap. You tell him he should go in. He shakes his head. He looks tired. 'There's going to be a few waves coming,' you tell him. He shakes his head again, but you get up and go to the boot and start taking everything out. You get ready. After a while your father gets ready and you both walk down to the water. You stand there. The waves seem so much bigger down here – you'd forgotten that. They seem to tower over you. They boom and snap. Neither of you move; you just stand there, holding your boards. 'Bit small,' your father says. You reach out and pass him his zip. He passes you your zip. When you step into the white water he carries on, paddling out until he's nothing more than a dark dot.

A wave comes in. You climb onto the board and slide off. You try again and it flips over and thwacks you across the arse. The wave knocks into you and pushes you down. The board jolts away and gets dragged in the backwash. Something sharp hits your ankle. Cold water flushes down your wetsuit. You look back at the beach. Everything onshore looks different and far away. You hardly recognise it. Another wave comes in, and then another. You climb onto the board and start to paddle. The white water is coming, it's pounding into you, and then the board is lifting and you're going, you're shooting forwards and you get one knee up, then half-stand, shakily. For a moment you're above the noise and the tumult. Everything is pushing and pulling but you are suspended, still: a force, for a moment, that is unacted upon. Then you fall off, roll, and come up retching. Your throat and stomach burn. You stand up and steady your board. You look for your father. He's still drifting out the back, even though a set has just washed in. The wind is picking up. Another, bigger set develops and breaks but he leaves those as well. With each wave that passes he disappears, appears, disappears, appears. 'Catch one then,' you say. 'Why don't you catch one?' He leaves another one. Someone next to him catches it, but the wave flattens out halfway in and leaves them floundering. Still your father is out there, waiting. He leaves another wave, and another. He's watching them as they form; none of them are right yet, none of them are exactly right. You don't know it yet but he's waiting for the best one, the one that will be perfect. The one that will bring him in right in front of you, finally, in triumph.

Standing Water

SO THERE'S THESE NEIGHBOURS that live out past the quarry, down a rough track that goes nowhere and then stops at the edge of a slushy field. It's low ground out there, and it rains more days than it doesn't, giving the place a bottom-of-the-well kind of feel. The nettles grow to neck height.

There's her house, and then, almost opposite, there's his. They can see each other easily enough – whose car is there, whose lights are on. They can see through each other's windows. There aren't any trees. There aren't any other houses. No one passes by. It's just the two of them, but they haven't spoken since the ditch started overflowing.

The ditch runs along the bottom of the track, and there's a drain in the middle that serves both their houses. The drain is always blocked. When it rains, the water fills the ditch and starts spilling over. It rushes along the track and over the grass and pools outside their front doors. It happens every month, every week. The drain spits and gurgles and the water gushes out, greasy and rabbit-coloured. It smells like a jug that's been holding flowers too long – that slick dark bit that gets left around the edges. Sometimes it seeps under their doors. Sometimes it seeps through their walls. In winter it freezes to a gristly crust. In summer midges spawn and dance over it.

But they never get it fixed. He thinks it's on her land and so she should be the one to do it – he remembers seeing some kind of clause in some kind of document relating to the boundary line, although he's misplaced the paperwork. She says it's closer to his house and so it's his responsibility – she's measured the distance and there's at least four inches in it.

She puts out sandbags. He buys a stiff broom and pushes the water away with sharp jabs. If they're ever out the front at the same time they carry on in silence. She swings her heavy grey plait down behind her back. It's like the pulley on a church bell except nothing chimes. He pulls the hood of his coat down low, so that only the frayed wires of his beard can be seen. Sometimes someone will shake their fist. When their doors slam, they echo across the fields.

The months pass, and then the years. They watch each other, they know each other's small routines – how, on Mondays, she leaves the house at eleven and comes back at two, carrying a plastic bag with bread and some kind of bottle in it. How he stays in every day of the week except Sundays, when he goes out early and comes back at midnight on the dot, with dark lines below his eyes. How she never watches TV. How he leaves his bedroom light on all night. How she crushes tins so hard for the recycling that they split in the middle. How he carefully cleans his spades. How she checks twice that she's locked the door behind her. How he checks that he's locked his door three times.

Once in a while she looks out and sees that all his curtains are shut. They can stay like that for weeks.

Once in a while he smells smoke and sees that she's having a bonfire; tearing out bits of paper from folders and feeding them into the flames. The cinders land on his van. They're as big as fists.

He knows what days she washes her hair.

She knows what day he changes his bed.

Sometimes, at night, he thinks he sees a torch glinting around the track.

Sometimes, at night, she thinks she sees a torch glinting around the field.

If a delivery comes for her when she isn't in, he doesn't take it. He asks for it to be left outside her door instead. Often it gets wet. Sometimes deliveries don't seem to arrive at all.

When he goes out on Sundays, she flattens the gravel outside her house, which his van has churned into divots. She picks up the sharpest bits and puts them on his drive.

The months pass and then the years. Still it rains most days. The drain blocks up and the ditch overflows and water pools in front of their houses.

One evening, at the tail end of winter, the rain is coming down as thick and heavy as a tap on full throttle. The gutters pour. The drops are fat and grimy and smear on the windows. She's inside slicing the skin off potatoes, when she hears something scraping, then a thud. She goes over to the window and glimpses a torch somewhere down the track. The torch goes out. She starts on the potatoes again. The rain drums even louder. She cuts each potato and throws the pieces into a pan of cold water. They sink to the bottom. Something moves in the pelting rain and when she looks up he's there, outside the window, staring in. His eyes are pale and watery. She puts the knife down slowly. She goes to the door and opens it a few inches. He's hunched by the wall, wearing his mac and carrying a spade. There's mud up his legs and his back, and along both sleeves. His hood is streaming. Water runs off the bones of his face.

'Have you got a spade?' he says.

She squints out past the door. The light outside is dark brown, almost green, mud-coloured. She can hardly see him through it. 'Why?' she says.

'It's blocked.'

'It's always blocked.'

He turns and looks down the track. 'It's pouring over.' His voice is strange, lower than she remembers; it catches in his throat as if there's water bubbling in it.

She closes the door an inch and stands behind it.

He pulls his hood down further. 'I can't do it by myself.'

She sees that his spade is clagged with dirt. 'You've been trying to fix it?' she says.

He turns again and looks down towards the ditch. He clutches the spade. The skin on his fingers is damp and crinkled. He opens his mouth and it looks, for a moment, as if a trickle of something dark comes out, but it must just be mud from his hood. He closes his mouth, swallows, and walks back onto the track. His spade scrapes on the ground. Water pours over the tops of his boots. He blurs into the rain and disappears.

She opens the door wider and stands there, looking out. She can't really see anything except the rain. The water looks higher than usual – there's a pool under the door already and it's still rising. She looks back at her kitchen. The pool of water starts spreading. She unhooks her coat from the peg, pulls on her boots and gets her spade from the garage – it's rusty and buckling but it will have to do. She steps into the water and follows him down to the ditch, bending her head against the force of the downpour. He's walking slowly, almost bowed over, with one leg dragging. There are slabs of mud under his boots.

The ditch is overflowing fast. It's gritty and thick and slopping out like bathwater going over the rim. He bends down and starts digging, bringing out spadefuls and throwing them over his shoulder. The ditch looks much wider than it was before. The bank has collapsed down one side, water is flooding out, and there's mud and weeds choking everything. She leans in and digs, scooping out stones and roots. Sludge smears up her legs and hands.

After a while she notices that he keeps stopping and stretching his back. He tries to straighten it but it's stooped, as if there's something heavy pressing down on it. Sometimes he turns and spits behind him. His spit looks muddy. There's mud in his teeth and under his nails. It looks like there's water brimming under his coat.

She digs deeper, scooping out spade after spade of dirt and crushed nettles. The rain is like hands pummelling. It roars in her ears. The pile of mud is growing behind her, but the ditch is so deep and there's so much slipped mud, it doesn't seem like she's even cleared half of it yet.

'When did this happen?' she says.

Another bit of the bank slides into the water – a chunk of clay and roots that calves off stickily.

She keeps digging and the shape of the ditch becomes clearer. It is bigger. The sides have been scraped and cleared and a channel has been dug through the bank. She digs again and finds a section of pipe that wasn't there before. It definitely wasn't there before. She digs around it. The pipe is long. It's been laid in the channel and comes out onto her side of the track. Water is pouring out of it and diverting straight onto her land. She digs again. The pipe isn't finished, but there are more sections in there ready to be joined together, and more channels that, once they're finished, will siphon all the water off in the same direction.

She looks at his muddy coat, the mud on his spade. Suddenly she knows exactly what he's been doing.

'You underhanded bastard,' she says. She clenches the spade, turns around, and raises it in the air.

He's gone. He was standing right next to her and now there's nothing. The rain is dark and cold. She scans the track, the field, but there's no one there. She looks at the bank – there are no footprints where he was standing. The mud is smooth and clear, the grass is untrampled. She curses and swings her spade down deep into the water. It hits something solid. It doesn't feel like rock, it feels softer than that, but harder than the mud.

She wades into the ditch to get a better look. The water sucks at her boots. She digs in again and hits the same thing. It's down underneath where the bank has collapsed. It's big. She prods carefully, loosening, lifting the mud around it. She gets her spade in one more time but can't prise it loose. She kneels down and puts her arms into the water. It's deep. It comes up past her elbows. The chill comes up past her neck. She reaches down and feels around and her fingers catch at something. She grabs it and pulls. It's heavy, and it seems to be caught down there, wedged in under the piles of slipped mud. She pulls harder, leaning back with all her weight. The thing shifts. She reaches down, untangles a root, and moves some hard lumps of clay. She has a handful of dark, greasy cloth. She pulls again and steps back and more of it slides out.

Boots appear, then legs, a mac that's sopping and daubed in mud, a hood that's pulled down low over a wiry beard.

She picks up her spade and stands there for a long time. She looks down at the pipe, at the new channels cut into the bank, and the water pouring out across her land. She looks down at him. There are three deep spade marks – one in the top of his head, one in his chest, one in his thigh. His skin is as waxy as when you dig up a potato. He's sodden and cold and his teeth are gritted in his jaw, almost, maybe, as if he's grinning up at her.

Well, that's what she says happened anyway.

A Year of Buryings

January

The first was Lily Ennis and she did not go peacefully. It was just as the old year slipped into the new one and she clung on, gripping the chair. She'd always thought that the next year would be different – her son would phone, people would visit, someone would beg her to stop smoking. After all, it wasn't up to her to do everything, was it? She clung on. She waited. She's still waiting, actually. When the phone rings, there's a dry intake of breath. In the mornings, there are dents in the arms of the chair.

Next was Riley – sometimes, out of the corner of your eye, you might see him wandering around, talking to himself, wringing his hands. Maybe it's because of the things he did, or maybe it's because of the things he didn't. There he goes now, with his head down, hurrying past.

It's hard to keep track sometimes, what with all the comings, what with all the goings. Every bloody year it's like this – everything changing, everything staying the same.

Violet always swore she'd be kinder.

Lenny always said he'd fix that loose rung on his ladder.

Selwyn made a thousand resolutions. He was going to see an iceberg. He was going to settle down. He would never be late again. You can see his car on its side in the hedge. In spring, it gets covered over when the leaves grow thick. In winter, it shows through again. Thistles are growing in the tyres. There's a bird's nest in the exhaust.

And then there was that man at the crossroads, what was his name? He didn't make any resolutions at all. For him it was better if nothing changed. He'd fallen in love once and look how that turned out. One year blurred into the next, keeping himself to himself, following his own strange routines. He lived by the bus shelter (there's no buses, the timetable's ten years old; there's just a mouldy armchair someone dragged in to help with the waiting). Every year the nettles come up, every year they die back down. He was keeled over in the verge for a month up there before anyone found him. He was up there for a month listening to a badger scratching up his bones.

February

That's the thing – there's a constant shifting of earth round here. It's hard to keep track of it. Bulldozers, cement lorries, gardens ripped out for parking. Houses go up, houses come down. I'm supposed to be remembering, I'm supposed to be getting it all down, but all I can hear is digging and hammering.

Now someone's tapping on windows. Who is it? It's Jameson with his stick, out in the rain again, trying to remember where he used to live.

Wanda tried not to remember. She lay in bed listening to the young people speeding along the top road, doing two-wheelers round the bends. She'd done it herself a long time ago. She'd clipped someone in the dark. God, who was it? She hadn't stopped. She hadn't told anyone. The beat of the engines kept her awake all night. Finally she got up and climbed out of the window. She stood in the road with her arm raised to slow them down. The headlights passed straight through her.

Bradley gritted his teeth and swerved. He crawled out of the windscreen without looking back. He was only nineteen. Now he hangs around on empty farms and outside pubs. No one wants to be the one to say it, but things haven't turned out too differently for him after all.

It turned out Acer had a flair for making enemies.

It turned out everyone was a little bit in love with Annie.

Franklin, meanwhile, died of heartbreak. Don't let anyone try to tell you it was a stroke, or angina. He had all the years mapped out, and then he woke up and his girlfriend's shoes had gone, and her coat, and that suitcase she'd been keeping under the bed. He wrote letters and phoned her about a hundred times a day. He hung around outside her house until they made him stop. There's a holly tree growing out of him now. It's already covered in dark berries. When you walk past it reaches out and tries to snag your clothes.

In the next row across, there's a couple who'd been married over seventy years. They were buried side by side, as close as it was possible to get. If you look carefully you might see a gap appearing, hardly anything, a millimetre, then another millimetre – the kind of thing the gravedigger, a man of geography, would put down to shifting bedrock, or the damn moles, or one of those sinkholes that may one day open up and swallow everything.

March

The cliffs crumbled. A pockmark eroded into a crack, the crack widened until a small overhang broke off and shattered on the beach, exposing smooth shale underneath; the disgruntled fossil of something that remembered better days, when this place was a vast lake next to a mountain range. No one really noticed.

No one noticed the woman going in for a swim either. It was the last day of her holiday and she had that not-quite-left, not-quite-here-any-more feeling. Her bags were packed in the car. The cleaner was sweeping her away. The rip took her out almost to Lundy.

It finally gave back Mr Edwards, though, about forty years too late. There he is in the shallows – washing in and out, thin as a strand of kelp, practically see-through actually. Not a kind man. No one wants to be the one to say it. But maybe a changed man, what with the way the water has smoothed out his bones, the way those barnacles have found the gaps between his fingers.

When Jamie Silver washed off his boat in a storm he made sure he wouldn't be found. He was a claustrophobic man – he didn't want a wooden box, or the hard, packed earth. He'd survived a war and three divorces. Someone had once hit him in the dark with their car. He knew the currents very well; he knew the deepest channels and the rocks he could snag on to. He knew how far he needed to go so that no one could bring him in, or fix him down, or say his life had been this or his life had been that. As far as he was concerned people could look, the meddling, good-hearted bastards, but thank God they'd never find him.

April

In April there was no one.

A dolphin was stranded on the stones. There was no one around apart from the fossil and the fossil didn't give a shit. It had seen it all before. The dolphin started drying out. It forgot what being a dolphin was, until the tide came back in and it swam away and remembered.

A gun went off but I don't think anything's come of it yet.

Someone else came down to retire but I don't think anything's come of that yet either.

It's hard to keep track though. Another supermarket went up practically overnight. Someone built a house in their back garden. The mud got churned, there was nowhere for the rain to go. The last lumps of snow melted and mixed with the soil to create a sort of slush – not liquid exactly, not exactly solid ground.

Lyn slipped into the river on her way home at night. Water poured into her shoes. It was very dark, very cold, and something heavy came down, like a lid closing over her. But she knew she'd get out. She'd had her palm read once and they'd said that her life would be full of tall, exciting men, and she hadn't met any of them yet.

Lenny's nephew almost used the same bloody ladder.

Yardley woke up during his funeral, banging etc.

No one knows what's happened to Lonnie.

Lizzie Wheeler found out about the radioactive properties of granite. How the radon leaked out every day, every minute. How it was odourless and colourless. She started seeing clouds of gas hanging in the air. When she found a lump in her side, nestled under her ribs, she knew exactly what it was. The doctor referred her to the hospital for tests. She felt the lump every day, got to know its precise dimensions, its peculiar firmness. Then, one day, it disappeared. Reabsorbed, the doctor said, shaking her head. On her way home, Lizzie saw a cloud of gas right in front of her. She shivered and walked the long way around it.

And did you hear about Pinky Rowe, who walked unscathed in that lightning storm, carrying a metal umbrella?

May

In May the rain came. You know the kind – warmish, dampish, turning everything into pulpy paper. There's nowhere to go in rain like that. Nothing to do. You suddenly realise the long distances. There aren't many trees to hide under.

In Mikey's house they found washing stolen off the neighbour's line – socks, a nightie, three soft pillowcases.

In Sal's there was a mace under the bed.

June

Someone's watching out of that window. There's a face, too blurry to properly see. It's sort of mottled, sort of furtive-looking. Whoever it is they're just standing there, watching, listening, even though someone's come in and taken away the furniture, even though the curtains are down and the electricity's been switched off. Look – their face is pressed right up to the glass, even though the carpets have been rolled, even though someone's taken away the plates, and the chairs, and letters are banking up on the floor.

Ira didn't take any notice of anyone. Other people and their troubles dripped off him like rain from a waterproof coat. No one's judging – it would probably be blissful really, to not hear things, to not know things. Apart from the fact that he missed what everyone was saying about the eroding cliff path, but there are swings, aren't there, and there are roundabouts.

Look, there goes Maurice again, circling the fields, ripping up handfuls of daisies. He does that every day without stopping. See the way he's rushing, see the way his hands are shaking. It makes you think he must have done something terrible, the way he leaves flowers like that shoved under his wife's door.

Leslie empties a bottle of Scotch every week on the ground where her father's buried. The one time she forgot the grass withered up and turned as dry as ashes.

Which reminds me of something – what is it? Why's it always up to me to remember everything?

That's it. Mrs Edwards is still in the glove compartment of her daughter's car, bumping softly every time it turns. She's waiting to be let go of, waiting, finally, for when there'll be no one orbiting around her like strange, lost planets.

Whereas someone who shall not be named (that old lech) has been shoved unceremoniously, and for all time, into the back of a cupboard.

We should spare a thought for Fuller too, shouldn't we, who never got around to arranging his affairs. He always thought he had more time. He could have been scattered over the water, drifting with the seagulls, being pulled in all directions by the wind. Instead he's on two different mantelpieces in hot front rooms, getting polished every day. It's not so bad actually, except for the way the clocks tick slightly out of sync.

July

The rain turned into a heatwave overnight. The sand roasted. Mud dried for the first time in years. A gorse fire broke out. There was a landslip and the fossil fell onto the beach and was covered over again by stones, just as it had started to get used to things.

Jack Gilbert and Mitch Mitchell were old enemies. Neither of them could remember why. Their farms shared a border and they let out each other's animals and lit dirty bonfires. So when smoke came up from Mitch's place, Jack didn't do anything about it. He went back to his house and watched from the window. The smoke thickened and hung in the air. It didn't smell like one of his normal bonfires – it smelled like there was petrol in it. Mitch had probably left an old can lying around in the yard. He was, and always had been, a complacent man. Jack went to bed. He woke up the next morning with the taste of smoke in his mouth. It was in his hair and his clothes and now, according to him, it won't come out.

Someone got stabbed with a kitchen knife. Someone else got poisoned or something. My God, it's not like the old days round here is it; it's not like those times you could keep your door unlocked. What with people wandering around with blood in their mouths and did you hear that shouting, that choking? I'm just saying.

Two sisters lived in flats one above the other. The smallest noises set them off. They heard each other's TVs. They heard curtains opening and scraping across their runners. They heard each other's breathing and the sounds they made when they were eating. The sister upstairs banged her feet on the floor. The sister downstairs climbed a ladder and banged with her hands. Their hearts went on practically the same day, but they're still doing it – banging and cursing – and neither will be the one that stops first.

There was that woman who put a curse on her mother-in-law, wasn't there? Of course, no one can prove it.

No one can prove who started the rift between the Randalls either. They're a big, sprawling family and even they keep forgetting who's aligned with who. There are cousins who don't speak to cousins, aunts who don't speak to nieces. The grandfather can't go to the cinema any more because his grandson works there, even though he used to love the maroon seats, the musty smell of popcorn. One of the nephews was tiling his roof early in the morning, before the heat set in, when he fell off. There was no one around except his uncle, who was in the next field. The nephew didn't shout to him for help. He couldn't bring himself to do it. If only he'd just shouted. But you know what these things are like between families, don't you – the way they're deep and gleaming, like tin lodes.

August

Here are the last words of Yana: Can somebody …?

Here are the last words of Fletcher: Can anybody …?

No one heard the last words of Dina.

September

The cliffs down the coastline go like this: Mussel Rock, Pigsback Rock, Squench Rock, Cow and Calf, Tense Rocks. The sea chips away at them bit by bit.

A twelve-year-old boy called Rowan fell. His friends called him Yo-Yo. He tried to jump the gap that split a headland into two jutting points. He was wearing a red jumper and the red moved slowly down through the last dried pinks of the thrift and the heather. A group of oystercatchers went up like flares.

In their condolences everyone said he was too young, too full of potential. But the boy's father, stunned and silent, couldn't stop thinking of the boy's wildness: he'd run away from home when he was eight, at ten had broken both wrists falling off a stolen quad. He remembered the hot night back in July when he saw the boy slip into his room, red-eyed and reeking of petrol. That's when he should have done something, that's when he should have stepped in. 'But there was no proof,' he murmured along to the eulogy. After the service, people kept asking why they could smell petrol in the church; did someone's car have a leak that needed seeing to?

Daisy could see herself lying in hospital (she was somewhere at the back, towards the ceiling). Look at all the flowers and cards they'd given her. Listen to the nice things they were saying about her. She shook her head and tried to speak, to correct them – they didn't know about the time she'd left her ailing, fractious mother in a room alone all day, did they? Or the twenty pounds she'd taken from that collection bucket, or how, when her daughter was four and wouldn't stop biting, she'd bitten her right back.

Deano was noted for his wit. You should have seen his impressions. Everyone wanted to stand behind him in a queue; it made the time go so much quicker, it made the day a little bit easier. No one knew that every morning he had to roll his loneliness, his disappointments, down to his feet and step out of them, like a layer of old skin.

For Iris they sang 'Abide With Me' when she'd expected 'Ain't No Sunshine'.

For Peter they bought lilies, forgetting about his allergies.

Only five people came to Radley's – since then he's been passing through walls, knocking on doors, trying to make sure that no one forgets him.

No one's forgotten Greta. She married all three of the Randall brothers and once beat Lonnie in an arm-wrestling match. She had a kind word for everyone, even if it was just the drink talking.

Clement, on the other hand, was an arsehole.

October

More houses went up. More gardens were ripped out for parking. It's endless round here, isn't it? There's a constant shifting of absolutely bloody everything.

A boot washed up on the town beach. People said it was Jamie Silver's. There was the mark of his needle and thread at the soles – no one else in town repaired their own shoes.

A hand washed up as well, but how am I meant to know who that belongs to?

Selma Richards left her body at the bottom of the stairs and got up feeling so much lighter. That thing had always been a burden, what with its aches and its twinges, its cracked veins and its dry skin, its ridged nails and its gammy eyes, its stiff hips and its dropped arches, its swollen glands and its knotty hair and its wisdom teeth only ever half coming through.

Giles left a house full of things he couldn't throw away – damp newspapers, tins, plastic bottles. It was stacked up to the ceilings. He had his own elaborate systems. If you squint you might see him over there in the corner, slowly reordering it, folding bags, flattening cardboard boxes, dust in his hair, dust on his clothes. When he was nine, he'd thrown away his favourite toy – a bear with warm, coppery fur – because someone said he was too old for it. He'd looked for it for days but never found it.

There were a few more sloes, a few less people. The cliffs turned almost burgundy. Geese flew over, creaking like trains. There was one tangled in the weeds by the road, well, just its feathers – the whole thing sort of broken open the way a thistle-head opens.

Dooley, the thatcher, left his signature twist of straw in every roof in the area.

Gregory left nothing worth mentioning.

Myrtle left five sons, twelve grandchildren, eighteen great-grandchildren: a whole bloody dynasty.

November

Pinky Rowe went out in the lightning again. No one wants to be the one to say it, but people never really learn do they.

Farley, for example, never learned to say no to anything.

Hazel never gave up those burnt bits – the black edges of toast, the dark crackling, the scrapings at the side of the pot. The doctor said they'd take five years off her. It was probably worth it.

December

The problem is, there's always too much happening. There's one thing and then there's always another. How am I supposed to get it all down, what am I supposed to say exactly?

I could tell you, I suppose, that Kenny's last thought was the sound of the leaves on the poplar tree in his first garden.

And the last thought of Ikey involved that bright sauce from the Chinese takeaway. (He wished it had been something more highfalutin.)

Or maybe I should say that Vanda wished for one more kiss of the lobe of her husband's ear.

Or that Opal's wish was so small, so secret, that no one could hear?

Then there's Davey, who could have.

And Bunny, who should have.

And what about Floyd, who'd done everything he wanted? When he'd worked at the pub on the cliffs collecting empty glasses, he held the record for a stack of fifty. He'd given money to charity, been a godfather to a baby. He'd stroked a tiger at the carnival, had witnessed a glittering sweep of phosphorescence out across the water. And once, he'd caught the cleanest wave at Salthouse and ridden it right in, the dark green of it arching beautifully, smoothly, behind his back.

What the hell am I meant to do with that?

Thank God the year's almost ending. Although that just means there's another one coming.

It's hard to keep track, what with all the comings, what with all the goings. And I suppose it's up to me to see what happens next, is it? The cliffs are still crumbling, the fossil's getting buried deeper in the stones. Houses are still going up. Houses are still coming down. All that earth's still shifting around. Every bloody year it's like this – everything changing, everything staying the same.

Cables

'THERE'S THOSE HOLES AGAIN,' Morrie says. He leans back against the bench and puts his hands in his coat pockets.

'They weren't there yesterday,' Fran says. She watches a gull that is watching her. The concrete under the bench is sandy. There are bits of dry seaweed and cigarette butts that roll in the wind.

'It must have happened in the night.'

'That's when it usually happens.'

A woman walks towards them speaking into a phone. They stop talking and tilt their heads until she's gone past.

'There's a lot of them this time,' Morrie says. 'They're right across the beach.'

'Down to the water.'

'The sand's been flung everywhere.'

'Look how deep that one is.'

'Look how deep that one is.'

'You could climb a ladder down that one.'

Their eyes follow the holes across the sand.

'Apparently there's a pattern,' Morrie says.

'Is there?'

'That's what I heard.'

'Who said that?'

'I heard it.'

'What pattern?'

'It's systematic.'

They look from left to right along the sand, then from the top of the beach to the bottom. The gull edges closer and Fran claps her hands. The gull stops, but still watches them.

'The tide will come in soon,' Morrie says. 'It'll cover them over.'

'It's halfway in now.'

They look out at the sea. Small waves surge. There is a deeper, darker line of water where a current moves. The shadows of clouds skim across like boats.

Morrie sighs and rests his hands on his stomach. 'Then he'll come back with his spade and dig more.'

Fran zips her coat up tighter. 'That's what he always does.'

'Listening.'

'Digging.'

'Listening.'

The waves break then drag back across the stones. The stones clatter as they turn.

'He didn't always hear it,' Morrie says. 'He was just going along, doing what he always did.'

'Listening.'

'Finding things out.'

'He knew everything around here.'

'About everyone.'

'You couldn't tell him anything – he'd already heard it.'

'Who stole that statue.'

'What happened to Lonnie.'

They fold their arms and hunch against the wind.

'He had to know everything,' Morrie says.

'He did.'

'He always had to know more.'

'He did.'

'Then he started thinking about the cables.' Morrie leans back and stretches out his legs. 'How they come in under the beach. How they're passing by, right under his feet. With all that information. All those communications.'

'I heard it's telephone calls.'

'I heard it's emails.'

'Financial transactions.'

'The stock exchange.'

'Internet searches.'

'Messages.'

'All of it.'

'Everything.'

'Right here, under the sand.'

They look out across the beach. The gull takes a few running steps forward, then stops and sidles back.

'He couldn't stop thinking about it all.'

'He wouldn't stop thinking about it all.'

'Coming in every minute.'

'Every second.'

'Then, one day, he heard buzzing.'

'It was only faint at first.'

'He hardly noticed it,' Morrie says. 'He told himself it was just a fly, or machinery in the distance.'

'Like when someone's vacuuming in another house up the street.'

'Or you walk past the tattoo place.'

'Or when a bulb's about to go.'

'It was faint. But it carried on. After three days he went and got checked. He thought it could be something inside his ears.'

'Maybe it was.'

'They said there was nothing wrong.'

'Maybe there wasn't.'

'But then it started getting louder.'

'How loud?'

'Like a lawnmower.'

'I'd have said a Strimmer.'

'He started hearing it every day. It would suddenly start up when he was at the pub, and he'd glance around, seeing if anyone else could hear it. He'd shake his head and rub over the top of his jaw. Gradually it would fade and he'd tell himself he was imagining it. Then, later, when he was walking down the street, he'd notice it again.'

'What about at work?'

'I heard he doesn't work any more.'

They stop and their heads slowly tilt. Two people walk past talking quietly to each other.

Fran looks back at the road. There is sand, fields, a few scattered houses. 'He lives somewhere up there, doesn't he?'

'He does now.'

'By himself?'

'He is now.'

Fran sighs and shakes her head. The gull walks a bit closer, then stops just out of reach of their feet. It stares at something.

'He couldn't sleep. He stopped reading the paper. He stopped watching TV. You'd be trying to have a conversation with him and he'd just be staring out towards the beach.'

'I wouldn't like that,' Fran says.

'He put earplugs in. He walked around with his hands over his ears. It helped at first, but then the noise changed.'

'How did it change?'

'It got closer. It was very close, like it was inside the ear itself. It was more high-pitched, a sort of rushing sound.'

'Constant?'

'All the time.'

Fran sighs again. The tide presses in. Water and sand pour into one of the holes. When the wave pulls back, the hole is full and level again with the beach.

'I heard he's got maps,' Morrie says. 'Where he thinks they come in. The different beaches. Working out where to dig. How to get to them.'

'Who said that?'

'I heard it.'

'They must be deep.'

'They've got to be deep, cables like that.'

'And he reckons he can hear them?'

They lean forward slowly. There is the wind and the low beat of the waves. The bench creaks under them.

'I can't hear them,' Fran says.

Morrie leans further forward. 'I don't know if I just heard something.'

'What?'

'I don't know. Something.'

Fran tilts her head and frowns. 'Maybe I heard something too.'

'What?'

They listen. The sand blows lightly over their feet. The gull scrapes one of its claws against the concrete and tenses.

'What if you did start thinking about it?' Fran says.

'All of it.'

'Everything.'

'Passing you by.'

'Every minute.'

'Every second.'

They glance at each other, then sit upright on the bench. The gull rushes forward, grabs a cold chip from between Fran's feet and flies away.

Morrie leans back. 'I don't think I can hear it any more.'

'Neither can I.'

They watch the tide fill in another hole. The waves break then drag back against the stones. The stones clatter as they turn.

Morrie shifts again on the bench. 'Did you hear about what happened with those neighbours?'

'The ones with the ditch?'

'The ones with the ditch.'

Fran reaches down and brushes the sand off her shoes. 'I might have heard something about it,' she says.

The Sing of the Shore

HE'D BEEN DRIVING ALL day and his eyes were dry, his shoulders cracking like pipes. Three hours, maybe four, that's how long he'd thought it would take, but he'd been driving for over eight. The roads had narrowed the closer he got, and now they were single track, with clumps of grass down the middle and flanked by bulky hedges. Beyond them were ridged fields, pylons, a few barns with collapsed roofs, the wet wind dousing everything.

The road turned stony. Potholes made the car jump. The road narrowed again and Bryce stopped, tried to see where he was, then kept going. He was sure he'd missed it. Nothing around here looked exactly as he remembered – that farmhouse wasn't there before, was it? And that dark mass of trees? He stopped again and got out. Daffodils lit the bank like torches. He climbed up and looked over, could just glimpse the sea at the bottom of the fields. He stood there for a long time. There was the same old wind above and the same old waves below, knuckling together like they were shaping loaves of bread.

He drove forward again, then stopped suddenly at a gate, which was open and hanging off its hinges. He turned in and parked on the long grass. There were the campsite's corrugated huts – the kitchen, the laundry, the shower block. There was the office – a caravan at the bottom of the slope – and the swing; but it was all rusted out, overgrown, and one of the swing's chains had snapped. There was no one around. It was early spring and there should have been people staying by now; the fields scattered with tents and campfires, the roar of gas from stoves.

He took his bag out and crossed over to the bungalow to find Kensa. Skylarks rose up from the grass, their songs tangling together. He could smell clover, gorse, the mucky, shitty smell from the next field over. A tractor was ploughing in the distance, gulls following behind like a reel of cotton unspooling.

He was almost at the house when he looked up and saw a woman standing in the window, talking on a phone. He was about to wave then stopped, almost stumbling in the furrowed mud. It wasn't his sister. He turned and scanned the fields, then turned back to the house. The woman was staring at him and pointing at something over his shoulder. It took him a moment to realise that she meant the caravan. He nodded, pulled his bag higher onto his shoulder, and made his way down the slope.

There was a low sound coming from somewhere – almost too low to notice. The further down the field he went, the louder it got. It was a sort of booming. He stopped and looked around, but couldn't see anything. It wasn't the waves; he could hear those breaking slowly against the rocks. This was deeper, more like an echo, or a murmur behind a wall. He kept going through the long grass. After a while he told himself he couldn't hear it any more.

The caravan's door was shut and there were curtains across the windows. He went up the step and knocked. He waited a moment then knocked again, and when there was no answer he pushed gently on the door. Inside, the room was cramped and stale. He was expecting the desk and the swivel chair, but there was also a mattress on the floor with a blanket on it he recognised. There was a gas heater, and a pan on the hob. Kensa must have rigged the caravan up to the mains because the fridge under the sink was humming and there was a dim lamp in the corner.

He put his bag down and went over to the desk. All the office stuff was there – the money box, the check-in forms, the accounts book. He opened the accounts and looked at the figures. They were low. No one had stayed over the winter; hardly anyone the summer before. There were no bookings for the months coming up either.

There was a noise outside and suddenly Kensa was in the doorway. They stared at each other for a moment, then she came in, sat on the mattress and started pulling off her boots. 'You're back then,' she said.

Bryce closed the book. 'You sold the house,' he said. He moved away from the desk and knocked into a box of clothes. He pulled out the chair and sat in it, rubbing his fingers into the corners of his eyes. When their parents died, Kensa had taken over the place. Bryce had already gone. He remembered the day she'd moved back into the bungalow – it was the last time he'd been here.

'I saved your share,' Kensa said. 'Of the money.' She got up, opened one of the cupboards, closed it, then opened it again. She brought out a few tins, emptied one into the pan and lit the flame. 'Are you hungry?' She seemed smaller somehow; there was a stoop to the top of her back. She kept running her hand through her hair, which she'd cut short. There were the same three hoops in each ear. She was past forty; he could hardly believe it, Christ, he was almost forty himself. He felt too big for the space – he was suddenly aware of how bulky he'd let his waist get, the extra weight around his hips. There was still the same wiriness about Kensa, or maybe rigidity, like she was holding herself away from something.

'I just need a few days,' he said, gesturing to his bag. 'Maybe a week.' There was nowhere else he could go. A few things hadn't worked out, a few things needed waiting out, and then they'd be OK again, like a piece of glass battered into smoothness by time and the sea.

Kensa stirred the pan. 'Are beans alright?' she asked.

She'd never liked beans, and neither had Bryce. It was the way the skins peeled off and crumbled. As kids they'd gone round to an aunt's for dinner and fed them to the dog under the table.

He leaned back in the chair and it cracked softly. 'Yeah,' he said. 'Sounds good.'

~

Summers on the campsite were long and empty. There was nothing nearby – no shop, no park, no other houses. Their parents were too busy to take them anywhere so Bryce and Kensa would hang around by the kitchen block, light matches from the spare box, bet which tent would collapse first, or who would trip over a guy rope. They went through the left-behind clothes in the laundry and put on whatever they found. Sometimes Kensa wore rolled-up overalls; sometimes a velvet dress which slipped off her shoulders. Bryce wore a Hawaiian shirt that smelled of aftershave.

At first, they made friends with other kids who came to stay – there was that girl with the head-brace, and that boy who could burp the alphabet – but after a while Kensa decided they wouldn't do it any more. They didn't need anyone else. The other kids always left. They never turned round to wave from the backseats of their cars; they probably forgot about them as soon as they went past the gate. All that would be left were the yellow squares of grass where their tents had been, and a few charred sticks from their fires.

Instead, Kensa stole bright soaps from the showers and chocolate out of the communal fridge. At night, they would crouch by the tents, listening. Sometimes there would be arguing, sometimes singing. Sometimes there would be strange noises in there that Bryce didn't recognise, and Kensa would put her hands over his ears.

She could hold her breath until her lips went grey, and throw the peeling knife into the door so hard that it quivered. She would push back her stringy fringe and stare out at things Bryce couldn't see. Once, she found a chunk of ice on the grass outside the front door. It was about the size of a grapefruit and it was just there suddenly, one day, in the heat. They had no idea where it had come from. She picked it up and kept it in the freezer, behind the bread and the bag of peas. Sometimes they would take it out and look for a long time at the blue-tinged crystals. Bryce followed her everywhere.

They got sunburn, grass rash, nettle stings, bites from mosquitoes and horseflies. Kensa would find dock leaves and spit on Bryce's bites. She picked her own into scars. When a group of boys crushed a patch of strawberries he'd been growing, she went out in the night and undid their guy ropes, so that their tent collapsed on them in the rain.

Then, one summer, Nate came. It was a dark, muggy summer, the kind that always seems to be brewing storms, but no storm ever hits. Flies banged into the windows and lay twitching against the glass. Mushrooms bloomed and disappeared overnight.

Bryce was nine and Kensa was twelve. Nate arrived late one afternoon and set up a small tent in the corner of the field, far away from everyone. He was seventeen. No one knew where he'd come from. He paid by the week and said he didn't know how long he'd be staying. At night, a small torch would shine out from his tent, and stay on until morning.

~

Kensa cleared up the food and put the plates in the sink. She glanced at Bryce, then around the cramped room. 'I guess I should give you the mattress,' she said.

Bryce's eyes were closing and he forced them open. He zipped his jacket up to his chin – he could almost see his breath in front of him. 'Have you still got that tent?' he said. 'The spare one?' They always used to keep a spare in case any of the visitors' tents broke.

'Maybe,' Kensa said. 'I'll go and look.' She put her boots back on and went out into the dark. The wind came in the door and blew the papers across the desk.

Bryce's eyes closed again. He must have slept for a moment because he suddenly jerked awake. He didn't know where he was, and he stood up, took a step forward, felt the small walls pressing in. There was a low noise, deep and almost regular, as if there was too much pressure in his ears. He crossed the room and opened the door, almost walking into Kensa.

'Can you hear that?' Bryce said.

'What?'

'That.' He went down the step and onto the grass.

'The wind's picking up,' Kensa said. She was holding a tent, a sleeping bag and a torch. 'We should put this up.'

Bryce listened again. The wind was thudding against the caravan, making the loose glass in the window clunk. He'd forgotten the way the gales careened over the cliffs like that: head first, with nothing in their way. 'I'll do it,' he said. He went inside and got his bag, then took all the camping stuff from Kensa. 'See you in the morning, OK?'

A bat brushed past his face, then swung low over the grass. Kensa went inside. He could see her through a gap in the curtain – she sat down on the mattress, got up again, took out a bottle of something and poured it into a mug.

Bryce walked down the field until he found a flat pitch to put up the tent. He unpacked it from the bag and shone the torch on the poles and the rusty pegs. The wind dragged the material out of his hands. It was soft and mouldy and there were dead flies studded in the netting.

He put up the buckling poles, spread the canvas over them, then realised it was inside out. He took the canvas off, turned it over, fastened it back down, then hammered in the pegs, hammering his finger by mistake and swearing into the wind. The pegs bent against the stony ground but finally went in.

Kensa's light was still on when he zipped up the door.

He shifted on the ground, turned onto his side, then his back. He pulled the sleeping bag higher, then dropped it back down. He didn't want to know what the smell was in there.

An hour passed, and then another. It started to rain and a few drops splashed onto his legs. He moved over to the other side of the tent. He needed a mat, maybe some cushions. There was a stone jutting into his hip, and another one in his shoulder. He turned over again.

The sky lightened slowly and a chill came up from the ground. Finally he gave up, pulled on a jumper and jeans, found his towel at the bottom of his bag, and went over the wet grass to the shower block. It had been a long time since he'd walked outside with bare feet – he'd forgotten the sponginess of it, the bristle of dandelion leaves, the way daisies snapped off between his toes.

He took a piss then turned on the shower. There were cracks everywhere, clumps of mud and dried grass, brittle spiders that must have died years ago. Red gunk dribbled between the tiles and there were yellow flakes of old soap on the floor.

Whenever he'd imagined Kensa, it hadn't been like this. He'd thought of her walking between bright tents, letting down guy ropes in the night if someone had been a jerk. He'd thought it would all be the same, suspended somehow, like a point on a map, even when Kensa's face became blurred, even when the campsite faded like paint running over wet paper. They hadn't spoken for a long time. He'd meant to phone, to get in touch, but he'd always been moving from place to place, from job to job, always trying to find the next thing, waiting for when he could say, finally, here I am.

The water ran lukewarm, almost cold. He came out shivering.

~

That summer they would get up early, slipping out before their parents made them do any chores. What they hated was cleaning out the showers, emptying bins, looking for mouldy bread or bottles of green milk in the kitchen. So they disappeared, taking handfuls of dry cereal with them. They weren't meant to go far – beyond the campsite there were cliffs, rips, caves, long drops to gravelly beaches – so they stayed around the lanes and the fields, walking back and forth along the dark hedges. The fields had barley in them, which moved in the wind like muscles under a horse's back. Kensa would get Bryce to hide in the stalks and then she would try to find him. She'd start counting and Bryce would crawl away through the damp soil, his heart pummelling in his throat, trying not to snap any of the stalks and give himself away. Kensa always found him.

One morning he'd been hiding for a long time. He'd found a gap big enough to sit in, and he was waiting, the skin on his hands tingling. Minutes passed, then maybe half an hour. An ant crawled up his leg. It was hot down there and the clouds were getting thick and murky.

'Kensa?' he said quietly. Another ant went up his leg. Voices drifted over the field towards him. He stood up, but he couldn't see anything except the stalks rippling.

He walked towards the voices. One of them sounded like Kensa's but he couldn't tell who the other person was. He crossed the field and came out at the edge, his cheeks dusty, seeds knotted in his hair. It was Kensa and that boy, Nate, from the campsite. They were standing by the gate, talking. It took them a moment to see Bryce.

'Hello there,' Nate said. 'Who are you?'

'That's just Bryce,' Kensa said. 'My brother.'

Bryce brushed another ant off his leg. 'You were meant to find me,' he said.

Kensa picked at a barley stalk. 'This is Nate,' she said. 'Remember?'

'You were supposed to find me,' Bryce told her.

Kensa bent the stalk around her arm and left it there, like a bracelet. 'You were saying about the glow-worms,' she said to Nate.

Nate pushed at his glasses and sniffed. He sounded blocked up. 'I thought I'd go and look for them. There's meant to be a lot round here.' He was short and his face was pale and creased, like a pillow in the morning. He had a shaved head and bare feet.

'Glow-worms?' Bryce said.

'You can come if you want,' Nate said. He spoke quietly, almost with a lisp, which made him sound even more out of breath.

'I'll come,' Kensa said. She wound another stalk round her wrist.

'I was hiding for ages,' Bryce said. The clouds got even darker and a few drops of rain fell onto the dusty ground. He stood there, picking the seeds out of his hair, then he turned and went back into the field, found his hiding place and crouched down, muttering to himself and digging in the soil with his fingers. He didn't come out when Kensa called him.

That evening, Kensa didn't eat all her dinner. She kept some back and put it on another plate in secret. When she went out of the back door, Bryce followed. She crossed over the main site and went down to Nate's tent, then sat outside with him while he ate.

~

Bryce was alone on the campsite. He didn't know where Kensa had gone. It was late afternoon and just starting to get dark. He'd spent most of the day reinforcing his tent, patching the leaks with tape and trying to find a mat to lie on. Now he got a bucket of water, disinfectant, a cloth, crusty rubber gloves, and started cleaning the shower block. He turned on all the strip lights, which buzzed and clanked, then gritted his teeth while he did the plugholes and the sink and between the tiles. He washed away the spiders and the strung-up midges, then took the casing off the lights and tipped out the husks of wasps. He poured bleach everywhere and came out coughing, his eyes red around the rims.

He did the same with the laundry room and the kitchen. He emptied the bins, threw away old socks, swept up onion skins and peeled desiccated teabags from the floor. He mopped and threw down more bleach. Then he washed his hands a thousand times and went down to his tent.

Kensa was back. He knocked at the caravan and went in. She was sitting at the desk with a book in front of her, squinting in the dim light.

'I cleaned the blocks,' Bryce said. The bottle of whisky was out on the sink and he poured some into a mug for himself, topped up Kensa's, then sat down on the edge of the mattress. 'It doesn't look like many people have been using them.'

Kensa closed the book slowly. Bryce thought he recognised the cover from the pile of left-behind books in the kitchen. It was some kind of musty out-of-date travel guide, the edges yellow and curled. The lamp cast shadows below her eyes.

'When did the last person stay?' he said.

The light flickered and Kensa frowned and tapped the bulb. 'I don't remember.'

Bryce shifted on the mattress. He took a long drink, and then another.

'Other sites have opened up,' she said. 'Around.'

'Maybe you should …'

The bulb flickered again. 'Don't do this now,' Kensa said.

Bryce shifted again on the mattress. The blanket was rucked up under him. He picked it up. It was tiny and fraying, and there was a wobbly K written on the label. He looked around at the bare walls, the boxes of clothes, the bottle of drink. This was what he'd just come from – except that, for him, it was because he was always just on the cusp of leaving.

'Kensa,' he said.

The lamp pinged and snapped off.

'Shit,' Kensa said. The fridge stopped humming. The lights at the edge of the campsite went black and they were plunged into the dark.

Bryce got up and stepped forward, stumbling over Kensa's foot.

'Stay still,' Kensa said. 'I need to find the torch.'

'I'll find it.'

'Just stay still.'

There was the sound of cupboards and drawers opening, and a moment later the torch was on, the door had opened and she was zipping up her coat.

Bryce followed her down the step and onto the grass. The darkness was almost solid – it seemed to press outwards, filling everything like gas expanding to fill a space.

Kensa's voice came from behind him. 'It's the trip switch. I need to re-jig the wires,' she said. 'It always bloody does this.' She started walking over to the kitchen block.

He watched her go. He'd thought, for a moment, that when he'd stumbled in the caravan she was about to say, have a nice trip.

In the distance, the bungalow was still lit up. There was a family in there, sitting round a table, steam rising slowly from their plates.

Kensa spent more and more time with Nate. Instead of waiting for Bryce in the morning, she would leave before he was up, disappearing on long walks out to the cliffs and down to the rocks and the beaches.

When Bryce tried to follow, he always got caught by his parents. When they asked where Kensa was he pursed his lips and said he didn't know. He had to do the chores alone: mopping the floors, smearing the mud around the tiles, wiping the stained mirrors. The coins they gave him afterwards rattled like stones in his pocket.

A family came and set up a big tent with an awning that flapped like a broken wing.

There was a boy who looked about the same age as Bryce, so Bryce went over there and stood by the door until the boy finally came out and started kicking a ball around by himself.

'Kick it to me,' Bryce said.

'Why?'

'Kick it to me.'

The boy kept the ball under his foot so Bryce kicked it away, then picked it up and ran off with it. The boy followed him.

'Let's go and listen at that tent,' Bryce said.

The boy stared at him.

'Let's crouch down behind it and listen.'

'Why?'

Bryce picked a flake of rubber off the ball. It was muddy and wet. 'I don't know.'

The other boy went back into his tent. He didn't come back for the ball.

Bryce went down to the bottom of the campsite and looked out. He could see Kensa and Nate far below, on the rocks. They were just walking. They never did anything except walk and talk quietly about things, their heads bent together. Bryce could never catch what they were saying. He would listen, but their voices would blur and hum, and get tugged away by the wind.

Kensa took Nate bits of food from the fridge and loose change she found down the back of the sofa. She'd told him how to work the washing machine for free. She'd shown him the chunk of ice and Nate had held it up, looked at it for a long time, then said it was probably the remains of something someone had flushed out of a toilet on an aeroplane. After that Kensa didn't take it out any more.

Bryce watched them until they disappeared round the headland, then he started walking back. He passed Nate's tent, and slowed down. A thin sheet of rain was billowing in across the fields. The edges of the clouds were orange, almost smouldering. He looked around, then unzipped the tent and went in.

There was hardly anything in there: a mat, a sleeping bag, a damp towel, a jumper rolled up for a pillow. There was a rucksack by the door and Bryce took everything out and laid it on the ground. There was a book, a torch, a few spare clothes, a pair of fraying socks that had been repaired with neat stitches. Half a bar of chocolate. Allergy tablets. An old bus ticket.

He looked it all over carefully, then put every item back exactly as he'd found it, except he ripped a small corner of the book before closing it.

Bryce was taking his rubbish up to the bins when the woman from the bungalow came out. She was wearing a long, baggy jumper that went past her knees, and rubber gloves which dripped water on the path. A small boy was watching from the window, his hands pushed up against the glass.

'Are you Bryce?' she asked.

Bryce opened the bin and threw his bag in. He nodded.

'We bought the house, a couple of years back?'

'I saw that,' Bryce told her.

The woman glanced over to check on the boy. 'Kensa's talked about you.'

As soon as she'd turned away again, the boy pressed his face against the window, squashing his cheeks and lips into fat white shapes.

'She has?'

The woman pulled at a loose thread on her jumper, but couldn't hold on to it with the rubber gloves. 'We sometimes worry about her, living out in that caravan – we …' She trailed off and pulled again at the thread. 'What I wanted to ask you was, would you both like to come over for dinner later?'

'Dinner?' Bryce said.

'If you're free.'

'Later?' He looked behind him, as if that would determine the answer. He looked at the house – the kitchen, the hallway. That kid was probably sleeping in his old room. He rubbed a finger over his eye. 'We're actually busy tonight,' he said. 'Sorry about that.'

He waited for her to suggest another night, but she didn't. She just shrugged and smiled and said that it was OK.

He went back and told Kensa.

'Bollocks,' she said. 'Now we have to go somewhere.'

'Where can we go?'

'There isn't anywhere.'

'We have to go somewhere.' He glanced outside, saw the grass, the swing, his tent buckling in the wind. 'We'll drive to the pub.'

'It closed.'

'Bollocks.'

Kensa took two mugs out of the cupboard and sat down. 'We'll have to stay in here with all the lights off.'

'We're not doing that.'

'Why?'

Bryce went out of the caravan, scanned the fields, his car, then went up to the blocks. He paced around the kitchen. There was a pile of old driftwood in the corner, and a box of matches. He carried it all back down to the caravan, then found some tins of food, bread, a few cans of beer. 'Come on,' he said.

They crossed the fields and took the path that sloped down the cliff. There was a ledge, and then another, lower ledge – a wide outcrop ringed with thrift, which trembled in the wind. The sea was still far below them, the tops of the waves cutting across like torn edges of paper. A flock of gulls glided out towards the deeper water.

Bryce piled the wood up, rolled some newspaper and lit it. The wind blew the flame straight out. He tried again, shielding it with his back.

'Let me try,' Kensa said. She crouched down and blew into the middle of the wood. A flame sputtered and spread and a line of smoke twisted out.

Bryce rested the tins on the fire and soon they were scalding. They waited for them to cool, then pulled up the lids and ate, scooping out meatballs and folding them into bread, drinking the dregs of sauce at the bottom. The driftwood spat out salt into the dark.

Kensa sat forwards with her arms wrapped around her knees. Her head was slightly to one side, as if she was listening for something. She'd gone out again all that morning; he didn't know where she went, what she did all day. Bryce started to speak, stopped, shifted on the stones. Sparks spat and went out.

'When do you think we can go back up?' he said.

Kensa watched the fire. She opened the beers and passed one over to him. 'Not yet,' she said. She settled back against a rock.

Bryce threw another bit of wood on the fire. Smoke billowed like a sheet. It crossed his mind that Kensa had probably sat in the caravan with all the lights off before, to avoid other invitations.

'Remember when you almost stabbed yourself with that knife?' Kensa said.

Bryce looked up. 'I thought it was that fake one.'

'It wasn't the fake one,' Kensa said.

'Where the blade slid into the handle.'

'It wasn't the fake one.'

'I'd started pushing it into my stomach.'

'I had to knock it out of your hand.'

They drank their beers and watched the fire. Minutes passed, or maybe hours. Bryce could still feel the sharp point of the knife – there was a scar there somewhere, below his belly button, hidden now by a line of wiry hair.

'Do you remember Nate?' he asked suddenly.

'What?'

'Nate, that guy who stayed here on his own, do you remember him?'

Kensa put her can down slowly. The fire was almost out. She got up and stood on a rock, looking over towards the campsite. 'It's probably safe to go back up now.' She used her boot to scrape ash over the last few embers.

~

Bryce's bedroom was small and tidy. It had a desk and a globe and his shoes were lined up by the door. There was a stain on the carpet in the corner, which he'd covered with a cushion, from when he and Kensa had mixed together baking powder and vinegar to make a bomb, but it had gone off too quickly. He could still smell the vinegar on hot mornings.

He was just getting up – the long, empty day stretching ahead of him – when Kensa flung the door open, knocking over the globe. She was breathing hard, and her shoes and legs were flecked with wet grass.

'He's gone,' she said.

Bryce folded the top of his duvet down carefully.

'Did you hear what I said? Nate's gone.'

'What do you mean gone?'

'I went out there and he's gone.'

Bryce smoothed the duvet, then straightened and smoothed his pillow.

'His tent's there,' Kensa said. 'But none of his stuff. His bag's gone.' She was pacing the room now. The tops of her cheeks had gone very pale, almost white. 'I told Mum and Dad but they're not worried at all. They said he'd paid up yesterday and must have moved on. I said what about his tent, he wouldn't leave his tent, and they said it was a crummy old tent and people leave crummy tents all the time. They were annoyed because now they have to take it down and get rid of it themselves.' She paced over to the window and looked out. 'Say something.'

'He just left,' Bryce told her.

'But he didn't say he was going,' Kensa said. 'He wasn't meant to go.'

Bryce went over to the window and stood next to her. He scratched at the paint on the frame. 'Let's go and hide in the field,' he said. The day suddenly seemed not so long, not so empty. 'Let's go,' he said.

Kensa stayed by the window.

It was late morning and Bryce came down from the kitchen eating a handful of dry cereal and drinking from a mug of thick black coffee. He knocked on the caravan and waited for Kensa. He was going out to buy food and wanted to know if she needed anything.

He knocked again, then went in. Kensa had already gone. There was a bowl and mug in the sink and a pan on the hob, which was still warm when he touched it. He went outside and stood on the step, thought he just caught a glimpse of her walking across the field. He closed the door, turned, and followed her.

He took the same paths they used to take, past the edges of the fields, where the barley was only just starting to come up. When he used to crawl through the stalks, the fields had seemed vast, stretching for miles in all directions, the rows like corridors that never ended, but now he crossed them in a moment, remembering the feeling of warm dust on his cheeks, the scratchy earth under his knees. He almost heard the sound of Kensa counting down, almost felt the old fearful tingle that meant she'd got to zero.

He climbed the gate and turned towards the headland. The wind had dropped overnight and the air was warmer, denser. The clouds had a dark tinge to them, like damp behind a wall. The path edged down and he scanned the rocks below. The tide was in and the sea was gnashing at them, the white water roiling like a cauldron.

He pushed on further. His knees ached and his T-shirt was sticking to his skin. He should have caught up with her by now, or at least be able to see her somewhere further along the path.

He skidded on a gritty slope and stopped. He looked around again. There was nothing, no one, just a buzzard keening overhead, a swathe of blue flowers like stitches in the grass, and then the low, dull booming, so low he almost couldn't hear it, echoing across the rocks, the sea, the sky, as if it was coming from everywhere.

~

Sometimes, when their parents had to go away for the day, Bryce and Kensa would be in charge of the office, answering the phone and taking down bookings. There was a pile of forms and pens, and an old chipped phone. The caravan was cool and musty. Thin spiders hung in the corners. There was a new chair in there that swivelled and Kensa sat on it behind the desk. Bryce sat at the top of the step, in a wedge of sun that came in the doorway. Clouds moved across and the caravan went from dark to light, dark to light, until the skin on his arms turned to goosebumps.

Kensa spun slowly in the chair, staring out of the window. The phone rang but she didn't move.

'You have to answer that,' Bryce said.

'You do it.'

'It's your turn.' They always took it in turns. Bryce hated answering the phone. He could never remember how much they charged each night, or if they had electric hook-ups. The voices on the other end sounded impatient and far away. They asked him how old he was and where his parents were. Sometimes Kensa answered all the calls. She would use a funny accent and make him almost retch with laughing.

Bryce watched the phone. Eventually it stopped ringing.

Kensa was still looking out of the window. 'He wouldn't have left his tent,' she said. 'If he didn't have his tent, where would he sleep? He didn't have anywhere else to go. He didn't have anyone.'

'We should have taken that booking,' Bryce told her.

Kensa spun the chair back towards the desk. She opened the bookings folder and flipped back through the pages, running her finger down the columns. 'Here's where he checked in,' she said. 'And here …' She looked closer. 'See, he didn't actually check out.'

'He paid the full amount,' Bryce said.

'He didn't officially check out.'

The phone rang again but Kensa didn't look up.

Bryce's throat felt dry. It wasn't his turn. What was he meant to say? Hello, you're through to bookings? How can I help you? Welcome to …

Kensa didn't move. The phone kept ringing. Bryce picked it up and held it to his ear, and forgot to say anything at all.

A car drove into the campsite and a woman got out. She was about fifty, and she was wearing wellies and a leather jacket. There were a lot of silver bracelets on one of her wrists. She stood by the car for a moment, looking round at the site, then started walking down to the office.

Bryce had been washing the outside of the caravan, which was coated in a rind of mildew. The car's radio was blaring as it came in and he recognised the song but couldn't place it – the music sounded strange, too loud, like something half-familiar from a long time ago. He put the cloth down and dried his hands.

The woman nodded at him, and looked around again at the site, following the slope of grass up to the blocks, down to the sea.

'It's warming up,' Bryce said. He went inside and found a check-in form and a pen. He knew that, eventually, people would start coming. 'We've got a lot of flat pitches. How big's your tent?'

'I don't have a tent,' the woman said.

'Are you bringing a caravan?'

'I'm not staying.' She put her hand up as if to shield her eyes against a glare, even though it wasn't bright. Her bracelets jangled. 'I'm actually here to …'

Just then Kensa came back from the laundry room carrying a bag of washing. When she saw the woman she frowned and shook her head.

'You said you'd think about it again,' the woman said.

Kensa went inside with the washing, slammed a cupboard, then came back out. She took the cloth out of Bryce's bucket and started thumping it against the caravan. Soap ran down the metal and onto the grass. 'I told you last time,' she said.

'I'm offering good money,' the woman said. 'Take it off your hands. Like I said, I'd do the place up, look after it.'

Kensa scrubbed hard at a thick patch of green. She plunged the cloth in the bucket and slopped it out again. The water turned grey.

'You're living in the office,' the woman said.

'I told you already.'

The woman glanced at Bryce, then took one more look over the site. Her hair kept blowing across her face. She pushed it back behind her ears, holding the rest in her hand. A few staticky strands lifted, as if a balloon had been rubbed over it. 'I'll come back in a few weeks,' she said. 'Give you some more time.' She looked once more at Bryce, then started walking back to her car. A daisy she'd stepped on sprang back up slowly.

Bryce found another cloth and started cleaning around the back of the caravan. Grit worked its way in under his nails, and wet spiderwebs wrapped around his fingers. He stopped and picked them off – they felt tough but they were so thin they were almost impossible to see. He could hear Kensa banging and muttering to herself.

'What offer was it?' he said.

'I've told her it's a waste of time.' Kensa started on the side window, thumping the cloth over the glass in wide arcs. Bits of dirt flew across onto Bryce's feet.

'You don't have to stay,' he said. 'You could do something else. Go.'

'What?'

'You could go.'

The banging stopped for a moment.

'Something would come up,' Bryce said. 'You could figure it out as you went.'

'Like you?' Kensa said. Her cloth thumped again at the window.

~

'Stop following me.' Kensa turned back and waved her arms at Bryce. She was wearing her rolled-up overalls, and there were scabs and freckles like paint spatters up her thin legs. 'Go home.'

Bryce slowed down but didn't stop. He kicked at the dusty path. 'You're going too far,' he said.

Kensa crossed over to the cliffs and looked down. The sea was very dark and very grey. A mass of tangled wood and netting drifted past.

'You're going too far,' Bryce said again. 'We're not meant to.'

'We used to go down here all the time,' Kensa told him.

'No we didn't,' Bryce said. He realised too late that she meant her and Nate. She'd been circling their old routes for days now – skirting the fields, the path round the headland, the rocks below. She'd been staying out later and later, coming back just in time for dinner, with mud and bits of stone stuck to her hands. She would avoid their parents' questions, bend her head down to her plate and eat. She wouldn't look at Bryce.

Bryce stayed where he was. A seal dipped in the water and made a crying sound. It was hot and his T-shirt stuck to his skin. He waited until Kensa climbed back up. The next day she slipped out again before he was awake.

The spring wind blew in strange, sporadic gusts, like it was working itself up to something. The sky was leaden and low, but a shaft of sun broke through, sweeping across like a searchlight.

Bryce came out of the shower, an old towel wrapped around his waist, the smell of rusty pipes and chlorine in his wet hair. He'd managed to work the controls now so that the water came out mostly warm, apart from the last freezing jet at the end. He'd glanced in the mirror, realised his hair was long and tangled round his ears, his eyes bloodshot from lack of sleep. He needed to shave too – his cheeks and jaw were thick with dark bristles. He thought of himself aged nine; his arms were so thin he could reach behind the tumble dryer in the laundry block and find any dropped money.

He was just in the doorway when the woman from the bungalow came past. When she saw him she jumped, but she tried to hide it.

'Sorry,' he said. He kept hold of the towel. The wind went up there like a bastard.

'I wasn't expecting … I was looking for Kensa. Is she back yet?'

'Back?'

'Last night. I saw the torch. She's usually back by now.'

Bryce turned and looked at the caravan. The curtains were still shut.

'It's just something I need to ask her, about the house.' The woman was looking everywhere except at him. She studied a crack in the wall, the way a dandelion was bursting out of it. 'Tell her she doesn't have to come in, it's just those boxes she left in the loft – old clothes and household things, some of your stuff, I think – she said she'd clear them but she hasn't yet and I sort of need the space.' One of her hands was resting on her stomach, which curved out under her jumper. He hadn't noticed it before.

He shut the shower-block door. 'I'll tell her,' he said. He crossed the field. He'd thought Kensa was still asleep. Her curtains were across, her door was shut, how was he meant to know she wasn't in there? What did the woman say? She's usually back by now. He didn't know anything.

He was almost at his tent when Kensa came up from the path. She looked tired. Her boots were wet and she was carrying the torch.

Bryce unzipped his door but stayed where he was. 'Been out?' he said.

'I couldn't sleep.' She crouched down and started to tighten one of his ropes.

'Where'd you go?' he asked.

She pushed a peg further into the ground with her boot. 'This thing looks like it's going to blow away any second.' She worked her boot so the peg was right in. 'I could see it bending as I came up.'

'Where'd you go?' Bryce said.

Kensa went back round the tent, checking each peg, each rope. 'You have to be careful pitching here,' she said. 'Because of the rocks. The pegs don't catch. You think the tent's secure but in a gale it'll just skid right across the field.' She banged at another peg with her heel until it disappeared into the ground.

Bryce nodded. When she was crouched down like that, he could see how sloped her shoulders were – it looked as if she was hunching against cold weather. The hoops of her earrings clinked softly against each other.

'You must have gone pretty far,' he said.

'I guess.'

'Out past the fields?'

'I guess.'

'Is that burnt-out barn still there?'

'What barn?'

'Further out that way, past the fields.'

'I don't know.'

'You would have seen it, if you went that way,' Bryce said. An ant started to crawl up his leg and he leaned down and brushed it off. 'We used to go there. The roof was collapsing. It would make these cracking noises, where the wood was about to give in.'

'I don't remember.'

'It was over that way.'

Kensa frowned. She pulled the canvas so that it was taut over the frame. 'Are you sure?'

'What do you mean am I sure?' Bryce said. There was a dull ache behind his eyes. He needed coffee, or a drink, maybe both. 'We used to go there.' As soon as he said it, he remembered the barn was somewhere else; by the road near the first place he'd lived when he moved away. He'd climbed onto the roof one night and felt the soft wood almost give way under him.

'Maybe,' Kensa said. 'I think I remember. Out past the fields?'

Bryce nodded. 'Yeah,' he said. 'Over there.'

Kensa finished with his tent and Bryce went inside to get dressed. He lay on his sleeping bag and put his head on the rolled-up jumper he was using as a pillow. The tent was so thin he could almost see through it. He took everything out of his bag and looked over it: jeans, socks, spare shirts, a phone with no battery, no signal. His wallet. A receipt for petrol. He put it all back carefully.

~

Bryce's bedroom door opened and a crack of light from the hallway grew and spread over the floor, and across his bed. There were footsteps, the stifled sound of breathing, and then Kensa was standing over him.

'I know where he went,' she said.

Bryce opened one eye. His clock read midnight. It was dark and quiet. He closed his eye and tried to tell Kensa to go away, but his mouth wouldn't work properly. He pulled the covers over his head.

'Come on,' Kensa whispered. She opened Bryce's wardrobe, pulled out some clothes for him and threw them onto the bed.

'Whatnma?' Bryce said.

'We have to go.'

Bryce sat up and rubbed over his eyes. His chin dropped onto his chest and he tried to lift it back up, but couldn't do it.

'I don't know why I didn't think of this before,' Kensa was saying. 'He kept talking about going down there. He wanted to go right to the back; he thought there might be bats, or maybe those glow-worms he'd read about. He couldn't find them anywhere else and he really wanted to see them. All he wanted to do was see them.'

Bryce watched as she moved around the room, pushing at her fringe and adjusting the batteries in the torch.

'Where?' he said. 'Where did he talk about?'

'The sea caves. I already said that. We have to go there now.'

'We're not allowed.' The caves were meant to be huge and pitch-black – you could walk in deeper and deeper and never come out. No one knew how far they stretched back.

'We have to go,' Kensa said.

Bryce sat back on the bed and folded his arms. 'You told me to go home.'

'When?'

'You told me to go home.' He lay back down and rolled himself in the duvet, leaving just enough of a gap so he could see what Kensa was doing.

She came over to the bed and stood right in front of him. She pressed again at the batteries in the torch and the light came on, the beam tilting upwards into her chin, making her eyes look huge and roving. 'You have to come,' she said.

Bryce didn't move. The duvet muffled his voice. 'Why?'

Kensa moved over to the window and looked out. 'Because.'

Bryce rolled himself tighter into the covers.

'I need you to come,' she said. 'OK?' She wouldn't turn around.

Bryce got up and put on his shoes.

Kensa opened the window and started climbing out. She balanced her feet on the windowsill then jumped over the spiky bushes in the flower bed.

Bryce followed behind. 'He took his sleeping bag though,' he whispered. 'And his mat. Why would he have taken those down there?' He tried to jump from the window, but slipped, and grabbed at the bottom of the palm tree, scratching his hands on the gristly bark.

'Ssshh,' Kensa said. She turned the torch off and threaded her way through the campsite. It was full and they had to go between tents and ropes, past the sounds of people sleeping and awnings lifting in the wind. There were hushed, gurgling snores, as if a plug was loose in a bath. A little kid called something out in his sleep. A dog barked and they froze, waiting for someone to come out and see them, but no one came. They kept going. Once they were past the tents and into the first field, Kensa turned the torch back on.

There was a thin moon and the clouds crowded around it like moths. The barley bent in the wind. They walked in silence, Kensa first, Bryce behind, his legs heavy, his mouth dry, trying to stop himself turning back with every step.

They crossed the edge of the field, then climbed the gate. Something rustled on the ground, then darted away. Kensa turned round to look at Bryce. The buckles on her sandals rattled softly and the moon striped her face with silver. She looked different somehow, like his sister but also not like his sister at all. As they carried on along the path he reached out to touch her, to check, but just as he was about to do it, his hand fell away.

The path turned stony and started to drop down towards the sea. The gritty dust scraped with each step, waves cracked against the rocks like beaten rugs, and there was something else as well – a strange, low noise, that Bryce had never noticed before – a sort of deep booming that echoed through the cliff and up into his feet. He stumbled on the stony path, righted himself, then stumbled again.

'Kensa?' he said.

'We're almost there.'

'What's that noise?'

'What?'

'That.'

'It's the caves,' Kensa said.

The sound got louder until it was all Bryce could hear. It beat in his ears like a sail. He skidded and stones rolled; he couldn't find anywhere stable to put his feet so he stopped and stood very still. He couldn't see Kensa. He couldn't move forward. It was so dark. He couldn't tell where the path was any more, where anything was.

'Kensa?' he whispered.

There was no answer.

The sea was booming in the caves, knocking against the walls. It sounded like his heart against his chest. It was dangerous to go in; it was too dark, the tide was too high. Water might be pushing in through the tunnels. He turned and looked back. There were a few tiny glints of light from the campsite. He took another step forward, then turned again. There was a scrabbling noise from the path below and the torch's beam swept up across the rocks.

'Wait for me,' Bryce called. 'Wait there. I'm coming down, OK?' He waited until he was sure Kensa had stopped, then he turned and ran back to the campsite to get their parents.

~

The storm came in suddenly. Bryce had just slipped into an uneasy sleep – dreaming of stones rolling, the moon, Kensa's muffled cry of surprise, her eyes narrowing, her face turning away. He woke with the side of his tent pushing against his mouth, water sluicing down his legs, the tent poles bowing like they were about to snap.

He sat up, got dressed, then tried to unzip his tent to go out, but the force of the wind and the rain drove him back in. He lay back down, felt the storm wrenching at the tent, trying to drag it across the grass. The poles strained. There was a tightness in the air, and then the lightning started, fast and bright, scattering across the sky like gunshots. The thunder came straight after, pealing like huge bells, and below it all was the relentless booming of the caves – he could almost feel them reverberating up the earth and into his back, the sea pummelling at the stone, hurling itself around the hollow tunnels, right under the campsite, under his tent, under everything.

He didn't know how long the storm lasted. Gradually the wind eased, gradually the rain thinned to mizzle. Everything in the tent was drenched: his sleeping bag, his wallet, his clothes. He unzipped the door and went out. It was just getting light.

The grass was flattened. There were leaves and twigs everywhere, bits of wood, a rusty hinge that had been bowled down from the gate. His tent hadn't moved – the ropes were still tight, the pegs still deep in the ground – but the main pole had snapped and one of the walls had ripped, making the sides crumple inwards like old fruit. There was a fine layer of sand along the roof.

He looked over at the caravan. The curtains were shut, the door was open and swinging in the wind. Bryce looked at his tent, his stuff, his car. Maybe he should just go. Maybe it would be easier if he just went.

He packed his sodden bag, walked to the car, and put it in the boot. He opened the door and sat in the driver's seat. He didn't turn on the engine. He sat for a long time. Then he got out and started walking.

He took the path down to the caves. Halfway there he looked down and saw Kensa. She was sitting on the rocks at the side of the path, staring at the sea. The tide was out. The water was creased and battered after the storm, brown with churned sand and teeming with choppy waves. The path was wet, the stones stained with rain. They rolled under his feet as he made his way down.

'Did you hear them?' Kensa said. 'They were so loud.'

'What are they like inside?'

Kensa zipped her coat tighter and watched the waves. 'I don't know.'

Bryce looked down at the rocks and the beach. He thought of Kensa crossing the fields, going down to the rocks, standing outside the caves, but never going in. He thought of her on the path that night, in the dark, waiting for him.

'Come on,' he said.

He made his way down, slipping on grit, clutching at wet rocks, Kensa following behind until they were on the beach. The caves were in front of them – there was a deep gap in the cliff that widened out into the dark, the stone shattered and polished by the sea as it shouldered its way further in.

Bryce walked up the beach and stood at the caves' mouth. Tunnels arched ahead of him, echoing and gleaming like a cathedral.

Kensa stood next to him. She had one hand deep in her pocket, the other was clutching the torch. 'What if we …' she said, but she didn't finish.

The longer Bryce stared into the cave, the darker it looked. He took a step forward, his boots rattling on stones and bits of slate. He took another step and the slate became smooth, pale sand. He took a breath and walked in.

The air was cool and musty. After a moment, Kensa came in and turned on the torch, shining it on the dripping walls, which glistened black and red as if a flame had passed across them.

They walked forwards slowly. The walls dripped, the waves broke on the rocks very far away. Something moved above them, then a bat dropped down, circled the tunnel, and flew back up into the dark. Kensa moved ahead. Bryce picked his way through carefully, thinking of Nate's small torch, the way he used to keep it on all night, the worn straps on his rucksack, the look he had that Bryce now recognised, of someone who'd got used to moving on and not looking back.

The caves went deeper and the silence seemed to grow and thicken. There was no way of knowing which direction he was going – he just kept going, following the caves as they dipped and turned, breathing in the thin air. Sometimes the tunnels narrowed, sometimes they opened out like rooms. After a while he realised he couldn't hear Kensa any more. He couldn't see the torch. He stopped. There was no sound, no movement.

He waited in the tunnel. He didn't know how many turns he'd taken, how far or how deep he'd gone. He reached out and touched the wall, tried not to think of the miles of cliff all around him, the sea slowly making its way back in.

He held onto the cold stone. He called out. He waited. A stone clattered down and landed by his boot. Another bat dropped and circled. Then, finally, he heard footsteps in the distance. He let go of the wall and made his way towards the sound, stretching out his hands. He called again. His heart was pounding like the tide against the caves; the skin on his palms was damp and tingling. He half-expected to hear Kensa counting slowly down to zero. Any minute now she would stretch out her hands and find him.

Kensa called out, telling him to wait, to stay where he was, she'd be there in a minute, but he kept going. They could hear each other's footsteps, their breathing, they were getting closer, it was so dark, there were so many twists and bends, but any moment now they would find each other, any moment now they would know exactly where they were.

By-the-Wind Sailors

Marine organisms with an internal float and sail. The sail is either angled to the right or to the left. The sailor has no control over where it is blown by the wind – if the wind changes direction, the sailor may get pushed inland and stranded.

At the end of winter they move to a caravan. The site is on a cliff and easterlies cut across like scythes. There is a reception and a laundry area and a row of slot machines. On Saturday nights the disco floods music over the sea. There is the sound of waves constantly, and gulls, and the two-a.m. couple who chase each other around the fields full of wrath, then make up again inside, shouting out their remorse, bottles and tins rattling against the window. Some mornings there is frost crackling the grass. Other mornings the sun lays out warmth in copper sheets. The thrift is just starting to bloom. The sand martins arrive back to nest in the crumbling cliffs.

The family are Ruby and Nathan Tulley and their daughter Lacey. Ruby is short and gaunt and never keeps still – she bites her lips, pulls out eyelashes, gnaws at the skin around her nails.

Her hair is dyed maroon but it has faded to dusty purple, like the colour of sloes before they are ripe. She met Nathan at a garage sale, both of them buying someone else's chipped plates and flat cushions. Nathan is a sleepy man, and so shy that he hides if he sees anyone he knows. His skin is dry and red but he doesn't like to use the cream that Ruby got for him – it stings – so every night he gently squeezes some into the sink and turns on the tap.

He fixes fences and gates and anything else that comes his way. Ruby can sew better than anyone and she takes in dresses and shirts and works on them on her second-hand Singer, a thistle-head of pins stored in the corner of her mouth. Lacey sits very still under chairs and tables, sticks her tongue between the gap in her teeth, and doesn't say much at all.

There are a hundred static caravans and it's not the busy season, but somehow they end up in one right on the edge of the site, a mile from the facilities and half-tilted into a sloppy furrow. The front window is cracked and the curtains and walls are adorned with mould and midges. All the shine has been scoured off the plyboard. But it's dry enough inside and there's a small, neat table with three wicker chairs, which turn out to be a lot more comfortable than they look. Ruby and Nathan sit on theirs, Lacey sits under hers. There is a pattern of white leaves and cigarette burns on the carpet.

There was a fire at the flat they'd been renting in town, so they only have a few singed bags with them. The whole building went up overnight and by morning was nothing more than crumpled bricks, plaster and melted plastic. The family can't remember how they got out. Sometimes they go back over it but none of them can remember anything about getting out. One moment there was the reek and pressure of smoke, and the next they were standing on the street watching reams of yellow tape twisting in the wind, and blue lights flashing. They didn't know where to go. Then Ruby remembered how she and Nathan had stayed in a caravan for a few months after they'd got married. It had been a good time. There was no heating, so they would do their laundry in the evenings and then sit with the warm bags on their laps, drinking beer by the mugful.

But in this caravan they never seem to get warm. Nathan can't shake off a rattling cough and tightness in his lungs. Ruby scrubs the mould off the curtains but it keeps creeping back. There is always a strange, smoky smell on their clothes that won't wash out. Lacey takes to lying flat underneath the loose carpet. She's so small that Ruby sometimes stands on her by accident. Everything is broken and it's impossible to get hold of the site's owner, so Nathan fixes the beds and the overhead cupboards and Ruby prises out the black gunge between the tiles in the shower. Slowly, they find a routine. The chairs are the best thing in the whole place and they look forward to the end of the day when they can sit quietly in them. They lean back against the creaking wicker and close their eyes. I guess you can get used to anywhere, they say to each other. Nathan goes out early to the farm where he is repairing the fences. Ruby watches Lacey and mends the caravan's torn bedding and curtains. She's hoping that once they can get hold of the owner, he might give her the work of the whole site.

When the wind squalls, the caravan rocks from side to side. One godawful night it feels as if it's sliding down the field towards the sea. They run out into the dark and throw bricks in front of it. Nathan digs a deep trench and goes back to bed muddy. In the morning, there is no sign that the caravan has moved at all. Lacey strings a row of things she's found along the back window. There are bits of coal, wet feathers, and a clutch of plastic key rings, salt-scrubbed but still bright, covered in writing no one can read because it's in Mandarin.

In the middle of summer, more tourists come and the park is full. Nathan says that people look over at their caravan kind of funny but Ruby knows what Nathan is like. Really, he's glad that no one tries to speak to them. The hotplate is working better now, and the damp is drying out. In the mornings, the skylarks rise up into the air as if on ladders.

Then, one afternoon, they come back and find their door flung wide open. There's a man inside painting the walls. He has headphones on, and he whistles, glances round, then gets back to work. He paints straight over the mould and the midges, so that there are tiny bumps where they've been sealed in. It seems that their caravan is now needed as overflow for the holiday season, and they realise with a jolt that in the panic of finding somewhere to live they'd signed no contracts, been given no guarantees. Ruby goes to find the owner to demand an explanation but it's impossible to get hold of him.

They move to an annexe in town. It's almost the end of summer. The geese fly back from their breeding grounds inland. The annexe belongs to a man who is recently separated. At first his wife slept in the spare bedroom, then she lived in the annexe and then she moved out completely. She obviously wanted to do it in stages. When the family move in, the annexe is, as Ruby says, cold enough to freeze the balls off a swinging cat. Northerlies blow straight through the thin walls. There are still some of the wife's things lying around. Next to the bed, there's a shelf with her glasses on, a half-open book and a mug of freezing tea. Ruby tips the tea away, then puts everything in a bag and puts the bag under the bed. Under the bed there is a white shoe and a tightly rolled-up newspaper. Nathan finds her toothbrush by the sink. Lacey puts on a pair of silver clip-on earrings.

It's not a bad place: the window looks out over a chestnut tree and they find a brittle pack of cards in a drawer and teach Lacey to play blackjack and shithead. But sometimes the husband wanders in without knocking and paces through the small rooms. Or he sits on the sofa for a long time, doing nothing but staring at the wall. The family stay very quiet and retreat into the kitchen. Then only Nathan's coughing seems to startle him into getting up and going away.

The bag under the bed fills up. They find her hairbrush under the sofa, a T-shirt folded against the back of a drawer. The bathroom smells of her spicy perfume. Ruby finds Lacey under the bed, looking through the wife's book and frowning. The husband starts coming in more and more but now he is busy doing things: he paints the walls and does something to stop the damp coming through. The house smells of paint and clean carpets. They don't like it; they don't like the smell of the paint or the way he whistles as he's working. Why didn't he do all this last autumn, when they first moved in? They watch and wait. One day draught-proofing appears round the windows; another day there's a picture hanging above the sofa. Ruby shakes her head and bites at the skin around her nails. The picture is too much. They are already half-packed when the wife comes in, laden with bags. The husband is right behind her. The wife puts down her suitcase and looks around the room. It was never going to be for ever, she says.

The next place is a cabin at the bottom of an overrun garden. Bindweed chokes the windows. There are spiderwebs over the door. Every morning a gull taps rhythmically on the glass with its beak. It wants the flash of silver on the sink's plug, but can never seem to get any closer. The cabin was built for children who now prefer to spend their time drinking White Lightning in pool clubs and sand dunes. It's almost as big as the annexe, but the beds are narrow built-in bunks – something they didn't notice at first and now it's too late. At night, Ruby leans down from her bunk and watches Nathan and Lacey sleeping. Sometimes she reaches down and touches them gently on their shoulders.

There are dark shapes on the walls where pictures used to be stuck – shapes of birds and rainbows imprinted on the fading wood. There is a crap dolphin that looks more like a canoe. On good days, the cabin is warm and there is the sweet tang of pine trees. On bad days, the pine trees block out the sun, the windows steam up and drip onto the bunks, the tiny fridge shudders and stops, and there is the uneasy sound of gunshots from nearby fields. There are more bad days than good ones.

What follows is a dire stint in a shoddily divided house, crammed in with two families who seem to thrive on cacophony: yowling dogs, labyrinthine disputes, endless music broadcast from phones and laptops. The house has been split into three but to get to their rooms upstairs they have to go along the hallway of the downstairs flat, and across the kitchen of the middle one. The dogs go off whenever they open the kitchen door, and there are often parties that are so sudden and surging that it's impossible for Ruby and Nathan to get through them. Once they are trapped in the hall for two hours, until Ruby works out a way of getting in via a half-open upstairs window. Lacey stops going to bed, and instead lies on the floor, watching the parties through a crack in the boards. It's a dismal few months, and they feel more embroiled in the lives of the two families than they do in their own; but they've learned the words to all the songs by Elvis; how to calm an agitated greyhound; how to pass through rooms and doorways without being noticed at all.

They spend a few weeks in an empty bed and breakfast, with a landlady who cannot sleep and who soothes herself by walking up and down the stairs, her knees clicking like dice. Then another week in a dilapidated roadside hotel; another in a dim room above a pub, with a strange oil painting of a stag on the wall, which vibrates to the jukebox and the mirth and the brawls.

Living out of bags that must be packed and unpacked constantly is, they soon realise, a complete bastard, so they decide to go back to the caravan. The one they stayed in before is available again; for some reason no one else wants it. The geese move back inland for the summer. Inside, the paint is still almost fresh. The crack in the front window has been fixed. A single hotplate that is theirs and theirs alone is a sudden luxury, a miracle even. It's warm – the wind has shifted and is coming from the south. Lacey sits outside and plays solitaire with the cards she took from the annexe. Ruby sticks up adverts for her sewing. Work has been slow for her recently but Nathan is always busy – he goes out early in the mornings and works his way back and forth along the fences on the outer farms. There is always broken fencing. At night his cough is no louder than the breeze that pushes in around the door.

Then winter comes. At first they think they'll be able to manage. They get hold of an extra duvet and some thick material to line the curtains. In the evenings, they warm their hands over the hotplate. I guess you can get used to anything, they say to each other. But it's a ruthless winter. The windows freeze on the inside. The three of them huddle in one bed. Lacey keeps waking up and saying she can hear a gull tapping on the window. Nathan's cough gets worse. When Ruby finds mould growing on Lacey's jumper there's nothing left to do but pack up their bags.

It's not that hard to find a new place. Ruby has heard about a village just outside town, where the houses are full all summer, then empty all winter. Some of the houses are advertised as winter lets. The family move into a small cottage in a row of other small cottages. It's the kind of place they used to talk about living in one day. The doors are painted green and there's wisteria or clematis or whatever the hell it is growing all over them. The village is very quiet apart from the thrum of the sea in the distance. There's no sound from either of the houses next door: no footsteps on the stairs, no doors opening and closing. The curtains stay half-drawn. At night there are two lights from windows in the distance. Empty dustbins clatter in the wind.

For some reason they can't settle. A gate scrapes and Nathan goes out and fixes it. Now it's really quiet. Lacey wakes up in the night and thinks she's in the caravan. She goes into the hall and pees on the doormat. The fridge hums softly. The furniture and plates and curtains are almost too nice to touch and the family try not to touch them too much. They use the same three plates over and over. When Lacey breaks hers, Ruby shouts: shitting hell, Lacey, those are good plates. She tries to stick it together but it doesn't stick. Lacey goes and sits under the table and Nathan sits with her.

One evening, the house alarm goes off when Nathan comes in from work. They try to stop it but the electrics are complicated. After twenty minutes it turns off, but for a long time afterwards Ruby and Nathan pace and look out of the windows. It happens again the next night, and the next. They sit in the kitchen with their hands over their ears as the alarm blares over the village.

After six weeks there's a message on the answerphone. A woman's voice bellows a greeting to the house's owners: she's glad that they've finally decided to come down for Christmas, they should let her know as soon as they arrive so that she can drop round and see them. Should she bring trifle or mousse? The red light on the phone blinks. Nathan plays the message again. He doesn't like mousse or trifle. It's the first they've heard of the owners coming down. There were no timeframes when they moved in, but surely winter lasts until at least February. Ruby curses the owners in long and complicated ways. Nathan says that maybe they'll have to stay in a bed and breakfast and then come back once the house is empty again. But all the bed and breakfasts are booked and double the price around this time of year. They don't have to rush – they've got over four weeks to work something out, but they want to go as quickly as possible; they don't want to leave it so late that they have to witness the owner's arrival, the awkward crossover in the doorway.

They pack up. They can't remember why they used to imagine living in such a place; they can't connect that old dream with themselves at all. The alarm wails out one more time as they go. Lacey sticks a finger up at it and doesn't get told off.

Nathan has found a place they can stay. A shop in town has a flat above it that's available. The shop sells CDs and records and rents out films. Ruby and Nathan used to go there a lot. They've lived above shops before and had a good time. In the day there is bustle but at night it's mostly quiet. Shops are always heated and the heat rises into the flats. There is one bedroom and the sofa in the living room folds out for Lacey. It's snug and grungy, with the smell of old cigarettes and cooking – just what they like. There's no worrying about expensive plates here. There are always footsteps and voices and cars, and bright lights along the road at night. The roof is porous and lets in westerlies, dandelion seeds, hibernating butterflies, the sound of the Friday-night drunks calling up the street like mournful geese.

But the shop gets steadily quieter. A few months pass, and then a year. One afternoon Ruby notices a sign outside that says 'Clearance Sale'. No one has mentioned anything to them. Apparently people aren't renting out films from shops any more. Ruby and Nathan can't help remembering all the times they used to go in and pick out a film, then post it back through the door on Sunday mornings. They don't understand why people would rather click a button and stay indoors. Sometimes it seems like the world is moving on without them. The stock in the shop empties and a 'For Sale' sign goes up. They spend a sad hour packing their bags. They fold the bed up into the sofa. Lacey rips a corner of wallpaper, writes something behind it and sticks it back down with spit.

There is nowhere else to go except back to the winter-let cottage. They unpack their bags. The snowglobe and the ornament of a shepherdess they had in the old flat don't look right here and end up banished to the back of a cupboard. The owners have left a half-eaten box of chocolates from their last Christmas visit and Ruby tries one. It's strawberry and stale. The gate Nathan fixed before seems to have bust again, and creaks quietly in the wind.

One night Lacey comes into their room and says she can smell fire. Nathan and Ruby rush downstairs and there's a smouldering ember from the stove on the carpet. They stamp it out for a long time, then stay up until morning, clutching each other's hands, watching the carpet for smoke. The ember has burned a black mark right through to the floorboards.

Winter ends and they wait for word from the landlords. But no word comes. Spring turns into summer and then winter again. Ruby spends an afternoon ringing up to find their post – they are due final bills from old rentals, bank statements and God knows what else. Automated messages tell her the same thing each time: her details can't be located and will need to be found and looked into. The messages always promise that someone will get back to her. They start receiving supermarket offer sheets almost every day, addressed to old tenants and old owners of the house. Lacey draws circles around all the things she wants: half-price lemonade, bin liners, pasta shaped into thin, contorted faces.

They stop thinking of moving. Nathan puts up a shelf in the bathroom. Ruby paints one wall of the kitchen blue. Lacey lines up snail shells along her bedroom window. At night there are no other lights, and it's the calling of owls, not beer lorries or people or discos, that makes its way through the windows. The gate scrapes in the wind. They sleep lighter and wake much earlier than usual. The days become much longer. One night they see a meteor shower that rends the sky with silver. I guess you can get used to anywhere, they say to each other. After all, it's the most beautiful place they have ever lived – surely it's ridiculous to feel unsure about it, to miss the thump of music on Saturday nights, the smell of old cigarettes, the strange darkness of pine trees.

Overnight, the cottage sells. There hasn't even been a 'For Sale' sign outside but Nathan says he supposes it was all done over the internet. They've never got round to owning a computer or an expensive phone. Sometimes it seems like the world is moving on without them. Estate agents and surveyors prowl around. Nathan takes down the shelf and they paint over the blue wall in the kitchen. The wooden counters and draining board are bleached with watermarks and they spend a terrified few hours trying to scrub them off. What kind of sadist has a draining board you can't get wet? Ruby says. She had learned to love that stupid wood. They rub a damp teabag along it to stain the watermarks brown and make them blend back in. It seems to work. Slowly and systematically they erase themselves from the house. Nathan unpins Lacey's drawings and fills in the holes. Ruby patches up the burn mark in the carpet and does it so well that it looks as if they have never been there at all.

They move back to the caravan. It's the end of winter. They miss stairs and, for a while, a single hotplate does not seem like enough. The wicker chairs are as comfortable as ever. Nathan adjusts the table so they can all fit round it and play with the deck of cards. They are teaching Lacey whist, and arseholes and presidents. The wind cuts across the cliff like a scythe. Some days it brings with it the coconut smell of the gorse. Some days it rocks the caravan like a bloody cradle. They put bricks either side to balance it out. Really, you can get used to anywhere, they say to each other.

Nathan goes out and mends the fences on a farm he worked on only a year before. It seems that fence posts nowadays are rotting out sooner and sooner. Ruby mends the frayed seats in the caravan and hums a song by Elvis. Lacey digs out the chunk of coal she buried three summers ago. It's just the same as she left it: glossy and heavy and exactly the same size as her palm.

The mildew on the caravan is creeping back, and the corner of the window has a hairline crack that looks to be spreading. The two-a.m. couple fight and make up again. The thrift is just beginning to bloom. The sand martins come back to nest in the crumbling cliffs.

Acknowledgements

THANK YOU TO MY agent Elizabeth Sheinkman and my editor Helen Garnons-Williams for their encouragement, advice and enthusiasm. Thank you to everyone at 4th Estate. Thank you to the Roger and Laura Farnworth Residency at Warleggan, which gave me two weeks to write in such peaceful and beautiful surroundings. Thank you to Jos Smith and Anneliese Mackintosh for their help and advice. Thank you to Mum. Thank you to Ben, as always, for everything.

Grateful acknowledgement is also due to Arts Council England, who supported this book with a writer's grant.